W9-BXW-459

Introductory Digital Electronics

Nigel P. Cook

PRENTICE HALL
Upper Saddle River, New Jersey   Columbus, Ohio

Library of Congress Cataloging-in-Publication Data
Cook, Nigel P.
Introductory digital electronics / Nigel P. Cook.
p. cm.
Includes index.
ISBN 0-675-21334-7
1. Digital electronics. I. Title.
TK7868.D5C597 1997
621.39'5—dc21 97-10713
CIP

Cover Photo: Pasieka/Zefa/H. Armstrong Roberts
Editor: Linda Ludewig
Production Editor: Sheryl Glicker Langner
Project Management: Elm Street Publishing Services, Inc.
Design Coordinator: Karrie M. Converse
Cover Designer: Rod Harris
Production Manager: Pamela D. Bennett
Marketing Manager: Debbie Yarnell

Upper Saddle River, New Jersey 07458

Printed in the United States of America
10 9 8 7 6 5 4

ISBN: 0-675-21334-7

Prentice-Hall International (UK) Limited, *London*
Prentice-Hall of Australia Pty. Limited, *Sydney*
Prentice-Hall of Canada, Inc., *Toronto*
Prentice-Hall Hispanoamericana, S. A., *Mexico*
Prentice-Hall of India Private Limited, *New Delhi*
Prentice-Hall of Japan, Inc., *Tokyo*
Editora Prentice-Hall do Brasil, Ltda., *Rio de Janeiro*

To my father

Whom the gods favor die young.

Platus

Other Titles by This Author

Introductory DC/AC Electronics
Introductory DC/AC Circuits
Introductory Semiconductor Electronics
Introductory Mathematics
Practical Electricity
Practical Electronics
Microwave Principles and Systems

For more information on any of the other textbooks by Nigel Cook, see his web page at: www.prenhall.com/cook or ask your local Prentice Hall representative.

Preface

Following the success of *Introductory DC/AC Electronics* and *Introductory Semiconductor Electronics,* Nigel Cook was besieged with requests to follow with an *Introductory Digital Electronics* text using the same student-friendly writing style, practical treatment, real-world applications, and troubleshooting that made his other books come to life for technician-level students.

ORGANIZATION OF THIS TEXTBOOK

This textbook has been divided into three basic parts: Digital Basics (Chapters 1 through 6), Digital Circuits (Chapters 7 through 16), and Digital Systems (Chapter 17).

PART I: Digital Basics

Chapter 1 Electronics—Analog and Digital
Chapter 2 Number Systems and Codes
Chapter 3 Logic Gates
Chapter 4 The Bipolar and MOS Transistors Within Digital ICs
Chapter 5 Troubleshooting Logic Gate Circuits
Chapter 6 Logic Gate Circuit Simplification

PART II: Digital Circuits

Chapter 7 Decoders and Encoders
Chapter 8 Multiplexers, Demultiplexers, Comparators, and Parity Circuits
Chapter 9 Set-Reset and Data-Type Flip-Flops
Chapter 10 *J-K* Flip-Flops and Timer Circuits
Chapter 11 Registers
Chapter 12 Counters
Chapter 13 Arithmetic Operations and Circuits
Chapter 14 Semiconductor Memories
Chapter 15 Programmable Logic Devices
Chapter 16 Analog and Digital Signal Converters

PART III: Digital Systems

Chapter 17 Introduction to Microprocessors

DEVELOPMENT, CLASS TESTING, AND REVIEWING

The first phase of development was conducted in the classroom with students and instructors as critics. Each topic was class-tested by videotaping each lesson, and the results were then evaluated and implemented. This feedback was invaluable and enabled me to fine-tune my presentation of topics and instill understanding and confidence in the students.

The second phase of development was to forward a copy of the revised manuscript to several instructors at schools throughout the country. These technical and topical critiques helped to mold the text into a more accurate form.

The third and final phase was to class-test the final revised manuscript and then commission the last technical review in the final stages of production.

ILLUSTRATED TOUR OF TEXTBOOK FEATURES

The following are just a few of the features you'll find in *Introductory Digital Electronics.*

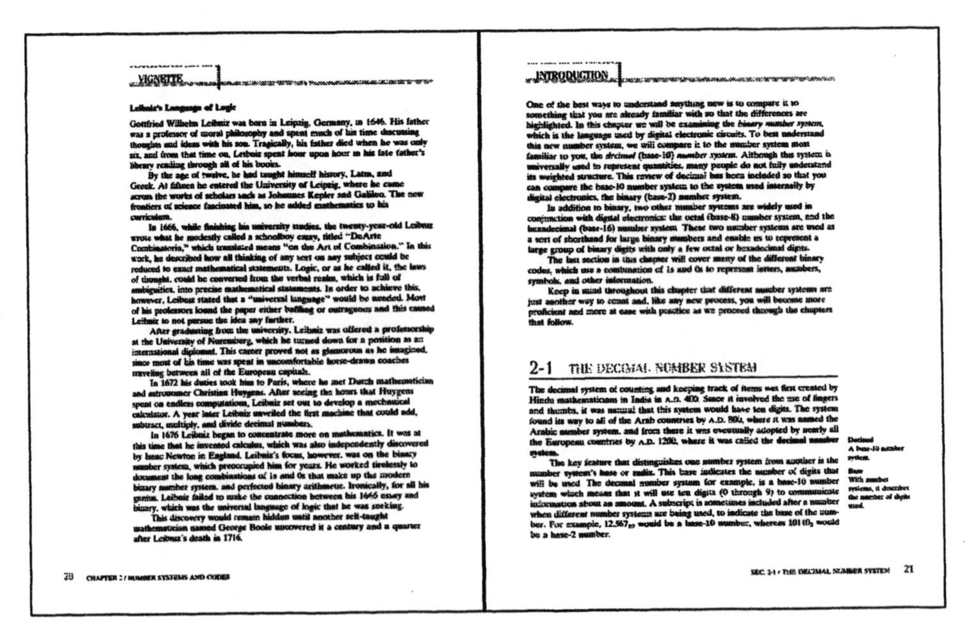

VIGNETTE

Leibniz's Language of Logic

Gottfried Wilhelm Leibniz was born in Leipzig, Germany, in 1646. His father was a professor of moral philosophy and spent much of his time discussing thoughts and ideas with his son. Tragically, his father died when he was only six, and from that time on, Leibniz spent hour upon hour in his late father's library reading through all of his books.

By the age of twelve, he had taught himself history, Latin, and Greek. At fifteen he entered the University of Leipzig, where he came across the works of scholars such as Johannes Kepler and Galileo. The new frontiers of science fascinated him, so he added mathematics to his curriculum.

In 1666, while finishing his university studies, the twenty-year-old Leibniz wrote what he modestly called a schoolboy essay, titled "DeArte Combinatoria," which translated means "on the Art of Combination." In this work, he described how all thinking of any sort on any subject could be reduced to exact mathematical statements. Logic, or as he called it, the laws of thought, could be converted from the verbal realm, which is full of ambiguities, into precise mathematical statements. In order to achieve this, however, Leibniz stated that a "universal language" would be needed. Most of his professors found the paper either baffling or outrageous and this caused Leibniz to not pursue the idea any further.

After graduating from the university, Leibniz was offered a professorship at the University of Nuremberg, which he turned down for a position as an international diplomat. This career proved not as glamorous as he imagined, since most of his time was spent in uncomfortable horse-drawn coaches traveling between all of the European capitals.

In 1672 his duties took him to Paris, where he met Dutch mathematician and astronomer Christian Huygens. After seeing the hours that Huygens spent on endless computations, Leibniz set out to develop a mechanical calculator. A year later Leibniz unveiled the first machine that could add, subtract, multiply, and divide decimal numbers.

In 1676 Leibniz began to concentrate more on mathematics. It was at this time that he invented calculus, which was also independently discovered by Isaac Newton in England. Leibniz's focus, however, was on the binary number system, which preoccupied him for years. He worked tirelessly to document the long combinations of 1s and 0s that make up the modern binary number system, and perfected binary arithmetic. Ironically, for all his genius, Leibniz failed to make the connection between his 1666 essay and binary, which was the universal language of logic that he was seeking.

This discovery would remain hidden until another self-taught mathematician named George Boole uncovered it a century and a quarter after Leibniz's death in 1716.

20 CHAPTER 2 / NUMBER SYSTEMS AND CODES

INTRODUCTION

One of the best ways to understand anything new is to compare it to something that you are already familiar with so that the differences are highlighted. In this chapter we will be examining the *binary number system*, which is the language used by digital electronic circuits. To best understand this new number system, we will compare it to the number system most familiar to you, the *decimal* (base-10) *number system*. Although this system is universally used to represent quantities, many people do not fully understand its weighted structure. This review of decimal has been included so that you can compare the base-10 number system to the system used internally by digital electronics, the binary (base-2) number system.

In addition to binary, two other number systems are widely used in conjunction with digital electronics: the octal (base-8) number system, and the hexadecimal (base-16) number system. These two number systems are used as a sort of shorthand for large binary numbers and enable us to represent a large group of binary digits with only a few octal or hexadecimal digits.

The last section in this chapter will cover many of the different binary codes, which use a combination of 1s and 0s to represent letters, numbers, symbols, and other information.

Keep in mind throughout this chapter that different number systems are just another way to count and, like any new process, you will become more proficient and more at ease with practice as we proceed through the chapters that follow.

2-1 THE DECIMAL NUMBER SYSTEM

The decimal system of counting and keeping track of items was first created by Hindu mathematicians in India in A.D. 400. Since it involved the use of fingers and thumbs, it was natural that this system would have ten digits. The system found its way to all of the Arab countries by A.D. 800, where it was named the Arabic number system, and from there it was eventually adopted by nearly all the European countries by A.D. 1200, where it was called the **decimal number system**.

The key feature that distinguishes one number system from another is the number system's **base** or **radix**. This base indicates the number of digits that will be used. The decimal number system for example, is a base-10 number system which means that it will use ten digits (0 through 9) to communicate information about an amount. A subscript is sometimes included after a number when different number systems are being used, to indicate the base of the number. For example, 12.567_{10} would be a base-10 number, whereas 1010_2 would be a base-2 number.

Decimal A base-10 number system.

Base With number systems, it describes the number of digits used.

SEC. 2-1 / THE DECIMAL NUMBER SYSTEM 21

Chapter opening vignettes, featuring electronic industry entrepreneurs, motivate students to read and understand chapter material, and conversational introductions review what has been previously covered and what is about to be covered.

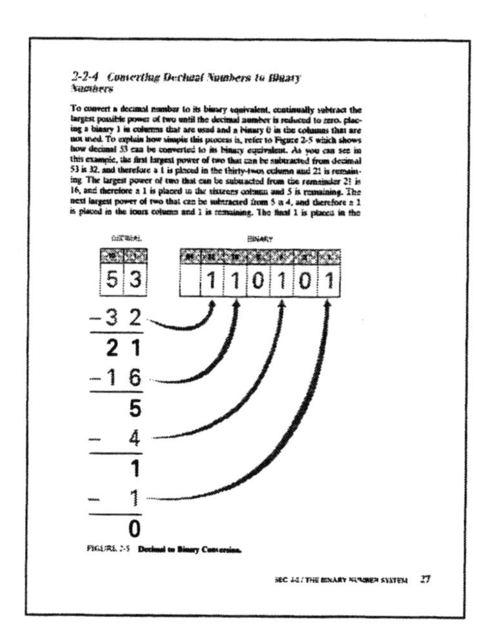

2-2-4 Converting Decimal Numbers to Binary Numbers

To convert a decimal number to its binary equivalent, continually subtract the largest possible power of two until the decimal number is reduced to zero, placing a binary 1 in columns that are used and a binary 0 in the columns that are not used. To explain how simple this process is, refer to Figure 2-5 which shows how decimal 53 can be converted to its binary equivalent. As you can see in this example, the first largest power of two that can be subtracted from decimal 53 is 32, and therefore a 1 is placed in the thirty-twos column and 21 is remaining. The largest power of two that can be subtracted from the remainder 21 is 16, and therefore a 1 is placed in the sixteens column and 5 is remaining. The next largest power of two that can be subtracted from 5 is 4, and therefore a 1 is placed in the fours column and 1 is remaining. The final 1 is placed in the

FIGURE 2-5 Decimal to Binary Conversion.

SEC 2-2 / THE BINARY NUMBER SYSTEM 27

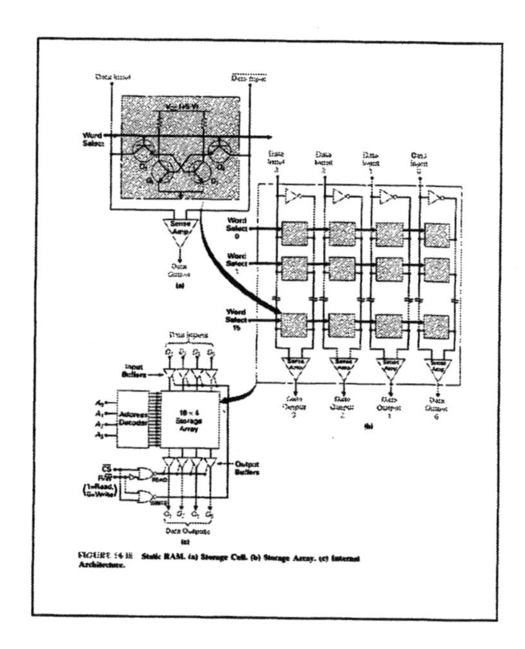

FIGURE 14-18 Static RAM. (a) Storage Cell. (b) Storage Array. (c) Internal Architecture.

A student-friendly writing style coupled with dynamic diagrams enables the student to comfortably master the material.

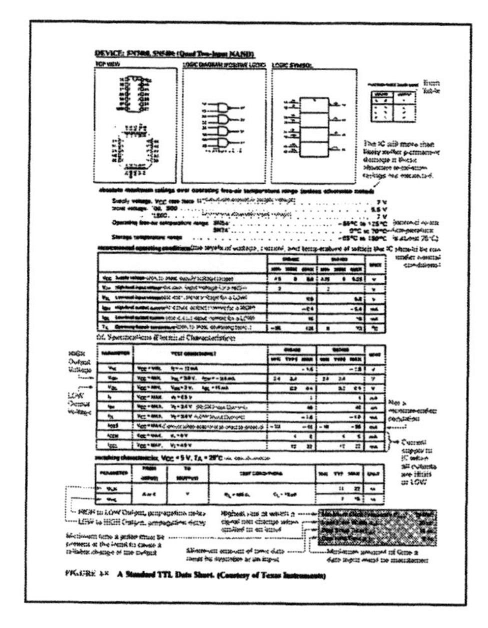

FIGURE [illegible] A Standard TTL Data Sheet. (Courtesy of Texas Instruments)

A large number of data sheets are integrated into their appropriate positions with highlighted annotations included to explain the meaning of key characteristics and symbols.

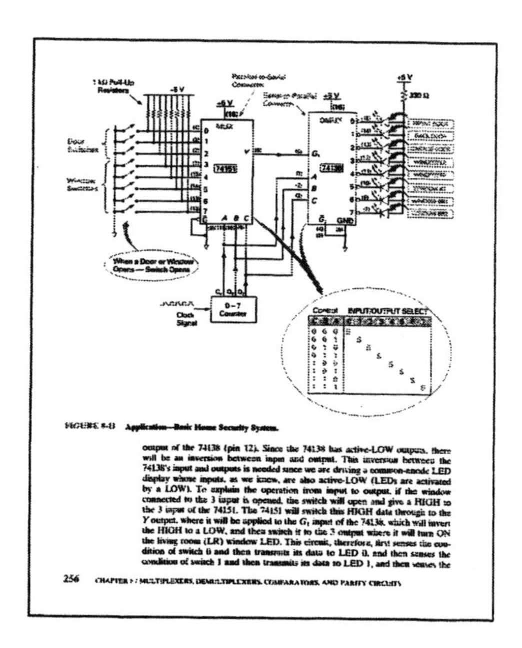

FIGURE 8-13 Application—Basic Home Security System.

output of the 74138 (pin 12). Since the 74138 has active-LOW outputs, there will be an inversion between input and output. This inversion between the 74138's input and outputs is needed since we are driving a common-anode LED display whose inputs, as we know, are also active-LOW (LEDs are activated by a LOW). To explain the operation from input to output, if the window connected to the 3 input is opened, the switch will open and give a HIGH to the 3 input of the 74151. The 74151 will switch this HIGH data through to the Y output, where it will be applied to the G_1 input of the 74138, which will invert the HIGH to a LOW, and then switch it to the 3 output where it will turn ON the living room (LR) window LED. This circuit, therefore, first senses the condition of switch 0 and then transmits its data to LED 0, and then senses the condition of switch 1 and then transmits its data to LED 1, and then senses the

256 CHAPTER 8 / MULTIPLEXERS, DEMULTIPLEXERS, COMPARATORS, AND PARITY CIRCUITS

Actual circuit examples are included to help the student connect a topic to a practical application.

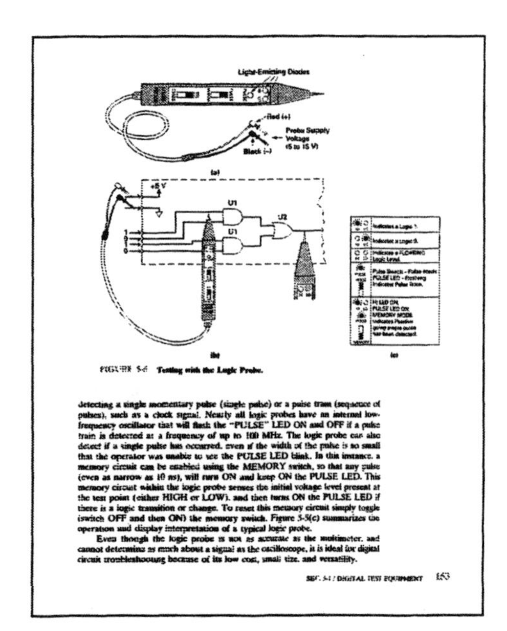

FIGURE 5-6 Testing with the Logic Probe.

detecting a single momentary pulse (single pulse) or a pulse train (sequence of pulses), such as a clock signal. Nearly all logic probes have an internal low-frequency oscillator that will flash the "PULSE" LED ON and OFF if a pulse train is detected at a frequency of up to 100 MHz. The logic probe can also detect if a single pulse has occurred, even if the width of the pulse is so small that the operator was unable to see the PULSE LED blink. In this instance, a memory circuit can be enabled using the MEMORY switch, so that any pulse (even as narrow as 10 ns), will turn ON and keep ON the PULSE LED. This memory circuit within the logic probe senses the initial voltage level present at the test point (either HIGH or LOW), and then turns ON the PULSE LED if there is a logic transition or change. To reset this memory circuit simply toggle (switch OFF and then ON) the memory switch. Figure 5-5(c) summarizes the operation and display interpretation of a typical logic probe.

Even though the logic probe is not as accurate as the multimeter, and cannot determine as much about a signal as the oscilloscope, it is ideal for digital circuit troubleshooting because of its low cost, small size, and versatility.

SEC. 5-1 / DIGITAL TEST EQUIPMENT 153

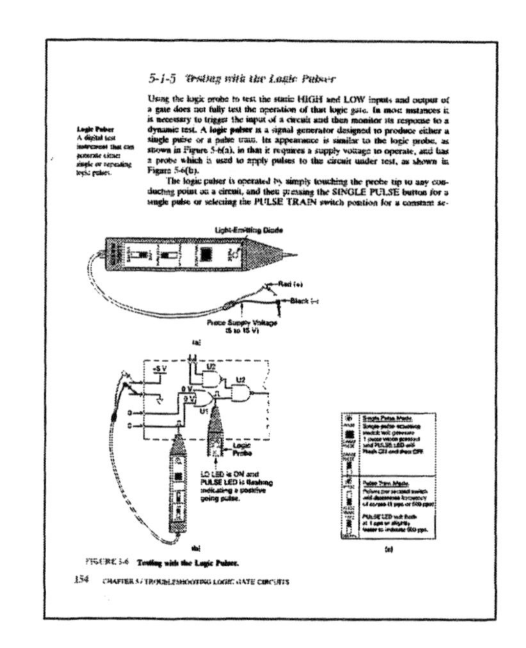
5-1-5 Testing with the Logic Pulser

Logic Pulser A digital test instrument that can generate either single or repeating logic pulses.

Using the logic probe to test the static HIGH and LOW inputs and output of a gate does not fully test the operation of that logic gate. In most instances it is necessary to trigger the input of a circuit and then monitor its response to a dynamic test. A **logic pulser** is a signal generator designed to produce either a single pulse or a pulse train. Its appearance is similar to the logic probe, as shown in Figure 5-6(a), in that it requires a supply voltage to operate, and has a probe which is used to apply pulses to the circuit under test, as shown in Figure 5-6(b).

The logic pulser is operated by simply touching the probe tip to any conducting point on a circuit, and then pressing the SINGLE PULSE button for a single pulse or selecting the PULSE TRAIN switch position for a constant se-

FIGURE 5-6 Testing with the Logic Pulser.

154 CHAPTER 5 / TROUBLESHOOTING LOGIC GATE CIRCUITS

A strong testing, test equipment, and troubleshooting emphasis prepares the technician student for the working world.

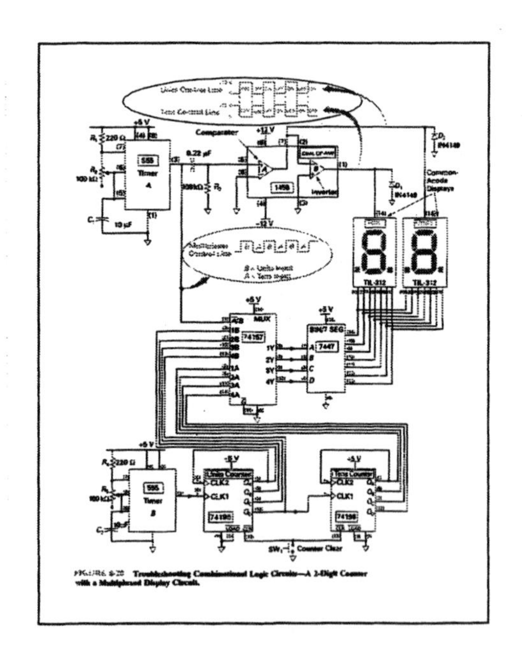

FIGURE 8-20 Troubleshooting Combinational Logic Circuits—A 2-Digit Counter with a Multiplexed Display Circuit.

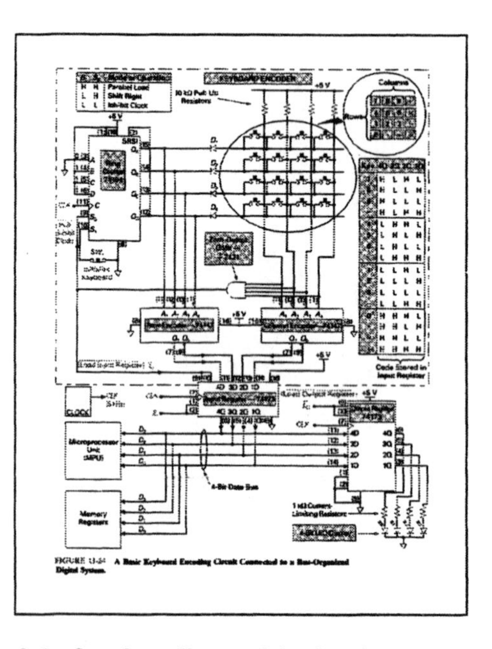

FIGURE 11-34 A Basic Keyboard Encoding Circuit Connected to a Bus-Organized Digital System.

An end-of-chapter application circuit combines all of the functions discussed in the chapter, so the student can see how all of the building blocks can be put to practical use.

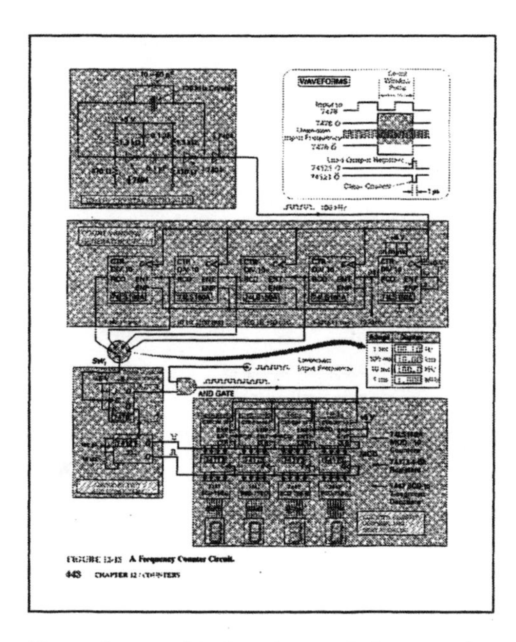

FIGURE 12-18 A Frequency Counter Circuit.

448 CHAPTER 12 / COUNTERS

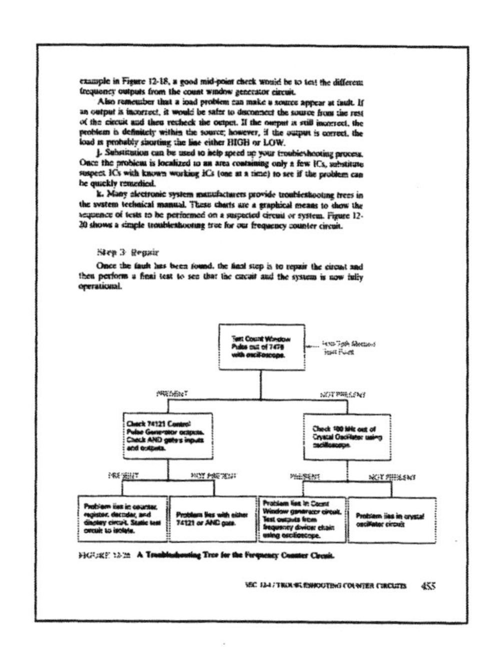

example in Figure 12-18, a good mid-point check would be to test the different frequency outputs from the count window generator circuit.

Also remember that a load problem can make a source appear at fault. If an output is incorrect, it would be safer to disconnect the source from the rest of the circuit and then recheck the output. If the output is still incorrect, the problem is definitely within the source; however, if the output is correct, the load is probably shorting the line either HIGH or LOW.

j. Substitution can be used to help speed up your troubleshooting process. Once the problem is localized to an area containing only a few ICs, substitute suspect ICs with known working ICs (one at a time) to see if the problem can be quickly remedied.

k. Many electronic system manufacturers provide troubleshooting trees in the system technical manual. These charts are a graphical means to show the sequence of tests to be performed on a suspected circuit or system. Figure 12-20 shows a simple troubleshooting tree for our frequency counter circuit.

Step 3: Repair

Once the fault has been found, the final step is to repair the circuit and then perform a final test to see that the circuit and the system is now fully operational.

FIGURE 12-20 A Troubleshooting Tree for the Frequency Counter Circuit.

SEC 12-4 / TROUBLESHOOTING COUNTER CIRCUITS 455

Extensive troubleshooting techniques and procedures are applied to all combined chapter application circuits, preparing the student for the working technician environment.

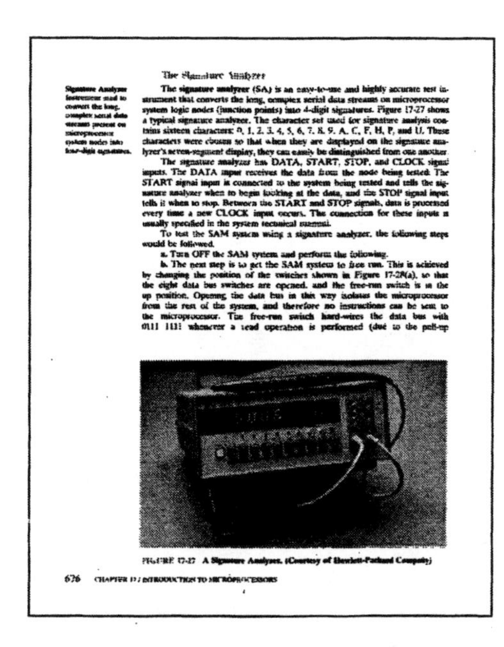

The Signature Analyzer

Signature Analyzer Instrument used to convert the long, complex serial data streams present on microprocessor system nodes into four-digit signatures.

The **signature analyzer (SA)** is an easy-to-use and highly accurate test instrument that converts the long, complex serial data streams on microprocessor system logic nodes (junction points) into 4-digit signatures. Figure 17-27 shows a typical signature analyzer. The character set used for signature analysis contains sixteen characters: 0, 1, 2, 3, 4, 5, 6, 7, 8, 9, A, C, F, H, P, and U. These characters were chosen so that when they are displayed on the signature analyzer's seven-segment display, they can easily be distinguished from one another.

The signature analyzer has DATA, START, STOP, and CLOCK signal inputs. The DATA input receives the data from the node being tested. The START signal input is connected to the system being tested and tells the signature analyzer when to begin looking at the data, and the STOP signal input tells it when to stop. Between the START and STOP signals, data is processed every time a new CLOCK input occurs. The connection for these inputs is usually specified in the system technical manual.

To test the SAM system using a signature analyzer, the following steps would be followed.

a. Turn OFF the SAM system and perform the following.

b. The next step is to get the SAM system to free run. This is achieved by changing the position of the switches shown in Figure 17-28(a), so that the eight data bus switches are opened, and the free-run switch is in the up position. Opening the data bus in this way isolates the microprocessor from the rest of the system, and therefore no instructions can be sent to the microprocessor. The free-run switch hard-wires the data bus with 0111 1111 whenever a read operation is performed (due to the pull-up

FIGURE 17-27 A Signature Analyzer. (Courtesy of Hewlett-Packard Company)

676 CHAPTER 17 / INTRODUCTION TO MICROPROCESSORS

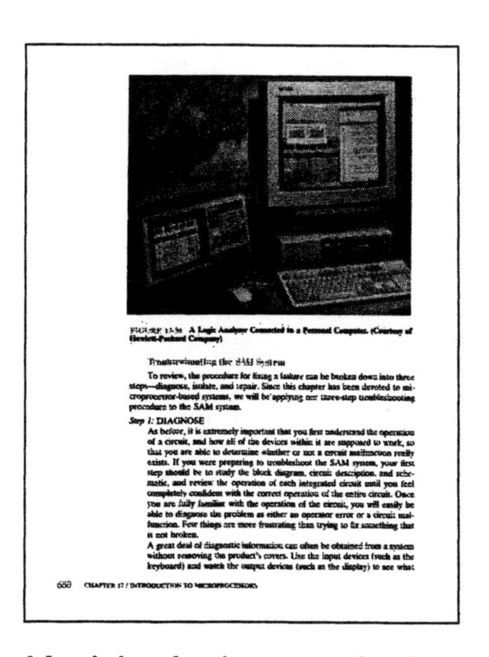

FIGURE 17-30 A Logic Analyzer Connected to a Personal Computer. (Courtesy of Hewlett-Packard Company)

Troubleshooting the SAM System

To review, the procedure for fixing a failure can be broken down into three steps—diagnose, isolate, and repair. Since this chapter has been devoted to microprocessor-based systems, we will be applying our three-step troubleshooting procedure to the SAM system.

Step 1: DIAGNOSE

As before, it is extremely important that you first understand the operation of a circuit, and how all of the devices within it are supposed to work, so that you are able to determine whether or not a circuit malfunction really exists. If you were preparing to troubleshoot the SAM system, your first step should be to study the block diagram, circuit description, and schematic, and review the operation of each integrated circuit until you feel completely confident with the correct operation of the entire circuit. Once you are fully familiar with the operation of the circuit, you will easily be able to diagnose the problem as either an operator error or a circuit malfunction. Few things are more frustrating than trying to fix something that is not broken.

A great deal of diagnostic information can often be obtained from a system without removing the product's covers. Use the input devices (such as the keyboard) and watch the output devices (such as the display) to see what

680 CHAPTER 17 / INTRODUCTION TO MICROPROCESSORS

The final chapter contains a detailed component-level description of a microprocessor-based system, including schematic diagram interpretation, troubleshooting procedures, and typical system faults.

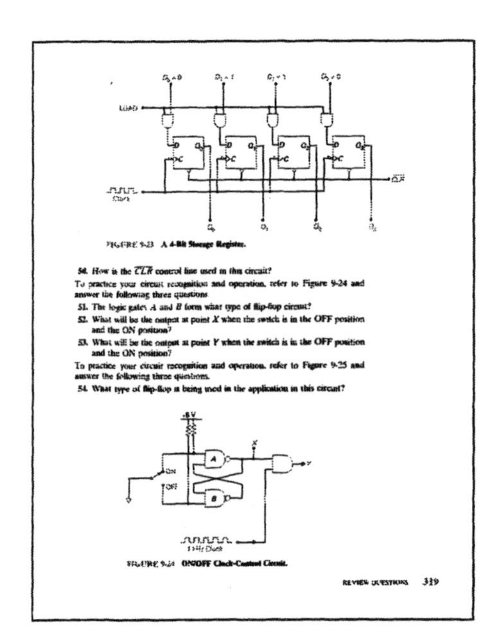
FIGURE 9-23 A 4-Bit Storage Register.

50. How is the $\overline{CLR}$ control line used in this circuit?

To practice your circuit recognition and operation, refer to Figure 9-24 and answer the following three questions.

51. The logic gates A and B form what type of flip-flop circuit?
52. What will be the output at point X when the switch is in the OFF position and the ON position?
53. What will be the output at point Y when the switch is in the OFF position and the ON position?

To practice your circuit recognition and operation, refer to Figure 9-25 and answer the following three questions.

54. What type of flip-flop is being used in the application in this circuit?

FIGURE 9-24 ON/OFF Clock-Control Circuit.

REVIEW QUESTIONS 319

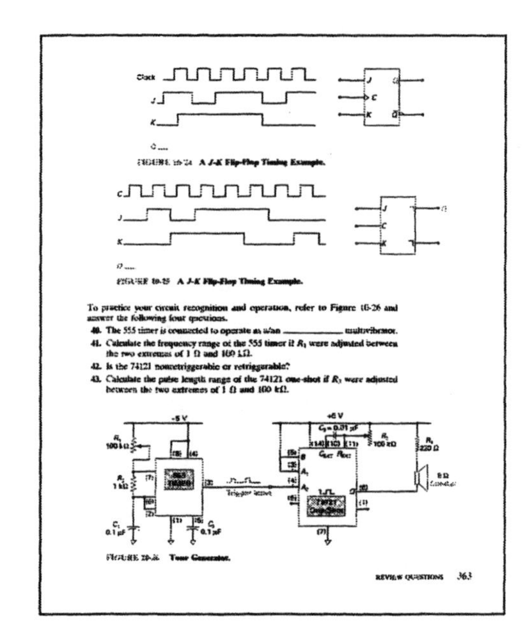
FIGURE 10-24 A J-K Flip-Flop Timing Example.

FIGURE 10-25 A J-K Flip-Flop Timing Example.

To practice your circuit recognition and operation, refer to Figure 10-26 and answer the following four questions.

40. The 555 timer is connected to operate as a/an ____________ multivibrator.
41. Calculate the frequency range of the 555 timer if R_1 were adjusted between the two extremes of 1 Ω and 100 kΩ.
42. Is the 74121 nonretriggerable or retriggerable?
43. Calculate the pulse length range of the 74121 one-shot if R_3 were adjusted between the two extremes of 1 Ω and 100 kΩ.

FIGURE 10-26 Tone Generator.

REVIEW QUESTIONS 363

An extensive end-of-chapter test bank tests the student's understanding with multiple-choice, essay, practice-problem, and troubleshooting questions.

ANCILLARIES ACCOMPANYING THIS TEXT

A comprehensive *lab manual,* co-authored by Nigel Cook and Gary Lancaster, contains numerous experiments that are designed to translate all of the textbook's theory into practical experimentation.

A free *electronics workbench data disk* containing all of the lab manual's computer simulation circuits.

A free *interactive student study disk* for Windows. This program will test a student's comprehension of each chapter, give the instructor a printed summary of each student's performance, and list the text sections that should be reviewed in order to improve student understanding.

A free *test item file manual* is available. This test bank includes additional questions that can be used for tests and examinations, or to give students additional practice problems. These questions are also available on a *test item file data disk* that will operate with the Prentice Hall custom test generator.

A free *instructor's reference manual with transparency masters* is available from your local Prentice Hall representative. This manual includes the answers to all the self-test review and end-of-chapter questions, quick-reference summary illustrations, and some additional resource information.

ACKNOWLEDGMENTS

I would first like to thank my wife, Dawn, my daughter, Candy, and my son, Jon, for their ever-enthusiastic love and support.

Secondly, a special thank-you goes to Steve Howe, who worked tirelessly to help ensure that this text had a high level of quality and accuracy. In addition,

I would like to thank Gary Lancaster for all his work on the lab manual that accompanies this text.

I would also like to thank Elene and Michael Rendler for the development of the student study disk.

A debt of thanks is extended also to my marketing director, Debbie Yarnell, for her special efforts with all my books; my production editors, Sheryl Langner and Martha Beyerlein, who always take a special interest in my projects; my supplements editor, Judy Casillo; and my editor, Linda Ludewig.

In addition, I would like to thank the following family and friends for their support: Frances Ford, Julia Cook, Gina and Bill Hoggetts, Frank and Audrey Ford, Brian and Liz Ford, Anita Gilmour, David and Vera Oliphant, and Brad and Rebecca Ross.

Last but not least, I would like to acknowledge all of the innovative, energetic, and motivational instructors who go out of their way to prove that mathematics, science, and technology are within everyone's grasp and can be made easy, interesting, and fun.

Nigel P. Cook

Contents

PART I
Digital Basics *2*

Chapter 15

Chapter 16

PART III

Chapter 17

Appendixes

PART I

Chapter 1

OBJECTIVES

After completing this chapter, you will be able to:

1. Define the differences between analog and digital signals, devices, and circuits.
2. Explain why the ten-state digital system was replaced by the two-state digital system.
3. Describe the differences between an analog and digital electrical circuit.
4. Explain why analog circuits are linear circuits.
5. Describe the evolution of digital electronics from its early beginnings to the present day.
6. Describe how the integrated circuit made the digital electronics revolution possible.

Electronics—Analog and Digital

OUTLINE

VIGNETTE

Moon Walk

In July of 1969 almost everyone throughout the world was caught at one stage or another gazing up to the stars and wondering. For American astronaut Neil Armstrong, the age-old human dream of walking on the moon was close to becoming a reality. On earth, millions of people and thousands of newspapers and magazines waited anxiously to celebrate a successful moon landing. Then the rather brief but eloquent message from Armstrong was transmitted 240,000 miles to Houston, Texas, where it was immediately retransmitted to a waiting world. This message was, "That's one small step for a man, one giant leap for mankind." These words were received by many via television, but for magazines and newspapers, this entire mission, including the speech from Armstrong, was converted into a special code made up of ON-OFF pulses that traveled from computer to computer. Every letter of every word was converted into a code which used the two symbols of the binary number system—zero and one. This code is still used extensively by all modern computers and is called the American Standard Code for Information Interchange, or ASCII (pronounced "askey").

It was only fitting that these codes made up of zeros and ones conveyed the finale to this historic mission since they played such an important role throughout. Commands encoded into zeros and ones were used to control almost everything, from triggering the takeoff to keeping the spacecraft at the proper angle for re-entry into the earth's atmosphere.

No matter what its size or application, a digital electronic computer is quite simply a system that manages the flow of information in the form of zeros (0) and ones (1). Referring to the ASCII code table in Figure 1-3 on p. 9, see if you can decode the following famous Armstrong message.

0100010	1010100	1101000	1100001	1110100	0100111	1110011	0100000	1101111	1101110
1100101	0100000	1110011	1101101	1100001	1101100	1101100	0100000	1110011	1110100
1100101	1110000	0100000	1100110	1101111	1110010	0100000	1100001	0100000	1101101
1100001	1101110	0101100	0100000	1101111	1101110	1100101	0100000	1100111	1101101
1100001	1101110	1110100	0100000	1101100	1100101	1100001	1110000	0100000	1100110
1101111	1110010	0100000	1101101	1100001	1101110	1101011	1101001	1101110	1100100
0101110	0100010	0100000	0100000	0101101	1001110	1100101	1101001	1101100	0100000
1000001	1110010	1101101	1110011	1110100	1110010	1101111	1101110	1100111	0101100
0100000	1000001	1110000	1101111	1101100	1101100	1101111	0100000	0110001	0110001

INTRODUCTION

At this point you have reached a milestone in your study of electricity and electronics. You have completed your courses in dc/ac and semiconductor electronics, and you are about to begin the next phase entitled "Introductory Digital Electronics."

Digital electronics is so widely used today that it is almost impossible to find an electronic system that does not make use of digital electronics. In fact, a recent statistic stated that, on average, 85 percent of the circuitry within electronic systems is digital and only 15 percent is analog. Future predictions indicate that more and more, electronic information will be "digitized" because information in digital form is easier to process and is more accurate because it is less sensitive to noise. It is easy, therefore, to see why this course in digital electronic concepts, terminology, components, circuits, applications, and troubleshooting is so vital to your understanding of electronics.

In this chapter, we will first define the differences between analog and digital, and then examine the applications and advantages of digital electronics.

1-1 REVIEW—ANALOG AND DIGITAL

Before we begin, let us review where we are and where we are going. Figure 1-1(a) illustrates the "tree of electricity and electronics," which is an easy way for us to connect all of the different pieces in the electronics puzzle. Working from the bottom up, you can see that everything rests on the four basic electrical roots; current, voltage, resistance, and power. Electrical and electronic components were developed to generate and control these four basic electrical phenomena. When a group of components or devices are interconnected, they form a circuit, and just as components are the building blocks for circuits, circuits are in turn the building blocks for electronic equipment or systems.

1-1-1 Your Course in Electronics

Figure 1-1(b) breaks up your electronics course into four basic steps, which correlate to the basic blocks shown in the tree. In your previous "Introductory DC/AC Electronics" course, you covered all of Step 1 (Basics of Electricity) and the first half of Step 2 (DC/AC Devices). In the "Introductory Semiconductor Electronics" course, you covered the second half of Step 2 (Semiconductor Devices) and the first half of Step 3 (Analog Circuits). In this "Introductory Digital Electronics" course, we will be completing the second half of Step 3 (Digital Circuits), which is shown shaded in Figure 1-1(b).

After completing this course, you should have a thorough knowledge of electrical and electronic devices and circuits (Steps 1, 2, and 3), and be ready to apply this understanding to communications, data processing, consumer, industrial, test and measurement, and biomedical system applications. These six different branches or classifications of electrical and electronic systems are shown at the top of the tree in Figure 1-1(a), and listed under Step 4 in Figure 1-1(b).

1-1-2 Analog and Digital

The two terms analog and digital need to be further explained since all of the electrical and electronic circuits listed in Step 3 of Figure 1-1(b) are classified in one of these two categories. To begin, let us examine analog signals, components, and circuits.

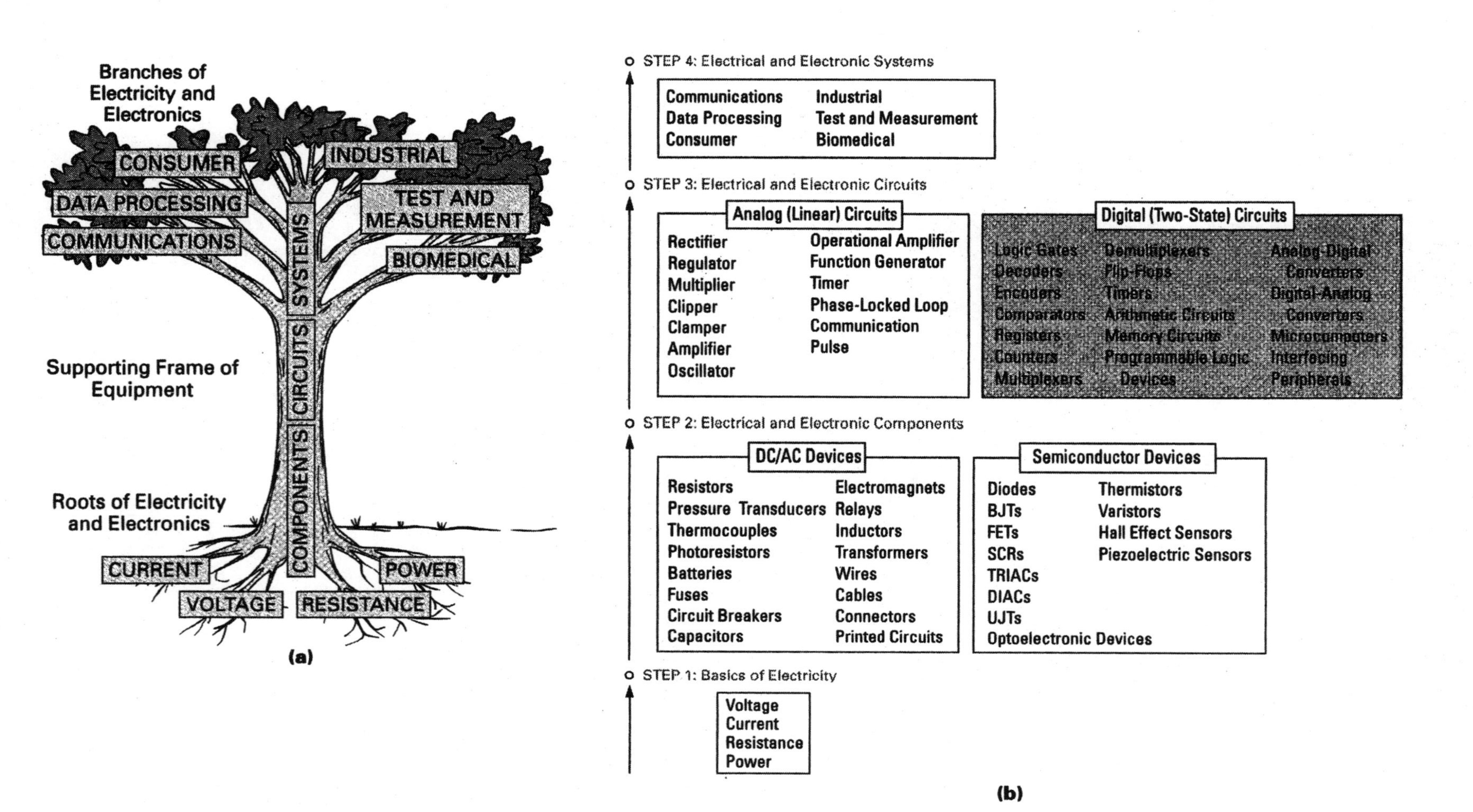

FIGURE 1-1 The Tree of Electricity and Electronics.

Analog Signals, Components, and Circuits

Figure 1-2(a) shows an electronic circuit designed to amplify speech information detected by a microphone. One of the easiest ways to represent information is to have a voltage change in direct proportion to the information it is representing. In the example in Figure 1-2(a), the *pitch* and *loudness* of the sound waves applied to the microphone should control the *frequency* and *amplitude* of the voltage signal from the microphone. The output voltage signal from the microphone is said to be an analog of the input speech signal. The word **analog** means "similar to," and in Figure 1-2(a) the electronic signal produced by the microphone is an analog (or is similar) to the speech signal, since a change in speech "loudness or pitch" will cause a corresponding change in signal voltage "amplitude or frequency."

In Figure 1-2(b), a light detector or solar cell converts light energy into an electronic signal. This information signal represents the amount of light present since changes in voltage amplitude result in a change in light-level intensity. Once again, the output electronic signal is an analog (or is similar) to the sensed light input.

Figure 1-2, therefore, indicates two analog electronic (information) circuits. The microphone in Figure 1-2(a) generates an **AC analog signal** that is then amplified by an **AC amplifier circuit.** The microphone is considered an analog device or component and the amplifier an analog circuit. The light detector would also be an analog component; however, in this example it generates a **DC analog signal** that is then amplified by a **DC amplifier circuit.** Both of the

Analog
The representation of physical properties by a proportionally varying signal.

AC Analog Signal
An analog signal that alternates positive and negative.

AC Amplifier
An amplifier designed to increase the magnitude of an AC signal.

DC Analog Signal
An analog signal that is always either positive or negative.

DC Amplifier
An amplifier designed to increase the magnitude of a DC signal.

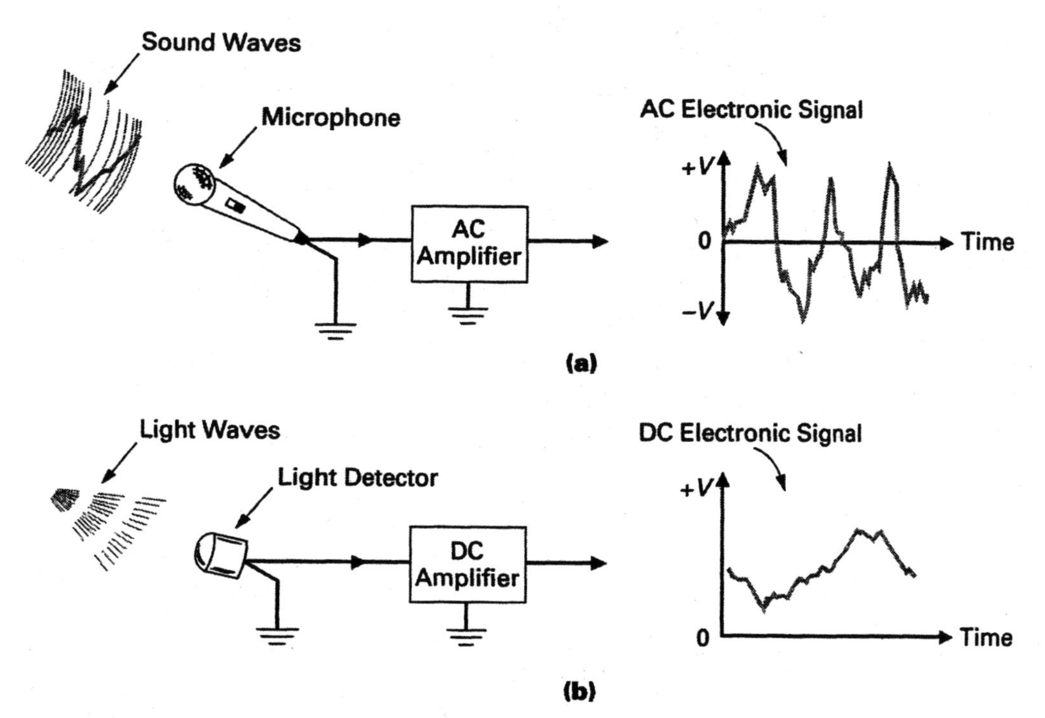

FIGURE 1-2 **Analog Electronic Signals, Components, and Circuits. (a) AC Information Signal. (b) DC Information Signal.**

information signals in Figure 1-2 vary smoothly and continuously in accordance with the natural properties they represent (sound and light).

Digital Signals, Components, and Circuits

Digital Electronic Circuit
A circuit designed to manage digital information signals.

With **digital electronic circuits,** information is first converted into a group of pulses, as can be seen in the example shown in Figure 1-3(a). This code consists of a series of HIGH and LOW voltages in which the HIGH voltages are called "1s" (ones) and the LOW voltages are called "0s" (zeros). Figure 1-3(b) lists the American Standard Code for Information Interchange (abbreviated ASCII, pronounced "askey"), which is one example of a digital code. Referring to Figure 1-3(a), you will notice that the "1101001" information, or data stream code, corresponds to the lower case "i" in the ASCII table shown highlighted in Figure 1-3(b). Computer keyboards are one of many devices that make use of the digital ASCII code. In Figure 1-3(c), you can see how the lower case "i" ASCII code is generated whenever the "i" key is pressed, encoding the information "i" into a group of pulses (1101001).

The next question you may have is: Why do we go to all of this trouble to encode all of our data or information into these two-state codes? The answer can best be explained by examining history. The early digital systems constructed in the 1950s made use of a decimal code that used ten levels or voltages, with each of these voltages corresponding to one of the ten digits in the decimal number system (0 = 0 V, 1 = 1 V, 2 = 2 V, 3 = 3 V, up to 9 = 9 V). The circuits that had to manage these decimal codes, however, were very complex since they had to generate one of ten voltages and sense the difference between all ten voltage levels. This complexity led to inaccuracy since some circuits would periodically confuse one voltage level for a different voltage level. *The solution to the problem of circuit complexity and inaccuracy was solved by adopting a two-state system instead of a ten-state system.*

Using a two-state or two-digit system, you can generate codes for any number, letter, or symbol, as we have seen in the ASCII table in Figure 1-3. The electronic circuits that manage these two-state codes are less complex since they only have to generate and sense either a HIGH or LOW voltage. In addition, two-state circuits are much more accurate since there is little room for error between the two extremes of ON and OFF or HIGH voltage and LOW voltage.

Abandoning the ten-state system and adopting the two-state system for the advantages of circuit simplicity and accuracy meant that we were no longer dealing with the decimal number system. As a result, having only two digits (0 and 1) means that we are now operating in the two-state number system, which is called **binary.**

Binary
Having only two alternatives, two-state.

Figure 1-4(a) shows how the familiar decimal system has ten digits or levels labeled 0 through 9, and Figure 1-4(b) shows how we could electronically represent each decimal digit. With the base-2, or binary number system, we only have two digits in the number scale. Therefore, only the digits 0 (zero) and 1 (one) exist in binary, as shown in Figure 1-4(c). These two states are typically represented in an electronic circuit as two different values of voltage (binary 0 = LOW voltage, binary 1 = HIGH voltage), as shown in Figure 1-4(d). Using combinations of **binary digits** (abbreviated as bits) we can represent information

Binary Digit
Abbreviated bit; it is either a 0 or 1 from the binary number system.

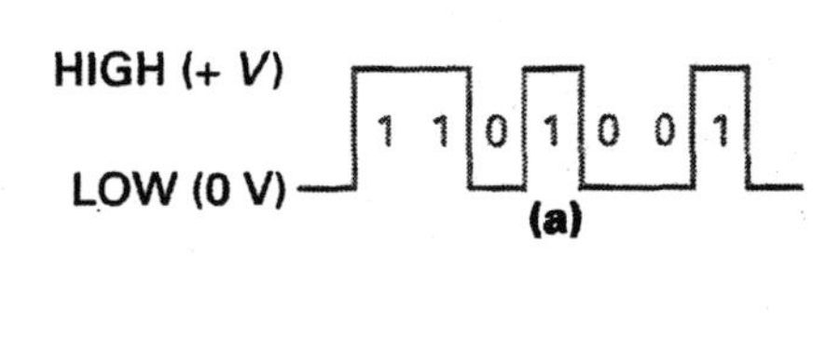

The American Standard Code for Information Interchange (ASCII)

Character	Code	Character	Code
SP	0100000	P	1010000
!	0100001	Q	1010001
"	0100010	R	1010010
#	0100011	S	1010011
$	0100100	T	1010100
%	0100101	U	1010101
&	0100110	V	1010110
'	0100111	W	1010111
(	0101000	X	1011000
)	0101001	Y	1011001
*	0101010	Z	1011010
+	0101011	[	1011011
,	0101100	/	1011100
–	0101101	]	1011101
.	0101110	^	1011110
/	0101111	_	1011111
0	0110000	'	1100000
1	0110001	a	1100001
2	0110010	b	1100010
3	0110011	c	1100011
4	0110100	d	1100100
5	0110101	e	1100101
6	0110110	f	1100110
7	0110111	g	1100111
8	0111000	h	1101000
9	0111001	i	1101001
:	0111010	j	1101010
;	0111011	k	1101011
<	0111100	l	1101100
=	0111101	m	1101101
>	0111110	n	1101110
?	0111111	o	1101111
@	1000000	p	1110000
A	1000001	q	1110001
B	1000010	r	1110010
C	1000011	s	1110011
D	1000100	t	1110100
E	1000101	u	1110101
F	1000110	v	1110110
G	1000111	w	1110111
H	1001000	x	1111000
I	1001001	y	1111001
J	1001010	z	1111010
K	1001011	(	1111011
L	1001100	⁝	1111100
M	1001101	)	1111101
N	1001110	~	1111110
O	1001111	DEL	1111111

(b)

(c)

FIGURE 1-3 Two-State Digital Information. (a) Information or Data Code for "i." (b) ASCII Code. (c) Keyboard ASCII Generator.

Note: See Figure 2-20 for a complete ASCII listing.

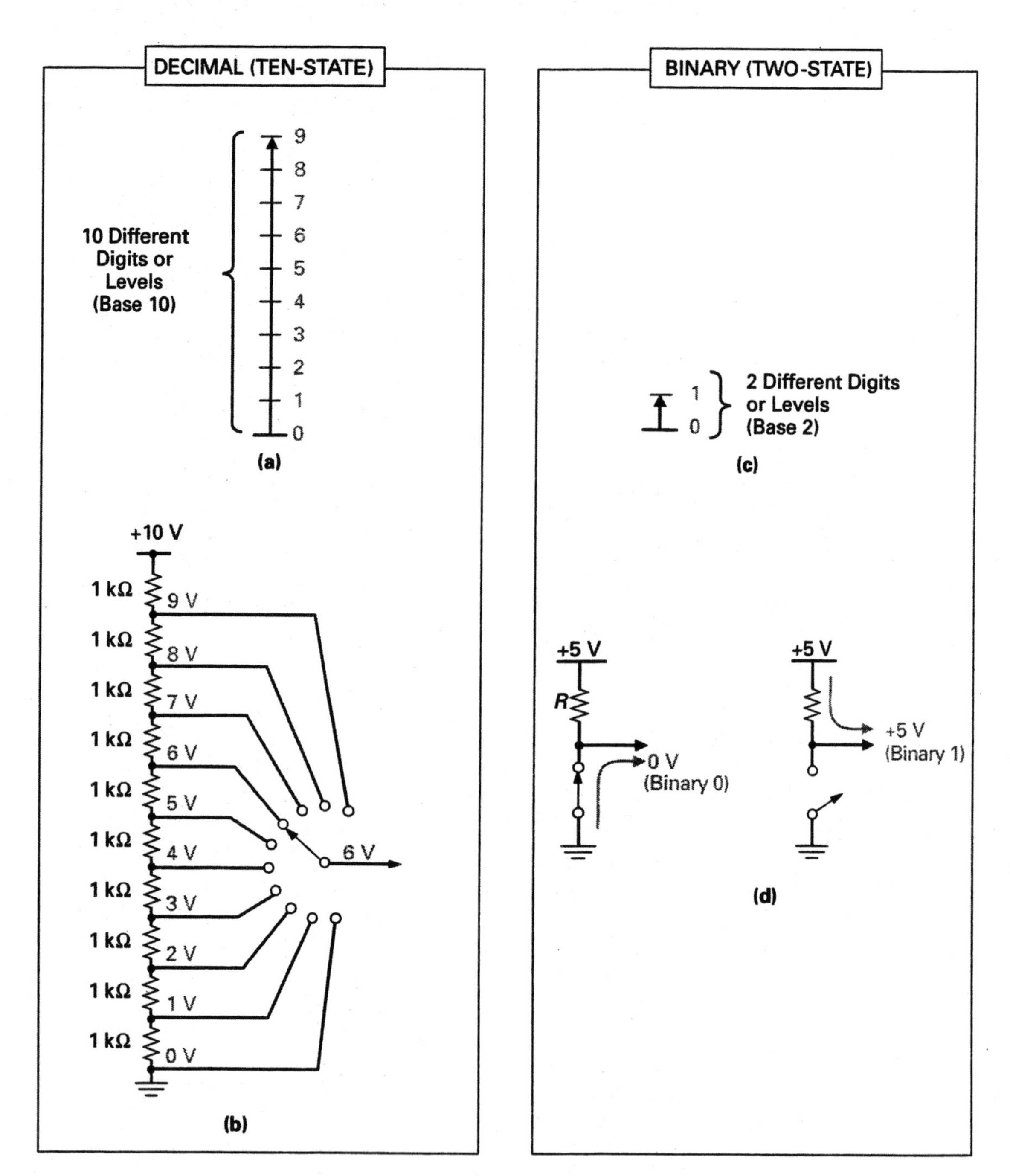

FIGURE 1-4 Electronically Representing the Digits of a Number System.

as a binary code. This code is called a **digital signal** because it is an *information signal* that makes use of *binary digits*. Today, almost all information, from your voice telephone conversations to the music on your compact disks, is **digitized** or converted to binary data form.

Digital Signal
An electronic signal made up of binary digits.

Digitize
To convert an analog signal to a digital signal.

The Analog and Digital Electrical Circuit

To further define the difference between analog and digital terms and devices, let us examine the difference in relation to an electrical circuit. To review, both electrical and electronic components, circuits, and systems control electron flow; however, their applications are distinctly different. The key difference is that ELECTRICAL components, circuits, and systems manage the flow of POWER, while ELECTRONIC components, circuits, and systems manage the flow of INFORMATION.

Figure 1-5(a) shows a simple example of an **analog electrical (power) circuit.** The amount of current through the light bulb in Figure 1-5(a), and therefore the lamp's brightness, can be varied by the dimmer. The brightness of the bulb is, therefore, an analog (or is similar) to the angular position of the dimmer control. When used in this way, the light bulb and dimmer are called analog components, and this analog electrical circuit can be used to produce an infinite number of brightness levels.

Analog Electrical Circuit
An analog circuit designed to manage power.

In contrast, Figure 1-5(b) shows a simple **digital electrical (power) circuit.** In this two-state power control circuit, the switch will turn the light bulb ON or OFF and generate one of the two states: switch open = light bulb OFF (binary 0), or switch closed = light bulb ON (binary 1).

Digital Electrical Circuit
A digital circuit designed to manage power.

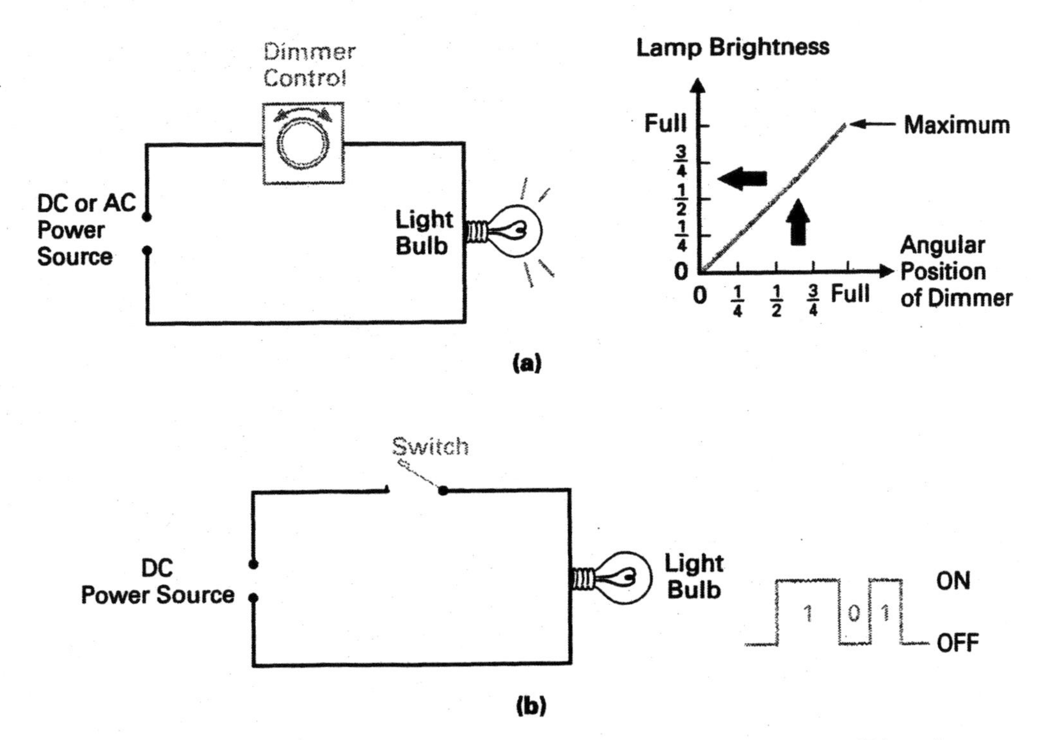

FIGURE 1-5 Analog and Digital Electrical Circuit. (a) Analog (Linear). (b) Digital (Two-State).

Analog Circuits Are Linear

Linear Circuit
A circuit in which the output varies in direct proportion to the input.

Analog circuits are often called **linear circuits** since linear by definition occurs *when an output varies in direct proportion to the input.* This linear circuit response is evident in both the analog electronic and electrical circuits shown in Figures 1-2 and 1-5(a). For example, referring to the electronic circuits in Figure 1-2, you can see that in both the dc and ac circuits the output signal voltage varies in direct proportion to the sound or light signal input. In relation to the electrical circuit in Figure 1-5(a), you can see by looking at the graph that the bulb's brightness (output) varies in direct proportion to the angular position of the dimmer (input).

Analog and Digital Readouts

As another example, let us consider the analog and digital readouts shown in Figure 1-6. With the analog readout watch shown in Figure 1-6(a), the movement of the hands of the watch on the calibrated scale is an analog (or is similar) to the amount of time that has passed. In contrast, the digital readout watch shown in Figure 1-6(b) indicates the passage of time using decimal digits.

Similarly, the amount of the pointer deflection across the scale of the analog readout multimeter shown in Figure 1-6(c) is an analog (or is similar) to the magnitude of the electrical property being measured, whereas the digital readout multimeter shown in Figure 1-6(d) indicates the magnitude of the electrical property being measured using decimal digits.

SELF-TEST REVIEW QUESTIONS FOR SECTION 1-1

1. If a temperature sensor generates a dc voltage that can be anywhere between 0 V and +15 V depending on the ambient heat, this signal would be considered a/an
   a. analog electronic signal.
   b. digital electronic signal.
   c. analog electrical signal.
   d. digital electrical signal.
2. A house thermostat basically contains two temperature-controlled switches that either turn on HEAT or COOL depending on the temperature settings of both and also the ambient temperature. The ON/OFF information from these temperature-controlled switches is considered a/an
   a. analog electronic signal.
   b. digital electronic signal.
   c. analog electrical signal.
   d. digital electrical signal.
3. Why are analog circuits also referred to as linear circuits?
4. Which types of displays need an operator to interpret the pointer's position on a calibrated scale?

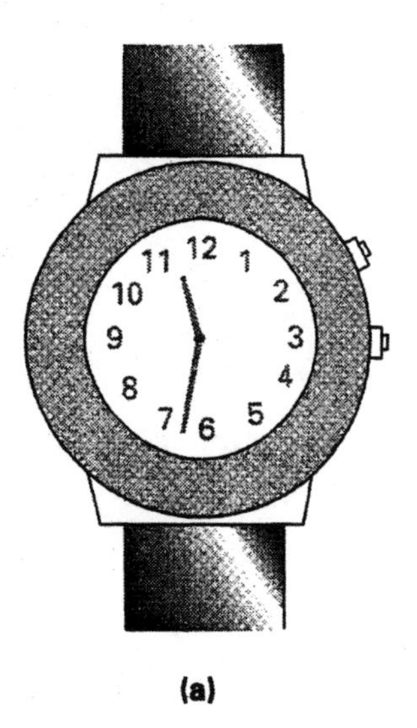

(a)

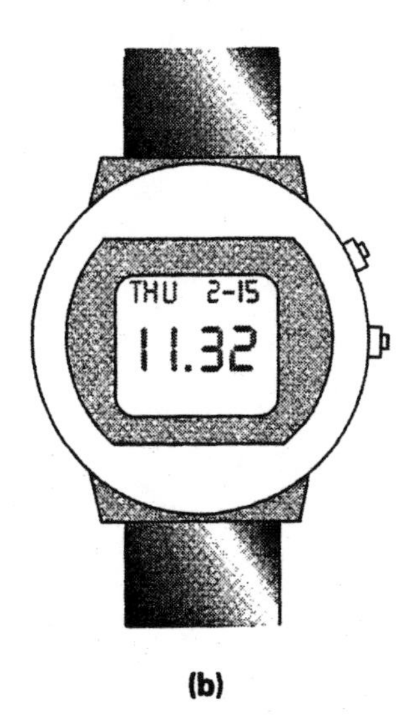

(b)

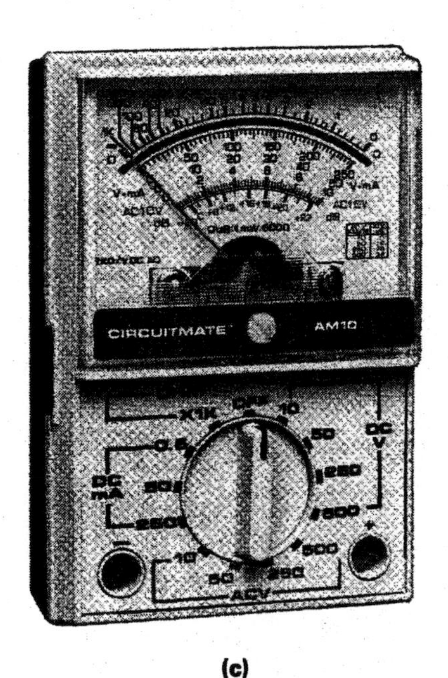

(c)

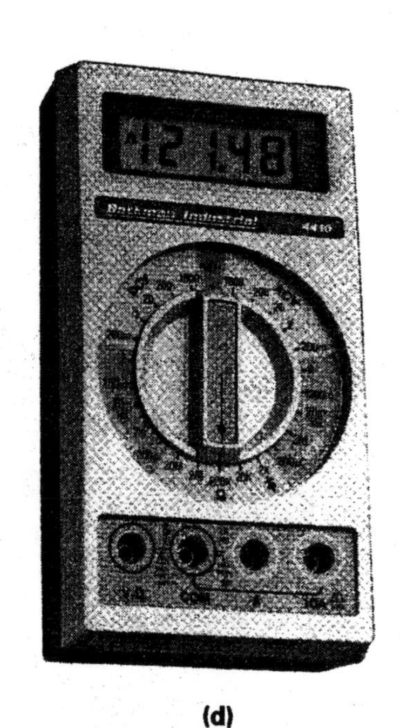

(d)

FIGURE 1-6 Analog and Digital Readouts.

1-2 DIGITAL ELECTRONICS

Integrated Circuit (IC)
A circuit in which all of the components are all interconnected and then encapsulated in a single package.

It was the **integrated circuit (IC),** like the one shown in Figure 1-7(a), that led to the digital electronics revolution. Before the invention of the IC in 1959, each component of a digital electronic circuit had to be manufactured separately and then wired together. With the IC, many components are constructed and interconnected on a single piece or "chip" of semiconductor material. Throughout

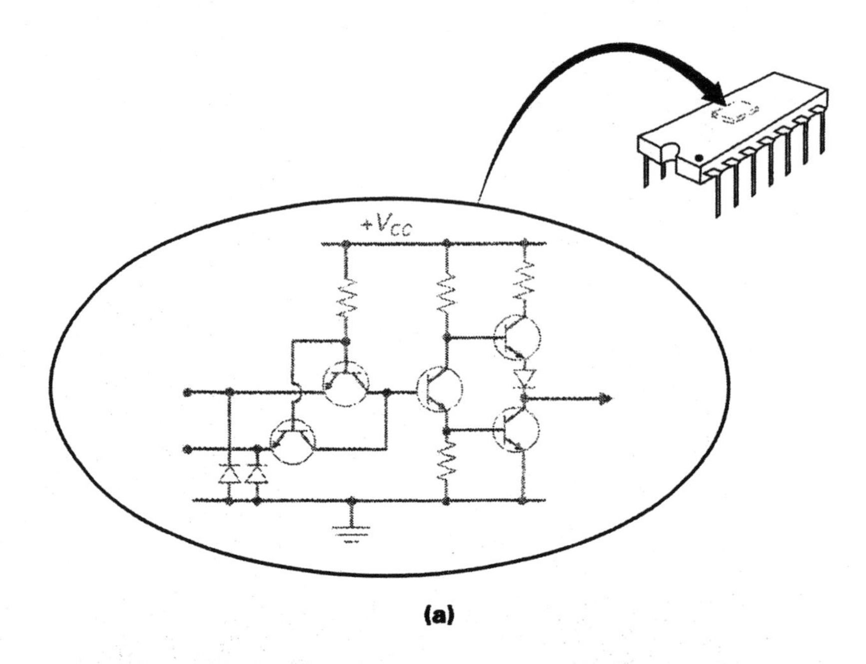

(a)

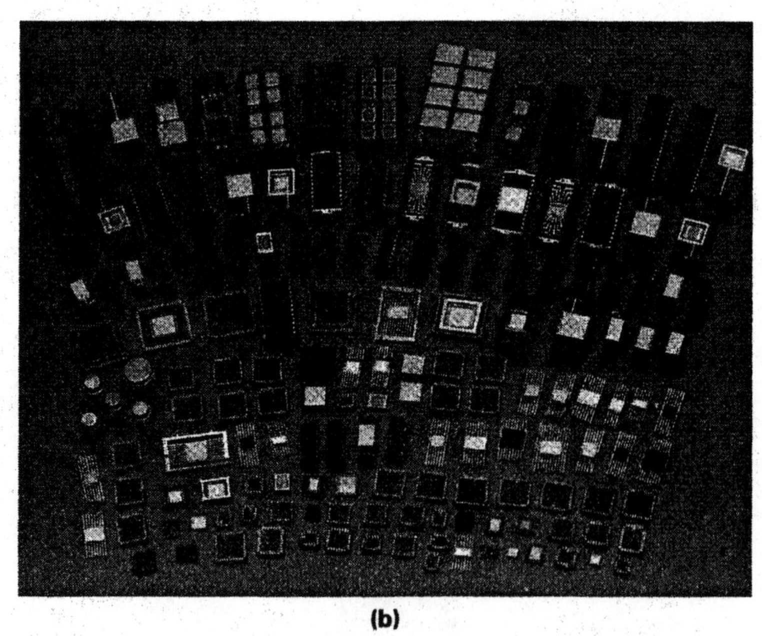

(b)

FIGURE 1-7 The Digital Integrated Circuit. (Photo courtesy of Harris Semiconductor.)

the 1960s, as the size of each component on the semiconductor chip shrank due to advances in miniaturization techniques, the incorporated components within the IC increased at an astounding rate. For example, in 1964 a chip measuring a tenth of an inch square contained ten transistors and their associated components. By 1970, almost a thousand components were integrated onto the same-sized chip, and the IC was available at the same cost. ICs saved space, weighed less, cost less, consumed less power, operated faster, and were more reliable than hard-wired discrete circuits. Slowly, ICs began to replace circuits containing discrete transistor circuits. The expression "yesterday's circuit is today's component" described the revolution that was occurring both then and now. For example, only a few years ago the average cellular telephone contained about four hundred components. Now, all of the discrete and integrated components have been replaced by only four ICs, and in the very near future these four ICs will be replaced by a single IC operating the entire phone. Figure 1-7(b) shows a variety of integrated circuit types.

In 1971 the incredible shrinkage of components had enabled engineers to fabricate a computer's central processing unit (CPU) on a single IC. This opened up a whole new arena since these units could process and store information for almost any task imaginable, from personal computers, video games, and home appliances to controlling robots on an assembly line. Initially, the four main digital circuits of a computer (CPU, memory, control, and input/output) were available as four separate ICs. However, as microelectronics techniques advanced, these four units were eventually all incorporated into a single microcomputer IC. In some applications such as home appliances, wrist watches, and cameras, for example, a single microcomputer IC is all that is needed, as shown in the example in Figure 1-8. These microcomputers are "specialists," with each one programmed to carry out a limited set of tasks. The personal computer, on the other hand, like the one shown in Figure 1-9, is a "generalist" that can switch from simulating the flight of an aircraft to rearranging a business report.

Digital electronic circuits are used in almost every electronic system. Their influence has greatly improved the ability of electronic equipment to perform its primary function, which is to manage the flow of information. In fact, there are probably very few people today who have not been affected in some way by the biggest application of digital electronics—the computer—or, as it is more

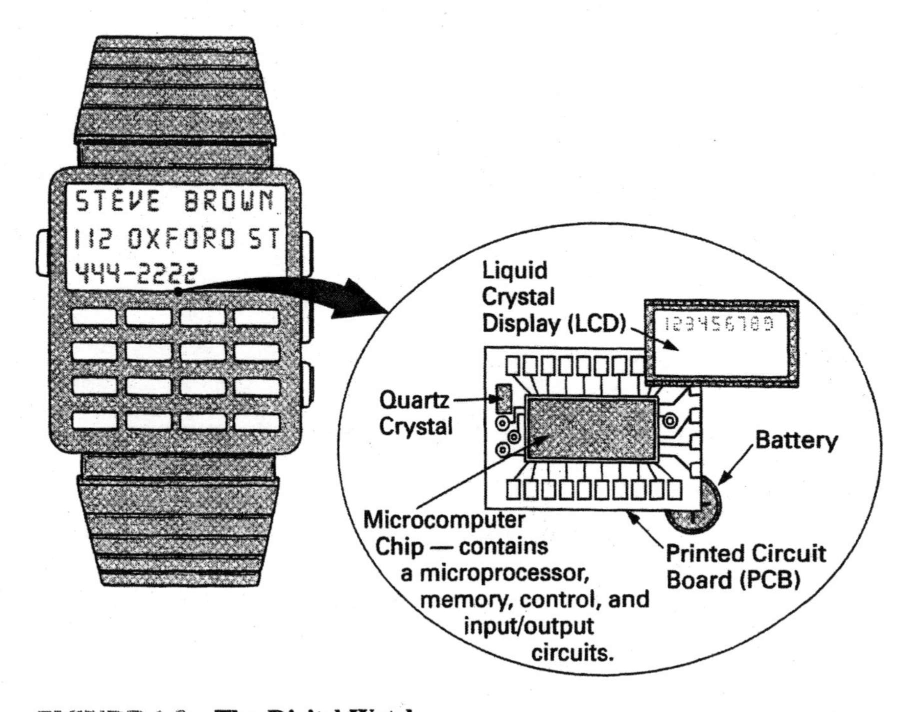

FIGURE 1-8 **The Digital Watch.**

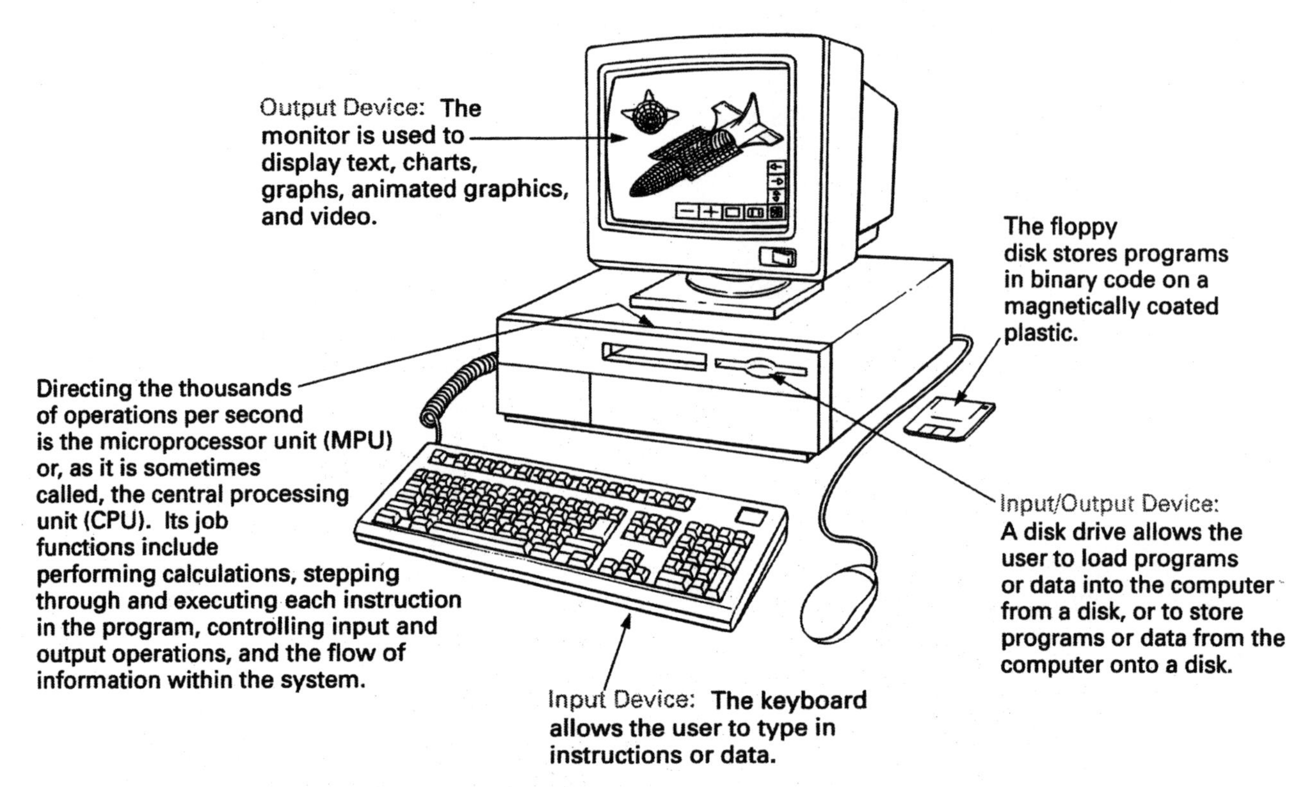

FIGURE 1-9 The Personal Computer or PC.

accurately named, the "data processor." In the following chapters, we will examine in detail digital electronic concepts, terminology, components, circuits, applications, and troubleshooting.

SELF-TEST REVIEW QUESTIONS FOR SECTION 1-2

1. Which of the following digital electronic systems is usually used to "compute" or perform mathematical calculations?
   a. Calculator **b.** Computer **c.** Amplifier **d.** Op-amp
2. Which of the following digital electronic systems is usually used to "process data"?
   a. Calculator **b.** Computer **c.** Amplifier **d.** Op-amp

REVIEW QUESTIONS

Multiple-Choice Questions

1. ____________ components, circuits, and systems manage the flow of POWER.
   a. Electronic **b.** Electrical **c.** Digital **d.** Analog
2. ____________ components, circuits, and systems manage the flow of INFORMATION.
   a. Electronic **b.** Electrical **c.** Digital **d.** Analog

3. ____________ devices perform such functions as generating, sensing, storing, retrieving, amplifying, transmitting, receiving, and displaying information.

   a. Electronic **b.** Electrical **c.** Digital **d.** Analog

4. ____________ devices perform such functions as generating, distributing, and converting electrical power.

   a. Electronic **b.** Electrical **c.** Digital **d.** Analog

5. Which of the following is an example of a digital electronic circuit?

   a. Amplifier **b.** CPU **c.** Power supply **d.** ON/OFF lighting circuit

6. Which of the following is an example of an analog electrical circuit?

   a. Amplifier **b.** CPU **c.** Power supply **d.** ON/OFF lighting circuit

7. Since digital circuits are based on the binary number system, they are often referred to as ____________ circuits. On the other hand, the output of analog circuits varies in direct proportion to the input, which is why analog circuits are often called ____________ circuits.

   a. ten-state, linear **b.** linear, two-state
   c. digital, ten-state **d.** two-state, linear

8. A digital multimeter displays the measured quantity using ____________.

   a. binary digits **b.** a pointer and scale
   c. decimal digits **d.** an abacus

9. Digital electronic systems use the two-state system instead of the ten-state system for the following reason(s):

   a. Circuit simplicity **b.** Accuracy
   c. Both a. and b. **d.** None of the above

10. The increased use of digital electronics resulted primarily due to the development of the ____________.

    a. binary number system **b.** operational amplifier
    c. integrated circuit **d.** none of the above

Essay Questions

11. Define the following: (1-1-2)

    a. Analog electronic circuit **b.** Analog electrical circuit
    c. Digital electronic circuit **d.** Digital electrical circuit

12. Why do digital systems have information coded into binary codes instead of decimal codes? (1-1-2)

13. Why are analog circuits also called linear circuits? (1-1-2)

14. Describe the differences between an analog and digital readout display. (1-1-2)

15. What device was primarily responsible for the increased use of digital electronics? (1-2)

Chapter 2

OBJECTIVES

After completing this chapter, you will be able to:

1. Explain how different number systems operate.
2. Describe the differences between the decimal and binary number systems.
3. Explain the positional weight and the reset and carry action for decimal.
4. Explain the positional weight, reset and carry action and conversion for binary.
5. Explain the positional weight, reset and carry action and conversion for hexadecimal.
6. Explain the positional weight, reset and carry action and conversion for octal.
7. Describe the following binary codes: binary coded decimal (BCD), the excess-3 code, the gray code, and the American Standard Code for Information Interchange (ASCII).

Number Systems and Codes

OUTLINE

VIGNETTE

Leibniz's Language of Logic

Gottfried Wilhelm Leibniz was born in Leipzig, Germany, in 1646. His father was a professor of moral philosophy and spent much of his time discussing thoughts and ideas with his son. Tragically, his father died when he was only six, and from that time on, Leibniz spent hour upon hour in his late father's library reading through all of his books.

By the age of twelve, he had taught himself history, Latin, and Greek. At fifteen he entered the University of Leipzig, where he came across the works of scholars such as Johannes Kepler and Galileo. The new frontiers of science fascinated him, so he added mathematics to his curriculum.

In 1666, while finishing his university studies, the twenty-year-old Leibniz wrote what he modestly called a schoolboy essay, titled "DeArte Combinatoria," which translated means "on the Art of Combination." In this work, he described how all thinking of any sort on any subject could be reduced to exact mathematical statements. Logic, or as he called it, the laws of thought, could be converted from the verbal realm, which is full of ambiguities, into precise mathematical statements. In order to achieve this, however, Leibniz stated that a "universal language" would be needed. Most of his professors found the paper either baffling or outrageous and this caused Leibniz to not pursue the idea any further.

After graduating from the university, Leibniz was offered a professorship at the University of Nuremberg, which he turned down for a position as an international diplomat. This career proved not as glamorous as he imagined, since most of his time was spent in uncomfortable horse-drawn coaches traveling between all of the European capitals.

In 1672 his duties took him to Paris, where he met Dutch mathematician and astronomer Christian Huygens. After seeing the hours that Huygens spent on endless computations, Leibniz set out to develop a mechanical calculator. A year later Leibniz unveiled the first machine that could add, subtract, multiply, and divide decimal numbers.

In 1676 Leibniz began to concentrate more on mathematics. It was at this time that he invented calculus, which was also independently discovered by Isaac Newton in England. Leibniz's focus, however, was on the binary number system, which preoccupied him for years. He worked tirelessly to document the long combinations of 1s and 0s that make up the modern binary number system, and perfected binary arithmetic. Ironically, for all his genius, Leibniz failed to make the connection between his 1666 essay and binary, which was the universal language of logic that he was seeking.

This discovery would remain hidden until another self-taught mathematician named George Boole uncovered it a century and a quarter after Leibniz's death in 1716.

INTRODUCTION

One of the best ways to understand anything new is to compare it to something that you are already familiar with so that the differences are highlighted. In this chapter we will be examining the *binary number system,* which is the language used by digital electronic circuits. To best understand this new number system, we will compare it to the number system most familiar to you, the *decimal* (base-10) *number system.* Although this system is universally used to represent quantities, many people do not fully understand its weighted structure. This review of decimal has been included so that you can compare the base-10 number system to the system used internally by digital electronics, the binary (base-2) number system.

In addition to binary, two other number systems are widely used in conjunction with digital electronics: the octal (base-8) number system, and the hexadecimal (base-16) number system. These two number systems are used as a sort of shorthand for large binary numbers and enable us to represent a large group of binary digits with only a few octal or hexadecimal digits.

The last section in this chapter will cover many of the different binary codes, which use a combination of 1s and 0s to represent letters, numbers, symbols, and other information.

Keep in mind throughout this chapter that different number systems are just another way to count and, like any new process, you will become more proficient and more at ease with practice as we proceed through the chapters that follow.

2-1 THE DECIMAL NUMBER SYSTEM

The decimal system of counting and keeping track of items was first created by Hindu mathematicians in India in A.D. 400. Since it involved the use of fingers and thumbs, it was natural that this system would have ten digits. The system found its way to all of the Arab countries by A.D. 800, where it was named the Arabic number system, and from there it was eventually adopted by nearly all the European countries by A.D. 1200, where it was called the **decimal number system.**

Decimal
A base-10 number system.

The key feature that distinguishes one number system from another is the number system's **base** or **radix.** This base indicates the number of digits that will be used. The decimal number system for example, is a base-10 number system which means that it will use ten digits (0 through 9) to communicate information about an amount. A subscript is sometimes included after a number when different number systems are being used, to indicate the base of the number. For example, $12{,}567_{10}$ would be a base-10 number, whereas 10110_2 would be a base-2 number.

Base
With number systems, it describes the number of digits used.

2-1-1 *Positional Weight*

The position of each digit of a decimal number determines the weight of that digit. A 1 by itself, for instance, is only worth 1, whereas a 1 to the left of three 0s makes the 1 worth 1,000.

In decimal notation, each position to the left of the decimal point indicates an increased positive power of ten, as seen in Figure 2-1(a). The total quantity or amount of the number is therefore determined by the size of each digit and the weighted position each digit is in. For example, the value shown in Figure 2-1(a) has six thousands, zero hundreds, one ten, and nine ones, which combined makes a total of $6{,}019_{10}$.

Most Significant Digit (MSD) The left-most, largest-weight digit in a number.

Least Significant Digit (LSD) The right-most, smallest-weight digit in a number.

In the decimal number system, the left-most digit is called the **most significant digit (MSD)** while the right-most digit is called the **least significant digit (LSD).** Applying this to the example in Figure 2-1(a), the 6 would be the MSD since its position carries the most weight, while the 9 would be the LSD since its position carries the least weight.

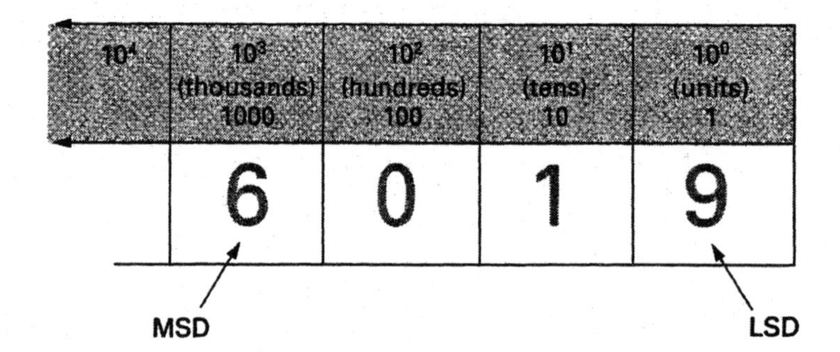

$$\text{Value of number} = (6 \times 10^3) + (0 \times 10^2) + (1 \times 10^1) + (9 \times 10^0)$$
$$= (6 \times 1000) + (0 \times 100) + (1 \times 10) + (9 \times 1)$$
$$= 6000 + 0 + 10 + 9 = 6019_{10}$$

(a)

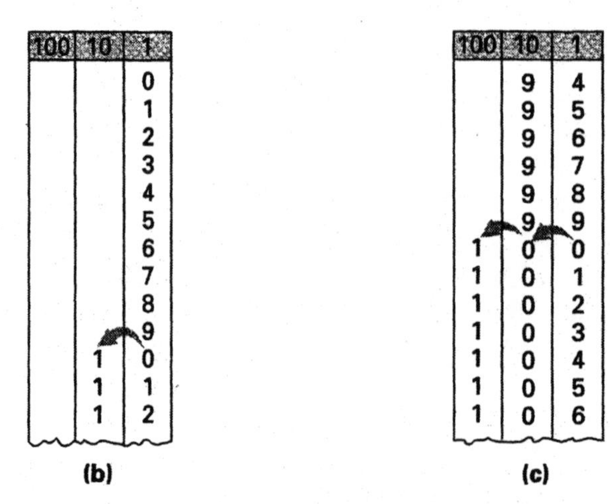

FIGURE 2-1 **The Decimal (Base-10) Number System.**

2-1-2 *Reset and Carry*

Before proceeding to the binary number system, let us review one other action that occurs when counting in decimal. This action, which is familiar to us all, is called **reset and carry.** Referring to Figure 2-1(b), you can see that a reset-and-carry operation occurs after a count of 9. The units column, which has reached its maximum count, resets to 0 and carries a 1 to the tens column resulting in a final count of 10.

Reset and Carry An action that occurs when a column has reached its maximum count.

This reset-and-carry action will occur in any column that reaches its maximum count. For example, Figure 2-1(c) shows how two reset-and-carry operations will take place after a count of 99. The units column resets and carries a 1 to the tens column, which in turn resets and carries a 1 to the hundreds column, resulting in a final count of 100.

SELF-TEST REVIEW QUESTIONS FOR SECTION 2-1

1. What is the difference between the Arabic number system and the decimal number system?
2. What is the base or radix of the decimal number system?
3. Describe the positional weight of each of the digits in the decimal number 2,639.
4. What action occurs when a decimal column advances beyond its maximum count?

2-2 THE BINARY NUMBER SYSTEM

As in the decimal system, the value of a binary digit is determined by its position relative to the other digits. In the decimal system, each position to the left of the decimal point increases by a power of 10. This case is also true with the binary number system; however, since it is a base-2 (bi) number system, each place to the left of the **binary point** increases by a power of 2. Figure 2-2 shows how columns of the binary and decimal number systems have different weights. For example, with binary, the columns are weighted so that 2^0 is one, 2^1 is two, 2^2 is four, 2^3 is eight ($2 \times 2 \times 2 = 8$), and so on.

Binary Point A symbol used to separate the whole from the fraction in a binary number.

As we know, the base or radix of a number system also indicates the maximum number of digits used by the number system. The base-2 binary number system only uses the first two digits on the number scale, 0 and 1. The 0s and 1s in binary are called "binary digits," or bits, for short.

2-2-1 *Positional Weight*

As in the decimal system, each column in binary carries its own weight, as seen in Figure 2-3(a). With the decimal number system, each position to the left increases 10 times. With binary, the weight of each column to the left increases

DECIMAL

10^2	10^1	10^0
100	10	1
		0
		1
		2
		3
		4
		5
		6
		7
		8
		9
	1	0
	1	1
	1	2

BINARY

2^3	2^2	2^1	2^0
8	4	2	1
			0
			1
		1	0
		1	1
	1	0	0
	1	0	1
	1	1	0
	1	1	1
1	0	0	0
1	0	0	1
1	0	1	0
1	0	1	1
1	1	0	0

FIGURE 2-2 A Comparison between Decimal and Binary.

2 times. The first column, therefore, has a weight of 1, the second column has a weight of 2, the third column 4, the fifth 8, and so on. The value or quantity of a binary number is determined by the digit in each column and the positional weight of the column. For example, in Figure 2-3(a), the binary number 101101_2 is equal in decimal to 45_{10}, since we have 1×32, 1×8, 1×4, and 1×1 ($32 + 8 + 4 + 1 = 45$). The 0s in this example are not multiplied by their weights (16 and 2) since they are "0" and of no value. The left-most binary digit is called the most significant bit (MSB) since it carries the most weight, while the right-most digit is called the least significant bit (LSB) since it carries the least weight. Applying this to the example in Figure 2-3(a), the 1 in the thirty-twos column is the MSB, while the 1 in the units column is the LSB.

2-2-2 *Reset and Carry*

The reset-and-carry action will occur in binary in exactly the same way it did in decimal. However, since binary has only two digits, a column will reach its maximum digit much sooner, and therefore the reset-and-carry action in the binary number system will occur much more frequently.

Referring to Figure 2-3(b), you can see that the binary counter begins with 0 and advances to 1. At this stage, the units column has reached its maximum, and therefore the next count forces the units column to reset and carry into the next column, producing a count of 0010_2 (2_{10}). The units column then advances to a count of 0011_2 (3_{10}). At this stage, both the units column and twos column have reached their maximums. As the count advances by one, it will cause the units column to reset and carry a 1 into the twos column, which will also have to reset and carry a 1 into the fours column. This will result in a final count of 0100_2 (4_{10}). The count will then continue to 0101_2 (5_{10}), 0110_2 (6_{10}), 0111_2 (7_{10}), and then 1000_2 (8_{10}) which, as you can see in Figure 2-3(b), is a result of three reset and carries.

Comparing the binary reset and carry to the decimal reset and carry in Figure 2-3(b), you can see that since the binary number system uses only two

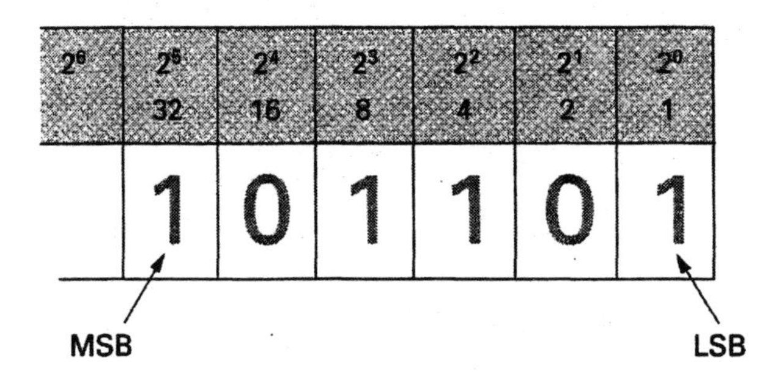

$$\begin{aligned}\text{Value of number} &= (1\times 2^5) + (0\times 2^4) + (1\times 2^3) + (1\times 2^2) + (1\times 2^1) + (1\times 2^0)\\ &= (1\times 32) + (0\times 16) + (1\times 8) + (1\times 4) + (0\times 2) + (1\times 1)\\ &= 32 + 0 + 8 + 4 + 0 + 1\\ &= 45_{10}\end{aligned}$$

(a)

DECIMAL

	10^1 10	10^0 1
		0
		1
		2
		3
		4
		5
		6
		7
		8
		9
	1	0
	1	1
	1	2
	1	3
	1	4
	1	5

BINARY

2^3 8	2^2 4	2^1 2	2^0 1
			0
			1
		1	0
		1	1
	1	0	0
	1	0	1
	1	1	0
	1	1	1
1	0	0	0
1	0	0	1
1	0	1	0
1	0	1	1
1	1	0	0
1	1	0	1
1	1	1	0
1	1	1	1

(b)

FIGURE 2-3 The Binary (Base-2) Number System.

digits, binary numbers quickly turn into multi-digit figures. For example, a decimal eight (8) uses only one digit, while a binary eight (1000) uses four digits.

2-2-3 Converting Binary Numbers to Decimal Numbers

Binary numbers can easily be converted to their decimal equivalent by simply adding together all of the column weights that contain a binary 1, as we did previously in Figure 2-3(a).

■ **EXAMPLE:**

Convert the following binary numbers to their decimal equivalents.

a. 1010
b. 101101

■ ***Solution:***

a.

Binary Column Weights	32	16	8	4	2	1
Binary Number			1	0	1	0

Decimal Equivalent $= (1 \times 8) + (0 \times 4) + (1 \times 2) + (0 \times 1)$
$= 8 + 0 + 2 + 0 = 10_{10}$

b.

Binary Column Weights	32	16	8	4	2	1
Binary Number	1	0	1	1	0	1

Decimal Equivalent $= (1 \times 32) + (0 \times 16) + (1 \times 8) + (1 \times 4) + (0 \times 2) + (1 \times 1)$
$= 32 + 0 + 8 + 4 + 0 + 1 = 45_{10}$

■ **EXAMPLE:**

The LEDs in Figure 2-4 are being used as a 4-bit (four binary digit) display. When the LED is OFF, it indicates a binary 0, and when the LED is ON, it indicates a binary 1. Determine the decimal equivalent of the binary displays shown in Figure 2-4(a), (b), and (c).

■ ***Solution:***

a. $0101_2 = 5_{10}$
b. $1110_2 = 14_{10}$
c. $1001_2 = 9_{10}$

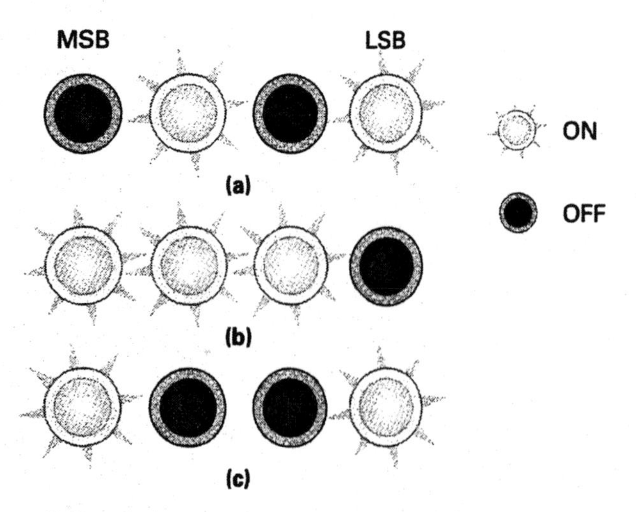

FIGURE 2-4 Using LEDs to Display Binary Numbers.

2-2-4 Converting Decimal Numbers to Binary Numbers

To convert a decimal number to its binary equivalent, continually subtract the largest possible power of two until the decimal number is reduced to zero, placing a binary 1 in columns that are used and a binary 0 in the columns that are not used. To explain how simple this process is, refer to Figure 2-5 which shows how decimal 53 can be converted to its binary equivalent. As you can see in this example, the first largest power of two that can be subtracted from decimal 53 is 32, and therefore a 1 is placed in the thirty-twos column and 21 is remaining. The largest power of two that can be subtracted from the remainder 21 is 16, and therefore a 1 is placed in the sixteens column and 5 is remaining. The next largest power of two that can be subtracted from 5 is 4, and therefore a 1 is placed in the fours column and 1 is remaining. The final 1 is placed in the

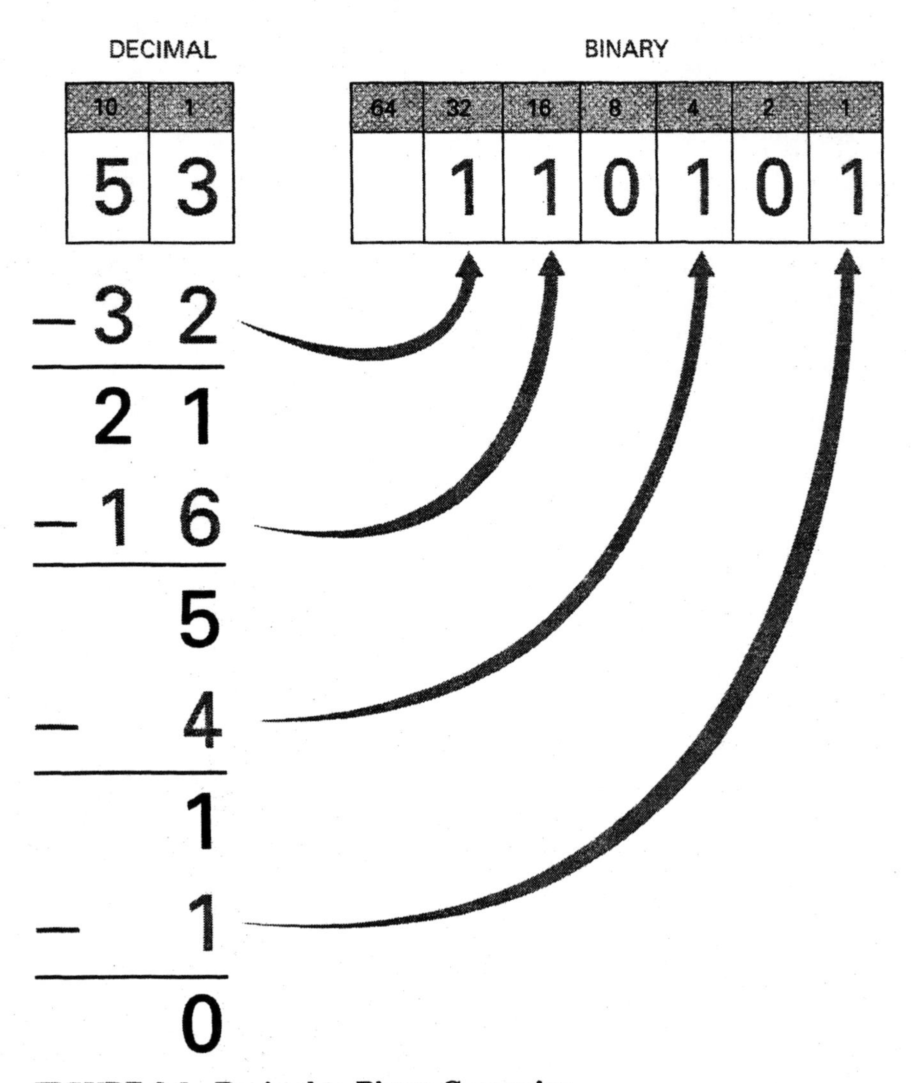

FIGURE 2-5 Decimal to Binary Conversion.

units column, and therefore decimal 53 is represented in binary as 110101, which indicates that the value is the sum of 1 × 32, 1 × 16, 1 × 4, and 1 × 1.

■ **EXAMPLE:**

Convert the following decimal numbers to their binary equivalents.

a. 25
b. 55

■ ***Solution:***

a. Binary Column Weights	32	16	8	4	2	1
Binary Number		1	1	0	0	1
Decimal Equivalent	25 − 16 = 9, 9 − 8 = 1, 1 − 1 = 0					
b. Binary Column Weights	32	16	8	4	2	1
Binary Number	1	1	0	1	1	1
Decimal Equivalent	55 − 32 = 23, 23 − 16 = 7, 7 − 4 = 3, 3 − 2 = 1, 1 − 1 = 0					

■ **EXAMPLE:**

Figure 2-6 shows a strip of magnetic tape containing binary data. The shaded dots indicate magnetized dots (which are equivalent to binary 1s) while the unshaded circles on the tape represent unmagnetized points (which are equivalent to binary 0s). What binary values should be stored in rows 4 and 5 so that decimal 19_{10} and 33_{10} are recorded?

■ ***Solution:***

a. Binary Column Weights	32	16	8	4	2	1
Binary Number		1	0	0	1	1
Decimal Equivalent	19 − 16 = 3, 3 − 2 = 1, 1 − 1 = 0					
Stored Magnetic Data	○●○○●●					
b. Binary Column Weights	32	16	8	4	2	1
Binary Number	1	0	0	0	0	1
Decimal Equivalent	33 − 32 = 1, 1 − 1 = 0					
Stored Magnetic Data	●○○○○●					

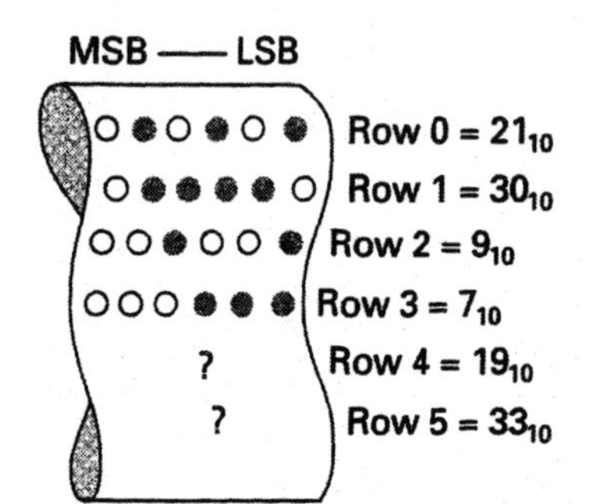

FIGURE 2-6 Binary Numbers on Magnetic Tape.

SELF-TEST REVIEW QUESTIONS FOR SECTION 2-2

1. What is the decimal equivalent of 11010_2?
2. Convert 23_{10} to its binary equivalent.
3. What are the full names for the abbreviations LSB and MSB?
4. Convert 110_{10} to its binary equivalent.

2-3 THE HEXADECIMAL NUMBER SYSTEM

If digital electronic circuits operate using binary numbers and we operate using decimal numbers, why is there any need for us to have any other system? The hexadecimal, or hex, system is used as a sort of shorthand for large strings of binary numbers, as will be explained in this section. To begin with, let us examine the basics of this number system and then look at its application.

Hexadecimal means "sixteen," and this number system has sixteen different digits as shown in Figure 2-7(a), which shows a comparison between decimal, hexadecimal, and binary. Looking at the first ten digits in the decimal and hexadecimal columns, you can see that there is no difference between the two columns; however, beyond 9, hexadecimal makes use of the letters A, B, C, D, E, and F. Digits 0 through 9 and letters A through F make up the total sixteen digits of the hexadecimal number system. Comparing hexadecimal to decimal once again, you can see that $A_{16} = 10_{10}$, $B_{16} = 11_{10}$, $C_{16} = 12_{10}$, $D_{16} = 13_{10}$, $E_{16} = 14_{10}$ and $F_{16} = 15_{10}$. Having these extra digits means that a column reset and carry will not occur until the count has reached the last and largest digit, F. The hexadecimal column in Figure 2-7(a) shows how reset and carry will occur whenever a column reaches its maximum digit, F, and this action is further illustrated in the examples in Figure 2-7(b).

Hexadecimal
A base-16 number system.

2-3-1 Converting Hexadecimal Numbers to Decimal Numbers

To find the decimal equivalent of a hexadecimal number, simply multiply each hexadecimal digit by its positional weight. For example, referring to Figure 2-8, you can see the positional weights of the hexadecimal columns which, as expected, are each progressively sixteen times larger as you move from right to left. The hexadecimal number 4C, therefore, indicates that the value has 4×16 ($4 \times 16 = 64$) and $C \times 1$. Since C is equal to 12 in decimal (12×1), the result of $12 + 64$ is 76, so hexadecimal 4C is equivalent to decimal 76.

■ EXAMPLE:

Convert hexadecimal 8BF to its decimal equivalent.

DECIMAL			HEXADECIMAL				BINARY					
10^2	10^1	10^0	16^3	16^2	16^1	16^0	2^5	2^4	2^3	2^2	2^1	2^0
100	10	1	4096	256	16	1	32	16	8	4	2	1
		0				0	0	0	0	0	0	0
		1				1	0	0	0	0	0	1
		2				2	0	0	0	0	1	0
		3				3	0	0	0	0	1	1
		4				4	0	0	0	1	0	0
		5				5	0	0	0	1	0	1
		6				6	0	0	0	1	1	0
		7				7	0	0	0	1	1	1
		8				8	0	0	1	0	0	0
		9				9	0	0	1	0	0	1
	1	0				A	0	0	1	0	1	0
	1	1				B	0	0	1	0	1	1
	1	2				C	0	0	1	1	0	0
	1	3				D	0	0	1	1	0	1
	1	4				E	0	0	1	1	1	0
	1	5				F	0	0	1	1	1	1
	1	6			1	0	0	1	0	0	0	0
	1	7			1	1	0	1	0	0	0	1
	1	8			1	2	0	1	0	0	1	0
	1	9			1	3	0	1	0	0	1	1
	2	0			1	4	0	1	0	1	0	0
	2	1			1	5	0	1	0	1	0	1
	2	2			1	6	0	1	0	1	1	0
	2	3			1	7	0	1	0	1	1	1
	2	4			1	8	0	1	1	0	0	0
	2	5			1	9	0	1	1	0	0	1
	2	6			1	A	0	1	1	0	1	0
	2	7			1	B	0	1	1	0	1	1
	2	8			1	C	0	1	1	1	0	0
	2	9			1	D	0	1	1	1	0	1
	3	0			1	E	0	1	1	1	1	0
	3	1			1	F	0	1	1	1	1	1
	3	2			2	0	1	0	0	0	0	0
	3	3			2	1	1	0	0	0	0	1
	3	4			2	2	1	0	0	0	1	0

(a)

A9	DE7	78	F9	10C
AA	DE8	79	FA	10D
AB	DE9	7A	FB	10E
AC	DEF	7B	FC	10F
AD	DF0	7C	FD	110
AE	DF1	7D	FE	111
AF	DF2	7E	FF	112
B0	DF3	7F	100	
B1		80	101	
		81		

(b)

FIGURE 2-7 Hexadecimal Reset and Carry. (a) Number System Comparison. (b) Hex Counting Examples.

16^4	16^3	16^2	16^1	16^0
65,536	4096	256	16	1
			4	C

Value of Number = (4 × 16) + (C × 1)
= (4 × 16) + (12 × 1)
= 64 + 12 = 76_{10}

FIGURE 2-8 Positional Weight of the Hexadecimal Number System.

■ ***Solution:***

	4096	256	16	1
Hexadecimal Column Weights		8	B	F

Decimal Equivalent

$$= (8 \times 256) + (B \times 16) + (F \times 1)$$
$$= 2048 + (11 \times 16) + (15 \times 1)$$
$$= 2048 + 176 + 15$$
$$= 2239$$

2-3-2 Converting Decimal Numbers to Hexadecimal Numbers

Decimal to hexadecimal conversion is achieved in the same way as decimal to binary. First subtract the largest possible power of sixteen, and then keep subtracting the largest possible power of sixteen from the remainder. Each time a subtraction takes place, add a 1 to the respective column until the decimal value has been reduced to zero. Figure 2-9 illustrates this procedure with an example showing how decimal 425 is converted to its hexadecimal equivalent. To begin with, the largest possible power of sixteen (256) is subtracted once from 425 and therefore a 1 is placed in the 256s column, leaving a remainder of 169. The next largest power of sixteen is 16, which can be subtracted ten times from 169, and therefore the hexadecimal equivalent of ten, which is A, is placed in the sixteens column, leaving a remainder of 9. Since nine 1s can be subtracted from the remainder of 9, the units column is advanced nine times, giving us our final hexadecimal result, 1A9.

■ **EXAMPLE:**

Convert decimal 4525 to its hexadecimal equivalent.

■ ***Solution:***

	4096	256	16	1
Hexadecimal Column Weights	1	1	A	D

Decimal Equivalent	= 4525 − 4096 = 429, 429 − 256 = 173, 173 − 16 − 16 − 16 − 16 − 16 − 16 − 16 − 16 − 16 − 16 = 13, 13 − 1 − 1 − 1 − 1 − 1 − 1 − 1 − 1 − 1 − 1 − 1 − 1 − 1 = 0

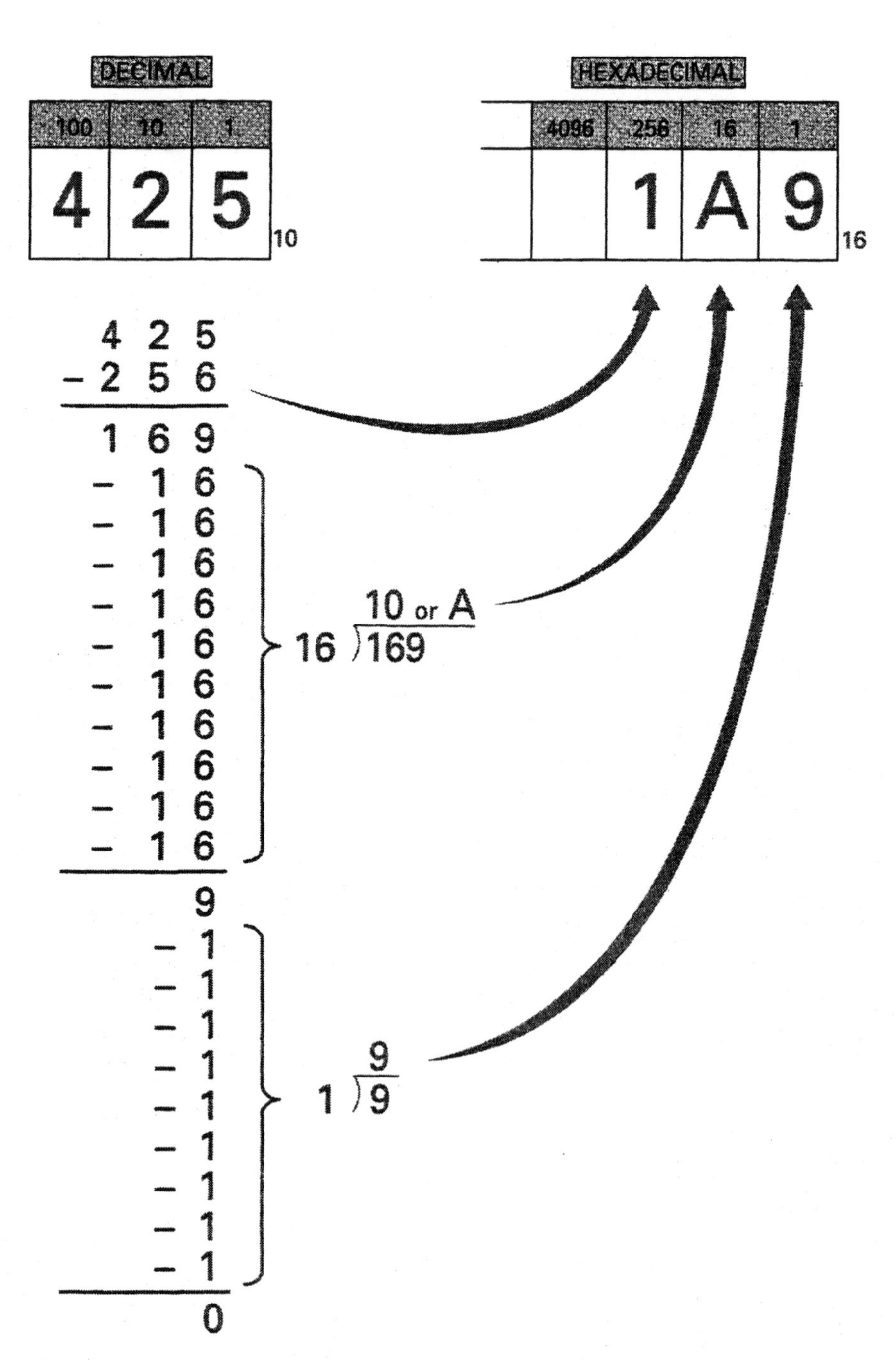

FIGURE 2-9 Decimal to Hexadecimal Conversion.

2-3-3 *Converting between Binary and Hexadecimal*

As mentioned in the beginning of this section, hexadecimal is used as a shorthand for representing large groups of binary digits. To illustrate this, Figure 2-10(a) shows how a 16-bit binary number, which is more commonly called a 16-bit binary **word,** can be represented by four hexadecimal digits. To explain this, Figure 2-10(b) shows how a 4-bit binary word can have any value from 0_{10} (0000_2) to 15_{10} (1111_2), and since hexadecimal has the same number of digits (0 through F), we can use one hexadecimal digit to represent four binary bits. As you can see in Figure 2-10(a), it is much easier to work with a number like $AE73_{16}$ than with 1010111001110011_2.

Word
An ordered set of characters that is treated as a unit.

To convert from hexadecimal to binary, we simply do the opposite, as shown in Figure 2-11. Since each hexadecimal digit represents four binary digits, a 4-digit hexadecimal number will convert to a 16-bit binary word.

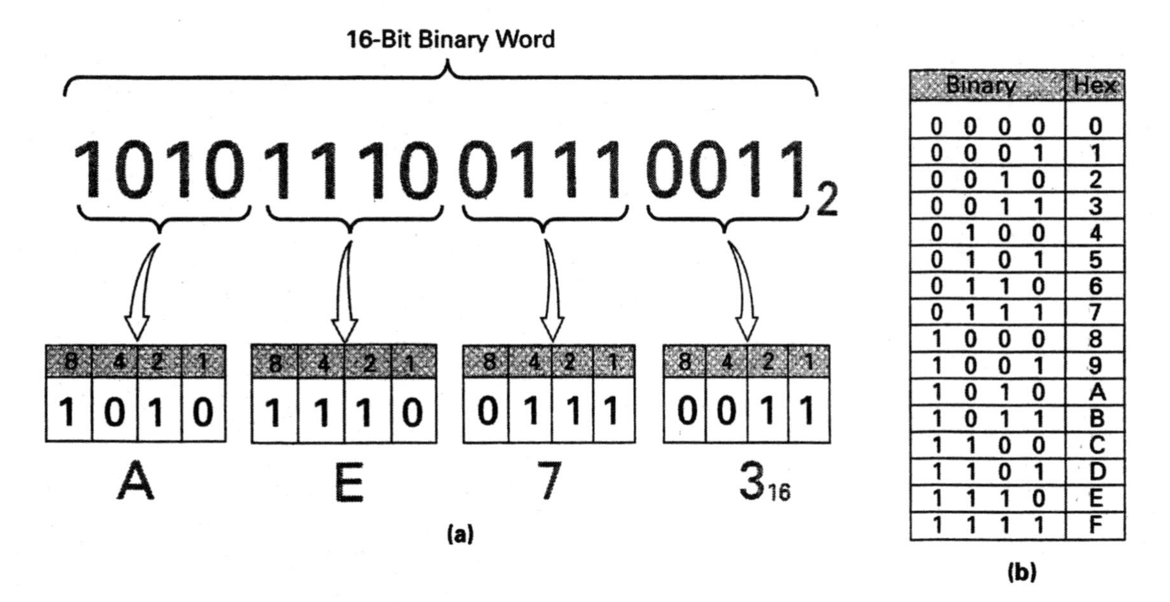

Binary	Hex
0 0 0 0	0
0 0 0 1	1
0 0 1 0	2
0 0 1 1	3
0 1 0 0	4
0 1 0 1	5
0 1 1 0	6
0 1 1 1	7
1 0 0 0	8
1 0 0 1	9
1 0 1 0	A
1 0 1 1	B
1 1 0 0	C
1 1 0 1	D
1 1 1 0	E
1 1 1 1	F

FIGURE 2-10 Representing Binary Numbers in Hexadecimal.

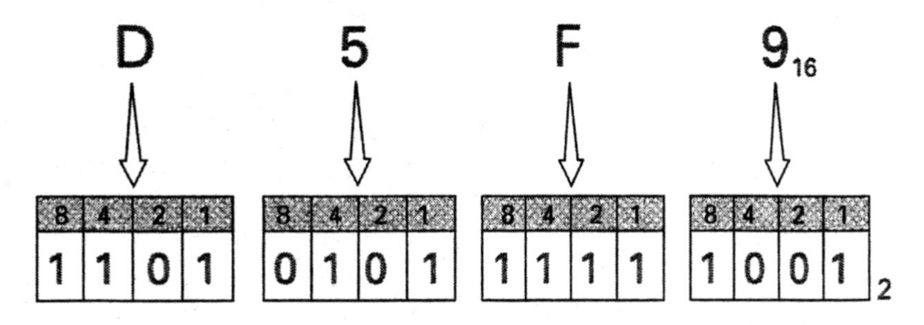

FIGURE 2-11 Hexadecimal to Binary Conversion.

■ **EXAMPLE:**

Convert hexadecimal 2BF9 to its binary equivalent.

■ ***Solution:***

Hexadecimal Number	2	B	F	9
Binary Equivalent	0010	1011	1111	1001

■ **EXAMPLE:**

Convert binary 110011100001 to its hexadecimal equivalent.

■ ***Solution:***

Binary Number	1100	1110	0001
Hexadecimal Equivalent	C	E	1

SELF-TEST REVIEW QUESTIONS FOR SECTION 2-3

1. What is the base of the
   a. decimal number system,
   b. binary number system, and
   c. hexadecimal number system?
2. What are the decimal and hexadecimal equivalents of 10101_2?
3. Convert $1011\ 1111\ 0111\ 1010_2$ to its hexadecimal equivalent.
4. What are the binary and hexadecimal equivalents of 33_{10}?

2-4 THE OCTAL NUMBER SYSTEM

Although not as frequently used as hexadecimal, the *octal number system* is also used as a shorthand for large binary words. It is very much like hexadecimal in that it allows for easy conversion from binary to octal and from octal to binary. Before we look at its application, however, let us examine the basics of this number system.

Octal
A base-8 number system.

Octal means eight, and the octal number system has eight different digits: 0 through 7. Figure 2-12(a) shows a comparison between decimal, octal, binary, and hexadecimal. Having only eight possible digits means that a column reset and carry will occur when the count has reached the last and largest digit of 7. At this point, the first octal column will reset and carry as shown in Figure 2-12(a). This action occurs when any of the octal columns reaches the maximum digit 7, as illustrated in the additional counting examples shown in Figure 2-12(b).

DECIMAL				OCTAL				BINARY					HEXADECIMAL			
1000	100	10	1	512	64	8	1	16	8	4	2	1	4096	256	16	1
			0				0	0	0	0	0	0				0
			1				1	0	0	0	0	1				1
			2				2	0	0	0	1	0				2
			3				3	0	0	0	1	1				3
			4				4	0	0	1	0	0				4
			5				5	0	0	1	0	1				5
			6				6	0	0	1	1	0				6
			7				7	0	0	1	1	1				7
			8			1	0	0	1	0	0	0				8
			9			1	1	0	1	0	0	1				9
		1	0			1	2	0	1	0	1	0				A
		1	1			1	3	0	1	0	1	1				B
		1	2			1	4	0	1	1	0	0				C
		1	3			1	5	0	1	1	0	1				D
		1	4			1	6	0	1	1	1	0				E
		1	5			1	7	0	1	1	1	1				F
		1	6			2	0	1	0	0	0	0			1	0
		1	7			2	1	1	0	0	0	1			1	1
		1	8			2	2	1	0	0	1	0			1	2
		1	9			2	3	1	0	0	1	1			1	3
		2	0			2	4	1	0	1	0	0			1	4

(a)

44	75	474
45	76	475
46	77	476
47	100	477
50	101	500
51	102	501

(b)

FIGURE 2-12 **Octal Reset and Carry. (a) Number System Comparison. (b) Octal Counting Examples.**

2-4-1 Converting Octal Numbers to Decimal Numbers

To find the decimal equivalent of an octal number, simply multiply each octal digit by its positional weight. For example, referring to Figure 2-13, you can see

8^5	8^4	8^3	8^2	8^1	8^0
32,768	4096	512	64	8	1
			1	2	6

$$\text{Value of Number} = (1 \times 64) + (2 \times 8) + (6 \times 1)$$
$$= 64 + 16 + 6 = 86_{10}$$

FIGURE 2-13 Positional Weight of the Octal Number System.

the positional weight of the octal columns, which, as expected, are each progressively eight times larger as you move from right to left. The octal number 126 therefore indicates that the value is the sum of 1 × 64, 2 × 8, and 6 × 1. The result, therefore, of 64 + 16 + 6 is equivalent to decimal 86.

■ **EXAMPLE:**

Convert octal 2479 to its decimal equivalent.

■ ***Solution:***

Octal Column Weights	512	64	8	1
	2	4	7	9

Decimal Equivalent $= (2 \times 512) + (4 \times 64) + (7 \times 8) + (9 \times 1)$
$= 1024 + 256 + 56 + 9$
$= 1345$

2-4-2 Converting Decimal Numbers to Octal Numbers

Decimal to octal conversion is achieved in the same way as decimal to binary and decimal to hex. First subtract the largest possible power of eight, and then keep subtracting the largest possible power of eight from the remainder. Each time a subtraction takes place, add 1 to the respective column until the decimal value has been reduced to zero. Figure 2-14 illustrates this procedure with an example showing how decimal 139 is converted to its octal equivalent. To begin with, the largest possible power of eight (64) can be subtracted twice from 139, and therefore a 2 is placed in the 64 column. The next largest power of eight is 8 itself, and therefore the octal 8 column is advanced to 1, leaving a remainder of 3. Since 3 units can be subtracted from this final remainder of 3, a 3 is placed in the octal units column, giving us our final octal result, 213.

■ **EXAMPLE:**

Convert decimal 3724 to its octal equivalent.

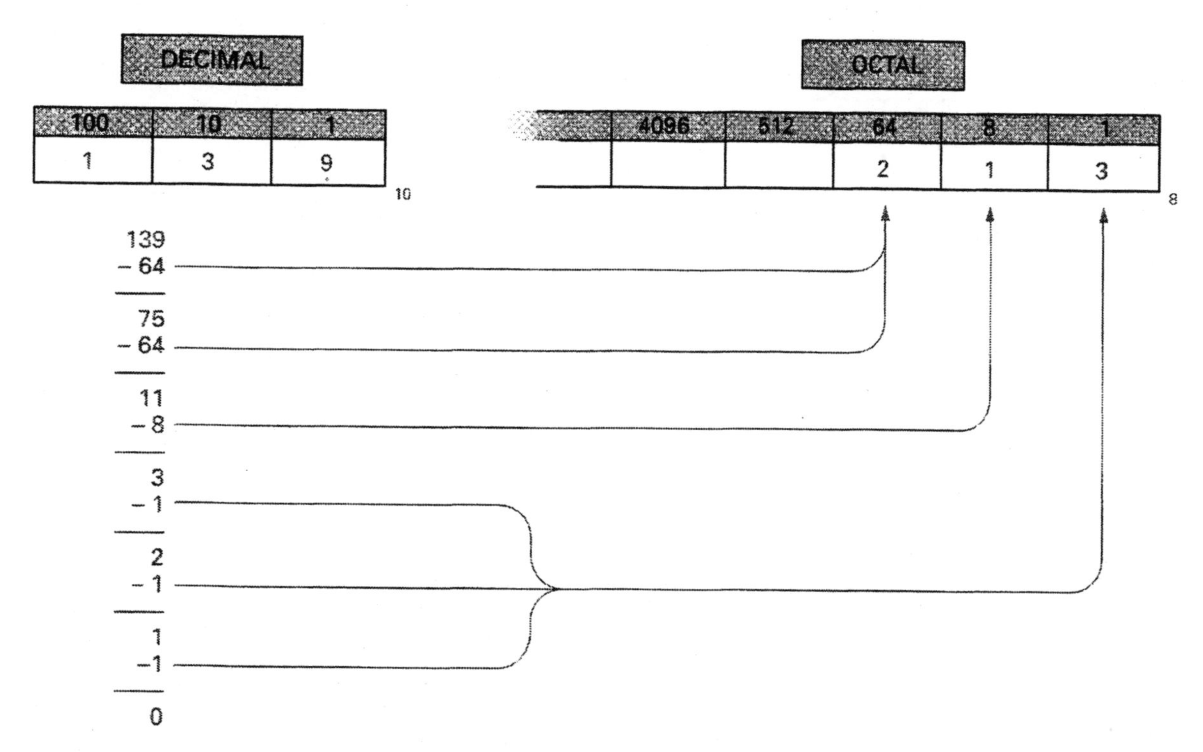

FIGURE 2-14 **Decimal to Octal Conversion.**

■ ***Solution:***

Octal Column Weights	512	64	8	1
	7	2	1	4

Decimal Equivalent = 3724 − 512 = 3212, 3212 − 512 = 2700, 2700 − 512 = 2188, 2188 − 512 = 1676, 1676 − 512 = 1164, 1164 − 512 = 652, 652 − 512 = 140, 140 − 64 = 76, 76 − 64 = 12, 12 − 8 = 4, 4 − 1 = 3, 3 − 1 = 2, 2 − 1 = 1, 1 − 1 = 0

2-4-3 *Converting between Binary and Octal*

Like hexadecimal, octal is used as a shorthand for representing large groups of binary digits. To illustrate this, Figure 2-15(a) shows how a 16-bit binary number, which is more commonly called a 16-bit binary word, can be represented by six octal digits. To explain this, Figure 2-15(b) shows how a 3-bit binary word can have any value from 0_{10} (000_2) to 7_{10} (111_2), and since octal has the same number of digits (0 through 7), we can use one octal digit to represent three binary bits. As you can see in the example in Figure 2-15(a), it is much easier to work with a number like 127163 than 1010111001110011.

To convert from octal to binary, we simply do the opposite, as shown in the example in Figure 2-16. Since each octal digit represents three binary digits, a 4-digit octal number will convert to a 12-bit binary word.

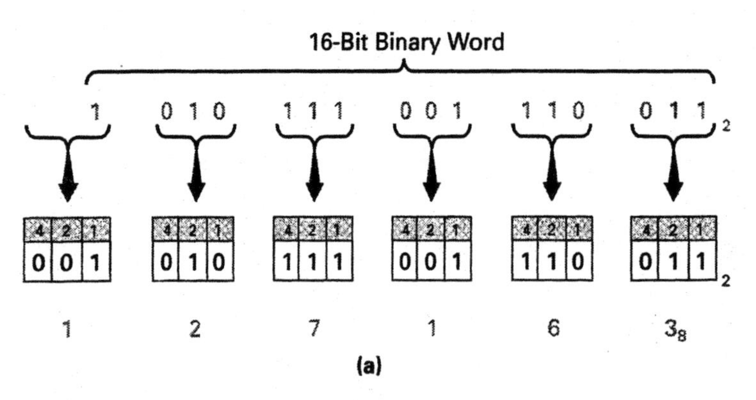

BINARY	OCTAL
0 0 0	0
0 0 1	1
0 1 0	2
0 1 1	3
1 0 0	4
1 0 1	5
1 1 0	6
1 1 1	7

(b)

FIGURE 2-15 Representing Binary Numbers in Octal.

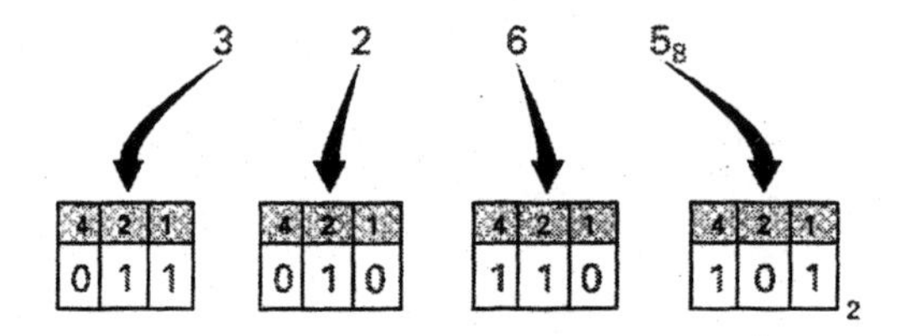

FIGURE 2-16 Octal to Binary Conversion.

■ **EXAMPLE:**

Convert octal number 2635 to its binary equivalent.

■ ***Solution:***

Octal Number	2	6	3	5
Binary equivalent	010	110	011	101

■ **EXAMPLE:**

Convert binary 101111001000010110 to its octal equivalent.

■ ***Solution:***

Binary Number	101	111	001	000	010	110
Octal Equivalent	5	7	1	0	2	6

SELF-TEST REVIEW QUESTIONS FOR SECTION 2-4

1. The octal number system has a base of ______.
2. Convert 154_{10} into octal.
3. A 12-bit binary word will convert into ______ octal digits.
4. What is the octal equivalent of 100010110111_2?

2-5 BINARY CODES

The process of converting a decimal number to its binary equivalent is called binary coding. The result of the conversion is a binary number or code that is called **pure binary.** There are, however, binary codes used in digital circuits other than pure binary, and in this section we will examine some of the most frequently used.

Pure Binary
Uncoded binary.

2-5-1 *The Binary Coded Decimal (BCD) Code*

No matter how familiar you become with binary, it will always be less convenient to work with than the decimal number system. For example, it will always take us a short time to convert 1111000_2 to 120_{10}. Designers realized this disadvantage early on and developed a binary code that had decimal characteristics and that was appropriately named **binary coded decimal (BCD).** Being a binary code, it has the advantages of a two-state system, and since it has a decimal format, it is also much easier for an operator to interface via a decimal keypad or decimal display to systems such as pocket calculators, wristwatches, and so on.

Binary Coded Decimal (BCD)
A code in which each decimal digit is represented by a group of 4 binary bits.

The BCD code expresses each decimal digit as a 4-bit word, as shown in Figure 2-17(a). In this example, decimal 1,753 converts to a BCD code of 0001 0111 0101 0011, with the first 4-bit code ($0001_2 = 1_{10}$) representing the 1 in the thousands column, the second 4-bit code ($0111_2 = 7_{10}$) representing the 7 in the hundreds column, the third 4-bit code ($0101_2 = 5_{10}$) representing the 5 in the tens column, and the fourth 4-bit code ($0011_2 = 3_{10}$) representing the 3 in the units column. As can be seen in Figure 2-17(a), the subscript "BCD" is often used after a BCD code to distinguish it from a pure binary number ($1753_{10} = 0001\ 0111\ 0101\ 0011_{BCD}$).

Figure 2-17(b) compares decimal, binary, and BCD. As you can see, the reset-and-carry action occurs in BCD at the same time it does in decimal. This is because BCD was designed to have only ten 4-bit binary codes, 0000, 0001, 0010, 0011, 0100, 0101, 0110, 0111, 1000, and 1001 (0 through 9), to make it easy to convert between this binary code and decimal. Binary codes 1010, 1011, 1100, 1101, 1110, and 1111 (A_{16} through F_{16}), are invalid codes that are not used in BCD because they are not used in decimal.

■ **EXAMPLE:**

Convert the BCD code 0101 1000 0111 0000 to decimal and the decimal number 369 to BCD.

■ ***Solution:***

BCD Code	0101	1000	0111	0000
Decimal Equivalent	5	8	7	0
Decimal Number	3	6	9	
BCD Code	0011	0110	1001	

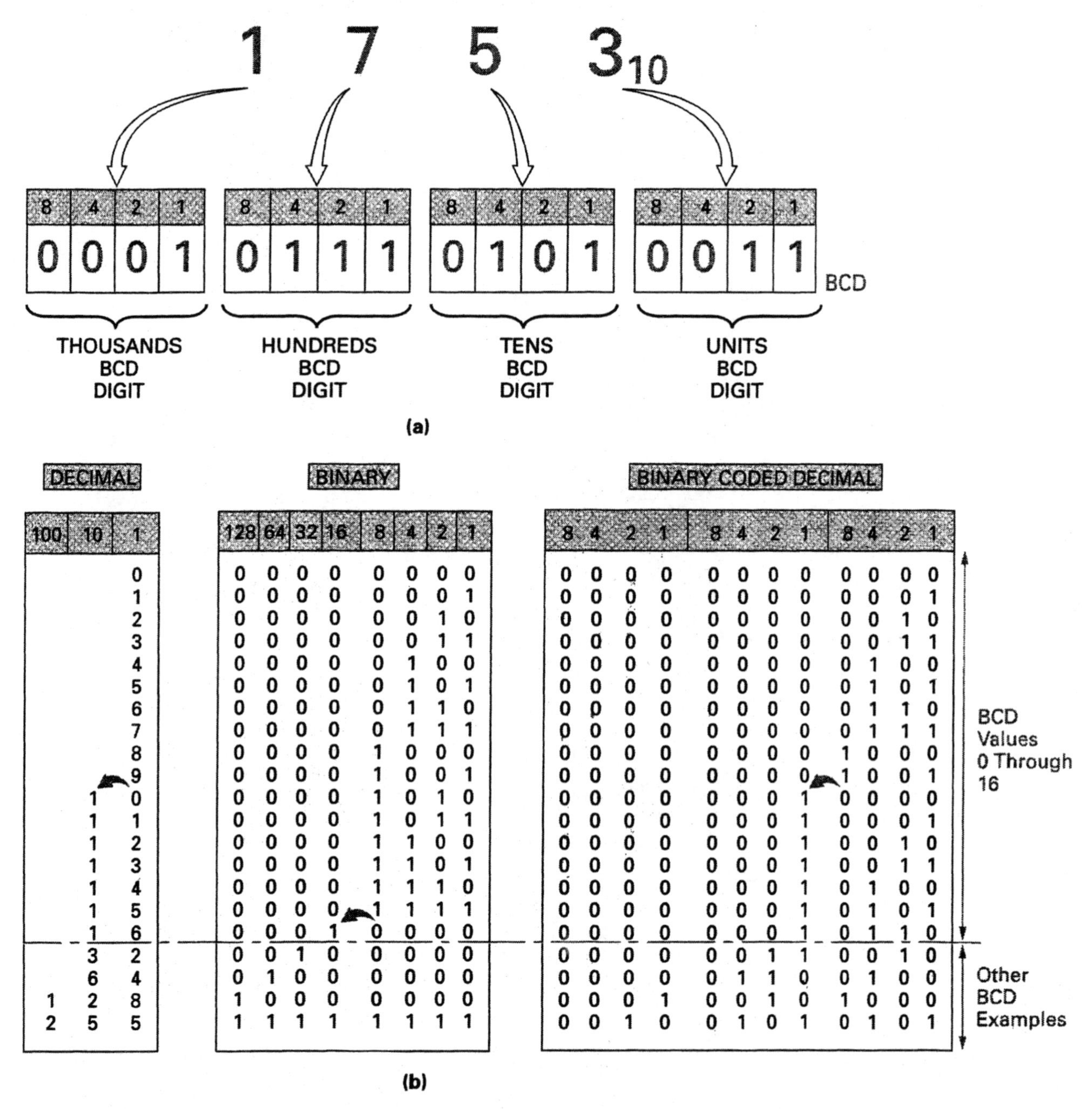

DECIMAL			BINARY								BINARY CODED DECIMAL											
100	10	1	128	64	32	16	8	4	2	1	8	4	2	1	8	4	2	1	8	4	2	1
		0	0	0	0	0	0	0	0	0	0	0	0	0	0	0	0	0	0	0	0	0
		1	0	0	0	0	0	0	0	1	0	0	0	0	0	0	0	0	0	0	0	1
		2	0	0	0	0	0	0	1	0	0	0	0	0	0	0	0	0	0	0	1	0
		3	0	0	0	0	0	0	1	1	0	0	0	0	0	0	0	0	0	0	1	1
		4	0	0	0	0	0	1	0	0	0	0	0	0	0	0	0	0	0	1	0	0
		5	0	0	0	0	0	1	0	1	0	0	0	0	0	0	0	0	0	1	0	1
		6	0	0	0	0	0	1	1	0	0	0	0	0	0	0	0	0	0	1	1	0
		7	0	0	0	0	0	1	1	1	0	0	0	0	0	0	0	0	0	1	1	1
		8	0	0	0	0	1	0	0	0	0	0	0	0	0	0	0	0	1	0	0	0
		9	0	0	0	0	1	0	0	1	0	0	0	0	0	0	0	0	1	0	0	1
	1	0	0	0	0	0	1	0	1	0	0	0	0	0	0	0	0	1	0	0	0	0
	1	1	0	0	0	0	1	0	1	1	0	0	0	0	0	0	0	1	0	0	0	1
	1	2	0	0	0	0	1	1	0	0	0	0	0	0	0	0	0	1	0	0	1	0
	1	3	0	0	0	0	1	1	0	1	0	0	0	0	0	0	0	1	0	0	1	1
	1	4	0	0	0	0	1	1	1	0	0	0	0	0	0	0	0	1	0	1	0	0
	1	5	0	0	0	0	1	1	1	1	0	0	0	0	0	0	0	1	0	1	0	1
	1	6	0	0	0	1	0	0	0	0	0	0	0	0	0	0	0	1	0	1	1	0
	3	2	0	0	1	0	0	0	0	0	0	0	0	0	0	0	1	1	0	0	1	0
	6	4	0	1	0	0	0	0	0	0	0	0	0	0	0	1	1	0	0	1	0	0
1	2	8	1	0	0	0	0	0	0	0	0	0	0	1	0	0	1	0	1	0	0	0
2	5	5	1	1	1	1	1	1	1	1	0	0	1	0	0	1	0	1	0	1	0	1

FIGURE 2-17 Binary Coded Decimal (BCD). (a) Decimal to BCD Conversion Example. (b) A Comparison of Decimal, Binary, and BCD.

2-5-2 The Excess-3 Code

Excess-3 Code
A code in which the decimal digit *n* is represented by the 4-bit binary code *n* + 3.

The **excess-3 code** is similar to BCD and is often used in some applications because of certain arithmetic operation advantages. Referring to Figure 2-18(a), you can see that, like BCD, each decimal digit converted into a 4-bit binary code. The difference with the excess-3 code, however, is that a value of 3 is added to each decimal digit before it is converted into a 4-bit binary code.

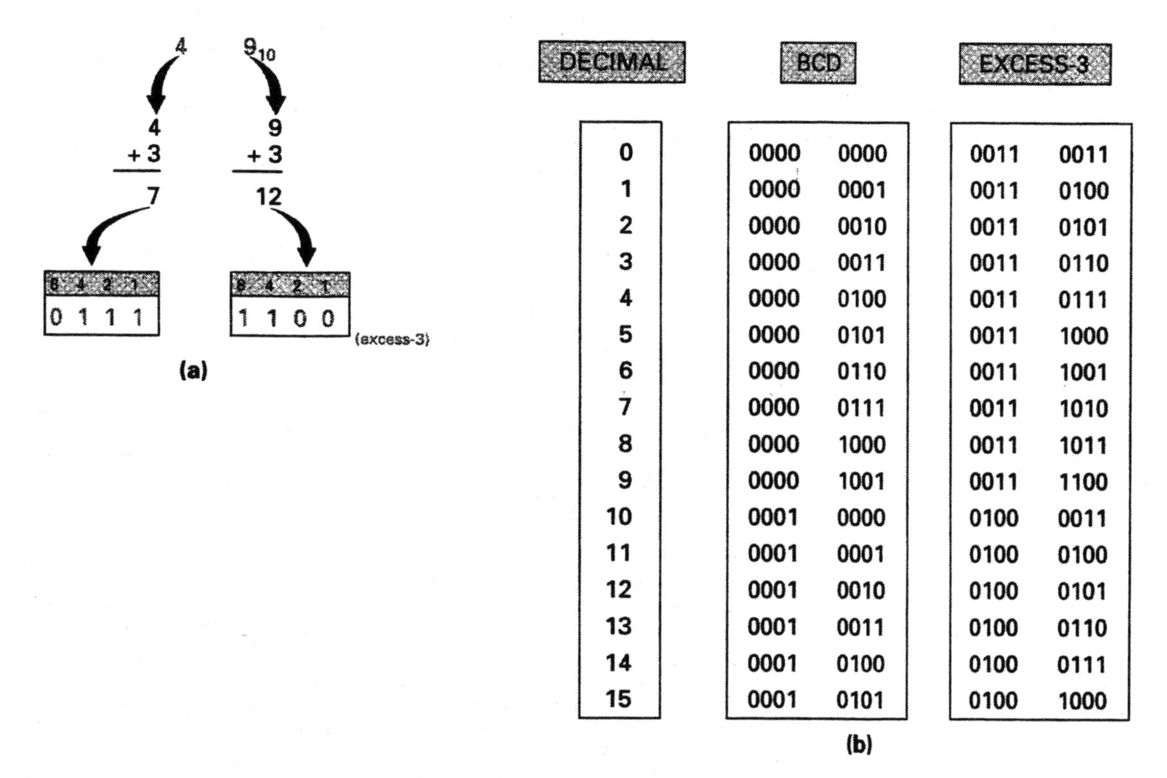

DECIMAL	BCD		EXCESS-3	
0	0000	0000	0011	0011
1	0000	0001	0011	0100
2	0000	0010	0011	0101
3	0000	0011	0011	0110
4	0000	0100	0011	0111
5	0000	0101	0011	1000
6	0000	0110	0011	1001
7	0000	0111	0011	1010
8	0000	1000	0011	1011
9	0000	1001	0011	1100
10	0001	0000	0100	0011
11	0001	0001	0100	0100
12	0001	0010	0100	0101
13	0001	0011	0100	0110
14	0001	0100	0100	0111
15	0001	0101	0100	1000

FIGURE 2-18 The Excess-3 Code. (a) Example. (b) Code Comparison.

Although this code has an offset or excess of 3, it still uses only ten 4-bit binary codes (0011 through 1100) like BCD. The invalid 4-bit values in this code are therefore 0000, 0001, 0010 (decimal 0 through 2), and 1101, 1110, 1111 (decimal 13 through 15). Figure 2-18(b) shows a comparison between decimal, BCD, and excess-3.

■ **EXAMPLE:**

Convert decimal 408 to its excess-3 code equivalent.

■ ***Solution:***

Decimal Number	4	0	8
+3 =	7	3	11
Excess-3 Binary	0111	0011	1011

2-5-3 The Gray Code

The **gray code,** shown in Figure 2-19, is a non-weighted binary code which means that the position of each binary digit carries no specific weight or value. This code, named after its inventor, was developed so that only one of the binary digits will change as you step from one code group to the next code group in sequence. For example, in pure binary, a change from 3 (0011) to 4 (0100) will result in a change in three bits, whereas in the gray code only one bit will change.

Gray Code
A non-weighted binary code in which sequential numbers are represented by codes that differ by one bit.

DECIMAL	BINARY	GRAY CODE
0	0000	0000
1	0001	0001
2	0010	0011
3	0011	0010
4	0100	0110
5	0101	0111
6	0110	0101
7	0111	0100
8	1000	1100
9	1001	1101
10	1010	1111
11	1011	1110
12	1100	1010
13	1101	1011
14	1110	1001
15	1111	1000

FIGURE 2-19 The Gray Code.

The minimum-change gray code is used to reduce errors in digital electronic circuitry. To explain this in more detail, when a binary count changes, it takes a very small amount of time for the bits to change from 0 to 1, or from 1 to 0. This transition time could produce an incorrect intermediate code. For example, if when changing from pure binary 0011 (3) to 0100 (4), the LSB were to switch slightly faster than the other bits, an incorrect code would be momentarily generated, as shown in the following example.

Binary	Decimal
0 0 1 1	3
0 0 1 0	Error
0 1 0 0	4

This momentary error could trigger an operation that should not occur, resulting in a system malfunction. By using the gray code, these timing errors are eliminated since only one bit changes at a time. The disadvantage of the gray code is that the codes must be converted to pure binary if numbers need to be added, subtracted, or used in other computations.

2-5-4 The American Standard Code for Information Interchange (ASCII)

To this point we have only discussed how we can encode numbers into binary codes. Digital electronic computers must also be able to generate and recognize binary codes that represent letters of the alphabet and symbols. The ASCII code, which was discussed in Chapter 1, is the most widely used **alphanumeric code** (alphabet and numeral code). As I write this book, the ASCII codes for each of these letters is being generated by my computer keyboard, and the computer is decoding these ASCII codes and then displaying the alphanumeric equivalent (letter, number, or symbol) on my computer screen.

Alphanumeric Code A code used to represent the letters of the alphabet and decimal numbers.

Figure 2-20 lists all of the 7-bit ASCII codes and the full names for the abbreviations used. This diagram also shows how the 7-bit ASCII codes are

Column		0	1	2	3	4	5	6	7
Row	Bits 4321 765	000	001	010	011	100	101	110	111
0	0000	NUL	DLE	SP	0	@	P	\	p
1	0001	SOH	DC1	!	1	A	Q	a	q
2	0010	STX	DC2	"	2	B	R	b	r
3	0011	ETX	DC3	#	3	C	S	c	s
4	0100	EOT	DC4	$	4	D	T	d	t
5	0101	ENQ	NAK	%	5	E	U	e	u
6	0110	ACK	SYN	&	6	F	V	f	v
7	0111	BEL	ETB	'	7	G	W	g	w
8	1000	BS	CAN	(	8	H	X	h	x
9	1001	HT	EM	)	9	I	Y	i	y
10	1010	LF	SUB	*	:	J	Z	j	z
11	1011	VT	ESC	+	;	K	[	k	{
12	1100	FF	FS	,	<	L	\	l	¦
13	1101	CR	GS	-	=	M	]	m	}
14	1110	SO	RS	.	>	N	⌒	n	~
15	1111	SI	US	/	?	O	—	o	DEL

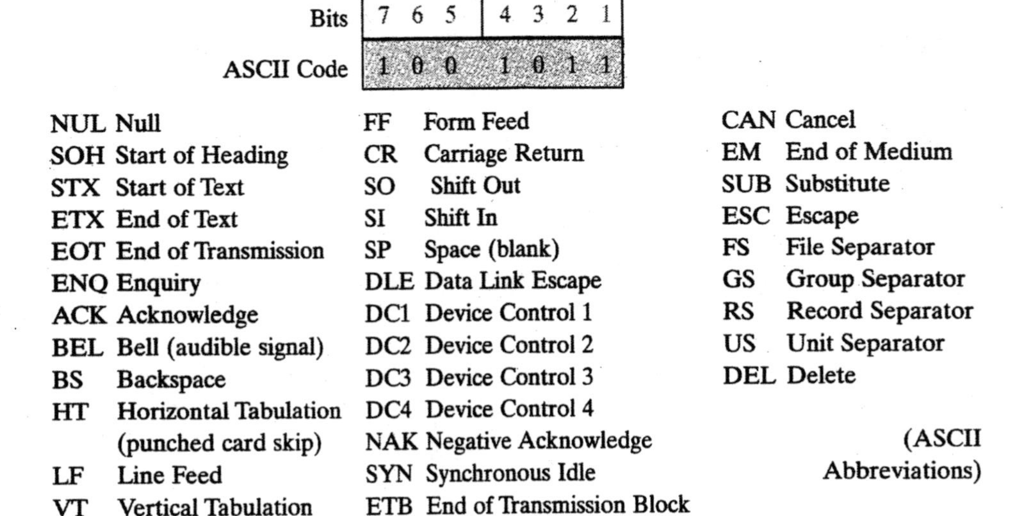

FIGURE 2-20 The ASCII Code.

each made up of a 4-bit group which indicates the row of the table, and a 3-bit group which indicates the column of the table. For example, the uppercase letter "K" is in column 100 (4_{10}) and in row 1011 (11_{10}), and is therefore represented by the ASCII code "1001011."

■ **EXAMPLE:**

List the ASCII codes for the message "Digital."

■ ***Solution:***

D = 1000100
i = 1101001
g = 1100111
i = 1101001
t = 1110100
a = 1100001
l = 1101100

■ **EXAMPLE:**

Figure 2-21 illustrates a 7-bit register. A register is a circuit that is used to store or hold a group of binary bits. If a switch is closed, the associated "Q output" will be grounded and 0 V (binary 0) will be applied to the output. On the other hand, if a switch is open, the associated "Q output" will be pulled HIGH due to the connection to +5 V via the 10 kΩ resistor, and therefore +5 V (binary 1) will be applied to the output. What would be the pure binary value, hexadecimal value, and ASCII code stored in this register?

■ ***Solution:***

	Q_6	Q_5	Q_4	Q_3	Q_2	Q_1	Q_0
Pure Binary	1	0	0	0	0	1	1_2
Hexadecimal				43_{16}			
ASCII				C (uppercase C)			

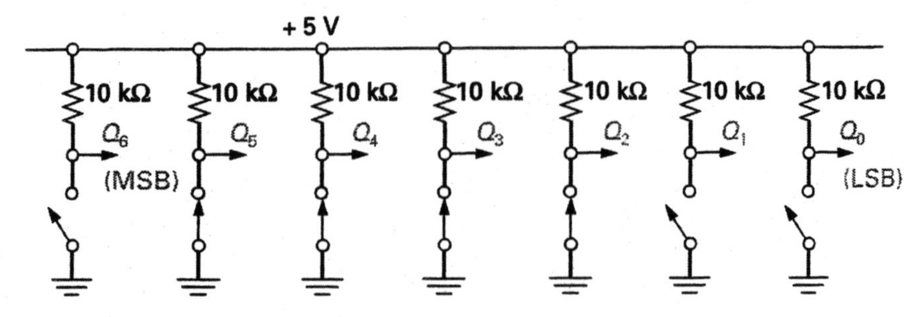

FIGURE 2-21 **A 7-Bit Switch Register.**

SELF-TEST REVIEW QUESTIONS FOR SECTION 2-5

1. Convert 7629_{10} to BCD.
2. What is the decimal equivalent of the excess-3 code 0100 0100?
3. What is a minimum-change code?
4. What does the following string of ASCII codes mean?
   1000010 1100001 1100010 1100010 1100001 1100111 1100101

REVIEW QUESTIONS

Multiple-Choice Questions

1. Binary is a base ____________ number system, decimal a base ____________ number system, and hexadecimal a base ____________ number system.

   a. 1, 2, 3 **b.** 2, 10, 16 **c.** 10, 2, 3 **d.** 8, 4, 2
2. What is the binary equivalent of decimal 11?

   a. 1010 **b.** 1100 **c.** 1011 **d.** 0111
3. Which of the following number systems is used as a shorthand for binary?

   a. Decimal **b.** Hexadecimal **c.** Binary **d.** Both a. and b.
4. What would be displayed on a hexadecimal counter if its count were advanced by one from $39FF_{16}$?

   a. 4000 **b.** 3A00 **c.** 3900 **d.** 4A00
5. What is the base of 101?

   a. 10 **b.** 2 **c.** 16 **d.** Could be any base
6. Which of the following is an invalid BCD code?

   a. 1010 **b.** 1000 **c.** 0111 **d.** 0000
7. What is the decimal equivalent of $1001\ 0001\ 1000\ 0111_{BCD}$?

   a. 9187 **b.** 9A56 **c.** 8659 **d.** 254,345
8. What would be the pure binary equivalent of 15_{16}?

   a. 0000 1111 **b.** 0001 0101 **c.** 0101 0101 **d.** 0000 1101
9. Which of the following is an invalid excess-3 code?

   a. 1010 **b.** 1000 **c.** 0111 **d.** 0000
10. Since the lower four bits of the ASCII code for the numbers 0 through 9 are the same as their pure binary equivalent, what is the 7-bit ASCII code for the number 6?

    a. 0111001 **b.** 0110111 **c.** 0110011 **d.** 0110110

Essay Questions

11. Give the full names of the following abbreviations. (2)

    a. Bit **b.** BCD **c.** ASCII

12. What is the Arabic number system? (2-1)
13. Why does reset and carry occur after 9 in decimal, and after 1 in binary? (2-1)
14. Briefly describe how to (2-2)
    a. convert a binary number to a decimal number.
    b. convert a decimal number to a binary number.
15. Why is the hexadecimal number system used in conjunction with digital circuits? (2-3)
16. Briefly describe how to (2-3)
    a. convert a hexadecimal number to a decimal number.
    b. convert a decimal number to a hexadecimal number.
    c. convert between binary and hexadecimal.
17. In what application is the octal number system used? (2-4)
18. Briefly describe how to (2-4)
    a. convert an octal number to a decimal number.
    b. convert a decimal number to an octal number.
    c. convert between binary and octal.
19. Why is the BCD code easier to decode than pure binary? (2-5)
20. What is binary coded decimal and why is it needed? (2-5)
21. Briefly describe how to convert a BCD code to its decimal equivalent. (2-5)
22. What is the excess-3 code? (2-5)
23. What is the advantage of the gray code? (2-5)
24. What is the ASCII code and in what applications is it used? (2-5)

Practice Problems

25. What is the pure binary output of the switch register shown in Figure 2-22 and what is its decimal equivalent?
26. Convert the following into their decimal equivalents.
    a. 110111_2 b. $2F_{16}$ c. 10110_{10}
27. What would be the decimal equivalent of the LED display in Figure 2-23 if it were displaying each of the following:
    a. Pure binary b. BCD c. Excess-3

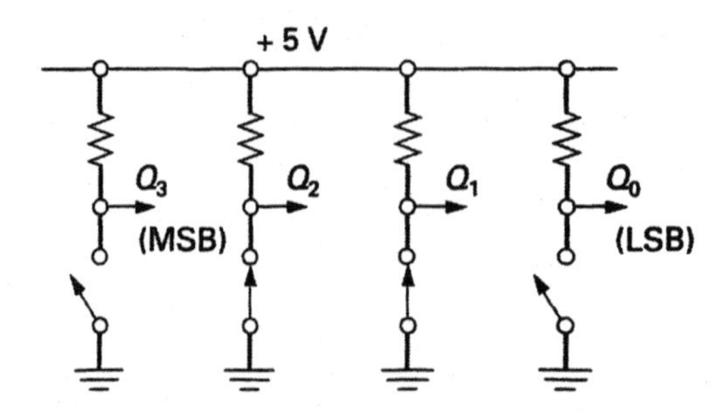

FIGURE 2-22 **4-Bit Switch Register.**

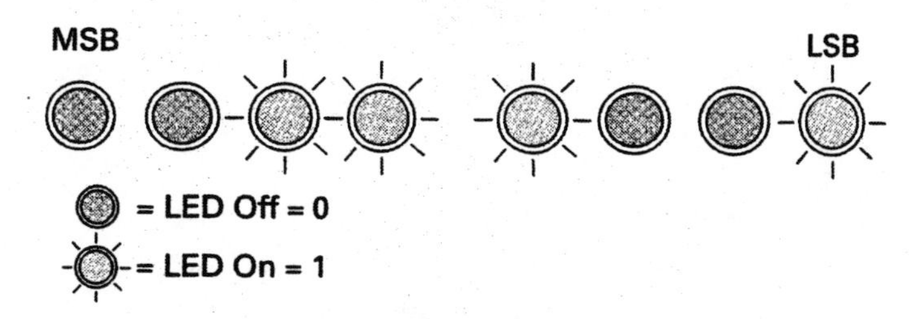

FIGURE 2-23 8-Bit LED Display.

28. Convert the following binary numbers to octal and hexadecimal.

a. 111101101001 **b.** 1011 **c.** 111000 **d.** 1111111

29. Convert the following decimal numbers to BCD.

a. 2,365 **b.** 24

30. Give the ASCII codes for the following:

a. ? **b.** $ **c.** 6

31. Identify the following codes.

DECIMAL	(a)	(b)	(c)	(d)
0	0000 0000	0000	0011 0011	011 0000
1	0000 0001	0001	0011 0100	011 0001
2	0000 0010	0011	0011 0101	011 0010
3	0000 0011	0010	0011 0110	011 0011
4	0000 0100	0110	0011 0111	011 0100
5	0000 0101	0111	0011 1000	011 0101

32. Convert the decimal number 23 into each of the following:

a. ASCII
b. Pure binary
c. BCD
d. Hexadecimal
e. Octal
f. Excess-3

33. If a flashlight is used to transmit ASCII codes, with a dot (short flash) representing a 0, and a dash (long flash) representing a 1, what is the following message?

0000010	1010000	1110010	1100001	1100011	1110100	1101001
1100011	1100101	0100000	1001101	1100001	1101011	1100101
1110011	0100000	1010000	1100101	1110010	1100110	1100101
1100011	1110100	0000011				

34. Convert the following:

a. $1110110_2 = \rule{3cm}{0.4pt}_{16}$
b. $175_{10} = \rule{3cm}{0.4pt}_{2}$
c. $ABC_{16} = \rule{3cm}{0.4pt}_{2}$
d. $00110110_2 = \rule{3cm}{0.4pt}_{8}$

35. Convert the following message to ASCII code: Easily said, not easily done.

Chapter 3

OBJECTIVES

After completing this chapter, you will be able to:

1. Describe the evolution of the two-state switch and how it is used to represent binary data.
2. Explain how the diode and transistor semiconductor switches can be used to construct digital electronic logic gates.
3. Describe the discrete circuit construction, truth table, operation, and an application for the following basic logic-gate types.
   a. The OR Gate
   b. The AND Gate
4. Describe the discrete circuit construction, truth table, operation, and an application for the following inverting logic-gate types.
   a. The NOT Gate
   b. The NOR Gate
   c. The NAND Gate
5. Describe the discrete circuit construction, truth table, operation, and an application for the following exclusive logic gate types.
   a. The XOR Gate
   b. The XNOR Gate
6. Explain the major differences between the traditional and IEEE/ANSI logic-gate symbols.

Logic Gates

OUTLINE

VIGNETTE

Back to the Future

Each and every part of a digital electronic computer is designed to perform a specific task. One would imagine that the function and operation of the computer's basic blocks were first thought of by some pioneer in the twentieth century; in fact, two of the key units were first described in 1833 by a man named Charles Babbage.

Born in England in 1791, Babbage became very well known for both his mathematical genius and eccentric personality. For many years he occupied the Cambridge Chair of Mathematics—once held by Isaac Newton—and although he never delivered a single lecture, he wrote a great number of papers on a variety of subjects ranging from politics to manufacturing techniques. He also helped develop several practical devices, including the tachometer and the railroad cowcatcher.

Babbage's ultimate pursuit, however, was that of mathematical accuracy. He delighted in spotting errors in everything from logarithmic tables to poetry. In fact, he once wrote to poet Alfred Lord Tennyson, pointing out an inaccuracy in his line "every moment dies a man—every moment one is born." Babbage explained to Tennyson that since the world population was actually increasing and not, as he indicated, remaining constant, the line should be rewritten to read "every moment dies a man—every moment one and one-sixteenth is born."

In 1822 Babbage described in a paper and built a model of what he called "a difference engine," which could be used to calculate mathematical tables. The Royal Society of Scientists described his machine as "highly deserving of public encouragement," and a year later the government awarded Babbage £1,500 for his project. Babbage originally estimated that the project would take three years; however, the design had its complications and after ten years of frustrating labor, in which the government grants increased to £17,000, Babbage was still no closer to completion. Finally the money stopped and Babbage reluctantly decided to let his brainchild go.

In 1833 Babbage developed an idea for a much more practical machine, which he named "the analytical engine." It was to be a machine of a more general nature that could be used to solve a variety of problems, depending on instructions supplied by the operator. It would include two units called the "mill" and the "store," both of which would be made of cogs and wheels. The store, which was equivalent to a modern-day computer memory, could hold up to 100 forty-digit numbers. The mill, which was equivalent to a modern computer's arithmetic and logic unit (ALU), could perform both arithmetic and logic operations on variables or numbers retrieved from the store, and the result could be stored in the store and then acted upon again or printed out. The program of instructions directing these operations would be fed into the analytical engine in the form of punched cards.

Sadly, the analytical engine was never built. All that remains are the volumes of descriptions and drawings and a section of the mill and printer built by Babbage's son, who also had to concede defeat. For Charles Babbage it was, unfortunately, a lifetime of frustration to have conceived the basic building blocks of the modern computer a century before the technology existed to build it.

INTRODUCTION

Within any digital electronic system you will find that diodes and transistors are used to construct **logic gate circuits.** Logic gates are in turn used to construct flip-flop circuits, and flip-flops are used to construct register, counter, and a variety of other circuits. These logic gates, therefore, are used as the basic building blocks for all digital circuits, and their purpose is to control the movement of binary data and binary instructions. Logic gates, and all other digital electronic circuits, are often referred to as **hardware** circuits. By definition, the hardware of a digital electronic system includes all of the electronic, magnetic, and mechanical devices of a digital system. In contrast, the **software** of a digital electronic system includes the binary data (like pure binary and ASCII codes) and binary instructions that are processed by the digital electronic system hardware. To use an analogy, we could say that a compact disk player is hardware, and the music information stored on a compact disk and processed by the player is software. Just as a CD player is useless without CDs, a digital system's hardware is useless without software. In other words, the information on the CD determines what music is played, and similarly the digital software determines the actions of the digital electronic hardware.

Logic Gate Circuits
Circuits containing predictable gate functions that either open or close their outputs.

Hardware
A term used to describe the electronic, magnetic, and mechanical devices of a digital system.

Software
A term used to describe the binary instructions and data processed by digital electronic system hardware.

Every digital electronic circuit uses logic-gate circuits to manipulate the coded pulses of binary language. These logic-gate circuits are constructed using diodes and transistors, and they are the basic decision-making elements in all digital circuits. In this chapter we will begin by discussing the evolution of the electronic switch, and then show how diode and transistor switches are used to construct these digital logic-gate hardware circuits.

3-1 HARDWARE FOR BINARY SYSTEMS

Digital circuits are often referred to as switching circuits because their control devices (diodes and transistors) are switched between the two extremes of ON and OFF. These digital circuits are also called **two-state circuits** because their control devices are driven into one of two states: either into the saturation state (fully ON), or cutoff state (fully OFF). These two modes of operation are used to represent the two binary digits of 1 and 0.

Two-state Circuits
Switching circuits that are controlled to be either ON or OFF.

To develop a digital electronic system that could manipulate binary information, inventors needed a two-state electronic switch. Early machines used mechanical switches and electromechanical relays—like the examples in Figure 3-1(a) and (b)—to represent binary data by switching current ON or OFF. These

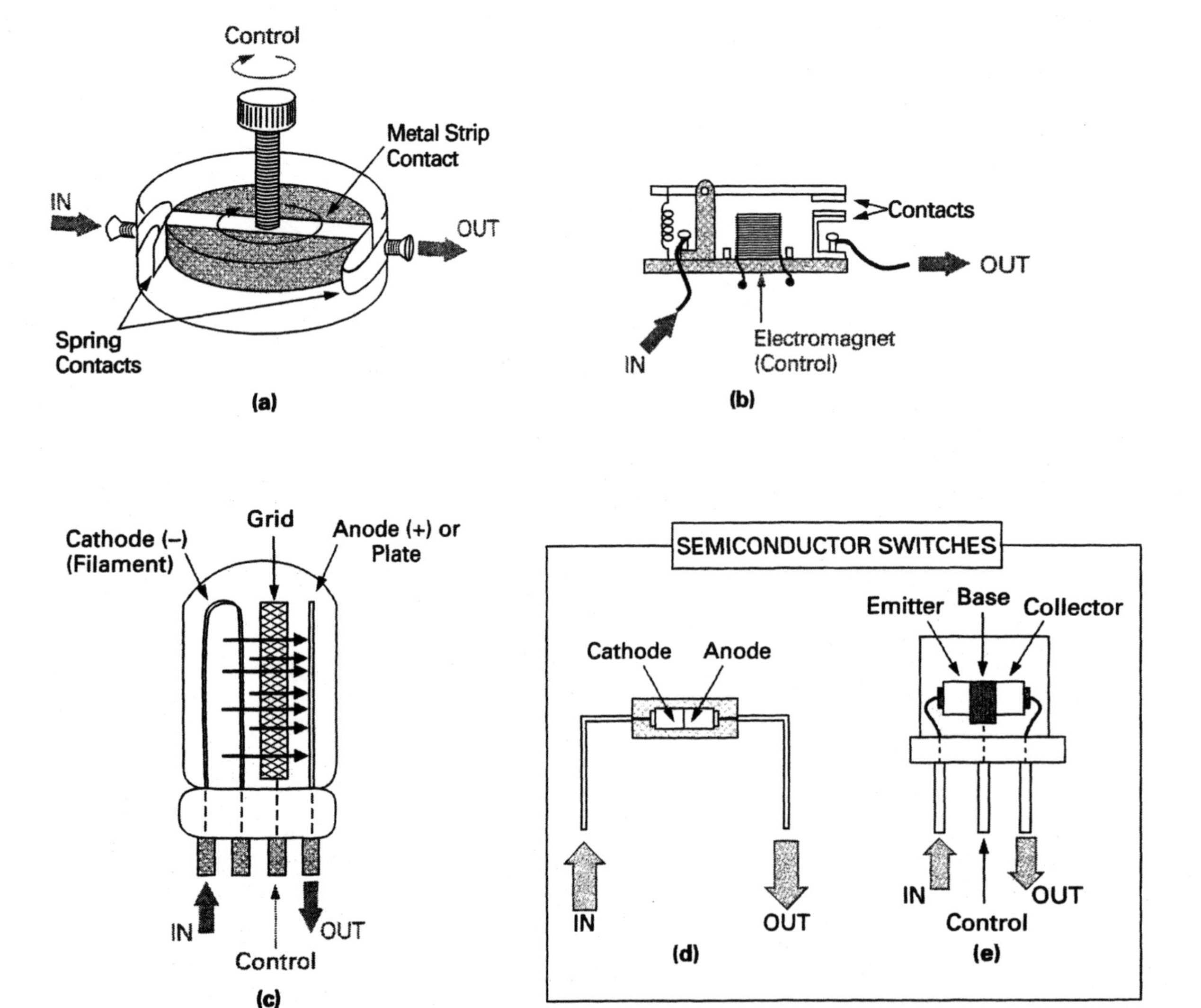

FIGURE 3-1 The Evolution of the Switch for Digital Electronic Circuits. (a) A 19th Century Mechanical Turn Switch. (b) An Electromechanical Relay. (c) A 1906 Vacuum Tube. (d) A 1939 Semiconductor Diode. (e) A 1948 Semiconductor Transistor.

mechanical devices were eventually replaced by the vacuum tube, shown in Figure 3-1(c), which, unlike switches and relays, had the advantage of no moving parts. The vacuum tube, however, was bulky, fragile, had to warm up, and consumed an enormous amount of power. Finally, compact and low-power digital electronic circuits and systems became a reality with the development of semiconductor diode and transistor switches, which are shown in Figures 3-1(d) and 3-1(e). Let us see how we can use these semiconductor switches to construct a logic gate.

3-1-1 *Using the Diode to Construct a Logic Gate*

A basic logic gate could be constructed using two junction diodes and a resistor, as shown in Figure 3-2(a). This circuit is called a logic gate because it will always

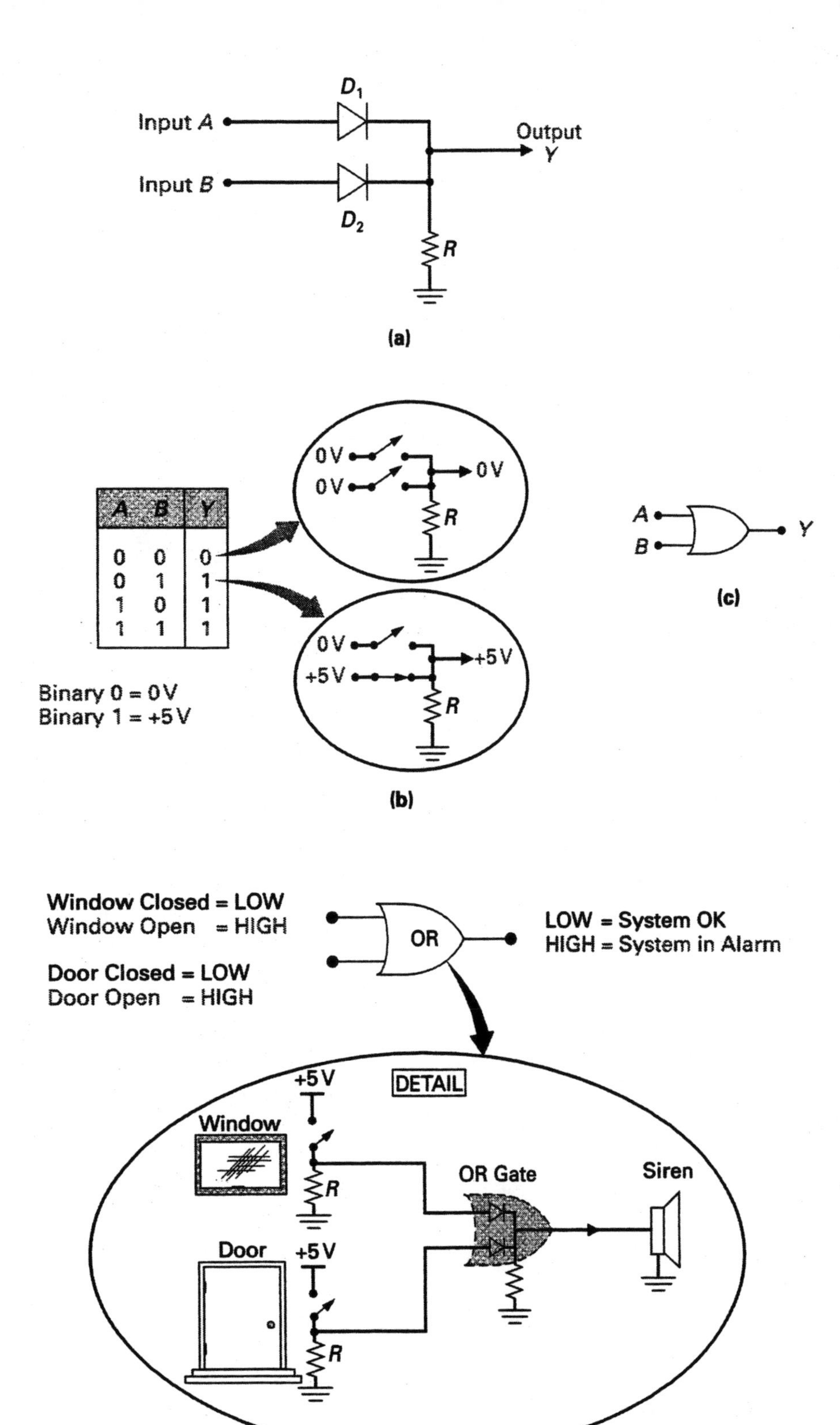

FIGURE 3-2 The Diode Being Used in a Digital Electronic Circuit. (a) Basic OR Gate Circuit. (b) OR Gate Function Table. (c) OR Gate Schematic Symbol. (d) OR Gate Security System Application.

produce a *logical* or *predictable* output, and this output will depend on the condition of its inputs. For example, Figure 3-2(b) shows a table which lists how this logic gate will react to all of the different input possibilities. The two binary states 0 and 1 are represented in the circuit as two voltages:

Binary 0 = 0 V (LOW voltage)

Binary 1 = 5 V (HIGH voltage)

If you study the table in Figure 3-2(b), you can see that when both inputs are LOW or at 0 V ($A = 0$, $B = 0$), both diodes will be OFF since the anodes are at 0 V (due to the inputs) and the cathodes of the diodes are at 0 V (due to the pull-down resistor, *R*). This input combination of $A = 0$ and $B = 0$ will therefore turn both diodes OFF and always produce an output at *Y* of 0 V (binary 0).

In all other combinations in the table in Figure 3-2(b), a HIGH or +5 V (logical 1) input is applied to either or both of the inputs *A* and *B*. Any HIGH input will always turn ON its associated diode since the anode will be at +5 V and the cathode will be at 0 V via *R*. When ON, a diode is equivalent to a closed switch, and therefore the +5 V input is switched through to the output, making *Y* equal to +5 V, or logical 1. Therefore, as far as summarizing the operation of this circuit, we could say that if A *input OR* B *input is HIGH, the output* Y *will be HIGH.* This circuit was named the OR gate because of this behavior.

The schematic symbol for the OR gate is shown in Figure 3-2(c). To show how the OR gate circuit could be used, consider the simple security system shown in Figure 3-2(d). In this application, if either the window OR the door were to open, a switch contact would close and deliver a HIGH input to the OR gate. As we know, any HIGH input to an OR gate will always generate a HIGH output, and in this circuit the +5 V out of the OR gate will activate the siren.

In summary, a logic gate accepts inputs in the form of HIGH or LOW voltages, judges these input combinations based on a predetermined set of rules, and then produces a single output in the form of a HIGH or LOW voltage. The term logic is used because the output is predictable or logical, and the term gate is used because only certain input combinations will "unlock the gate." For example, any HIGH input to an OR gate will unlock the gate and allow the HIGH at the input to pass through to the output.

3-1-2 ***Using the Transistor to Construct a Logic Gate***

NOT Gate
A logic gate with only one input and one output that will invert the binary input.

Figure 3-3(a) shows how the transistor can be used to construct another type of digital logic gate. This gate is called the **NOT gate** or, more commonly, the **INVERTER gate.** The basic NOT gate circuit is constructed using one NPN transistor and two resistors. This logic gate has only one input (*A*) and one output (*Y*), and its schematic symbol is shown in Figure 3-3(b). Figure 3-3(c) shows how this logic gate will react to the two different input possibilities.

When the input is 0 V (logic 0), the transistor's base-emitter PN diode will be reverse-biased and the transistor will turn OFF. Referring to the upper inset for this circuit condition in Figure 3-3(c), you can see that the OFF transistor

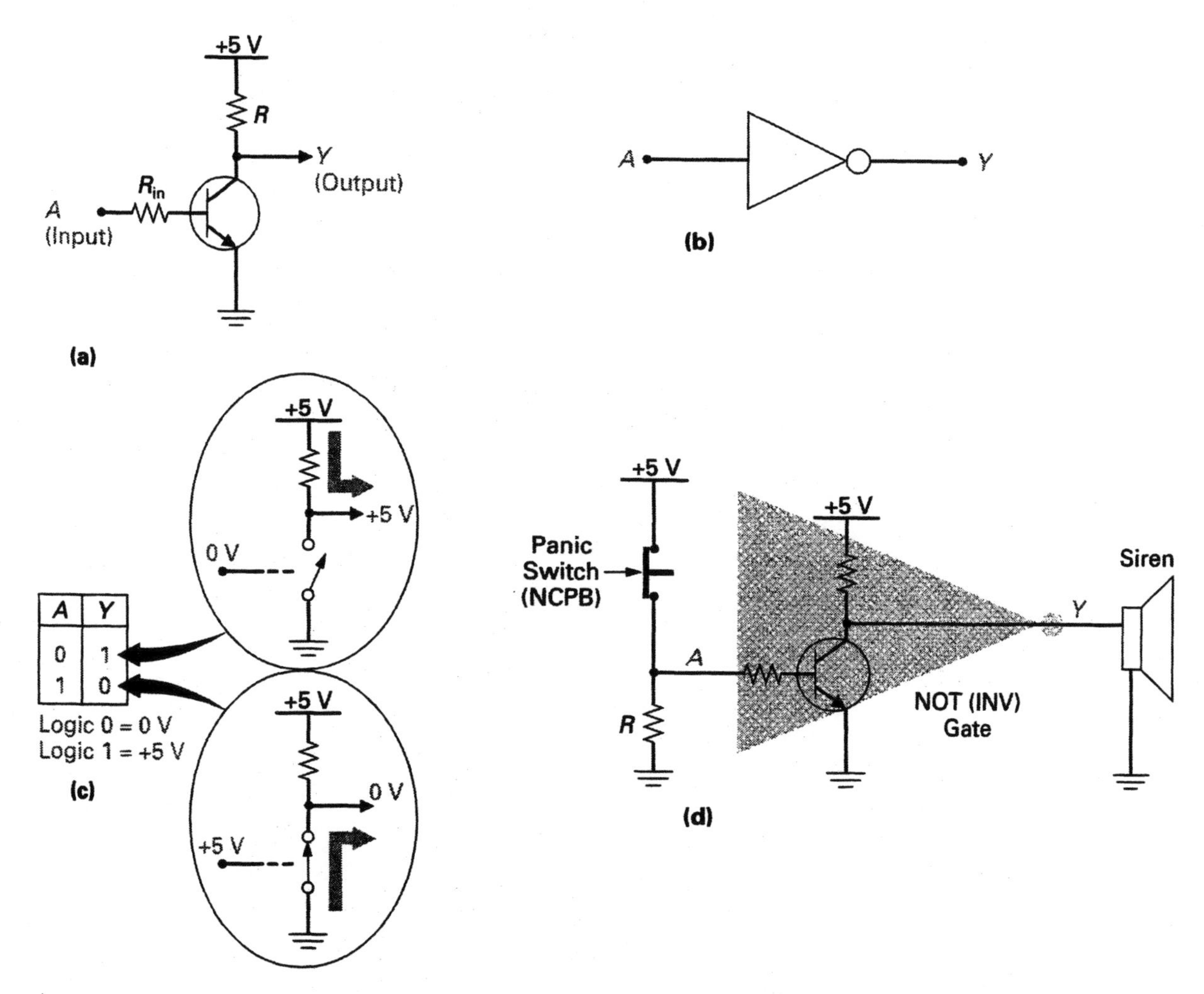

FIGURE 3-3 The Transistor Being Used in a Digital Electronic Circuit. (a) Basic NOT or INVERTER Gate Circuit. (b) NOT Gate Schematic Symbol. (c) NOT Gate Function Table. (d) NOT Gate Security System Application.

is equivalent to an open switch between emitter and collector, and therefore the +5 V supply voltage will be connected to the output. A logic 0 input (0 V) will therefore be converted or inverted to a logic 1 output (+5 V).

On the other hand, when the input is +5 V (logic 1), the transistor's base-emitter PN diode will be forward-biased and the transistor will turn ON. Referring to the lower inset for this circuit condition in Figure 3-3(c), you can see that the ON transistor is equivalent to a closed switch between emitter and collector, and therefore 0 V will be connected to the output. A logic 1 input (+5 V) will therefore be converted or inverted to a logic 0 output (0 V).

Referring to the function table in Figure 3-3(c), you can see why the operation of this gate can be summarized with the statement that *the output logic level is "NOT" the same as the input logic level* (a 1 input is NOT the same as the 0 output, or the 0 input is NOT the same as the 1 output), hence the name NOT gate.

As an application, Figure 3-3(d) shows how a NOT or INVERTER (INV) gate could be used to invert an input control signal. In this circuit, a normally closed push-button (NCPB) switch is used as a panic switch to activate a siren in a security system. Since the push button is normally closed, it will produce +5 V at *A* when it is not in an alarm condition. If this voltage were connected directly to the siren, the siren would be activated incorrectly. By including the NOT gate between the switch circuit and the siren, the normally HIGH output of the NCPB will be inverted to a LOW and not activate the siren when there is no alarm. When the panic switch is pressed, however, the NCPB contacts will open, producing a LOW input voltage to the NOT gate. This LOW input will be inverted to a HIGH output and activate the siren.

The OR gate and NOT gate are only two examples of logic gates used in digital electronics. In the remainder of this chapter we will examine in detail these two logic gates and other basic, inverting, and exclusive logic gates.

SELF-TEST REVIEW QUESTIONS FOR SECTION 3-1

1. The ____________ includes the electronic, magnetic, and mechanical devices of a digital electronic system.
2. The ____________ includes the binary data and binary instructions that are processed by the digital electronic system.
3. Which two-state switch was the first to have no moving parts?
4. Which two semiconductor switches are used to implement digital electronic circuits?

3-2 BASIC LOGIC GATES

The two basic types of logic gates used in digital circuits are the OR gate and the AND gate. We will begin with the OR gate, which was introduced in the previous section.

3-2-1 The OR Gate

OR Gate
A logic gate that will give a HIGH output if either of its inputs are HIGH.

Truth Table or Function Table
A table used to show the action of a device as it reacts to all possible input combinations.

The **OR gate** can have two or more inputs, but will always have a single output. Figure 3-4(a) shows how an OR gate can be constructed using two diodes and a resistor. Figure 3-4(b) shows a table listing all of the input possibilities for this two-input OR gate. This table is often referred to as a **truth table** or **function table** since it details the "truth" or the way in which this circuit will function. The insets in Figure 3-4(b) simplify how the circuit will basically produce one of two outputs. If both inputs are LOW, both diode switches are open and the output will be LOW. On the other hand, any HIGH input will cause the associated diode switch to close and connect the HIGH input through to the output. We could summarize the operation of this gate, therefore, by saying that *if either*

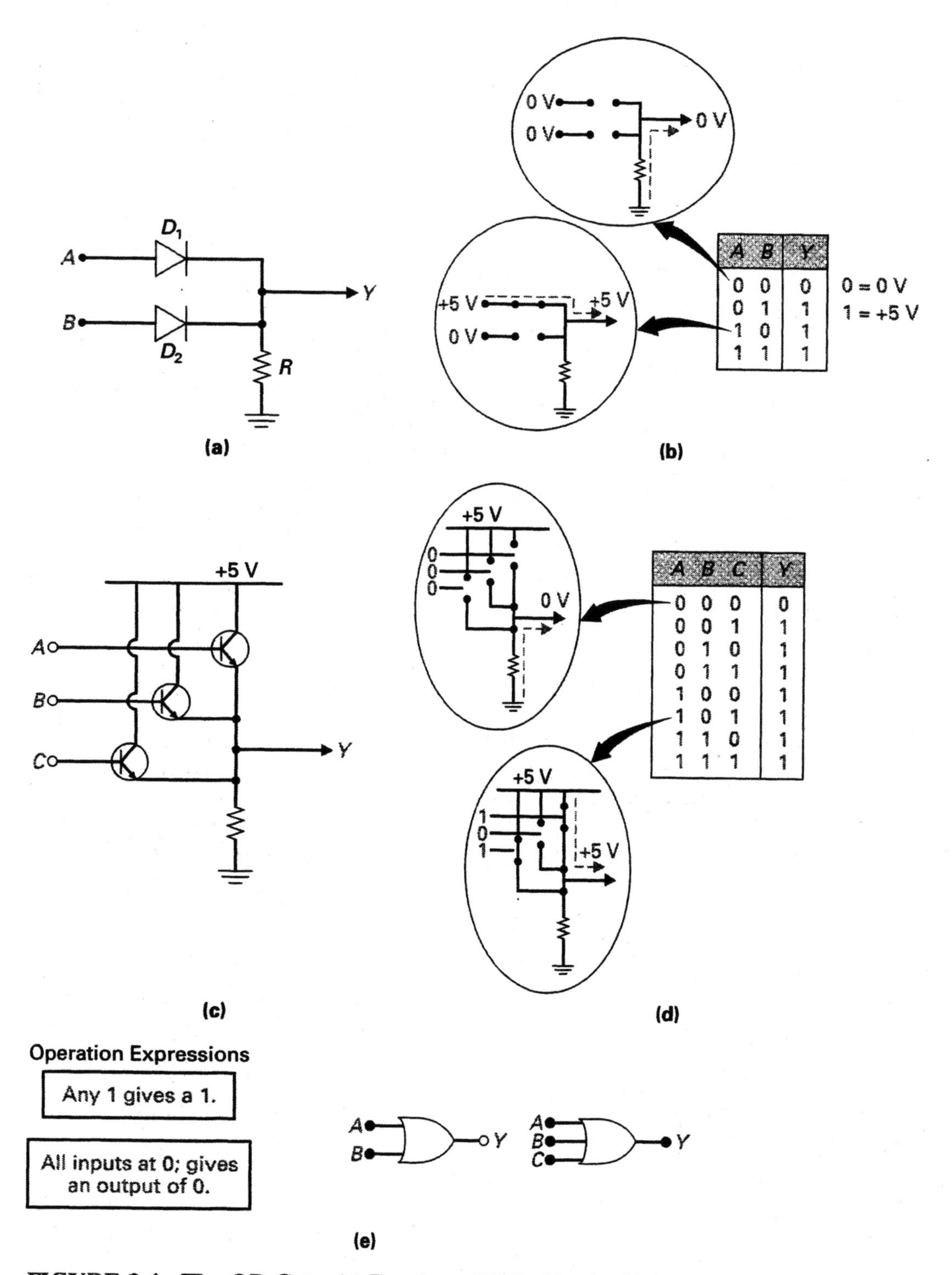

FIGURE 3-4 The OR Gate. (a) Two-Input Diode Circuit. (b) Two-Input Truth Table. (c) Three-Input Transistor Circuit. (d) Three-Input Truth Table. (e) OR Gate Symbols and Operation Expressions.

A *OR* B *is HIGH, the output* Y *will be HIGH, and only when both inputs are LOW will the output be LOW.*

Figure 3-4(c) shows how an OR gate can be constructed using transistors as switches instead of diodes. In this OR gate circuit, three two-state transistor switches are included to construct a 3-input OR gate. The truth table for this OR gate circuit is shown in Figure 3-4(d). Although another input has been added to this circuit, the action of the gate still remains the same since any HIGH input will still produce a HIGH output, and only when all of the inputs are LOW will the output be LOW. To explain the operation of this transistor logic gate in more detail, the insets in Figure 3-4(d) show how the transistor switches operate to produce one of two output voltage levels. When all inputs are LOW, the transistor's base emitter junctions are reverse-biased, all transistors are OFF, and the pull-down resistor will take the Y output LOW. On the other hand, any HIGH input to the base of an NPN transistor will forward-bias the associated transistor and cause the transistor's collector-to-emitter junction to be equivalent to a closed switch, and therefore switch the HIGH ($+V_{CC}$) collector supply voltage through to the Y output.

Figure 3-4(e) shows the logic schematic symbols for a two- and three-input OR gate and the abbreviated operation expressions for the OR gate. Logic gates with a large number of inputs will have a large number of input combinations; however, the gate's operation will still be the same. To calculate the number of possible input combinations, you can use the following formula:

$$n = 2^x$$

n = number of input combinations
2 is included because we are dealing with a base-2 number system
x = number of inputs

■ **EXAMPLE:**

Construct a truth table for a four-input OR gate and show the output logic level for each and every input combination.

■ ***Solution:***

Figure 3-5(a) shows the truth table for a four-input OR gate. The number of possible input combinations will be

$n = 2^x$

$n = 2^4 = 16$

(Calculator Sequence: press keys, 2 y^x 4 =)

With only two possible digits (0 and 1) and four inputs, we can have a maximum of sixteen different input combinations. Whether the OR gate has 2, 3, 4, or 444 inputs, it will still operate in the same predictable or logical way: when all inputs are LOW, the output will be LOW, and when any input is HIGH, the output will be HIGH. This is shown in the Y column in Figure 3-5(a).

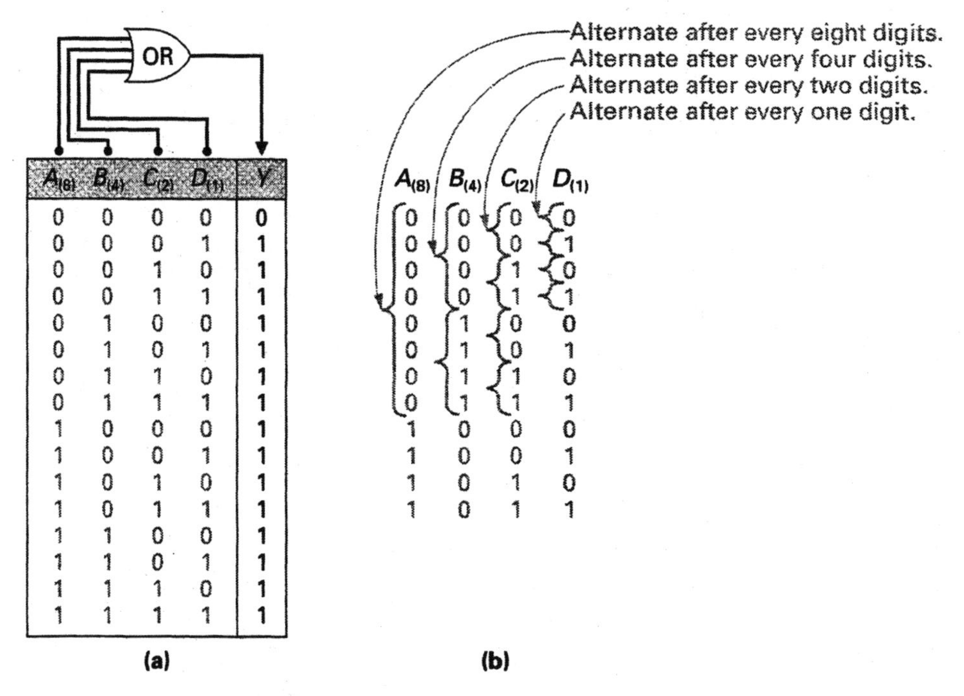

$A_{(8)}$	$B_{(4)}$	$C_{(2)}$	$D_{(1)}$	Y
0	0	0	0	0
0	0	0	1	1
0	0	1	0	1
0	0	1	1	1
0	1	0	0	1
0	1	0	1	1
0	1	1	0	1
0	1	1	1	1
1	0	0	0	1
1	0	0	1	1
1	0	1	0	1
1	0	1	1	1
1	1	0	0	1
1	1	0	1	1
1	1	1	0	1
1	1	1	1	1

$A_{(8)}$	$B_{(4)}$	$C_{(2)}$	$D_{(1)}$
0	0	0	0
0	0	0	1
0	0	1	0
0	0	1	1
0	1	0	0
0	1	0	1
0	1	1	0
0	1	1	1
1	0	0	0
1	0	0	1
1	0	1	0
1	0	1	1

FIGURE 3-5 A Four-Input OR Gate.

An easy way to construct these truth tables is to first calculate the maximum number of input combinations and then start in the units (1s) column (D column) and move down, alternating the binary digits after every single digit (0101010101, and so on) up to the maximum count, as shown in Figure 3-5(b). Then go to the twos (2s) column (C column), and move down, alternating the binary digits after every two digits (001100110011, and so on) up to the maximum count. Then go to the fours (4s) column (B column), and move down, alternating the binary digits after every four digits (000011110000, and so on) up to the maximum count. Finally, go to the eights (8s) column (A column), and move down, alternating the binary digits after every eight digits (0000000011111111, and so on) up to the maximum count.

These input combinations—or binary words—will start at the top of the truth table with binary 0 and then count up to a maximum value that is always one less than the maximum number of combinations. For example, with the four-input OR gate in Figure 3-5(a), the truth table begins with a count of 0000_2 (0_{10}) and then counts up to a maximum of 1111_2 (15_{10}). The maximum count within a truth table (1111_2 or 15_{10}) is always one less than the maximum number of combinations (16_{10}) because 0000_2 (0_{10}) is one of the input combinations (0000, 0001, 0010, 0011, 0100, 0101, 0110, 0111, 1000, 1001, 1010, 1011, 1100, 1101, 1110, 1111 is a total of sixteen different combinations, with fifteen being the maximum count). Stated with a formula:

$$Count_{max} = 2^X - 1$$

For example, in Figure 3-5: $Count_{max} = 2^4 - 1 = 16 - 1 = 15$
(Calculator Sequence: press keys, 2 y^X 4 − 1 =)

■ **EXAMPLE:**

Referring to the circuit in Figure 3-6(a), list the 4-bit binary words generated at the outputs of the four OR gates when each of the 0 through 9 push buttons are pressed.

■ ***Solution:***

An encoder circuit is a code generator, and the circuit in Figure 3-6(a) is a decimal-to-binary code generator circuit. Whenever a decimal push button is pressed, a 4-bit binary word equivalent to the decimal value is generated at the outputs of OR gates *D, C, B,* and *A.* The table in Figure 3-6(b) lists the pure binary codes generated for each of the decimal push buttons. For example, when push button 6 is pressed, a HIGH input is applied to OR gates *C* and *B,* and

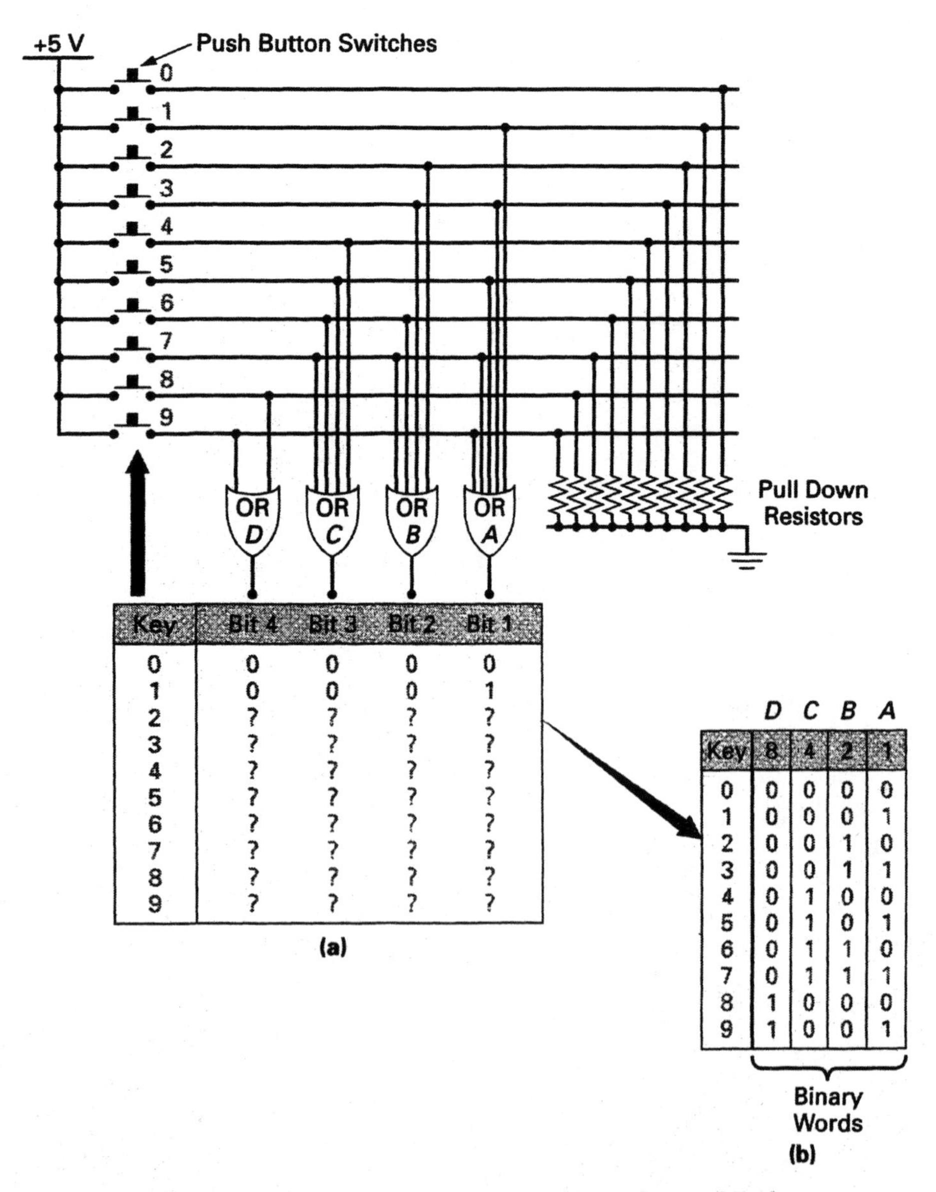

Key	Bit 4	Bit 3	Bit 2	Bit 1
0	0	0	0	0
1	0	0	0	1
2	?	?	?	?
3	?	?	?	?
4	?	?	?	?
5	?	?	?	?
6	?	?	?	?
7	?	?	?	?
8	?	?	?	?
9	?	?	?	?

Key	8	4	2	1
0	0	0	0	0
1	0	0	0	1
2	0	0	1	0
3	0	0	1	1
4	0	1	0	0
5	0	1	0	1
6	0	1	1	0
7	0	1	1	1
8	1	0	0	0
9	1	0	0	1

FIGURE 3-6 A Decimal-to-Binary Encoder Circuit Using OR Gates.

therefore their outputs will go HIGH, while the outputs of OR gates *C* and *A* will stay LOW because all of their inputs are LOW due to the pull-down resistors. In this instance, the 4-bit binary code 0110_2 (6_{10}) will be generated.

Rather than construct OR gates using discrete components such as the diode and transistor circuits shown in Figure 3-4(a) and (c), you can experiment with some of the digital integrated circuits (ICs) available from semiconductor manufacturers. These digital ICs contain complete logic-gate circuits, and in the following chapter we will be examining all of the different types of digital ICs in more detail. For now, however, Figure 3-7(a) introduces the "7432 digital IC" which contains four (quad) two-input OR gates within a single 14-pin package. All four OR gates within this IC will operate if power is connected to the IC by using a dc power supply to apply +5 V to pin 14 and ground to pin 7. To test the OR gate that has its inputs connected to pins 1 and 2 and its output at pin 3, we will apply a square wave from a function generator to the OR gate's pin 1 input, a HIGH/LOW control input to the OR gate's pin 2 input, and monitor the output of the OR gate at pin 3 with an oscilloscope.

Referring to the timing diagram in Figure 3-7(b), you can see that whenever the square wave input (pin 1) or the control input (pin 2) are HIGH (shown shaded), the output will also be HIGH. Only when both inputs to the OR gate are LOW will the output be LOW. You may want to try constructing this test circuit in a lab and test each one of the four OR gates in a 7432 IC.

3-2-2 The AND Gate

Like the OR gate, an **AND gate** can have two or more inputs, but will always have a single output. Figure 3-8(a) shows how an AND gate can be constructed using two diodes and a resistor. Figure 3-8(b) shows the truth table for this two-input AND gate. The insets in Figure 3-8(b) simplify how the circuit will basically produce one of two outputs. Since both of the diode anodes are connected to +5 V via *R*, any LOW input to the AND gate will turn ON the associated diode, as shown in the first inset in Figure 3-8(b). In this instance, the diode or diodes will be equivalent to a closed switch, and therefore switch the LOW input through to the output. On the other hand, when both *A* AND *B* inputs are HIGH, both diodes will be reverse-biased and equivalent to open switches, and the output *Y* will be pulled HIGH, as shown in the second inset in Figure 3-8(b). We could summarize the operation of this gate, therefore, by saying that *any LOW input will cause a LOW output, and only when both* A *AND* B *inputs are HIGH will the output* Y *be HIGH.*

AND Gate
A logic gate that will give a HIGH output only if all inputs are HIGH.

Figure 3-8(c) shows how a three-input AND gate can be constructed using three two-state transistor switches. The truth table for this AND gate circuit is shown in Figure 3-8(d). Although another input has been added to this circuit, the action of the gate still remains the same since any LOW input will still produce a LOW output, and only when all of the inputs are HIGH will the output be HIGH. To explain the operation of this transistor logic gate in more detail, the insets in Figure 3-8(d) show how the transistor switches operate to produce one of two output voltage levels. When any of the inputs are LOW,

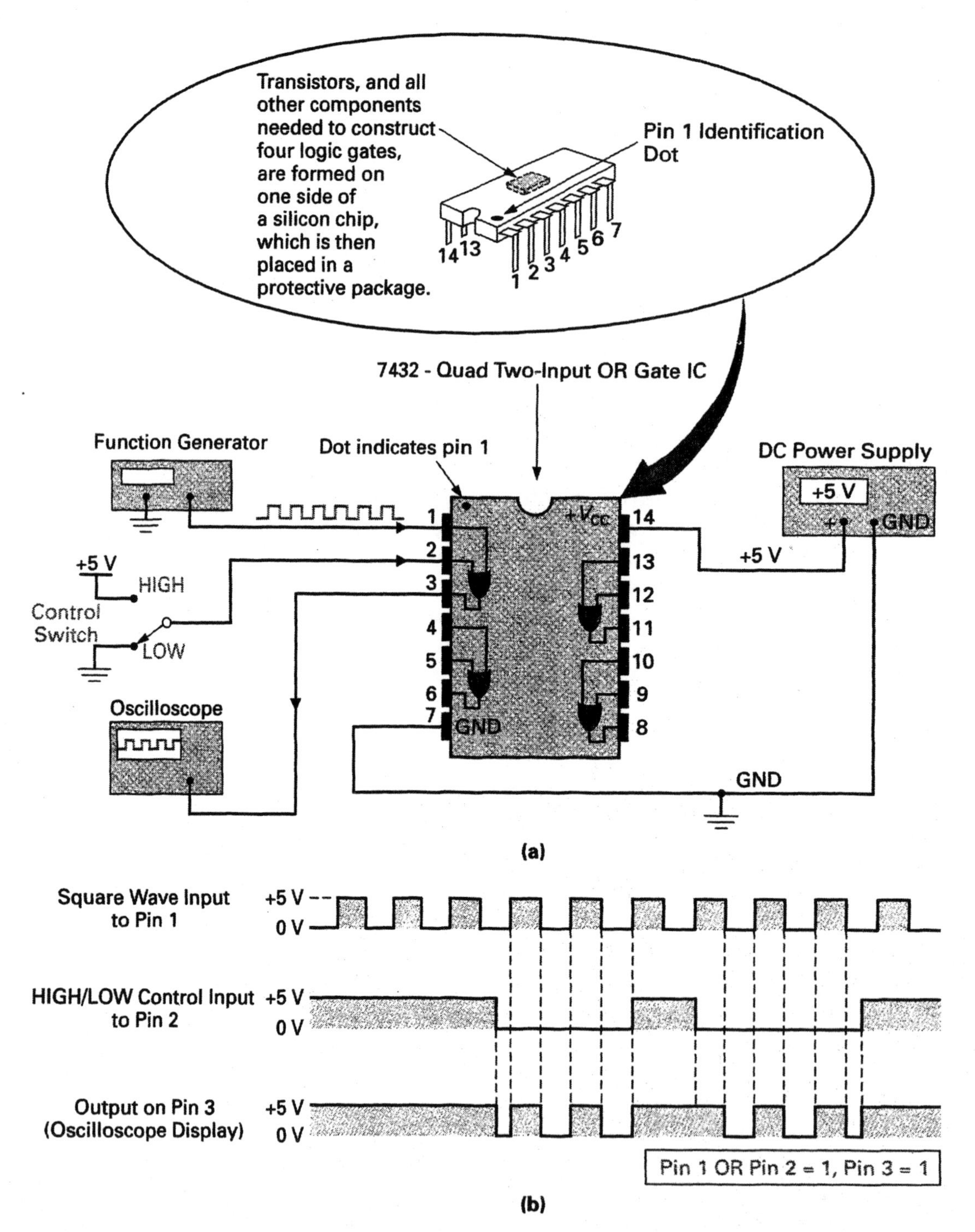

FIGURE 3-7 Testing the Operation of Digital IC Containing Four Two-Input OR Gates.

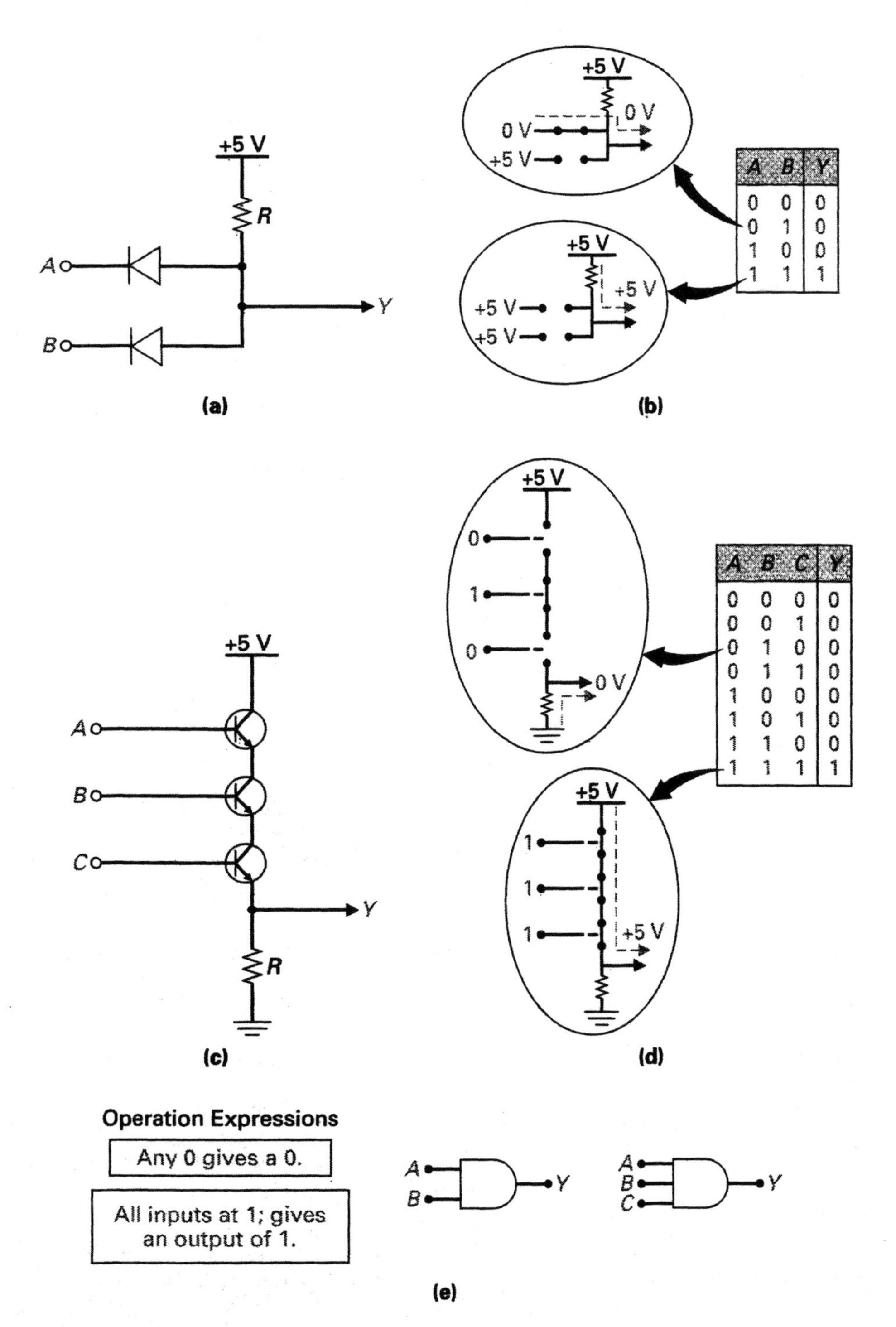

A	B	Y
0	0	0
0	1	0
1	0	0
1	1	1

A	B	C	Y
0	0	0	0
0	0	1	0
0	1	0	0
0	1	1	0
1	0	0	0
1	0	1	0
1	1	0	0
1	1	1	1

FIGURE 3-8 The AND Gate. (a) Two-Input Diode Circuit. (b) Two-Input Truth Table. (c) Three-Input Transistor Circuit. (d) Three-Input Truth Table. (e) AND Gate Symbols.

the transistor's base-emitter junction is reverse-biased, the OFF transistor is equivalent to an open switch between collector and emitter, and the pull-down resistor will take the Y output LOW. On the other hand, when all inputs are HIGH, all of the NPN transistor's base-emitter junctions will be forward-biased, causing all of the transistor's collector-to-emitter junctions to be equivalent to closed switches, and therefore switching the HIGH ($+V_{CC}$) collector supply voltage through to the Y output.

Figure 3-8(e) shows the logic schematic symbols for a two- and three-input AND gate, and the abbreviated operation expressions for the AND gate.

■ **EXAMPLE:**

Develop a truth table for a five-input AND gate. Show the output logic level for each and every input combination, and indicate the range of values within the truth table.

■ ***Solution:***

Figure 3-9 shows the truth table for a five-input AND gate. The number of possible input combinations will be

$$n = 2^x$$
$$n = 2^5 = 32$$

A five-input AND gate will still operate in the same predictable or logical way: when any input is LOW, the output will be LOW, and when all inputs are HIGH, the output will be HIGH. This is shown in the Y column in Figure 3-9.

The range of values within the truth table will be

$$Count_{\max} = 2^5 - 1 = 32 - 1 = 31$$

With the five-input OR gate in Figure 3-9, the truth table begins with a count of 00000_2 (0_{10}) and then counts up to a maximum of 11111_2 (31_{10}).

■ **EXAMPLE:**

Referring to the circuit in Figure 3-10(a), describe what would be present at the output of the AND gate for each of the CONTROL switch positions.

■ ***Solution:***

When the CONTROL switch is put in the ENABLE position, a HIGH is applied to the lower input of the AND gate, as shown in Figure 3-10(b). In this mode, the output Y will follow the square wave input, since when the square wave is HIGH, the AND gate inputs are both HIGH, and Y is HIGH. When the square wave is LOW, the AND gate will have a LOW input, and Y will be LOW. The Y output, therefore, follows the square wave input, and the AND gate is said to be equivalent to a closed switch, as shown in Figure 3-10(b). On the other hand, when the CONTROL switch is put in the DISABLE position, a LOW is applied to the lower input of the AND gate, as shown in Figure 3-10(c). In this mode, the output Y will always remain LOW, since any LOW

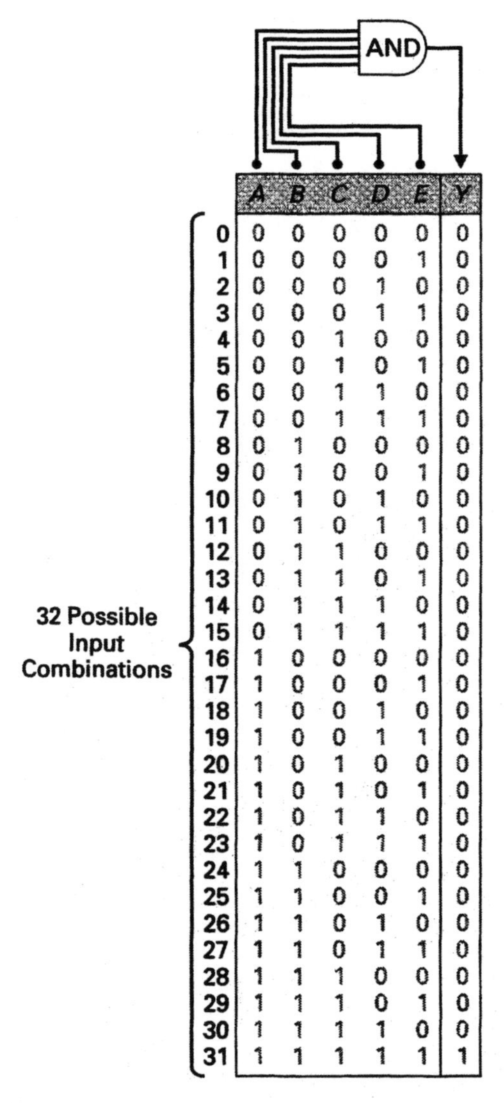

	A	B	C	D	E	Y
0	0	0	0	0	0	0
1	0	0	0	0	1	0
2	0	0	0	1	0	0
3	0	0	0	1	1	0
4	0	0	1	0	0	0
5	0	0	1	0	1	0
6	0	0	1	1	0	0
7	0	0	1	1	1	0
8	0	1	0	0	0	0
9	0	1	0	0	1	0
10	0	1	0	1	0	0
11	0	1	0	1	1	0
12	0	1	1	0	0	0
13	0	1	1	0	1	0
14	0	1	1	1	0	0
15	0	1	1	1	1	0
16	1	0	0	0	0	0
17	1	0	0	0	1	0
18	1	0	0	1	0	0
19	1	0	0	1	1	0
20	1	0	1	0	0	0
21	1	0	1	0	1	0
22	1	0	1	1	0	0
23	1	0	1	1	1	0
24	1	1	0	0	0	0
25	1	1	0	0	1	0
26	1	1	0	1	0	0
27	1	1	0	1	1	0
28	1	1	1	0	0	0
29	1	1	1	0	1	0
30	1	1	1	1	0	0
31	1	1	1	1	1	1

FIGURE 3-9 A Five-Input AND Gate.

input to an AND gate will always result in a LOW output. In this situation, the AND gate is said to be equivalent to an open switch, as shown in Figure 3-10(c).

When used in applications such as this, the AND gate is said to be acting as a **controlled switch.**

Controlled Switch An electronically controlled switch.

Like the OR gate digital IC, there are several digital ICs available containing AND gates. Figure 3-11(a) introduces the "7408 digital IC," which contains four (quad) two-input AND gates, and shows how to test the AND gates to see if they operate as controlled switches. As before, all four AND gates within this IC will operate if power is connected to the IC by using a dc power supply to apply +5 V to pin 14 and ground to pin 7. To test the AND gate that

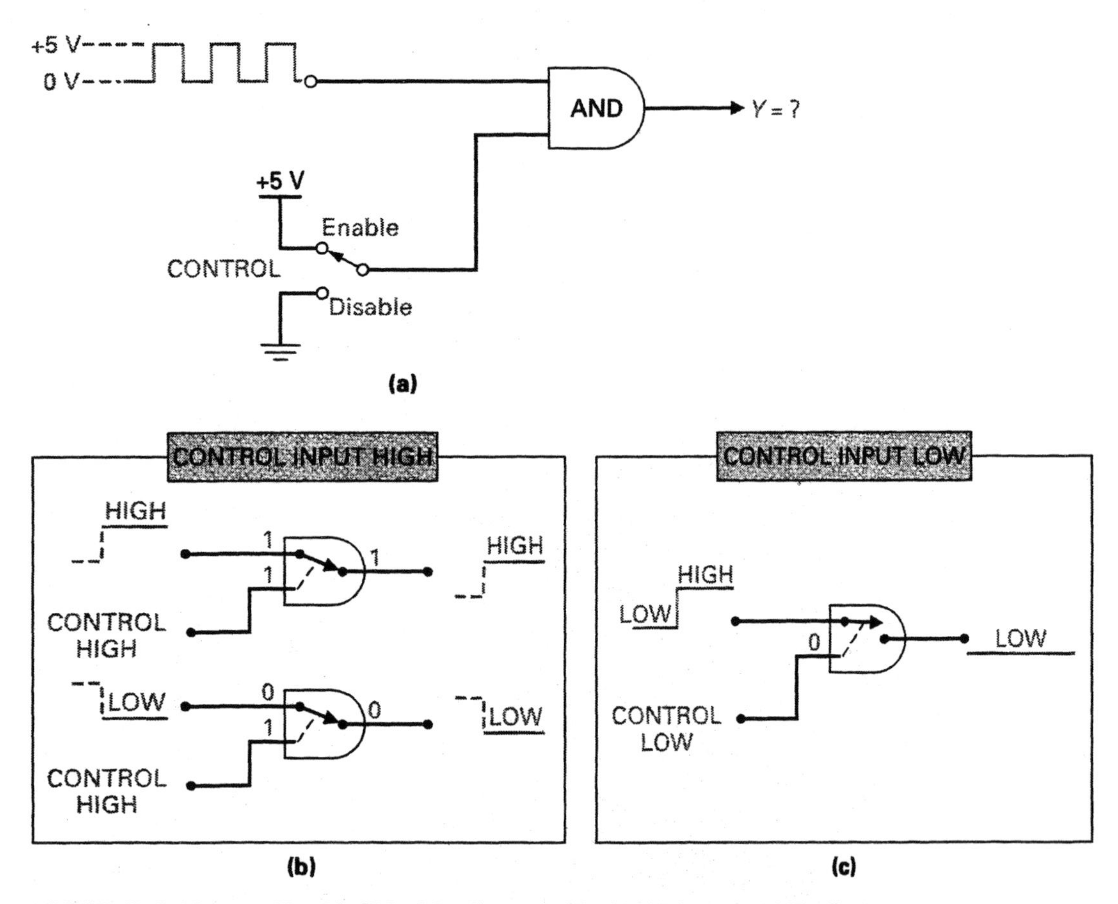

FIGURE 3-10 An Enable/Disable Control Circuit Using an AND Gate.

has its inputs connected to pins 1 and 2 and its output at pin 3, we will apply a square wave from a function generator to the AND gate's pin 1 input, a HIGH/LOW control input to the AND gate's pin 2 input, and monitor the output of the AND gate at pin 3 with an oscilloscope. Referring to the timing diagram in Figure 3-11(b), you can see that only when both inputs are HIGH (shown shaded), will the output be HIGH. You may want to try constructing this test circuit in a lab and testing each one of the four AND gates in a 7408 IC.

SELF-TEST REVIEW QUESTIONS FOR SECTION 3-2

1. The two basic logic gates are the __________ gate and the __________ gate.
2. With an OR gate, any binary 1 input will give a binary __________ output.
3. Which logic gate can be used as a controlled switch?
4. Only when all inputs are LOW will the output of an __________ gate be LOW.

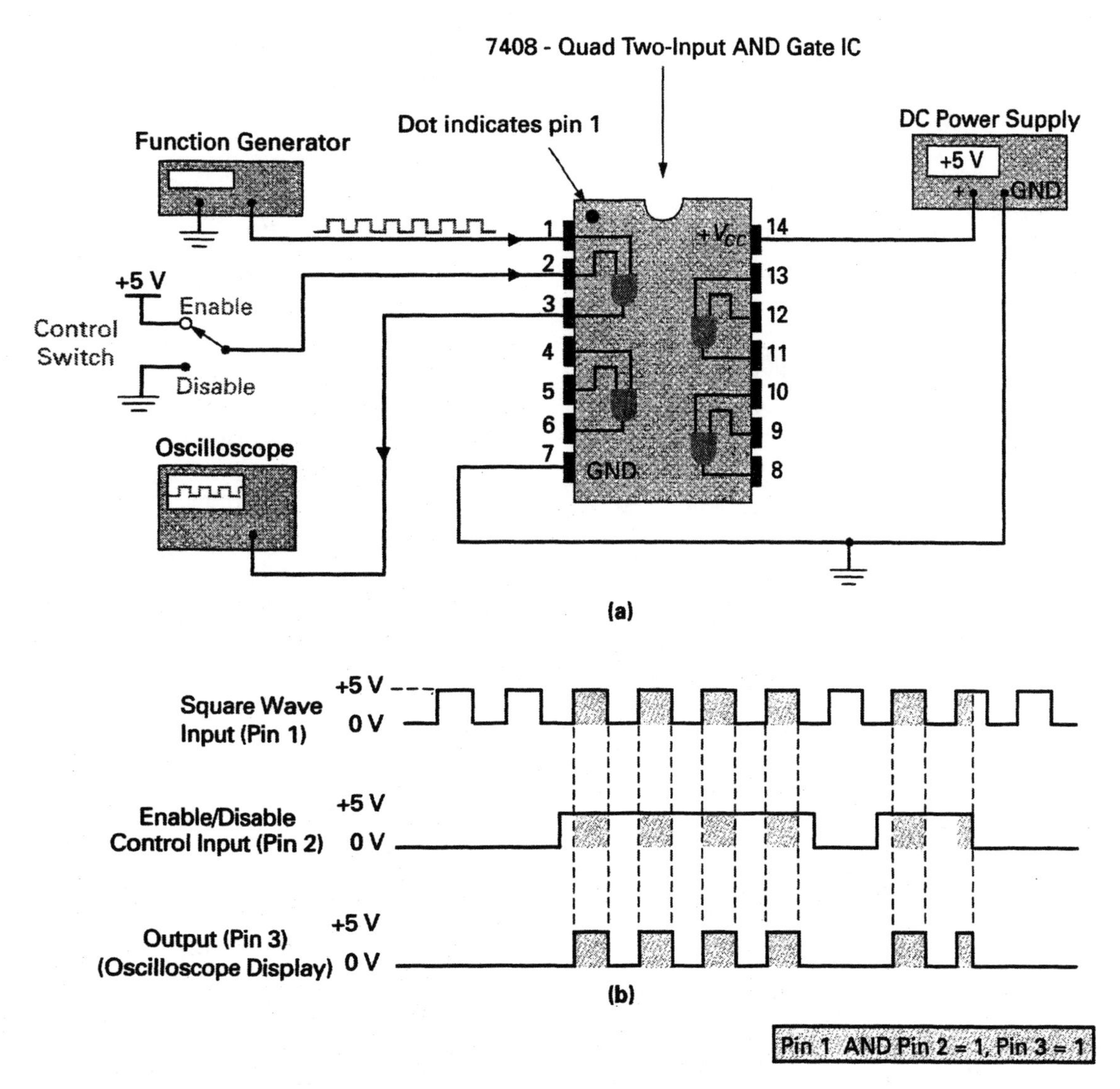

FIGURE 3-11 Testing the Operation of a Digital IC Containing Four Two-Input AND Gates.

5. With an AND gate, any binary 0 input will give a binary __________ output.
6. Only when all inputs are HIGH will the output of an __________ gate be HIGH.

3-3 INVERTING LOGIC GATES

Although the basic OR and AND logic gates can be used to construct many digital circuits, in some circuit applications other logic operations are needed. For this reason, semiconductor manufacturers made other logic-gate types available. In this section, we will examine three inverting-type logic gates, called the

NOT gate, the NOR gate, and the NAND gate. We will begin with the NOT or INVERTER gate, which was introduced earlier.

3-3-1 The NOT Gate

The NOT, or logic INVERTER gate, is the simplest of all the logic gates because it has only one input and one output. Figure 3-12(a) shows how a NOT gate can be constructed using two resistors and a transistor. Figure 3-12(b) shows the truth table for the NOT gate. The insets in Figure 3-12(b) simplify how the circuit will produce one of two outputs. Referring to the upper inset, you can see that when the input is LOW, the transistor's base-emitter diode will be reverse-biased and so the transistor will turn OFF. In this instance, the transistor is equivalent to an open switch between emitter and collector, and therefore the +5 V supply voltage will be connected to the output. A LOW input, therefore, will be converted or inverted to a HIGH output. On the other hand, referring to the lower inset in Figure 3-12(b), you can see that when the input is HIGH, the transistor's base-emitter diode will be forward-biased and the transistor will turn ON. In this circuit condition, the transistor is equivalent to a closed switch between emitter and collector, and the output will be switched

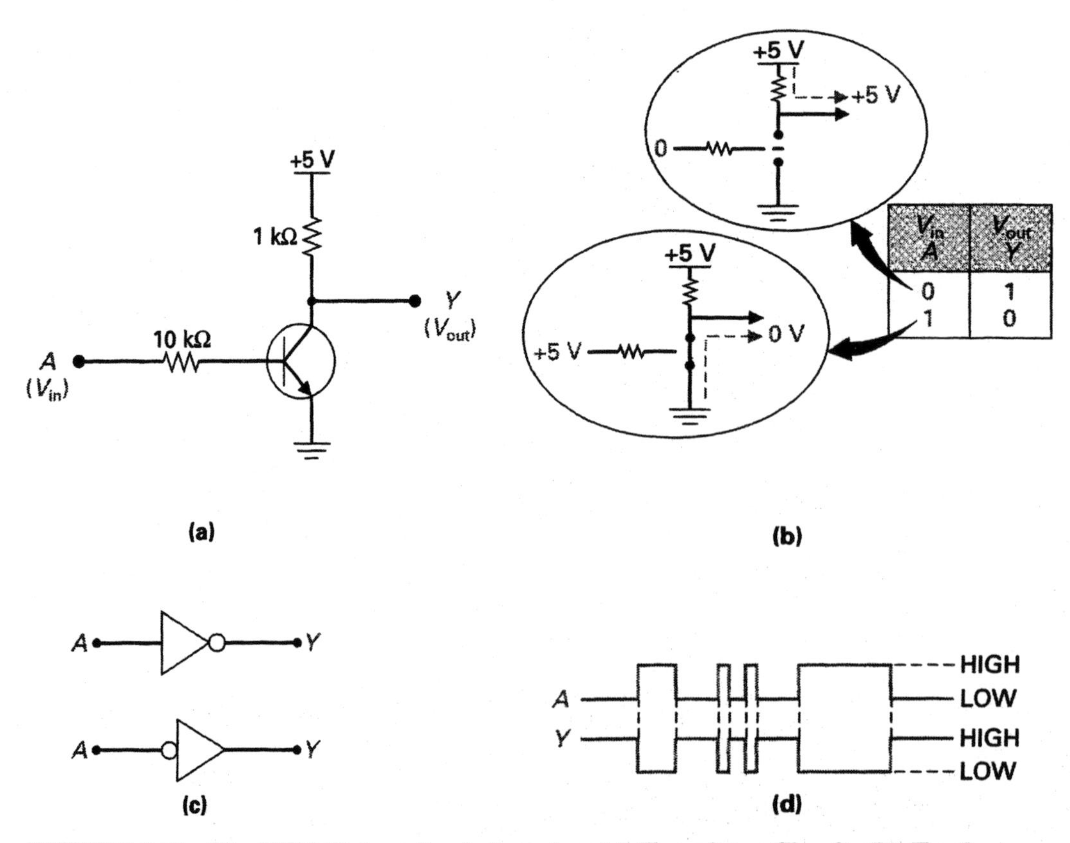

FIGURE 3-12 The NOT Gate or Logic Inverter. (a) Transistor Circuit. (b) Truth Table. (c) Logic Symbols for a NOT Gate. (d) Timing Diagram Showing Input-to-Output Inversion.

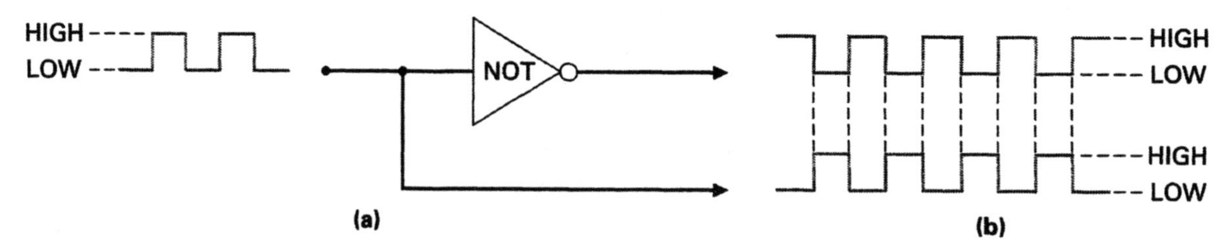

FIGURE 3-13 **A Two-Phase Signal Generator Circuit Using a NOT Gate.**

LOW. A HIGH input, therefore, will be converted or inverted to a LOW output. We could summarize the operation of this gate by saying that *the output logic level is "NOT" the same as the input logic level* (a 1 input is NOT the same as the 0 output, or a 0 input is NOT the same as the 1 output), hence the name NOT gate. The output is, therefore, always the **complement,** or opposite, of the input.

Complement
Opposite.

Figure 3-12(c) shows the schematic symbols normally used to represent the NOT gate. The triangle is used to represent the logic circuit, while the **bubble,** or small circle either before or after the triangle, is used to indicate the complementary or inverting nature of this circuit. Figure 3-12(d) shows a typical example of a NOT gate's input/output waveforms. As you can see from these waveforms, the output waveform is an inverted version of the input waveform.

Bubble
A small circle symbol used to signify an invert function.

■ EXAMPLE:

What will be the outputs from the circuit in Figure 3-13(a), assuming the square wave input shown?

■ ***Solution:***

As can be seen in Figure 3-13(b), the two-phase signal generator circuit will supply two square wave output signals that are 180° out of phase due to the inverting action of the NOT gate.

■ EXAMPLE:

What will be the outputs from each of the six NOT gates within the 7404 digital IC shown in Figure 3-14?

■ ***Solution:***

The NOT gate will always invert its input, and therefore:

	NOT Gate Input	NOT Gate Output
A	HIGH	LOW
B	LOW	HIGH
C	LOW	HIGH
D	HIGH	LOW
E	HIGH	LOW
F	LOW	HIGH

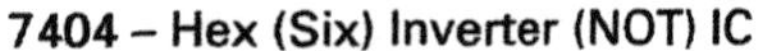

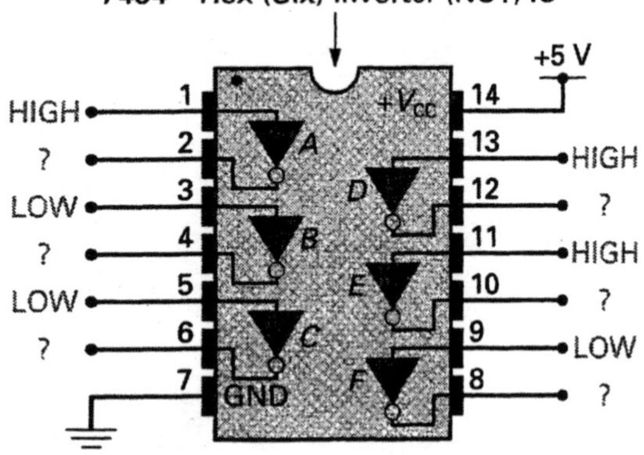

FIGURE 3-14 A Digital IC Containing Six NOT Gates.

3-3-2 The NOR Gate

NOR Gate
A NOT-OR gate that will give a LOW output if any of its inputs are HIGH.

The name "NOR" is a combination of the two words "NOT" and "OR," and is used to describe a type of logic gate that contains an OR gate followed by a NOT gate, as shown in Figure 3-15(a). The **NOR gate** could be constructed using the previously discussed transistor OR gate circuit followed by a transistor NOT gate circuit, as shown in the inset in Figure 3-15(a). The NOR gate should not be thought of as a new type of logic gate because it isn't. It is simply a combination of two previously discussed gates—an OR gate and a NOT gate.

Figure 3-15(b) shows the standard symbol used to represent the NOR gate. Studying the NOR gate symbol, you can see that the OR symbol is used to show how the two inputs are first "ORed," and that the bubble is used to indicate that the result from the OR operation is inverted before being applied to the output. As a result, the NOR gate's output is simply the opposite, or complement, of the OR gate, as can be seen in the truth table for a two-input NOR gate shown in Figure 3-15(c). If you compare the NOR gate truth table to the OR gate truth table, you can see that the only difference is that the output is inverted:

OR Gate: When all inputs are 0, the output is **0,** and any 1 input gives a **1** output.

NOR Gate: When all inputs are 0, the output is **1,** and any 1 input gives a **0** output.

The operation expressions for the NOR gate are listed in Figure 3-15(d), along with the standard symbols for a three-input and four-input NOR gate.

■ **EXAMPLE:**

What will be the outputs from each of the four NOR gates within the 7402 digital IC shown in Figure 3-16?

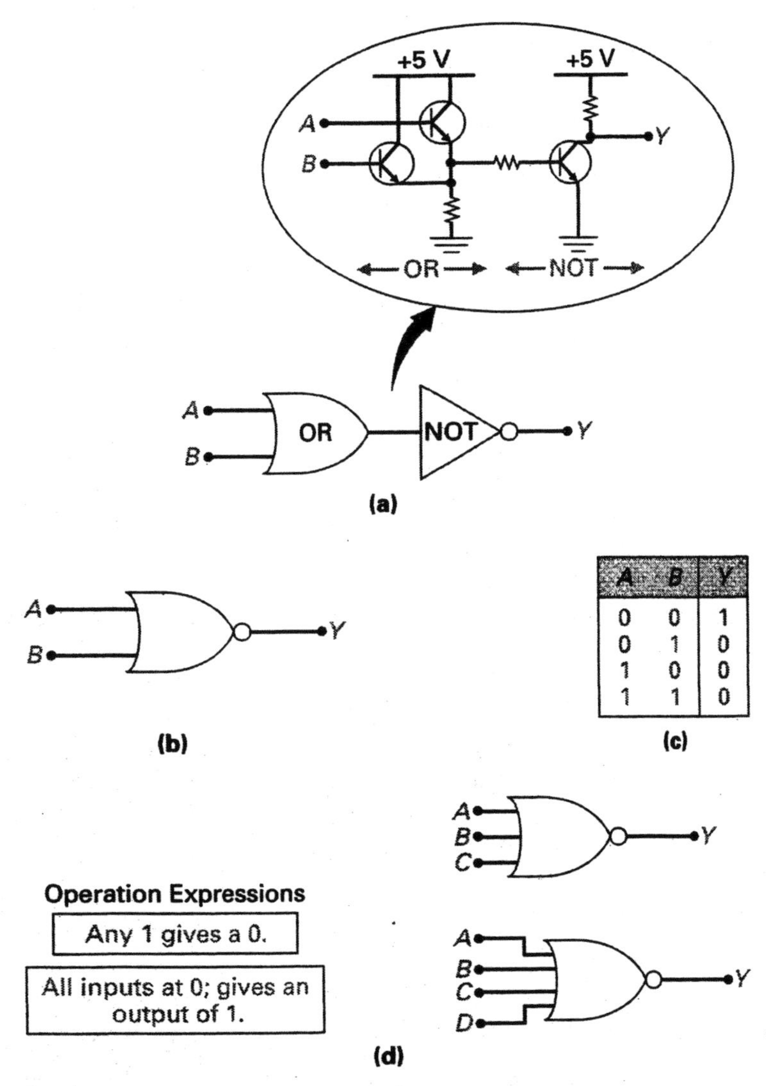

A	B	Y
0	0	1
0	1	0
1	0	0
1	1	0

FIGURE 3-15 The NOR Gate. (a) Two-Input Circuit. (b) Two-Input Logic Symbol. (c) Two-Input Truth Table. (d) NOR Gate Operation Expressions and Three- and Four-Input Logic Symbols.

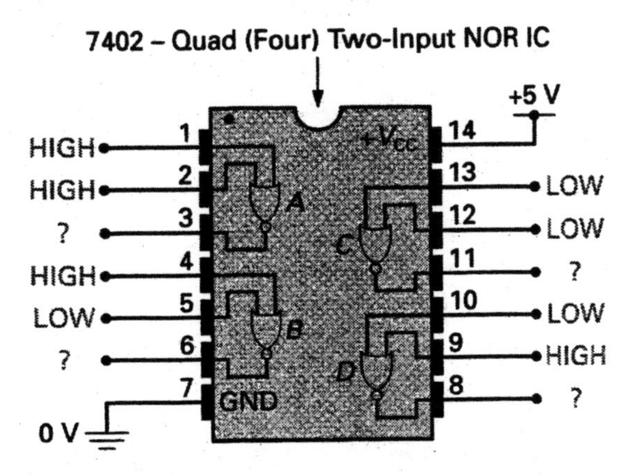

FIGURE 3-16 A Digital IC Containing Four NOR gates.

■ ***Solution:***

When all inputs are 0, the output is 1, and when any input is 1, the output is 0. Therefore:

	NOR Gate Inputs		NOR Gate Output
A	HIGH	HIGH	LOW
B	HIGH	LOW	LOW
C	LOW	LOW	HIGH
D	LOW	HIGH	LOW

■ **EXAMPLE:**

Referring to the circuit shown in Figure 3-17, determine whether the LED will flash at a rate of 1 Hz or 2 Hz for the inputs and switch position shown.

■ ***Solution:***

The circuit shown in Figure 3-17 is an example of a multiplexer circuit, which is a circuit that switches only one of its inputs through to a single output. In this circuit, two AND gates are operated as controlled switches, with the data select switch determining which of the AND gates is enabled. In this example, the upper AND gate is enabled, and so the *A*-data input (1 Hz signal) will be switched through to the NOR gate's upper input. The lower input to the NOR gate will be LOW, because the lower AND gate has been disabled by the data select switch. (AND: Any 0 input gives a 0 output.) Since any HIGH input to a NOR gate will give a LOW output, the NOR gate will basically act as an INVERTER and invert the 1 Hz square wave input. The anode of the LED is connected to +5 V, so when the output of the NOR gate goes LOW, the LED will turn ON, and when the output of the NOR gate goes HIGH, the LED will turn OFF. This multiplexer circuit will therefore switch the 1 Hz *A*-data input through to the output, and the LED will flash ON and OFF at a 1 Hz rate.

If the data select switch were placed in the *B* position, the 2 Hz *B*-data input would be switched through to the output and the LED would flash ON and OFF at a 2 Hz rate.

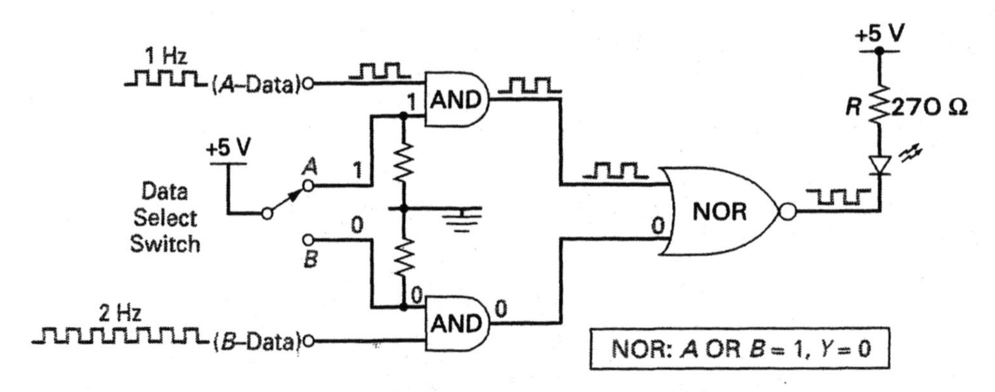

FIGURE 3-17 A Data Selector or Multiplexer Circuit Using a NOR Gate.

3-3-3 *The NAND Gate*

Like NOR, the name "NAND" is a combination of the two words "NOT" and "AND," and it describes another type of logic gate that contains an AND gate followed by a NOT gate, as shown in Figure 3-18(a). The **NAND gate** could be

NAND Gate
A NOT-AND logic gate circuit that will give a HIGH output if any of its inputs are LOW.

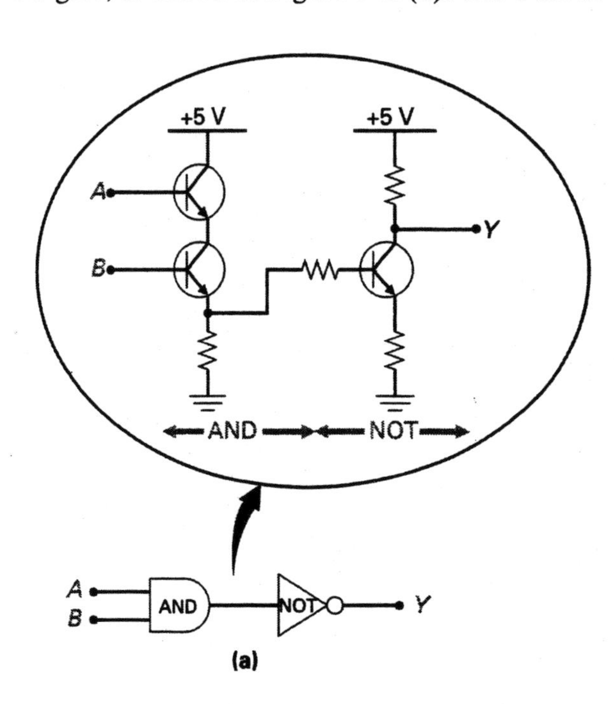

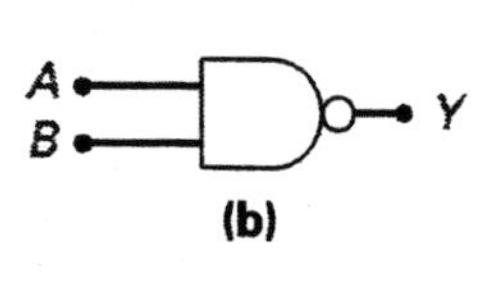

A	B	Y
0	0	1
0	1	1
1	0	1
1	1	0

(c)

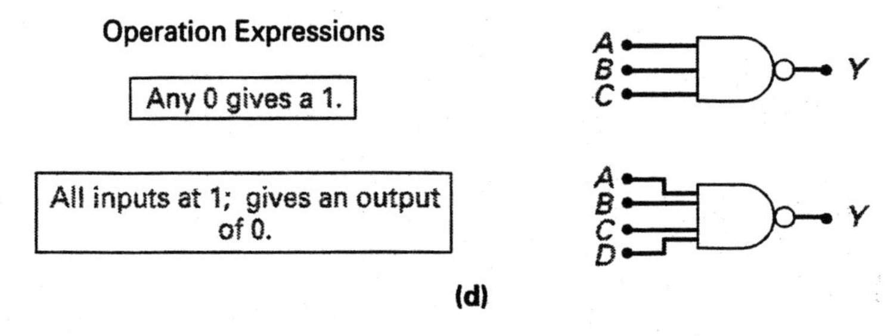

FIGURE 3-18 The NAND Gate. (a) Two-Input Circuit. (b) Two-Input Logic Symbol. (c) Two-Input Truth Table. (d) NAND Gate Operation Expressions and Three- and Four-Input Logic Symbols.

constructed using the previously discussed transistor AND gate circuit followed by a transistor NOT gate circuit, as shown in the inset in Figure 3-18(a). The NAND gate should also not be thought of as a new type of logic gate because it is simply an AND gate and a NOT gate combined to perform the NAND logic function.

Figure 3-18(b) shows the standard symbol used to represent the NAND gate. Studying the NAND gate symbol, you can see that the AND symbol is used to show how the two inputs are first "ANDed," and the bubble is used to indicate that the result from the AND operation is inverted before being applied to the output. As a result, the NAND gate's output is simply the opposite, or complement, of the AND gate, as can be seen in the truth table for a two-input NAND gate shown in Figure 3-18(c). If you compare the NAND gate truth table to the AND gate truth table, you can see that the only difference is that the output is inverted. To compare:

AND Gate: Any 0 input gives a **0** output, and only when both inputs are 1 will the output be **1**.

NAND Gate: Any 0 input gives a **1** output, and only when both inputs are 1 will the output be **0**.

The operation expressions for the NAND gate are listed in Figure 3-18(d), along with the standard symbols for a three-input and four-input NAND gate.

■ **EXAMPLE:**

What will be the outputs from each of the four NAND gates within the 7400 digital IC shown in Figure 3-19?

■ ***Solution:***

With the NAND gate, any 0 input gives a 1 output, and only when both inputs are 1 will the output be 0. Therefore:

	NAND Gate Inputs		NAND Gate Output
A	HIGH	HIGH	LOW
B	LOW	LOW	HIGH
C	HIGH	LOW	HIGH
D	LOW	HIGH	HIGH

■ **EXAMPLE:**

Which of the LEDs in Figure 3-20 would turn ON, if the binary input at *A* and *B* were both HIGH?

■ ***Solution:***

The circuit shown in Figure 3-20 is an example of a decoder circuit, which is a circuit that translates coded characters into a more understandable form. In this example, the circuit is used to decode a 2-bit binary code and then activate a decimal display. A set of four NAND gates is used to drive the decimal display, which contains four LEDs numbered 0 through 3. Since the NAND gate outputs

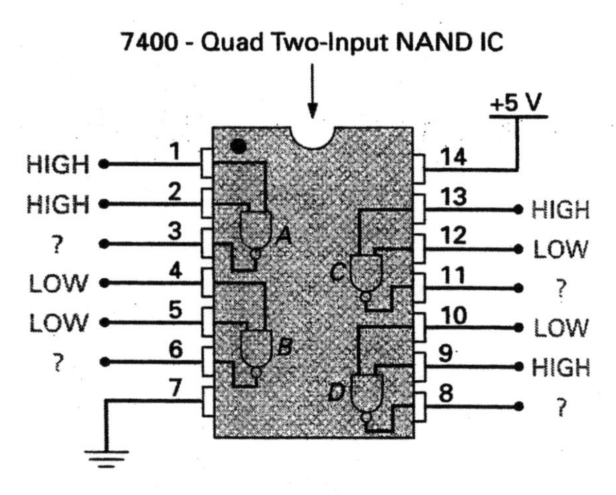

FIGURE 3-19 A Digital IC Containing Four NAND Gates.

are connected to the cathodes of the LEDs, and the anodes are connected to +5 V, a LOW output from any NAND gate will turn ON its associated LED. As we now know, a NAND gate will only give a LOW output when both of its inputs are HIGH; and the inputs in this circuit are controlled by the *A* and *B*—and inverted *A* and *B*—binary input lines. For example, when *A* and *B* inputs are both LOW, the inverted *A* and *B* lines will both be HIGH. Only NAND gate *D* will have both of its inputs HIGH, so its output will go LOW, causing the number 3 LED to turn ON. As a result, when binary 11_2 is applied to the input of the decoder, the output will display decimal 3_{10}. Therefore, each of the

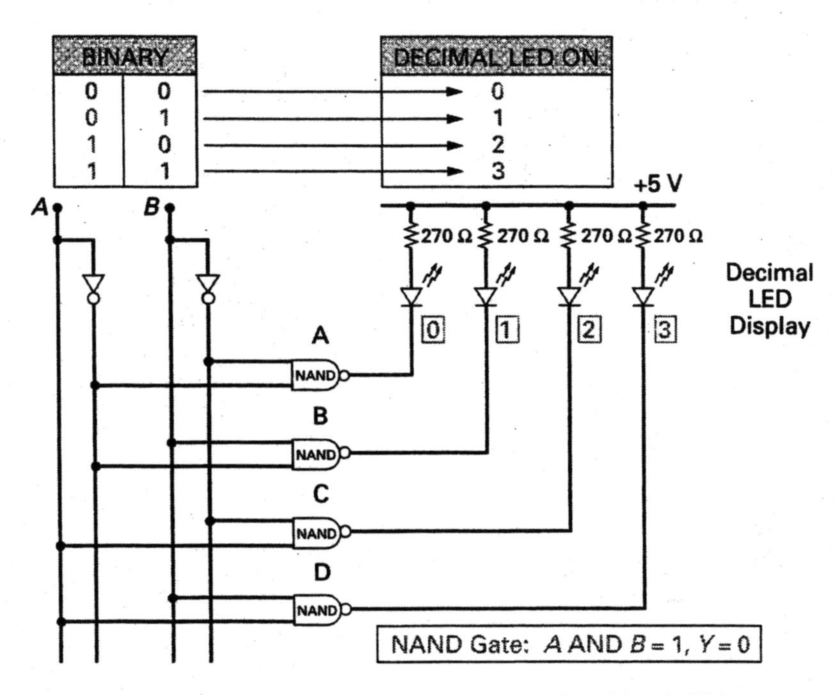

FIGURE 3-20 A Binary-to-Decimal Decoder Circuit Using NAND Gates.

four 2-bit binary input codes will activate one of the equivalent decimal output LEDs.

SELF-TEST REVIEW QUESTIONS FOR SECTION 3-3

1. Which of the logic gates has only one input?
2. With a NOT gate, any binary 1 input will give a binary ________ output.
3. Which logic gate is a combination of an AND gate followed by a logic INVERTER?
4. Only when all inputs are LOW, will the output of a/an ________ gate be HIGH.
5. With a NAND gate, any binary 0 input will give a binary ________ output.
6. Only when all inputs are HIGH, will the output of a/an ________ gate be LOW.

3-4 EXCLUSIVE LOGIC GATES

In this section, we will discuss the final two logic gate types: the exclusive-OR (XOR) gate and exclusive-NOR (XNOR) gate. Although these two logic gates are not used as frequently as the five basic OR, AND, NOT, NOR, and NAND gates, their function is ideal in some circuit applications.

3-4-1 The XOR Gate

Exclusive-OR Gate A logic gate circuit that will give a HIGH output if any odd number of binary 1s are applied to the input.

The **exclusive-OR (XOR) gate** logic symbol is shown in Figure 3-21(a), and its truth table is shown in Figure 3-21(b). Like the basic OR gate, the operation of the XOR is dependent on the HIGH or binary 1 inputs. With the basic OR, any 1 input would cause a 1 output. With the XOR, any odd number of binary 1s at the input will cause a binary 1 at the output. Looking at the truth table in Figure 3-21(b), you can see that when the binary inputs are 01 or 10, there is one binary 1 at the input (odd) and the output is 1. When the binary inputs are 00 or 11, however, there are two binary 0s or two binary 1s at the input (even) and the output is 0. To distinguish the XOR gate symbol from the basic OR gate symbol, you may have noticed that an additional curved line is included across the input.

Exclusive-OR logic gates are constructed by combining some of our previously discussed basic logic gates. Figure 3-21(c) shows how two NOT gates, two AND gates, and an OR gate can be connected to form an XOR gate. If inputs *A* and *B* are both HIGH or both LOW (even input), the AND gates will both end up with a LOW at one of their inputs, so the OR gate will have both

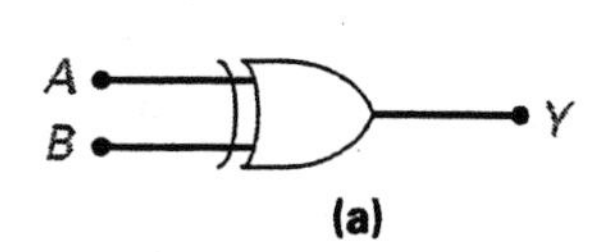

(a)

	A	B	Y
Even	0	0	0
Odd	0	1	1
Odd	1	0	1
Even	1	1	0

(b)

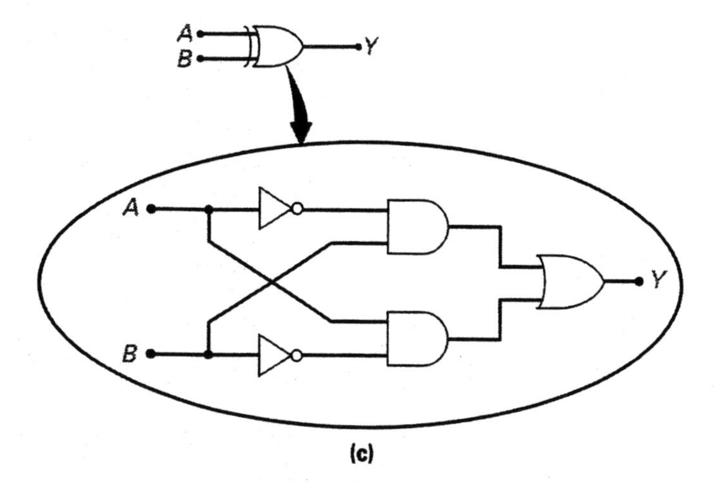

(c)

Operation Expression

Odd number of 1s at the input gives a 1 at the output.

(d)

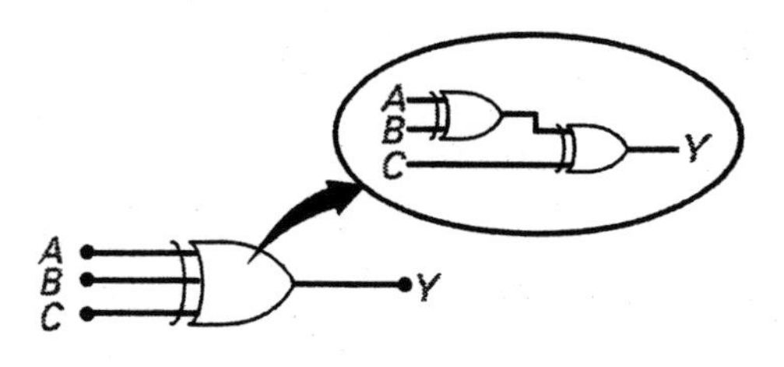

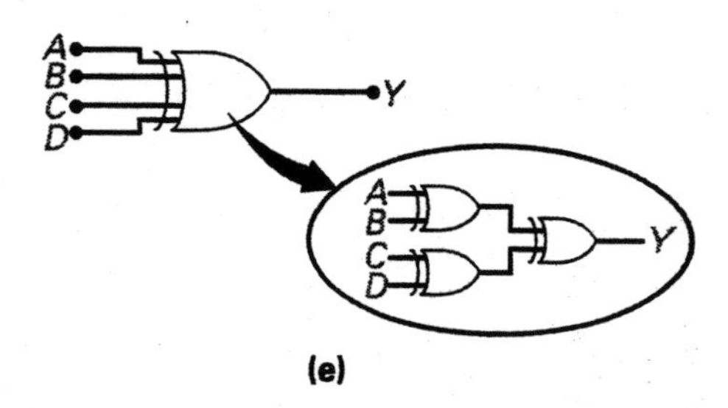

(e)

FIGURE 3-21 The Exclusive-OR (XOR) Gate. (a) Two-Input Logic Symbol. (b) Two-Input Truth Table. (c) Basic Gate Circuit. (d) Operation Expression. (e) Three- and Four-Input XOR Gates.

of its inputs LOW, and therefore the final output will be LOW. On the other hand, if A is HIGH and B is LOW or if B is HIGH and A is LOW (odd input), one of the AND gates will have both of its inputs HIGH, giving a HIGH to the OR gate and, therefore, a HIGH to the final output. We could summarize the operation of this gate therefore by saying that *the output is only HIGH when there is an odd number of binary 1s at the input.*

Figure 3-21(d) lists the operation expression for the XOR gate and Figure 3-21(e) shows how two-input XOR gates can be used to construct XOR gates with more than two inputs.

■ **EXAMPLE:**

List the outputs you would expect for all the possible inputs to a three-input XOR gate.

■ ***Solution:***

$$n = 2^x$$
$$n = 2^3 = 8$$

A	B	C	Y	
0	0	0	0	Even
0	0	1	1	Odd
0	1	0	1	Odd
0	1	1	0	Even
1	0	0	1	Odd
1	0	1	0	Even
1	1	0	0	Even
1	1	1	1	Odd

The output is only 1 when there is an odd number of binary 1s at the input.

■ **EXAMPLE:**

What will be the outputs from each of the four XOR gates within the 7486 digital IC shown in Figure 3-22?

■ ***Solution:***

With the XOR gate, the output is only HIGH when there is an odd number of binary 1s at the input. Therefore:

	XOR Gate Inputs		XOR Gate Output
A	HIGH	HIGH	LOW
B	HIGH	LOW	HIGH
C	LOW	LOW	LOW
D	LOW	HIGH	HIGH

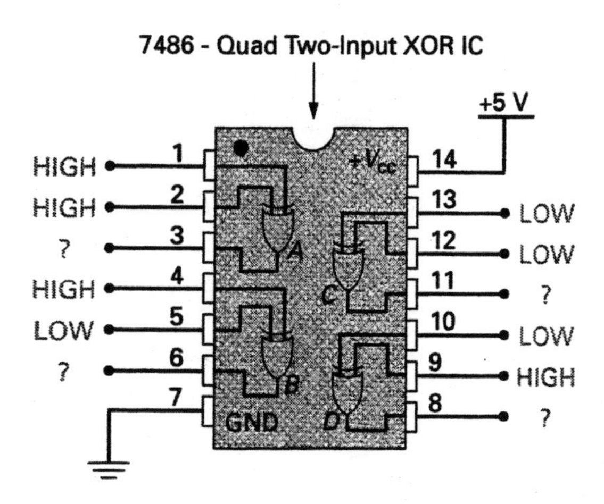

FIGURE 3-22 A Digital IC Containing Four XOR Gates.

3-4-2 *The XNOR Gate*

The **exclusive-NOR (XNOR) gate** is simply an XOR gate followed by a NOT gate, as shown in Figure 3-23(a). Its logic symbol, which is shown in Figure 3-23(b), is the same as the XOR symbol except for the bubble at the output, which is included to indicate that the result of the XOR operation is inverted before it appears at the output. Like the NAND and NOR gates, the XNOR is not a new gate, but simply a previously discussed gate with an inverted output.

Exclusive-NOR Gate
A NOT-exclusive OR gate that will give a HIGH output if any even number of binary 1s are applied to the input.

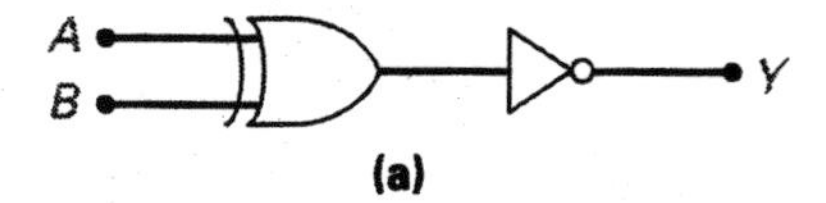

(a)

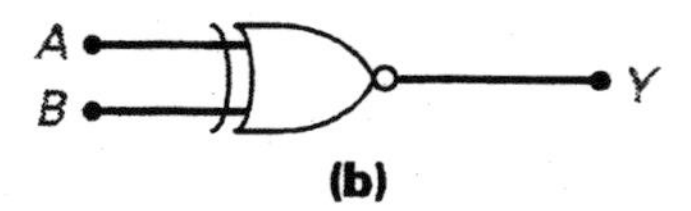

(b)

A	B	Number of 1s	Y
0	0	Even	1
0	1	Odd	0
1	0	Odd	0
1	1	Even	1

(c)

Operation Expression

Even number of 1s at the input gives a 1 at the output.

(d)

FIGURE 3-23 The Exclusive-NOR (XNOR) Gate. (a) Circuit. (b) Logic Symbol. (c) Truth Table. (d) Operation Expression.

The truth table for the XNOR gate is shown in Figure 3-23(c). Comparing the XOR gate to the XNOR gate, the XOR gate's output is only HIGH when there is an odd number of binary 1s at the input, whereas with the XNOR gate, the output is only HIGH when there is an even number of binary 1s at the input. The operation expression for the XNOR gate is listed in Figure 3-23(d).

■ **EXAMPLE:**

What will be the X and Y outputs from the circuit in Figure 3-24(a) if the A and B inputs are either out-of-phase or in-phase?

■ ***Solution:***

The circuit in Figure 3-24(a) is a phase detector circuit. The two square-wave inputs at A and B are applied to both the XOR and XNOR gate. If these two waveforms are out-of-phase, as shown in Figure 3-24(b), an odd number of binary 1s will be constantly applied to both gates, and the XOR output (X) will go HIGH and the XNOR output (Y) will go LOW. On the other hand, if the two waveforms are in-phase, as shown in Figure 3-24(c), an even number of

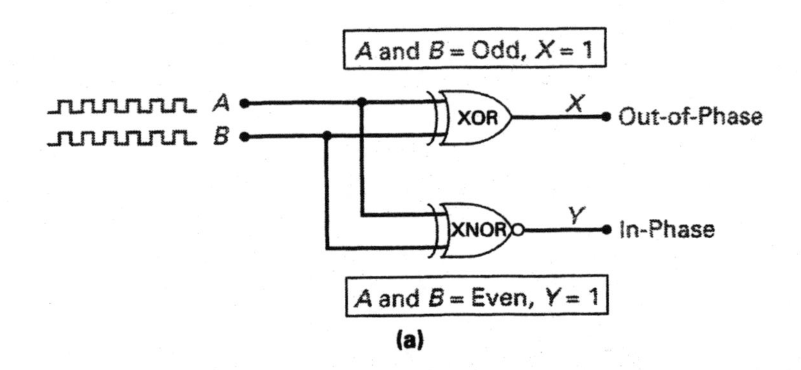

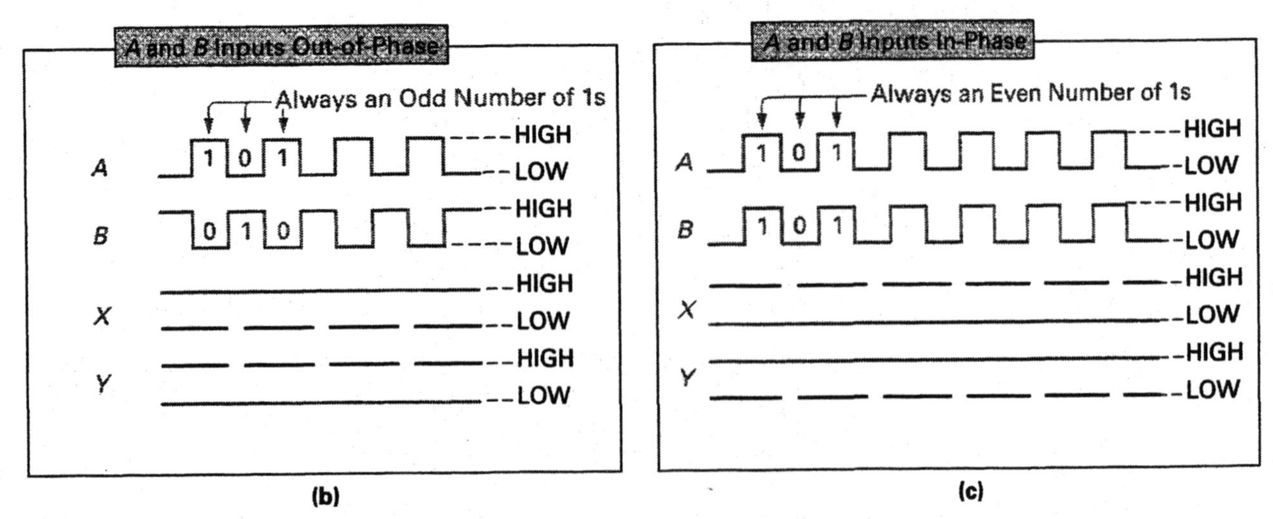

FIGURE 3-24 A Phase Detector Circuit Using an XOR and XNOR Gate.

binary 1s will constantly be applied to both gates, and the XOR output (X) will go LOW and the XNOR output (Y) will go HIGH.

■ **EXAMPLE:**

The four logic circuits within the 74135 digital IC, shown in Figure 3-25(a), can be made to operate as either XOR or XNOR logic gates. Their logic function is determined by the HIGH or LOW control input applied to pins 4 and 12.

Consider the logic circuit that has its inputs at pins 1 and 2, its output at pin 3, and a control input at pin 4. How will this logic circuit function if its control input is first made LOW, and then made HIGH?

■ ***Solution:***

Figure 3-25(b) follows the logic levels throughout the logic circuit when the control input is LOW, and an odd and even number of binary 1s are applied to the inputs A and B. Studying the Y output for these odd and even inputs, you can see that an odd input results in a HIGH output, and an even input results in a LOW output. When the control input is LOW, therefore, the logic circuits function as XOR gates.

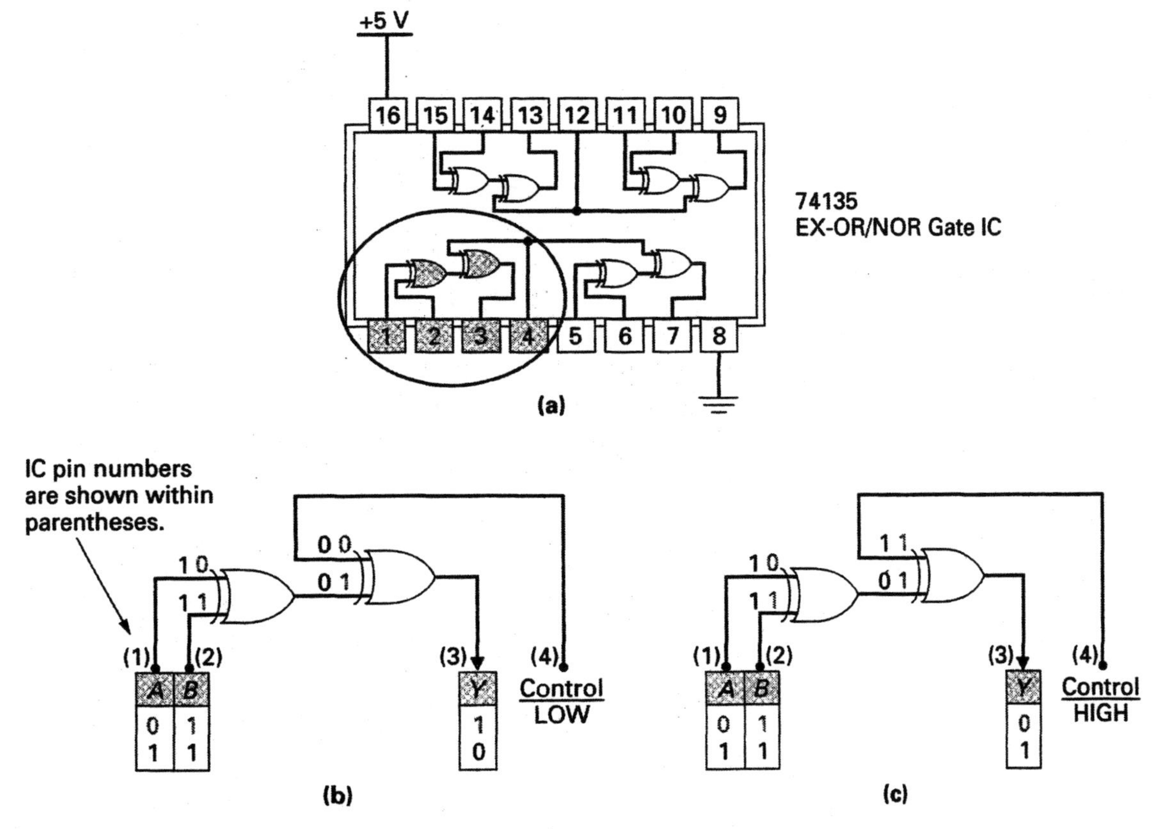

FIGURE 3-25 A Digital IC Containing Four Logic Gates that Can Be Controlled to Act as XOR Gates or XNOR Gates.

Figure 3-25(c) follows the logic levels throughout the logic circuit when the control input is HIGH, and an odd and even number of binary 1s are applied to the inputs *A* and *B*. Studying the *Y* output for these odd and even inputs, you can see that an odd input results in a LOW output, and an even input results in a HIGH output. When the control input is HIGH, therefore, the logic circuits function as XNOR gates.

SELF-TEST REVIEW QUESTIONS FOR SECTION 3-4

1. Which type of logic gate will always give a HIGH output when any of its inputs are HIGH?
2. Only when an odd number of 1s is applied to the input, will the output of an ________ gate be HIGH.
3. Only when an even number of 1s is applied to the input, will the output of an ________ gate be HIGH.
4. Which type of logic gate will always give a LOW output when an odd number of 1s is applied at the input?

3-5 IEEE/ANSI SYMBOLS FOR LOGIC GATES

The logic symbols presented so far in this chapter have been used for many years in the digital electronics industry. In 1984 the *Institute of Electrical and Electronic Engineers (IEEE)* and the *American National Standards Institute (ANSI)* introduced a new standard for logic symbols, which is slowly being accepted by more and more electronics companies. The advantage of this new standard is that instead of using distinctive shapes to represent logic gates, it uses a special **dependency notation system.** Simply stated, the new standard uses a *notation* (or note) within a rectangular or square block to indicate how the output is *dependent* on the input. Figure 3-26 compares traditional logic symbols with the newer IEEE/ANSI logic symbols and describes the meaning of the dependency notations.

Dependency Notation
A coding system used on schematic diagrams that uses notations to indicate how an output is dependent on inputs.

You should be familiar with both logic symbol types since you will come across both in industry schematics.

■ **EXAMPLE:**

Identify which of the logic circuits shown in Figure 3-27 is traditional and which is using the IEEE/ANSI standard.

■ ***Solution:***

Figure 3-27(a) shows the traditional logic symbol and Figure 3-27(b) shows the IEEE/ANSI logic symbol for a 7400 digital IC (quad two-input NAND).

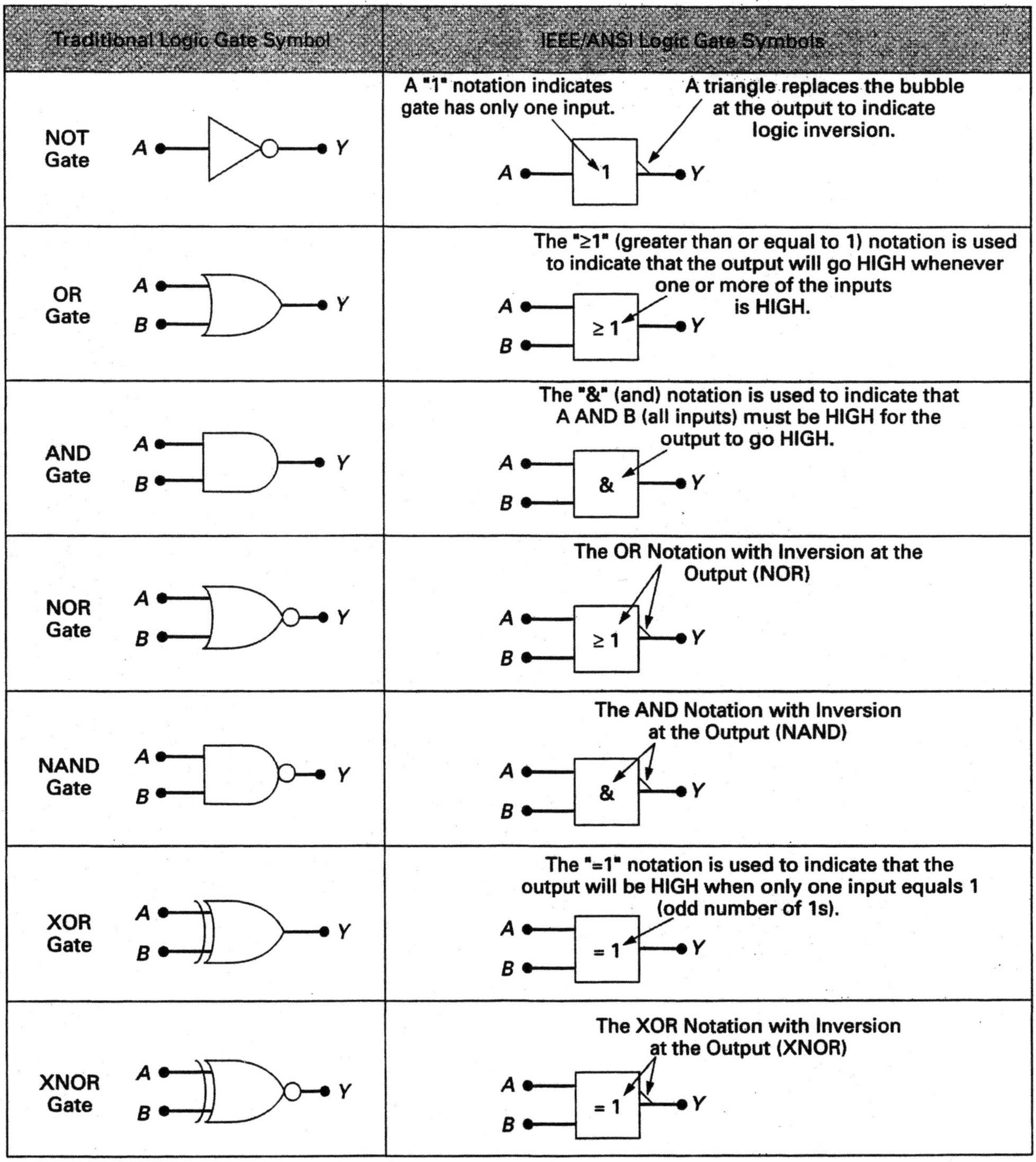

Traditional Logic Gate Symbol	IEEE/ANSI Logic Gate Symbols
NOT Gate (A, Y)	A "1" notation indicates gate has only one input. A triangle replaces the bubble at the output to indicate logic inversion. (A, 1, Y)
OR Gate (A, B, Y)	The "≥1" (greater than or equal to 1) notation is used to indicate that the output will go HIGH whenever one or more of the inputs is HIGH. (A, B, ≥ 1, Y)
AND Gate (A, B, Y)	The "&" (and) notation is used to indicate that A AND B (all inputs) must be HIGH for the output to go HIGH. (A, B, &, Y)
NOR Gate (A, B, Y)	The OR Notation with Inversion at the Output (NOR) (A, B, ≥ 1, Y)
NAND Gate (A, B, Y)	The AND Notation with Inversion at the Output (NAND) (A, B, &, Y)
XOR Gate (A, B, Y)	The "=1" notation is used to indicate that the output will be HIGH when only one input equals 1 (odd number of 1s). (A, B, = 1, Y)
XNOR Gate (A, B, Y)	The XOR Notation with Inversion at the Output (XNOR) (A, B, = 1, Y)

FIGURE 3-26 Traditional and IEEE/ANSI Symbols for Logic Gates.

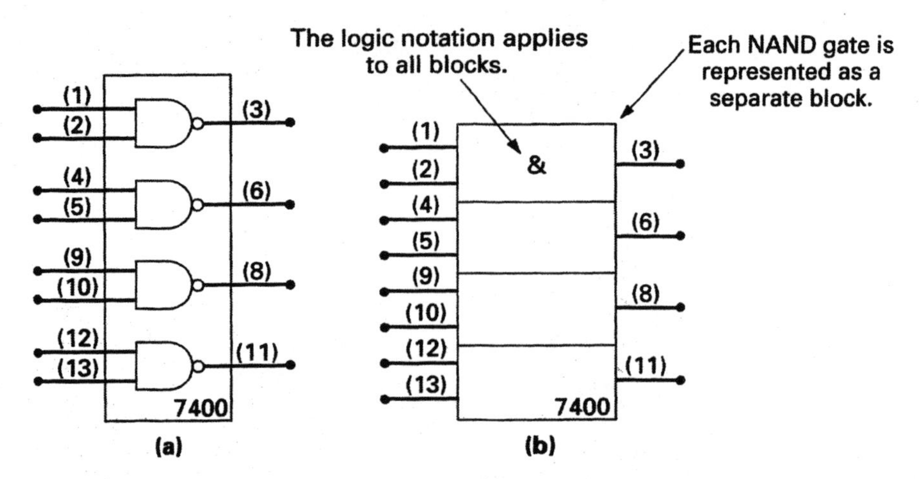

FIGURE 3-27 Traditional and IEEE/ANSI Logic Symbols for a 7400 Digital IC (Quad Two-Input NAND).

SELF-TEST REVIEW QUESTIONS FOR SECTION 3-5

1. Which type of logic-gate symbols make use of rectangular blocks?
2. What advantage does the IEEE/ANSI standard have over the traditional symbol standard?

REVIEW QUESTIONS

Multiple-Choice Questions

1. Which device is currently being used extensively as a two-state switch in digital electronic circuits?
   a. Transistor **b.** Vacuum tube **c.** Relay **d.** Toggle switch
2. How many input combinations would a four-input logic gate have?
   a. Eight **b.** Sixteen **c.** Thirty-two **d.** Four
3. What would be the maximum count within the truth table for a four-input logic gate?
   a. 7 **b.** 15 **c.** 17 **d.** 31
4. What would be the output from a three-input OR gate if its inputs were 101?
   a. 1 **b.** 0 **c.** Unknown **d.** None of the above
5. What would be the output from a three-input AND gate if its inputs were 101?
   a. 1 **b.** 0 **c.** Unknown **d.** None of the above
6. What would be the output from a three-input NAND gate if its inputs were 101?
   a. 1 **b.** 0 **c.** Unknown **d.** None of the above
7. What would be the output from a three-input NOR gate if its inputs were 101?
   a. 1 **b.** 0 **c.** Unknown **d.** None of the above

8. What would be the output from a three-input XOR gate if its inputs were 101?

   a. 1 **b.** 0 **c.** Unknown **d.** None of the above

9. What would be the output from a three-input XNOR gate if its inputs were 101?

   a. 1 **b.** 0 **c.** Unknown **d.** None of the above

10. What would be the output from a NOT gate if its input were 1?

    a. 1 **b.** 0 **c.** Unknown **d.** None of the above

11. Which of the following logic gates is ideal as a controlled switch?

    a. OR **b.** AND **c.** NOR **d.** NAND

12. Which of the following logic gates will always give a LOW output whenever a LOW input is applied?

    a. OR **b.** AND **c.** NOR **d.** NAND

13. Which of the following logic gates will always give a HIGH output whenever a LOW input is applied?

    a. OR **b.** AND **c.** NOR **d.** NAND

14. Which of the following logic gates will always give a HIGH output whenever a HIGH input is applied?

    a. OR **b.** AND **c.** NOR **d.** NAND

15. Which of the following logic gates will always give a LOW output whenever a HIGH input is applied?

    a. OR **b.** AND **c.** NOR **d.** NAND

16. An XOR gate will always give a HIGH output whenever an ____________ number of 1s is applied to the input.

    a. odd **b.** even **c.** unknown **d.** none of the above

17. An XNOR gate will always give a HIGH output whenever an ____________ number of 1s is applied to the input.

    a. odd **b.** even **c.** unknown **d.** none of the above

18. Which of the dependency notation logic symbols contains an "&" within the square block and no triangle at the output?

    a. OR **b.** AND **c.** NOR **d.** NAND

19. Which of the dependency notation logic symbols contains a "≥1" within the square block and no triangle at the output?

    a. OR **b.** AND **c.** NOR **d.** NAND

20. Which of the dependency notation logic symbols contains a "≥1" within the square block and a triangle at the output?

    a. OR **b.** AND **c.** NOR **d.** NAND

Essay Questions

21. List some of the two-state devices that have been used in digital electronic switches. (3-1)

22. What were the reasons for using two-state switches instead of ten-state switches? (Chapter 2)

23. Sketch and describe how the semiconductor diode can be used to construct a logic-gate circuit. (3-1)

24. What is the meaning of the term "logic gate"? (3-1)

25. Sketch and describe how the semiconductor transistor can be used to construct a logic-gate circuit. (3-1)

26. Sketch the traditional logic symbols used for the seven logic-gate types. (Chapter 3)

27. Describe the differences between the truth tables for the OR gate and the AND gate. (3-2)

28. Describe the basic differences between the following: (3-2 and 3-3)

a. The OR and the NOR gate. **b.** The AND and the NAND gate.

29. Give the full names for the following abbreviations. (Chapter 3)

a. XOR **b.** & **c.** =1 **d.** ≥1 **e.** NAND **f.** XNOR **g.** NOR

30. Sketch the truth table and the IEEE/ANSI logic gate symbols for the OR, AND, NOT, NOR, NAND, XOR, and XNOR gates. (Chapter 3)

Practice Problems

31. In the home security system shown in Figure 3-28(a), a two-input logic gate is needed to actuate an alarm (output Y) if either the window is opened (input A) or the door is opened (input B). Which decision-making logic gate should be used in this application?

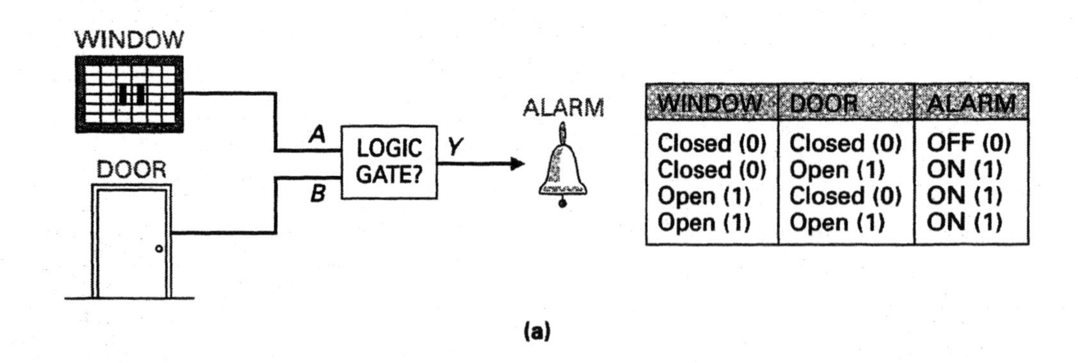

WINDOW	DOOR	ALARM
Closed (0)	Closed (0)	OFF (0)
Closed (0)	Open (1)	ON (1)
Open (1)	Closed (0)	ON (1)
Open (1)	Open (1)	ON (1)

(a)

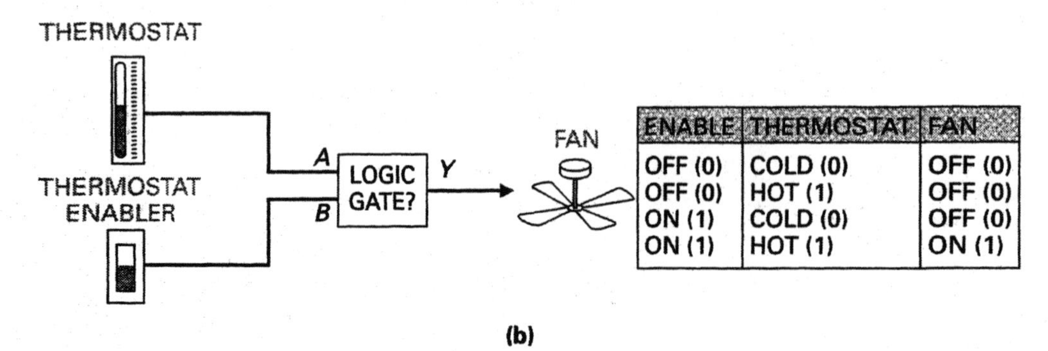

ENABLE	THERMOSTAT	FAN
OFF (0)	COLD (0)	OFF (0)
OFF (0)	HOT (1)	OFF (0)
ON (1)	COLD (0)	OFF (0)
ON (1)	HOT (1)	ON (1)

(b)

FIGURE 3-28 Decision-Making Logic Gates.

32. In the office temperature control system shown in Figure 3-28(b), a thermostat (input A) is used to turn ON and OFF a fan (output Y). A thermostat-enable switch is connected to the other input of the logic gate (input B), and it will either enable thermostat control of the fan or disable thermostat control of the fan. Which decision-making logic gate should be used in this application?
33. Briefly describe the operation of the photodiode receiver circuit in Figure 3-29. What will be the output at Y when light is being applied to the photodiode and when light is not being applied?
34. Briefly describe the operation of the gated LED flasher circuit shown in Figure 3-30.
35. Which logic gate would produce the output waveform shown in Figure 3-31?

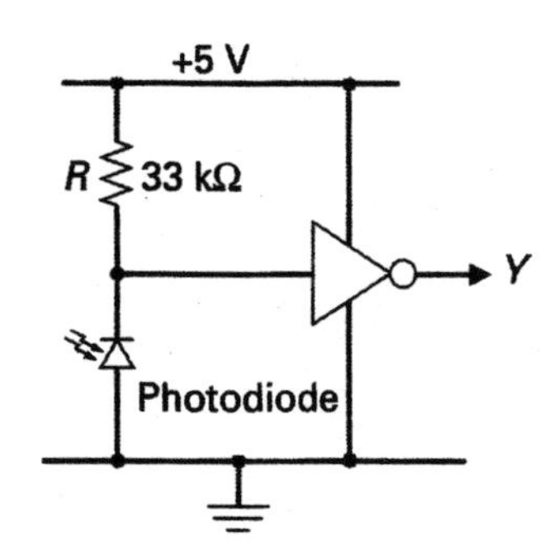

FIGURE 3-29
Photodiode Receiver.

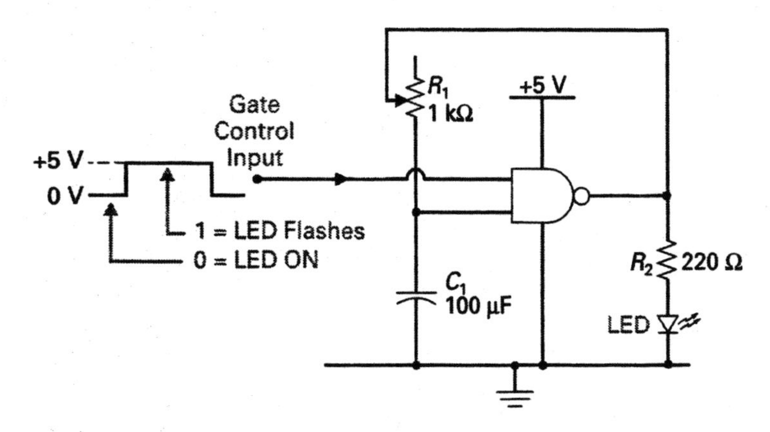

FIGURE 3-30 Gated LED Flasher Circuit.

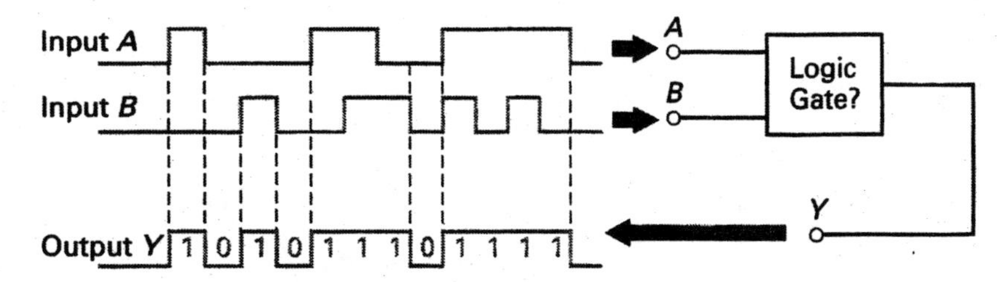

FIGURE 3-31 Timing Analysis of a Logic Gate.

36. Choose one of the six logic gates shown in Figure 3-32(a) and then sketch the output waveform that would result for the input waveforms shown in Figure 3-32(b).
37. Develop truth tables for each of the logic circuits shown in Figure 3-33.
38. State the logic operation and sketch the IEEE/ANSI logic symbol for each of the logic circuits shown in Figure 3-33.

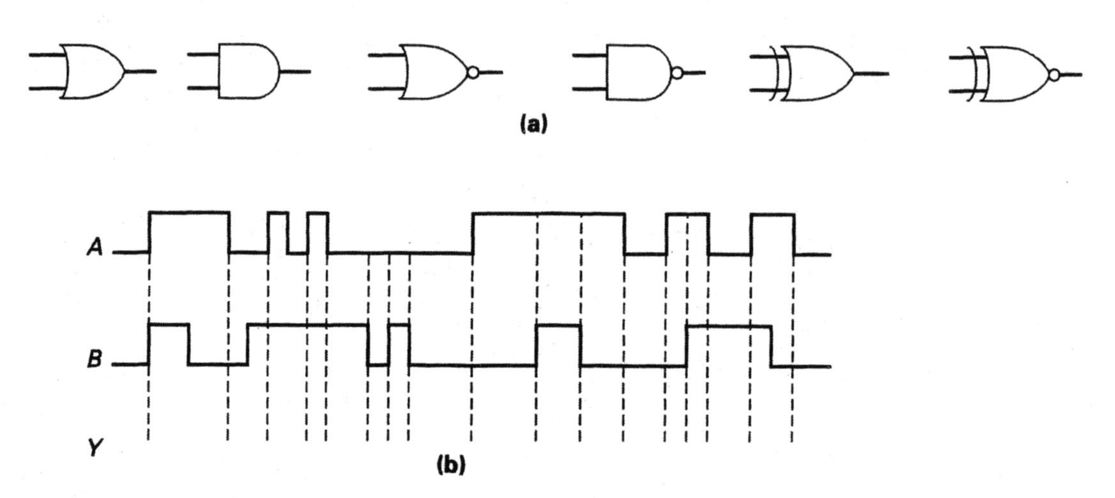

FIGURE 3-32 Timing Waveforms.

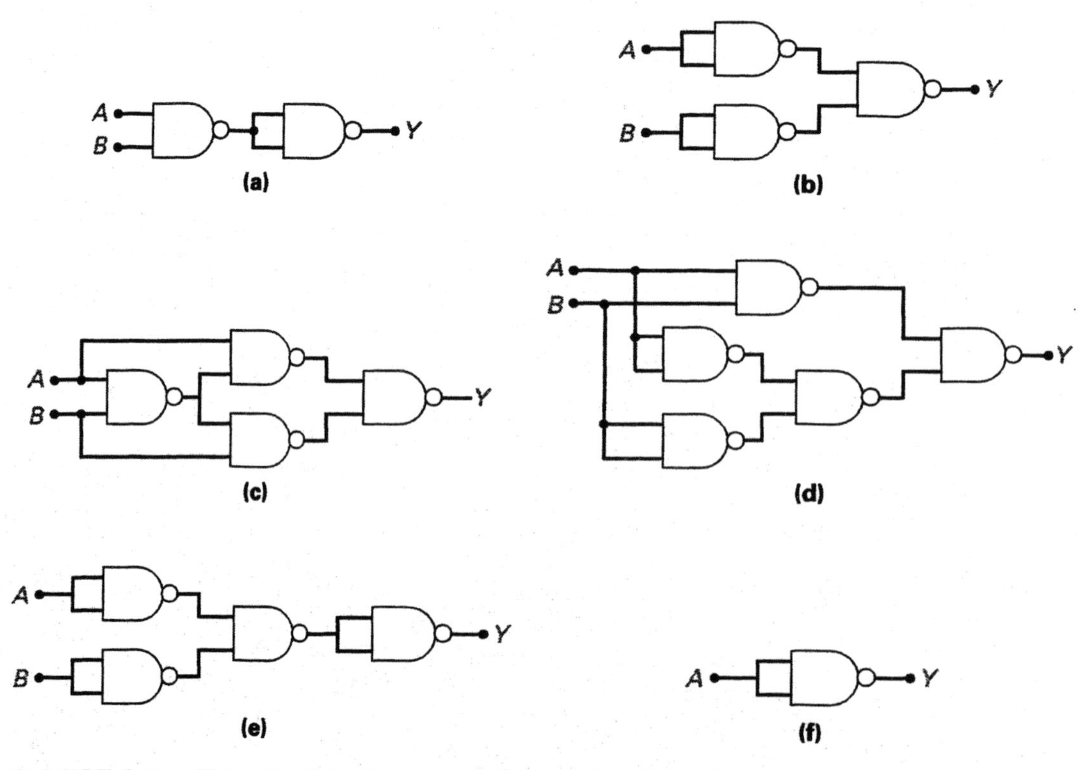

FIGURE 3-33 Using NAND Gates as Building Blocks for Other Gates.

39. Referring to the circuit in Figure 3-34, what will be the outputs at Y_0, Y_1, Y_2, and Y_3 if the control line input is

a. HIGH? **b.** LOW?

40. Referring to the circuit in Figure 3-34, what will the output logic levels be if the control line input is low?

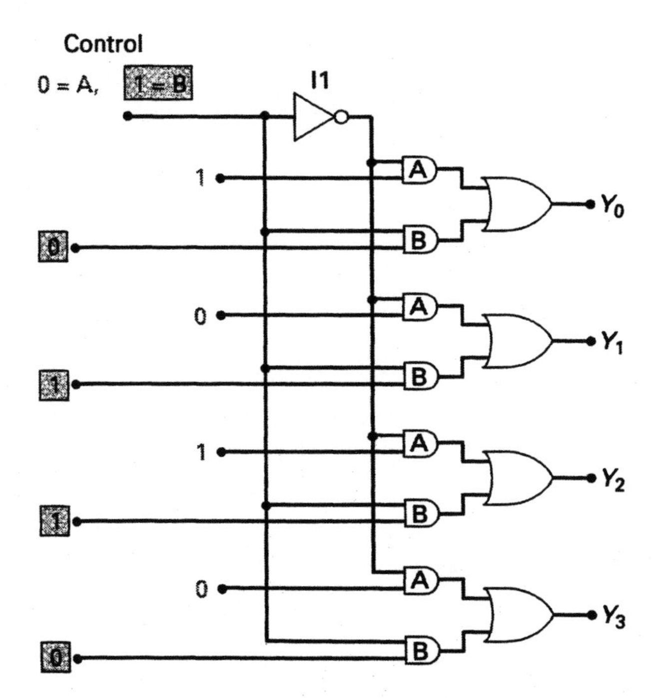

FIGURE 3-34 Four-Bit Word Multiplexer.

Chapter 4

OBJECTIVES

After completing this chapter, you will be able to:

1. Describe how the bipolar transistor is used in digital or two-state switching circuit applications.
2. Describe the operation of a bipolar transistor standard TTL (transistor-transistor logic) gate circuit.
3. List and describe the standard bipolar transistor TTL logic gate characteristics.
4. Describe the circuit operation, characteristics, and applications of other bipolar transistor TTL logic gate circuits, such as
   a. Low-power and high-speed TTL logic gates
   b. Schottky TTL logic gates
   c. Open collector TTL logic gates
   d. Three-state output TTL logic gates
   e. Buffer/driver TTL logic gates
   f. Schmitt-trigger TTL logic gates
5. Describe the circuit operation, characteristics, and applications of other non-standard bipolar transistor logic gate circuits, such as
   a. Emitter-coupled logic (ECL) gates
   b. Integrated-injection (I^2L) logic gates
6. Compare bipolar logic circuits to MOS logic circuits.
7. Explain the operation and characteristics of the following digital logic circuits:
   a. PMOS
   b. NMOS
   c. CMOS
8. List and describe the CMOS logic gate characteristics and the different CMOS series types available.
9. Describe the handling precautions that should be observed when handling MOS logic ICs.
10. Describe the different IC package types available and how ICs are classified based on circuit complexity.
11. Explain how to interface one logic family to another.
12. List some of the other different logic family types.

The Bipolar and MOS Transistors Within Digital ICs

OUTLINE

VIGNETTE

A Noyce Invention

On June 3, 1990, Robert Noyce, who was best known as the inventor of the silicon integrated circuit, or IC, died of a heart attack at the age of 62.

Noyce was the son of a Congregational minister and became fascinated by computers and electronics while studying to earn his bachelor's degree at Grinnell College in Grinnell, Iowa. Luckily, he was the student of Grant Gale, who in 1948 was given one of the first transistors by one of its inventors, John Bardeen. Using this device in his lectures, Gale taught one of the first courses in solid-state physics to a class of eighteen physics majors, one of whom was Noyce.

Wanting to pursue further interests in electronics, Noyce went on to obtain his doctorate in physics from the Massachusetts Institute of Technology in 1953. Later in that same year, Noyce took his first job as a research engineer at Philco Corp. in Philadelphia. He left in 1956 to join Shockley Semiconductor Laboratory in Mountain View, a company that was founded by another one of the inventors of the transistor, William Shockley.

A year later Noyce and seven other colleagues, whom Shockley called "the traitorous eight," resigned in what was to become a common pattern in Silicon Valley to join Fairchild Camera and Instruments Corp.'s semiconductor division. While at Fairchild, Noyce invented a process for interconnecting transistors on a single silicon chip, which was officially named an IC and was nicknamed "chip." This technological breakthrough, which brought the miniaturization of electronic circuits, was to be used in almost every electronic product, but, probably the first to make use of the process was the digital electronic computer.

The honor for this discovery is actually shared by Jack Kilbey, a staff scientist at Texas Instruments, who independently developed the same process and also holds patents for the invention.

In 1968 Noyce and More, another Fairchild scientist, founded Intel, a company which grew to become the nation's leading semiconductor company and a pioneer in the development of memory and microprocessor circuits.

In 1979 Noyce was awarded the National Medal of Science by President Jimmy Carter, and in 1987 he received the National Medal of Technology from President Ronald Reagan. He held more than a dozen patents, and in 1983 was inducted into the National Inventors Hall of Fame.

While Noyce's name never became widely known, his invention, the IC, is used in every electronic product today, and he was an instrumental figure in creating the $50 billion semiconductor industry, which is at the heart of the $500 billion electronics industry. Sadly, he did not possess a feeling of success or real accomplishment, and this was made clear when he once described his career as "the result of a succession of dissatisfactions."

INTRODUCTION

In the previous chapter, we saw how diodes and transistors could be used to construct digital logic-gate circuits, which are the basic decision making elements in all digital circuits. In reality, many more components are included in a typical logic gate circuit to obtain better input and output characteristics. In this chapter, we will see how the bipolar transistor and the MOS transistor are used to construct the two most frequently used digital IC logic families.

BIPOLAR LOGIC FAMILY	MOS LOGIC FAMILY
TTL (Transistor-Transistor Logic)	PMOS (P-Channel MOSFET)
Standard TTL	
Low-Power TTL	NMOS (N-Channel MOSFET)
High-Speed TTL	
Schottky TTL	CMOS (Complementary MOSFET)
Low-Power Schottky TTL	High-Speed CMOS
Advanced Low-Power Schottky TTL	High-Speed CMOS TTL Compatible
ECL (Emitter-Coupled Logic)	Advanced CMOS Logic
IIL (Integrated-Injection Logic)	Advanced CMOS TTL Compatible

A logic family is a group of digital circuits with nearly identical characteristics. Each of these two groups or families of digital ICs has its own characteristics and, therefore, advantages. For instance, as you will discover in this chapter, logic gates constructed using MOS transistors use less space due to their simpler construction, have a very high noise immunity, and consume less power than equivalent bipolar transistor logic gates; however, the high input impedance and input capacitance of the E-MOSFET transistor due to its insulated gate means that time constants are larger and, therefore, the transistor ON/OFF switching speeds are slower than equivalent bipolar gates. There are, therefore, trade-offs between the two logic families, and those are

a. Bipolar circuits are faster than MOS circuits, but their circuits are larger and consume more power.

b. MOS circuits are smaller and consume less power than bipolar circuits, but they are generally slower.

In this chapter, we will be examining the operation and characteristics of these bipolar and MOS circuits within digital ICs. This understanding is important if you are going to be able to determine the use of digital ICs within electronic circuits, interpret the characteristics of manufacturer data sheets, test digital ICs, and troubleshoot digital electronic circuits.

4-1 THE BIPOLAR FAMILY OF DIGITAL INTEGRATED CIRCUITS

In the previous chapter, diodes and bipolar transistors were used to construct each of the basic logic-gate types. These simplified logic-gate circuits were used to show how we could use the bipolar transistor to perform each logic gate function. In reality, many more components are included in a typical logic-gate circuit to obtain better input and output characteristics. For example, Figure 4-1 shows a typical digital IC containing four logic gates. Referring to the inset in this figure, you can see that a logic gate is actually constructed using a number of bipolar transistors, diodes, and resistors. All of these components are formed and interconnected on one side of a silicon chip. This single piece of silicon actually contains four logic-gate transistor circuits, with the inputs and output of each logic gate connected to the external pins of the IC package. For example, looking at the top view of the IC in Figure 4-1, you can see that one logic gate has its inputs on pins 1 and 2 and its output on pin 3. Since all four logic-gate circuits will need a $+V_{CC}$ supply voltage (typically +5 V), and a ground (0 V), two pins are assigned for this purpose (pin 14 = $+V_{CC}$, pin 7 = ground).

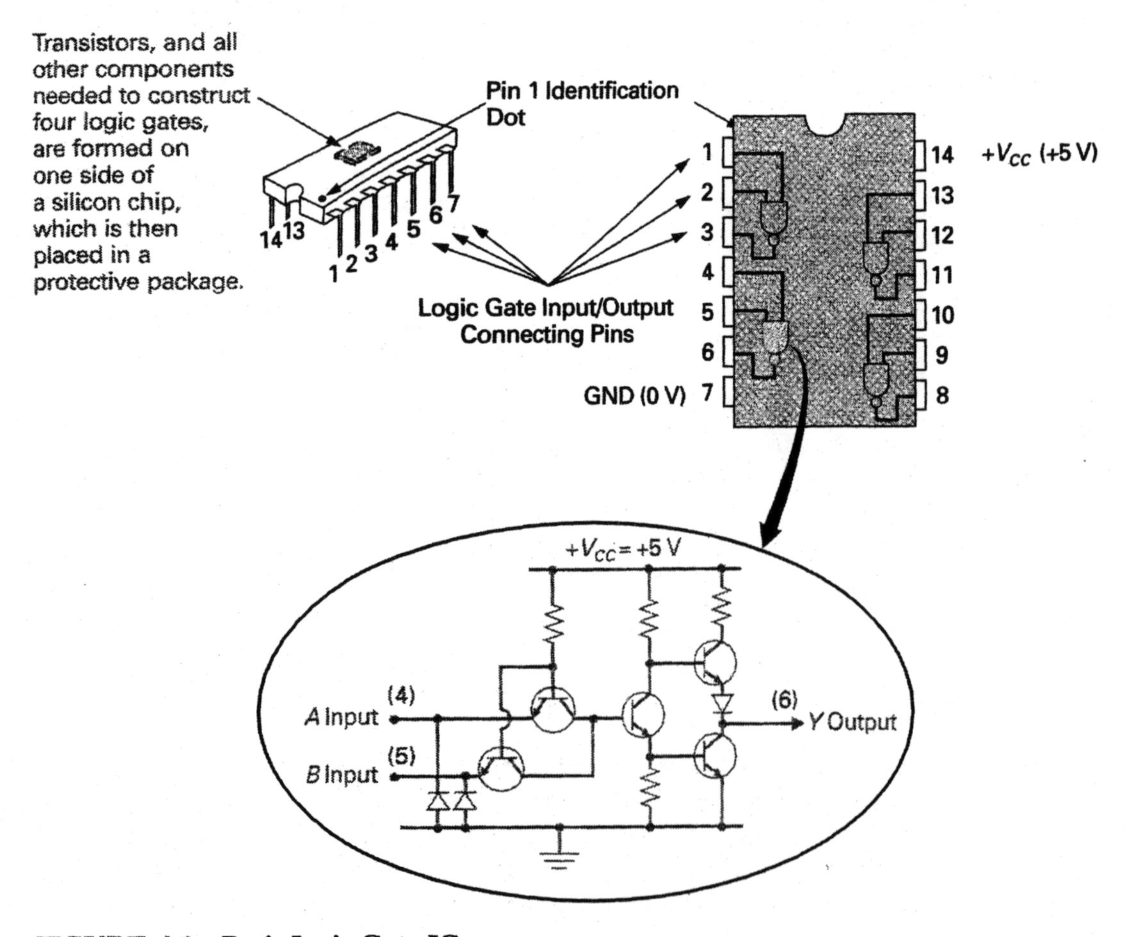

FIGURE 4-1 Basic Logic-Gate IC.

In the example in Figure 4-1, the IC contains four two-input NAND gates. This IC package type is called a **dual-in-line package (DIP)** because it contains two rows of connecting pins.

Dual-In-Line Package (DIP)
An electronic package having two rows of connecting pins.

In 1964 Texas Instruments introduced the first complete range of logic gate integrated circuits. They called this line of products **transistor-transistor logic (TTL) ICs** because the circuit used to construct the logic gates contained several interconnected transistors, as shown in the example in the inset in Figure 4-1. These ICs became the building blocks for all digital circuits. They were in fact, in the true sense of the word, building blocks, since no knowledge of electronic circuit design was necessary because the electronic circuits had already been designed and fabricated on a chip. All that had to be done was connect all of the individual ICs in a combination that achieved the specific or the desired circuit operation. All of these TTL ICs were compatible, which meant they all responded and generated the same logic 1 and 0 voltage levels, and they all used the same value of supply voltage; therefore, the output of one TTL IC could be directly connected to the input of one or more other TTL ICs in any combination.

Transistor-Transistor Logic (TTL)
A line of digital ICs that have logic gates containing several interconnected transistors.

By modifying the design of the bipolar transistor circuit within the logic gate, TTL IC manufacturers can change the logic function performed. The circuit can also be modified to increase the number of inputs to the gate. The most common type of TTL circuits are the 7400 series, which were originally developed by Texas Instruments but are now available from almost every digital IC semiconductor manufacturer. As an example, Table 4-1 lists all of the 7400 series of TTL ICs discussed in the previous chapter.

The TTL IC example shown in Figure 4-1, therefore, would have a device part number of "7400" since it contains four (quad) two-input NAND gates.

4-1-1 Standard TTL Logic Gate Circuits

It is important that you understand the inner workings of a logic gate circuit, since you will only be able to diagnose if a logic gate is working or not working if you know how a logic gate is supposed to operate. In this section, we will examine the bipolar transistor circuits within two typical **standard TTL logic gates.**

Standard TTL
The original transistor-transistor logic circuit type.

TABLE 4-1 A Few of the 7400 Standard TTL Series of Digital ICs

DEVICE NUMBER	LOGIC GATE CONFIGURATION
7400	Quad (four) two-input NAND gates
7402	Quad two-input NOR gates
7404	Hex (six) NOT or INVERTER gates
7408	Quad two-input AND gates
7432	Quad two-input OR gates
7486	Quad two-input XOR gates
74135	Quad-two-input XOR or XNOR gates

Bipolar Transistor NOT Gate Circuit

Figure 4-2(a) shows the pin assignment for the six (hex) NOT gates within a 7404 TTL IC. All six NOT gates are supplied power by connecting the $+V_{CC}$ terminal (pin 14) to +5 V and the ground terminal (pin 7) to 0 V. Connecting power and ground to the 7404 will activate all six of the NOT gate circuits so they are ready to function as inverters.

Coupling Transistor
A transistor connected to couple or connect one part of a circuit to another.

Totem Pole Circuit
A circuit arrangement in which two devices are stacked so that the operation of one affects the operation of the other.

Phase Splitter
A circuit designed to generate two out-of-phase outputs.

The inset in Figure 4-2(a) shows the circuit for a bipolar TTL NOT or logic INVERTER gate. As expected, the INVERTER has only one input which is applied to the **coupling transistor** Q_1, and one output which is developed by the **totem pole circuit** made up of Q_3 and Q_4. This totem pole circuit derives its name from the fact that Q_3 sits on top of Q_4 like the elements of an American Indian totem pole. Transistor Q_2 acts as a **phase splitter** since its collector and emitter outputs will be out of phase with one another (base-to-emitter = 180° phase shift, base-to-collector = 0° phase shift), and it is these two opposite outputs that will drive the bases of Q_3 and Q_4.

Let us now see how this circuit will respond to a LOW input. A 0 V (binary 0) input to the NOT gate circuit will apply 0 V to the emitter of Q_1. This will forward-bias the PN base-emitter junction of Q_1 (base is connected to +5 V, emitter has an input of 0 V), causing electron flow from the 0 V input through Q_1's base emitter, through R_1 to $+V_{CC}$. Since the base of Q_1 is only 0.7 V above the emitter voltage, the collector diode of Q_1 will be reverse-biased so the base of Q_2 will receive no input voltage and therefore turn OFF. With Q_2 cut OFF, all of the +5 V supply voltage will be applied via R_2 to the base of Q_3, sending Q_3 into saturation. With Q_3 saturated, its collector-to-emitter junction is equivalent to a closed switch between collector and emitter. As a result, the +5 V supply voltage ($+V_{CC}$) will be connected to the output via the low output resistance path of R_3, Q_3's collector-to-emitter, and D_2. A LOW input, therefore, will result in a HIGH output. In this condition, transistor Q_4 is cut OFF and equivalent to an open switch between collector and emitter because Q_2 is OFF and therefore the emitter of Q_2 and the base of Q_4 are at 0 V.

Let us now see how this circuit will respond to a logic 1 input. A HIGH input (+5 V) to the emitter of Q_1 will reverse bias the PN base-emitter junction of Q_1 (base is connected to +5 V, emitter has an input of +5 V). The +5 V at the base of Q_1 will forward bias Q_1's base collector junction, applying +5 V to the base of Q_2, sending it into saturation. In the last condition, when Q_2 was cut OFF, all of the $+V_{CC}$ supply voltage was applied to the base of Q_3. In this condition, Q_2 is heavily ON and therefore equivalent to a closed switch between collector and emitter, so the $+V_{CC}$ supply voltage will be proportionally developed across R_2, Q_2's collector-to-emitter, and R_4. Transistor Q_4 will turn ON due to the positive voltage drop across R_4 and switch its emitter voltage of 0 V to the collector and, therefore, the output. A HIGH input will therefore result in a LOW output. In this condition, Q_3 is kept OFF by a sufficiently low voltage level on Q_2's collector, and by diode D_2 which adds an additional diode to the emitter diode of Q_3, ensuring Q_3 stays OFF when Q_4 is ON.

The only other component that has not been discussed is the input diode D_1, which is included to prevent any negative input voltage spikes (negative

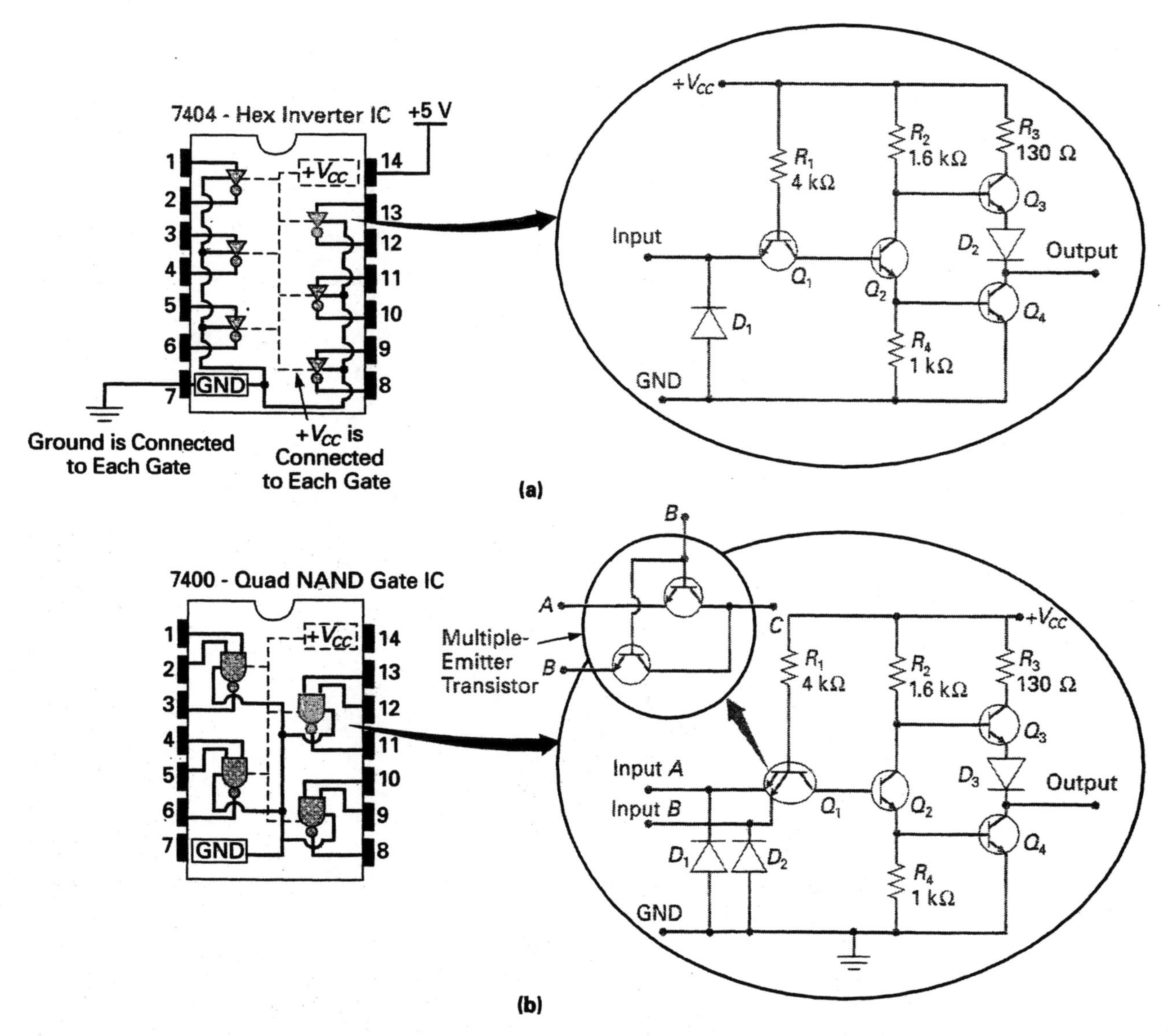

FIGURE 4-2 Standard TTL Logic-Gate Circuits. (a) NOT (Inverter) Gate. (b) NAND Gate.

transient) from damaging Q_1. D_1 will conduct these negative input voltages to ground.

Bipolar Transistor NAND Gate Circuit

Figure 4-2(b) shows the pin-out for the standard TTL 7400 IC and the bipolar transistor circuitry needed for each of the four NAND gates within this IC. The circuit is basically the same as the INVERTER or NOT gate circuit previously discussed, except for the **multiple-emitter input transistor,** Q_1. This transistor is used in all two-input logic gates, and if three inputs were needed, you would see a three-emitter input transistor. Q_1 can be thought of as having two transistors with separate emitters, but with the bases and collectors con-

Multiple-Emitter Transistor
A bipolar transistor having more than one emitter control terminal.

nected, as shown adjacent to the NAND circuit in Figure 4-2(b). Any LOW input to either A or B will cause the respective emitter diode to conduct current from the LOW input, through Q_1's base emitter and R_1, to $+V_{CC}$. The LOW voltage at the base of Q_1 will reverse bias Q_1's collector diode, preventing any voltage from reaching Q_2's base, and so Q_2 will cut OFF. The remainder of the circuit will operate in exactly the same way as the NOT gate circuit. With Q_2 OFF, Q_3 will turn ON and connect $+V_{CC}$ through to the output. Any LOW input, therefore, will generate a HIGH output, which is in keeping with the NAND gate's truth table. Only when both inputs are HIGH, will both of Q_1's emitter diodes be OFF and the HIGH at the base of Q_1 be applied via the forward-biased collector diode of Q_1 to the base of Q_2. With Q_2 ON, Q_4 will receive a positive base voltage and therefore switch ground, or a LOW voltage, through to the output.

TTL Logic Gate Characteristics

Almost every digital IC semiconductor manufacturer has its own version of the standard TTL logic gate family or series of ICs. For instance, Fairchild has its 9300 series of TTL ICs, Signetics has the 8000 series, and so on. No matter what manufacturer is used, all of the TTL circuits are compatible in that they all have the same input and output characteristics. To examine the compatibility of these bipolar transistor logic-gate circuits, we will have to connect two logic gates to see how they will operate when connected to form a circuit. Figure 4-3 shows a **TTL driver,** or logic source, connected to a **TTL load.** (In this example a NAND gate output is connected to the input of a NOT gate.)

TTL Driver A transistor-transistor logic source device.

TTL Load A transistor-transistor logic device that will encumber or burden the source.

Figure 4-4 lists all of the standard TTL logic-gate characteristics. Let us now step through the items mentioned in this list of characteristics and explain their meaning.

a. For TTL circuits, the valid output logic levels from a gate and valid input logic levels to a gate are shown in the blocks in Figure 4-4. Looking at the block on the left you can see that a gate must generate an output voltage that is between the minimum and maximum points in order for that output to be recognized as a valid logical 1 or 0 (valid LOW output = 0 V to +0.4 V, valid HIGH output = +2.4 V to +5 V). Referring to the block on the right of Figure 4-4, you can see that as long as the input voltages to a gate are between the minimum and maximum points, the gate will recognize the input as a valid logical 1 or 0 (valid LOW input = +0 V to +0.8 V, valid HIGH input = +2.0 V to +5 V).

Power Dissipation The power dissipated by a device or circuit.

b. The second item on our list of TTL characteristics in Figure 4-4 is **power dissipation.** A standard TTL circuit will typically dissipate 10 mW per gate.

c. The third item on our list of TTL characteristics in Figure 4-4 is the **propagation delay time** of a gate, which is the time it takes for the output of a

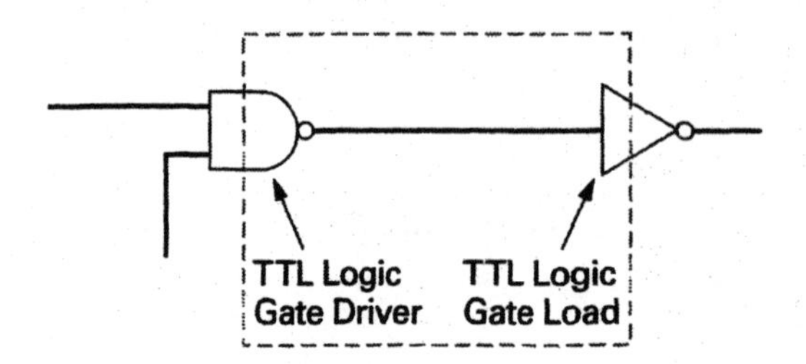

FIGURE 4-3 Connecting Logic Gates.

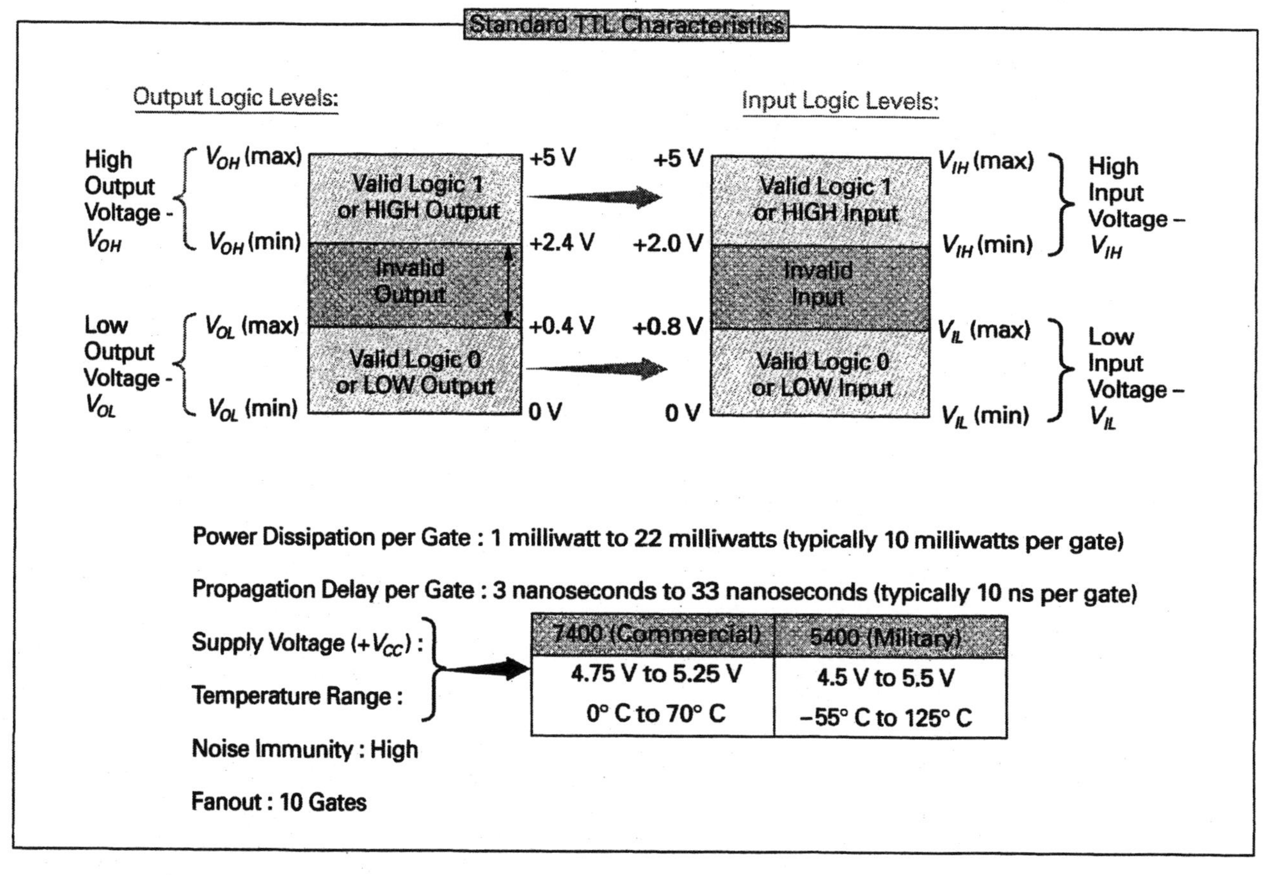

	7400 (Commercial)	5400 (Military)
Supply Voltage ($+V_{CC}$)	4.75 V to 5.25 V	4.5 V to 5.5 V
Temperature Range	0° C to 70° C	-55° C to 125° C

FIGURE 4-4 Standard TTL Logic-Gate Characteristics.

gate to change after the inputs have changed. This delay time is typically about 10 nanoseconds (ns) for a standard TTL gate. For example, it would normally take 10 ns for the output of an AND gate to go HIGH after all of its inputs go HIGH.

Propagation Delay Time
The time it takes for an output to change after a new input has been applied.

d. As you can see from the next two items mentioned in Figure 4-4, there are two standard TTL series of ICs available. The 7400 TTL series is used for all commercial applications and it will operate reliably when the supply voltage is between 4.75 V and 5.25 V, and when the ambient temperature is between 0°C and 70°C. The **5400 series** is a more expensive line used for military applications because of its increased supply voltage tolerance (4.5 V to 5.5 V) and temperature tolerance (−55°C to 125°C). A 5400 series TTL IC will have the same pin configuration and perform almost exactly the same logic function as a 7400 series device. For example, a 5404 (hex INVERTER) has the same gate input and output characteristics, pin numbers, $+V_{CC}$, and ground pin numbers as a 7404.

Noise Immunity
The ability of a device to not be affected by noise.

e. Noise is an unwanted voltage signal that is induced into conductors due to electromagnetic radiation from adjacent current-carrying conductors. All TTL logic circuits have a high **noise immunity,** which means they will be immune to most noise fluctuations at the gate's inputs. A standard TTL logic gate actually has a **noise margin** of 0.4 V between a TTL driver and TTL load, as is shown

Noise Margin
The range in which noise will not affect the operation of a circuit.

in Figure 4-4, if you compare the valid HIGH and LOW outputs to the HIGH and LOW inputs. This means that if a TTL driver were to output its worst case LOW output (0.4 V), and induced noise increased this signal further away from 0.4 V to 0.8 V, the TTL load would still accept this input as a valid LOW input since it is between 0 V and 0.8 V. This is shown in the simplified diagram in Figure 4-5(a). Similarly, if a TTL driver were to output its worst case HIGH

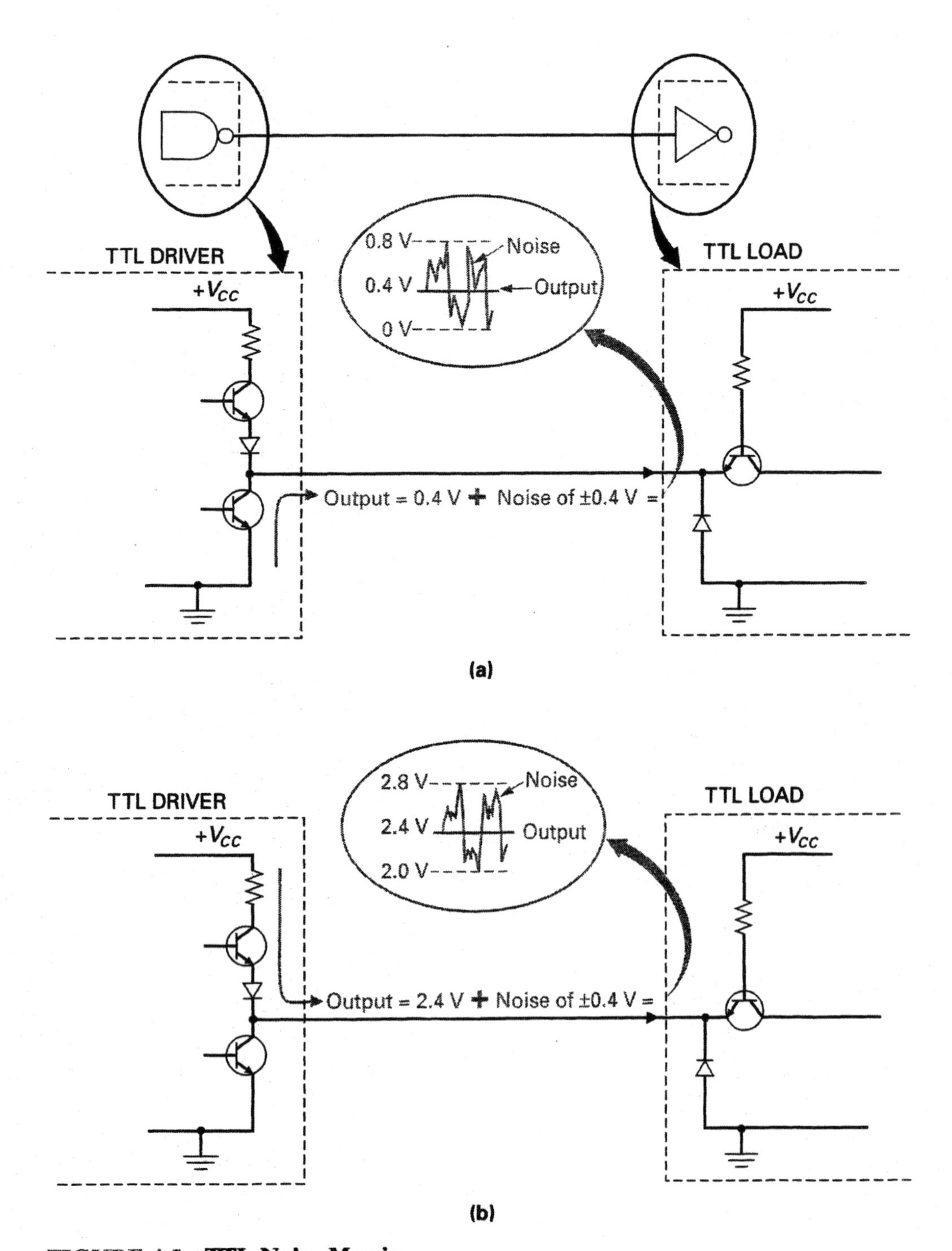

FIGURE 4-5 TTL Noise Margin.

output (2.4 V), and induced noise decreased this signal further away from 2.4 V to 2.0 V, the TTL load would still accept this input as a valid logic 1 since it is between 2.0 V and 5 V. This is shown in the simplified diagram in Figure 4-5(b).

f. The last specification that has to be discussed in Figure 4-4, is the listing called **fanout.** To explain this term, Figure 4-6 shows how a TTL driver has to act as both a current **sink** and a current **source.** Looking at the direction of current in this illustration, you can see that this characteristic is described using conventional current flow (current travels from $+V \rightarrow -V$) instead of electron current flow (current travels from $-V \rightarrow +V$). Whether the current flow is from ground to $+V_{CC}$ (electron flow), or from $+V_{CC}$ to ground (conventional flow), the value of current is always the same if the same circuit exists between these two points. The only reason we have changed to conventional at this point is that most semiconductor manufacturers' specifications for the terms "sinking" and "sourcing" are detailed using conventional current flow.

Fanout
The number of parallel loads that can be driven simultaneously by one logic source, keeping logic levels within spec.

Sink
A power-consuming device or circuit.

Source
A circuit or device that supplies power.

In Figure 4-6(a), a single logic gate load has been connected to a TTL logic gate driver, which is producing a LOW output (Q_4 in driver's totem pole is ON). In this instance, Q_4 will "sink" the single gate load current of 1.6 mA from Q_1 to ground. Manufacturer's data sheets state that Q_4 can sink a maximum of 16 mA and still produce a LOW output voltage of 0.4 V or less (the voltage drop across Q_4 will always be 0.4 V or less when it is sinking 16 mA). Since the maximum sinking current of 16 mA is ten times larger than the single gate load current of 1.6 mA, we can connect ten TTL loads to each TTL driver or source, as seen in Figure 4-6(b), and still stay "within spec" (within the parameters defined in the data sheet).

To examine both sides of the coin, Figure 4-6(c) shows a single TTL load connected to a TTL driver, which is producing a HIGH output (Q_3 in driver's totem pole is ON). In this instance, Q_3 will "source" the single gate load current of 40 μA. Manufacturers' data sheets state that Q_3 can source a maximum of 400 μA and still maintain a HIGH output voltage of 2.4 V or more. Since the maximum source current of 400 μA is ten times larger than the single gate load current of 40 μA, we can connect ten TTL loads to each TTL driver, as seen in Figure 4-6(d), and still stay within spec.

As we have seen in Figure 4-6, the maximum number of TTL loads that can be reliably driven by a single TTL driver is called the fanout. Referring to the standard TTL logic gate characteristics in Figure 4-4, you will now understand why a standard TTL gate has a fanout of 10.

g. Using Figure 4-6, we can describe another TTL circuit characteristic. When a TTL load receives a LOW input, as seen in Figure 4-6(a), a large current exists in the emitter of Q_1. On the other hand, when a TTL load receives a HIGH input, as seen in Figure 4-6(c), the emitter current in Q_1 is almost zero. This action explains why a TTL logic gate acts like it is receiving a HIGH input when its input lead is not connected. For example, if the input lead to a NOT gate was not connected to a circuit (an open input), and power was applied to the gate, the output of the NOT gate would be LOW. An unconnected input is called a **floating input.** In this condition, no emitter current will exist in Q_1 since the input is disconnected; so we can say *a floating TTL input is equivalent to a HIGH input.*

Floating Input
An input that is not connected.

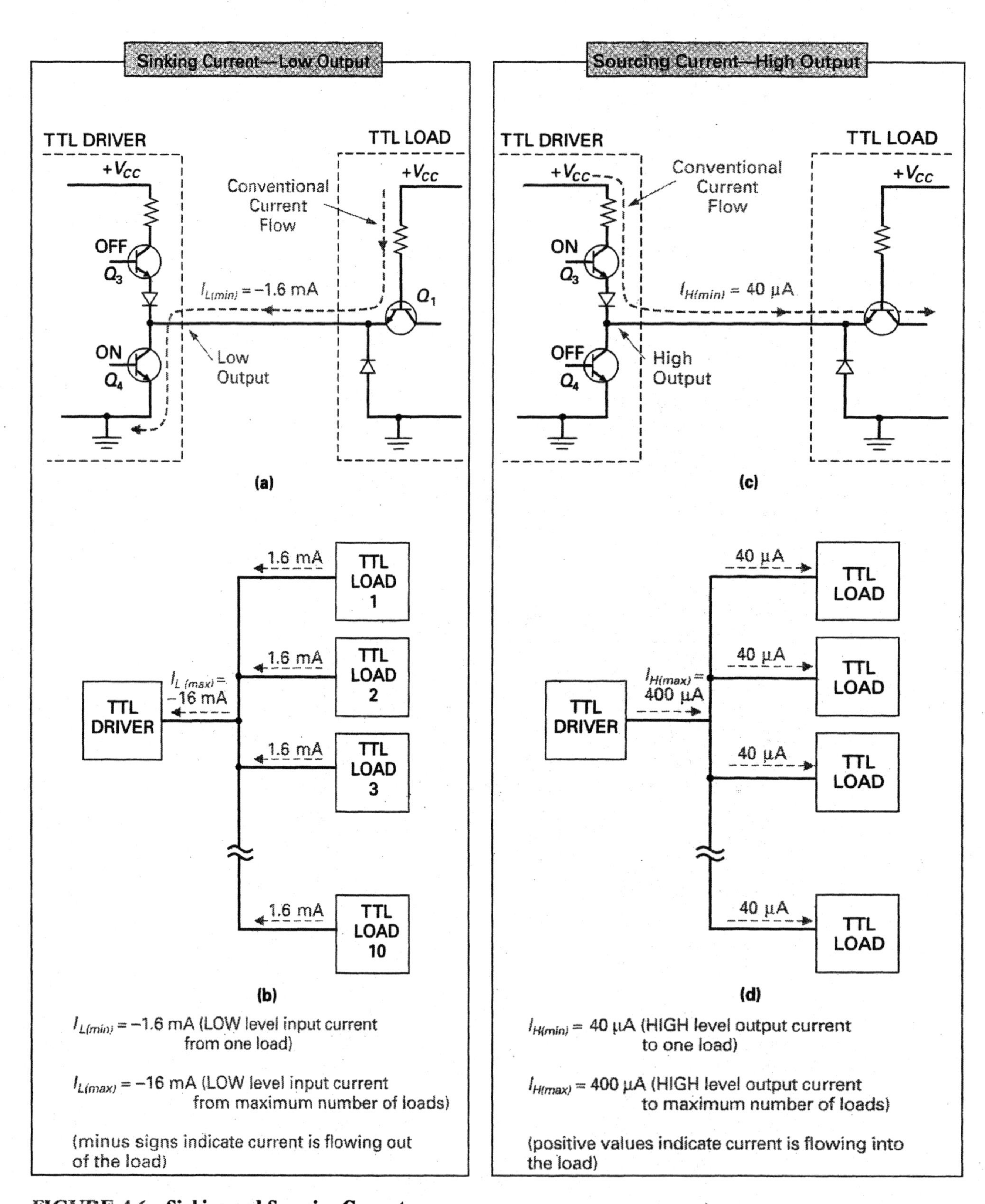

FIGURE 4-6 Sinking and Sourcing Current.

4-1-2 *Low-Power and High-Speed TTL Logic Gates*

Low-Power TTL
A line of transistor-transistor logic ICs that have a low power dissipation figure.

High-Speed TTL
A line of transistor-transistor logic ICs that have a high speed or low propagation delay speed figure.

A low-power TTL circuit is basically the same as the standard TTL circuit previously discussed except that all of the internal resistance values in the logic gate circuit are ten times larger. This means that the power dissipation of a low-power TTL circuit is one-tenth that of a standard TTL circuit (low-power TTL logic gates will typically have a power dissipation of about 1 mW per gate).

Increasing the internal resistances, however, will decrease the circuit's internal currents, and since the bipolar transistor is a current operated device, the response or switching time of these logic gate types will decrease. (Low-power TTL logic gates will typically have a propagation delay time of 33 ns per gate.) With low-power TTL circuits, therefore, high speed is sacrificed for low power consumption. Low-power logic gate ICs can be identified because the device number will have the letter "L" following the series number 74 or 54; for example, 74L00, 74L01, 74L02, and so on.

By decreasing the resistances within a standard logic gate, manufacturers can lower the logic gate circuit's internal time constants, and therefore decrease the gate's propagation delay time (typically 6 ns). The lower resistances, however, increase the gate's power dissipation (typically 22 mW), so the trade-off is therefore once again speed versus power. High-speed logic gate ICs can be identified because the device number will have the letter "H" following the series number 74 or 54; for example, 74H00, 74H01, 74H02, and so on.

4-1-3 *Schottky TTL Logic Gates*

Schottky TTL
A line of transistor-transistor logic ICs that use schottky transistors.

Saturation Delay Time
The additional time it takes a saturated transistor to turn off.

Schottky Transistor
A transistor having a schottky diode connected between its collector and base.

With standard TTL and low-power TTL logic gates, transistors go into saturation when they are switched ON and are flooded with extra carriers. When a transistor is switched OFF, it takes a short time for these extra carriers to leave the transistor, and therefore it takes a short time for the transistor to cut OFF. This **saturation delay time** accounts for the slow propagation delay times, or switching times, of standard and low-power TTL logic gates.

Schottky TTL logic gates overcome this problem by fabricating a schottky diode between the collector and base of each bipolar transistor, as seen in Figure 4-7(a). This transistor is called a **schottky transistor** and uses the symbol shown in Figure 4-7(b). Let us now see how this schottky diode will affect the operation of the transistor and decrease saturation delay time. When the transistor begins to turn ON, its collector voltage will fall. When the collector drops below 0.4 V of the base voltage, the schottky diode will conduct and shunt current away from the collector junction. This action effectively clamps the collector to 0.4 V of the base voltage, keeping the collector diode of the transistor reverse-biased. Keeping the collector diode reverse-biased prevents the transistor from slipping into heavy saturation, and therefore decreases the transistor's saturation delay time.

Schottky TTL devices are fast, with propagation delay times of typically 3 ns. This fast switching time means that the input signals can operate at extremely high frequencies of typically 100 MHz. Schottky TTL ICs are labeled with an "S" after the first two digits of the series number, for example, 74S04, 74S08, and so on.

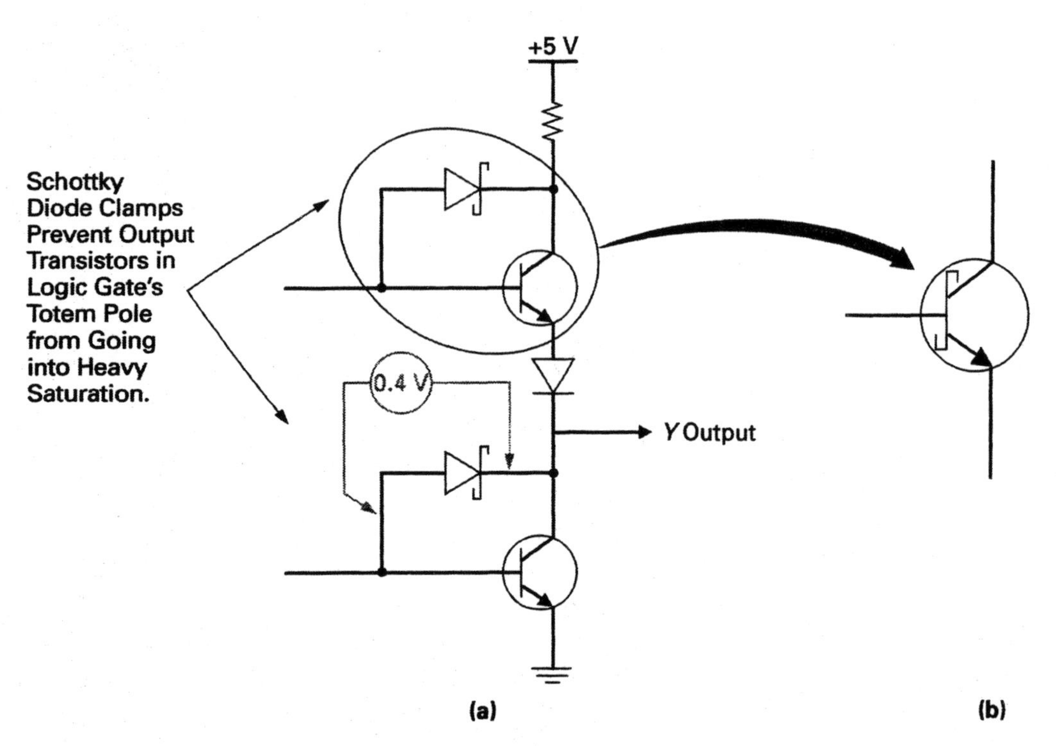

FIGURE 4-7 Schottky Clamped Bipolar Transistors. (a) Connecting a Schottky Diode across a Transistor Collector-Base Junction. (b) Schottky Transistor symbol.

By using schottky transistors and increasing the internal resistances in a logic gate circuit, manufacturers compromised the speed of schottky to obtain a lower dissipation rating. Schottky TTL devices will typically have a power dissipation of 19 mW per gate, whereas *low-power schottky TTL devices* typically have a power dissipation of 2 mW per gate and propagation delay times of 9.5 ns. These low-power schottky TTL devices include the letters "LS" in the device number; for example, 74LS32, 74LS123, and so on.

Special schottky TTL devices are available that use a new integration process called oxide isolation, in which the transistor's collector diode is isolated by a thin oxide layer instead of a reverse-bias junction. These *advanced schottky (AS) TTL devices* (labeled 74AS00) have a typical power dissipation of 8.5 mW per gate and a propagation delay time of 1.5 ns. By increasing internal resistances once again, manufacturers came up with an *advanced low-power schottky TTL* line of products (labeled 74ALS00) which have power dissipation and propagation time figures of 1 mW and 4 ns. Some manufacturers, such as Fairchild, use the letter "F" for fast, in their AS series of ICs; for example, 74F10, 74F14, and so on.

As you have seen, an IC contains quite a complex circuit, and this circuit has a detailed list of circuit characteristics. These characteristics or specifications are listed in a manufacturer's data sheet or spec sheet. As an example, Figure 4-8 shows a typical data sheet for a 7400 (quad two-input NAND) TTL IC.

DEVICE: SN7400, SN5400 (Quad Two-Input NAND)

TOP VIEW — LOGIC DIAGRAM (POSITIVE LOGIC) 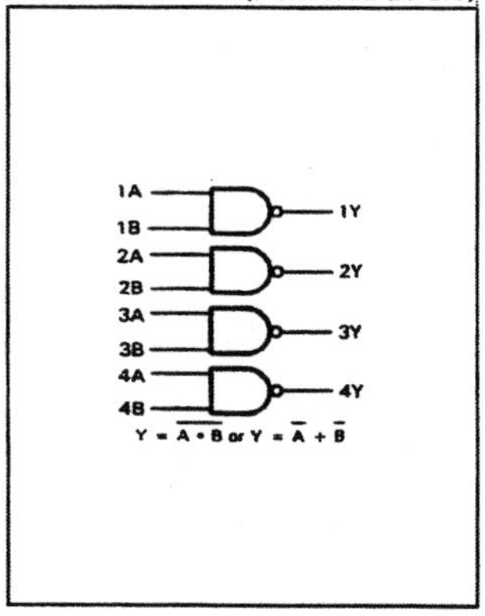 — LOGIC SYMBOL

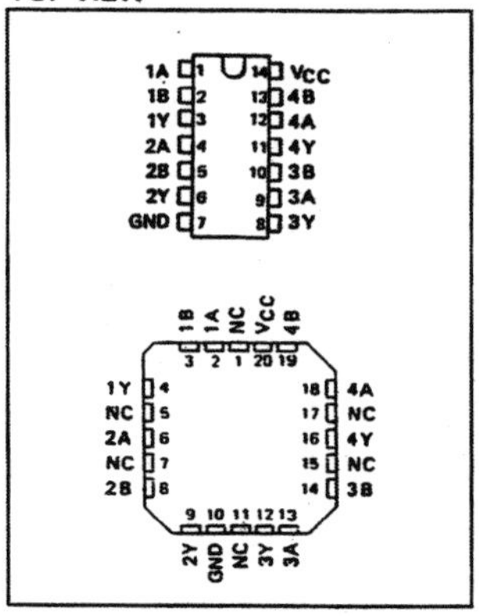

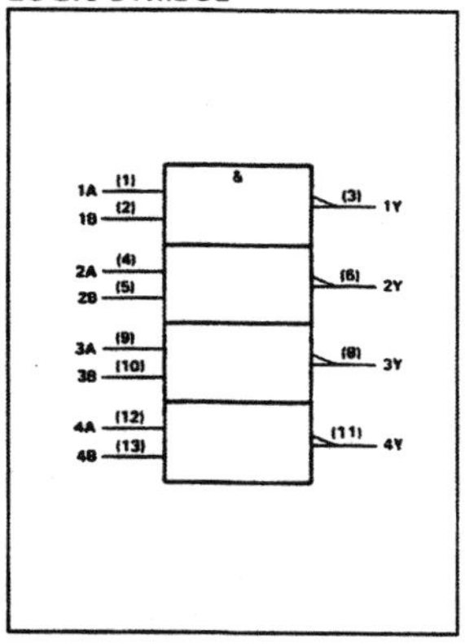

FUNCTION TABLE (each gate) ← Truth Table

INPUTS A	INPUTS B	OUTPUT Y
H	H	L
L	X	H
X	L	H

The IC will more than likely suffer permanent damage if these absolute maximum ratings are exceeded.

absolute maximum ratings over operating free-air temperature range (unless otherwise noted)

Supply voltage, V_{CC} (see Note 1) (maximum allowable supply voltage) .... 7 V
Input voltage: '00, 'S00 .... 5.5 V
'LS00 (maximum allowable input voltage) .... 7 V
Operating free-air temperature range: SN54' .... −55°C to 125°C
SN74' .... 0°C to 70°C ← (normal room temperature is about 25°C)
Storage temperature range .... −65°C to 150°C

recommended operating conditions (the levels of voltage, current, and temperature at which the IC should be run under normal conditions)

	SN5400 MIN	SN5400 NOM	SN5400 MAX	SN7400 MIN	SN7400 NOM	SN7400 MAX	UNIT
V_{CC} Supply voltage (min. to max. supply voltage range)	4.5	5	5.5	4.75	5	5.25	V
V_{IH} High-level input voltage (the min. input voltage for a HIGH)	2			2			V
V_{IL} Low-level input voltage (the max. input voltage for a LOW)			0.8			0.8	V
I_{OH} High-level output current (the max. output current for a HIGH)			−0.4			−0.4	mA
I_{OL} Low-level output current (the max. output current for a LOW)			16			16	mA
T_A Operating free-air temperature (min. to max. operating temp.)	−55		125	0		70	°C

DC Specifications (Electrical Characteristics)

PARAMETER	TEST CONDITIONS †	SN5400 MIN	SN5400 TYP‡	SN5400 MAX	SN7400 MIN	SN7400 TYP‡	SN7400 MAX	UNIT
V_{IK}	V_{CC} = MIN, I_I = −12 mA			−1.5			−1.5	V
V_{OH} (← HIGH Output Voltage)	V_{CC} = MIN, V_{IL} = 0.8 V, I_{OH} = −0.4 mA	2.4	3.4		2.4	3.4		V
V_{OL} (← LOW Output Voltage)	V_{CC} = MIN, V_{IH} = 2 V, I_{OL} = 16 mA		0.2	0.4		0.2	0.4	V
I_I	V_{CC} = MAX, V_I = 5.5 V			1			1	mA
I_{IH}	V_{CC} = MAX, V_I = 2.4 V (HIGH Input Current)			40			40	μA
I_{IL}	V_{CC} = MAX, V_I = 0.4 V (LOW Input Current)			−1.6			−1.6	mA
I_{OS}§	V_{CC} = MAX (Current when output is shorted to ground)	−20		−55	−18		−55	mA
I_{CCH}	V_{CC} = MAX, V_I = 0 V		4	8		4	8	mA
I_{CCL}	V_{CC} = MAX, V_I = 4.5 V		12	22		12	22	mA

Not a recommended condition (← I_{OS})

Current supply to IC when all outputs are HIGH or LOW (← I_{CCH}, I_{CCL})

switching characteristics, V_{CC} = 5 V, T_A = 25°C (AC specifications)

PARAMETER	FROM (INPUT)	TO (OUTPUT)	TEST CONDITIONS	MIN	TYP	MAX	UNIT
t_{PLH}	A or B	Y	R_L = 400 Ω, C_L = 15 pF		11	22	ns
t_{PHL}					7	15	ns

HIGH to LOW Output, propagation delay
LOW to HIGH Output, propagation delay

Maximum Clock Frequency (f_{max})	15 MHz
Input Pulse Width (t_w)	20 ns
Data Setup Time (t_{su})	15 ns
Data Hold Time (t_h)	5 ns

Highest rate at which a signal can change when applied to an input
Minimum time a pulse must be present at the input to cause a reliable change at the output
Minimum amount of time data must be available at an input
Minimum amount of time a data input must be maintained

FIGURE 4-8 A Standard TTL Data Sheet. (Courtesy of Texas Instruments)

Most data sheets can be broken down into five basic blocks:

a. pin-out and logic symbols b. maximum ratings
c. recommended ratings d. DC specifications or electrical characteristics
e. AC specifications or switching characteristics

Study all of the annotations included in this data sheet, because they explain the meaning the terms used and the characteristics.

■ **EXAMPLE:**

What differences can you notice between the 7400 and 5400 operating conditions and specifications listed in Figure 4-8?

■ ***Solution:***

Generally, the key differences between the commercial 7400 and the military 5400 series of TTL ICs are operating supply voltage range, operating temperature range, and output current when the output is shorted.

7400:	Supply Voltage Range	4.75 V to 5.25 V
	Operating Temperature Range	0°C to 70°C
	Current When the Output Is Shorted	−18 mA to −55 mA
5400:	Supply Voltage Range	4.5 V to 5.5 V
	Operating Temperature Range	−55°C to 125°C
	Current When the Output Is Shorted	−20 mA to −55 mA

4-1-4 Open Collector TTL Gates

Open-Collector Output
A logic circuit in which the final output, which is a transistor's collector, is open or floating.

Pull-up Resistor
The name given to a resistor that is connected between a power line and a signal line. Its function is to pull-up or make HIGH a signal line when it is not switched LOW.

Some TTL devices have an **open collector output** instead of the totem pole output circuit previously discussed. Figure 4-9(a) shows a standard TTL INVERTER circuit with an open collector output, and Figure 4-9(b) shows the logic symbol used to represent an open collector output gate.

In order to get a HIGH or LOW output from an open collector gate, an external **pull-up resistor** (R_P) must be connected between $+V_{CC}$ and the collector of Q_3 (which is the output) as seen in Figure 4-9(c). In this circuit, two open collector inverters are driving the same output Y. If two standard TTL logic gates with totem pole outputs were wired together in this way and the A INVERTER produced a HIGH output and the B INVERTER a LOW output, there would be a direct short between the two, causing both gates to burn out. With open collector gates, the outputs can be wired together or connected to a common output line, without causing any HIGH-LOW output conflicts between gates. With the circuit in Figure 4-9(c), only when both inputs are LOW will the output transistors of both inverters be floating and the output be pulled up by R_P to +5 V ($A = 0$, $B = 0$, $Y = 1$). If either A input or B input is HIGH, one of the output transistors will turn ON and pull down or ground the output Y ($A = 1$ or $B = 1$, $Y = 0$). You may recognize this logic operation as that of a NOR gate.

Open collector gates can sink around 40 mA, which is a vast improvement over standard TTL gates that only sink 16 mA (74ASXX gates can sink 20 mA).

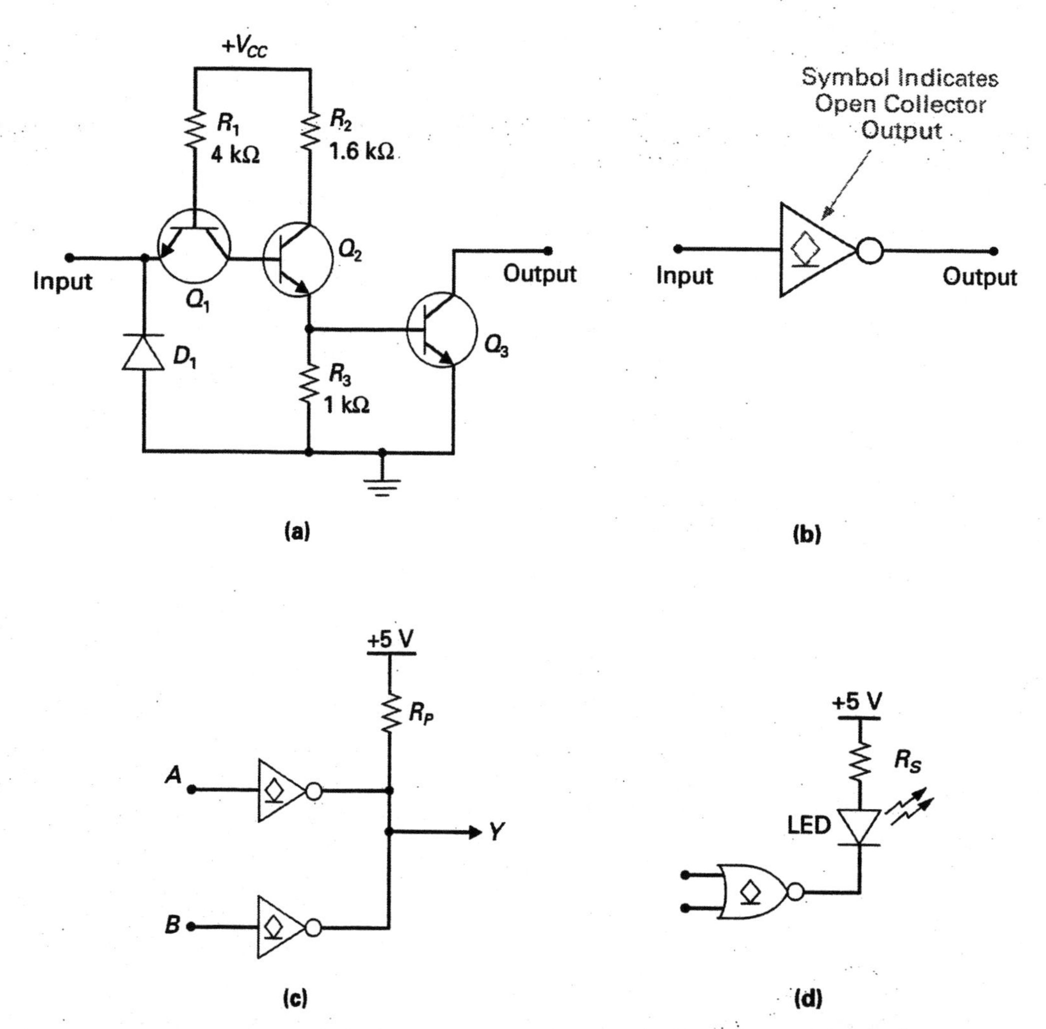

FIGURE 4-9 Open Collector Gates. (a) Bipolar Transistor Open Collector INVERTER Circuit. (b) Open Collector Symbol. (c) Wired INVERTER Open Collector Gates. (d) Open Collector NOR Driver Gate.

In Figure 4-9(d), you can see how the high sinking current ability of an open collector NOR gate can be used to drive the high current demand of a light emitting diode (LED) load. The current limiting resistor (R_S) is being used to limit the current through the LED to about 25 mA, which produces a good level of brightness from an LED. In many applications, open collector gates are used to drive external loads that require increased current such as LED displays, lamps, relays, and so on. As an example, Figure 4-10 shows typical open-collector current characteristics.

4-1-5 Three-State (Tri-State) Output TTL Gates

Three-State Output TTL
A line of transistor-transistor logic gates that can have three possible outputs: LOW, HIGH, or OPEN.

The **three-state output gate** has, as its name implies, three output states or conditions. These three output states are LOW, HIGH, and FLOATING, or a "high-impedance output" state. Before we discuss why we need this third output state, let us see how this circuit operates.

Symbol	Parameter	54/74	54/74 H	54/74 S	54/74 LS
I_{OH} (max)	Power Supply	12 mA	26 mA	19.8 mA	2.4 mA
I_{OL} (max)	Current	33 mA	58 mA	54 mA	6.6 mA
t_{PLH} (max)	Propogation Delay	55 ns	18 ns	7.5 ns	22 ns
t_{PHL} (max)		15 ns	15 ns	7.0 ns	18 ns

A Standard TTL open-collector gate can sink a maximum of 33 mA when supplying a LOW output, whereas a high-speed TTL open-collector gate can sink a maximum of 58 mA.

7401 - QUADRUPLE 2-INPUT
POSITIVE-NAND GATES
WITH OPEN-COLLECTOR OUTPUTS

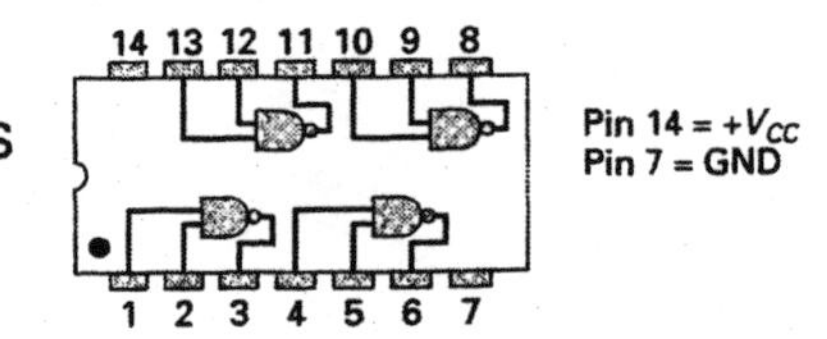

A
B
Y

FIGURE 4-10 **An Open Collector Output Data Sheet.**

Figure 4-11(a) illustrates the internal bipolar transistor circuit for a three-state NOT gate. The circuit is basically the same as the previously discussed INVERTER gate, except for the additional control input applied to transistor Q_4. When the enable input is LOW, Q_4 is OFF and has no effect on the normal inverting operation of the circuit. However, when the enable input is HIGH, Q_4 is turned ON, and the ground on its emitter will be switched through to the second emitter of Q_1. The ON Q_1 will switch ground from its emitter through to the base of Q_2, turning Q_2 OFF and therefore Q_5 OFF. Transistor Q_4 will also ground the base of Q_3 via D_1, and so Q_3 will turn OFF. The result will be that both totem pole transistors are cut OFF, as seen in the equivalent circuit shown in the inset in Figure 4-11(a). This high-impedance output state, therefore, is created by turning OFF both totem pole output transistors so that the output is isolated from ground and the supply voltage. With the output of the gate completely disconnected from the circuit, this lead is said to be a disconnected, or a FLOATING high-impedance line.

Active-LOW Control Line
A control line that goes LOW when activated.

Figure 4-11(b) shows the schematic symbol for this three-state output NOT gate. The bubble attached to the enable input is used to indicate an **active-LOW control line,** which means that if you want to "activate the gate" you must make this control line "LOW." To disable the gate therefore, we would not make this line active, so the enable line would be HIGH. Studying the truth table or function table in Figure 4-11(c), you can see this gate's operation summarized. When the enable control line is LOW, the circuit is enabled for normal INVERTER operation. When the enable control line is HIGH, however, the output is always FLOATING, no matter what input logic level is present. The label for active-LOW control lines will usually have a bar over the label's letters, as shown with the output enable $(\overline{OE})$ in Figure 4-11(b). This bar over the label is used to distinguish an active-LOW control line label from an active-HIGH control line label.

Like the previously discussed open collector gates, these devices are used in applications where two or more gate outputs are connected to a single output line. These common signal lines are called "signal buses," and they allow logic

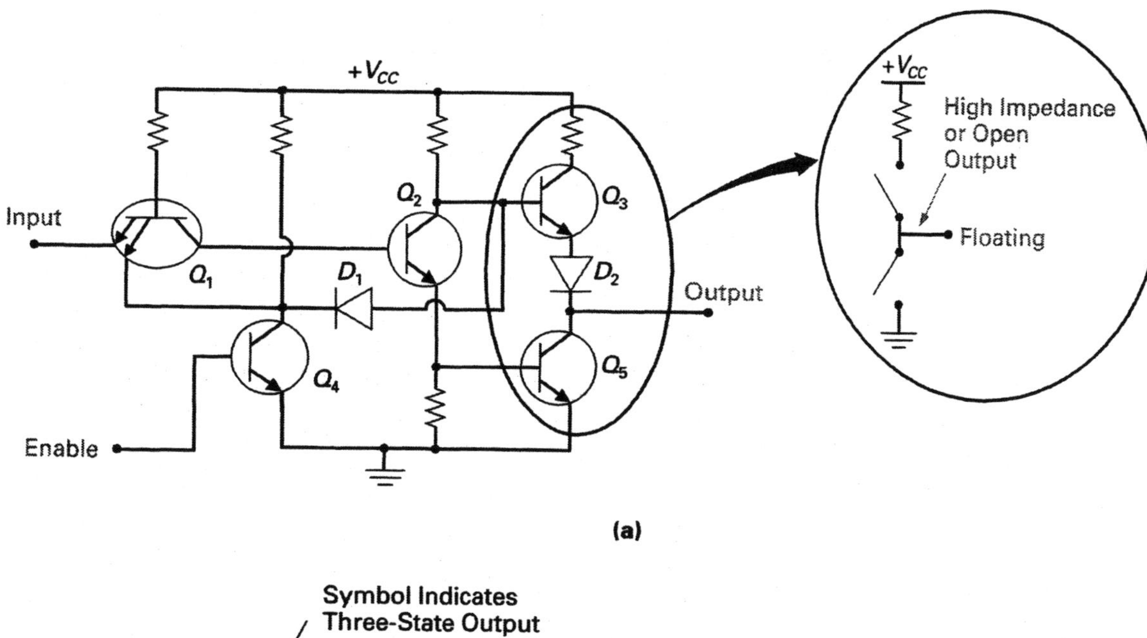

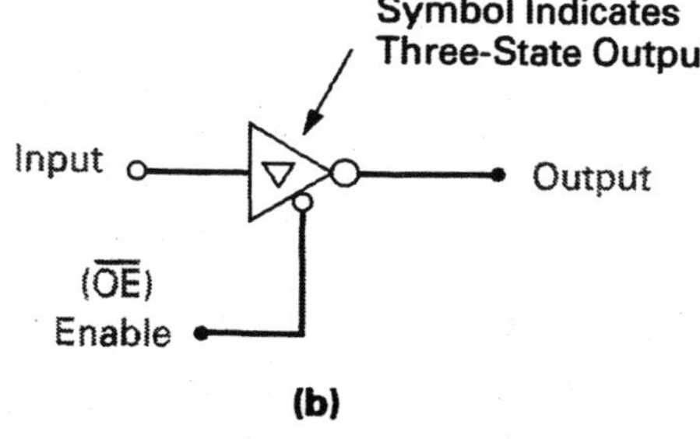

Enable	Input	Output
LOW (enabled)	LOW	HIGH
LOW (enabled)	HIGH	LOW
HIGH (disabled)	LOW	FLOATING
HIGH (disabled)	HIGH	FLOATING

(c)

FIGURE 4-11 Three-State Output Gates. (a) Bipolar Transistor Circuits. (b) Symbol. (c) Truth Table.

devices to individually take control of a common set of signal lines without causing signal interferences with other devices on the bus. To avoid HIGH/LOW output line conflicts between gates, only one gate is enabled at one time to drive the output line, with all other gates being disabled, and therefore their outputs FLOATING. These applications will be discussed in more detail in future chapters.

4-1-6 Buffer/Driver TTL Gates

A **buffer** is a device that isolates one circuit from another. In order to achieve this isolation, a buffer should have a high input impedance (low input current) and a low output impedance (high output current). A standard TTL gate provides a certain amount of isolation or buffering, since the output current is ten times that of the input current.

Buffer
A device that isolates one device from another. Its schematic symbol is a triangle with one input and one output.

	Input	Output	
Standard TTL	$I_{IL} = -1.6$ mA	$I_{OL} = -16$ mA	(output is ten times larger)
	$I_{IH} = 40\ \mu$A	$I_{OH} = 400\ \mu$A	(output is ten times larger)

A buffer or driver logic gate is a basic gate that has been slightly modified to increase the output current. For example, a 7428 is a "quad two-input NOR gate buffer" whose output current is thirty times greater than the input currents, enabling it to drive heavier loads.

	Input	Output	
Buffer/Driver TTL	$I_{IL} = -1.6$ mA	$I_{OL} = -48$ mA	(output is thirty times larger)
	$I_{IH} = 40\ \mu$A	$I_{OH} = 1.2$ mA	(output is thirty times larger)

As an example, Figure 4-12 shows a typical three-state output, buffer/driver data sheet. Annotations have been included to highlight and explain the meaning of some terms and characteristics discussed in the previous sections.

■ **EXAMPLE:**

What is the maximum sourcing and sinking output current for the 74244?

■ ***Solution:***

Referring to Figure 4-12, you can see that each output can source 15 mA (HIGH level output current, I_{OH}), and sink 24 mA (LOW level output current, I_{OL}).

4-1-7 Schmitt-Trigger TTL Gates

Schmitt-Trigger
A level-sensitive input circuit that has a two-state output.

The **schmitt-trigger circuit,** named after its inventor, is a two-state device that is used for pulse shaping. A typical schmitt-trigger circuit can be seen in Figure 4-13(a), while the schematic symbol for the schmitt-trigger logic symbol is shown in Figure 4-13(b). The waveforms shown in Figure 4-13(c) and (d) illustrate the two basic applications of the schmitt trigger, which are to convert a sine wave into a rectangular wave or to sharpen the rise and fall times of a rectangular wave.

Like all of the previously discussed digital switching circuits, the schmitt-trigger circuit may be constructed with discrete or individual components, or purchased as an IC containing several complete schmitt-trigger circuits. Using the circuit shown in Figure 4-13(a) and the example input/output waveforms shown in Figure 4-13(c) and (d), let us see how this bipolar transistor circuit will operate. The input signal is applied to the base of Q_1, and if this signal is below the ON voltage (V_{ON}), Q_1 will be cut OFF and its collector voltage will be HIGH. The HIGH Q_1 collector voltage is coupled to the base of Q_2 causing it to saturate (to turn heavily ON). As a result, Q_2 conducts a large current (shown as electron flow) through R_6, Q_2 emitter-to-collector, and R_2 to $+V_{CC}$. The voltage developed across R_6 (V_{R6}) establishes the V_{ON} voltage of the schmitt trigger, since the base of Q_1, and therefore the input, will have to be 0.7 V greater than the emitter of Q_1, or V_{R6}, if Q_1 is to turn ON.

When the input does reach a value that is 0.7 V above V_{R6}, Q_1 will turn ON and its collector voltage will fall. This decrease is coupled to the base of Q_2, turning it OFF and causing its collector voltage, and therefore the output, to rise. With Q_1 now ON and Q_2 OFF, the voltage across R_6 is now being

DEVICE: SN74S244 (Octal Three-State Buffer/Driver)

- **3-State Outputs Drive Bus Lines or Buffer Memory Address Registers**
- **PNP Inputs Reduce D-C Loading**
- **Hysteresis at Inputs Improves Noise Margins**

description

These octal buffers and line drivers are designed specifically to improve both the performance and density of three-state memory address drivers, clock drivers, and bus-oriented receivers and transmitters. The designer has a choice of selected combinations of inverting and noninverting outputs, symmetrical $\overline{G}$ (active-low output control) inputs, and complementary G and $\overline{G}$ inputs. These devices feature high fan-out, improved fan-in, and 400-mV noise-margin. The SN74LS' and SN74S' can be used to drive terminated lines down to 133 ohms.

TOP VIEW

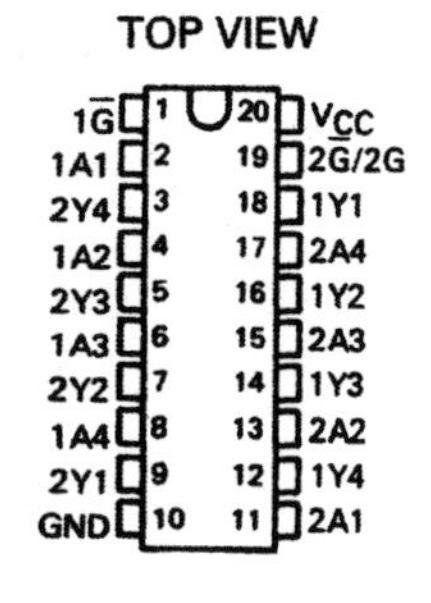

TOP VIEW

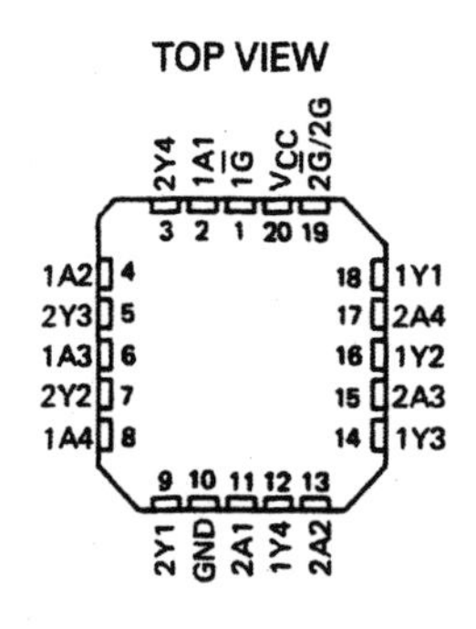

LOGIC DIAGRAM

'LS244, 'S244

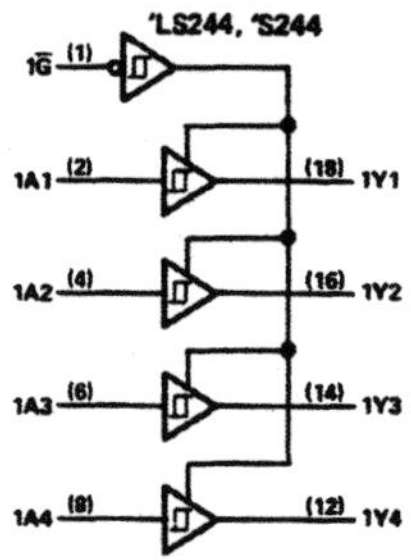

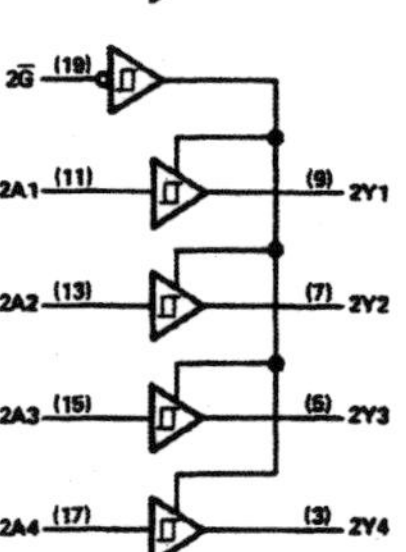

LOGIC SYMBOL

'LS244, 'S244

1G (1), 1A1 (2), 1A2 (4), 1A3 (6), 1A4 (8), 2G (19), 2A1 (11), 2A2 (13), 2A3 (15), 2A4 (17)

(18) 1Y1, (16) 1Y2, (14) 1Y3, (12) 1Y4, (9) 2Y1, (7) 2Y2, (5) 2Y3, (3) 2Y4

These two active-LOW output enable lines are used to control whether the inputs are switched through to the outputs ($\overline{OE}$ = LOW), or the outputs are open ($\overline{OE}$ = HIGH)

Three-State Output Symbol

Buffer-Driver Symbol

Outputs

Inputs

absolute maximum ratings over operating free-air temperature range (unless otherwise noted)

Supply voltage, V_{CC} (see Note 1) . . . . . 7 V
Input voltage: 'LS Circuits . . . . . 7 V
'S Circuits . . . . . 5.5 V
Off-state output voltage . . . . . 5.5 V
Operating free-air temperature range: SN54LS', SN54S' Circuits . . . . . −55°C to 125°C
SN74LS', SN74S' Circuits . . . . . 0°C to 70°C
Storage temperature range . . . . . −65°C to 150°C

NOTE 1: Voltage values are with respect to network ground terminal.

recommended operating conditions

PARAMETER		SN54LS'			SN74LS'			UNIT
		MIN	NOM	MAX	MIN	NOM	MAX	
V_{CC}	Supply voltage (see Note 1)	4.5	5	5.5	4.75	5	5.25	V
V_{IH}	High-level input voltage	2			2			V
V_{IL}	Low-level input voltage			0.7			0.8	V
I_{OH}	High-level output current			−12			−15	mA
I_{OL}	Low-level output current			12			24	mA
T_A	Operating free-air temperature	−55		125	0		70	°C

NOTE 1: Voltage values are with respect to network ground terminal.

Buffer/Driver Increased Output Currents.

$I_{OH} = -15$ mA

$I_{OL} = 24$ mA

FIGURE 4-12 A Three-State Output Buffer/Driver Data Sheet. (Courtesy of Texas Instruments)

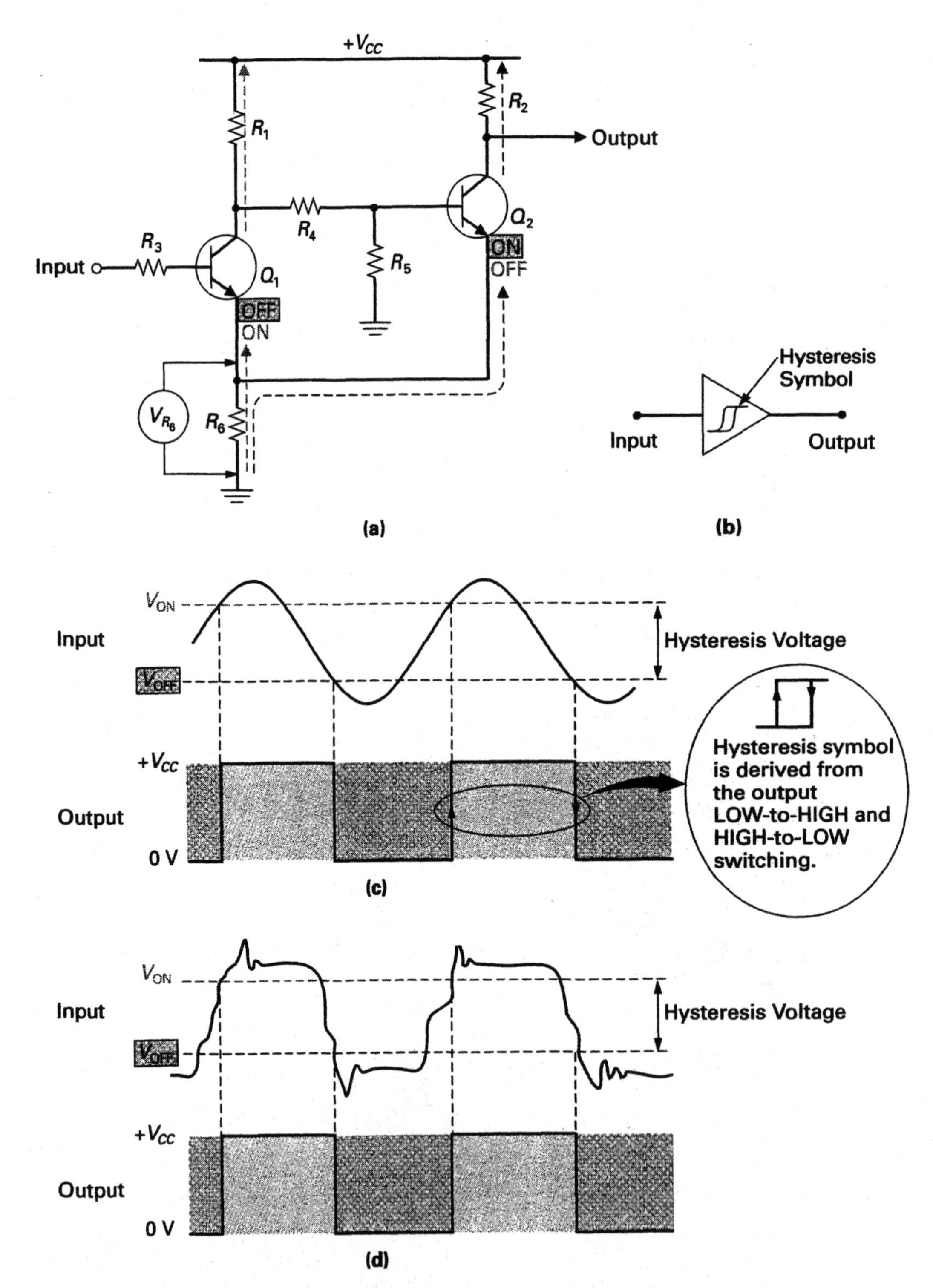

FIGURE 4-13 The Schmitt Trigger. (a) Bipolar Transistor Circuit. (b) Logic Gate Symbol. (c) Converting a Sine Wave into a Rectangular Wave. (d) Sharpening the Rise and Fall Times of a Rectangular Wave.

established by the current path through R_6, Q_1 emitter-to-collector, and R_1 to $+V_{CC}$. This will set up a different V_{R6} voltage, and therefore a different V_{OFF} voltage since the input must fall 0.7 V below V_{R6} for Q_1 to turn OFF. When the input falls below this voltage, Q_1 will turn OFF, its collector voltage will rise and turn ON Q_2, causing its collector voltage and the output voltage to fall. The circuit is now returned to its original condition, awaiting the input to once again exceed the V_{ON} threshold voltage.

The symbol for any schmitt-trigger gate will contain the distinctive hysteresis symbol, as shown in Figure 4-13(b). Hysteresis can be generally defined as the lag between cause and effect. Referring to the waveforms in Figure 4-13(c) and (d), you can see that there is in fact a lag between the input (cause) and output (effect). The input has to rise to an "ON voltage" (V_{ON}) before the output will rise to its positive peak, and must drop below an "OFF voltage" (V_{OFF}) before the output will drop to its negative peak. This lag is desirable since it produces a sharp, well defined, output signal. The difference between the V_{ON} and V_{OFF} voltage is called the **hysteresis voltage.**

Hysteresis Voltage The difference between two voltages.

Several schmitt-trigger TTL ICs are available, including 7413 (dual four-input NAND schmitt-trigger), 7414 (hex schmitt-trigger inverters), 74132 (quad two-input NAND schmitt-trigger), and so on. These logic circuits will have a schmitt-trigger circuit connected to the output of each logic gate circuit so that the gate will perform its normal function and then sharpen the rise and fall times of pulses that have been corrupted by noise and attenuation. As an example, Figure 4-14 shows a typical TTL schmitt-trigger data sheet.

4-1-8 Emitter-Coupled Logic (ECL) Gate Circuits

Schottky TTL circuits improved the switching speed of digital logic circuits by preventing the bipolar transistor from saturating. **Emitter-coupled logic,** or **ECL,** is another non-saturating bipolar transistor logic circuit arrangement; however, ECL is the fastest of all logic circuits. It is used in large digital computer systems that require fast propagation delay times (typically 0.8 ns per gate) and are able to sacrifice power dissipation (typically 40 mW per gate).

Emitter-Coupled Logic (ECL) A bipolar digital logic circuit family that uses emitter input coupling.

Figure 4-15(a) shows a typical ECL gate circuit. The two inputs to this circuit, *A* and *B,* are applied to the transistors Q_1 and Q_2, which are connected in parallel. If additional inputs are needed, more input transistors can be connected in parallel across Q_1 and Q_2, as shown in the dashed lines in Figure 4-15(a). The two collector outputs of Q_2 and Q_3 drive the two low output impedance emitter followers Q_5 and Q_6. These emitter follower outputs provide an OR output from Q_6, and its complement, or opposite, which is a NOR output from Q_5. This dual output is one of the advantages of ECL gates.

Pull-Down Resistor A name given to a resistor that is connected between a signal line and ground. Its function is to pull-down or make LOW a signal line when it is not switched HIGH.

External **pull-down resistors** are required, as seen in Figure 4-15(b), which also shows the logic symbol for this OR/NOR ECL gate. The open outputs of ECL gates means that two or more gate outputs can drive the same line without causing any HIGH/LOW voltage conflicts. The V_{EE} pin of an ECL IC is typically connected to −5.2 V, and the V_{CC} pin is connected to ground. Although these dc supply values seem to not properly bias the circuit, remember that the V_{CC} power line is still positive (since it is at ground) relative to the V_{EE} power

DEVICE: SN7414, SN74LS14 (Hex Inverter Schmitt Trigger)

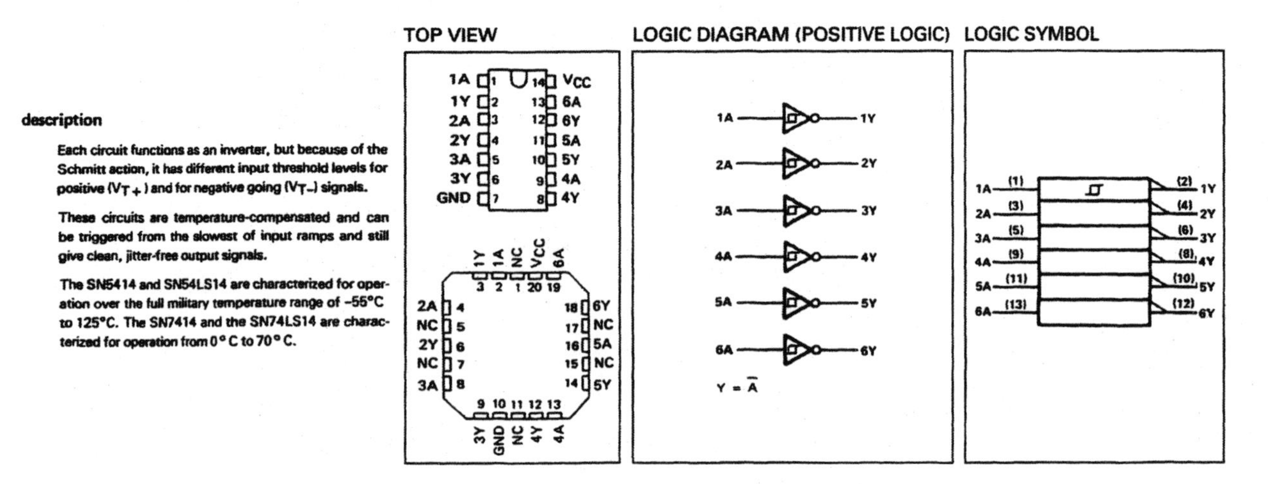

absolute maximum ratings over operating free-air temperature range (unless otherwise noted)

Supply voltage, V_{CC} (see Note 1) .... 7 V
Input voltage: '14 .... 5.5 V
'LS14 .... 7 V
Operating free-air temperature: SN54' .... −55°C to 125°C
SN74' .... 0°C to 70°C
Storage temperature range .... −65°C to 150°C

recommended operating conditions

	SN5414			SN74LS14			UNIT
	MIN	NOM	MAX	MIN	NOM	MAX	
V_{CC} Supply voltage	4.5	5	5.5	4.75	5	5.25	V
I_{OH} High-level output current			−0.8			−0.4	mA
I_{OL} Low-level output current			16			8	mA
T_A Operating free-air temperature	−55		125	0		70	°C

FIGURE 4-14 A Schmitt-Trigger Data Sheet. (Courtesy of Texas Instruments)

line (which is at −5.2 V). The transistor Q_4 has a constant base bias voltage and an emitter resistor and will produce a temperature stabilized constant voltage, or reference voltage, of approximately −1.3 V to the base of Q_3. The bipolar transistors in the circuit simply conduct more or conduct less depending on whether they represent a LOW or HIGH, ensuring that none of the transistors saturate.

Preventing transistors from going into saturation accounts for the fast speed of ECL logic gate circuits, and the fact that transistors, for the most part, are conducting and therefore drawing a current, accounts for the ECL's increase in power dissipation. The typical characteristics of ECL gates are listed in Figure 4-15(c). Notice that the binary 0 and binary 1 voltage levels are slightly different from TTL circuits; however, the binary 1 voltage level (−0.8 V) is still positive relative to binary 0 (−1.7 V).

Let us now examine this ECL circuit's operation. When either one or both of the inputs go to a HIGH (−0.8 V), that transistor will conduct more (since its base is now positive with respect to the emitter which is at −1.3 V). In so doing, the input transistor will make transistor Q_3 turn more OFF (since the

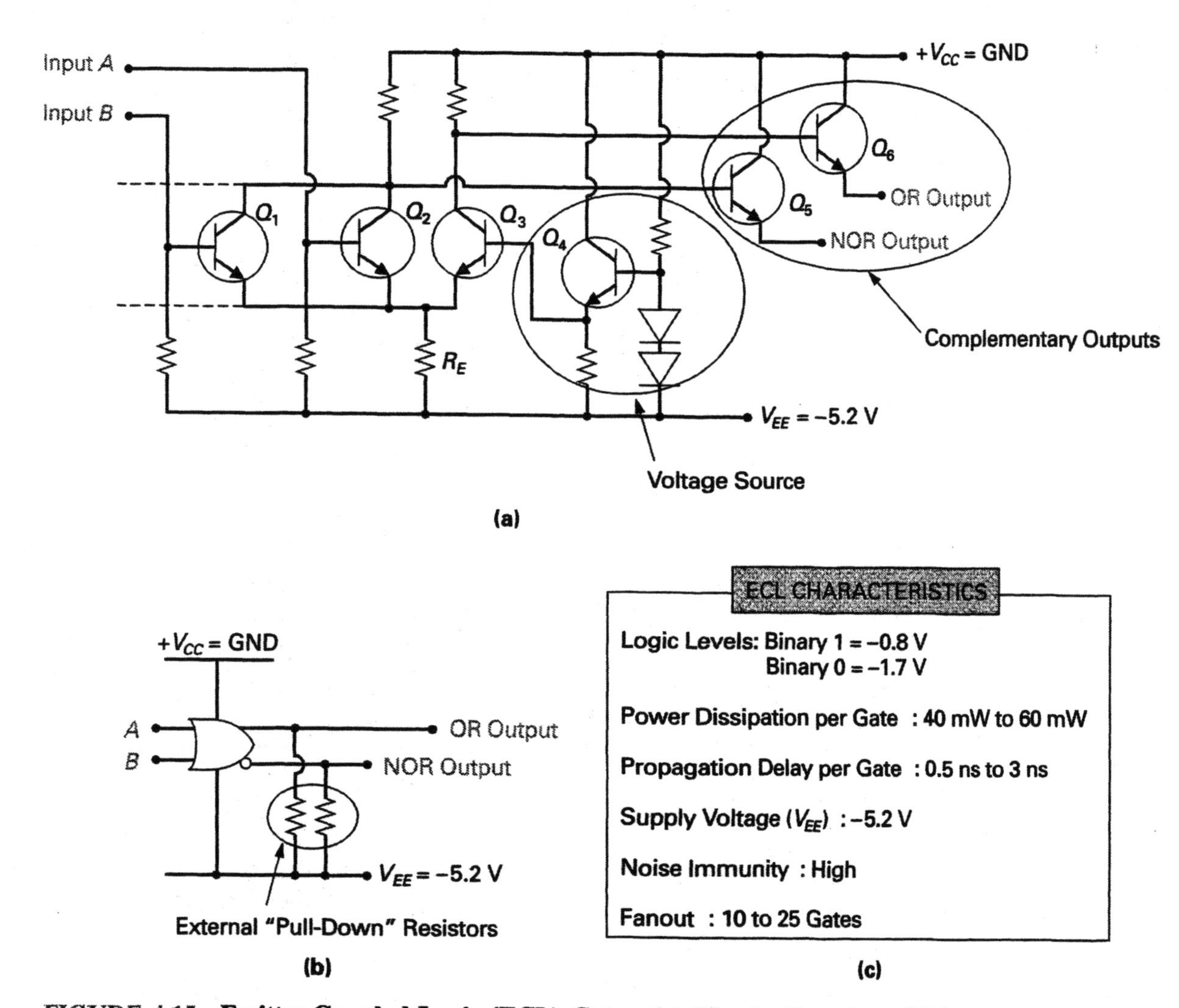

FIGURE 4-15 Emitter-Coupled Logic (ECL) Gates. (a) Bipolar Transistor OR/NOR Circuit. (b) Logic Symbol. (c) Characteristics.

base-emitter junction of the input transmitter will make Q_3's emitter positive with respect to its large negative base voltage). The conducting input transistor will switch its negative emitter voltage through to the collector and therefore to the base of Q_5. Q_5 will turn more OFF, and its output will be pulled LOW. For a NOR gate output, any binary 1 input will produce a binary 0 output.

On the other hand, with Q_3 more OFF, its collector will be more positive (closer to 0 V) and Q_6 will turn more ON and connect the more positive V_{CC} to the OR output. For an OR gate output, any binary 1 input will produce a binary 1 output. When all inputs are LOW (−1.7 V), all of the input transistors will be more OFF, producing a HIGH output to Q_5 (all inputs are 0, output from NOR is 1). With Q_1 and Q_2 more OFF, Q_3 will conduct more producing a LOW output to Q_6 (all inputs are 0, output of OR is 0).

The high speed, high power consumption, and high-cost logic of these circuits account for why they are only used when absolutely necessary. The most widely used ECL family of ICs is Motorola's ECL (MECL) 10K and 100K series, in which devices are numbered MC10XXX or MC100XXX.

4-1-9 *Integrated-Injection Logic (I^2L) Gate Circuits*

Integrated-Injection Logic (IIL)
A bipolar digital logic family that uses a current injector transistor.

Packing Density
A term used to describe the number of devices that can be packed into a given area.

Current Injector Transistor
A transistor connected to generate a constant current.

Integrated-injection logic (IIL), or as they are sometimes abbreviated, I^2L (pronounced "I squared L"), is a series of digital bipolar transistor circuits that have good speed, low power consumption and very good **circuit packing density.** In fact, one I^2L gate is one-tenth the size of a standard TTL gate, making it ideal for compact or high density circuit applications.

A typical I^2L logic NOT gate is shown in Figure 4-16(a). Transistor Q_1 acts as a current source and the multiple collector transistor Q_2 acts as the INVERTER with two outputs. Studying the circuit, you can see that the base of Q_1 is connected to the emitter of Q_2, and the collector of Q_1 is connected to the base of Q_2. These common connections enable the entire I^2L gate to be constructed as one transistor with two emitters and two collectors, and use the same space on a silicon chip as one standard TTL multiple-emitter transistor.

The circuit in Figure 4-16(a) operates in the following way. The emitter of Q_1 is connected to an external supply voltage ($+V_S$) which can be anywhere from 1 V to 15 V depending on the injector current required. The constant base-emitter bias of Q_1 means that a constant current is available at the collector of Q_1 (called the **current injector transistor**) and this current will either be applied to the base of Q_2 or out of the input, depending on the input voltage logic level. To be specific, a LOW input will pull the injector current out of the input and away from the base of Q_2. This will turn Q_2 OFF and its outputs will be open, which is equivalent to a HIGH output. On the other hand, a HIGH input will cause the injector current of Q_1 to drive the base of Q_2. This will turn Q_2 ON and switch the LOW on its emitter through to the collector outputs.

Figure 4-16(b) illustrates an I^2L NAND gate. As you can see, the only difference between this circuit and the INVERTER gate circuit in Figure 4-16(a) is the two inputs A and B. Transistors Q_3 and Q_4 are shown in this circuit; however, they are the multiple-collector transistors of the gates driving inputs A and B of this NAND gate circuit. If either one of these transistors is ON (a low input to the NAND), the injector current of Q_1 will be pulled away from Q_2's base, turning OFF Q_2 and causing a HIGH output. Only when both inputs are HIGH (Q_3 and Q_4 are OFF), will the injector current from Q_1 turn ON Q_2 and produce a LOW output.

Figure 4-16(c) lists the characteristics of I^2L logic gate circuits. These circuits are used almost exclusively in digital watches, cameras, and almost all compact, battery-powered circuit applications. As mentioned earlier, its good speed, low power consumption, and small size make it ideal for any portable system. Also, the simplicity of I^2L circuits means that it is easy to combine analog or linear circuits (such as amplifiers and oscillators) with digital circuits to create complete electronic systems on one chip. The circuit complexity of standard TTL digital circuits means that it is normally too difficult, and not cost effective (due to fabrication difficulties and therefore high cost), to place digital and linear circuits on the same IC chip.

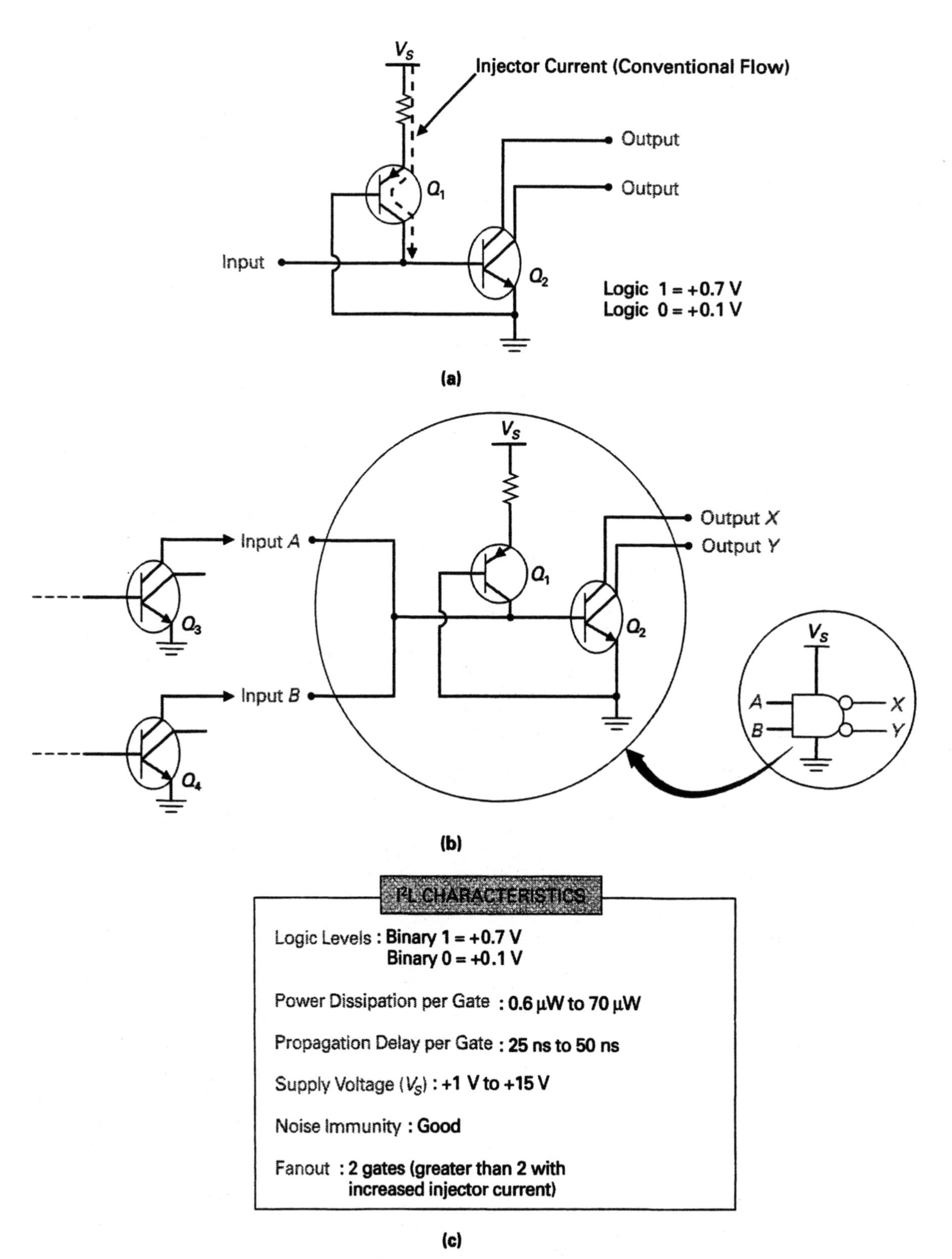

FIGURE 4-16 Integrated-Injection Logic (IIL or I^2L) Gates. (a) Bipolar Transistor INVERTER (NOT) Gate Circuit. (b) Bipolar Transistor NAND Gate Circuit. (c) Characteristics.

SELF-TEST REVIEW QUESTIONS FOR SECTION 4-1

1. TTL is an abbreviation for ____________.
2. What is the fanout of a standard TTL gate?
3. A floating TTL input is equivalent to a ____________ (HIGH/LOW) input.
4. The three output states of a tri-state logic gate are ____________, ____________, and ____________.
5. In what application would ECL circuits be used?
6. What is the advantage of I^2L logic circuits?

4-2 THE MOS FAMILY OF DIGITAL INTEGRATED CIRCUITS

In this section we will examine the three basic types of MOS ICs, which are called PMOS (pronounced "pea-moss"), NMOS (pronounced "en-moss") and CMOS (pronounced "sea-moss").

The E-MOSFET is ideally suited for digital or two-state circuit applications. One reason is that it naturally operates as a "normally-OFF voltage controlled switch," since it can be turned ON when the gate voltage is positive, and turned OFF when the gate voltage falls below a threshold level. This threshold level is a highly desirable characteristic because it prevents noise from false triggering, or accidentally turning ON, the device. The other E-MOSFET advantage is its extremely high input impedance which means that the device's circuit current, and therefore power dissipation, are low. This enables us to densely pack or integrate many thousands of E-MOSFETs onto one small piece of silicon, forming a high component density IC. These low-power and high-density advantages make the E-MOSFET ideal in battery-powered, small-sized (portable) applications such as calculators, wristwatches, notebook computers, hand-held video games, digital cellular phones, and so on.

In the following sections we will see how the E-MOSFET can be used to construct digital logic gate circuits, and then compare these circuits and their characteristics to the previously discussed bipolar logic gate circuit characteristics.

4-2-1 PMOS (P-Channel MOS) Logic Circuits

PMOS Logic
A digital logic family of ICs that use P-Channel MOSFETs.

The name **PMOS logic** is used to describe this logic circuitry because the logic gates are constructed using P-channel E-MOSFETs. Figure 4-17(a) shows how three P-channel E-MOSFETs can be used to construct a NOR logic gate. Figure 4-17(b) shows the logic levels used in PMOS circuits, and Figure 4-17(c) reviews the operation of a P-channel E-MOSFET.

The PMOS NOR gate circuit in Figure 4-17(a) will operate as follows. Transistor Q_3 can be thought of as a "current limiting resistor" between source and drain because its gate is constantly connected to $-V_{GG}$, and therefore Q_3 is always ON. Any HIGH input will turn OFF its respective E-MOSFET, resulting in a negative or LOW (−8 V) output due to the pull-down action of Q_3. Only when both inputs are LOW, will both Q_1 and Q_2 be ON, allowing $+V_{DD}$ (0 V, which is a HIGH) to be connected through Q_1 and Q_2 to the output.

4-2-2 *NMOS (N-Channel MOS) Logic Circuits*

The name **NMOS logic** is used to describe this logic circuitry because the logic gates are constructed using N-channel E-MOSFETs. Figure 4-18(a) shows how three N-channel E-MOSFETs can be used to construct a NAND gate. Figure 4-18(b) shows the logic levels used in NMOS circuits, and Figure 4-18(c) reviews the operation of an N-channel E-MOSFET.

NMOS Logic
A digital logic family of ICs that use N-channel MOSFETs.

The NMOS NAND gate circuit in Figure 4-18(a) will operate as follows. Transistor Q_1 functions as a "current limiting resistor" between source and drain because its gate is constantly connected to $+V_{DD}$, and therefore Q_1 is always ON. A LOW (0 V) on either or both of the inputs will turn OFF one or both of the transistors Q_2 and Q_3, causing the output to be pulled up to a HIGH (+3.5 V) by Q_1. Only when both inputs are HIGH, will Q_1 and Q_2 turn ON and switch the ground on Q_3's source through Q_3 and Q_2 to the output.

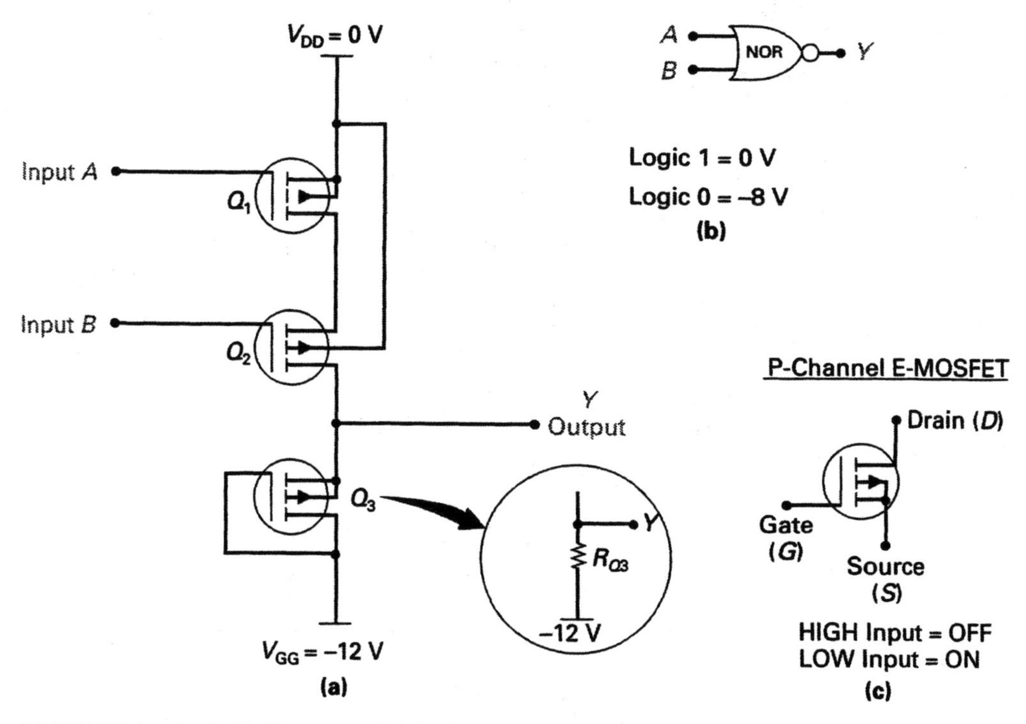

FIGURE 4-17 **A P-Channel MOS (PMOS) NOR Gate.**

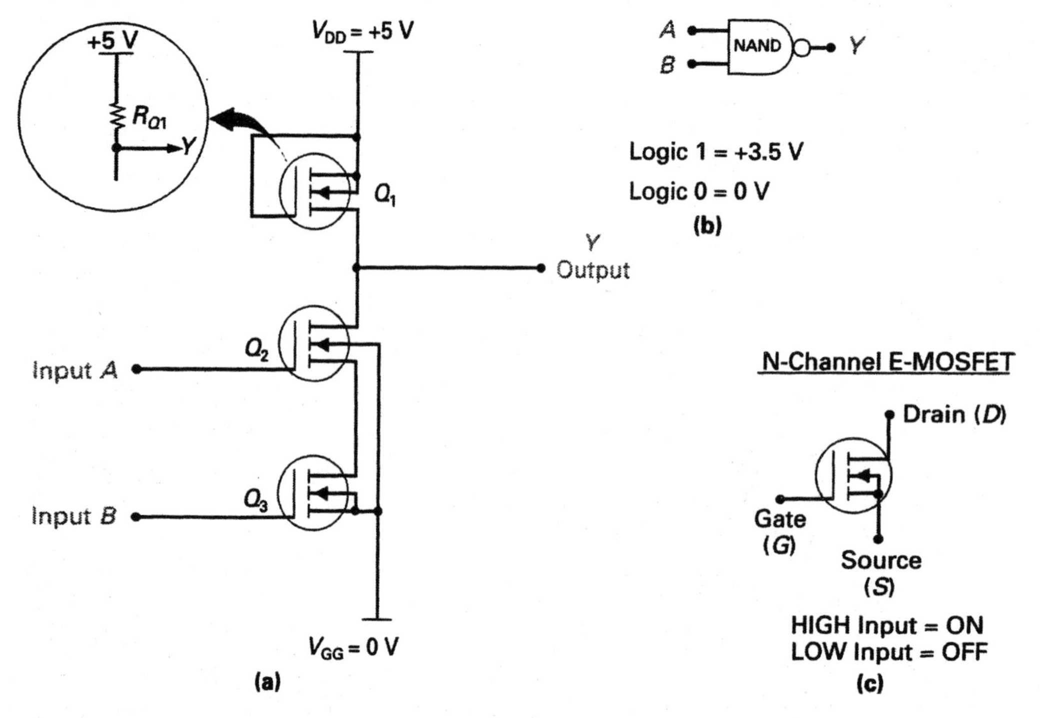

FIGURE 4-18 An N-Channel MOS (NMOS) NAND Gate.

4-2-3 *CMOS (Complementary MOS) Logic Circuits*

CMOS Logic
A digital logic family that makes use of both P-channel and N-channel MOSFETs.

The name **CMOS logic** is used to describe this logic circuitry because the logic gates are constructed using both a P-channel E-MOSFET (Q_1), and its complement, or opposite, an N-channel E-MOSFET (Q_2).

A CMOS NOT Gate Circuit

Figure 4-19(a) shows how a CMOS NOT or INVERTER gate can be constructed. Notice the N-channel E-MOSFET (Q_2) has its source connected to 0 V, and therefore its gate needs to be positive (relative to the 0 V source) for it to turn ON. On the other hand, the P-channel E-MOSFET has its source connected to +10 V, and therefore its gate needs to be negative (relative to the +10 V source) for it to turn ON. The table in Figure 4-19(a) shows how this circuit will function. When V_{IN} or A is LOW (0 V), Q_1 will turn ON (N-P gate-source is forward-biased; $G = 0$ V, $S = +10$ V) and Q_2 will turn OFF (P-N gate-source is reverse-biased; $G = 0$ V, $S = 0$ V). A LOW input will therefore turn ON Q_1 creating a channel between source and drain, connecting the +10 V supply voltage to the output, or Y. On the other hand, when V_{IN} or A is HIGH (+10 V), Q_1 will turn OFF (N-P gate-source is reverse-biased; $G = +10$ V, $S = +10$ V) and Q_2 will turn ON (P-N gate-source is forward-biased; $G = +10$ V, $S = 0$ V). A HIGH input will therefore turn ON Q_2, creating a channel between source and drain connecting the 0 V at Q_2's source to the output, or Y.

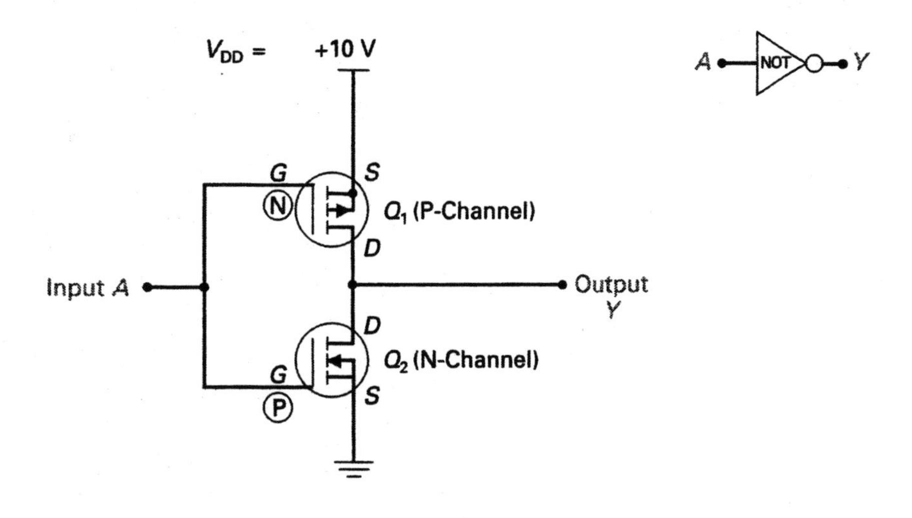

A	Y	Q_1	Q_2
0 V	+10 V	ON	OFF
+10 V	0 V	OFF	ON

(a)

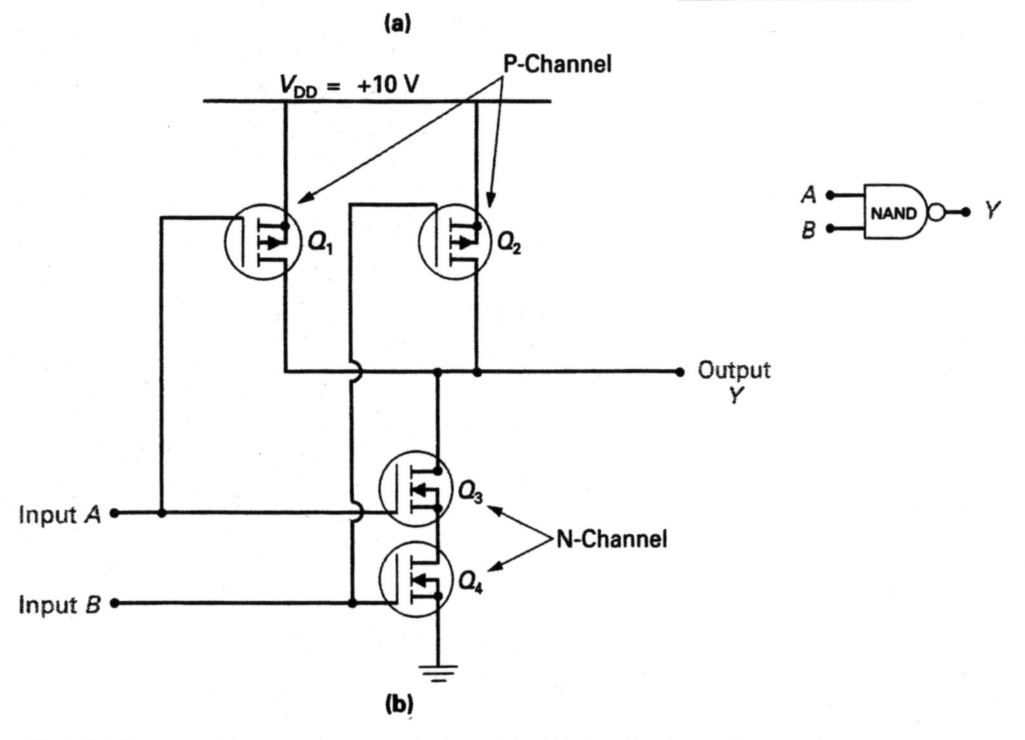

(b)

FIGURE 4-19 Complementary MOS (CMOS) Logic Gates. (a) An Inverter. (b) A NAND Gate.

A CMOS NAND Gate Circuit

Figure 4-19(b) shows how two complementary MOSFET pairs (two N-channel and two P-channel) could be connected to form a two-input CMOS NAND gate. If either input A or B is LOW, one or both of the P-channel MOSFETs (Q_1 and Q_2) will turn ON and switch $+V_{DD}$ or a HIGH through to the output. Only when both inputs A and B are HIGH will Q_1 and Q_2 be OFF, and Q_3 and Q_4 will be ON, switching ground or a LOW through to the output.

CMOS Logic Gate Characteristics

CMOS logic is probably the best all-round logic circuitry because of its low power consumption, very high noise immunity, large power supply voltage range, and high fanout. These characteristics are given in detail in Figure 4-20. The low power consumption achieved by CMOS circuits is due to the "complementary pairs" in each CMOS gate circuit. As was shown in Figure 4-19, during gate operation there is never a continuous path for current between the supply voltage and ground. When the P-channel device is OFF, $+V_{DD}$ is disconnected from the circuit, and when the N-channel device is OFF, ground is disconnected from the circuit. It is only when the output voltage switches from a HIGH to LOW, or LOW to HIGH, that the P-channel and N-channel E-MOSFET are momentarily ON at the same time. It is during this time of switching over from one state to the next that a small current will flow due to the complete path from $+V_{DD}$ to ground. This is why the power dissipation of CMOS logic gates increases as the operating frequency of the circuit increases.

The CMOS Series of ICs

There is a full line of CMOS ICs available that can provide almost all the same functions as their TTL counterparts. These different types are given different series numbers to distinguish them from others. Some of the more frequently used CMOS ICs are described in the following paragraphs.

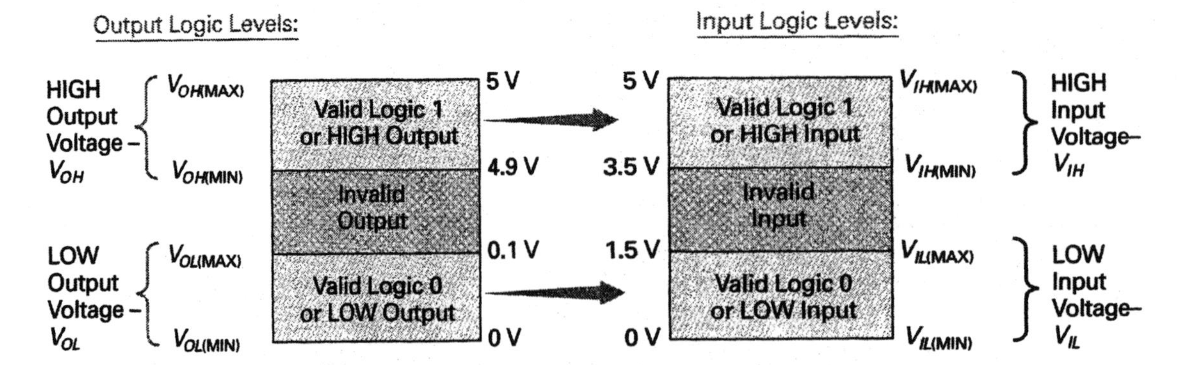

Power Dissipation per Gate: 2.7 nanowatts (static) to 170 μW (at 100 kHz)
Propagation Delay per Gate: 10 ns
Supply Voltage: +3 V to +6 V
Noise Immunity: Very High
Fanout: 10

FIGURE 4-20 **CMOS Logic Gate Characteristics (High Speed CMOS—74HCXX).**

a. *4000 Series* This first line of CMOS digital ICs was originally made by RCA and uses the numbers 40XX, with the number in the "XX" position indicating the circuit type. For example, the 4001 is a quad two-input NOR gate, while the 4011 is a "quad two-input NAND gate." This series was eventually improved and called the 4000B series, and the original was labeled the 4000A series. Both use a V_{DD} supply of between +3 V and +15 V, with a logic 0 (LOW) = ⅓ of V_{DD} and logic 1 (HIGH) = ⅔ of V_{DD}. Propagation delay times for this series are typically 40 ns to 175 ns.

b. *40H00 Series* This "high-speed" CMOS series improved the propagation delay to approximately 20 ns; however, it was still not able to match the speed of the bipolar TTL 74LS00 series.

c. *74C00 Series* This series was designed to be "pin-compatible" with its TTL counterpart. This meant that all of the ICs used the same input and output pin numbers. Supply voltages can be anywhere from 3 V_{dc} to 15 V_{dc}, and propagation delays are long (90 ns, typically) and so applications are limited to signal frequency applications below 2 MHz. Low frequency and dc input signals result in almost no power dissipation; however, at high frequencies the gate power dissipation can exceed 10 mW to 15 mW.

d. *74HC00 and 74HCT00 Series* This high-speed CMOS series of ICs matches the propagation delay times of the 74LS00 bipolar ICs, and still maintains the CMOS advantage of low power consumption. It was introduced to overcome some of the disadvantages that plagued older CMOS series. The HC and HCT series are both pin-compatible with TTL. The HCT (high-speed CMOS TTL-compatible) series is a truly compatible CMOS line of ICs since it uses the same input and output voltage levels for a HIGH and LOW. This advantage makes it easier to swap a TTL IC for a CMOS IC and not cause any interfacing problems. The 74HC series logic gates can operate from a supply voltage of 2 V_{dc} to 6 V_{dc} (typically 5 V_{dc} so that it is identical to TTL levels). They have a typical gate propagation delay of about 20 ns and a power dissipation of about 1 mW. At a supply voltage of 5 V_{dc}, a 74HCT gate will typically have a propagation delay of 40 ns and a power dissipation of 1 mW. As an example, Figure 4-21 shows a typical HCT CMOS data sheet.

e. *74AC00 and 74ACT00 Series* This "advanced CMOS logic" series and "advanced CMOS, TTL-compatible" series has even better characteristics than the 74HC00 and 74HCT00 series. When operated from a supply voltage of 5 V_{dc}, a 74AC gate has a 3 ns propagation delay, and a 74ACT gate has a 5 ns propagation delay, enabling both families to operate beyond 100 MHz yet dissipate only 0.5 mW of power per gate. The 74AC series is not completely compatible with TTL, however the 74ACT series is fully compatible.

■ **EXAMPLE:**

Referring to Figure 4-21, what would be the output HIGH and LOW worst-case voltage if the output current is 4 mA and the IC is being operated at room temperature?

■ ***Solution:***

Finding the 54/74HCT $T_A$column in the DC electrical characteristics block in Figure 4-21, you can see that when the output current is 4 mA, the minimum

DEVICE: SN74HCT00 (Quad Two–Input NAND)

TOP VIEW

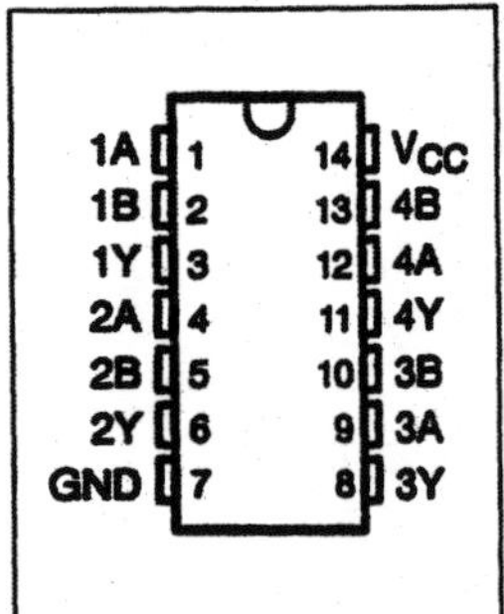

LOGIC DIAGRAM (POSITIVE LOGIC)

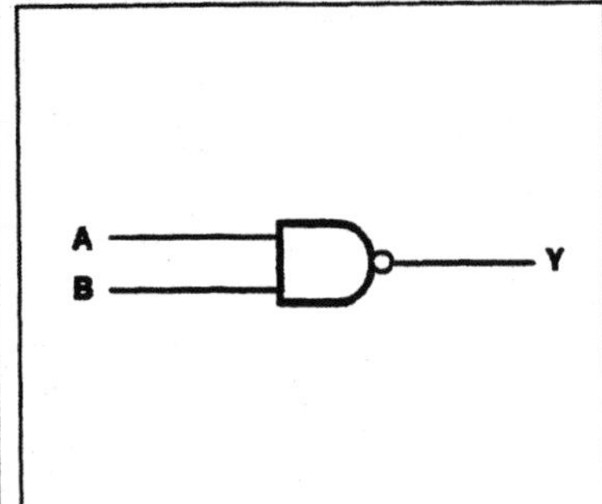

LOGIC SYMBOL

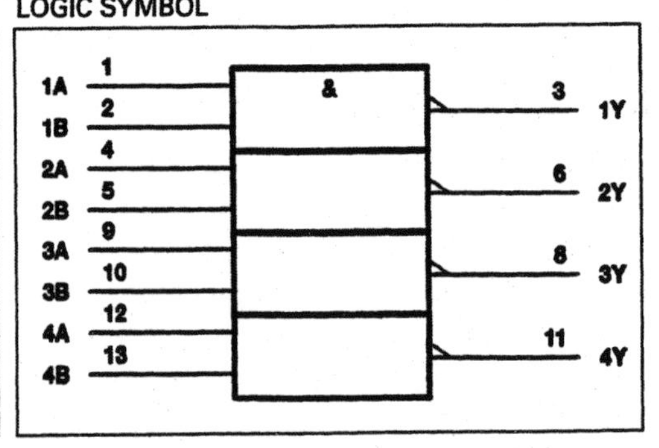

absolute maximum ratings over operating free-air temperature range

Supply voltage range, V_{CC} .... –0.5 V to 7 V
Input clamp current, I_{IK} ($V_I < 0$ or $V_I > V_{CC}$) (see Note 1) .... ±20 mA
Output clamp current, I_{OK} ($V_O < 0$ or $V_O > V_{CC}$) (see Note 1) .... ±20 mA
Continuous output current, I_O ($V_O = 0$ to V_{CC}) .... ±25 mA
Continuous current through V_{CC} or GND .... ±50 mA
Maximum power dissipation at $T_A = 55°C$ (in still air) (see Note 2): D package .... 1.25 W
N package .... 1.1 W
PW package .... 0.5 W
Storage temperature range, T_{stg} .... –65°C to 150°C

recommended operating conditions

			SN54HCT00			SN74HCT00			UNIT
			MIN	NOM	MAX	MIN	NOM	MAX	
V_{CC}	Supply voltage		4.5	5	5.5	4.5	5	5.5	V
V_{IH}	High-level input voltage	V_{CC} = 4.5 V to 5.5 V	2			2			V
V_{IL}	Low-level input voltage	V_{CC} = 4.5 V to 5.5 V	0		0.8	0		0.8	V
V_I	Input voltage		0		V_{CC}	0		V_{CC}	V
V_O	Output voltage		0		V_{CC}	0		V_{CC}	V
t_t	Input transition (rise and fall) time		0		500	0		500	ns
T_A	Operating free-air temperature		–55		125	–40		85	°C

electrical characteristics over recommended operating free-air temperature range (unless otherwise noted)

PARAMETER	TEST CONDITIONS		V_{CC}	$T_A = 25°C$			SN54HCT00		SN74HCT00		UNIT
				MIN	TYP	MAX	MIN	MAX	MIN	MAX	
V_{OH}	$V_I = V_{IH}$ or V_{IL}	$I_{OH} = -20\ \mu A$	4.5 V	4.4	4.499		4.4		4.4		V
		$I_{OH} = -4$ mA		3.98	4.3		3.7		3.84		
V_{OL}	$V_I = V_{IH}$ or V_{IL}	$I_{OL} = 20\ \mu A$	4.5 V		0.001	0.1		[illegible]		0.1	V
		$I_{OL} = 4$ mA			0.17	0.26		0.4		0.33	
I_I	$V_I = V_{CC}$ or 0		5.5 V		±0.1	±100		±1000		±1000	nA
I_{CC}	$V_I = V_{CC}$ or 0,	$I_O = 0$	5.5 V			2		40		20	µA
ΔI_{CC}‡	One input at 0.5 V or 2.4 V, Other inputs at 0 or V_{CC}		5.5 V		1.4	2.4		3		2.9	mA
C_i			4.5 V to 5.5 V		3	10		10		10	pF

‡ This is the increase in supply current for each input that is at one of the specified TTL voltage levels rather than 0 or V_{CC}.

FIGURE 4-21 A High-Speed CMOS, TTL Compatible (HCT) Data Sheet. (Courtesy of Texas Instruments)

HIGH-level output voltage (V_{OH}) is 3.98 V, and the maximum LOW-level output voltage (V_{OL}) is 0.26 V.

4-2-4 *MOSFET Handling Precautions*

Certain precautions must be taken when handling any MOSFET devices. The very thin insulating layer between the gate and the substrate of a MOSFET can easily be punctured if an excessive voltage is applied. Your body can build up extremely large electrostatic charges due to friction. If this charge were to come in contact with the pins of a MOSFET device, an electrostatic-discharge (ESD) would occur, resulting in a possible arc across the MOSFET's thin insulating layer, causing permanent damage. Most MOSFETs presently manufactured have zeners internally connected between the gate and source to bypass high voltage static or in-circuit potentials to protect the MOSFET; however, it is important that:

a. All MOS devices are shipped and stored in a "conductive foam" when not in use so that all of the IC pins are kept at the same potential, and therefore electrostatic voltages cannot build up between terminals.

b. When MOS devices are removed from the conductive foam, be sure not to touch the pins, since your body may have built up an electrostatic charge.

c. When MOS devices are removed from the conductive foam, always place them on a grounded surface such as a metal tray.

d. When continually working with MOS devices, use a wrist-grounding strap, which is a length of cable with a 1 MΩ resistor in series to prevent electrical shock if you were to come in contact with a voltage source.

e. All test equipment, soldering irons, and work benches should be properly grounded.

f. All power in equipment should be off before MOS devices are removed or inserted into printed circuit boards.

g. Any unused MOSFET terminals must be connected, since an unused input left open can build up an electrostatic charge and float to high voltage levels.

h. Any boards containing MOS devices should be shipped or stored with the connection side of the board in conductive foam.

SELF-TEST REVIEW QUESTIONS FOR SECTION 4-2

1. True or false: the E-MOSFET naturally operates as a voltage controlled switch.
2. List three advantages that MOS logic gates have over bipolar logic gates.
3. Which type of E-MOSFETs are used in CMOS logic circuits?
4. What is the key advantage of CMOS logic?

4-3 DIGITAL IC PACKAGE TYPES AND COMPLEXITY CLASSIFICATION

There are four basic methods of packaging integrated circuits, and these are illustrated in Figure 4-22.

4-3-1 *Early Digital IC Package Types*

Transistor Outline
A package type used to house a transistor or circuit that resembles a small can.

Figure 4-22(a) shows the first type of package used for ICs, called the **transistor outline** or **TO can** package. This transistor package was modified from the basic three-lead transistor package to have the extra leads needed, and is still used to house linear ICs due to its good heat dissipation ability, but is rarely used for digital ICs. Figure 4-22(b) illustrates another early IC package called the flat pack. The leads of the package were designed to be soldered directly to the tracks of the printed circuit board. This package allowed circuits to be placed close together for small size applications, such as avionics and military systems. The package is generally made of ceramic to withstand the high temperatures due to the densely packaged circuits.

4-3-2 *Present Day Digital IC Package Types*

Figure 4-22(c) shows the dual-in-line package, or DIP, which has been a standard for many years. Called a DIP because of its two (dual) lines of parallel connecting pins, its leads feed through holes punched in the printed circuit board. Connection pads circle the holes both on the top and the bottom of the printed circuit board (PCB) so that when the DIP package is soldered in place it makes a connection.

Surface Mount Technology
A package type that has connections that connect to the surface of a printed circuit board.

Figure 4-22(d) illustrates the three types of **surface-mount technology (SMT)** packages. The key advantage of SMT over DIP is that a DIP package needs both a hole and a connecting pad around the hole. With SMT, no holes are needed since the package is mounted directly onto the printed circuit board. To illustrate this point further, Figure 4-22(e) shows the space that can be saved when a surface mount resistor is used instead of a through-hole axial lead resistor. Without the need for holes, pads can be placed closer together, resulting in a considerable space savings. Figure 4-22(d) shows the three basic types of SMT packages. The "small outline IC" (SOIC) package uses L-shaped leads, the "plastic-leaded chip carrier" (PLCC) uses J-shaped leads, and the "leadless ceramic chip carrier" (LCCC) has metallic contacts molded into its ceramic body. Figure 4-22(f) shows a selection of IC package types.

4-3-3 *Digital IC Circuit Complexity Classification*

Although ICs are generally classified as being either linear (analog) or digital, another method of classification groups ICs based on their internal circuit complexity. The four categories are called small scale integration (SSI),

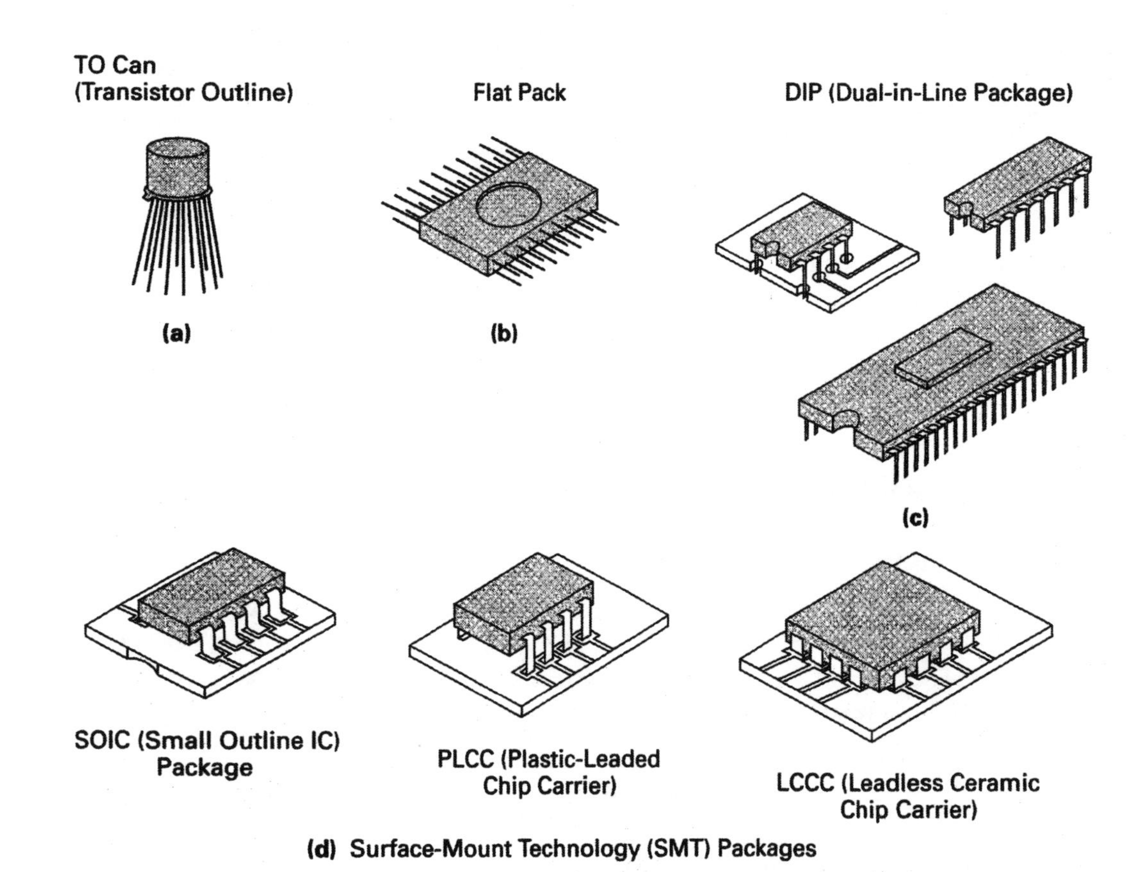

(f)

FIGURE 4-22 IC Package Types. (Photo courtesy Harris Semiconductor)

medium scale integration (MSI), large scale integration (LSI) and very large scale integration (VLSI).

Small Scale Integration (SSI) ICs

These circuits are the simplest form of IC, such as single-function digital logic gate circuits. They contain less than 180 interconnected components on a single chip.

Medium Scale Integration (MSI) ICs

These ICs contain between 180 and 1,500 interconnected components, and their advantage over an SSI IC is that the number of ICs needed in a system is less, resulting in a reduced cost and assembly time.

Large Scale Integration (LSI) ICs

These circuits contain several MSI circuits all integrated on a single chip to form larger functional circuits or systems such as memories, microprocessors, calculators, and basic test instruments. These ICs contain between 1,500 and 15,000 interconnected components.

Very Large Scale Integration (VLSI) ICs

VLSI circuits contain extremely complex circuits such as large microprocessors, memories, and single-chip computers. These ICs contain more than 15,000 interconnected components.

SELF-TEST REVIEW QUESTIONS FOR SECTION 4-3

1. Give the full names for the following abbreviations.
   a. TO Can **b.** DIP **c.** SMT
2. Which package type saves the most printed circuit board space?
3. An IC containing four digital bipolar logic gates with about forty-four components would be classified as a/an ____________ IC.
   a. SSI **b.** MSI **c.** LSI **d.** VLSI
4. The low circuit current and therefore low heat dissipation of ____________ circuits makes it possible for small size circuits, and this is why ____________ technology dominates the VLSI circuit market.

4-4 COMPARING AND INTERFACING BIPOLAR AND MOS LOGIC FAMILIES

Figure 4-23(a) summarizes the differences between the two main digital IC logic families.

4-4-1 *The Bipolar Family*

The bipolar family of digital ICs has three basic types, called TTL (transistor-transistor logic), ECL (emitter-coupled logic), and IIL, or I^2L, (integrated-injection logic). The characteristics of these three basic bipolar transistor logic gate types are summarized in the upper table in Figure 4-23(a). Generally, the power dissipation of TTL and ECL logic circuits limits their packing density and therefore they are only used for SSI and MSI circuit applications. On the other hand, I^2L digital circuits are easy to combine with linear circuits (such as op-amps and oscillators) to create LSI and VLSI systems on a single chip.

4-4-2 *The MOS Family*

The lower table in Figure 4-23(a) shows the three basic E-MOSFET digital IC types, which are more frequently called the MOS family of digital ICs. The three basic types of MOS ICs are called PMOS, NMOS, and CMOS. Comparing the

	Bipolar Family		
	LS TTL	ECL	IIL
Cost	Low	High	Medium
Fanout	20	10 to 25	2
Power Dissipation/Gate	2 mW	40 to 60 mW	0.06 to 70 μW
Propagation Delay/Gate	8 ns	0.5 to 3 ns	25 to 50 ns
Typical Input Signal Frequency	15 to 120 MHz	200 to 1000 MHz	1 to 10 MHz
External Noise Immunity	Good	Good	Fair–Good
Typical Supply Voltage	+5 V	−5.2 V	1 V to 15 V
Temperature Range	−55 to 125°C 0 to 70°C	−55 to 125°C	0 to 70°C
Internally Generated Noise	Medium–High	Low–Medium	Low

	MOS Family		
	PMOS	NMOS	HCMOS
Cost	High	High	Medium
Fanout	20	20	10
Power Dissipation/Gate	0.2 to 10 mW	0.2 to 10 mW	2.7 nW to 170 μW
Propagation Delay/Gate	300 ns	50 ns	10 ns
Typical Input Signal Frequency	2 MHz	5 to 10 MHz	5 to 100 MHz
External Noise Immunity	Good	Good	Very Good
Typical Supply Voltage	−12 V	+5 V	+3 V to +6 V
Temperature Range	−55 to 125°C 0 to 70°C	−55 to 125°C 0 to 70°C	−55 to 125°C −40 to 85°C
Internally Generated Noise	Medium	Medium	Low–Medium

(a)

FIGURE 4-23 Comparing Logic Types (a) Logic Families (b) TTL and CMOS Series.

Series	74	74L	74S	74LS	74AS	74ALS	74F	74C	4000B	74HC	74HCT	74AC	74ACT
Supply Voltage Range (V_{CC})	4.75 VDC to 5.25 VDC	4.75 VDC to 5.25 VDC	4.75 VDC to 5.25 VDC	4.75 VDC to 5.25 VDC	4.5 VDC to 5.5 VDC	4.5 VDC to 5.5 VDC	4.5 VDC to 5.5 VDC	3.0 VDC to 15.0 VDC	3.0 VDC to 18.0 VDC	2.0 VDC to 6.0 VDC	4.5 VDC to 5.5 VDC	3.0 VDC to 5.5 VDC	4.5 VDC to 5.5 VDC
Min. Logic-1 Input Voltage (V_{IH})	2.0 VDC	2.0 VDC	2.0 VDC	2.0 VDC	2.0 VDC	2.0 VDC	2.0 VDC	2.0 VDC	2.3 V_{cc}	3.15 VDC	2.0 VDC	3.15 VDC	2.0 VDC
Max. Logic-0 Input Voltage (V_{IL})	0.8 VDC	0.8 VDC	0.8 VDC	0.8 VDC	0.8 VDC	0.8 VDC	0.8 VDC	1.5 VDC	1.3 V_{cc}	0.9 VDC	0.8 VDC	1.35 VDC	0.8 VDC
Min. Logic-1 Output Voltage (V_{OH})	2.4 VDC	2.4 VDC	2.7 VDC	2.7 VDC	2.7 VDC	2.7 VDC	2.7 VDC	4.5 VDC	$\sim V_{cc}$	4.4 VDC	3.84 VDC	4.2 VDC	3.8 VDC
Max. Logic-0 Output Voltage (V_{OL})	0.4 VDC	0.4 VDC	0.5 VDC	0.4 VDC	0.4 VDC	0.4 VDC	0.5 VDC	0.5 VDC	$\sim$0 VDC	0.1 VDC	0.33 VDC	0.5 VDC	0.5 VDC
Max Logic-0 Input Current (I_{IL})	−1.6 mA	−0.18 mA	−2.0 mA	−0.36 mA	−0.5 mA	−0.1 mA	−0.6 mA	−0.5 nA	±1.0 μA	±1.0 μA	±1.0 μA	±1.0 μA	±1.0 μA
Max Logic-1 Input Current (I_{IH})	40.0 μA	10.0 μA	50.0 μA	20.0 μA	20.0 μA	20.0 μA	20.0 μA	5.0 nA	±1.0 μA	±1.0 μA	±1.0 μA	±1.0 μA	±1.0 μA
Max. Logic-0 Output Current (I_{OL})	16.0 mA	3.6 mA	20.0 mA	4.0 mA	20.0 mA	8.0 mA	20.0 mA	0.4 mA	3.0 mA	20.0 μA	20.0 μA	24.0 mA	24.0 mA
Max Logic-1 Output Current (I_{OH})	−400.0 μA	−200.0 μA	−1.0 mA	−400.0 μA	−2.0 mA	−400.0 μA	−1.0 mA	−0.36 mA	−3.0 mA	−20.0 μA	−20.0 μA	−24.0 mA	−24.0 mA
Propagation Delay (High to Low) (t_{PHL})	8.0 ns	30.0 ns	5.0 ns	8.0 ns	1.5 ns	7 ns	3.7 ns	90.0 ns	50.0 ns	20.0 ns	40.0 ns	3.0 ns	5.0 ns
Propagation Delay (Low to High) (t_{PLH})	13.0 ns	60.0 ns	5.0 ns	8.0 ns	1.5 ns	5 ns	3.2 ns	90.0 ns	65.0 ns	20.0 ns	40.0 ns	3.0 ns	5.0 ns
Max. Operating Frequency (f_{max})	35 MHz	3 MHz	125 MHz	45 MHz	80 MHz	35 MHz	100 MHz	2 MHz	6 MHz	20 MHz	24 MHz	125 MHz	125 MHz
Power Dissipation Per Gate (mW)	10.0 mW	1.0 mW	20.0 mW	2.0 mW	4.0 mW	1.0 mW	4.0 mW	Note 1	Note 1	25.0 μW	80.0 μW	440 μW/device	440 μW/device

NOTE 1: Value is Frequency-Dependent

(b)

FIGURE 4-23 Continued

characteristics, you can see that MOS logic gates use less space due to their simpler construction, have a very high noise immunity, and consume less power than equivalent bipolar logic gates. However, the high input impedance and capacitance due to the E-MOSFET's insulated gate means that time constants are larger and therefore transistor ON/OFF switching speeds are slower than equivalent bipolar gates. Both PMOS and NMOS ICs can be used in LSI and VLSI applications such as large memory circuits, single-chip microcomputers, and test instruments. On the other hand, CMOS ICs are generally used in SSI and MSI circuit applications.

The trade-offs, therefore, between the two logic families are

a. Bipolar circuits are faster than MOS circuits, but their circuits are larger and consume more power.

b. MOS circuits are smaller and consume less power than bipolar circuits, but they are generally slower.

4-4-3 *Interfacing Logic Families*

Interface
A device or circuit that connects an output to an input.

The table in Figure 4-23(b) lists the key differences between the TTL series and CMOS series of digital ICs. In some circuit applications, there is a need to connect or **interface** one logic gate family type to another logic gate family type.

For example, in some circuit applications we may need to interface a TTL logic gate to a CMOS logic gate, or connect a CMOS gate to a TTL gate. In these instances, it is important that both devices are **compatible,** meaning that the HIGH and LOW logic levels at the output of the source logic gate will be of the correct amplitude and polarity so that they will be interpreted correctly at the input of the load logic gate. If we were to use the 74HCT00 CMOS series, there would be no interfacing problems at all, since this series generates and recognizes the TTL HIGH and LOW voltage levels (it is TTL-compatible). Other CMOS series, however, are not voltage- and current-compatible, and therefore an interfacing device will have to be included between the two different types of logic gates to compensate for the differences.

Compatible
In relation to ICs, it describes a circuit family that has input/output logic levels and operating characteristics which are compatible with other logic IC families.

Interfacing Standard TTL and 4000B Series CMOS

The voltage and current differences for the key TTL and CMOS series of logic gates are listed in Figure 4-24(a). Figure 4-24(b) shows the incompatibility between the HIGH and LOW voltage levels of standard TTL series and 4000B CMOS series logic gates. A LOW logic level output voltage from the TTL logic gate will be recognized at the CMOS input as a LOW, since even a worst-case LOW output of 0.8 V is still less than the worst-case LOW level CMOS input of 1.67 V. The problem arises with a HIGH output from the TTL gate, which even if it is at 2.4 V, will still be less than the minimum acceptable CMOS input voltage of 3.33 V. To compensate for this problem, a 10 KΩ pull-up resistor is connected between the TTL and CMOS gate, as shown in Figure 4-24(c). This pull-up resistor will pull a HIGH output from the TTL gate up towards 5 V so that it will be recognized at the input as a HIGH by the CMOS gate.

Figure 4-24(d) shows the interfacing device that needs to be included when connecting a CMOS gate to a TTL gate. Referring to Figure 4-24(b), you can see that a HIGH output voltage of 4.95 V will be easily recognized as a HIGH by the TTL gate, and a LOW of 0.05 V will also easily be recognized as a LOW by the TTL gate. The voltage levels in this instance are not the problem, but the current levels are because CMOS devices are low-current devices. For example, referring to Figure 4-24(d) you can see that the LOW output current of the 4069 CMOS gate (0.51 mA) is not sufficient to drive the TTL LOW input current requirement (1.6 mA). Similarly, the HIGH output current of the 4069 CMOS gate (0.51 mA) is not sufficient to drive the TTL HIGH input current requirement (40 μA). Including the 40508 CMOS buffer/driver will ensure current compatibility, as shown in Figure 4-24(d), since the output LOW and HIGH current values from this buffer are more than sufficient for two standard TTL loads.

Interfacing TTL and 4000B CMOS Using Level-Shifters

In the previous example, the 4000B CMOS gates and the standard TTL gates were using the same supply voltage (5 V). The 4000B CMOS series can use a supply voltage that is anywhere between +3 V and +15 V. To interface two different logic gate types with different supply voltages, we will have to include a **level-shifter** or **translator,** as shown in Figure 4-25. The 4050B level-shifter will convert 0 V to 15 V CMOS logic to 0 V to 5 V TTL logic, as shown in Figure 4-25(a), enabling us to interface CMOS to TTL. The 4504B level-

Level-Shifter
A circuit that will shift or change logic levels.

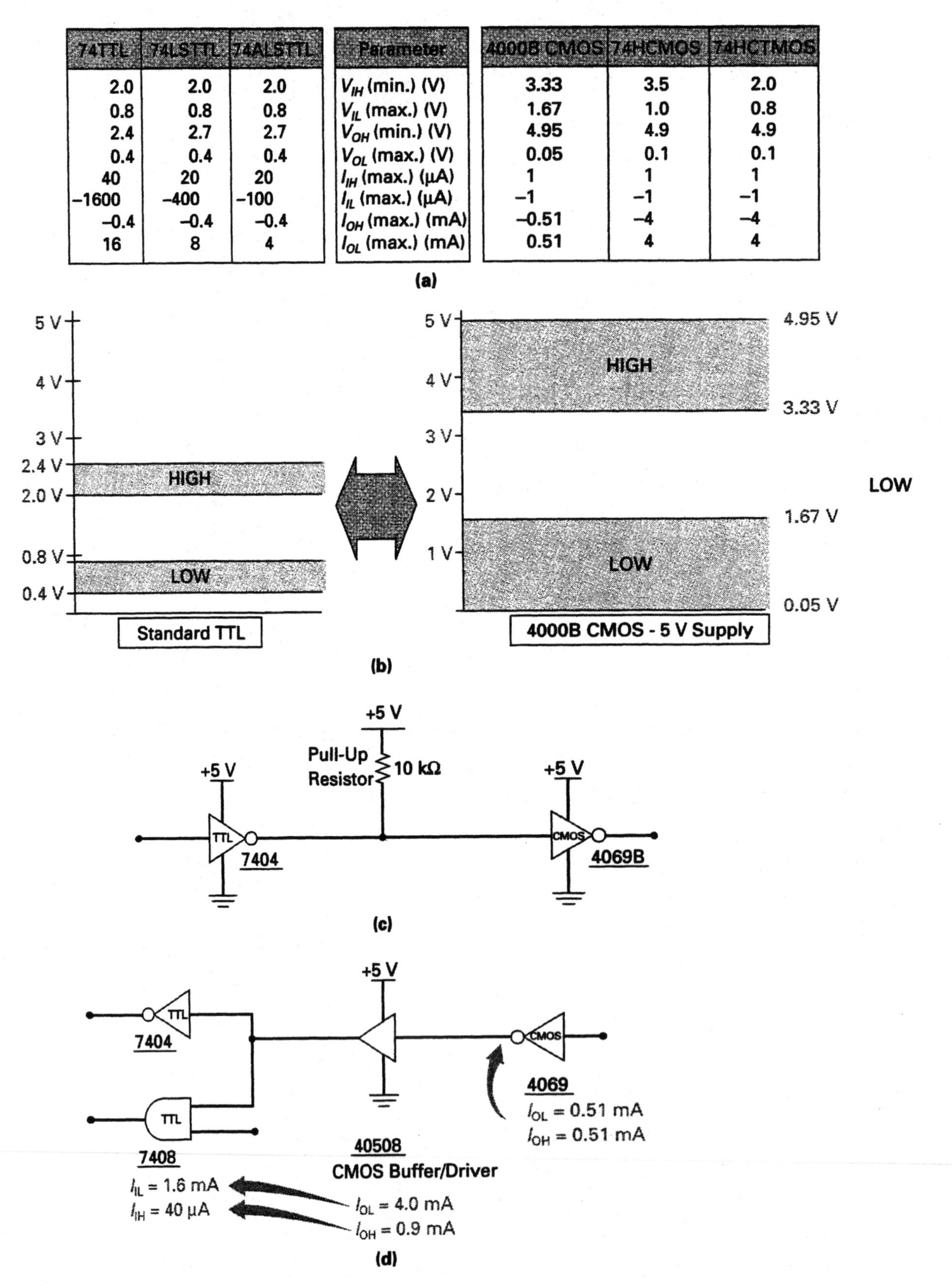

74TTL	74LSTTL	74ALSTTL	Parameter	4000B CMOS	74HCMOS	74HCTMOS
2.0	2.0	2.0	V_{IH} (min.) (V)	3.33	3.5	2.0
0.8	0.8	0.8	V_{IL} (max.) (V)	1.67	1.0	0.8
2.4	2.7	2.7	V_{OH} (min.) (V)	4.95	4.9	4.9
0.4	0.4	0.4	V_{OL} (max.) (V)	0.05	0.1	0.1
40	20	20	I_{IH} (max.) (μA)	1	1	1
−1600	−400	−100	I_{IL} (max.) (μA)	−1	−1	−1
−0.4	−0.4	−0.4	I_{OH} (max.) (mA)	−0.51	−4	−4
16	8	4	I_{OL} (max.) (mA)	0.51	4	4

FIGURE 4-24 **Interfacing a TTL Gate and a 4000B CMOS Gate that are Using the Same Supply Voltage Value.**

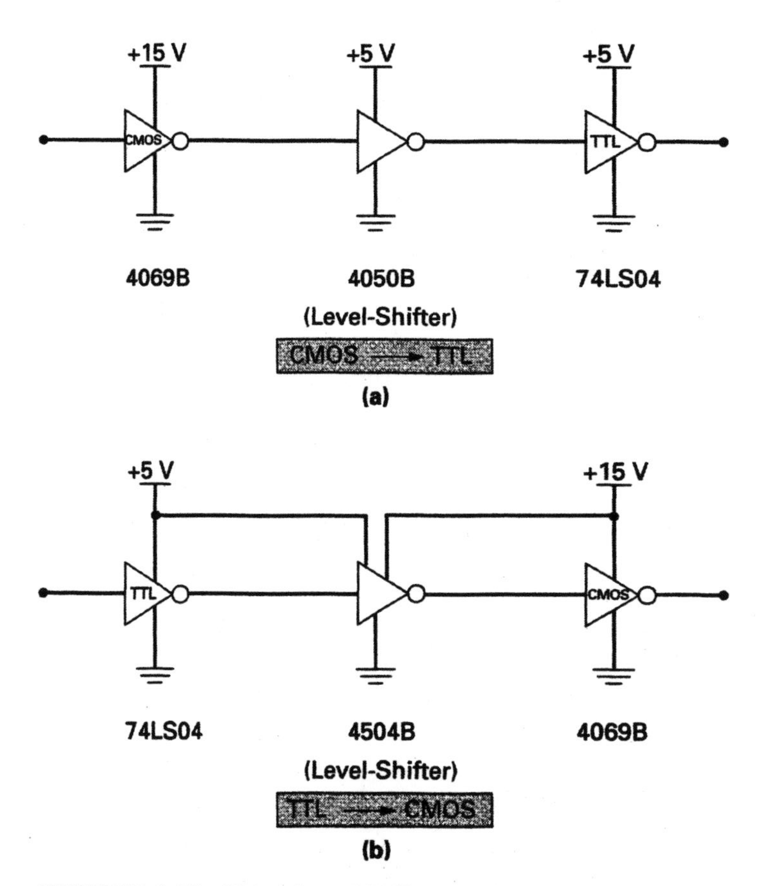

FIGURE 4-25 **Using Level-Shifters or Translators to Interface TTL and 4000B CMOS Gates with Different Supply Voltages.**

shifter will convert 0 V to 5 V TTL logic to 0 V to 15 V CMOS logic, as shown in Figure 4-25(b), enabling us to interface TTL to CMOS.

Interfacing ECL and TTL Using Level-Shifters

The 0 V to +5 V logic levels of TTL are very much different than the −5.2 V to 0 V logic levels of ECL. Figure 4-26(a) shows how a 10124 level-shifter can be used to interface a TTL logic gate to an ECL logic gate, and Figure 4-26(b) shows how a 10125 level-shifter can be used to interface an ECL logic gate to a TTL logic gate.

4-4-4 *Other Logic-Gate Families*

Semiconductor manufacturers are continually searching for new semiconductor circuits that have faster propagation delay times (since this enables them to operate at higher frequencies), lower power consumption, better noise immunity, and increased packing density. Some of the new hopefuls are gallium arsenide (GaAs) circuits, BiCMOS circuits (which is a combination of bipolar and

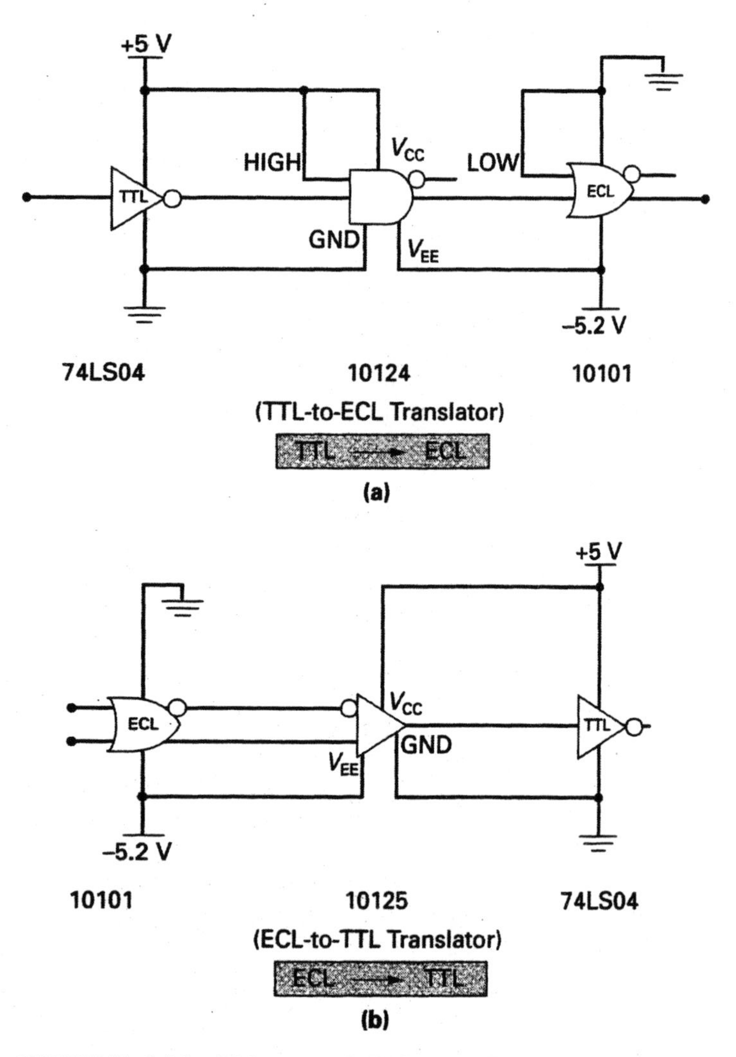

FIGURE 4-26 Using Level-Shifters or Translators to Interface ECL and TTL.

CMOS on the same chip, an example of which was featured in Figure 4-12), silicon-on-sapphire (SOS) circuits, and josephen junction circuits. It is therefore important in this ever-changing semiconductor industry that you read electronics magazines and manufacturers' technical publications to stay abreast of the latest technological advances. Most of the semiconductor manufacturers have their own home page on the Internet with access to all data sheets, circuit applications, technical support, and new product information.

SELF-TEST REVIEW QUESTIONS FOR SECTION 4-4

1. What is the speed/power trade-off, in regards to the selection of TTL logic gates?

2. Which logic-gate type is probably the best all-round logic because of its low power consumption, very high noise immunity, wide power supply voltage range, and high fanout?
3. What interfacing problems would occur between a 74HCT00 gate and a 74ALS00 gate?
4. What is a level-shifter gate?
5. Which logic family combines both bipolar and CMOS on a single chip?
6. Which logic gate series would be best suited if a circuit is processing input signal frequencies in the 110 MHz range?

REVIEW QUESTIONS

Multiple-Choice Questions

1. Which stage within a standard TTL gate provides a low output impedance for both a HIGH and LOW output?

   a. Multiple-emitter **b.** Phase splitter
   c. Totem pole **d.** Injection emitter

2. A floating TTL input has exactly the same effect as a/an

   a. LOW input. **b.** HIGH input. **c.** invalid input. **d.** both (a) and (c).

3. A standard TTL logic gate can sink a maximum of __________, and source a maximum of __________ .

   a. 400 μA, 1.6 mA **b.** 40 μA, 16 mA
   c. 16 mA, 400 μA **d.** 1.6 mA, 40 μA

4. Which TTL logic gates require an external pull-up resistor?

   a. ECL **b.** Open Collector **c.** Schottky TTL **d.** Both (a) and (b)

5. The output current of a standard TTL gate is __________ times greater than the input current, while the output current of a buffer/driver TTL gate is typically __________ times greater than the input.

   a. 10, 30 **b.** 10, 15 **c.** 30, 10 **d.** 15, 30

6. __________ logic gates are generally incorporated in digital circuits to sharpen the rise and fall times of input pulses.

   a. Buffer/driver **b.** Schmitt trigger **c.** Open collector **d.** Tri-state

7. The fanout of a standard TTL logic gate is __________.

   a. 10 **b.** 20 **c.** 2 **d.** 40

8. The output current of a standard TTL logic gate is __________ times greater than the input current.

   a. 10 **b.** 20 **c.** 2 **d.** 40

9. Which is the fastest of all the logic gate types?

   a. CMOS **b.** Open Collector **c.** Schottky TTL **d.** ECL

10. Which of the following CMOS series types is both pin- and voltage-compatible with TTL?

    a. 4000 **b.** 74C00 **c.** 74HCT00 **d.** 74HC00

11. Which logic gate type contains both a P-channel and N-channel E-MOSFET?

a. CMOS **b.** NMOS **c.** PMOS **d.** TREEMOS

12. A __________ __________ gate is used to convert a sine wave into a rectangular wave, or to sharpen the rise and fall times of a rectangular wave.

a. open collector **b.** high speed **c.** schmitt trigger **d.** mos logic

13. __________ circuits have a combination of bipolar and CMOS logic circuits on the same chip.

a. GaAs **b.** SOS **c.** Josephen junction **d.** BiCMOS

14. __________ circuits can combine linear circuits (such as amplifiers and oscillators) with digital circuits to create complete electronic systems on one chip.

a. CMOS **b.** IIL **c.** ECL **d.** NMOS

15. __________ circuits are faster, but their circuits are larger and consume more power, whereas __________ circuits are smaller and consume less power, but they are generally slower.

a. Bipolar, MOS **b.** CMOS, ECL **c.** MOS, CMOS **d.** MOS, bipolar

Essay Questions

16. What is "transistor-transistor logic"? (4-1)
17. Sketch a standard TTL bipolar transistor NAND gate circuit and describe the operation of the circuit for all input combinations. (4-1-1)
18. Define the purpose of the following: (4-1-1)
    - **a.** A bipolar multiple-emitter input transistor
    - **b.** A totem-pole circuit
    - **c.** A bipolar phase splitter transistor
19. List and describe the following standard TTL logic gate characteristics: (4-1-1)
    - **a.** A valid logic 1 output voltage range
    - **b.** A valid logic 1 input voltage range
    - **c.** A valid logic 0 output voltage range
    - **d.** A valid logic 0 input voltage range
    - **e.** The typical power dissipation for each gate
    - **f.** The typical propagation delay for each gate
    - **g.** The logic gate fanout
    - **h.** The $+V_{CC}$ commercial supply voltage range
20. What is the maximum and minimum current a standard TTL logic gate can source and sink? (4-1-1)
21. Why is a floating input equivalent to a HIGH input? (4-1-1)

22. Describe the basic operation, characteristics and applications of the following bipolar transistor logic gate circuits.
    a. Low-power and high-speed TTL (4-1-2)
    b. Schottky TTL (4-1-3)
    c. Open collector TTL (4-1-4)
    d. Three-state output TTL (4-1-5)
    e. Buffer/driver TTL (4-1-6)
    f. Schmitt trigger (4-1-7)
    g. Emitter-coupled logic (4-1-8)
    h. Integrated-injection logic (4-1-9)
23. What advantages does the MOS family of digital ICs have over the bipolar family of digital ICs? (4-2)
24. Briefly describe the basic circuit construction and operation of a typical PMOS and NMOS logic gate. (4-2-1 and 4-2-2)
25. What type of switching transistors are used within CMOS logic gate circuits? (4-2-3)
26. List and describe the following CMOS logic gate characteristics. (4-2-3)
    a. A valid logic 1 output voltage range
    b. A valid logic 1 input voltage range
    c. A valid logic 0 output voltage range
    d. A valid logic 0 input voltage range
    e. The typical power dissipation for each gate
    f. The typical propagation delay for each gate
    g. The logic gate fanout
    h. The $+V_{DD}$ supply voltage range
27. What precaution should be observed when storing MOS logic gates? (4-2-4)
28. Describe what you know about the following CMOS series types. (4-2-3)
    a. 4000 series
    b. 40H00 series
    c. 74C00 series
    d. 74HC00 and 74HCT00 series
    e. 74AC00 and 74ACT00 series
29. What precautions should be observed when handling MOS logic gate types? (4-2-4)
30. What key advantage does an SMT package have over a DIP package? (4-3-2)
31. Fill in the following blanks. (4-3-3)
    a. ____________ ICs. These circuits contain extremely complex circuits such as large microprocessors, memories, and single-chip computers. These ICs contain more than 15,000 interconnected components.
    b. ____________ ICs. These ICs contain between 180 and 1,500 interconnected components, and their advantage over an SSI IC is that the number of ICs needed in a system is less, resulting in a reduced cost and assembly time.

c. ____________ ICs. These circuits are the simplest form of integrated circuit, such as single function digital logic gate circuits. They contain less than 180 interconnected components on a single chip.

d. ____________ ICs. These circuits contain several MSI circuits all integrated on a single chip to form larger functional circuits or systems such as memories, microprocessors, calculators, and basic test instruments. These ICs contain between 1,500 and 15,000 interconnected components.

32. Fill in the following blanks. (4-4-1 and 4-4-2)
    a. Bipolar circuits are ____________ (faster/slower) than MOS circuits, but their circuits are ____________ (slower/larger) and consume ____________ (less/more) power.
    b. MOS circuits are ____________ (larger/smaller) and consume ____________ (less/more) power than bipolar circuits, but they are generally ____________ (faster/slower).
33. Define the term interfacing. (4-4-3)
34. What is a level-shifter? (4-4-3)
35. What is a BiCMOS logic gate? (4-4-4)

Practice Problems

36. Referring to the TTL logic circuit shown in Figure 4-27, complete the truth table.
37. Sketch the logic symbol and describe the logic function being performed by the logic circuit in Figure 4-27.
38. What circuit arrangement is formed by Q_5 and Q_6 in Figure 4-27, and what is the circuit's purpose?
39. What does the symbol within the logic gate in Figure 4-28 indicate?

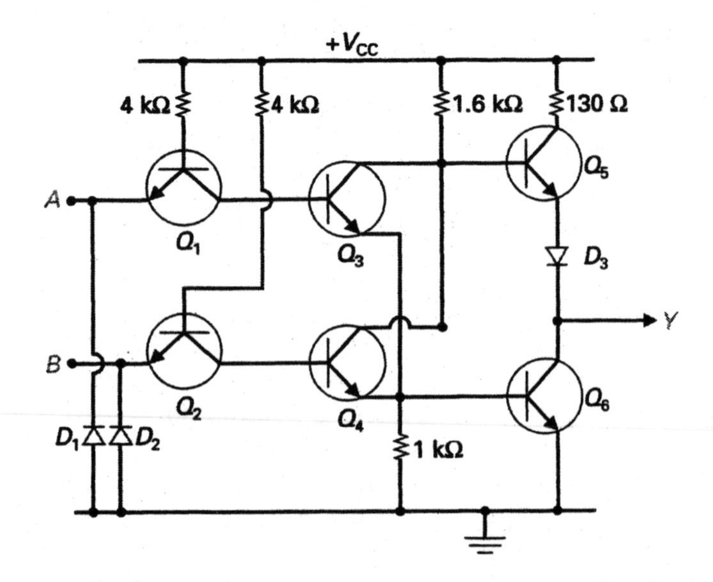

A	B	Y
LOW	LOW	?
LOW	HIGH	?
HIGH	LOW	?
HIGH	HIGH	?

FIGURE 4-27 **A Standard TTL Logic Gate Circuit.**

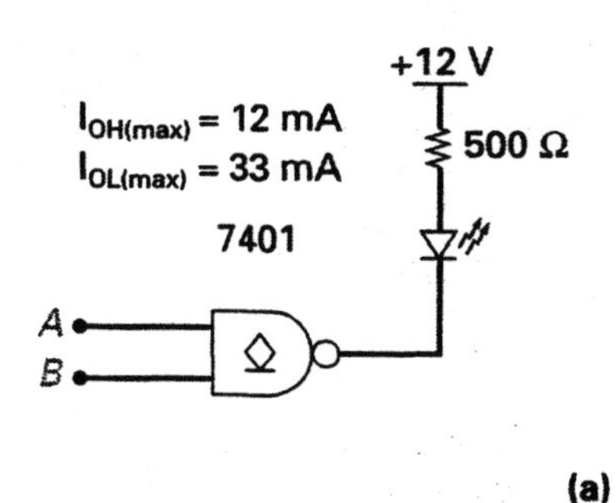

A	B	LED
0	0	?
0	1	?
1	0	?
1	1	?

(a)

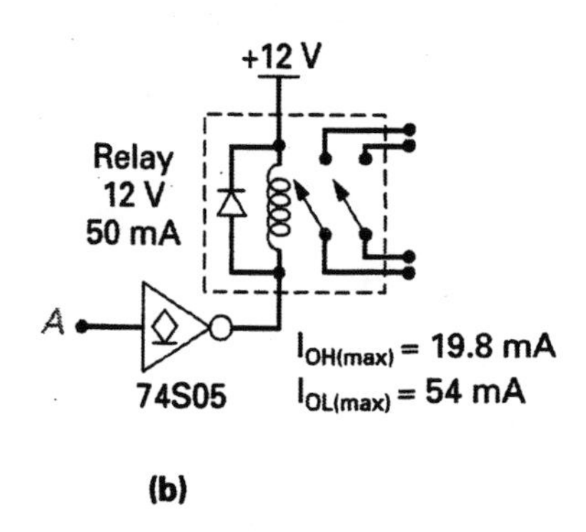

(b)

FIGURE 4-28 Controlling an LED with a Logic Gate.

40. Referring to Figure 4-28(a), complete the truth table.

41. Referring to Figure 4-28(a), determine whether the 7401 can handle the LOW output sinking current for this circuit (assuming a 2 V LED drop).

42. Will the 74LS05 be able to handle the LOW output sinking current needed to energize the 12 V, 50 mA relay?

43. What open collector output gate in Figure 4-10 would be ideal for a greater-than-50 mA load?

44. Identify the logic symbol used within the logic gate in Figure 4-29(a).

45. Referring to Figure 4-29(a), complete the truth table given.

46. Referring to Figure 4-29(b), complete the truth table given.

47. What is the difference between the 74LS125 shown in Figure 4-29(a) and the 74LS126 shown in Figure 4-29(b)?

48. Identify the logic symbol used within the logic gate in Figure 4-30(a).

49. Describe the operation of the complete circuit shown in Figure 4-30(a).

50. Comparing the inputs to the outputs in Figure 4-30(b), (c), (d), and (e), describe the circuit function being performed.

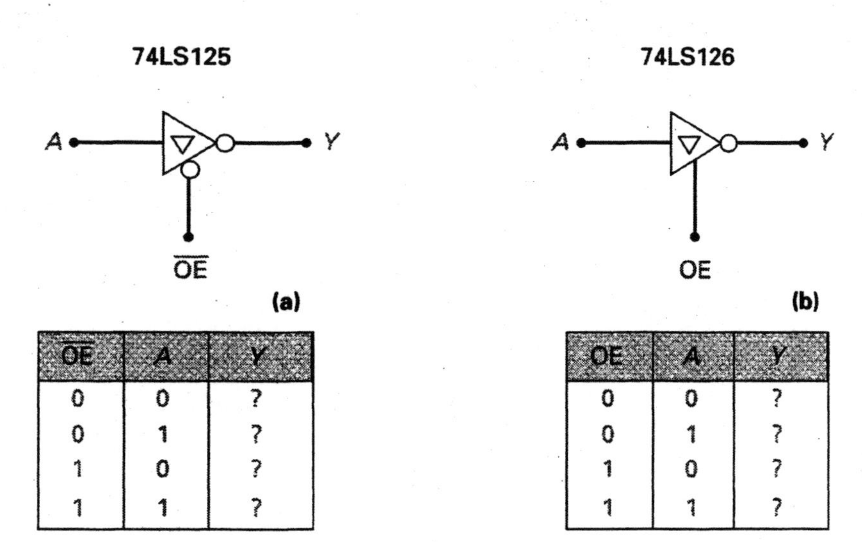

$\overline{OE}$	A	Y
0	0	?
0	1	?
1	0	?
1	1	?

OE	A	Y
0	0	?
0	1	?
1	0	?
1	1	?

FIGURE 4-29 Logic Gates With Control Inputs.

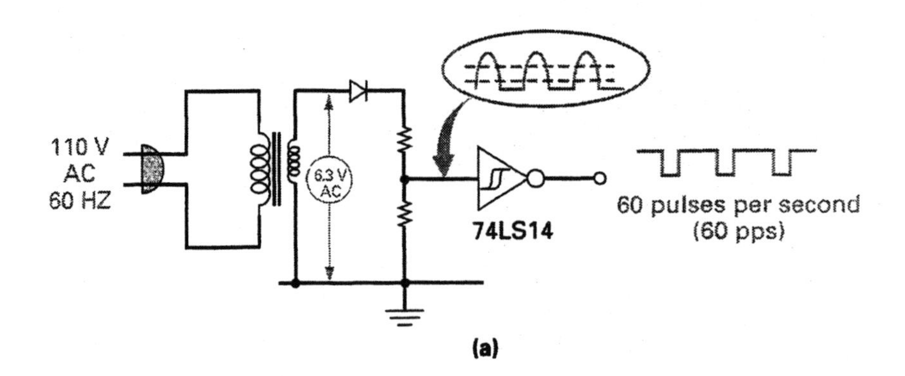

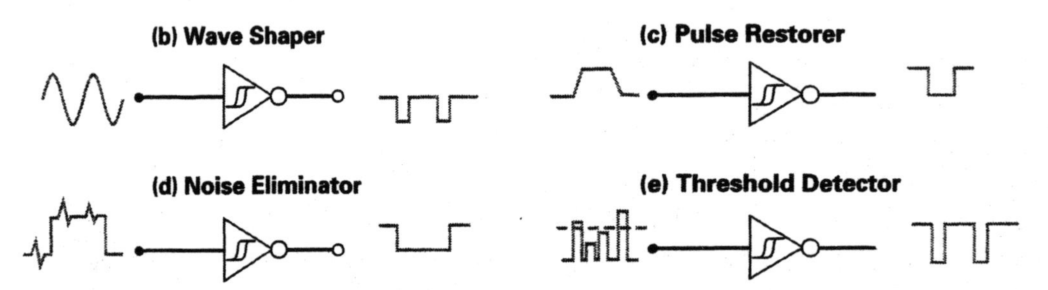

FIGURE 4-30 Logic Gate Circuit Applications.

51. What type of E-MOSFET is being used in the logic circuit in Figure 4-31?
52. Complete the truth table for the logic gate shown in Figure 4-31.
53. Sketch the logic symbol and describe the logic function being performed by the logic circuit in Figure 4-31.
54. Referring to Figure 4-32, what would be the output at pin 6 if the inputs at pins 1, 2, 4, and 5 were 1101?
55. Referring to Figure 4-32, what is the maximum propagation delay of a 74HC20 if $V_{CC} = 6$ V?

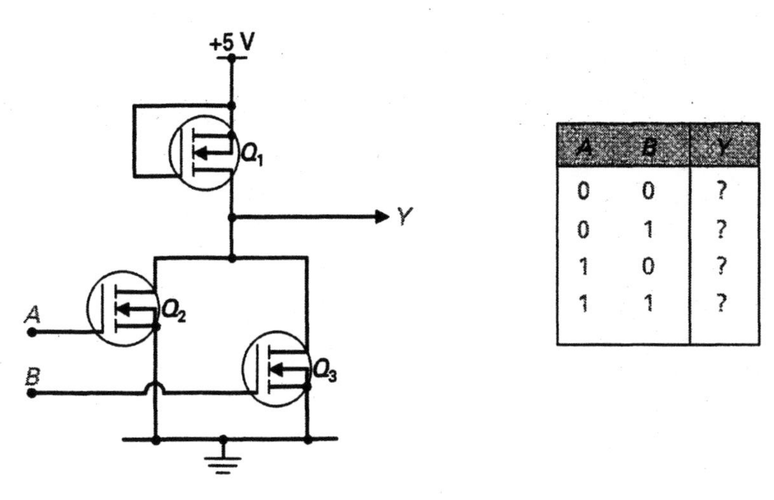

A	B	Y
0	0	?
0	1	?
1	0	?
1	1	?

FIGURE 4-31 A MOS Logic Gate.

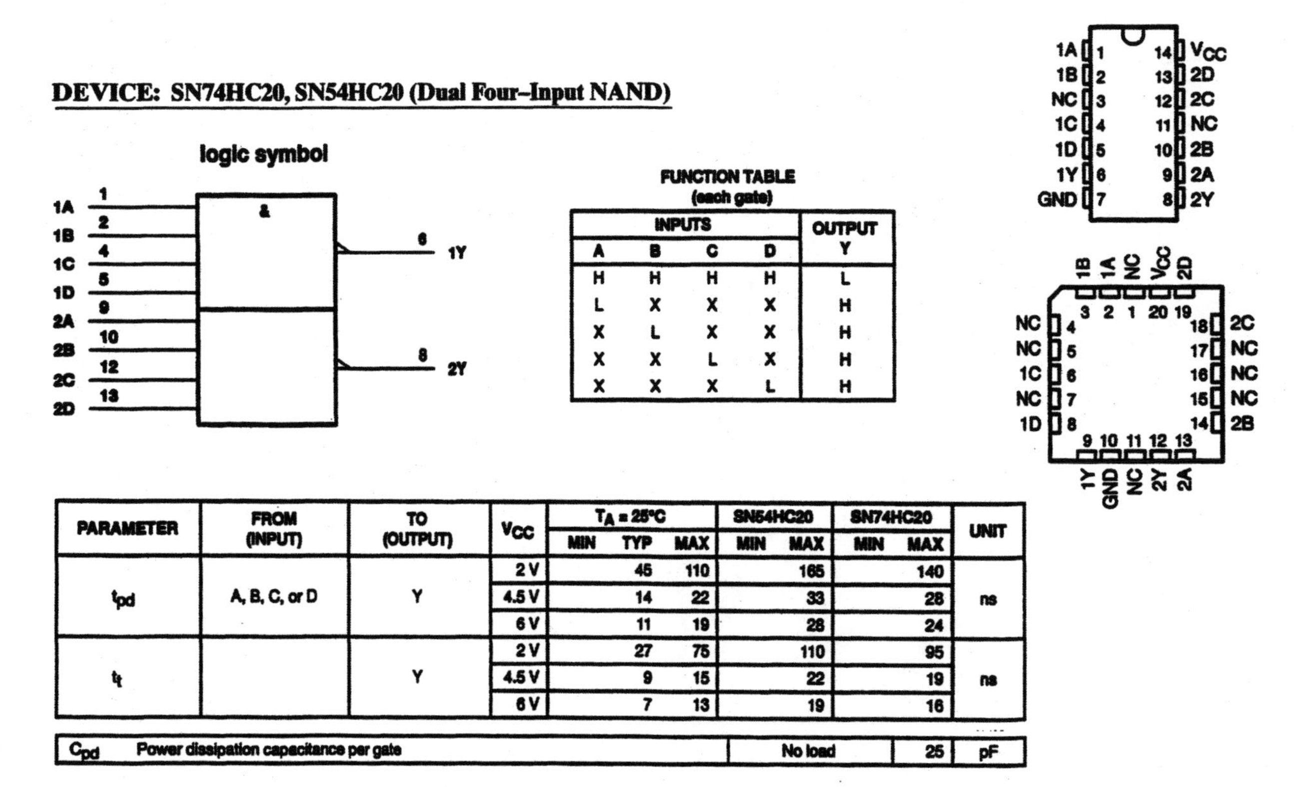

DEVICE: SN74HC20, SN54HC20 (Dual Four–Input NAND)

FUNCTION TABLE
(each gate)

INPUTS				OUTPUT
A	B	C	D	Y
H	H	H	H	L
L	X	X	X	H
X	L	X	X	H
X	X	L	X	H
X	X	X	L	H

PARAMETER	FROM (INPUT)	TO (OUTPUT)	V_{CC}	T_A = 25°C MIN	T_A = 25°C TYP	T_A = 25°C MAX	SN54HC20 MIN	SN54HC20 MAX	SN74HC20 MIN	SN74HC20 MAX	UNIT
t_{pd}	A, B, C, or D	Y	2 V		45	110		165		140	ns
			4.5 V		14	22		33		28	
			6 V		11	19		28		24	
t_t		Y	2 V		27	75		110		95	ns
			4.5 V		9	15		22		19	
			6 V		7	13		19		16	

C_{pd}	Power dissipation capacitance per gate	No load	25	pF

FIGURE 4-32 A High-Speed CMOS Data Sheet. (Courtesy of Texas Instruments)

Chapter 5

OBJECTIVES

After completing this chapter, you will be able to:

1. Define the term troubleshooting.
2. Describe the three-step troubleshooting process: diagnose, isolate, and repair.
3. Describe the basic operation of the following test instruments and explain how they can be used to troubleshoot digital electronic circuits.
   a. Multimeter
   b. Oscilloscope
   c. Logic clip
   d. Logic probe
   e. Logic pulser
   f. Current tracer
4. Explain the difference between a static test and a dynamic test.
5. List and describe the more common internal and external IC failures.
6. Describe how to isolate internal and external IC shorts and opens.
7. Describe how to repair a circuit once a faulty component has been identified.

Troubleshooting Logic Gate Circuits

OUTLINE

VIGNETTE

Space, the Final Frontier

On May 24, 1962 a deep silence and tension filled the Mercury Control Center at Cape Canaveral, Florida. The only voice to be heard was that of Gus Grissom as he repeatedly tried to make contact with fellow astronaut Scott Carpenter aboard the *Aurora 7,* which had just completed three orbits around the earth and was now attempting a re-entry. The normal procedure at this stage of the mission is to fire the thrusters to correctly orient the craft to the proper angle for returning to earth. The next step is to slow down the capsule by applying the brakes or retrorockets, which causes the spacecraft to fall into the atmosphere and to earth. The final step is to pop open the parachutes at about 25,000 feet, and lower the capsule into the sea. All of these steps are crucial, and if any one of the steps is not followed exactly, it can spell disaster. For example, if the spacecraft is pitched at the wrong angle, or if it is traveling at too great a speed, the craft and astronaut will burn up due to the friction of the earth's atmosphere.

This was the second American orbital flight; however, unlike the first, this mission had not gone at all according to plan. The problems began when Carpenter was asked to position *Aurora 7* for the return and the capsule's automatic stabilization system had failed. Carpenter responded immediately by switching to manual control, but due to the malfunction, the craft was misaligned by 25 degrees. To make matters worse, the retrorockets fired three seconds too late, failing to produce the expected braking power.

Nine minutes had now passed and Gus Grissom at the Cape urgently repeated the same message into his microphone, not knowing whether Carpenter was dead or alive: "*Aurora 7, Aurora 7,* do you read me? Do you read me?"

Hundreds of miles away, a pair of mainframe computers at Goddard Space Flight Center in Maryland knew exactly where he was. These machines had continually digested the direction and altitude data from a radar ground station in California that had tracked the capsule from its time of re-entry. Eleven minutes after the last radar contact, the mainframe displayed an incredibly accurate estimate of the capsule's splashdown position. Aircraft were immediately dispatched and homed in on the capsule's signal beacon, where Carpenter was found riding in an inflatable life raft alongside his spacecraft. Luckily, the combined effects of the errors had brought Carpenter safely through the atmosphere, but some 250 miles away from the planned splashdown point, and well out of range of Grissom's radio transmission and the awaiting naval recovery task force.

At this time in history, the computer was not completely accepted as a valued partner in the space business. Astronauts, aeropace engineers, and ground control personnel were wary of putting too much trust in a machine that at that time would sometimes fail to operate for no apparent reason.

Today, the computer has proved itself indispensable, lending itself to almost every phase of our exploration of the final frontier. Computers are used by aerospace engineers to design space vehicles and to simulate the designs to determine their flight characteristics. These digital electronic computer systems can also simulate missions into deep space; an astronaut can take a flight simulation to almost any place in the galaxy, developing good flight skills on the way. Computers are also used to continually monitor the thousands of tests preceding a launch and will stop the countdown for any abnormality. Once the mission is underway, the on-board computer supplies the astronaut quickly and accurately with all information needed, enabling mankind to explore new worlds and new civilizations, and to boldly go where no one has gone before!

INTRODUCTION

This chapter will introduce you to digital test equipment and digital troubleshooting techniques. An effective electronics technician or troubleshooter must have a thorough knowledge of electronics, test equipment, troubleshooting techniques, and equipment repair. Like analog circuits, digital circuits occasionally fail, and in most cases a technician is required to quickly locate the problem within the system and then make the repair. The procedure for fixing a failure can be broken down into three basic steps.

Step 1: DIAGNOSE

The first step is to determine whether a problem really exists. To carry out this step, a technician must collect as much information about the system, the circuit, and the components used, and then diagnose the problem.

Step 2: ISOLATE

The second step is to apply a logical and sequential reasoning process to isolate the problem. In this step, a technician will operate, observe, test, and apply troubleshooting techniques in order to isolate the malfunction.

Step 3: REPAIR

The third and final step is to make the actual repair, and final test the circuit.

Troubleshooting, by definition, is the process of locating and diagnosing breakdowns in equipment by means of systematic checking and analysis. To troubleshoot, you will need a thorough knowledge of troubleshooting techniques, a very good understanding of test equipment, documentation in the form of technical and service manuals, and experience. Troubleshooting experience can only be acquired with practice; therefore, in future chapters, complete troubleshooting sections will be included and applied to all new devices and circuits. As far as practical experience, very few of your lab circuits will work perfectly the first time, so you will gain troubleshooting experience each and every time you experiment in lab. Although it seems very frustrating when a circuit is not operating correctly, remember that the more problems you have with a circuit, the better.

In this chapter, you will first be introduced to digital test equipment, and then in the following section you will see how these test instruments can be used to isolate internal and external digital IC failures. In the final section we will discuss how to make repairs to an electronics circuit once the problem has been isolated.

5-1 DIGITAL TEST EQUIPMENT

A variety of test equipment is available to help you troubleshoot digital circuits and systems. Some are standard, such as the multimeter and oscilloscope, which can be used for either analog or digital circuits. Other test instruments, such as the logic clip, logic probe, logic pulser, and current tracer have been designed specifically to test digital logic circuits. In this section we will be examining each of these test instruments, beginning with the multimeter.

5-1-1 Testing with the Multimeter

Multimeter
Test instrument able to perform multiple tasks in that it can be used to measure voltage, current, or resistance.

Both analog and digital type **multimeters** can be used to troubleshoot digital circuits. In most cases you will only use the multimeter to test voltage and resistance. The current settings are rarely used since a path needs to be opened to measure current, and as most circuits are soldered onto a printed circuit board, making current measurements is impractical. Some of the more common digital circuit tests using the multimeter include the following:

a. You can use the multimeter to test power supply voltages. On the AC VOLTS setting, you can check the 120 V ac input into the system's dc power supply, and using the DC VOLTS setting you can check all of the dc power supply voltages out of the power supply circuit, as seen in Figure 5-1(a). The multimeter can also be used to check that these dc supply voltages are present at each printed circuit board within the system, as in the case of the 5 V supply to the display board shown in Figure 5-1(b).

b. You can also use the multimeter on the DC VOLTS setting to test the binary 0 and binary 1 voltages throughout a digital circuit, as shown in Figure 5-1(b).

c. The multimeter is also frequently used on the OHMS setting to test components such as resistors, fuses, diodes, transistors, wires, cables, printed circuit board tracks, and other disconnected devices, as shown in Figure 5-1(c).

5-1-2 Testing with the Oscilloscope

Pulse Wave
A repeating wave that only alternates between two levels or values and remains at one of these values for a small amount of time relative to the other.

Since digital circuits manage two-state information, most of the signals within a digital system will be either square or rectangular. To review, the **pulse, or rectangular, wave** alternates between two peak values; however, unlike the square wave, the pulse wave does not remain at the two peak values for equal lengths of time. The *positive pulse waveform* seen in Figure 5-2(a), for instance, remains at its negative value for long periods of time and only momentarily

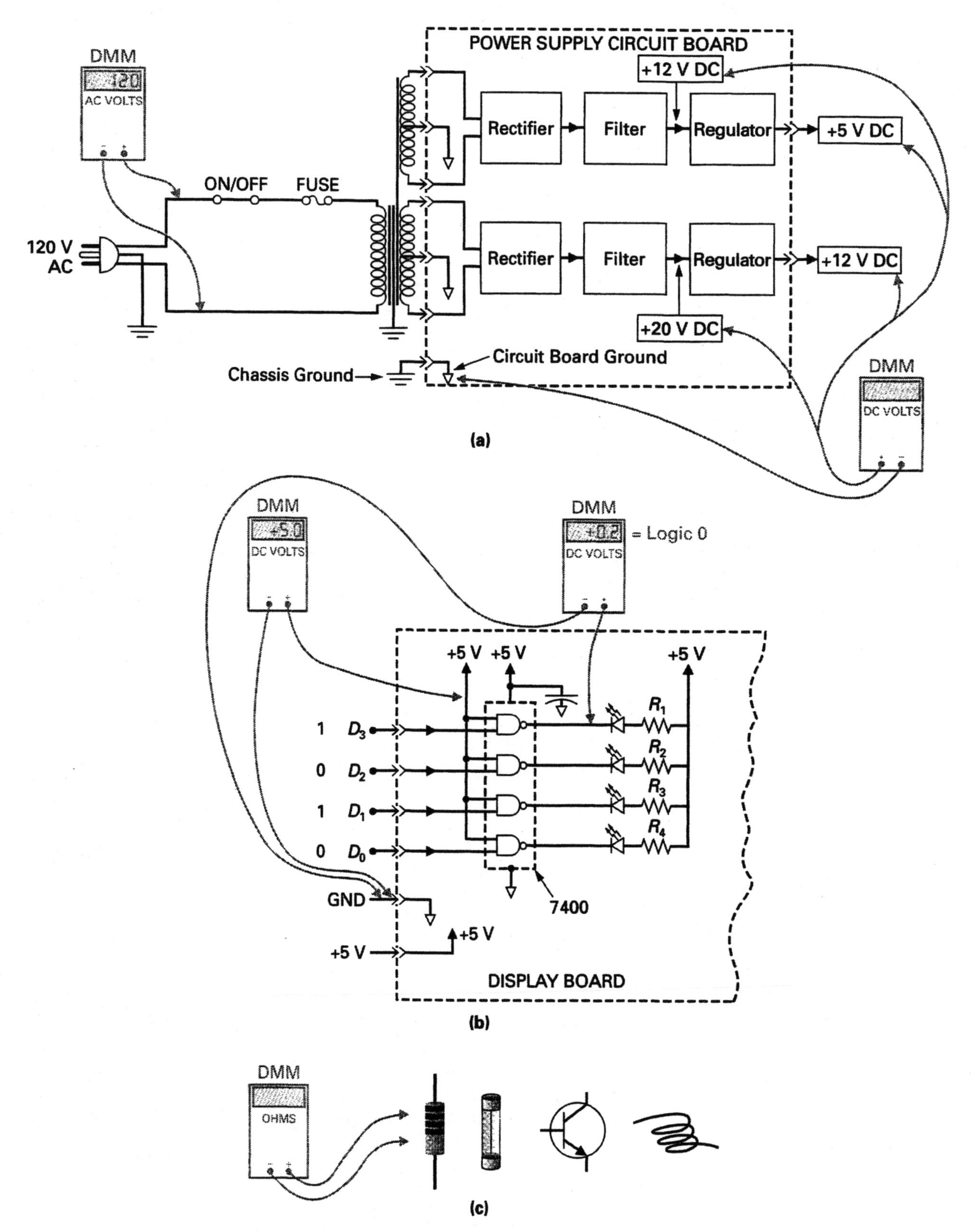

FIGURE 5-1 Testing with the Multimeter.

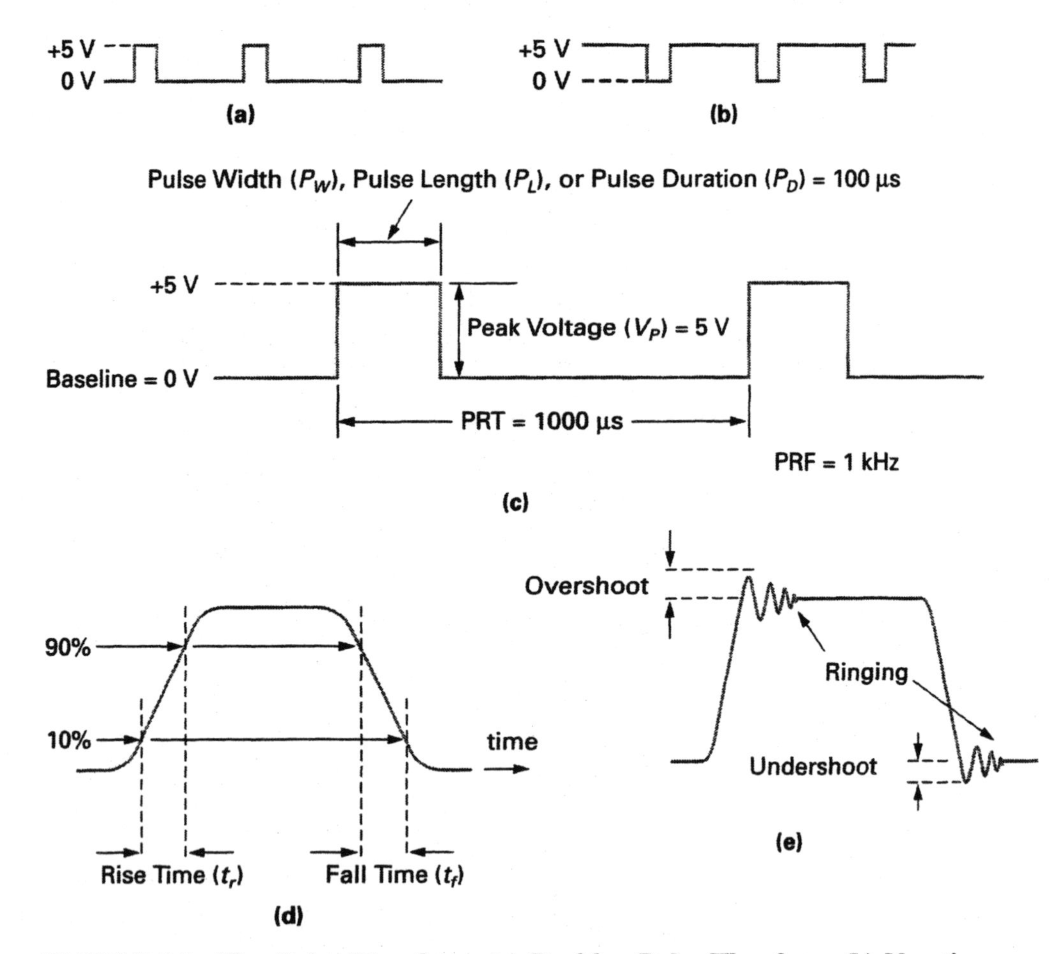

FIGURE 5-2 The Pulse Waveform. (a) Positive Pulse Waveform. (b) Negative Pulse Waveform. (c) Peak Voltage, Pulse Width, PRF, and PRT. (d) Rise Time and Fall Time. (e) Overshoot, Undershoot, and Ringing.

pulses positive. On the other hand, the *negative pulse waveform* seen in Figure 5-2(b) remains at its positive value for long periods of time and momentarily pulses negative.

Pulse Repetition Frequency
The number of times a second a pulse is generated.

Pulse Repetition Time
The time interval between the start of two consecutive pulses.

When referring to pulse waveforms, the term **pulse repetition frequency** (PRF) is used to describe the frequency or rate of the pulses, and the term **pulse repetition time** (PRT) is used to describe the period or time of one complete cycle, as seen in Figure 5-2(c). For example, if 1000 pulses are generated every second (PRF = 1000 pulses per second, pps or 1 kHz), each cycle will last 1/1000 of a second or 1 ms (PRT = 1/PRF = 1/1 kHz = 1 ms or 1000 μs).

The duty cycle of a pulse waveform indicates the ratio of pulse width time to the complete cycle time (it compares ON time to OFF time), and is calculated with the formula

$$\text{Duty Cycle} = \frac{\text{Pulse Width } (P_W)}{\text{Period } (t)} \times 100\%$$

■ **EXAMPLE:**

Calculate the pulse width, PRT, PRF, and duty cycle for the waveform shown in Figure 5-2(c).

■ ***Solution:***

In the example in Figure 5-2(c), a pulse width of 100 μs and a period of 1000 μs will produce a duty cycle of

$$\text{Duty Cycle} = \frac{\text{Pulse Width } (P_W)}{\text{Period } (t)} \times 100\%$$

$$= \frac{100\ \mu\text{s}}{1000\ \mu\text{s}} \times 100\% = 10\%$$

A duty cycle of 10% means that the positive pulse lasts for 10% of the complete cycle time, or PRT.

Since the pulse repetition time is 1000 μs, the pulse repetition frequency will equal the reciprocal of this value, or 1 kHz (PRF = 1/PRT = 1/1000 μs = 1 kHz).

Another area that needs to be reviewed in regards to digital signals is rise and fall time. The pulse waveforms seen in Figure 5-2(a), (b), and (c) all rise and fall instantly between their two states or peak values. In reality, there is a time lapse referred to as the *rise time* and *fall time* as seen in Figure 5-2(d). The rise time is defined as the time needed for the pulse to rise from 10% to 90% of the peak amplitude, while the fall time is the time needed for the pulse to fall from 90% to 10% of the peak amplitude.

When carefully studying pulse waveforms on the oscilloscope, you will be able to measure pulse width, PRT, rise time, and fall time. You will also see a few other conditions known as *overshoot, undershoot,* and *ringing.* These unwanted conditions tend to always accompany high-frequency pulse waveforms due to imperfections in the circuit, as seen in Figure 5-2(e). As you can see, the positive rising edge of the waveform "overshoots" the peak, and then a series of gradually decreasing sine-wave oscillations occur, called ringing. At the end of the pulse this ringing action reoccurs, as the negative falling edge "undershoots" the negative peak value.

The graticules on an oscilloscope's display can be used to measure amplitude and time, as seen in Figure 5-3(a). This time/amplitude measuring ability of the oscilloscope enables us to measure almost every characteristic of a digital signal. To begin with, the vertical scale can be used to measure HIGH/LOW logic levels, as shown in Figure 5-3(b). On the other hand, the horizontal scale can be used to measure the period, rise time, fall time, and pulse width of a digital signal as seen in Figure 5-3(c). Once the period of a cycle has been calculated, the frequency can be determined ($f = 1/t$).

Some of the more common digital circuit tests using the oscilloscope include the following:

a. You can use the oscilloscope to check the clock or master timing signal within a digital system to see that it is not only present at every point in the circuit, but also that its frequency, wave shape, and amplitude are correct.

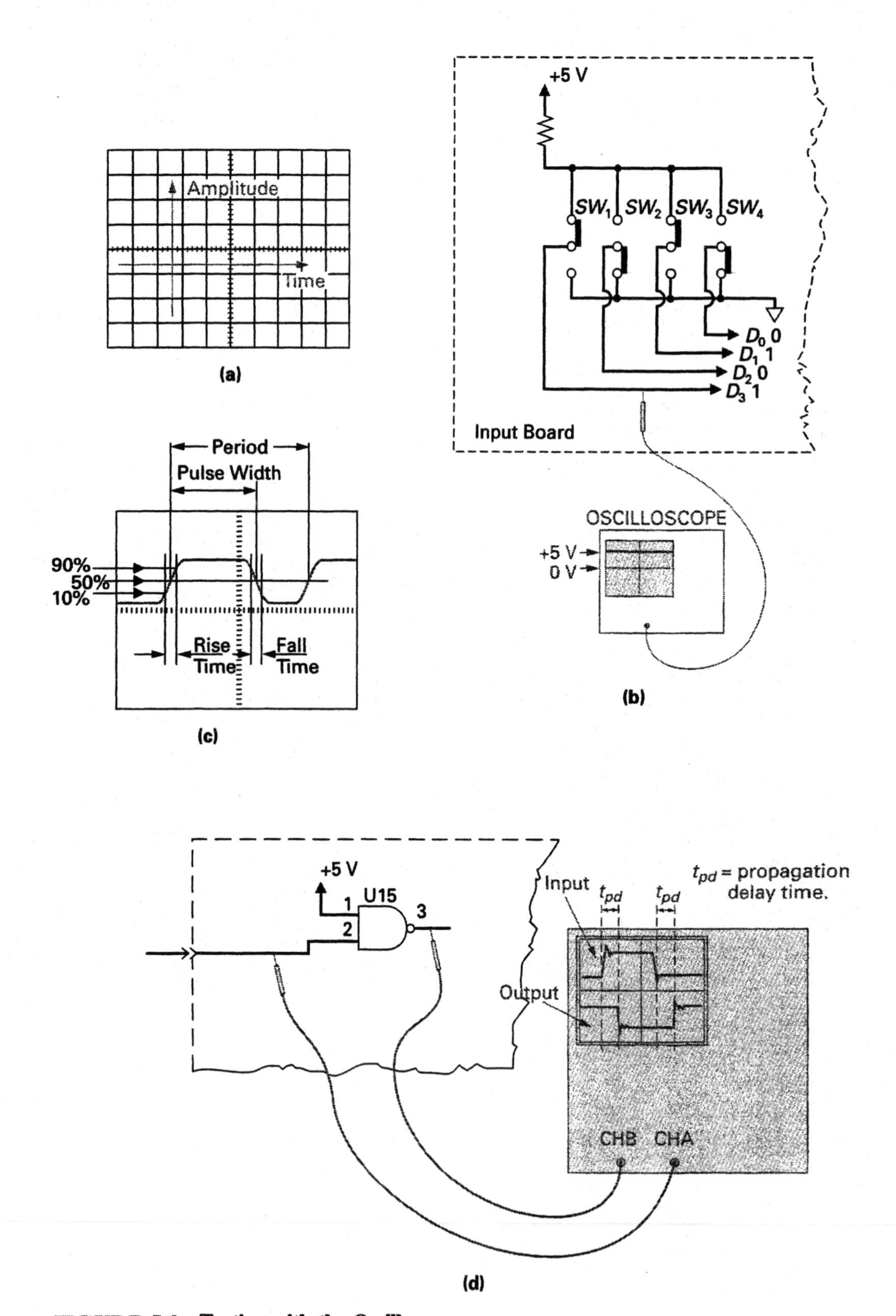

FIGURE 5-3 **Testing with the Oscilloscope.**

b. You will also use the oscilloscope frequently to monitor two or more signals simultaneously, as shown in Figure 5-3(d). In this instance, an input and output waveform are being compared to determine the NAND gate's propagation delay time.

c. The oscilloscope can also be used to measure waveform distortions, such as the ringing shown in Figure 5-3(d), to see if these undershoots or overshoots are causing any false operations, such as incorrectly triggering a logic gate. A noise problem in a digital circuit is often called a **glitch,** which stands for "gremlins loose in the computer housing."

Glitch
An acronym for "gremlins loose in the computer housing." The term is used to describe a problem within an electronic system.

From the previous discussion you can see that the oscilloscope is a very versatile test instrument, able to measure almost every characteristic of a digital signal. Its accuracy, however, does not match the multimeter for measuring logic level voltages, or the frequency counter for measuring signal frequency. The scope is a more visual instrument, able to display a picture image of signals at each point in a circuit, and enabling its operator to see phase relationships, wave shape, pulse widths, rise and fall times, distortion, and other characteristics.

When using an oscilloscope to troubleshoot a digital circuit, try to remember the following:

a. Always use a *multi-trace oscilloscope* when possible so you can compare input and output waveforms. Remember that a chopped or multiplexed display will not be accurate at high frequencies since one trace is trying to display two waveforms. A dual-beam oscilloscope is best for high-frequency digital circuit troubleshooting.

b. Trigger the horizontal sweep of the oscilloscope with the input signal, or a frequency related signal. Having a *triggered sweep* will ensure more accurate timing measurements and enable the oscilloscope to lock on to your input more easily.

c. Use an oscilloscope with a large *bandwidth.* The input signals are fed to vertical amplifiers within the oscilloscope, and these circuits need to have a high upper-frequency limit. Since all scopes can measure dc signals (0 Hz), this lower frequency is generally not listed. The upper-frequency limit, however, is normally printed somewhere on the front panel of the scope and should be 40 MHz for most digital measurements, 60 MHz for high-speed TTL, and between 100 and 1000 MHz for high-speed ECL circuits.

5-1-3 Testing with the Logic Clip

The **logic clip,** which is shown in Figure 5-4(a), was specially designed for troubleshooting digital circuits. It consists of a spring-loaded clip that clamps on to a standard IC package, where it makes contact with all the pins of the IC, as shown in Figure 5-4(b). These contacts are then available at the end of the clip, so a multimeter or oscilloscope probe can easily be connected.

Logic Clip
A spring loaded digital testing clip that can be clamped onto an IC package so that the signal present on any pin is available at the end of the clip. Some include LEDs to display logic levels present.

Some logic clips simply have connections on the top of the clip, while others have light-emitting diodes (LEDs) and test points on the top of the clip, as seen in Figure 5-4(a). The LEDs give a quick indication of the binary level on the pins of the IC, with a logic 0 turning an LED OFF and a binary 1 turning an LED ON.

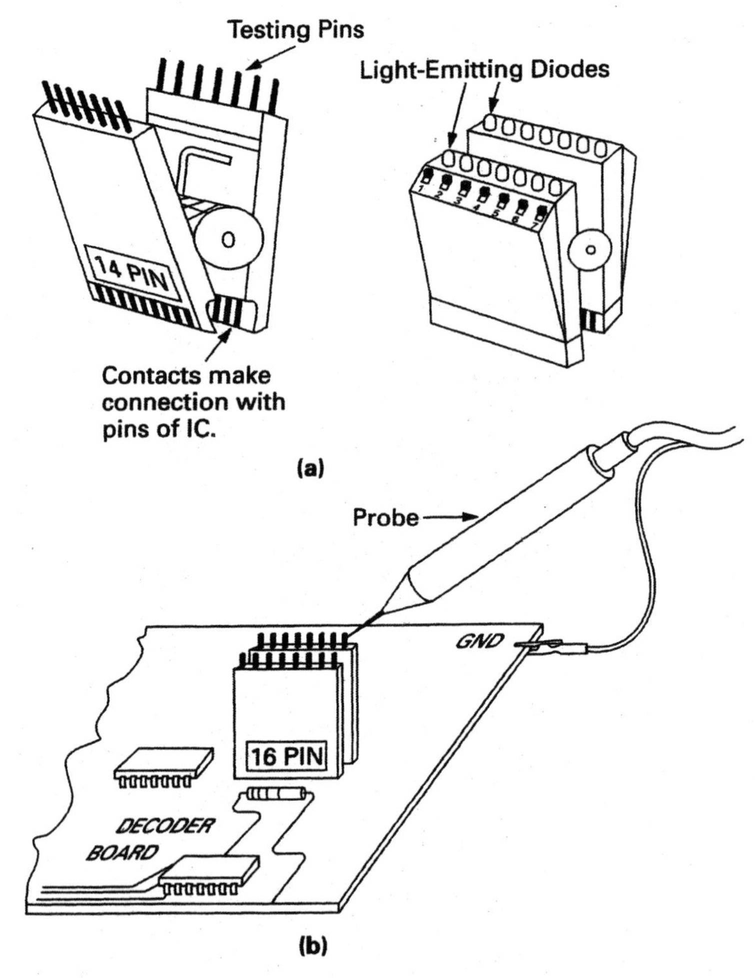

FIGURE 5-4 Testing with the Logic Clip.

Logic Probe
A digital test instrument that can indicate whether a HIGH, LOW, or PULSE is present at a circuit point.

Static Test
A test performed on a circuit that has been held constant in one condition.

Dynamic Test
A test performed on a circuit that is running at its normal operating speed.

5-1-4 *Testing with the Logic Probe*

Another test instrument designed specifically for digital circuit troubleshooting is the **logic probe,** shown in Figure 5-5(a). In order for the probe to operate, its red and black power leads must be connected to a power source. This power source can normally be obtained from the circuit being tested, as shown in Figure 5-5(b). Once the power leads have been connected, the logic probe is ready to sense the voltage at any point in a digital circuit and give an indication as to whether that voltage is a valid binary 1 or 0. The HI and LO LEDs on the logic probe are used to indicate the logic level, as seen in Figure 5-5(b). Since the voltage thresholds for a binary 0 and 1 can be different, most logic probes have a selector switch to select different internal threshold detection circuits for TTL, ECL, or CMOS.

In addition to being able to perform **static tests** (constant or non-changing HIGH and LOW signals) like those of the multimeter, the logic probe can also perform **dynamic tests** (changing signals) like those of the oscilloscope, such as

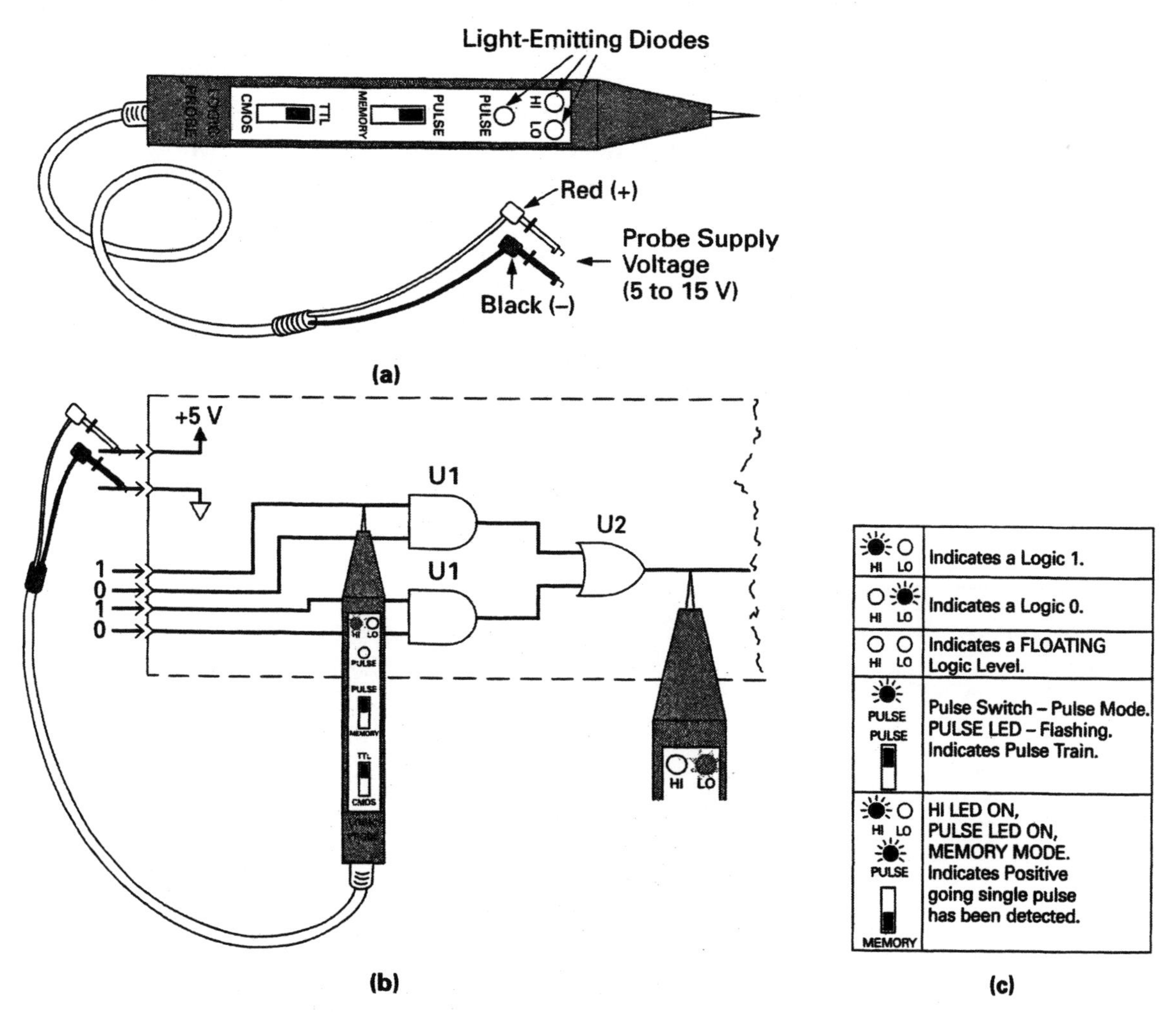

FIGURE 5-5 Testing with the Logic Probe.

detecting a single momentary pulse (single pulse) or a pulse train (sequence of pulses), such as a clock signal. Nearly all logic probes have an internal low-frequency oscillator that will flash the "PULSE" LED ON and OFF if a pulse train is detected at a frequency of up to 100 MHz. The logic probe can also detect if a single pulse has occurred, even if the width of the pulse is so small that the operator was unable to see the PULSE LED blink. In this instance, a memory circuit can be enabled using the MEMORY switch, so that any pulse (even as narrow as 10 ns), will turn ON and keep ON the PULSE LED. This memory circuit within the logic probe senses the initial voltage level present at the test point (either HIGH or LOW), and then turns ON the PULSE LED if there is a logic transition or change. To reset this memory circuit simply toggle (switch OFF and then ON) the memory switch. Figure 5-5(c) summarizes the operation and display interpretation of a typical logic probe.

Even though the logic probe is not as accurate as the multimeter, and cannot determine as much about a signal as the oscilloscope, it is ideal for digital circuit troubleshooting because of its low cost, small size, and versatility.

5-1-5 Testing with the Logic Pulser

Using the logic probe to test the static HIGH and LOW inputs and output of a gate does not fully test the operation of that logic gate. In most instances it is necessary to trigger the input of a circuit and then monitor its response to a dynamic test. A **logic pulser** is a signal generator designed to produce either a single pulse or a pulse train. Its appearance is similar to the logic probe, as shown in Figure 5-6(a), in that it requires a supply voltage to operate, and has a probe which is used to apply pulses to the circuit under test, as shown in Figure 5-6(b).

Logic Pulser
A digital test instrument that can generate either single or repeating logic pulses.

The logic pulser is operated by simply touching the probe tip to any conducting point on a circuit, and then pressing the SINGLE PULSE button for a single pulse or selecting the PULSE TRAIN switch position for a constant se-

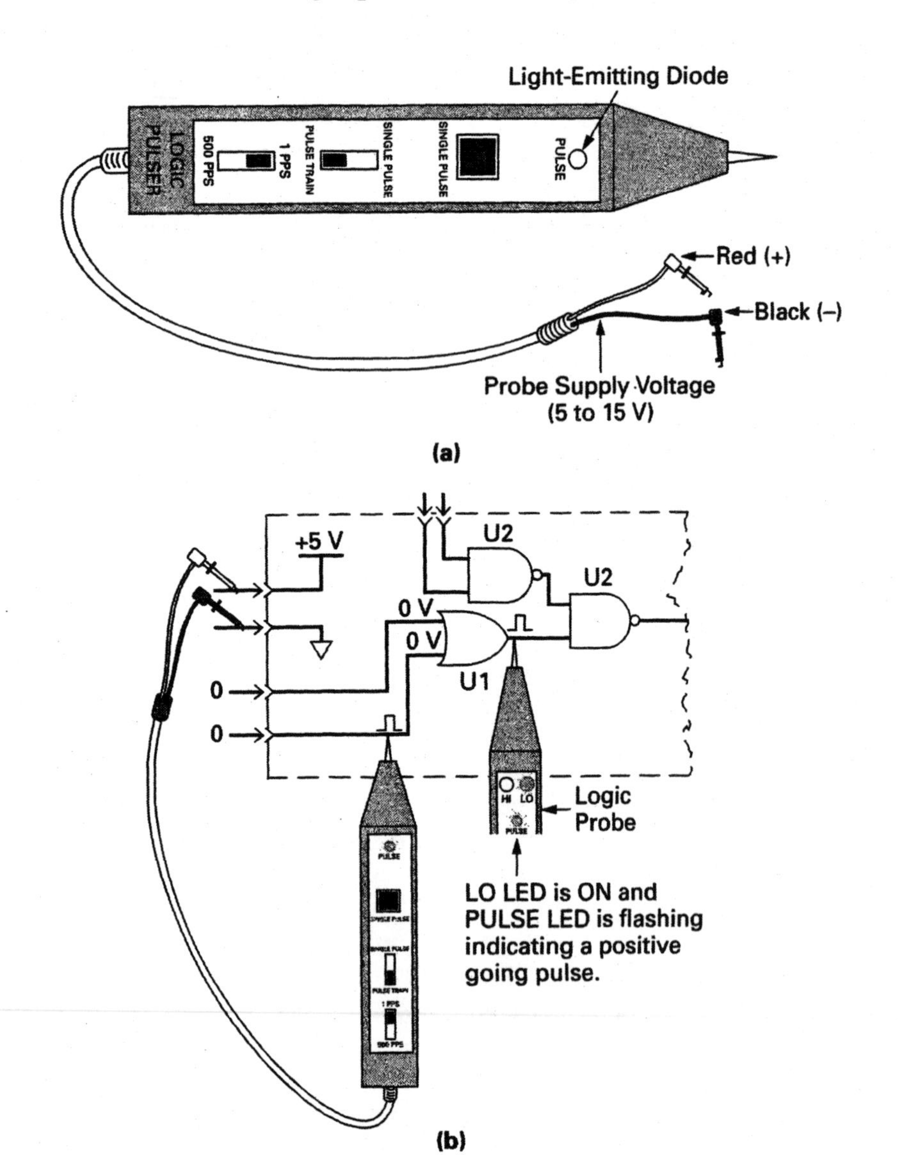

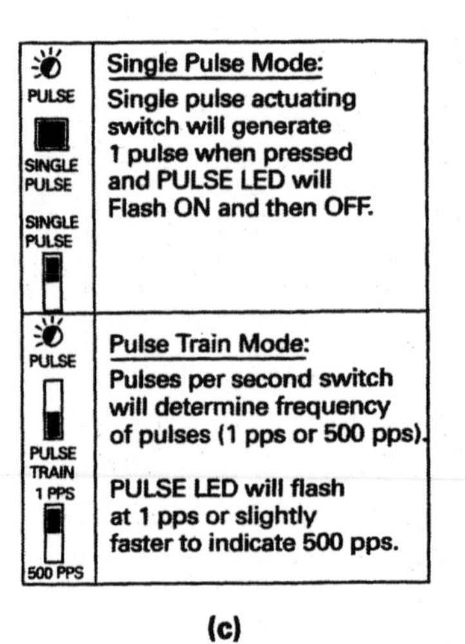

FIGURE 5-6 Testing with the Logic Pulser.

quence of pulses. The frequency of the pulse train, with this example model, can be either one pulse per second (1 pps) or 500 pulses per second (500 pps), based on the pps selector switch.

The logic probe is generally used in conjunction with the logic pulser to sense the pulse or pulses generated by the logic pulser, as shown in Figure 5-6(b). In this example, the logic pulser is being used to inject a single pulse into the input of an OR gate, and a logic probe is being used to detect or sense this pulse at the output of the OR gate. If the input to the OR gate was HIGH, the logic pulser would have automatically sensed the binary 1 voltage level, and then pulsed the input to the OR gate LOW when the pulse button was pressed. In this example, however, the input logic level to the OR gate was LOW, so the logic pulser sensed the binary 0 voltage level, and then pulsed the OR gate input HIGH. This ability of the logic pulser to sense the logic level present at any point and then pulse the line to its opposite state means that the operator does not have to determine these logic levels before a point is tested, making the logic pulser a fast and easy instrument to use. The logic pulser achieves this feature with an internal circuit that will override the logic level in the circuit under test, and either source current or sink current when a pulse needs to be generated. This complete in-circuit testing ability means that components do not need to be removed and input and output paths do not need to be opened in order to carry out a test. Figure 5-6(c) summarizes the operation and display interpretation of a typical logic pulser.

Since TTL, ECL, and CMOS circuits all have different input and output voltages and currents, logic pulsers will need different internal circuits for each family of logic ICs. Typical pulse widths for logic pulsers are between 500 ns and 1 μs for TTL logic pulsers, and 10 μs for CMOS logic pulsers.

5-1-6 Testing with the Current Tracer

The logic pulser can also be used in conjunction with another very useful test instrument called a **current tracer,** which is shown in Figure 5-7(a). Like the logic probe, the current tracer is also a sensing test instrument, but unlike the logic probe, the current tracer senses the relative values of current in a conductor. It achieves this by using an insulated inductive pick-up tip, which senses the magnetic field generated by the current in a conductor. By adjusting the sensitivity control on the current tracer and observing the lamp's intensity when the probe is placed on a pulsating logic signal line, a shorted path can be found by simply tracing the path of high current.

Current Tracer
A digital test instrument used to sense relative values of current in a conductor using an inductive pick-up tip.

To troubleshoot a short in an IC's input or output circuit, a power supply, or a printed circuit board (PCB) track or cable, we would traditionally have to cut PCB tracks, snip pins, or open component leads. This would isolate the short, since the excessive current path or short would be broken when the path was opened. The current tracer allows the troubleshooter to isolate the shorted path without tampering with the circuit. Figure 5-7(b) illustrates how the current tracer should be correctly aligned on a conducting PCB track. Like the logic probe and logic pulser, the current tracer can use the same supply voltage as the circuit under test. Since the probe's tip is insulated, it can be placed directly on the track; however, be sure that the probe is always perpendicular to the

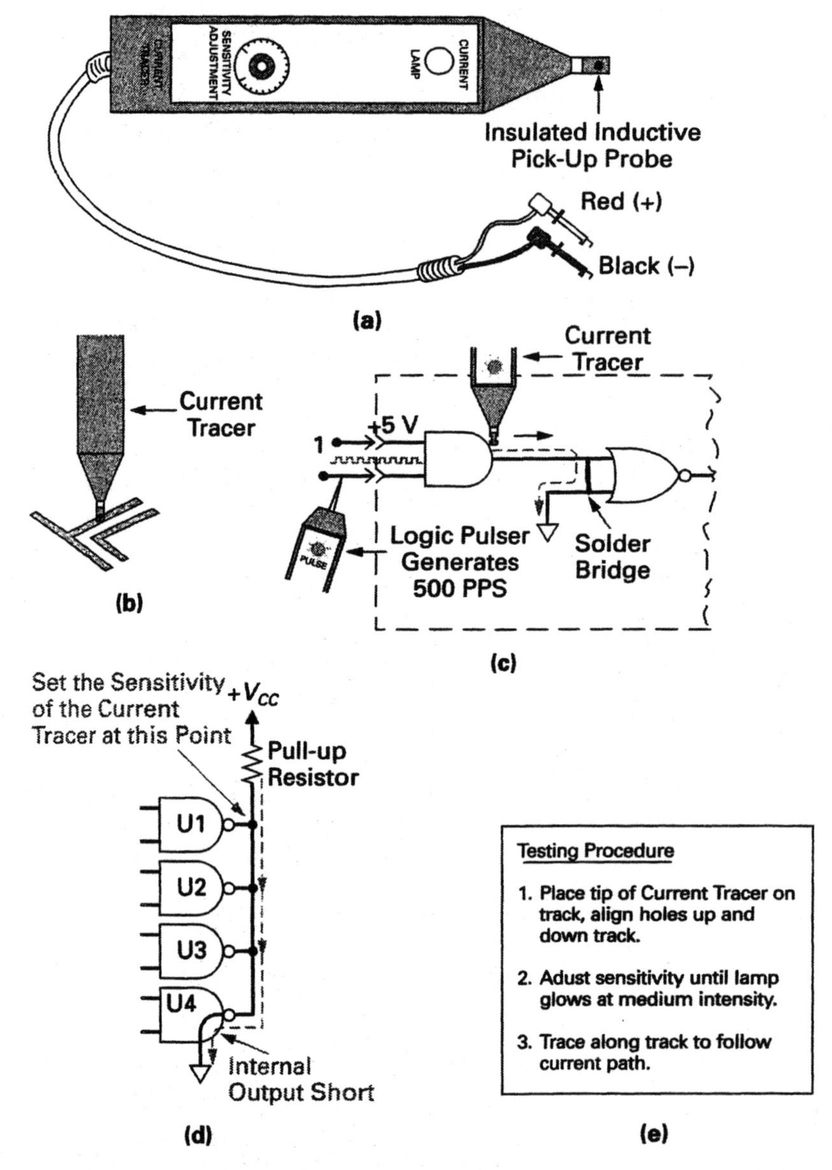

FIGURE 5-7 Testing with the Current Tracer.

board. The other important point to remember is to ensure that the small holes or dots on either side of the tip are aligned so they face up and down the track being traced at all times. This is so that the inductive coil in the insulated tip is oriented to pick up the maximum amount of magnetic flux.

Figure 5-7(c) shows an example of how a current tracer and logic pulser can be used to locate a short caused by a solder bridge. In this example, we will first connect the logic pulser so that it will generate a pulsating current, and therefore a changing magnetic field, which can be detected by the inductive tip of the current tracer. The next step is to touch the tip of the current tracer on the output of the AND gate and adjust the sensitivity level of the current tracer to light the current lamp (the sensitivity adjustment allows the current tracer to sense a wide range of current levels, from 1 mA to 1 A, typically). Once this reference level of brightness has been set, trace along the path between the output of the AND gate and the input of the OR gate. Since the current value should be the same at all points along this track, the lamp should glow at the same level of brightness. Once the current tracer moves past the solder bridge, the lamp will go out, since no current is present beyond this point. At this point you should, after a visual inspection, be able to locate the short.

The current tracer is also an ideal tool for troubleshooting a circuit in which many outputs are tied to a common point (such as open collector gates), as seen in the example in Figure 5-7(d). If any of these gates were to develop an internal short, it would be difficult to isolate which of the parallel paths is causing the problem. In the past, you would have to remove each gate from the circuit (which is difficult with a soldered PCB), or clip the output IC pin of each gate until the short was removed and the faulty gate isolated. The current tracer can quickly isolate faults of this nature since it will highlight the current path provided by the short, and therefore lead you directly to the faulty gate.

SELF-TEST REVIEW QUESTIONS FOR SECTION 5-1

1. What would be the most accurate instrument for measuring dc voltages?
2. Which test instrument would be best for measuring gate propagation delay time?
3. A logic __________ is used to generate a single or train of test pulses.
4. A __________ detects current pulses, while a __________ detects voltage levels and pulses.

5-2 DIGITAL CIRCUIT PROBLEMS

Since ICs account for almost 85 percent of the components within digital systems, it is highly likely that most digital circuit problems will be caused by a faulty IC. Digital circuit failures can be basically divided into two categories: failures within ICs, or failures outside of ICs. When troubleshooting digital circuits, therefore, you will have to first isolate the faulty IC and then determine whether the problem is internal or external to the IC. An internal IC problem cannot be fixed and therefore the IC will have to be replaced, whereas an external IC problem can be caused by electronic devices, connecting devices, mechanical or magnetic devices, power-supply voltages, system clock signals, and so on. To begin with, let us discuss some of the typical internal IC failures.

5-2-1 ***Digital IC Problems***

Digital IC logic-gate failures can basically be classified as either opens or shorts in either the inputs or output, as summarized in Figure 5-8. In all cases, these failures will result in the IC having to be discarded and replaced. The following examples will help you to isolate an internal IC failure.

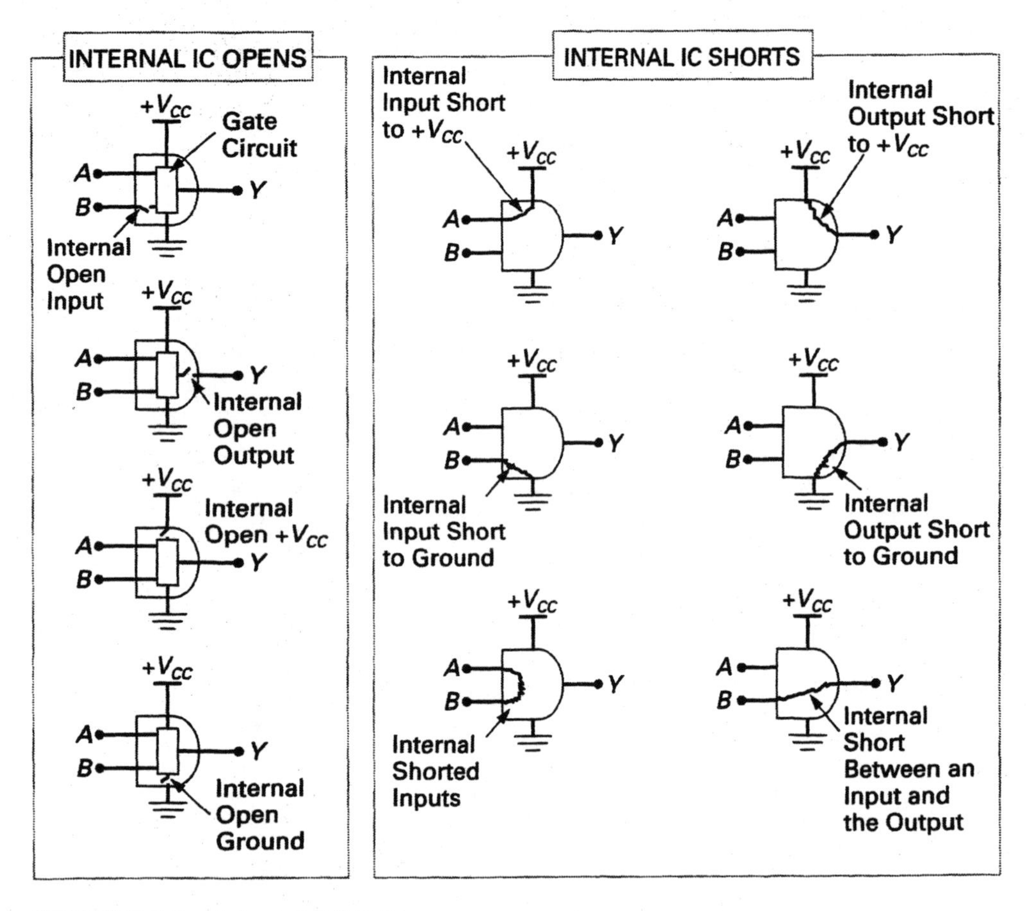

FIGURE 5-8 **Internal IC Failures.**

Internal IC Opens

An internal open gate input or an open gate output are very common internal IC failures. These failures are generally caused by the very thin wires connecting the IC to the pins of the package coming loose or burning out due to excessive values of current or voltage.

Figure 5-9 illustrates how a logic probe and logic pulser can be used to isolate an internal open NAND gate input. In Figure 5-9(a) the open input on pin 10 has been jumper connected to $+V_{CC}$, and the other gate input (pin 9) is being driven by the logic pulser. As you know, any LOW input to a NAND gate will produce a HIGH output, and only when both inputs are HIGH will the output be LOW, as shown in the inset in Figure 5-9(a). The logic probe, which is monitoring the gate on pin 8, indicates that a negative going pulse waveform is present at the output, and therefore the gate seems to be functioning normally. When the jumper and pulser inputs are reversed, however, as seen in Figure 5-9(b), the output remains permanently LOW. These conditions could only occur if the pin 10 input was permanently HIGH. As discussed previously, an open or floating input is equivalent to HIGH input, since both conditions have next-to-no input current. The cause, therefore, must be either an internal open at pin 10 (since this is equivalent to a HIGH input), or a $+V_{CC}$ short to the pin 10 input of this gate

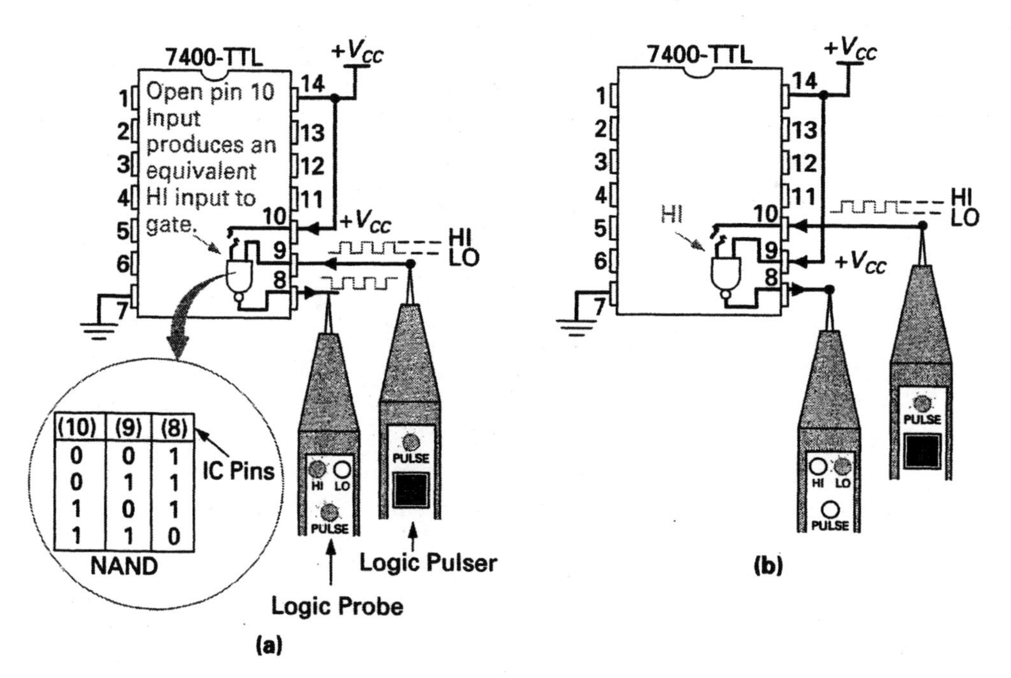

FIGURE 5-9 **An Internal Open Gate Input.**

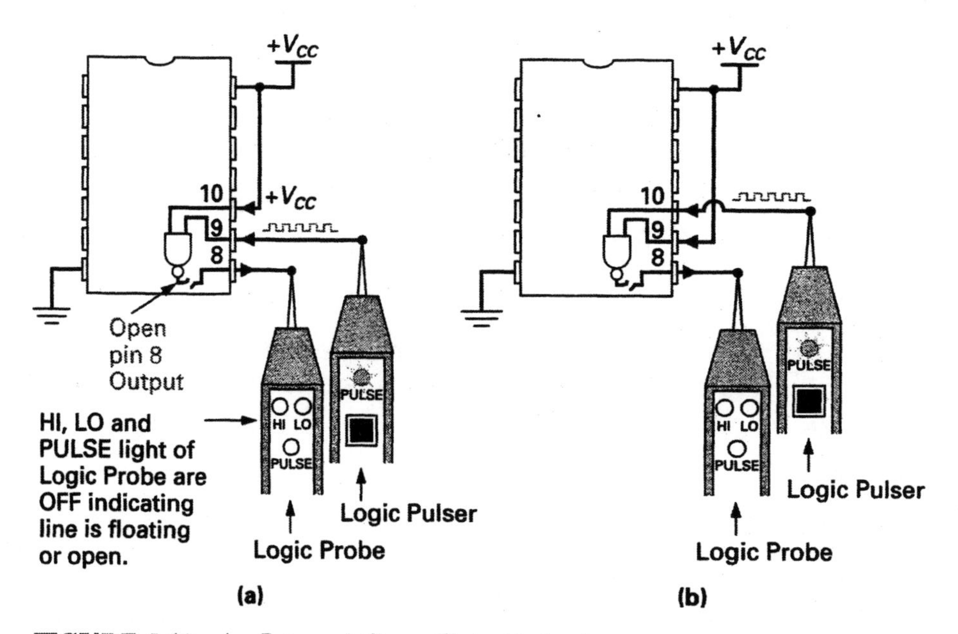

FIGURE 5-10 **An Internal Open Gate Output.**

(which is less common than an open). In either case, however, the logic gate IC has an internal failure and will have to be replaced.

Figure 5-10 illustrates how a logic pulser and logic probe can be used to isolate an internal open gate output. In Figure 5-10(a), the logic gate's input at pin 10 has been jumper connected to $+V_{CC}$, and the other input at pin 9 is being driven by the logic pulser. The logic probe, which is monitoring the output on pin 8, has none of its indicator LEDs ON, which indicates that the logic level is neither a valid logic 1 or a valid logic 0 (line is probably floating). In Figure 5-10(b), the inputs have been reversed; however, the logic probe is still indicating the same effect. This test highlights the failure, which is an open gate output. Once again, the logic gate IC has an internal failure and will have to be replaced.

Internal IC Shorts

Internal logic gate shorts to either $+V_{CC}$ or ground will cause the inputs or output to be stuck either HIGH or LOW. These internal shorts are sometimes difficult to isolate, so let us examine a couple of typical examples.

Figure 5-11(a) shows the effect an internal input lead short to ground will have on a digital logic gate circuit. The input to gate *B* is shorted to ground, so this input is being pulled LOW. This LOW between *A* output and *B* input makes it appear as though gate *A* is malfunctioning, since it has two HIGH inputs and therefore should be giving a LOW output. The problem is solved by disconnecting gate *A* from gate *B*. Since gate *A*'s output will go HIGH when it is no longer being pulled down by gate *B*, the fault will be isolated to a short at the input of gate *B*.

In Figure 5-11(b), the output of gate *A* has shorted to $+V_{CC}$. This circuit condition gives the impression that gate *A* is malfunctioning; however, if the *B* gate input had shorted to $+V_{CC}$ we would get the same symptoms. Once again, we will have to isolate, by disconnecting the *A* output from the *B* input to determine which gate is faulty.

Troubleshooting Internal IC Failures

Let us now try a few examples to test our understanding of digital test equipment and troubleshooting techniques.

■ **EXAMPLE:**

Determine whether a problem exists in the circuit in Figure 5-12(a) based on the logic probe readings indicated. If a problem does exist, indicate what you think could be the cause.

■ ***Solution:***

Gates *B* and *C* seem to be producing the correct outputs based on their input logic levels. Gate *A*'s output, on the other hand, should be LOW, and it is reading a HIGH. To determine whether the problem is the source gate *A* or the load gate *B*, the two gates have been isolated as shown in Figure 5-12(b).

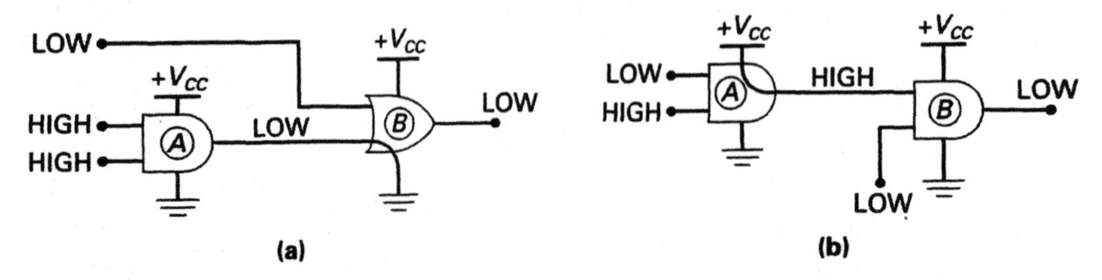

FIGURE 5-11 Internal IC Shorts.

7408 (Quad Two-Input AND Gate)

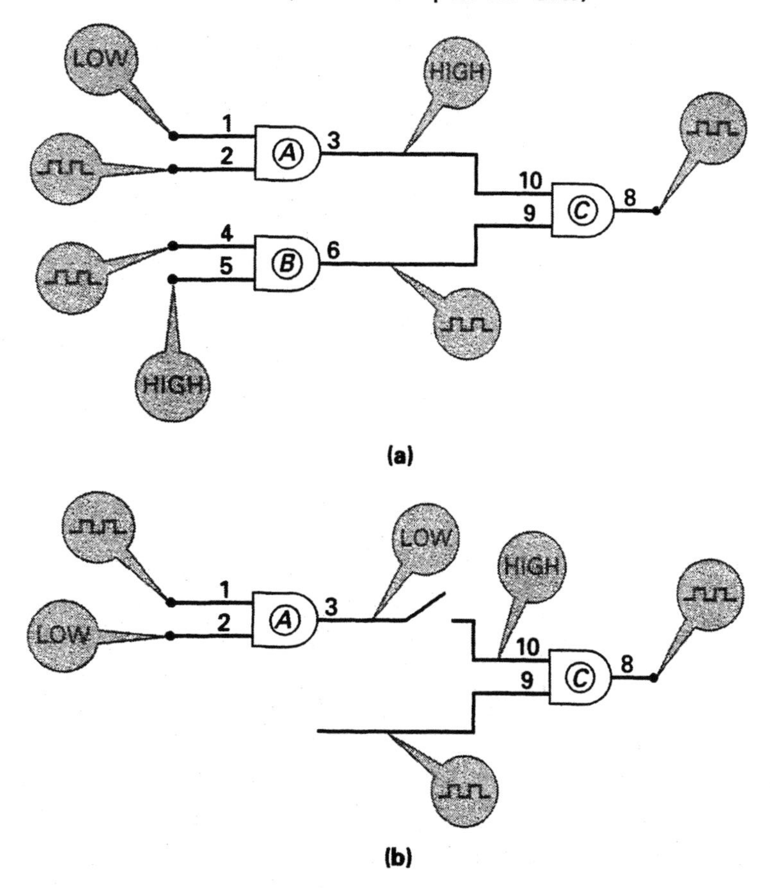

FIGURE 5-12 Troubleshooting Gate Circuits.

After isolating gate *A* from gate *B*, you can see that the pin 10 input to gate *C* seems to be stuck HIGH, probably due to an internal short to $+V_{CC}$.

Even though gate *C* seems to be the only gate malfunctioning, the whole 7408 IC will have to be replaced. In situations like this, it is a waste of time isolating the problem to a specific gate when the problem has already been isolated down to a single component. Use your understanding of internal logic gate circuits to isolate the problem to a specific component and not to a device within that component.

■ **EXAMPLE:**

What could be the possible fault in the circuit shown in Figure 5-13?

■ ***Solution:***

The output of NOT gate *A* should be HIGH; however, it is pulsating. This pulsating signal can only have come from pin 2 of the NAND gate. By disconnecting the pulsating input at pin 2 of the NAND gate, and then single pulsing this input using the logic pulser, you can use the logic probe to see if the pin 1 input of the NAND gate always follows what is on pin 2. Once an input short

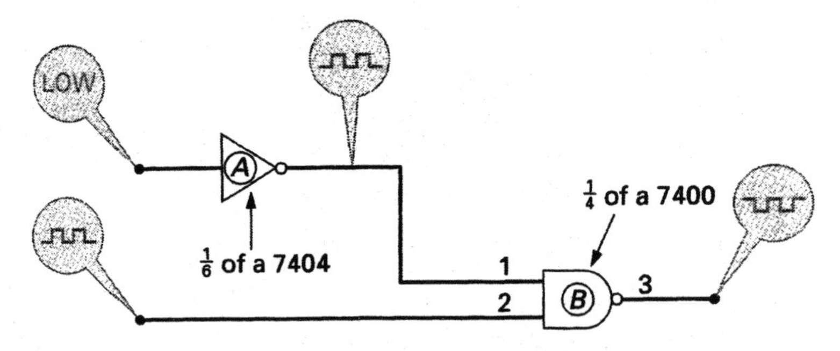

FIGURE 5-13 **Troubleshooting Gate Circuits.**

has been determined, visually inspect the NAND gate IC to be sure that there is no external short between the pins 1 and 2.

5-2-2 *Other Digital Circuit Device Problems*

Failures external to ICs can produce almost exactly the same symptoms as internal IC failures, so your task is to determine whether the problem is internal or external to the IC. The external IC circuit problems can be wide and varied; however, let us discuss some of the more typical failures.

External IC Shorts and Opens

Most digital signal line shorts can be isolated by injecting a signal onto the line using the logic pulser, and then using the current tracer to find the shorted path. Here are some examples of typical external IC short circuit problems.

a. A short between the pins of the IC package due to a solder bridge, sloppy wiring, wire clippings, or an improperly etched printed circuit board.

b. A bending in or out of the IC pins as they are inserted into a printed circuit board.

c. The shorting of an externally connected component, such as a shorted capacitor, resistor, diode, transistor, LED, switch, and so on.

d. A shorted connector such as an IC socket, circuit board connector, or cable connector.

Most digital signal line opens are easily located since the digital signal does not arrive at its destination. The logic probe is ideal for tracing these open paths. Here are some examples of typical external IC open circuit problems.

a. An open in a signal line due to an improperly soldered pin, a deep scratch in the etched printed circuit board, a bent or broken IC pin, a broken wire, and so on.

b. An IC that is inserted into the printed circuit board backwards.

c. The failure of externally connected components such as resistors, diodes, transistors, and so on.

d. An open connector such as an IC socket, circuit board connector, or cable connector.

In summary, the logic pulser and current tracer are ideal for tracing shorts and the logic probe is ideal for tracing opens.

Power Supply Problems

Like analog circuits, digital circuits suffer from more than their fair share of power-supply problems. When a digital circuit seems to have a power-supply problem, remember to use the half-split method of troubleshooting first, as seen in Figure 5-14(a). First check the dc supply voltages from the dc power-supply

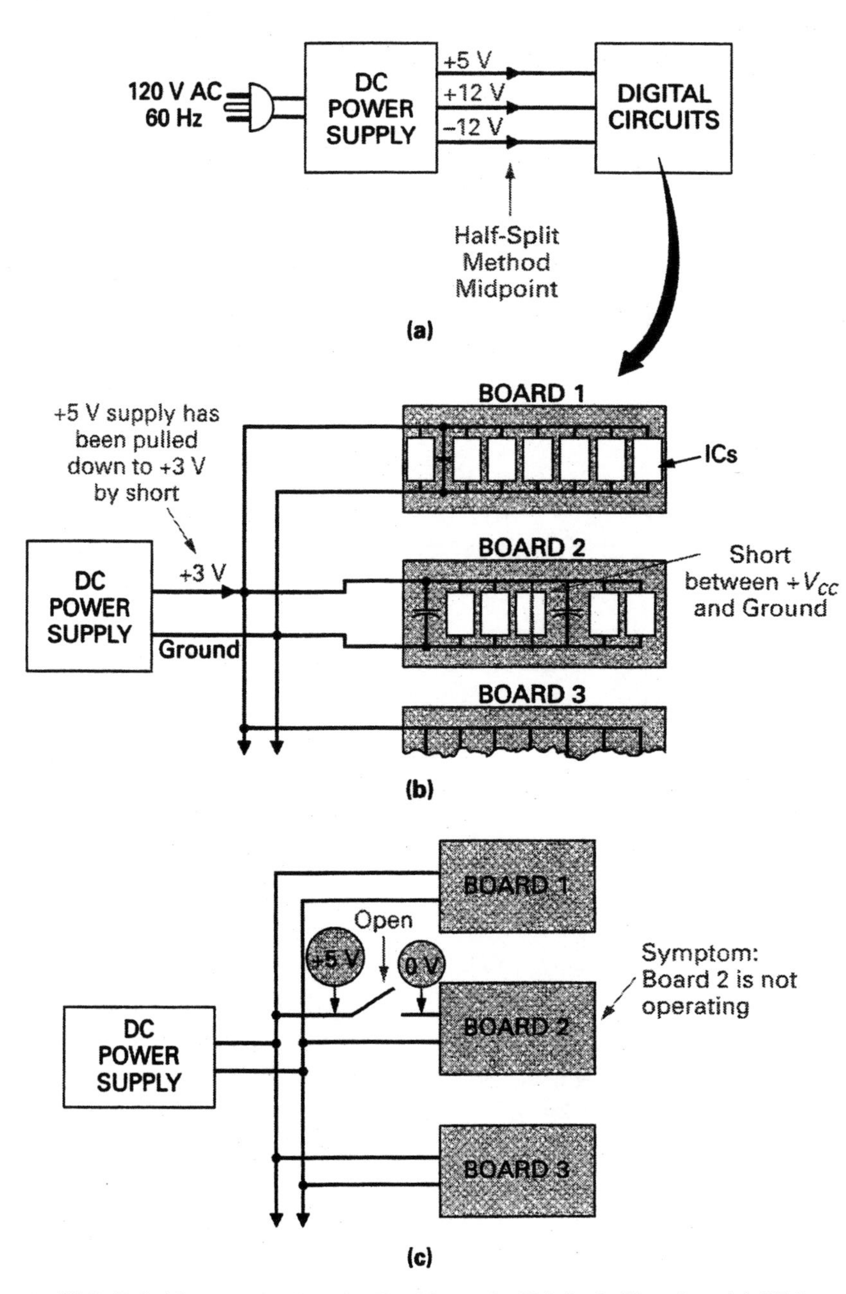

FIGURE 5-14 **Power Supply Problems in Digital Circuits. (a) Using the Half-Split Method. (b) Tracing Shorts. (c) Tracing Opens.**

unit to determine whether the power problem is in the power supply (source) or in one of the digital circuits (load).

If the dc output voltages are not present at the output of the dc power supply, then the fault probably exists before this test point. In this instance, you should test the dc power-supply circuit; however, a short in the digital circuits can pull down the dc supply voltage and make it appear as though the problem is in the dc power supply, as shown in Figure 5-14(b). Once again, the current tracer is ideal for tracing the faulty board and then the faulty device in that printed circuit board. If you cannot get power to the digital circuits because the short causes the dc power supply fuse to continually blow, use the logic pulser to inject a current into the supply line and then use the current tracer to follow the current path to the short.

Open power-supply lines are easy to locate since power is prevented from reaching its destination. The logic probe is ideal for tracking down power line opens, as shown in Figure 5-14(c).

Clock Signal Problems

Clock
A two-state timing signal that is distributed throughout a digital electronic system to ensure that all circuits operate in sync, or at the correct time.

All digital systems use a master timing signal called a **clock.** This two-state timing signal is generated by a digital oscillator circuit. This clock or timing signal is distributed throughout the digital electronic system to ensure that all

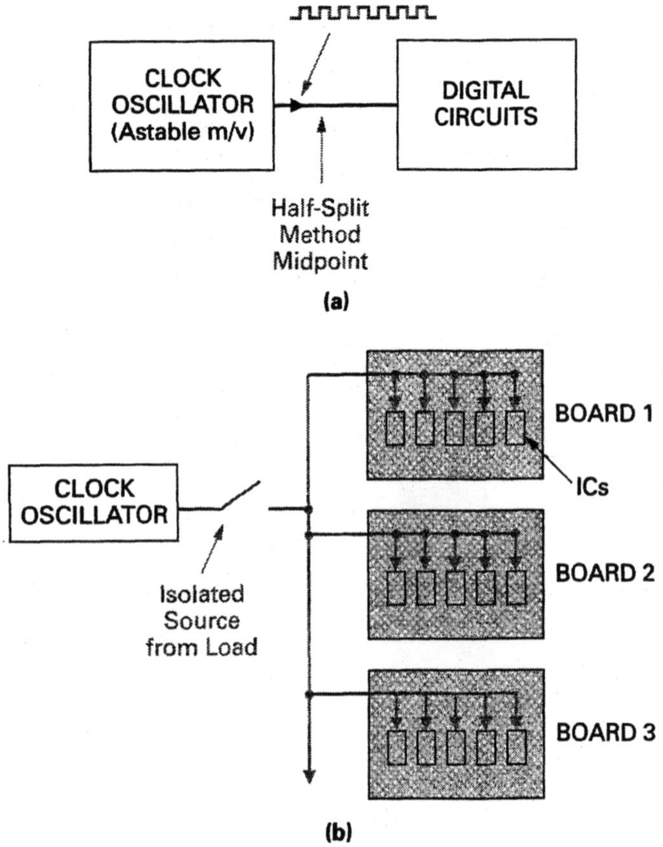

FIGURE 5-15 Clock Signal Problems.

circuits operate in sync, or at their correct time. The presence or absence of this clock signal can be checked with a logic probe; however, in most cases it is best to check the quality of the signal with an oscilloscope. Even small variations in frequency, pulse width, amplitude, and other characteristics can cause many problems throughout the digital system.

To isolate a clock signal problem, first apply the half-split method to determine whether the fault is in the oscillator circuit (source), or in one of the digital circuits (load), as shown in Figure 5-15(a). If the clock signal from the oscillator is incorrect, or if no signal exists, the problem will generally be in the oscillator clock circuit; however, like the dc supply voltage, the clock signal is distributed throughout the digital system, and therefore you may have to disconnect the clock signal output from the oscillator to determine whether a short is pulling down the signal, or whether the oscillator clock circuit is not generating the proper signal, as shown in Figure 5-15(b). If there is a short somewhere in one of the digital circuits, the current tracer can be used to find the shorted path, whereas an open in a clock signal line can easily be traced with a logic probe.

SELF-TEST REVIEW QUESTIONS FOR SECTION 5-2

1. How are most internal IC failures repaired?
2. An ____________ is easier to isolate and will generally not affect other components while a ____________ is harder to isolate and can often damage other components (short or open).
3. What is the half-split method?
4. What is a clock signal?
5. The ____________ and ____________ are ideal for tracing shorts, and the ____________ is ideal for tracing opens.

5-3 CIRCUIT REPAIR

Isolating the faulty component will always take a lot longer than repairing the circuit once the problem has been found. Some repairs are simple, such as replacing a cable, removing a solder bridge from a PCB, reseating a connector, or adjusting a variable resistor. In most instances, however, the repair will involve replacing a component such as an IC, transistor, diode, resistor, or capacitor. This will mean that you will need to use the soldering iron to remove the component from the PCB. When soldering and desoldering components to and from a PCB, follow the techniques discussed in your dc/ac electronics course, and also remember the following:

a. Always make a note of the component's orientation in the PCB since certain components such as ICs, diodes, transistors, and electrolytic capacitors have to be correctly oriented.

b. When desoldering components, use either a vacuum bulb or solder wick to remove the solder.

c. Be extremely careful not to overheat the PCB or the component being soldered or desoldered.

d. Avoid using too much or too little solder because this may cause additional problems.

e. Use a grounded soldering iron and a grounding strap when working with MOS devices to prevent any damage to the IC from static discharge.

Once the equipment has been repaired, always "final test" the equipment to see if it is now fully operational. Check that the system operates correctly in all aspects, especially in the area that was previously malfunctioning, and then reassemble the system.

SELF-TEST REVIEW QUESTIONS FOR SECTION 5-3

1. List the steps you would follow to remove an IC from a PCB.
2. What is a "final test"?

REVIEW QUESTIONS

Multiple-Choice Questions

1. Which test instrument would most accurately measure logic voltage levels?

   a. Logic probe b. Multimeter c. Oscilloscope d. Logic pulser

2. Which test instrument would be best suited for testing a clock signal's duty cycle?

   a. Logic probe b. Multimeter c. Oscilloscope d. Logic pulser

3. Which test instrument would be best suited for testing the HIGH and LOW logic levels at the inputs and outputs of a series of several logic gates?

   a. Logic probe b. Multimeter c. Oscilloscope d. Logic pulser

4. Which test instrument would be best suited for generating a test signal?

   a. Logic probe b. Multimeter c. Oscilloscope d. Logic pulser

5. The voltmeter can be used to make ____________ tests, while the oscilloscope can be used to make ____________ tests.

   a. Dynamic, static b. Static, dynamic

6. The logic pulser

   a. has an insulated tip.
   b. senses current pulses at different points in a circuit.
   c. is a signal generator.
   d. is a signal sensing instrument.

7. Which digital test instrument can be used to sense the logic levels at different points in a circuit?

   a. Multimeter **b.** Logic probe **c.** Oscilloscope **d.** Current tracer

8. The current tracer is ideal for locating what type of circuit faults?

   a. Shorts
   b. Opens
   c. Invalid logic levels
   d. Rise/fall signal problems

9. A scratch in a PCB track will more than likely cause a/an __________.

   a. short **b.** open **c.** no problem **d.** any of the above

10. Which of the following is most likely to cause a short?

    a. Broken IC pin
    b. Low current
    c. Solder bridge
    d. IC pin not in an IC socket

Essay Questions

11. List some of the tests you would typically use the multimeter for in digital electronic systems. (5-1-1)
12. Define the following, as they relate to two-state waveforms. (5-1-2)

    a. Rise time **b.** Fall time **c.** Pulse width
    d. Duty cycle **e.** PRT **f.** PRF

13. What is the difference between a pulse waveform and a square waveform? (5-1-2)
14. What is a logic clip, and how is it used to test digital ICs? (5-1-3)
15. What is a logic probe? (5-1-4)
16. How can the logic probe be used for both static and dynamic testing? (5-1-4)
17. If the multimeter and oscilloscope will measure digital signals more accurately than a logic probe, why would anyone test digital circuits with a logic probe? (5-1-4)
18. How are the logic probe and logic pulser used together to test digital circuits? (5-1-5)
19. How are the logic pulser and current tracer used together to test digital circuits? (5-1-6)
20. List some of the typical internal IC failures. (5-2-1)
21. Why is it necessary to isolate a source from a load in order to localize a fault? (5-2-2)
22. List some of the typical external IC shorts and opens. (5-2-2)
23. If a digital circuit seems to have some timing problems, with one gate being enabled too early and another being enabled too late, what would you check? (5-2-2)
24. How are opens and shorts in power lines isolated? (5-2-2)
25. List some of the precautions you should observe when desoldering an IC from a PCB. (5-3)

Practice Problems

26. Identify the waveform shown in Figure 5-16.

27. What test instrument would you use to check the rise and fall time of the waveform shown in Figure 5-16?

28. Referring to Figure 5-16, calculate the following characteristics for the waveform:

a. Pulse width **b.** Period **c.** Frequency **d.** Duty cycle

29. Calculate the PRF and PRT of the waveform shown in Figure 5-16.

30. Could a logic probe be used to identify whether the waveform in Figure 5-16 was a square or rectangular wave?

Troubleshooting Questions

31. Why is it important to be able to recognize different types of internal IC failures if the ICs themselves cannot be repaired?

32. In Figure 5-17(a), a logic probe has been used to check the circuit since it is not functioning as it should. In Figure 5-17(b), you can see what logic levels were obtained when the output of the NAND gate was isolated. Which IC should be replaced?

33. What do you suspect is wrong with the faulty IC in Figure 5-17?

34. In Figure 5-18, an oscilloscope has been used to test several points on the digital circuit. Which IC should be replaced, and what do you think is the problem?

35. Are the logic gates in Figure 5-19 operating as they should?

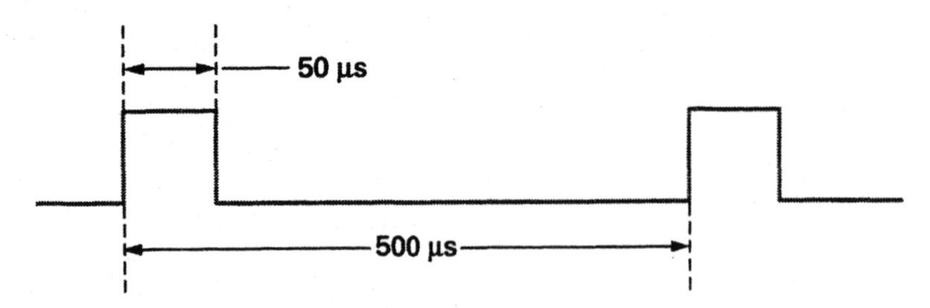

FIGURE 5-16 A Repeating or Pulse Waveform.

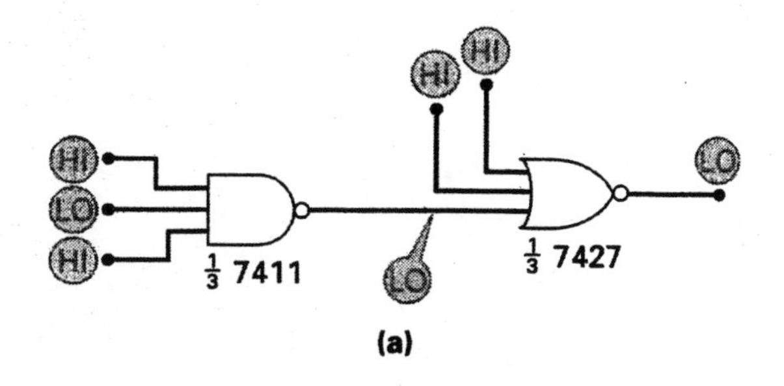

(a)

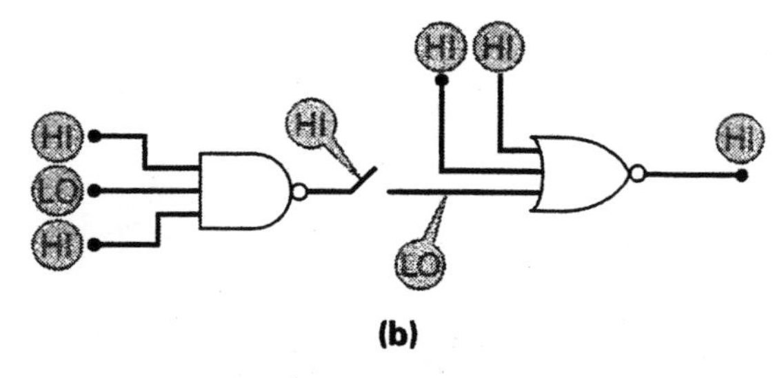

(b)

FIGURE 5-17 Troubleshooting Exercise 1.

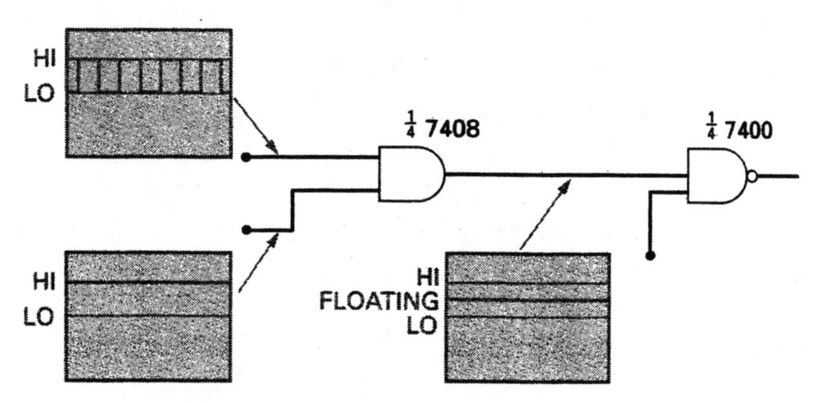

FIGURE 5-18 Troubleshooting Exercise 2.

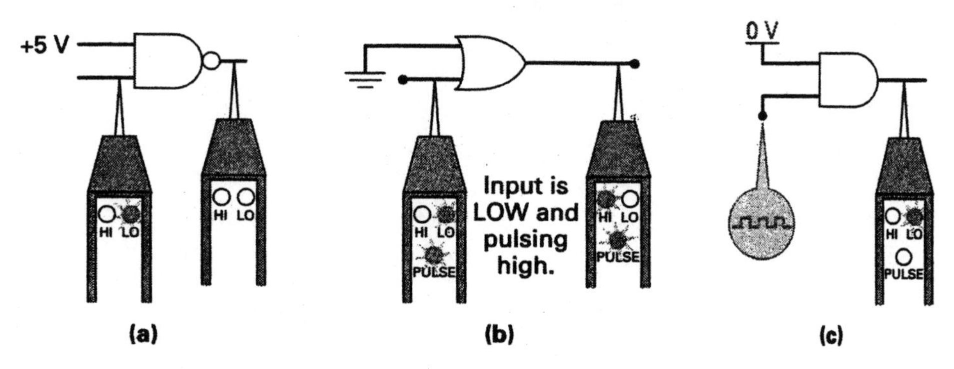

(a) (b) (c)

FIGURE 5-19 Troubleshooting Exercise 3.

Chapter 6

OBJECTIVES

After completing this chapter, you will be able to:

1. Give the Boolean expressions for each of the seven basic logic gates.
2. Describe how to write an equivalent Boolean expression for a logic circuit and how to sketch an equivalent logic circuit for a Boolean expression.
3. List and explain the laws and rules that apply to Boolean algebra.
4. Describe the logic circuit design process from truth table to gate circuit.
5. Explain the difference between a sum-of-products equation and a product-of-sums equation.
6. Describe how Boolean algebra and Karnaugh maps can be used to simplify logic circuits.

Logic Gate Circuit Simplification

OUTLINE

VIGNETTE

From Folly to Foresight

George Boole was born in the industrial town of Lincoln in eastern England in 1815. His parents were poor trade people, and even though there was a school for boys in Lincoln, there is no record of him ever attending. In those hard times, children of the working class had no hope of receiving any form of education, and their lives generally followed the familiar pattern of their parents. George Boole, however, was to break the mold. He would rise up from these humble beginnings to become one of the most respected mathematicians of his day.

Boole's father had taught himself a small amount of mathematics, and since his son of six seemed to have a thirst for learning, he began to pass on all of his knowledge. At eight years old, George had surpassed his father's understanding and craved more. He quickly realized that his advancement was heavily dependent on his understanding Latin. Luckily, a family friend who owned a local book shop knew enough about the basics of Latin to get him on his way, and once he had taught Boole all he knew, Boole continued with the books at his disposal. By the age of twelve he had conquered Latin, and by fourteen he had added Greek, French, German, and Italian to his repertoire.

At the age of sixteen, however, poverty stood in his way. Since his parents could no longer support him, he was forced to take a job as a poorly paid teaching assistant. After studying the entire school system, he left four year laters and opened his own school in which he taught all subjects. It was in this role that he discovered that his mathematics was weak, so he began studying the mathematical journals at the local library in an attempt to stay ahead of his students. He quickly discovered that he had a talent for mathematics. As well as mastering all the present-day ideas, he began to develop some of his own, which were later accepted for publication. After a stream of articles, he became so highly regarded that he was asked to join the mathematics faculty at Queens College in 1849.

After accepting the position, Boole concentrated more on his ideas, one of which was to develop a system of symbolic logic. In this system he created a form of algebra which had its own set of symbols and rules. Using this system, Boole could encode any statement that had to be proven (a proposition) into his symbolic language, and then manipulate it to determine whether it was true or false. Boole's algebra had three basic operations that are often called logic functions—AND, OR, and NOT. Using these three operations, Boole could perform such operations as add, subtract, multiply, divide, and compare. These logic functions were binary in nature, and therefore dealt with only two entities—TRUE or FALSE, YES or NO, OPEN or CLOSED, ZERO or ONE, and so on. Boole's theory was that if all logical arguments could be reduced to one of two basic levels, the questionable middle ground would be removed, making it easier to arrive at a valid conclusion.

At the time, Boole's system, which was later called "Boolean algebra," was either ignored or criticized by colleagues who called it a folly with no practical purpose. Almost a century later, however, scientists would combine George Boole's Boolean algebra with binary numbers and make possible the digital electronic computer.

INTRODUCTION

In 1854 George Boole invented a symbolic logic which linked mathematics and logic. Boole's logical algebra, which today is known as **Boolean algebra,** states that each variable (input or output) can assume one of two values or states—true or false.

Boolean Algebra
A form of algebra invented by George Boole that deals with classes, propositions, and ON/OFF circuit elements associated with operators such as NOT, OR, NOR, AND, NAND, XOR, and XNOR.

Up until 1938 Boolean algebra had no practical application until Claude Shannon used it to analyze telephone switching circuits in his MIT thesis. In his paper he described how the two variables of the Boolean algebra (true and false) could be used to represent the two states of the switching relay (open and closed).

Today the mechanical relays have been replaced by semiconductor switches. However, Boolean algebra is still used to express both simple and complex two-state logic functions in a convenient mathematical format. These mathematical expressions of logic functions make it easier for technicians to analyze digital circuits, and are a primary design tool for engineers. By using Boolean algebra, circuits can be made simpler, less expensive, and more efficient.

In this chapter we will discuss the basics of Boolean algebra and then see how it can be applied to logic gate circuit simplification.

6-1 BOOLEAN EXPRESSIONS FOR LOGIC GATES

The operation of each of the seven basic logic gates can be described with a Boolean expression. To explain this in more detail, let us first consider the most basic of all the logic gates, the NOT gate.

6-1-1 *The NOT Expression*

In Figure 6-1 you can see the previously discussed NOT gate, and because of the inversion that occurs between input and output, Y is always equal to the opposite, or complement, of input A.

FIGURE 6-1 The Boolean Expression for a NOT Gate.

$$Y = \text{NOT } A$$

Therefore, if the input were 0, the output would be 1.

$$Y = \text{NOT } 0 = 1$$

If, on the other hand, the A input were 1, the output would be 0.

$$Y = \text{NOT } 1 = 0$$

In Boolean algebra, this inversion of the A input is indicated with a bar over the letter A as follows:

$Y = \overline{A}$ (pronounced "Y equals not A")

Using this Boolean expression, we can easily calculate the output Y for either of the two input A conditions. For example, if $A = 0$,

$$Y = \overline{A} = \overline{0} = 1$$

If, on the other hand, the A input is 1,

$$Y = \overline{A} = \overline{1} = 0$$

6-1-2 *The OR Expression*

The operation of the OR gate can also be described with a Boolean expression, as seen in Figure 6-2(a). In Boolean algebra, the "+" sign is used to indicate the OR operation.

$Y = A + B$ (pronounced "Y equals A or B")

The expression $Y = A + B$ (Boolean equation), is the same as $Y = A$ OR B (word equation). If you find yourself wanting to say "plus" instead of "OR,"

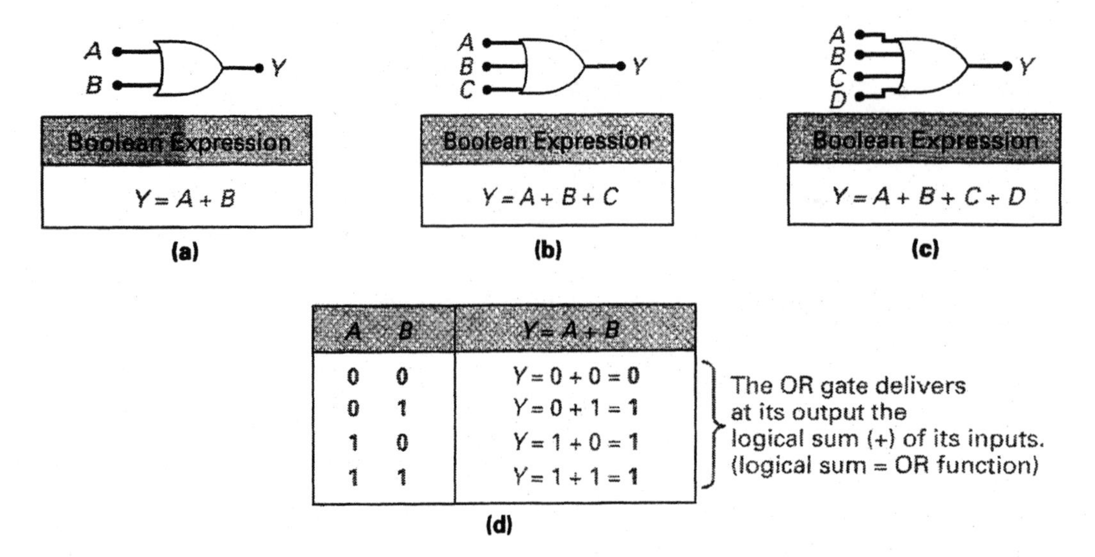

FIGURE 6-2 The Boolean Expression for an OR Gate.

don't worry: it will require some practice to get used to this traditional symbol
application. Just as the word "wind" can be used in two
y wind" or "to wind a clock"), then so can the "+"
either addition or the OR operation. With practice you
with this symbol's dual role.
re than two inputs, as seen in Figure 6-2(b) or (c),
extended, as follows.

put OR gate: $Y = A + B + C$

put OR gate: $Y = A + B + C + D$

wo-input OR gate has been repeated in Figure
pression for this gate, we can determine the output
possible 1 and 0 input combinations for the input
this truth table you can see that the output of the
er A OR B is 1. George Boole described the OR
n, which is why the "plus" (addition) symbol was
logical addition, is different from normal addition,
roduce at its output the *logical sum* of the inputs.

ression

D operation is indicated with a "·", and therefore
the AND gate shown in Figure 6-3(a), would be

B (pronounced "Y equals A and B")

ndicate multiplication, and therefore like the "+" sym-
two meanings. In multiplication, the "·" is often
aning is still known to be multiplication. For example:

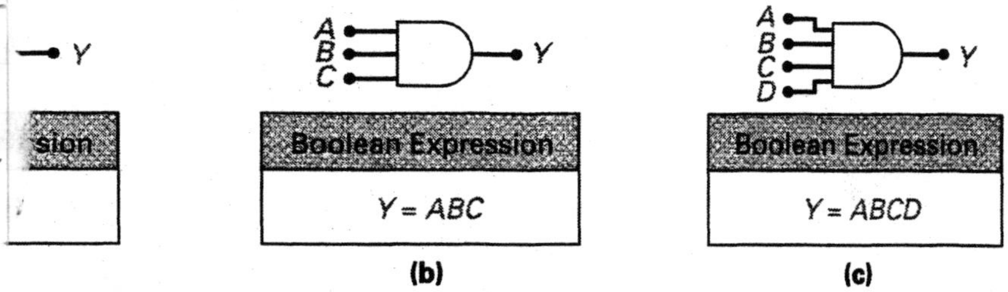

Y

sion

A B C Y

Boolean Expression
$Y = ABC$

(b)

A B C D Y

Boolean Expression
$Y = ABCD$

(c)

A	B	$Y = A \cdot B$
0	0	$Y = 0 \cdot 0 = 0$
0	1	$Y = 0 \cdot 1 = 0$
1	0	$Y = 1 \cdot 0 = 0$
1	1	$Y = 1 \cdot 1 = 1$

The AND gate delivers at its output the logical product (•) of its inputs. (logical product = AND function)

(d)

FIGURE 6-3 The Boolean Expression for an AND Gate.

$$V = I \times R \text{ is the same as } V = I \cdot R \text{ or } V = IR$$

In Boolean algebra the same situation occurs, in that

$$Y = A \cdot B \text{ is the same as } Y = AB$$

If AND gates have more than two inputs, as seen in Figure 6-3(b) and (c), the AND expression is simply extended.

The truth table for a two-input AND gate is shown in Figure 6-3(d). Using the Boolean expression, we can calculate the output Y by substituting all of the possible 1 and 0 input combinations for the input variables A and B. Looking at the truth table, you can see that the output of the AND gate is a 1 only when both A and B are 1. George Boole described the AND function as *Boolean multiplication* or *logical multiplication,* which is why the period (multiplication) symbol was used. Boolean multiplication is different from standard multiplication as you have seen, and the AND gate is said to produce at its output the *logical product* of its inputs.

■ **EXAMPLE:**

What would be the Boolean expression for the circuit shown in Figure 6-4, and what would be the output if $A = 0$ and $B = 1$?

■ ***Solution:***

The A input in Figure 6-4 is first inverted before it is ANDed with the B input. The Boolean equation would be, therefore

$$Y = \overline{A} \cdot B$$

Using this expression, we can substitute the A and B for the 0 and 1 inputs, and then determine the output Y.

$$Y = \overline{A} \cdot B = \overline{0} \cdot 1 = 1 \cdot 1 = 1$$

(Y equals NOT A and B = NOT 0 AND 1 = 1 AND 1 = 1)

■ **EXAMPLE:**

Give the Boolean expression and truth table for the logic circuit shown in Figure 6-5(a).

■ ***Solution:***

To begin with, there are three inputs or variables to this circuit: A, B, and C. Inputs A and B are ANDed together ($A \cdot B$) and the result is then ORed with input C ($+ C$). The resulting equation will be

$$Y = (A \cdot B) + C \qquad \text{or} \qquad Y = AB + C$$

As you can see in the truth table in Figure 6-5(b), the Y output is HIGH when A AND B are HIGH, OR if C is HIGH.

In the two previous examples, we have gone from a logic circuit to a Boolean equation. Now let us reverse the procedure and see how easy it is to go from a Boolean equation to a logic circuit.

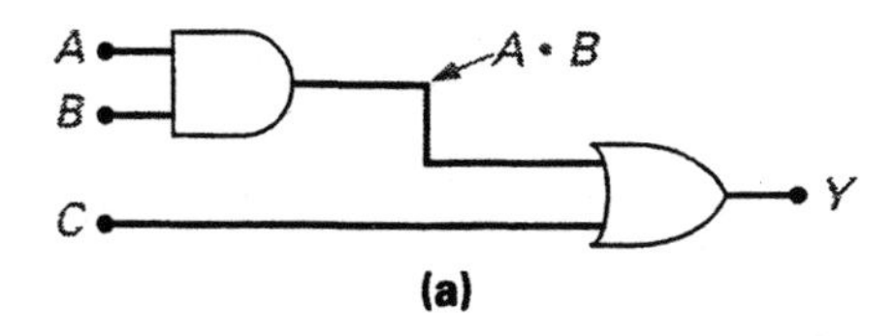

(a)

A	B	C	Y = AB + C
0	0	0	$Y = (0 \cdot 0) + 0 = 0 + 0 = 0$
0	0	1	$Y = (0 \cdot 0) + 1 = 0 + 1 = 1$
0	1	0	$Y = (0 \cdot 1) + 0 = 0 + 0 = 0$
0	1	1	$Y = (0 \cdot 1) + 1 = 0 + 1 = 1$
1	0	0	$Y = (1 \cdot 0) + 0 = 0 + 0 = 0$
1	0	1	$Y = (1 \cdot 0) + 1 = 0 + 1 = 1$
1	1	0	$Y = (1 \cdot 1) + 0 = 1 + 0 = 1$
1	1	1	$Y = (1 \cdot 1) + 1 = 1 + 1 = 1$

(b)

FIGURE 6-5 (*A* AND *B*) OR (*C*) Logic Circuit.

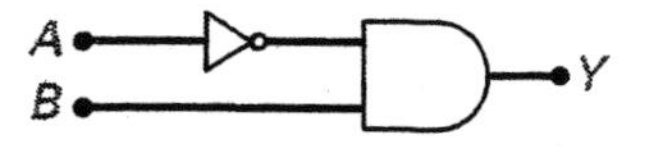

FIGURE 6-4 Single, Bubble Input AND Gate.

■ EXAMPLE:

Sketch the logic-gate circuit for the following Boolean equation.

$$Y = AB + CD$$

■ ***Solution:***

Studying the equation, we can see that there are four variables: *A, B, C,* and *D*. Input *A* is ANDed with input *B*, giving $A \cdot B$ or *AB*, as seen in Figure 6-6(a). Input *C* is ANDed with input *D*, giving *CD*, as seen in Figure 6-6(b). The ANDed *AB* output and the ANDed *CD* output are then ORed to produce a final output *Y*, as seen in Figure 6-6(c).

$Y = \overline{A + B}$ (pronounced "Y equals not A or B")

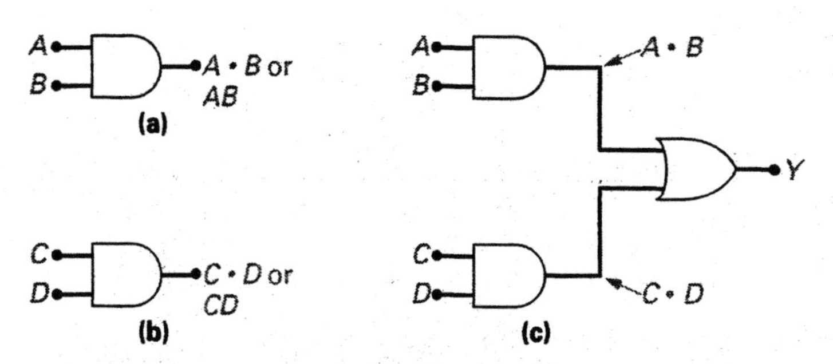

FIGURE 6-6 (*A* AND *B*) OR (*C* AND *D*) Logic Circuit.

6-1-4 *The NOR Expression*

The Boolean expression for the two-input NOR gate shown in Figure 6-7(a) will be

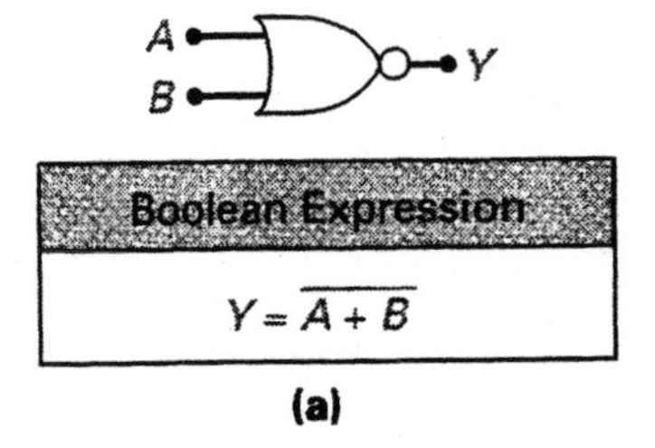

A	B	$Y = \overline{A + B}$
0	0	$Y = \overline{0 + 0} = \overline{0} = 1$
0	1	$Y = \overline{0 + 1} = \overline{1} = 0$
1	0	$Y = \overline{1 + 0} = \overline{1} = 0$
1	1	$Y = \overline{1 + 1} = \overline{1} = 0$

(b)

FIGURE 6-7 The Boolean Expression for a NOR Gate.

To explain why this expression describes a NOR gate, let us examine each part in detail. In this equation the two input variables A and B are first ORed (indicated by the $A + B$ part of the equation), and then the result is complemented, or inverted (indicated by the bar over the whole OR output expression).

The truth table in Figure 6-7(b) tests the Boolean expression for a NOR gate for all possible input combinations.

While studying Boolean algebra, Augustus DeMorgan discovered two important theorems. The first theorem stated that a bubbled input AND gate was logically equivalent to a NOR gate. These two logic gates are shown in Figure 6-8(a) and (b), respectively, with their associated truth tables. Comparing the two truth tables you can see how any combination of inputs to either gate circuit will result in the same output, making the two circuits interchangeable.

■ **EXAMPLE:**

Give the two Boolean equations for the two logic-gate circuits shown in Figure 6-8(a) and (b).

■ ***Solution:***

With the bubbled input AND gate shown in Figure 6-8(a), you can see that input A is first inverted by a NOT gate, giving $\overline{A}$. Input B is also complemented by a NOT gate, giving $\overline{B}$. These two complemented inputs are then ANDed, giving the following final function.

$$Y = \overline{A} \cdot \overline{B}$$

With the NOR gate shown in Figure 6-8(b), inputs A and B are first ORed and then the result is complemented, giving

$$Y = \overline{A + B}$$

DeMorgan's first theorem, therefore, is as follows:

DeMorgan's First Theorem: $\overline{A} \cdot \overline{B} = \overline{A + B}$

Studying the differences between the left equation and the right equation, you can see that there are actually two basic changes: the line is either broken or

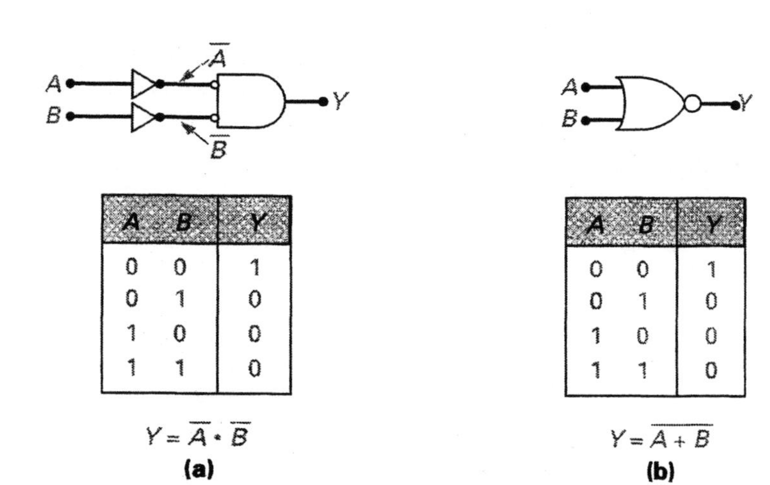

A	B	Y
0	0	1
0	1	0
1	0	0
1	1	0

A	B	Y
0	0	1
0	1	0
1	0	0
1	1	0

FIGURE 6-8 DeMorgan's First Theorem.

solid, and the logical sign is different. Therefore, if we wanted to convert the left equation into the right equation, all we would have to do is follow this simple procedure:

$$Y = \overline{A} \cdot \overline{B} \longrightarrow Y = \overline{A + B}$$

"Mend the line, and change the sign."

To convert the right equation into the left equation, all we would have to do is follow this simple procedure:

"Break the line, and change the sign."

$$Y = \overline{A} \cdot \overline{B} \longleftarrow Y = \overline{A + B}$$

This ability to interchange an AND function with an OR function, and to interchange a bubbled input with a bubbled output, will come in handy when we are trying to simplify logic circuits, as you will see in this chapter. For now, remember the rules on how to "DeMorganize" an equation.

Mend the line, and change the sign.
Break the line, and change the sign.

6-1-5 The NAND Expression

The Boolean expression for the two-input NAND gate shown in Figure 6-9(a) will be

$Y = \overline{A \cdot B}$ or $Y = \overline{AB}$ (pronounced "Y equals not A and B")

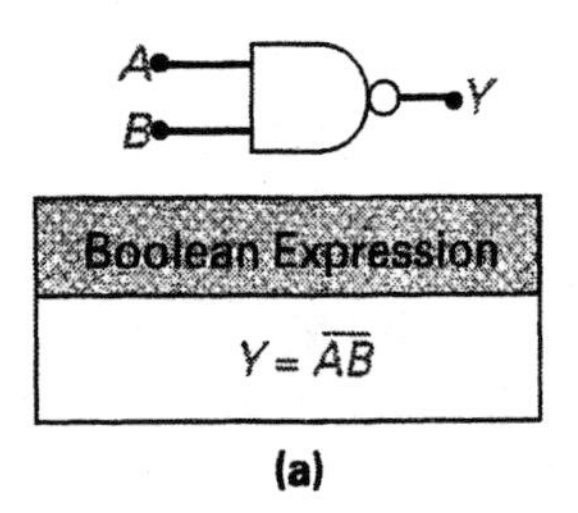

(a)

A	B	$Y = \overline{A \cdot B}$
0	0	$Y = \overline{0 \cdot 0} = \overline{0} = 1$
0	1	$Y = \overline{0 \cdot 1} = \overline{0} = 1$
1	0	$Y = \overline{1 \cdot 0} = \overline{0} = 1$
1	1	$Y = \overline{1 \cdot 1} = \overline{1} = 0$

(b)

FIGURE 6-9 The Boolean Expression for a NAND Gate.

This expression states that the two inputs A and B are first ANDed (indicated by the $A \cdot B$ part of the equation), and then the result is complemented (indicated by the bar over the AND expression).

The truth table shown in Figure 6-9(b) tests this Boolean equation for all possible input combinations.

■ **EXAMPLE:**

DeMorganize the following equation, and then sketch the logically equivalent circuits with their truth tables and Boolean expressions.

$$Y = \overline{A} + \overline{B}$$

■ ***Solution:***

To apply DeMorgan's theorem to the equation $Y = \overline{A} + \overline{B}$, we simply apply the rule "mend the line, and change the sign," as follows:

$$\overline{A} + \overline{B} = \overline{A \cdot B}$$

The circuits for these two equations can be seen in Figure 6-10(a) and (b), and if you study the truth tables you can see that the outputs of both logic gates are identical for all input combinations.

This observation—that a bubbled input OR gate is interchangeable with a bubbled output AND gate, or NAND gate—was first made by DeMorgan and is referred to as DeMorgan's second theorem.

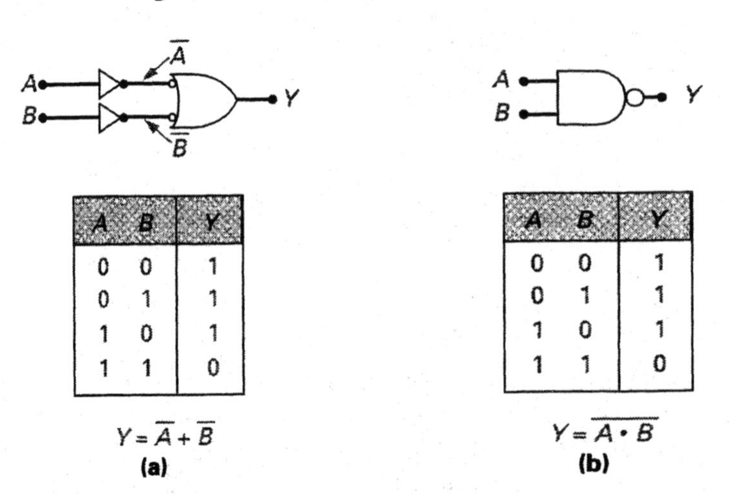

A	B	Y
0	0	1
0	1	1
1	0	1
1	1	0

$Y = \overline{A} + \overline{B}$

(a)

A	B	Y
0	0	1
0	1	1
1	0	1
1	1	0

$Y = \overline{A \cdot B}$

(b)

FIGURE 6-10 DeMorgan's Second Theorem.

DeMorgan's Second Theorem: $\overline{A} + \overline{B} = \overline{A \cdot B}$

6-1-6 *The XOR Expression*

In Boolean algebra, the "⊕" symbol is used to describe the exclusive-OR action. This means that the Boolean expression for the two-input XOR gate shown in Figure 6-11(a) will be

$Y = A \oplus B$ (pronounced "Y equals A exclusive-OR B")

The truth table in Figure 6-11(b) shows how this Boolean expression is applied to all possible input combinations.

■ **EXAMPLE:**

Give the Boolean expression and truth table for the circuit shown in Figure 6-12(a).

■ ***Solution:***

Looking at the circuit in Figure 6-12(a), you can see that the upper AND gate has an inverted A input ($\overline{A}$) and a B input, giving a result of $\overline{A} \cdot B$. This $\overline{A} \cdot B$

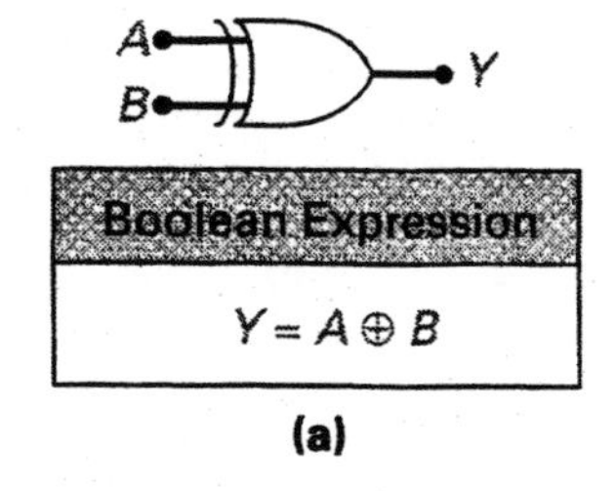

A	B	$Y = A \oplus B$
0	0	$Y = 0 \oplus 0 = 0$
0	1	$Y = 0 \oplus 1 = 1$
1	0	$Y = 1 \oplus 0 = 1$
1	1	$Y = 1 \oplus 1 = 0$

(b)

FIGURE 6-11 The Boolean Expression for an XOR Gate.

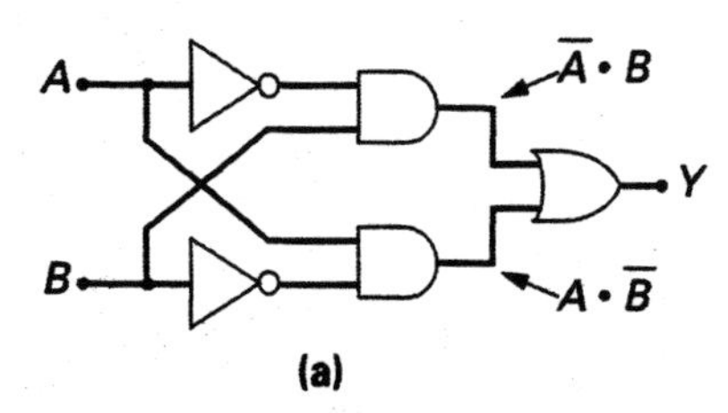

(a)

A	B	$Y = \overline{A}B + A\overline{B}$
0	0	$Y = (\overline{0} \cdot 0) + (0 \cdot \overline{0}) = (1 \cdot 0) + (0 \cdot 1) = 0 + 0 = 0$
0	1	$Y = (\overline{0} \cdot 1) + (0 \cdot \overline{1}) = (1 \cdot 1) + (0 \cdot 0) = 1 + 0 = 1$
1	0	$Y = (\overline{1} \cdot 0) + (1 \cdot \overline{0}) = (0 \cdot 0) + (1 \cdot 1) = 0 + 1 = 1$
1	1	$Y = (\overline{1} \cdot 1) + (1 \cdot \overline{1}) = (0 \cdot 1) + (1 \cdot 0) = 0 + 0 = 0$

(b)

FIGURE 6-12 (NOT *A* AND *B*) OR (*A* AND NOT *B*).

output is then ORed with the input from the lower AND gate, which has an inverted B input and A input giving a result of $A \cdot \overline{B}$. The final equation will therefore be

$$Y = (\overline{A} \cdot B) + (A \cdot \overline{B}) \quad \text{or} \quad Y = \overline{A}\,B + A\,\overline{B}$$

The truth table for this logic circuit that combines five logic gates is shown in Figure 6-12(b). Looking at the input combinations and the output, you can see that the output is only 1 when an odd number of 1s appears at the input. This circuit is therefore acting as an XOR gate, and so

$$\overline{A}\,B + A\,\overline{B} = A \oplus B$$

6-1-7 The XNOR Expression

The Boolean expression for the two-input XNOR gate shown in Figure 6-13(a) will be

$$Y = \overline{A \oplus B} \text{ (pronounced “Y equals not A exclusive-or B”)}$$

From this equation, we can see that the two inputs A and B are first XORed and then the result is complemented.

The truth table in Figure 6-13(b) tests this Boolean expression for all possible input combinations.

■ **EXAMPLE:**

Give the Boolean expression and truth table for the circuit shown in Figure 6-14(a).

■ ***Solution:***

Studying the circuit in Figure 6-14(a), you can see that the upper AND gate has an inverted A input and an inverted B input, giving $\overline{A} \cdot \overline{B}$ at its output. The lower AND gate simply ANDs the A and B input, giving $A \cdot B$ at its output. The two AND gate outputs are then ORed, so the final equation for this circuit will be

$$Y = \overline{A} \cdot \overline{B} + A \cdot B \quad \text{or} \quad Y = \overline{A}\,\overline{B} + A\,B$$

The truth table for this circuit is shown in Figure 6-14(b). Comparing this truth table to an XNOR gate's truth table, you can see that the two are equivalent since a 1 output is only present when an even number of 1s are applied to the input; therefore

$$\overline{A}\,\overline{B} + A\,B = \overline{A \oplus B}$$

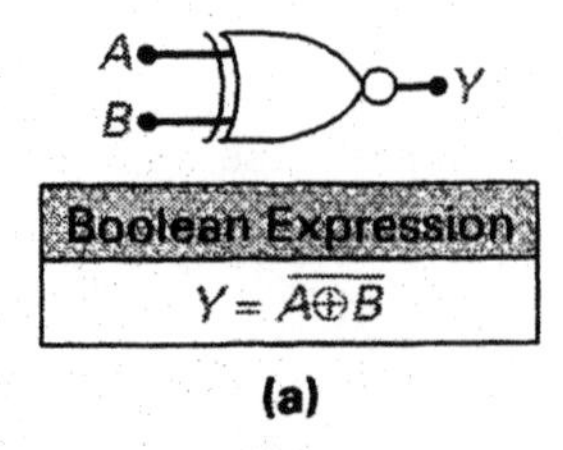

(a)

A	B	$Y = \overline{A \oplus B}$
0	0	$Y = \overline{0 \oplus 0} = \overline{0} = 1$
0	1	$Y = \overline{0 \oplus 1} = \overline{1} = 0$
1	0	$Y = \overline{1 \oplus 0} = \overline{1} = 0$
1	1	$Y = \overline{1 \oplus 1} = \overline{0} = 1$

(b)

FIGURE 6-13 **The Boolean Expression for an XNOR Gate.**

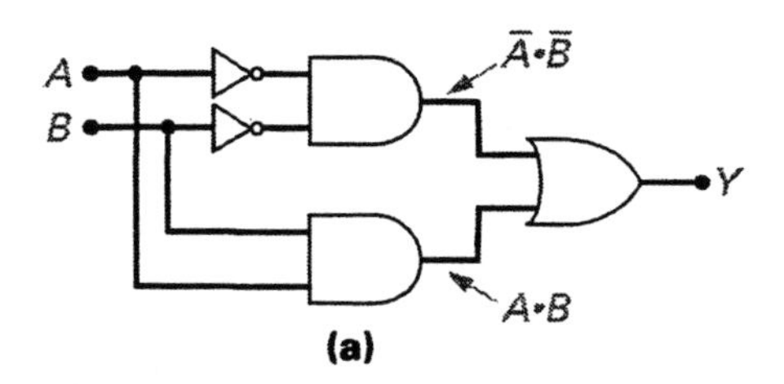

(a)

A	B	$Y = \overline{A}\,\overline{B} + AB$
0	0	$Y = (\overline{0}\cdot\overline{0}) + (0\cdot 0) = (1\cdot 1) + (0\cdot 0) = 1 + 0 = 1$
0	1	$Y = (\overline{0}\cdot\overline{1}) + (0\cdot 1) = (1\cdot 0) + (0\cdot 1) = 0 + 0 = 0$
1	0	$Y = (\overline{1}\cdot\overline{0}) + (1\cdot 0) = (0\cdot 1) + (1\cdot 0) = 0 + 0 = 0$
1	1	$Y = (\overline{1}\cdot\overline{1}) + (1\cdot 1) = (0\cdot 0) + (1\cdot 1) = 0 + 1 = 1$

(b)

FIGURE 6-14 (NOT *A* AND NOT *B*) OR (*A* AND *B*).

SELF-TEST REVIEW QUESTIONS FOR SECTION 6-1

1. Which logic gate would be needed to perform each of the following Boolean expressions?
   a. $A\,B$ **b.** $A \oplus B$ **c.** $N + M$ **d.** $\overline{X \cdot Y}$ **e.** $\overline{S + T}$ **f.** $\overline{GH}$
2. State DeMorgan's first theorem using Boolean expressions.
3. Apply DeMorgan's theorem to the following:
   a. $\overline{A} + \overline{B}$ **b.** $Y = \overline{(A + B) \cdot C}$
4. Sketch the logic gates or circuits for the following expressions.
   a. ABC **b.** $\overline{L \oplus D}$ **c.** $\overline{A + B}$ **d.** $(\overline{A} \cdot \overline{B}) + \overline{C}$ **e.** $A \cdot (\overline{L \oplus D})$
   f. $\overline{(A + B) \cdot C}$

6-2 BOOLEAN ALGEBRA LAWS AND RULES

In this section we will discuss some of the rules and laws that apply to Boolean algebra. Many of these rules and laws are the same as ordinary algebra and, as you will see in this section, are quite obvious.

6-2-1 The Commutative Law

Commutative Law
Combining elements in such a manner that the result is independent of the order in which the elements are taken.

The word "commutative" is defined as "combining elements in such a manner that the result is independent of the order in which the elements are taken." This means that the order in which the inputs to a logic gate are ORed or ANDed for example, is not important since the result will be the same, as shown in Figure 6-15.

In Figure 6-15(a), you can see that ORing *A* and *B* will achieve the same result as reversing the order of the inputs and ORing *B* and *A*. As stated in the

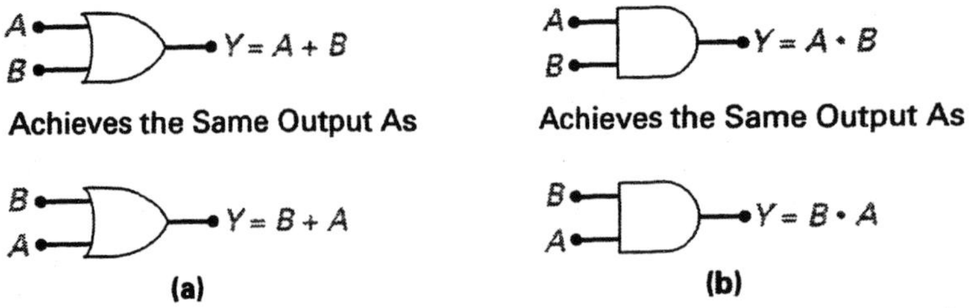

FIGURE 6-15 **The Commutative Law. (a) Logical Addition. (b) Logical Multiplication.**

previous section, the OR function is described in Boolean algebra as logical addition, so the *commutative law of addition* can be algebraically written as

Commutative Law of Addition: $A + B = B + A$

Figure 6-15(b) shows how this law relates to an AND gate, which in Boolean algebra is described as logical multiplication. The *commutative law of multiplication* can be algebraically written as

Commutative Law of Multiplication: $A \cdot B = B \cdot A$

6-2-2 *The Associative Law*

Associative Law Combining elements such that when the order of the elements is preserved, the result is independent of the grouping.

The word "associative" is defined as "combining elements such that when the order of the elements is preserved, the result is independent of the grouping." To explain this in simple terms, Figure 6-16 shows that how you group the inputs in an ORing process or ANDing process will have no effect on the output.

In Figure 6-16(a), you can see that ORing B and C and then ORing the result with A, will achieve the same result as ORing A and B and then ORing the result with C. This *associative law of addition* can be algebraically written as

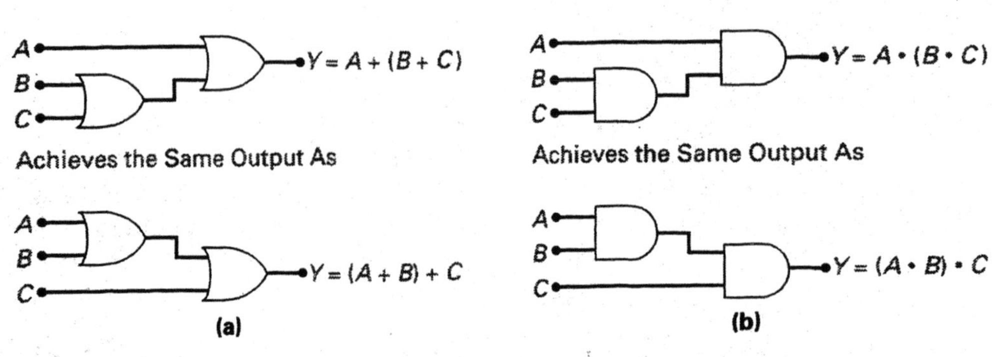

FIGURE 6-16 **The Associative Law. (a) Logical Addition. (b) Logical Multiplication.**

Associative Law of Addition: $A + (B + C) = (A + B) + C$

Figure 6-16(b) shows how this law relates to an AND gate, which in Boolean algebra is described as logical multiplication. The *associative law of multiplication* can be algebraically written as

Associative Law of Multiplication: $A \cdot (B \cdot C) = (A \cdot B) \cdot C$

6-2-3 The Distributive Law

Distributive Law Producing the same element when operating on a whole as when operating on each part, and collecting the results.

By definition, the word "distributive" means "producing the same element when operating on a whole, as when operating on each part, and collecting the results." It can be algebraically stated as

Distributive Law: $A \cdot (B + C) = (A \cdot B) + (A \cdot C)$

Figure 6-17 illustrates this law by showing that ORing two or more inputs and then ANDing the result, as shown in Figure 6-17(a), achieves the same output as ANDing the single variable (A) with each of the other inputs (B and C) and then ORing the results, as shown in Figure 6-17(b). To help reinforce this algebraic law, let us use an example involving actual values for A, B, and C.

■ **EXAMPLE:**

Prove the distributive law by inserting the values $A = 2$, $B = 3$ and $C = 4$.

■ ***Solution:***

$$
\begin{aligned}
A \cdot (B + C) &= (A \cdot B) + (A \cdot C) \\
2 \times (3 + 4) &= (2 \times 3) + (2 \times 4) \\
2 \times 7 &= 6 + 8 \\
14 &= 14
\end{aligned}
$$

In some instances, we may wish to reverse the process performed by the distributive law to extract the common factor. For example, consider the following equation:

$$Y = \overline{A}B + AB$$

Since "$\cdot B$" (AND B) seems to be a common factor in this equation, by *factoring,* we could obtain

$$
\begin{aligned}
Y &= \overline{A}B + AB && \text{Original Expression} \\
Y &= (\overline{A} + A) \cdot B && \text{Factoring AND } B
\end{aligned}
$$

6-2-4 *Boolean Algebra Rules*

Now that we have covered the three laws relating to Boolean algebra, let us now concentrate on these Boolean algebra rules, beginning with those relating to OR operations.

OR Gate Rules

The first rule can be seen in Figure 6-18(a). This shows what happens when one input to an OR gate is always 0 and the other input (A) is a variable. If

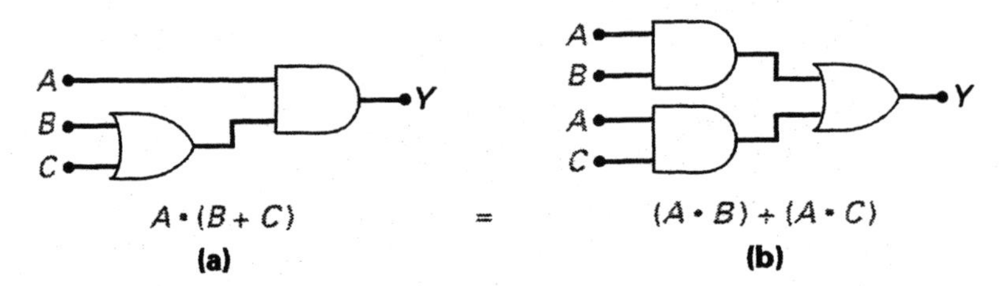

FIGURE 6-17 The Distributive Law.

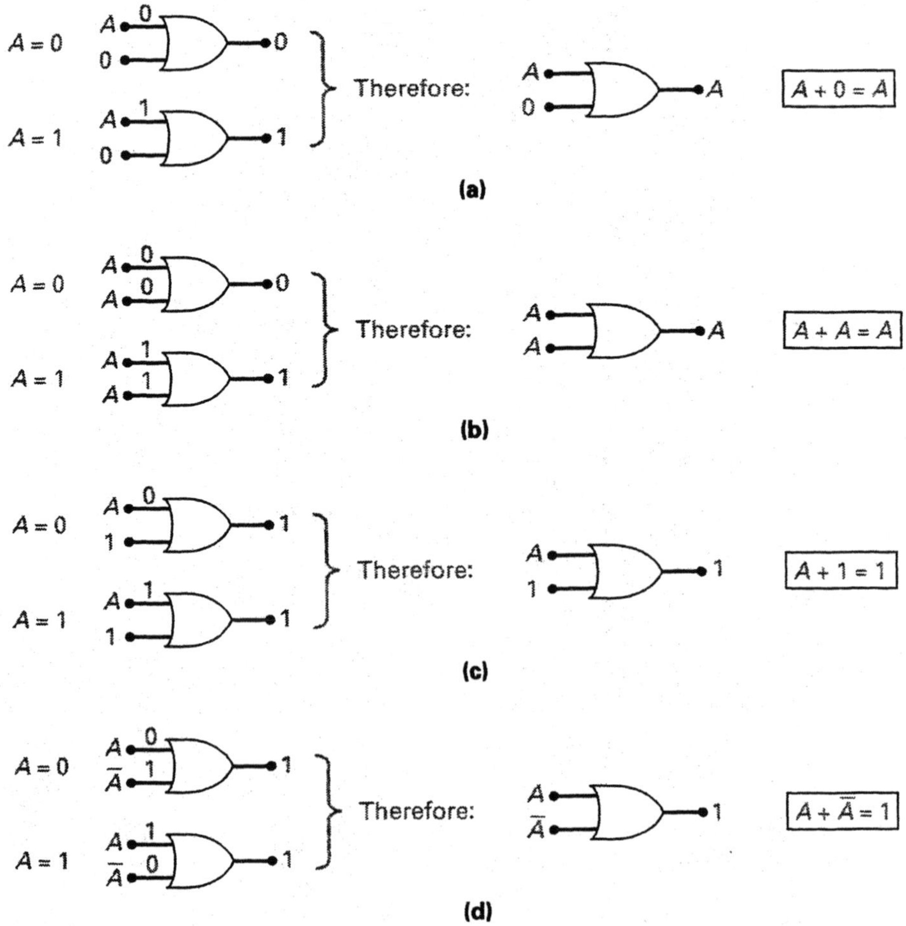

FIGURE 6-18 Boolean Rules for OR Gates.

$A = 0$, the output equals 0, and if $A = 1$, the output equals 1. Therefore, a variable input ORed with 0 will always equal the variable input. Stated algebraically:

$$A + 0 = A$$

Another OR gate Boolean rule is shown in Figure 6-18(b), and states that

$$A + A = A$$

As you can see in Figure 6-18(b), when a variable is ORed with itself, the output will always equal the logic level of the variable input.

Figure 6-18(c) shows the next OR gate Boolean rule. This rule states that any 1 input to an OR gate will result in a 1 output, regardless of the other input. Stated algebraically:

$$A + 1 = 1$$

The final Boolean rule for OR gates is shown in Figure 6-18(d). This rule states that when any variable (A) is ORed with its complement ($\overline{A}$) the result will be a 1.

$$A + \overline{A} = 1$$

AND Gate Rules

Like the OR gate, there are four Boolean rules relating to AND gates. The first is illustrated in Figure 6-19(a) and states that if a variable input A is ANDed with a 0, the output will be 0 regardless of the other input.

$$A \cdot 0 = 0$$

The second Boolean rule for AND gates is shown in Figure 6-19(b). In this illustration you can see that if a variable is ANDed with a 1, the output will equal the variable.

$$A \cdot 1 = A$$

Another AND gate Boolean rule is shown in Figure 6-19(c). In this instance, any variable that is ANDed with itself will always give an output that is equal to the variable.

$$A \cdot A = A$$

The last of the AND gate Boolean rules is shown in Figure 6-19(d). In this illustration you can see that if a variable (A) is ANDed with its complement ($\overline{A}$), the output will always equal 0.

$$A \cdot \overline{A} = 0$$

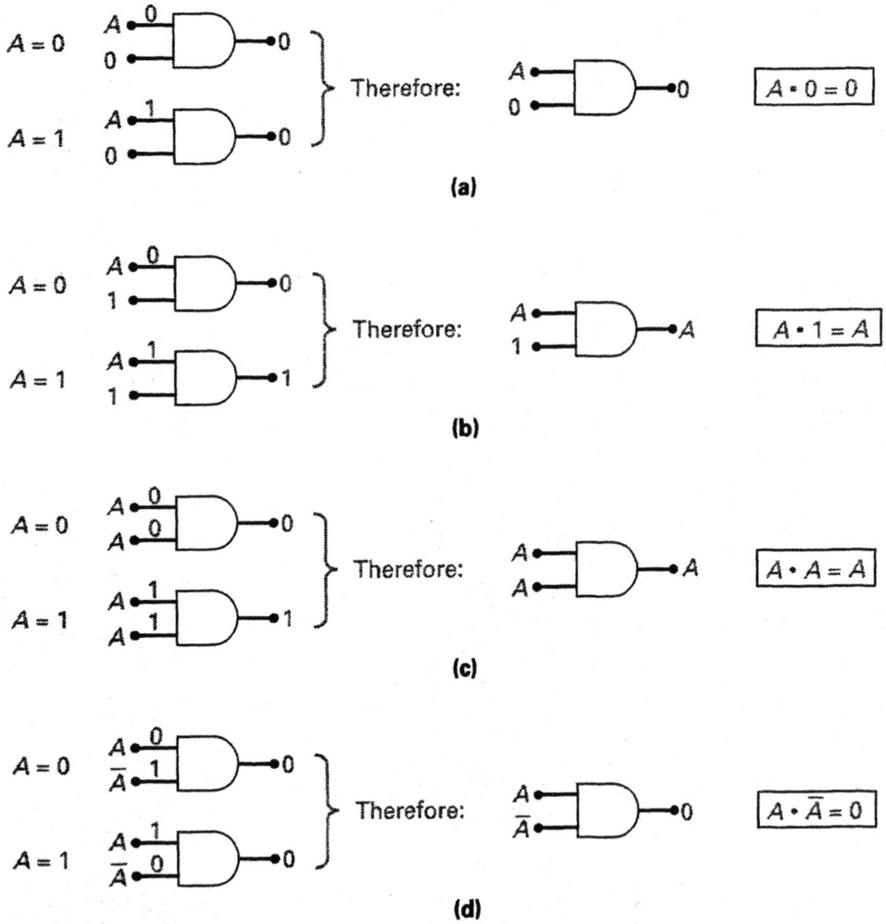

$A \cdot 0 = 0$

(a)

$A \cdot 1 = A$

(b)

$A \cdot A = A$

(c)

$A \cdot \overline{A} = 0$

(d)

FIGURE 6-19 Boolean Rules for AND Gates.

Double-Inversion Rule

The *double-inversion rule* is illustrated in Figure 6-20(a) and states that if a variable is inverted twice, then the variable will be back to its original state. To state this algebraically, we use the double bar as follows:

$$\overline{\overline{A}} = A$$

In Figure 6-20(b), you can see that if the NOR gate were replaced with an OR gate, the INVERTER would not be needed since the double inversion returns the logic level to its original state.

DeMorgan's Theorems

DeMorgan's first and second theorems were discussed earlier in the previous section, and are repeated here since they also apply as Boolean rules.

DeMorgan's First Theorem: $\overline{A} \cdot \overline{B} = \overline{A + B}$

DeMorgan's Second Theorem: $\overline{A} + \overline{B} = \overline{A \cdot B}$

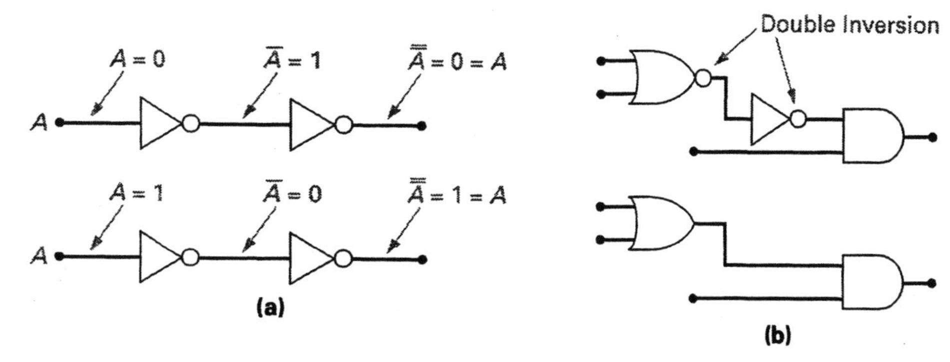

FIGURE 6-20 Double-Inversion Rule.

■ **EXAMPLE:**

Apply DeMorgan's theorem and the double-inversion rule to find equivalent logic circuits of the following:

$$\text{a. } \overline{\bar{A} + B} \qquad \text{b. } \overline{A \cdot \bar{B}}$$

■ ***Solution:***

With both equations we simply "break the line and change the sign," and then cancel any double inversions to find the equivalent logic circuits, as follows:

$$\text{a. } \overline{\bar{A} + B} = \bar{\bar{A}} \cdot \bar{B} = A \cdot \bar{B}$$

In the first example we found that a NOR gate with a bubbled A input is equivalent to an AND gate with a bubbled B input.

$$\text{b. } \overline{A \cdot \bar{B}} = \bar{A} + \bar{\bar{B}} = \bar{A} + B$$

In the second example we found that a NAND gate with a bubbled B input is equivalent to an OR gate with a bubbled A input.

The Duality Theorem

The *duality theorem* is very useful since it allows us to change Boolean equations into their *dual,* and to produce new Boolean equations. The two steps to follow for making the change are very simple.

a. First, change each OR symbol to an AND symbol and each AND symbol to an OR symbol.

b. Second, change each 0 to a 1 and each 1 to a 0.

To see if this works, let us apply the duality theorem to a simple example.

■ **EXAMPLE:**

Apply the duality theorem to $A + 1 = 1$, and then state whether the resulting rule is true.

■ ***Solution:***

Changing the OR symbol to an AND symbol, and changing the 1s to 0s, we arrive at the following:

Original Rule:	$A + 1 = 1$
Dual Rule:	$A \cdot 0 = 0$

OR symbol is changed to AND symbol. — 1s are changed to 0s.

The dual Boolean rule states that any variable ANDed with a 0 will always yield a 0 output, which is true.

■ **EXAMPLE:**

Determine the dual Boolean equation for the following:

$$A \cdot (B + C) = (A \cdot B) + (A \cdot C)$$

Once you have determined the new Boolean equation, sketch circuits and give the truth tables for both the left and right parts of the equation.

■ ***Solution:***

Applying the duality theorem to $A \cdot (B + C) = (A \cdot B) + (A \cdot C)$ means that we will have to change all the AND symbols to OR symbols, and vice versa.

Original Equation:	$A \cdot (B + C) = (A \cdot B) + (A \cdot C)$
Dual Equation:	$A + (B \cdot C) = (A + B) \cdot (A + C)$

Figure 6-21 shows the dual equation and its circuits and truth tables. As you can see by comparing the truth tables, the left and right parts of the equation are equivalent.

As a summary, here is a list of the Boolean laws and rules described in this section, with their dual equations.

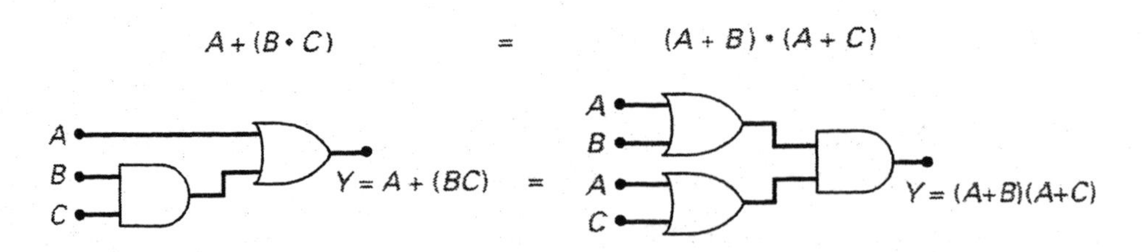

A	B	C	Y = A + (B • C)
0	0	0	Y = 0 + (0 • 0) = 0
0	0	1	Y = 0 + (0 • 1) = 0
0	1	0	Y = 0 + (1 • 0) = 0
0	1	1	Y = 0 + (1 • 1) = 1
1	0	0	Y = 1 + (0 • 0) = 1
1	0	1	Y = 1 + (0 • 1) = 1
1	1	0	Y = 1 + (1 • 0) = 1
1	1	1	Y = 1 + (1 • 1) = 1

=

A	B	C	Y = (A + B) • (A + C)
0	0	0	Y = (0 + 0) • (0 + 0) = 0
0	0	1	Y = (0 + 0) • (0 + 1) = 0
0	1	0	Y = (0 + 1) • (0 + 0) = 0
0	1	1	Y = (0 + 1) • (0 + 1) = 1
1	0	0	Y = (1 + 0) • (1 + 0) = 1
1	0	1	Y = (1 + 0) • (1 + 1) = 1
1	1	0	Y = (1 + 1) • (1 + 0) = 1
1	1	1	Y = (1 + 1) • (1 + 1) = 1

FIGURE 6-21 **A OR (B AND C) Equals (A OR B) AND (A OR C).**

	ORIGINAL EQUATION	DUAL EQUATION
1. Commutative Law	$A + B = B + A$	$A \cdot B = B \cdot A$
2. Associative Law	$A \cdot (B \cdot C) = (A \cdot B) \cdot C$	$A + (B + C) = (A + B) + C$
3. Distributive Law	$A + (B \cdot C) = (A + B) \cdot (A + C)$	$A \cdot (B + C) = (A \cdot B) + (A \cdot C)$
4. OR-AND Rules	$A + 0 = A$	$A \cdot 1 = A$
	$A + A = A$	$A \cdot A = A$
	$A + 1 = 1$	$A \cdot 0 = 0$
	$A + \overline{A} = 1$	$A \cdot \overline{A} = 0$
5. DeMorgan's Laws	$\overline{A} \cdot \overline{B} = \overline{A + B}$	$\overline{A} + \overline{B} = \overline{A \cdot B}$

SELF-TEST REVIEW QUESTIONS FOR SECTION 6-2

1. Give the answers to the following Boolean algebra rules.
   a. $A + 1 = ?$ **b.** $A + \overline{A} = ?$ **c.** $A \cdot 0 = ?$ **d.** $A \cdot A = ?$
2. What is the dual equation of $A + (B \cdot C) = (A + B) \cdot (A + C)$?
3. Apply the associative law to the equation $A \cdot (B \cdot C)$.
4. Apply the distributive law to the equation $(A \cdot B) + (A \cdot C) + (A \cdot D)$.

6-3 FROM TRUTH TABLE TO GATE CIRCUIT

Now that you understand the Boolean rules and laws for logic gates, let us put this knowledge to some practical use. If you need a logic-gate circuit to perform a certain function, the best way to begin is with a truth table that details what input combinations should drive the output HIGH and what input combinations should drive the output LOW. As an example, let us assume that we need a logic circuit that will follow the truth table shown in Figure 6-22(a). This means that the output (Y) should only be 1 when

$$A = 0, B = 0, C = 1$$
$$A = 0, B = 1, C = 1$$
$$A = 1, B = 0, C = 1$$
$$A = 1, B = 1, C = 1$$

In the truth table in Figure 6-22(a), you can see the **fundamental products** listed for each of these HIGH outputs. A fundamental product is a Boolean expression that describes what the inputs will need to be in order to generate a HIGH output. For example, the first fundamental product ($Y = \overline{A} \cdot \overline{B} \cdot C$) states that A must be 0, B must be 0, and C must be 1 for the output Y to be 1 ($Y = \overline{A} \cdot \overline{B} \cdot C = \overline{0} \cdot \overline{0} \cdot 1 = 1 \cdot 1 \cdot 1 = 1$). The second fundamental product ($Y = \overline{A} \cdot B \cdot C$) states that A must be 0, B must be 1, and C must be 1 for the output Y to be 1 ($Y = \overline{A} \cdot B \cdot C = \overline{0} \cdot 1 \cdot 1 = 1 \cdot 1 \cdot 1 = 1$). The third fundamental product ($Y = A \cdot \overline{B} \cdot C$) states that A must be 1, B must be 0, and C must be 1 for Y to be 1 ($Y = A \cdot \overline{B} \cdot C = 1 \cdot \overline{0} \cdot 1 = 1 \cdot 1 \cdot 1 = 1$). The fourth and

Fundamental Products
The truth table input combinations that are essential since they produce a HIGH output.

A	B	C	Y	Fundamental Product
0	0	0	0	
0	0	1	1	→ $\overline{A}\overline{B}C$
0	1	0	0	
0	1	1	1	→ $\overline{A}BC$
1	0	0	0	
1	0	1	1	→ $A\overline{B}C$
1	1	0	0	
1	1	1	1	→ ABC

(a)

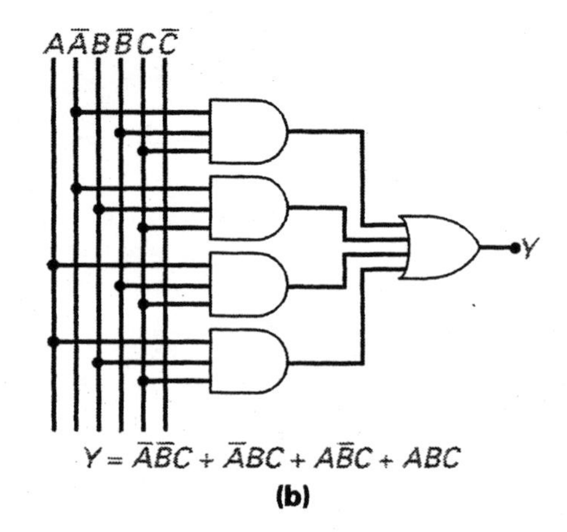

(b)

FIGURE 6-22 **Sum-of-Products (SOP) Form.**

final fundamental product ($Y = A \cdot B \cdot C$) states that A must be 1, B must be 1, and C must be 1 for Y to be 1 ($Y = A \cdot B \cdot C = 1 \cdot 1 \cdot 1 = 1$). The fundamental products for each of the HIGH outputs are listed in the truth table in Figure 6-22(a). By ORing all of these fundamental products, a Boolean equation can be derived, as follows:

$$Y = \overline{A}\ \overline{B}\ C + \overline{A}\ B\ C + A\ \overline{B}\ C + A\ B\ C$$

From this Boolean equation we can create a logic network that is the circuit equivalent of the truth table. To complete this step, we will need to study in detail the different parts of the Boolean equation. First, the Boolean equation states that the outputs of four three-input AND gates are connected to the input of a four-input OR gate. The first AND gate is connected to inputs $\overline{A}$ $\overline{B}$ and C, the second to inputs $\overline{A}$ B and C, the third to inputs A $\overline{B}$ and C, and finally the fourth to inputs A B and C. The resulting logic circuit equivalent for the truth table in Figure 6-22(a) is shown in Figure 6-22(b).

Sum-of-Products (SOP) Form
A Boolean expression that describes the ORing of two or more AND functions.

The Boolean equation for the logic circuit in Figure 6-22 is in the **sum-of-products (SOP) form.** To explain what this term means, we must recall once again that in Boolean algebra, an AND gate's output is the *logical product* of its inputs. On the other hand, the OR gate's output is the *logical sum* of its inputs. A sum-of-products expression, therefore, describes the ORing together of (sum of) two or more AND functions (products). Here are some examples of sum-of-product expressions.

$$Y = AB + \overline{A}B$$
$$Y = A\overline{B} + C\overline{D}E$$
$$Y = \overline{A}BCD + AB\overline{C}\overline{D} + A\overline{B}C\overline{D}$$

Product-of-Sums (POS) Form
A Boolean expression that describes the ANDing of two or more OR functions.

The other basic form for Boolean expressions is the **product-of-sums (POS) form.** This form describes the ANDing of (product of) two or more OR functions (sums). Here are some examples of product-of-sums expressions.

$$Y = (A + B + \overline{C}) \cdot (C + D)$$
$$Y = (\overline{A} + \overline{B}) \cdot (A + B)$$
$$Y = (A + B + C + \overline{D}) \cdot (\overline{A} + B + \overline{C} + D) \cdot (A + \overline{B} + \overline{C} + \overline{D})$$

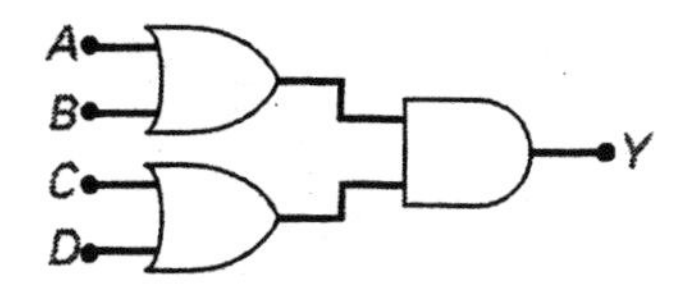

FIGURE 6-23 SOP or POS?

■ **EXAMPLE:**

Is the circuit shown in Figure 6-23 an example of sum-of-products or product-of-sums, and what would this circuit's Boolean expression be?

■ ***Solution:***

The circuit in Figure 6-23 is a product-of-sums equivalent circuit, and its Boolean expression is

$$Y = (A + B) \cdot (C + D)$$

■ **EXAMPLE:**

Determine the logic circuit needed for the truth table shown in Figure 6-24(a).

■ ***Solution:***

The first step is to write the fundamental product for each HIGH output in the truth table, as seen in Figure 6-24(a). By ORing (+) these fundamental products, we can obtain the following sum-of-products Boolean equation.

$$Y = \bar{A}\bar{B}\bar{C}D + \bar{A}B\bar{C}\bar{D} + A\bar{B}\bar{C}D + AB\bar{C}D$$

The next step is to develop the circuit equivalent of this equation. The equation states that the outputs of four four-input AND gates are connected to a four-

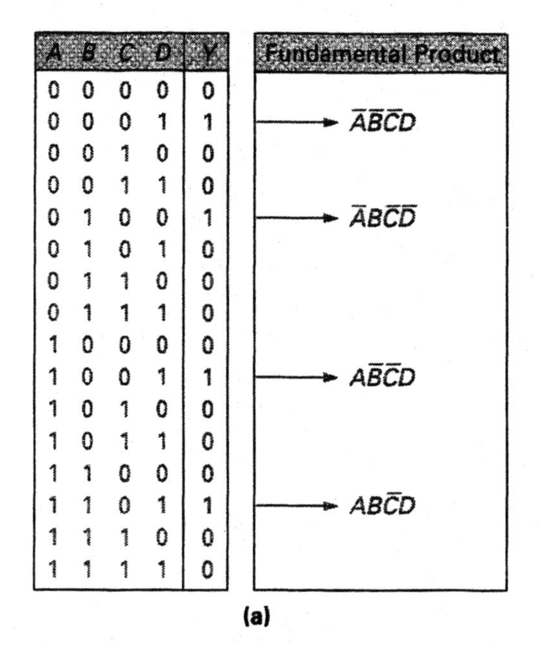

A	B	C	D	Y	Fundamental Product
0	0	0	0	0	
0	0	0	1	1	→ $\bar{A}\bar{B}\bar{C}D$
0	0	1	0	0	
0	0	1	1	0	
0	1	0	0	1	→ $\bar{A}B\bar{C}\bar{D}$
0	1	0	1	0	
0	1	1	0	0	
0	1	1	1	0	
1	0	0	0	0	
1	0	0	1	1	→ $A\bar{B}\bar{C}D$
1	0	1	0	0	
1	0	1	1	0	
1	1	0	0	0	
1	1	0	1	1	→ $AB\bar{C}D$
1	1	1	0	0	
1	1	1	1	0	

(a)

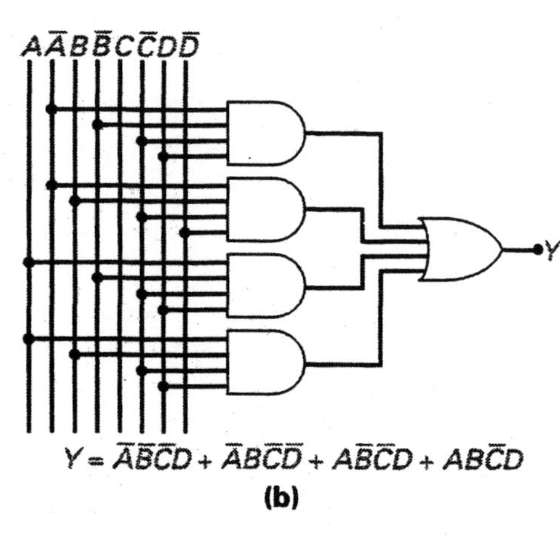

(b)

FIGURE 6-24 Sum-of-Products Example.

input OR gate. The first AND gate had inputs $\overline{A}\overline{B}\overline{C}D$, the second $\overline{A}B\overline{C}\overline{D}$, the third has $A\overline{B}\overline{C}D$, and the fourth has $AB\overline{C}D$. The circuit equivalent for this Boolean expression can be seen in Figure 6-24(b). This circuit will operate in the manner detailed by the truth table in Figure 6-24(a), and generate a 1 output only when

$$A = 0, B = 0, C = 0, D = 1$$
$$A = 0, B = 1, C = 0, D = 0$$
$$A = 1, B = 0, C = 0, D = 1$$
$$A = 1, B = 1, C = 0, D = 1$$

In all other instances, the output of this circuit will be 0.

■ **EXAMPLE:**

Apply the distributive law to the following Boolean expression in order to convert it into a sum-of-products form.

$$Y = A\overline{B} + C(D\overline{E} + F\overline{G})$$

■ ***Solution:***

$Y = A\overline{B} + C(D\overline{E} + F\overline{G})$	Original Expression
$Y = A\overline{B} + CD\overline{E} + CF\overline{G}$	Distributive Law Applied

---

SELF-TEST REVIEW QUESTIONS FOR SECTION 6-3

1. A two-input AND gate will produce at its output the logical ____________ of A and B, while an OR gate will produce at its output the logical ____________ of A and B.
2. What are the fundamental products of each of the following input words?
   a. $A, B, C, D = 1011$ **b.** $A, B, C, D = 0110$
3. The Boolean expression $Y = AB + \overline{A}B$ is an example of a/an ____________ (SOP/POS) equation.
4. Describe the steps involved to develop a logic circuit from a truth table.

6-4 GATE CIRCUIT SIMPLIFICATION

In the last section we saw how we could convert a desired set of conditions in a truth table to a sum-of-products equation, and then to an equivalent logic circuit. In this section we will see how we can simplify the Boolean equation using Boolean laws and rules, and therefore simplify the final logic circuit. As to why we would want to simplify a logic circuit, the answer is: A simplified circuit performs the same function, but is smaller, easier to construct, consumes less power, and costs less.

6-4-1 *Boolean Algebra Simplification*

Generally, the simplicity of one circuit compared to another circuit is judged by counting the number of logic-gate inputs. For example, the logic circuit in Figure 6-25(a) has a total of six inputs (two on each of the AND gates and two on the OR gate). The Boolean equation for this circuit would be

$$Y = \overline{A}B + AB$$

Since "$\cdot\ B$" (AND B) seems to be a common factor in this equation, by factoring we could obtain the following:

$$Y = \overline{A}B + AB \qquad \text{Original Expression}$$
$$Y = (\overline{A} + A) \cdot B \qquad \text{Factoring AND } B$$

This equivalent logic circuit is shown in Figure 6-25(b), and if you count the number of logic-gate inputs in this circuit (two for the OR and two for the AND, a total of four), you can see that the circuit has been simplified from a six-input circuit to a four-input circuit.

This, however, is not the end of our simplification process, since another rule can be applied to further simplify the circuit. Remembering that any variable ORed with its complement results in a 1, we can replace $\overline{A} + A$ with a 1, as follows:

$$Y = (\overline{A} + A) \cdot B$$
$$Y = (1) \cdot B$$
$$Y = 1 \cdot B$$

The resulting equation $Y = 1 \cdot B$ can be further simplified since any variable ANDed with a 1 will always cause the output to equal the variable. Therefore, $Y = 1 \cdot B$ can be replaced with B, as follows:

$$Y = 1 \cdot B$$
$$Y = B$$

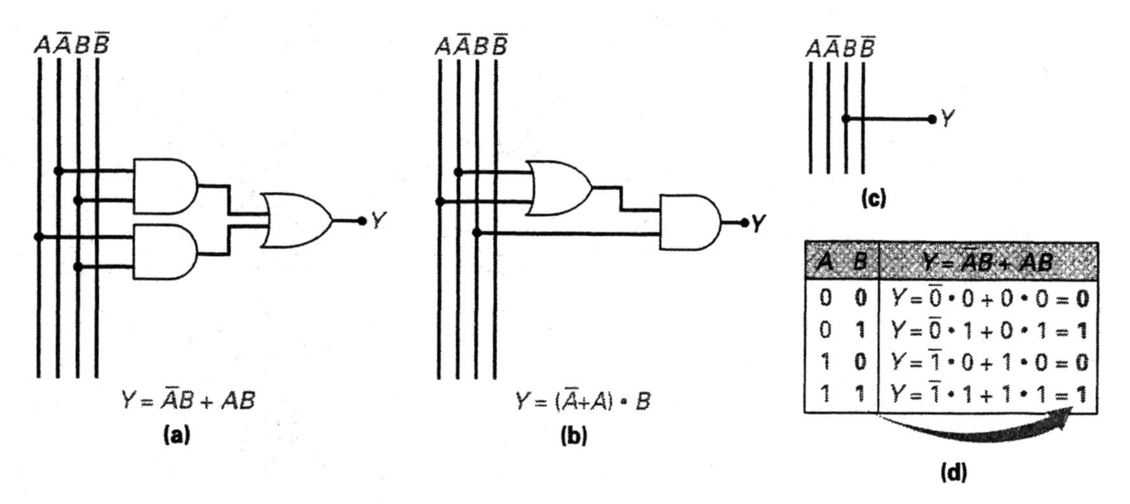

A	B	$Y = \overline{A}B + AB$
0	0	$Y = \overline{0} \cdot 0 + 0 \cdot 0 = 0$
0	1	$Y = \overline{0} \cdot 1 + 0 \cdot 1 = 1$
1	0	$Y = \overline{1} \cdot 0 + 1 \cdot 0 = 0$
1	1	$Y = \overline{1} \cdot 1 + 1 \cdot 1 = 1$

FIGURE 6-25 Simplifying Gate Circuits.

This means that the entire circuit in Figure 6-25(a) can be replaced with a single wire connected from the B input to the output, as seen in Figure 6-25(c). This fact is confirmed in the truth table in Figure 6-25(d), which shows that output Y exactly follows the input B.

■ **EXAMPLE:**

Simplify the Boolean equation shown in Figure 6-26(a).

■ ***Solution:***

The Boolean equation for this circuit is

$$Y = A\bar{B}\bar{C} + \bar{A}B\bar{C} + A\bar{B}C + \bar{A}\bar{B}\bar{C}$$

The first step in the simplification process is to try and spot common factors in each of the ANDed terms, and then rearrange these terms as shown in Figure 6-26(b). The next step is to factor out the common expressions, which are $A\bar{B}$ and $\bar{A}\bar{C}$, resulting in

$$Y = A\bar{B}\bar{C} + \bar{A}B\bar{C} + A\bar{B}C + \bar{A}\bar{B}\bar{C} \qquad \text{Original Equation}$$
$$Y = A\bar{B}\bar{C} + A\bar{B}C + \bar{A}B\bar{C} + \bar{A}\bar{B}\bar{C} \qquad \text{Rearranged Equation}$$
$$Y = A\bar{B}(\bar{C} + C) + \bar{A}\bar{C}(B + \bar{B}) \qquad \text{Factored Equation}$$

Studying the last equation you can see that we can apply the OR rule $(\bar{A} + A) = 1$ and simplify the terms within the parentheses to obtain the following:

$$Y = A\bar{B}(\bar{C} + C) + \bar{A}\bar{C}(B + \bar{B})$$
$$Y = A\bar{B} \cdot (1) + \bar{A}\bar{C} \cdot (1)$$

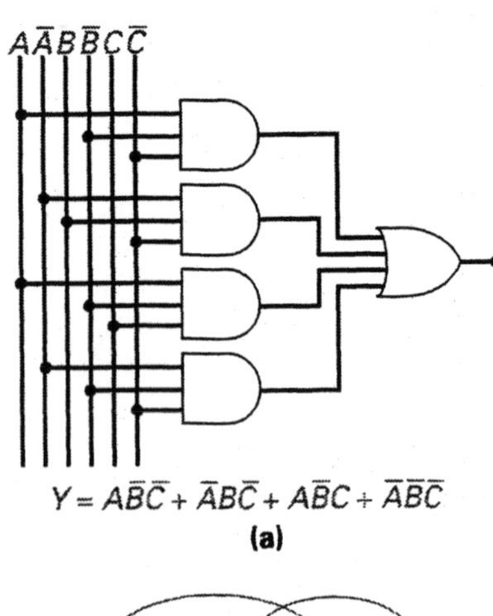

(a)

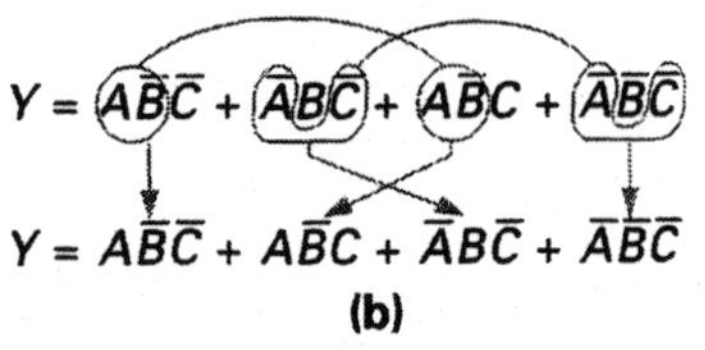

(b)

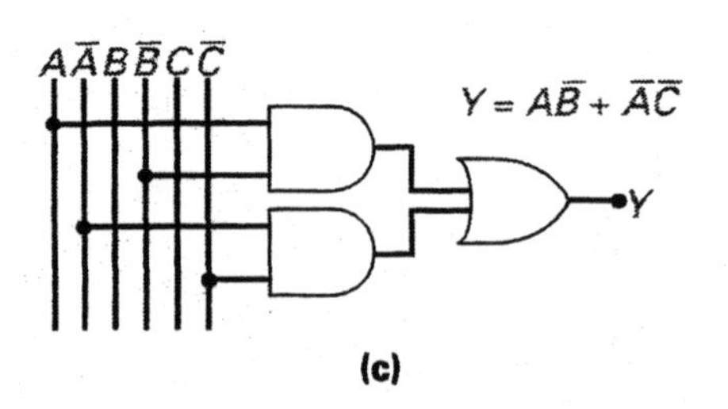

(c)

A	B	C	$Y = A\bar{B}\bar{C} + \bar{A}B\bar{C} + A\bar{B}C + \bar{A}\bar{B}\bar{C}$	$Y = A\bar{B} + \bar{A}\bar{C}$
0	0	0	$Y = 0\bar{0}\bar{0} + \bar{0}0\bar{0} + 0\bar{0}0 + \bar{0}\bar{0}\bar{0} = \mathbf{1}$	$Y = 0\bar{0} + \bar{0}\bar{0} = \mathbf{1}$
0	0	1	$Y = 0\bar{0}\bar{1} + \bar{0}0\bar{1} + 0\bar{0}1 + \bar{0}\bar{0}\bar{1} = \mathbf{0}$	$Y = 0\bar{0} + \bar{0}\bar{1} = \mathbf{0}$
0	1	0	$Y = 0\bar{1}\bar{0} + \bar{0}1\bar{0} + 0\bar{1}0 + \bar{0}\bar{1}\bar{0} = \mathbf{1}$	$Y = 0\bar{1} + \bar{0}\bar{0} = \mathbf{1}$
0	1	1	$Y = 0\bar{1}\bar{1} + \bar{0}1\bar{1} + 0\bar{1}1 + \bar{0}\bar{1}\bar{1} = \mathbf{0}$	$Y = 0\bar{1} + \bar{0}\bar{1} = \mathbf{0}$
1	0	0	$Y = 1\bar{0}\bar{0} + \bar{1}0\bar{0} + 1\bar{0}0 + \bar{1}\bar{0}\bar{0} = \mathbf{1}$	$Y = 1\bar{0} + \bar{1}\bar{0} = \mathbf{1}$
1	0	1	$Y = 1\bar{0}\bar{1} + \bar{1}0\bar{1} + 1\bar{0}1 + \bar{1}\bar{0}\bar{1} = \mathbf{1}$	$Y = 1\bar{0} + \bar{1}\bar{1} = \mathbf{1}$
1	1	0	$Y = 1\bar{1}\bar{0} + \bar{1}1\bar{0} + 1\bar{1}0 + \bar{1}\bar{1}\bar{0} = \mathbf{0}$	$Y = 1\bar{1} + \bar{1}\bar{0} = \mathbf{0}$
1	1	1	$Y = 1\bar{1}\bar{1} + \bar{1}1\bar{1} + 1\bar{1}1 + \bar{1}\bar{1}\bar{1} = \mathbf{0}$	$Y = 1\bar{1} + \bar{1}\bar{1} = \mathbf{0}$

(d)

FIGURE 6-26 **Boolean Equation Simplification.**

Now, we can apply the $A \cdot 1 = A$ rule to further simplify the terms, since $A\overline{B} \cdot 1 = A\overline{B}$ and $\overline{A}\overline{C} \cdot 1 = \overline{A}\overline{C}$; therefore:

$$Y = A\overline{B} + \overline{A}\overline{C}$$

This equivalent logic circuit is illustrated in Figure 6-26(c), and has only six logic gate inputs compared to the original circuit shown in Figure 6-26(a), which has sixteen logic gate inputs. The truth table in Figure 6-26(d) compares the original equation with the simplified equation and shows how the same result is obtained.

6-4-2 Karnaugh Map Simplification

In most instances, engineers and technicians will simplify logic circuits using a **Karnaugh map, or K-map.** The Karnaugh map, named after its inventor, is quite simply a rearranged truth table, as shown in Figure 6-27. With this map the essential gating requirements can be more easily recognized and reduced to their simplest form.

Karnaugh Map
Also called K-map, it is a truth table that has been rearranged to show a geometrical pattern of functional relationships for gating configurations. Using this map, essential gating requirements can be more easily recognized and reduced to their simplest form.

The total number of boxes or cells in a K-map depends on the number of input variables. For example, in Figure 6-27, only two inputs (A and B) and their complements ($\overline{A}$ and $\overline{B}$) are present, and therefore the K-map contains (like a two-variable truth table) only four combinations (00, 01, 10, and 11).

Each cell of this two-variable K-map represents one of the four input combinations. In practice, the input labels are placed outside of the cells, as shown in Figure 6-28(a), and apply to either a column or row of cells. For example, the row label $\overline{A}$ applies to the two upper cells, while the row label A applies to the two lower cells. Running along the top of the K-map, the label $\overline{B}$ applies to the two left cells, while the label B applies to the two right cells. As an example, the upper right cell represents the input combination $\overline{A}B$. Figures 6-28(b) and (c) show the formats for a three-variable ($2^3 = 8$ cells) and four-variable ($2^4 = 16$ cells) K-map.

Now that we have an understanding of the K-map, let us see how it can be used to simplify a logic circuit. As an example, imagine that we need to create an equivalent logic circuit for the truth table given in Figure 6-29(a). The first step is to develop a sum-of-products Boolean expression. This is achieved by writing the fundamental product for each HIGH output in the truth table and then ORing all of the fundamental products, as shown in Figure 6-29(b). The equivalent logic circuit for this equation is shown in Figure 6-29(c). The next step is to plot this Boolean expression on a two-variable K-map, as seen

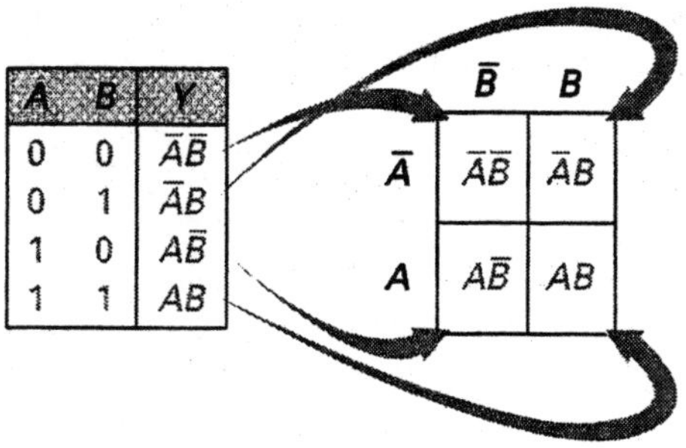

A	B	Y
0	0	$\overline{A}\overline{B}$
0	1	$\overline{A}B$
1	0	$A\overline{B}$
1	1	AB

	$\overline{B}$	B
$\overline{A}$	$\overline{A}\overline{B}$	$\overline{A}B$
A	$A\overline{B}$	AB

FIGURE 6-27 The Karnaugh Map—A Rearranged Truth Table.

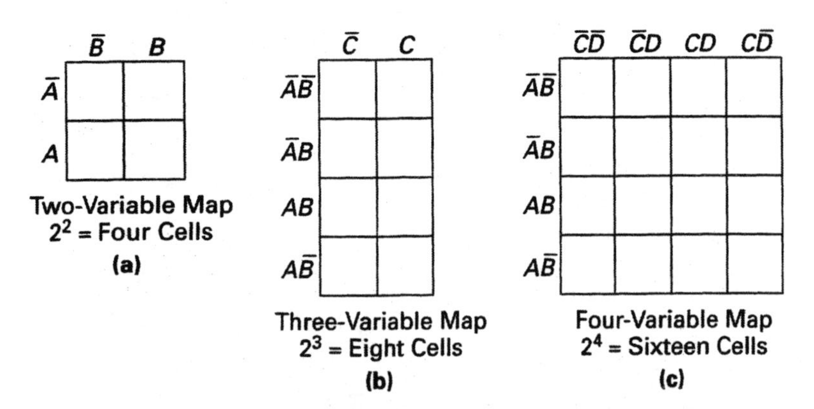

FIGURE 6-28 Karnaugh Maps. (a) Two-Variable. (b) Three-Variable. (c) Four-Variable.

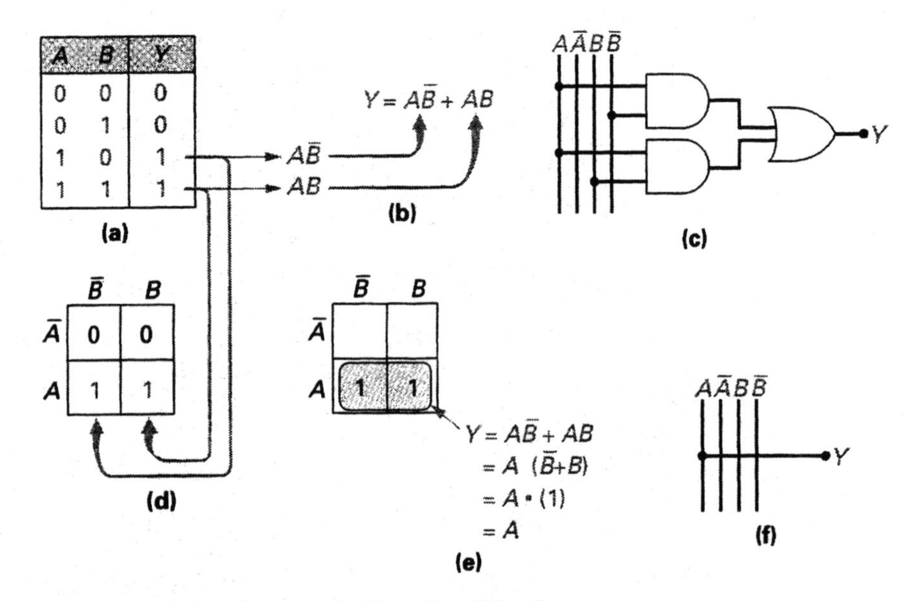

FIGURE 6-29 Karnaugh Map Simplification.

in Figure 6-29(d). When plotting a sum-of-products expression on a K-map, remember that each cell corresponds to each of the input combinations in the truth table. A HIGH output in the truth table should appear as a 1 in its equivalent cell in the K-map, and a LOW output in the truth table should appear as a 0 in its equivalent cell. A 1, therefore, will appear in the lower left cell (corresponding to $A\overline{B}$) and in the lower right cell (corresponding to AB). The other input combinations ($\overline{A}\overline{B}$ and $\overline{A}B$) both yield a 0 output, and therefore a 0 should be placed in these two upper cells.

Reducing Boolean equations is largely achieved by applying the rule of complements, which states that $A + \overline{A} = 1$. Now that the SOP equation has been plotted on the K-map shown in Figure 6-29(d), the next step is to group terms and then factor out the common variables. If you study the K-map in Figure 6-29(d), you will see that adjacent cells differ by only one input variable. This means that if you move either horizontally or vertically from one cell to an adjacent cell, only one variable will change. By grouping adjacent cells containing a 1, as seen in Figure 6-29(e), cells can be compared and simplified (using the rule of complements) to create one-product terms. In this example, cells $A\overline{B}$ and AB contain B and $\overline{B}$, so these opposites or complements cancel, leaving A, as follows:

$Y = A\overline{B} + AB$	Grouped Pair
$Y = A \cdot (\overline{B} + B)$	Factoring A AND
$Y = A \cdot 1$	
$Y = A$	

This procedure can be confirmed by studying the original truth table in Figure 6-29(a) in which you can see that output Y exactly follows input A. The equivalent circuit, therefore, is shown in Figure 6-29(f).

■ **EXAMPLE:**

Determine the simplest logic circuit for the truth table shown in Figure 6-30(a). Illustrate each step in the process.

■ ***Solution:***

Since this example has three input variables, the first step is to draw a three-variable K-map, as shown in Figure 6-30(b). The next step is to look for the HIGH outputs in the truth table in Figure 6-30(a) and plot these 1s in their equivalent cells in the K-map, as shown in Figure 6-30(b). After inserting 0s in the remaining cells, group the 1s into pairs, as shown in Figure 6-30(b), and then study the row and column variable labels associated with that grouped pair to see which variable will drop out due to the rule of complements. In the upper group, $\overline{A}$ and A will cancel, leaving $B\overline{C}$, and in the lower group $\overline{C}$ and C will cancel, leaving $A\overline{B}$. These reduced products will form an equivalent Boolean equation and logic circuit, as seen in Figure 6-30(c). In this example, the original sixteen-input equation in Figure 6-30(b) has been simplified to the equivalent six-input logic gate circuit shown in Figure 6-30(c).

The 1s in a K-map can be grouped in pairs (groups of two), quads (groups of four), octets (groups of eight), and all higher powers of two. Figure 6-31

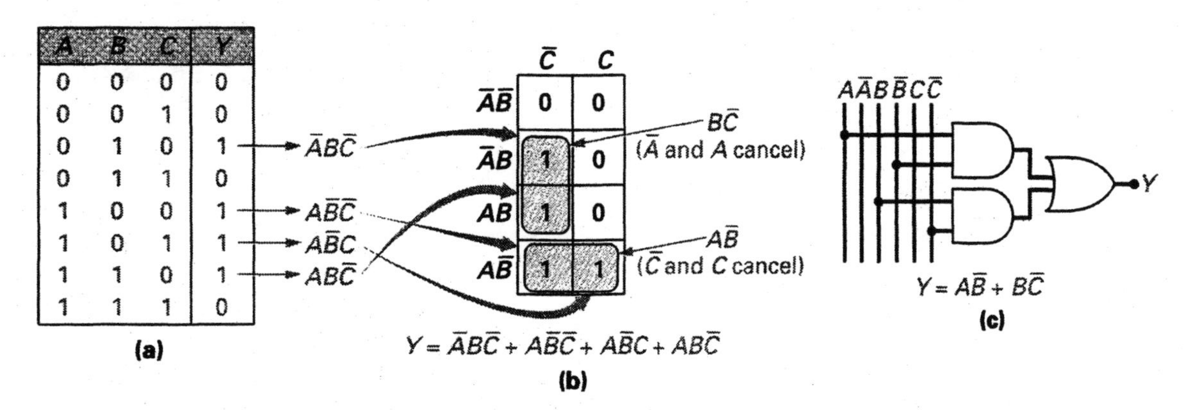

FIGURE 6-30 Three-Variable K-Map Simplification.

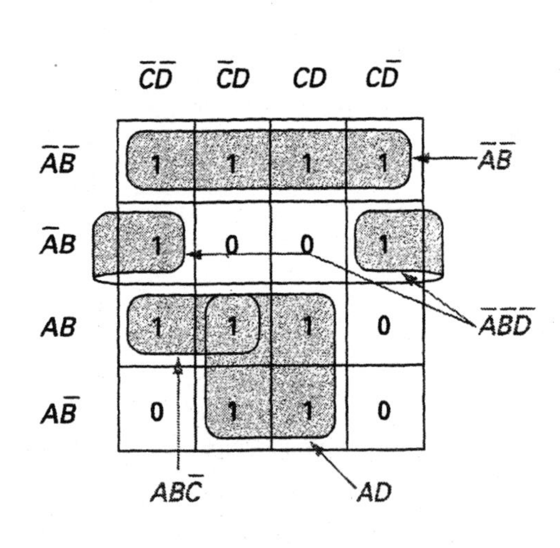

Before : $Y = \overline{A}\overline{B}\overline{C}\overline{D} + \overline{A}\overline{B}\overline{C}D + \overline{A}\overline{B}CD + \overline{A}\overline{B}C\overline{D} +$
$\overline{A}B\overline{C}\overline{D} + \overline{A}BC\overline{D} + AB\overline{C}\overline{D} + AB\overline{C}D +$
$ABCD + A\overline{B}\overline{C}D + A\overline{B}CD$

After : $Y = AB\overline{C} + AD + \overline{A}B\overline{D} + \overline{A}\overline{B}$

(a)

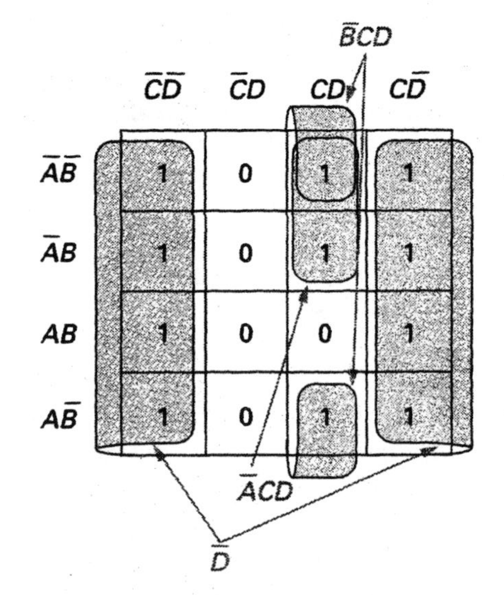

Before : $Y = \overline{A}\overline{B}\overline{C}\overline{D} + \overline{A}\overline{B}CD + \overline{A}\overline{B}C\overline{D} +$
$\overline{A}B\overline{C}\overline{D} + \overline{A}BCD + \overline{A}BC\overline{D} +$
$AB\overline{C}\overline{D} + ABC\overline{D} + A\overline{B}\overline{C}\overline{D} +$
$A\overline{B}CD + A\overline{B}C\overline{D}$

After : $Y = \overline{A}CD + \overline{B}CD + \overline{D}$

(b)

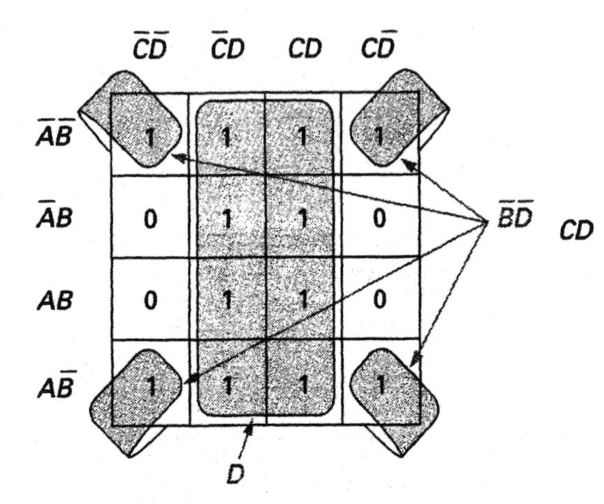

Before : $Y = \overline{A}\overline{B}\overline{C}\overline{D} + \overline{A}\overline{B}\overline{C}D + \overline{A}\overline{B}CD + \overline{A}\overline{B}C\overline{D} +$
$\overline{A}B\overline{C}D + \overline{A}BCD + AB\overline{C}D + ABCD +$
$A\overline{B}\overline{C}\overline{D} + A\overline{B}\overline{C}D + A\overline{B}CD + A\overline{B}C\overline{D}$

After : $Y = \overline{B}\overline{D} + D$

(c)

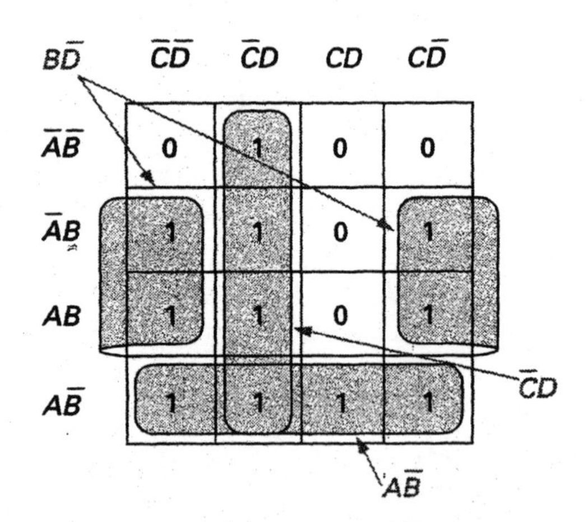

Before : $Y = \overline{A}\overline{B}\overline{C}D + \overline{A}B\overline{C}\overline{D} + \overline{A}B\overline{C}D + \overline{A}BC\overline{D} +$
$AB\overline{C}\overline{D} + AB\overline{C}D + ABC\overline{D} + A\overline{B}\overline{C}\overline{D} +$
$A\overline{B}\overline{C}D + A\overline{B}CD + A\overline{B}C\overline{D}$

After : $Y = A\overline{B} + B\overline{D} + \overline{C}D$

(d)

FIGURE 6-31 K-Map Grouping Examples.

shows some examples of grouping and how the K-map has been used to reduce large Boolean equations. Notice that the larger groups will yield smaller terms and, therefore, gate circuits with fewer inputs. For this reason, you should begin by looking for the largest possible groups, and then step down in group size if none are found (for example, begin looking for octets, then quads, and finally pairs). Also notice that you can capture 1s on either side of a map by wrapping a group around behind the map.

■ **EXAMPLE:**

What would be the sum-of-products Boolean expression for the truth table in Figure 6-32(a)? After determining the Boolean expression, plot it on a K-map to see if it can be simplified.

■ ***Solution:***

The first step in developing the sum-of-products Boolean expression is to write down the fundamental products for each HIGH output in the truth table, as seen in Figure 6-32(a). From this, we can derive an SOP equation, which will be

$$Y = \overline{A}\,\overline{B}\,\overline{C}D + \overline{A}\,\overline{B}CD + \overline{A}B\overline{C}D + \overline{A}BCD + A\overline{B}CD + ABCD$$

The next step is to draw a four-variable K-map, as seen in Figure 6-32(b), and then plot the HIGH outputs of the truth table in their equivalent cells in the K-map. Looking at Figure 6-32(b), you can see that the 1s can be grouped into two quads. With the square-shaped quad, row labels B and $\overline{B}$ will cancel and column labels C and $\overline{C}$ will cancel, leaving $\overline{A}D$. With the rectangular-shaped quad, all of its row labels will cancel, leaving CD. The simplified Boolean expression, therefore, will be

$$Y = \overline{A}D + CD$$

Referring to the truth table in Figure 6-33(a), you can see that input words 0000 through 1001 have either a 0 or 1 at the output Y. Input words 1010 through 1111 on the other hand, have an "X" written in the output column to

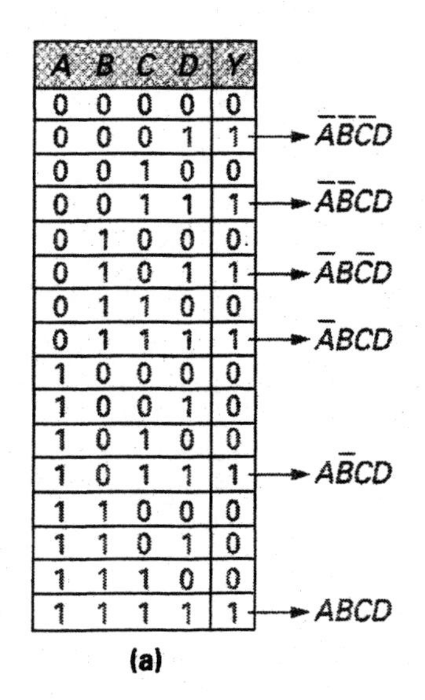

A	B	C	D	Y	
0	0	0	0	0	
0	0	0	1	1	→ $\overline{A}\,\overline{B}\,\overline{C}D$
0	0	1	0	0	
0	0	1	1	1	→ $\overline{A}\,\overline{B}CD$
0	1	0	0	0	
0	1	0	1	1	→ $\overline{A}B\overline{C}D$
0	1	1	0	0	
0	1	1	1	1	→ $\overline{A}BCD$
1	0	0	0	0	
1	0	0	1	0	
1	0	1	0	0	
1	0	1	1	1	→ $A\overline{B}CD$
1	1	0	0	0	
1	1	0	1	0	
1	1	1	0	0	
1	1	1	1	1	→ $ABCD$

(a)

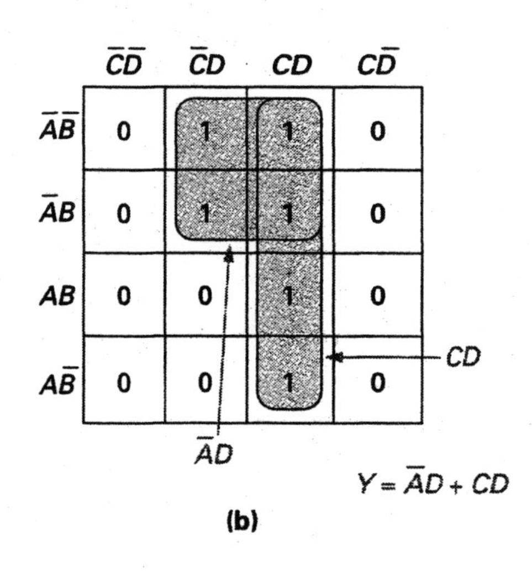

(b)

FIGURE 6-32 Example of Four-Variable K-Map.

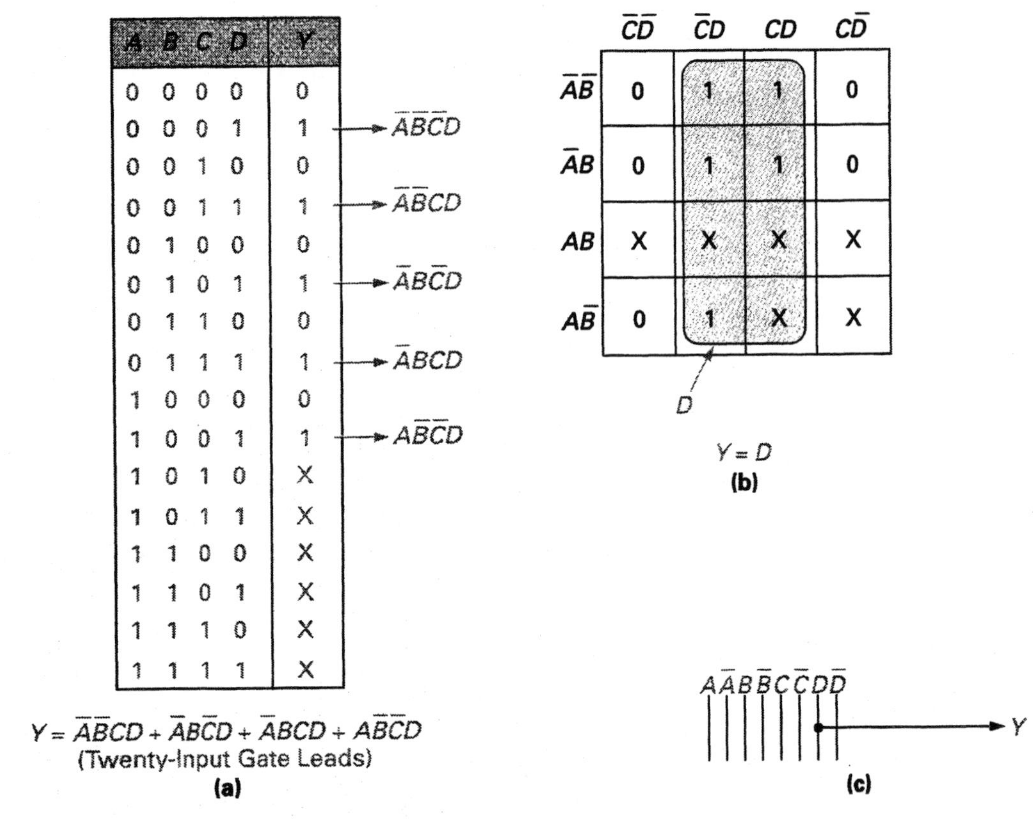

FIGURE 6-33 Four-Variable K-Map Example with Don't-Care Conditions.

indicate that the output Y for these input combinations is unimportant (output can be either a 0 or 1). These Xs in the output are appropriately called "don't-care conditions," and can be treated as either a 0 or 1. In K-maps, these Xs can be grouped with 1s to create larger groups, and therefore a simpler logic circuit, as we will see in the next example.

■ **EXAMPLE:**

Sketch an equivalent logic circuit for the truth table given in Figure 6-33(a) using the least amount of inputs.

■ ***Solution:***

Figure 6-33(b) shows a four-variable K-map with the 0s, 1s, and Xs (don't-care conditions) inserted into their appropriate cells. Since a larger group will always yield a smaller term, visualize the Xs as 1s, and then try for as big a group as possible. In this example, the Xs have allowed us to group an octet, resulting in the single product D. As a result, the twenty-input Boolean equation in Figure 6-33(a) can be replaced with a single line connected to D, as shown in Figure 6-33(c).

SELF-TEST REVIEW QUESTIONS FOR SECTION 6-4

1. A two-variable K-map will have ____________ cells.
2. Simplify the equation $Y = \overline{A}\,B + A\,B$.
3. When grouping 1s in a K-map, the ____________ (smaller/larger) the groups, the smaller the terms.
4. Can the don't-care conditions in a K-map be grouped with 0s or with 1s, or doesn't it matter?

REVIEW QUESTIONS

Multiple-Choice Questions

1. In Boolean, the OR function is described as logical ____________ and the AND function is described as logical ____________.

   a. addition, division **b.** multiplication, addition
   c. addition, multiplication **d.** multiplication, subtraction

2. The OR gate is said to produce at its output the logical ____________ of its inputs, while the AND gate is said to produce at its output the logical ____________ of its inputs.

   a. sum, product **b.** product, quotient
   c. sum, subtrahend **d.** product, sum

3. If inputs A and B are ANDed together and then the result is ORed with C, the Boolean expression equation describing this operation would be

   a. $Y = AB + BC$ **b.** $Y = (A + B)C$
   c. $Y = AB + C$ **d.** $Y = AB + AC$

4. The Boolean expression $Y = (A \cdot B) + (A \cdot C)$ describes a logic circuit that has

   a. six inputs **b.** two OR and one AND gate
   c. two AND and one OR gate **d.** four inputs

5. The Boolean expression $Y = \overline{A + B}$ describes a two-input ____________.

   a. NAND gate **b.** NOR gate **c.** AND gate **d.** OR gate

6. DeMorgan's first theorem states that

   a. $\overline{A} \cdot \overline{B} = \overline{A + B}$ **b.** $\overline{A} + \overline{\overline{B}} = \overline{A} + B$
   c. $\overline{A} + \overline{B} = \overline{A \cdot B}$ **d.** $\overline{\overline{A}} \cdot \overline{B} = A \cdot \overline{B}$

7. DeMorgan's second theorem states that

   a. $\overline{A} \cdot \overline{B} = \overline{A + B}$ **b.** $\overline{A} + \overline{\overline{B}} = \overline{A} + B$
   c. $\overline{A} + \overline{B} = \overline{A \cdot B}$ **d.** $\overline{\overline{A}} \cdot \overline{B} = A \cdot \overline{B}$

8. In Boolean algebra which symbol is used to describe the exclusive OR action?

   a. "+" **b.** "·" **c.** "×" **d.** "⊕"

9. Which of the following algebraically states the commutative law?

   a. $A + B = B + A$
   b. $A \cdot (B \cdot C) = (A \cdot B) \cdot C$
   c. $A + (B \cdot C) = (A + B) \cdot (A + C)$
   d. None of the above

10. $A + (B \cdot C) = (A + B) \cdot (A + C)$ is the algebraic definition of the ____________.

    a. associative law
    b. distributive law
    c. commutative law
    d. complements law

11. The four Boolean rules for an OR gate are $A + 0 = ?$, $A + A = ?$, $A + 1 = ?$, $A + \overline{A} = ?$.

    a. $A, 0, 1, 1$ b. $A, A, 0, 0$ c. $A, A, 1, 1$ d. $A, A, 0, 1$

12. The four Boolean rules for an AND gate are $A \cdot 1 = ?$, $A \cdot A = ?$, $A \cdot 0 = ?$, $A \cdot \overline{A} = ?$.

    a. $A, 0, 1, 1$ b. $A, A, 0, 0$ c. $A, A, 1, 1$ c. $A, A, 0, 1$

13. If you applied DeMorgan's theorem and the double-inversion rule to the equation $Y = \overline{\overline{A} + B\overline{C}}$, what would be the result?

    a. $A\overline{B} + C$ b. $A + B\overline{C}$ c. $\overline{A}B + C$ d. $AB + \overline{C}$

14. What would be the dual equation of $A + (B \cdot C) = (A + B) \cdot (A + C)$?

    a. $A \cdot (B + C) = (A \cdot B) + (A \cdot C)$
    b. $\overline{\overline{A} + B} = AB + \overline{C}$
    c. $A \cdot (B \cdot C) = (A \cdot B) + (A \cdot C)$
    d. $A + (B \cdot C) = (A \cdot B) \cdot (A + C)$

15. $Y = AB + BC + AC$ is an example of a ____________.

    a. product-of-sums expression
    b. sum-of-products expression
    c. quotient-of-sums expression
    d. sum-of-quotients expression

16. What would be the fundamental product of the input word $ABCD = 1101$?

    a. $ABCD$ b. $\overline{A}\overline{B}C\overline{D}$ c. $AB\overline{C}D$ d. $ABC\overline{D}$

17. How many cells would a three-variable Karnaugh map have?

    a. Four b. Six c. Eight d. Sixteen

18. What would be the Boolean expression for a four-variable logic circuit that NANDs inputs A and B and also NANDs inputs C and D, and then NANDs the results.

    a. $\overline{\overline{AB} + \overline{CD}}$
    b. $(\overline{AB}) \cdot (\overline{CD})$
    c. $\overline{(\overline{AB}) \cdot (\overline{CD})}$
    d. $\overline{(\overline{A + B}) \cdot (\overline{C + D})}$

19. If you applied DeMorgan's theorem and the double-inversion rule to the answer in question 18, you will obtain which of the following equations?

    a. $AB + CD$
    b. $A + B + C + D$
    c. $(A + B) \cdot (C + D)$
    d. $(A + B) + (CD)$

20. In reference to questions 18 and 19, you can see that the DeMorgan equivalent of a NAND-NAND circuit is a/an ____________.

    a. OR-AND **b.** AND-NOR **c.** AND-OR **d.** NOR-AND

Essay Questions

21. List the Boolean expression for the seven basic logic gates. (6-1)
22. Describe briefly, with an example, the process of converting a desired set of output conditions into an equivalent logic circuit. (6-3)
23. What is the difference between an SOP and a POS equation? (6-3)
24. What is a Karnaugh map? (6-4-2)
25. Briefly describe how K-maps can be used to simplify a logic circuit. (6-4-2)
26. How is the simplicity of a logic circuit generally judged? (6-4)
27. Why is it necessary to simplify logic circuits? (6-4)
28. What is meant by the terms logical product and logical sum? (6-1)
29. How can DeMorgan's theorems be stated using Boolean algebra? (6-1)
30. Briefly describe why larger groups of 1s on a K-map yield smaller terms, and why smaller terms are desired. (6-4-2)

Practice Problems

31. Determine the Boolean expressions for the logic circuits shown in Figure 6-34.
32. Which of the expressions in Question 31 would be considered POS, and which SOP?
33. Sketch the equivalent logic circuit for the following Boolean equations.

    a. $Y = (AB) + (\overline{AB}) + (\overline{A + B}) + (A \oplus B)$ **b.** $Y = (\overline{\overline{A} + B}) \oplus ABC$

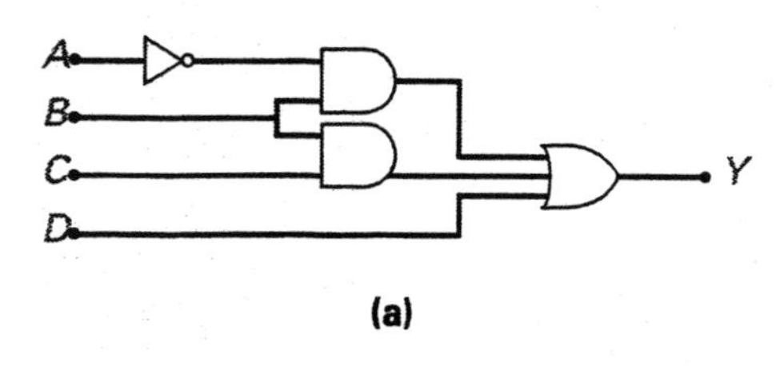

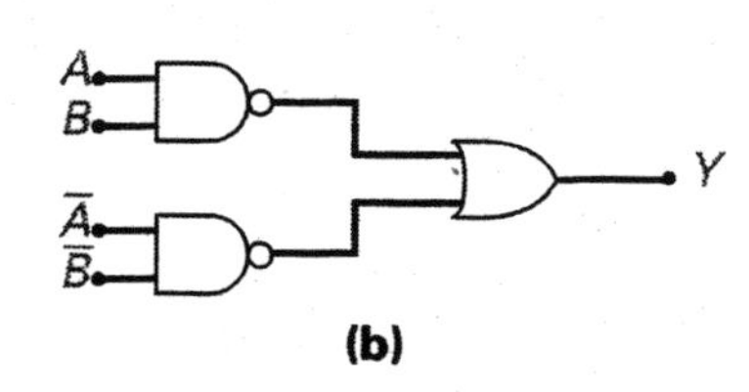

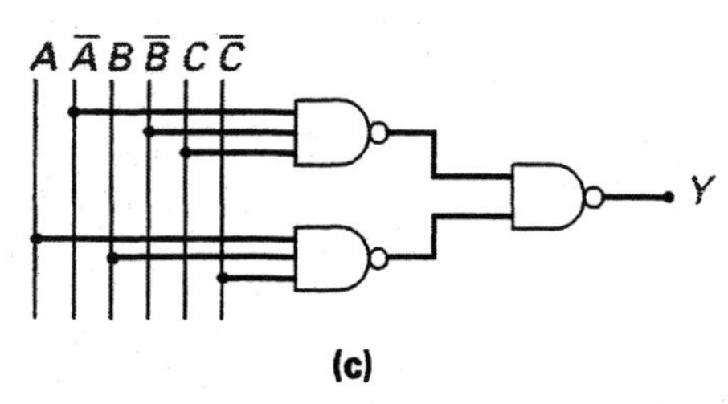

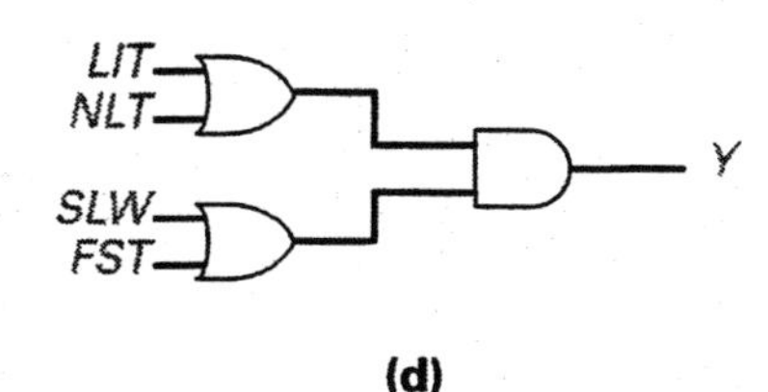

FIGURE 6-34 From Logic Circuit to Boolean Equation.

34. Determine the Boolean expression for the Y_2 output in Table 6-1.

35. Use a K-map to simplify the Y_2 expression from Question 34.

36. How much was the Y_2 logic circuit simplified in Question 35?

37. Referring to Table 6-1, (a) determine the Boolean expression for the Y_1 output. (b) Simplify the expression using a K-map. (c) Describe how much the logic circuit was simplified.

38. Determine the Boolean expression for the Y_3 output in Table 6-1, simplify the expression using a K-map, and then describe how much the logic circuit was simplified.

39. Sketch the simplified logic circuits for the Y_1, Y_2, and Y_3 outputs listed in Table 6-1.

40. If the don't-care condition in the Y_3 output in Table 6-1 were 0s, what would the simplified expression be? Is the expression more or less simplified with don't-care conditions?

TABLE 6-1 The Y_1, Y_2, and Y_3 Truth Table

A	B	C	D	Y_1	Y_2	Y_3
0	0	0	0	1	1	1
0	0	0	1	1	1	0
0	0	1	0	1	0	1
0	0	1	1	1	0	0
0	1	0	0	0	0	1
0	1	0	1	0	0	1
0	1	1	0	0	1	1
0	1	1	1	0	1	1
1	0	0	0	0	1	0
1	0	0	1	1	1	0
1	0	1	0	0	0	0
1	0	1	1	1	0	0
1	1	0	0	1	0	1
1	1	0	1	1	0	X
1	1	1	0	1	0	X
1	1	1	1	1	0	X

Chapter 7

OBJECTIVES

After completing this chapter, you will be able to:

1. Define the term "combinational logic circuits" and list some of the different types.
2. Describe the function and explain the basic operation of a decoder circuit.
3. Describe the operation, data sheet characteristics, and typical application for the following digital decoder ICs.
   a. 7442—BCD-to-decimal decoder
   b. 74154—four-line to sixteen-line decoder
   c. 7447—binary-to-seven-segment decoder/driver
   d. 4511—binary-to-seven-segment decoder/driver
4. Explain how digital decoders are used to drive both light-emitting diode (LED) displays and liquid-crystal displays (LCD).
5. Describe the function and explain the basic operation of a decoder circuit.
6. Describe the operation, data sheet characteristics, and typical application for the following digital encoder IC.
   a. 74147—ten-line decimal to four-line BCD priority encoder
7. Explain the three-step troubleshooting procedure and then show how it can be applied to a typical encoder/decoder circuit.
8. Describe some frequently used troubleshooting procedures.

Decoders and Encoders

OUTLINE

VIGNETTE

Working with Wang

Born in China in 1920, An Wang was a hard-working and diligent student who at the young age of sixteen entered the prestigious Shanghai Chiao Tung University to begin working on his electrical engineering degree. At twenty-five he immigrated to the United States to continue his studies, and three years later earned his doctorate in physics from Harvard. Around this time, Wang invented and patented a type of computer memory which utilized a doughnut-shaped ring of iron. In his autobiography titled *Lessons,* Wang described how he eventually sold his patent for the memory core to IBM, but only after an episode that he described as corporate thuggery. Apparently, while negotiating the price with IBM, his patent was challenged by another inventor whose claim later proved insubstantial. This uncertainty, however, weakened Wang's bargaining position and he settled for a reduced price. Later Wang heard from the family of the man who challenged his patent shortly after he had died, and they told him that IBM had put him up to challenging his patent.

In 1951 Wang Laboratories, a one-man electrical fixtures store above a garage, was founded. Through Wang's continued inventions, the company prospered and in 1964 he released his first desktop calculator. In this same year Wang began developing what would become the mainstay of the company—office computer systems. After developing a string of successful calculators and office word-processing systems, Wang Laboratories reached its peak in 1984 with earnings at $2 billion, and Wang equipment could be found in eighty percent of the nation's largest companies.

With an initial investment of $600, Wang's worth in 1984 was estimated by Forbes magazine at $1.6 billion, making him the fifth richest person in America. This slim, shy inventor, who was usually dressed in a well-tailored suit and a bow tie, was very generous with his wealth, donating millions of dollars to the arts and scholarships. On one occasion in 1980, when the roof of the Boston Performing Arts Center was literally about to collapse, Wang promptly wrote a check for $4 million the moment he was told of the situation, saving the theater.

An Wang was indeed one of the pioneering giants of the computer industry. The holder of forty patents and twenty-three honorary degrees, he was one of only sixty-nine members in the National Inventors Hall of Fame, and his company was vital to the "Massachusetts Miracle" of the 1980s when high technology industries brought thousands of jobs to aging American factory towns.

In March 1990 An Wang, the founder of Wang Laboratories and inventor of the iron ring that served as the computer memory until the microchip, died of cancer at the age of 70.

INTRODUCTION

The first digital ICs contained only four logic gates, so if you wanted to decode a digital signal, you would have to construct a digital decoder circuit on a printed circuit board (PCB) using several logic gate ICs. As semiconductor manufacturers improved their miniaturization techniques, they were able to incorporate larger circuits within a single IC package. Studying the evolving digital electronics industry, semiconductor manufacturers noticed that several circuit functions are always needed in digital electronic systems, so they formed these complete functional circuits within a single IC. They called these digital ICs **combinational logic circuits** since logic gates were combined on a single chip of semiconductor to create functional logic circuits, such as decoders, encoders, multiplexers, demultiplexers, code converters, comparators, and parity generators and checkers. As time has passed, this practice of reducing a frequently needed digital PCB circuit to a single IC has continued, in step with the ability of semiconductor manufacturers to continuously improve their circuit miniaturization techniques.

Combinational Logic Circuits
Digital logic circuits in which the outputs are dependent only on the present input states and the logic path delays.

In this chapter we will take a close look at two types of combinational logic circuits—decoders and encoders.

7-1 DECODERS

The term decode means to translate coded characters into a more understandable form, or to reverse a previous encoding process. A **digital decoder circuit** is a logic circuit that responds to one specific input word or code, while rejecting all others.

Decoder Circuit
A circuit that responds to a particular coded signal while rejecting all others.

7-1-1 *Basic Decoder Circuits*

Figure 7-1(a) shows how an AND gate and two INVERTERS can be connected to act as a basic decoder. As you know, all inputs to an AND gate need to be 1 for the output to be 1. Referring to the function table for this circuit, you can see that only when the input code is 001 will the AND gate have all of its inputs HIGH, and therefore produce a HIGH output.

While some decoders are designed to recognize only one input word combination, other decoders are designed to activate one output for every input word combination. For example, Figure 7-1(b) shows how four AND gates and two INVERTERS can be connected to activate one of four outputs based on the binary input word applied to inputs A and B. The two INVERTERS are included to invert the A and B inputs to create two additional input lines called "NOT A" and "NOT B" (symbolized $\overline{A}$ and $\overline{B}$). Looking at the function table for this circuit, you can see that when inputs A and B are 00_2 (0), only output 0 will be activated or made HIGH. This is because both of the inputs to AND gate 0 are connected to the $\overline{A}$ and $\overline{B}$ input lines, so when A and B are LOW, $\overline{A}$ and $\overline{B}$ are both HIGH, and therefore the output of AND gate 0 will be

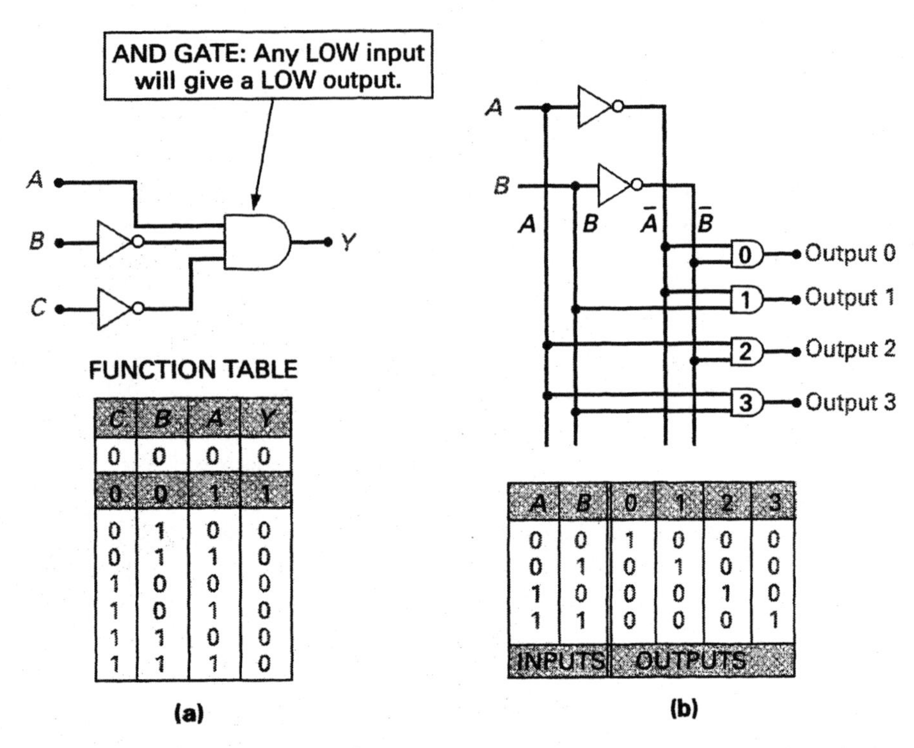

FUNCTION TABLE

C	B	A	Y
0	0	0	0
0	0	1	1
0	1	0	0
0	1	1	0
1	0	0	0
1	0	1	0
1	1	0	0
1	1	1	0

A	B	0	1	2	3
0	0	1	0	0	0
0	1	0	1	0	0
1	0	0	0	1	0
1	1	0	0	0	1
INPUTS		OUTPUTS			

FIGURE 7-1 Basic Decoders.

HIGH. The inputs of the other AND gates are connected to the input lines so that a 01_2 (1) input will make output 1 HIGH, a 10_2 (2) input will make output 2 HIGH, and a 11_2 (3) input will make output 3 HIGH. This circuit would be described as a *one-of-four decoder* since an input code is used to select or activate one of four outputs. If the four outputs were connected to light-emitting diodes (LEDs) that were labeled 0 through 3, the decoder would function as a **binary-to-decimal decoder** since the circuit is translating a binary coded input into a decimal output display.

Binary-to-Decimal Decoder
A decoder circuit that selects one of ten outputs (decimal), depending on the 4-bit binary input applied.

7-1-2 Decimal Decoders

Binary-to-decimal conversion is one of the most common applications for digital decoder circuits. One widely used decimal decoder IC is the 7442. This IC is a BCD-to-decimal decoder, and its data sheet is shown in Figure 7-2, with annotations describing some of the details of the device. Looking first at this ICs internal schematic diagram, you can see that the decoder has a 4-bit BCD input labeled A (pin 15), B (pin 14), C (pin 13), and D (pin 12). As you know, a 4-bit binary input can have a maximum of sixteen possible input combinations; however, this BCD decoder uses only ten NAND gates to decode the ten BCD input codes 0000_2 (0), through 1001_2 (9). Referring to the function table for the 7442, which is shown in the data sheet in Figure 7-2, you can see that the binary value of the BCD input will determine which one of the decimal outputs is enabled. For example, a binary input of 0010_2 (2) will enable the 2 output, whereas a binary input of 0111_2 (7) will enable the 7 output.

DEVICE: SN74LS42—BCD to Decimal Decoder

TOP VIEW

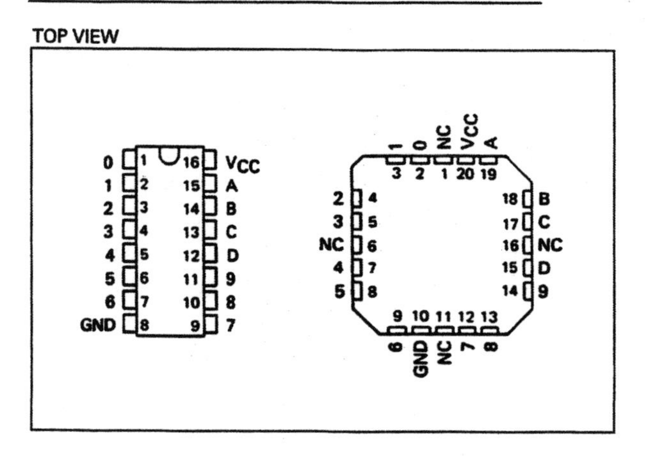

LOGIC SYMBOL

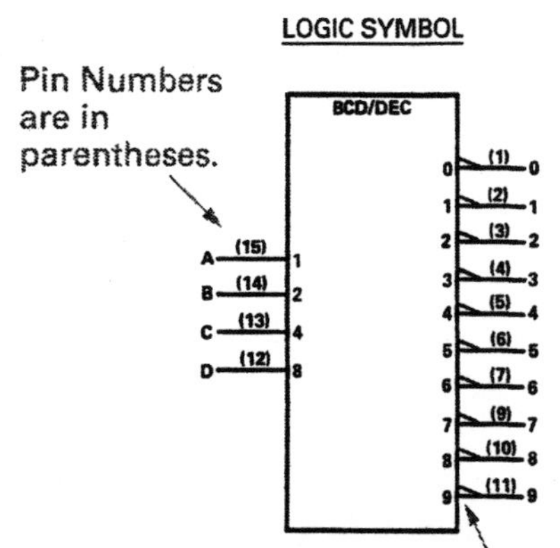

Triangle on outputs denotes an **active LOW output** meaning that the output goes low when active.

Schematic of Inputs and Outputs

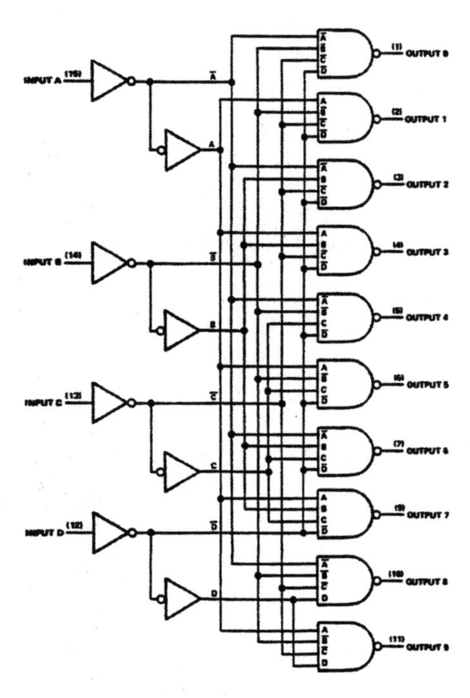

FUNCTION TABLE

NO.	BCD INPUT				DECIMAL OUTPUT									
	D	C	B	A	0	1	2	3	4	5	6	7	8	9
0	L	L	L	L	L	H	H	H	H	H	H	H	H	H
1	L	L	L	H	H	L	H	H	H	H	H	H	H	H
2	L	L	H	L	H	H	L	H	H	H	H	H	H	H
3	L	L	H	H	H	H	H	L	H	H	H	H	H	H
4	L	H	L	L	H	H	H	H	L	H	H	H	H	H
5	L	H	L	H	H	H	H	H	H	L	H	H	H	H
6	L	H	H	L	H	H	H	H	H	H	L	H	H	H
7	L	H	H	H	H	H	H	H	H	H	H	L	H	H
8	H	L	L	L	H	H	H	H	H	H	H	H	L	H
9	H	L	L	H	H	H	H	H	H	H	H	H	H	L
INVALID	H	L	H	L	H	H	H	H	H	H	H	H	H	H
	H	L	H	H	H	H	H	H	H	H	H	H	H	H
	H	H	L	L	H	H	H	H	H	H	H	H	H	H
	H	H	L	H	H	H	H	H	H	H	H	H	H	H
	H	H	H	L	H	H	H	H	H	H	H	H	H	H
	H	H	H	H	H	H	H	H	H	H	H	H	H	H

H = high level, L = low level

FIGURE 7-2 A BCD-to-Decimal Decoder IC. (Courtesy of Texas Instruments)

■ **EXAMPLE:**

Which of the 7442 outputs would be enabled if 1100_2 were applied to the input?

■ ***Solution:***

Referring to the 7442 function table in Figure 7-2, you can see that whenever an invalid BCD code such as 1100_2 or HHLL (decimal 12) is applied, none of the outputs is made active.

Application—BCD-to-Decimal Counter Circuit

Figure 7-3 shows how a 7442 can be used in conjunction with a 7490 decade (ten) counter, to produce a BCD-to-decimal counter/decoder circuit. The 7490 counter, and other counters, will be discussed in more detail in Chapter 12; for now, however, simply think of the counter as a circuit that counts the number of clock pulses applied to its input (pin 14) and then generates a 4-bit binary equivalent value at its outputs (pins 12, 9, 8, 11). In this circuit, the input clock pulses are continuous, so the output binary count from the 7490 will continually advance or cycle through all ten BCD codes—0000, 0001, 0010, 0011, 0100, 0101, 0110, 0111, 1000, 1001—and then reset to 0000 and repeat the cycle. This incrementing 4-bit BCD count from the 7490 is applied to the 4-bit input of the 7442 decoder. An input of 0000 to the 7442 will cause it to make its 0 output LOW, an input of 0001 to the 7442 will cause it to make its 1 output LOW, an input of 0010 to the 7442 will cause it to make its 2 output LOW, and so on. Since the LEDs are arranged as a common-anode display, a LOW at any of the 7442 outputs will turn ON the respective LED. As a result, the LED decimal display will turn ON each LED in sequence (0 through 9), and then repeat the cycle.

■ **EXAMPLE:**

Imagine you had constructed the circuit in Figure 7-3 in lab, and when you switched on the power the decimal display did not flash sequentially. What tests would you make to try and isolate the problem?

■ ***Solution:***

The following tests should help you isolate a problem with the circuit shown in Figure 7-3.

a. Verify +5 V is applied to the 7490, 7442, and LED anodes using the logic probe.

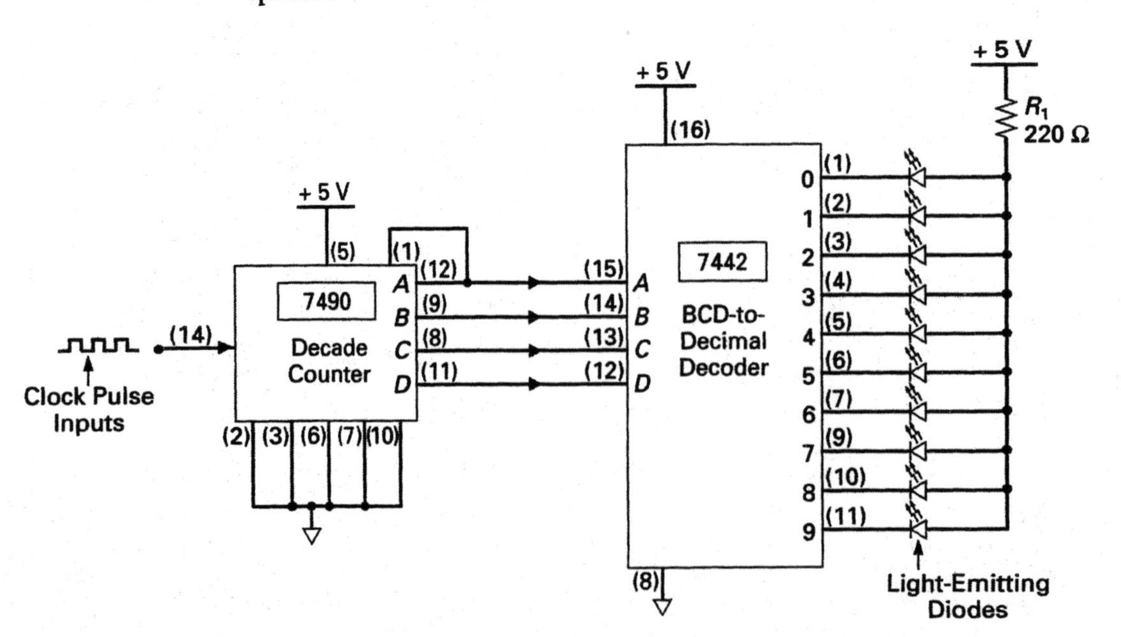

FIGURE 7-3 **Application—BCD-to-Decimal Counter Circuit.**

b. Verify ground is applied to all necessary IC pins using the logic probe.
c. Check clock pulse input at pin 14 of 7490 using the logic probe (dynamic test). You can also use the logic pulser to single pulse the counter input (static test).
d. Apply half split method to circuit—isolate outputs from 7490, and then test to see if counter is operating as it should. This will determine whether problem is in counter half of circuit or decoder half of circuit.
e. Further isolate to determine whether problem is internal or external to suspected IC.

7-1-3 *Hexadecimal Decoders*

The 74154 is a good example of a **hexadecimal decoder,** and its data sheet is shown in Figure 7-4. This widely used decoder will select one of sixteen (hexadecimal) outputs, depending on the 4-bit binary input value. Referring to the function table in Figure 7-4, you can see how each of the sixteen possible binary input combinations (0000 through 1111, or 0 through 15) will select or make LOW one of the sixteen outputs (0–15). This function table also shows how the gate enable inputs ($\overline{G}1$ and $\overline{G}2$) have to be LOW (active-LOW control lines) in order for the logic gates within the IC to be enabled or operate. If these two control inputs are LOW, the input binary word will determine which one of the sixteen outputs is made LOW.

Hexadecimal Decoder
A decoder circuit that selects one of sixteen outputs (hexadecimal) depending on the 4-bit binary input applied.

Application—Sixteen-LED Back-and-Forth Flasher Circuit

Figure 7-5 shows how the 74154 can be used in conjunction with a 74193 up/down counter to produce a sixteen-LED back-and-forth flasher circuit. The circuit in Figure 7-5 operates in the following way. Let us begin by assuming that *B* NAND gate has a HIGH on pin 5, and *A* NAND gate has a LOW on pin 1. These two points are always at opposite logic levels due to the cross-coupling between NAND gates *C* and *D*. With *B* NAND gate pin 5 HIGH, the clock pulses on pin 4 appear at the COUNT-UP input of the counter, and therefore the counter's count will increase in binary (0000, 0001, 0010, and so on toward 1111). This binary count appears at the *ABCD* outputs of the 74193, and is applied to the *ABCD* inputs of the 74154 decoder. The decoder will respond to the binary input and make LOW the equivalent output, and consequently turn ON the associated LED (due to the LOW outputs). When the count reaches 1111 (15), the 15 output (pin 17) goes LOW, turning ON the left-most LED. This LOW output at pin 17 of the 74154 also appears at pin 10 of NAND gate *C,* which produces a HIGH output to drive pin 1 of NAND gate *A,* and allows the clock pulses through to the COUNT-DOWN input of the 74193. The HIGH output of NAND gate *C* also drives pin 12 of NAND gate *D,* whose other input on pin 13 is also HIGH since the 0 output of the decoder is not active at this time. The two HIGH inputs into NAND gate *D* therefore produce a LOW output, which is applied to NAND gate *B* to block the clock pulses from driving the COUNT-UP input of the 74193. The LOW output of NAND gate *D* is also applied to pin 9 of NAND gate *C* to keep the COUNT-

DEVICE: SN74154—Four-Line to Sixteen-Line Decoder

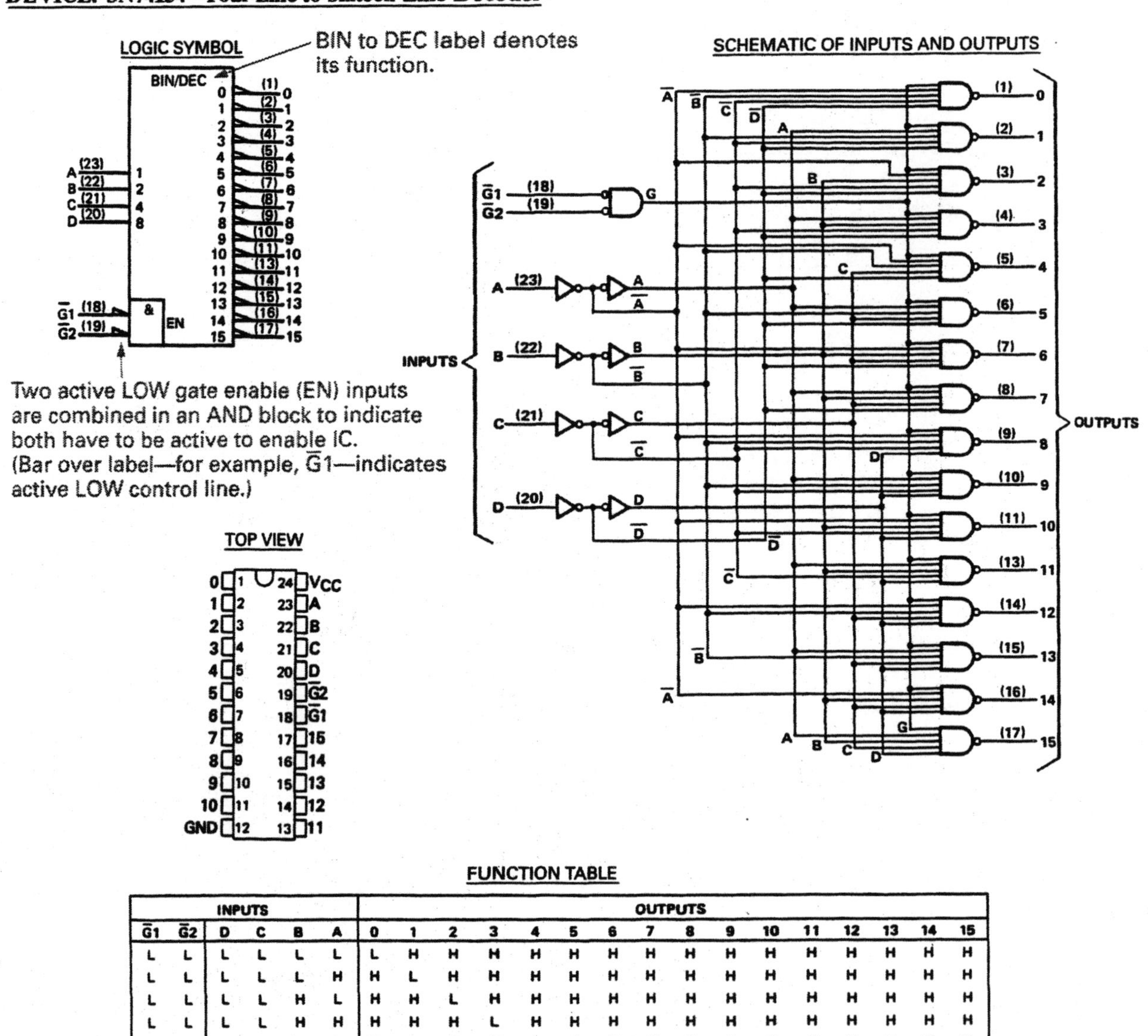

FUNCTION TABLE

INPUTS						OUTPUTS															
$\overline{G1}$	$\overline{G2}$	D	C	B	A	0	1	2	3	4	5	6	7	8	9	10	11	12	13	14	15
L	L	L	L	L	L	L	H	H	H	H	H	H	H	H	H	H	H	H	H	H	H
L	L	L	L	L	H	H	L	H	H	H	H	H	H	H	H	H	H	H	H	H	H
L	L	L	L	H	L	H	H	L	H	H	H	H	H	H	H	H	H	H	H	H	H
L	L	L	L	H	H	H	H	H	L	H	H	H	H	H	H	H	H	H	H	H	H
L	L	L	H	L	L	H	H	H	H	L	H	H	H	H	H	H	H	H	H	H	H
L	L	L	H	L	H	H	H	H	H	H	L	H	H	H	H	H	H	H	H	H	H
L	L	L	H	H	L	H	H	H	H	H	H	L	H	H	H	H	H	H	H	H	H
L	L	L	H	H	H	H	H	H	H	H	H	H	L	H	H	H	H	H	H	H	H
L	L	H	L	L	L	H	H	H	H	H	H	H	H	L	H	H	H	H	H	H	H
L	L	H	L	L	H	H	H	H	H	H	H	H	H	H	L	H	H	H	H	H	H
L	L	H	L	H	L	H	H	H	H	H	H	H	H	H	H	L	H	H	H	H	H
L	L	H	L	H	H	H	H	H	H	H	H	H	H	H	H	H	L	H	H	H	H
L	L	H	H	L	L	H	H	H	H	H	H	H	H	H	H	H	H	L	H	H	H
L	L	H	H	L	H	H	H	H	H	H	H	H	H	H	H	H	H	H	L	H	H
L	L	H	H	H	L	H	H	H	H	H	H	H	H	H	H	H	H	H	H	L	H
L	L	H	H	H	H	H	H	H	H	H	H	H	H	H	H	H	H	H	H	H	L
L	H	X	X	X	X	H	H	H	H	H	H	H	H	H	H	H	H	H	H	H	H
H	L	X	X	X	X	H	H	H	H	H	H	H	H	H	H	H	H	H	H	H	H
H	H	X	X	X	X	H	H	H	H	H	H	H	H	H	H	H	H	H	H	H	H

H = high level, L = low level, X = irrelevant

FIGURE 7-4 A Four-Input to Sixteen-Output Decoder IC. (Courtesy of Texas Instruments)

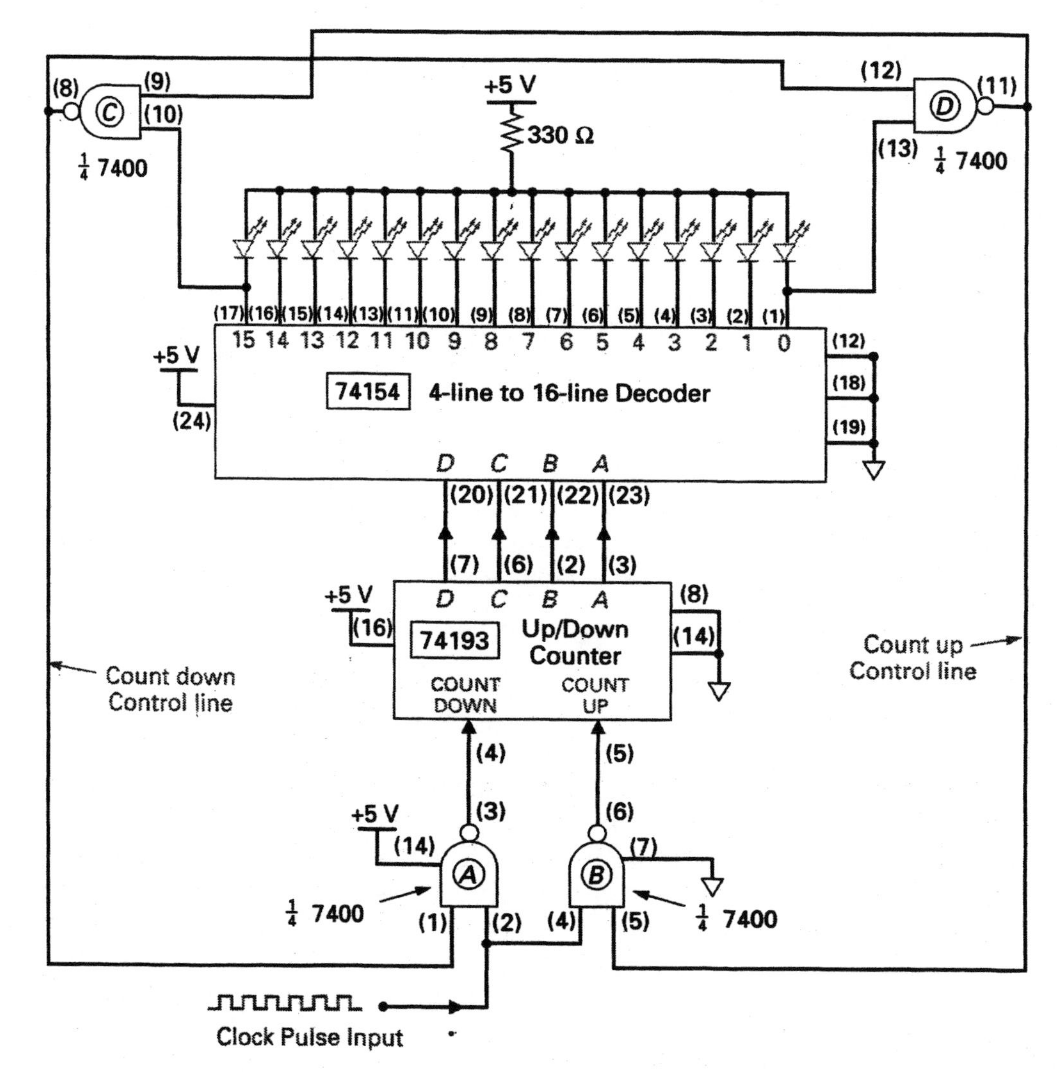

FIGURE 7-5 Application—Sixteen-LED Back-and-Forth Flasher Circuit.

DOWN control line active (HIGH), even after the LOW is no longer present at the 74154's pin 17. The counter, therefore, counts down from 15 to 0, causing the LED's ON/OFF sequence to move from left to right. When the counter reaches a count of 0, the NAND gates will switch the clock pulses once again to the COUNT-UP input of the counter, and the cycle will repeat.

■ **EXAMPLE:**

What modification would you make to the circuit in Figure 7-5 to increase the back-and-forth flashing rate?

■ ***Solution:***

The back-and-forth flashing rate can be increased by simply increasing the clock pulse input frequency.

7-1-4 *Display Decoders*

Many systems such as digital watches, calculators, pagers, and cellular phones make use of a multi-segment display. Decoders are needed in these systems to decode the binary data into the multi-segment data needed to drive the display.

Light-Emitting Diode (LED) Display Decoders

The 7447 IC, shown in Figure 7-6, is one example of a display decoder/driver. Looking at this IC's logic symbol, you can see that it has four inputs and seven outputs. The "numerical designations and resultant displays" section in Figure 7-6 shows what display will result for each of the possible BCD inputs. The function table gives more detail as to which segments (*a* through *g*) will be ON and which will be OFF for each of the 4-bit binary inputs. To review, Figure 7-7(a) shows how the seven segments of an LED display are labeled—*a, b, c, d, e, f,* and *g*—and Figure 7-7(b) shows how certain ON/OFF segment combinations can be used to display the decimal digits numbers 0 through 9.

Application—Seven-Segment LED Display Decoder/Driver

BCD-to-Seven-Segment Decoder A decoder circuit that converts a BCD input code into an equivalent seven-segment output code.

Figure 7-7(c) shows how a 7447 **BCD-to-seven-segment decoder** IC can be connected to drive a common-anode seven-segment LED display. The function table for this digital IC has been repeated in the upper inset in Figure 7-7(c). Looking at the first line of this table, for example, you can see that if a binary input of 0000 (0) is applied to the *ABCD* inputs, the 7447 will make LOW outputs *abcdef,* and make HIGH output *g.* These outputs are applied to the cathodes of the LEDs within the common-anode display (TIL 312), as seen in the lower inset, and since all of the anodes are connected to +5 V, a LOW output will turn ON a segment and a HIGH output will turn OFF a segment. The 7447, therefore, generates active-LOW outputs, which means that a "LOW" output signal will "activate" its associated output LED. In this example the active-LOW outputs from the 7447 will turn ON segments *abcdef,* while the HIGH output will turn OFF segment *g,* causing the decimal digit 0 to be displayed. The remaining lines in the function table indicate which segments are made active for the other binary count inputs 0001 (1) through 1001 (9).

■ **EXAMPLE:**

What would be the *abcdefg* HIGH/LOW code generated by a 7447 if 0101 were applied to the input?

■ ***Solution:***

Referring to the 7447 function table in Figure 7-6, you can see that when 0101, or LHLH (decimal 5), is applied at the input, the 7447 will turn ON (by generating a LOW output) all of the segments except *b* and *e.* These two segment outputs will therefore be HIGH, and the rest will be LOW, so the output code will be ON, OFF, ON, ON, OFF, ON, ON (LHLLHLL).

The circuit shown in Figure 7-7(c) operates in the following way. A 1 Hz clock input signal is applied via a count ON/OFF switch to the input of a 7490

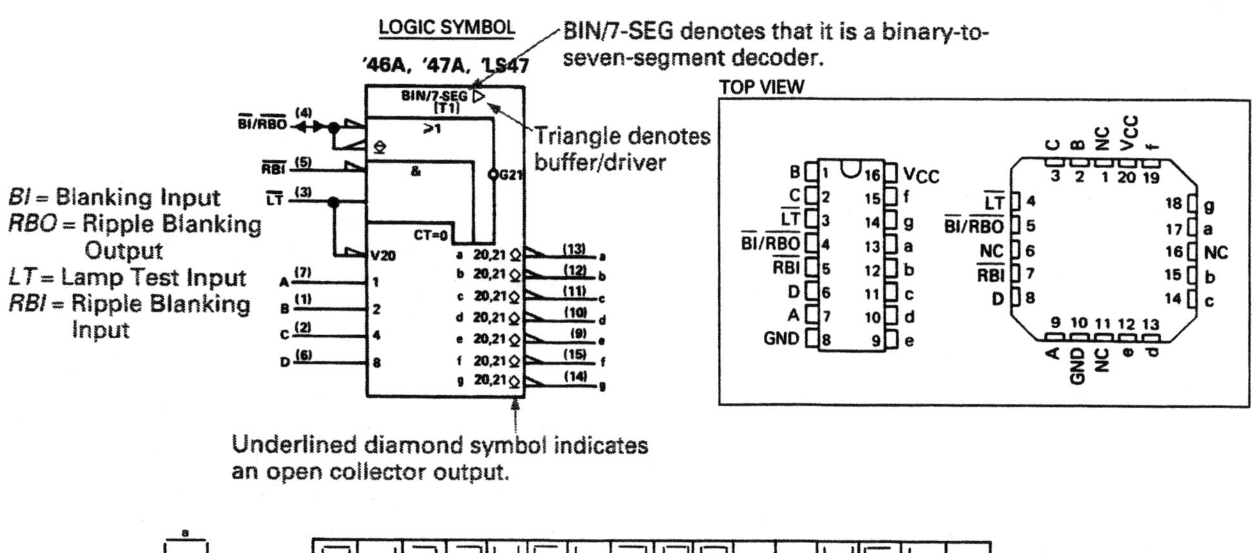

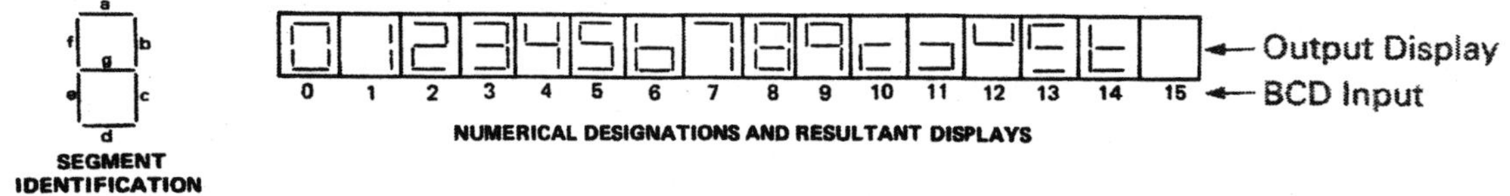

'46A, '47A, 'LS47 FUNCTION TABLE (T1)

DECIMAL OR FUNCTION	INPUTS						$\overline{BI}/\overline{RBO}$†	OUTPUTS							NOTE
	$\overline{LT}$	$\overline{RBI}$	D	C	B	A		a	b	c	d	e	f	g	
0	H	H	L	L	L	L	H	ON	ON	ON	ON	ON	ON	OFF	
1	H	X	L	L	L	H	H	OFF	ON	ON	OFF	OFF	OFF	OFF	
2	H	X	L	L	H	L	H	ON	ON	OFF	ON	ON	OFF	ON	
3	H	X	L	L	H	H	H	ON	ON	ON	ON	OFF	OFF	ON	
4	H	X	L	H	L	L	H	OFF	ON	ON	OFF	OFF	ON	ON	
5	H	X	L	H	L	H	H	ON	OFF	ON	ON	OFF	ON	ON	
6	H	X	L	H	H	L	H	OFF	OFF	ON	ON	ON	ON	ON	
7	H	X	L	H	H	H	H	ON	ON	ON	OFF	OFF	OFF	OFF	
8	H	X	H	L	L	L	H	ON	ON	ON	ON	ON	ON	ON	1
9	H	X	H	L	L	H	H	ON	ON	ON	OFF	OFF	ON	ON	
10	H	X	H	L	H	L	H	OFF	OFF	OFF	ON	ON	OFF	ON	
11	H	X	H	L	H	H	H	OFF	OFF	ON	ON	OFF	OFF	ON	
12	H	X	H	H	L	L	H	OFF	ON	OFF	OFF	OFF	ON	ON	
13	H	X	H	H	L	H	H	ON	OFF	OFF	ON	OFF	ON	ON	
14	H	X	H	H	H	L	H	OFF	OFF	OFF	ON	ON	ON	ON	
15	H	X	H	H	H	H	H	OFF	OFF	OFF	OFF	OFF	OFF	OFF	
BI	X	X	X	X	X	X	L	OFF	OFF	OFF	OFF	OFF	OFF	OFF	2
RBI	H	L	L	L	L	L	L	OFF	OFF	OFF	OFF	OFF	OFF	OFF	3
LT	L	X	X	X	X	X	H	ON	ON	ON	ON	ON	ON	ON	4

FIGURE 7-6 A Binary-to-Seven-Segment Decoder/Driver IC. (Courtesy of Texas Instruments)

digital counter IC. This counter will produce a 4-bit binary word at its outputs *ABCD* based on the number of input clock pulses received at it input (pin 14). This 4-bit binary word at the output of the 7490 is applied to the 7447 seven-segment decoder/driver, where it will be decoded into a seven-segment code that is applied to a seven-segment common-anode display. When the count ON/OFF switch is closed, the 7490 will count, the 7447 will decode the count, and the display will constantly cycle through the digits 0 through 9. When the count

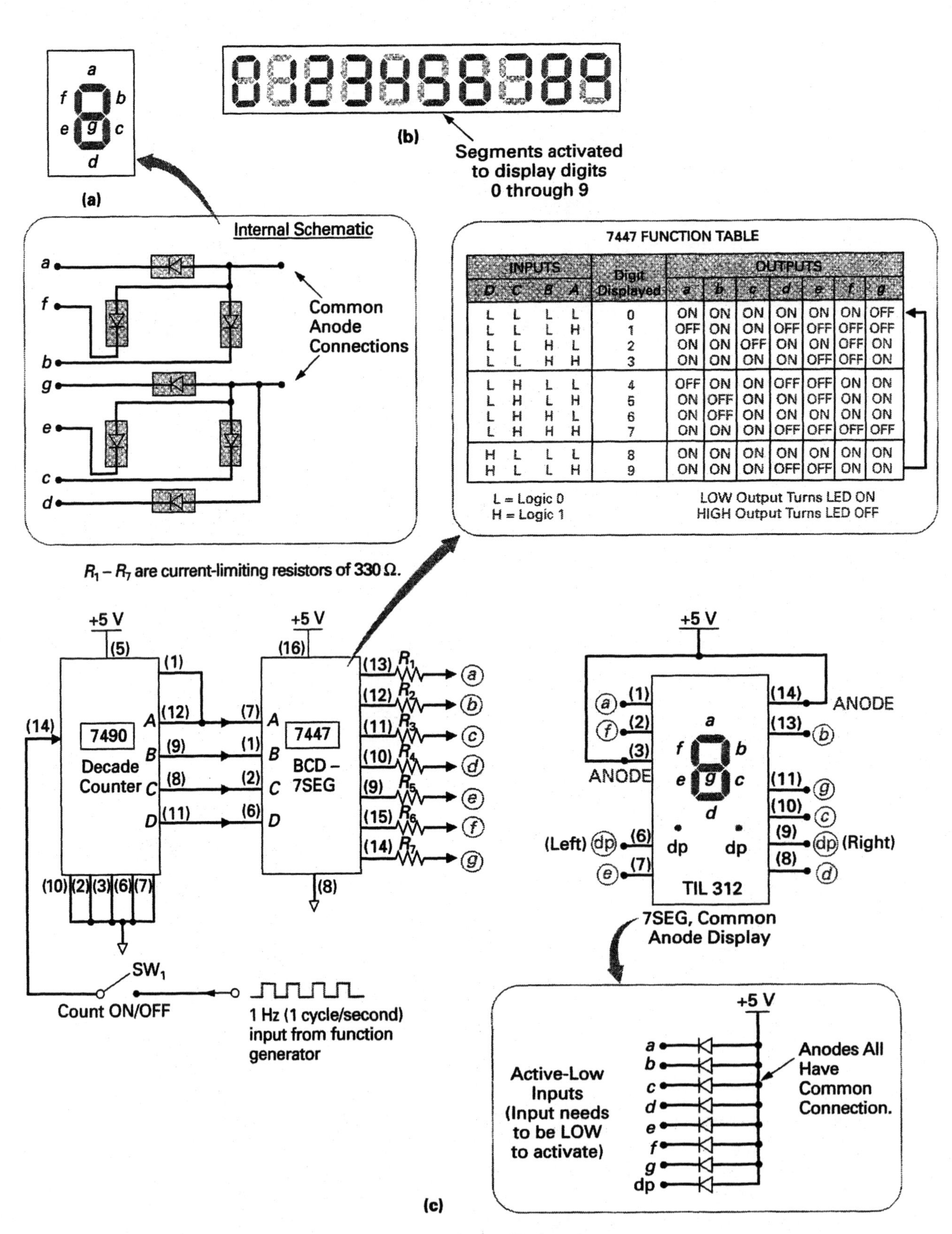

7447 FUNCTION TABLE

INPUTS				Digit Displayed	OUTPUTS						
D	*C*	*B*	*A*		*a*	*b*	*c*	*d*	*e*	*f*	*g*
L	L	L	L	0	ON	ON	ON	ON	ON	ON	OFF
L	L	L	H	1	OFF	ON	ON	OFF	OFF	OFF	OFF
L	L	H	L	2	ON	ON	OFF	ON	ON	OFF	ON
L	L	H	H	3	ON	ON	ON	ON	OFF	OFF	ON
L	H	L	L	4	OFF	ON	ON	OFF	OFF	ON	ON
L	H	L	H	5	ON	OFF	ON	ON	OFF	ON	ON
L	H	H	L	6	ON	OFF	ON	ON	ON	ON	ON
L	H	H	H	7	ON	ON	ON	OFF	OFF	OFF	OFF
H	L	L	L	8	ON	ON	ON	ON	ON	ON	ON
H	L	L	H	9	ON	ON	ON	OFF	OFF	ON	ON

L = Logic 0
H = Logic 1

LOW Output Turns LED ON
HIGH Output Turns LED OFF

FIGURE 7-7 Application—Seven-Segment LED Display/Driver.

ON/OFF switch is opened, the counter will freeze at its present count, and, therefore, so will the number shown on the display.

The 7447 is an example of a BCD-to-seven-segment decoder/driver that has active-LOW outputs for a common-anode seven-segment display. Also available is a 7448 IC, which is a BCD-to-seven-segment decoder/driver that has active-HIGH outputs for a common-cathode seven-segment display. In most applications the 7447 is more commonly used because a 7447 can sink more output current from a common-anode display than a 7448 can source current to a common-cathode display. The 7447 and 7448 are also referred to as **code converters,** because they convert a BCD code into a seven-segment code.

Code Converter A decoder circuit that converts an input code into a different but equivalent output code.

Referring to the 7447 logic symbol in the data sheet in Figure 7-6, you can see that the 7447 has three other inputs that should be explained. They are called lamp test input ($\overline{LT}$), blanking input ($\overline{BI}$), and ripple blanking input ($\overline{RBI}$). All three of these input control lines are active-LOW inputs, as indicated by the triangle on the input lines and the NOT bar over the abbreviated letter designations.

When the lamp test input ($\overline{LT}$) is made active (taken LOW), all of the 7447 will be pulled LOW, and therefore all of the seven-segment LEDs will go ON. This feature is good for quickly testing the decoder and all of the LEDs within the seven-segment display. The blanking input ($\overline{BI}$) can be used to blank the display (turn OFF all of the segments), overriding the present binary input being applied. This feature is used when several seven-segment displays are grouped together to form a multi-digit display and the most significant displays need to be blanked when they are zero. For example, if we had to display the value "003456," there would be no need to display the first two zeros, so they could be blanked so the display shows "3456." The third input, the ripple blanking input ($\overline{RBI}$), is an active-LOW control input that allows us to dim or brighten the display without having to adjust the display voltage applied to the LEDs. The ripple blanking input is usually driven by a pulse waveform, as shown in Figure 7-8. The duty cycle, or ratio of the pulse waveform's HIGH time to LOW time, is made variable under the control of a dimmer control. When the pulse waveform is LOW, the display is blanked, and when the pulse waveform is HIGH, the display is ON and controlled by the 4-bit input word applied to the 7447's *ABCD* inputs. By using a pulse frequency that is greater than 25 Hz so no display flicker is visible, the brightness of the display can be varied by changing the pulse waveform's duty cycle, and therefore the display ON-to-OFF ratio, as seen in the examples in the inset in Figure 7-8.

Liquid-Crystal Display (LCD) Decoders

Liquid-crystal displays are also frequently used in electronic systems to display information. Figure 7-9(a) shows a seven-segment LCD. The sealed liquid crystal segments are normally transparent, but when an ac voltage (typically 3 V to 15 V) is applied between a contact and the back electrode or plate, the crystal molecules become disorganized, resulting in a darkening of the segment. In some instances, edge- or back-lighting is used to improve the display's visibility. Displays using LCDs draw much less current than LED displays, which is why you will often see LCDs used in battery-operated devices such as watches and calculators.

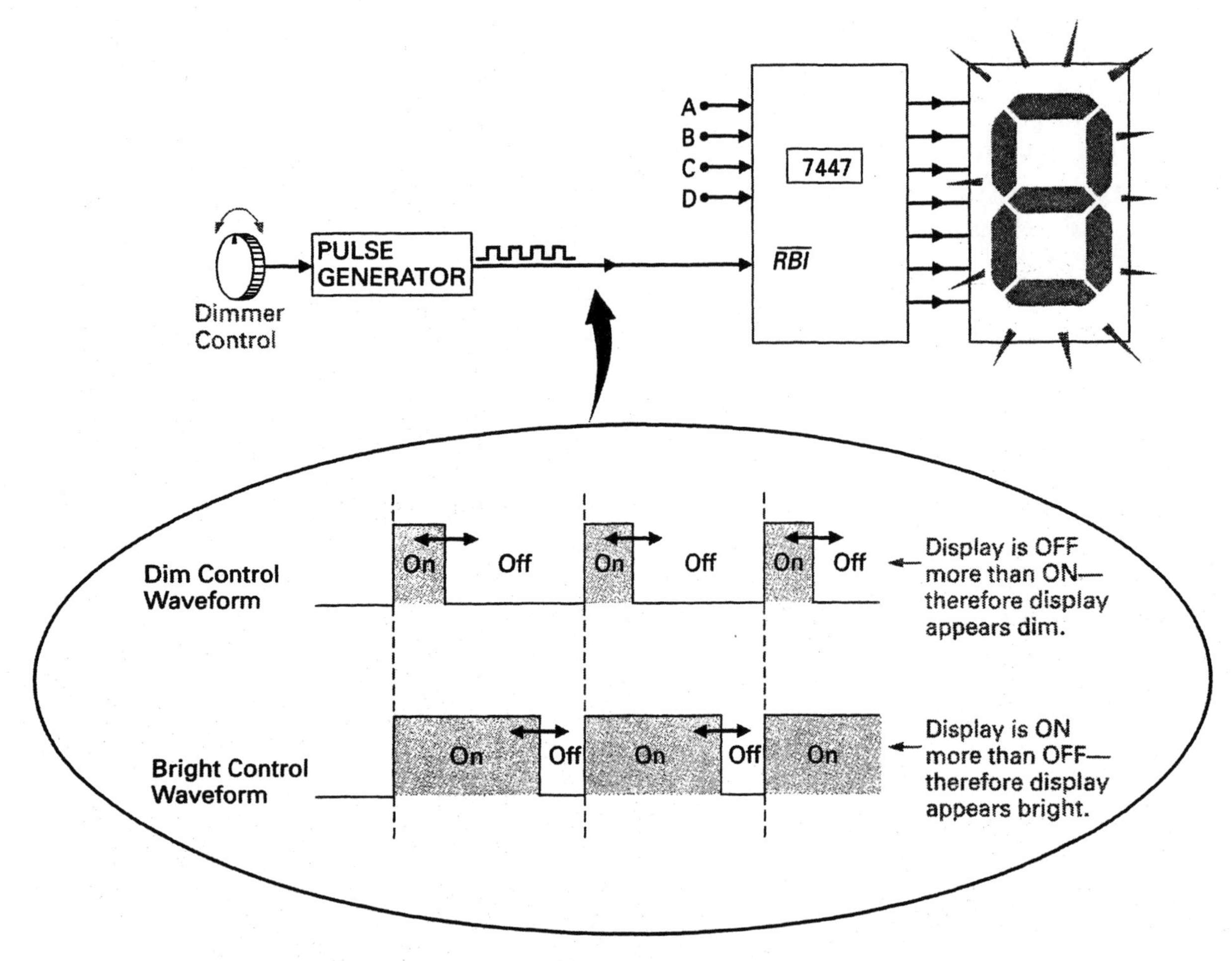

FIGURE 7-8 Ripple Blanking of a Display.

Application—Seven-Segment LCD Display Decoder/Driver

Figure 7-9(b) shows how a CMOS 4511 (BCD-to-seven-segment decoder/driver) can be used in conjunction with two CMOS 4070 ICs (quad exclusive-OR gates) to drive a seven-segment liquid crystal display. The XOR gates act as controlled inverters, and the circuit operates as follows. To turn ON a segment, the 4511 generates a HIGH output which is applied to an XOR gate input. This HIGH input to the XOR will cause the 30 Hz square wave applied to the other input to be inverted due to the action of the XOR, and therefore the 30 Hz being applied to the LCD's segment will be out of phase with the 30 Hz being applied to the LCD's backplate. This potential difference between the backplate and a specific segment will cause the respective segment to darken since an electric field will disorganize the segment molecules, as shown in the inset in Figure 7-9(b). On the other hand, when a 4511 output is LOW, there is no inversion of the 30 Hz square wave, so the 30 Hz being applied to the LCD's segment will be in phase with the 30 Hz being applied to the LCD's backplate. In this instance, there is no potential difference between the backplate and segment, so the molecules remain in their normally organized condition and appear transparent, as shown in the inset in Figure 7-9(b).

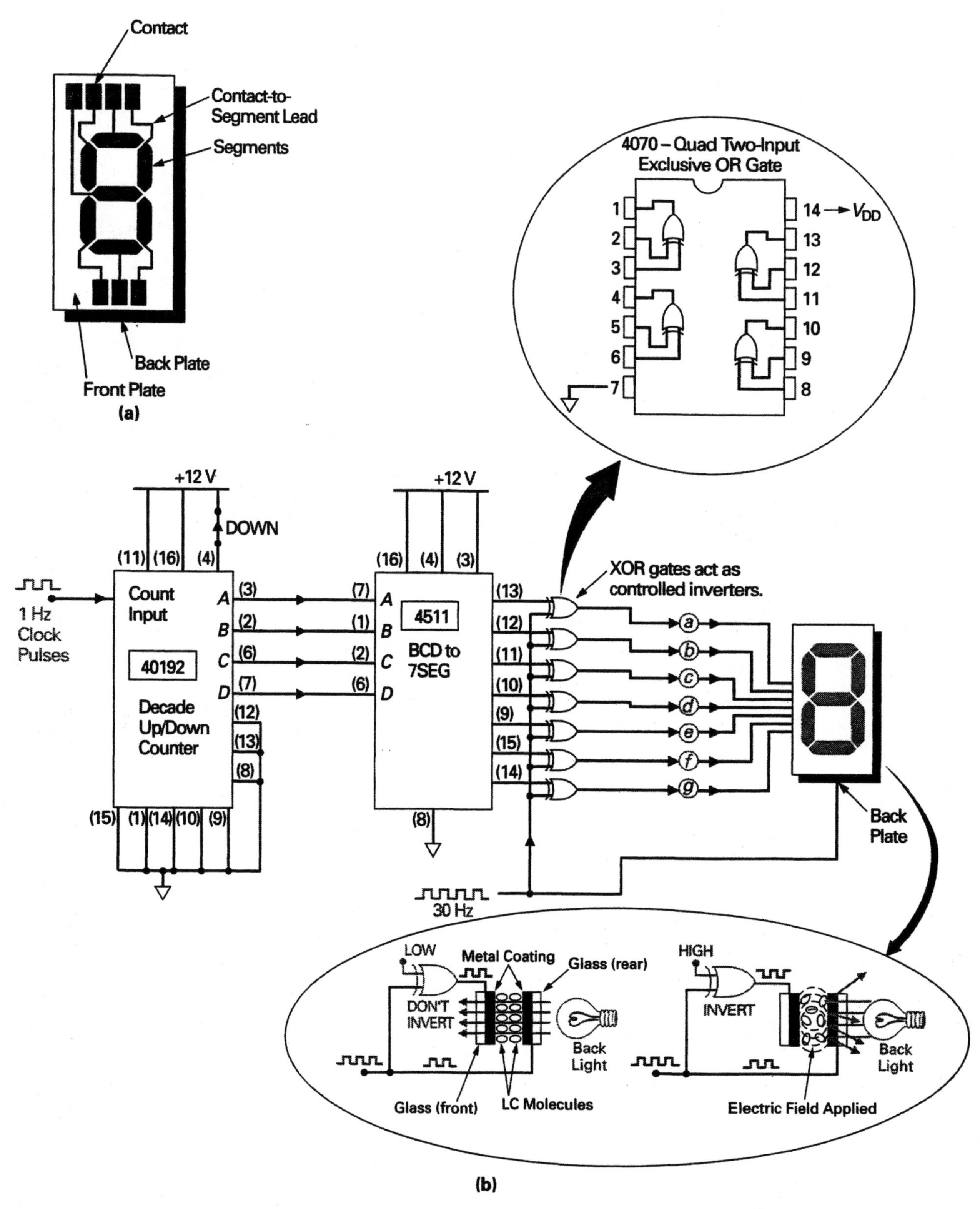

FIGURE 7-9 Application—A Liquid Crystal Display (LCD) Decoder Circuit.

■ **EXAMPLE:**

Would the 4511 BCD-to-seven-segment decoder/driver in Figure 7-9(b) have active-HIGH or active-LOW outputs?

■ ***Solution:***

A HIGH output from the 4511 in Figure 7-9(b) will cause the XOR gate to invert the square wave and turn ON the associated segment. A HIGH output therefore activates the segment, so the outputs of the 4511 are active-HIGH.

In most applications CMOS devices are used to drive LCDs because, like the LCD, they also have a very low power consumption, making them ideal for battery-operated systems. The second reason for using CMOS instead of TTL to drive LCD displays is that the LOW output voltage of the TTL gate can be as much as 0.4 V, and this dc level reduces the life span of the LCD display. This condition does not occur with CMOS logic since its worst-case LOW output is 0.1 V.

SELF-TEST REVIEW QUESTIONS FOR SECTION 7-1

1. What would be the 0 through 9 output from a 7442 BCD-to-decimal decoder if the input were 1001?
2. How many outputs would the following decoders have?
   a. BCD-to-decimal
   b. Binary-to-hexadecimal
3. What would be the *abcdefg* logic levels out of a 7447 if 0110 were applied at the input?
4. Why are CMOS decoder/drivers normally used to drive liquid crystal displays?

7-2 ENCODERS

Encoder Circuit A circuit that generates specific output codes in response to certain input conditions.

An **encoder circuit** performs the opposite function of a decoder. A decoder is designed to *detect* specific codes, whereas an encoder is designed to *generate* specific codes.

7-2-1 ***Basic Encoder Circuits***

Figure 7-10(a) shows how a simple decimal-to-binary encoder circuit can be constructed using three push-buttons, three pull-up resistors, and two NAND gates. The pull-up resistors are included to ensure that the input to the NAND gates are normally HIGH. When button 1 is pressed, the upper input of NAND gate *A* will be pulled LOW, and since any LOW input to a NAND gate will produce a HIGH output, A_0 will be driven HIGH. The two inputs to NAND gate *B* are unaffected by SW_1, so the two HIGH inputs to this NAND gate will

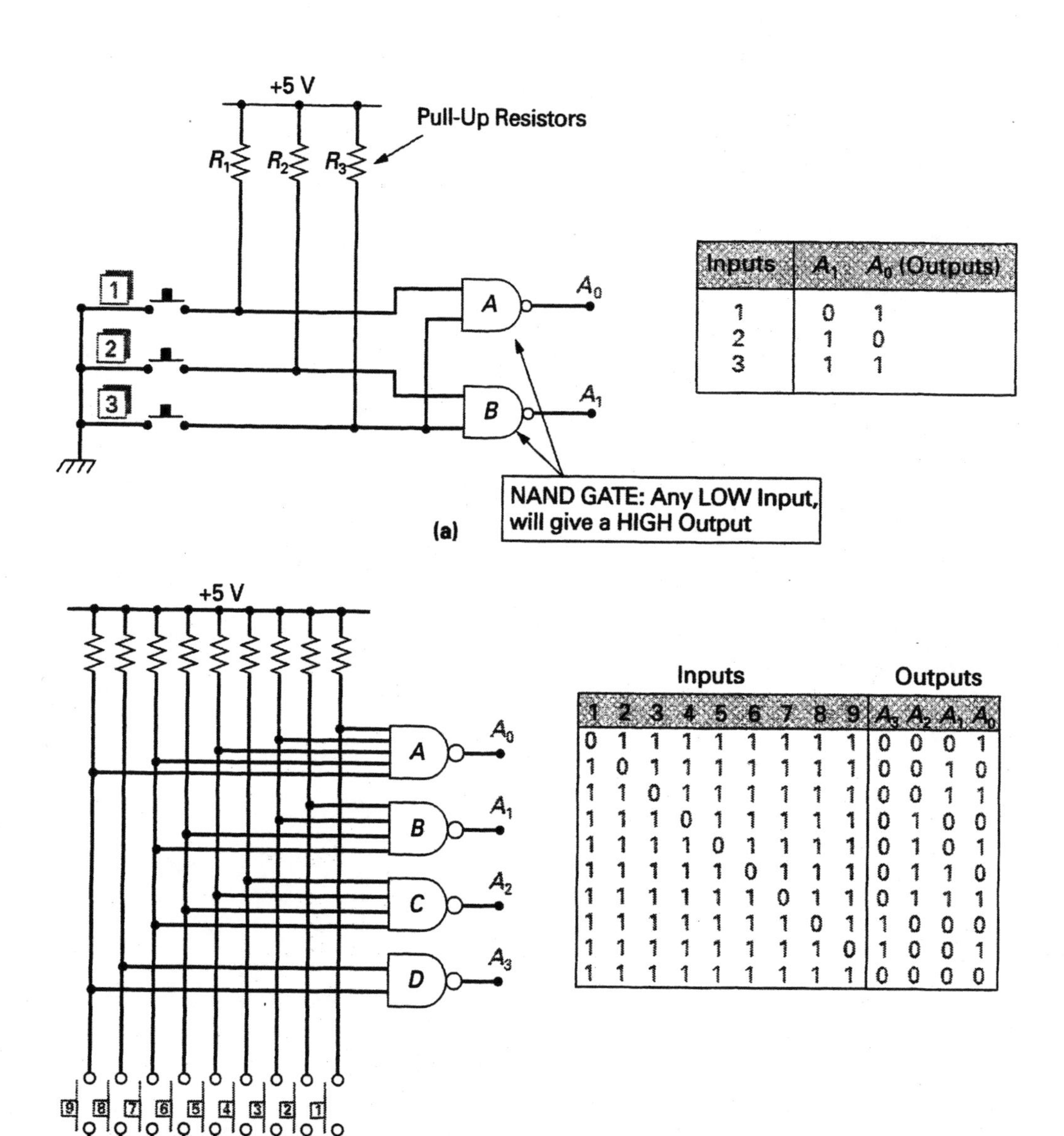

Inputs	A_1	A_0 (Outputs)
1	0	1
2	1	0
3	1	1

Inputs									Outputs			
1	2	3	4	5	6	7	8	9	A_3	A_2	A_1	A_0
0	1	1	1	1	1	1	1	1	0	0	0	1
1	0	1	1	1	1	1	1	1	0	0	1	0
1	1	0	1	1	1	1	1	1	0	0	1	1
1	1	1	0	1	1	1	1	1	0	1	0	0
1	1	1	1	0	1	1	1	1	0	1	0	1
1	1	1	1	1	0	1	1	1	0	1	1	0
1	1	1	1	1	1	0	1	1	0	1	1	1
1	1	1	1	1	1	1	0	1	1	0	0	0
1	1	1	1	1	1	1	1	0	1	0	0	1
1	1	1	1	1	1	1	1	1	0	0	0	0

FIGURE 7-10 Basic Encoder Circuits.

produce a LOW output at A_1. Referring to the truth table in Figure 7-10(a), you can see that the two-bit code generated when switch 1 is pressed is 01 (binary 1). Studying the circuit and the rest of the function table shown in Figure 7-10(a), you can see that the code 10 (binary 2) will be generated when switch 2 is pressed, and the code 11 (binary 3) will be generated when switch 3 is pressed. This basic decimal-to-binary encoder circuit, which has three inputs and two outputs, would be described as a three-line to two-line encoder.

Figure 7-10(b) expands on the basic encoder circuit discussed in Figure 7-10(a) by including extra push-button switches and NAND gates so we can generate a 4-bit BCD code that is equivalent to the decimal key pressed. This circuit, therefore, is a *decimal-to-BCD encoder* circuit, and it will operate in the same manner as the previously discussed basic encoder. For example, when the decimal 2 key is pressed, it grounds an input to NAND gate *B*, so only the A_1 output goes HIGH, giving a 0010 (binary 2) output. As another example, when the decimal 7 key is pressed, it grounds inputs to NAND gates *A, B,* and *C,* so the A_0, A_1, A_2 outputs go HIGH, giving a 0111 (binary 7) output.

■ **EXAMPLE:**

Referring to the encoder circuit in Figure 7-10(b), determine what the output code would be if push-buttons 2 and 4 were pressed simultaneously.

■ ***Solution:***

If push-buttons 2 and 4 in Figure 7-10(b) were pressed simultaneously, NAND gates *B* and *C* would both have a LOW input, so both would generate a HIGH output. The output codes for buttons 2 (0010) and 4 (0100) therefore would be combined, producing a 0110 output code.

In most applications there is no need to build an encoder circuit using discrete components, since a variety of combinational logic ICs are available. Let us now examine the operation and application of a frequently used digital encoder IC.

Decimal-to-BCD Encoder
An encoder circuit that in response to decimal input conditions generates equivalent BCD output codes.

Negative-Logic
1. An active-LOW signal or code. 2. Digital logic in which the more negative logic level represents a binary 1.

Positive-Logic
1. An active-HIGH signal or code. 2. Digital logic in which the more positive logic level represents a binary 1.

Priority Encoder
An encoder circuit that in response to several simultaneous inputs will generate an output code that is equivalent to the highest priority input.

7-2-2 Decimal-to-BCD Encoders

Figure 7-11 shows the data sheet for the 74147 IC, which is a ten-line to four-line (decimal-to-BCD) priority encoder. The bars over the 74147's $\overline{A_1}$ through $\overline{A_9}$ input labels and the small triangle on the input lines are used to indicate that these are active-LOW inputs. This means that the encoder's inputs will be activated whenever these inputs are made LOW. There are also bars over the 74147's $\overline{Q_0}$ through $\overline{Q_3}$ outputs. This is again used to indicate that the output is an active-LOW code. This active-LOW output code is also known as **negative-logic,** and if you need to convert the output to an active-HIGH code, or **positive-logic,** you will need to invert the output from the 74147. For example, looking at the bottom line of the function table in Figure 7-11, you can see that if the $\overline{A_1}$ input is made LOW (active), the 74147 will generate the negative-logic output code HHHL (1110), which will have to be inverted, or reversed, to the positive-logic code LLLH (0001 or binary 1) in order to be equivalent to the activated input $\overline{A_1}$.

As another example, looking at the second line of the function table, you can see that if the $\overline{A_9}$ input is made LOW (active), the 74147 will generate the negative-logic code LHHL (0110), which will have to be inverted to the positive-logic code HLLH (1001 or binary 9) in order to be equivalent to the activated input $\overline{A_9}$. The rest of the function table shows how a LOW on any of the $\overline{A_1}$ through $\overline{A_9}$ inputs will generate an equivalent negative-logic code. To explain one other detail, the Xs in the function table in Figure 7-11 indicate a don't-care condition. This is because the 74147 is a **priority encoder,** or high priority encoder (HPRI), which means that if more than one key is pressed at the same time, the 74147 will only generate the code for the larger digit, or higher priority. For example, if the 9 and 6 keys are pressed at the same time, you can see on

DEVICE: SN74LS147—Ten-Line Decimal to Four-Line BCD Priority Encoder

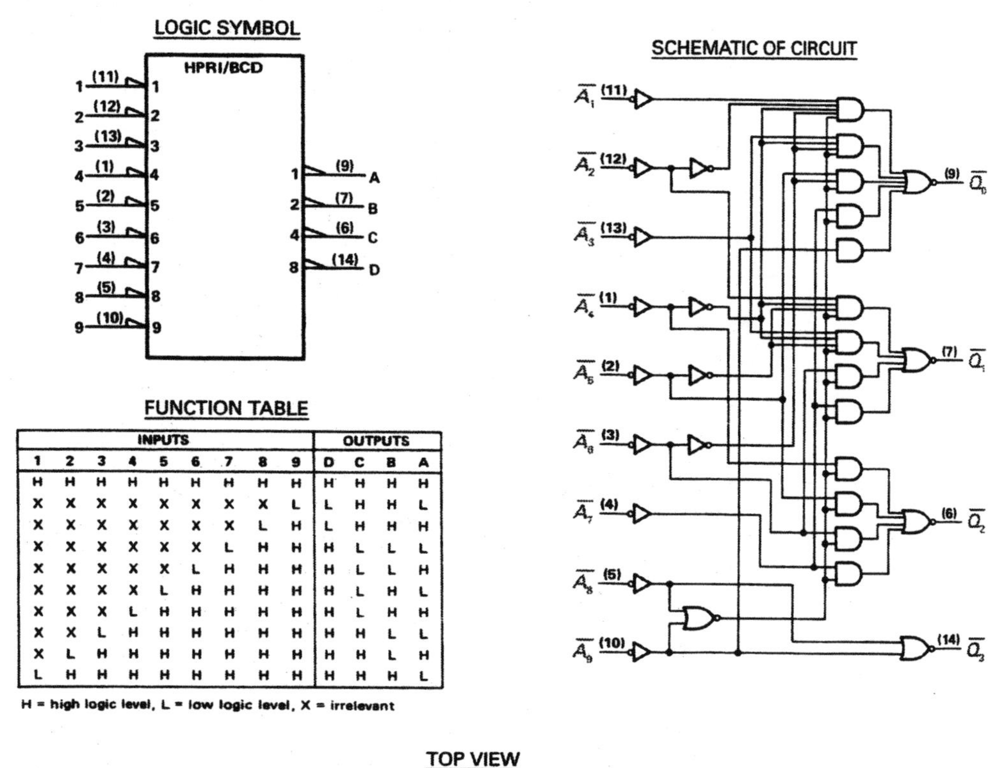

INPUTS									OUTPUTS			
1	2	3	4	5	6	7	8	9	D	C	B	A
H	H	H	H	H	H	H	H	H	H	H	H	H
X	X	X	X	X	X	X	X	L	L	H	H	L
X	X	X	X	X	X	X	L	H	L	H	H	H
X	X	X	X	X	X	L	H	H	H	L	L	L
X	X	X	X	X	L	H	H	H	H	L	L	H
X	X	X	X	L	H	H	H	H	H	L	H	L
X	X	X	L	H	H	H	H	H	H	L	H	H
X	X	L	H	H	H	H	H	H	H	H	L	L
X	L	H	H	H	H	H	H	H	H	H	L	H
L	H	H	H	H	H	H	H	H	H	H	H	L

H = high logic level, L = low logic level, X = irrelevant

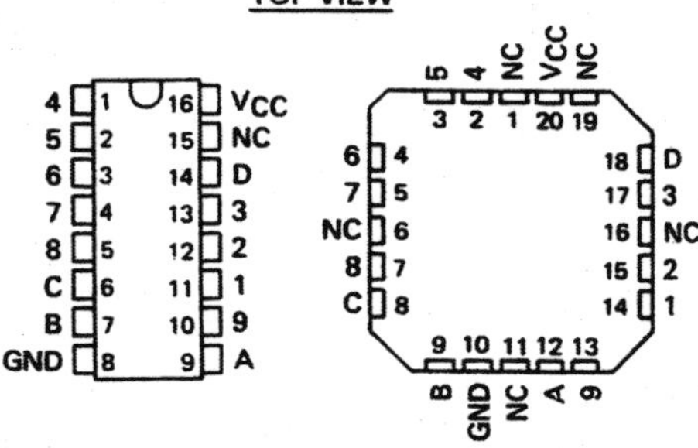

FIGURE 7-11 A Ten-Input (Decimal) to Four-Output (BCD) Priority Encoder IC. (Courtesy of Texas Instruments)

the second line of the function table that the 74147 will generate a nine output-code, responding only to the LOW on the $\overline{A_9}$ input and ignoring the LOW on the $\overline{A_6}$ input. As another example, looking at the third line of the function table you can see that if input $\overline{A_8}$ is taken LOW, the output code will be LHHH which is the inverse of 1000, the BCD code for 8. This BCD code of 8, however, will only be present at the output if the $\overline{A_9}$ input is inactive, since the $\overline{A_9}$ input would have a higher priority than the $\overline{A_8}$ input. As you move down the truth table, you can see how a lower priority input will produce a corresponding BCD output code as long as a higher priority input is not active.

Application—A 0-to-9 Keypad Encoder Circuit

As an application, Figure 7-12 shows how a 74147 can be used to encode a calculator's keypad. Looking first at the push-button switches, which are con-

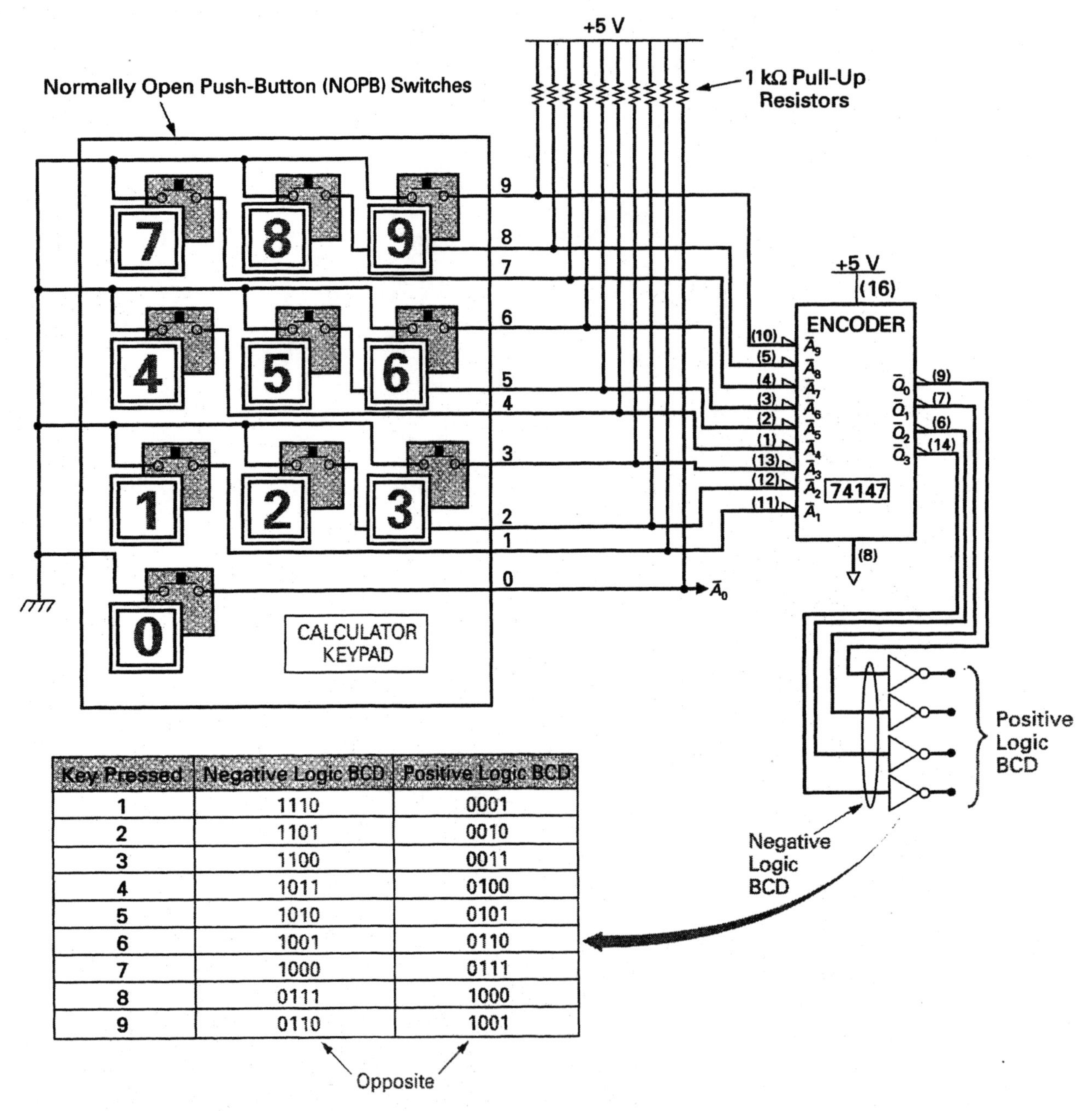

Key Pressed	Negative Logic BCD	Positive Logic BCD
1	1110	0001
2	1101	0010
3	1100	0011
4	1011	0100
5	1010	0101
6	1001	0110
7	1000	0111
8	0111	1000
9	0110	1001

FIGURE 7-12 Application—A 0-to-9 Keypad Encoder Circuit.

trolled by the 1 through 9 key pads, you can see that pull-up resistors are included to ensure that the 74147's inputs are normally HIGH. When a key is pressed, the respective input is taken LOW, making that input to the 74147 active. The table shown in the inset in Figure 7-12 shows how this encoder circuit will respond to different inputs. For example, when the 1 key is pressed, the 74147 will receive a LOW at the $\overline{A_1}$ input, so it will generate a 1110 negative-logic output code, which will be inverted by a set of four INVERTER logic gates to the positive-logic code 0001. Therefore, a binary 1 (0001) output code is generated whenever the 1 key is pressed. This circuit, therefore, will generate a positive-logic BCD code at the output of the INVERTER gates that is equivalent to the value of the keypad key pressed.

■ **EXAMPLE:**

What would be the negative-logic and positive-logic code generated by the circuit in Figure 7-12, if none of the keys were pressed?

■ ***Solution:***

Referring to the first line of the function table for the 74147 data sheet shown in Figure 7-11, you can see that when none of the inputs is active (LOW), a 1111 negative-logic output code is generated at the output of the 74147 which, after the inversion, will be a positive-logic code 0000 (0).

SELF-TEST REVIEW QUESTIONS FOR SECTION 7-2

1. How many inputs and outputs would a decimal-to-BCD encoder have?
2. How can the negative BCD logic output of a 74147 be converted to a positive BCD logic output?
3. If the 3, 4 and 7 inputs of a priority encoder are all activated at the same time, what would be the binary output code?
4. Convert the following negative-logic codes to positive-logic codes.
   a. 0010 **b.** 1110 **c.** LHLL **d.** ON, OFF, ON, OFF

7-3 TROUBLESHOOTING DECODERS AND ENCODERS

To be an effective electronics technician or troubleshooter, you must have a thorough knowledge of electronics, test equipment, troubleshooting techniques, and equipment repair. In most cases, a technician is required to quickly locate the problem within an electronic system and then make the repair. The procedure for fixing a failure can be broken down into three steps.

Step 1: DIAGNOSE

The first step is to determine whether a circuit problem really exists, or if it is simply an operator error. To carry out this step, a technician must collect as much information about the system, circuit, and components used, and then diagnose the problem.

Step 2: ISOLATE

The second step is to apply a logical and sequential reasoning process to isolate the problem. In this step a technician will operate, observe, test, and apply troubleshooting techniques in order to isolate the malfunction.

Step 3: REPAIR

The third and final step is to make the actual repair, and then final test the circuit.

As far as troubleshooting is concerned, practice really does make perfect. In this and each of the following digital circuit chapters, therefore, we will complete the chapter by applying this three-step troubleshooting process to an application circuit that combines all of the circuits discussed previously in the chapter. Since this chapter was devoted to digital encoder and decoder circuits, let us first examine the operation of a typical encoder/decoder circuit, and then apply our three-step troubleshooting process to this circuit.

7-3-1 An Encoder-and-Decoder Circuit

Figure 7-13 combines two application circuits discussed previously in this chapter. This circuit makes use of a 74147 HPRI decimal-to-BCD encoder to generate a negative-logic BCD code that is equivalent to the value of the keypad key pressed. The 4-bit negative-logic BCD code from the output of the 74147 is inverted by four NOT gates from a 7404 hex INVERTER IC, and the resulting positive-logic BCD code is applied to a 7447 BCD-to-seven-segment decoder/driver. The 7447 decodes the BCD input code and generates a seven-segment active-LOW output code to drive a common-anode seven-segment LED display. Whenever a key is pressed and held, therefore, the decimal equivalent of that key is displayed on the seven-segment display.

Let us now apply our three-step troubleshooting procedure to this encoder/decoder circuit so that we can practice troubleshooting procedures and methods.

Step 1: Diagnose

It is extremely important that you first understand the operation of a circuit and how all of the devices within it are supposed to work so that you are able to determine whether or not a circuit malfunction really exists. If you were preparing to troubleshoot the encoder/decoder circuit in Figure 7-13, your first step should be to read through the circuit description and review the operation of each integrated circuit until you feel completely confident with the correct operation of the entire circuit. The circuit description, or theory of operation, for an electronic circuit can generally be found in a service or technical manual, along with troubleshooting guides. As far as the circuit's ICs are concerned, manufacturers' digital data books contain a full description of the IC's operation, characteristics, and pin allocation. Referring to all of this documentation before you begin troubleshooting will generally speed up and simplify the isolation step. Many technicians bypass this "data collection" step and proceed directly to the isolation step. If you are completely familiar with the circuit's operation, this shortcut would not hurt your performance. However, if you are not completely familiar with the circuit, keep in mind the following expression and let it act as a brake to stop you from racing past the problem: Less haste, more speed.

Once you are fully familiar with the operation of the circuit, you will easily be able to diagnose the problem as either an *operator error* or a *circuit malfunction.*

Distinguishing an operator error from an actual circuit malfunction is an important first step, and a wrong diagnosis can waste much time and effort. For

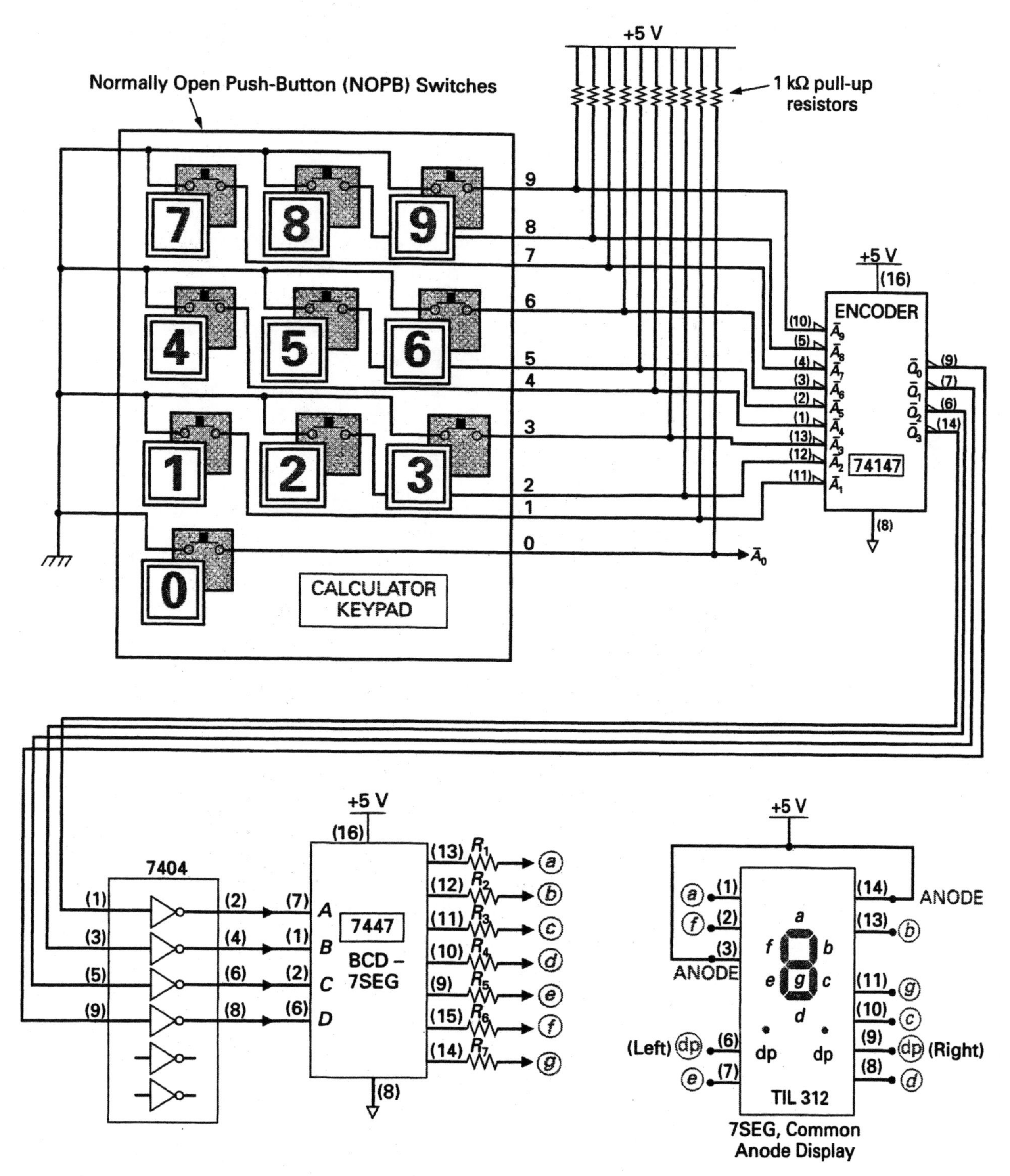

FIGURE 7-13 Troubleshooting Decoder and Encoder Circuits.

example, the following could be interpreted as circuit malfunctions, when in fact they are simply operator errors.

Example 1.
Symptom: The display shows a 0 when I am not pressing any of the keys.
Diagnosis: Operator error—this is normal, since a 0000 positive-logic BCD code will be generated when none of the keys are being pressed.

Example 2.
Symptom: The decimal points on the seven-segment display are not working.
Diagnosis: Operator error—this circuit is not designed to activate the decimal points, only to display digit 0 when no keys are pressed, and digits 1 through 9 whenever keys 1 through 9 are pressed.

Once you have determined that the problem is not an operator error, but is in fact a circuit malfunction, proceed to step 2 and isolate the circuit failure.

Step 2: Isolate

No matter what circuit or system failure has occurred, you should always follow a logical and sequential troubleshooting procedure. Let us review some of the isolating techniques and apply them to our example circuit in Figure 7-13.

a. Use a cause-and-effect troubleshooting process, which means study the effects you are getting from the faulty circuit and then logically reason out what could be the cause.

b. Check first for obvious errors before leaping into a detailed testing procedure. Is the power OFF or not connected to the circuit? Are there wiring errors? Are all of the ICs correctly oriented?

Many of your protoboard circuit problems will be caused by construction errors. These problems include wiring errors, using incorrect ICs or other devices, and incorrectly oriented components. A single construction error is not too difficult to find; however, multiple construction errors can be extremely difficult to find since the combined symptoms will send you in all directions. To limit these construction errors, remember that "prevention is better than cure," so try to spend a little more time during the construction of a circuit, double-checking your circuit along the way.

c. Using a logic probe or voltmeter, test that power and ground are connected to the circuit and are present at all points requiring power and ground. If the whole circuit—or a large section of the circuit—is not operating, the problem is normally power. Using a multimeter, check that all of the dc voltages for the circuit are present at all IC pins that should have power or a HIGH input, and are within tolerance. Secondly, check that 0 V or ground is connected to all IC ground pins and all inputs that should be tied LOW.

d. Use your senses to check for broken wires, loose connectors, overheating or smoking components, pins not making contact, and so on.

e. Apply the half-split method of troubleshooting first to a circuit, and then to a section of a circuit, to help speed up the isolation process. With our encoder/decoder circuit, a good mid-point check would be to test the negative logic output codes from the encoder using the logic probe.

- If the output codes are correct when several keys are pressed but the digit is not shown on the display, you know that the keypad and encoder section of the circuit is okay, and therefore the problem must be somewhere in the INVERTER, decoder, or display section of the circuit.
- If the output codes from the encoder are incorrect when several keys are pressed, then the problem is probably within the keypad and encoder section of the circuit. Remember, however, that a load problem can make a source appear at fault, so it would be safer to disconnect the encoder from the rest of the circuit, and then recheck its output codes. If these codes are still incorrect, the problem is definitely within the keypad and encoder section; however, if the encoder's codes are now correct, the chances are that a short in one of the data lines or NOT gates is pulling down the output of the encoder.

f. Substitution can be used to help speed up your troubleshooting process. Once the problem is localized to an area containing only a few ICs, substitute suspect ICs with known working ICs (one at a time) to see if the problem can be quickly remedied.

g. Many electronic system manufacturers provide troubleshooting trees in the system technical manual. These charts are a graphical means to show the sequence of tests to be performed on a suspected circuit or system.

Step 3: Repair

Once the fault has been found, the final step is to repair the circuit, which could involve simply removing an excess piece of wire, re-soldering a broken connection, reconnecting a connector, or adjusting the power supply voltage. In most instances, however, the repair will involve the replacement of a faulty component. For a circuit that has been constructed on a prototyping board or bread board, the removal and replacement of a component is simple; however, when a printed circuit board is involved, you should make a note of the component's orientation and observe good soldering and de-soldering techniques. Also be sure to handle any MOS ICs with care to prevent any damage due to static discharge.

When the circuit has been repaired, always perform a final test to see that the circuit and the system is now fully operational.

7-3-2 ***Sample Problems***

Once you have constructed a circuit like the encoder/decoder circuit in Figure 7-13, introduce a few errors to see what effect or symptoms they produce. Then logically reason out why a particular error or cause has a particular effect on the circuit. Never short any two points together unless you have carefully thought out the consequences, but generally it is safe to open a path and see the results. Here are some examples for the encoder/decoder circuit in Figure 7-13.

a. Disconnect power to the 74147.

b. Disconnect power from the keypad's pull-up resistors.

c. Open one of the NOT gate inputs.
d. Disconnect an input to a NOT gate and connect it directly to the 7447 input.
d. Disconnect power to the 7447.
e. Open one of the leads from the 7447 to the TIL312 by removing one of the current-limiting resistors.
f. Disconnect power to the TIL312.
g. Connect one of the 7447 outputs to the "left dp" input.

SELF-TEST REVIEW QUESTIONS FOR SECTION 7-3

1. List the steps of the three-step troubleshooting procedure.
2. What is the basic difference between an operator error and a circuit malfunction?
3. What is the half-split troubleshooting technique?
4. What symptom would you get from the encoder/decoder circuit in Figure 7-13 if the *c* segment LED in the seven-segment display had burned out?

REVIEW QUESTIONS

Multiple-Choice Questions

1. A ___________ translates coded characters into a more understandable form.

   a. decoder b. encoder c. multiplexer d. code converter

2. How many inputs and outputs would a BCD-to-decimal decoder have?

   a. three inputs, sixteen outputs b. ten inputs, four outputs
   c. ten inputs, one output d. four inputs, ten outputs

3. A common-cathode seven-segment display would need active-___________ inputs, while a common-anode seven-segment display would need active-___________ inputs.

   a. HIGH, LOW b. LOW, LOW c. LOW, HIGH d. HIGH, HIGH

4. Which segments of a seven-segment display would be active when the decimal digit seven is being displayed?

   a. abfg b. abde c. ace d. abc

5. The ripple blanking feature of a seven-segment display decoder IC is used to ___________.

   a. blank all segments permanently
   b. remove or blank any ac ripple on the power line
   c. adjust the brightness of the display
   d. turn ON all of the segments

6. What code conversion is performed by a 7448 IC?

   a. BCD to gray b. Hex to BCD
   c. BCD to seven-segment d. Gray to seven-segment

7. ____________ are designed to detect specific codes, while ____________ are designed to generate specific codes.

   a. Code converters, decoders b. Decoders, encoders
   c. Encoders, code converters d. Encoders, decoders

8. A decimal-to-BCD encoder will have ____________ inputs and ____________ outputs.

   a. ten, four b. three, ten c. four, ten d. ten, three

9. The negative-logic BCD code for 9 is ____________.

   a. 1001 b. 1010 c. 0111 d. 0110

10. What is the positive-logic code for the negative-logic code 1010?

   a. 0110 b. 1100 c. 0101 d. 1001

Essay Questions

11. Briefly describe the function and applications of a decoder. (7-1)
12. Sketch an example of a decoder circuit. (7-1)
13. What is a decimal decoder? (7-1)
14. Briefly describe the pin-out and function table of a 7442 IC. (7-1)
15. What is a hexadecimal decoder? (7-1)
16. Briefly describe the pin-out and function table of a 74154 IC. (7-1)
17. What is a seven-segment decoder? (7-1)
18. Briefly describe the pin-out and function table of a 7447 IC. (7-1)
19. What is the difference between a 7447 decoder and a 7448 decoder? (7-1)
20. Why are CMOS decoder/drivers generally used in conjunction with liquid crystal displays? (7-1)
21. Briefly describe the function and applications of an encoder. (7-2)
22. Sketch an example of a encoder circuit. (7-2)
23. What is a priority encoder? (7-2)
24. Briefly describe the pin-out and function table of a 74147 IC. (7-2)
25. What is negative-logic, and how does it compare to positive-logic? (7-2)

Practice Problems

26. What binary codes would have to be applied to the decoder circuits in Figure 7-14 in order to generate a HIGH output?
27. Sketch a decoder circuit, like the ones shown in Figure 7-14, that will generate a HIGH output for the following input codes.

   a. 1110101 b. 110011 c. 1011101101

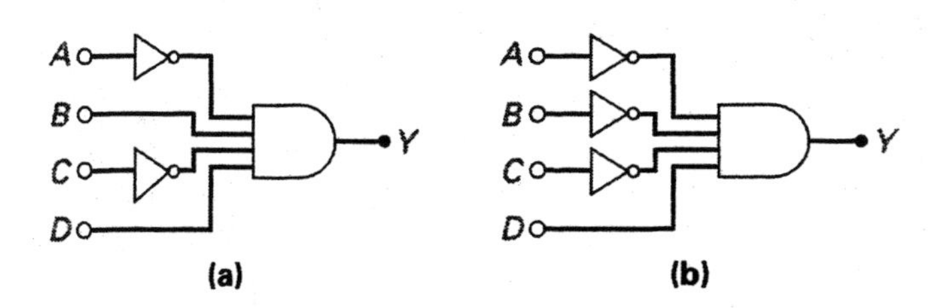

FIGURE 7-14 Basic Decoders.

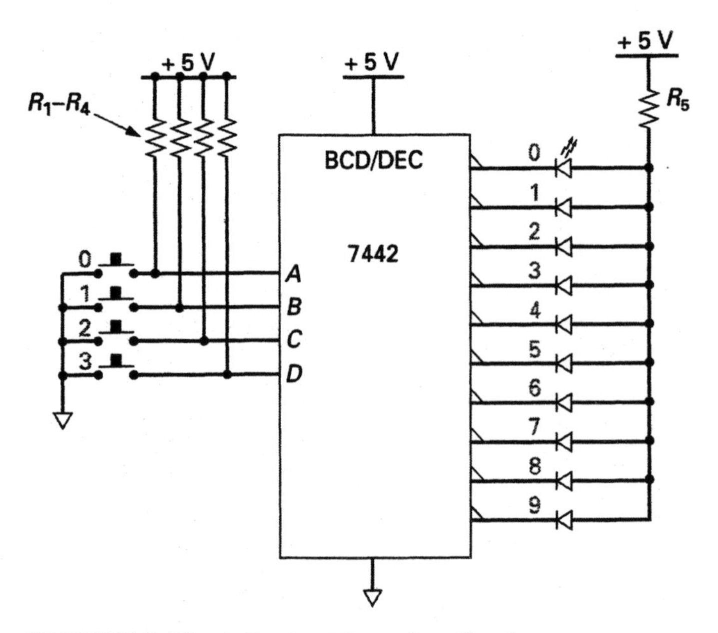

FIGURE 7-15 A Decimal Decoder Circuit.

To practice your circuit recognition and operation, refer to Figure 7-15 and answer the following six questions.

28. What is the function of the 7442 in this circuit?

29. What is the purpose of R_1 through R_4?

30. Do the LEDs in this circuit form a common-anode or common-cathode display?

31. Are the outputs from the 7442 active-LOW or active-HIGH?

32. What would happen to the output display if the following input switches were pressed simultaneously?

a. 3 and 1 **b.** 0, 2, and 3 **c.** 2 and 1

33. If the number 4 LED were ON, what binary logic levels would you expect to find on the *ABCD* inputs of the 7442, if you were to check using the logic probe?

To practice your circuit recognition and operation, refer to Figure 7-16 and answer the following six questions.

34. What is the function of the 7448 in this circuit?

35. Is the seven-segment display in this circuit a common-anode or common-cathode display?

36. Are the outputs of the 7448 in this circuit active-LOW or active-HIGH?

37. What is the purpose of the push-button switch in this circuit?

38. What would be displayed if the start/stop push-button were held down permanently?

39. If decimal 8 were displayed on the seven-segment display, what binary logic levels would you expect to find on the 7448's input and output pins if you were to check using a logic probe?

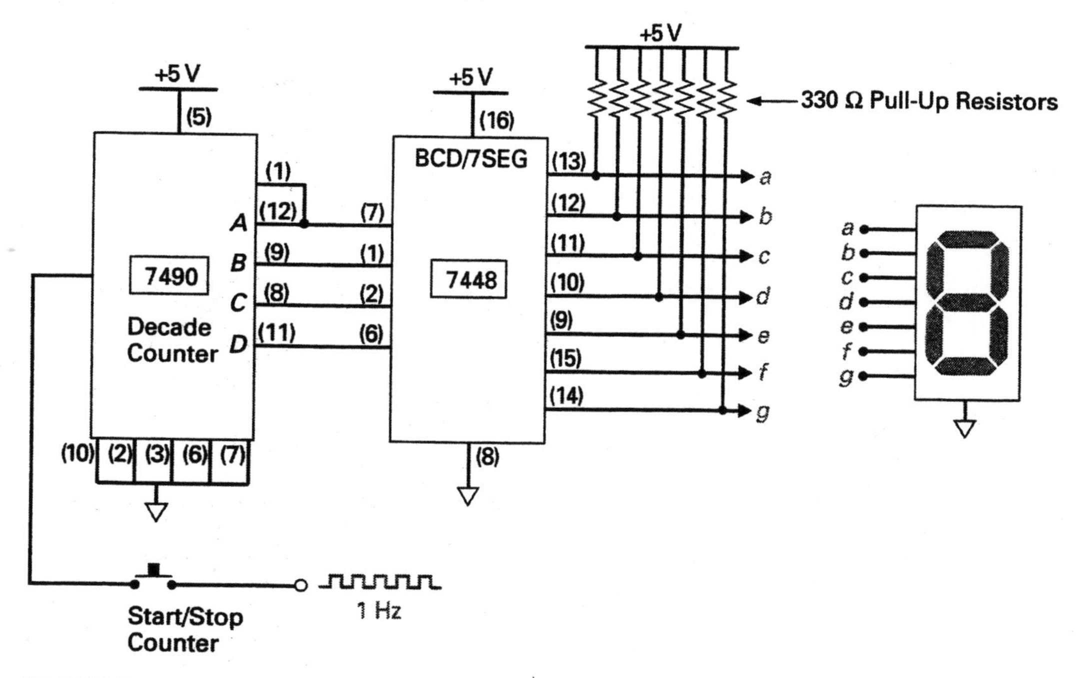

FIGURE 7-16 A 0–9 Second Timer Circuit.

To practice your circuit recognition and operation, refer to Figure 7-17, and answer the following three questions.

40. Is this circuit a decoder or encoder?

41. Briefly describe the operation of this circuit.

42. Determine the *A*, *B*, and *C* outputs for each of the rotary switch positions and insert the results in the function table alongside the circuit.

To practice your circuit recognition and operation, refer to Figure 7-18 and answer the following eight questions.

43. Briefly describe the operation of this circuit.

44. What would be the output logic levels from the 74147 when each of the input switches are activated?

45. Are the inputs to the 74147 active-LOW or active-HIGH?

46. Are the outputs from the 74147 active-LOW or active-HIGH?

47. The _____________-logic (positive/negative) output of the 74147 is compatible with the common-_____________ (anode/cathode) display, since a LOW output will turn the respective LED _____________ (ON/OFF).

48. What would be the output logic levels from the 74147 if switch 5 were pressed?

49. What would be the output logic levels from the 74147 if switches 5 and 7 were pressed simultaneously?

50. What would be the condition of the display if none of the input switches were being pressed?

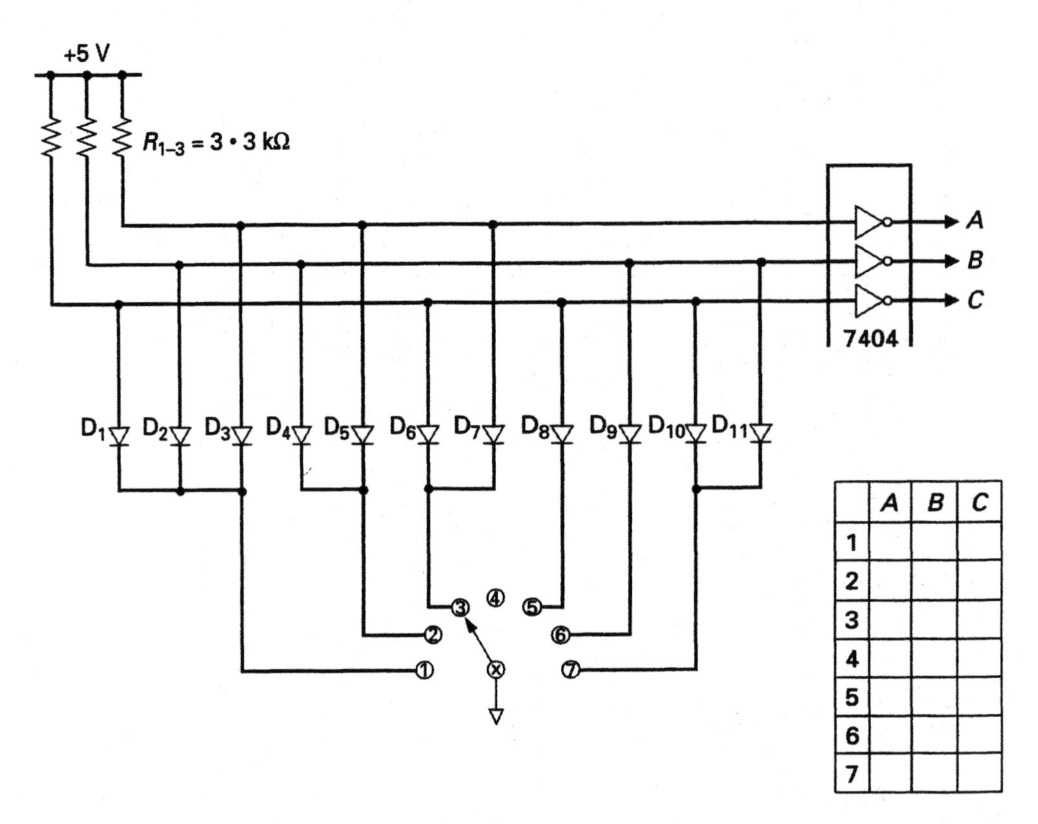

	A	B	C
1			
2			
3			
4			
5			
6			
7			

FIGURE 7-17 A Switch Encoder Circuit.

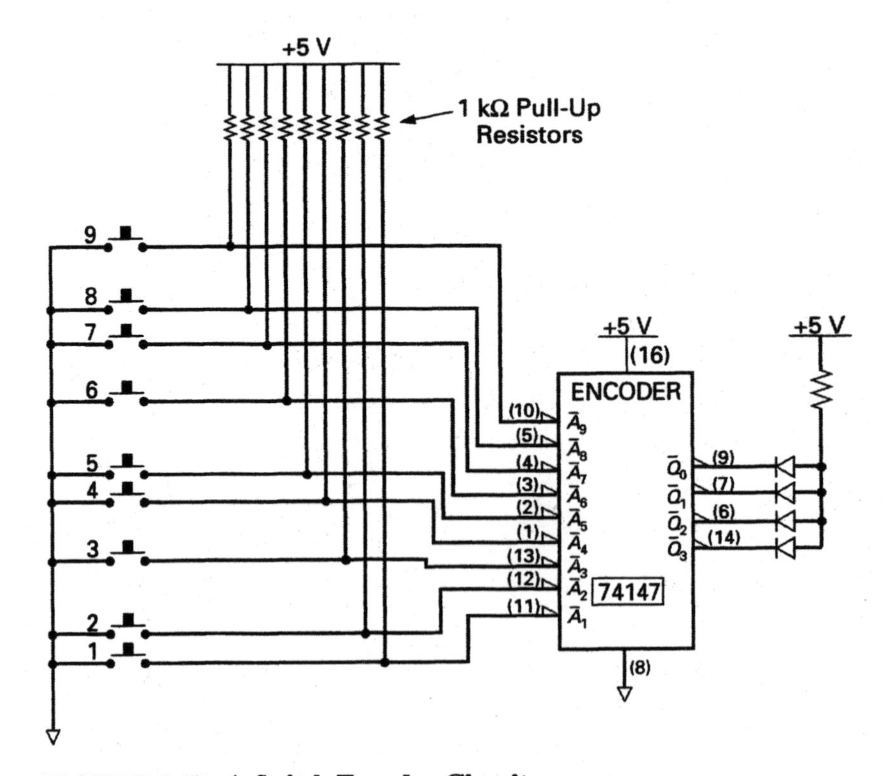

FIGURE 7-18 A Switch Encoder Circuit.

Troubleshooting Questions

51. Briefly describe each of the three steps in the three-step troubleshooting procedure.

52. What symptoms would you expect from the circuit shown in Figure 7-15 if the following malfunctions were to occur?

a. R_5 was open.
b. Switch 2 was stuck down.
c. LED 8 burned out.
d. The 7442 output 4 was shorted to ground.

53. What symptoms would you expect from the circuit shown in Figure 7-16 if the following malfunctions were to occur?

a. There was no ground connection to the seven-segment display.
b. The *g* output of the 7448 was shorted to ground.
c. The 7490 output remained locked at 1001.

54. What symptoms would you expect from the circuit shown in Figure 7-17 if the following malfunctions were to occur?

a. D_1 was to open.
b. R_1 was to open.
c. The center wiper position of the rotary switch was disconnected from ground.
d. The NOT gate *A* output was stuck HIGH.

55. What symptoms would you expect from the circuit shown in Figure 7-18 if the following malfunctions were to occur?

a. One of the LEDs was to burn out.
b. The 8 input switch was stuck down.
c. One of the 74147 outputs was permanently HIGH.
d. One of the 74147 outputs was permanently LOW.

Chapter 8

OBJECTIVES

After completing this chapter, you will be able to:

1. Define the function of a digital multiplexer.
2. Describe the operation and give an application for the following digital multiplexer ICs.
   a. 74151—one-of-eight data multiplexer/selector
   b. 74157—quadruple two-line to one-line data multiplexer
3. Define the function of a digital demultiplexer.
4. Describe the operation and give an application for the following digital demultiplexer ICs.
   a. 7442—BCD-to-decimal decoder/demultiplexer
   b. 74154—four-line to sixteen-line decoder/demultiplexer
   c. 74138—three-line to eight-line decoder/demultiplexer
5. Define the function of a digital comparator.
6. Describe the operation and give an application for the following digital comparator IC
   a. 7485—4-bit magnitude comparator
7. Explain the differences between odd and even parity.
8. Describe the operation of a simplified 4-bit parallel data transmission system with even parity error detection.
9. Describe the operation and give an application for the following digital parity generator/checker IC.
   a. 74180—9-bit odd/even parity generator/checker
10. Explain the three-step troubleshooting procedure and then show how it can be applied to a typical combinational logic circuit.
11. Describe some frequently used troubleshooting procedures.

Multiplexers, Demultiplexers, Comparators, and Parity Circuits

OUTLINE

VIGNETTE

Keeping it BASIC

The complexity and frustration endured by John Kemeny and Thomas Kurtz in the early days of computer programming inspired them to develop a programming language that was simple and easy to use. Kemeny was a Hungarian immigrant who had a gift for mathematics and frequently used this gift on many occasions to, as he put it, "beat the system." On one occasion in 1942 the sixteen-year-old Hungarian, who knew very little English, was given an aptitude test by his teachers at Washington Heights High School in Manhattan. Recalling this time, he said, "I had no vocabulary and could only understand a few words of each question, but it was a multiple-choice test and I understood enough to see there was a pattern to the answers. I cracked the code and got one of the highest scores in New York City."

As an undergraduate working on his doctorate at Princeton in 1945, Kemeny was exposed to some important people. He worked with fellow Hungarian John Von Neumann for a while on the top-secret Manhattan project (constructing an atomic bomb), and he also worked as a research assistant for Albert Einstein toward the end of the war.

Although Kemeny and Kurtz both received their doctorates from Princeton, they did not meet until they went to work as mathematics professors at Dartmouth College. In those days both Kemeny and Kurtz had to travel 135 miles by train to MIT in Boston to run their math programs on an IBM 704 computer. All programs had to be keypunched on IBM cards and then handed to an operator who fed the cards into the computer. After processing, which could take twenty-four hours, the computer would typically print out a cryptic error message. Once this error had been corrected and new cards punched, the process was repeated. Sometimes even experienced programmers would take two weeks to work out all of the bugs in their programs.

In 1959 Dartmouth College finally acquired their own small computer. At about the same time, a team of IBM programmers developed a programming language called FORTRAN (formula translator) which could express scientific and mathematical formulas. Like everyone else, Kemeny and Kurtz found FORTRAN extremely hard to master and decided to develop two ideas of their own. The first was a simple programming language students could use, and the second was a system whereby several teletypewriters, and therefore programmers, could be connected to the same computer. At 4 A.M. on 1 May, 1964, Kemeny and Kurtz sat down at two separate terminals connected to one computer and started generating programs in their new simple language. What made this day different was that everything worked and both professors were able to simultaneously "timeshare" the computer. Their new programming language was called BASIC, which stands for Beginners All-Purpose Symbolic Instruction Code. BASIC used simple English words such as SAVE, LIST, PRINT, and RUN, and was definitely all-purpose since it could be used by any student whose studies ranged from sociology to engineering. BASIC went on to become a universal language that was built into almost every microcomputer in the late 1970s. In the late 1980s millions of people would learn BASIC, making this computer language more internationally understood than all of the Scandinavian languages combined.

INTRODUCTION

In the previous chapter we discussed the operation, characteristics, applications, and troubleshooting of digital decoder and encoder circuits. Decoders and encoders are two examples of combinational logic circuits, which are single-chip ICs that have logic gates combined internally to form a functional logic circuit. In this chapter we will continue our coverage of combinational logic circuits and discuss the operation, characteristics, applications, and troubleshooting of digital multiplexers, demultiplexers, comparators, and parity generators and checkers.

8-1 MULTIPLEXERS

A **digital multiplexer** (MUX, pronounced "mucks"), or selector, is a circuit that is controlled to switch one of several inputs through to one output. The most basic of all multiplexers is the rotary switch, shown in Figure 8-1(a), which will switch any of the six inputs through to the one output based on the switch position selected. In most high-speed applications, the multiplexer will need to be electronically rather than mechanically controlled. **Analog multiplexer** circuits use relays or transistors to electronically switch analog signals, while **digital multiplexer** circuits use logic gates to switch one of several digital input signals through to one output. Figure 8-1(b) illustrates a simple digital multiplexer designed to switch one of its two inputs through to the output (a one-of-two multiplexer). The logic level applied to the SELECT input determines whether AND gate A or B is enabled, and therefore which of the data inputs (D_0 or D_1) is switched to the output. If SELECT = 0, AND gate A is enabled while

Multiplexer (MUX) Also called a data selector, it is a circuit that is controlled to switch one of several inputs through to one output.

Analog Multiplexer An analog signal multiplexer circuit.

Digital Multiplexer A digital signal multiplexer circuit.

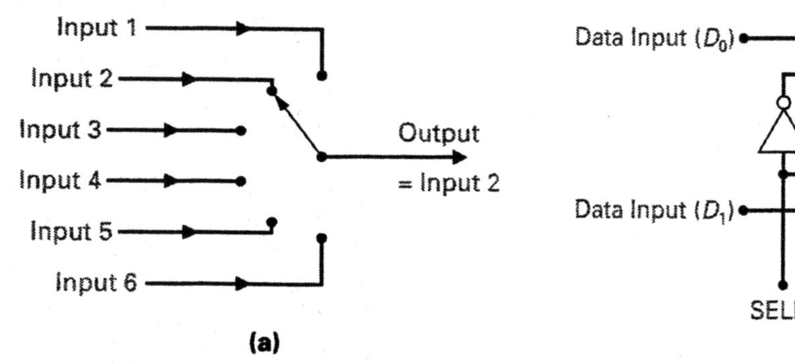

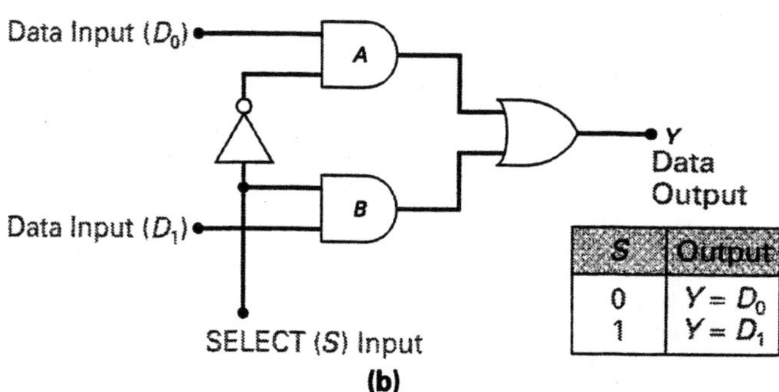

FIGURE 8-1 **Basic Multiplexers.**

AND gate B is disabled, and therefore output Y will follow input D_0 ($Y = D_0$). On the other hand, if SELECT = 1, AND gate B is enabled while AND gate A is disabled, and output Y will follow input D_1 ($Y = D_1$).

8-1-1 *One-of-Eight Data Multiplexer/Selector*

Figure 8-2 shows the data sheet for the 74151 IC, which is an eight-input data multiplexer, or selector. Referring to the IC's internal schematic and function table, you can see that the binary value applied to the three DATA SELECT control inputs will determine which one of the eight inputs is switched through to the output. For example, when ABC = 000 (binary 0) data input D_0 is

DEVICE: SN74LS151 – One-of-Eight Data Multiplexer/Selector

Description:

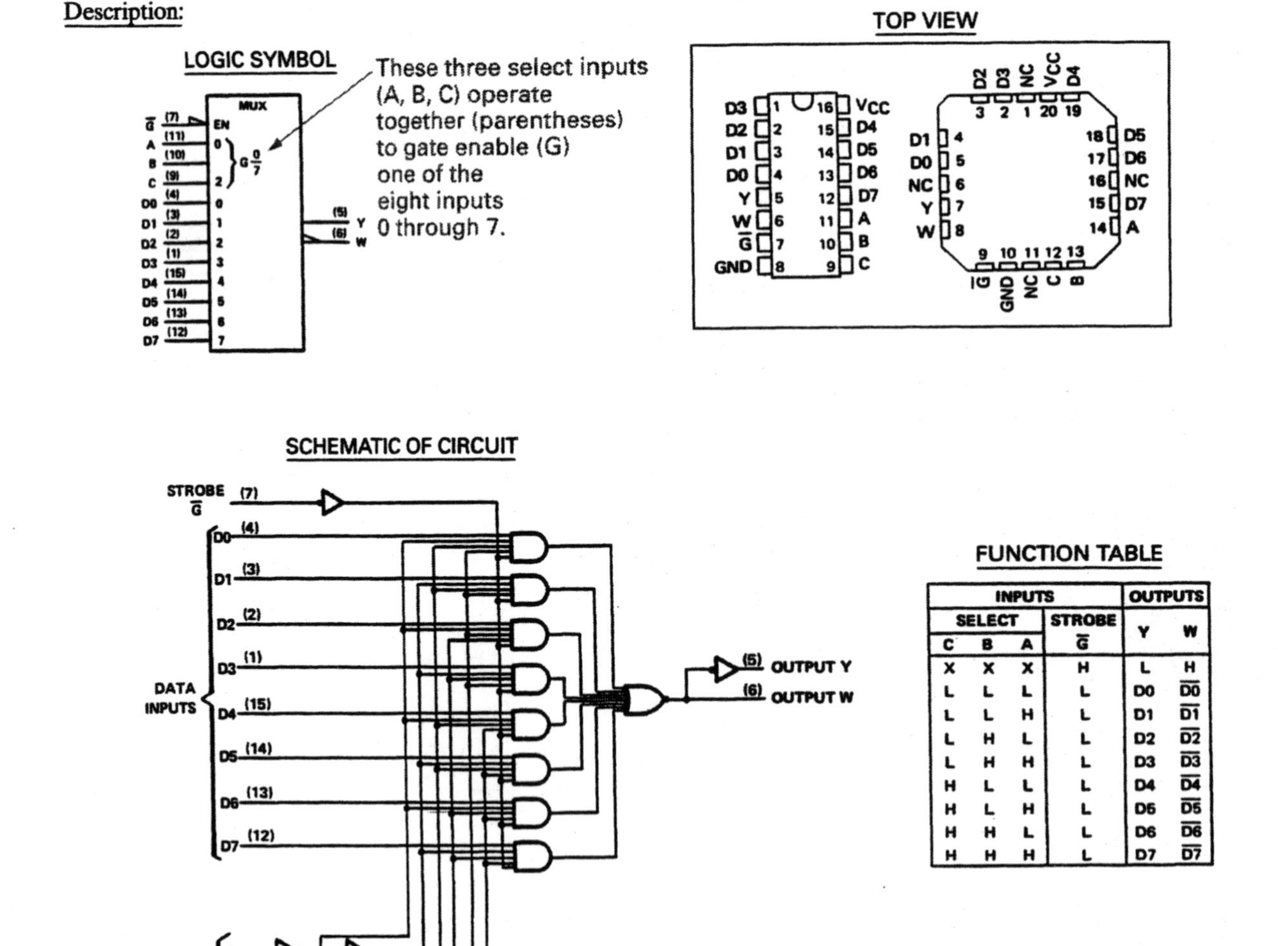

INPUTS				OUTPUTS	
SELECT			STROBE		
C	B	A	$\overline{G}$	Y	W
X	X	X	H	L	H
L	L	L	L	D0	$\overline{D0}$
L	L	H	L	D1	$\overline{D1}$
L	H	L	L	D2	$\overline{D2}$
L	H	H	L	D3	$\overline{D3}$
H	L	L	L	D4	$\overline{D4}$
H	L	H	L	D5	$\overline{D5}$
H	H	L	L	D6	$\overline{D6}$
H	H	H	L	D7	$\overline{D7}$

FIGURE 8-2 A Multiplexer IC That Will Connect One of Eight Inputs through to the Output. (Courtesy of Texas Instruments)

switched through to the Y output, or when ABC = 110 (binary 6) data input D_6 is switched through to the Y output. An eight-input MUX will therefore need three data select input control lines so that eight binary combinations can be obtained ($2^3 = 8$), and each of these combinations can be used to select one of the eight data inputs. Referring again to the data sheet of the 74151 in Figure 8-2, let us discuss a few other details regarding this IC. The 74151 has a direct output Y (pin 5) and a complemented, or inverted, output W (pin 6). The active-LOW strobe enable or gate enable ($\overline{G}$) is used to either enable ($\overline{G} = 0$) or disable ($\overline{G} = 1$) all of the internal AND select gates, as detailed in the function table.

■ **EXAMPLE:**

Which input would be switched through to the Y output if the following logic levels were applied to the 74151?

a. $A = 0$, $B = 1$, $C = 0$, $D_0 = 1$, $D_2 = 1$, $D_3 = 0$, $D_4 = 1$, $D_5 = 1$, $D_6 = 1$, $D_7 = 0$, $\overline{G} = 1$
b. $A = 1$, $B = 1$, $C = 1$, $D_0 = 1$, $D_2 = 1$, $D_3 = 0$, $D_4 = 1$, $D_5 = 1$, $D_6 = 1$, $D_7 = 0$, $\overline{G} = 0$.

■ ***Solution:***

a. Y = LOW because the gate enable or strobe is not active.
b. $Y = D_7 = 0$ because select inputs ABC = 111, or decimal 7.

Application—Parallel-to-Serial Data Conversion

Parallel-to-Serial Converter A circuit that will convert a parallel data word input into a serial data stream output.

Figure 8-3(a) shows how a 74151 can be used as a **parallel-to-serial converter.** Binary data can be transferred either in parallel form or serial form between two digital circuits. The circuit in Figure 8-3 shows the differences between these two data formats by showing how we can convert an 8-bit parallel-data input into an 8-bit serial-data output. Two 4-bit storage registers hold the 8-bit parallel word that has to be converted to serial form. This 8-bit parallel binary word (0110 1001) is applied to the data inputs of the 8-bit multiplexer. A clock or timing signal is applied to the input of a number 0 through 7 counter whose three outputs (Q_0, Q_1 and Q_2) are applied to the three multiplexer data select inputs (A, B and C). Referring to the waveforms in Figure 8-3(b), you can see that each positive-going edge of the clock signal causes the counter to advance by one count. As the 3-bit counter is incremented through each of its eight output states (000, 001, 010, 011, 100, 101, 110, and 111), the ABC data-select MUX inputs are sequenced in order, causing each of the eight data inputs (D_0 through D_7) to be selected one after the other. As a result, the 8-bit parallel binary word at the input of the 74151 is switched, one bit at a time, producing an 8-bit serial binary word at the output.

Parallel data transfer is generally not used for the long distance transmission of binary data (for example, over a telephone line) since it requires several data lines connected in parallel. Parallel data is therefore converted to serial data before it is transmitted, since serial data transmission only requires a single data line. The serial data transfer of binary digits is, however, a lot slower than parallel data transmission. To explain this, let us assume that the time interval

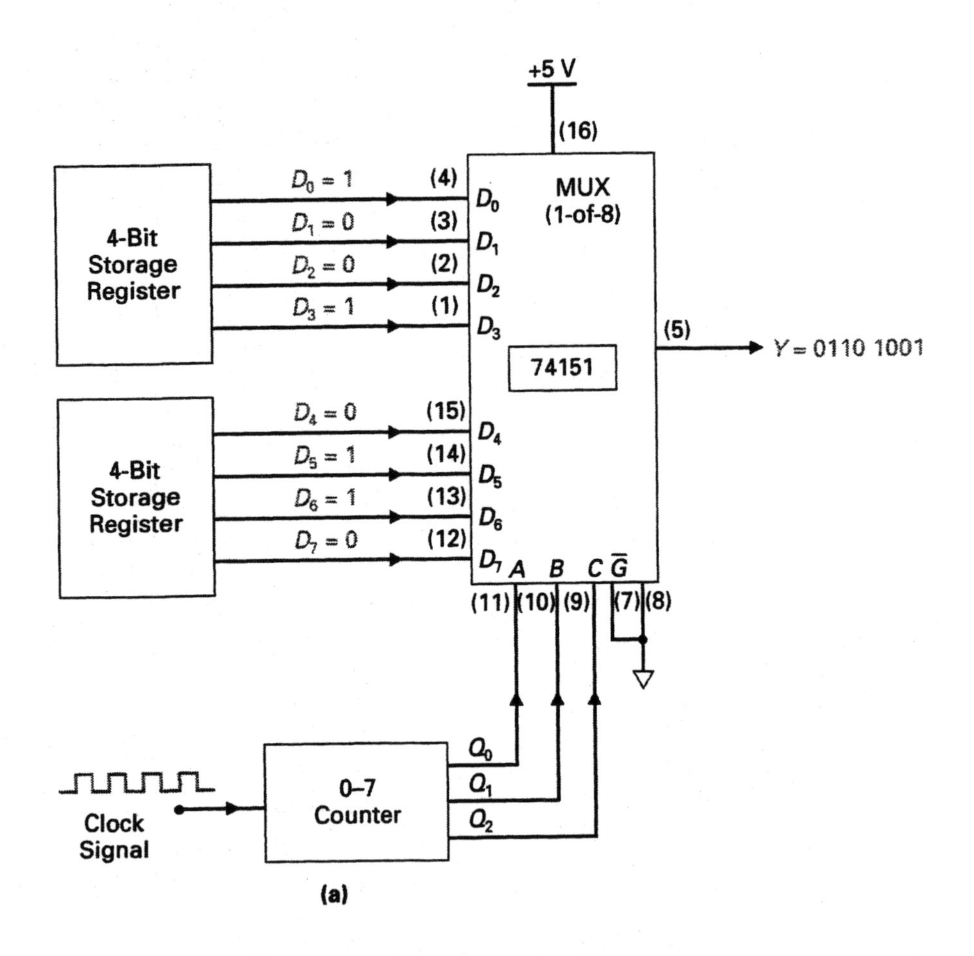

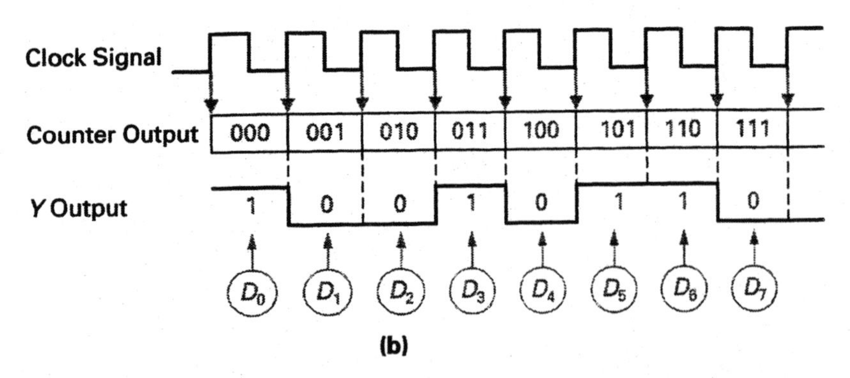

FIGURE 8-3 Application—Parallel-to-Serial Data Conversion.

allotted to each bit in the example in Figure 8-3 is one microsecond. It will therefore take eight microseconds for the 8-bit word to be transmitted in serial form; however, all of the binary bits in a parallel word are transferred simultaneously in one microsecond over a group of parallel lines. The difference between these two data transfer methods therefore seems to come down to the standard speed-versus-cost trade-off. Serial data transfer needs only one transmitter circuit, one receiver circuit, and one data line to transfer any number of binary bits. Parallel data transfer, on the other hand, requires a separate transmitter, receiver, and line for each bit of the word. The transmission and reception of parallel data is therefore more expensive; however, when high-speed data transfer is required, the data handling time for a parallel transfer of data is extremely fast compared to the data handling time needed for a serial transfer of data.

Application—A Logic-Function Generator Circuit

Logic-Function Generator A circuit that will generate the desired logic level at the output for each input combination.

Multiplexers are also very useful as **logic-function generators.** As an example, Figure 8-4(a) shows a truth table and sum-of-products expression for a desired logic-function. By connecting the multiplexer's data inputs either HIGH or LOW, as shown in Figure 8-4(b), we can generate the desired HIGH or LOW

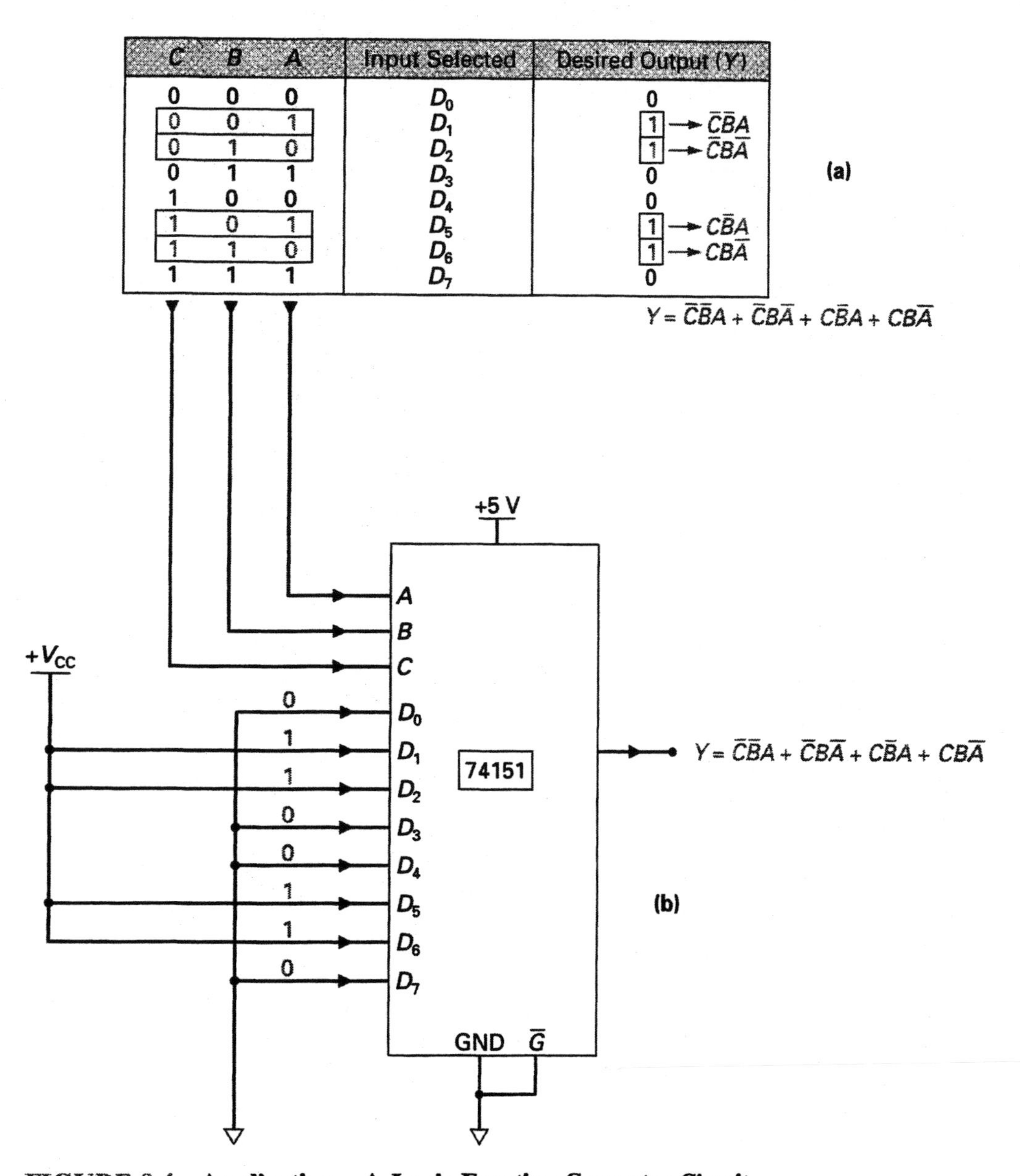

C	B	A	Input Selected	Desired Output (Y)
0	0	0	D_0	0
0	0	1	D_1	1 → $\overline{C}\,\overline{B}A$
0	1	0	D_2	1 → $\overline{C}B\overline{A}$
0	1	1	D_3	0
1	0	0	D_4	0
1	0	1	D_5	1 → $C\overline{B}A$
1	1	0	D_6	1 → $CB\overline{A}$
1	1	1	D_7	0

$$Y = \overline{C}\,\overline{B}A + \overline{C}B\overline{A} + C\overline{B}A + CB\overline{A}$$

FIGURE 8-4 Application—A Logic-Function Generator Circuit.

output at Y for each of the ABC input combinations. Using a MUX instead of discrete logic gates will greatly reduce the number of ICs used, and therefore the circuit cost, size, and power consumption. For example, to implement the sum-of-products expression in Figure 8-4 using discrete logic gates would require four three-input AND gates, one four-input OR gate, and three INVERTERS—a total of four ICs compared to one 74151 IC.

8-1-2 *Four-of-Eight Data Multiplexer/Selector*

Figure 8-5 shows the data sheet for the 74157 IC which is an eight-input four-output multiplexer. This multiplexer will switch either the A 4-bit word (1A, 2A, 3A, and 4A) or the B 4-bit word (1B, 2B, 3B, and 4B) through to the

DEVICE: SN74157 – Quadruple Two-Line to One-Line Data Multiplexer

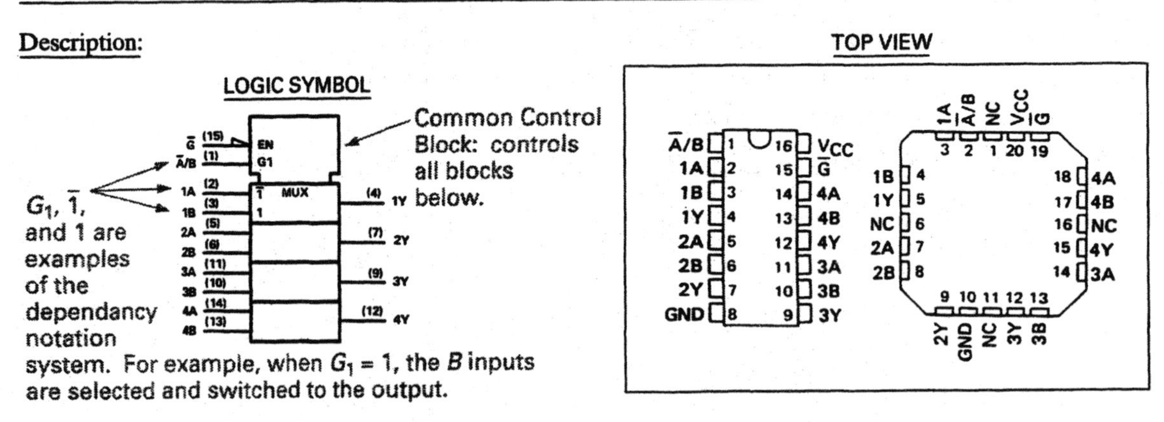

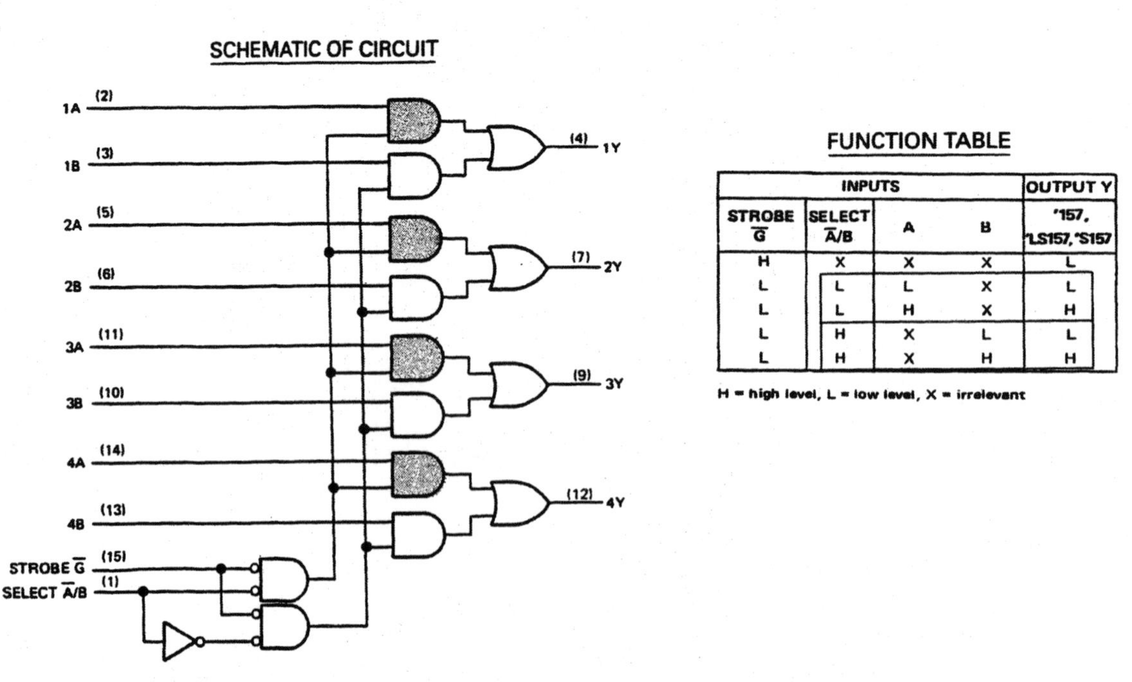

INPUTS				OUTPUT Y
STROBE $\overline{G}$	SELECT $\overline{A}/B$	A	B	'157, 'LS157, 'S157
H	X	X	X	L
L	L	L	X	L
L	L	H	X	H
L	H	X	L	L
L	H	X	H	H

H = high level, L = low level, X = irrelevant

FIGURE 8-5 A Multiplexer IC That Will Switch One of Two 4-Bit Input Words through to the Output. (Courtesy of Texas Instruments)

4-bit output (1Y, 2Y, 3Y, and 4Y). The active-LOW gate enable input ($\overline{G}$) is used to either enable ($\overline{G} = 0$) or disable ($\overline{G} = 1$) all of the logic gates within the IC. When the gate enable line is active, the Y outputs will follow whichever inputs have been selected by the $\overline{A}/B$ select control input, as shown in the function table in Figure 8-5. For example, when $\overline{A}/B = 0$, the Y outputs will follow whatever is applied to the A inputs, whereas when $\overline{A}/B = 1$, the Y outputs will follow whatever is applied to the B inputs.

■ **EXAMPLE:**

Which input would be switched through to the output if the following logic levels were applied to the 74157?

a. $\overline{A}/B = 0, \overline{G} = 0, A = 1001, B = 0110.$
b. $\overline{A}/B = 1, \overline{G} = 0, A = 1001, B = 0110.$

■ ***Solution:***

a. $Y = A = 1001$ because the $\overline{G}$ input is low, the $\overline{A}/B$ input is LOW and so the A inputs are selected.
b. $Y = A = 0110$ because the $\overline{G}$ input is low, $\overline{A}/B$ input is HIGH and so the B inputs are selected.

Application—A Multiplexed Display

Most calculators have an $\boxed{X \longleftrightarrow M}$ key that is used to select whether the calculator's seven-segment display shows the current value being operated on (X) or the value stored in memory (M). Each time the $\boxed{X \longleftrightarrow M}$ key is pressed, the display switches between these two values. Figure 8-6 shows how

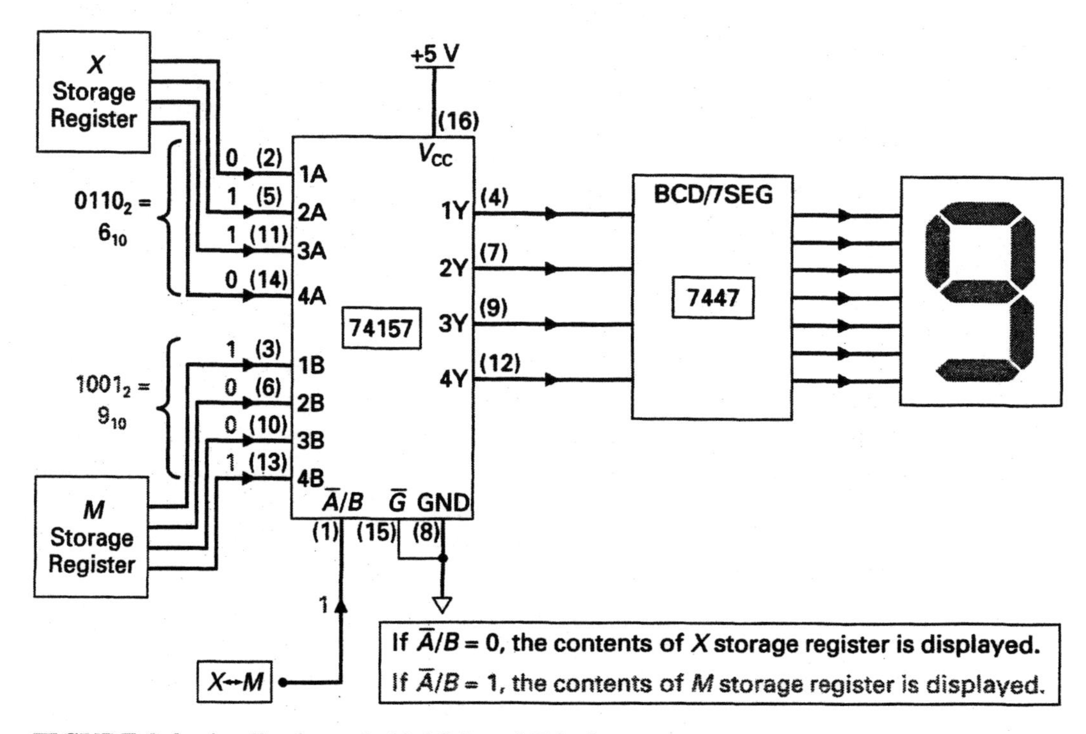

FIGURE 8-6 Application—A Multiplexed Display.

this function could be implemented using a 74157, 7447, and a seven-segment display. The calculator's $\boxed{X \longleftrightarrow M}$ key controls the $\overline{A}/B$ input of the 74157. When $\overline{A}/B = 0$, the contents of the X storage register (which is applied to the 74157's A inputs) is switched through to the Y outputs ($Y = 0110$), decoded by the 7447, and displayed on the seven-segment display (display will show decimal 6). On the other hand, when $\overline{A}/B = 1$, the contents of the M storage register (which is applied to the 74157's B inputs) is switched through to the Y outputs ($Y = 1001$), decoded by the 7447, and displayed on the seven-segment display (display will show decimal 9).

Multiplexed Display
A display circuit that timeshares the display or the display drive circuitry.

A **multiplexed display** uses one display to present two or more different pieces of information. This "timesharing" of the decoder driver, display, and interconnect wiring greatly reduces the number of ICs needed, power consumption, size, and printed circuit board interconnections. The big disadvantage of using a single multiplexed display instead of two separate displays is that you cannot see both values at the same time.

SELF-TEST REVIEW QUESTIONS FOR SECTION 8-1

1. A multiplexer is a circuit that through a control input will switch __________ input(s) through to __________ output(s).
   a. one, one of several
   b. one of several, one
   c. one, one
   d. one of several, one of several
2. How many data select inputs would a one-of-sixteen multiplexer need?
3. True or false: multiplexers are often used for serial-to-parallel data conversion.
4. Which data transmission method is considered the fastest: serial or parallel?
5. Which data transmission method requires more hardware: serial or parallel?
6. Would the 74151 or 74157 best be described as a 4-bit word multiplexer?

8-2 DEMULTIPLEXERS

Demultiplexer (DMUX)
Also called a data distributor, it is a circuit that is controlled to switch a single input through to one of several outputs.

A digital multiplexer (MUX), or selector, is a circuit that is controlled to switch one of several inputs through to one output. On the other hand, a **digital demultiplexer** (DMUX, pronounced "demucks"), or data distributor, operates in exactly the opposite way since it is a circuit that is controlled to switch a single input through to one of several outputs. The demultiplexer is also known as a *data distributor* since the data applied to the single input is distributed to one of the many outputs.

Figure 8-7(a) shows how a rotary switch can be made to act as a mechanically controlled demultiplexer. In this circuit the switch is controlled to connect a single input through to one of several outputs. Like the multiplexer, mechanical devices are too slow for high-speed applications, and therefore most demultiplexers are also electronically rather than mechanically controlled.

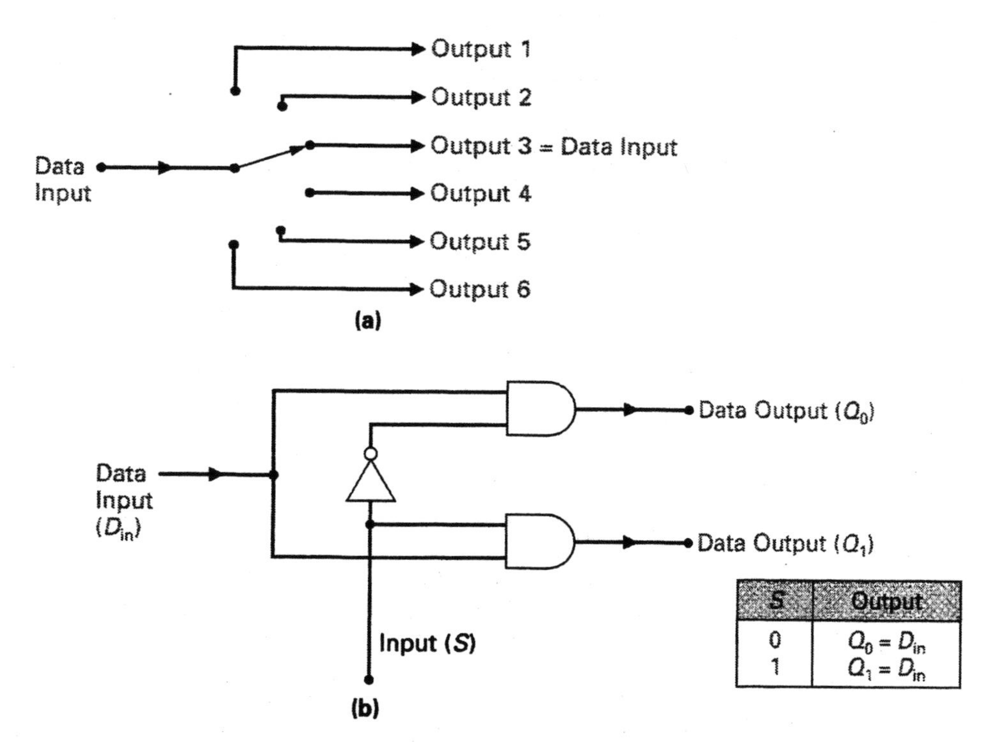

S	Output
0	$Q_0 = D_{in}$
1	$Q_1 = D_{in}$

FIGURE 8-7 Basic Demultiplexers/Data Distributors.

Analog demultiplexers will switch a single analog signal input through to one of several outputs, while **digital demultiplexers** will switch a single digital signal input through to one of several outputs. Figure 8-7(b) shows a simple digital demultiplexer circuit designed to switch its single input through to one of two outputs. This one-line to two-line demultiplexer is controlled by the SELECT INPUT which will determine which AND gate is enabled and which AND gate is disabled, as described in the function table.

Analog Demultiplexer An analog signal demultiplexer circuit.

Digital Demultiplexer A digital signal demultiplexer circuit.

In the following sections you will be introduced to three typical demultiplexer ICs. Two of these ICs were discussed previously in the decoder section of Chapter 7, and they are reintroduced in this section since most decoder ICs can be made to function as either decoders or demultiplexers.

8-2-1 A One-Line to Eight-Line Demultiplexer

The 7442 BCD-to-decimal decoder, which was discussed previously in Chapter 7 (Figure 7-2), can be used as a demultiplexer, as shown in Figure 8-8. The internal schematic of the 7442 IC is shown in Figure 8-8(a). As a decoder, any one of the 7442's ten outputs could be driven LOW by applying an equivalent *BCD* input code to the *ABCD* inputs. As a demultiplexer, the 7442 can be made to function as a one-line to eight-line data distributor. To operate in this way, inputs *A, B,* and *C* are used to select one of the internal logic gates 0 through 7 (three select lines will give a total of eight combinations). For example, if 010 is applied to inputs *ABC,* internal gate 2 will be enabled. The serial data to be distributed is applied to input *D,* and will be applied to all of the lower

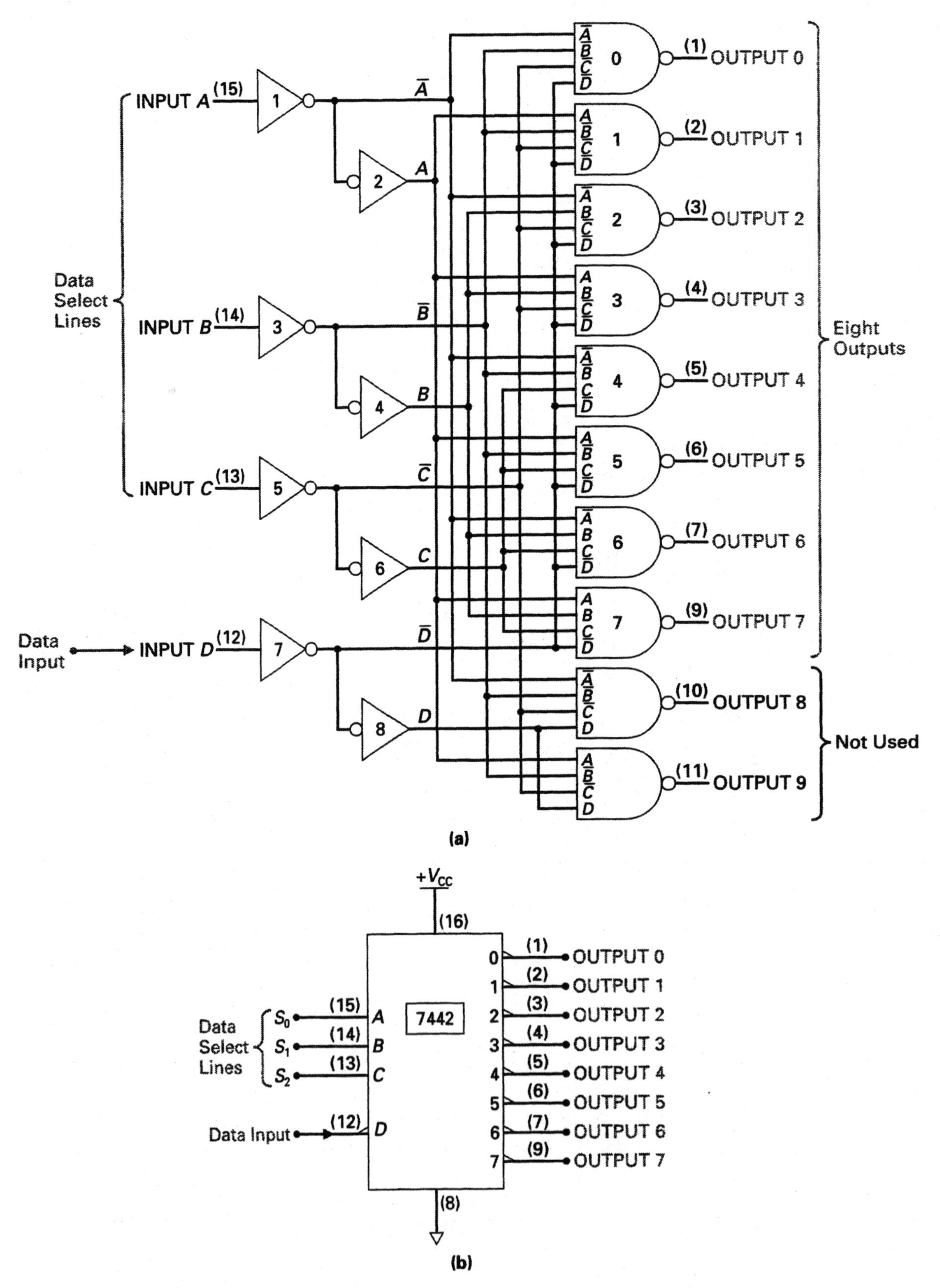

FIGURE 8-8 The 7442 Decoder as a Demultiplexer.

inputs of NAND gates 0 through 7. INVERTER 7 will invert the input data, but this process will be reversed by the inversion of the NAND gates. The output of NAND gate 2 will therefore be under the control of the *D* input, so a HIGH input on *D* will appear as a HIGH on output 2 and a LOW input on *D* will appear as a LOW on output 2. When a 7442 IC is used as a demultiplexer, its logic symbol is drawn differently to that of a decoder logic symbol, as shown in Figure 8-8(b).

■ **EXAMPLE:**

The *D* data input would be switched through to which output if the following logic levels were applied to the 7442?

a. $CBA = 011$ b. $CBA = 101$

■ ***Solution:***

a. Output 3 b. Output 5

8-2-2 A One-Line to Sixteen-Line Demultiplexer

Figure 8-9 shows how a 74154 decoder IC, which was also discussed previously in Chapter 7 (Figure 7-4), can be made to function as a one-line to sixteen-line demultiplexer. With this IC all four of the decoder OUTPUT SELECT inputs

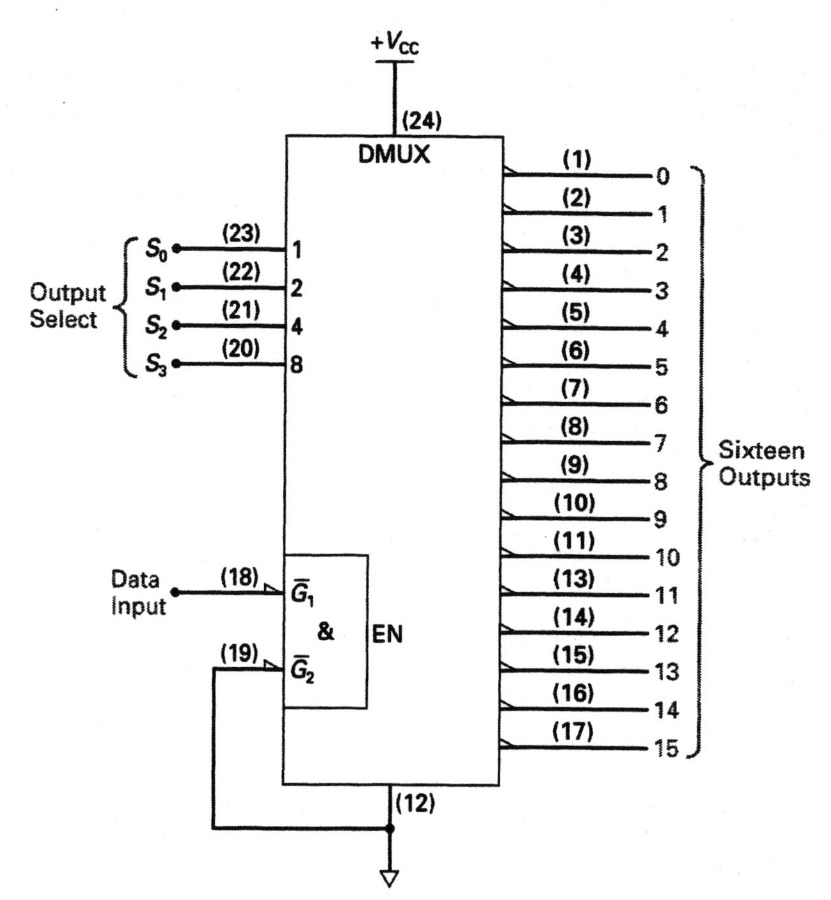

FIGURE 8-9 The 74154 as a Four-line to Sixteen-line Demultiplexer.

are used to select one of the sixteen available outputs ($2^4 = 16$), and the DATA input is applied to one of the gate enable inputs. When the DATA input is LOW, all of the 74154's internal gates are enabled, so the selected output (based on the OUTPUT SELECT code applied) is driven LOW. On the other hand, when the DATA input is HIGH, all of the 74154's internal gates are disabled, so the selected output is HIGH.

■ **EXAMPLE:**

The D input would be switched through to which output if the following logic levels were applied to the 74154?

a. $S = 1011$ b. $S = 1110$

■ ***Solution:***

a. Output 11 b. Output 14

8-2-3 *A Three-Line to Eight-Line Decoder/Demultiplexer*

Like the 7442 and 74154 ICs, the 74138 is listed as a decoder/demultiplexer in the IC data books. Data sheet details for this three-line to eight-line decoder or demultiplexer IC are shown in Figure 8-10. Referring to the logic symbol, you will notice that the 74138 has three inputs. The G_1 input is active-HIGH, while the $\overline{G_{2A}}$ and the $\overline{G_{2B}}$ inputs are both active-LOW. All three of these gate enable inputs must be active in order for the IC to be enabled. This factor is confirmed in the first two lines of the function table, which shows that all of the outputs will remain HIGH if G_1 or G_2 (G_{2A} and G_{2B}) are inactive. Lines three through ten of the function table show how the 74138 operates once it is enabled. The three SELECT inputs (ABC) are used to select one of the eight ($2^3 = 8$) active-LOW outputs (Y_0 through Y_7). For example, when $CBA = 101$ (decimal 5) the Y_5 output is made active (LOW) while all the other outputs remain inactive (HIGH).

■ **EXAMPLE:**

What would be the outputs from a 74138 for the following input conditions?

a. $CBA = 011$, $G_1 =$ LOW, $\overline{G_2} =$ LOW
b. $CBA = 101$, $G_1 =$ HIGH, $\overline{G_2} =$ LOW

■ ***Solution:***

a. All outputs will be HIGH because G_1 is not active.
b. All outputs will be HIGH, except for Y_5 (101), which will be LOW.

Application—A Basic Home Security System

Figure 8-11 shows how a 74151 eight-line to one-line multiplexer (MUX), and a 74138 one-line to eight-line demultiplexer (DMUX) could be connected to form a basic home security system. A binary 0 (000) to 7 (111) counter drives the ABC select or control inputs of both the MUX and DMUX. As the counter counts from 0 to 7, it "selects" in turn each of the 74151 inputs and switches

DEVICE: SN74LS138 – Three-Line to Eight-Line Decoder/Demultiplexer

Description:

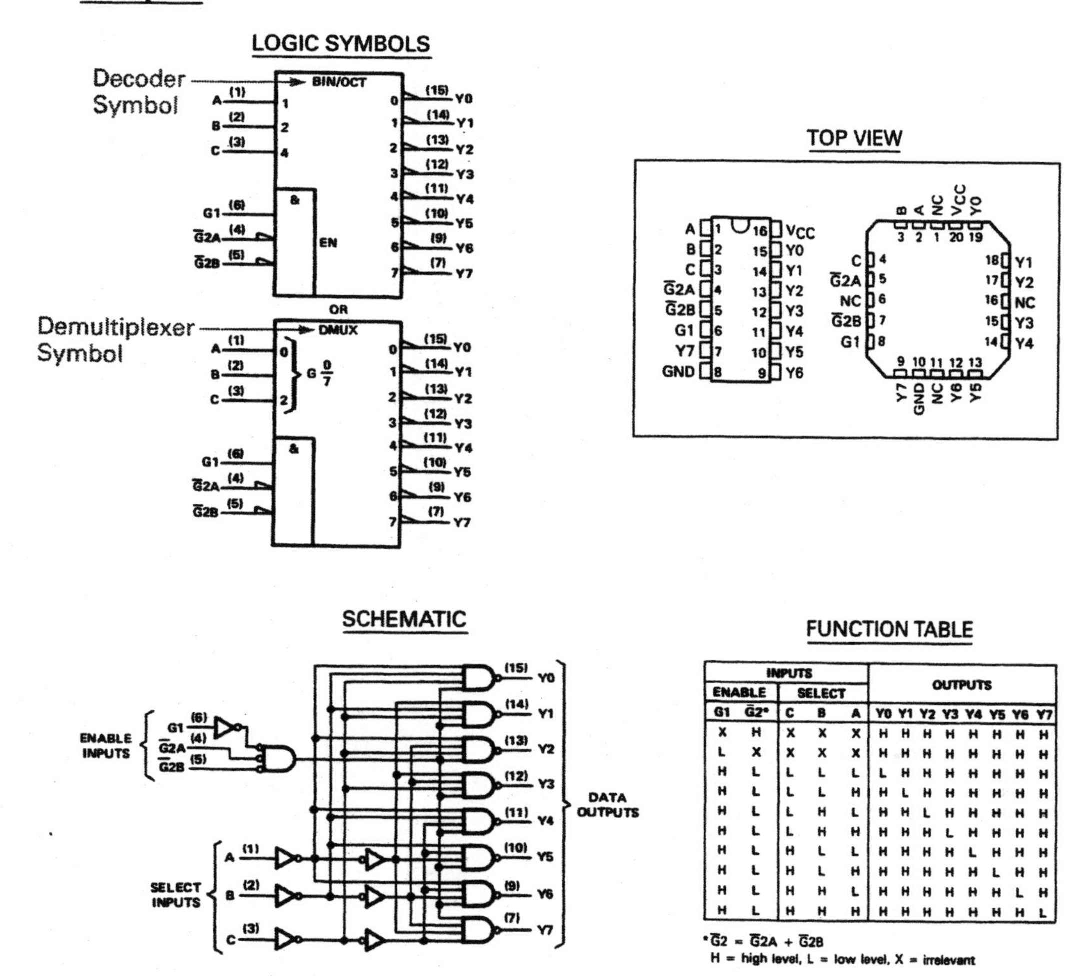

INPUTS					OUTPUTS							
ENABLE		SELECT										
G1	$\overline{G}2$*	C	B	A	Y0	Y1	Y2	Y3	Y4	Y5	Y6	Y7
X	H	X	X	X	H	H	H	H	H	H	H	H
L	X	X	X	X	H	H	H	H	H	H	H	H
H	L	L	L	L	L	H	H	H	H	H	H	H
H	L	L	L	H	H	L	H	H	H	H	H	H
H	L	L	H	L	H	H	L	H	H	H	H	H
H	L	L	H	H	H	H	H	L	H	H	H	H
H	L	H	L	L	H	H	H	H	L	H	H	H
H	L	H	L	H	H	H	H	H	H	L	H	H
H	L	H	H	L	H	H	H	H	H	H	L	H
H	L	H	H	H	H	H	H	H	H	H	H	L

*$\overline{G}2 = \overline{G}2A + \overline{G}2B$
H = high level, L = low level, X = irrelevant

FIGURE 8-10 An IC That Can Function as Either a Decoder or Demultiplexer. (Courtesy of Texas Instruments)

the HIGH or LOW data through to the *Y* output, as seen in the function table in the inset in Figure 8-11. For example, when the counter is at a count of 011 (3), the HIGH or LOW data present at the 3 input of the 74151 (pin 1) is switched through to the *Y* output. In this example circuit, the 3 input of the 74151 is connected to a window switch. If the window is open, its associated switch will also be open and the input to the 74151 will be HIGH due to the pull-up resistors. On the other hand, if the window and its associated switch are closed, ground or a LOW logic level will be applied to the input of the 74151.

The 74138 DMUX in this circuit application operates in exactly the opposite way of the 74151. For example, as we mentioned earlier, when the counter is at a count of 011 (3), the HIGH or LOW data present at the 3 input of the 74151 (pin 1) is switched through to the *Y* output. This HIGH or LOW data is applied to the G_1 input of the 74138, which is also receiving a 011 (3) select input from the counter, so the data at G_1 is switched through to the 3

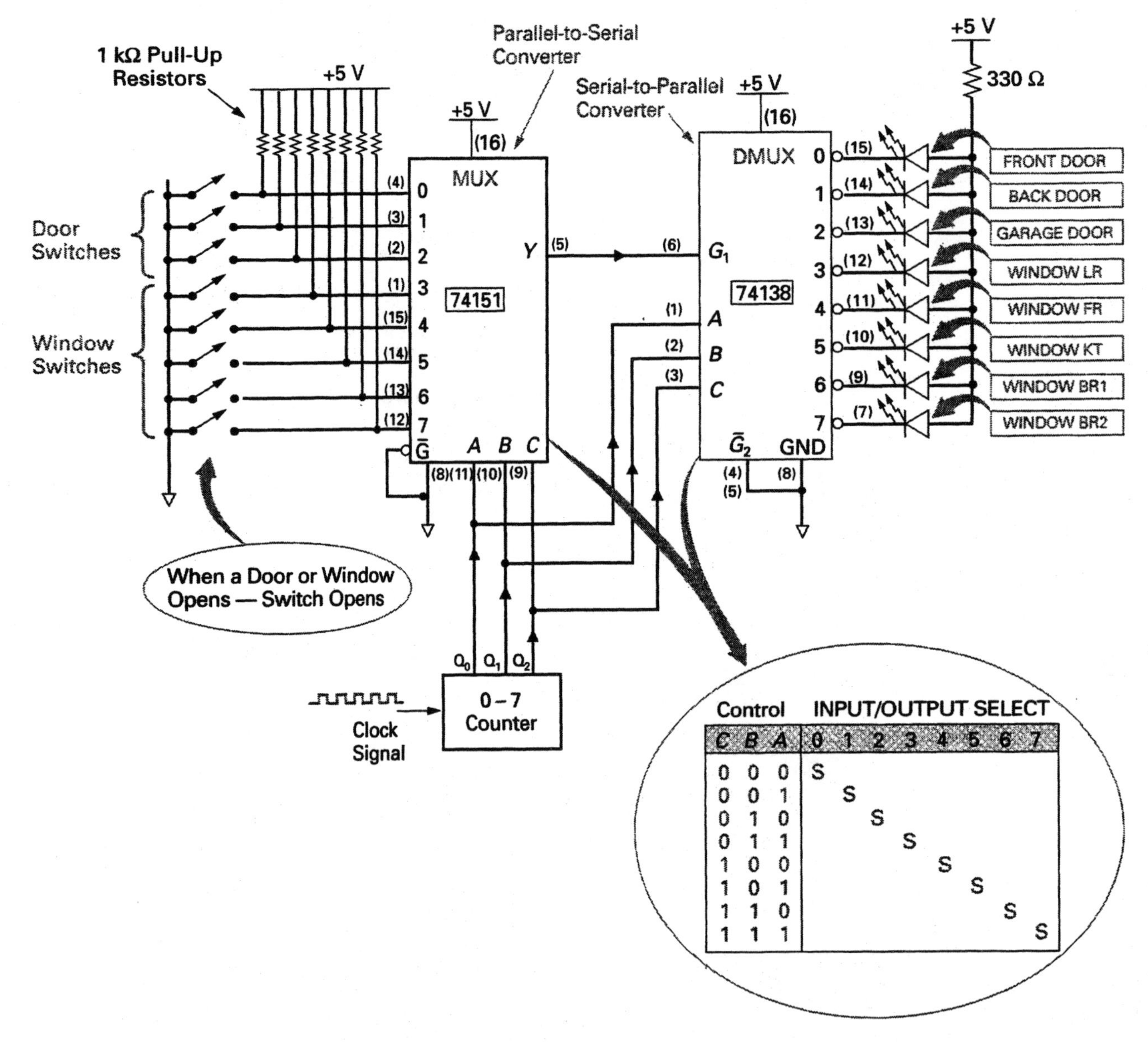

C	B	A	0	1	2	3	4	5	6	7
0	0	0	S							
0	0	1		S						
0	1	0			S					
0	1	1				S				
1	0	0					S			
1	0	1						S		
1	1	0							S	
1	1	1								S

FIGURE 8-11 Application—Basic Home Security System.

output of the 74138 (pin 12). Since the 74138 has active-LOW outputs, there will be an inversion between input and output. This inversion between the 74138's input and outputs is needed since we are driving a common-anode LED display whose inputs, as we know, are also active-LOW (LEDs are activated by a LOW). To explain the operation from input to output, if the window connected to the 3 input is opened, the switch will open and give a HIGH to the 3 input of the 74151. The 74151 will switch this HIGH data through to the Y output, where it will be applied to the G_1 input of the 74138, which will invert the HIGH to a LOW, and then switch it to the 3 output where it will turn ON the living room (LR) window LED. This circuit, therefore, first senses the condition of switch 0 and then transmits its data to LED 0, and then senses the condition of switch 1 and then transmits its data to LED 1, and then senses the condition of switch 2 and then transmits its data to LED 2, and so on until it reaches 7 (111). The counter then resets and the process is repeated.

The 74151 functions as a parallel-to-serial converter since it converts the parallel 8-bit word applied to its inputs into an 8-bit serial data stream. Conversely, the 74138 functions as a **serial-to-parallel converter,** since it converts an 8-bit serial data stream applied to its input into a parallel 8-bit output word. One question you may have about this circuit is, why don't we simply connect the switches straight to the LEDs and bypass the 74151 and 74138? The answer lies in the basic difference between serial data transmission and parallel data transmission. With this circuit, the 74151 can be placed in close proximity to all the windows and doors, and then only four wires (three for the counter and one for data) need to be run to the 74138 and its display. If the switches were connected directly to the display, eight wires would be needed to connect the switch sensors to the LED display. The advantage of serial data transmission, therefore, is that only one data line is needed. On the other hand, the advantage of parallel data transmission is speed, since a parallel connection would mean that all of the data is transmitted at one time.

Serial-to-Parallel Converter
A circuit that will convert serial data into a parallel format.

SELF-TEST REVIEW QUESTIONS FOR SECTION 8-2

1. A demultiplexer or data distributor is a circuit that through a control input will switch __________ input(s) through to __________ output(s).
   a. one, one of several
   b. one, one
   c. one of several, one
   d. one of several, one of several
2. A/an __________ can also be used as a demultiplexer. (encoder/decoder)
3. Which 3-bit code must be applied to the *CBA* inputs of a 74138 to select the Y_4 output?
4. The main difference between a decoder and a demultiplexer is that a decoder has only SELECT inputs, while a demultiplexer has both SELECT inputs and a DATA input. (true/false)

8-3 COMPARATORS

A digital **comparator** compares two binary input words to see if they are equal. The exclusive NOR (XNOR) gate is a comparator logic gate since it compares the two binary input digits applied, as seen in Figure 8-12(a). When the two input bits are the same, the output of the XNOR gate is a 1, however when the input bits are different, the output of the XNOR gate is 0. Let us now see how we could use a group of XNOR gates to create a 4-bit binary comparator circuit.

Comparator
A circuit that compares two inputs to determine whether or not the two inputs are equal.

8-3-1 A 4-Bit Binary Comparator

Figure 8-12(b) shows a 4-bit binary comparator circuit that will compare two 4-bit words being applied by two storage registers. Four XNOR gates are used to compare WORD *A* (A_0, A_1, A_2, and A_3) with WORD *B* (B_0, B_1, B_2, and B_3).

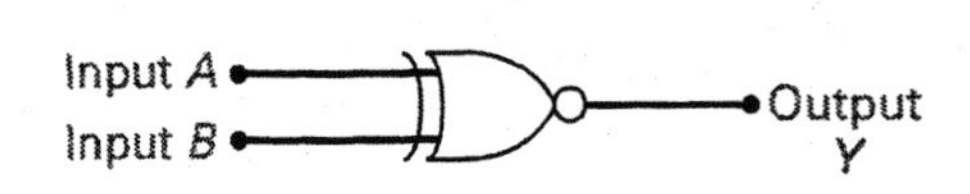

A	B	Output (Y)	
0	0	1	$A = B$
0	1	0	$A \neq B$
1	0	0	$A \neq B$
1	1	1	$A = B$

(= equal, ≠ not equal)

(a)

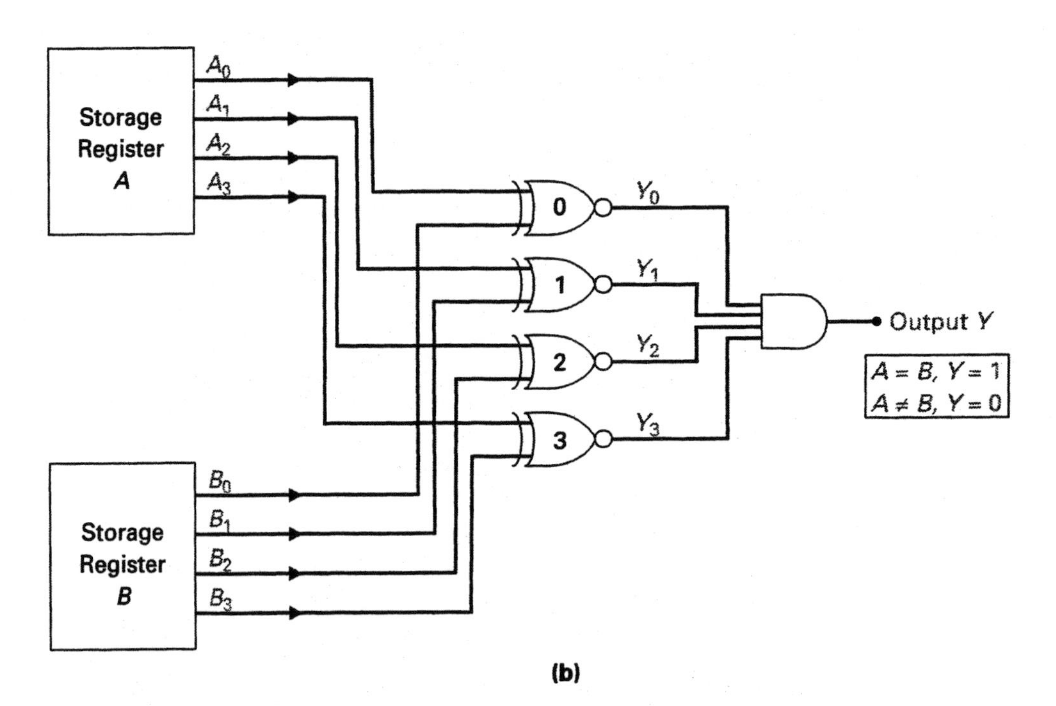

(b)

FIGURE 8-12 **Basic Comparators.**

Logic gate XNOR 0 compares A_0 and B_0, and if they are the same, Y_0 is a 1. Logic gate XNOR 1 compares A_1 and B_1, and if they are the same, Y_1 is a 1. In turn, XNOR 2 compares A_2 and B_2, and XNOR 3 compares A_3 and B_3, producing a 1 output if the bits are equal (=) and a 0 output if the bits are not equal (≠). If WORD *A* is identical to WORD *B*, all of the XNOR outputs will be HIGH, and therefore the output of the AND gate will also be HIGH. If WORD *A* differs from WORD *B* in any way, the AND gate will receive a LOW input, and will produce a LOW output. Therefore, if WORD *A* equals WORD *B*, the output $Y = 1$, whereas if WORD *A* is not equal to WORD *B*, the output $Y = 0$.

8-3-2 A 4-Bit Magnitude Comparator

A wide variety of digital comparator ICs are available that contain a complete internal binary comparator circuit. Let us now examine the operation and application of a frequently used digital comparator IC, the TTL 7485. Data sheet details for this 4-bit magnitude comparator IC are shown in Figure 8-13. The 7485 compares the 4-bit word inputs (WORD A and WORD B) and provides an active-HIGH output that indicates whether the two input words are equal ($A = B$, pin 6). In addition, the 7485 provides two other active-HIGH outputs indicating whether WORD A is greater than WORD B ($A > B$, pin 5), and whether WORD A is less than WORD B ($A < B$, pin 7). The greater than and less than magnitude outputs account for why this IC is called a magnitude comparator.

The three **cascading inputs** (pins 2, 3, and 4) provide a means for stringing together, or cascading, two or more 7485s so you can compare words that are larger than 4 bits in size. Figure 8-14(a) shows how the cascading inputs are connected so that they do not interfere with a comparison of two 4-bit words. Figure 8-14(b) illustrates how two 7485s would be connected to compare two 8-bit words. Comparator X compares the 4 low-order bits of the two 8-bit words

Cascading Inputs Connecting in series or stringing together two or more devices.

DEVICE: SN7485 – 4-Bit Magnitude Comparator

Description:

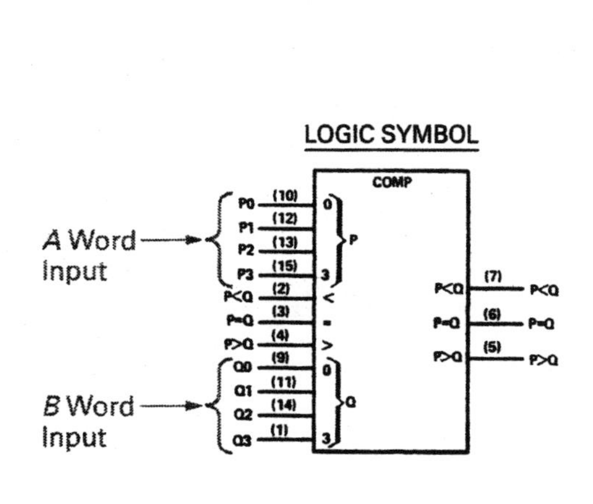

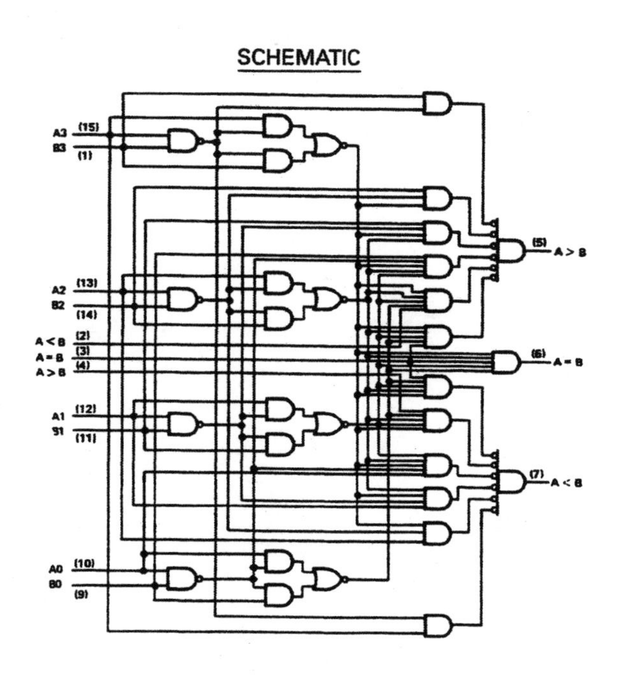

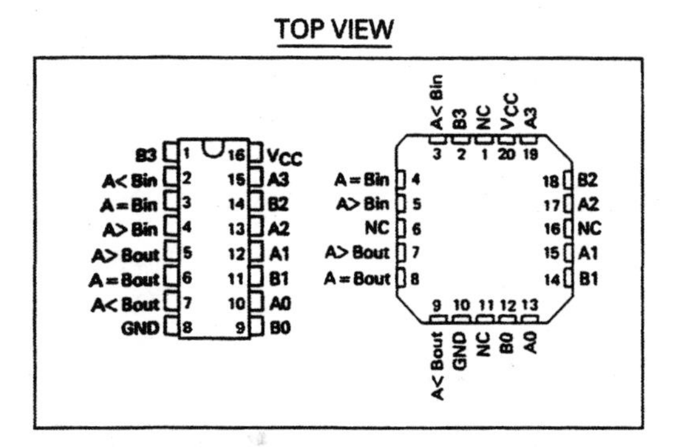

FUNCTION TABLE

COMPARING INPUTS				CASCADING INPUTS			OUTPUTS		
A3, B3	A2, B2	A1, B1	A0, B0	A > B	A < B	A = B	A > B	A < B	A = B
A3 > B3	X	X	X	X	X	X	H	L	L
A3 < B3	X	X	X	X	X	X	L	H	L
A3 = B3	A2 > B2	X	X	X	X	X	H	L	L
A3 = B3	A2 < B2	X	X	X	X	X	L	H	L
A3 = B2	A2 = B2	A1 > B1	X	X	X	X	H	L	L
A3 = B3	A2 = B2	A1 < B1	X	X	X	X	L	H	L
A2 = B3	A2 = B2	A1 = B1	A0 > B0	X	X	X	H	L	L
A3 = B3	A2 = B2	A1 = B1	A0 < B0	X	X	X	L	H	L
A3 = B3	A2 = B2	A1 = B1	A0 = B0	H	L	L	H	L	L
A3 = B3	A2 = B2	A1 = B1	A0 = B0	L	H	L	L	H	L
A3 = B3	A2 = B2	A1 = B1	A0 = B0	X	X	H	L	L	H
A3 = B3	A2 = B2	A1 = B1	A0 = B0	H	H	L	L	L	L
A3 = B3	A2 = B2	A1 = B1	A0 = B0	L	L	L	H	H	L

FIGURE 8-13 A Comparator IC That Compares Two 4-bit Input Words. (Courtesy of Texas Instruments)

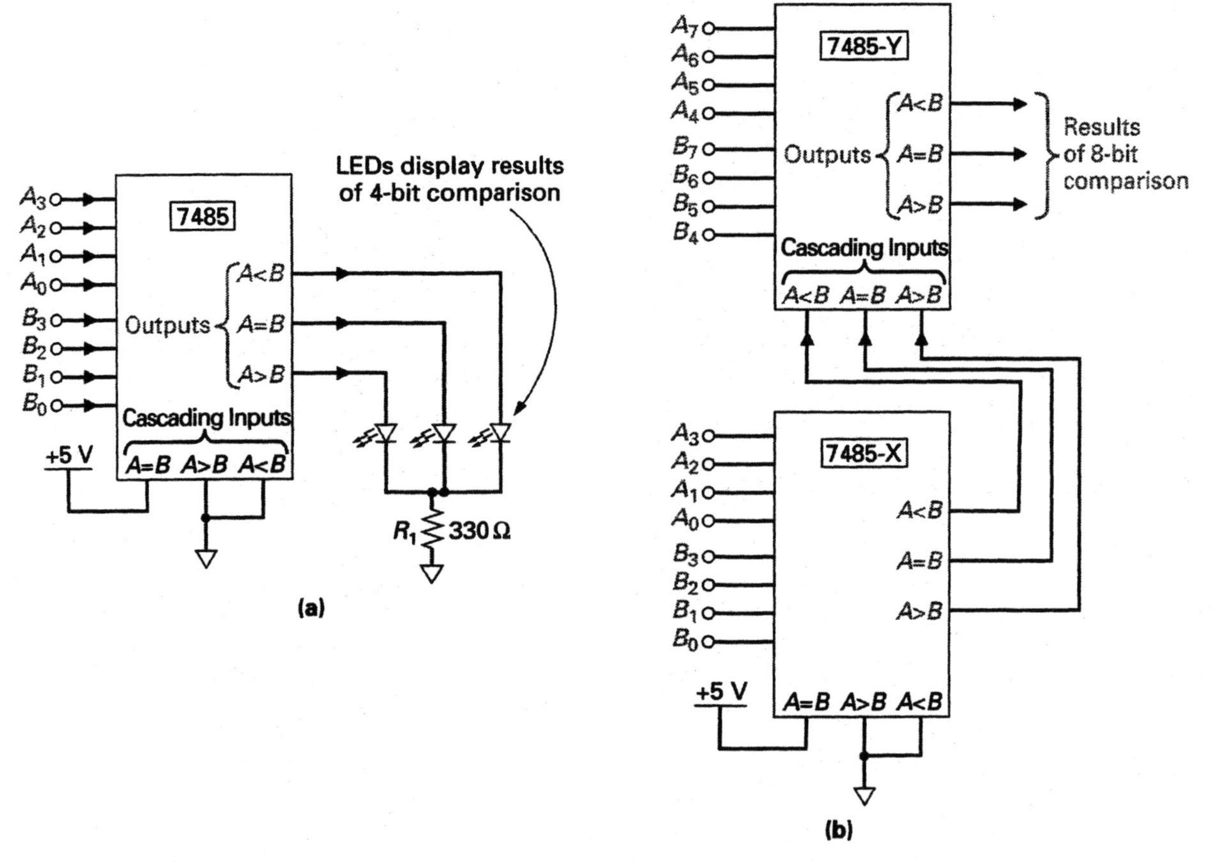

FIGURE 8-14 Using the 7485 to Compare. (a) Two 4-Bit Words. (b) Two 8-Bit Words.

(X compares A_0, A_1, A_2, and A_3 with B_0, B_1, B_2, and B_3) and its outputs are fed into the cascade inputs of the Y comparator. The Y comparator compares the 4 high-order bits of the two 8-bit words (Y compares A_4, A_5, A_6, and A_7 with B_4, B_5, B_6, and B_7) and the results from the X comparator at its cascade inputs, to produce the final outputs indicating the result of the 8-bit comparison.

■ **EXAMPLE:**

What would be the outputs from a 7485 if the following inputs were applied?

a. $A = 1010$, $B = 1101$ b. $A = 0011$, $B = 0010$

■ ***Solution:***

a. See the fourth line of the function table in Figure 8-13.
   $A > B$ output will be LOW
   $A < B$ output will be HIGH
   $A = B$ output will be LOW
b. See the last line of the function table in Figure 8-13.
   $A > B$ output will be HIGH
   $A < B$ output will be LOW
   $A = B$ output will be LOW

Let us now see how the 7485 magnitude comparator can be used in a typical control application.

Application—A Photocopier Control Circuit

Figure 8-15 shows how a 7485 comparator could be used in a photocopier to control the number of copies made. A storage register (A) holds the 4-bit binary word 0010 (2), indicating the number of copies that have been made so far. The storage register (B) holds the 4-bit binary word 1000 (8), indicating the number of copies requested. The 7485 compares these two inputs and generates three output control signals. If the 4-bit input A is less than the 4-bit input B, the 7485 pin 7 output will be HIGH, signaling that the copier should continue copying. When input A is equal to input B, the 7485 pin 6 output will be HIGH, signaling that the requested number of copies have been made and the copier should stop copying.

SELF-TEST REVIEW QUESTIONS FOR SECTION 8-3

1. Which logic gate type is considered a basic comparator?
2. What are the three outputs generally present at the output of a magnitude comparator IC, such as the 7485?
3. Why would a comparator IC have cascading inputs?
4. What would be the logic levels at the three outputs of a 7485 if WORD A = 1010 and WORD B = 1011?

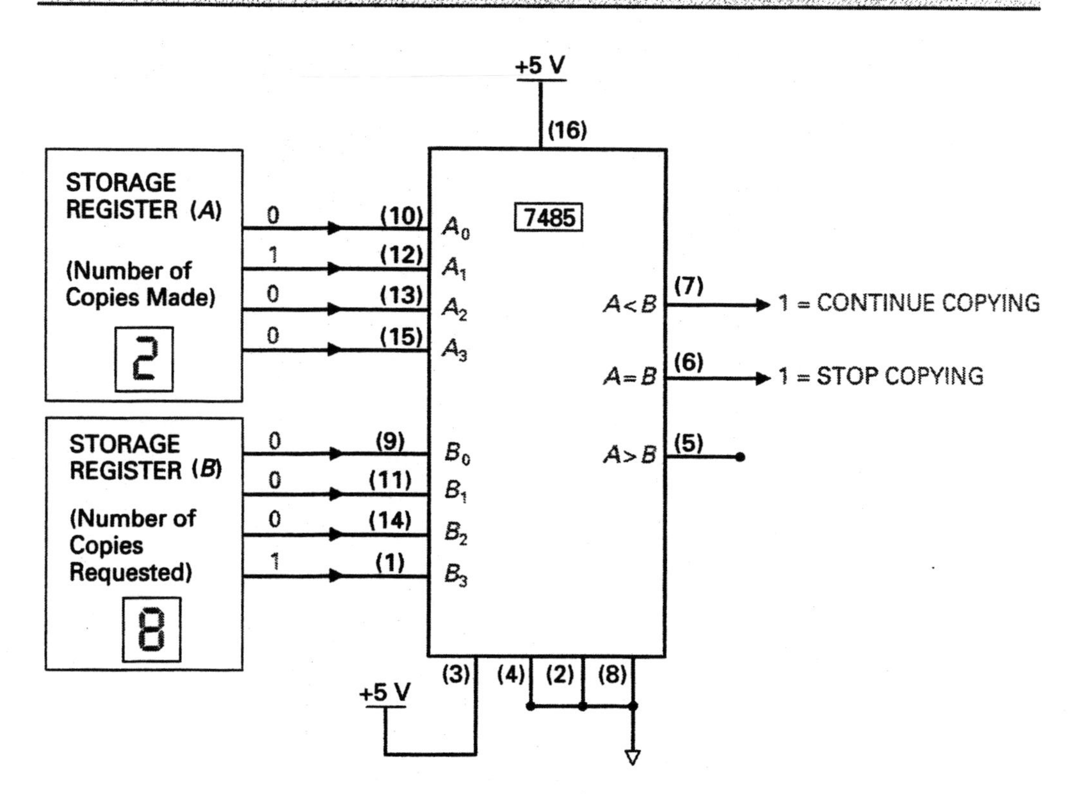

FIGURE 8-15 **Application—Comparator Control of a Photocopier.**

8-4 PARITY GENERATORS AND CHECKERS

The main objective of a digital electronic system is to manage the flow of digital information or data. In all digital systems, therefore, it is no surprise that a large volume of binary data is continually being transferred from one point to another. The accuracy of this data is susceptible to electromagnetic noise, interference, and circuit fluctuations that can cause a binary 1 to be misinterpreted as a binary 0, or a binary 0 to be misinterpreted as a binary 1. Although the chance of a data transfer error in a two-state system is very remote, it can occur, and if it does, it is important that some checking system is in place so that a data error can be detected. The **parity check** system uses circuitry to check whether the odd or even number of binary 1s present in the transmitted code matches the odd or even number of binary 1s present in the received code.

Parity Check
An odd-even check that makes the total number of 1s in a binary word either an odd or even number.

8-4-1 *Even or Odd Parity*

Parity Generator
A circuit that generates the parity bit that is added to a binary word so that a parity check can later be performed.

Parity Checker
A circuit that tests the parity bit that has been added to a binary word so that a parity check can be performed.

Parity Bit
A binary digit that is added to a binary word to make the sum of the bits either odd or even.

Figure 8-16(a) shows a simplified 4-bit parallel data transmission system that has a **parity generator** circuit and a **parity checker** circuit included for data error detection. The parity generator circuit monitors the 4-bit code being transmitted from the keyboard storage register and generates a **parity bit** based on the total number of binary 1s present in the 4-bit parallel word. The table in Figure 8-16(b) shows what EVEN-PARITY BIT (P_E) or ODD-PARITY BIT (P_O) would be generated for each of the possible sixteen 4-bit codes from the keyboard. For example, let us assume that we wish to transmit an even-parity check bit. The P_E will be made a 1 or 0 so that the number of binary 1s in the final transmitted 5-bit code (D_0, D_1, D_2, D_3, and P_E) will always equal an even number (which means that there are zero, two, or four binary 1s in the transmitted 5-bit code). If, on the other hand, we wish to transmit an odd-parity check bit, the P_O will be made a 1 or 0 so that the number of binary 1s in the final transmitted 5-bit code (D_0, D_1, D_2, D_3, and P_O) will always equal an odd number (which means that there are one, three, or five binary 1s in the transmitted 5-bit code).

Referring to the circuit in Figure 8-16(a), you can see that XOR gates are used in both the parity generator and parity checker circuits. In this example circuit, an even-parity check system is being used, and therefore an even-parity bit is generated by XOR A and transmitted along with the 4-bit code from keyboard to computer. Within the computer, the 4-bit code is stored in a register, and at the same time it is monitored by XOR gates *B* and *C*, which will produce an active-HIGH PARITY ERROR output. This HIGH error signal could be used to turn on a DATA ERROR LED, signal the computer to display a keyboard data transmission error, or signal the computer to reject the transmitted code and signal the keyboard to retransmit the code for the key pressed.

8-4-2 *A 9-Bit Parity Generator/Checker*

Parity generator and checker ICs, such as the 74180 detailed in Figure 8-17, are available for use in parity check applications. The 74180, like all parity generators and checkers, assumes that if an error were to occur, it would more than

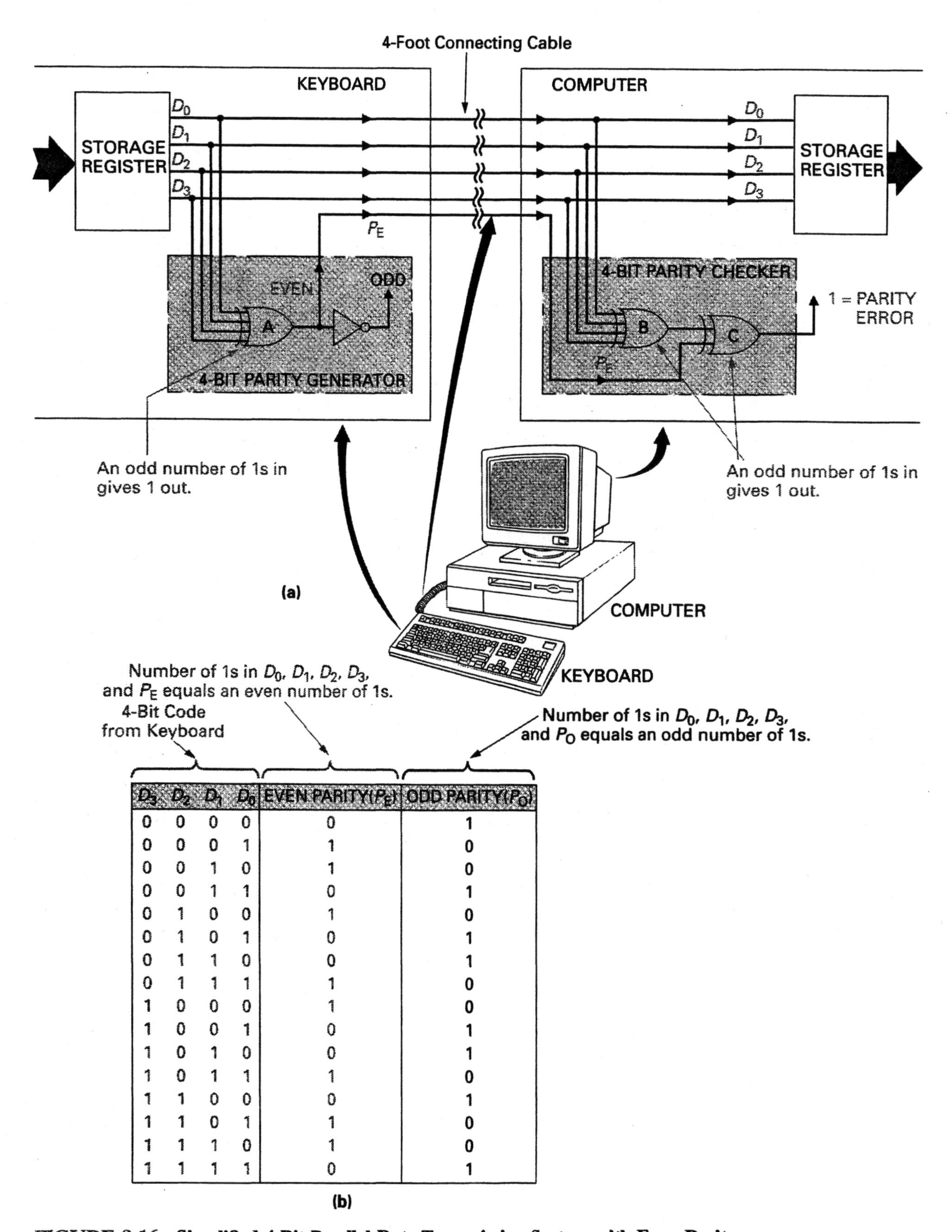

D_3	D_2	D_1	D_0	EVEN PARITY (P_E)	ODD PARITY (P_O)
0	0	0	0	0	1
0	0	0	1	1	0
0	0	1	0	1	0
0	0	1	1	0	1
0	1	0	0	1	0
0	1	0	1	0	1
0	1	1	0	0	1
0	1	1	1	1	0
1	0	0	0	1	0
1	0	0	1	0	1
1	0	1	0	0	1
1	0	1	1	1	0
1	1	0	0	0	1
1	1	0	1	1	0
1	1	1	0	1	0
1	1	1	1	0	1

FIGURE 8-16 Simplified 4-Bit Parallel Data Transmission System with Even-Parity Error Detection.

DEVICE: SN74180 – 9-Bit Odd/Even Parity Generator/Checker

Description:

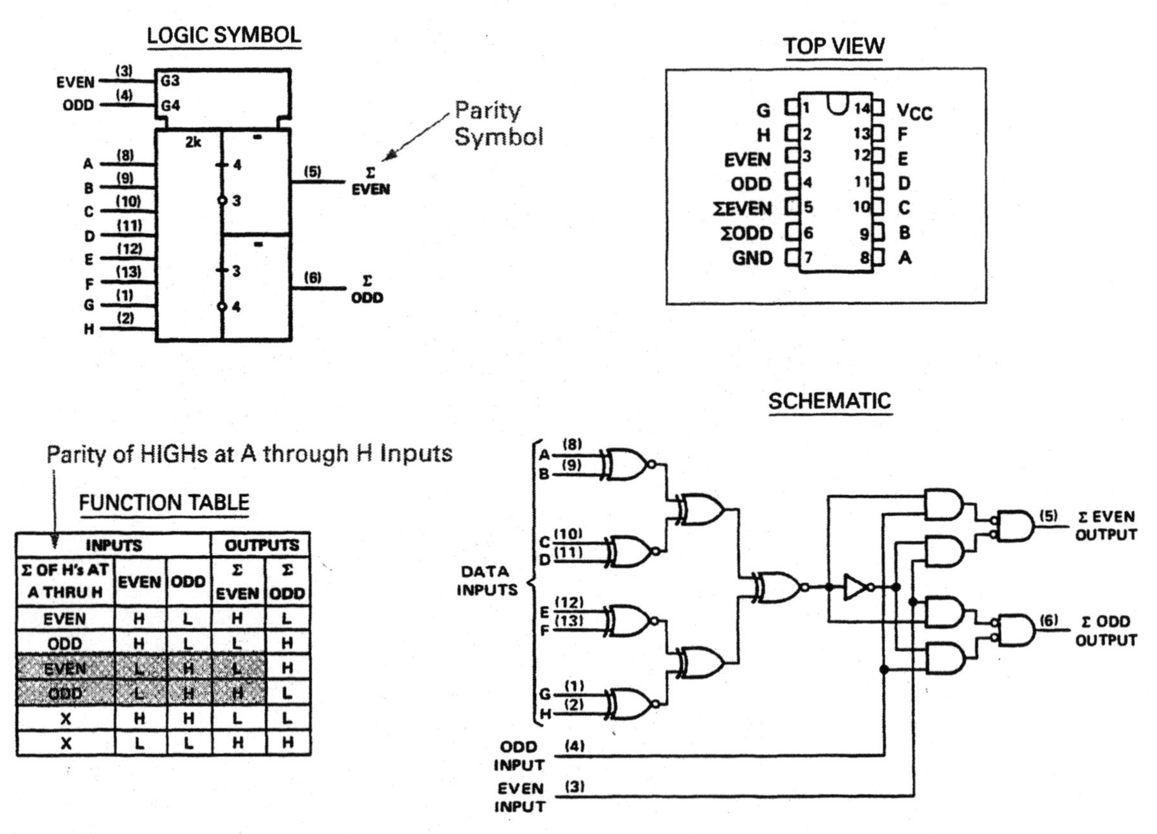

INPUTS			OUTPUTS	
Σ OF H's AT A THRU H	EVEN	ODD	Σ EVEN	Σ ODD
EVEN	H	L	H	L
ODD	H	L	L	H
EVEN	L	H	L	H
ODD	L	H	H	L
X	H	H	L	L
X	L	L	H	H

FIGURE 8-17 A Parity Generator and Checker IC. (Courtesy of Texas Instruments)

likely occur in only one of the bits within the transmitted word. If an error were to occur in two bits of the transmitted word, it is possible that an incorrect word will pass the parity test (if a 1 is degraded by noise to a 0, and at the same time a 0 is upgraded by noise to a 1). Multi-bit errors, however, are extremely rare due to the reliability and accuracy of two-state digital electronic systems.

The 74180 is able to function as either a parity generator or as a parity checker. The best way to fully understand the operation of this IC is to see it in an application.

Application—8-Bit Parallel Data Transmission System with Even-Parity Error Detection

Figure 8-18 shows how two 74180 ICs can be used in an 8-bit parallel data transmission system with even-parity error detection. The parity generator (74180 *A*) has its EVEN control input LOW and its ODD control input HIGH, so it will operate in the manner shown shaded in the function table in Figure 8-17. To explain all of the possibilities, Figure 8-19 lists each condition.

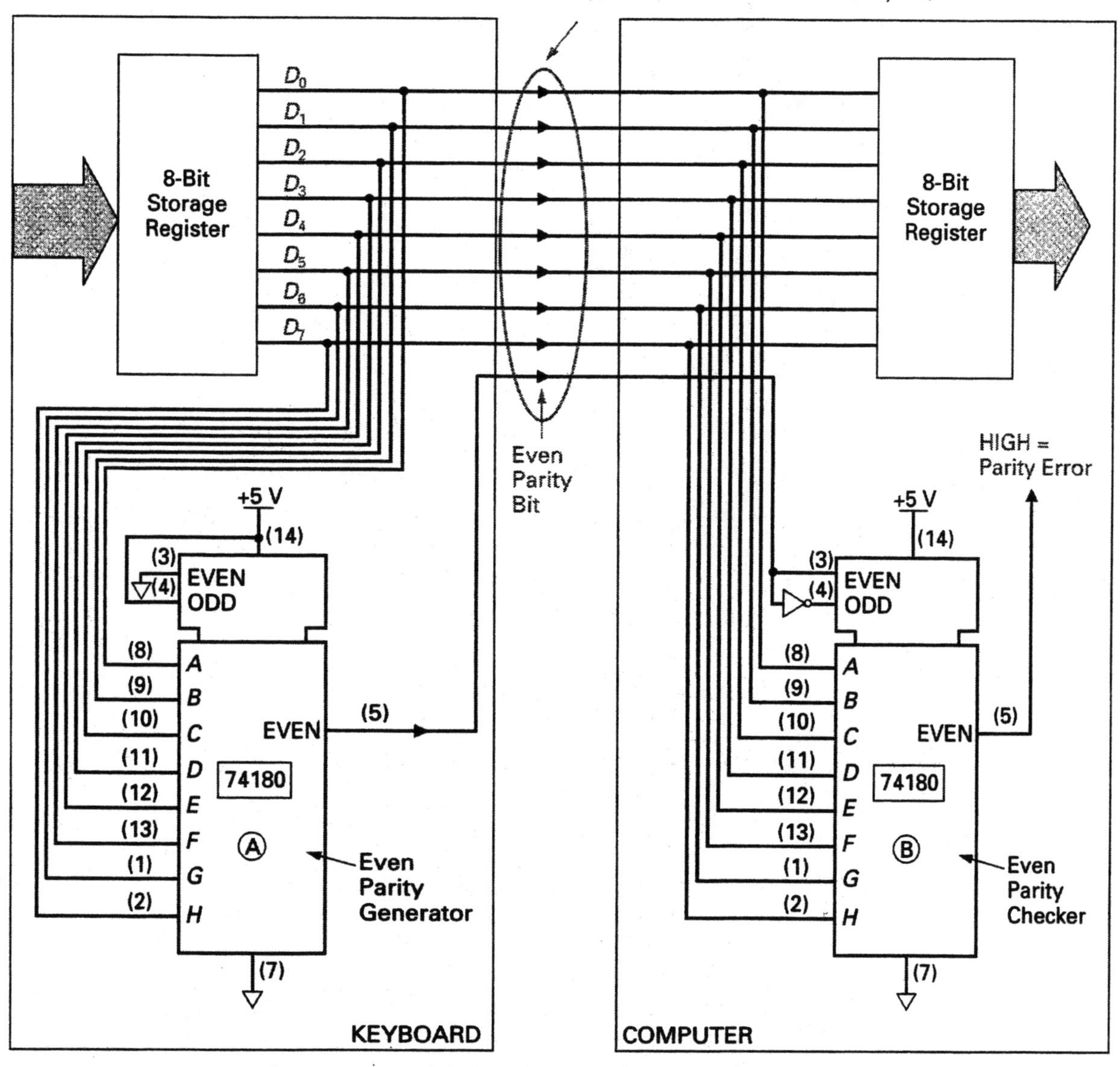

FIGURE 8-18 Application—Simplified 8-Bit Parallel Data Transmission System with Even-Parity Error Detection.

a. In the first condition, the 8-bit code to be transmitted from the keyboard contains an even number of 1s, and therefore the parity generator (74180 *A*) generates a LOW, even-parity bit output. The parity checker IC (74180 *B*) receives an even number of 1s at its *A* through *H* inputs, and the same control inputs as the parity generator IC (EVEN input = LOW, ODD input = HIGH), so the even output is LOW, indicating no transmission error has occurred.

b. In the second condition, the 8-bit code to be transmitted from the keyboard contains an even number of 1s, and therefore the parity generator (74180 *A*) generates a LOW, even-parity bit output. In this condition, however, a bit error occurs due to electrical noise so an ODD number of 1s is received at the

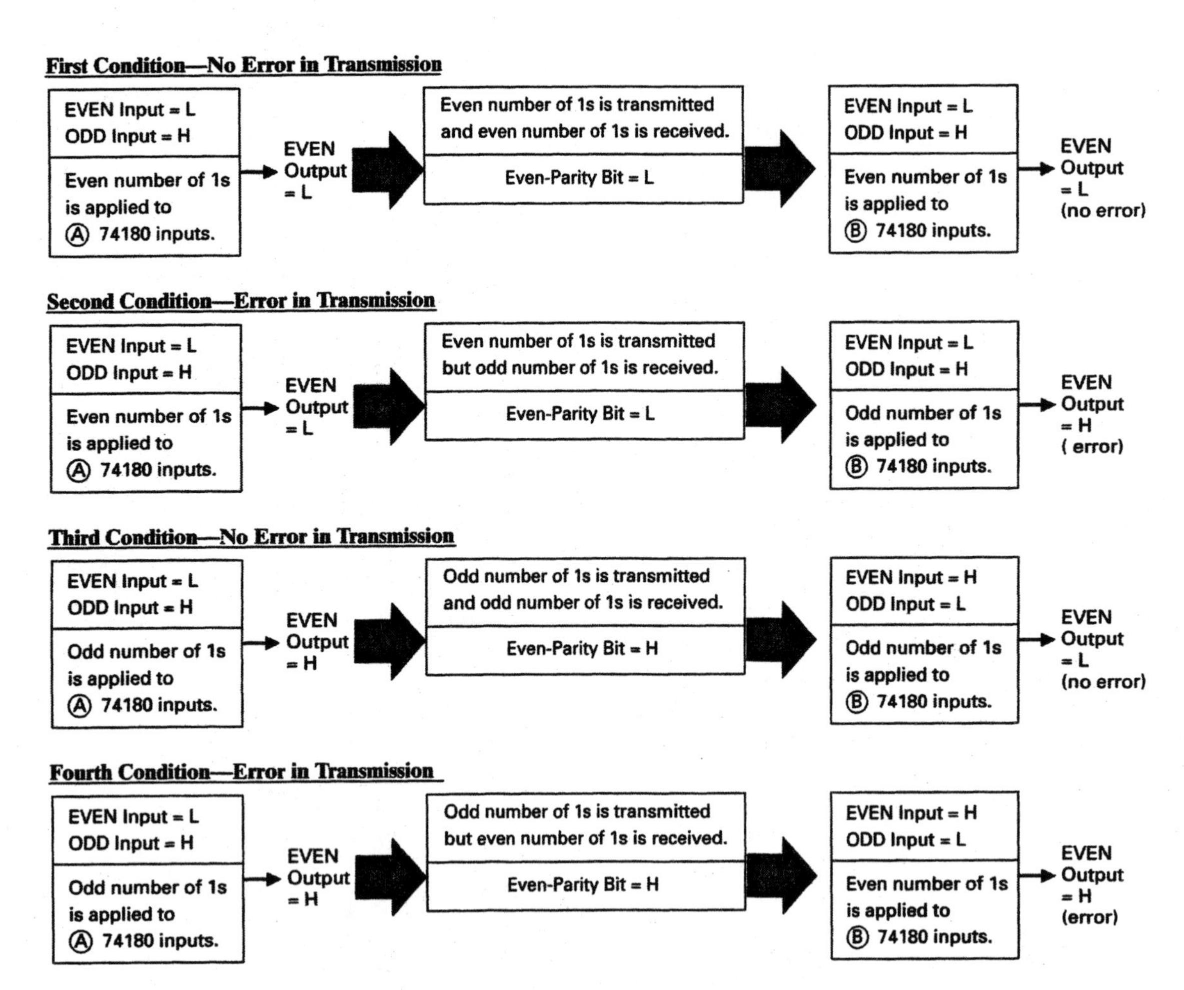

FIGURE 8-19 Four Basic Conditions of 74180 Even-Parity Circuit.

computer. The parity checker IC (74180 *B*) receives an ODD number of ones at its *A* through *H* inputs, and the same control inputs as the parity generator IC (EVEN input = LOW, ODD input = HIGH), so the even output is HIGH, indicating a transmission error has occurred.

c. In the third condition, the 8-bit code to be transmitted from the keyboard contains an odd number of 1s, and therefore the parity generator (74180 *A*) generates a HIGH, even-parity bit output. The parity checker IC (74180 *B*) receives an odd number of 1s at its *A* through *H* inputs, and opposite control inputs to the parity generator IC (EVEN input = HIGH, ODD input = LOW), so the even output is LOW, indicating no transmission error has occurred.

d. In the fourth condition, the 8-bit code to be transmitted from the keyboard contains an odd number of 1s, and therefore the parity generator (74180 *A*) generates a HIGH, even-parity bit output. In this condition, however, a bit error occurs due to electrical noise so an even number of 1s is received at the computer. The parity checker IC (74180 *B*) receives an even number of ones at its *A* through *H* inputs, and the opposite control inputs to the parity generator IC (EVEN input = HIGH, ODD input = LOW), so the even output is HIGH, indicating a transmission error has occurred.

SELF-TEST REVIEW QUESTIONS FOR SECTION 8-4

1. Include an odd-parity bit for the following 4-bit gray codes.

DECIMAL	GRAY CODE	ODD-PARITY BIT
0	0000	?
1	0001	?
2	0011	?
3	0010	?
4	0110	?
5	0111	?

2. The parity system can only be used to detect single-bit errors. (true/false)
3. Is the 74180 used as a parity generator or as a parity checker?
4. If an even-parity bit were added to the ASCII codes being generated by a keyboard, what would be the size of the words being transmitted?

8-5 TROUBLESHOOTING COMBINATIONAL LOGIC CIRCUITS

To be an effective electronics technician or troubleshooter, you must have a thorough knowledge of electronics, test equipment, troubleshooting techniques, and equipment repair. In most cases a technician is required to quickly locate the problem within an electronic system and then make the repair. To review, the procedure for fixing a failure can be broken down into three steps.

Step 1: DIAGNOSE
The first step is to determine whether a circuit problem really exists, or if it is simply an operator error. To carry out this step, a technician must collect as much information about the system, circuit, and components used, and then diagnose the problem.

Step 2: ISOLATE
The second step is to apply a logical and sequential reasoning process to isolate the problem. In this step, a technician will operate, observe, test, and apply troubleshooting techniques in order to isolate the malfunction.

Step 3: REPAIR
The third and final step is to make the actual repair, and then final test the circuit.

As far as troubleshooting is concerned, practice really does make perfect. Since this chapter has been devoted to additional combinational logic circuits, let us first examine the operation of a typical combinational logic circuit, and then apply our three-step troubleshooting process to this circuit.

8-5-1 A Combinational Logic Circuit

Figure 8-20 shows a 2-digit counter with a multiplexed display circuit. The 555 timer *A* functions as a square wave oscillator with its frequency controlled by R_2. The clock signal generated by this 555 timer serves as a multiplexing clock signal, and is applied to both the 74157 multiplexer and the 1458 dual op-amp. The *A* op-amp of the 1458 is connected to operate as a comparator, while the *B* op-amp is connected to operate as an INVERTER. These two op-amps provide the two complementary, or opposite, waveforms shown at the top of Figure 8-20. Since both the units and tens seven-segment displays are common-anode types, only when the anode pin (14) is made HIGH will the display be activated. Since these anode control signals are out of phase with one another, the displays will alternate between ON and OFF, with the units ON and the tens OFF during one-half cycle, and then the units OFF and the tens ON during the following half-cycle.

As mentioned previously, the multiplexer control signal from 555 timer *A* is also applied to the $\overline{A}/B$ switching control input of the 74157 multiplexer, making this multiplexer synchronized with the ON/OFF switching of the seven-segment displays. When the $\overline{A}/B$ multiplexer control line is HIGH, the 74157 switches the 4-bit *B*-input word from the units counter through to the 74157 *Y*-outputs. This unit count is decoded by the seven-segment decoder (7447) and the seven-segment code is applied to both of the displays; however, only the units display is enabled at this time and therefore the units count will be displayed on the units display. When the $\overline{A}/B$ multiplexer control line is LOW, the 74157 switches the 4-bit *A*-input word from the tens counter through to the 74157 *Y*-outputs. This tens count is decoded by the seven-segment decoder (7447), and the seven-segment code is applied to both of the displays; however, only the tens display is enabled at this time, and therefore the tens count will be displayed on the tens display.

The 555 timer *B* provides a clock signal for the counters. By varying this timer's clock frequency (adjusting R_5), you can vary the rate at which the counters count. Switch 1 is used to clear both counters to 0 when the normally open push-button switch is pressed.

Let us now apply our three-step troubleshooting procedure to this combinational logic circuit so that we can practice troubleshooting procedures and methods.

Step 1: Diagnose

It is extremely important that you first understand the operation of a circuit and how all of the devices within it are supposed to work, so that you are able to determine whether or not a circuit malfunction really exists. If you were preparing to troubleshoot the circuit in Figure 8-20, your first step should be to read through the circuit description and review the operation of each integrated circuit until you feel completely confident with the correct operation of the entire circuit. The circuit description, or theory of operation, for an electronic circuit can generally be found in a service or technical manual, along with troubleshooting guides. As far as the circuit's ICs are concerned, manufacturer's

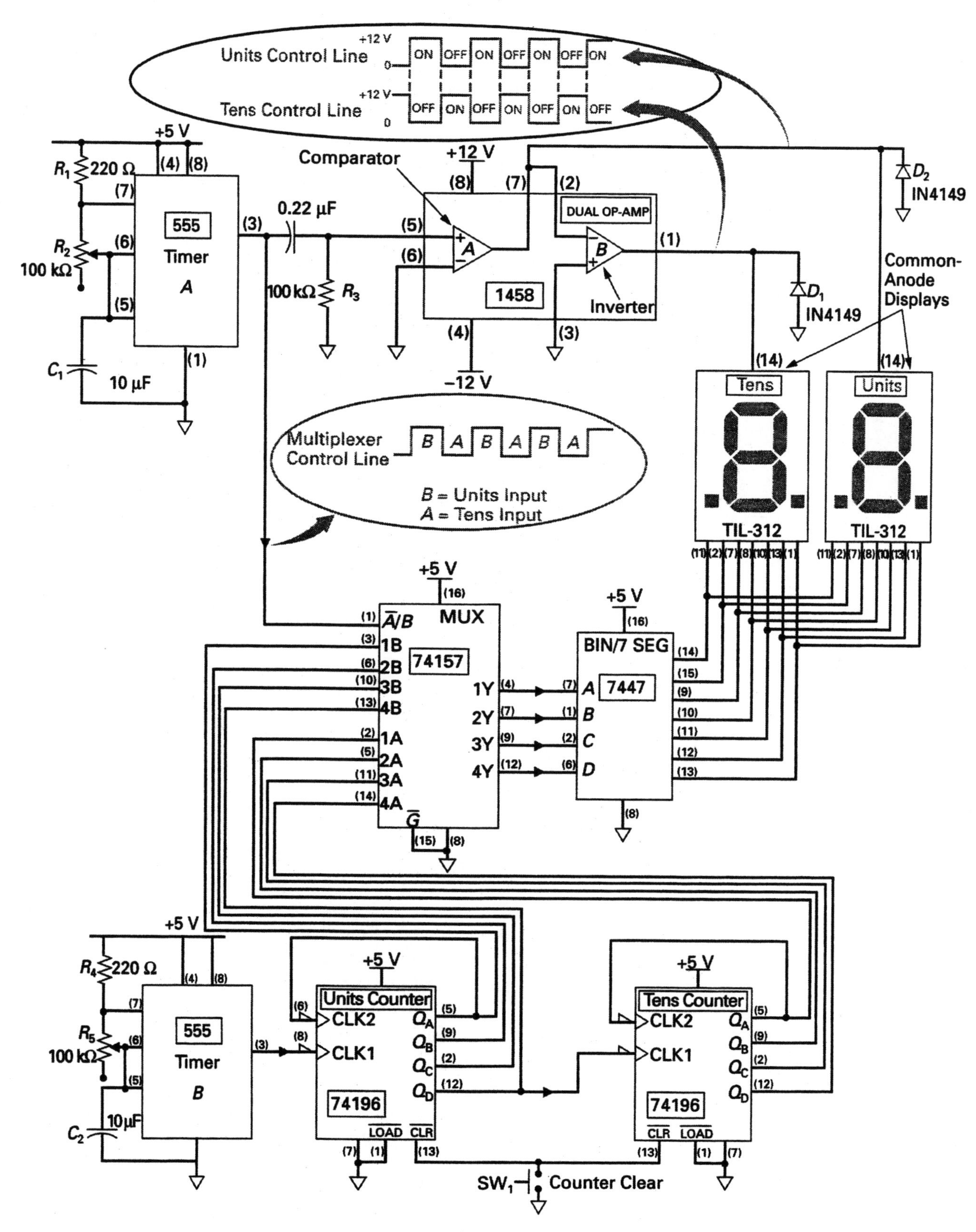

FIGURE 8-20 Troubleshooting Combinational Logic Circuits—A 2-Digit Counter with a Multiplexed Display Circuit.

digital data books contain a full description of the IC's operation, characteristics, and pin allocation. Referring to all of this documentation before you begin troubleshooting will generally speed up and simplify the isolation step. Many technicians bypass this data collection step and proceed directly to the isolation step. If you are completely familiar with the circuit's operation, this shortcut would not hurt your performance. However, if you are not completely familiar with the circuit, keep in mind the following expression and let it act as a brake to stop you from racing past the problem: *Less haste, more speed.*

Once you are fully familiar with the operation of the circuit, you will easily be able to diagnose the problem as either an *operator error* or a *circuit malfunction.* Distinguishing an operator error from an actual circuit malfunction is an important first step, and a wrong diagnosis can waste a lot of time and effort. For example, the following could be interpreted as circuit malfunctions, when in fact they are simply operator errors.

Example 1.
Symptom: If switch 1 is pressed, both displays go to 0.
Diagnosis: Operator Error—This is normal since switch 1 is a display reset switch.

Example 2.
Symptom: Only one display is on at a time. The circuit seems to be switching between the units and tens.
Diagnosis: Operator error—The 555 timer *A* clock frequency is set too low. Adjust R_2 until the display switching is so fast that the eye sees both displays as being constantly ON.

Example 3.
Symptom: The circuit is malfunctioning because the display is counting up in half-seconds instead of seconds.
Diagnosis: Operator error—There is not a circuit malfunction, the 555 timer *B* clock frequency is set too high. Adjust R_5 until the display counts up in seconds.

Once you have determined that the problem is not an operator error, but is in fact a circuit malfunction, proceed to Step 2 and isolate the circuit failure.

Step 2: Isolate

No matter what circuit or system failure has occurred, you should always follow a logical and sequential troubleshooting procedure. Let us review some of the isolating techniques and apply them to our example circuit in Figure 8-20.

a. Use a cause-and-effect troubleshooting process, which means study the effects you are getting from the faulty circuit and then logically reason out what could be the cause.

b. Check first for obvious errors before leaping into a detailed testing procedure. Is the power OFF or not connected to the circuit? Are there wiring errors? Are all of the ICs correctly oriented?

c. Using a logic probe or voltmeter test that power and ground are connected to the circuit and are present at all points requiring power and ground. If the whole circuit, or a large section of the circuit, is not operating, the problem is normally power. Using a multimeter, check that all of the dc voltages for the circuit are present at all IC pins that should have power or a HIGH input, and are within tolerance. Secondly, check that 0 V or ground is connected to all IC ground pins and all inputs that should be tied LOW.

d. Use your senses to check for broken wires, loose connectors, overheating or smoking components, pins not making contact, and so on.

e. Test the clock signals using the logic probe or the oscilloscope. Although clock signals can be easily checked with a logic probe, the oscilloscope is best for checking the signal's amplitude, frequency, pulse width, and so on. The oscilloscope will also display timing problems and noise spikes (glitches) which could be false-triggering a circuit into operation at the wrong time.

f. Perform a static test on the circuit. With our circuit example in Figure 8-20, you could static test the circuit by disconnecting the clock input to the units counter, and then use a logic pulser's SINGLE PULSE feature to clock the counter in single steps. A logic probe could then be used to test that valid logic levels are appearing at the outputs of the units and tens counters. For example, after resetting the counter and then applying nineteen clock pulse inputs from the logic pulser, the logic probe should indicate the following logic levels at the counter outputs:

	TENS	UNITS
Counter	0001	1001_{BCD}
Display	1	9_{10}

By also disconnecting the clock output from 555 timer *A,* and then switching the multiplexer control line first HIGH and then LOW, you can trace the 4-bit values through the 74157 multiplexer and 7447 decoder to the displays. Holding the circuit stationary at different stages in its operation (static testing), and using the logic pulser to single-step the circuit and the logic probe to detect logic levels at the inputs and outputs during each step, will enable you to isolate any timing problems within the circuit.

g. With a dynamic test, the circuit is tested while it is operating at its normal clock frequency, and therefore all of the inputs and outputs are continually changing. Although a logic probe can detect a pulse waveform, the oscilloscope will display more signal detail, and since it can display two signals at a time, it is ideal for making signal comparisons and looking for timing errors. A *logic analyzer,* which will be discussed in more detail in Chapter 17, is a type of oscilloscope that can display typically eight to sixteen digital signals at one time.

In some instances you will discover that some circuits will operate normally when undergoing a static test, and yet fail a dynamic test. This effect usually points to a timing problem involving the clock signals and/or the propagation delay times of the ICs used within the circuit.

h. "Noise" due to electromagnetic interference (EMI) can false-trigger an IC, such as the counters in Figure 8-20. This problem can be overcome by not leaving any of the IC's inputs unconnected, and therefore, floating. Connect unneeded active-HIGH inputs to ground and active-LOW inputs to $+V_{CC}$.

i. Apply the half-split method of troubleshooting first to a circuit, and then to a section of a circuit, to help speed up the isolation process. With our circuit example in Figure 8-20, a good mid-point check would be to first static-test the inputs to the seven-segment displays. If the units control signal or the tens control signal to the seven-segment displays are not as they should be, then the problem is more than likely in the upper section of the circuit (555 timer *A* and the 1458). If, on the other hand, the seven-segment code from the 7447 is not as it should be, then the problem more than likely lies in the lower section of the circuit (555 timer *B,* the counters, the 74157, and the 7447).

Remember that a load problem can make a source appear at fault. If an output is incorrect, it would be safer to disconnect the source from the rest of the circuit, and then recheck the output. If the output is still incorrect, the problem is definitely within the source; however, if the output is correct, the load is probably shorting the line either HIGH or LOW.

j. Substitution can be used to help speed up your troubleshooting process. Once the problem is localized to an area containing only a few ICs, substitute suspect ICs with known working ICs (one at a time) to see if the problem can be quickly remedied.

Step 3: Repair

Once the fault has been found, the final step is to repair the circuit, which could involve simply removing an excess piece of wire, re-soldering a broken connection, reconnecting a connector, adjusting the power supply voltage or clock frequency. In most instances, however, the repair will involve the replacement of a faulty component. For a circuit that has been constructed on a prototyping board or bread board, the removal and replacement of a component is simple; however, when a printed circuit board is involved, you should make a note of the component's orientation and observe good soldering and de-soldering techniques. Also be sure to handle any MOS ICs with care to prevent any damage due to static discharge.

When the circuit has been repaired, always perform a final test to see that the circuit and the system is now fully operational.

8-5-2 Sample Problems

Once you have constructed a circuit like the 2-digit counter with a multiplexed display shown in Figure 8-20, introduce a few errors to see what effect or symptoms they produce. Then logically reason out why a particular error or cause has a particular effect on the circuit. Never short any two points together unless you have carefully thought out the consequences. It is, however, generally safe to open a path and see the results. Here are some problems for our example circuit in Figure 8-20.

a. Disconnect the output of 555 timer *A* from the circuit.

b. Disconnect the output of 555 timer *A* from the 1458 input.

c. Open the pin 14 connection to the tens display.

d. Disconnect an input to the 7447.
e. Disconnect the output of 555 timer B from the counter.
f. Open the connection between 555 timer A and the 74157 $\overline{A}/B$ input.
g. Disconnect power to the 7447.
h. Disconnect one of the 7447 outputs from the 7447 and connect it to ground.

SELF-TEST REVIEW QUESTIONS FOR SECTION 8-5

1. What are the three basic troubleshooting steps?
2. What is a static test?
3. What is the half-split troubleshooting technique?
4. What symptom would you get from the circuit in Figure 8-20 if the units and tens control signals from the 1458 were in phase with one another?

REVIEW QUESTIONS

Multiple-Choice Questions

1. A ____________ is controlled to switch a single input through to one of several outputs.
   a. multiplexer b. comparator c. demultiplexer d. encoder
2. A ____________ is controlled to switch one of several inputs through to one output.
   a. multiplexer b. comparator c. demultiplexer d. encoder
3. Multiplexers can be used for ____________ data conversion, while demultiplexers can be used for ____________ data conversion.
   a. parallel-to-serial, serial-to-parallel
   b. parallel-to-serial, parallel-to-serial
   c. serial-to-parallel, serial-to-parallel
   d. serial-to-parallel, parallel-to-serial
4. Demultiplexers are often referred to as ____________.
   a. encoders b. data selectors c. comparators d. data distributors
5. Multiplexers are often referred to as ____________.
   a. encoders b. data selectors c. comparators d. data distributors
6. ____________ ICs can also be used as ____________ ICs.
   a. Encoder, decoders b. Decoder, multiplexer
   c. Decoder, demultiplexer d. Demultiplexer, encoder
7. A digital ____________ is a circuit used to determine whether two parallel binary input words are equal or unequal.
   a. multiplexer b. comparator c. decoder d. parity generator
8. What would the even-parity bit be for the code 0011011?
   a. HIGH b. LOW c. Either 1 or 0 d. None of the above

9. What would be the odd-parity bit for the code 0011011?

   a. HIGH **b.** LOW **c.** Either 1 or 0 **d.** None of the above

10. The parity check system can be used to detect single or multiple-bit errors.

    a. True **b.** False

11. Which basic logic gate functions as an even-parity generator?

    a. XNOR **b.** NAND **c.** NOR **d.** XOR

12. What are the three steps in the three-step troubleshooting procedure?

    a. Isolate, diagnose, repair **b.** Repair, diagnose, isolate
    c. Diagnose, isolate, repair **d.** Isolate, repair, diagnose

13. Disconnecting the clock signal to a circuit and then point-to-point testing all the logic levels throughout the circuit is an example of __________ testing.

    a. cause and effect **b.** dynamic **c.** half-split **d.** static

14. Running a circuit at its normal operating frequency and testing different points throughout the circuit is an example of __________ testing.

    a. cause and effect **b.** dynamic **c.** half-split **d.** static

15. The __________ method of troubleshooting tests a mid-point in the faulty circuit to determine which half section of the circuit contains the fault.

    a. half-split **b.** substitution **c.** cause and effect **d.** dynamic

Essay Questions

16. What is the key difference between a multiplexer and a demultiplexer? (8-2)
17. Using a sketch show how a rotary switch operates as a basic

    a. Multiplexer (8-1) **b.** Demultiplexer (8-2)

18. Why is a multiplexer also referred to as a data selector? (8-1)
19. How many DATA SELECT (binary) inputs would be needed for each of the following: (8-1)

    a. One-of-eight multiplexer
    b. One-of-sixteen multiplexer
    c. Two 4-bit inputs to one 4-bit output multiplexer

20. What is the difference between parallel data and serial data, and what are the advantages of each data transmission method? (8-1)
21. What is parallel-to-serial data conversion? (8-1)
22. How can a multiplexer be used as a logic-function generator? (8-1)
23. What is a multiplexed display, and what advantages does it have over a standard display? (8-1)
24. Why is a demultiplexer also referred as a data distributor? (8-2)
25. Why is it that decoders can also function as demultiplexers? (8-2)
26. What is serial-to-parallel data conversion? (8-2)
27. What is a digital comparator circuit? (8-3)

28. Which type of logic gate can determine whether the applied binary input word is equal or unequal? (8-3)
29. What is a magnitude comparator? (8-3)
30. What is the parity check system? (8-4)
31. Briefly describe the difference between an even-parity check bit and an odd-parity check bit. (8-4)
32. What is the function of a parity generator IC? (8-4)
33. What is the function of a parity check IC? (8-4)
34. Can the parity check system be used to detect multi-bit data transmission errors? (8-4)
35. If a parity generator IC were configured to generate an even-parity bit, would the parity check IC be configured to detect even or odd parity? (8-4)

Practice Problems

36. Add an even-parity bit to the following ASCII codes.

    a. 1100001 **b.** 0111010 **c.** 1010110

To practice your circuit recognition and operation, refer to Figure 8-21 and answer the following two questions.

37. Is the circuit operating as a serial-to-parallel or parallel-to-serial data transmission circuit?
38. What binary data will appear at the output as the counter cycles from 0 through 7?

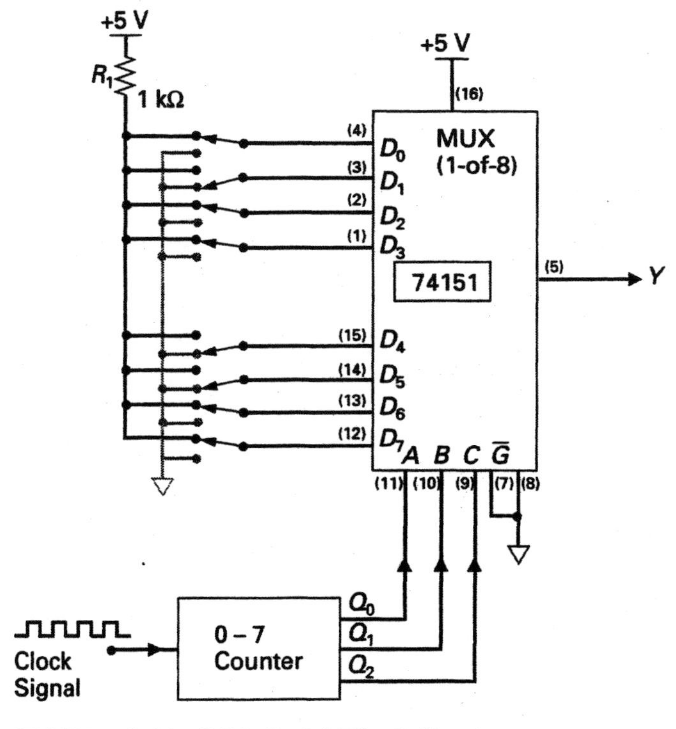

FIGURE 8-21 8-Bit Serial Word Generator

To practice your circuit recognition and operation, refer to Figure 8-22 and answer the following five questions.

39. Briefly describe the operation of the circuit, describing how the display will show the binary value of whichever push-button key is pressed.

40. If push-button 4 is pressed, the output of the 74151 will output a LOW when the counter has reached a count of ____________, and this LOW output from the 74151 will cause the NAND gate to ____________ (block/pass) the clock pulses to the 74193. The binary display will therefore display ____________.

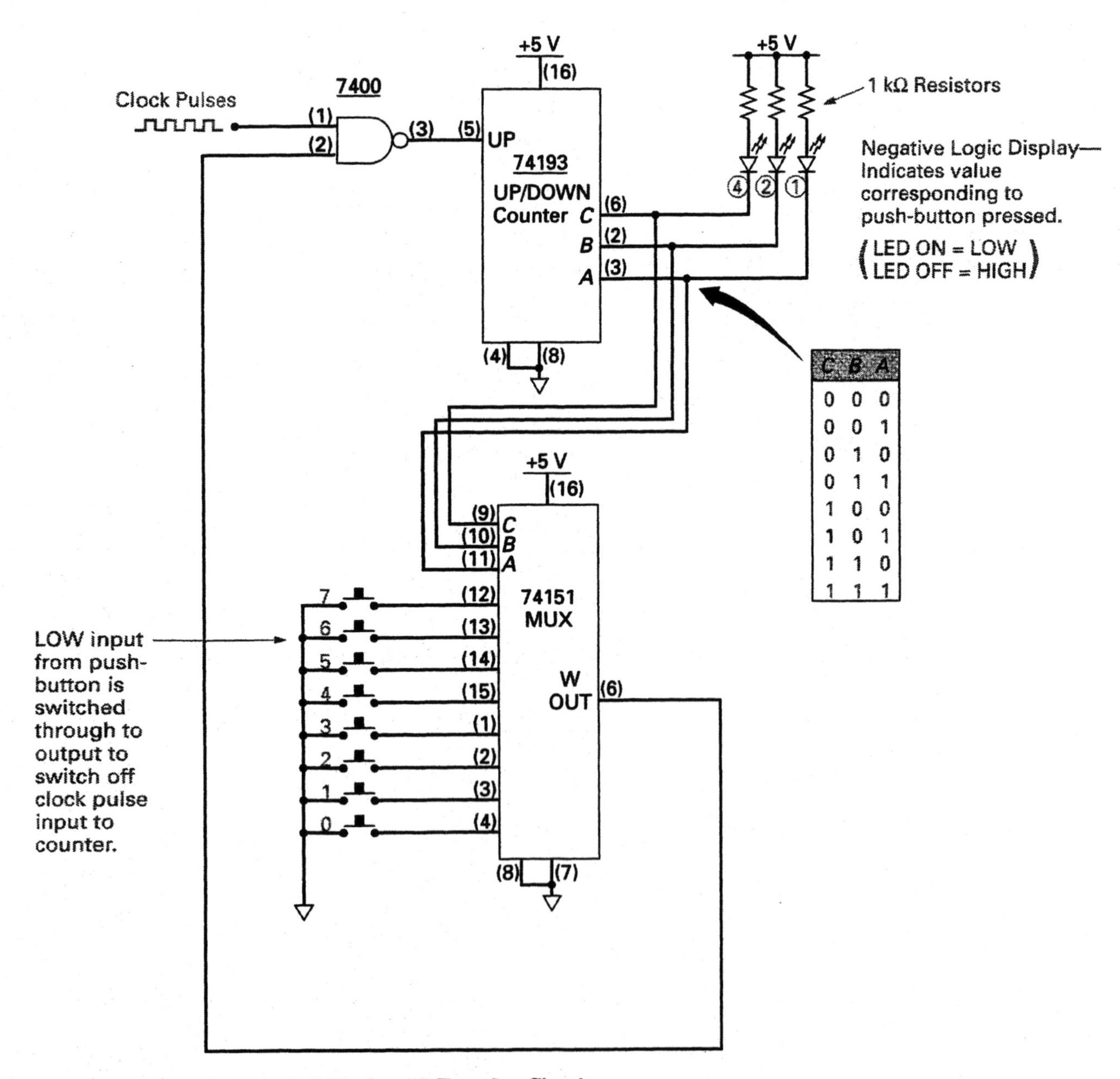

FIGURE 8-22 **0 through 7 Keyboard Encoder Circuit**

41. The display in this circuit is a common-____________ display, and therefore a LED ON indicates a ____________ (LOW/HIGH) logic level.

42. What binary value would the display finally indicate if the counter were at 000 and it began its upward count cycle with push-buttons 2 and 5 pressed?

43. What would be the condition of the display if none of the input switches were being pressed?

To practice your circuit recognition and operation, refer to Figure 8-23 and answer the following six questions.

44. What is the purpose of the 74157 IC in this circuit?

45. What is the purpose of the 7485 IC in this circuit?

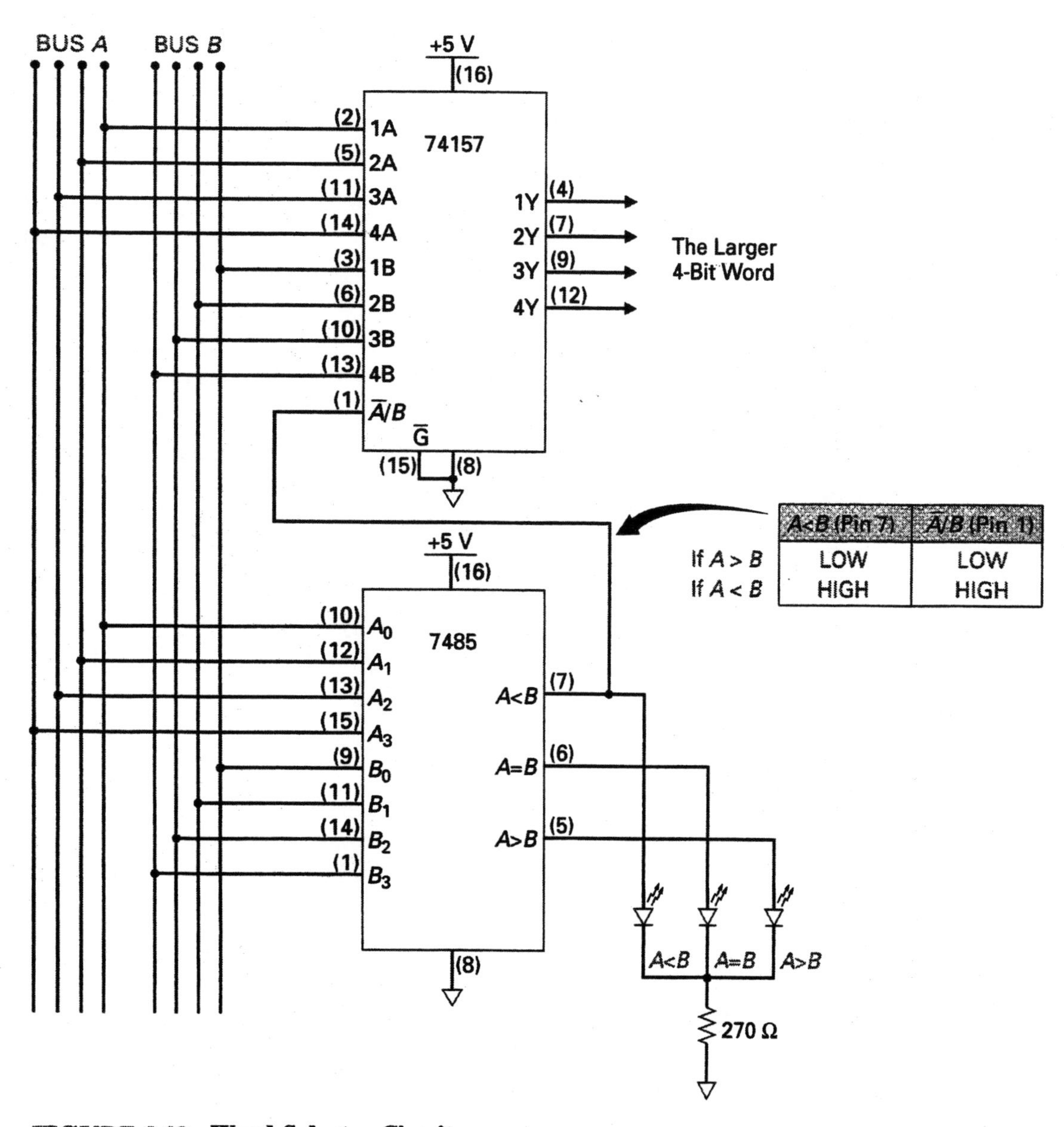

	A<B (Pin 7)	$\overline{A}/B$ (Pin 1)
If $A > B$	LOW	LOW
If $A < B$	HIGH	HIGH

FIGURE 8-23 Word Selector Circuit.

46. If the A input word were 1010 and the B input word were 1001:

   a. What would the outputs of the 7485 be?
   b. What would the LED display indicate?
   c. What word would be present at the output of the 74157?

47. Briefly describe the purpose of this circuit.

48. Is the display in this circuit a common-anode or common-cathode type?

49. What would the 7485 and 74157 outputs be if the A input word were the same as the B input word?

To practice your circuit recognition and operation, refer to Figure 8-24 and answer the following six questions.

50. What is the purpose of the 74193 IC in this circuit?

51. What is the purpose of the 74138 IC in this circuit?

52. Briefly describe the operation of this circuit.

53. How would this circuit operate if the jumper were connected to the 5 output of the 74138?

54. Is the display in this circuit a common-anode or common-cathode type?

55. What would the display show if the jumper were connected to the 3 output of the 74138?

Troubleshooting Questions

56. Briefly describe the difference(s) between a static test and a dynamic test.

57. What symptoms would you expect from the circuit shown in Figure 8-21 if pin 7 of the 74151 were connected permanently HIGH instead of LOW?

58. What symptoms would you expect from the circuit shown in Figure 8-22 if the following malfunctions were to occur?

   a. The C output of the 74193 was permanently HIGH.
   b. The LSB LED were burned out.
   c. Power was disconnected from the LED display.
   d. The output of the NAND gate was connected to the 74193 DOWN input instead of the UP input.

59. What symptoms would you expect from the circuit shown in Figure 8-23 if the following malfunctions were to occur?

   a. Pin 1 of the 74157 was connected to pin 6 of the 7485.
   b. Pin 1 of the 74157 was connected to pin 5 of the 7485.
   c. Pin 15 of the 74157 was connected HIGH instead of LOW.
   d. Power was not connected to the 74157.

60. What symptoms would you expect from the circuit shown in Figure 8-24 if the following malfunctions were to occur?

   a. The jumper was permanently connected LOW.
   b. The data input to the 74138 was connected LOW instead of HIGH.
   c. The clock input was disconnected from the 74193.
   d. Pin 5 of the 74138 was connected HIGH instead of LOW.

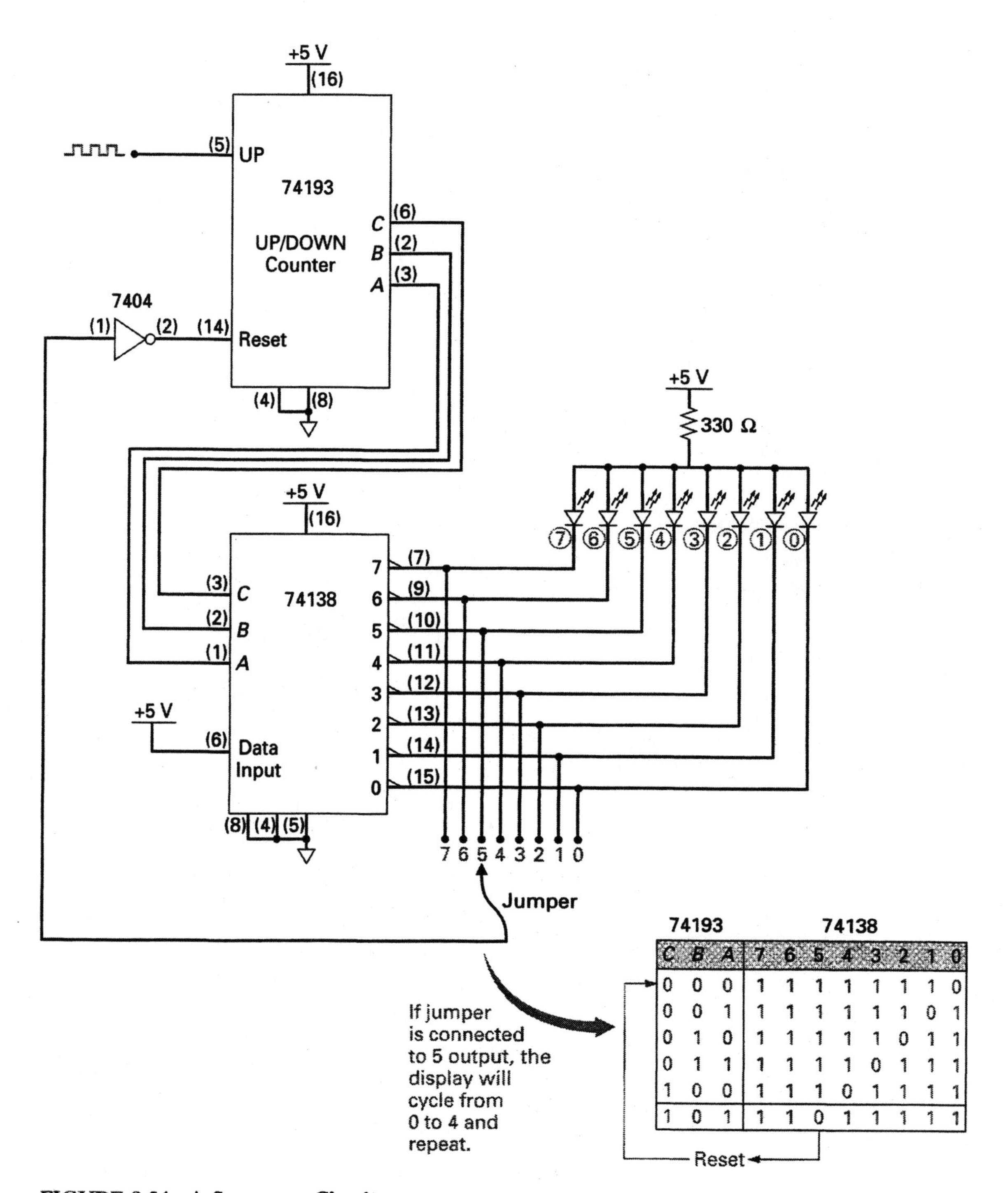

C	B	A	7	6	5	4	3	2	1	0
0	0	0	1	1	1	1	1	1	1	0
0	0	1	1	1	1	1	1	1	0	1
0	1	0	1	1	1	1	1	0	1	1
0	1	1	1	1	1	1	0	1	1	1
1	0	0	1	1	1	0	1	1	1	1
1	0	1	1	1	0	1	1	1	1	1

FIGURE 8-24 A Sequencer Circuit.

Chapter 9

OBJECTIVES

After completing this chapter, you will be able to:

1. Describe the basic function of a flip-flop or latch circuit.
2. List the three basic types of flip-flops.
3. Describe the function and explain the basic operation of a Set-Reset (*S-R*) flip-flop circuit.
4. Explain the differences between a level-triggered, edge-triggered, and pulse-triggered Set-Reset flip-flop.
5. Describe the operation, characteristics, and typical application, for the digital 74L71—AND Gated *S-R* master-slave flip-flop.
6. Describe the function and explain the basic operation of a Data-type flip-flop circuit.
7. Explain the differences between a level-triggered, edge-triggered, and pulse-triggered D-type flip-flop.
8. Describe the operation, data sheet characteristics, and typical application, for the following digital set-reset flip-flop ICs.
   a. 7475—4-bit bistable latches
   b. 7474—Dual D-type positive edge-triggered flip-flop
   c. 74175—Quadruple D-type positive edge-triggered flip-flop
9. Explain the three-step troubleshooting procedure and then show how it can be applied to a typical flip-flop circuit.
10. Describe some frequently used troubleshooting procedures.

Set-Reset and Data-Type Flip-Flops

OUTLINE

VIGNETTE

The Persistor

In the late 1930s, experimenters had successfully demonstrated that semiconductors could act as rectifiers and be used in place of the then-dominant vacuum tube. The only problem was that nobody could work out how to control semiconductors to make them predictable so they could function like the vacuum tube as an amplifier or a switch. It was not until the beginning of World War II that researchers and physicists began seriously investigating semiconductors in hopes that they could be used to create superior components for radar applications.

Convinced that semiconductor devices would be playing a major role in the future of electronics, AT&T's Bell Laboratories launched an extensive research program. Their goal was to develop a component to replace the vacuum-tube amplifier and mechanical-relay switch that were being used extensively throughout the nation's telephone system. The three men leading the assault were Walter Brattain, a sixteen-year Bell experimenter, John Bardeen, a young theoretician, and the team leader, William Shockley.

Shockley had been fascinated by semiconductors for the past ten years and had a talent for simplifying a research problem to its basic elements and then directing the project into its next avenue of investigation. He had a dominant, intensely serious, and competitive personality and on many occasions forged ahead without waiting to evaluate all of the details of the present situation. On one occasion, after several months of extensive research on a semiconductor amplifier, one of Shockley's ideas which worked perfectly in theory failed completely when tested. Shockley forged off in another direction, but Bardeen and Brattain became fascinated by the reasons for the failure.

On December 23, 1947, after three years of research and at a cost of $1 million, Brattain and Bardeen conducted an experiment on their newly constructed semiconductor prototype. Applying an audio signal to the input of the device, a replica was viewed on an oscilloscope at the output, which was fifty times larger in size—they had achieved amplification. The input signal seemed to control the resistance between the output terminals and this controlled the amount of current flowing through the resistive section of the device. Since the input controlled the amount of current transferring through the resistive section of the device, the component was called a *transresistor,* a name which was later shortened to *transistor.* Six months later the device was released on the market, but its unpredictable behavior and cost of eight dollars, compared to its seventy-five-cent vacuum tube counterpart, resulted in it not making much of an impression.

To perhaps make up for his lack of involvement in the final design, Shockley immediately began work on refining the design and removing what he called "the mysterious witchcraft" that the device seemed to possess. In a matter of days Shockley had much of the theory outlined, but perfecting the design was going to be a long and arduous task. One colleague working with Shockley jokingly referred to the device as a "persistor" because of its stubbornness to work. Shockley's persistence, however, eventually won out, and in 1951 the first reliable commercial transistor was released. It could do everything that its vacuum-tube counterpart could do and it was smaller, had no fragile glass envelope, did not need a heating filament to warm it up, and only consumed a fraction of the power.

In 1956 Shockley, Brattain, and Bardeen were all recognized by the world's scientific community when they shared the Nobel Prize in physics. Today this point in history is recognized as the most significant milestone in the history of electronics. Every electronic system manufactured today has an abundance of transistors within its circuitry—not bad for a device that was originally ridiculed by many as "a flash in the pan."

INTRODUCTION

In the two previous chapters we discussed how logic gates could be used in combination to decode, encode, select, distribute, and check binary data. In this chapter you will see how logic gates can be used in combination to form a memory circuit that is able to store binary information. This memory circuit is called a flip-flop because it can be flipped into its set condition in which it stores a binary 1, or flopped into its reset condition in which it stores a binary 0.

There are three basic types of flip-flop circuits available:

the Set-Reset (*S-R*) flip-flop

the Data-type (D-type) flip-flop

the *J-K* flip-flop

In this chapter we will cover the set-reset flip-flop and the data-type flip-flop; in the following chapter we will discuss the *J-K* flip-flop.

9-1 SET-RESET (*S-R*) FLIP-FLOPS

The flip-flop is a digital logic circuit that is capable of storing a single bit of data. It is able to store either a binary 1 or a binary 0 because of the circuit's two stable operating states—SET and RESET. Once the flip-flop has been flipped into its set condition (in which it stores a binary 1) or flopped into its reset condition (in which it stores a binary 0), the output of the circuit remains latched or locked in this state as long as power is applied to the circuit. This latching or holding action accounts for why the SET or RESET (*S-R*) flip-flop is also known as a SET or RESET latch. Let us now examine this circuit in detail.

9-1-1 *Basic* S-R *Flip-Flop or* S-R *Latch*

Set-Reset (*S-R*) Flip-Flop
A bistable multivibrator or two-state flip-flop circuit that can have its output latched HIGH (set) or LOW (reset).

The bistable multivibrator has two (bi) stable states, and its bipolar transistor circuit is illustrated in Figure 9-1(a). The circuit has two inputs called the SET and RESET inputs, and these inputs drive the base of Q_1 and Q_2. The two outputs from this circuit are taken from the collectors of Q_1 and Q_2, and are called Q and $\overline{Q}$ (pronounced "queue" and "not queue"). The $\overline{Q}$ output derives its name from the fact that its voltage level is always the opposite of the Q output. For example, if Q is HIGH, $\overline{Q}$ will be LOW, and if Q is LOW, $\overline{Q}$ will be HIGH. The bistable multivibrator circuit is often called an ***S-R* (set-reset) flip-flop,** since *a pulse on the SET input will flip the circuit into the set state (Q output is set HIGH), while a pulse on the RESET input will flop the circuit into its reset stage (Q output is reset LOW).*

Figure 9-1(b) shows the logic symbol for a *S-R* or *R-S* flip-flop, or bistable multivibrator circuit.

To fully understand the operation of the circuit, refer to the waveforms in Figure 9-1(c). When power is first applied, one of the transistors will turn ON first, and because of the cross-coupling, turn the other transistor OFF. Let us assume that the circuit in Figure 9-1(a) starts with Q_1 ON and Q_2 OFF. The low voltage (approximately 0.3 V) on Q_1's collector will be coupled to Q_2's base, keeping it OFF, while the high voltage on Q_2's collector (approximately +5 V) will be coupled to the base of Q_1, keeping it ON. This condition is called the *RESET state* since the primary output (output 1, or Q) has been reset to binary 0 (LOW), or 0 V. The cross-coupling action between the transistors will keep the transistors in the reset state (output 1 or $Q = 0$ V, and output 2 or $\overline{Q} = 5$ V) until an input appears.

Following the waveforms in Figure 9-1(c), you can see that the first input to go active is the SET input at time t_1. This positive pulse will be applied to the base of Q_2, and forward bias its base-emitter junction. As a result, Q_2 will go ON and its LOW collector voltage will be applied to output 2 ($\overline{Q}$). This LOW on Q_2's collector will also be cross-coupled to the base of Q_1, turning it OFF and making output 1 (Q) HIGH. When a pulse appears on the SET input, the circuit will therefore be put in its *SET state,* which means that the primary output (output 1 or Q) will be set to binary 1 (HIGH), or +5 V.

Studying the waveforms in Figure 9-1(c) once again, you will notice that after the positive SET input pulse has ended, the bistable circuit will still remain *latched* or held in its last state due to the cross-coupling between the transistors. This ability of the bistable multivibrator to remain in its last condition or state, and therefore remember what was last stored in the circuit, accounts for why the *S-R* flip-flop is also called a ***S-R* latch.**

Set-Reset Latch
A set-reset flip-flop.

Continuing with the waveforms in Figure 9-1(c), you can see that a RESET pulse occurs at time t_2, and resets the primary output (output 1 or Q) LOW. The flip-flop then remains latched in its reset state until a SET pulse is applied to the set input at time t_3, setting Q HIGH. Finally, a positive RESET pulse is applied to the reset input at time t_4, and Q is reset LOW.

The operation of the *S-R* flip-flop is summarized in the truth table shown in Figure 9-1(d). When only the R input is pulsed HIGH (reset condition), the

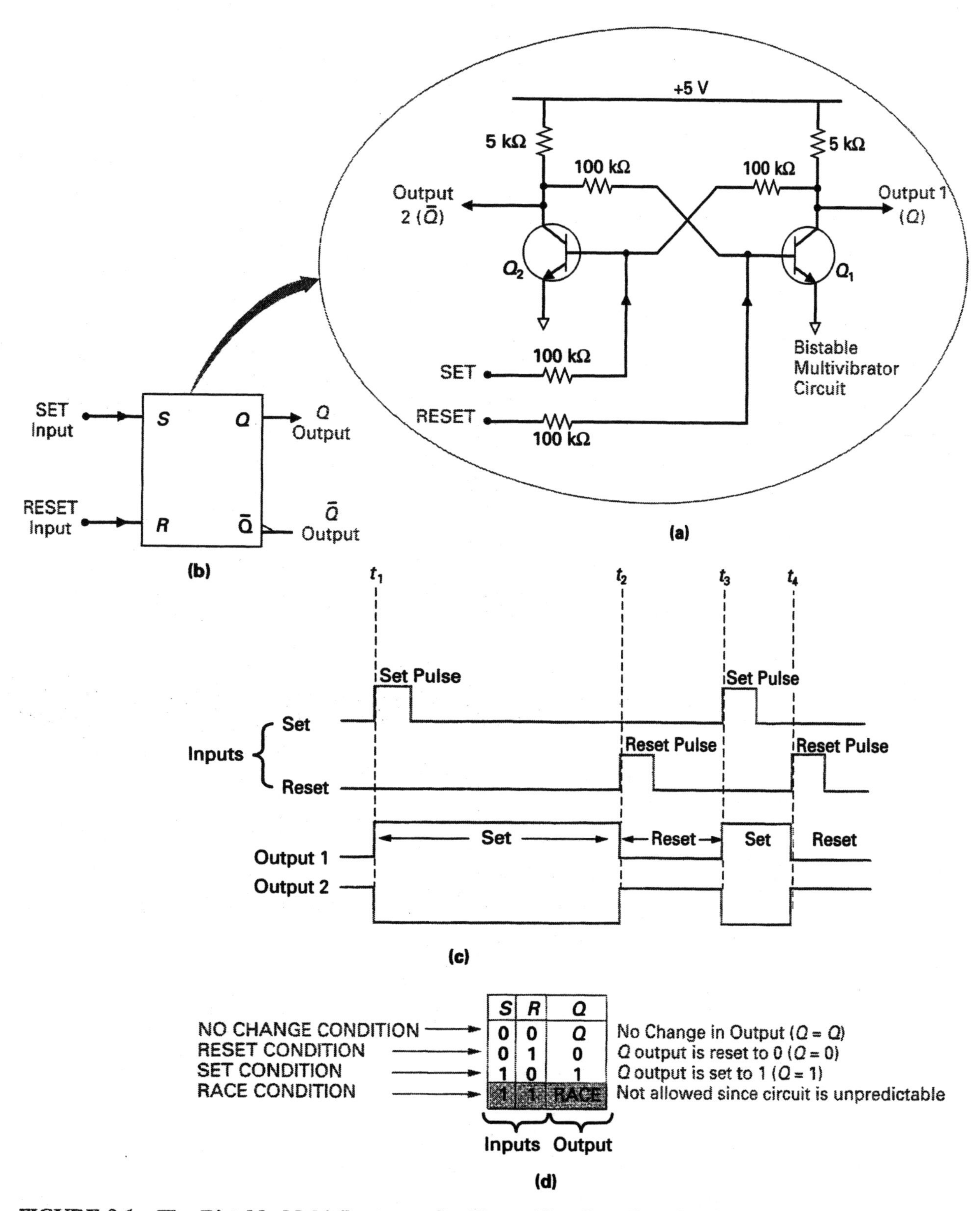

S	R	Q
0	0	Q
0	1	0
1	0	1
1	1	RACE

FIGURE 9-1 The Bistable Multivibrator or Set-Reset ($S = \bar{R}$ or $R = \bar{S}$) Flip-Flop. (a) Bipolar Transistor Circuit. (b) S-R Symbol. (c) Timing Waveforms. (d) Truth Table.

Q output is reset to a binary 0, or reset LOW ($\overline{Q}$ will be the opposite, or HIGH). On the other hand, when only the S input is pulsed HIGH (set condition), the Q output is set to a binary 1, or set HIGH ($\overline{Q}$ will be the opposite, or LOW). When both the S and R inputs are LOW, the S-R flip-flop is said to be in the **NO CHANGE,** or latch, condition, since there will be no change in the output Q. For example, if the output Q is SET, and then the S and R inputs are made LOW, the Q output will remain SET, or HIGH. On the other hand, if the output Q is RESET, and then the S and R inputs are made LOW, the Q output will remain RESET, or LOW.

NO CHANGE Condition
An input combination or condition that when applied to a flip-flop circuit will cause no change in the output logic level.

The external circuits driving the S and R inputs will be designed so that these inputs are never both HIGH, as shown in the last condition in the table in Figure 9-1(d). This is called the **RACE condition,** since both bipolar transistors will have their bases made positive, and therefore they will race to turn ON and then shut the other transistor OFF via the cross-coupling. This input condition is not normally applied since the output condition is unpredictable.

RACE Condition
An input combination or condition that when applied to a flip-flop circuit will cause the internal circuit to race to turn on and result in an unpredictable logic level output.

The bistable multivibrator S-R flip-flop, or S-R latch, has become one of the most important circuits in digital electronics. It is used in a variety of applications, including data storage, counting, and frequency division.

9-1-2 NOR *S-R* Latch and NAND *S-R* Latch

Figure 9-2 shows how set-reset flip-flop circuits can be constructed using logic gates instead of discrete components. Figure 9-2(a) shows how two NOR gates could be connected to form a set-reset NOR latch, or flip-flop. This circuit operates in exactly the same way as the previously discussed S-R flip-flop in Figure 9-1. Looking at the truth table for the NOR latch in Figure 9-2(b), you can see that when both the S and R inputs are LOW, there will be NO CHANGE in the Q's output logic level. In the second line of the truth table, you can see that a HIGH on the R input will RESET the Q output to 0, while the third line of the truth table shows how a HIGH on the S input will SET the Q output to 1. The last line of the truth table shows that when the S and the R inputs are both HIGH, the circuit will race and the Q output will be unpredictable. As mentioned previously, the S-R latch is only operated in the NO CHANGE, RESET, and SET conditions. The RACE condition is not normally used because the output cannot be controlled.

The timing diagram in Figure 9-2(c) shows how the Q output can be either SET HIGH by momentarily pulsing the S input HIGH, or RESET LOW by momentarily pulsing the R input HIGH. Let us study a few examples to see how this circuit operates.

■ **EXAMPLE:**

Remembering that any HIGH into a NOR gate will give a LOW output, and only when both inputs are LOW will the output be HIGH, study Figure 9-3, which shows a variety of input-output conditions for the set-reset NOR latch. Determine the Q output for each of the examples in Figure 9-3.

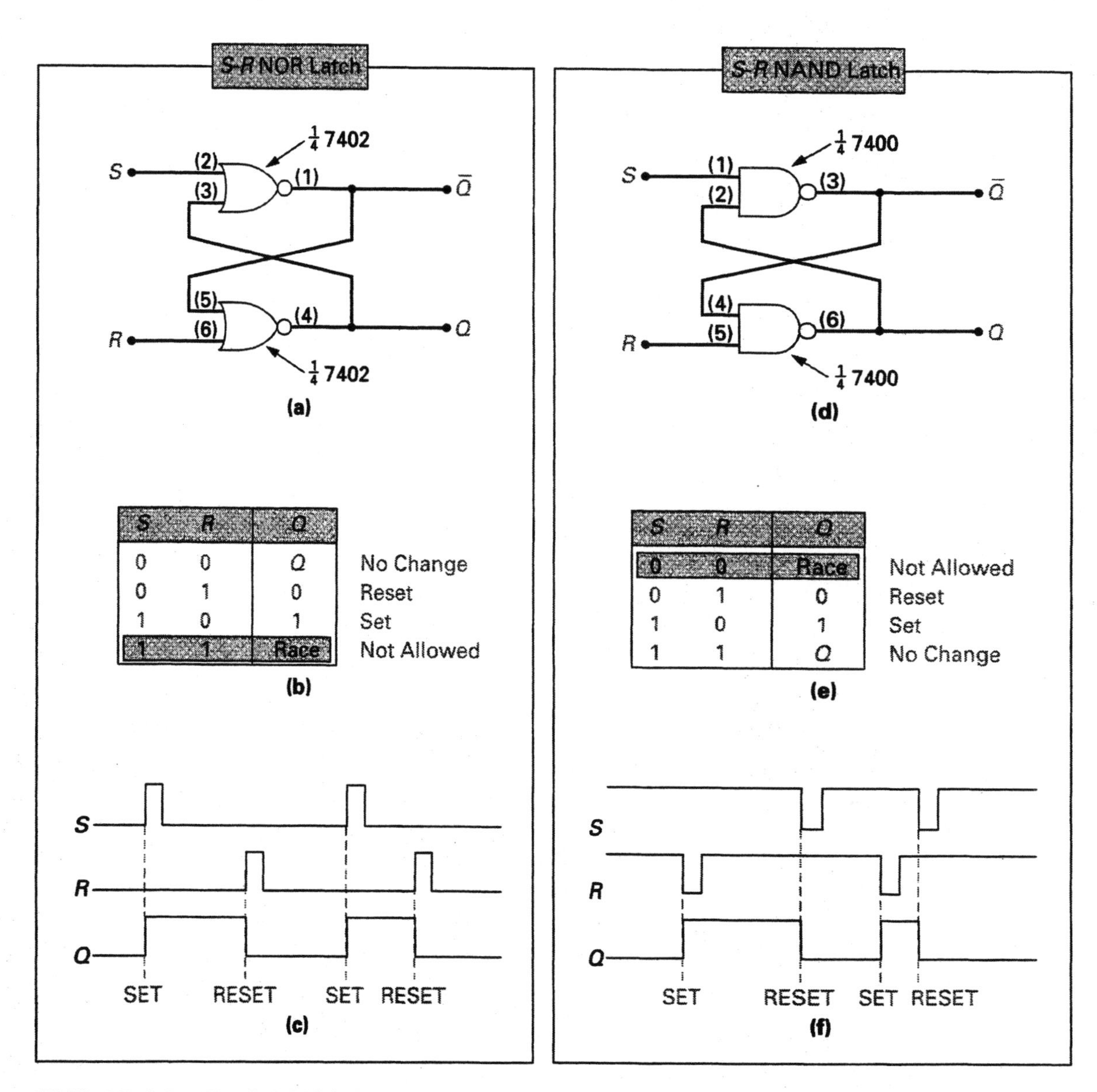

FIGURE 9-2 The *S-R* NOR Latch and the *S-R* NAND Latch.

■ ***Solution:***

By considering the S and R inputs to the NOR gates and the Q and $\overline{Q}$ outputs that are cross-coupled to the other NOR gate inputs, we can determine the final NOR gate outputs.

a. In Figure 9-3(a), NOR gate X inputs = 0 and 0, therefore $\overline{Q}$ will be 1. NOR gate Y inputs = 0 and 1, therefore Q will be 0. In this instance, the Q and $\overline{Q}$ outputs will not change ($Q = 0$, $\overline{Q} = 1$). This ties in with the truth table given in Figure 9-2(b), which states that when $S = 0$ and $R = 0$, there will be no change in Q ($Q = Q$).

b. Moving to the example in Figure 9-3(b), you can see that the only change that has occurred is that the S input is now HIGH. This HIGH input to NOR

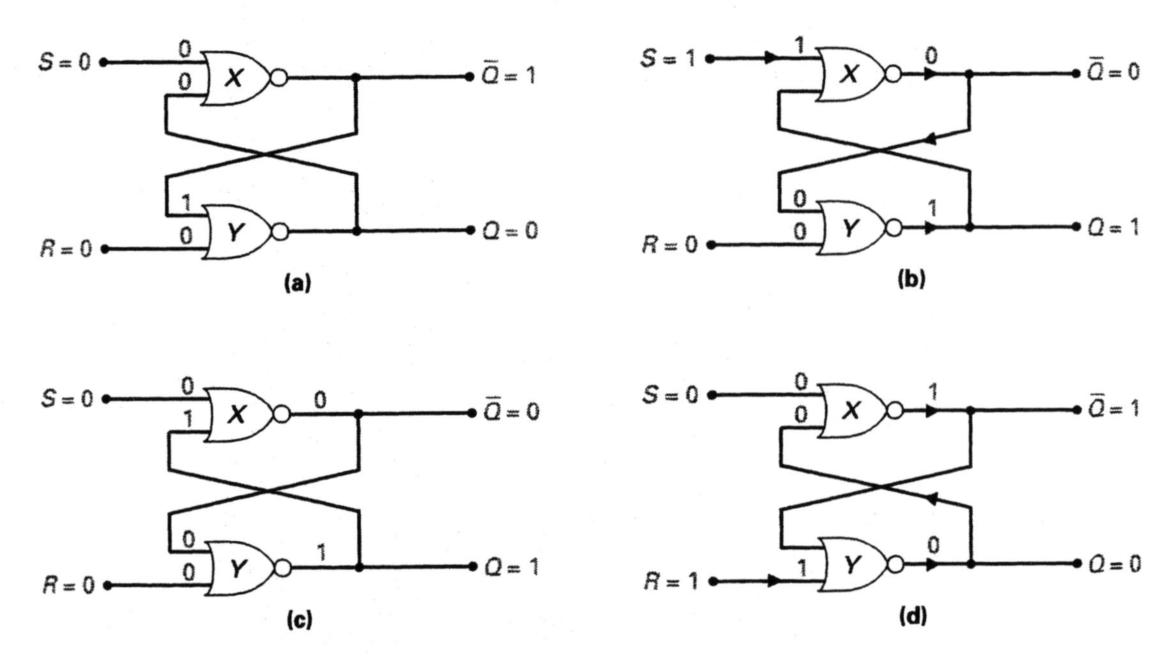

FIGURE 9-3 NOR Latch Input Conditions.

gate X will produce a LOW output at $\overline{Q}$ which will be coupled to the input of NOR gate Y. With both inputs LOW, the output of NOR gate Y will go HIGH, and Q will therefore change to a 1 and $\overline{Q}$ to a 0. In this example, a SET condition was applied to the input to the NOR latch and the Q output was therefore SET to 1.

c. Moving to the example in Figure 9-3(c), you can see that the Q output is now SET HIGH. The S input, which was taken HIGH momentarily, has now gone LOW and therefore the NOR latch inputs are in the NO CHANGE condition. NOR gate X has inputs of 0 and 1, and therefore $\overline{Q} = 0$ (no change). NOR gate Y has inputs 0 and 0, and therefore $Q = 1$ (no change).

d. Moving to the example in Figure 9-3(d), you can see that the R input is now being pulsed HIGH. This HIGH input to NOR gate Y will produce a LOW output at Q, which will be cross-coupled back to the input of NOR gate X. With two LOW inputs to NOR gate X, its $\overline{Q}$ output will be HIGH. This HIGH level at $\overline{Q}$ will be coupled to the input of NOR gate Y and will keep Q at 0 even when R returns to a LOW. This RESET input condition ($S = 0$, $R = 1$) will therefore reset the Q output to 0.

From these examples you may have noticed that it is the cross-coupling that keeps the circuit outputs latched in either the SET or RESET condition.

An S-R latch can also be constructed using two NAND gates, as shown in Figure 9-2(d). Referring to the truth table in Figure 9-2(e), you can see that the NO CHANGE and RACE conditions for a NAND latch are opposite that of the NOR latch. With the NAND latch, a LOW on both inputs will cause the circuit to race and result in an unpredictable output, while a HIGH on both inputs will result in no change at the outputs. Since the NO CHANGE condition occurs when both inputs are HIGH, the S and R inputs will normally be HIGH, as seen in the timing diagram in Figure 9-2(f). To SET the Q output to a 1, therefore, the R input must be pulsed LOW so that $S = 1$ and $R = 0$ (SET condition). On the other hand, to reset the Q output to a 0, the S input must be pulsed LOW so that $S = 0$ and $R = 1$ (RESET condition). Let us study a few examples to see how this circuit operates.

■ **EXAMPLE:**

Remembering that any 0 into a NAND gate gives a 1 output, and only when both inputs are 1 will the output be 0, study Figure 9-4, which shows a variety of input-output conditions for the set-reset NAND latch. Determine the Q output for each of the examples in Figure 9-4.

■ ***Solution:***

By considering the S and R inputs to the NAND gates and the Q and $\overline{Q}$ outputs that are cross-coupled to the other NAND gate inputs, we can determine the final NAND gate outputs.

a. In Figure 9-4(a), NAND gate X inputs = 1 and 0 and $\overline{Q}$ will therefore be 1, and NAND Y inputs = 1 and 1 and Q will therefore be 0. With both S and R inputs HIGH, the NAND latch is in the NO CHANGE condition, and the outputs Q and $\overline{Q}$ therefore remain in their last condition.

b. Moving to the example in Figure 9-4(b), you can see that the R input has been brought LOW. This LOW to the input of NAND gate Y will produce a HIGH output, which will change the Q to a 1. The HIGH Q will be cross-coupled back to the input of NAND gate X, which will now have both inputs HIGH, and therefore produce a LOW output. In this example, a SET condition was applied to the input of the NAND latch ($S = 1$, $R = 0$), and therefore the Q output was set HIGH.

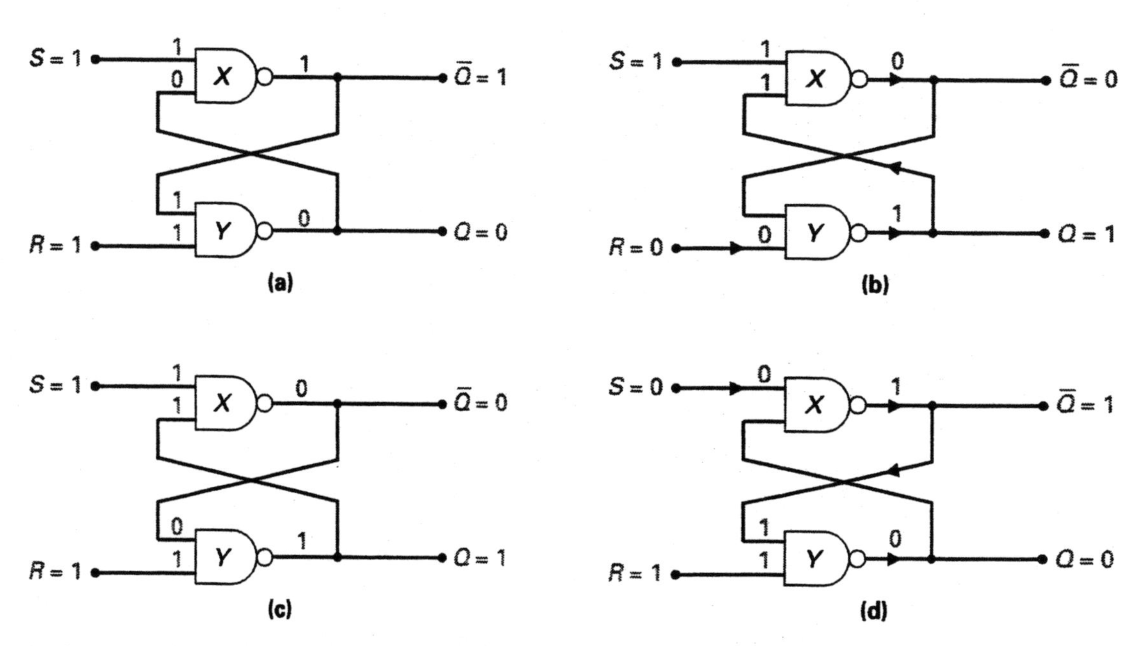

FIGURE 9-4 NAND Latch Input Conditions.

c. Moving to the example in Figure 9-4(c), you can see that the R input has returned to a HIGH level, causing the NAND latch to return to a NO CHANGE condition. NAND gate X inputs = 1 and 1, and therefore $\overline{Q} = 0$ (no change), and NAND gate Y inputs = 0 and 1, and therefore $Q = 1$ (no change).

d. Moving to the example in Figure 9-4(d), you will notice that the S input is now being momentarily pulsed LOW. This 0 applied to NAND gate X will produce a HIGH output, and therefore $\overline{Q} = 1$. This HIGH at $\overline{Q}$ will be cross-coupled to the input of NAND gate Y, and since both of its inputs are now HIGH, it will produce a LOW output at Q. This RESET input condition ($S = 0$, $R = 1$) will therefore reset Q to a 0.

Like the NOR latch, it is the cross-coupling between the NAND gates that keeps the circuit latched into either the SET or RESET condition.

Application—Contact-Bounce Eliminator Circuit

Switch Debouncer
A circuit designed to eliminate the contact bounce that occurs whenever a switch is thrown into a new position.

Set-reset flip-flops or latches are often used as **switch debouncers,** as shown in Figure 9-5. Whenever a switch is thrown into a new position, the moving and stationary contacts of the switch tend to bounce, causing the switch to make and break for a few milliseconds, before finally settling in the new position. This effect is illustrated in Figure 9-5(a) and (b), which shows the before and after conditions. In Figure 9-5(a), the ON/OFF is in the OFF position, and due to the action of the pull-up resistor R_1, the input to the INVERTER gate is HIGH and the final output is therefore LOW. In Figure 9-5(b), the switch has been thrown from the OFF to the ON position, and due to the bouncing of the switch's moving contact (pole) on the stationary ON contact, the output vibrates HIGH and LOW several times before it settles down permanently in the ON position. If the output of this circuit were connected to the input of a counter, the counter would erroneously assume that several OFF-to-ON transitions had occurred, instead of just one. It is therefore vital in any digital system that contact bounce be eliminated.

Figure 9-5(c) shows how a S-R NAND latch can be used as a *contact-bounce eliminator.* With the switch in the OFF position, the S input is LOW and the R input is HIGH. This RESET input condition ($S = 0$, $R = 1$) resets the Q output to 0, as seen in the timing diagram in Figure 9-5(c) at time t_0. When the switch is thrown to the ON position (time t_1), the S input is pulled permanently HIGH and the R input will bounce between LOW and HIGH for a few milliseconds. The first time that the switch's pole makes contact and the R input goes LOW, the latch will be SET since $S = 1$ and $R = 0$. Subsequent bounces will have no effect on the SET or HIGH Q output, since the alternating R input is only causing the latch to switch between its SET ($S = 1$, $R = 0$) and NO CHANGE ($S = 1$, $R = 1$) input conditions.

Similarly, when the switch is thrown into its OFF position, the R input will be pulled permanently HIGH and the S input will bounce between LOW and HIGH for a few milliseconds. The first time that the switch's pole makes contact and the S input goes LOW (t_2), the latch will be RESET ($S = 0$, $R = 1$). Subsequent bounces will have no effect on the RESET or LOW Q output, since the alternating S input is only causing the latch to switch between its RESET ($S = 0$, $R = 1$) and NO CHANGE ($S = 1$, $R = 1$) input conditions.

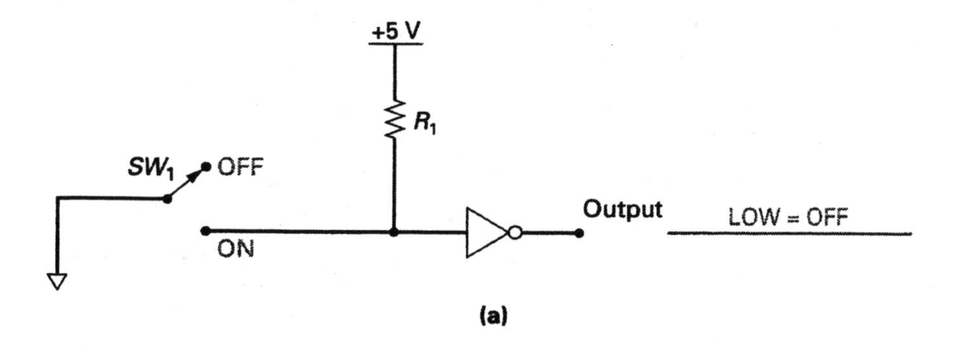

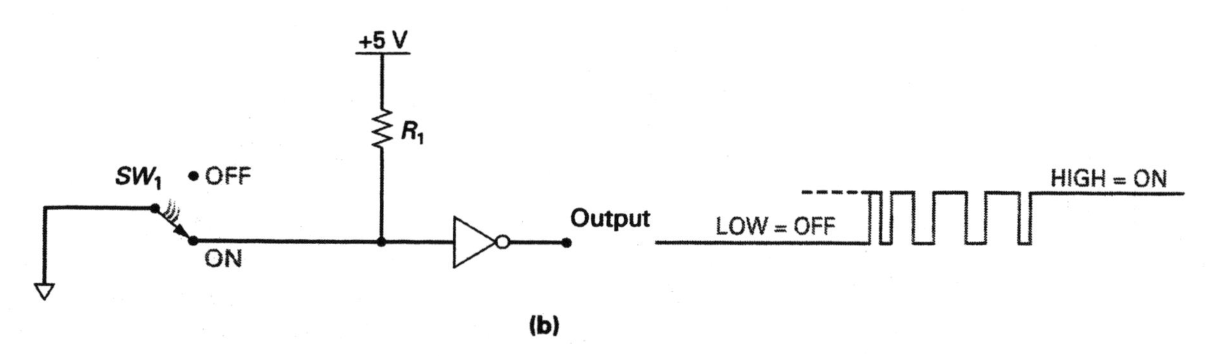

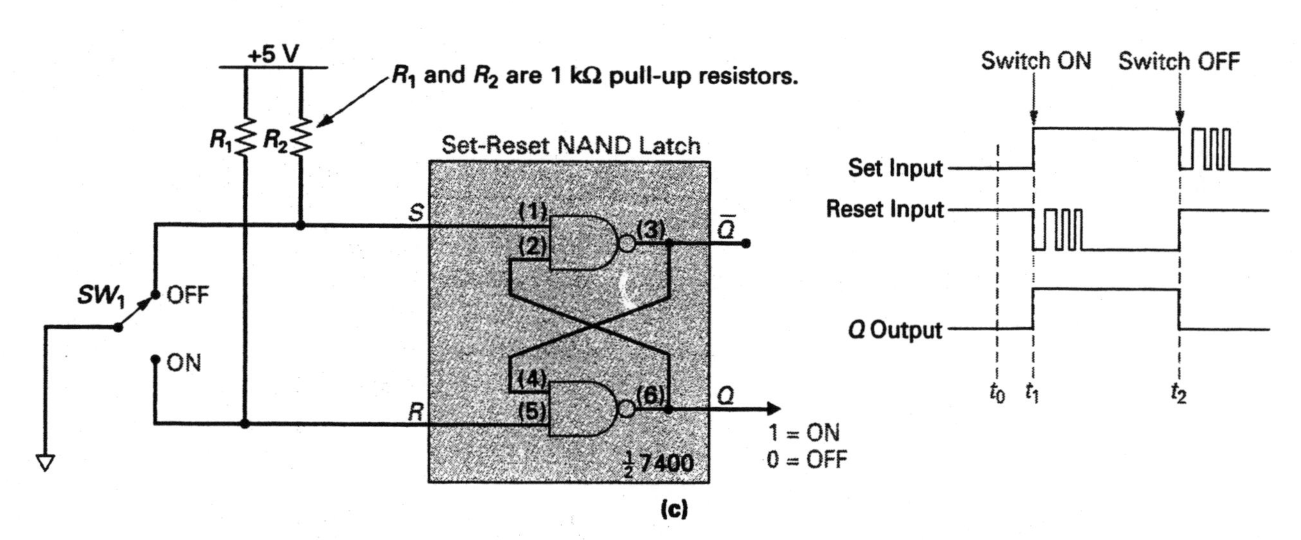

FIGURE 9-5 The *S-R* Latch as a Contact-Bounce Eliminator.

9-1-3 *Level-Triggered* S-R *Flip-Flops*

Triggered
A device or circuit that initiates an action in response to a trigger pulse.

Level-Triggered
A device or circuit that initiates an action in response to a trigger's logic level.

A typical digital electronic system contains many thousands of flip-flops. To coordinate the overall operation of a digital system, a clock signal is applied to each flip-flop to ensure that each device is **triggered** into operation at the right time. A clock signal, therefore, controls when a flip-flop is enabled or disabled, and when its outputs change state.

Figure 9-6(a) illustrates how a *S-R* NAND latch can be controlled by the HIGH or positive **level** of a clock signal. NAND gates *A* and *B* act as controlled switches, only allowing the *S* and *R* inputs through to the inputs of the latch when the clock signal is HIGH. This operation is summarized in the truth table in Figure 9-6(b). Whenever the clock signal is 0, the latch is in the NO CHANGE condition; however, when the clock signal is 1, the *S-R* latch functions.

Referring to the waveforms in Figure 9-6(c), you can see that this *S-R* latch is only enabled when the clock signal is positive. This circuit is normally referred to as a *positive level-triggered S-R flip-flop* since a positive clock level will trigger the circuit to operate. A *negative level-triggered S-R flip-flop* circuit will simply include an INVERTER between the clock input and the *A* and *B* NAND gates so that the latch will only be triggered into operation when the clock signal is 0 or LOW. Figure 9-6(d) shows the logic symbols for a positive level-triggered and negative level-triggered *S-R* flip-flop.

A level-triggered *S-R* flip-flop is also referred to as a *gated S-R flip-flop or latch* since NAND gates *A* and *B* act as a gate to either pass or block the SET and RESET inputs through to the latch.

9-1-4 *Edge-Triggered* S-R *Flip-Flops*

Edge-Triggered
A device or circuit that initiates an action in response to a trigger's positive or negative edge.

The problem with a level-triggered flip-flop is that the *S* and *R* inputs have to be held in the desired input condition (SET, RESET, or NO CHANGE) for the entire time that the clock signal is enabling the flip-flop. With **edge-triggered flip-flops,** the device is only enabled during the positive edge of the clock signal *(positive edge-triggering)* or the negative edge of the clock signal *(negative edge-triggering).*

Figure 9-7(a) shows how the previously discussed level-triggered *S-R* flip-flop can be modified so that it will only be triggered into operation when the clock is changing from LOW to HIGH (positive transition or positive edge of the clock signal). The modification to this circuit involves including a *positive-transition pulse generator* in the internal circuitry of the *S-R* flip-flop. The inset in Figure 9-7(a) shows a basic positive edge pulse generator circuit. As you can see from the waveforms in the inset, one input of the AND gate receives the clock signal directly (input *Y*) while the other input receives a delayed and inverted version of the clock input (input *X*). The two inputs to the AND gate are both HIGH for only a short period of time, starting at the positive edge of the clock input and lasting as long as the propagation delay time introduced by the INVERTER gate. The final output from this positive edge pulse generator circuit is one narrow positive pulse for every LOW to HIGH transition of the clock input signal. This positive pulse will enable NAND gates *A* and *B* during the time this signal is HIGH, and connect the *S* and *R* inputs through to the *S-R* latch section of the circuit to control the *Q* and $\overline{Q}$ outputs.

The operation of a positive edge-triggered *S-R* flip-flop is summarized by the truth table in Figure 9-7(b). The up and down arrows in the clock column represent the rising (positive transition) and falling (negative transition) edges of the clock. The first three rows of this table indicate that if the clock is at a LOW level (0), HIGH level (1), or on a negative edge (↓), the output *Q* will remain unchanged. The last four rows of this truth table indicate how the *S-R* flip-flop will operate as expected whenever a positive edge of the clock is applied (↑).

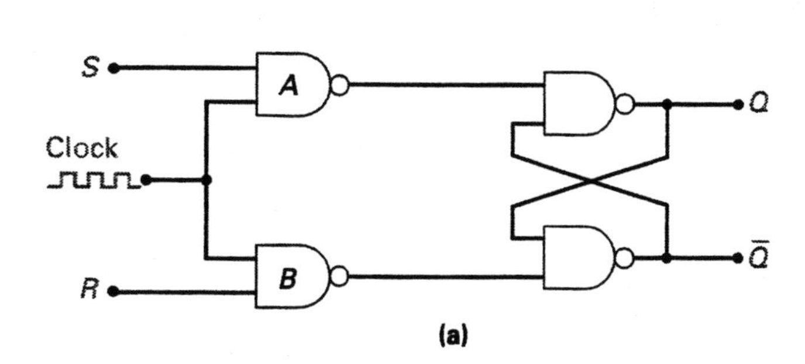

Clock	S	R	Q
0	0	0	No Change
0	0	1	No Change
0	1	0	No Change
0	1	1	No Change
1	0	0	No Change
1	0	1	0-Reset
1	1	0	1-Set
1	1	1	Race

(b)

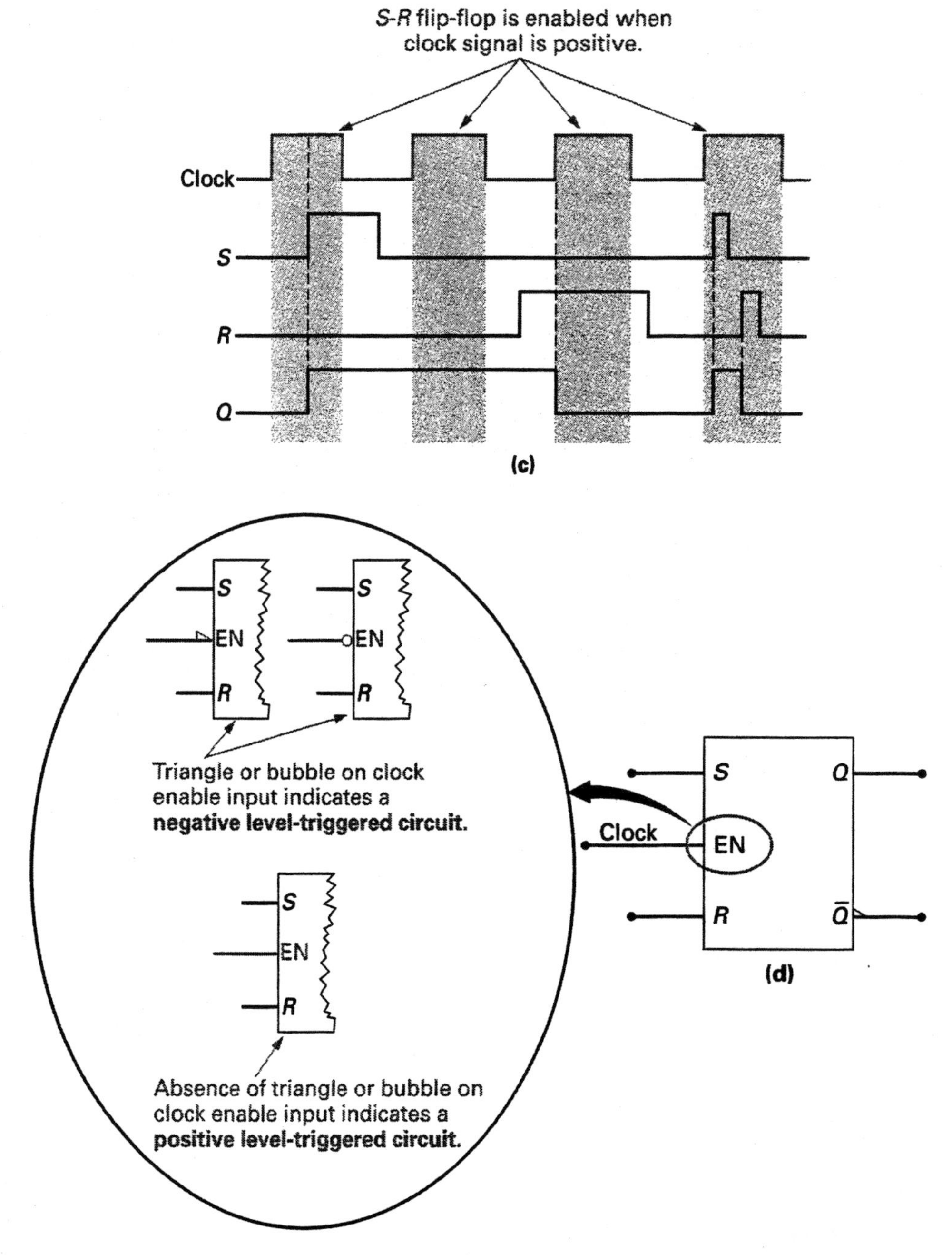

FIGURE 9-6 Level-Triggered or Gated Set-Reset Flip-Flop.

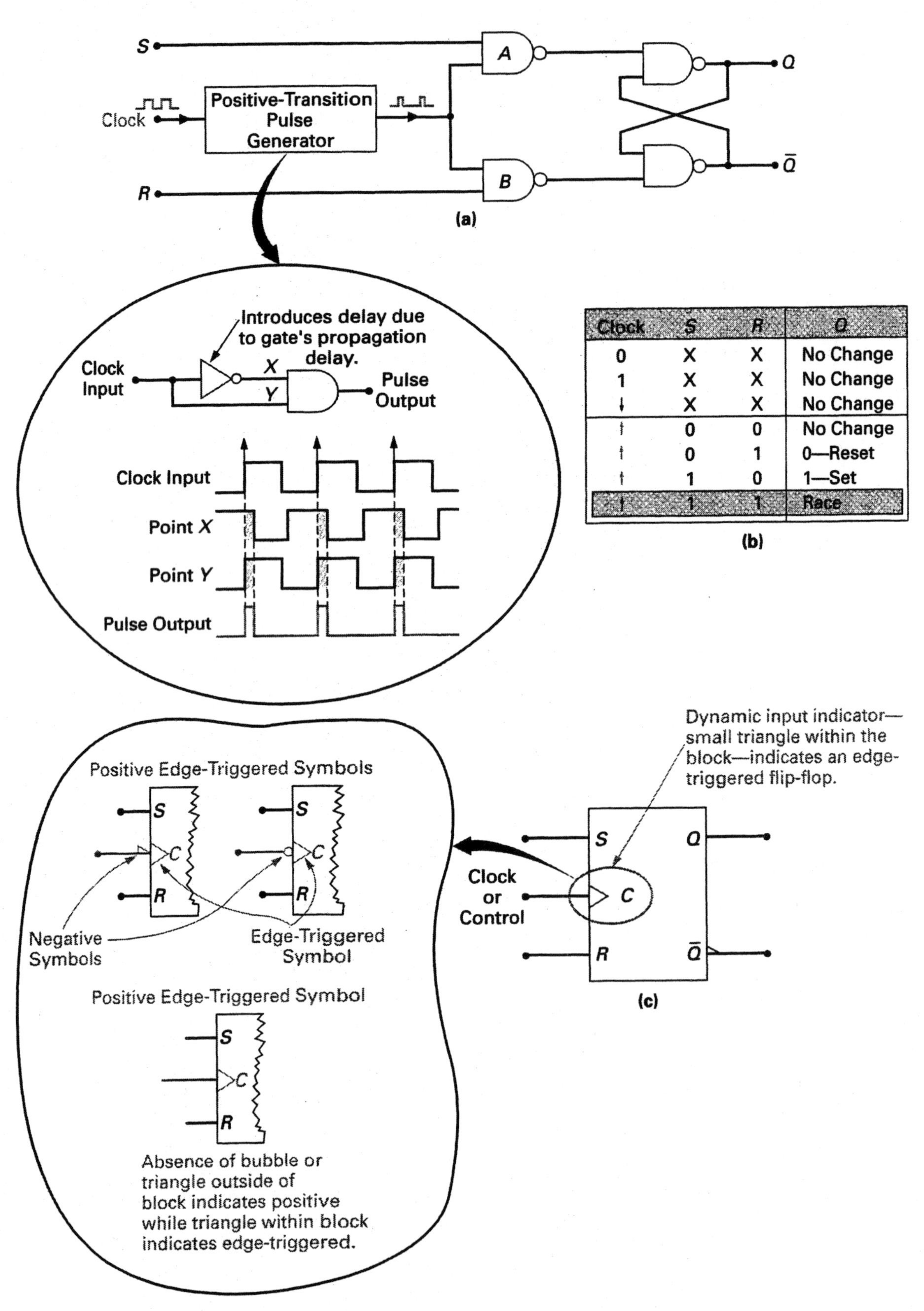

Clock	S	R	Q
0	X	X	No Change
1	X	X	No Change
↓	X	X	No Change
↑	0	0	No Change
↑	0	1	0—Reset
↑	1	0	1—Set
↑	1	1	Race

FIGURE 9-7 Edge-Triggered Set-Reset Flip-Flops.

The logic symbol for the positive edge-triggered *S-R* flip-flop circuit discussed in Figure 9-7(a) is shown in Figure 9-7(c). The "*C*" input stands for clock, or control, and the triangle within the block next to the *C* is called the dynamic (changing) input indicator and is used to identify an edge-triggered flip-flop. As you can see in the inset in Figure 9-7(c), a bubble or triangle outside of the block on the clock input indicates a negative edge-triggered flip-flop. A negative edge-triggered flip-flop circuit would be exactly the same as the circuit in Figure 9-7(a) except that the clock input would have to be inverted before it was applied to the transition pulse generator.

9-1-5 *Pulse-Triggered* S-R *Flip-Flops*

Up to this point, we have seen how set-reset flip-flops can be controlled by the level or edge of a clock signal. A **pulse-triggered** *S-R* flip-flop is a level-clocked flip-flop; however, both the HIGH and LOW levels of the input clock are needed before the *Q* outputs will reflect the *S* and *R* input condition. With this circuit, therefore, a complete cycle of the input clock (a complete clock pulse) is needed to trigger the circuit into operation, which is why this circuit is called a pulse-triggered flip-flop.

Pulse-Triggered
A device or circuit that initiates an action in response to a complete trigger pulse.

Figure 9-8(a) shows a simplified circuit of a pulse-triggered *S-R* flip-flop. Pulse-triggered circuits are also referred to as **master-slave flip-flops** since they contain two clocked *S-R* latches called the master latch and the slave latch. The first-stage master latch receives the clock directly *(C),* while the second-stage slave latch receives an inverted version of the clock $(\overline{C})$. Referring to the *C* and $\overline{C}$ waveforms in Figure 9-8(c), you can see that the master section is enabled while the clock *(C)* is positive. The slave section, on the other hand, is enabled when the clock is negative, because it is at this time that the inverted clock $(\overline{C})$ is positive. There are therefore two distinct steps involved before the *Q* and $\overline{Q}$ outputs reflect the input condition applied to the SET and RESET inputs.

Master-Slave Flip-Flop
A two-latch flip-flop circuit in which the master latch is loaded on the leading edge of the clock pulse, and the slave latch is loaded on the trailing edge of the clock pulse.

Step 1: During the HIGH level of the clock input, the master SET-RESET latch is enabled and its output *Y* is either SET, RESET, or left unchanged (NO CHANGE).

Step 2: During the LOW level of the clock input, the slave SET-RESET latch is enabled and its output *Q* will follow whatever logic level is present on the *Y* input.

This two-step action is sometimes referred to as *cocking and triggering.* The master latch is cocked during the positive level of the clock and the slave latch is triggered during the negative level of the clock.

The truth table in Figure 9-8(b) summarizes the operation of the pulse-triggered (master-slave) *S-R* flip-flop. The last four lines of the truth table reflect the normal operating conditions (NO CHANGE, RESET, SET, and RACE) of a *S-R* flip-flop. A complete clock pulse is shown in the clock column for these conditions since both the HIGH and LOW levels of the clock input are needed to enable both sections of the flip-flop. The first three lines of the truth table describe how the active-LOW **PRESET** $(\overline{PRE})$ and active-LOW **CLEAR**

Preset
A control input that is independent of the clock and is used to set the output HIGH.

Clear
A control input that is independent of the clock and is used to reset the output LOW.

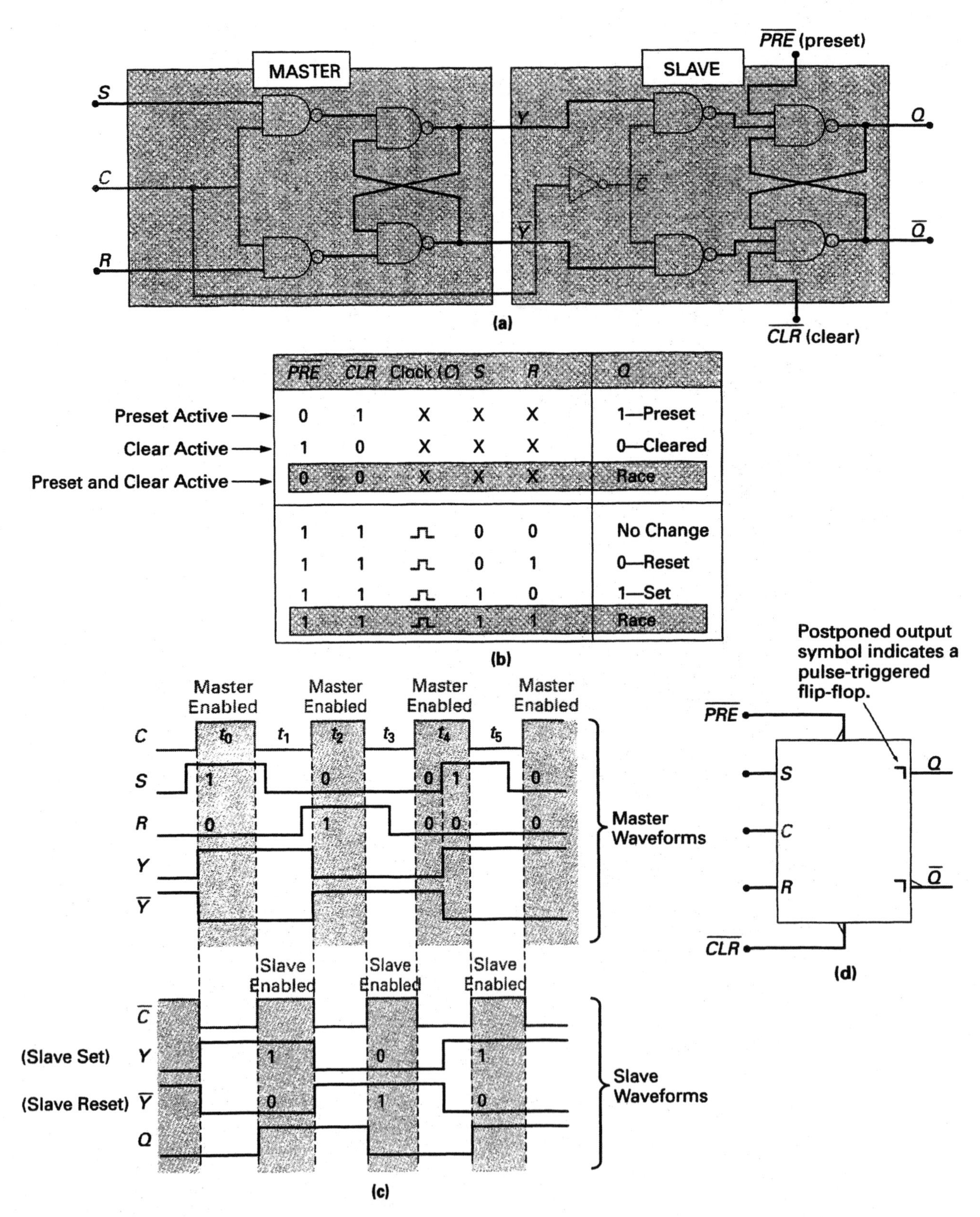

$\overline{PRE}$	$\overline{CLR}$	Clock (C)	S	R	Q
0	1	X	X	X	1—Preset
1	0	X	X	X	0—Cleared
0	0	X	X	X	Race
1	1	⎍	0	0	No Change
1	1	⎍	0	1	0—Reset
1	1	⎍	1	0	1—Set
1	1	⎍	1	1	Race

FIGURE 9-8 Pulse-Triggered (Master-Slave) Set-Reset Flip-Flop.

($\overline{CLR}$) input functions. When power is first applied to a digital system, flip-flops may start up in either the SET ($Q = 1$) or RESET ($Q = 0$) condition. This could be both dangerous and damaging if these Q outputs were controlling external devices. For this reason the direct PRESET (synonymous with SET) and direct CLEAR (synonymous with RESET) inputs are generally always available with most commercially available flip-flop ICs. They are called direct inputs because they do not need the clock signal to be active in order to PRESET the Q to a 1, or CLEAR the Q to a 0. One application for these direct PRESET and CLEAR inputs would be to control the outputs of any flip-flops controlling machinery in an automated manufacturing plant so that when power is first applied, all devices start off in their OFF state.

Returning to the truth table in Figure 9-8(b), you can see that when the CLEAR is HIGH and the PRESET is LOW (active), the output is PRESET to 1 regardless of the *C, S,* and *R* inputs. In the second line of the table, the PRESET is HIGH and the CLEAR input is LOW (active), causing the Q to be cleared to 0, regardless of the *C, S,* and *R* inputs. These active-LOW inputs are therefore normally pulled HIGH to make them inactive. Referring to the circuit in Figure 9-8(a) you can see that the PRESET and CLEAR inputs control the final NAND gates of the slave latch. As with any NAND gate, any LOW input will produce a HIGH output, and therefore a LOW PRESET will produce a HIGH Q (LOW $\overline{Q}$), while a LOW CLEAR will produce a LOW Q (HIGH $\overline{Q}$). Bringing both the PRESET and CLEAR LOW at the same time will produce a RACE condition as summarized in the truth table in Figure 9-8(b), since both Q and $\overline{Q}$ will try to go HIGH.

Figure 9-8(c) shows a timing diagram for a pulse-triggered (master-slave) set-reset flip-flop. Referring to the clock input (*C*), let us step through times t_0 through t_5 and see how this circuit responds to a variety of input conditions.

At time t_0, the master latch section is enabled by the positive level of *C*. Since the *S* input is HIGH and the *R* input is LOW at this time (SET condition), the *Y* output of the master latch will be SET to 1 ($\overline{Y} = 0$).

At time t_1, the master latch section is disabled by a LOW *C*, while the slave latch section is enabled by a HIGH $\overline{C}$. Since the *Y* and $\overline{Y}$ inputs to the slave latch section are in the SET condition ($Y = 1, \overline{Y} = 0$), the Q output is SET to 1. This action demonstrates how the slave section simply switches its inputs through to the outputs when enabled by the clock (if $Y = 1$ and $\overline{Y} = 0$, $Q = 1$ and $\overline{Q} = 0$, when $\overline{C}$ is HIGH). It can therefore be said that the second latch is a slave to the master latch.

At time t_2, the master latch section is enabled by a HIGH *C*. Since its SET input is LOW and its RESET input is HIGH at this time, the *Y* output will be reset to 0.

At time t_3, the master latch section is disabled and the slave latch section is enabled. The reset output of the master latch is clocked into the enabled slave section and transferred through to the output, resetting Q LOW (if $Y = 0$ and $\overline{Y} = 1$, $Q = 0$ and $\overline{Q} = 1$, when $\overline{C}$ is HIGH).

At time t_4, the master-slave *S-R* flip-flop is a level-clocked circuit, which means that the inputs have to remain constant for the entire time that the clock is enabling the latch. At the beginning of time t_4, the *S* and *R* inputs are both LOW and therefore *Y* remains in its last state, which was RESET ($Y = 0$).

Half-way through time t_4, the S input goes HIGH and the Y output is therefore SET HIGH.

At time t_5, the master latch is disabled while the slave latch is enabled, connecting the HIGH Y input through to the Q output.

The logic symbol for a pulse-triggered master-slave flip-flop is shown in Figure 9-8(d). The active-LOW PRESET *($\overline{PRE}$)* and active-LOW CLEAR *($\overline{CLR}$)* inputs are included, with the traditional triangle being used to represent the fact that they are activated by a LOW logic level. The *postponed output symbol* (⏋) at the Q and $\overline{Q}$ outputs indicates that these outputs do not change state until the clock input has fallen from a HIGH to a LOW level. This is due to the slave section, which, as you know, is only enabled—and the output updated—when the clock signal drops LOW. If the postponed output were turned the other way, it would indicate that the master latch would be enabled by a LOW clock input, and the slave latch would be enabled by a HIGH clock input. The output with this type of S-R master-slave flip-flop would be updated whenever the clock rose to a HIGH level.

Application: A Memory Latch/Unlatch Circuit

The inset in Figure 9-9 shows the function table for a 74L71, which is a pulse-triggered set-reset master-slave flip-flop IC with three ANDed SET inputs and three ANDed RESET inputs. By ANDing the SET and RESET inputs in this way, the IC has more control versatility since all inputs will have to be HIGH for the control line to be HIGH, and any LOW input will pull the corresponding control line LOW. The first three lines of the function table show

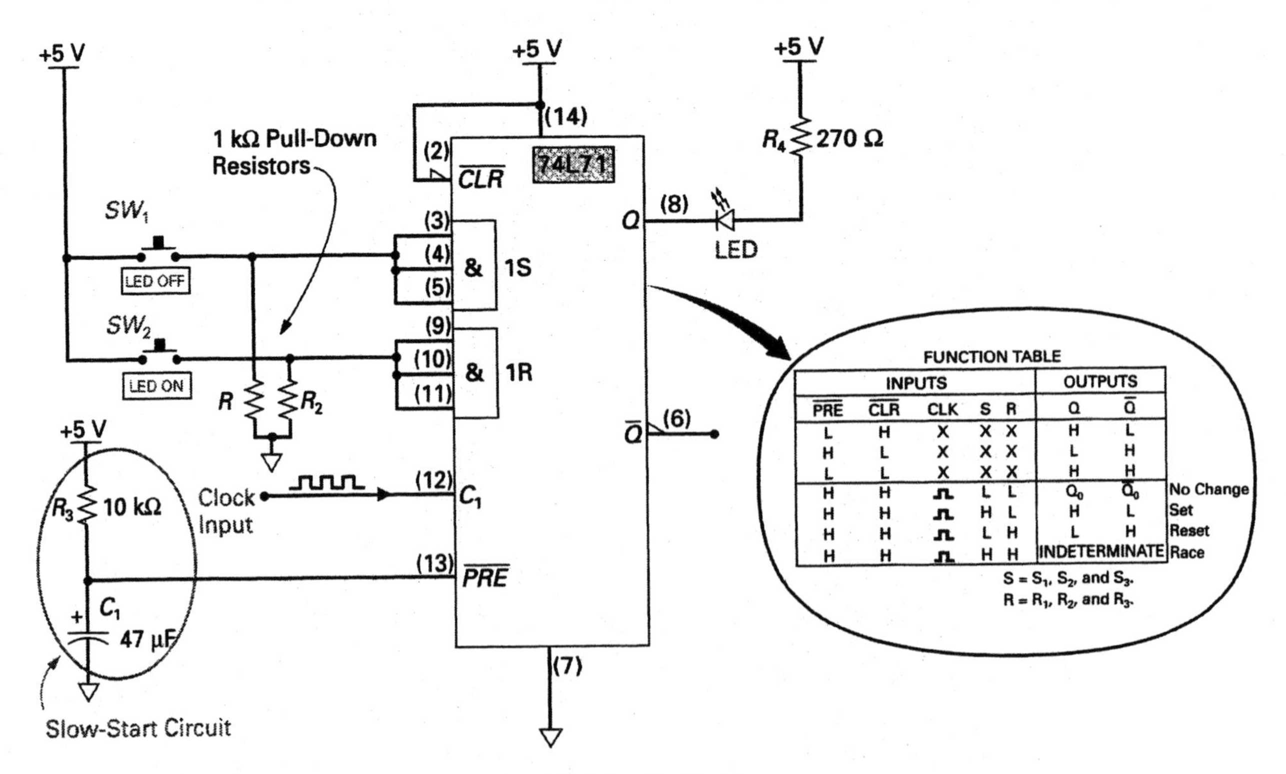

INPUTS					OUTPUTS		
$\overline{PRE}$	$\overline{CLR}$	CLK	S	R	Q	$\overline{Q}$	
L	H	X	X	X	H	L	
H	L	X	X	X	L	H	
L	L	X	X	X	H	H	
H	H	⎍	L	L	Q_0	$\overline{Q}_0$	No Change
H	H	⎍	H	L	H	L	Set
H	H	⎍	L	H	L	H	Reset
H	H	⎍	H	H	INDETERMINATE		Race

S = S_1, S_2, and S_3.
R = R_1, R_2, and R_3.

FIGURE 9-9 Application—A Memory Latch/Unlatch Circuit.

how the active-LOW PRESET and CLEAR inputs will override the outputs, and the last four lines of the function table show the NO CHANGE, SET, RESET, and RACE input conditions.

■ **EXAMPLE:**

What would be the Q output from the 7471 for the following input conditions?

a. $S_1 = 1$, $S_2 = 1$, $S_3 = 1$, $R_1 = 0$, $R_2 = 0$, $R_3 = 0$, CLK = complete pulse, PRE = 0, CLR = 1.
b. $S_1 = 1$, $S_2 = 1$, $S_3 = 0$, $R_1 = 0$, $R_2 = 0$, $R_3 = 0$, CLK = complete pulse, PRE = 1, CLR = 1.

■ ***Solution:***

a. Q would be HIGH since PRESET is active.
b. Q would not change, since $S = 0$ (due to S_3) and $R = 0$.

To understand the operation of this IC better, let us see how it could be used in an application.

The circuit in Figure 9-9 shows how a pulse-triggered set-reset master-slave flip-flop can be used to latch or unlatch a control output. The output Q is connected to +5 V via an LED and current-limiting resistor (R_4), so a HIGH output (SET) will turn OFF the LED and a LOW output (RESET) will turn ON the LED.

All of the SET inputs are tied together and connected to the LED OFF push-button switch, SW_1. The RESET inputs are also all tied together and connected to the LED ON push-button switch, SW_2. The SET and RESET inputs are normally both LOW (NO CHANGE) due to the pull-down resistors R_1 and R_2; however, when a push-button switch is pressed, the corresponding input is switched HIGH. For example, if SW_1 were momentarily pressed, the Q output would be SET HIGH after one clock pulse and the LED would turn OFF. On the other hand, if SW_2 were momentarily pressed, the Q output would be RESET LOW after one clock pulse and the LED would turn ON. A momentary press of SW_2, therefore, will lock or ON the LED, while a momentary press of SW_1 will lock or OFF the LED.

As described earlier, a flip-flop's output may start out either HIGH or LOW when power is first applied to a circuit. To ensure that a flip-flop's output always starts off in the desired condition, a start circuit is normally included and connected to either the PRESET or CLEAR input. In Figure 9-9, R_3 and C_1 form a *slow-start circuit* that is connected to the flip-flop's PRESET input to ensure that the LED is initially OFF when the circuit is powered up. This circuit operates in the following way. When power is first applied to the circuit, there is no charge on C_1 and therefore the PRESET input to the 7471 will be active, so the Q output will be PRESET to a 1, causing the LED to turn OFF. After a small delay (equal to five time constants of R_3 and C_1), C_1 will have charged to 5 V and the PRESET input will be inactive (HIGH). This slow-start circuit therefore delays the PRESET input from going immediately HIGH to ensure that the Q output starts off SET or HIGH, causing the LED to be OFF when power is first applied. If you wished the Q output to be 0 at power-up, you would simply connect the PRESET input to +5 V and connect the CLEAR input to the slow-start circuit.

SELF-TEST REVIEW QUESTIONS FOR SECTION 9-1

1. Once triggered, the SET-RESET flip-flop's output will remain locked in either the SET condition or RESET condition, which is why a flip-flop is also known as a ________.
2. List the three different ways a flip-flop can be triggered.
3. If $S = 0$ and $R = 0$, the NOR latch would be in the ________ input condition and the NAND latch would be in the ________ input condition.
4. Which form of triggering makes use of a
   a. transition pulse generator?
   b. master latch and slave latch?
5. What is a contact-bounce eliminator circuit?
6. What is a slow-start circuit?

9-2 DATA-TYPE (D-TYPE) FLIP-FLOPS

D-Type Flip-Flop
A data-type flip-flop circuit that can latch or store a binary 1 or 0.

The **D-type flip-flop** or **D-type latch** is basically a SET-RESET flip-flop with a small circuit modification. This modification was introduced so that the data-type flip-flop could not RACE like the *S-R* flip-flop.

The basic D-type flip-flop logic circuit is shown in Figure 9-10(a). The modification to the circuit is the inclusion of an INVERTER gate, which ensures that the *R* and *S* inputs to the NAND latch are never at the same logic level. The single data input bit *(D)* appears at the *S* input of the latch, and its complement $(\overline{D})$ appears at the *R* input of the latch. Therefore, when the *D* input is 1, the *Q* output is SET to 1, and when the *D* input is 0, the *Q* output is RESET to 0, as summarized in the truth table in Figure 9-10(b). The logic symbol for this unclocked D-type flip-flop is shown in Figure 9-10(c). As can be seen in the timing waveforms in Figure 9-10(d), the *Q* output is either SET or RESET as soon as the *D* input goes HIGH or LOW.

To coordinate the overall action of a digital system, devices are triggered into operation at specific times. This synchronization is controlled by a master timing signal called a clock. Like the *S-R* flip-flop, the D-type flip-flop can be either level-triggered, edge-triggered, or pulse-triggered. Let us now examine each of these three types in more detail.

9-2-1 Level-Triggered D-Type Flip-Flops

Figure 9-11(a) shows the basic logic circuit for a *level-triggered* or *gated D-type flip-flop*. A LOW clock level will disable the input gates *A* and *B* and prevent the latch from changing states, as indicated in the first line of the truth table in

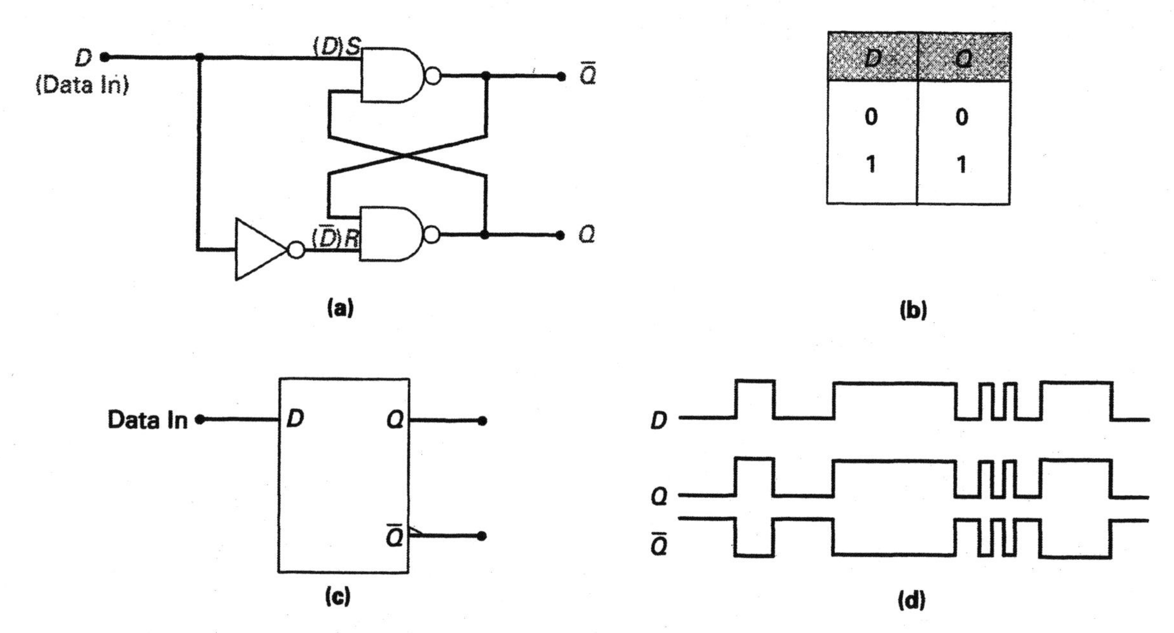

FIGURE 9-10 The Basic Data-Type (D-type) Flip-Flop or Latch.

Figure 9-11(b). When the clock level is HIGH, however, gates *A* and *B* are enabled and the *D* input controls the *Q* output, as seen in the second and third lines of the truth table. In the timing diagram in Figure 9-11(c), the operation of the level-clocked or level-triggered D-type flip-flop is further reinforced. This circuit's operation can be summarized by saying that when the clock input is HIGH, the *Q* output equals the *D* input.

The logic symbol for both the positive level-triggered and negative level-triggered D-type flip-flop is shown in Figure 9-11(d).

Data sheet details for the 7475 IC are shown in Figure 9-12. This device contains four level-triggered D-type latches. Referring to the logic symbol, you can see that the four data inputs are labeled 1D, 2D, 3D, and 4D. There are only two enable or control inputs (*C*); pin 13 will enable the first and second latches, and pin 4 will enable the third and fourth latches. The function table for each of the latches shows that when the control (*C*) input is HIGH, *Q* will equal *D*. When the *C* input is LOW, however, there will be no change in the output, regardless of the *D* input.

■ **EXAMPLE:**

What would be the *Q* output from the 7475 for the following input conditions?

1D = 1, 2D = 0, 1C, 2C = H, 3D = 1, 4D = 1, 3C, 4C = L

■ ***Solution:***

Outputs 1Q = 1 and 2Q = 0, since the enables (1C and 2C = HIGH) for these two flip-flops are active. On the other hand, 3Q = NO CHANGE and 4Q = NO CHANGE, since these two flip-flops are disabled (3C and 4C = LOW).

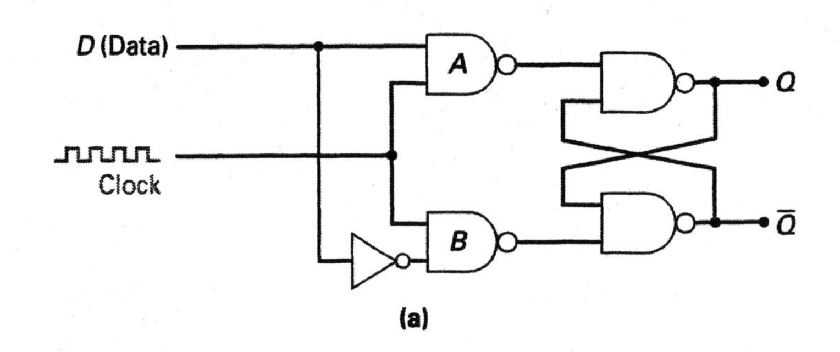

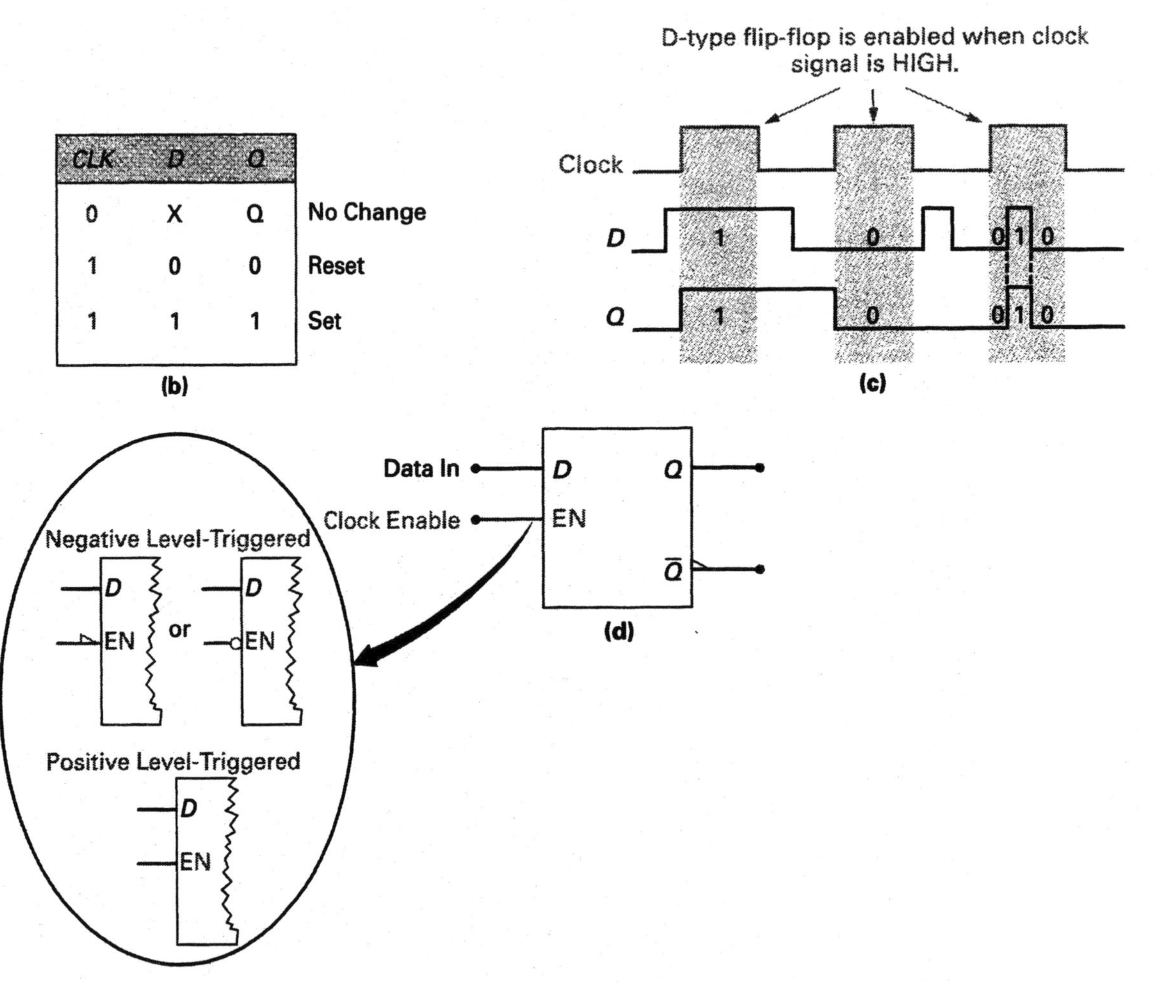

FIGURE 9-11 **Level-Triggered or Gated D-Type Flip-Flop.**

Application—A Decimal Counting Unit with Freeze Control

The four latches of a 7475 are ideally suited for use as temporary storage of binary information between a processing unit and an input-output, or indicator, unit. Figure 9-13 shows how a 7475 can be connected to operate in such an application. The two enable inputs for the 7475 (pins 4 and 13) are tied together and connected to switch 2 (SW_2). When this switch is placed in the LATCH position, all four of the 7475 D-type latches are enabled and the BCD

IC TYPE: SN7475 – 4-Bit Bistable Latches

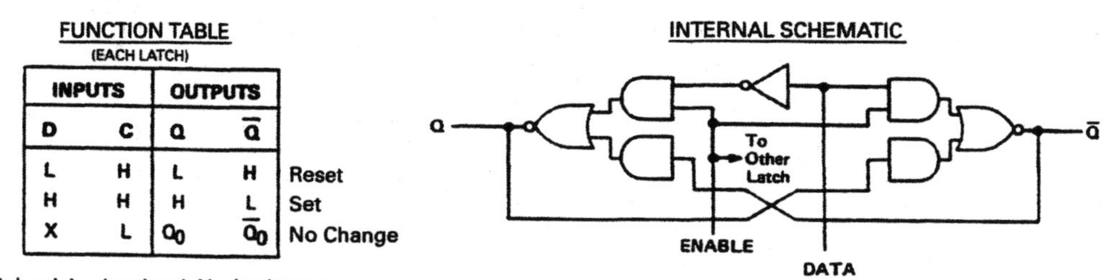

FUNCTION TABLE
(EACH LATCH)

INPUTS		OUTPUTS		
D	C	Q	$\overline{Q}$	
L	H	L	H	Reset
H	H	H	L	Set
X	L	Q_0	$\overline{Q}_0$	No Change

H = high level, L = low level, X = irrelevant
Q_0 = the level of Q before the high-to-low transition of C

FIGURE 9-12 An IC Containing Four D-Type Flip-Flops. (Courtesy of Texas Instruments)

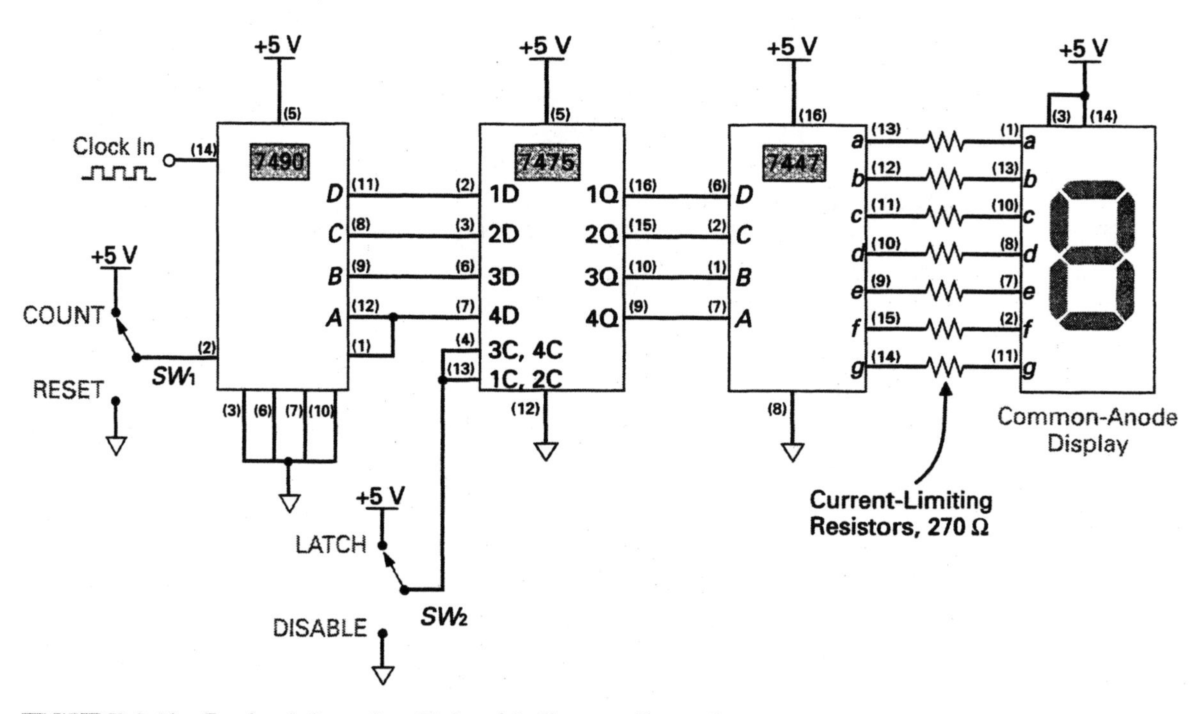

FIGURE 9-13 Decimal Counting Unit with Freeze Control.

count from the 7490 will pass through the 7475 to the 7447 and finally be displayed on the common-anode seven-segment display. When SW_2 is placed in the DISABLE position, however, all four of the 7445 D-type latches are disabled, but the previously latched BCD code will remain at the output and, therefore, on the display, regardless of the BCD code variations at the D inputs. The effect is to freeze the display when SW_2 disables the 7445, but allow the display to reflect the 7490 count when SW_2 enables the 7475.

9-2-2 Edge-Triggered D-Type Flip-Flops

To be useful in fast-acting digital circuits, a flip-flop needs to respond to the edge of a clock signal rather than the level of a clock. The advantage of edge-triggered devices is that the inputs do not need to be held stable for the entire time of the HIGH or LOW level of the clock.

Figure 9-14(a) shows how an edge-triggered D-type flip-flop incorporates the same positive transition pulse generator as the edge-triggered *S-R* flip-flop. Referring to the truth table in Figure 9-14(b), you can see that the active-LOW PRESET and CLEAR inputs are not dependent on the clock input or data input, and can be used to PRESET or CLEAR the Q output. The last three lines of the truth table summarize how a positive edge-triggered D-type flip-flop will operate. If the clock input is at a HIGH level, LOW level, or making a negative transition from LOW to HIGH (↓), the D-type flip-flop will not be enabled and therefore NO CHANGE will occur at the output. On a positive transition of the clock (↑), however, the Q output will follow the D input.

The logic symbol for a positive edge-triggered D-type flip-flop is shown in Figure 9-14(c). The inset illustrates how positive edge-triggered and negative edge-triggered symbols will differ.

Data sheet details for the 7474 IC are shown in Figure 9-15. This device contains two independent positive edge-triggered D-type flip-flops with individual PRESET and CLEAR inputs. The logic symbol for this IC is a single block (representing the IC) split into two sections (representing each of the two D-type flip-flops).

■ **EXAMPLE:**

What would be the Q outputs from the 7474 for the following input conditions?

a. 1D = 1, 1CLK = ↓, $\overline{1PRE}$ = 1, $\overline{1CLR}$ = 1
b. 2D = 1, 2CLK = ↑, $\overline{2PRE}$ = 1, $\overline{2CLR}$ = 1

■ ***Solution:***

a. 1Q = NO CHANGE (since the clock, preset, and clear are not active).
b. 2Q = SET (since D = 1, the *CLK* is active, and the *PRE* and *CLR* are not active).

Application—A Storage Register Circuit

Figure 9-16(a) shows how the two D-type flip-flops within a 7474 IC can be used as a 2-bit storage register. In this example, a combinational logic circuit produces a Y and Z data output, which is only present for a short space of time.

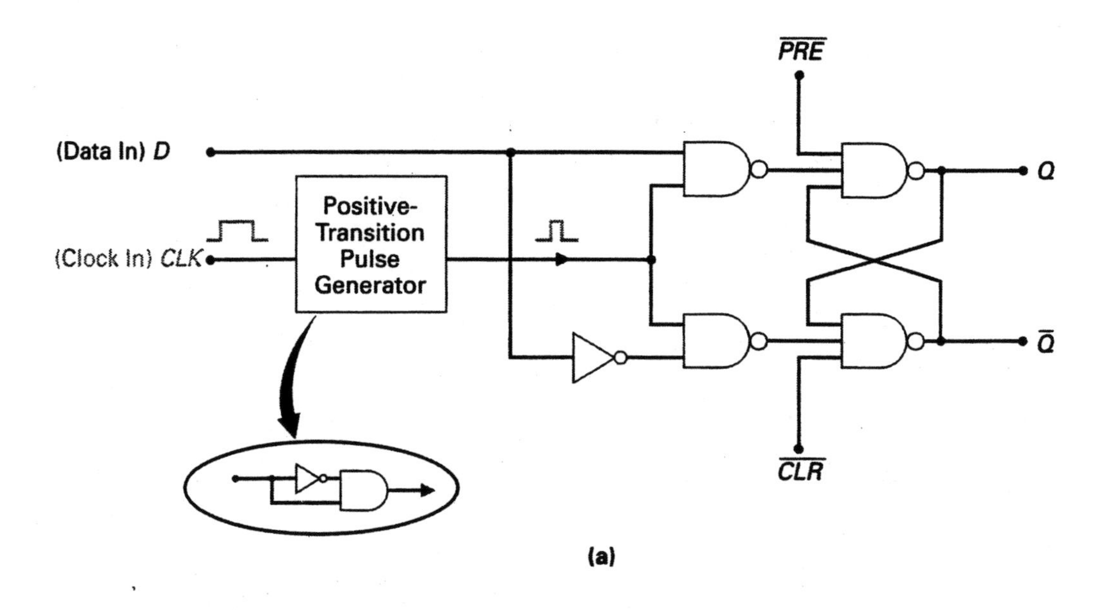

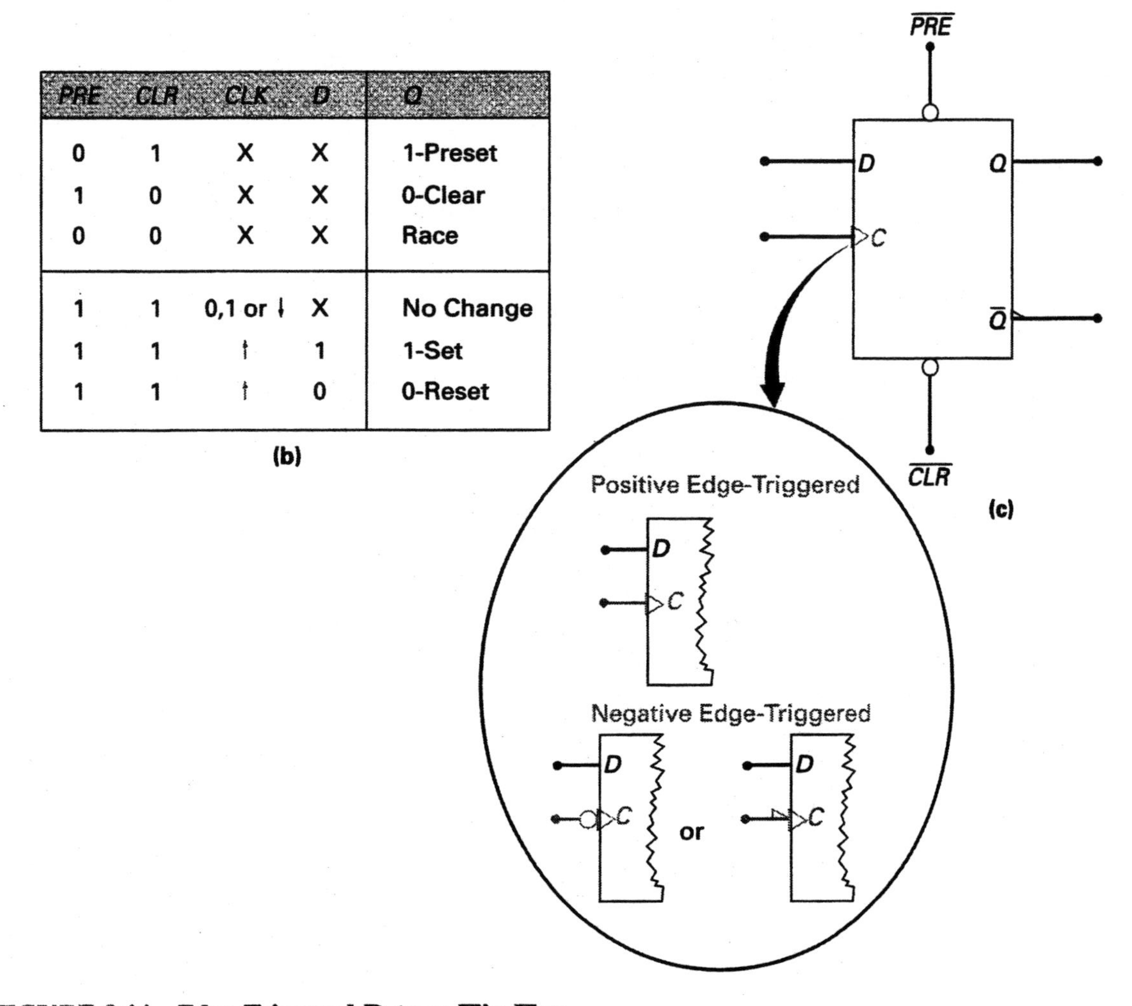

PRE	CLR	CLK	D	Q
0	1	X	X	1-Preset
1	0	X	X	0-Clear
0	0	X	X	Race
1	1	0,1 or ↓	X	No Change
1	1	↑	1	1-Set
1	1	↑	0	0-Reset

FIGURE 9-14 **Edge-Triggered D-type Flip-Flop.**

IC TYPE: SN7474 – Dual D-Type Positive Edge-triggered Flip-Flop with Present and Clear

LOGIC SYMBOL

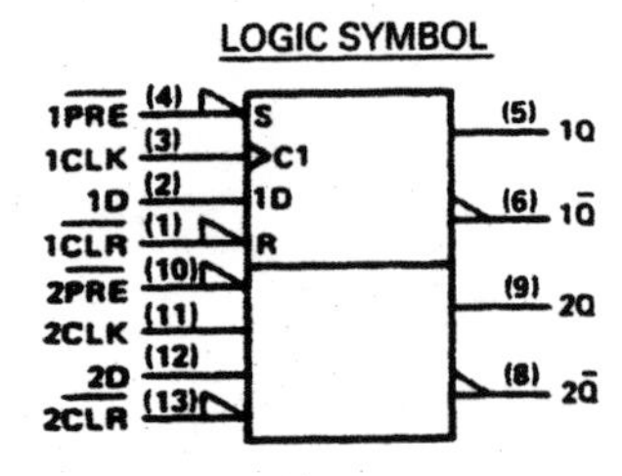

FUNCTION TABLE

INPUTS				OUTPUTS		
$\overline{PRE}$	$\overline{CLR}$	CLK	D	Q	$\overline{Q}$	
L	H	X	X	H	L	Preset
H	L	X	X	L	H	Clear
L	L	X	X	H†	H†	Race
H	H	↑	H	H	L	Set
H	H	↑	L	L	H	Reset
H	H	L	X	Q_0	$\overline{Q}_0$	No Change

TOP VIEW

Pin		Pin	
1	$1\overline{CLR}$	14	V_{CC}
2	1D	13	$2\overline{CLR}$
3	1CLK	12	2D
4	$1\overline{PRE}$	11	2CLK
5	1Q	10	$2\overline{PRE}$
6	$1\overline{Q}$	9	2Q
7	GND	8	$2\overline{Q}$

LOGIC DIAGRAM

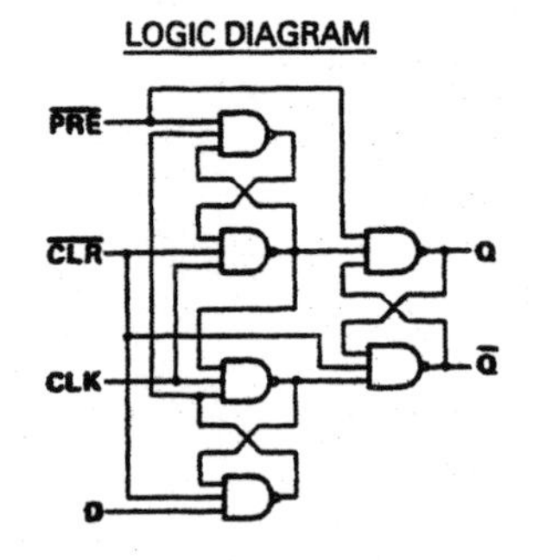

FIGURE 9-15 An IC Containing Two Positive Edge-Triggered D-Type Flip-Flops. (Courtesy of Texas Instruments)

The two D-type flip-flops are included to store the 2-bit output from the combinational logic circuit on the positive edge of the clock signal. Once this information is safely stored in the D-type latches, the combinational logic circuit can begin working on another task, since the 2-bit word is permanently stored and appears at the Q outputs of the D-type latches.

Application—A Divide-by-Two and Counter Circuit

Figure 9-16(b) shows how a D-type latch can function as a *divide-by-two* and *counter circuit.* To make a D-type latch divide the clock input by two, you need to connect the $\overline{Q}$ output back to the D input. Referring to the thick lines in the waveform in Figure 9-16(b), you can see that two cycles of the clock input are needed to produce one cycle at the Q output. The D-type latch has therefore divided the clock input by two by making use of the phase reversal between the D input and the $\overline{Q}$.

Referring to the binary values within the thick lines of the waveforms, you may have noticed that the divide-by-two D-type latch is also counting down in binary from 3 to 0. For example, at time t_0, the binary count present on the clock input and the $\overline{Q}$ output is 11_2, or 3_{10}. At time t_1, the count has dropped to 10_2, or 2_{10}, and then at time t_2 the count has decreased to 01_2, or 1_{10}. For each clock pulse following, the count decreases by one, starting at the maximum 2-bit count of 3_{10} (11_2), then decreasing to 0 (00), and then repeating the cycle.

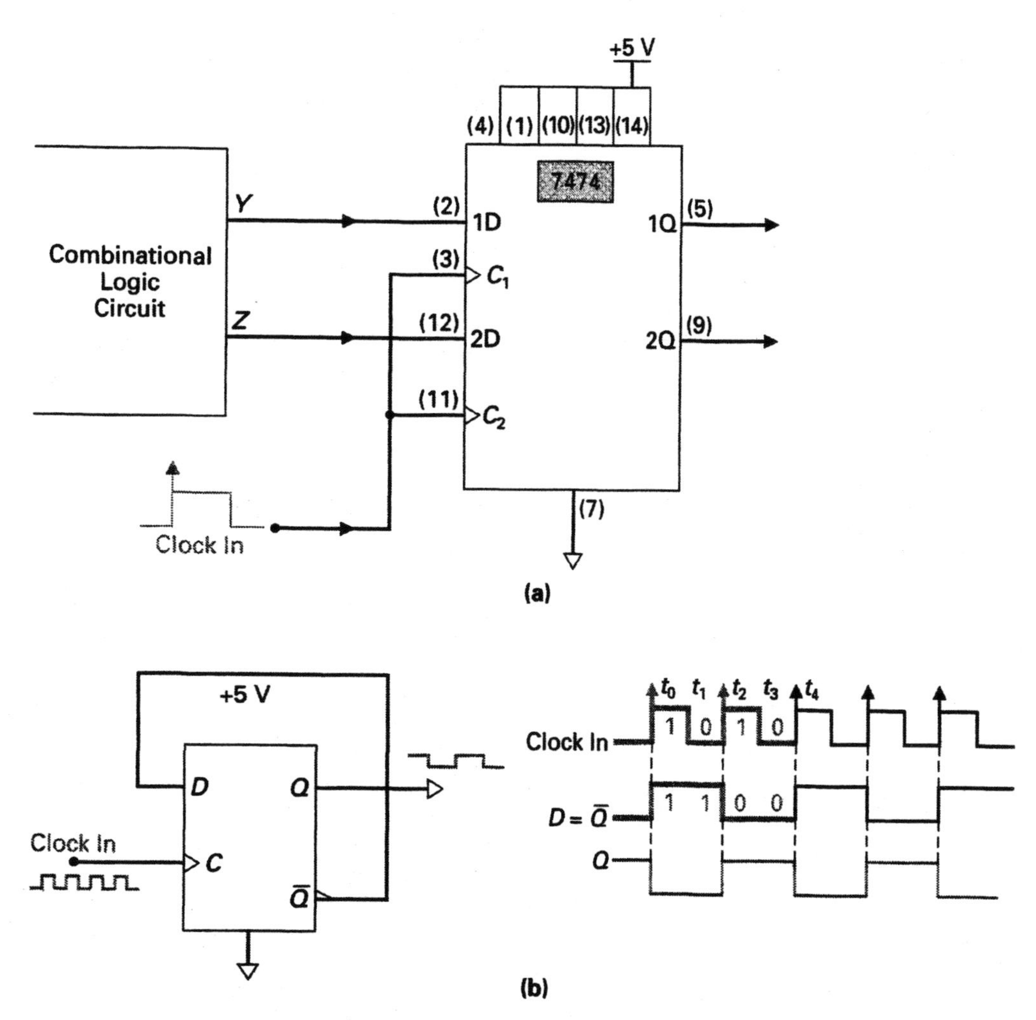

FIGURE 9-16 D-type Flip-Flop Applications. (a) Storage Register. (b) Divide-by-Two Counter.

The 74175, shown in Figure 9-17, is another example of an IC containing positive edge-triggered D-type flip-flops. The clock input drives all four D-type latches within the 74175, and on the positive edge of the clock signal, Q will equal D for each of the flip-flops. An active LOW CLEAR input drives all four D-type latches and can be used to clear all of the Q outputs to 0. Like the 7474, the 74175 can be used in storage register, counter, and divider applications.

■ **EXAMPLE:**

What would be the Q outputs from the 74175 for the following input conditions?

a. 1D = 1, 2D = 0, 3D = 0, 4D = 1, $CLK = \downarrow$, $CLR = 1$.
b. 1D = 1, 2D = 1, 3D = 0, 4D = 1, $CLK = \uparrow$, $CLR = 1$.
c. 1D = 0, 2D = 1, 3D = 0, 4D = 1, $CLK = \uparrow$, $CLR = 0$.

IC TYPE: SN74175 – Quadruple D-Type Positive Edge-triggered Flip-Flop with Clear

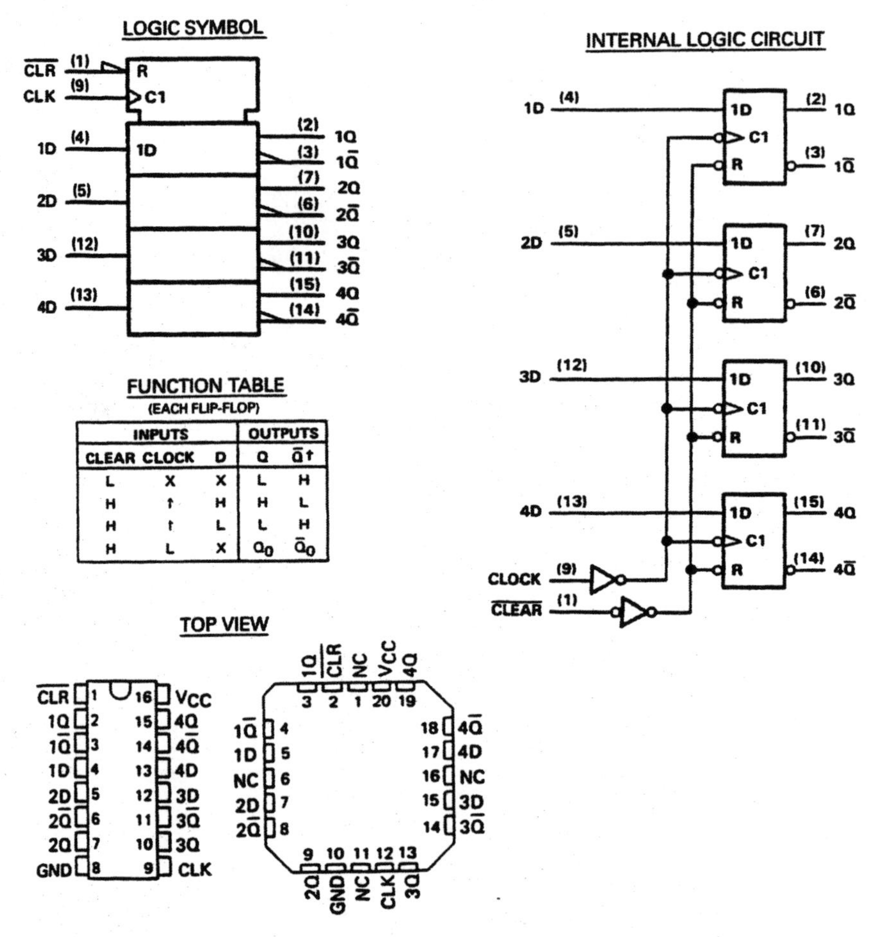

INPUTS			OUTPUTS	
CLEAR	CLOCK	D	Q	Q̄†
L	X	X	L	H
H	↑	H	H	L
H	↑	L	L	H
H	L	X	Q_0	$\bar{Q}_0$

FIGURE 9-17 An IC Containing Four Positive Edge-Triggered D-Type Flip-Flops. (Courtesy of Texas Instruments)

■ ***Solution:***

a. There will be NO CHANGE in the outputs because the clock and clear are not active.

b. 1Q = 1, 2Q = 1, 3Q = 0, 4Q = 1 (outputs will equal inputs since the clock is active and the clear is not active).

c. All outputs will be 0 because the clear is active.

9-2-3 Pulse-Triggered D-Type Flip-Flops

The logic circuit, truth table, and logic symbol for a pulse-triggered (master-slave) D-type flip-flop is shown in Figure 9-18. Like the pulse-triggered S-R flip-flop, the pulse-triggered D-type flip-flop requires a complete clock pulse input before the Q output will reflect the D input.

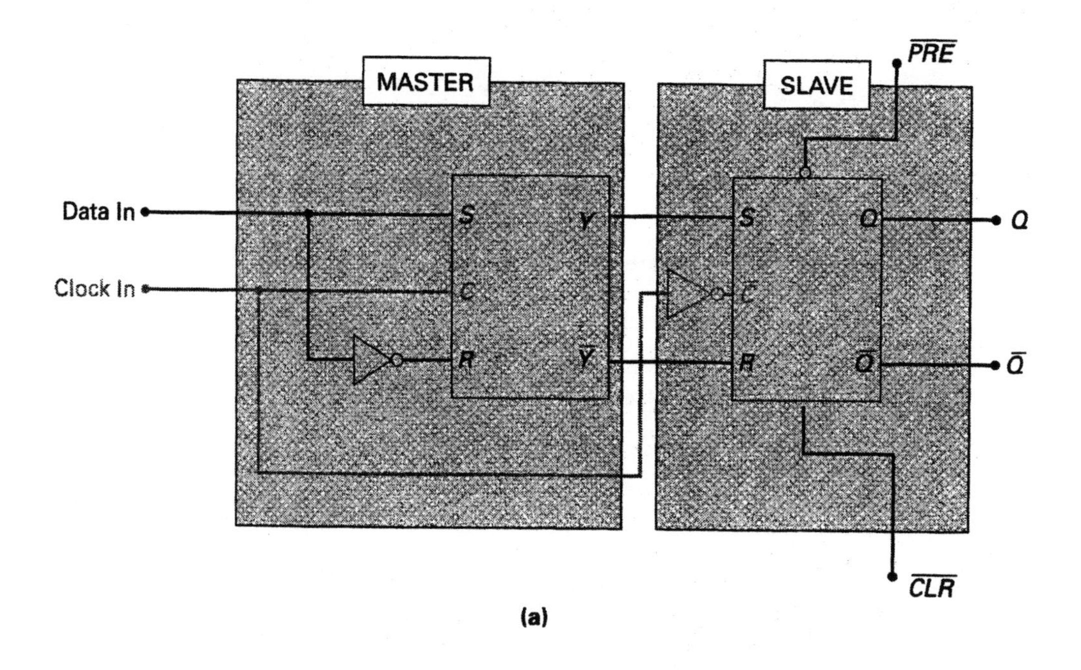

(a)

$\overline{PRE}$	$\overline{CLR}$	Clock	D	Q
0	1	X	X	1—Preset
1	0	X	X	0—Cleared
0	0	X	X	Race
1	1	⎍	0	0—Reset
1	1	⎍	1	1—Set

(b)

(c)

FIGURE 9-18 Pulse-Triggered (Master-Slave) D-Type Flip-Flop.

■ **EXAMPLE:**

What would be the Q output for the pulse-triggered D-type flip-flop shown in Figure 9-19, assuming the D and C inputs shown in the waveforms?

■ ***Solution:***

All pulse-triggered flip-flops are level-clocked devices, which means that the master section is enabled for the entire time that the C input is HIGH and the slave section is enabled during the LOW input of C. The D input is therefore transferred through the master section when the clock is HIGH, and then transferred through the slave section, and to the Q output, when the clock input is LOW, as shown in Figure 9-19.

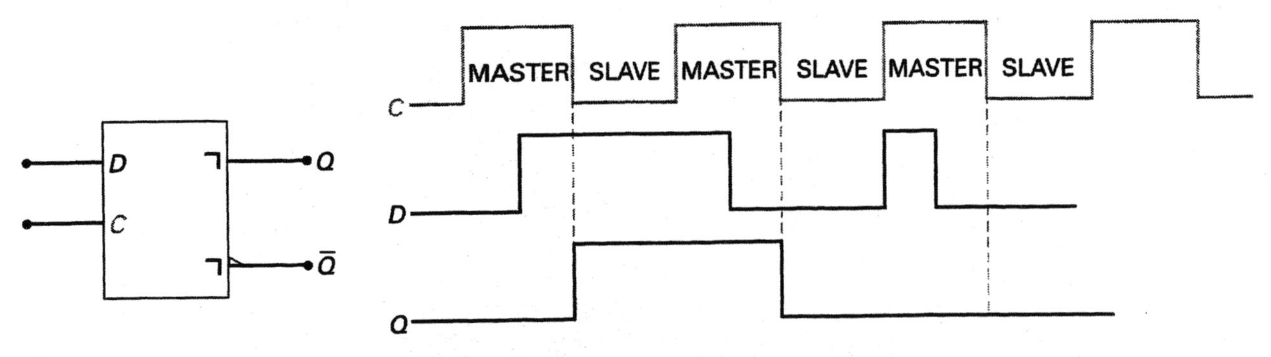

FIGURE 9-19 Input/Output Waveforms for a Pulse-Triggered D-Type Flip-Flop.

SELF-TEST REVIEW QUESTIONS FOR SECTION 9-2

1. The "D" in D-type stands for ________.
2. Can the D-type flip-flop RACE?
3. What are the input and output differences between an *S-R* flip-flop and a D-type flip-flop?
4. A D-type flip-flop could be used as a/an ________.
   a. binary counter
   b. frequency divider
   c. storage register
   d. all of the above

9-3 TROUBLESHOOTING SET-RESET AND DATA-TYPE FLIP-FLOP CIRCUITS

To review, the procedure for fixing a failure can be broken down into the three steps diagnose, isolate, and repair. Since this chapter has been devoted to *S-R* and D-type flip-flop logic circuits, let us first examine the operation of a typical flip-flop circuit, and then apply our three-step troubleshooting process to this circuit.

9-3-1 *A Flip-Flop Logic Circuit*

As an example, Figure 9-20 shows a four-stage frequency divider circuit with single-step or continuous-clock control. This circuit operates in the following way. As discussed previously, the D-type flip-flop will divide the input clock frequency by two if the $\overline{Q}$ output is connected back to the *D* input. Referring to the four-stage frequency divider circuit (using two 74LS74s) and its associated waveforms in the lower half of the circuit, you can see that each D-type flip-flop will halve its applied input frequency (÷ 2). For example, if a 16 Hz clock signal is applied to the input:

> The first stage will divide the 16 Hz input by two, giving an 8 Hz output at test point 1 (16 Hz ÷ 2 = 8 Hz).

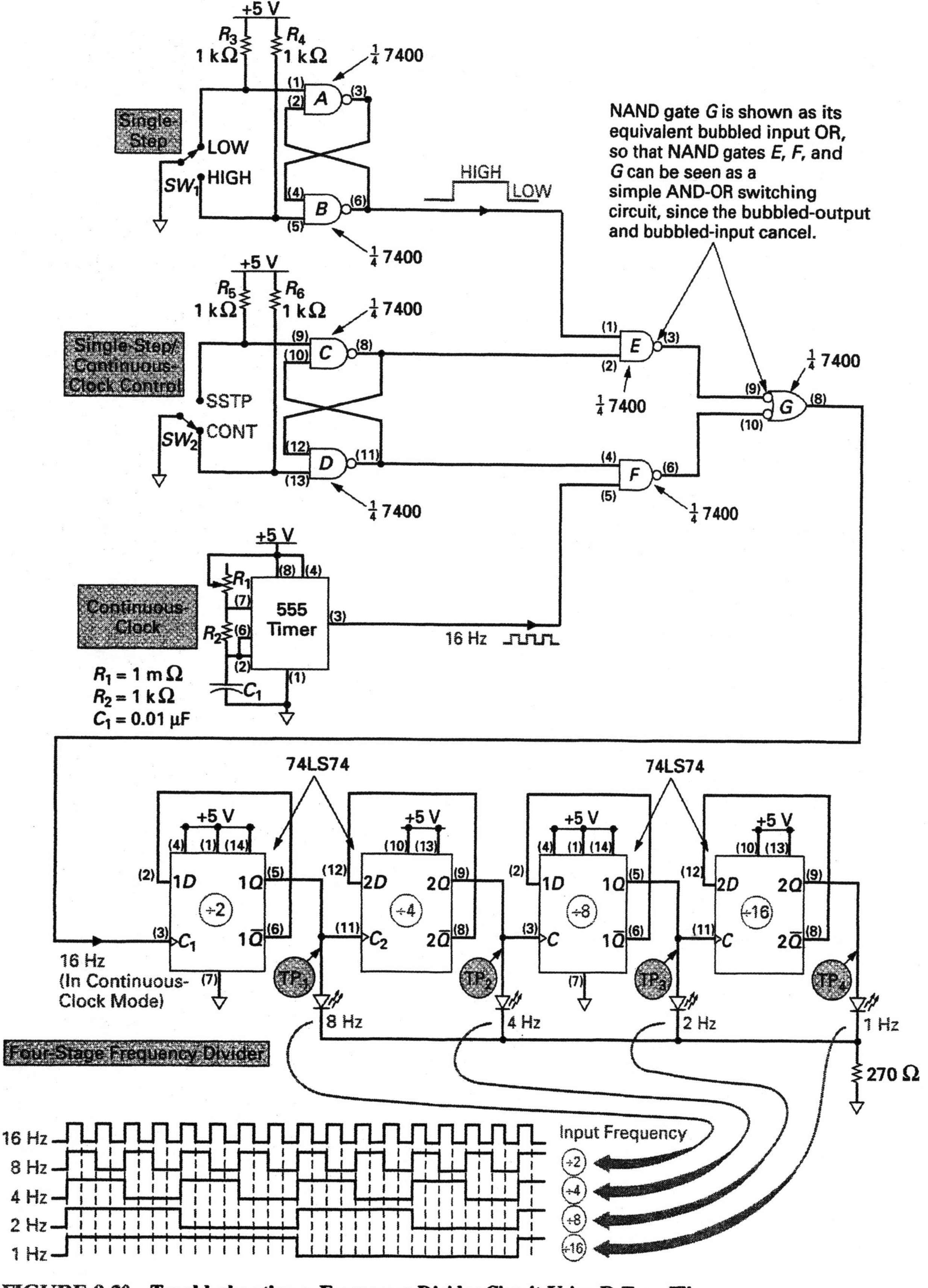

FIGURE 9-20 Troubleshooting a Frequency Divider Circuit Using D-Type Flip-Flops.

The first and second stages will divide the 16 Hz input by four, giving a 4 Hz output at test point 2 (16 Hz ÷ 2 = 8 Hz, 8 Hz ÷ 2 = 4 Hz).

The first, second, and third stages will divide the 16 Hz input by eight, giving a 2 Hz output at test point 3 (16 Hz ÷ 2 = 8 Hz, 8 Hz ÷ 2 = 4 Hz, 4 Hz ÷ 2 = 2 Hz).

The first, second, third, and fourth stages will divide the 16 Hz input by sixteen, giving a 1 Hz output at test point 4 (16 Hz ÷ 2 = 8 Hz, 8 Hz ÷ 2 = 4 Hz, 4 Hz ÷ 2 = 2 Hz, 2 Hz ÷ 2 = 1 Hz).

These divided outputs at test points 1, 2, 3, and 4 can be seen on a four-LED display connected to each of the D-type flip-flop outputs.

The clock input applied to the four-stage frequency divider is derived from the single-step or continuous-clock control circuit in the upper half of the circuit in Figure 9-20. With this circuit, SW_2 is a single-pole double-throw (SPDT) switch that is used to select either a single-step or continuous-clock output for the frequency divider circuit. The NAND latch, made up of NAND gates C and D, acts as a switch debouncer circuit for SW_2. When SW_2 is in the single-step position, NAND C will produce a HIGH output, which will enable NAND gate E and allow the single clock pulses, generated by SW_1 and debounced by NAND gates A and B, through to NAND G, and on to the frequency divider. When SW_2 is in the continuous-clock position, NAND D will produce a HIGH output which will enable NAND gate F and allow the 16 Hz clock pulses, generated by the 555 timer circuit, through to NAND G, and on to the frequency divider.

Let us now apply our three-step troubleshooting procedure to this flip-flop logic circuit so that we can practice troubleshooting procedures and methods.

9-3-2 Step 1: Diagnose

As before, it is extremely important that you first understand the operation of a circuit and how all of the devices within it are supposed to work, so that you are able to determine whether or not a circuit malfunction really exists. If you were preparing to troubleshoot the circuit in Figure 9-20, your first step should be to read through the circuit description and review the operation of each IC until you feel completely confident with the correct operation of the entire circuit. Once you are fully familiar with the operation of the circuit, you will easily be able to diagnose the problem as either an *operator error* or a *circuit malfunction*.

Distinguishing an operator error from an actual circuit malfunction is an important first step, and a wrong diagnosis can waste a lot of time and effort. For example, the following could be interpreted as circuit malfunctions, when in fact they are simply operator errors.

Example 1.
Symptom: The LEDs remain frozen when SW_1 and SW_2 are in the upper position.
Diagnosis: Operator Error—Single-step mode has been selected; however, SW_1 has to be switched HIGH and LOW in order to generate single clock pulses.

Example 2.
Symptom: When SW_2 is placed in the lower position, the single-step clock switch SW_1 will not operate.
Diagnosis: Operator Error—In the continuous-clock mode, the single-step clock from switch SW_1 is blocked.

Example 3.
Symptom: The circuit is malfunctioning because the display at test point 4 is operating at a frequency that is higher than 1 Hz.
Diagnosis: Operator Error—There is not a circuit malfunction; the 555 timer clock frequency is set too high. Adjust R_1 until the LED at TP_4 flashes ON and OFF in one-second intervals.

Once you have determined that the problem is not an operator error, but is in fact a circuit malfunction, proceed to Step 2 and isolate the circuit failure.

9-3-3 Step 2: Isolate

No matter what circuit or system failure has occurred, you should always follow a logical and sequential troubleshooting procedure. Let us review some of the isolating techniques and apply them to our example circuit in Figure 9-20.

a. Use a cause and effect troubleshooting process, which means study the effects you are getting from the faulty circuit and then logically reason out what could be the cause.

b. Check first for obvious errors before leaping into a detailed testing procedure. Is the power OFF or not connected to the circuit? Are there wiring errors? Are all of the ICs correctly oriented?

c. Using a logic probe or voltmeter test that power and ground are connected to the circuit and are present at all points requiring power and ground. If the whole circuit, or a large section of the circuit, is not operating, the problem is normally power. Using a multimeter, check that all of the dc voltages for the circuit are present at all IC pins that should have power or a HIGH input, and are within tolerance. Secondly, check that 0 V or ground is connected to all IC ground pins and all inputs that should be tied LOW.

d. Use your senses to check for broken wires, loose connectors, overheating or smoking components, pins not making contact, and so on.

e. Test the clock signals using the logic probe or the oscilloscope. Although clock signals can be easily checked with a logic probe, the oscilloscope is best for checking the signal's amplitude, frequency, pulse width, and so on. The oscilloscope will also display timing problems and noise spikes (glitches) which could be false-triggering a circuit into operation at the wrong time.

f. Perform a static test on the circuit. With our circuit example in Figure 9-20, you can static test the frequency divider circuit by switching the circuit to the single-step mode, assuming that the clock section of the circuit is operating correctly. You can also check the frequency divider section of the circuit by disconnecting the clock input to the circuit, and then using a logic pulser's single pulse feature to clock the divider in single steps. A logic probe could then be used to test that valid logic levels are appearing at the outputs of the D-type flip-flops.

g. With a dynamic test, the circuit is tested while it is operating at its normal clock frequency and all of the inputs and outputs are continually changing. Although a logic probe can detect a pulse waveform, the oscilloscope will display more signal detail, and since it can display more than one signal at a time, it is ideal for making signal comparisons and looking for timing errors. Once again, you can dynamic test the frequency divider circuit in Figure 9-20 by switching the circuit to the continuous-clock mode, assuming that the clock section of the circuit is operating correctly.

A logic analyzer, which will be discussed in more detail in the digital system chapter, is a type of oscilloscope that can display typically eight to sixteen digital signals at one time.

In some instances, you will discover that some circuits will operate normally when undergoing a static test, and yet fail a dynamic test. This effect usually points to a timing problem involving the clock signals and/or the propagation delay times of the ICs used within the circuit.

h. Noise due to electromagnetic interference (EMI) can false-trigger an IC, such as the D-type flip-flops in Figure 9-20. This problem can be overcome by not leaving any of the ICs' inputs unconnected and, therefore, floating. Connect unneeded active-HIGH inputs to ground and active-LOW inputs to $+V_{CC}$.

i. Apply the half-split method of troubleshooting first to a circuit, and then to a section of a circuit, to help speed up the isolation process. With our circuit example in Figure 9-20, a good mid-point check would be to test the clock signal at the output of NAND *G* to determine whether the problem is in the clock section or divider section of the circuit.

Also remember that a load problem can make a source appear at fault. If an output is incorrect, it would be safer to disconnect the source from the rest of the circuit and then recheck the output. If the output is still incorrect, the problem is definitely within the source; however, if the output is correct, the load is probably shorting the line either HIGH or LOW.

j. Substitution can be used to help speed up your troubleshooting process. Once the problem is localized to an area containing only a few ICs, substitute suspect ICs with known working ICs (one at a time) to see if the problem can be quickly remedied.

9-3-4 Step 3: Repair

Once the fault has been found, the final step is to repair the circuit and then perform a final test to see that the circuit and the system are now fully operational.

9-3-5 Sample Problems

Once you have constructed a circuit, like the frequency divider circuit in Figure 9-20, introduce a few errors to see what effect or symptoms they produce. Then logically reason out why a particular error or cause has a particular effect on the circuit. Never short any two points together unless you have carefully thought out the consequences, but generally it is safe to open a path and see the results. Here are some examples for our example circuit in Figure 9-20.

a. Disconnect the output of the 555 timer from the circuit.

b. Bypass the NAND *A* and *B* debounce circuit so that SW_1 feeds directly to NAND *E*.

c. Bypass the NAND *C* and *D* debounce circuit so that SW_2 feeds directly to NAND *E* and *F*.

d. Switch the Q and $\overline{Q}$ outputs of a D-type flip-flop.

e. Connect pin 13 of a 7474 LOW instead of HIGH.

f. Connect the LEDs to the $\overline{Q}$ outputs of the 7474s instead of the Q outputs.

SELF-TEST REVIEW QUESTIONS FOR SECTION 9-3

1. What are the three basic troubleshooting steps?
2. What is a static test?
3. What is the half-split troubleshooting technique?
4. What symptom would you get from the circuit in Figure 9-20 if the switch control input to NAND gates *C* and *D* were both held LOW?

REVIEW QUESTIONS

Multiple-Choice Questions

1. A flip-flop is a ____________ circuit that can store 1 bit of binary data.

   a. decoder **b.** comparator **c.** encoder **d.** memory

2. Like the bistable multivibrator, the flip-flop is a ____________ device.

   a. three-state **b.** monostable **c.** two-state **d.** astable

3. A NOR latch has ____________ inputs.

   a. active-HIGH **b.** active-LOW

4. A NAND latch has ____________ inputs.

   a. active-HIGH **b.** active-LOW

5. A NOR latch will ____________ when $S = 0$ and $R = 0$, whereas a NAND latch will ____________ when $S = 1$ and $R = 1$.

   a. RACE, RACE
   b. have NO CHANGE, RACE
   c. have NO CHANGE, have NO CHANGE
   d. RACE, have NO CHANGE

6. A NAND latch will ____________ when $S = 1$ and $R = 0$, and a NOR latch will ____________ when $S = 1$ and $R = 0$.

   a. SET, RESET **b.** SET, SET **c.** RESET, SET **d.** RESET, RESET

7. Digital systems contain thousands of flip-flop circuits that are all synchronized by a ____________.

   a. PRESET input **b.** clock signal **c.** CLEAR input **d.** all of the above

8. A gated S-R flip-flop is ____________.

   a. level-triggered
   b. positive edge-triggered
   c. pulse-triggered
   d. negative edge-triggered

9. With a positive edge-triggered SET-RESET flip-flop, the inputs will control the output:

   a. when the clock rises from LOW to HIGH.
   b. when a complete pulse of the clock signal is applied.
   c. when a HIGH level is applied to the clock.
   d. all of the above.

10. A master-slave flip-flop is ____________.

    a. level-triggered
    b. positive edge-triggered
    c. pulse-triggered
    d. negative edge-triggered

11. With a D-type latch, the data input drives the flip-flop's internal ____________ input, while the complement of the data input drives the flip-flop's internal ____________ input.

    a. R, S **b.** R, CLR **c.** S, R **d.** S, PRE

12. If a 1 MHz clock signal were applied to the C input of a D-type latch, and the $\overline{Q}$ output were connected to the D input, what would be the output frequency at Q?

    a. 2 MHz **b.** 0.5 MHz **c.** 4 MHz **d.** 1.5 MHz

13. With a pulse-triggered flip-flop, the clock is applied directly to the ____________ latch, while the complement of the clock is applied to the ____________ latch.

    a. SET, RESET **b.** SET, master **c.** master, slave **d.** RESET, slave

14. If a LOW Q output turns ON an output LED, what input should be connected to a slow-start circuit to ensure the LED is OFF when circuit power is first applied?

    a. PRESET **b.** SET **c.** CLEAR **d.** RESET

15. Since the S and R inputs of a D-type latch are always opposite, it is impossible for this type of flip-flop to have the ____________ input condition.

    a. SET **b.** RESET **c.** NO CHANGE **d.** RACE

Essay Questions

16. What are the three basic types of flip-flops? (Introduction)
17. What is a flip-flop circuit? (Introduction)
18. What is the relationship between a bistable multivibrator and a SET-RESET flip-flop? (9-1)
19. List and describe the four SET-RESET flip-flop input conditions. (9-1)
20. Why is a flip-flop circuit also called a latch? (9-1)
21. Briefly describe the differences between a NOR latch and a NAND latch. (9-1)
22. What are the three basic triggering methods used for flip-flop circuits? (9-1)

23. Briefly describe why latch circuits are used in connection with control switches. (9-1)

24. What advantage does an edge-triggered flip-flop have over a level-triggered flip-flop? (9-1)

25. What is the purpose of the PRESET and CLEAR inputs applied to flip-flop circuits? (9-1)

26. Briefly describe the operation of a pulse-triggered *S-R* flip-flop circuit. (9-1)

27. What is the basic difference between a pulse-triggered flip-flop and a master-slave flip-flop? (9-1)

28. How is a SET-RESET flip-flop used as a memory latch circuit? (9-1)

29. What circuit modification was made to a SET-RESET flip-flop in order to create a D-type flip-flop? (9-2)

30. What key advantage does the D-type latch have over the SET-RESET latch? (9-2)

31. Why does the D-type flip-flop not have a RACE input condition? (9-2)

32. Briefly describe how a set of D-type flip-flops can be used as a storage register. (9-2)

33. Briefly describe how a D-type flip-flop can be used as a frequency divider circuit. (9-2)

34. Briefly describe how a D-type flip-flop can be used as a binary counter. (9-2)

35. Sketch a frequency divider circuit using three D-type flip-flops that will divide the input clock frequency by eight. (9-2)

Practice Problems

36. Identify the latch circuits shown in Figure 9-21(a) and (b).

37. Sketch the *Q* output waveform from the latch circuit shown in Figure 9-21(a).

38. What would have to be applied to the latch circuit in Figure 9-21(a) to cause the output to:

    a. have NO CHANGE? **b.** RACE?

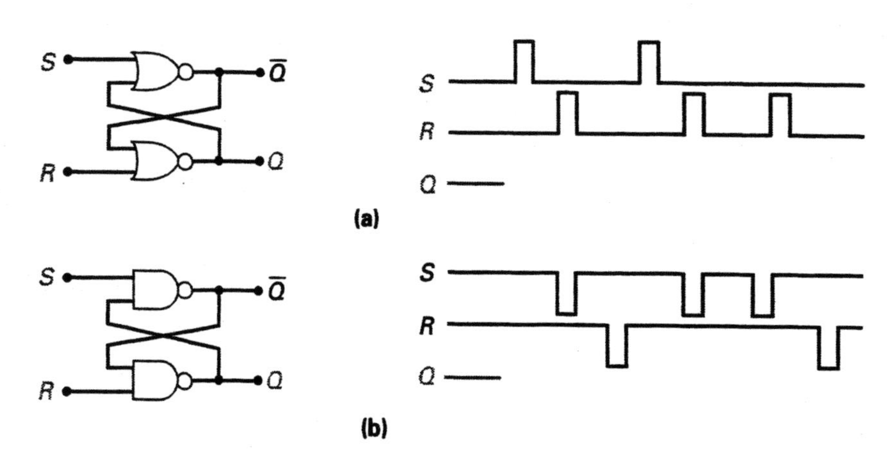

FIGURE 9-21 Input and Output NAND and NOR Waveforms.

39. Sketch the Q output waveform from the latch circuit shown in Figure 9-21(b).

40. What would have to be applied to the latch circuit in Figure 9-21(b) to cause the output to:

    a. have NO CHANGE? **b.** RACE?

41. Using the four NAND gates within a 7400 IC, sketch a circuit showing how the gates would need to be connected to form a level-triggered *S-R* flip-flop (show IC pin numbers).

42. Sketch a circuit and include waveforms to show how an AND gate and INVERTER gates could be used to construct a

    a. positive-transition pulse generator.
    b. negative-transition pulse generator.

43. Sketch a circuit showing how 7400 NAND gates would need to be connected to form a positive-transition pulse generator circuit and a positive edge-triggered *S-R* flip-flop (show IC pin numbers).

44. Identify the flip-flop logic symbols shown in Figure 9-22.

To practice your circuit recognition and operation, refer to Figure 9-23 and answer the following six questions.

45. What type of flip-flop circuit is being used in this application?

46. If the LOAD control line were LOW, what would be applied to the D inputs of the flip-flops?

47. If the LOAD control line were HIGH, what would be applied to the D inputs of the flip-flops?

48. What type of triggering is used in this circuit?

49. Describe the basic steps involved in loading a 4-bit word into this storage register circuit.

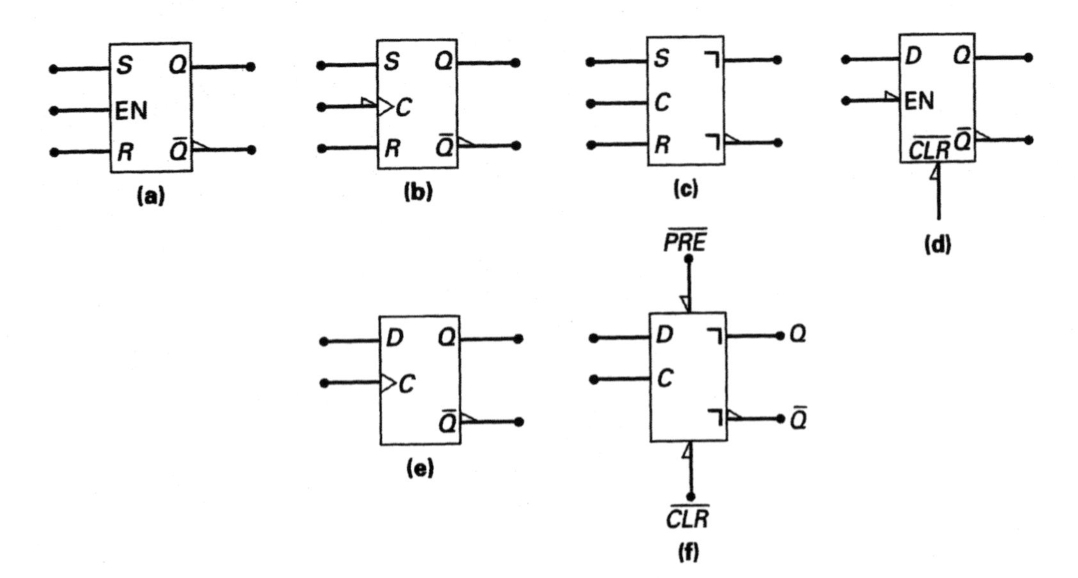

FIGURE 9-22 Flip-Flop Logic Symbols.

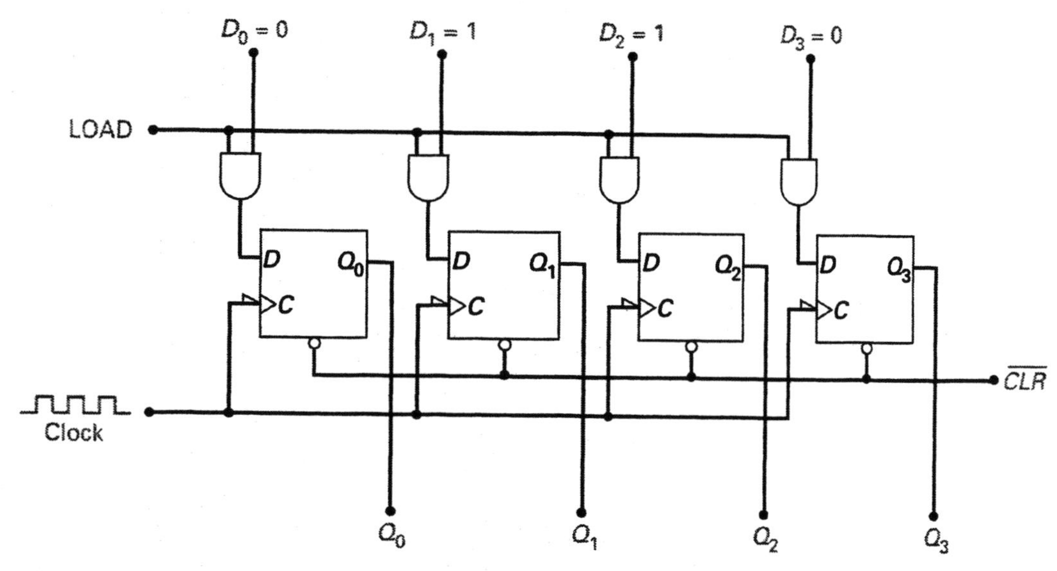

FIGURE 9-23 A 4-Bit Storage Register.

50. How is the $\overline{CLR}$ control line used in this circuit?

To practice your circuit recognition and operation, refer to Figure 9-24 and answer the following three questions.

51. The logic gates A and B form what type of flip-flop circuit?
52. What will be the output at point X when the switch is in the OFF position and the ON position?
53. What will be the output at point Y when the switch is in the OFF position and the ON position?

To practice your circuit recognition and operation, refer to Figure 9-25 and answer the following three questions.

54. What type of flip-flop is being used in the application in this circuit?

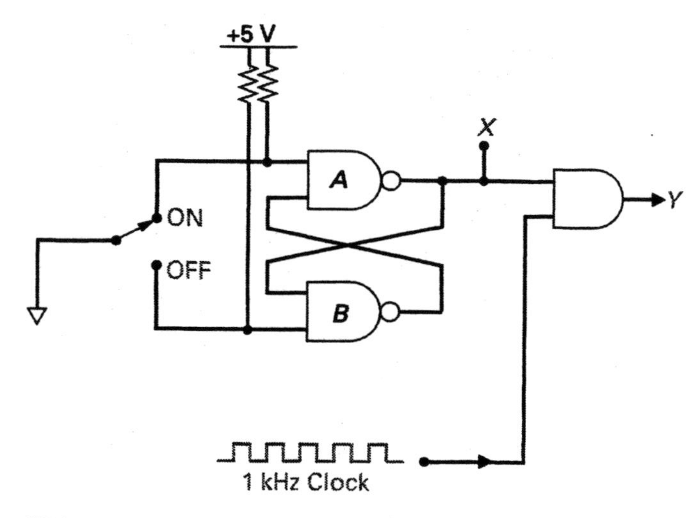

FIGURE 9-24 ON/OFF Clock-Control Circuit.

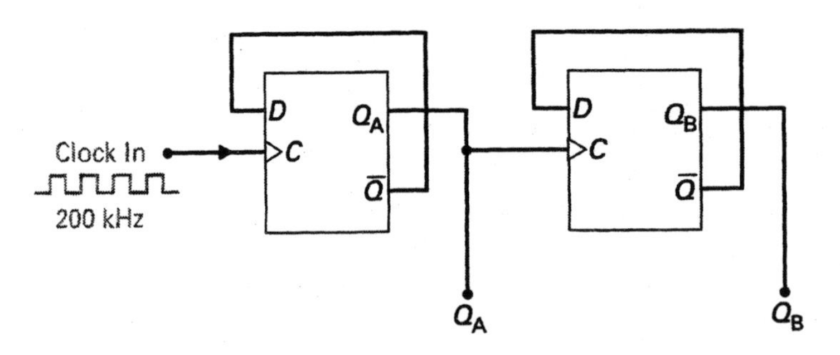

FIGURE 9-25 **Clock-Division Circuit.**

55. What will be the output frequency at:

   a. Q_A? b. Q_B?

56. Sketch the input/output waveforms for this circuit.

To practice your circuit recognition and operation, refer to Figure 9-26 and answer the following four questions.

57. What type of flip-flop is being used in this circuit?
58. What type of triggering is being used in this circuit?
59. Describe how this circuit will operate when inputs A and B are out-of-phase with one another.
60. Describe how this circuit will operate when inputs A and B are in-phase with one another.

Troubleshooting Questions

61. How could you use a logic pulser and logic probe to static test the circuit in Figure 9-23?
62. What symptoms would you expect from the circuit shown in Figure 9-23 if the following malfunctions were to occur?

   a. The clock input was permanently HIGH.
   b. The LOAD input was permanently HIGH.
   c. Power was disconnected from one of the D-type flip-flops.
   d. The CLEAR input was permanently LOW.
   e. One of the D-type flip-flop's D inputs was shorted to $+V_{CC}$.

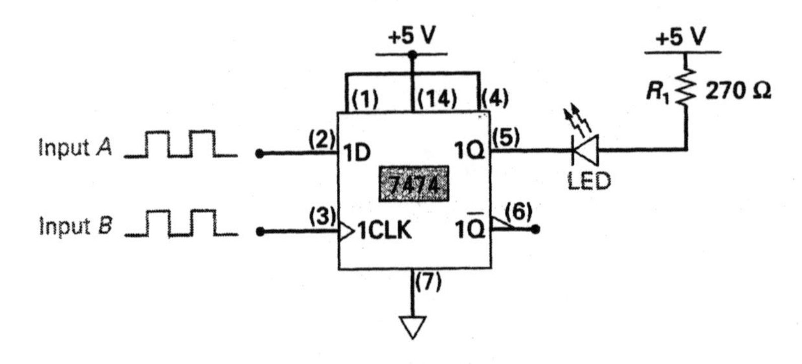

FIGURE 9-26 **A Phase-Detector Circuit.**

63. What symptoms would you expect from the circuit shown in Figure 9-24 if the following malfunctions were to occur?

a. The switch's pole was not connected to ground.
b. The control input to the AND gate was shorted LOW.

64. What symptoms would you expect from the circuit shown in Figure 9-25 if the following malfunctions were to occur?

a. The D and C inputs of the first flip-flop were switched.
b. The Q and $\overline{Q}$ outputs of the first flip-flop were switched.

65. What symptoms would you expect from the circuit shown in Figure 9-26 if the following malfunctions were to occur?

a. The D and CLK inputs of the flip-flop were switched.
b. The LED was connected to the $\overline{Q}$ output instead of the Q.
c. Pin 1 was connected to ground.

Chapter 10

OBJECTIVES

After completing this chapter, you will be able to:

1. Describe the differences between the *J-K* flip-flop, the *S-R* flip-flop, and the D-type flip-flops.
2. Explain the operation of the *J-K* flip-flop.
3. Describe the differences between an edge-triggered and a pulse-triggered *J-K* flip-flop.
4. Sketch the output waveform from a *J-K* flip-flop for any given input timing diagram.
5. Describe the operation of a basic transistor and logic-gate astable multivibrator circuit.
6. Explain how astable multivibrators can be used as digital clock oscillators.
7. Describe the operation of a basic transistor and logic-gate monostable multivibrator circuit.
8. List and explain the different applications of monostable multivibrators or one-shot ICs.
9. Describe the block diagram and operation of the 555 timer IC.
10. Explain how the 555 timer can be used as an astable or a monostable multivibrator.
11. Apply the three-step troubleshooting procedure to a *J-K* flip-flop and timer circuit.

J-K Flip-Flops and Timer Circuits

OUTLINE

VIGNETTE

Go and Do "Something More Useful"

Vladamir Zworykin was born in Mourom, Russia, on July 30, 1889. He graduated in 1912 from the St. Petersburg Institute of Technology and then did his postgraduate work at the College de France in Paris. After World War I, he immigrated to the United States and joined the research staff of Westinghouse Electric Corporation in Pittsburgh. Here he invented the iconoscope (electronic camera) and the kinescope (CRT picture tube). In 1923 he proudly demonstrated to his employers what he considered his biggest achievement—a cloudy image of boats on the river outside his lab on a small screen. Management was not at all impressed, and Zworykin later said that he was told, "to spend my time on something more useful."

In 1929 David Sarnoff, the founder of the Radio Corporation of America (RCA), asked Zworykin what it would take to develop television for commercial use. Zworykin, who at that time was not at all sure, answered confidently. "Eighteen months and $100,000." Sarnoff liked what he heard and hired Zworykin, putting him in charge of RCA's electronic research laboratory at Camden, New Jersey. Years later Sarnoff loved pointing out the slight difference in development cost and time, which finally worked out to be twenty years and $50 million.

While at RCA, Zworykin also developed the electron microscope (in only three months), the electron multiplier tube, and the first infrared sniper scopes which were used in World War II.

In 1967 Zworykin received the National Medal of Science for his contributions to the instruments of science, engineering, and television, and for his stimulation of the application of engineering to medicine. Recognized as the "father of television," he lived long enough to see all of his inventions flourish before he died in 1982. When asked in a television interview in 1980 what he thought of television today he answered, "The technique is wonderful. It is beyond my expectation. But the programs! I would never let my children even close to this thing."

INTRODUCTION

In the previous chapter we examined two of the three types of flip-flop circuits—the set-reset (*S-R*) flip-flop, and the data-type flip-flop. In the first section of this chapter, we will be discussing the third type of flip-flop circuit which is called the *J-K* flip-flop. The *J-K* flip-flop evolved from the D-type flip-flop in the same way that the D-type flip-flop evolved from the *S-R* flip-flop. The *J-K* flip-flop can outperform both the *S-R* and D-type flip-flops, making it the most versatile of all the flip-flops. Its internal circuit is more complex, however, and therefore more expensive, which is why the low-cost *S-R* and D-type flip-flops are still ideal for simple circuit applications.

Flip-flop circuits are often referred to as bistable multivibrator circuits because they have two stable (bistable) states. In the second section of this chapter, we will examine the astable (unstable) multivibrator circuit which has no stable states, and the monostable multivibrator circuit which has one stable (monostable) state. As you will see in this section, these multivibrator circuits are often used to generate timing signals in digital circuits.

10-1 *J-K* FLIP-FLOPS

With the previously discussed *S-R* flip-flop and D-type flip-flop, the letters "S," "R," and "D" were abbreviations for SET, RESET, and DATA. With the ***J-K* flip-flop,** the letters "J" and "K" are not abbreviations, they are arbitrarily chosen letters. As you will see in the following section, the *J-K* flip-flop operates in almost exactly the same way as the *S-R* flip-flop in that there will be NO CHANGE at the output when the *J* and *K* inputs are both LOW, the output will be SET HIGH when the *J* input (set input) is HIGH, and the output will be RESET LOW when the *K* input (reset input) is HIGH. The distinctive difference with the *J-K* flip-flop is that it will not RACE when both its SET (*J*) and RESET (*K*) inputs are HIGH. When this input condition is applied ($J = 1$ and $K = 1$), the *J-K* flip-flop will **toggle,** which means that it will simply switch or reverse the present logic level at its *Q* output. This toggle feature of the *J-K* flip-flop is achieved by modifying the basic *S-R* flip-flop's internal logic circuit to include two cross-coupled feedback lines between the output and the input. This circuit modification, however, means that the *J-K* flip-flop cannot be level-triggered, it can only be edge-triggered or pulse-triggered.

***J-K* Flip-Flop**
A circuit that will operate in the same way as a set-reset flip-flop, except that it will toggle its output instead of race when both inputs are HIGH.

Toggle
An input combination or condition that causes the output to toggle or switch its output to the opposite logic level.

10-1-1 ***Edge-Triggered* J-K *Flip-Flops***

Figure 10-1(a) shows the internal logic circuit for an edge-triggered *J-K* flip-flop circuit, and Figure 10-1(b) shows this circuit's function table. The first three lines of the function table show that if the clock input line is LOW, HIGH, or making a transition from LOW to HIGH (a positive edge), there will be no change at the *Q* output regardless of the *J* and *K* inputs. This is because this *J-K* flip-flop is a negative edge-triggered device, which means it will only "wake up" and perform the operation applied to the *J* and *K* inputs when the clock input is making a transition from HIGH to LOW (a negative edge). The negative-transition pulse generator will generate a positive output pulse every time the clock pulse input drops from a HIGH to a LOW level. In the last four lines of the function table, you can see that when a negative edge is applied to the clock input, the *J-K* flip-flop will react to its *J* (SET) and *K* (RESET) inputs. For example, when negative edge-triggered, the *J-K* flip-flop will operate in almost the same way as a *S-R* flip-flop, since it will not change its *Q* output when *J* and *K* are LOW, RESET its *Q* output when the reset input (*K*) is HIGH, and SET its *Q* output when the set input (*J*) is HIGH. In the last line of the function table, you can see why the *J-K* flip-flop is an improvement over the

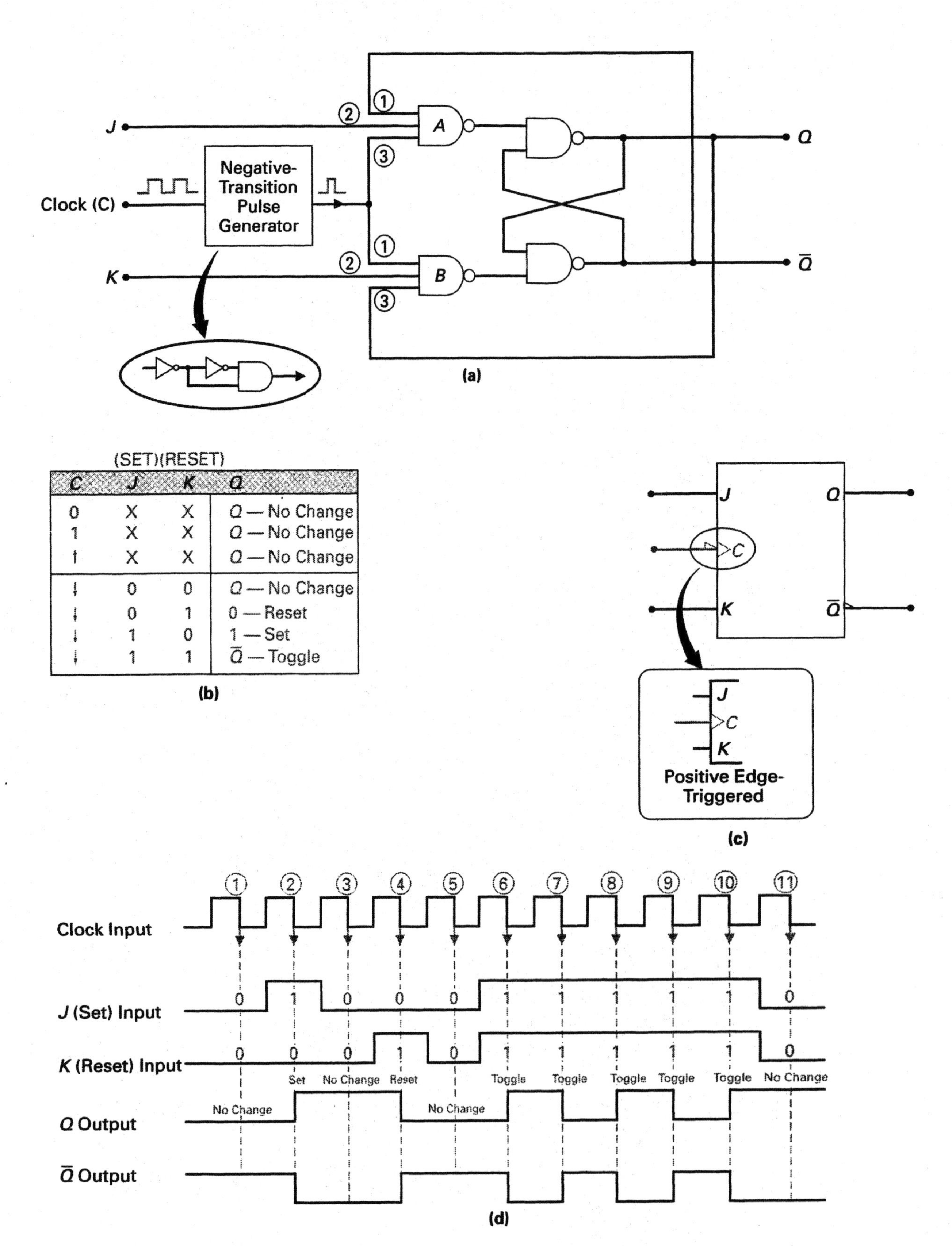

FIGURE 10-1 **Edge-Triggered *J-K* Flip-Flops.**

S-R flip-flop. When both the *S* and *R* inputs of a *S-R* flip-flop are HIGH, it would race and generate an unpredictable *Q* output, which is why this input condition is never used. When both the *J* and *K* inputs of a *J-K* flip-flop are made HIGH, on the other hand, the *Q* output will toggle, or switch, to the opposite state. This means that if *Q* is HIGH, it will switch to a LOW, and if *Q* is LOW, it will switch to a HIGH. The function table shows that if the *J-K* flip-flop is in the toggle condition and the clock input makes a transition from HIGH to LOW, the *Q* output will switch to the opposite logic level, or the logic level of the $\overline{Q}$ output.

Figure 10-1(c) shows the logic symbol for a negative edge-triggered *J-K* flip-flop. A full triangle at the clock input, inside the rectangular block, is used to indicate that this device is edge-triggered, and the smaller right triangle (or bubble) outside the rectangular block is used to indicate that this input is active-LOW, or negative edge-triggered. The inset in Figure 10-1(c) shows the logic symbol for a positive edge-triggered *J-K* flip-flop.

To reinforce your understanding of the *J-K* flip-flop, Figure 10-1(d) shows how a negative edge-triggered *J-K* flip-flop will respond to a variety of input combinations. Since this *J-K* flip-flop is negative edge-triggered, the device will only respond to the *J* and *K* inputs and change its outputs when the clock signal makes a transition from HIGH to LOW, as indicated by the negative arrows shown on the square-wave clock input waveform. At negative clock edge 1, $J = 0$ and $K = 0$, so there is no change in the *Q* output (which stays LOW), and the $\overline{Q}$ output (which stays HIGH). At negative clock edge 2, $J = 1$ and $K = 0$ (set condition), so the *Q* output is set HIGH ($\overline{Q}$ is switched to the opposite, LOW). At negative clock edge 3, both *J* and *K* are logic 0, so once again the output will not change (*Q* will remain latched in the set condition). At negative clock edge 4, $J = 0$ and $K = 1$ (reset condition), so the *Q* output will be reset LOW. At negative edge 5, both *J* and *K* are again LOW, so there will be no change in the *Q* output (*Q* will remain latched in the reset condition). At negative clock edge 6, both *J* and *K* are HIGH (toggle condition), so the output will toggle, or switch, to its opposite state (since *Q* is LOW it will be switched HIGH). At negative clock edges 7, 8, 9, and 10, the *J* and *K* inputs remain HIGH, and therefore the *Q* output will continually toggle or switch to its opposite logic level. At negative edge 11, both *J* and *K* return to 0, so the *Q* output remains in its last state, which in this example is HIGH.

Because of the toggle condition, the *J-K* flip-flop cannot be level-triggered. If the *J* and *K* inputs were both HIGH, and the clock were connected directly to the input gates without a transition pulse generator, the output would toggle continuously for as long as the clock input was HIGH. To explain this in a little more detail, when the clock is at its HIGH level, the HIGH *J* and *K* inputs would be passed through to the flip-flop outputs, and because of the cross-coupling, the outputs would toggle. These new outputs would then be fed back to the input gates producing a new input condition, and the cycle would repeat. The result would be a continual change in the output or oscillations during the time the clock signal was active. For this reason, the *J-K* flip-flop is only either edge-triggered (so that the input gates are only enabled momentarily) or pulse-triggered (master is first enabled while slave—and therefore the output—is disabled). With edge-triggered *J-K* flip-flops, the pulse from the transition generator is always less than the propagation delay of the flip-flop, so that output oscillation will not occur when the flip-flop is in the toggle condition.

Figure 10-2 shows the data sheet for the 74LS76A IC, which contains two negative edge-triggered *J-K* flip-flops. Let us see how this flip-flop could be used as a frequency divider or as a frequency counter.

Application: Frequency Divider/Binary Counter Circuit

Figure 10-3(a) shows how a 555 timer and a 74LS76A *J-K* flip-flop can be connected to form a frequency divider or binary counter circuit. The 555 timer, which will be discussed in detail in the second section of this chapter, is connected to operate as an astable multivibrator, and will generate the square-wave output waveform shown in Figure 10-3(b). The 74LS76A digital IC actually contains two complete *J-K* flip-flop circuits, and since only one is needed for this circuit, the flip-flop is labeled "½ of 74LS76A." Notice that the *J* and *K* inputs of the 74LS76A flip-flop are both connected HIGH, so the *Q* output will continually toggle for each negative edge of the clock signal input from the 555 timer, as seen in the waveforms in Figure 10-3(b). Since two negative edges are needed at the clock input to produce one cycle at the output, the *J-K* flip-flop is in fact dividing the input frequency by two. An input frequency of 2 kHz

IC TYPE: SN74LS76A – Dual Edge-Triggered *J-K* Flip-Flop with Preset and Clear

LOGIC SYMBOL

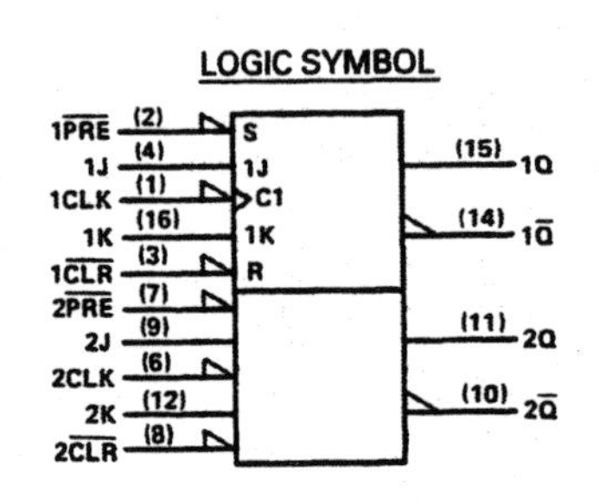

FUNCTION TABLE

INPUTS					OUTPUTS	
PRE	CLR	CLK	J	K	Q	Q̄
L	H	X	X	X	H	L
H	L	X	X	X	L	H
L	L	X	X	X	H†	H†
H	H	↓	L	L	Q_0	$\overline{Q}_0$
H	H	↓	H	L	H	L
H	H	↓	L	H	L	H
H	H	↓	H	H	TOGGLE	
H	H	H	X	X	Q_0	$\overline{Q}_0$

TOP VIEW

Pin		Pin	
1CLK	1	16	1K
1 PRE	2	15	1Q
1 CLR	3	14	1Q̄
1 J	4	13	GND
Vcc	5	12	2K
2CLK	6	11	2Q
2 PRE	7	10	2Q̄
2 CLR	8	9	2J

INTERNAL LOGIC CIRCUIT

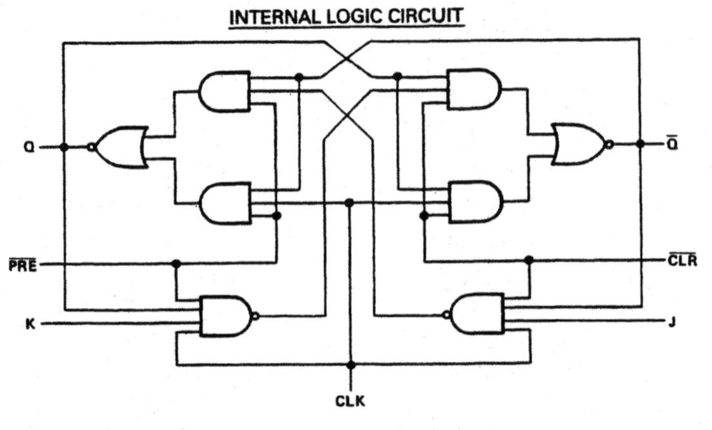

FIGURE 10-2 An IC Containing Two Negative Edge-Triggered *J-K* Flip-Flops. (Courtesy of Texas Instruments)

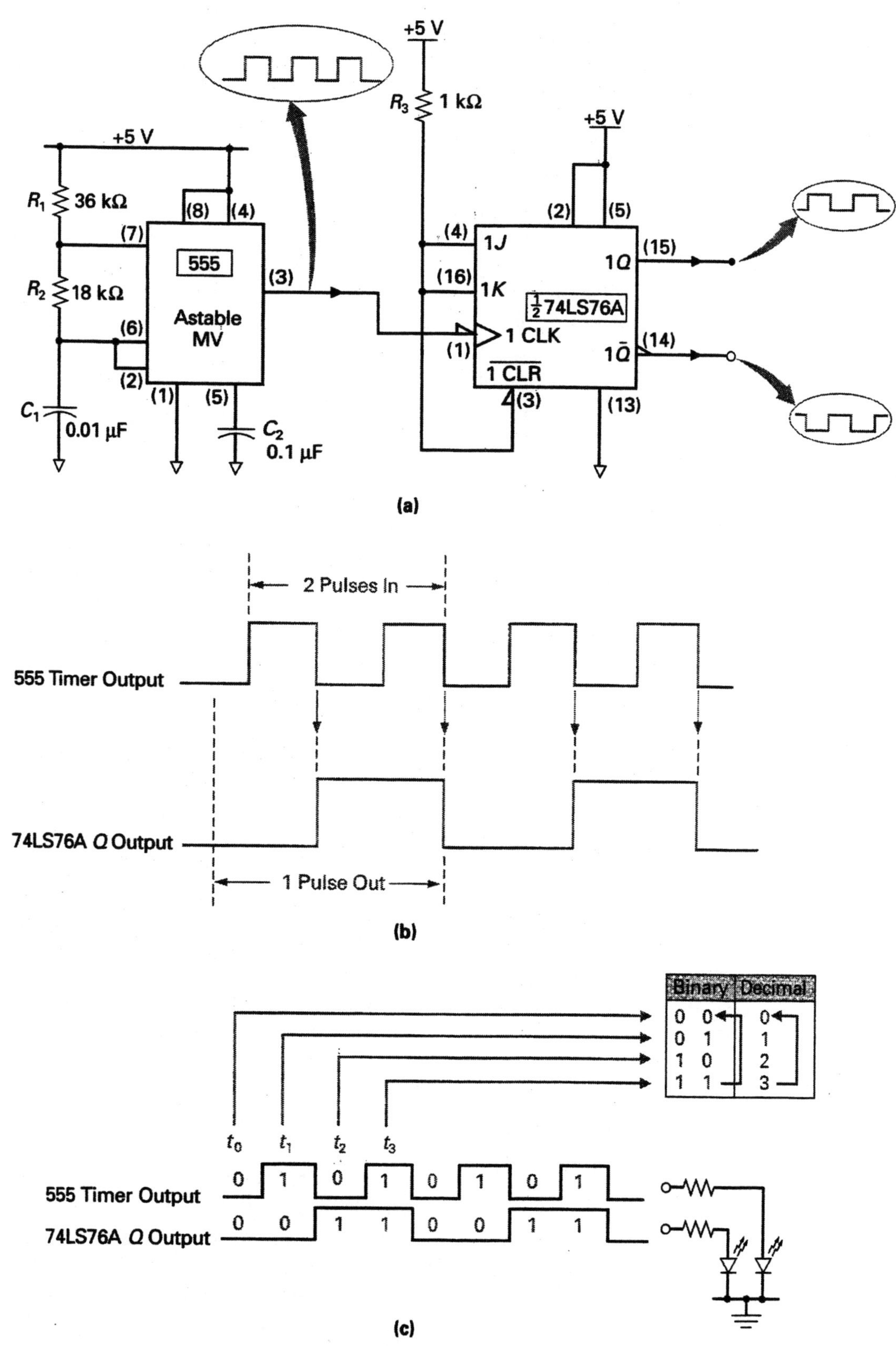

FIGURE 10-3 Application—The Edge-Triggered *J-K* Flip-Flop Acting as a Frequency Divider and Binary Counter.

from the 555 timer will therefore appear as 1 kHz at the *J-K* flip-flop's *Q* output, since two pulses in to the 74LS76A will produce one pulse out.

Now that we have seen how the *J-K* flip-flop can be used as a frequency divider, let us see how it can function as a binary counter. Figure 10-3(c) repeats the output waveforms from the 555 timer and the 74LS76A *Q* output. If these two outputs were connected to two LEDs, the LEDs would count up in binary as shown in the table in Figure 10-3(c). To explain this in more detail, at time t_0, both the 555 and 74LS76A *Q* output are LOW, so our LED binary display shows a count of 00 (binary 0). At time t_1, the binary display will be driven by a HIGH from the 555 timer output, and a LOW from the 74LS76A *Q* output, and therefore will display 01 (binary 1). At time t_2, the display will show 10 (binary 2), and at time t_3 the display will show 11 (binary 3). Combined, the 555 timer and the divided-by-two output of the 74LS76A can be used to generate a 2-bit word which will count from binary 0 (00) to binary 3 (11), and then continuously repeat the count.

Figure 10-4(a) shows how two 74LS76A ICs can be connected to produce a 4-bit binary up-counter. This and other counters will be discussed in more detail in Chapter 12; however, for now let us understand how this circuit operates. Since the *J* and *K* inputs of all the flip-flops in this circuit are tied HIGH, the flip-flops will permanently be in the toggle input condition. The timing diagram for this circuit is shown in Figure 10-4(b). In this circuit, the clock input triggers flip-flop 1 (FF1), the output of FF1 (1Q) triggers FF2, the output of FF2 triggers FF3, and the output of FF3 triggers FF4. Since all of the flip-flops are negative edge-triggered, each negative edge of the clock input will toggle FF1, each negative edge of the FF1 output will toggle FF2, each negative edge of the FF2 output will toggle FF3, and each negative edge of the FF3 output will toggle FF4. Four active-HIGH light-emitting diodes are connected to display the logic level at each of the outputs. As far as the binary count is concerned, the display is in its reverse order, since LED1 is the display's LSB (2^0) and LED4 is the display's MSB (2^3). As a result, we will see a 1248 binary display instead of the customary 8421 binary display.

Referring to the decimal count line in Figure 10-4(b), you can see that at first all outputs are LOW and therefore all LEDs are OFF (a binary count of 0000 or decimal 0). The count then advances by one for every clock pulse input, until it reaches its maximum 4-bit count with all LEDs ON (a binary count of 1111 or decimal 15).

10-1-2 Pulse-Triggered J-K Flip-Flops

The basic circuit for a pulse-triggered (master-slave) *J-K* flip-flop is shown in Figure 10-5(a). The circuit is almost identical to the master-slave *S-R* flip-flop except for the distinctive *J-K* cross-coupled feedback connections from the *Q* and $\overline{Q}$ slave outputs back to the master input gates. Figure 10-5(b) summarizes the operation of the pulse-triggered *J-K* flip-flop. The operation of the PRESET and CLEAR inputs are not included in this table, since they will operate in exactly the same way as any other flip-flop. The logic symbol for this flip-flop is shown in Figure 10-5(c). The postponed output symbol (˥) is used to indicate

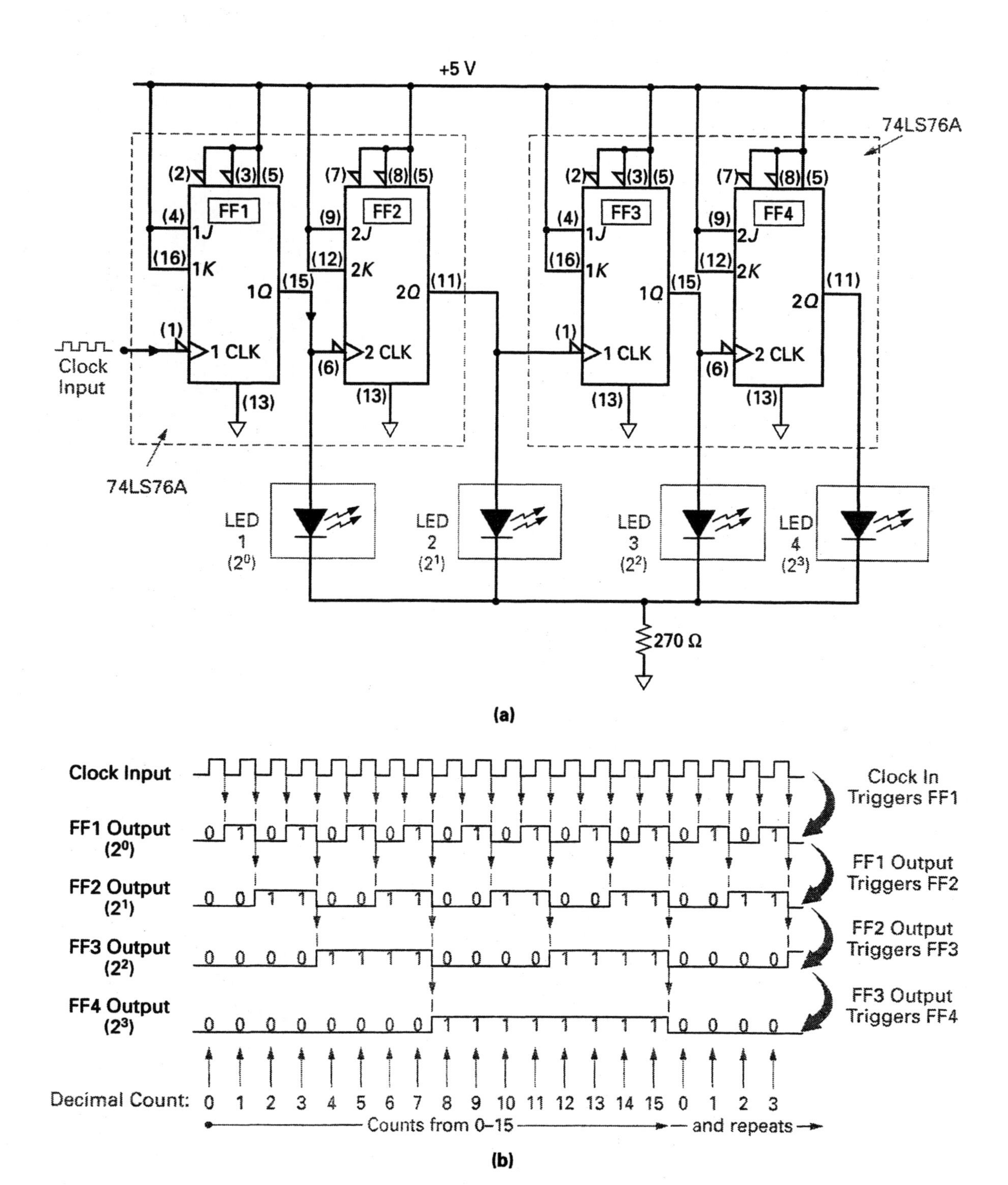

FIGURE 10-4 Application—A 4-Bit Binary Counter.

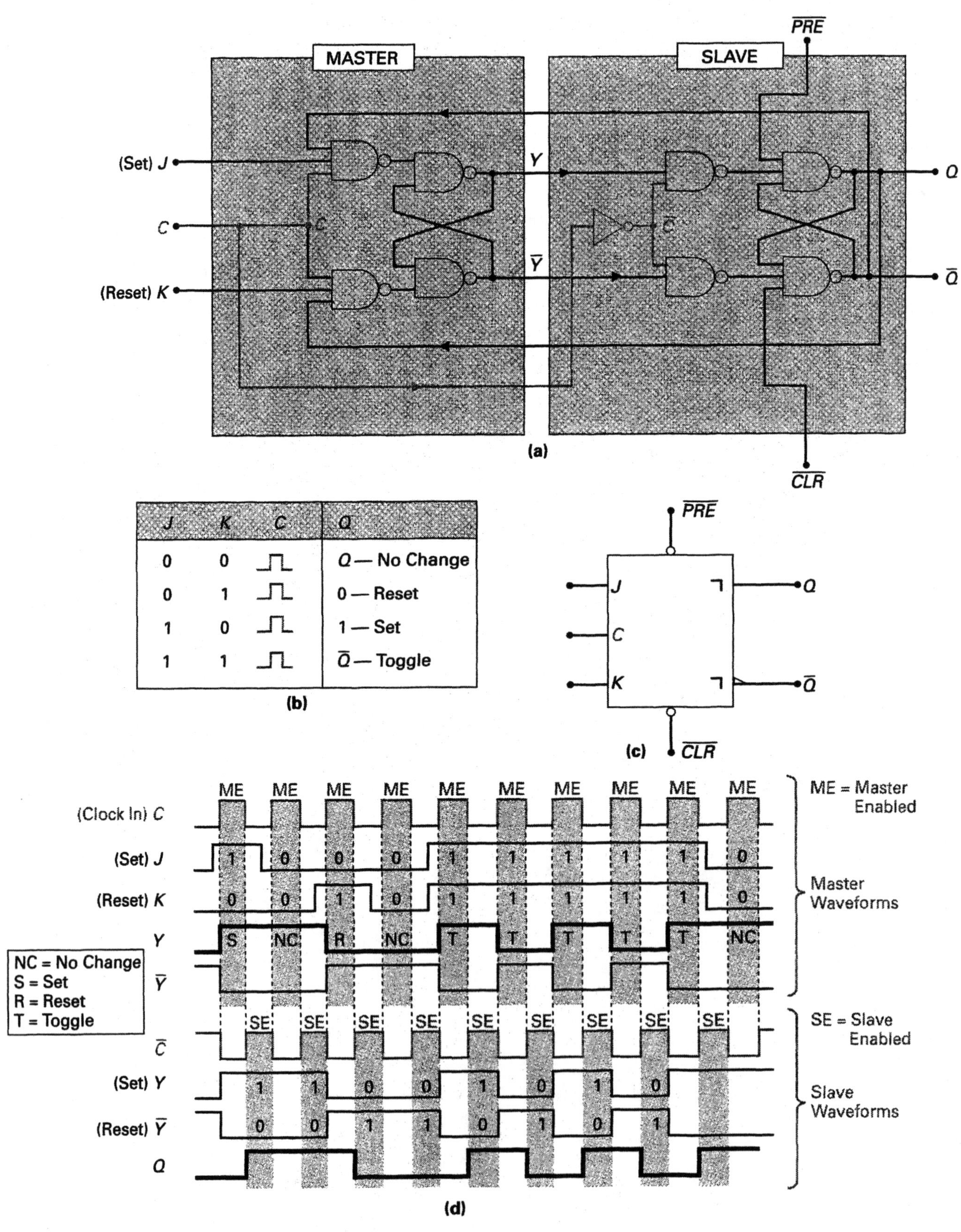

FIGURE 10-5 **Pulse-Triggered (Master-Slave) *J-K* Flip-Flop.**

that the Q and $\overline{Q}$ outputs will only change when the clock input (C) falls from a HIGH to a LOW level (which is when the slave is enabled).

The timing diagram in Figure 10-5(d) serves as a visual summary of the master-slave J-K flip-flop's operation. When the clock input (C) goes HIGH, the master section is enabled (ME). If $J = 1$ and $K = 0$, the Y output will be SET to 1 (S), whereas if $J = 0$ and $K = 1$, the Y output will be RESET to 0 (R). On the other hand, if $J = 0$ and $K = 0$ there will be NO CHANGE at the output (NC), whereas if $J = 1$ and $K = 1$ the output will toggle (T) based on the feedback control inputs from the Q and $\overline{Q}$ outputs. When the clock input (C) goes LOW and disables the master section, the inverted clock input ($\overline{C}$) goes HIGH and enables the slave section. Looking at the thick Y and Q output waveforms in Figure 10-5(d), you can see that the Q output is a replica of the Y output except that the Q lags Y by half a clock pulse. This is because the Q output will only follow the Y output when the slave section is enabled, and this occurs when the clock input (C) falls from a HIGH to a LOW (⌉).

Figure 10-6 shows the data sheet for the 74107 IC, which contains two pulse-triggered (master-slave) J-K flip-flops. An active-LOW CLEAR input is available for each of the flip-flops so that the Q outputs can be reset by a slow-

IC TYPE: SN74107 – Dual Pulse-Triggered *J-K* Flip-Flops with Clear

LOGIC SYMBOL

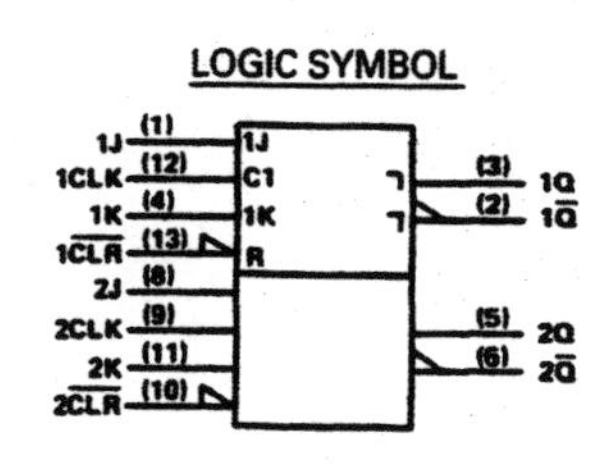

FUNCTION TABLE

INPUTS				OUTPUTS	
$\overline{CLR}$	CLK	J	K	Q	$\overline{Q}$
L	X	X	X	L	H
H	⎍	L	L	Q_0	$\overline{Q}_0$
H	⎍	H	L	H	L
H	⎍	L	H	L	H
H	⎍	H	H	TOGGLE	

TOP VIEW

1J 1 — 14 V_{CC}
1Q̄ 2 — 13 1CLR̄
1Q 3 — 12 1CLK
1K 4 — 11 2K
2Q 5 — 10 2CLR̄
2Q̄ 6 — 9 2CLK
GND 7 — 8 2J

INTERNAL LOGIC SCHEMATIC

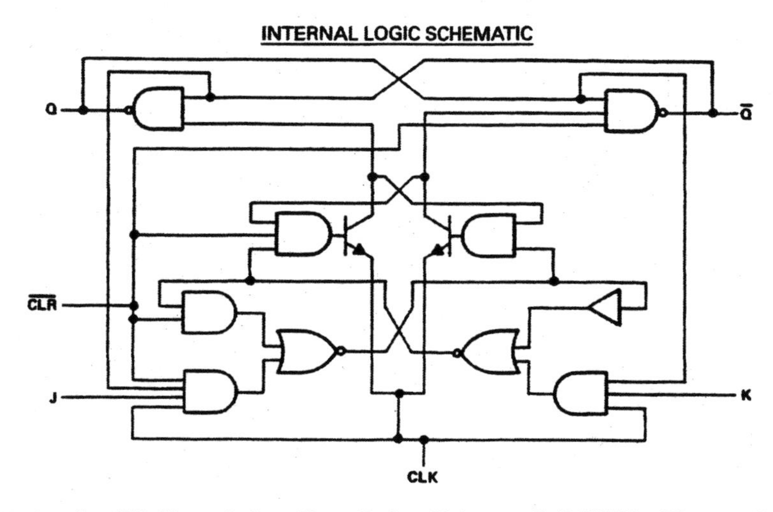

FIGURE 10-6 An IC Containing Two Pulse-Triggered *J-K* Flip-Flops. (Courtesy of Texas Instruments)

start circuit, for example, when power is first applied. Let us see how this flip-flop could be used as a frequency divider.

Application: Frequency Divider

Figure 10-7(a) shows how the two pulse-triggered *J-K* flip-flops within a 74107 could be connected as a divide-by-four circuit. Referring to the waveforms in Figure 10-7(b), you can see that two complete cycles of the clock input are needed to produce one complete cycle at the output of flip-flop 1 (FF_1 therefore divides by two). Comparing the clock input to the output of FF_2, you can see that four complete cycles of the clock input are needed to produce one complete cycle at the output of FF_2 (FF_1 divides by two and then FF_2 divides by two, resulting in a final divide-by-four output.

Like the 74107, the 74111, shown in Figure 10-8, contains two pulse-triggered (master-slave) *J-K* flip-flops. The 74111, however, has an additional data lockout feature that enables it to operate at extremely fast clock speeds (typically 25 MHz). Referring to the logic symbol, you can see that the dynamic indicator is present on the clock input and the postponed output indicator is

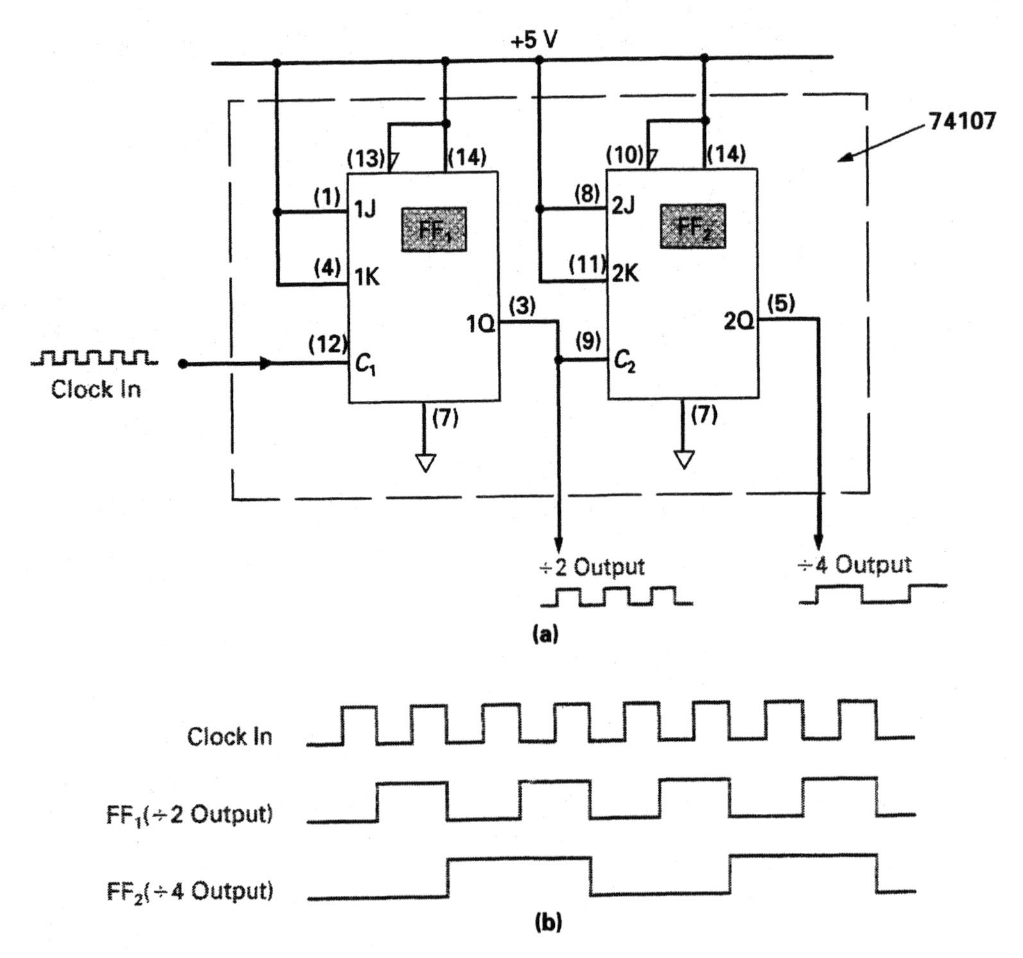

FIGURE 10-7 Divide-by-Four Circuit.

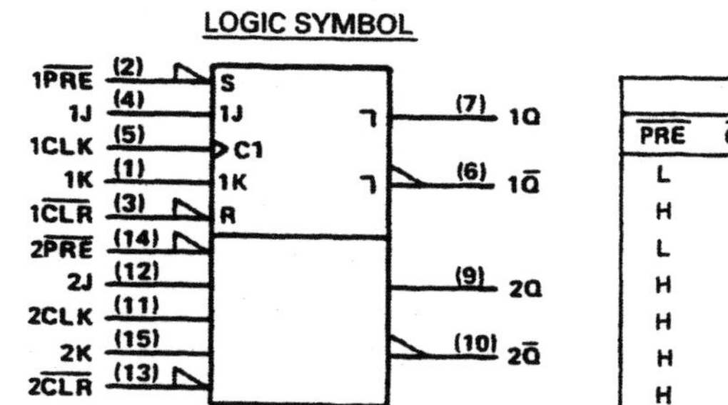

FUNCTION TABLE

INPUTS					OUTPUTS	
$\overline{PRE}$	$\overline{CLR}$	CLK	J	K	Q	$\overline{Q}$
L	H	X	X	X	H	L
H	L	X	X	X	L	H
L	L	X	X	X	H‡	H‡
H	H	⎍	L	L	Q_0	$\overline{Q}_0$
H	H	⎍	H	L	H	L
H	H	⎍	L	H	L	H
H	H	⎍	H	H	TOGGLE	

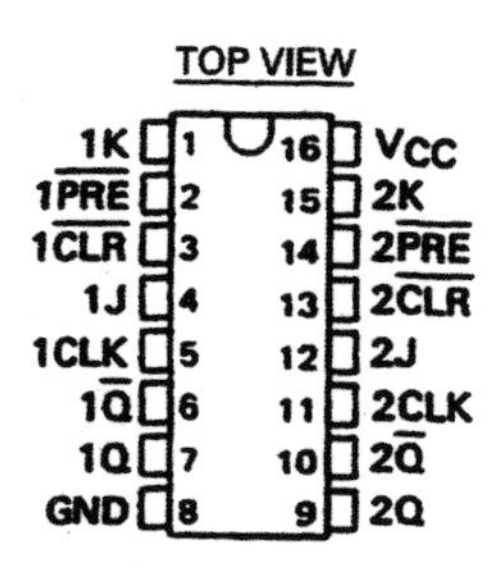

FIGURE 10-8 An IC Containing Two Pulse-Triggered *J-K* Flip-Flops. (Courtesy of Texas Instruments)

present at the Q and $\overline{Q}$ outputs. These symbols are used to indicate that the master section of each flip-flop is edge-triggered, while the slave section of each flip-flop is level-triggered. Since no bubble or right triangle exists outside of the block on the C input, the master section is positive edge-triggered. Therefore, the J and K inputs are enabled during the rising edge of the clock pulse (for about 30 ns). After this, the J and K inputs are disabled and their data is locked-out, making design easier since these inputs do not have to be held constant while the clock input is HIGH. When the clock falls from HIGH to LOW level, the slave section is enabled and the Q and $\overline{Q}$ outputs are updated.

SELF-TEST REVIEW QUESTIONS FOR SECTION 10-1

1. Which of the following input conditions will set HIGH the Q output of a negative edge-triggered *J-K* flip-flop?
   a. $J = 1, K = 1, C = \uparrow, \overline{PRE} = 1, \overline{CLR} = 1.$
   b. $J = 1, K = 0, C = \downarrow, \overline{PRE} = 1, \overline{CLR} = 0.$
   c. $J = 1, K = 1, C = \downarrow, \overline{PRE} = 1, \overline{CLR} = 1.$
   d. $J = 1, K = 0, C = \downarrow, \overline{PRE} = 1, \overline{CLR} = 1.$
2. When both the inputs of a *J-K* flip-flop are __________, the Q output will __________, or switch to the opposite state.
3. If the clock input to a *J-K* flip-flop was a 126 kHz square wave and $J = 1$, $K = 1$, $\overline{PRE} = 1$, $\overline{CLR} = 1$, what would the Q output be?
   a. A 126 kHz square wave
   b. A 63 kHz square wave
   c. A 31.5 kHz square wave
   d. A 252 kHz square wave
4. The __________ flip-flop can perform all the functions of the __________ flip-flop and __________ flip-flop.
   a. *S-R*, *J-K*, D-type
   b. D-type, *J-K*, *S-R*
   c. *J-K*, D-type, *S-R*
   d. *S-R*, D-type, *J-K*

10-2 DIGITAL TIMER AND CONTROL CIRCUITS

Timing is everything in digital logic circuits. To control the timing of digital circuits, a clock signal is distributed throughout the digital system. This square-wave clock signal is generated by a clock oscillator, and its sharp positive (leading) and negative (trailing) edges are used to control the sequence of operations in a digital circuit.

The *S-R*, D-type, and *J-K* flip-flops are all examples of bistable multivibrators, since they have two (bi) stable states (SET and RESET). In this section we will discuss the astable, or unstable, multivibrator which has no stable states, and is commonly used as a clock oscillator. The third type of multivibrator is the monostable multivibrator which has only one (mono) stable state, and when triggered will generate a rectangular pulse of a fixed duration.

10-2-1 *The Astable Multivibrator Circuit*

Astable Multivibrator
A free-running oscillator circuit that can be used to generate either a square or rectangular output waveform.

The **astable multivibrator** circuit is used to produce an alternating two-state square or rectangular output waveform. This circuit is often called a *free-running multivibrator* since the circuit requires no input signal to start its operation. It will simply begin oscillating the moment the dc supply voltage is applied.

Bipolar Transistor Astable Multivibrator Circuits

The circuit seen in Figure 10-9(a) consists of two *cross-coupled bipolar transistors,* which means that there is a cross connection between the base and collector of the two transistors Q_1 and Q_2. This circuit also contains two *RC* timing networks—R_1/C_1 and R_2/C_2.

Let us now examine the operation of this astable multivibrator circuit. When no dc supply voltage is present ($V_{CC} = 0$ V), both transistors are OFF and therefore there is no output. When a V_{CC} supply voltage is applied to the circuit (for example, +5 V), both transistors will receive a positive-bias base voltage via R_1 and R_2. Although both Q_1 and Q_2 are matched bipolar transistors—which means that their manufacturer ratings are identical—no two transistors are ever the same. This difference, and the differences in R_1 and R_2 due to resistor tolerances, means that one transistor will turn ON faster than the other. Let us assume that Q_1 turns on first, as seen in Figure 10-9(b). As Q_1 conducts, its collector voltage decreases since it is like a closed switch between collector and emitter, and this decrease in collector voltage is coupled through C_1 to the base of Q_2, causing it to conduct less and eventually turn OFF. With Q_2 OFF, its collector voltage will be high (+5 V), since Q_2 is equivalent to an open switch between collector and emitter. This increase in collector voltage is coupled through C_2 to the base of Q_1, causing it to conduct more and eventually turn fully ON. The cross-coupling between these two bipolar transistors will reinforce this condition with the LOW Q_1 collector voltage keeping Q_2 OFF and the HIGH Q_2 collector voltage keeping Q_1 ON. With Q_1 equivalent to a closed switch, a current path now exists for C_1 to charge, as seen in Figure 10-9(b). As soon as the charge on C_1 reaches about 0.7 V, Q_2 will conduct because its base-emitter junction will be forward-biased. This condition is shown in

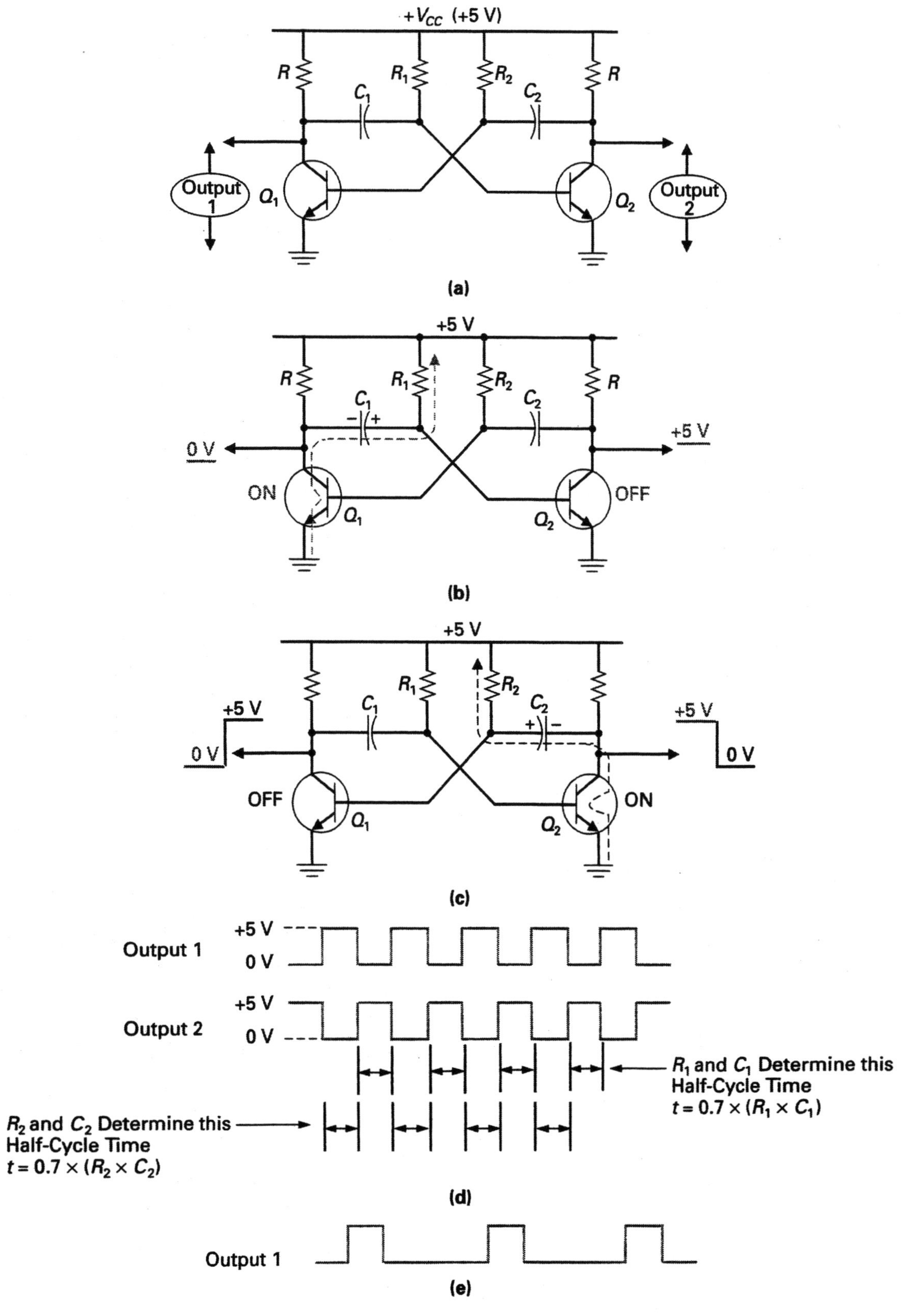

FIGURE 10-9 The Bipolar Transistor Astable Multivibrator. (a) Basic Circuit. (b) Q_1 ON Condition. (c) Q_2 ON Condition. (d) Square-Wave Mode. (e) Rectangular-Wave Mode.

Figure 10-9(c). When Q_2 conducts, its collector voltage will drop, cutting OFF Q_1 and creating a charge path for C_2. As soon as the charge on C_2 reaches 0.7 V, Q_1 will conduct again, and the cycle will repeat.

The output waveforms switch between the supply voltage ($+V_{CC}$) when a transistor is cut off and zero volts when a transistor is saturated (ON). The result is two square-wave outputs that are out-of-phase with one another, as seen in the waveforms in Figure 10-9(d). Referring to the output waveforms in Figure 10-9(d), you can see that the time constants of R_1 and C_1, and R_2 and C_2 determine the complete cycle time. If the R_1/C_1 time constant is equal to the R_2/C_2 time constant, both halves of the cycle will be equal (50% duty cycle) and the result will be a square wave. Referring to Figure 10-9(d), you can see that the R_1/C_1 time constant will determine the time of one half-cycle, while the R_2/C_2 time constant will determine the time of the other half-cycle. The formula for calculating the time of one half-cycle is equal to

$$t = 0.7 \times (R_1 \times C_1) \text{ or } t = 0.7 \times (R_2 \times C_2)$$

The frequency (f) of this square wave can be calculated by taking the reciprocal of both half-cycles, which will be

$$f = \frac{1}{1.4 \times RC}$$

■ **EXAMPLE:**

Calculate the positive and negative cycle time and the circuit frequency of the astable multivibrator circuit in Figure 10-9, if

$$R_1 \text{ and } R_2 = 100\ \text{k}\Omega \qquad C_1 \text{ and } C_2 = 1\ \mu\text{F}$$

■ ***Solution:***

Since the time constants of R_1/C_1 and R_2/C_2 are the same, each half-cycle time will be the same and equal to

$$\begin{aligned} t &= 0.7 \times (R \times C) \\ &= 0.7 \times (100\ \text{k}\Omega \times 1\mu\text{F}) = 0.07 \text{ s or } 70 \text{ ms} \end{aligned}$$

The frequency of the astable circuit will be equal to the reciprocal of the complete cycle, or the reciprocal of twice the half-alternation time.

$$f = \frac{1}{1.4 \times RC} = \frac{1}{1.4 \times (100\ \text{k}\Omega \times 1\mu\text{F})} = 7.14 \text{ Hz}$$

$$\text{or } f = \frac{1}{2 \times t} = \frac{1}{2 \times 70 \text{ ms}} = \frac{1}{0.14} = 7.14 \text{ Hz}$$

If the time constants of the two RC timing networks in the astable circuit are different, however, the result will be a rectangular or pulse waveform, as seen in Figure 10-9(e). In this instance, the same formula can be used to calculate the time for each alternation. The frequency will be equal to the reciprocal of the time for both alternations.

Logic Gate Astable Multivibrator Circuits

The astable multivibrator can also be constructed using logic gates, as seen in Figure 10-10. In Figure 10-10(a), a schmitt-trigger INVERTER is connected to operate as a clock oscillator. When power is first applied to this circuit, the capacitor will have no charge and this LOW input is inverted by the NOT gate giving a HIGH output (seen as black in illustration). The capacitor (*C*) will begin to charge via the resistor (*R*), and the increasing positive charge across the capacitor will be felt at the input of the INVERTER. After a time (which is dependent on the values of *R* and *C*), the capacitor charge will be large enough to apply a valid HIGH to the INVERTER input. This HIGH input will cause the output of the INVERTER to go LOW, so the capacitor will begin to discharge (seen as blue in illustration). When the capacitor's charge falls to a valid LOW logic level, the INVERTER will generate a HIGH output, and the cycle will repeat.

Figure 10-10(b) shows another basic astable multivibrator circuit using two INVERTERS. The operation of this circuit is similar to the one in Figure 10-10(a). A continual capacitor charge and discharge causes the astable or free-running multivibrator to switch back and forth between its two unstable states,

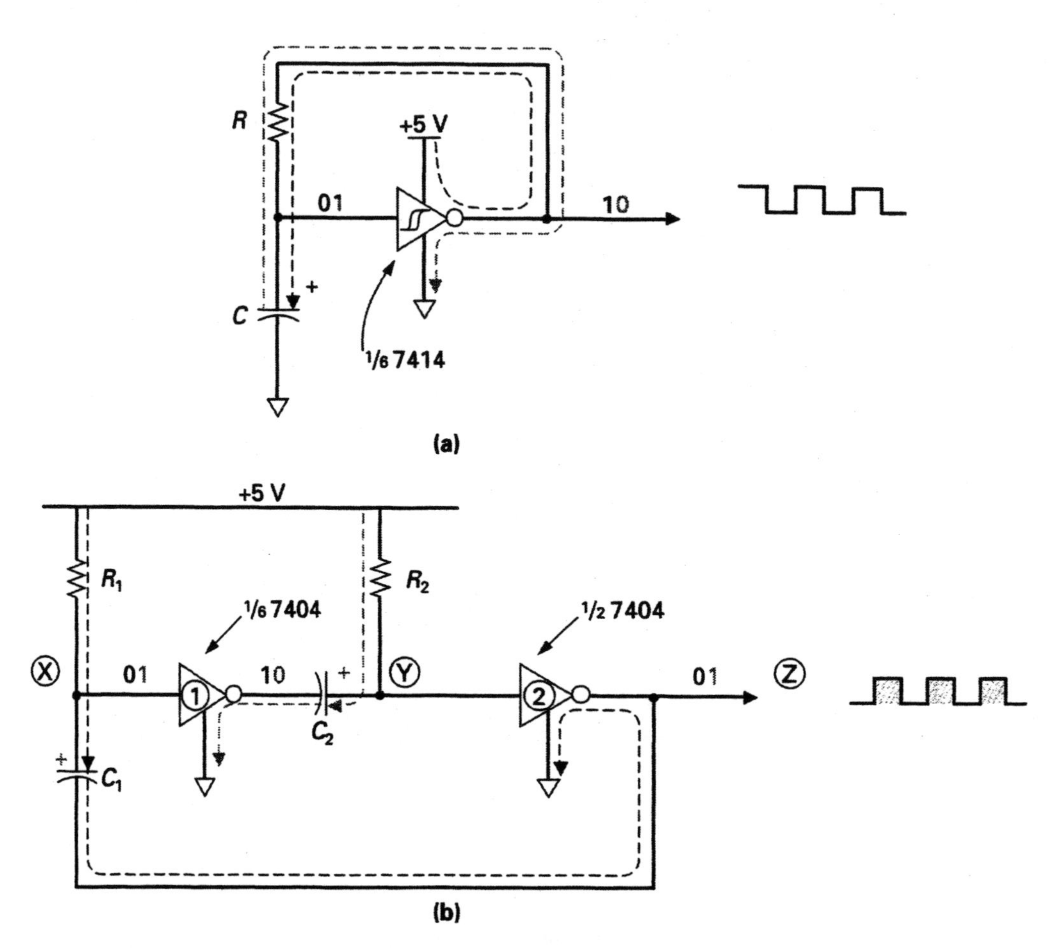

FIGURE 10-10 **Logic Gate Astable Multivibrators.**

producing a repeating rectangular wave at the output (two conditions are shown in the illustration as black and blue).

10-2-2 The Monostable Multivibrator Circuit

Monostable Multivibrator Also called a one-shot multivibrator, it is a circuit that will generate an output pulse for each input trigger.

The astable multivibrator is often referred to as an unstable multivibrator, because it is continually alternating or switching back and forth, and therefore it has no stable state. The **monostable multivibrator,** on the other hand, has as its name implies one (mono) stable state. The circuit will remain in this stable state indefinitely, until a trigger is applied and forces the monostable multivibrator into its unstable state. It will then remain in its unstable state for a small period of time, and then switch back to its stable state and await another trigger. The monostable multivibrator is often compared to a gun and called a *one-shot multivibrator,* since it will produce one output pulse or shot for each input trigger.

Bipolar Transistor Monostable Multivibrator Circuits

Referring to the bipolar transistor monostable multivibrator circuit in Figure 10-11(a), you can see that the monostable multivibrator is similar to the astable except for the trigger input circuit, and for the fact that it has only one *RC* timing network. To begin with, let us consider the stable state of the monostable. Components R_2, D_1, and R_5 form a voltage divider, the values of which are chosen to produce a large positive Q_2 base voltage. This large positive-base bias voltage will cause Q_2 to saturate (turn heavily ON), which in turn will produce a LOW Q_2 collector voltage, which will be coupled via R_4 to the base of Q_1, cutting it OFF. The circuit remains in this stable state (Q_2 ON, Q_1 OFF) until a *trigger input* is received.

Referring to the timing waveforms in Figure 10-11(b), you can see how the circuit reacts when a positive input trigger is applied. The pulse is first applied to a differentiator circuit (C_2 and R_5) that converts the pulse into a positive and a negative spike. These spikes are then applied to the positive clipper diode D_1, which only allows the negative spike to pass to the base of Q_2. This negative spike will reverse-bias Q_2's base-emitter junction, turning Q_2 OFF, and causing its collector voltage to rise to $+V_{CC}$, as seen in the waveforms. This increased Q_2 collector voltage will be coupled to the base of Q_1, turning it ON. The monostable multivibrator is now in its unstable state, which is indicated in blue in Figure 10-11(a). In this condition, C_1 will charge as shown by the dashed current line; however, as soon as the voltage across C_1 reaches 0.7 V (which is dependent on the R_2/C_1 time constant), it will force Q_2 to conduct, which in turn will cause Q_1 to cut OFF and the monostable to return to its stable state. The output pulse width or pulse time (t), seen in Figure 10-11(b), can be calculated with the same formula used for the astable multivibrator:

$$t = 0.7 \times (R_2 \times C_1)$$

The one-shot multivibrator is sometimes used in *pulse-stretching* applications. For example, referring to the waveforms in Figure 10-11(b), imagine the input positive trigger pulse were 1 μs in width and the *RC* time constant of R_2 and C_1 were such that the output pulse width (t) were 500 μs. In this example,

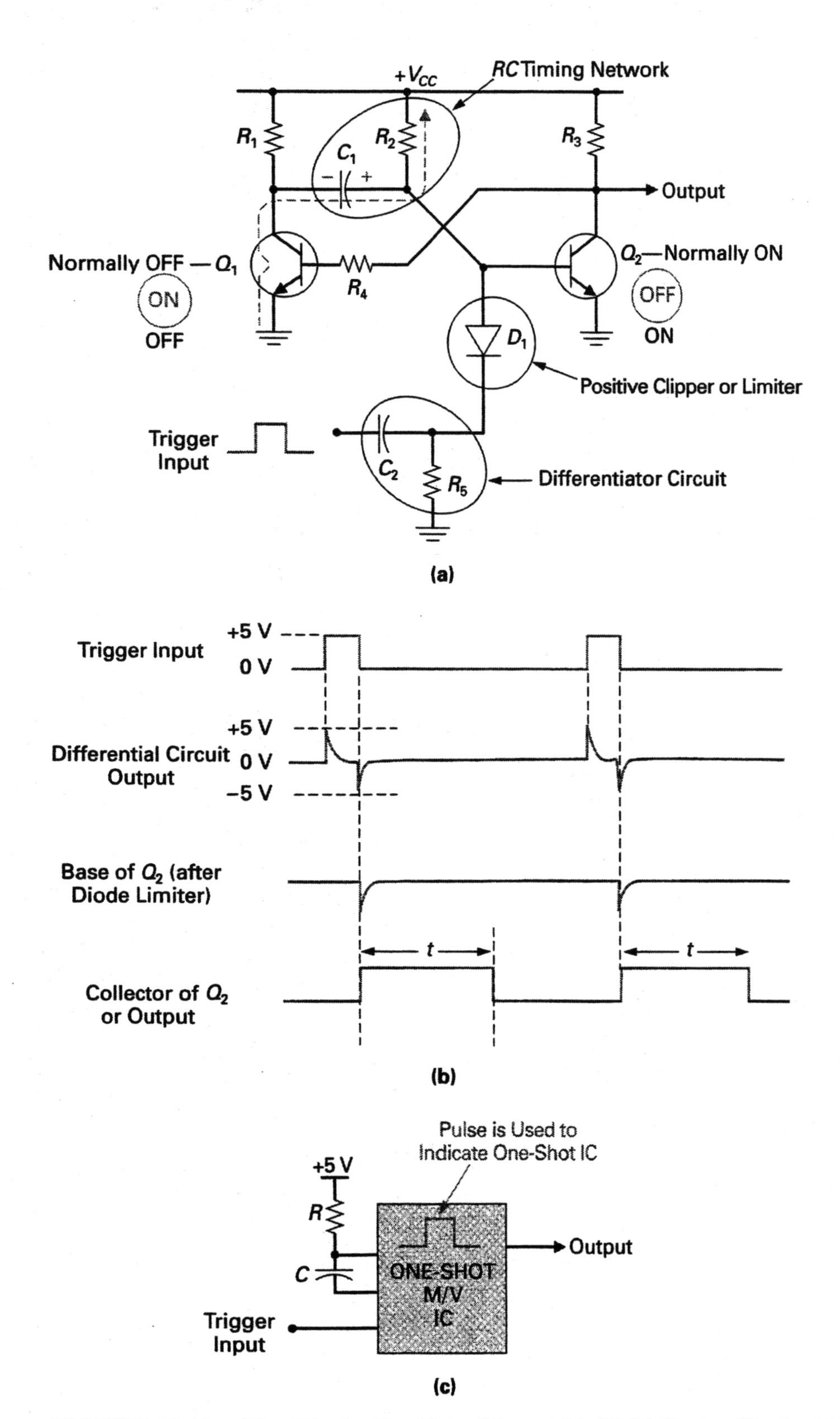

FIGURE 10-11 The Bipolar Transistor Monostable Multivibrator Circuit. (a) Bipolar Transistor Circuit. (b) Input/Output Timing Waveforms. (c) Symbol.

the input pulse would have been effectively stretched from 1 μs to 500 μs. The monostable multivibrator, or one-shot timer circuit, is also used to introduce a *time delay.* Referring again to the waveforms in Figure 10-11(b), imagine a differentiator circuit connected to the output of the monostable circuit. If the output pulse width were again set to 500 μs, there would be a 500 μs delay between the differentiated negative edge of the input pulse and the differentiated negative edge of the output pulse. Figure 10(c) shows the schematic symbol for a monostable multivibrator.

Logic Gate Monostable Multivibrator Circuits

Like the astable multivibrator, the monostable multivibrator can be constructed using logic gates, as seen in Figure 10-12(a). When the trigger input is LOW and the Q output is LOW, the output from the NOR gate is HIGH, and therefore the output from the INVERTER is LOW, keeping the circuit in its stable state. When the trigger input pulses HIGH, it causes the output of the

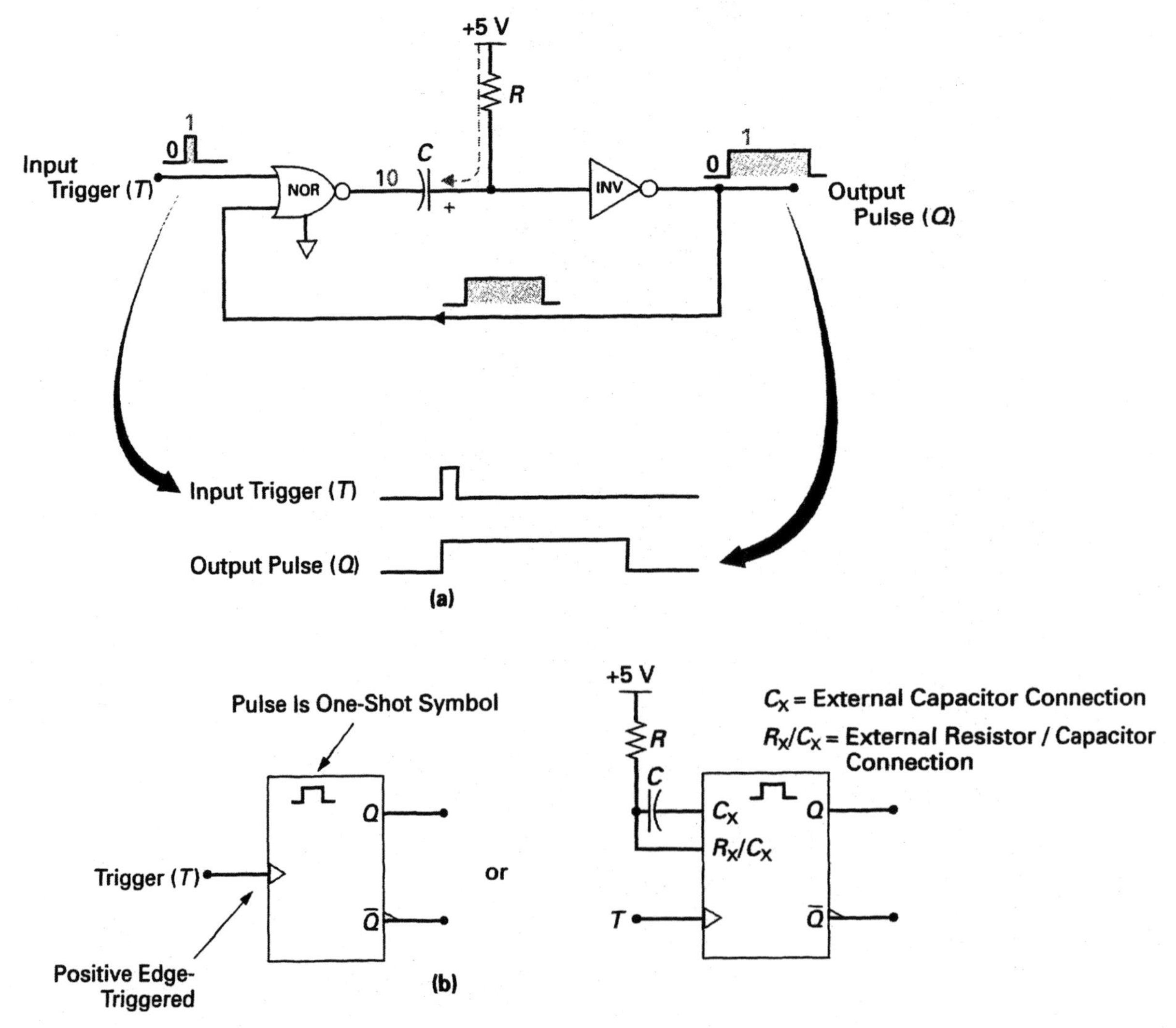

FIGURE 10-12 **Logic Gate Monostable (One-Shot) Multivibrator.**

NOR gate to go LOW. This HIGH-to-LOW transition is coupled through the capacitor to the INVERTER, which produces a HIGH Q output, as can be seen in the timing diagram below the circuit. This HIGH Q output is fed back to the NOR gate's other input and keeps the NOR gate output LOW even after the trigger pulse has ended. A LOW NOR gate output will produce a potential difference across the resistor-capacitor network, and therefore the capacitor will begin to charge. After a time, dependent on the values of R and C, the charge on C is large enough for the INVERTER to recognize it as a valid HIGH input, and therefore it generates a LOW Q output ending the pulse.

The typical block logic symbols for a monostable multivibrator, or one-shot, are shown in Figure 10-12(b). Some logic symbols group the entire circuit within a block, while others block only the logic gates and show the time-determining capacitor and resistor separately.

Today, one-shot circuits are very rarely constructed with discrete transistors or logic gates since integrated circuits are available. These one-shot ICs are classified as being either *nonretriggerable one-shots* or *retriggerable one shots.* The difference between these two types is best described by comparing the waveforms in Figure 10-13.

Figure 10-13(a) shows the action of the nonretriggerable one-shot. When triggered, an output pulse of a certain width (P_W) is generated, as shown when triggers 1 and 2 are applied. If another trigger is applied before the output pulse has ended, it will be ignored. For example, trigger 3 starts an output pulse, but trigger 4 is ignored since it occurs before the output pulse has *timed out.* As another example, trigger 5 *fires* the one-shot, while triggers 6 and 7 are ignored because the one-shot is in its unstable state. Once triggered, therefore, this type

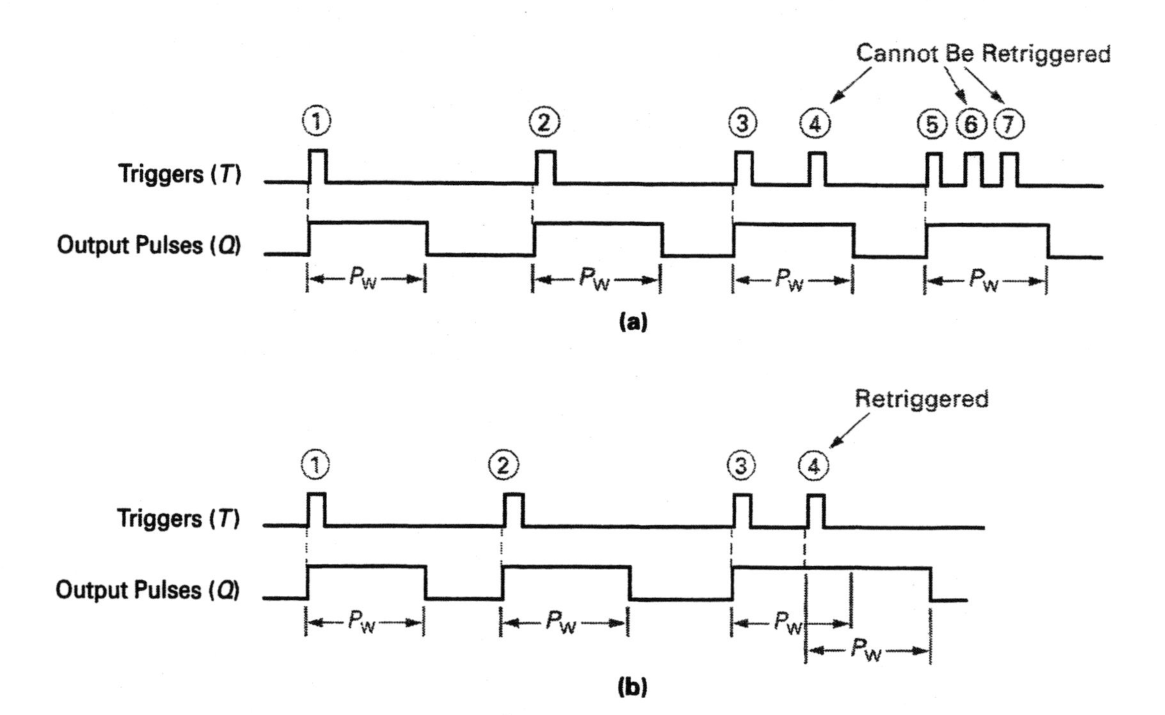

FIGURE 10-13 One-Shot IC Types. (a) Nonretriggerable Action. (b) Retriggerable Action.

of one-shot cannot be retriggered until its output pulse has ended (the one-shot is nonretriggerable).

Figure 10-13(b) shows the action of the retriggerable one-shot. When triggered, an output pulse is produced of a certain pulse width (P_W), as shown when triggers 1 and 2 are applied. The difference with this type of one-shot can be seen with triggers 3 and 4. Trigger 3 fires the one-shot, so it begins to produce an output pulse; however, before this pulse has timed out, trigger 4 retriggers the one-shot. The output pulse will now be extended for a time equal to a pulse width starting at the time trigger 4 occurred. Once triggered, therefore, this type of one-shot can be retriggered (the one-shot is retriggerable). Let us now examine in more detail a nonretriggerable one-shot IC (74121) and a retriggerable one-shot (74122).

Figure 10-14 shows the data sheet details for the 74121 nonretriggerable monostable multivibrator IC. Inputs A_1, A_2, and B are used to trigger the 74121 one-shot. Referring to the internal logic circuit, you can see that a positive edge or LOW to HIGH transition is needed to trigger this one-shot. Due to the bubbled input OR gate and schmitt-trigger AND gate, this positive edge can be

IC TYPE: SN74121 – Nonretriggerable Monostable Multivibrator

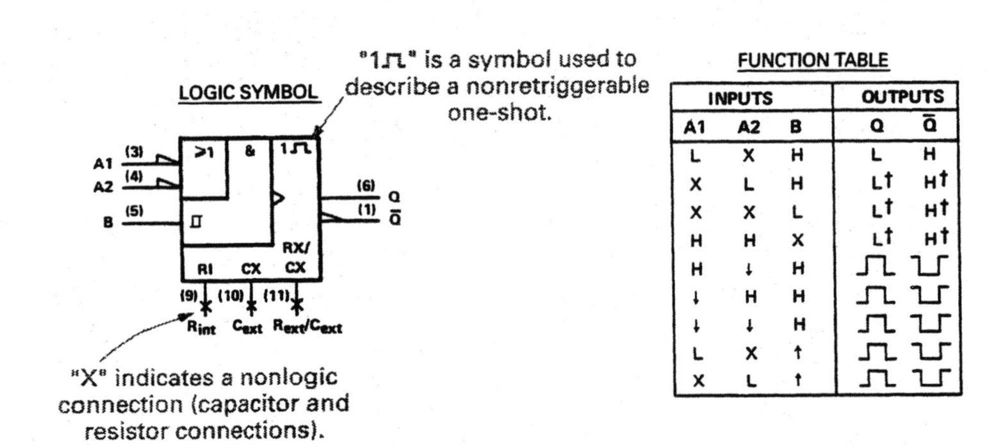

INPUTS			OUTPUTS	
A1	A2	B	Q	Q̄
L	X	H	L	H
X	L	H	L†	H†
X	X	L	L†	H†
H	H	X	L†	H†
H	↓	H	⊓	⊔
↓	H	H	⊓	⊔
↓	↓	H	⊓	⊔
L	X	↑	⊓	⊔
X	L	↑	⊓	⊔

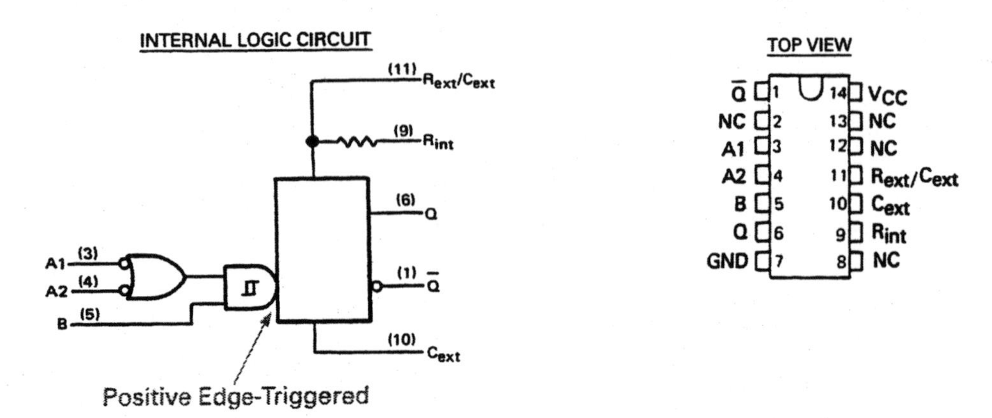

FIGURE 10-14 An IC Containing A Nonretriggerable Monostable Multivibrator. (Courtesy of Texas Instruments)

obtained in one of five ways, as listed in the last five lines of the function table. Complementary output pulses can be obtained from the Q and $\overline{Q}$ output. The pulse width of the output pulse is determined by the values of R and C, which are connected to the three nonlogic inputs labeled R_{INT} (R internal), C_{EXT} (C external), and R_{EXT}/C_{EXT} (R external and C external). Using these three inputs, there are three ways to control the output pulse width as shown in Figure 10-15.

1. To not connect any external components and leave all three inputs open, as shown in Figure 10-15(a). An output pulse width of approximately 30 ns will be generated for each input trigger. This is due to the internal (2 kΩ) timing resistor, which is connected between pins 9 and 11, and is shown in the 74121 internal logic circuit and top view within the data sheet in Figure 10-14.
2. To use R_{INT} (2 kΩ) in conjunction with an external capacitor, as shown in Figure 10-15(b). Connecting R_{INT} (pin 9) to $+V_{CC}$ and then connecting an external capacitor between C_{EXT} (pin 10) and R_{EXT}/C_{EXT} (pin 11), the output pulse width will be equal to

$$P_W = 0.7 \times R_{INT} \times C_{EXT}$$

3. To connect an external resistor between pin 11 and $+V_{CC}$ and an external capacitor between pins 10 and 11, as shown in Figure 10-15(c). The pulse width in this instance can be set anywhere between 40 ns and 28 s, and can be calculated with the formula

$$P_W = 0.7 \times R_{EXT} \times C_{EXT}$$

One-shots are used basically in one of three applications—pulse generation, timing and sequencing, or delay.

Figure 10-16(a) shows how a 74121 can be connected to function as a variable output pulse width generator. The normally open push-button (NOPB) switch is used to trigger the one-shot, and the external 100 kΩ variable resistor is used to adjust the output pulse width. The $\overline{Q}$ output is used to drive an output

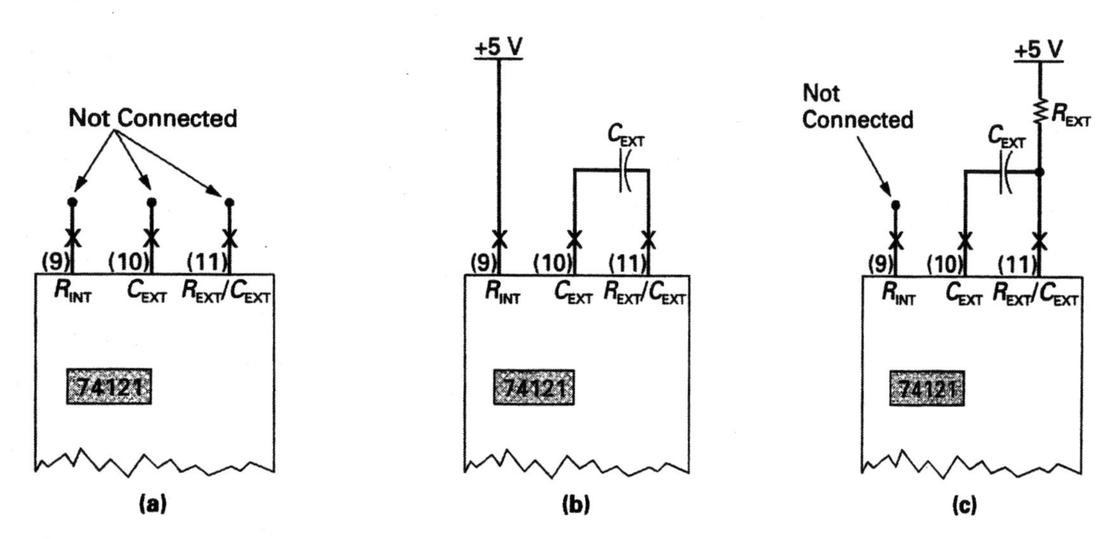

FIGURE 10-15 **Controlling the Output Pulse Width of a 74121 IC.**

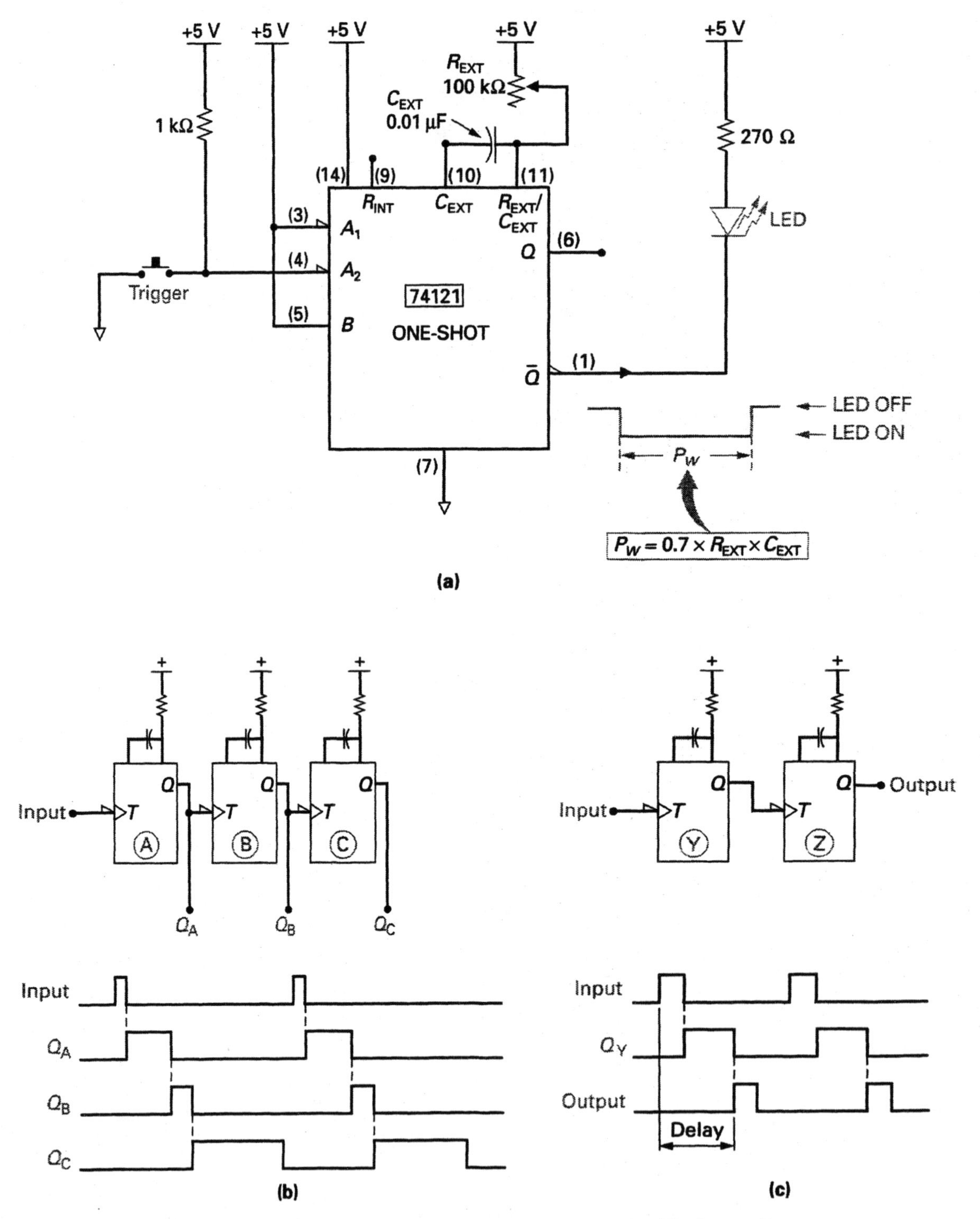

FIGURE 10-16 One-Shot Applications. (a) Pulse Generation. (b) Timing and Sequencing. (c) Time Delay.

LED instead of the Q output, since a common-anode display is being used for a higher current and therefore a higher light-level display.

Figure 10-16(b) shows how three one-shots can be used to generate a sequence of timing pulses. Since all inputs are negative edge-triggered, the trailing edge of each output is used to trigger the next one-shot, as can be seen in the associated waveforms in Figure 10-16(b).

Figure 10-16(c) shows how two one-shots can be used to introduce a delay. An input trigger fires one-shot Y, which introduces a certain time delay, and its trailing edge triggers one-shot Z which then produces an output pulse. The result is that the input pulse is delayed from appearing at the output for a time equal to the pulse width time of the Y one-shot. This time delay function is shown in the accompanying waveforms in Figure 10-16(c).

Figure 10-17 shows some of the data sheet details for the 74122 IC, which is a retriggerable one-shot IC with a clear control. The inputs A_1, A_2, B_1, and B_2 are gated inputs used to trigger the monostable into operation. The output pulse width (P_W) is controlled by the external resistor connected between V_{CC} and pin 11, and the external capacitor connected between pin 10 and pin 11. The pulse width formula for this IC is as follows:

$$P_W = 0.32 \times R_{EXT} \times C_{EXT} \left(1 + \frac{0.7}{R_{EXT}}\right)$$

10-2-3 *The 555 Timer Circuit*

One of the most frequently used low-cost integrated circuit timers is the **555 Timer.** Its IC package consists of eight pins, as seen in Figure 10-18(a), and derives its number identification from the distinctive voltage divider circuit seen in Figure 10-18(b), consisting of three 5 kΩ resistors. It is a highly versatile timer that can be made to function as an astable multivibrator, monostable multivibrator, frequency divider, or modulator, depending on the connection of external components.

555 Timer
A versatile IC timer that can, for example, be configured to function as a modulator, multivibrator, or frequency divider.

IC TYPE: SN74122 – Retriggerable Monostable Multivibrator with Clear

LOGIC SYMBOL

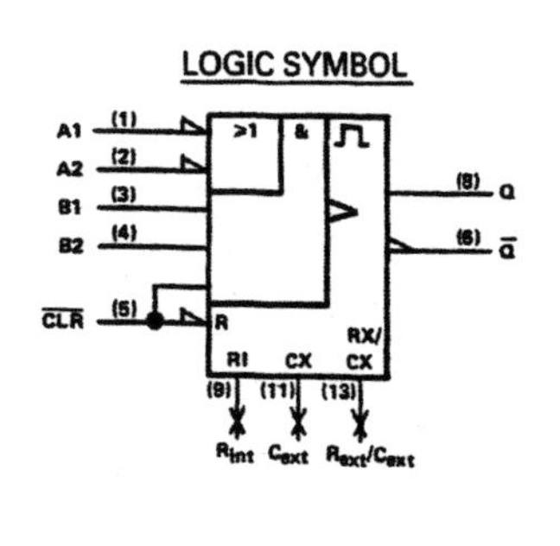

FUNCTION TABLE

INPUTS					OUTPUTS	
CLEAR	A1	A2	B1	B2	Q	Q̄
L	X	X	X	X	L	H
X	H	H	X	X	L†	H†
X	X	X	L	X	L†	H†
X	X	X	X	L	L†	H†
H	L	X	↑	H	⊓	⊔
H	L	X	H	↑	⊓	⊔
H	X	L	↑	H	⊓	⊔
H	X	L	H	↑	⊓	⊔
H	H	↓	H	H	⊓	⊔
H	↓	↓	H	H	⊓	⊔
H	↓	H	H	H	⊓	⊔
↑	L	X	H	H	⊓	⊔
↑	X	L	H	H	⊓	⊔

TOP VIEW

Pin		Pin	
A1	1	14	VCC
A2	2	13	Rext/Cext
B1	3	12	NC
B2	4	11	Cext
CLR	5	10	NC
Q̄	6	9	Rint
GND	7	8	Q

FIGURE 10-17 An IC Containing a Retriggerable Monostable Multivibrator. (Courtesy of Texas Instruments)

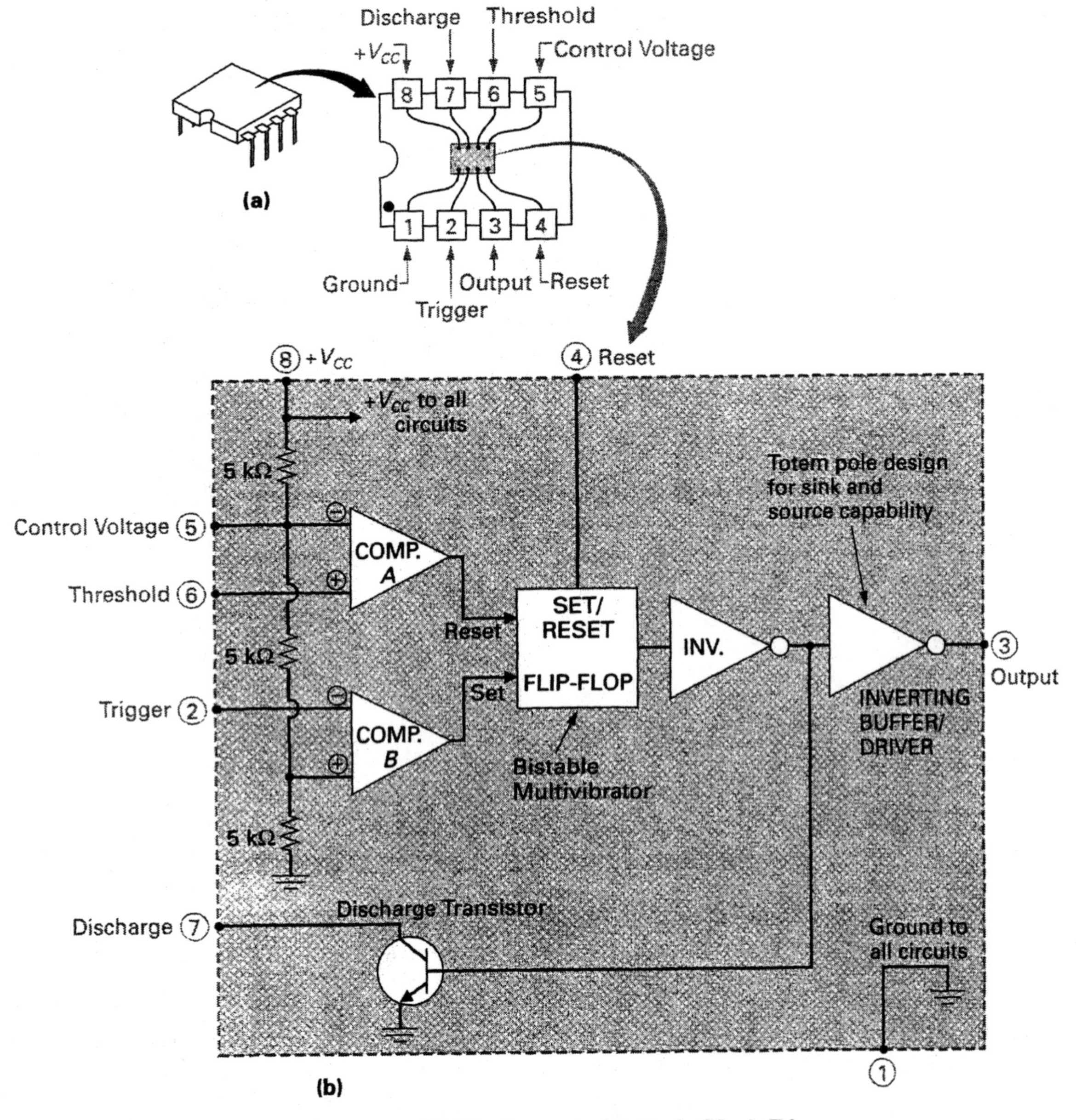

FIGURE 10-18 The 555 Timer. (a) IC Pin Layout. (b) Basic Block Diagram.

Nearly all the IC manufacturers produce a version of the 555 timer, which can be labeled in different ways, such as SN72 555, MC14 555, SE 555, and so on. Two 555 timers are also available in a 16-pin dual IC package that is labeled with the numbers "556."

Basic 555 Timer Circuit Action

Referring to the basic block diagram in Figure 10-18(b), let us examine the basic action of all the devices within a 555 timer circuit. The three 5 kΩ resistors develop reference voltages at the inputs of the two comparators *A* and *B*. A comparator is a circuit that compares an input signal voltage to an input reference voltage and then produces a YES/NO or HIGH/LOW decision output. The negative input of comparator A will have a reference that is ⅔ of V_{CC}, and therefore the positive input (pin 6—threshold) will have to be more positive than ⅔ of V_{CC} for the output of comparator A to go HIGH. With comparator B, the positive input has a reference that is ⅓ of V_{CC}, and therefore the negative input (pin 2—trigger) will have to be more negative, or fall below, ⅓ of V_{CC} for the output of comparator B to go HIGH. If the output of comparator A were to go HIGH, the set-reset flip-flop output would be RESET LOW to 0 V. This LOW output would be inverted by the INVERTER to a HIGH, and then inverted and buffered (boosted in current) by the final INVERTER to appear as a LOW at the output pin 3. If the output of comparator B were to go HIGH, the set-reset flip-flop would be SET HIGH to +5 V. This HIGH output would be inverted by the INVERTER to a LOW, and then inverted and buffered by the final INVERTER, to appear as a HIGH at the output pin 3. When the output of the set-reset flip-flop is LOW (reset), the input at the base of the discharge transistor will be HIGH. The transistor will therefore turn ON, and its low emitter-to-collector resistance will ground pin 7. When the output of the set-reset flip-flop is HIGH (set), the input at the base of the discharge transistor will be LOW. The transistor will therefore turn OFF and its high emitter-to-collector resistance will disconnect pin 7 from ground (it will be floating).

The 555 Timer as an Astable Multivibrator

Figure 10-19(a) shows how the 555 timer can be connected to operate as an astable or free-running multivibrator. The waveforms in Figure 10-19(b) show how the externally-connected capacitor C will charge and discharge, and how the output will continually switch between its positive ($+V_{CC}$) and negative (0 V) supply voltage peaks.

To begin with, let us assume that the output of the S-R flip-flop was previously SET HIGH (blue condition, time T_1). The HIGH output of the S-R flip-flop will be inverted to a LOW and turn OFF the 555's internal discharge transistor. With this transistor OFF, the external capacitor C can begin to charge towards $+V_{CC}$ via R_A and R_B. At time T_2, the capacitor's charge has increased beyond ⅔ of V_{CC}, and therefore the output of comparator A will go HIGH, and RESET the S-R flip-flop's output LOW (black condition). This will cause the output (pin 3) of the 555 to go LOW; however, the discharge transistor's base will be HIGH, so it will turn ON. With the discharge transistor ON, the capacitor can begin to discharge as indicated in Figure 10-19(a) and (b). At time T_3, the capacitor's charge has fallen below ⅓ of V_{CC}, or the trigger level of comparator B. As a result, comparator B's output will go HIGH and SET the output of the S-R flip-flop HIGH, or back to its original state. The discharge transistor will once again be cut OFF, allowing the capacitor to charge, and the cycle to repeat.

As you can see in Figure 10-19(a), the capacitor charges through R_A and R_B to ⅔ of V_{CC}, and then discharges through R_B to ⅓ of V_{CC}. As a result, the positive half-cycle time (t_p) can be calculated with the formula

$$t_p = 0.7 \times C \times (R_A + R_B)$$

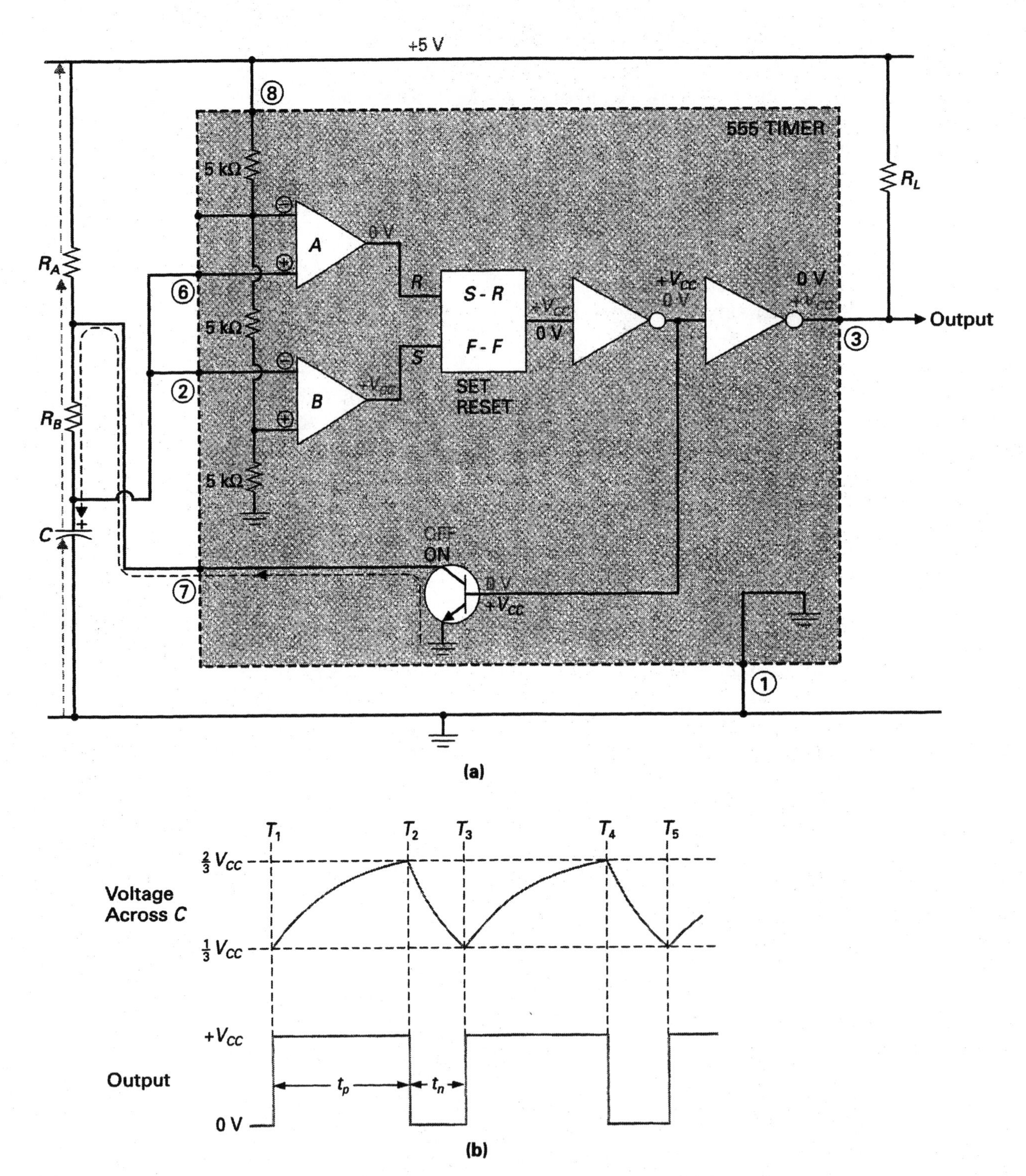

FIGURE 10-19 The 555 Timer as an Astable Multivibrator. (a) Circuit. (b) Waveforms.

The negative half-cycle time (t_n) can be calculated with the formula

$$t_n = 0.7 \times C \times R_B$$

The total cycle time will equal the sum of both half-cycles ($t = t_p + t_n$) and the frequency will equal the reciprocal of time ($f = 1/t$).

■ **EXAMPLE:**

Calculate the positive half-cycle time, negative half-cycle time, complete cycle time, and frequency of the 555 astable multivibrator circuit in Figure 10-19 if

$$R_A = 1\text{ k}\Omega \qquad R_B = 2\text{ k}\Omega \qquad C = 1\ \mu\text{F}$$

■ ***Solution:***

The positive half-cycle will last for

$$\begin{aligned} t_p &= 0.7 \times C \times (R_A + R_B) \\ &= 0.7 \times 1\ \mu\text{F} \times (1\text{ k}\Omega + 2\text{ k}\Omega) = 2.1\text{ ms} \end{aligned}$$

The negative half-cycle will last for

$$\begin{aligned} t_n &= 0.7 \times C \times R_B \\ &= 0.7 \times 1\ \mu\text{F} \times 2\text{ k}\Omega = 1.4\text{ ms} \end{aligned}$$

The complete cycle time will be

$$\begin{aligned} t &= t_p + t_n \\ &= 2.1\text{ ms} + 1.4\text{ ms} = 3.5\text{ ms} \end{aligned}$$

The frequency of this 555 astable multivibrator will be

$$\begin{aligned} f &= \frac{1}{t} \\ &= \frac{1}{3.5\text{ ms}} = 285.7\text{ Hz} \end{aligned}$$

By combining the above steps we can obtain the following formula, which can be used to calculate frequency directly.

$$\begin{aligned} f &= \frac{1.43}{(R_1 + 2R_2) \times C_1} \\ &= \frac{1.43}{[1\text{ k}\Omega + (2 \times 2\text{ k}\Omega)] \times 1\ \mu\text{F}} = \frac{1.43}{0.005} = 286\text{ Hz} \end{aligned}$$

■ **EXAMPLE:**

Figure 10-20(a) shows a typical clock-oscillator circuit using a 555 timer and a *J-K* flip-flop. Perform the following:

a. Calculate the output frequency from the 555 timer.
b. Determine the HIGH time and LOW time of the 555 timer's output.
c. Determine the output frequency from the *J-K* flip-flop.
d. Sketch the 555 timer output and the 74LS76A's Q output in time relation.
e. Explain the purpose of the RUN/ $\overline{\text{HALT}}$ switch.

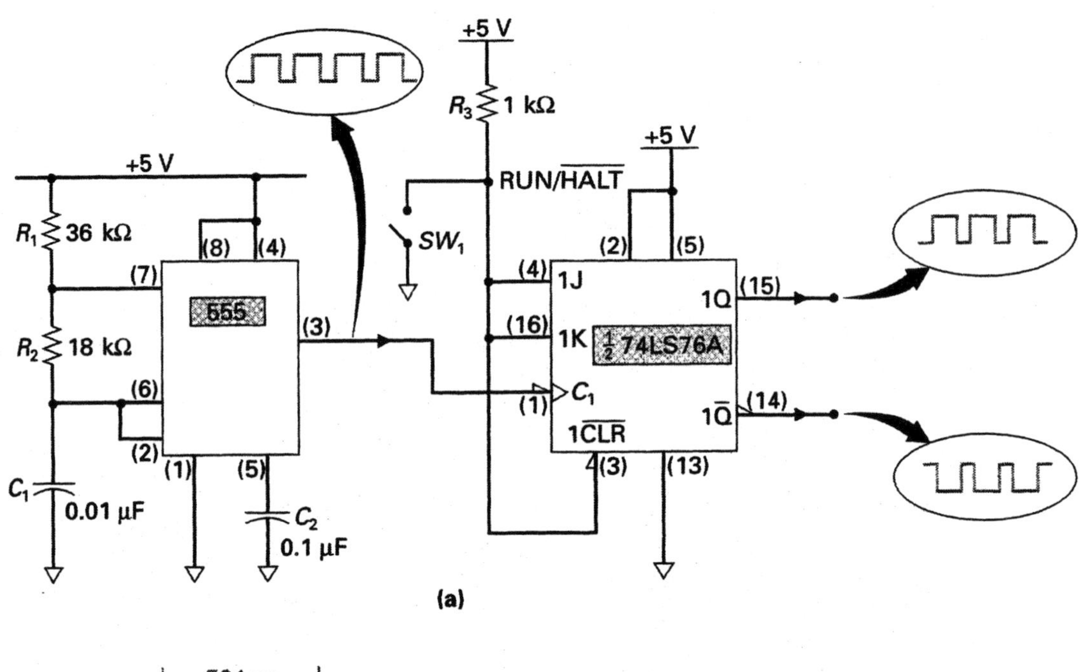

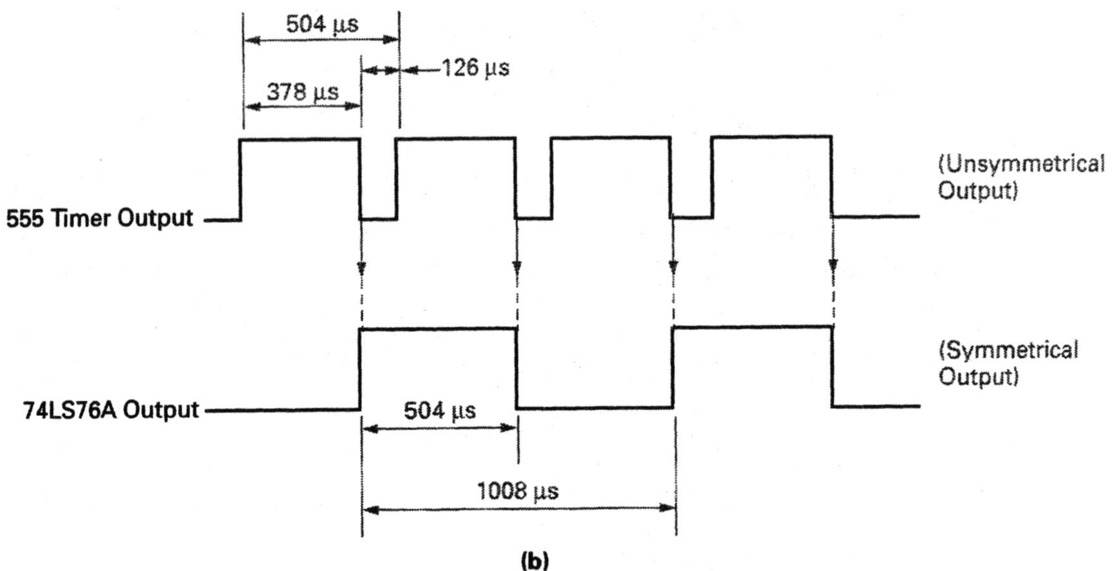

FIGURE 10-20 1 kHz Clock Oscillator with RUN/$\overline{\text{HALT}}$ Control and Low-Level Start.

■ ***Solution:***

a.
$$f = \frac{1.43}{(R_1 + 2R_2) \times C_1}$$
$$= \frac{1.43}{[36\ \text{k}\Omega + (2 \times 18\ \text{k}\Omega)] \times 0.01\ \mu\text{F}} = 2\ \text{kHz}$$

b. The positive half-cycle will last for
$$t_p = 0.7 \times C \times (R_A + R_B)$$
$$= 0.7 \times 0.01\ \mu\text{F} \times (36\ \text{k}\Omega + 18\ \text{k}\Omega) = 378\ \mu\text{s}$$

The negative half-cycle will last for

$$t_p = 0.7 \times C \times R_B$$
$$= 0.7 \times 0.01\ \mu F \times 18\ k\Omega = 126\ \mu s$$

c. Since the J and K inputs of the flip-flop are normally tied HIGH, the output will toggle and therefore divide the input frequency by two. Since we know that the 555 timer's output is 2 kHz, the J-K flip-flop's output will be 1 kHz (2 kHz ÷ 2 = 1 kHz).

d. The 555 timer output and the 74LS76A's Q output in time relation are shown in Figure 10-20(b). The 555 timer output drives the clock input (C) of the negative edge-triggered 74LS76A J-K flip-flop. Each negative edge of the unsymmetrical 2 kHz clock input will cause the output to toggle, or change state, producing a symmetrical 1 kHz output.

e. With SW_1 open, the RUN/ $\overline{\text{HALT}}$ line will be HIGH, so the J and K inputs of the flip-flop will also be HIGH (toggle) condition. If SW_1 is closed, the RUN/$\overline{\text{HALT}}$ line will be LOW, forcing the J-K flip-flop into the NO CHANGE condition. This will freeze the flip-flop's outputs and therefore halt or stop the clock signal division. The RUN/$\overline{\text{HALT}}$ line is also connected to the active-LOW clear input of the J-K flip-flop. This will ensure that when we switch from HALT to the RUN mode, the clock output at Q will always start LOW (in the HALT mode Q is reset LOW, and therefore when we switch to the RUN mode, Q will start LOW and then switch HIGH. Having both a Q and $\overline{Q}$ output means that this circuit can be used to supply either a single-phase clock signal (Q only) or a two-phase clock signal (Q and $\overline{Q}$).

The 555 Timer as a Monostable Multivibrator

Figure 10-21(a) shows how the 555 timer can be connected to operate as a monostable or one-shot multivibrator. The waveforms in Figure 10-21(b) show the time relationships between the input trigger, the charge and discharge of the capacitor, and the output pulse. The width of the output pulse (P_W) is dependent on the values of the external timing components R_A and C.

At time T_1 in Figure 10-21(b), the set-reset flip-flop is in the reset condition and is therefore producing a LOW output. This LOW from the S-R F-F is inverted by the INVERTER, and then inverted and buffered by the final stage, to produce a LOW (0 V) output from the 555 timer at pin 3. The LOW output from the S-R F-F will be inverted and appear as a HIGH at the base of the discharge transistor, turning it ON, and therefore providing a discharge path for the capacitor to ground.

At time T_2, a trigger is applied to pin 2 of the 555 monostable multivibrator. This negative trigger will cause negative input of comparator B to fall below ⅓ of V_{CC}, so the output of comparator B will go HIGH and SET the output of the S-R F-F HIGH. This HIGH from the S-R F-F will send the output of the 555 timer (pin 3) HIGH and turn OFF the discharge transistor. Once the path to ground through the discharge transistor has been removed from across the capacitor, the capacitor can begin to charge via R_A to + V_{CC}, as seen in the waveforms in Figure 10- 21(b). The output of the 555 timer remains HIGH until

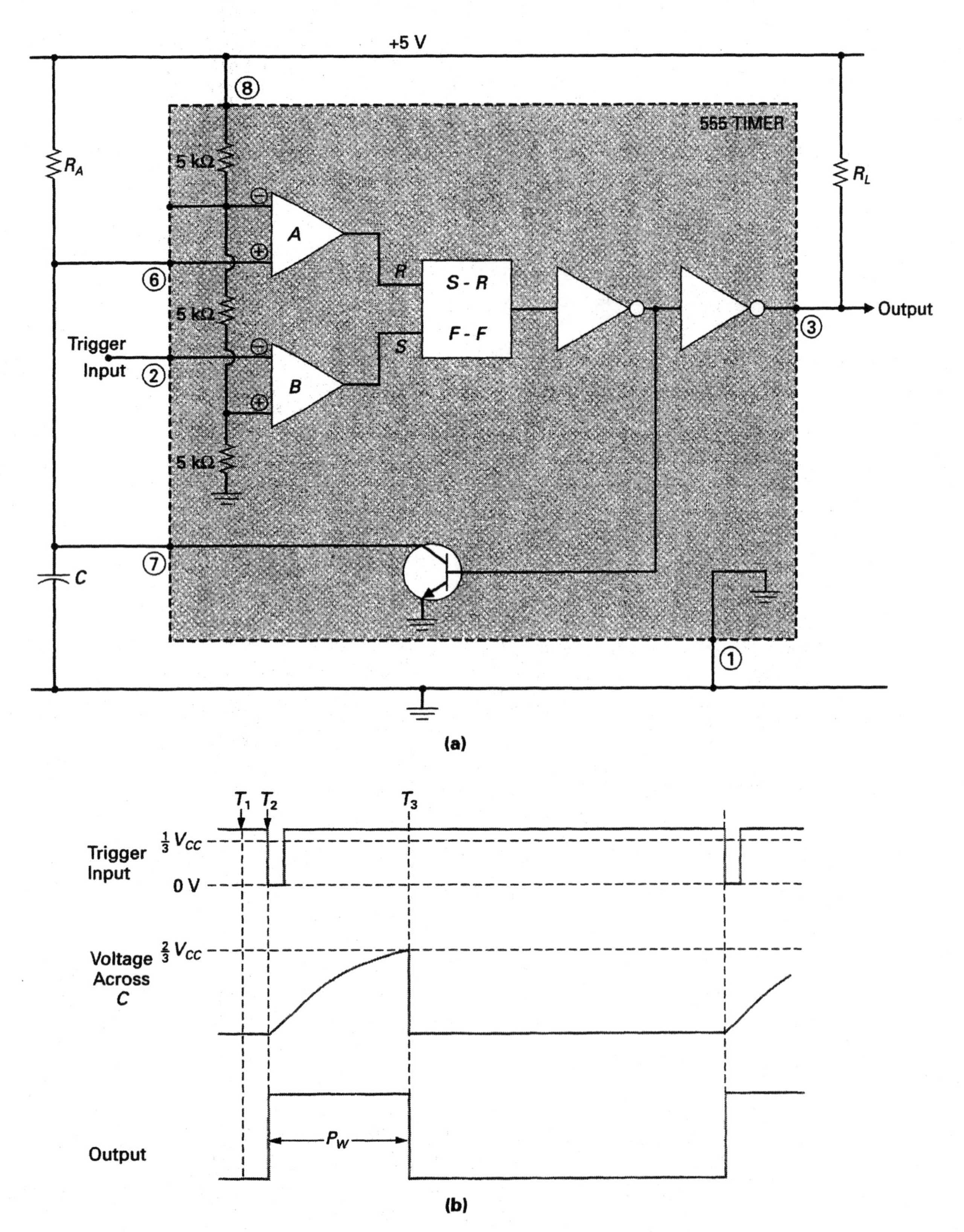

FIGURE 10-21 The 555 Timer as a Monostable Multivibrator. (a) Circuit. (b) Waveforms.

the charge on the capacitor exceeds ⅔ of V_{CC}. At this time (T_3), the output of comparator A will go HIGH, resetting the S-R F-F and causing the output of the 555 timer to go LOW, and also turning ON the discharge transistor to discharge the capacitor. The circuit will then remain in this stable condition until a new trigger arrives to initiate the cycle once again.

The leading edge of the positive output pulse is initiated by the input trigger, while the trailing edge of the output pulse is determined by the R_A and C charge time, which is dependent on their values. Since the capacitor can charge to ⅔ of V_{CC} in a little more than one time constant (1 time constant = 0.632, ⅔ = 0.633), the following formula can be used to calculate the pulse width (P_W).

$$P_W = 1.1 \times (R_A \times C)$$

■ EXAMPLE:

Calculate the pulse width of the 555 monostable multivibrator if R_A = 2 MΩ and C = 1 μF.

■ *Solution:*

The width of the output pulse will be

$$P_W = 1.1 \times (R_A \times C)$$
$$= 1.2 \times (1\ \text{M}\Omega \times 1\ \mu\text{F}) = 2.2\ \text{s}$$

SELF-TEST REVIEW QUESTIONS FOR SECTION 10-2

1. The astable multivibrator is also called the ____________ multivibrator.
2. Which of the multivibrator circuits has only one stable state?
3. Which multivibrator is also called a set-reset flip-flop?
4. In what applications are one-shots used?
5. The 555 timer derived its number identification from ____________
6. List two applications of the 555 timer.

10-3 TROUBLESHOOTING *J-K* FLIP-FLOP AND TIMER CIRCUITS

To review, the procedure for fixing a failure can be broken down into three steps—diagnose, isolate, and repair. Since this chapter has been devoted to *J-K* flip-flop and timer circuits, let us first examine the operation of a typical *J-K* flip-flop and timer circuit, and then apply our three-step troubleshooting process to this circuit.

10-3-1 A Flip-Flop Logic Circuit

As an example, Figure 10-22 shows a 4-bit binary counter circuit with single-step or continuous-clock control. The four-stage binary counter in the upper half of the circuit was covered previously in Figure 10-4. The only addition to this circuit is SW_1 which, when pressed, will clear the counter to 0000.

The clock input applied to the four-stage counter is derived from the single-step or continuous-clock control circuit in the lower half of the circuit in Figure 10-22. In this circuit, SW_3 is a single-pole double-throw (SPDT) switch that is used to select either a single-step or continuous-clock input for the binary counter circuit. The 74121 one-shot circuit was covered previously in Figure 10-16(a). When SW_3 is in the single-step position, NAND *A* will produce a HIGH output which will enable NAND gate *C* and allow the single clock pulses, generated by the 74121 when SW_2 is pressed, through to NAND *E* and onto the counter. When SW_3 is in the continuous-clock position, NAND *B* will produce a HIGH output which will enable NAND gate *D* and allow the 1 kHz clock pulses, generated by the 555 timer and divide-by-two circuit covered previously in Figure 10-20, through to NAND *E* and onto the counter.

Let us now apply our three-step troubleshooting procedure to this flip-flop logic circuit so that we can practice our troubleshooting procedures and methods.

Step 1: Diagnose

As before, it is extremely important that you first understand the operation of a circuit and how all of the devices within it are supposed to work, so that you are able to determine whether or not a circuit malfunction really exists. If you were preparing to troubleshoot the circuit in Figure 10-22, your first step should be to read through the circuit description and review the operation of each integrated circuit until you feel completely confident with the correct operation of the entire circuit. Once you are fully familiar with the operation of the circuit, you will easily be able to diagnose the problem as either an *operator error* or a *circuit malfunction.*

Distinguishing an operator error from an actual circuit malfunction is an important first step, and a wrong diagnosis can waste a lot of time and effort. For example, the following could be interpreted as circuit malfunctions, when in fact they are simply operator errors.

Example 1.
Symptom: The LEDs remain frozen, even though SW_3 is in the upper position.
Diagnosis: Operator Error—Single-step mode has been selected; however, SW_2 will have to be pressed in order to generate single clock pulses.

Example 2.
Symptom: When SW_3 is placed in the lower position, all of the LEDs are only partly ON.
Diagnosis: Operator Error—In the continuous-clock mode, a 1 kHz clock signal is applied, so the counting will be too rapid for the eye to see the changes.

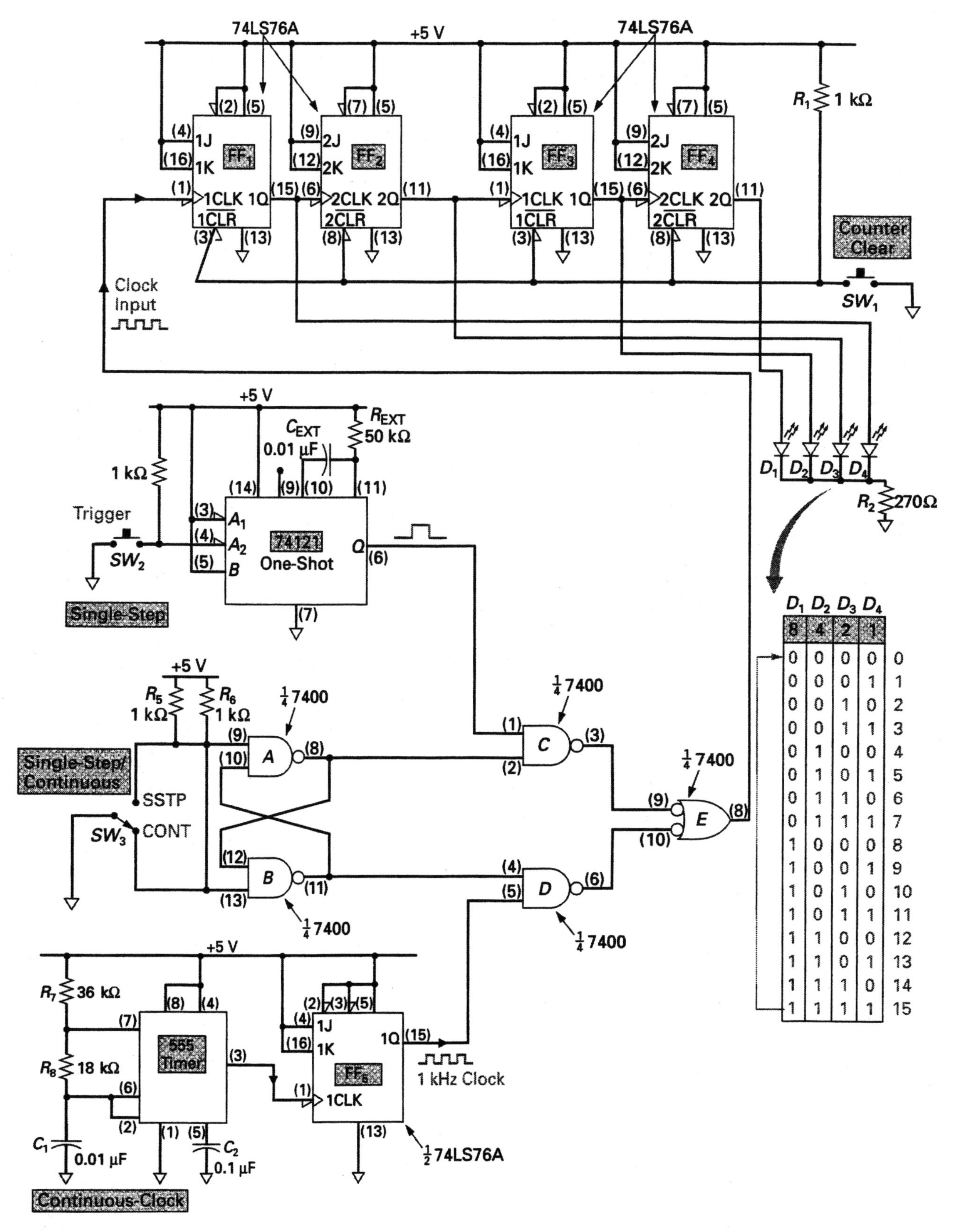

D_1	D_2	D_3	D_4	
8	4	2	1	
0	0	0	0	0
0	0	0	1	1
0	0	1	0	2
0	0	1	1	3
0	1	0	0	4
0	1	0	1	5
0	1	1	0	6
0	1	1	1	7
1	0	0	0	8
1	0	0	1	9
1	0	1	0	10
1	0	1	1	11
1	1	0	0	12
1	1	0	1	13
1	1	1	0	14
1	1	1	1	15

FIGURE 10-22 Four-Bit Binary Up-Counter with Single-Step or Continuous-Clock Control.

Example 3.
Symptom: The circuit is malfunctioning because the display is dead, even when SW_3 is in the upper position and SW_1 is pressed to generate triggers.
Diagnosis: Operator Error—There is not a circuit malfunction. SW_2 should be pressed to generate single-step trigger pulses. The switch SW_1 is used to clear the display.

Once you have determined that the problem is not an operator error, but is in fact a circuit malfunction, proceed to Step 2 and isolate the circuit failure.

Step 2: Isolate

No matter what circuit or system failure has occurred, you should always follow a logical and sequential troubleshooting procedure. Let us review some of the isolating techniques, and then apply them to our example circuit in Figure 10-22.

a. Use a cause and effect troubleshooting process, which means study the effects you are getting from the faulty circuit and then logically reason out what could be the cause.

b. Check first for obvious errors before leaping into a detailed testing procedure. Is the power OFF or not connected to the circuit? Are there wiring errors? Are all of the ICs correctly oriented?

c. Using a logic probe or voltmeter test that power and ground are connected to the circuit and are present at all points requiring power and ground. If the whole circuit, or a large section of the circuit, is not operating, the problem is normally power. Using a multimeter, check that all of the dc voltages for the circuit are present at all IC pins that should have power or a HIGH input, and are within tolerance. Second, check that 0 V or ground is connected to all IC ground pins and all inputs that should be tied LOW.

d. Use your senses to check for broken wires, loose connectors, overheating or smoking components, pins not making contact, and so on.

e. Test the clock signals using the logic probe or the oscilloscope. Although clock signals can be easily checked with a logic probe, the oscilloscope is best for checking the signal's amplitude, frequency, pulse width, and so on. The oscilloscope will also display timing problems and noise spikes (glitches) which could be false-triggering a circuit into operation at the wrong time.

f. Perform a static test on the circuit. With our circuit example in Figure 10-22, you can static test the counter circuit by switching the circuit to the single-step mode, assuming that the clock section of the circuit is operating correctly. If the clock section is suspected, you can check the counter section of the circuit by disconnecting the clock input to the circuit, and then using a logic pulser's single pulse feature to clock the counter in single steps. A logic probe could then be used to test that valid logic levels are appearing at the outputs of the *J-K* flip-flops.

g. With a dynamic test, the circuit is tested while it is operating at its normal clock frequency, and therefore all of the inputs and outputs are continually changing. Although a logic probe can detect a pulse waveform, the oscilloscope will display more signal detail, and since it can display more than one signal at a time, it is ideal for making signal comparisons and looking for timing errors.

Once again, you can dynamic test the counter circuit in Figure 10-22 by switching the circuit to the continuous-clock mode, assuming that the clock section of the circuit is operating correctly. If the clock section is suspected, you can check the counter section of the circuit by disconnecting the clock input to the circuit, and then using a logic pulser's pulse train feature to continuously clock the counter.

A logic analyzer, which will be discussed in more detail in the digital system chapter, is a type of oscilloscope that can display typically eight to sixteen digital signals at one time.

In some instances, you will discover that some circuits will operate normally when undergoing a static test, and yet fail a dynamic test. This effect usually points to a timing problem involving the clock signals and/or the propagation delay times of the ICs used within the circuit.

h. Noise due to electromagnetic interference (EMI) can false-trigger an IC, such as the *J-K* flip-flops in Figure 10-22. This problem can be overcome by not leaving any of the IC's inputs unconnected and, therefore, floating. Connect unneeded active-HIGH inputs to ground and active-LOW inputs to $+V_{CC}$.

i. Apply the half-split method of troubleshooting first to a circuit, and then to a section of a circuit, to help speed up the isolation process. With our circuit example in Figure 10-22, a good mid-point check would be to test the clock signal at the output of NAND *E* to determine whether the problem is in the clock section or divider section of the circuit.

Also remember that a load problem can make a source appear at fault. If an output is incorrect, it would be safer to disconnect the source from the rest of the circuit and then recheck the output. If the output is still incorrect, the problem is definitely within the source; however, if the output is correct, the load is probably shorting the line either HIGH or LOW.

j. Substitution can be used to help speed up your troubleshooting process. Once the problem is localized to an area containing only a few ICs, substitute suspect ICs with known working ICs (one at a time) to see if the problem can be quickly remedied.

Step 3: Repair

Once the fault has been found, the final step is to repair the circuit and then perform a final test to see that the circuit and the system is now fully operational.

10-3-2 Sample Problems

Once you have constructed a circuit, like the frequency divider circuit in Figure 10-22, introduce a few errors to see what effect or symptoms they produce. Then logically reason out why a particular error or cause has a particular effect on the circuit. Never short any two points together unless you have carefully thought out the consequences, but generally it is safe to open a path and see the results. Here are some examples for our example circuit in Figure 10-22.

a. Disconnect the output of the 555 timer.

b. Connect the *J* and *K* inputs of FF5 LOW instead of HIGH.

c. Disconnect C_{EXT} and R_{EXT} from the 74121.
d. Connect the LED display of the counter to the $\overline{Q}$ outputs instead of the Q outputs.
e. Connect pin 5 of the 74121 LOW instead of HIGH.

SELF-TEST REVIEW QUESTIONS FOR SECTION 10-3

1. What are the three basic troubleshooting steps?
2. What is a static test?
3. What is the half-split troubleshooting technique?
4. What symptom would you get from the circuit in Figure 10-22, if SW_1 were connected to pins 2 and 7 of the 74LS76A instead of pins 3 and 8?

REVIEW QUESTIONS

Multiple-Choice Questions

1. What is another name for a flip-flop circuit?
   a. Astable multivibrator b. Monostable multivibrator
   c. Bistable multivibrator d. Horstable multivibrator
2. Which two multivibrator types require trigger inputs?
   a. Astable and monostable b. Monostable and bistable
   c. Bistable and astable d. Both a. and c. are true
3. What input condition does the *J-K* flip-flop have that the *S-R* and D-type flip-flops don't have?
   a. SET b. RESET c. NO CHANGE d. TOGGLE
4. What would be the output from a negative edge-triggered *J-K* flip-flop if $J = 1$, $K = 0$, $C =$ LOW to HIGH, and $Q = 0$?
   a. NO CHANGE b. $Q = 1$ c. $Q = 0$ d. Both a. and c. are true
5. A four-stage *J-K* flip-flop can be made to function as a ________ binary counter.
   a. 0-to-7 b. 0-to-3 c. 0-to-15 d. 0-to-16
6. If $J = 0$ and $K = 1$, a *J-K* flip-flop when clocked is said to be in the ________ condition.
   a. SET b. RESET c. NO CHANGE d. TOGGLE
7. If $J = 1$, $K = 1$, and the input clock frequency is 2 MHz, what would be the Q output frequency?
   a. 2 MHz b. 1 MHz c. 0.5 MHz d. 4 MHz
8. Which triggering system is used to control a master-slave latch?
   a. Level-triggering b. Edge-triggering
   c. Pulse-triggering d. Either a. or c.
9. The *J-K* flip-flop responds to only ________.
   a. Level-triggering b. Edge-triggering
   c. Pulse-triggering d. Both b. and c.

10. Which multivibrator type is generally used as a clock oscillator?
   a. Astable multivibrator **b.** Monostable multivibrator
   c. Bistable multivibrator **d.** Tristable multivibrator
11. Which multivibrator type is used as a pulse generator?
   a. Astable multivibrator **b.** Monostable multivibrator
   c. Bistable multivibrator **d.** Tristable multivibrator
12. Monostable or one-shot multivibrators are generally used as a
   a. pulse generator. **b.** timer and sequencer.
   c. delay. **d.** all of the above.
13. Which multivibrator type has two stable states?
   a. Astable multivibrator **b.** Monostable multivibrator
   c. Bistable multivibrator **d.** Tristable multivibrator
14. Which multivibrator is also called a set-reset flip-flop?
   a. Astable **b.** Monostable **c.** Bistable **d.** Schmitt
15. The output pulse width of a 555 timer is determined by the externally connected ____________ and ____________.
   a. power supply, resistor **b.** load resistance, capacitor
   c. capacitor, load resistor **d.** input resistor, capacitor
16. The 555 timer consists of a ____________ -resistor voltage divider, ____________ comparator(s), ____________ *R/S* flip-flop(s), an INVERTER, an output stage, and a discharge transistor on a single IC.
   a. two, two, two **b.** three, two, one
   c. one, two, three **d.** three, one, two
17. A monostable multivibrator will generally make use of a(n) ____________ and ____________ circuit on the trigger input.
   a. integrator, clipper **b.** differentiator, schmitt
   c. schmitt, clipper **d.** differentiator, clipper
18. Which of the two-state circuits could be used to convert a sine wave into a square wave?
   a. Schmitt **b.** Bistable **c.** Astable **d.** Monostable
19. Which of the multivibrators could be used to stretch the width of a pulse?
   a. Schmitt **b.** Bistable **c.** Astable **d.** Monostable
20. How many *J-K* flip-flop stages will be needed to achieve a divide-by-eight frequency division?
   a. Two **b.** Three **c.** Four **d.** Five

Essay Questions

21. What makes the *J-K* flip-flop different from the *S-R* flip-flop? (10-1)
22. Define the term toggle. (10-1)
23. List the two basic applications for *J-K* flip-flops. (10-1-1)
24. Sketch a 0-to-7 binary counter circuit using negative edge-triggered *J-K* flip-flops. (10-1-1)
25. What is the difference between an edge-triggered and pulse-triggered *J-K* flip-flop? (10-1-2)

26. Give a brief description of the following circuits. (10-2)
   a. Astable multivibrator
   b. Monostable multivibrator
   c. Bistable multivibrator
27. Which type of multivibrator circuit is normally used as a clock oscillator? (10-2-1)
28. Which type of multivibrator is often referred to as a one-shot? (10-2-2)
29. Define the difference between a nonretriggerable and retriggerable one-shot. (10-2-2)
30. List the three key applications for one-shots. (10-2-2)

Practice Problems

To practice your circuit recognition and operation, refer to Figure 10-23 and answer the following three questions.

31. How is the *J-K* flip-flop in Figure 10-23 triggered?
32. Identify the circuit formed by NAND gates *A* and *B*.
33. What will be the output frequency at *Q* when
   a. the switch is in the ON position? b. the switch is on the OFF position?

To practice your circuit recognition and operation, refer to Figure 10-24 and answer the following three questions.

34. Is this *J-K* flip-flop edge-triggered or pulse-triggered?
35. Sketch the *Q* output waveform for the input timing waveforms shown.
36. Sketch the $\overline{Q}$ output waveform for the input timing waveforms shown.

To practice your circuit recognition and operation, refer to Figure 10-25 and answer the following three questions.

37. Is this *J-K* flip-flop edge-triggered or pulse-triggered?
38. Sketch the *Q* output waveform for the input timing waveforms shown.
39. Sketch the $\overline{Q}$ output waveform for the input timing waveforms shown.

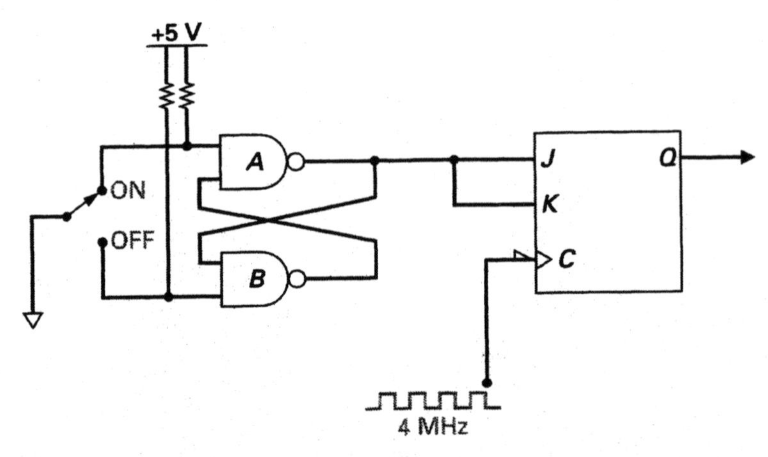

FIGURE 10-23 Divide-by-Two Clock Circuit with ON/OFF Control.

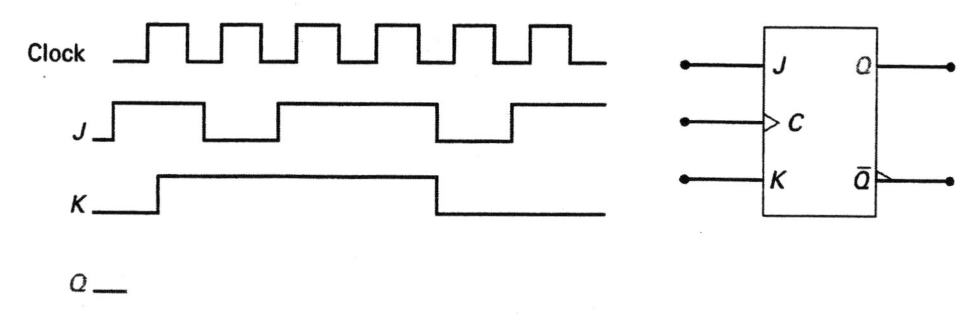

FIGURE 10-24 **A *J-K* Flip-Flop Timing Example.**

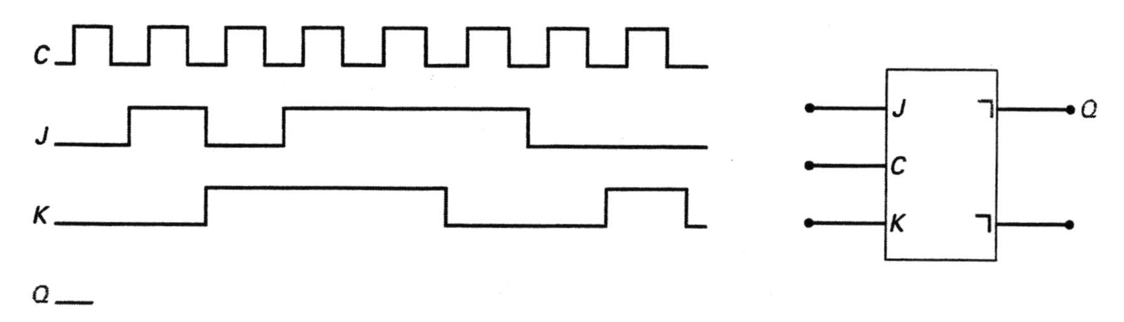

FIGURE 10-25 **A *J-K* Flip-Flop Timing Example.**

To practice your circuit recognition and operation, refer to Figure 10-26 and answer the following four questions.

40. The 555 timer is connected to operate as a/an ____________ multivibrator.

41. Calculate the frequency range of the 555 timer if R_1 were adjusted between the two extremes of 1 Ω and 100 kΩ.

42. Is the 74121 nonretriggerable or retriggerable?

43. Calculate the pulse length range of the 74121 one-shot if R_3 were adjusted between the two extremes of 1 Ω and 100 kΩ.

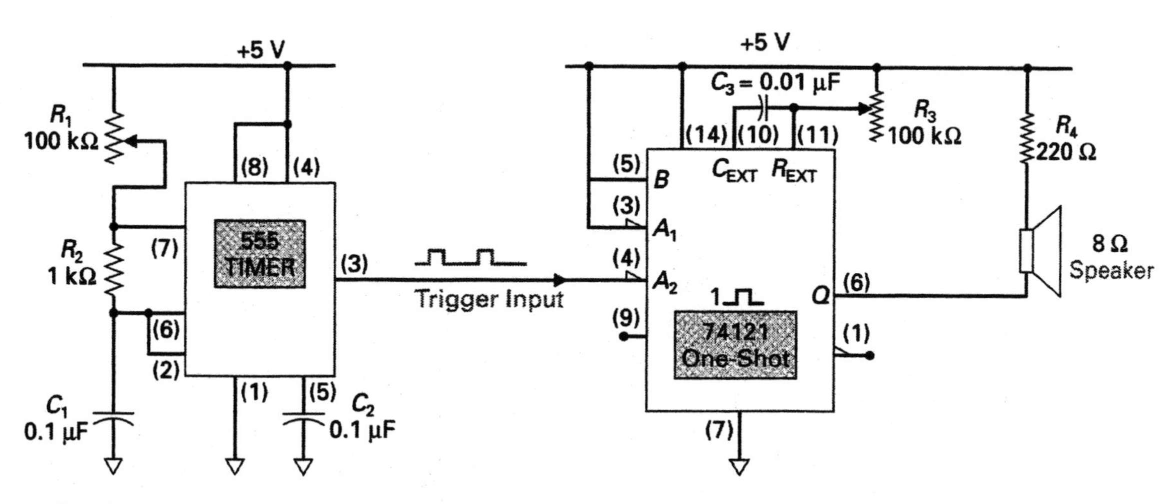

FIGURE 10-26 **Tone Generator.**

To practice your circuit recognition and operation, refer to Figure 10-27 and answer the following four questions.

44. Which type of *J-K* flip-flops are being used?
45. Which of the *J-K* flip-flop circuits will function as a divide-by-two circuit?
46. Which of the *J-K* flip-flop circuits will function as a divide-by-three circuit?
47. Which of the *J-K* flip-flop circuits will function as a divide-by-four circuit?

To practice your circuit recognition and operation, refer to Figure 10-28 and answer the following three questions.

48. What would be at the *Q* output for the circuit condition shown in Figure 10-28(a)?
49. What would be at the *Q* output for the circuit condition shown in Figure 10-28(b)?
50. What would be at the *Q* output for the circuit condition shown in Figure 10-28(c)?

Troubleshooting Questions

51. How could you use a logic pulser and logic probe to static test the circuit in Figure 10-23?
52. How would you apply the half-split troubleshooting method to the circuit in Figure 10-23?

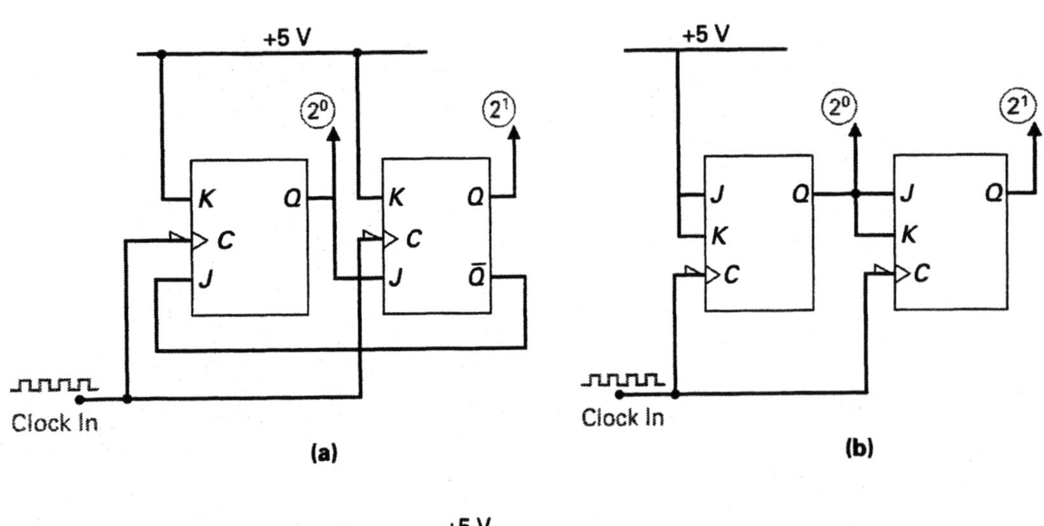

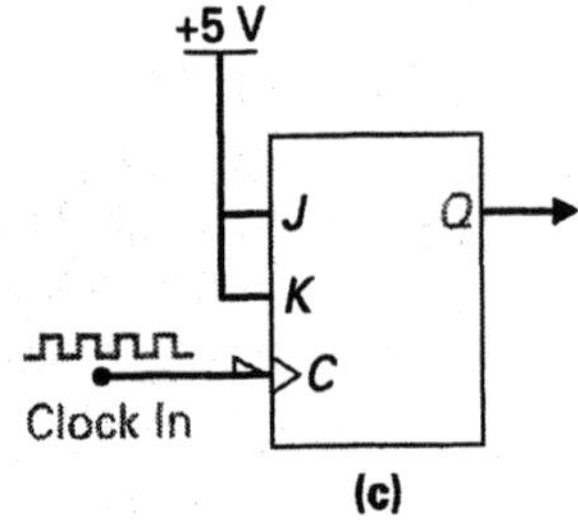

FIGURE 10-27 Divider Circuits.

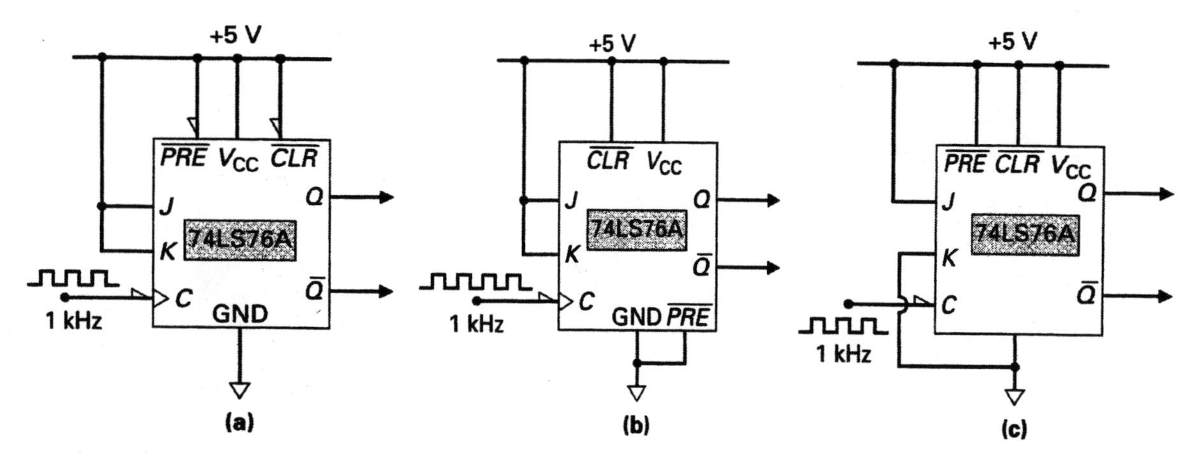

FIGURE 10-28 74LS76A IC Connections.

53. What symptoms would you expect from the circuit shown in Figure 10-23 if the following malfunctions were to occur?

a. The clock input was permanently HIGH.
b. The J and K inputs were permanently HIGH.
c. Power was disconnected from one J-K flip-flop.
d. The output of NAND A was shorted LOW.
e. The switch debounce circuit was bypassed.

54. How could you use a logic pulser and logic probe to static test the circuit in Figure 10-26?

55. What symptoms would you expect from the circuit shown in Figure 10-26 if the following malfunctions were to occur?

a. R_1 was to short.
b. Pin 4 of the 555 was connected LOW.
c. R_1 and R_2 were connected in the reverse order.
d. The speaker was connected to the $\overline{Q}$ instead of the Q output.
e. The trigger inputs A_1 and A_2 were reversed.
f. R_{INT} was used in conjunction with C_{EXT} instead of R_3.

Chapter 11

OBJECTIVES

After completing this chapter, you will be able to:

1. Explain the basic function of a register.
2. Describe the operation and characteristics of the parallel-in, parallel-out register or buffer register.
3. List the different types of shift registers.
4. Explain the operation and characteristics of the following types of shift registers, along with a representative IC and an application for each of the following:
   a. Serial-in, serial-out (SISO) shift register
   b. Serial-in, parallel-out (SIPO) shift register
   c. Parallel-in, serial-out (PISO) shift register
   d. Bidirectional universal shift register
5. Describe the operation of three-state output registers as well as how three-state output control is used in bus-organized systems.
6. Explain how registers are used in the following applications.
   a. Memory register
   b. Serial-to-parallel and parallel-to-serial data conversion
   c. Arithmetic operations
   d. Shift register counters/sequencers
7. Describe how to use the three-step method to troubleshoot digital logic circuits containing registers.

Registers

OUTLINE

VIGNETTE

Trash

In 1976 a Texas-based chain of stores known as Tandy Radio Shack, which up to that time had only been involved with electronic kits and parts for citizen band (CB) radios and stereo equipment, sensed a sudden demand amongst its hobbyists for a personal computer. In July of that year the company recruited Steven Leininger, who had been working for National Semiconductor since graduating from college. He seemed to possess the right qualifications in that he was an electronics engineer and a computer buff. In a double feat, Leininger designed both the computer's architecture and its built-in software, and used at the heart of his computer the brand-new Z80 microprocessor. This new microprocessor from the newly formed company, Zilog, resembled Intel's 8080 microprocessor, but was exceedingly more powerful. The reason it resembled Intel's processor was because Zilog was founded by a group of disgruntled Intel engineers; a situation that would later lead to some legal problems—but then, that's another story.

On February 2, 1977, after Leininger had not left the plant for the last two weeks and had been working almost completely around the clock, the TRS-80-Model 1 was demonstrated faultlessly to company owner Charles Tandy. Blowing smoke from his ever-present large cigar, Tandy said that it was "quite a nice little gadget." An executive then asked, "How many should be built—one thousand, two thousand?" The company's financial controller answered by saying, "We've got 3,500 company-owned stores. I think we can build that many. If nothing else, we can use them in the back for accounting." For Leininger, his baby had not received the welcome he expected from the general management, who did not share his confidence in its ability or attraction.

Not even Leininger was prepared for the TRS-80 (nicknamed Trash 80) splash that was going to take place. Going on sale in September of 1977, ten thousand orders flooded in almost overnight and the demand remained so strong that the company's production plant took over a year to catch up with the orders. A year later, in 1978, Tandy Computers took an impressive lead in the personal computer race, leaving behind their chief competitor, Commodore, and its PET computer.

INTRODUCTION

In Chapters 9 and 10 we examined the operation of the *S-R* flip-flop, the D-type flip-flop, and the *J-K* flip-flop. These flip-flops, or memory elements, are the building blocks we will use to construct sequential logic circuits, which can perform such functions as storing, sequencing, and counting. To help define the term sequential logic circuits, let us compare them to combinational logic circuits, which perform such functions as decoding, encoding, and comparing. The output from a combinational logic circuit is determined by the present state of the inputs, whereas the output of a sequential logic circuit, because of its memory ability, is determined by the previous state of the inputs.

Like combinational logic circuits, there is a large variety of sequential logic circuits; however, two types predominate in digital systems: registers and counters. Both of these circuit types have been briefly discussed in previous applications; however, in this chapter we will examine registers in more detail, while in the following chapter we will concentrate on counters.

A **register** is a group of flip-flops or memory elements that work together to store and shift a group of bits or a binary word. The most basic type of register is called the buffer register, and its function is simply to store a binary word. Other registers, such as the shift register, modify the stored word by shifting the stored bits to the left or to the right. In this chapter we will discuss the operation characteristics, applications, and troubleshooting of these and other registers along with some of the available register-integrated circuits.

11-1 BUFFER REGISTERS

The **buffer register** simply stores a digital word. Figure 11-1(a) shows how a buffer register could be constructed using positive edge-triggered D-type flip-flops. This buffer register has actually been constructed using the four D-type flip-flops within a 74175 IC, which was covered in Chapter 9. The 4-bit data input word (1D, 2D, 3D, and 4D) will be stored in the register and appear at the Q output (Q_1, Q_2, Q_3, and Q_4) when the first positive clock edge occurs at the 74175 clock input (pin 9). The 4-bit Q output word drives a 4-bit LED display which will display the contents of the 4-bit buffer register. Referring to the waveforms in Figure 11-1(b), you can see that the 4-bit data input, which is only momentarily present on the input data lines (1D, 2D, 3D, and 4D), is stored, or retained, in the buffer register when a rising clock edge occurs. This data is permanently present at the output, and therefore displayed on the LED display.

Buffer Register
A temporary data storage circuit able to store a digital word.

Since a 4-bit parallel word is delivered to the input of the register, and a 4-bit parallel word is present at the output, buffer registers are often referred to as **parallel-in, parallel-out (PIPO) registers.** The active-LOW clear input of the 74175 can be used to clear all of the flip-flops (erase the stored word) and therefore turn OFF all of the LEDs.

Parallel-In, Parallel-Out (PIPO) Register
A register that will parallel-load an input word and parallel-output its stored contents.

SELF-TEST REVIEW QUESTIONS FOR SECTION 11-1

1. Buffer registers are also referred to as __________ registers.
   a. SISO b. PISO c. PIPO d. SIPO
2. What is the main function of a buffer register?

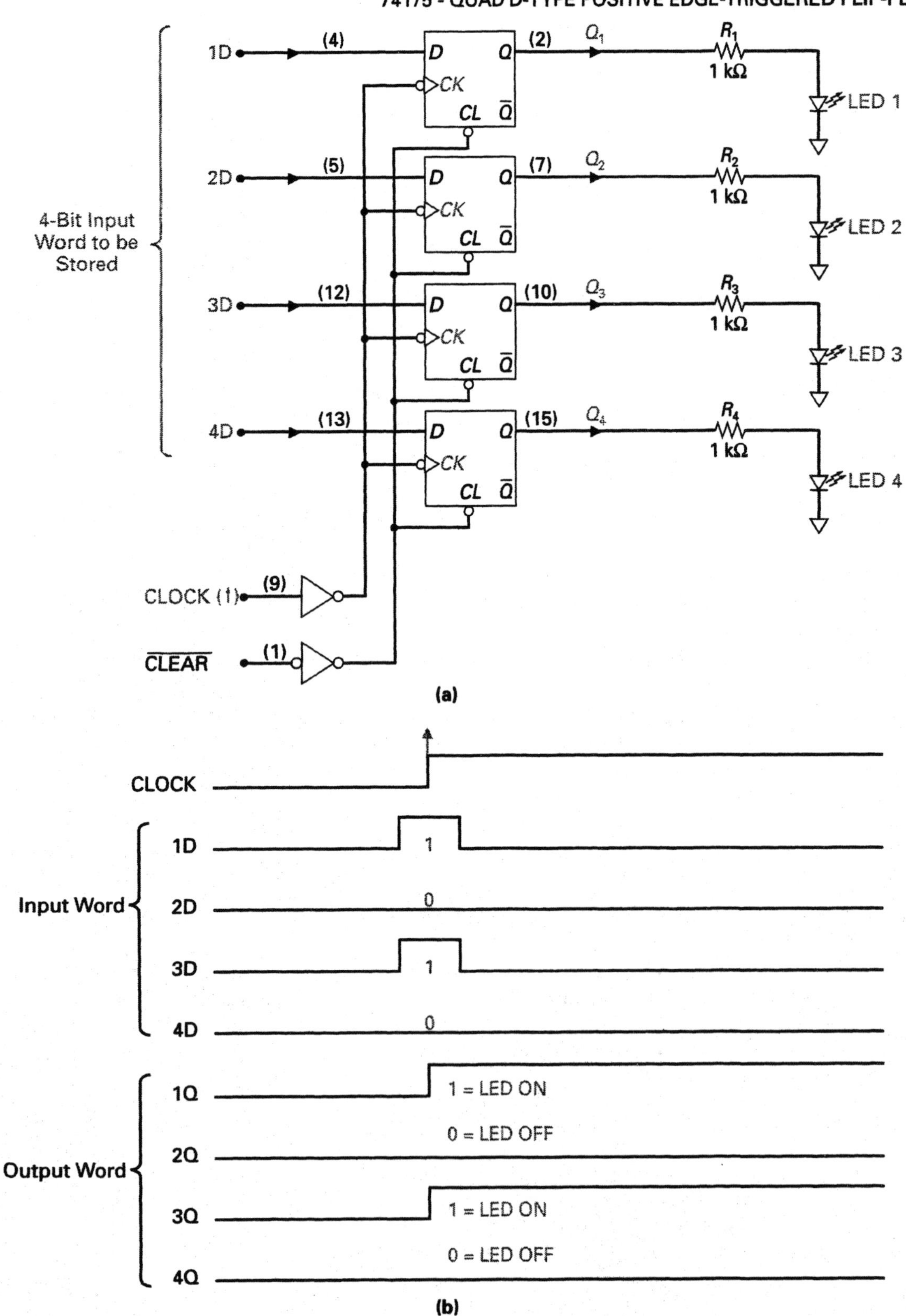

FIGURE 11-1 A Buffer Register. (a) 74175 Parallel-In, Parallel-Out Register. (b) Timing Waveforms.

11-2 SHIFT REGISTERS

A **shift register** is a storage register that will move or shift the bits of the stored word either to the left or to the right. The three basic types of shift registers are

Shift Register
A temporary data storage circuit able to shift or move the stored word either left or right.

a. Serial-in, serial-out (SISO) shift registers
b. Serial-in, parallel-out (SIPO) shift registers
c. Parallel-in, serial-out (PISO) shift registers

Each of these shift register classifications is shown in Figure 11-2. The names given to these three types actually describes how data is entered, or inputted, into the shift register for storage, and how data exits, or is outputted from, the shift register. To understand these operations in more detail, let us take a close look at each of these three shift registers in the following sections.

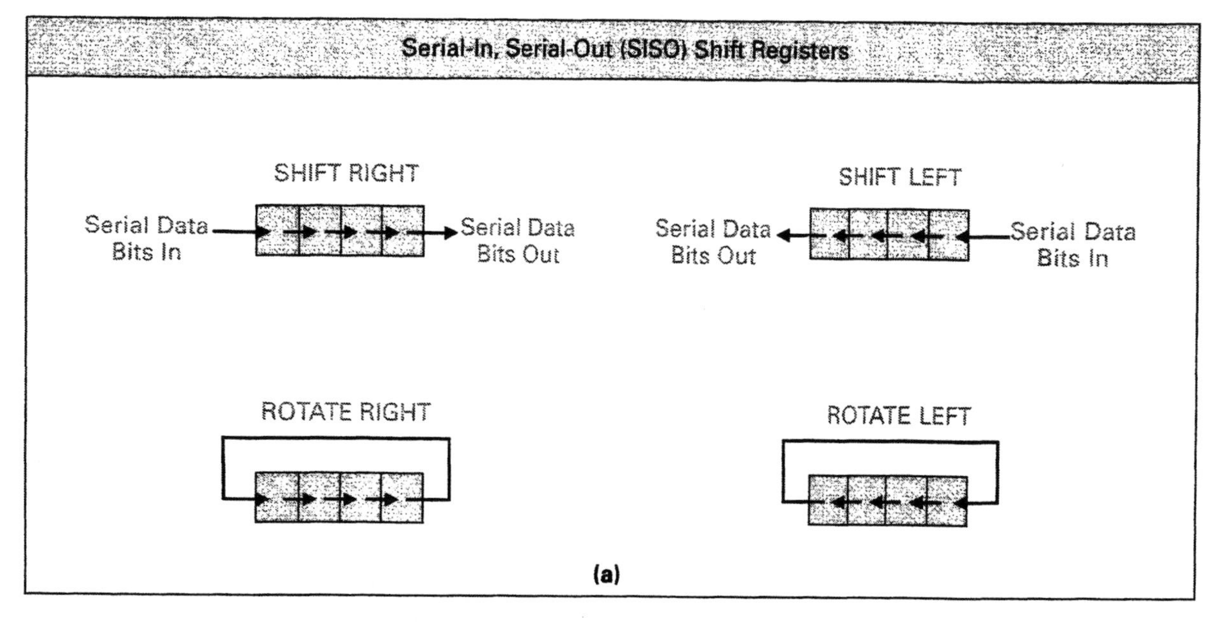

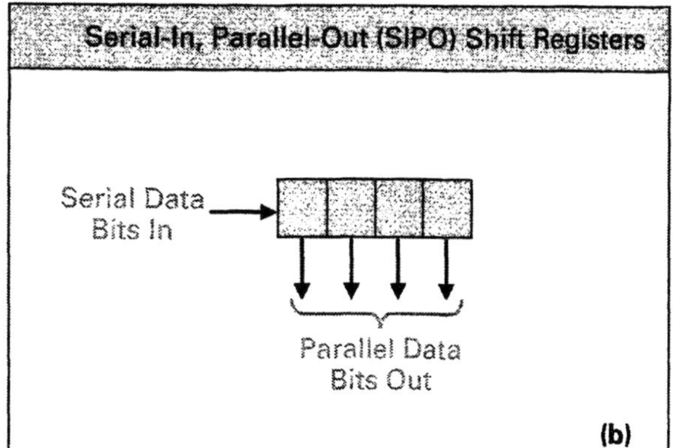

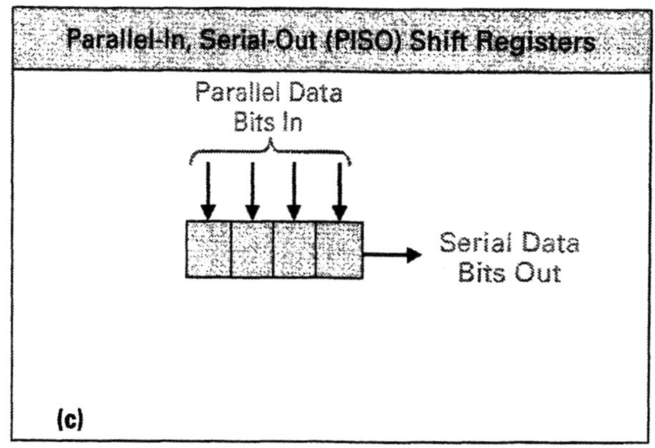

FIGURE 11-2 **Shift Register Classifications. (a) SISO. (b) SIPO. (c) PISO.**

11-2-1 Serial-In, Serial-Out (SISO) Shift Registers

To begin with, Figure 11-3(a) shows how a shift register operates. In this example, the initial condition shows that the 4-bit word 0110 is stored within the 4-bit shift register, while the external serial word 1001 is being applied to the shift register's input. After the first clock pulse, the data stored within the shift register is shifted one position to the right, and the first bit of the applied serial word is shifted into the first position of the shift register. After the second clock pulse, two bits of the stored 0110 word have been shifted out of the shift register while two bits of the applied 1001 word have been shifted into the shift register. After the third clock pulse, three shift-right operations have occurred. After four clock pulses the originally stored 0110 word has been completely shifted out of the register while the applied 1001 input word has been completely shifted into the register and is now being stored.

Now that the basic operation of the shift register is understood, let us see how we can use flip-flops to construct a shift register circuit. Figure 11-3(b) shows how a 4-bit shift register circuit can be constructed using four D-type flip-flops. The serial data input is applied to the D input of flip-flop 0 (FFO). The output from FFO (Q_0) is applied to the D input of FF1, the output of FF1 (Q_1) is applied to the D input of FF2, the output of FF2 (Q_2) is applied to the D input of FF3, and the output of FF3 (Q_3) is the final serial data output of the 4-bit shift register. The clock input is applied simultaneously to all of the D-type flip-flops, with each positive edge of the clock causing the stored 4-bit word to shift one position to the right. With this **serial-in, serial-out shift register,** data is serially fed into the register for storage under the control of the clock (one bit is stored per clock pulse, so four clock pulses will be needed to shift in a serially-applied 4-bit word). To extract a stored 4-bit word from the shift register, four clock pulses will have to be applied to serially shift the stored word out of the register.

Serial-In, Serial-Out (SISO) Register
A register that will serial-load an input word, and serial-output its stored contents.

In summary, the circuit in Figure 11-3(b) shows how four D-type flip-flops should be connected to form a *SISO shift-right shift register.* Figure 11-3(c) shows how the *D* and *Q* connections can be altered to form a *SISO shift-left shift register.* In some applications, the serial data outputs of the circuits shown in Figure 11-3(b) and (c), are connected directly back to the serial data inputs so that the outputted data is immediately inputted. These SISO operations are called *SISO rotate-right* and *SISO rotate-left,* and are shown in Figure 11-3(d).

Figure 11-4 shows data sheet details for the 7491A, which is an 8-bit SISO shift-right shift register IC. Looking at this IC's internal logic diagram, you can see that *S-R* flip-flops are used to form an 8-bit SISO shift register, which is positive edge-triggered, and has a gated *A* and *B* data input. Since the *A* and *B* data inputs are ANDed, both inputs must be HIGH to shift in a binary 1, while either input can be LOW to shift in a binary 0.

■ **EXAMPLE:**

Referring to the circuit shown in Figure 11-5, describe how you would

a. input or store the word "11001010" within the 7491A SISO shift register.
b. output the stored 8-bit word so that it can be displayed on the LED.

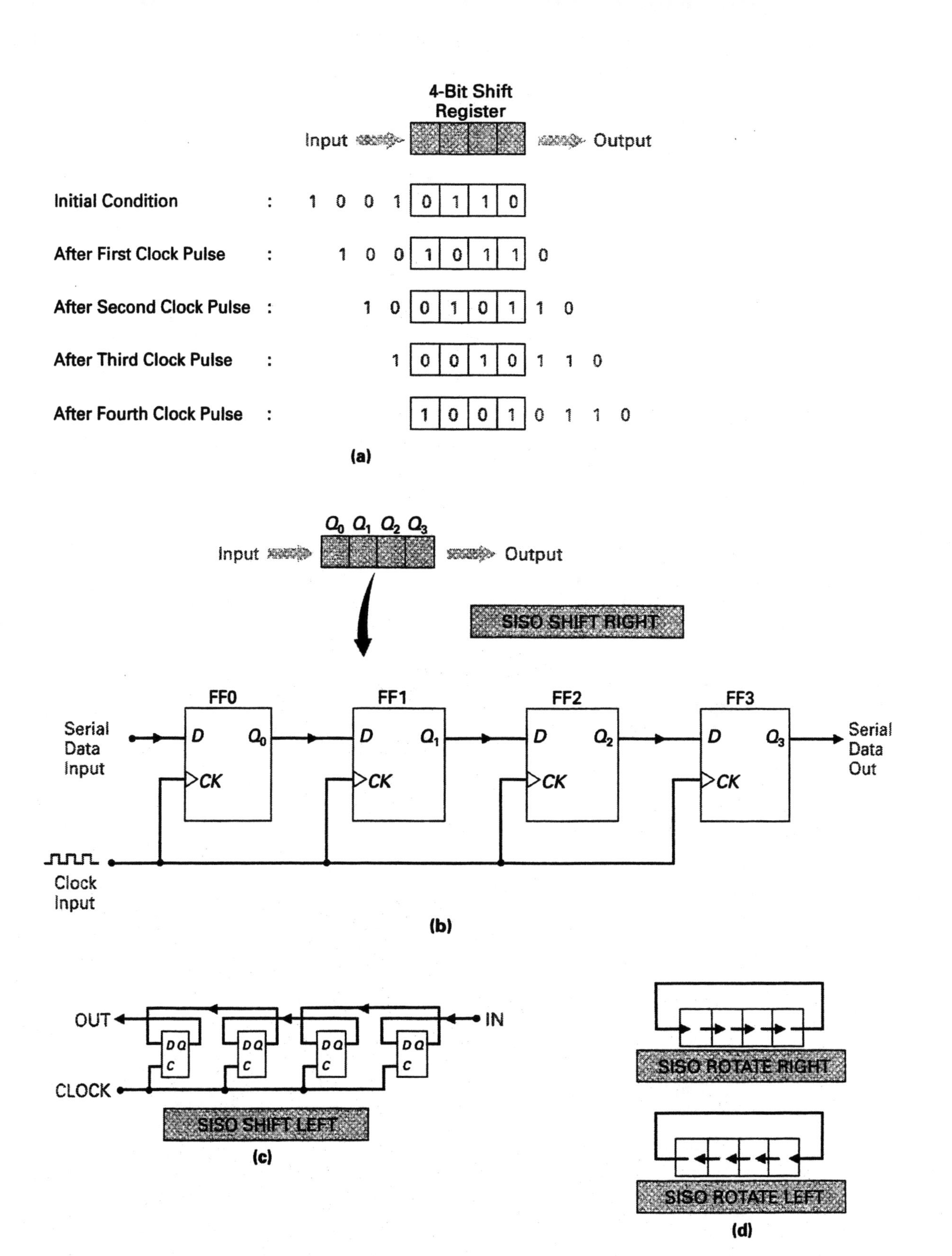

FIGURE 11-3 Serial-In, Serial-Out (SISO) Shift Registers.

IC TYPE: SN7491A– 8-Bit Serial-In, Serial-Out Shift Register

SRG8 = Shift Register with an 8-bit Capacity.

LOGIC SYMBOL

SRG8
CLK (9) C1/→
B (11)
A (12)
&
1D
(13) Q_H
(14) $\overline{Q}_H$

ANDed *A* and *B* Serial Inputs

FUNCTION TABLE

INPUTS AT t_n		OUTPUTS AT t_{n+8}	
A	B	Q_H	$\overline{Q}_H$
H	H	H	L
L	X	L	H
X	L	L	H

t_n = Reference bit time, clock low

t_{n+8} = Bit time after 8 low-to-high clock transitions.

TOP VIEW

Pin		Pin	
NC	1	14	$\overline{Q}_H$
NC	2	13	Q_H
NC	3	12	A
NC	4	11	B
V_{CC}	5	10	GND
NC	6	9	CLK
NC	7	8	NC

LOGIC DIAGRAM

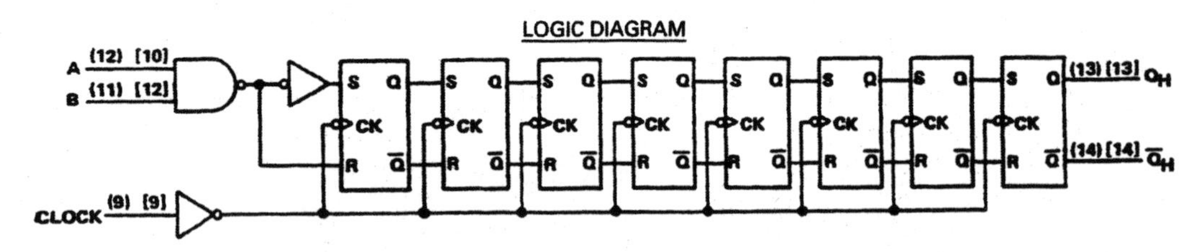

FIGURE 11-4 An IC Containing an Eight-Stage Serial-In, Serial-Out (SISO) Shift Register. (Courtesy of Texas Instruments)

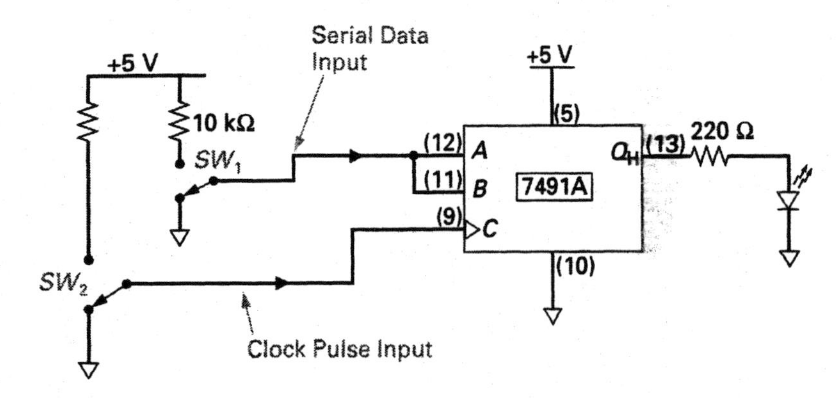

FIGURE 11-5 Test Circuit for a 7491A Shift Register.

■ ***Solution:***

This test circuit operates as follows. SW_1 controls the data input to the IC, while SW_2 controls the clock input to the IC. If SW_1 is in the upper position, a HIGH will be applied to the IC's serial data input, whereas if SW_1 is in the lower position, a LOW will be applied to the IC's serial data input. To shift-right the data stored within the 7491A, and therefore load the input data applied by SW_1, SW_2 should be momentarily moved into its upper position and then placed back in its lower position. Switching SW_2 up and then down in this way will apply a positive edge to the clock input of the 7491A, and therefore all of the *S-R* flip-flops within the 7491A.

a. To store the word "11001010" within the 7491A SISO shift register, first apply a LOW (LSB) from SW_1 and then clock the IC using SW_2, then apply a HIGH and clock, then a LOW and clock, and then a HIGH and clock, and so on until all eight bits are inputted.

b. To display the previously stored 8-bit word on the LED, simply clock the 7491A eight times and each bit of the word 11001010 will be shifted right and appear in turn at the 7491A output (pin 13). Since the 8-bit word's LSB was the first to be applied to the input of the shift register, it will be the first to appear at the output of the shift register.

11-2-2 Serial-In, Parallel-Out (SIPO) Shift Registers

Figure 11-6 shows the second shift register type, which is called a **serial-in, parallel-out (SIPO) shift register.** As can be seen in the simplified diagram in Figure 11-6(a), data is entered into this shift register type in serial form, and the stored data is available at the output as a parallel word.

Serial-In, Parallel-Out (SIPO) Register A register that will serial-load an input word and parallel-output its stored contents.

Figure 11-6(b) shows how a 4-bit SIPO shift register can be constructed using D-type flip-flops. To input data into this shift register, a 4-bit serial word is applied at the serial data input and shifted in under the control of the clock input (one shift-right/clock pulse). To input or store a 4-bit serial word into this shift register, therefore, would require four clock pulses. The data stored within the shift register is available at the four *Q* outputs (Q_0, Q_1, Q_2, and Q_3) as a 4-bit parallel data output.

Figure 11-7 shows data sheet details for the 74164, which is an 8-bit SIPO shift-right shift register IC. Looking at the logic diagram, you can see that this device has two ANDed serial inputs (*A* and *B*), an active-LOW clear input, and an 8-bit parallel output (Q_A through Q_H).

Application—Serial-to-Parallel Data Converter

Figure 11-8(a) on page 378 shows how a 74164 can be used to convert a serial data bit stream into a parallel data word output. Referring to the waveforms in Figure 11-8(b), you can see that one bit of serial data is loaded into the 74164 for each positive transition of the clock. The clock signal is also fed to a divide-by-eight circuit which divides the clock input by eight and loads the 8-bit data from the 74164 into an 8-bit buffer register (made up of two 74175s) at 8-bit intervals.

11-2-3 Parallel-In, Serial-Out (PISO) Shift Registers

Parallel-In, Serial-Out (PISO) Register A register that will parallel-load an input word and serial-output its stored contents.

The simplified diagram in Figure 11-9(a) on page 379 shows the basic operation of a **parallel-in, serial-out shift register.** With this shift register type, the data bits are entered simultaneously, or in parallel, into the register, and then the binary word is shifted out of the register bit by bit as a serial data stream.

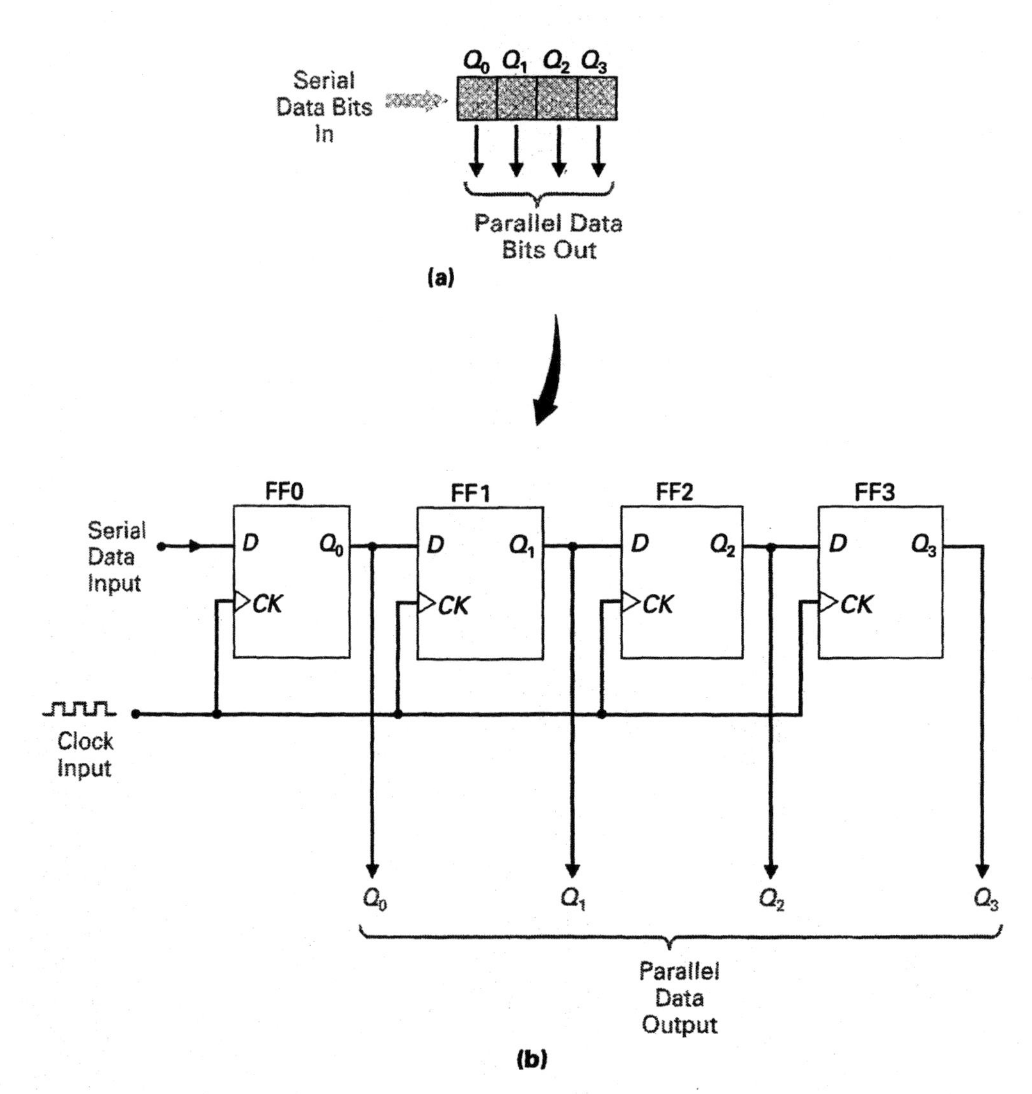

FIGURE 11-6 Serial-In, Parallel-Out (SIPO) Shift Registers.

Figure 11-9(b) shows how a 4-bit PISO shift register can be constructed using D-type flip-flops. The circuit is controlled by the SHIFT/$\overline{\text{LOAD}}$ control input. When the SHIFT/$\overline{\text{LOAD}}$ control line is LOW, all of the shaded AND gates are enabled, due to the inversion of this control signal by the shaded INVERTER. These enabled AND gates connect the 4-bit parallel input word on the data input lines (D_0, D_1, D_2, and D_3) to the data inputs of the flip-flops. When a clock pulse occurs, this 4-bit parallel word will be latched simultaneously into the four flip-flops and appear at the Q outputs (Q_0, Q_1, Q_2, and Q_3). When the SHIFT/$\overline{\text{LOAD}}$ control line is HIGH, all of the unshaded AND gates are enabled. These enabled AND gates connect the Q_0 output to the D input of FF_1, the Q_1 output to the D input of FF_2, and the Q_2 output to the D input of FF_3. In this mode, therefore, the data stored in the shift register will be shifted to the right by one position for every clock pulse applied.

IC TYPE: SN74164– 8-bit Serial-In, Parallel-Out Shift Register

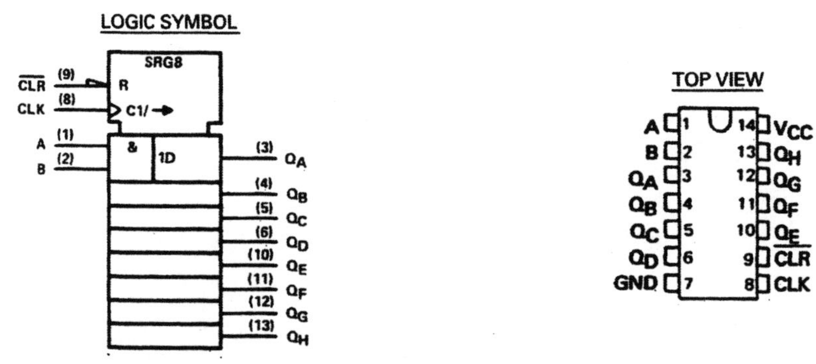

FUNCTION TABLE

INPUTS				OUTPUTS		
$\overline{CLEAR}$	CLOCK	A	B	Q_A	Q_B ...	Q_H
L	X	X	X	L	L	L
H	L	X	X	Q_{A0}	Q_{B0}	Q_{H0}
H	↑	H	H	H	Q_{An}	Q_{Gn}
H	↑	L	X	L	Q_{An}	Q_{Gn}
H	↑	X	L	L	Q_{An}	Q_{Gn}

H = high level (steady state), L = low level (steady state)
X = irrelevant (any input, including transitions)
↑ = transition from low to high level.
Q_{A0}, Q_{B0}, Q_{H0} = the level of Q_A, Q_B, or Q_H, respectively, before the indicated steady-state input conditions were established.
Q_{An}, Q_{Gn} = the level of Q_A or Q_G before the most recent ↑ transition of the clock; indicates a one-bit shift.

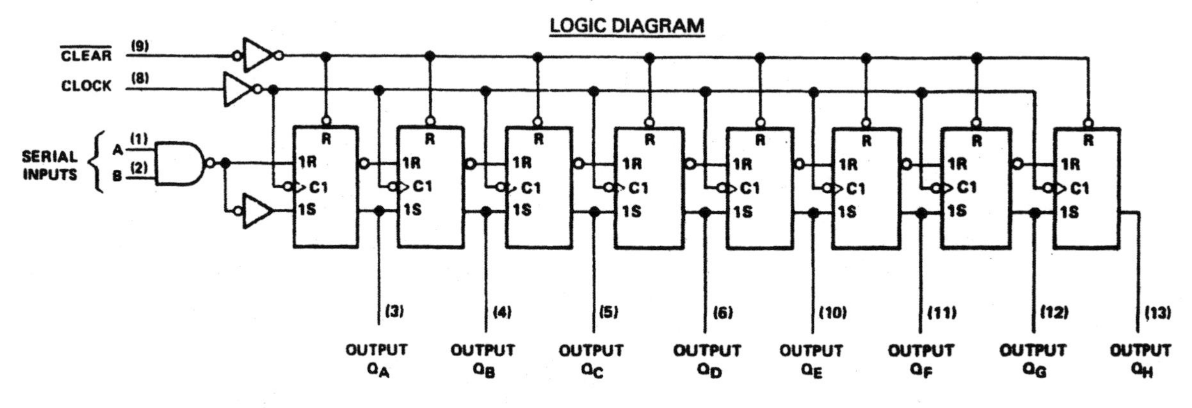

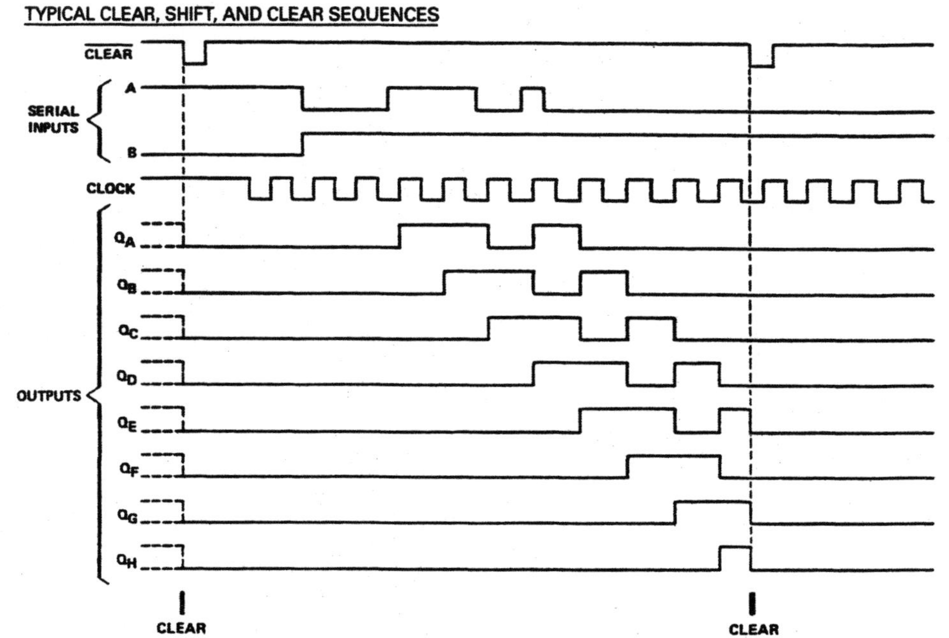

FIGURE 11-7 An IC Containing an Eight-Stage Serial-In, Parallel-Out (SIPO) Shift Register. (Courtesy of Texas Instruments)

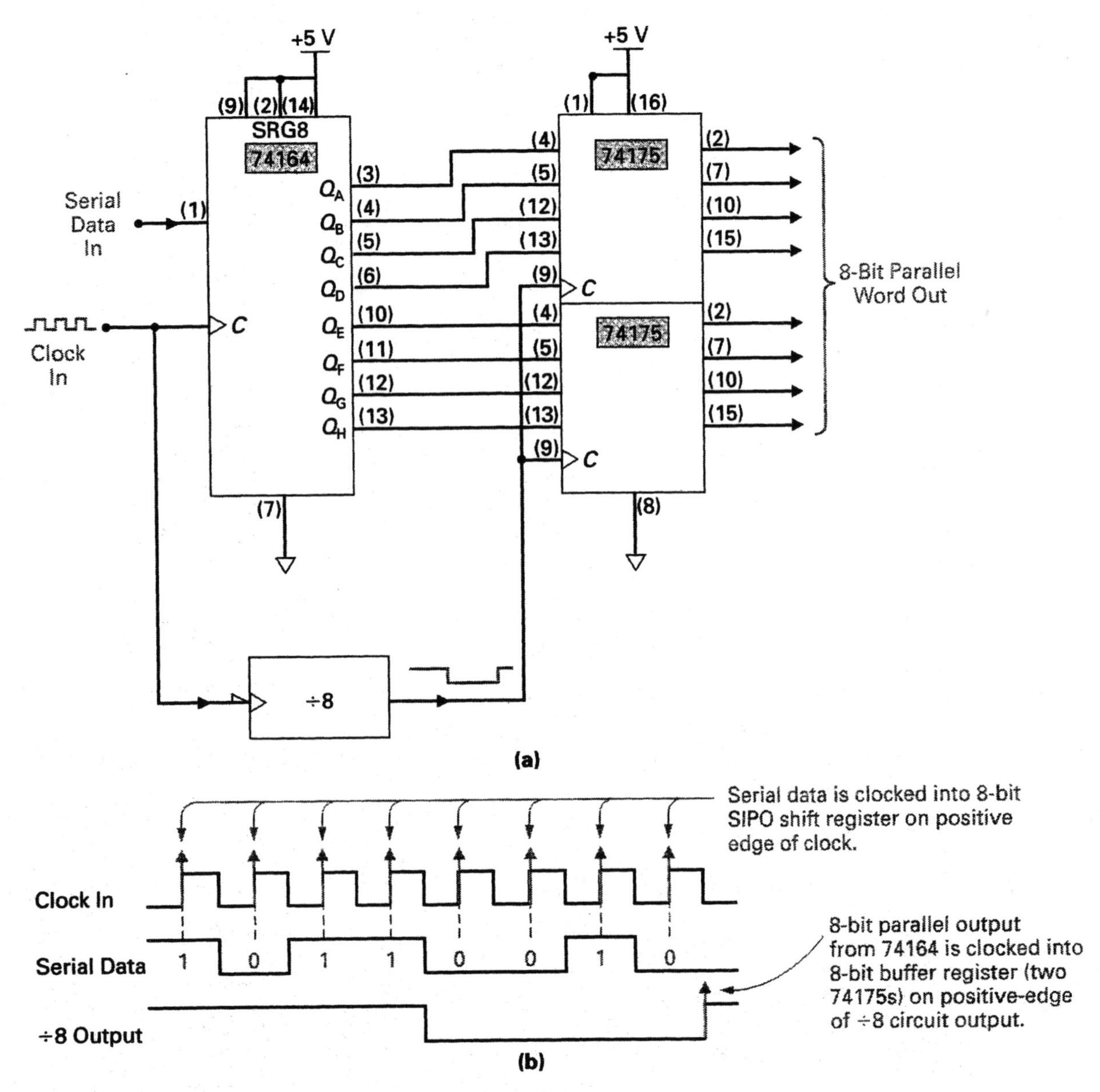

FIGURE 11-8 Serial-to-Parallel Data Converter. (a) Circuit. (b) Waveforms.

Figure 11-10 shows data sheet details for the 74165. This IC can function as either an 8-bit parallel-in (inputs *A*–*H*), serial-out (output Q_H), or an 8-bit serial-in (input SER) shift register. A LOW on the $SH/\overline{LD}$ (pin 1) control line will enable all of the NAND gates and allow simultaneous loading of the eight *S-R* flip-flops. A LOW on any of the parallel *A* through *H* inputs will RESET the respective flip-flop, while a HIGH on any of the parallel *A* through *H* inputs will SET the respective flip-flop. The loading of the register is called an *asynchronous operation* since the loading operation will not be in step with the occurrence of an active clock signal (the loading operation and clock are not synchronized). A HIGH on the $SH/\overline{LD}$ control line will disable all of the NAND gates and shift the stored 8-bit word to the right at a rate of one position for

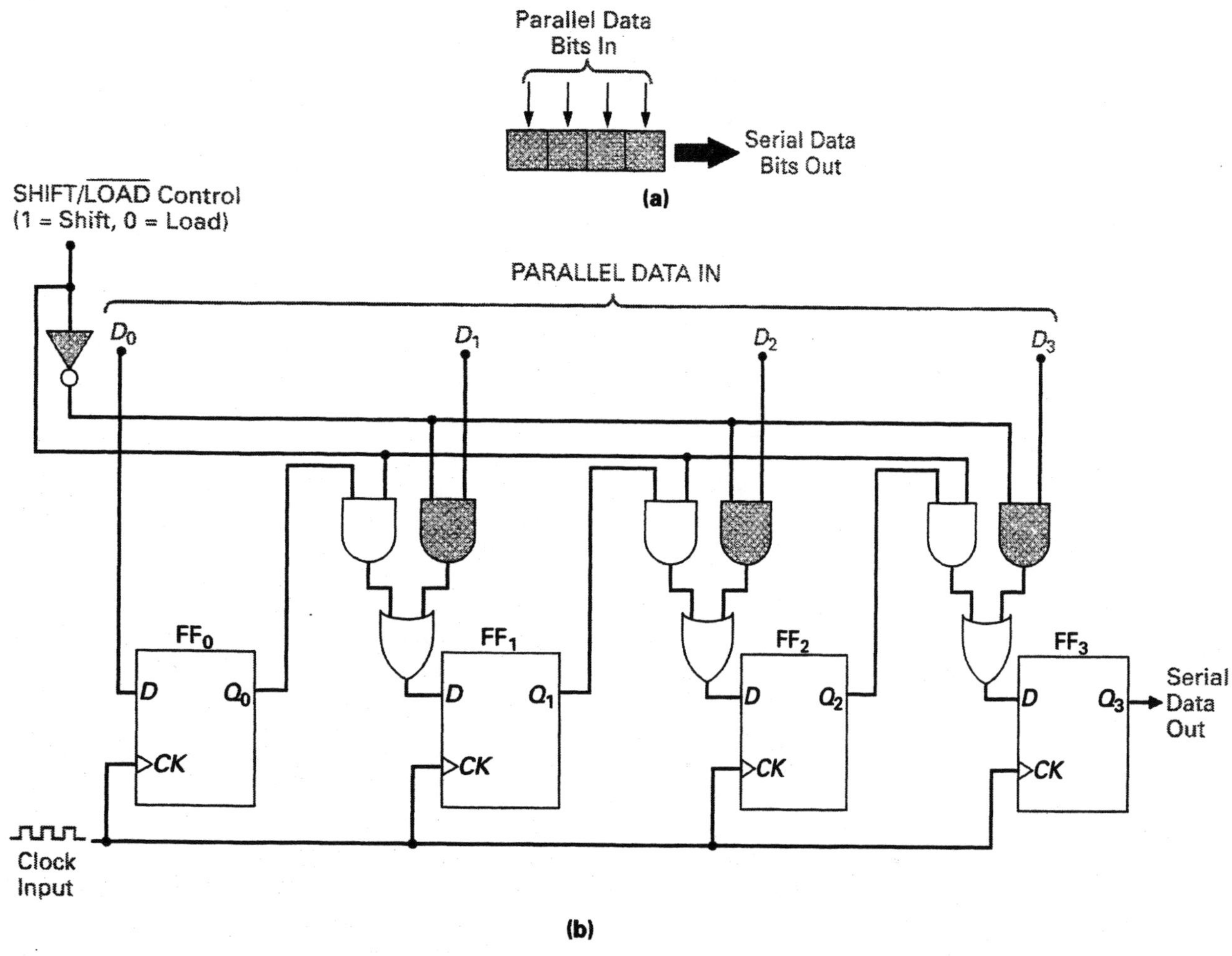

FIGURE 11-9 Parallel-In, Serial-Out (PISO) Shift Registers.

every one clock pulse. The shifting of data within the shift register is called a *synchronous operation* since the shifting of data will be in step or in sync with the occurrence of an active clock signal (the shifting operation and clock are synchronized). The clock input for this IC can be inhibited, if desired, by applying a HIGH to the *CLK INH* (pin 15) input.

Application—Parallel-to-Serial Data Converter

Figure 11-11 shows how a 74165 can be connected to convert an 8-bit parallel data input into an 8-bit serial data output. In this example, switches have been used to supply an 8-bit parallel input of 11010110 to the *A* through *H* inputs of the 74165. Moving SW_1 to the LOAD position will make pin 1 of the 74165 ($SH/\overline{LD}$) LOW, so the 8-bit word applied to the parallel inputs *A* through *H* will be loaded into the 74165's eight internal flip-flops. Since the LSB *H* input is LOW, the output LED will initially be OFF since Q_H is LOW. The 74165 is positive edge-triggered, and therefore as SW_2 is toggled, the stored bits will be shifted to the output (in the order *H, G, F, E, D, C, B,* and then *A*) at a rate of one bit per HIGH/LOW switching of SW_2.

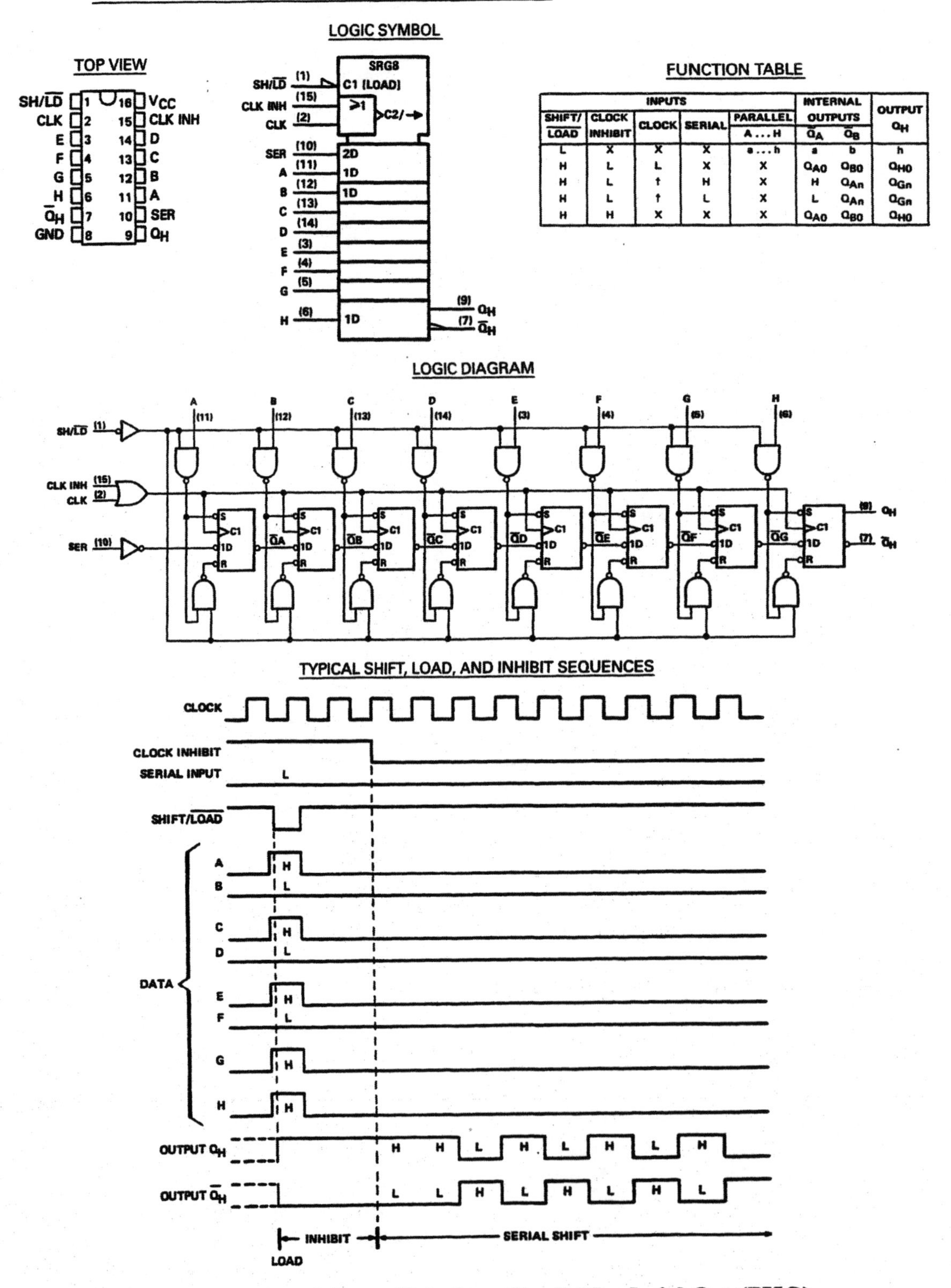

INPUTS					INTERNAL OUTPUTS		OUTPUT
SHIFT/$\overline{\text{LOAD}}$	CLOCK INHIBIT	CLOCK	SERIAL	PARALLEL A . . . H	$\overline{Q}_A$	$\overline{Q}_B$	Q_H
L	X	X	X	a . . . h	a	b	h
H	L	L	X	X	Q_{A0}	Q_{B0}	Q_{H0}
H	L	↑	H	X	H	Q_{An}	Q_{Gn}
H	L	↑	L	X	L	Q_{An}	Q_{Gn}
H	H	X	X	X	Q_{A0}	Q_{B0}	Q_{H0}

FIGURE 11-10 An IC Containing an Eight-Stage Parallel-In, Serial-Out (PISO) Shift Register or Serial-In, Serial-Out (SISO) Shift Register. (Courtesy of Texas Instruments)

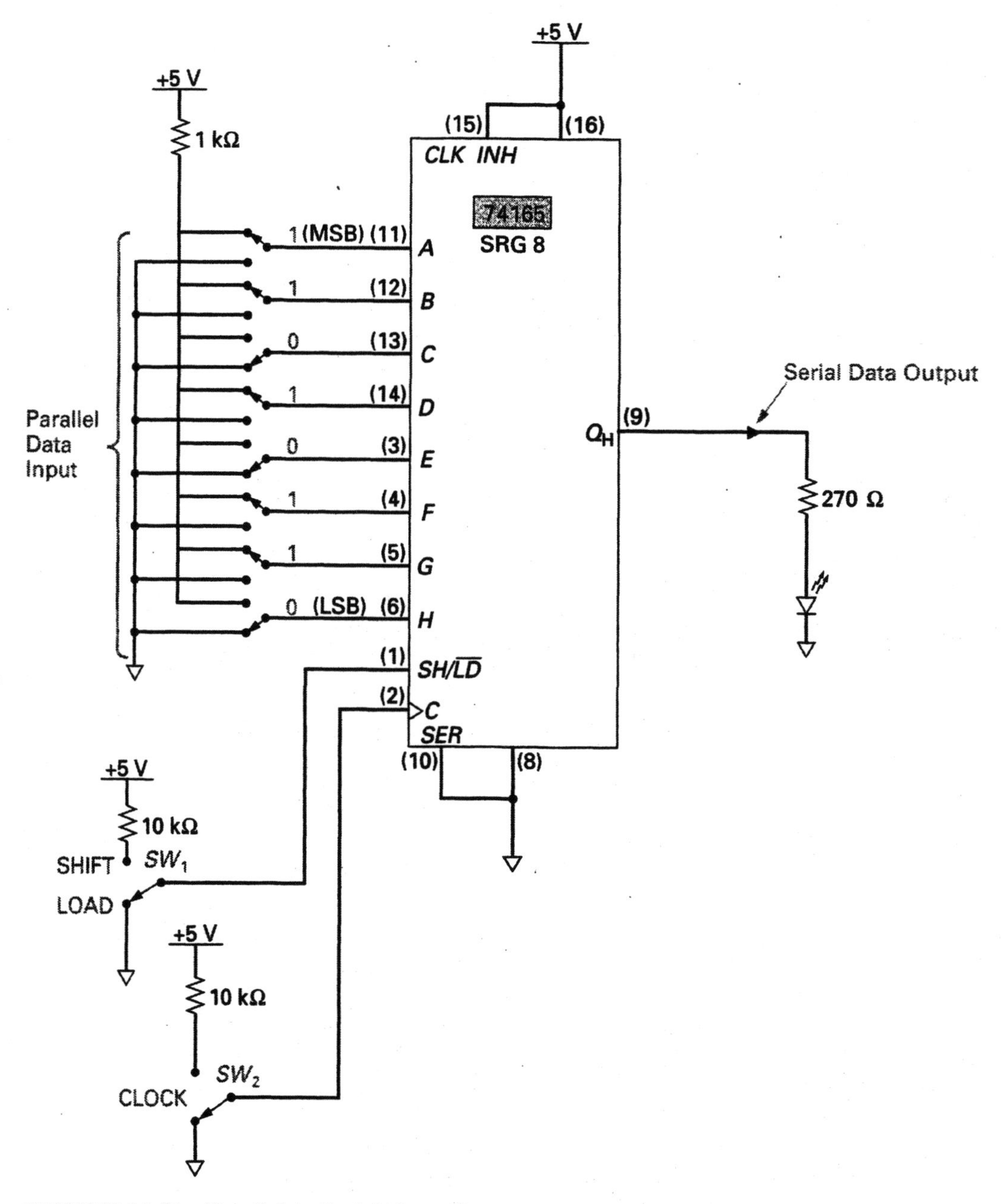

FIGURE 11-11 Parallel-to-Serial Data Converter.

11-2-4 Bidirectional Universal Shift Register

A bidirectional shift register can shift stored data either left or right depending on the logic level applied to a control line. The 74194 IC is frequently used in applications requiring bidirectional shifting of data. Figure 11-12 shows data sheet details for the 74194. This IC contains forty-six gates and can be made to operate in one of four modes by applying different logic levels to the mode control inputs S_0 and S_1. These modes of operation are listed on the next page.

IC TYPE: SN74194– 4-bit Bidirectional Universal Shift Register

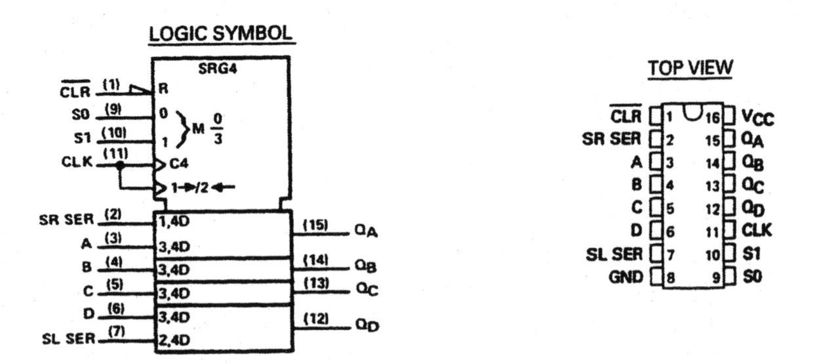

FUNCTION TABLE

INPUTS										OUTPUTS			
CLEAR	MODE		CLOCK	SERIAL		PARALLEL				Q_A	Q_B	Q_C	Q_D
	S1	S0		LEFT	RIGHT	A	B	C	D				
L	X	X	X	X	X	X	X	X	X	L	L	L	L
H	X	X	L	X	X	X	X	X	X	Q_{A0}	Q_{B0}	Q_{C0}	Q_{D0}
H	H	H	↑	X	X	a	b	c	d	a	b	c	d
H	L	H	↑	X	H	X	X	X	X	H	Q_{An}	Q_{Bn}	Q_{Cn}
H	L	H	↑	X	L	X	X	X	X	L	Q_{An}	Q_{Bn}	Q_{Cn}
H	H	L	↑	H	X	X	X	X	X	Q_{Bn}	Q_{Cn}	Q_{Dn}	H
H	H	L	↑	L	X	X	X	X	X	Q_{Bn}	Q_{Cn}	Q_{Dn}	L
H	L	L	X	X	X	X	X	X	X	Q_{A0}	Q_{B0}	Q_{C0}	Q_{D0}

H = high level (steady state)
L = low level (steady state)
X = irrelevant (any input, including transitions)
↑ = transition from low to high level
a, b, c, d = the level of steady-state input at inputs A, B, C, or D, respectively.
Q_{A0}, Q_{B0}, Q_{C0}, Q_{D0} = the level of Q_A, Q_B, Q_C, or Q_D, respectively, before the indicated steady-state input conditions were established.
Q_{An}, Q_{Bn}, Q_{Cn}, Q_{Dn} = the level of Q_A, Q_B, Q_C, respectively, before the most-recent ↑ transition of the clock.

FIGURE 11-12 An IC Containing a 4-Bit Universal Shift Register. (Courtesy of Texas Instruments)

S_1	S_0	MODE OF OPERATION
H	H	Parallel (broadside) Load
L	H	Shift Right (Q_A is shifted towards Q_D)
H	L	Shift Left (Q_D is shifted towards Q_A)
L	L	Inhibit Clock (do nothing)

To parallel-load a 4-bit word, therefore, the S_1 and S_0 mode control inputs must both be HIGH, and the 4-bit word to be loaded must be applied to inputs *A*, *B*, *C*, and *D*. Parallel loading is accomplished synchronously (in step or in sync) with an active clock signal, which for the 74194 is a positive clock transition. Once loaded, the stored 4-bit word will appear at the outputs Q_A, Q_B, Q_C and Q_D. During a parallel loading operation, serial data flow is inhibited.

To shift data to the right, the mode control input S_1 must be LOW and the mode control input S_0 must be HIGH. In this mode, data within the shift register will be shifted right synchronously from Q_A to Q_D, with new data being inputted from the shift-right serial input (pin 2).

To shift data to the left, the mode control input S_1 must be HIGH and the mode control input S_0 must be LOW. In this mode, data within the shift register will be shifted left synchronously from Q_D to Q_A, with new data being inputted from the shift-left serial input (pin 7).

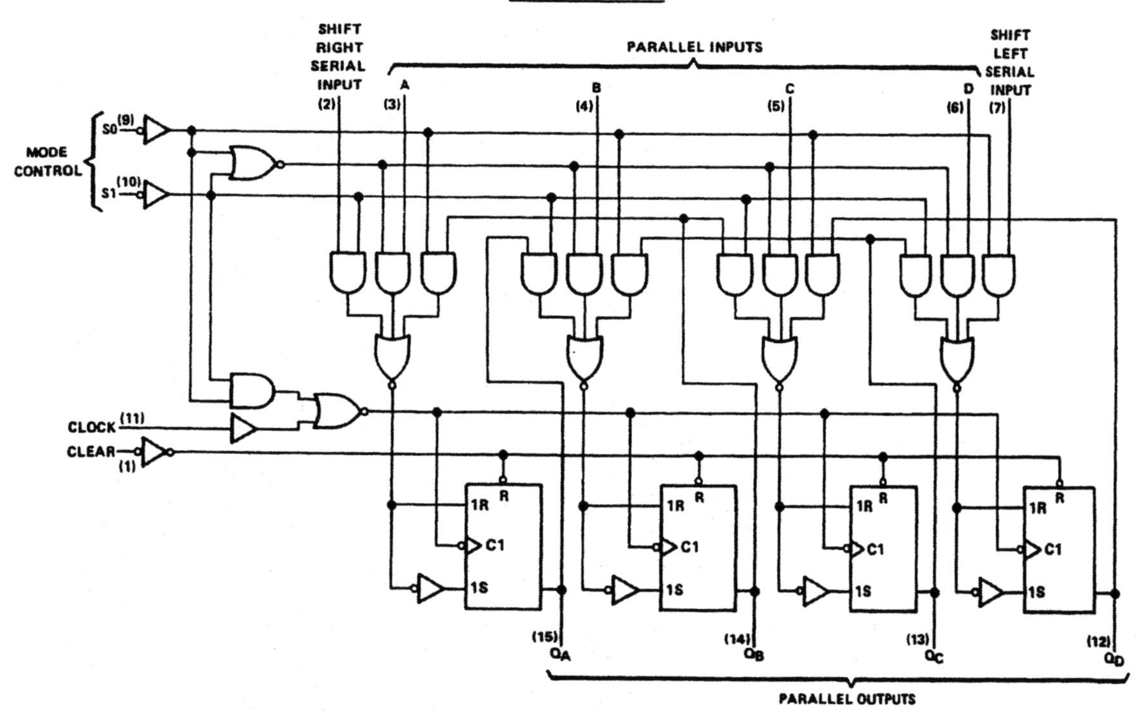

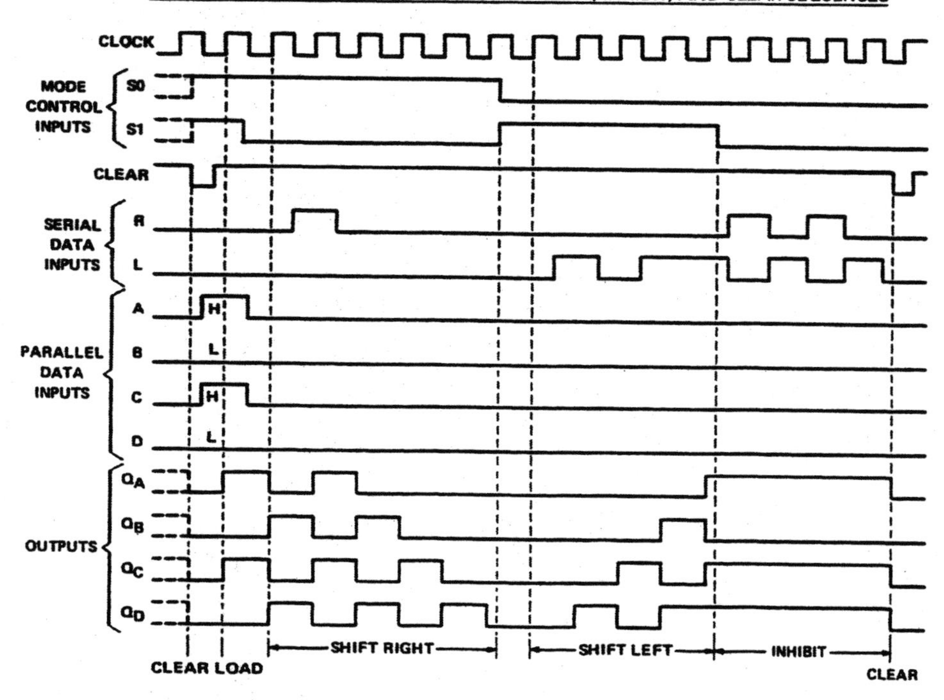

FIGURE 11-12 Continued.

When both the mode control inputs S_1 and S_0 are LOW, the clocking of the shift register is inhibited. The mode control inputs S_1 and S_0 should only be changed when the clock input is HIGH (inactive).

Application—Shift Register Sequence Generator

Figure 11-13 shows how two 74194 ICs could be connected to form an 8-bit shift-right sequence generator. Switches SW_0 through SW_7 can be set so that they apply any 8-bit pattern or word value (open switch = HIGH, closed switch = LOW). This 8-bit word can be loaded into the shift register by pressing the normally closed push-button (NCPB) SW_8. SW_8 controls the logic level applied to the mode control input S_1 (pin 10), as shown in the table in Figure 11-13. With S_0 connected permanently HIGH, an open SW_8 (switch pressed) will apply an equivalent HIGH input to S_1, which will instruct the two 74194s to parallel-load the 8-bit word which, when loaded by a positive transition of the clock, will appear at the Q_A, Q_B, Q_C, and Q_D outputs. When SW_8 is closed (switch released) the S_1 mode control input will be pulled LOW instructing both 74194s to shift right the data stored synchronously with each positive edge of the clock.

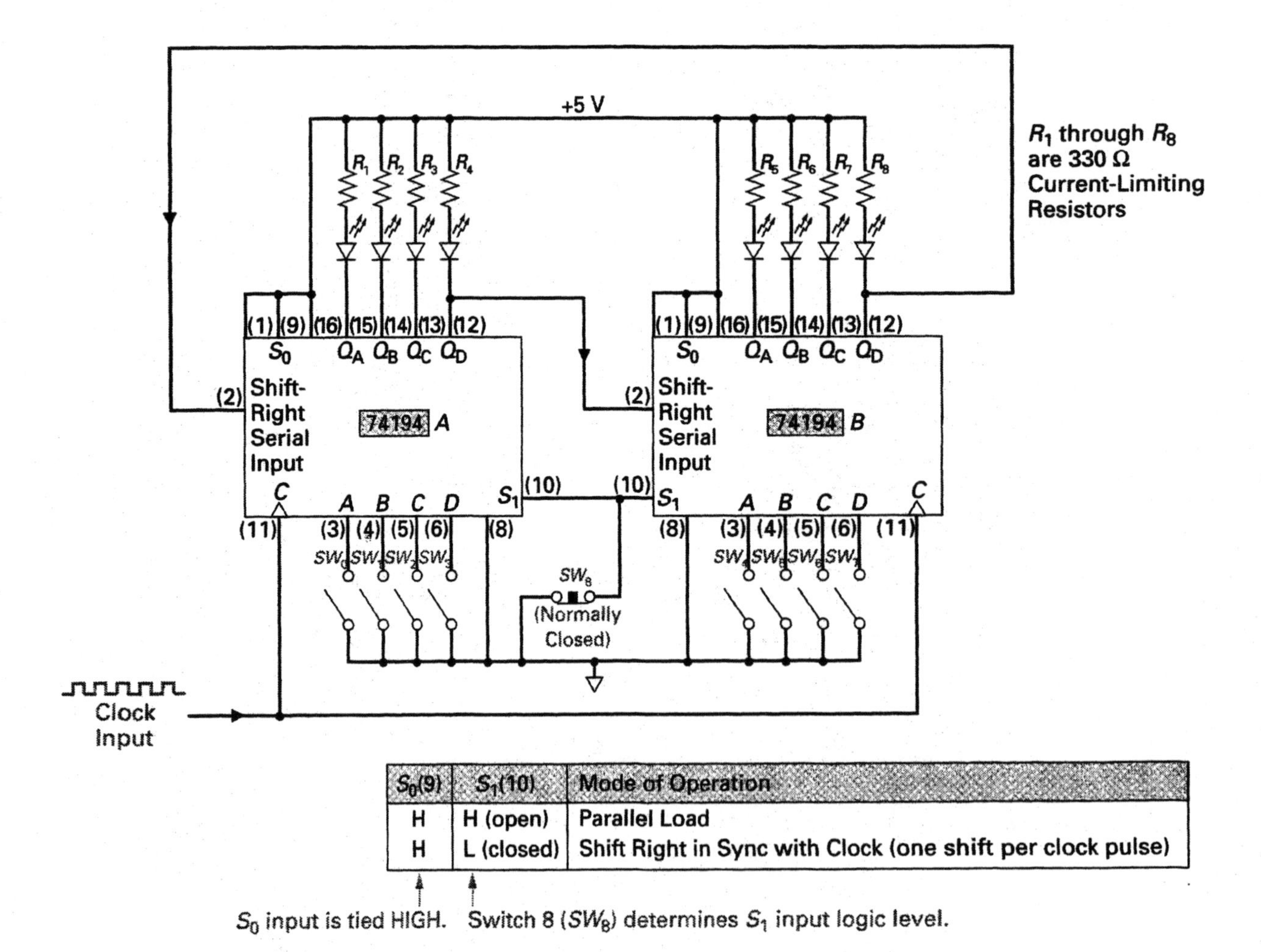

S_0(9)	S_1(10)	Mode of Operation
H	H (open)	Parallel Load
H	L (closed)	Shift Right in Sync with Clock (one shift per clock pulse)

S_0 input is tied HIGH. Switch 8 (SW_8) determines S_1 input logic level.

FIGURE 11-13 Shift-Right Sequence Generator.

In this shift-right mode, the Q_D output of 74194 *B* on the right of Figure 11-13 is connected back to the shift-right serial input of the 74194 *A*, and therefore the data is continually rotated right.

The LEDs in Figure 11-13 form a common-anode display, which means that a HIGH output will turn OFF an LED, and a LOW output will turn ON an LED.

SELF-TEST REVIEW QUESTIONS FOR SECTION 11-2

1. How many clock pulses would be needed to serially load an 8-bit SISO shift register with an 8-bit word?
2. The ____________ shift register could be used for serial-to-parallel conversion, while the ____________ shift register could be used for parallel-to-serial conversion. (SISO, SIPO, PISO)
3. How many clock pulses would be needed to parallel-load an 8-bit PISO shift register with an 8-bit word?
4. The ____________ shift register can parallel-load, shift right, shift left, input serially, and output serially.

11-3 THREE-STATE OUTPUT REGISTERS

Many registers have three-state or tri-state output gates connected between the flip-flop's *Q* outputs and the IC's output pins, as illustrated in Figure 11-14(a). The three-state output gate, which was discussed previously in Section 4-1-5, has as its name implies three output states, which are LOW (logic 0), HIGH (logic 1), and float (an open output or HIGH impedance state). With the active-HIGH three-state gates shown in Figure 11-14(a), a HIGH on the enable output control line (*E*) will connect the *Q* outputs from the D-type flip-flops directly through to the *Y* outputs of the IC. On the other hand, when the enable output control line is LOW, the three-state gates are disabled and the *Y* outputs of the IC will float since they have effectively been disconnected from the *Q* outputs. Figure 11-14(b) shows how a small inverted triangle on each output is used to indicate that this 4-bit controlled buffer register (PIPO) has *three-state output control.*

Referring to Figure 11-14(a), let us now see how this PIPO buffer register with three-state output control operates. Other than the enable output control line (*E*), this register has a load control line (*L*) and a clock input (*C*). When the load control line is LOW (inactive), the contents of each of the D-type flip-flops within the register remain unchanged since the data is recirculated from each *Q* output back to its respective *D* input for each positive edge of the clock input via the enabled unshaded AND gates. When the load control input is HIGH, however, all of the shaded AND gates are enabled, so the parallel input word on the D_0, D_1, D_2, and D_3 inputs are connected to the flip-flop data inputs and latched into the flip-flops on the next positive edge of the clock.

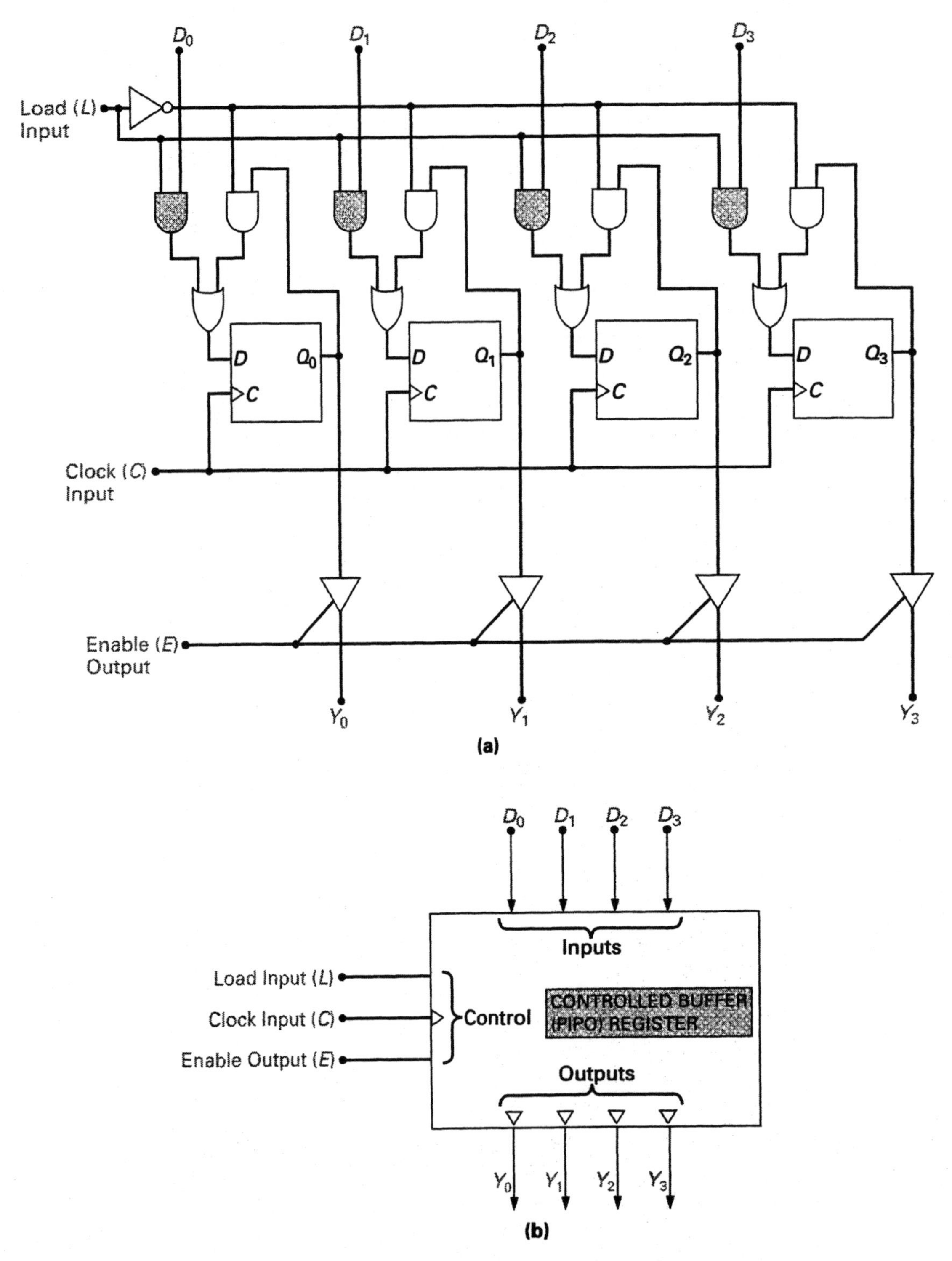

FIGURE 11-14 A Controlled Buffer Register with Three-State Output.

The three-state gate was originally developed by National Semiconductor in the early 1970s, and its purpose was to reduce the number of wires in a digital system. Three-state gates made possible a **bus-organized digital system.** To explain this concept, Figure 11-15 shows a simple circuit consisting of four input switches, three registers, and four output LEDs. A group of wires used to carry binary data is called a **bus,** and in Figure 11-15 a 4-bit bus is used as a common transmission path between the input device, output device, and the registers. Since the input device and the registers all have three-state outputs, all of these devices (unless enabled) are effectively isolated or disconnected from the bus. If they were not disconnected, a *bus conflict* would occur since one register could be driving a bus line HIGH while another register could be trying to drive that same bus line LOW.

Bus-Organized Digital System
A digital system interconnection technique in which a common bus is time-shared by all the devices connected to the bus.

Bus
A group of conductors used as a path for transmitting digital information from one of several sources to one of several destinations.

To transfer a 4-bit word from register A to register B, we must connect the contents of register A on to the bus making E_A HIGH, and then load this data into register B by making L_B HIGH. Since the clock (CLK) is connected to all registers, the next positive edge of the clock will load the 4-bit word from register A into register B.

If we now wanted to transfer the contents of register B to the output, we would simply make E_B HIGH so that contents of the B register would now be connected to the bus, and L_O HIGH so that the data on the bus will be loaded into the output register. On the next positive clock edge, the contents of the B register will be stored in the output register and displayed immediately on the output LEDs, since the output of the output register has no three-state control. The output register does not connect to the common bus, so its output can therefore be permanently enabled.

Before three-state outputs were available, separate groups of wires were needed between all of the devices in a circuit. For example, without three-state control, the circuit in Figure 11-15 would need a set of wires connecting register A to register B, another set connecting A to C, another set connecting B to C, another set connecting the input to A, and so on. As you can easily imagine, even the smallest of digital electronic systems containing only several devices would need a mass of connecting wires. With a bus-organized digital system, only one set of connecting wires is needed, and these can be connected to all devices as long as those devices that transmit data onto the common bus have three-state output control.

In Figure 11-15, only a 4-bit bus was used. In many digital systems 8-bit (8 wires), 16-bit (16 wires) and 32-bit (32 wires) buses are common, and are used to transfer 8-bit, 16-bit, or 32-bit parallel words from one device to another. In schematic diagrams, a solid bar with arrows is generally used to represent a bus of any size, as shown in the inset in Figure 11-15.

The 74173 IC is a 4-bit (PIPO) buffer register with three-state output control. Data sheet details describing this IC are given in Figure 11-16. This IC contains four D-type flip-flops with an active-HIGH clear input (pin 15). The three-state output buffers are controlled by the active-LOW inputs M and N (pins 1 and 2). When both of these output control inputs (M and N) are LOW, the HIGH or LOW logic levels stored in the four flip-flops will be available at the four outputs (pins 3, 4, 5, and 6). The parallel loading of data into the flip-flops is controlled by the active-LOW ANDed inputs $\overline{G}_1$ and $\overline{G}_2$ (pins 9 and

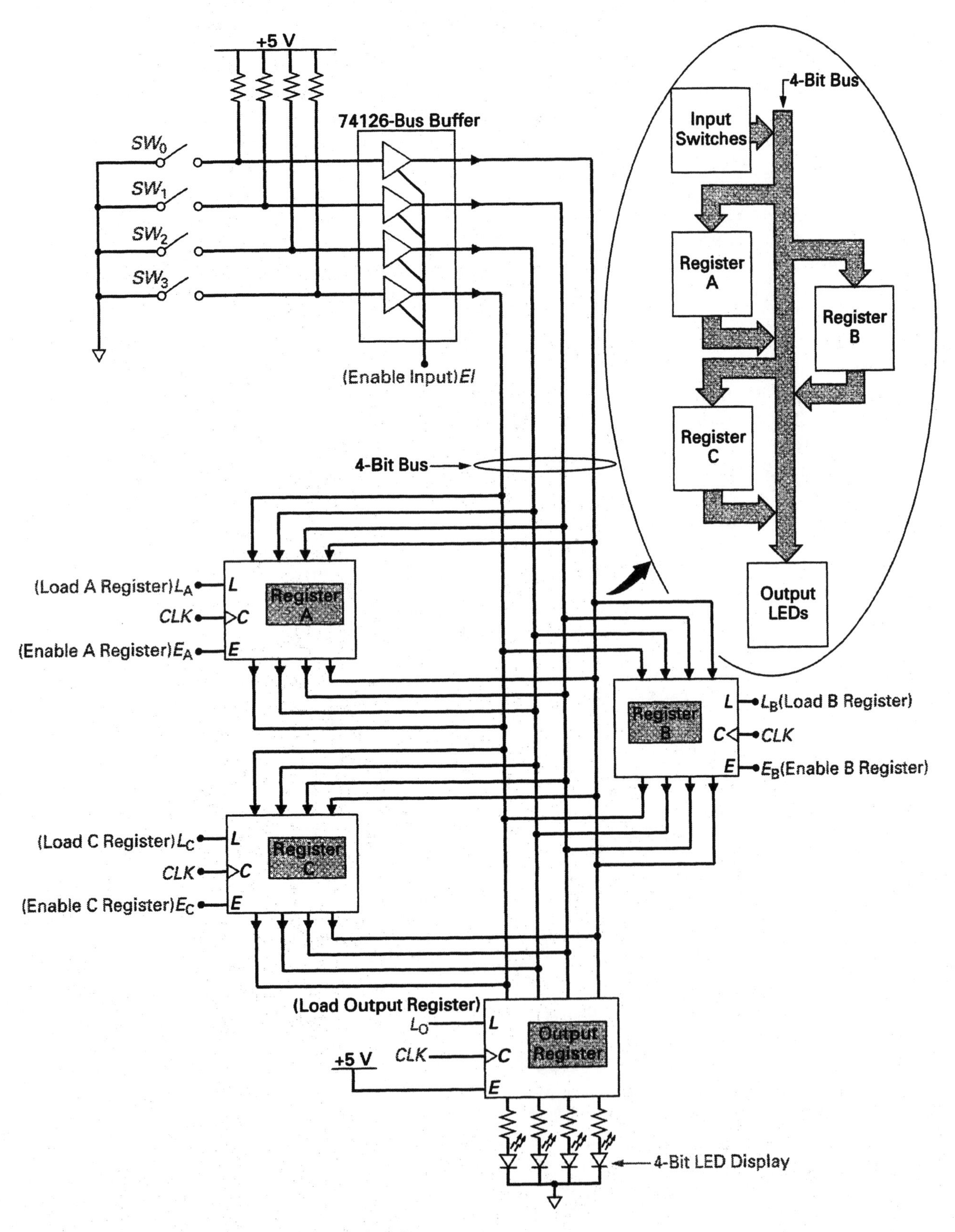

FIGURE 11-15 A Bus-Organized Digital System.

IC TYPE: SN74173– 4-bit D-Type Register with Three-State Outputs

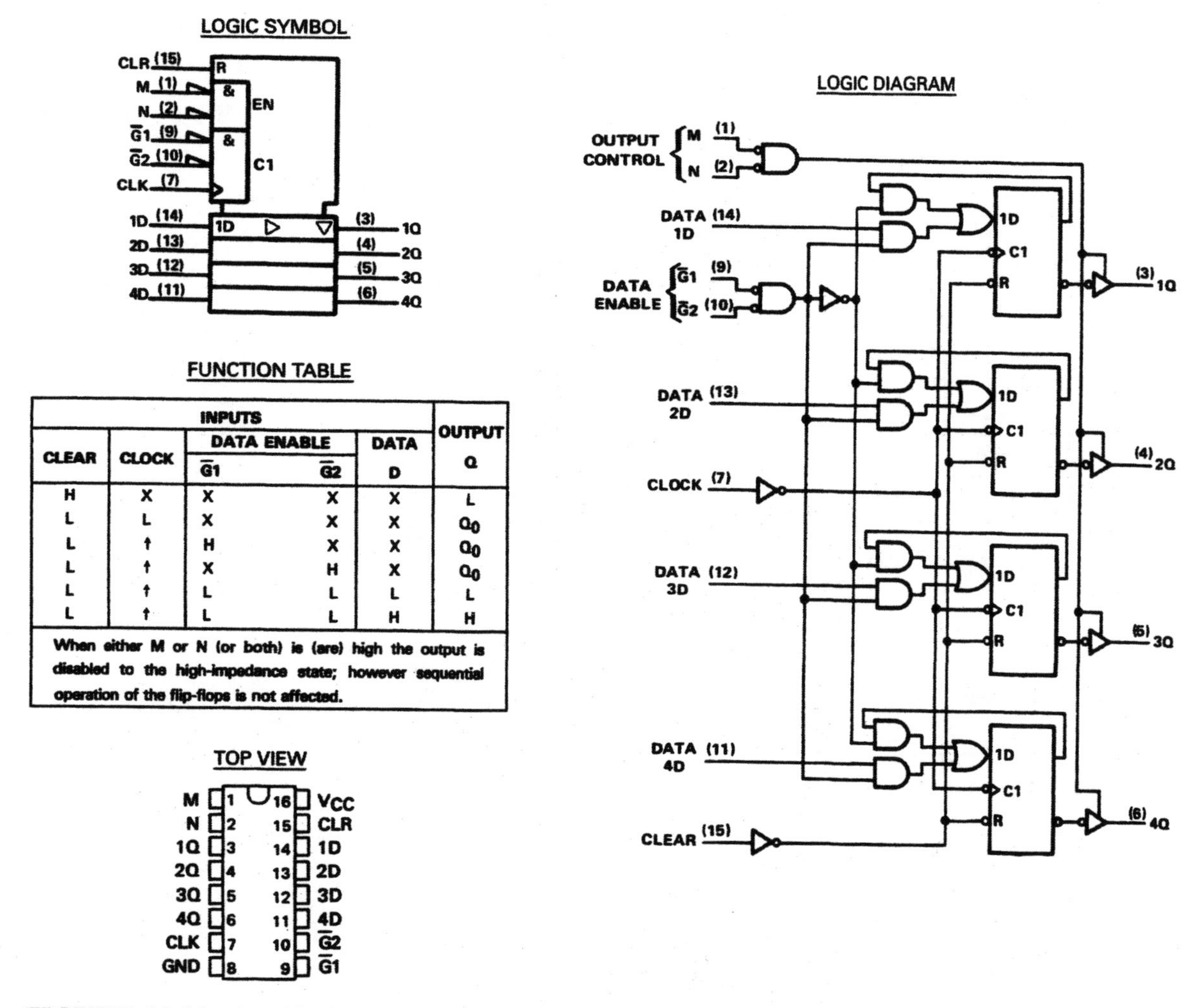

FUNCTION TABLE

INPUTS					OUTPUT
		DATA ENABLE		DATA	
CLEAR	CLOCK	$\overline{G1}$	$\overline{G2}$	D	Q
H	X	X	X	X	L
L	L	X	X	X	Q_0
L	↑	H	X	X	Q_0
L	↑	X	H	X	Q_0
L	↑	L	L	L	L
L	↑	L	L	H	H

When either M or N (or both) is (are) high the output is disabled to the high-impedance state; however sequential operation of the flip-flops is not affected.

FIGURE 11-16 An IC Containing a 4-Bit D-Type Register With Three-State Outputs. (Courtesy of Texas Instruments)

10). When both of these data enable inputs are LOW, the data at the D-inputs (pins 14, 13, 12, and 11) are loaded into their respective flip-flops on the net positive transmission clock input.

Application—A Bus-Organized Digital System

Figure 11-17 shows how four 74173s and one 74126 (quad buffer gates with three-state outputs) could be connected to construct the bus-organized digital system discussed previously in Figure 11-15.

■ EXAMPLE:

Referring to the circuit in Figure 11-17, list which control lines need to be made active in order to perform the following operations.

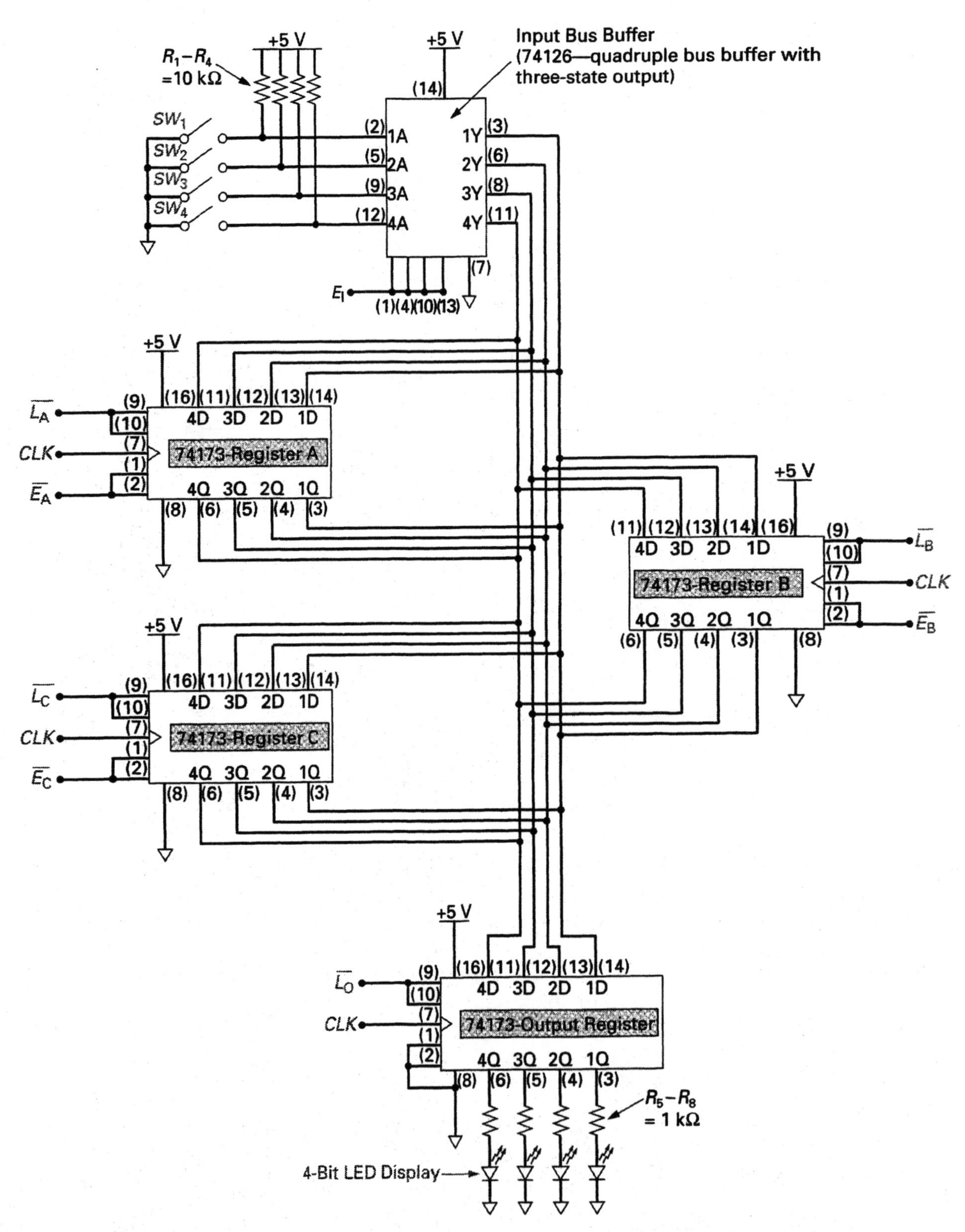

FIGURE 11-17 A 4-Bit Bus-Organized Digital System Containing Four Input Switches and an Input Buffer, Three PIPO Storage Registers, and an Output Register Driving a Four-LED display.

a. Load the 4-bit word from the input bus buffer into register A.
b. Copy the contents of register B into register C.
c. Display the contents of register C.

■ ***Solution:***

a. E_I(HIGH), $\overline{L}_A$(LOW), $CLK\uparrow$
b. E_B(HIGH), $\overline{L}_C$(LOW), $CLK\uparrow$
c. E_C(HIGH), $\overline{L}_O$(LOW), $CLK\uparrow$

SELF-TEST REVIEW QUESTIONS FOR SECTION 11-3

1. A three-state output can be either HIGH, LOW, or ____________.
2. A group of parallel wires designed to carry binary data is called a ____________.
3. What is bus conflict, and how is it avoided?
4. What are the three main inputs to a controlled buffer register with three-state output control?

11-4 REGISTER APPLICATIONS

In this section we will review the register applications discussed so far in this chapter, and examine other register uses.

11-4-1 Memory Registers

Since a register is basically a storage or memory element used to store a binary word, it is often referred to as a **memory register.** A memory unit may contain one memory register capable of storing only one binary word, or many memory registers capable of storing many binary words. Figure 11-18 shows how a controlled buffer register PIPO with three-state output control can be configured to function as a memory register. In order to perform as a memory register, a register must be able to achieve two basic operations, which are referred to as **read** (which is an input operation, since you are reading the contents of the register in the same way that we read or input words in a book), and **write** (which is an output operation, since you are writing data into the register in the same way that you write or output words onto a pad). These two operations are controlled by a READ/$\overline{\text{WRITE}}$ ($R/\overline{W}$) control line, which is used to control the buffer register's load and enable control inputs, as shown in Figure 11-18. When the $R/\overline{W}$ control line is LOW, the load control input is enabled while the enable control input is disabled, so new data is stored or written into the memory register from the bus (a write operation). When the $R/\overline{W}$ control line is HIGH, the enable control input is active while the load control input is inactive, so the 8 bits of stored data within the register are retrieved, or read, from the

Memory Register
A register capable of storing several binary words.

Read
An operation in which data is acquired from some source.

Write
An operation in which data is transferred to some destination.

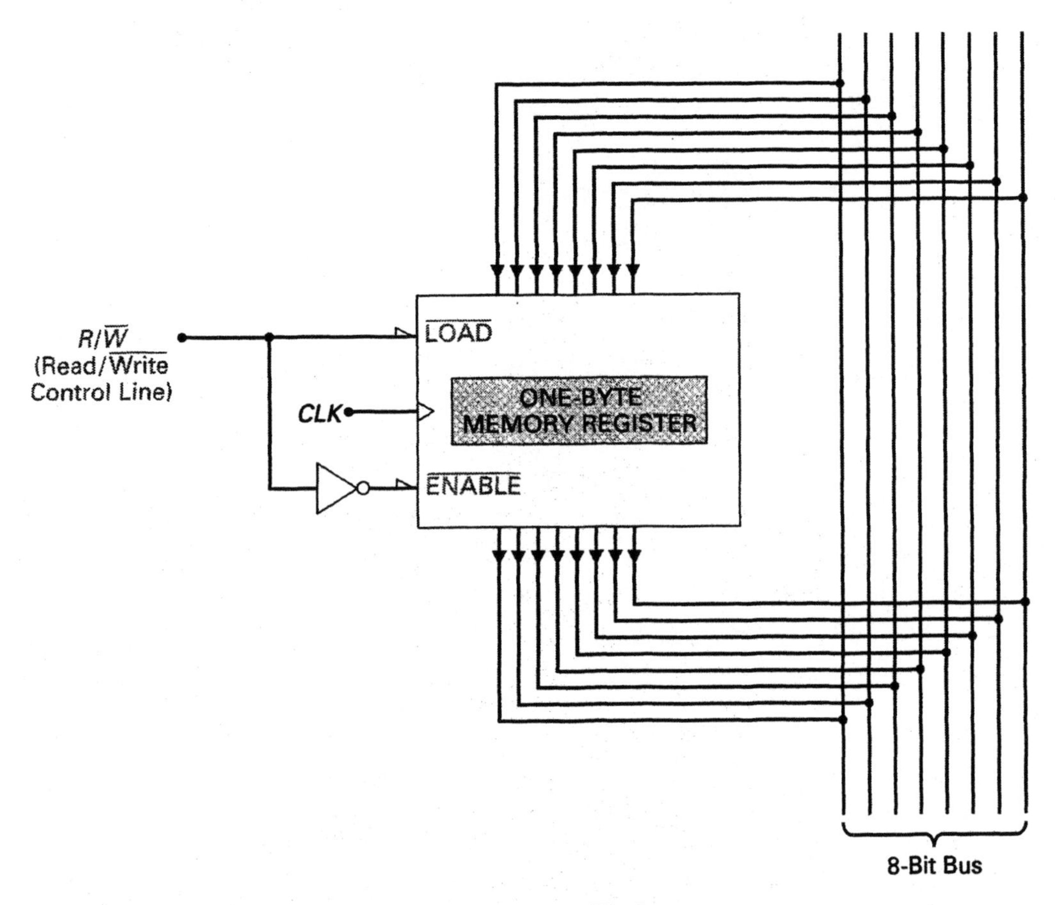

FIGURE 11-18 A One-Byte (8-Bit) Memory Register.

register on to the bus (read operation). This basic memory circuit, as well as other complete memory circuits, will be covered in more detail in Chapter 14.

11-4-2 Serial-to-Parallel and Parallel-to-Serial Conversions

Figure 11-19(a) shows how a SIPO shift register can be used for serial-to-parallel conversion. This shift register shifts in an 8-bit serial input at a rate of one bit per clock pulse, and after eight clock pulses all of the byte is stored and all bits are simultaneously available as an 8-bit parallel output word. Figure 11-19(b) shows how a PISO shift register can be used for parallel-to-serial conversion. This shift register simultaneously loads all 8-bits of a parallel input, and then shifts the data out serially at a rate of one bit per clock pulse, and therefore after eight clock pulses the byte has been outputted.

UART
An abbreviation for universal asynchronous receiver/transmitter. A communication device that can interface a parallel-word system to a serial communication network.

The ability of a shift register to convert binary data from serial-to-parallel or parallel-to-serial is made use of in an interfacing (data transfer) device known as a **UART,** which is an abbreviation for **universal asynchronous receiver and transmitter.** Data within a digital system is generally transferred from one device to another in parallel format on a bus, as shown in Figure 11-19(c). In many

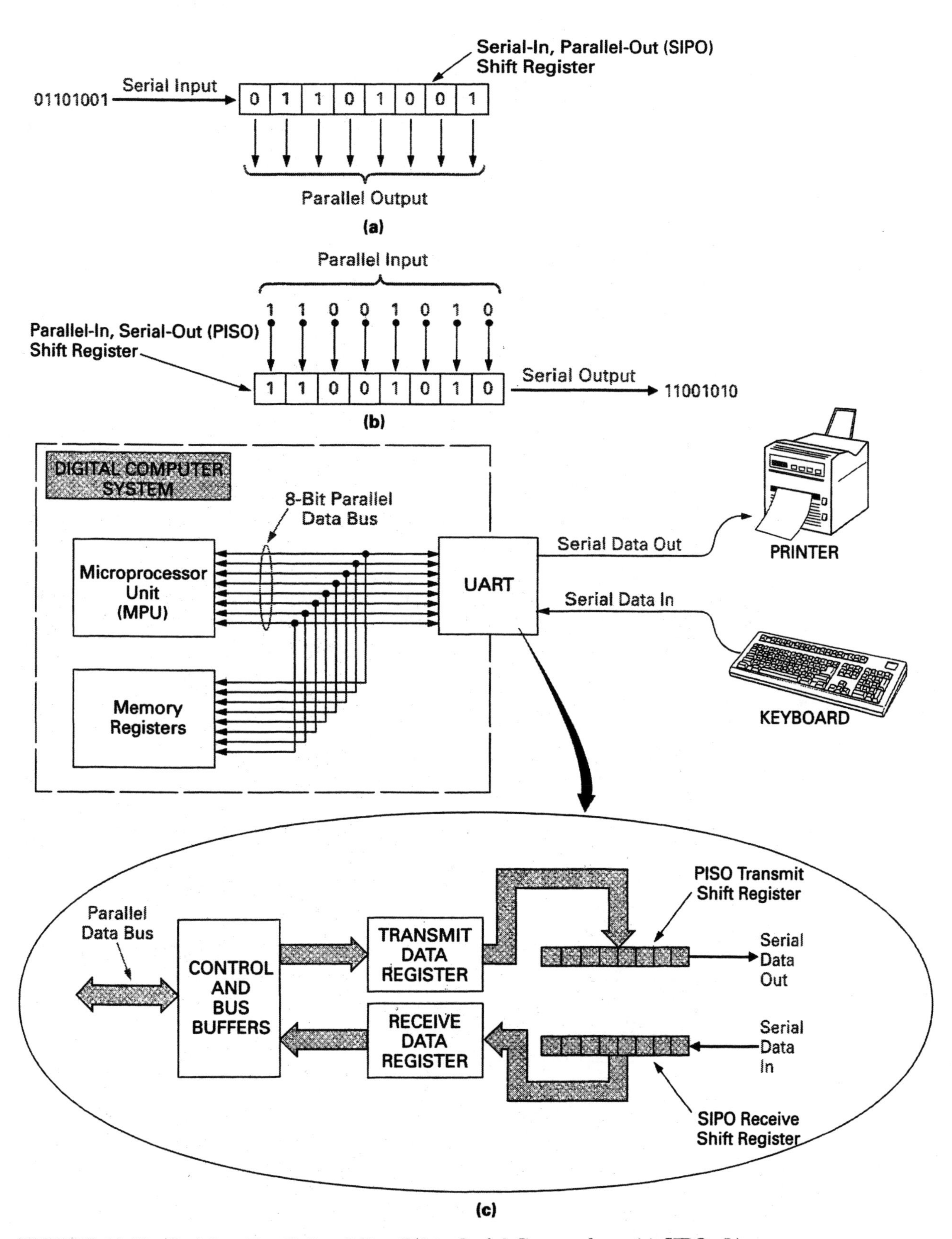

FIGURE 11-19 Serial-to-Parallel and Parallel-to-Serial Conversions. (a) SIPO. (b) PISO. (c) UART.

cases these digital electronic systems must communicate with external equipment that handles data in serial format. Under these circumstances, a UART can be used to interface the parallel data format within a digital electronics system to the serial data format of external equipment. The inset in Figure 11-19(c) shows the basic block diagram of a UART's internal circuit. In this particular example, the UART will be receiving data in parallel format from the microprocessor in the digital system via an 8-bit data bus. This parallel data will be converted to serial data and transmitted to a printer. The serial data received by the UART from a keyboard is converted to parallel data and then transmitted to the digital system microprocessor via the parallel 8-bit data bus. To carry out these operations, you can see from the inset in Figure 11-19(c) that the UART makes use of a PISO shift register to transmit serial data and a SIPO shift register to receive serial data in.

11-4-3 *Arithmetic Operations*

The shift register can also be used to perform arithmetic operations such as multiplication and division. Since binary is a base-2 number system, binary numbers can be changed by some power of two by simply moving the position of the number. For example, the contents of the upper shift register in Figure 11-20(a) is 00000111 (decimal 7). By shifting the contents once to the left and shifting a zero into the least significant bit (LSB) position, we can multiply the initial number of 7 by 2 to obtain 14 ($7 \times 2 = 14$). If we continue to shift to the left and add zeros, the value within the shift register in Figure 11-20(a) is continually doubled. The multiplying factor is equal to 2^n where n is the number of times the contents are shifted. In the example in Figure 11-20(a), the contents of the shift register are shifted three times, and therefore the original number is multiplied by $2^3 = 8$ ($7 \times 8 = 56$).

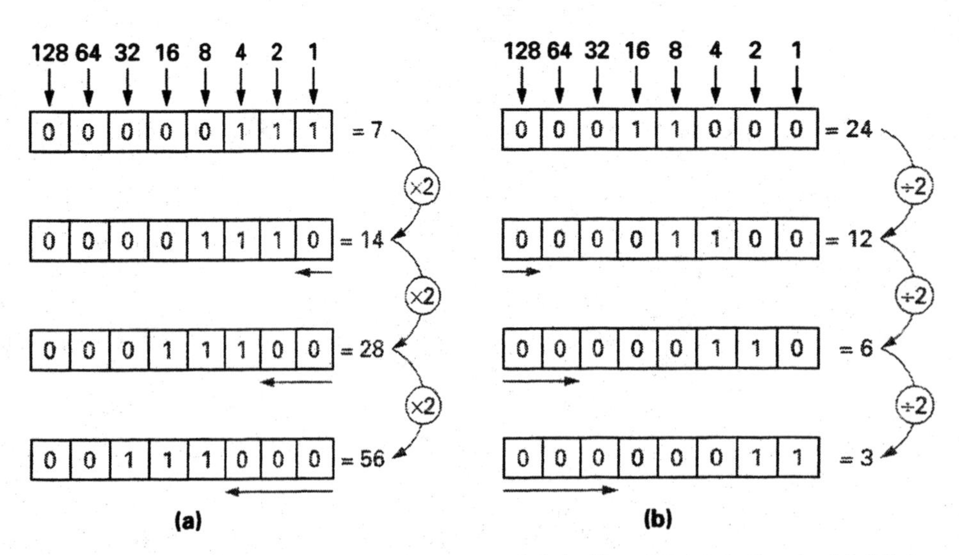

FIGURE 11-20 Shifting Binary Numbers. (a) Left to Multiply by 2. (b) Right to Divide by 2.

If shifting a binary value to the left causes it to double (× 2), then shifting a binary value to the right should cause it to halve (÷ 2). As an example, the contents of the upper shift register in Figure 11-20(b) is 00011000 (decimal 24). By shifting the contents once to the right, and shifting a zero into the most significant bit (MSB) position, we can divide the initial number of 24 by 2 to obtain 12 (24 ÷ 2 = 12). If we continue to shift to the right and add zeros, the value within the shift register in Figure 11-20(b) is continually halved. The divide ratio is also equal to 2^n where n is the number of times the contents is shifted. In the example in Figure 11-20(b), the contents of the shift register is shifted three times, and therefore the original number is divided by $2^3 = 8$ ($24 \div 8 = 3$).

11-4-4 *Shift Register Counters/Sequencers*

One common application of shift registers is as a **sequencer,** in which the parallel outputs of the shift register are used to generate a sequence of equally spaced timing pulses. When connected to operate in this way, shift registers are often referred to as counters, since they generate a specific sequence of logic levels at their parallel outputs. Two of the most widely used shift register *counters/sequencers* are the ring counter and the johnson counter, both of which we will now discuss.

Sequencer
A circuit that generates a sequence of equally spaced timing pulses.

The Ring Counter

Figure 11-21(a) shows how four D-type flip-flops could be connected to form a **ring counter** or **shift register sequencer.** All shift register counters have a feedback connection from the output of the last flip-flop (Q_3) to the input of the first flip-flop (D_0), as shown in Figure 11-21(a). This characteristic circle or "ring" that causes the stored data to be constantly cycled through the shift register accounts for the name ring counter.

Ring Counter
A shift register sequencer circuit in which the position of a binary 1 moves through the counter in an ordered sequence.

The circuit shown in Figure 11-21(a) operates as follows. When the active-LOW $\overline{\text{START}}$ control input is brought LOW, Q_0 will be preset HIGH and Q_1, Q_2, and Q_3 will be cleared LOW, as shown in the waveforms in Figure 11-21(b). The stored word (1000) is then shifted right as clock pulses are applied, and since a feedback loop exists between Q_3 and D_0, the single HIGH "start bit" will be continually recirculated. This ring counter action is shown in the waveforms in Figure 11-21(b) and the logic state table in Figure 11-21(c).

Figure 11-22 shows how two 74194s (4-bit bidirectional universal shift registers) could be connected to form an 8-bit ring counter. The ring or feedback connection is made by connecting Q_D of 74194 B back to the shift-right serial input of 74194 A. This ring counter circuit is initiated by pressing SW_1, which will put a HIGH on the mode control S_1. The HIGH applied to the S_1 input will instruct both 74194s to parallel-load their *A, B, C,* and *D* inputs, and since only the *A* input of 74194 A is HIGH, the word 10000000 will be stored. When SW_1 is released, the normally closed switch will connect a LOW to the S_1 mode inputs, instructing them to shift right the stored data. The stored 8-bit value 10000000 will be shifted to the right in sync with the positive edges of the clock input, and the result is shown in the table on page 397.

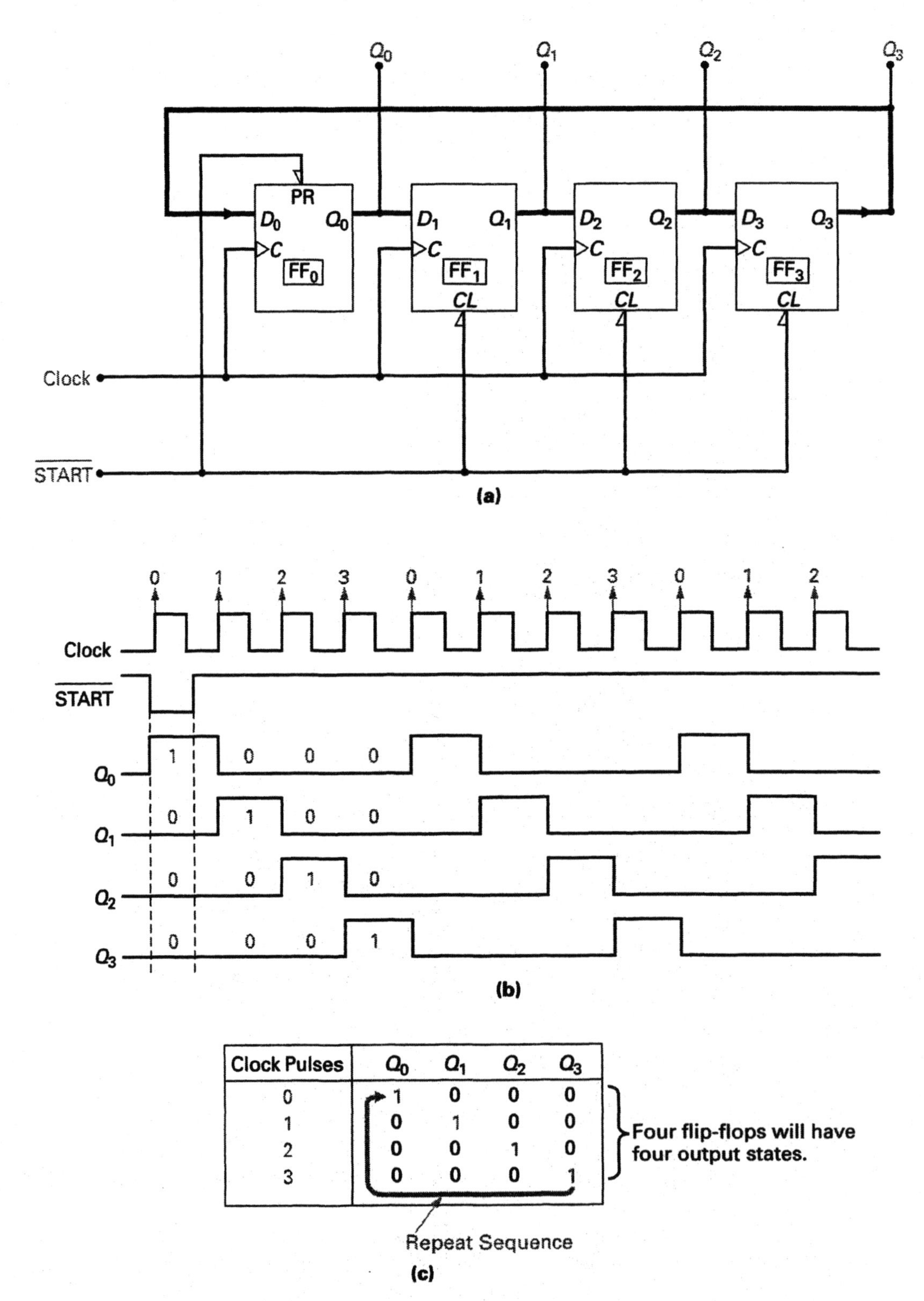

Clock Pulses	Q_0	Q_1	Q_2	Q_3
0	1	0	0	0
1	0	1	0	0
2	0	0	1	0
3	0	0	0	1

FIGURE 11-21 The Ring Counter. (a) Basic Circuit. (b) Timing Waveforms. (c) Logic State Table.

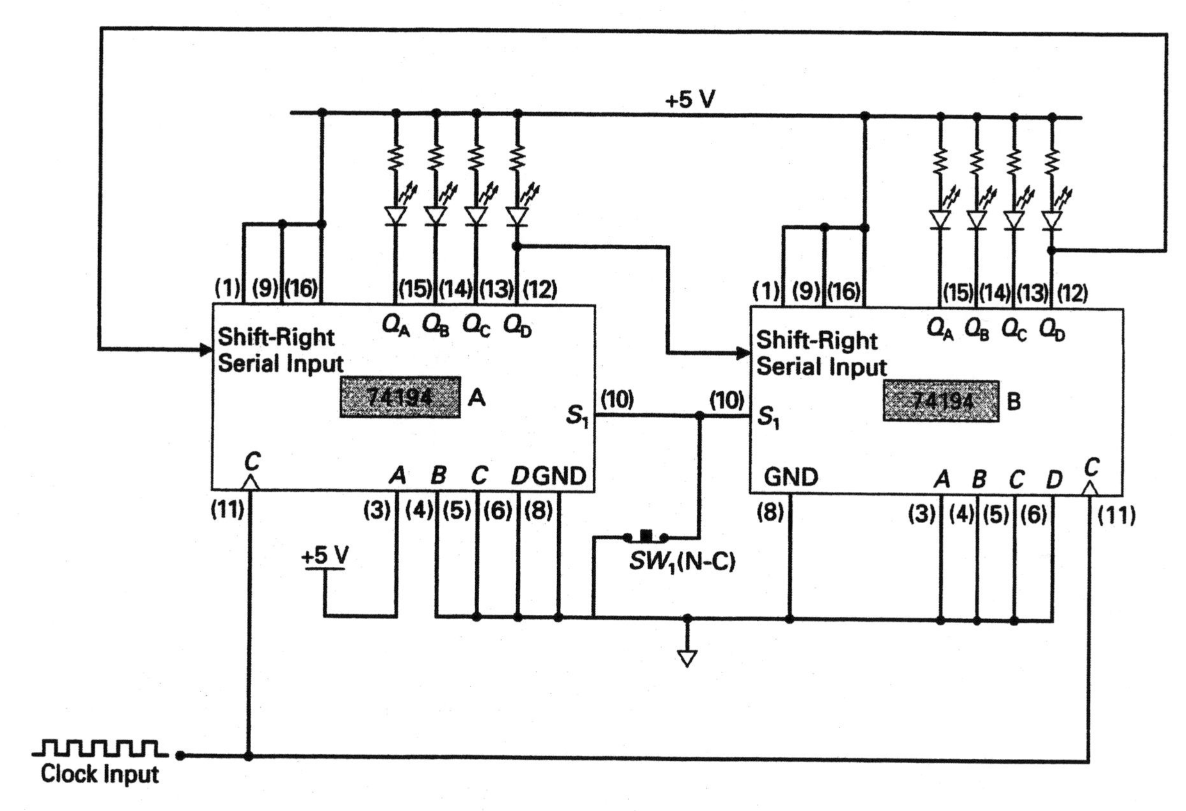

FIGURE 11-22 8-Bit Ring Counter Using Two 74194s.

CLOCK PULSES	74194 A Q_A	Q_B	Q_C	Q_D	74194 B Q_A	Q_B	Q_C	Q_D
0	1	0	0	0	0	0	0	0
1	0	1	0	0	0	0	0	0
2	0	0	1	0	0	0	0	0
3	0	0	0	1	0	0	0	0
4	0	0	0	0	1	0	0	0
5	0	0	0	0	0	1	0	0
6	0	0	0	0	0	0	1	0
7	0	0	0	0	0	0	0	1

In this example circuit, the outputs Q_A, Q_B, Q_C, and Q_D of both 74194s are used to sequence ON and OFF LEDs (common-anode display, HIGH = LED OFF, LOW = LED ON). In a real digital circuit application, the outputs would be connected to logic circuits and the ON/OFF sequence generated by the ring counter would control when each of these logic circuits is sequenced ON and then OFF. Since the frequency of the clock pulses is fixed, a sequence of equally spaced timing pulses would be generated to control the external logic circuits.

Johnson Counter
Also known as a twisted ring counter because of its inverted feedback path, it is a circuit that first fills with 1s and then fills with 0s and then repeats.

The Johnson Counter

Figure 11-23(a) shows how four D-type flip-flops could be connected to form a **johnson counter.** As you can see from this illustration, the johnson

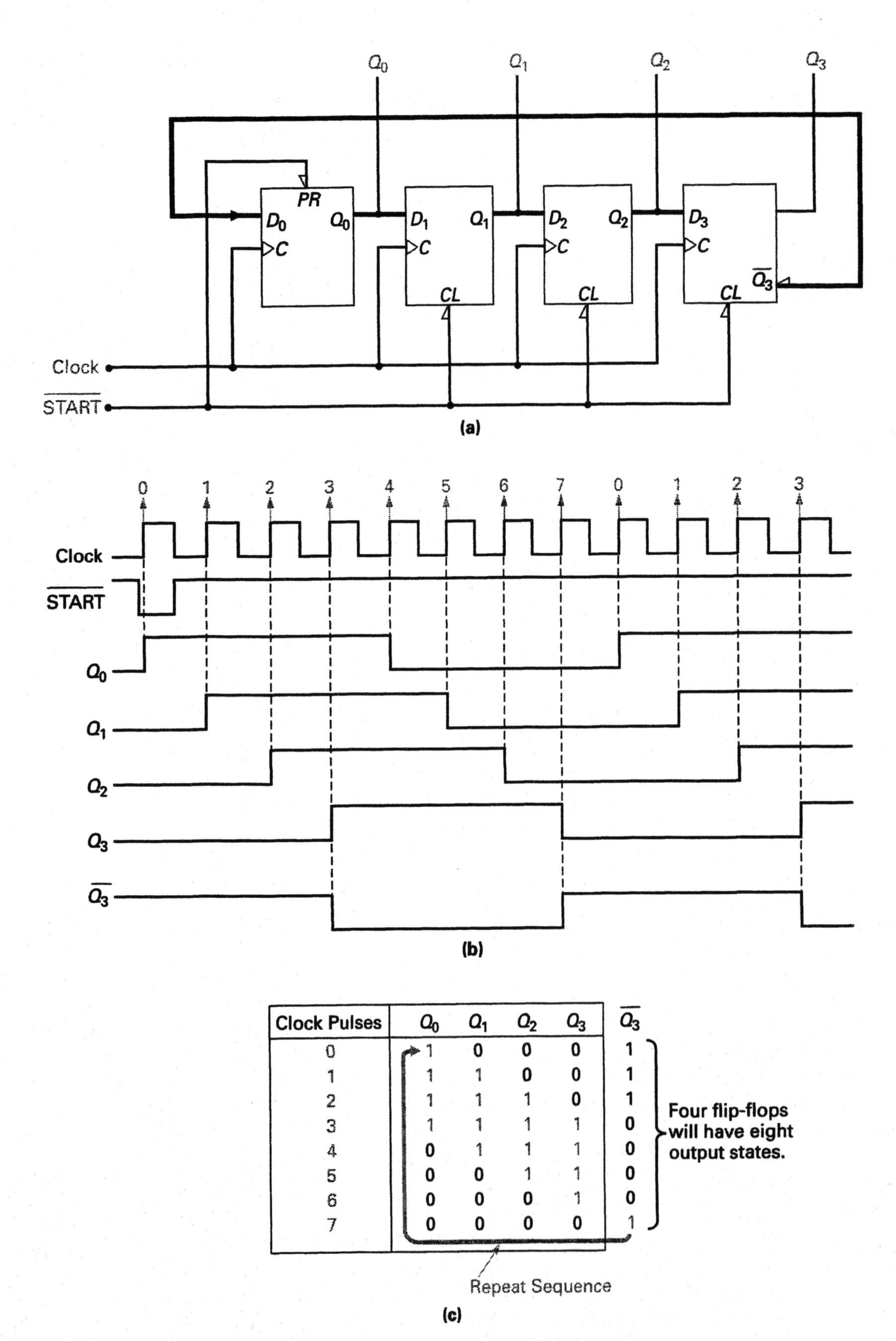

Clock Pulses	Q_0	Q_1	Q_2	Q_3	$\overline{Q_3}$
0	1	0	0	0	1
1	1	1	0	0	1
2	1	1	1	0	1
3	1	1	1	1	0
4	0	1	1	1	0
5	0	0	1	1	0
6	0	0	0	1	0
7	0	0	0	0	1

FIGURE 11-23 The Johnson Counter. (a) Basic Circuit. (b) Timing Waveforms. (c) Logic State Table.

counter is constructed in almost exactly the same way as the ring counter except that the inverted output of the last flip-flop ($\overline{Q}_3$) is connected to the input of the first flip-flop (D_0). This physical difference in the feedback path of the johnson counter accounts for why this circuit is often referred to as a *twisted ring counter,* instead of being named after its inventor.

Like the ring counter, the johnson counter needs a start circuit to initialize the first output sequence which, as you can see in the timing waveforms in Figure 11-23(b) and the logic state diagram in Figure 11-23(c), is 1000. Since Q_3 is LOW at the start, $\overline{Q}_3$ will be HIGH and this HIGH will be fed back to the D_0 input, so HIGH inputs will be clocked into the shift register in a left-to-right motion until the shift register is full of 1s. When Q_3 finally goes HIGH (after the third clock pulse), $\overline{Q}_3$ will go LOW, and so will D_0. The shift register will now shift in LOW inputs in a left-to-right motion until it is fully loaded with 0s. When Q_3 finally goes LOW (after the seventh clock pulse), $\overline{Q}_3$ will go HIGH and so therefore will D_0, causing the entire cycle to repeat.

With the ring counter, the number of different output states is governed by the number of flip-flops in the register and therefore a four-stage ring counter would have four different output states. Since the johnson counter first fills up with 1s (in four clock pulse inputs) and then fills up with 0s (in the next four clock pulses), the number of different output states is equal to twice the number of flip-flops. In the example circuit in Figure 11-23, the four-stage johnson counter will have eight different output states (2×4 flip-flops $= 8$), as listed in the logic state table in Figure 11-23(c).

SELF-TEST REVIEW QUESTIONS FOR SECTION 11-4

1. What is the function of the load and enable inputs to a controlled buffer register with three-state outputs?
2. Define the following terms.
   a. Bus **b.** Bus-organized digital system **c.** Bus conflict **d.** 16-bit bus
3. Give the full name of the abbreviation UART, and briefly describe the device's purpose.
4. If a binary number is shifted six positions to the left, the resulting number will be ____________ (16, 32, 64, 128) times ____________ (larger/smaller).
5. Describe the basic circuit difference between a ring counter and a johnson counter.

11-5 TROUBLESHOOTING REGISTER CIRCUITS

To review, the procedure for fixing a failure can be broken down into three steps—diagnose, isolate, and repair. Since this chapter has been devoted to registers, let us first examine the operation of a typical register circuit, and then apply our three-step troubleshooting process to this circuit.

11-5-1 A Register Circuit

Figure 11-24 shows a keyboard encoding circuit connected to a bus-organized digital system. A 4-bit data bus (D_0, D_1, D_2, and D_3) interconnects a microprocessor unit (which contains control, arithmetic, and logic circuits), a group of memory registers, an input register, and an output register. Every digital computer system from the simple to the complex is always made up of these three basic blocks—a microprocessor, memory, and input-output. Although the microprocessor and memory unit are shown in this circuit, we will only be concentrating on the keyboard encoder circuit and the input and output registers.

The output display is made up of four LEDs, which are driven by a PIPO control buffer register. There is no need for three-state output control for this register, since the output does not connect to the data bus, so its active-LOW enable inputs (pins 1 and 2) are permanently tied LOW, resulting in a two-state output to the LEDs. To load the contents of the data bus into the output register, the active-LOW load control ($\overline{L}_O$) must be made LOW synchronous with a positive transition of the clock input.

The input register, which is also a 74173, interfaces or connects the keyboard encoder circuit to the 4-bit system data bus. Once again, the 74173 will load its 4-bit input (1D, 2D, 3D, and 4D) into the internal flip-flops when the active-LOW load input control ($\overline{L}_I$) is taken LOW and a positive clock edge occurs. The 4-bit stored code will be delivered on to the data bus when the active-LOW enable input control ($\overline{E}_I$) is taken LOW. This 4-bit code will indicate which key in the sixteen-key matrix in the table in Figure 11-24 has been pressed and, as usual, each code will be different. (This is a 4-bit code, therefore there are sixteen combinations.)

The key matrix contains four rows and four columns, with 10 kΩ pull-up resistors connected to each column keeping these lines normally HIGH. The normally closed SW_1 needs to be pressed when the system is first turned on to initialize the keyboard. When pressed, the S_1 input to the 74194 will go HIGH and since S_0 will also be HIGH (since no key has yet been pressed) the ring counter will parallel-load the 4-bit input 0111 at inputs A, B, C, and D. The zero stored in the first flip-flop is the LOW output that will be shifted through the ring counter to scan the rows of the key matrix. The 74194 ring counter will shift right and rotate this LOW output at each of its outputs Q_A, Q_B, Q_C, and Q_D. When a key is pressed, the switch will make a connection between a row and a column, so when the ring counter is outputting a LOW, the associated column line will be switched LOW and so will one of the inputs to the column encoder. In addition, an input to the zero-detect gate will also go LOW, and this four-input AND gate will produce a LOW output when any of the column lines go LOW (when any switch is pressed). This LOW will activate two IC operations:

a. The LOW into S_0 (pin 9 of the ring counter) will put the 74194 in the inhibit clock mode (S_1 = LOW, S_0 = LOW) and prevent the ring counter from scanning while the row and column data is being encoded.

b. The same LOW on the load input register control ($\overline{L}_I$) will load the 4-bit code for the key pressed into the input register.

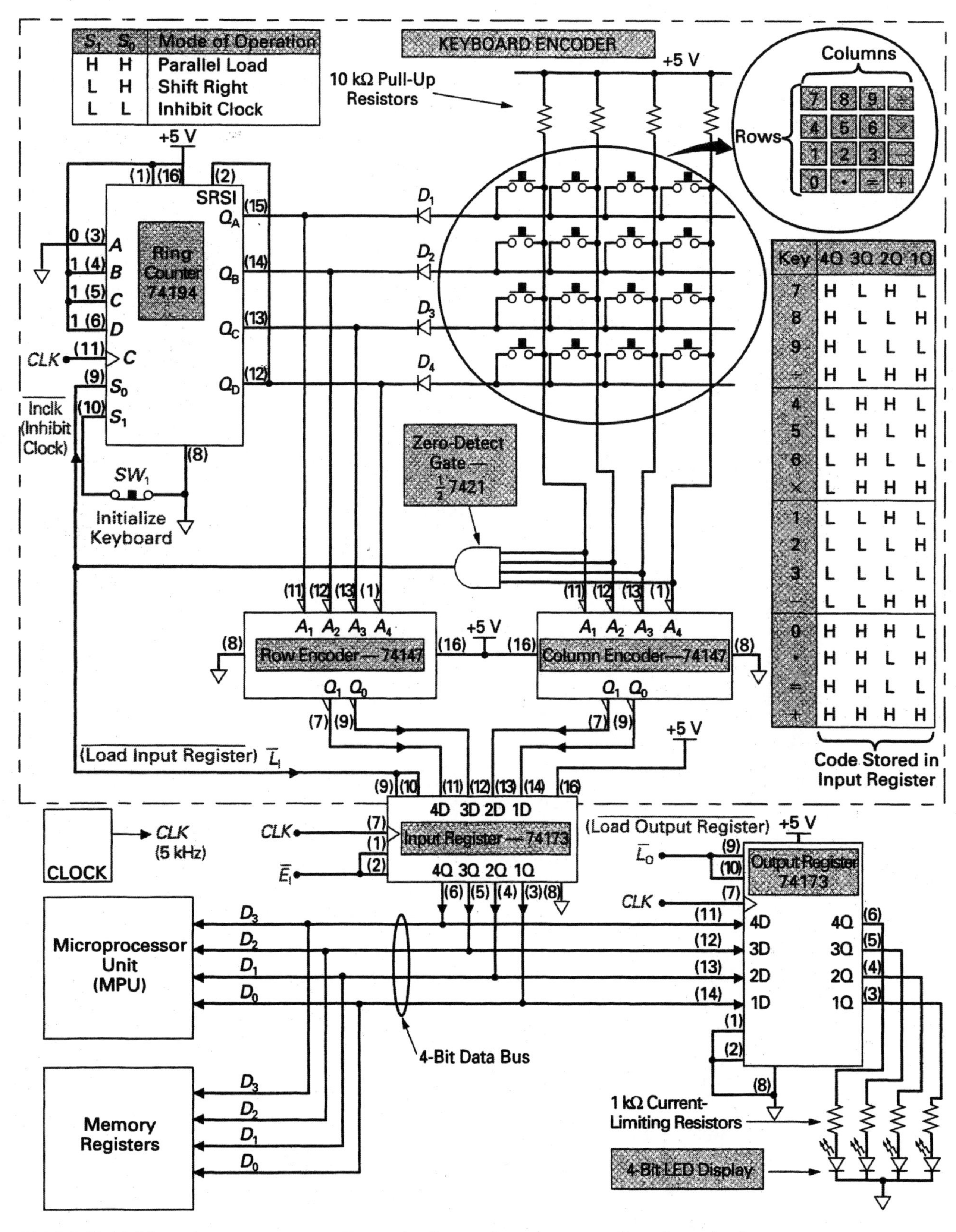

FIGURE 11-24 A Basic Keyboard Encoding Circuit Connected to a Bus-Organized Digital System.

Two keyboard encoders (74147s) are used to generate a 2-bit code that indicates the row in which the key has been pressed (row encoder), and a 2-bit code that indicates the column in which the key has been pressed (column encoder).

The clock frequency of 5 kHz (t = 200 μs) is chosen so that all four rows are scanned in 800 μs (0.8 ns), and since even the fastest key closure will only last several milliseconds, a keyboard input will not be missed.

The diodes on the four row lines (D_1 to D_4) prevent two outputs of the ring counter from being short-circuited if two keys in the same column are pressed simultaneously by accident. Let us now apply our three-step troubleshooting procedure to this flip-flop logic circuit so that we can practice our troubleshooting procedures and methods.

Step 1: Diagnose

As before, it is extremely important that you first understand the operation of a circuit and how all of the devices within it are supposed to work, so that you are able to determine whether or not a circuit malfunction really exists. If you were preparing to troubleshoot the circuit in Figure 11-24, your first step should be to read through the circuit description and review the operation of each integrated circuit until you feel completely confident with the correct operation of the entire circuit. Once you are fully familiar with the operation of the circuit, you will easily be able to diagnose the problem as either an *operator error* or a *circuit malfunction.*

Distinguishing an operator error from an actual circuit malfunction is an important first step, and a wrong diagnosis can waste a lot of time and effort. For example, the following could be interpreted as circuit malfunctions, when in fact they are simply operator errors.

Example 1.
Symptom: When a key is pressed on the keypad, its code is not being displayed on the LED display.
Diagnosis: Operator Error—The LED display does not automatically display the code for the key pressed. To transfer the code from the input register to the output register, you must make $\overline{E}_I$ LOW and $\overline{L}_O$ LOW.

Example 2.
Symptom: Keyboard encoder circuit is malfunctioning since it is not scanning keypad.
Diagnosis: Operator Error—Keyboard encoder circuit has not been initialized (SW_1).

Once you have determined that the problem is not an operator error, but is in fact a circuit malfunction, proceed to Step 2 and isolate the circuit failure.

Step 2: Isolate

No matter what circuit or system failure has occurred, you should always follow a logical and sequential troubleshooting procedure. Let us review some of the isolating techniques, and then apply them to our example circuit in Figure 11-24.

a. Use a cause and effect troubleshooting process, which means study the effects you are getting from the faulty circuit and then logically reason out what could be the cause.

b. Check first for obvious errors before leaping into a detailed testing procedure. Is the power OFF or not connected to the circuit? Are there wiring errors? Are all of the ICs correctly oriented?

c. Using a logic probe or voltmeter test that power and ground are connected to the circuit and are present at all points requiring power and ground. If the whole circuit, or a large section of the circuit, is not operating, the problem is normally power. Using a multimeter, check that all of the dc voltages for the circuit are present at all IC pins that should have power or a HIGH input, and are within tolerance. Second, check that 0 V or ground is connected to all IC ground pins and all inputs that should be tied LOW.

d. Use your senses to check for broken wires, loose connectors, overheating or smoking components, pins not making contact, and so on.

e. Test the clock signals using the logic probe or the oscilloscope. Although clock signals can be easily checked with a logic probe, the oscilloscope is best for checking the signal's amplitude, frequency, pulse width, and so on. The oscilloscope will also display timing problems and noise spikes (glitches) which could be false-triggering a circuit into operation at the wrong time.

f. Perform a static test on the circuit. With our circuit example in Figure 11-24, you will need to disconnect the clock input to the circuit and then trigger the circuit with the logic pulser's single pulse feature to clock the circuit. A logic probe could then be used to test that valid logic levels are appearing at all IC outputs. Some of the important points to test when single-stepping this circuit are as follows:

1. Is the ring counter parallel-loading $ABCD$ with 0001 when SW_1 is pressed?
2. Is the ring counter shifting the LOW output to Q_A, Q_B, Q_C, and Q_D and then repeating the cycle, when it is in the shift-right mode ($S_1 = L$, $S_0 = H$)?
3. Are the Q outputs of the ring counter locked in one state when it is in the inhibit clock mode ($S_1 = L$, $S_0 = L$), when a key is pressed?
4. Are both the row and column encoders functioning correctly and producing the 4-bit code for each key? Is this 4-bit code being loaded into the input register?
5. Is the zero detect AND gate generating a LOW inhibit clock and load input register output when a key is pressed?
6. Is the input register transmitting its contents on to the bus when $\overline{E}_I$ is made LOW?
7. Are the contents of the data bus being loaded into the output register when $\overline{L}_O$ is made LOW?
8. Are the contents of the output register being displayed on the 4-bit LED display?

g. With a dynamic test, the circuit is tested while it is operating at its normal clock frequency (which in Figure 11-24 is 5 kHz), and therefore all of the inputs and outputs are continually changing. Although a logic probe can detect a pulse waveform, the oscilloscope will display more signal detail, and since it can display more than one signal at a time, it is ideal for making signal comparisons and looking for timing errors. If the clock section is suspected, you can check the rest of the circuit by disconnecting the clock input to the circuit and then using a function generator to generate a 5 kHz clock.

In some instances, you will discover that some circuits will operate normally when undergoing a static test, and yet fail a dynamic test. This effect usually points to a timing problem involving the clock signals and/or the propagation delay times of the ICs used within the circuit.

h. Noise due to electromagnetic interference (EMI) can false-trigger an IC, such as the registers in Figure 11-24. This problem can be overcome by not leaving any of the IC's inputs unconnected and, therefore, floating. Connect unneeded active-HIGH inputs to ground and active-LOW inputs to $+V_{CC}$.

i. Apply the half-split method of troubleshooting first to a circuit, and then to a section of a circuit, to help speed up the isolation process. With our circuit example in Figure 11-24, a good mid-point check would be to test the code loaded into the input register to determine whether the problem is within the keyboard encoder circuit or outside the keyboard encoder circuit.

Also remember that a load problem can make a source appear at fault. If an output is incorrect, it would be safer to disconnect the source from the rest of the circuit and then recheck the output. If the output is still incorrect, the problem is definitely within the source; however, if the output is correct, the load is probably shorting the line either HIGH or LOW.

j. Substitution can be used to help speed up your troubleshooting process. Once the problem is localized to an area containing only a few ICs, substitute suspect ICs with known working ICs (one at a time) to see if the problem can be quickly remedied.

Step 3: Repair

Once the fault has been found, the final step is to repair the circuit and then perform a final test to see that the circuit and the system is now fully operational.

11-5-2 Sample Problems

Once you have constructed a circuit, like the circuit in Figure 11-24, introduce a few errors to see what effect or symptoms they produce. Then logically reason out why a particular error or cause has a particular effect on the circuit. Never short any two points together unless you have carefully thought out the consequences, but generally it is safe to open a path and see the results. Here are some examples for our example circuit in Figure 11-24.

a. Disconnect the clock input to the ring counter.

b. Disconnect one of the column encoder inputs (A_1, A_2, A_3, or A_4).

c. Disconnect the ground connection to the cathodes of the LEDs.

d. Permanently ground the S_1 input of the ring counter.

e. Open one of the inputs to the input register (4D, 3D, 2D, or 1D).

f. Disconnect the output from the zero-detect AND gate.
g. Disconnect the +5 V supply to the row encoder IC.

SELF-TEST REVIEW QUESTIONS FOR SECTION 11-5

1. What are the three basic troubleshooting steps?
2. Describe how you would static test the output register in Figure 11-24.
3. How would you apply the half-split troubleshooting method to only the keyboard encoder circuit?
4. What symptom would you get from the circuit in Figure 11-24 if SW_1 were permanently jammed open?

REVIEW QUESTIONS

Multiple-Choice Questions

1. The output of a combinational logic circuit is determined by the ________ state of the inputs, whereas the output of a sequential logic circuit is determined by the ________ state of the inputs.
   a. present, past **b.** past, past **c.** present, present **d.** past, present
2. Which of the following are examples of sequential logic circuits?
   a. Decoders and encoders **b.** Parity generators and checkers
   c. Registers and counters **d.** Both a. and b.
3. Which of the following are examples of combinational logic circuits?
   a. Decoders and encoders **b.** Parity generators and checkers
   c. Registers and counters **d.** Both a. and b.
4. Which of the following abbreviations best describes a buffer register?
   a. SISO **b.** PISO **c.** SIPO **d.** PIPO
5. Which of the following abbreviations best describes a shift register that inputs the data serially and outputs the stored data serially?
   a. SISO **b.** PISO **c.** SIPO **d.** PIPO
6. Which of the following abbreviations best describes a shift register that inputs the data serially and outputs the stored data as a parallel word?
   a. SISO **b.** PISO **c.** SIPO **d.** PIPO
7. Which of the following abbreviations best describes a shift register that inputs the data as a parallel word and then outputs the stored data serially?
   a. SISO **b.** PISO **c.** SIPO **d.** PIPO
8. A 4-bit SISO shift register will take ________ clock pulse(s) to store a 4-bit word.
   a. one **b.** two **c.** three **d.** four
9. A 4-bit PISO shift register will take ________ clock pulse(s) to store a 4-bit word.
   a. one **b.** two **c.** three **d.** four

10. Which of the following shift registers could be used for serial-to-parallel data conversion?

   a. SISO **b.** PISO **c.** SIPO **d.** PIPO **e.** All of the above

11. Which of the following shift registers could be used for parallel-to-serial data conversion?

   a. SISO **b.** PISO **c.** SIPO **d.** PIPO **e.** All of the above

12. Which of the following is considered a serial-in, serial-out shift register operation?

   a. Shift-right **b.** Shift-left **c.** Rotate-right **d.** Rotate-left
   e. All of the above

13. When an operation is in step with the clock signal, it is called a(an) ____________ operation.

   a. asynchronous **b.** UART **c.** sequenced **d.** synchronous

14. When an operation is not in step with the clock signal, it is called a(an) ____________ operation.

   a. asynchronous **b.** UART **c.** sequenced **d.** synchronous

15. A buffer register with three-state output control will generally have three control line inputs called ____________, ____________ and ____________.

   a. V_{CC}, clock, S_1 **b.** clock, enable, load
   c. SRI, load, S_0 **d.** shift-left, shift-right, inhibit

16. New data is ____________ into a memory register, while stored data is ____________ out of a memory register.

   a. read, written **b.** written, read

17. A universal asynchronous receiver/transmitter (UART) has a ____________ and a ____________ internally for interfacing between a bus and a single line.

   a. serial-to-parallel shift register, parallel-to-serial shift register
   b. memory register, arithmetic shift register
   c. ring counter, johnson counter
   d. buffer register, PIPO register

18. What will happen to the value of a binary number stored in a shift register if it is shifted to the left by three positions and zeros are shifted in through the LSB position?

   a. Value will be divided by 8.
   b. Value will be multiplied by 6.
   c. Value will be multiplied by 8.
   d. Value will be divided by 4.

19. With a ring counter, the ____________ output of the last flip-flop is connected back to the input of the first flip-flop, whereas with a johnson counter the ____________ output of the last flip-flop is connected back to the input of the first flip-flop.

   a. Q, CLEAR **b.** $\overline{Q}$, Q **c.** PRESET, $\overline{Q}$ **d.** Q, $\overline{Q}$

20. A five flip-flop ring counter will have ____________ different output combinations, while a five flip-flop johnson counter will have ____________ output combinations.

    a. ten, sixteen **b.** five, eight **c.** five, ten **d.** five, sixteen

Essay Questions

21. What is a buffer register, and why is it called a PIPO? (11-1)
22. Briefly describe the basic operation of a shift register, and describe the differences between the three types. (11-2)
23. What is a bidirectional universal shift register? (11-2-4)
24. What is a bus? (11-3)
25. Why are devices given three-state outputs, and what is a bus-organized digital system? (11-3)
26. What is a memory register, and how is its read and write operation controlled? (11-4-1)
27. Briefly describe the function and operation of a UART. (11-4-2)
28. How can shift registers be used to perform arithmetic operations? (11-4-3)
29. Sketch a ring counter and johnson counter, and briefly describe the differences between the two. (11-4-4)
30. Describe how a ring counter could be used in a keyboard encoder circuit. (11-5).

Practice Problems

31. Using an illustration, show the difference between a buffer register and a shift register.
32. Sketch three simplified block diagrams showing each of the three shift register types.

To practice your circuit recognition and operation, refer to Figure 11-25 and answer the following four questions.

33. Describe the 74175 IC in one sentence.
34. Which type of flip-flop is being used to form these two register circuits?
35. The circuit configuration in Figure 11-25(a) is a ____________.

    a. PIPO **b.** SISO **c.** SIPO **d.** PISO

36. The circuit configuration in Figure 11-25(b) is a ____________.

    a. PIPO **b.** SISO **c.** SIPO **d.** PISO

To practice your data sheet interpretation, refer to Figure 11-26 and answer the following ten questions.

37. Briefly describe the function of the following lines.

    a. Line 1 **b.** Line 2 **c.** Line 3 **d.** Line 4 **e.** Line 23

38. What are the parallel load input pins for this IC?
39. What is the purpose of the SHIFT/$\overline{\text{LOAD}}$ input control line?
40. What are the serial data input pins for this IC?
41. Is the shifting of the stored data synchronous or asynchronous?

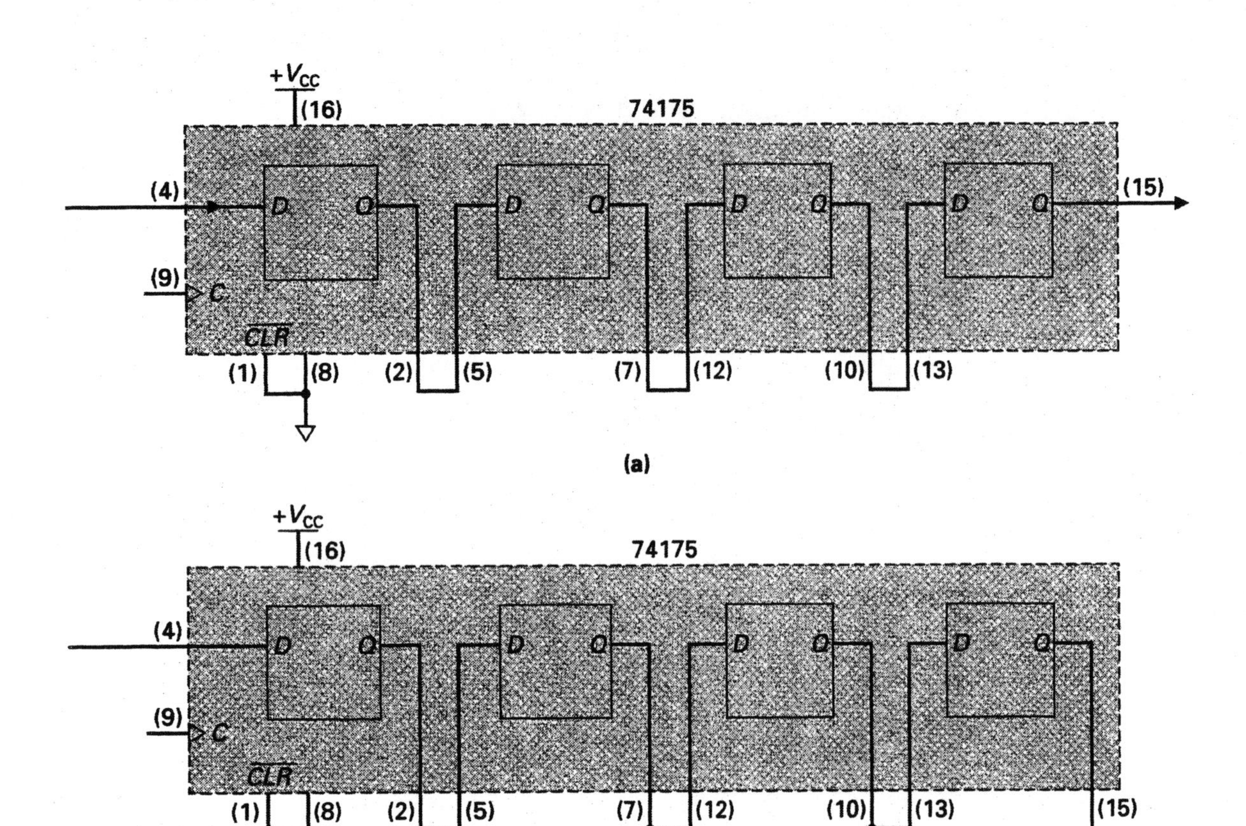

FIGURE 11-25 74175 Register Configurations.

42. Referring to the circuit in Figure 11-27, describe how and when this *RC* network is controlling the 74199.
43. Is the clear input for this IC active-HIGH or active-LOW?
44. Is this IC level-triggered, edge-triggered, or pulse-triggered?
45. If the stored word is 10001111 (Q_A–Q_H) and $SH/\overline{LD}$ is LOW, what will be the Q_A through Q_H outputs after three clock pulses (J and $\overline{K}$ are LOW)?
46. If the stored word is 11000000 (Q_A–Q_H) and $SH/\overline{LD}$ is HIGH, what will be the Q_A through Q_H outputs after four clock pulses (J and $\overline{K}$ are LOW)?

To practice your circuit recognition and operation, refer to Figure 11-28 and answer the following four questions.

47. The 74199 in this circuit is functioning as what type of register?
48. What is the purpose of the 74147 A and 74147 B in this circuit?
49. Briefly describe what will happen when
    a. power is first applied to the circuit.
    b. a key in the matrix is pressed.
50. Write a complete circuit description for this 8-by-8 keyboard encoder circuit.

IC TYPE: SN74199– 8-bit Shift Register

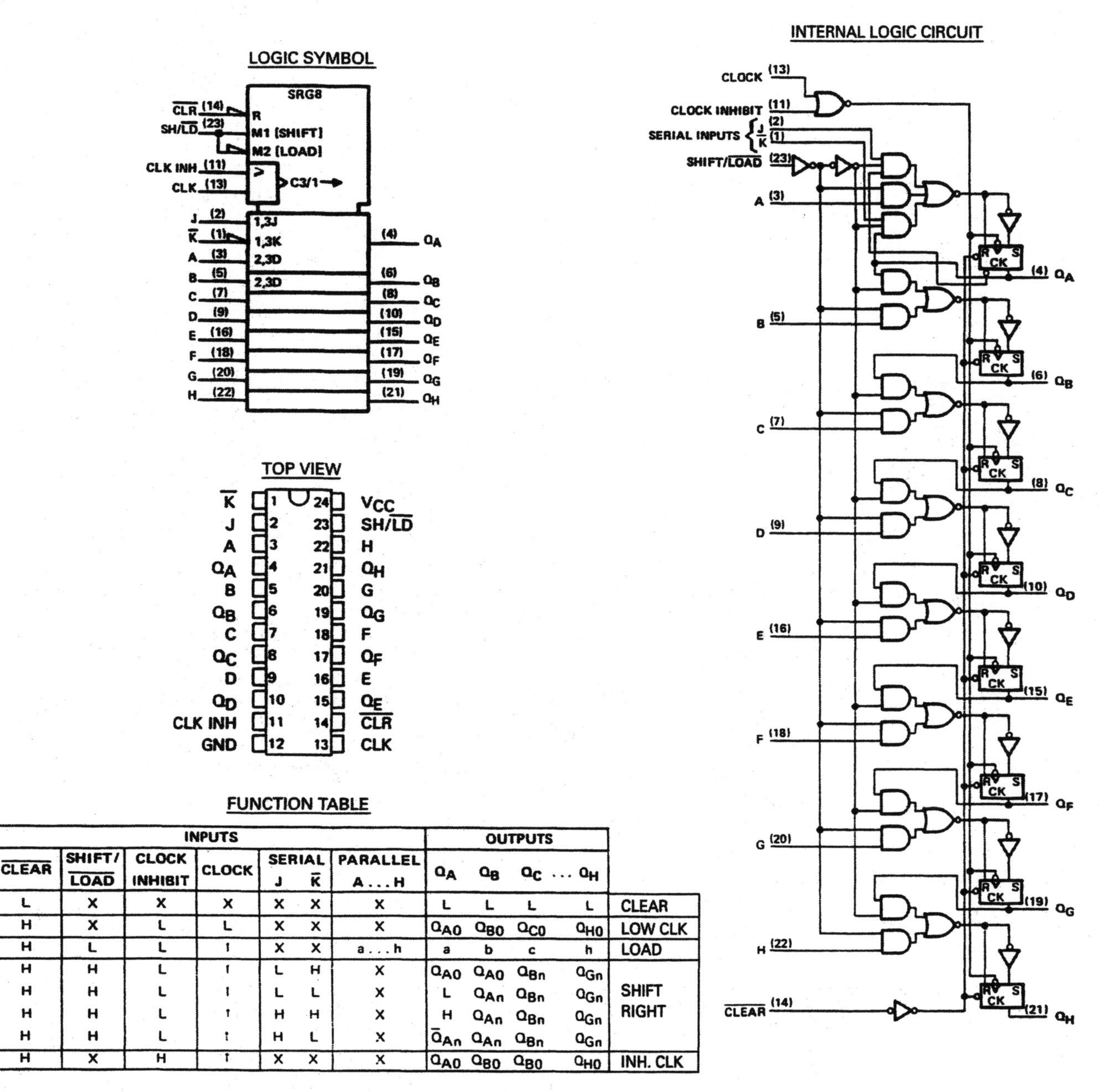

FUNCTION TABLE

INPUTS							OUTPUTS				
$\overline{\text{CLEAR}}$	SHIFT/$\overline{\text{LOAD}}$	CLOCK INHIBIT	CLOCK	SERIAL J	SERIAL $\overline{K}$	PARALLEL A . . . H	Q_A	Q_B	Q_C	. . . Q_H	
L	X	X	X	X	X	X	L	L	L	L	CLEAR
H	X	L	L	X	X	X	Q_{A0}	Q_{B0}	Q_{C0}	Q_{H0}	LOW CLK
H	L	L	↑	X	X	a . . . h	a	b	c	h	LOAD
H	H	L	↑	L	H	X	Q_{A0}	Q_{A0}	Q_{Bn}	Q_{Gn}	SHIFT RIGHT
H	H	L	↑	L	L	X	L	Q_{An}	Q_{Bn}	Q_{Gn}	
H	H	L	↑	H	H	X	H	Q_{An}	Q_{Bn}	Q_{Gn}	
H	H	L	↑	H	L	X	$\overline{Q}_{An}$	Q_{An}	Q_{Bn}	Q_{Gn}	
H	X	H	↑	X	X	X	Q_{A0}	Q_{B0}	Q_{B0}	Q_{H0}	INH. CLK

FIGURE 11-26 An IC Containing an 8-Bit Shift Register. (Courtesy of Texas Instruments)

Troubleshooting Questions

51. How could you use a logic pulser and logic probe to static test the circuit in Figure 11-28?

52. How would you apply the half-split troubleshooting method to the circuit in Figure 11-28?

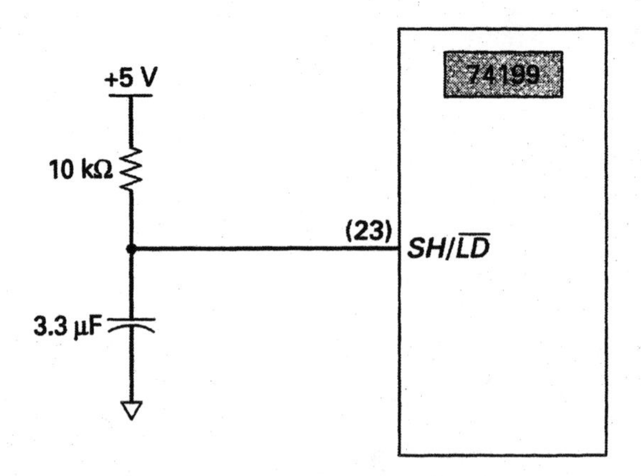

FIGURE 11-27 Shift/Load Control.

53. What symptoms would you expect from the circuit shown in Figure 11-28 if the following malfunctions were to occur?
    - **a.** C_1 was to short.
    - **b.** R_1 was open.
    - **c.** C_1 was open.
    - **d.** One of the diodes, D_1 through D_8, was to open.
    - **e.** Pin 13 of the 74199 was shorted to pin 12.
    - **f.** Pin 1 of the 74199 was shorted to pin 2.
    - **g.** There was an open in the path between pin 21 and pins 1 and 2 of the 74199.
54. Sketch a simple troubleshooting tree for the circuit shown in Figure 11-28.
55. Determine the 6-bit code generated by the shaded key in Figure 11-28. What would happen if this switch were a permanent short?

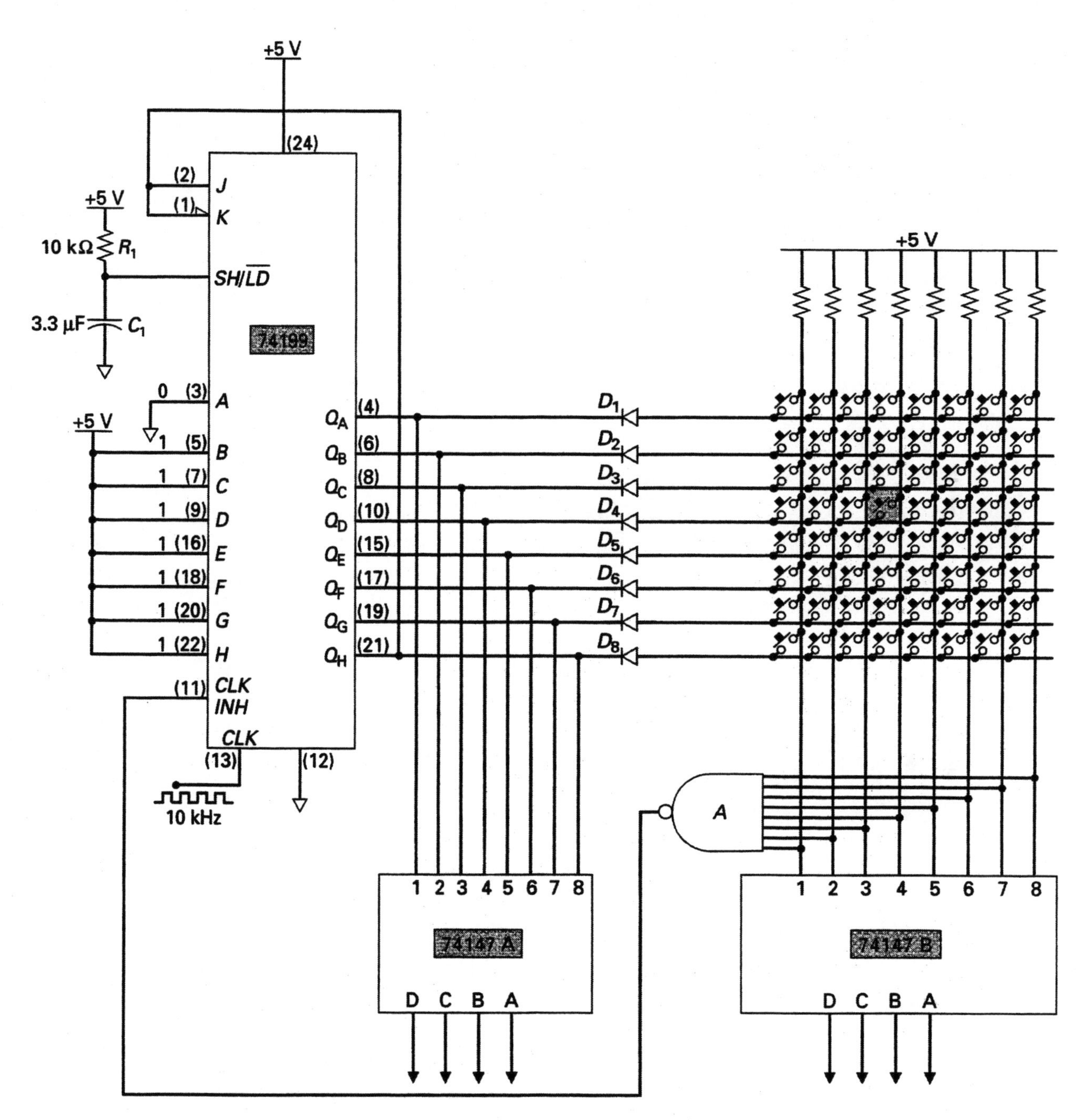

FIGURE 11-28 8-by-8 Keyboard Encoder.

Chapter 12

OBJECTIVES

After completing this chapter, you will be able to:

1. Define the difference between an asynchronous counter and a synchronous counter.
2. Describe how flip-flops can be connected to form a binary counter circuit or a frequency divider circuit.
3. Explain the following counter terms.
   a. Maximum count
   b. Modulus
   c. Frequency division
   d. Propagation delay time
4. Describe the operation of the following asynchronous counters.
   a. Binary up counter
   b. Binary down counter
   c. Binary up/down counter
   d. Decade counter
   e. Presettable or programmable counter
5. Explain the purpose of a state transition diagram.
6. List the advantages and disadvantages of synchronous and asynchronous counters.
7. Describe how a synchronous counter can function as a
   a. binary up counter.
   b. presettable counter.
   c. decade counter.
   d. up/down counter.
8. Explain how counters can be cascaded to obtain a higher modulus.
9. Describe how counter circuits can be used in the following applications.
   a. Digital clock circuit
   b. Frequency counter circuit
   c. Multiplexed display circuit
10. Describe how to troubleshoot digital logic circuits containing counters using the three-step method.

Counters

OUTLINE

VIGNETTE

The Great Experimenter

Michael Faraday was born to James and Margaret Faraday on September 22, 1791. James Faraday, a blacksmith in failing health, had brought his family to London in hopes of finding work; however, his poor health meant that the Faradays were often penniless and hungry. At the age of twelve, the young Faraday left school to work as an errand boy for a local bookseller, and was later apprenticed to the bookbinder's trade. At twenty-two, the gifted and engaging Faraday was at the right place at the right time, and with the right talents. He impressed the brilliant professor Humphry Davvy, who made him his research assistant at Count Rumford's Royal Institution. After only two years, Faraday was given a promotion and an apartment at the Royal Institution, and in 1821 he married. He lived at the Royal Institution for the rest of his active life, working in his laboratory, giving very dynamic lectures, and publishing over 450 scientific papers. Unlike many scientific papers, Faraday's papers would never use a calculus formula to explain a concept or idea. Instead, Faraday would explain all of his findings using logic and reasoning, so that a person trying to understand science did not need to be a scientist. It was this gift of being able to make even the most complex areas of science easily accessible to the student, coupled with his motivational teaching style, that made him so popular.

In 1855 he had written three volumes of papers on electromagnetism, the first dynamo, the first transformer, and the foundations of electrochemistry, a large amount on dielectrics and even some papers on plasma. The unit of capacitance is measured in *Farads,* in honor of his work in these areas of science.

Faraday began two series of lectures at the Royal Institution, which have continued to this day. Michael and Sarah Faraday were childless, but they both loved children, and in 1860 Faraday began a series of Christmas lectures expressly for children, the most popular of which was called "The Chemical History of a Candle." The other lecture series was the "Friday Evening Discourses," of which he himself delivered over a hundred. These very dynamic, enlightening, and entertaining lectures covered areas of science or technology for the lay person, and were filled with demonstrations. On one evening in 1846, an extremely nervous and shy speaker ran off just moments before he was scheduled to give the Friday Evening Discourse. Faraday had to fill in, and to this day a tradition is still enforced whereby the lecturer for the Friday Evening Discourse is locked up for half an hour before the presentation with a single glass of whiskey.

Faraday was often referred to as "the great experimenter," and it was this consistent experimentation that led to many of his findings. He was fascinated by science and technology, and was always exploring new and sometimes dangerous horizons. In fact, in one of his reports he states, "I have escaped, not quite unhurt, from four explosions." When asked to comment on experimentation, his advice was to "Let your imagination go, guiding it by judgment and principle, but holding it in and directing it by experiment. Nothing is so good as an experiment which, while it sets an error right, gives you as a reward for your humility an absolute advance in knowledge."

INTRODUCTION

A *counter* is like a register in that it is a sequential logic circuit made up of several flip-flops. A register, however, is designed to store a binary word, whereas the binary word stored in a counter represents the number of clock pulses that have occurred at the clock input. The input clock pulses, therefore, cause the flip-flops of the counter to change state, and by monitoring the outputs of the flip-flops you can determine how many clock pulses have been applied.

There are basically two types of counters—*asynchronous counters* and *synchronous counters*. The key difference between these two counter types is whether or not the operation of the counter is in synchronism with the clock signal. Most of the flip-flops of an asynchronous counter are not connected to the clock signal, and so the operation of this counter is not in sync with the master clock signal. On the other hand, all of the flip-flops of a synchronous counter are connected to the clock signal, and so the operation of this counter is in sync with the master clock signal.

In this chapter we will examine the different types of asynchronous and synchronous counters, several commonly used counter ICs, counter applications, and counter circuit troubleshooting.

12-1 ASYNCHRONOUS COUNTERS

As mentioned in the introduction, most of the flip-flops in an **asynchronous counter** are not connected to the clock signal, and therefore the flip-flops do not change state in sync with the master clock signal. To better understand this distinction, let us examine some of the different types of asynchronous counters.

Asynchronous Counter
A counter in which an action starts in response to a signal generated by a previous operation, rather than in response to a master clock signal.

12-1-1 Asynchronous Binary Up Counters

Figure 12-1(a) shows how a 4-bit or 4-stage asynchronous binary up counter can be constructed with four *J-K* flip-flops. In this circuit you can see that the *J-K* flip-flops are cascaded, or strung together, in series so that the output of one flip-flop will drive the clock input of the next flip-flop. Since the *J* and *K* inputs of all four flip-flops are tied HIGH, the output of each flip-flop will toggle, or change state, each time a negative edge occurs at the flip-flop's clock input.

The clock input to this circuit and the Q output waveforms for each of the four flip-flops are shown in Figure 12-1(b). The Q_0, Q_1, Q_2, and Q_3 outputs make up a 4-bit word which we will initially assume is 0000, as shown at the far

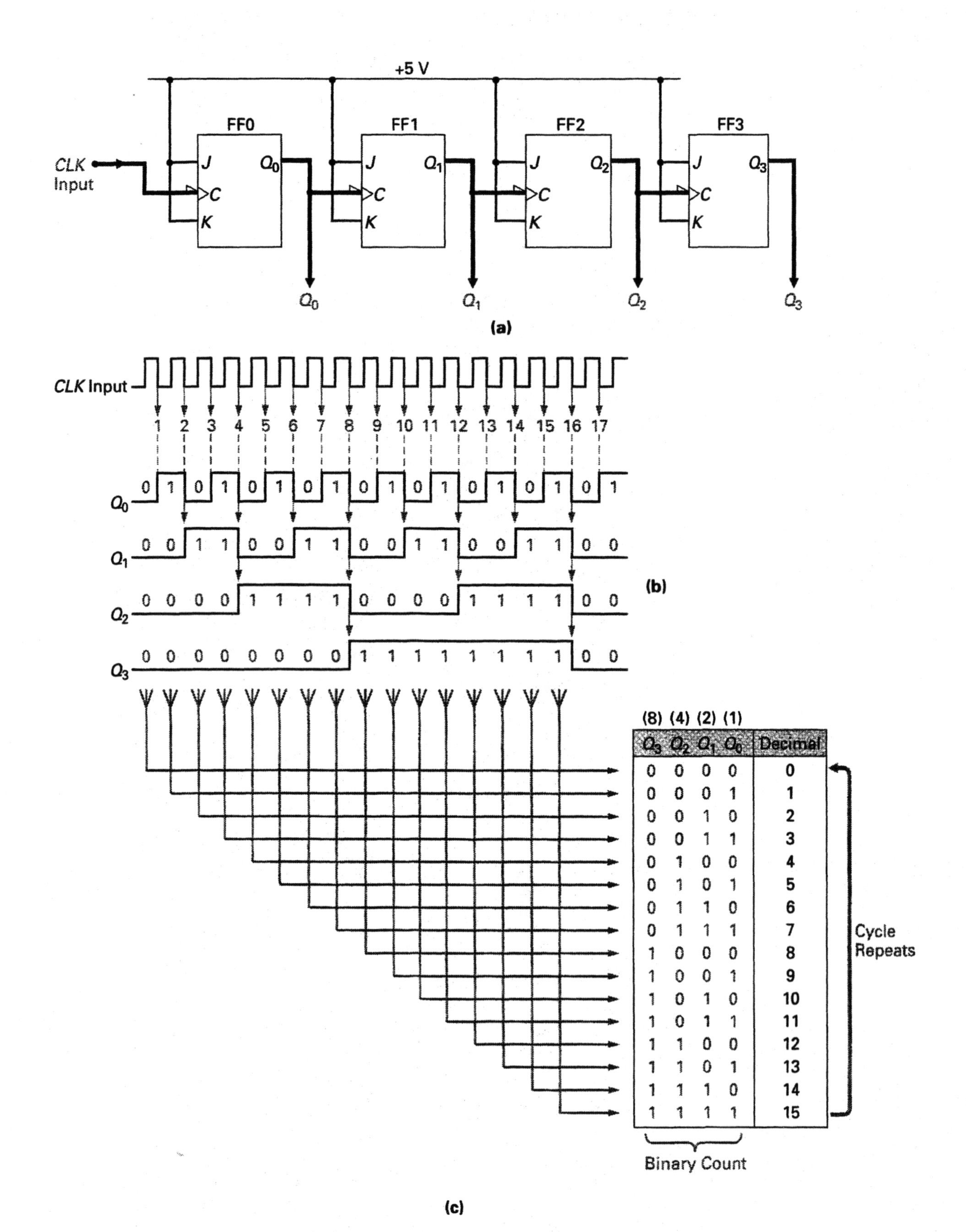

(8) Q_3	(4) Q_2	(2) Q_1	(1) Q_0	Decimal
0	0	0	0	0
0	0	0	1	1
0	0	1	0	2
0	0	1	1	3
0	1	0	0	4
0	1	0	1	5
0	1	1	0	6
0	1	1	1	7
1	0	0	0	8
1	0	0	1	9
1	0	1	0	10
1	0	1	1	11
1	1	0	0	12
1	1	0	1	13
1	1	1	0	14
1	1	1	1	15

FIGURE 12-1 An Asynchronous Binary Up Counter. (a) Logic Circuit. (b) Waveforms. (c) Truth Table.

left of the waveforms, and on the first line of the truth table in Figure 12-1(c). The output of FF0 (Q_0) generates the least significant bit (LSB) of the 4-bit output word, while the output of FF3 (Q_3) supplies the most significant bit (MSB) of the 4-bit output word. Since FF0 is driven by the clock input, Q_0 will toggle once for every negative edge of this clock input, as shown in the Q_0 timing waveform. This means that clock input negative edge 1 will cause Q_0 to change from a 0 to a 1, clock input negative edge 2 will cause Q_0 to change from a 1 to a 0, clock input negative edge 3 will cause Q_0 to change from a 0 to a 1, and so on. Since Q_0 is connected to the clock input of FF1, each negative edge of Q_0 will cause the Q_1 output to toggle. Similarly, a negative edge at the Q_1 output will cause the output of FF2 to toggle, and a negative edge at the Q_2 output will cause the output of FF3 to toggle.

The Maximum Count (N) of a Counter

Referring to the truth table in Figure 12-1(c), you may have noticed that the 4-bit binary value appearing at the Q_0, Q_1, Q_2, and Q_3 outputs indicated the number of clock pulses applied to this circuit. For example, before the clock pulse, the output was 0000 (decimal 0). After the first clock pulse, the count was 0001 (decimal 1); after the second clock pulse, the count was 0010 (decimal 2); after the third clock pulse, the count was 0011 (decimal 3); and so on. The **maximum count (N)** of a binary counter is governed by the number of flip-flops in the counter.

Maximum Count
For a counter, the maximum count is $2^n - 1$ or one less than the counter's modulus.

$$N = 2^n - 1$$

N = maximum count before cycle repeats.
n = number of flip-flops in the counter circuit.

With the counter circuit illustrated in Figure 12-1, the maximum count will be

$$\begin{aligned} N &= 2^n - 1 \\ &= 2^4 - 1 \\ &= 16 - 1 \\ &= 15_{10}\ (1111_2) \end{aligned}$$

This can be verified by referring to the waveforms in Figure 12-1(b) and the last line of the truth table in Figure 12-1(c), which shows the final maximum count of the 4-bit binary output is 1111 (decimal 15). You may have noticed that the maximum binary count that can be output from a binary counter before its cycle repeats is calçulated in the same way that we determine the maximum count of a binary word containing a certain number of bits.

■ **EXAMPLE:**

Calculate the maximum count of a

a. 1-bit counter
b. 2-bit counter
c. 3-bit counter
d. 4-bit counter
e. 8-bit counter

■ ***Solution:***

a. A 1-bit counter will have a maximum count of

$$N = 2^n - 1 = 2^1 - 1 = 2 - 1 = 1$$

b. A 2-bit counter will have a maximum count of

$$N = 2^n - 1 = 2^2 - 1 = 4 - 1 = 3$$

c. A 3-bit counter will have a maximum count of

$$N = 2^n - 1 = 2^3 - 1 = 8 - 1 = 7$$

d. A 4-bit counter will have a maximum count of

$$N = 2^n - 1 = 2^4 - 1 = 16 - 1 = 15$$

e. An 8-bit count will have a maximum count of

$$N = 2^n - 1 = 2^8 - 1 = 256 - 1 = 255$$

The Modulus (*MOD*) of a Counter

Modulus
The maximum number of output combinations generated by a counter, equal to 2^n.

The **modulus** of a counter describes the number of different output combinations generated by the counter. As an example, the 4-bit counter in Figure 12-1 has a modulus of 16, since the counter generates sixteen different output words or combinations. These sixteen different output combinations (0000 through 1111) are listed in the truth table in Figure 12-1(c). The modulus of a counter, therefore, can be calculated in the same way that we determine the number of different combinations generated by a binary word with a certain number of bits.

$$MOD = 2^n$$

MOD = modulus of the counter.
n = number of flip-flops in the counter.

With the counter circuit illustrated in Figure 12-1, the modulus of the counter will be

$$MOD = 2^n = 2^4 = 16$$

■ **EXAMPLE:**

Calculate the modulus of a

a. 1-bit counter
b. 2-bit counter
c. 3-bit counter
d. 4-bit counter
e. 8-bit counter

■ ***Solution:***

a. A 1-bit counter will have a modulus of $MOD = 2^n = 2^1 = 2$
b. A 2-bit counter will have a modulus of $MOD = 2^n = 2^2 = 4$
c. A 3-bit counter will have a modulus of $MOD = 2^n = 2^3 = 8$

d. A 4-bit counter will have a modulus of $MOD = 2^n = 2^4 = 16$
e. An 8-bit counter will have a modulus of $MOD = 2^n = 2^8 = 256$

The Frequency Division of a Counter

Referring again to the waveforms in Figure 12-1(b), you can see how a binary counter is also a **frequency divider.** Each of the flip-flops divides its input frequency by two. For example, two clock pulse cycles (or two negative edges) are needed at the C input of FF0 to produce one cycle at the Q_0 output. Likewise, two complete cycles of the clock are needed at the input of FF1 to produce one cycle at the Q_1 output, and two clock cycles are needed at the input of FF2 to produce one cycle at the Q_2 output. Finally, two clock cycles have to be applied at the input of FF3 in order to produce one cycle at the Q_3 output. Each flip-flop, therefore, acts as a **divide-by-two circuit.** If two flip-flops were connected together, the input clock frequency would be divided by two, and then divided by two again. This overall action would divide the input clock frequency by four, as shown in Figure 12-1(b), which shows that four negative edges are needed at the clock input in order to produce one complete cycle at the Q_1 output. Therefore, one flip-flop divides by two, two flip-flops will divide by four, three flip-flops will divide by eight, four flip-flops will divide by sixteen, and so on. The frequency division performed by a counter circuit can be calculated with the following formula.

Frequency Divider
A counter circuit that will divide the input signal frequency.

Divide-by-Two Circuit
A counter circuit that will divide the input signal frequency by 2.

$$\textit{Division Factor} = 2^n$$

n = number of flip-flops in the counter circuit.

■ **EXAMPLE:**

Calculate the following for a 3-bit asynchronous binary counter.
a. Maximum count
b. Modulus
c. Division factor

■ ***Solution:***

a.
$$N = 2^n - 1$$
$$= 2^3 - 1 = 8 - 1 = 7$$

b.
$$MOD = 2^n$$
$$= 2^3 = 8$$

c.
$$\textit{Division Factor} = 2^n$$
$$= 2^3 = 8$$

A 3-bit counter, therefore, has eight different output combinations (a modulus of 8), a maximum count of seven (111_2), and will divide the input frequency by eight.

The Propagation Delay Time (t_p) of a Counter

The asynchronous counter is also known as a **ripple counter.** This term is used because the clock pulses are applied to the input of only the first flip-flop,

Ripple Counter
An asynchronous counter in which the first flip-flop signal change affects the second flip-flop, which in turn affects the third, and so on until the last flip-flop in the series is changed.

and as the count is advanced, the effect "ripples" through to the other flip-flops in the circuit. Since each flip-flop in the counter circuit is triggered by the preceding flip-flop, it can take a certain amount of time for a pulse to ripple through all of the flip-flops and update the outputs to their new value. For example, when the eighth negative clock pulse edge occurs, all four Q outputs need to be changed from 0111 to 1000. If each flip-flop has a propagation delay time (t_p) of 10 ns, it will take 40 ns (4 flip-flops × 10 ns) to update the counter's count from 0111 to 1000. The counting speed or clock pulse frequency is therefore limited by the propagation delay time of all the flip-flops in the counter circuit. The HIGH input clock frequency limit can be calculated with the following formula.

$$f = \frac{1 \times 10^9}{n \times t_p}$$

f = upper clock pulse frequency limit.
n = number of flip-flops in the counter circuit.
t_p = propagation delay time of each flip-flop in nanoseconds.

■ **EXAMPLE:**

If a 4 MHz clock input were applied to a 4-bit asynchronous (ripple) binary counter, and each flip-flop had a propagation delay time of 32 ns,

a. would this input frequency be too fast for the counter?
b. what would be the frequency at the Q_0, Q_1, Q_2, and Q_3 outputs?

■ ***Solution:***

a.

$$f = \frac{1 \times 10^9}{n \times t_p}$$

$$= \frac{1 \times 10^9}{4 \times 32} = 7.8 \text{ MHz (upper frequency limit)}$$

A 4 MHz clock input is less than the 7.8 MHz upper frequency limit, and so this input is not too fast.

b. Frequency at Q_0 = 4 MHz ÷ 2 = 2 MHz
Frequency at Q_1 = 4 MHz ÷ 4 = 1 MHz
Frequency at Q_2 = 4 MHz ÷ 8 = 500 kHz
Frequency at Q_3 = 4 MHz ÷ 16 = 250 kHz

12-1-2 Asynchronous Binary Down Counters

With the asynchronous binary up counter just described, each input clock pulse would cause the parallel output binary word to be increased by one, or incremented. By making a small modification to the up counter circuit, it is possible to produce a down counter circuit which will cause the parallel output word to be decreased by one, or decremented, for each input clock pulse.

Figure 12-2(a) shows how a 4-bit or four-stage asynchronous binary down counter circuit can be constructed with four *J-K* flip-flops. Once again, the *J-K*

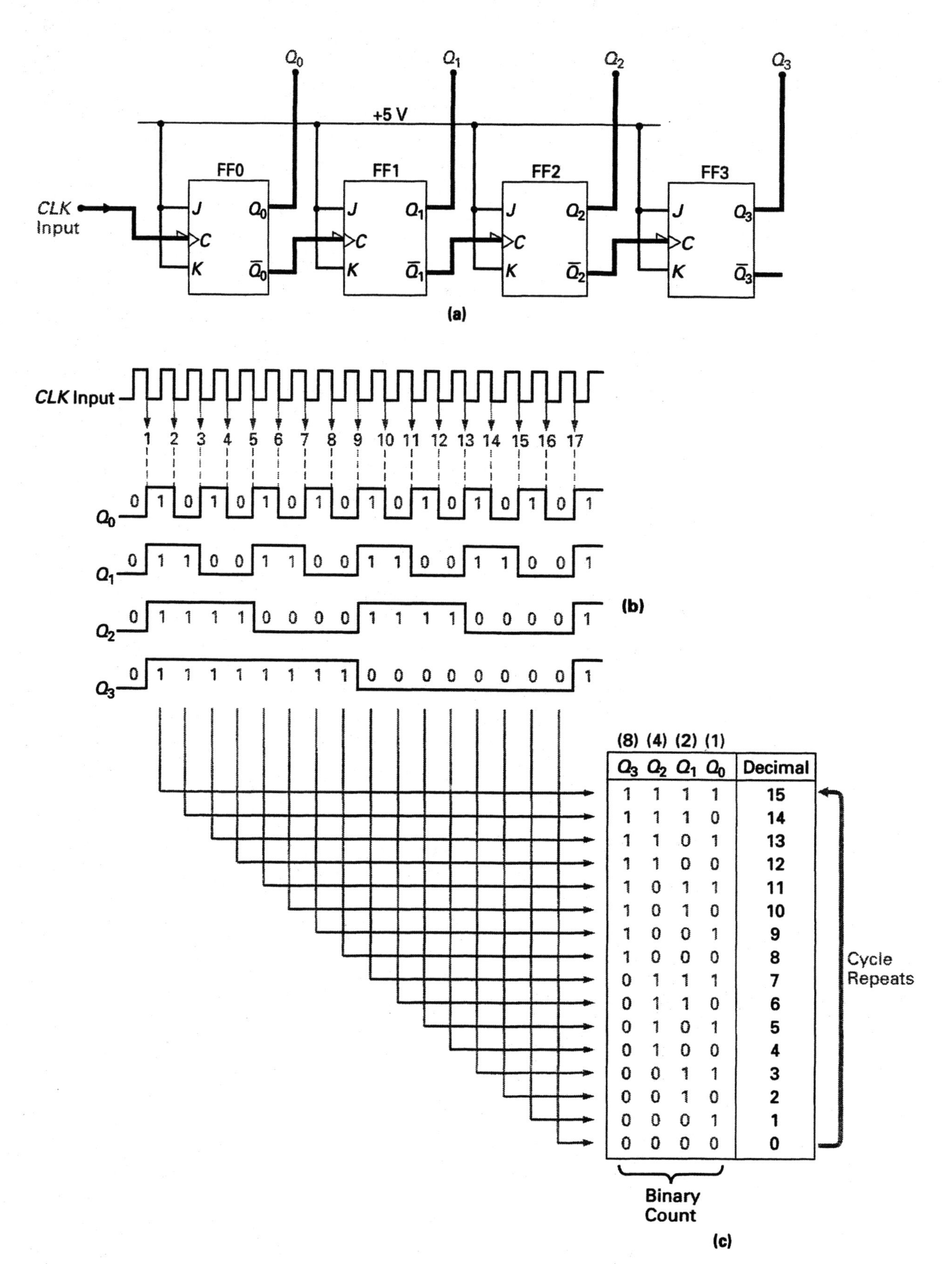

Q_3 (8)	Q_2 (4)	Q_1 (2)	Q_0 (1)	Decimal
1	1	1	1	15
1	1	1	0	14
1	1	0	1	13
1	1	0	0	12
1	0	1	1	11
1	0	1	0	10
1	0	0	1	9
1	0	0	0	8
0	1	1	1	7
0	1	1	0	6
0	1	0	1	5
0	1	0	0	4
0	0	1	1	3
0	0	1	0	2
0	0	0	1	1
0	0	0	0	0

FIGURE 12-2 An Asynchronous Binary Down Counter. (a) Logic Circuit. (b) Waveforms. (c) Truth Table.

flip-flops are cascaded, or strung together. In this instance, however, the $\overline{Q}$ outputs of each stage are connected to the clock inputs of the next stage. This small circuit difference causes the count sequence to be the complete opposite of the previously described up counter.

The clock input and Q output waveforms for this 4-bit binary down counter are shown in Figure 12-2(b). Referring to the far left side of the waveforms, you can see that we will begin with all the flip-flops initially RESET, and so the Q_0, Q_1, Q_2, and Q_3 outputs are 0000. If all the Q outputs are LOW, then all of the $\overline{Q}$ outputs must be HIGH, and therefore so are all of the C inputs to FF1, FF2, and FF3. Since the J and K inputs of all four flip-flops are tied HIGH, the output of each flip-flop will toggle each time a negative edge occurs at the circuit's clock (CLK) input. When the first clock pulse negative edge is applied to FF0, the Q_0 output will toggle from a binary 0 to a 1. This will cause the $\overline{Q}_0$ output to change from a binary 1 to a 0, and this negative edge or transition will clock the input of FF2, causing it to toggle, sending Q_2 HIGH and $\overline{Q}_2$ LOW. This HIGH-to-LOW transition at $\overline{Q}_2$ will clock the input of FF3 which will toggle FF3, sending its Q_3 output HIGH and its $\overline{Q}_3$ LOW. After the first clock pulse, therefore, the Q_0, Q_1, Q_2, and Q_3 outputs from the down counter will be 1111 (decimal 15), as shown in the waveforms in Figure 12-2(b) and on the first line of the truth table in Figure 12-2(c).

The down counter circuit will then proceed to count down by one for every clock pulse input applied. Referring again to the waveforms in Figure 12-2(b), you can see that FF0 will toggle for each negative edge of the CLK input, producing a Q_0 output that is half the input frequency (since each flip-flop divides by two). Referring to the Q_1, Q_2, and Q_3 waveforms, you can see that these outputs seem to toggle in response to each positive edge of the previous flip-flop's Q output. For instance, Q_1 will toggle for every positive edge of the Q_0; Q_2 will toggle for every positive edge of the Q_1; and Q_3 will toggle for every positive edge of the Q_2. This is because a positive edge at any of the Q outputs is a negative edge at the associated $\overline{Q}$ output, and it is the $\overline{Q}$ outputs that clock each following stage.

Referring again to the truth table in Figure 12-2(c), you can see that this MOD-16 (a 4-bit counter has sixteen possible output combinations) down counter will count down from 1111 (decimal 15) to 0000 (decimal 0), and then repeat the down-count cycle.

12-1-3 Asynchronous Binary Up/Down Counters

Comparing the asynchronous binary up counter in Figure 12-1(a) to the asynchronous down counter in Figure 12-2(a), you can see that the only difference between the two circuits is whether we clock FF1, FF2, and FF3 with the Q or $\overline{Q}$ outputs from the previous flip-flops.

Figure 12-3 shows how three AND-OR arrangements can be controlled by an UP/$\overline{\text{DOWN}}$ control line to create an asynchronous up/down binary counter. If the UP/$\overline{\text{DOWN}}$ control line is made HIGH, all of the shaded AND gates will be enabled, connecting the Q outputs through to the flip-flop's C inputs, causing the counter circuit to count up. On the other hand, if the UP/$\overline{\text{DOWN}}$ control

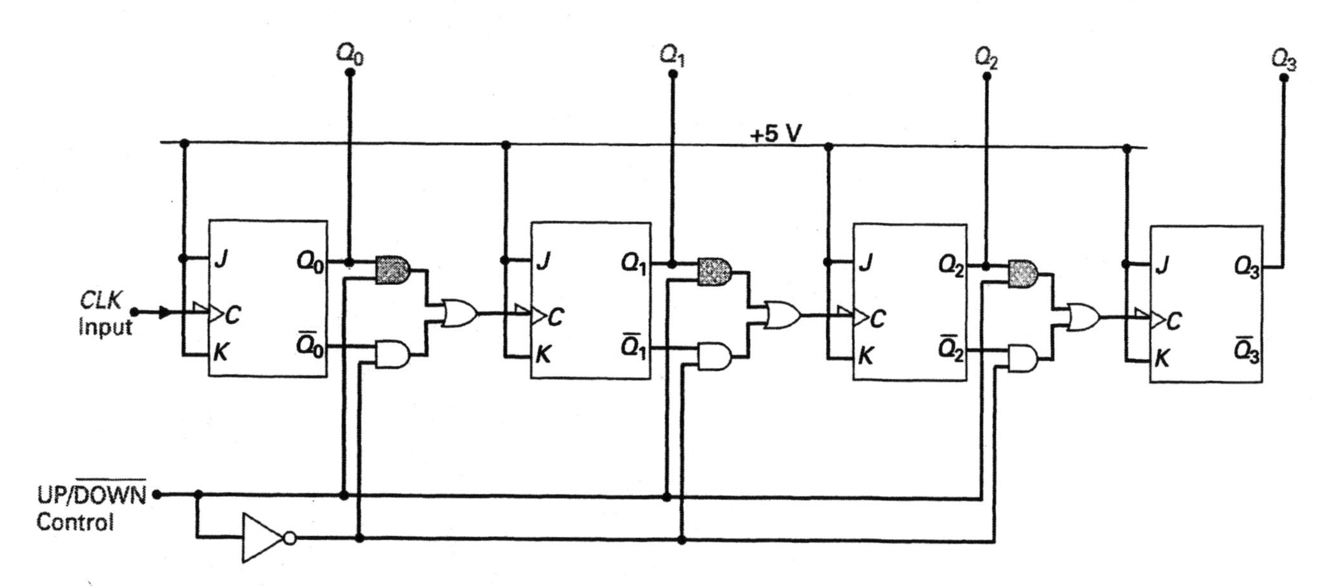

FIGURE 12-3 An Asynchronous Binary Up/Down Counter.

line is made LOW, all of the shaded AND gates will be disabled and all of the unshaded AND gates will be enabled, connecting the $\overline{Q}$ outputs through to the flip-flop's C inputs, causing the counter circuit to count down.

12-1-4 Asynchronous Decade (MOD-10) Counters

Figure 12-4(a) shows how the previously discussed asynchronous MOD-16 up counter circuit could be modified to construct a MOD-10 or **decade counter** circuit. This counter will count from 0000 (decimal 0) to 1001 (decimal 9), and then the cycle will repeat, as shown in the timing waveforms in Figure 12-4(b) and the truth table in Figure 12-4(c). The reason this counter skips the counts 1010 through 1111 (decimal 10 through 15) is due to the action of the NAND gate, which controls the active-LOW clear inputs ($\overline{CLR}$) of all four flip-flops. Since the two inputs to this NAND gate are connected to the Q_1 and Q_3 outputs, when the counter reaches a count of 1010 (decimal 10), both Q_1 and Q_3 will go HIGH, and therefore the output of the NAND gate will go LOW and CLEAR the counter. Referring to the timing waveforms in Figure 12-4(b), you can see that the $\overline{CLR}$ line is inactive for the counts 0000 through 1001. However, when the tenth clock pulse is applied, both Q_1 and Q_3 are HIGH. This condition of having both Q_1 and Q_3 HIGH is only temporary since the active $\overline{CLR}$ will almost immediately go LOW, and therefore CLEAR or RESET all of the flip-flop outputs, producing a 0000 counter output.

Decade Counter
A counter that has a modulus of 10, and therefore counts from 0 to a maximum of 9.

The truth table in Figure 12-4(c) summarizes the operation of this decade counter by showing how it counts from 0 to 9 and then repeats the cycle.

Figure 12-4(d) illustrates another way to show how a counter cycles through its different output states. In this *state transition diagram,* each circled binary number represents one of the output states, and the solid arrow indicates how one state changes to another as each clock pulse is received. For example, if we start at the count of 0000, after one clock pulse, the counter will go to

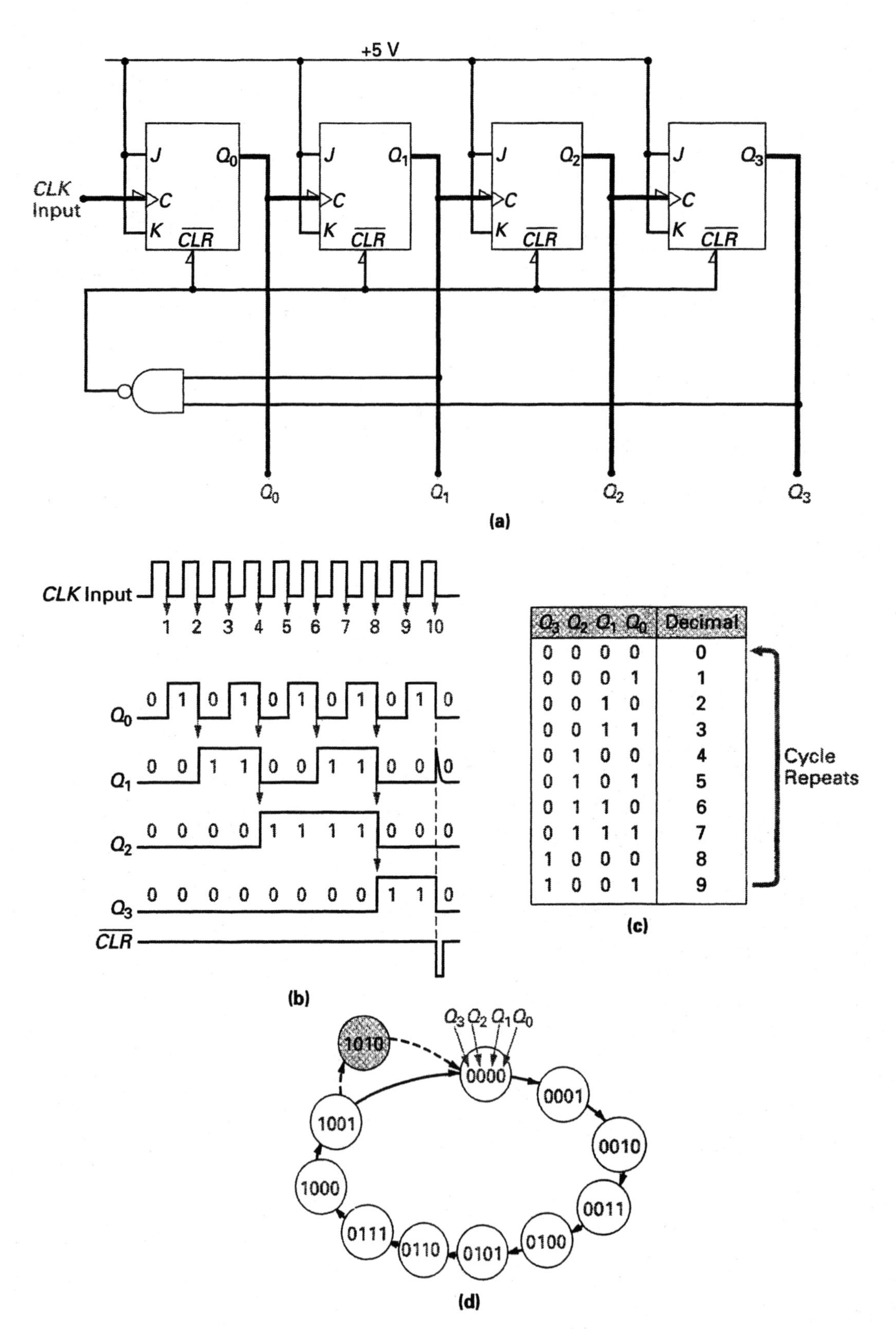

Q_3 Q_2 Q_1 Q_0	Decimal
0 0 0 0	0
0 0 0 1	1
0 0 1 0	2
0 0 1 1	3
0 1 0 0	4
0 1 0 1	5
0 1 1 0	6
0 1 1 1	7
1 0 0 0	8
1 0 0 1	9

FIGURE 12-4 An Asynchronous Decade Counter. (a) Logic Circuit. (b) Timing Waveforms. (c) Truth Table. (d) State Transition Diagram.

0001; after the next clock pulse, the counter will go to 0010; after the next clock pulse, the counter will go to 0011; and so on. The dotted line and shaded circle in this diagram indicate a temporary state that is only momentarily reached. For most purposes, this temporary state is ignored and the counter is said to go directly from 1001 to 0000, as indicated by the solid arrow.

In summary, a decade counter counts from 0 up to a maximum count of 9, a total of ten different output states (MOD-10). Since the counter requires ten input clock pulses before the output of the counter is reset, the Q_3 output frequency will be one-tenth that of the input clock frequency (*CLK* input). A decade counter, therefore, can also function as a divide-by-ten circuit. Remember that the only difference between a counter circuit and frequency divider circuit is how many of the outputs are used. For example, with a binary counter circuit, all of the flip-flops' outputs are used to deliver a parallel output word that represents the count, whereas with a frequency divider circuit, the clock input is applied to the first flip-flop, and the last flip-flop delivers serial output that is some fraction of the input clock frequency.

In the counter applications section of this chapter, you will see how the decade counter is frequently chosen in applications that have a decimal display, such as digital clocks, digital voltmeters, and frequency counters. This is because the decade counter allows us to interface the binary information within a digital system to a decimal display that you and I understand.

Figure 12-5 shows data sheet details for the 74293 counter IC. This device contains four master-slave J-*K* flip-flops that for versatility are connected as a single divide-by-two (÷2) stage and a ripple counter divide-by-eight (÷8) stage. To achieve the maximum count (0 to 15) or the maximum frequency division (÷16), simply connect the output of the ÷2 stage (Q_A) to the input of the ÷8 stage (input *B* or *CLK B*). The Reset/Count Table shows how both reset inputs ($R_{0(1)}$ and $R_{0(2)}$) must be HIGH in order to RESET the outputs Q_D, Q_C, Q_B, and Q_A. If either of these inputs is inactive (LOW), the counter will count up as shown in the Count Sequence Table.

Application—Binary Counter/Frequency Divider Circuits

Figures 12-6(a) through (c) are examples of how a 74293 could be configured to have a different modulus.

In Figure 12-6(a), the Q_A output has been connected to the CK_B input and therefore the ÷2 flip-flop is followed by the ÷8 flip-flops, producing a ÷16 frequency divider circuit or a MOD-16 binary counter circuit (0000 through 1111).

In Figure 12-6(b), the Q_B and Q_D outputs have been connected to the reset inputs ($R_{0(1)}$ and $R_{0(2)}$), so the counter will be RESET when both of these Q outputs go HIGH (at a count of 1010). The circuit will therefore function as a MOD-10 counter circuit (0000 through 1001) in which all four outputs will be used, or as a ÷10 frequency divider circuit in which the Q_D output is one-tenth that of the clock input frequency applied to CK_A.

In Figure 12-6(c), only the ÷8 ripple counter section of the 74293 is being used to create a MOD-6 counter or ÷6 frequency divider circuit. Since the Q_C and Q_D outputs have been connected to the reset inputs ($R_{0(1)}$ and $R_{0(2)}$), the counter will RESET the moment the counter reaches a count of 110 (decimal

IC TYPE: SN74293– Decade and 4-bit Binary Counter

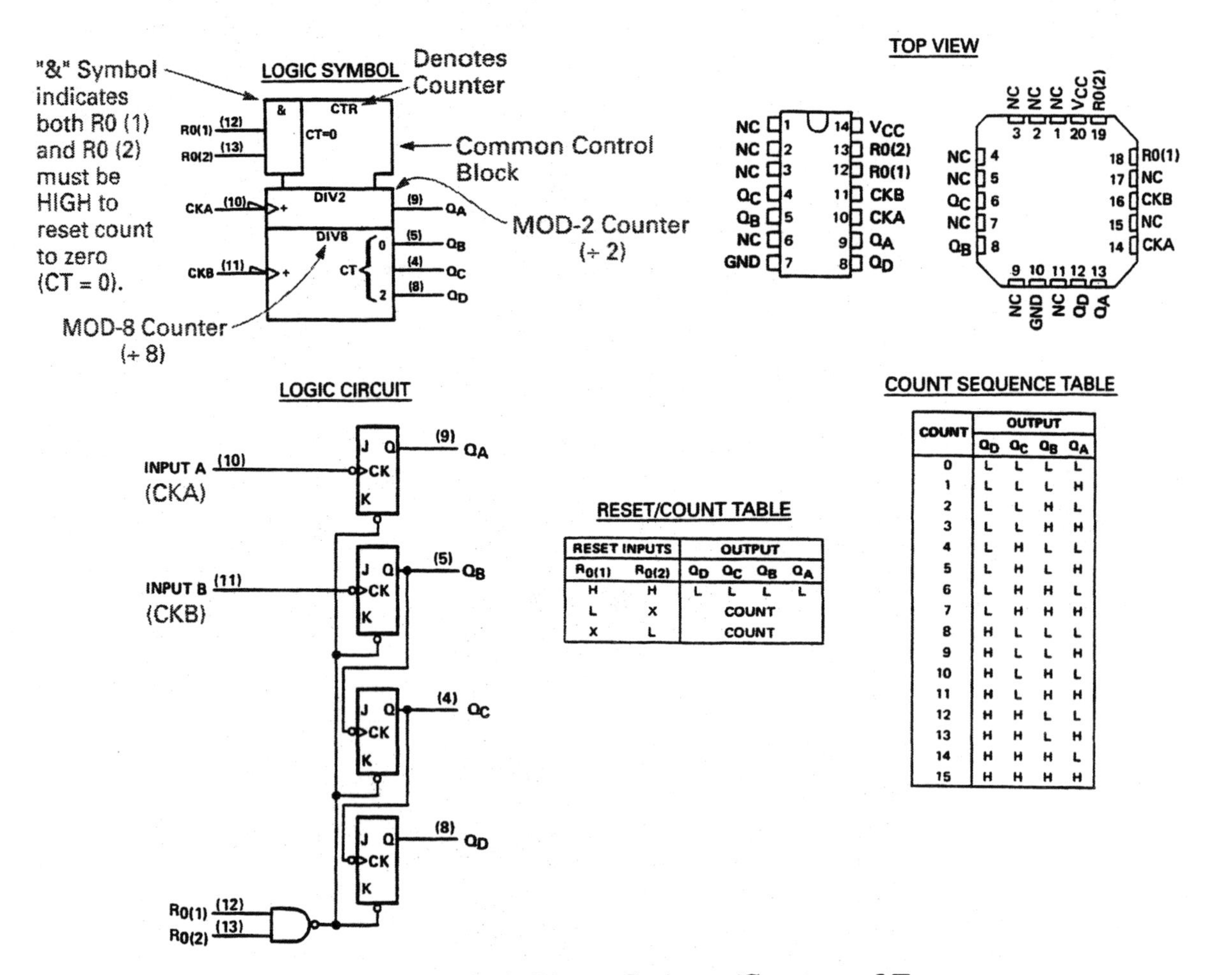

RESET/COUNT TABLE

RESET INPUTS		OUTPUT			
$R_{0(1)}$	$R_{0(2)}$	Q_D	Q_C	Q_B	Q_A
H	H	L	L	L	L
L	X	COUNT			
X	L	COUNT			

COUNT SEQUENCE TABLE

COUNT	OUTPUT			
	Q_D	Q_C	Q_B	Q_A
0	L	L	L	L
1	L	L	L	H
2	L	L	H	L
3	L	L	H	H
4	L	H	L	L
5	L	H	L	H
6	L	H	H	L
7	L	H	H	H
8	H	L	L	L
9	H	L	L	H
10	H	L	H	L
11	H	L	H	H
12	H	H	L	L
13	H	H	L	H
14	H	H	H	L
15	H	H	H	H

FIGURE 12-5 An IC Containing a 4-Bit Binary Counter. (Courtesy of Texas Instruments)

6). The circuit will therefore function as a MOD-6 counter circuit (000 through 101) in which outputs Q_B, Q_C, and Q_D will be used, or as a ÷6 frequency divider circuit in which the Q_D output is one-sixth that of the clock input frequency applied to CK_B.

In Figure 12-6(d), only the ÷2 section of the 74293 is being used to create a MOD-2 counter or ÷2 frequency divider.

Asynchronous Presettable Counter
A counter that will count from its loaded preset value up to its maximum value. By changing the preset value, the modulus of the counter can be changed.

12-1-5 Asynchronous Presettable Counters

Figure 12-7(a) shows how four *J-K* flip-flops and several gates could be connected to construct an **asynchronous presettable counter.** As an example, this circuit is loading a preset data value of 1001 (decimal 9) into the counter directly following the maximum count of 1111 (decimal 15), as shown in the state transition diagram in Figure 12-7(b). Once the value of 1001 has been loaded, the

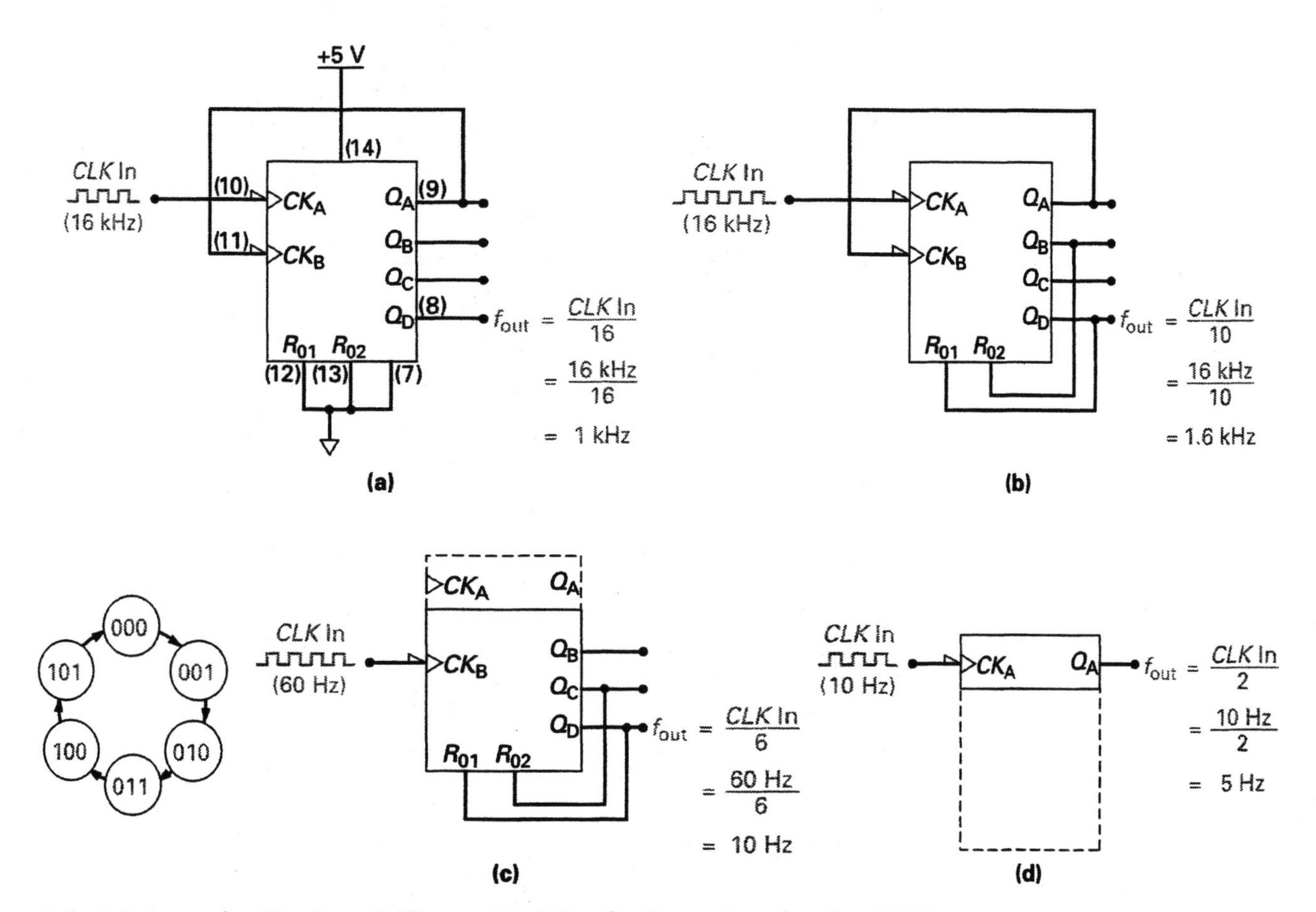

FIGURE 12-6 Application—Different Modulus Configurations for the 74293.

counter will again count up to a maximum of 1111, reload the preset value of 1001 once again, and then repeat the cycle.

The preset value, which in this example is 1001, is applied to the data inputs D_0, D_1, D_2, and D_3. The logic levels applied to these inputs are normally prevented from reaching the *J-K* flip-flop's $\overline{PRE}$ and $\overline{CLR}$ inputs because all of the NAND gates are disabled by a LOW on the LOAD control line. The logic level on the LOAD control line is determined by a NOR, whose four inputs are connected to the counter's four outputs Q_0, Q_1, Q_2, and Q_3.

To understand the operation of the circuit in Figure 12-7(a), let us begin by assuming that the counter is at its maximum count of 1111. On the next negative edge of the clock, the counter will increment to 0000, and only this output will cause the NOR gate, and therefore the LOAD control line, to go HIGH. With the LOAD control line HIGH, the data inputs and their complements are allowed to pass through the NAND gates to activate either the PRESET or CLEAR inputs of the flip-flops. In this example, the unshaded NAND gates will produce a HIGH output while the shaded NAND gates will produce a LOW output. Since the flip-flop's $\overline{PRE}$ and $\overline{CLR}$ inputs are both active-LOW, Q_0 will be preset to 1, Q_1 will be cleared to 0, Q_2 will be cleared to 0, and Q_3 will be preset to 1. The 4-bit data value of 1001 is therefore asynchronously

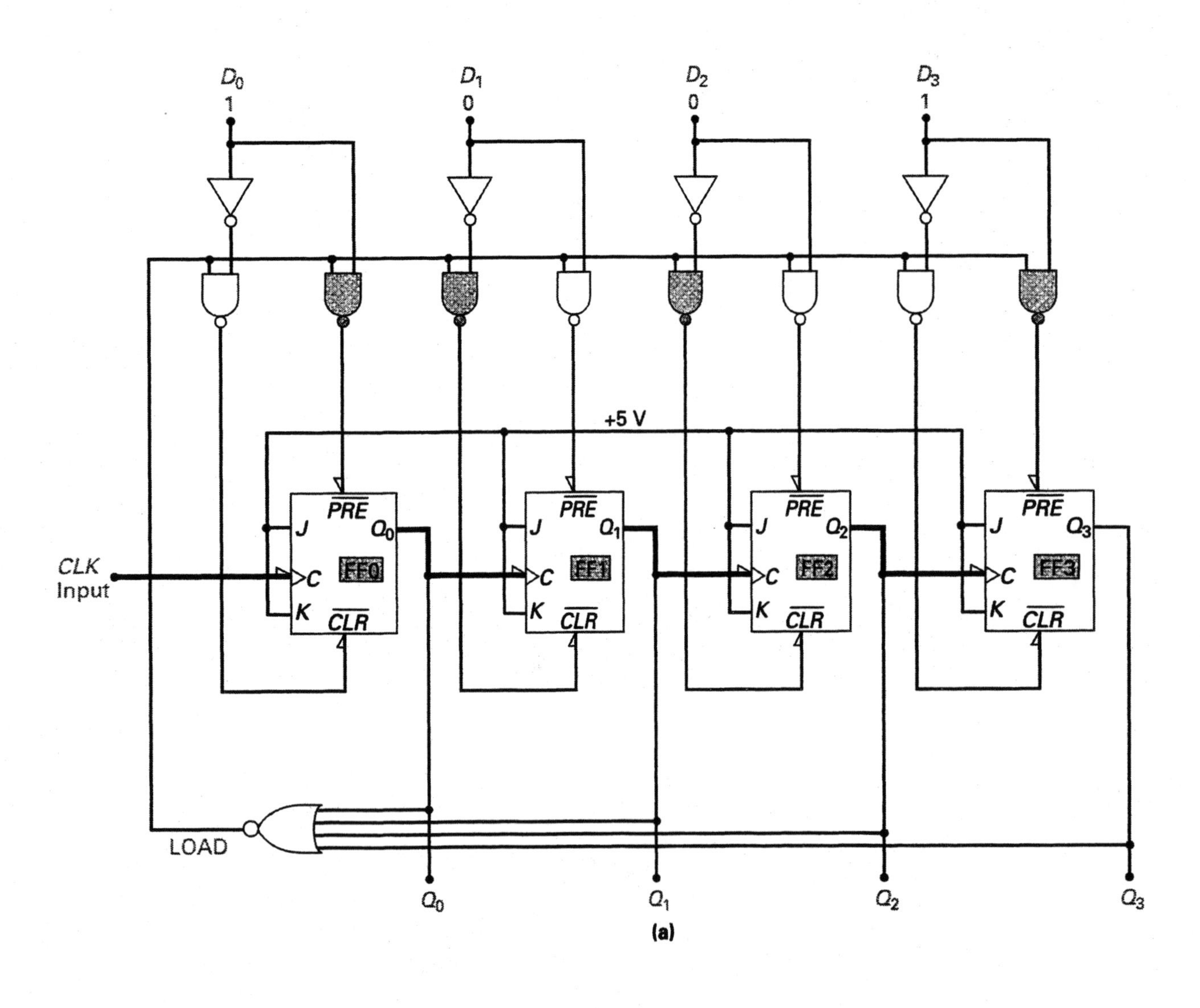

(a)

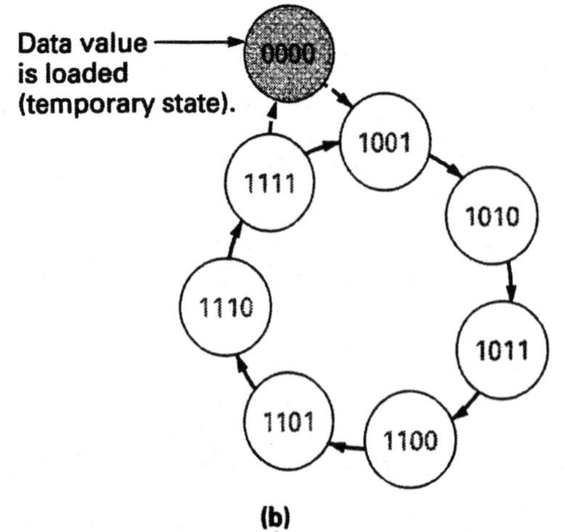

(b)

FIGURE 12-7 An Asynchronous Presettable Counter. (a) Logic Circuit. (b) State Transition Diagram for a Preset Value of 1001.

loaded into the counter (independent of the clock signal) the moment the counter goes to 0000. This 0000 state is only a temporary state, as shown in the state transition diagram in Figure 12-7(b), since it is almost immediately changed to the loaded state. Successive clock pulses at the input will cause the counter to count from 1001 to 1010, to 1011, to 1100, to 1101, to 1110, and then to the maximum count of 1111. The next clock pulse will reset the counter's count to 0000, enable the LOAD control line, load the data value, and then repeat the cycle.

One of the big advantages with a presettable counter is that it has a *programmable modulus.* For example, in Figure 12-7 we loaded a data value of 1001 (decimal 9) and then counted to 1111 (decimal 15), resulting in a total of seven different output combinations (1001, 1010, 1011, 1100, 1101, 1110, and 1111). As a result, this circuit will function as a MOD-7 binary counter or ÷7 frequency divider.

By changing the data value loaded into a presettable counter, we can create a counter of any modulus. The simple formula for calculating the modulus of a presettable counter is

$$M_P = MOD - D$$

M_P = modulus of the presettable counter.
MOD = natural modulus of the counter.
D = preset data value.

In our example circuit in Figure 12-7,

$$\begin{aligned} M_P &= MOD - D \\ &= 16 - 9 \quad \text{(counter skips states 0 to 8)} \\ &= 7 \end{aligned}$$

■ **EXAMPLE:**

Assuming a 6-bit asynchronous presettable counter, calculate

a. its natural modulus.
b. its preset modulus if 001100_2 is loaded.

■ ***Solution:***

a.
$$\begin{aligned} MOD &= 2^n \\ &= 2^6 = 64 \end{aligned}$$

b.
$$\begin{aligned} M_P &= MOD - D \\ &= 64 - 12 = 52 \end{aligned}$$

SELF-TEST REVIEW QUESTIONS FOR SECTION 12-1

1. Most of the flip-flops within an asynchronous counter __________ (are/are not) connected to the clock input and therefore they __________ (do/do not) change state in sync with the master clock signal.

2. What is the natural modulus and maximum count of an 8-bit counter?
3. What would be the upper frequency limit of an 8-bit counter if each flip-flop within an asynchronous counter had a propagation delay time of 30 ns?
4. What would be the natural modulus and preset modulus of an 8-bit counter if 0001 1110 was the load value?

12-2 SYNCHRONOUS COUNTERS

As mentioned previously, asynchronous counters are often referred to as ripple counters. The term ripple is used because the flip-flops making up the asynchronous counter are strung together with the output of one flip-flop driving the input of the next. As a result, the flip-flops do not change state simultaneously or in sync with the input clock pulses, since the new count has to ripple through and update all of the flip-flops. This ripple-through action leads to propagation delay times that limit the counter's counting speed. This limitation can be overcome with a **synchronous counter** in which all of the flip-flops within the counter are triggered simultaneously and therefore synchronized to the master timing clock signal.

Synchronous Counter
A counter in which all operations are controlled by the master clock signal.

12-2-1 Synchronous Binary Up Counters

Figure 12-8(a) shows how four *J-K* flip-flops and two AND gates can be connected to form a 4-bit synchronous MOD-16 up counter. The clock input signal line has been drawn as a heavier line in this illustration to show how all of the flip-flops in a synchronous counter circuit are triggered simultaneously by the common clock input signal. This parallel connection makes the counter synchronous since all of the flip-flops will now operate in step with the input clock signal.

The synchronous binary up counter shown in Figure 12-8(a) will operate as follows. The J and K inputs of FF0 are tied HIGH, and therefore the Q_0 output will continually toggle, like a single-stage asynchronous counter. This fact is verified by looking at the Q_0 column in Figure 12-8(b), which shows how the Q_0 output continually alternates HIGH and LOW. The J and K inputs of FF1 are controlled by the divide-by-two output of FF0. This means that when Q_0 is LOW, the Q_1 output of FF1 will not change; however, when Q_0 is HIGH, the Q_1 output of FF1 will toggle. This action can be confirmed by referring to the Q_1 column in Figure 12-8(b), which shows that when the Q_0 output goes HIGH, the Q_1 output will toggle or change state on the next active edge of the clock signal. The J and K inputs of FF2 are controlled by the ANDed outputs of Q_0 and Q_1. This means that only when Q_0 and Q_1 are both HIGH will the output of AND gate A be HIGH, and this HIGH will enable FF2 to toggle. Referring to the Q_2 column in Figure 12-8(b), you can see that only when Q_0 and Q_1 are HIGH will Q_2 toggle or change state on the next active edge of the clock. The J and K inputs of FF3 are controlled by the ANDed outputs of Q_0,

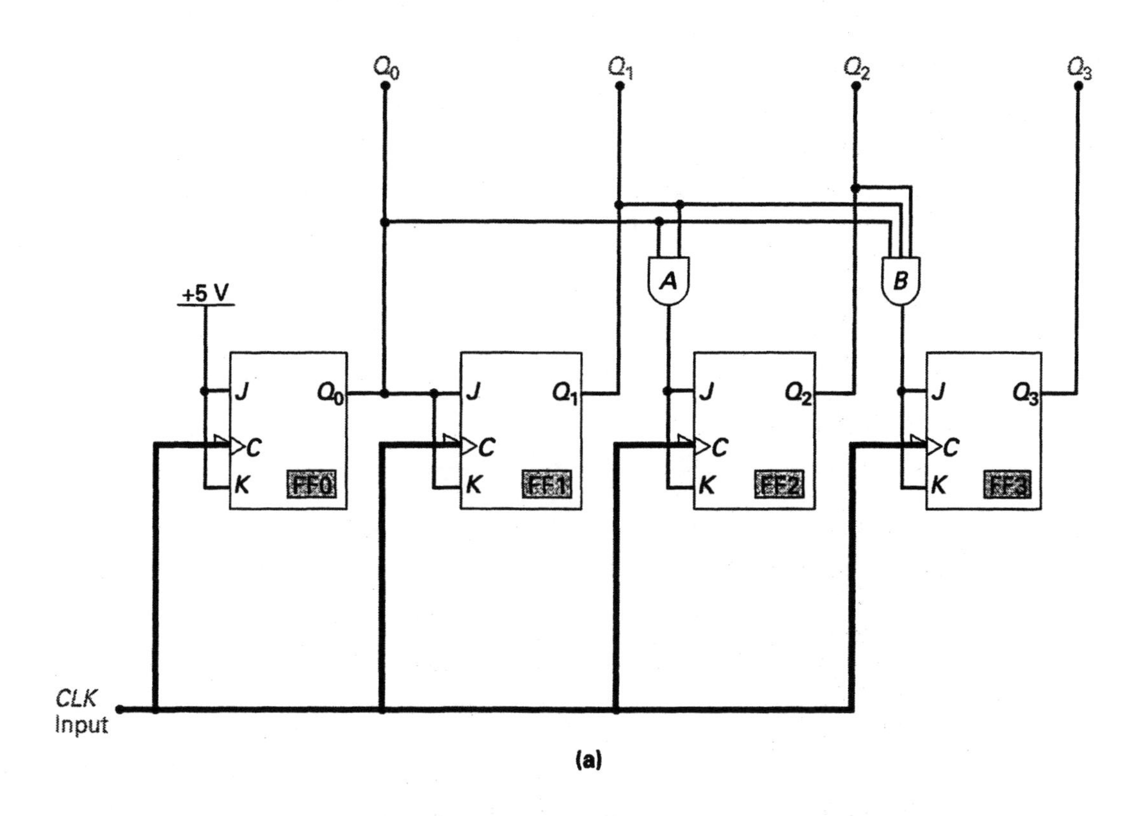

Q_3	Q_2	Q_1	Q_0	Decimal
0	0	0	0	0
0	0	0	1	1
0	0	1	0	2
0	0	1	1	3
0	1	0	0	4
0	1	0	1	5
0	1	1	0	6
0	1	1	1	7
1	0	0	0	8
1	0	0	1	9
1	0	1	0	10
1	0	1	1	11
1	1	0	0	12
1	1	0	1	13
1	1	1	0	14
1	1	1	1	15

(b)

FIGURE 12-8 A Synchronous Binary Up Counter. (a) Logic Circuit. (b) Truth Table.

Q_1, and Q_2. This means that only when Q_0, Q_1, and Q_2 are HIGH, will the output of AND gate *B* be HIGH, enabling FF3 to toggle. If you look at the Q_3 column in Figure 12-8(b), you will notice that only when Q_0, Q_1, and Q_2 are all HIGH, will Q_3 toggle or change state on the next active edge of the clock.

12-2-2 *Synchronous Counter Advantages*

The key advantage of the asynchronous or ripple counter is its circuit simplicity, as can be seen by comparing the synchronous binary up counter in Figure 12-8(a) to the asynchronous binary up counter in Figure 12-1(a). The key disadvantage of the asynchronous counter is its frequency of operation or counting speed limitation. Since the input is applied only to the first flip-flop, it takes a certain time for all of the outputs to be updated to reflect the new count. This propagation delay time of the counter is equal to the sum of all the flip-flop propagation delay times. This limitation means that we cannot trigger or clock the input of an asynchronous counter before all of the outputs have settled into their new state, and therefore the clock input frequency (or pulses to be counted) has a speed or frequency limitation. Asynchronous counters are sometimes constructed using ECL flip-flops which have an extremely small propagation delay time, and therefore a high input clock frequency can be applied. These ECL flip-flops, however, are more expensive and have a higher power consumption.

Synchronous counters are a direct solution to the asynchronous counter limitations, since they have small propagation delay times and are less expensive and consume less power than ECL asynchronous counters. The key to the synchronous counter's small propagation delay time is that all of its flip-flops are triggered or clocked simultaneously by a common clock input signal. This means that all of the flip-flops will be changing their outputs at the same time, so the total counter propagation delay time is equal to the delay of only one flip-flop. In reality, we should also take into account the time it takes for a Q output to propagate through a single AND gate and reach the J-K inputs of the following stage. Taking these two factors into consideration, we arrive at the following formula for calculating the delay time for a synchronous counter.

$$t_p = \text{single flip-flop } t_p + \text{single AND gate } t_p$$

■ **EXAMPLE:**

Assuming a J-K flip-flop propagation delay time of 30 ns and an AND gate propagation delay time of 10 ns, calculate the counter propagation time for a four-stage asynchronous counter and four-stage synchronous counter.

■ ***Solution:***

$$t_{p(asynchronous)} = \text{single flip-flop } t_p \times \text{number of flip-flops}$$
$$= 30 \text{ ns} \times 4 = 120 \text{ ns}$$
$$t_{p(synchronous)} = \text{single flip-flop } t_p + \text{single AND gate } t_p$$
$$= 30 \text{ ns} + 10 \text{ ns} = 40 \text{ ns}$$

In summary, the two advantages of a synchronous counter are that no transient output states occur due to the ripple through additive propagation delays, and that synchronous counters can operate at much higher clock input frequencies.

12-2-3 *Synchronous Presettable Binary Counters*

Like the presettable asynchronous counter, the presettable synchronous counter has a programmable modulus. The 74LS163A is an example of a presettable synchronous up counter IC, and its data sheet details are shown in Figure 12-9. Referring to the logic symbol, you can see that a 4-bit binary value can be loaded into this counter to obtain any desired modulus by applying a LOW to the $\overline{\text{LOAD}}$ control input (pin 9). When the $\overline{\text{LOAD}}$ input is made LOW, the Q outputs (Q_A, Q_B, Q_C, and Q_D) will reflect the data inputs (D, D, D, and D) after the next positive edge of the clock input (pin 2). There are two active-HIGH enable inputs (pins 7 and 10), and because of the ANDing of these inputs, both must be HIGH for the counter to count.

Referring to the timing diagram in Figure 12-9, you can see an example of how the counter is first loaded with a preset value of 1100 (decimal 12) and then counted up to its maximum count of 1111 (decimal 15). At this time, the ripple carry output (*RCO,* pin 15) generates an output pulse. This last stage output is used to enable the input of another counter, if two counters were needed in a circuit application. For example, when a counter has reached a count of

Q_3	Q_2	Q_1	Q_0	
1	1	1	1	$= 15_{10}$

a reset-and-carry action occurs from right to left as follows:

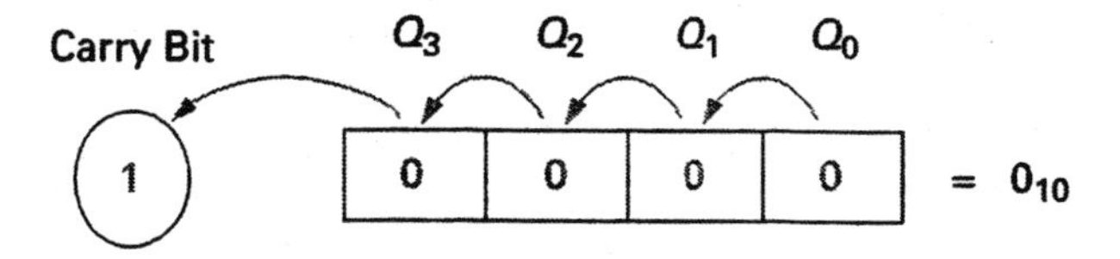

If only a 4-bit counter is needed, all of the Q outputs are reset to 0, the cycle repeats, and the carry bit is lost. If, however, two 4-bit counters are needed, the *RCO* of the LSB counter will be connected to the input of the MSB counter so that the Q_3 output will carry into the Q_4 flip-flop. In this situation the *carry bit* is not lost and the 8-bit counter counts beyond 15 as follows:

Most Significant 4 Bits					Least Significant 4 Bits				
Q_7	Q_6	Q_5	Q_4		Q_3	Q_2	Q_1	Q_0	
0	0	0	0		1	1	1	1	15
0	0	0	1	RCO	0	0	0	0	16
0	0	0	1		0	0	0	1	17
0	0	0	1		0	0	1	0	18

IC TYPE: SN74LS163A– Synchronous 4-bit Binary (MOD-16) Up Counter

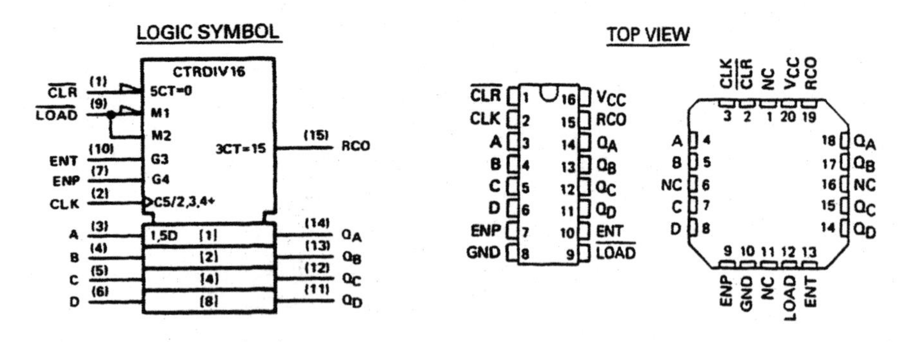

TYPICAL CLEAR, PRESET, COUNT, AND INHIBIT SEQUENCES

Illustrated below is the following sequence:

1. Clear outputs to zero
2. Preset to binary twelve
3. Count to thirteen, fourteen, fifteen, zero, one, and two
4. Inhibit

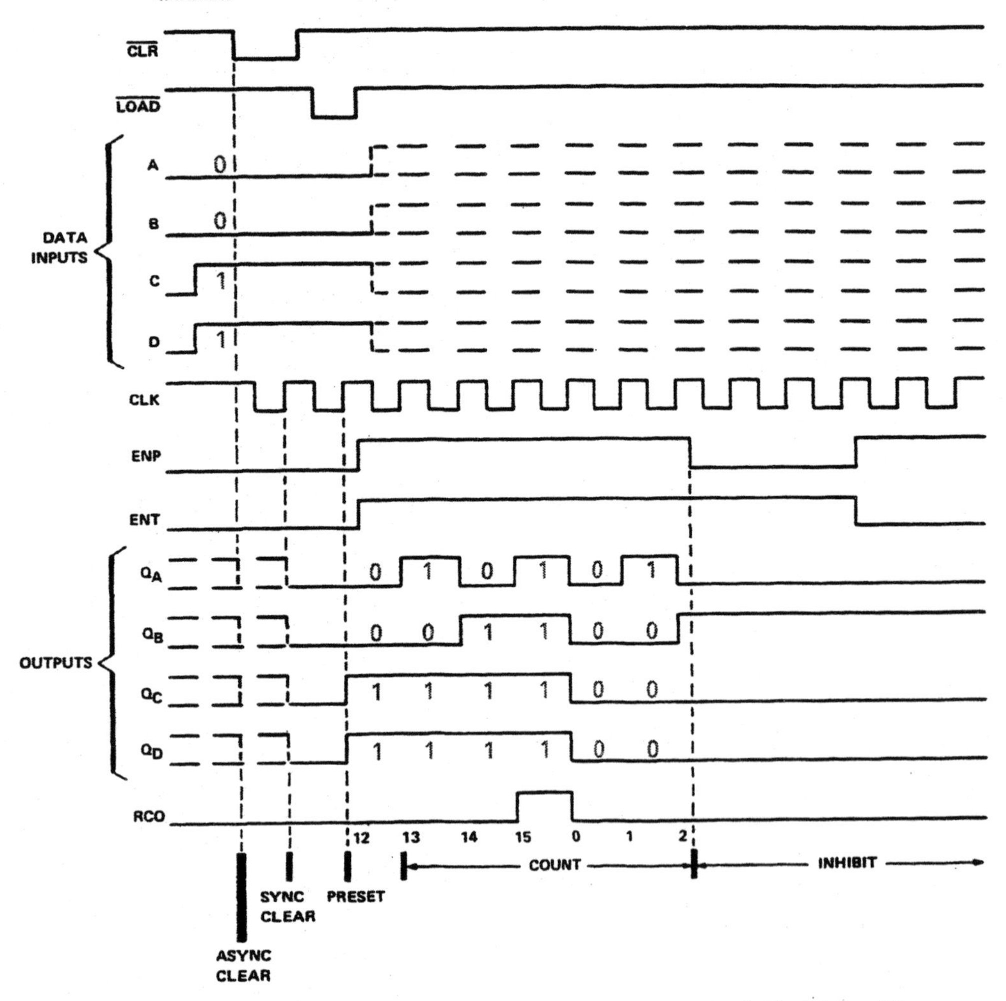

FIGURE 12-9 An IC Containing a Synchronous Presettable 4-Bit Binary Up Counter. (Courtesy of Texas Instruments)

INTERNAL LOGIC CIRCUIT

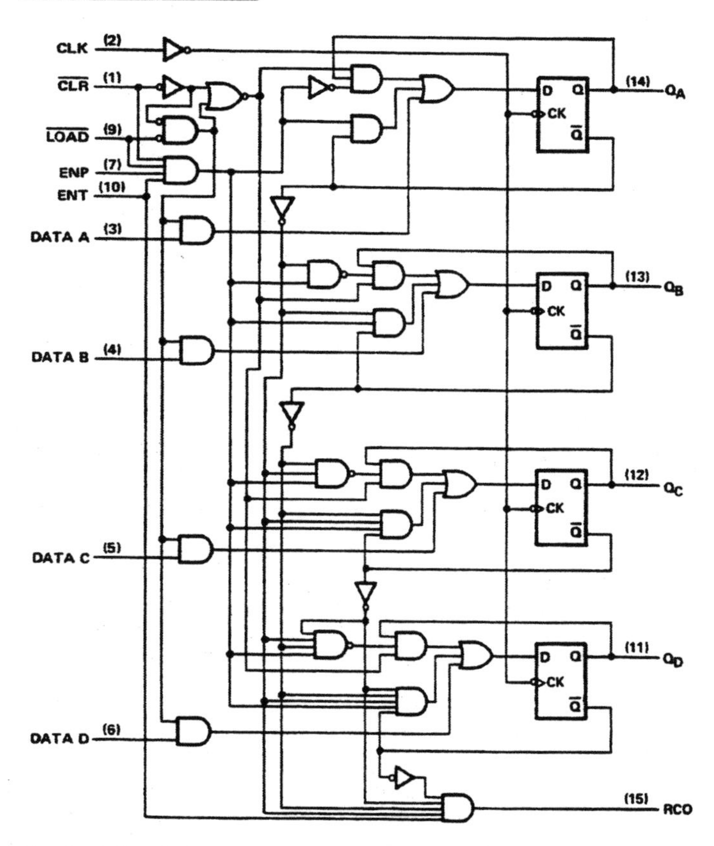

FIGURE 12-9 (Continued).

Application—Cascaded Counter Circuit

Figure 12-10(a) shows how two MOD-16 74163A ICs can be *cascaded,* or strung together, to obtain a MOD-256 counter. When connecting synchronous counters in cascade, the *RCO* (so called because it ripples out of the counter's last stage) is connected to the enable inputs of counter 2 (CNTR 2). As a result, when CNTR 1 reaches its maximum count (1111), the *RCO* from CNTR 1 will go HIGH and enable CNTR 2 so that on the next positive edge of the clock, CNTR 2 will have a count of 0001 and CNTR 1 will have reset to 0000.

Since the counter in Figure 12-10(a) contains eight outputs (Q_0 through Q_7), it will produce a total of 2^8, or 256, different output combinations, with a maximum count of 255 as shown in Figure 12-10(b). If this circuit were used as a frequency divider, it would produce a single output pulse at the Q_7 output for every 256 input clock pulses (a divide-by-256 circuit).

If some other counter modulus or frequency division were required (anywhere between 2 and 256), the preset inputs could be used to load in any value. For example, if 0001 1000 (decimal 24) were loaded into the 8-bit counter, the counter would have a modulus of

$$M_P = MOD - D$$
$$= 256 - 24 = 232$$

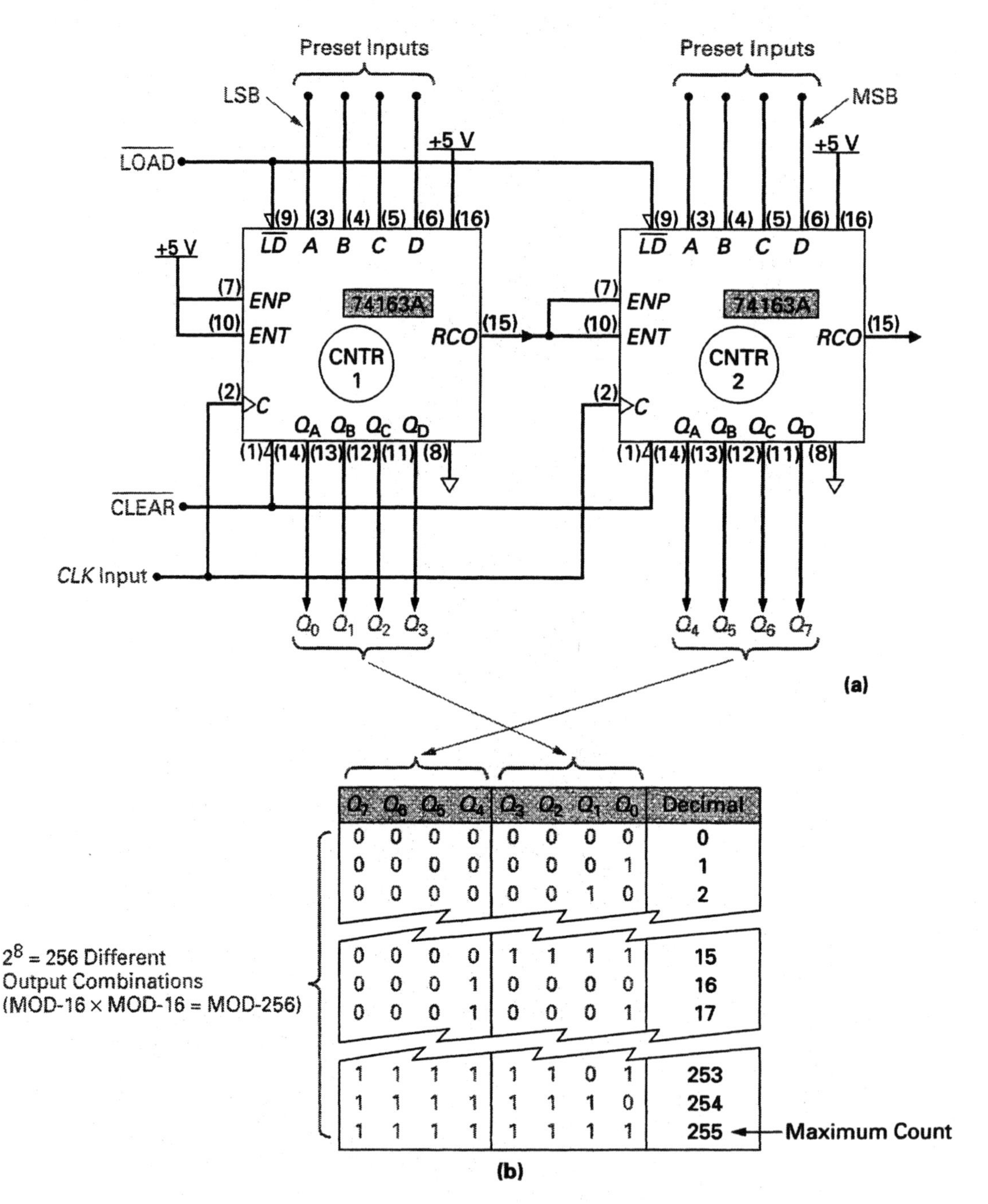

Q_7	Q_6	Q_5	Q_4	Q_3	Q_2	Q_1	Q_0	Decimal
0	0	0	0	0	0	0	0	0
0	0	0	0	0	0	0	1	1
0	0	0	0	0	0	1	0	2
0	0	0	0	1	1	1	1	15
0	0	0	1	0	0	0	0	16
0	0	0	1	0	0	0	1	17
1	1	1	1	1	1	0	1	253
1	1	1	1	1	1	1	0	254
1	1	1	1	1	1	1	1	255

FIGURE 12-10 Cascading Counters to Obtain a Higher Modulus.

This counter would now count from 24 to 255 (a total of 232 different output combinations), or it would divide the input clock frequency by 232.

Figure 12-11 shows how the *RCO* could be used to construct a 16-bit (MOD-65,536) synchronous up counter circuit. Other counter ICs could be added if needed, which is why this circuit is called an N-bit counter. For example, if only three ICs were used, the circuit would be a 12-bit synchronous counter (N = 3 ICs × 4-bits = 12-bit counter), whereas if six ICs were used,

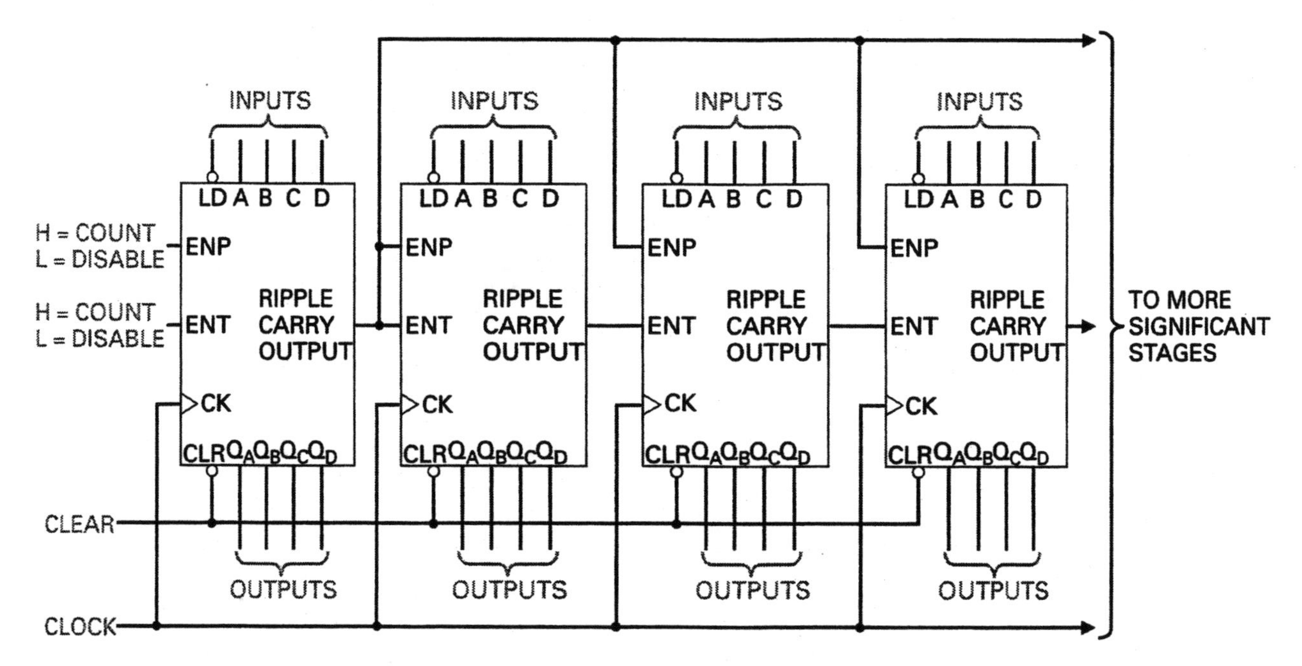

FIGURE 12-11 N-Bit Counter/Frequency Divider.

the circuit would be a 24-bit synchronous counter (N = 6 ICs × 4-bits = 24-bit counter).

12-2-4 Synchronous Decade (MOD-10) Counters

Like the asynchronous decade counter, a synchronous decade (MOD-10) counter generates only the BCD outputs 0000 though 1001 (decimal 0 through 9). In the counter applications section of this chapter, you will see how decade counters are frequently chosen in applications that have a decimal display, such as digital clocks, digital voltmeters, and frequency counters. This is because the decade counter allows us to interface the binary information within a digital system to a decimal display that you and I understand.

The 74LS160A is a good example of a synchronous decade counter IC. Data sheet details for this IC are shown in Figure 12-12. The 74LS160A has exactly the same input and output pin assignments as the 74LS163A covered in Figure 12-9; however, this counter has only ten different output combinations. Referring to the timing waveforms for this IC, you can see an example of how this BCD counter is first synchronously cleared and then synchronously loaded with a preset value of 0111 (decimal 7). The counter then counts clock pulse positive edges up to 1001 (decimal 9), at which time the *RCO* goes HIGH. On the next positive edge of the clock, the counter resets to 0000 and the *RCO* line goes LOW. The *RCO* pulse generated could be used to trigger the next counter IC in a cascaded decade counter circuit. As the clock pulse continues, the counter counts up from 0000 to 0011, at which time in these example waveforms the count is inhibited, as shown in Figure 12-12(b).

IC TYPE: SN74LS160A– Synchronous 4-bit Decade (MOD-10) Up Counter

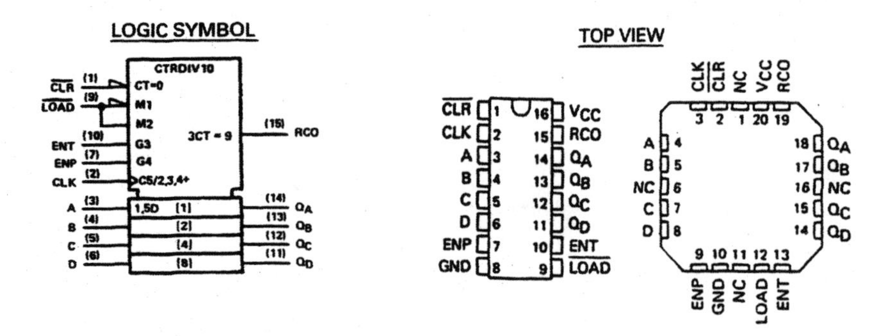

TYPICAL CLEAR, PRESET, COUNT, AND INHIBIT SEQUENCES

Illustrated below is the following sequence:

1. Clear outputs to zero
2. Preset to BCD seven
3. Count to eight, nine, zero, one, two, and three
4. Inhibit

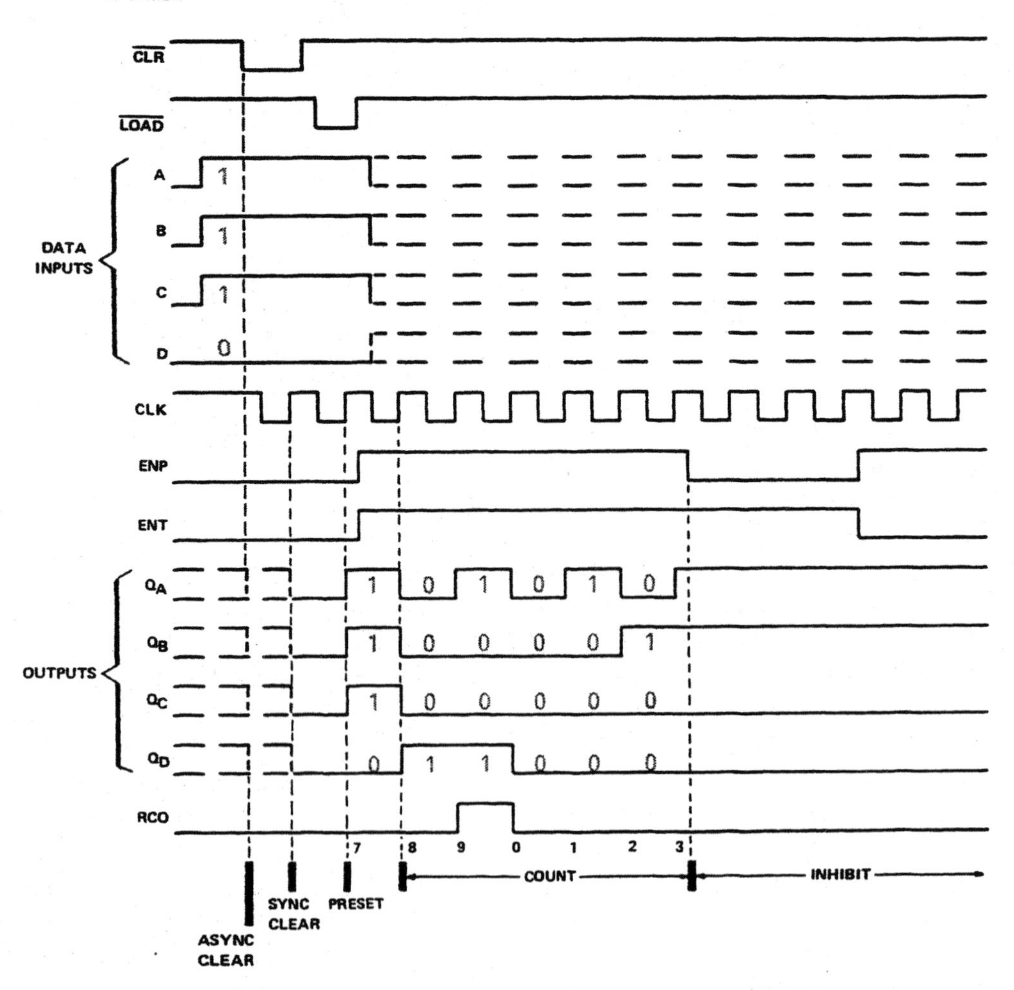

FIGURE 12-12 An IC Containing a Synchronous Decade Counter. (Courtesy of Texas Instruments)

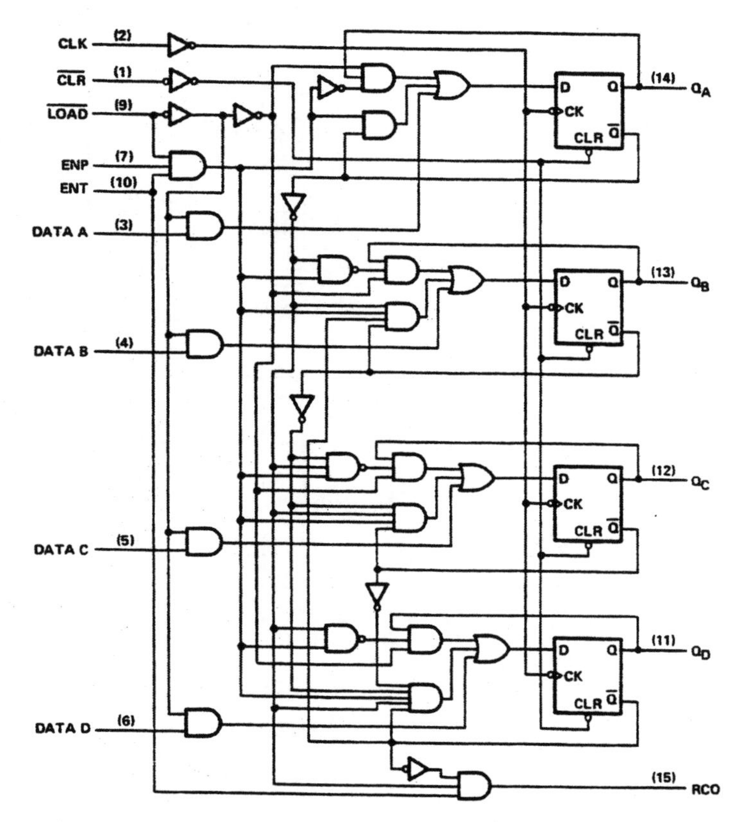

FIGURE 12-12 (Continued).

Application—A 0-to-99 Counter and Display Circuit

Figure 12-13 shows how two 74LS160A synchronous decade counter ICs, two 7447 BCD-to-seven-segment decoder ICs, and two common-anode seven-segment LED displays could be connected to construct a 0-to-99 counter and display circuit. Up until this point, we have always followed the industry schematics standard of having inputs on the left side of the page and outputs on the right side of the page. A difficulty arises, however, when we are showing the various digits of a display, since the units digit is normally always on the right side and digits of increasing weight (such as tens, hundreds, and thousands) are to the left of the units digit. In this illustration, we will have the input clock signal applied at the right and propagate to the left so that the digital displays will be in their correct position.

The operation of this circuit is as follows. The units counter counts the input clock pulses, and after every ten clock pulses it activates the *RCO*. This enables the tens counter, which counts up by one on the next positive edge of the clock. The two decade counters will produce BCD outputs that range from $0000\ 0000_{BCD}$ (decimal 0) to $1001\ 1001_{BCD}$ (decimal 99). The 4-bit BCD output codes from the decade counters are decoded by the 7447s and displayed as their decimal equivalents on the seven-segment displays.

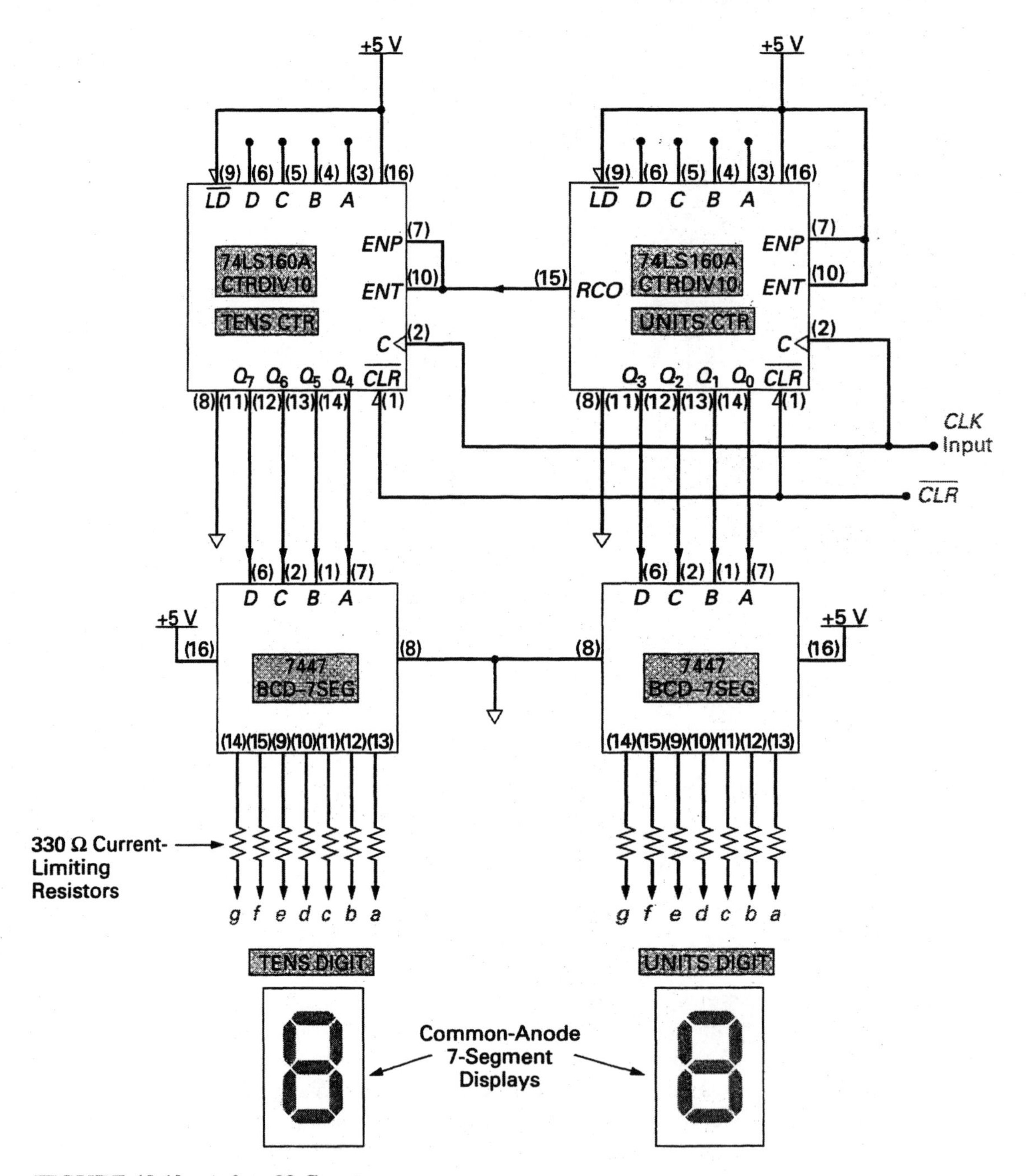

FIGURE 12-13 A 0-to-99 Counter.

12-2-5 *Synchronous Up/Down Counters*

Bidirectional Counter
A counter that can be controlled to either count up or count down.

All up/down counters are sometimes referred to as **bidirectional counters** since they can count in one of two directions—up or down. A control line is generally used to determine whether the count is incremented (advanced upward) or decremented (advanced downward). For example, if a 2-bit counter's DOWN/$\overline{\text{UP}}$ control line were made LOW, the counter would count up (0,1,2,3,0,1,2, . . .), whereas if the DOWN/$\overline{\text{UP}}$ control were made HIGH, the counter would count down (3,2,1,0,3,2,1, . . .).

The 74193 IC is described in data books as a presettable MOD-16 up/down synchronous counter with asynchronous loading and asynchronous master reset. Data sheet details for this IC are shown in Figure 12-14. This IC counter contains four *J-K* flip-flops and a variety of control gates. The counter can be used as a MOD-*N* counter, or divider, by applying any preset value to the data inputs (*A, B, C,* and *D*) and activating the active-LOW load input (pin 11). An asynchronous CLEAR or master reset input (pin 14) is available, and will force all outputs LOW when a HIGH level is applied. Synchronous operation is provided by having all flip-flops clocked simultaneously by a LOW-to-HIGH transition (positive edge-triggered). The direction of counting (up or down) is controlled by the two trigger inputs COUNT UP (pin 5) and COUNT DOWN (pin 4). To count up, simply apply the clock input to the COUNT UP input while the COUNT DOWN input is HIGH, whereas to count down, apply the clock input to the COUNT DOWN input while the COUNT UP input is HIGH.

The 74193 up/down counter was also designed to be cascaded, and therefore a CARRY output will pulse HIGH when the counter cycles beyond its maximum count, as follows:

	(8)	(4)	(2)	(1)		(10)	(1)
	1	1	1	0	=	1	4
Carry 1	1	1	1	1	=	1	5
1	0	0	0	0	=	1	6
1	0	0	0	1	=	1	7

Since the 74193 can also count down, this IC also has a BORROW output which will pulse HIGH when the counter cycles down to its lowest value, resets to maximum, and then repeats, as follows:

	(16)	(8)	(4)	(2)	(1)		(10)	(1)
	1	0	0	0	1	=	1	7
	1	0	0	0	0	=	1	6
Borrow 1								
	0	1	1	1	1	=	1	5
	0	1	1	1	0	=	1	4

Application—Back-and-Forth Flashing Circuit

Figure 12-15 shows how a 74193 up/down counter, a 74154 four-line to sixteen-line decoder and a 7400 could be connected to form a back-and-forth flashing circuit. This circuit was previously covered in Figure 7-5. In this circuit, the 74193 is made to count up from 0000 to 1111 and is then switched to count down from 1111 to 0000, and then switched again to repeat the cycle. The NAND gates provide the count-up and count-down control, while the 74154 enables (makes LOW) its 0 through 15 outputs in response to the binary count

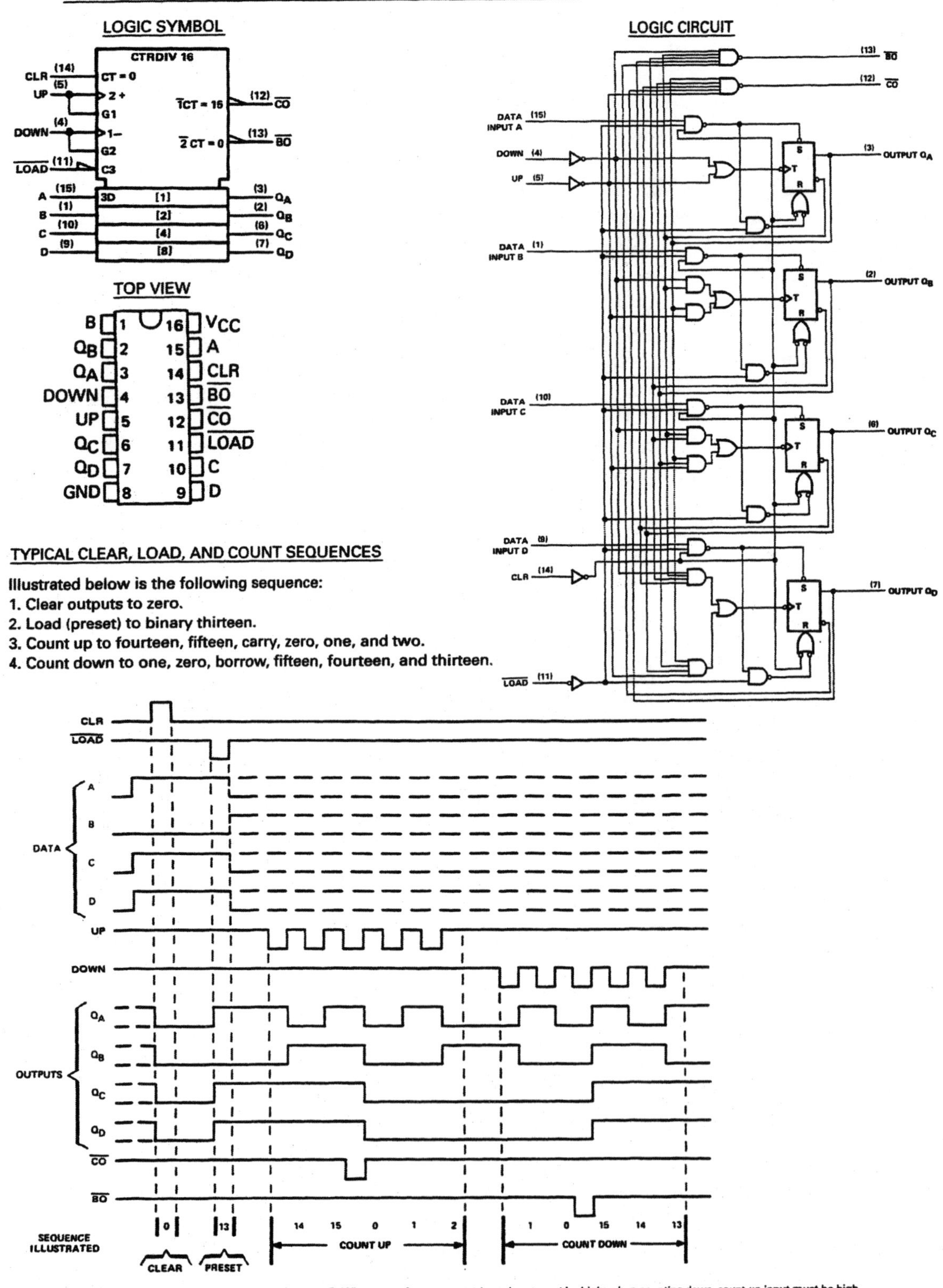

FIGURE 12-14 An IC Containing a Synchronous 4-Bit Up/Down Counter. (Courtesy of Texas Instruments)

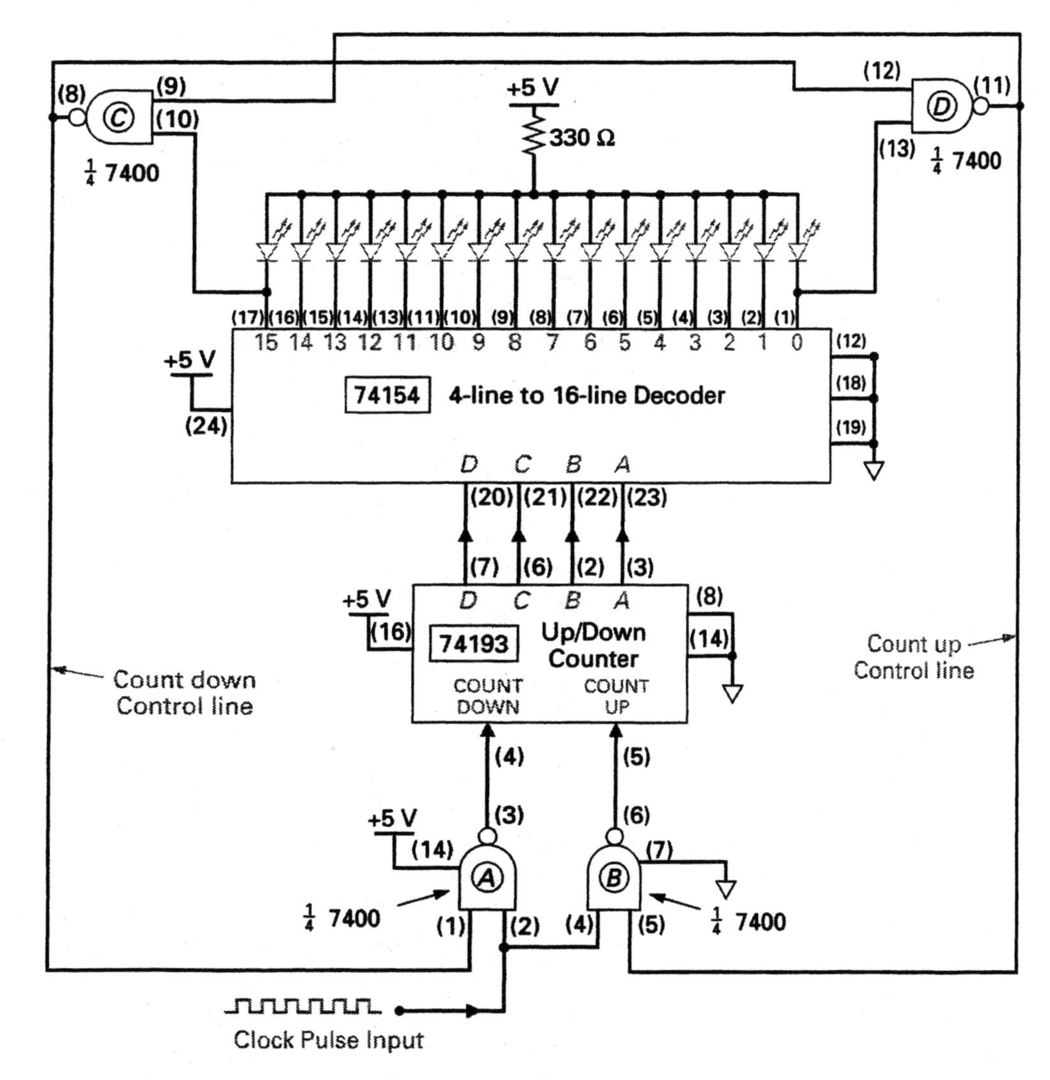

FIGURE 12-15 Application—Up/Down Flashing Circuit.

from the 74193 counter. As a result, the sixteen LEDs in the display are sequenced ON and OFF in a continual back-and-forth action.

SELF-TEST REVIEW QUESTIONS FOR SECTION 12-2

1. Synchronous counters are also referred to as ripple counters. (True/False)
2. What are the main advantages of the asynchronous counter and the synchronous counter?
3. The ____________ output from an up/down counter will go active when the counter counts up to its maximum count and then resets, while the ____________ output will go active when the counter goes down to 0 and then recycles to its maximum count.
4. By cascading counters we can obtain a higher ____________.

12-3 COUNTER APPLICATIONS

The counter is one of the most versatile of all the digital circuits and is used in a wide variety of applications. In most cases, however, it is basically used to count or to divide the input signal frequency. As a counter, it generates a parallel output that represents the number of clock pulses that have occurred. As a frequency divider, it generates a serial output that is a fraction of the input frequency.

In the three circuit applications that follow we will see how counters are used as a counter and as a frequency divider.

12-3-1 *A Digital Clock*

Counters are probably most frequently used to form a digital clock circuit, like the one seen in Figure 12-16. This digital time-keeping circuit displays the hours, minutes, and seconds of the day.

Starting at the top of this circuit you can see that the power supply unit is converting the 120 V ac input into a +5 V dc output, which is used to power all of the digital logic circuits. Since the 120 V ac input from the power company alternates at an extremely accurate 60 cycles per second (60 Hz), we can make use of this timing signal to generate a one-cycle-per-second (1 Hz) master timing, or clock signal. This is achieved by tapping off the 60 Hz ac from the secondary of the power supply transformer and then feeding this signal into a pulse-shaping circuit that will convert the 60 Hz sine-wave input into a sixty-pulses-per-second (60 pps) square-wave output. The 60 pps output from the pulse shaping circuit is then applied to a divide-by-ten counter and then to a divide-by-six counter, that in combination will divide the 60 pps input by sixty, to produce a one-pulse-per-second output from the divide-by-sixty circuit. This 1 pps clock signal from the divide-by-sixty circuit is the master timing signal for the seconds counter, minutes counter, and hours counter circuits.

The seconds counter circuit contains a MOD-10, or ÷10, counter followed by a MOD-6, or ÷6, counter. These two counters in combination will count the 1 pps clock input, and after every sixty seconds or sixty pulses will produce a 1 ppm output to the minutes counter circuit. The minutes counter circuit also contains a MOD-10, or ÷10, counter followed by a MOD-6, or ÷6, counter. These two counters in combination will count the 1 ppm clock input, and after every sixty minutes or sixty pulses will produce a 1 pph output to the hours counter circuit. The hours counter circuit contains a MOD-10, or ÷10, counter followed by a MOD-2, or ÷2, counter (which is a single flip-flop). These two counters in combination will count the 1 pph clock input. Once the count advances from 12 ($1\ 0010_{BCD}$) to 13 ($1\ 0011_{BCD}$), NAND gate A will almost immediately clear and preset load the hours counter to 1 ($0\ 0001_{BCD}$), ensuring that the hours count increments from 12:59 to 1:00.

The BCD outputs from the seconds counters, minutes counters, and hours counters are decoded by BCD-to-seven-segment 7447s, and these seven-segment codes drive the common-anode displays.

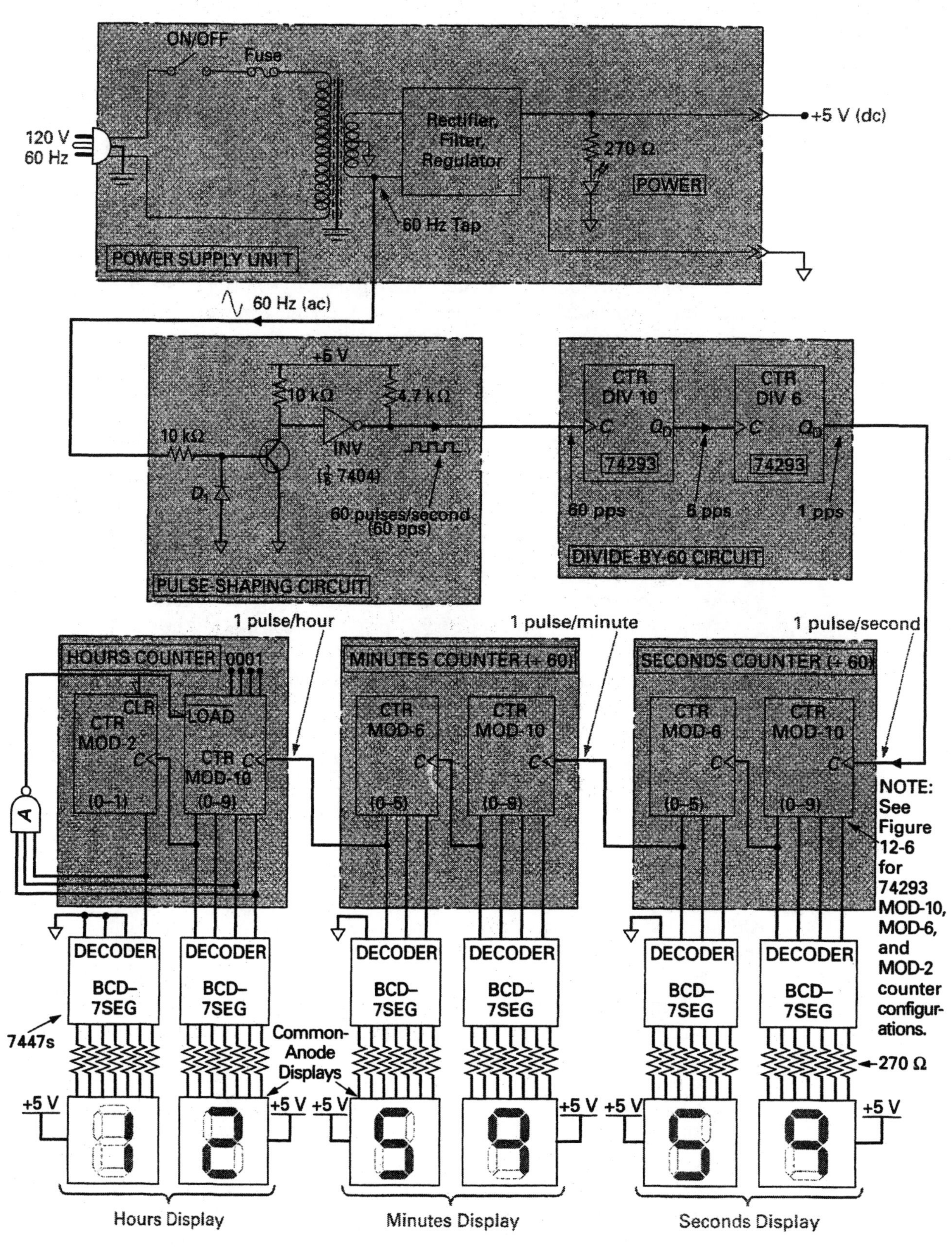

FIGURE 12-16 A Digital Clock Circuit.

12-3-2 A Frequency Counter

Frequency Counter A test instrument used to detect and display the frequency of a tested signal.

A **frequency counter** is a good example of a digital system that makes use of a clock oscillator, frequency dividers, counters, registers, decoders, and a seven-segment display.

Frequency Counter Block Diagram

Before we discuss the frequency counter circuit diagram, let us first examine the basic block diagram of the frequency counter, shown in Figure 12-17. To start at the far left of the timing diagram, you can see that the counter is cleared by a HIGH pulse just before the count window pulse begins. The count window pulse and the unknown input frequency are both applied to an AND gate, which acts as a controlled switch, as shown in the inset in Figure 12-17. When the count window pulse applies a LOW to the AND gate, it acts as an open switch since any LOW input to an AND gate produces a constant LOW out. On the other hand, when the count window pulse applies a HIGH to the AND gate, it acts as a closed switch connecting the unknown input frequency pulses through to the output. In the example shown in Figure 12-17, five pulses are switched through the AND gate to the clock input of the counter during the one second count window pulse. These five pulses will trigger the decade counter five times, causing it to count to 5 (0101). This BCD count of 5 from the counter will be decoded by the BCD-to-seven-segment decoder, and displayed on the seven-segment LED display. Since five pulses occurred in one second, the unknown input frequency is 5 pps.

The count window pulse is sometimes called the *sample pulse,* because the unknown input frequency is sampled during the time that this signal is active. The simple frequency counter circuit shown in Figure 12-17 is extremely limited, since it can only determine the unknown input frequency of inputs between 1 and 9 pps. To construct a more versatile frequency counter, we will have to include some additional digital ICs in the circuit.

Frequency Counter Circuit Diagram

Figure 12-18 shows a more complete frequency counter circuit diagram. Although at first glance this appears to be a complex circuit, it can easily be understood by examining one block at a time.

The accuracy of a frequency counter is directly dependent on the accuracy of the count window pulse. To ensure this accuracy, a 100 kHz crystal oscillator is used to generate a very accurate square wave. The output from the crystal oscillator is used to clock a chain of divide-by-ten ICs (74LS160As) in the count window generator circuit block. The 100 kHz input at the right is successively divided by ten to produce a 1 kHz, 100 Hz, 10 Hz, and 1 Hz output. Any one of these count window pulse frequencies can be selected by SW_1 and applied to a divide-by-two stage (1/2 7476) in the divide-by-two and monostable block. The waveforms shown in the inset in Figure 12-18 illustrate what occurs when SW_1 is placed in the 1 Hz position. The square-wave input to the 7476 completes one cycle in one second; however the positive half cycle of this waveform only lasts for half a second. By including the divide-by-two stage, the 1 Hz input is

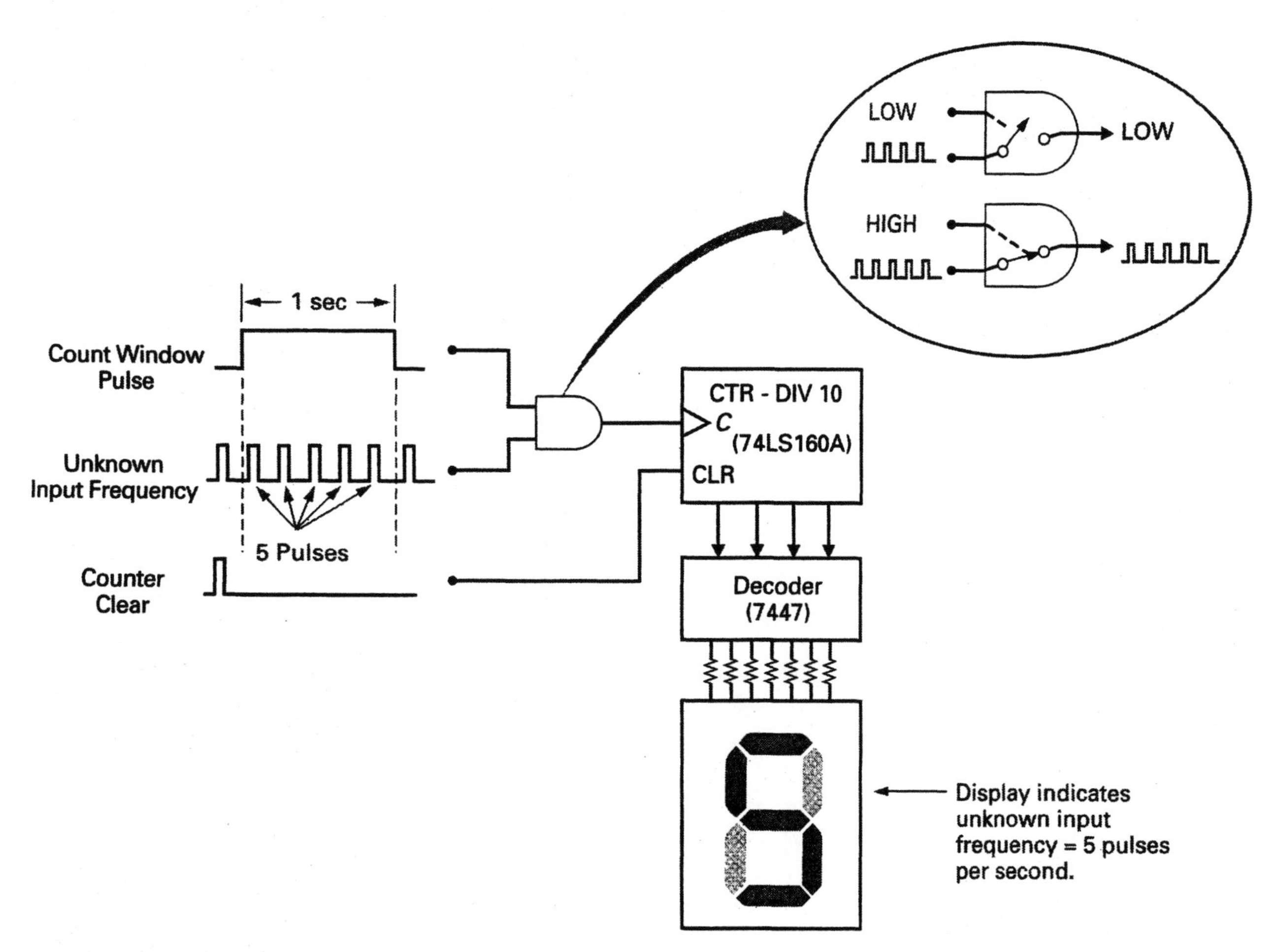

FIGURE 12-17 Simplified Frequency Counter Circuit.

divided by two to produce an output frequency of 0.5 Hz, which means one complete cycle lasts for two seconds, and therefore the positive alternation, or count window pulse, lasts for one second. Including a ÷2 stage in this way means that when SW_1 selects

a. 1 Hz, the count window pulse out of the ÷2 stage will have a duration of one second.

b. 10 Hz, the count window pulse out of the ÷2 stage will have a duration of 100 ms (0.1 s).

c. 100 Hz, the count window pulse out of the ÷2 stage will have a duration of 10 ms (0.01 s).

d. 1 kHz, the count window pulse out of the ÷2 stage will have a duration of 1 ms (0.001 s).

As we will see later, these smaller count window pulses will allow us to measure higher input frequencies.

The count window pulse out of the ÷2 stage is applied to an AND gate along with the unknown input frequency. In the example timing waveforms shown in the inset in Figure 12-18, you can see that ten pulses are switched

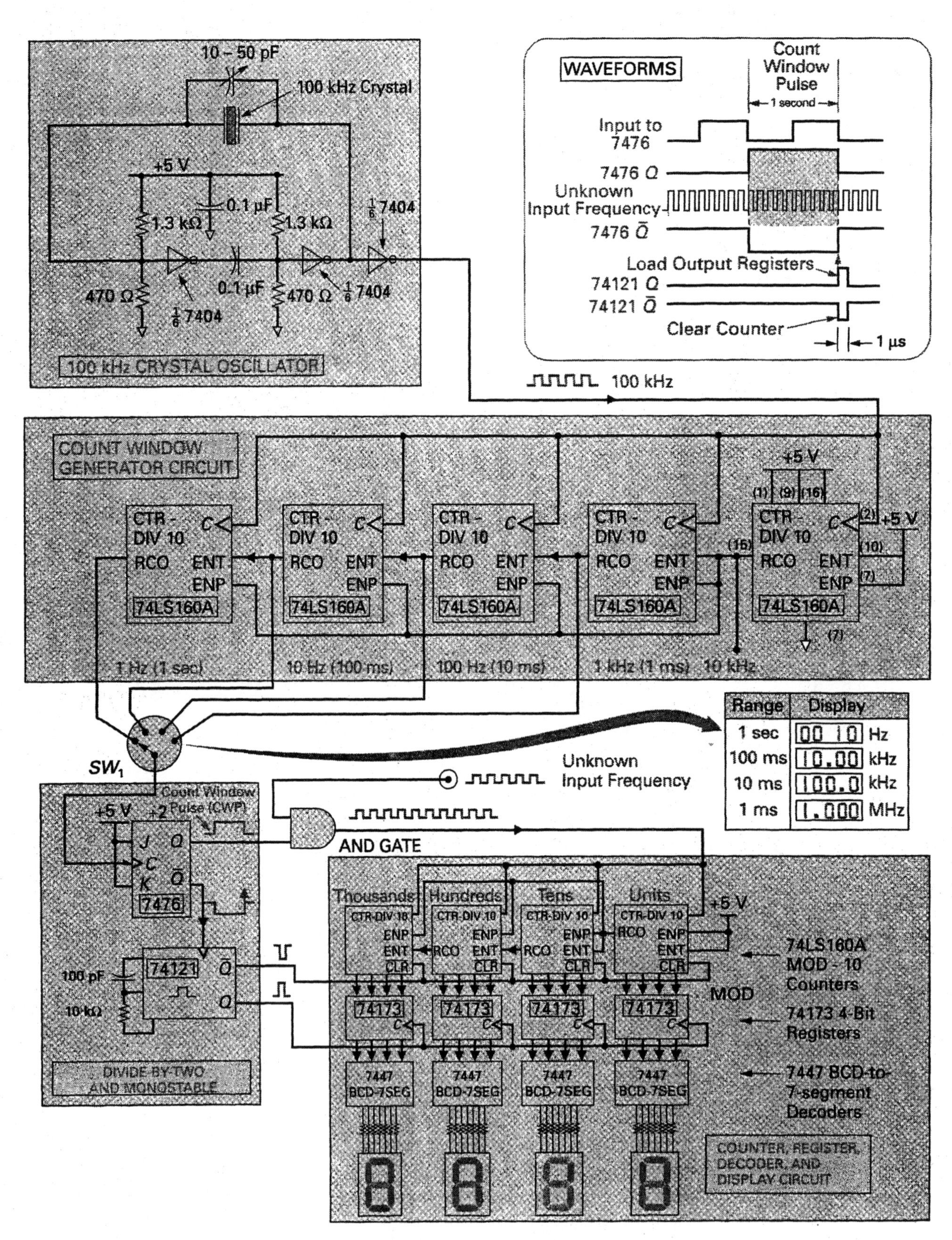

FIGURE 12-18 A Frequency Counter Circuit.

through the AND gate during the window of one second. These ten pulses are used to clock a four-stage counter in the counter, register, decoder, and display circuit block. The units counter on the right will count from 0 through 9, and if the count exceeds 9, it will reset to zero and carry a single tens pulse into the tens counter. In this example, this is exactly what will happen, and the final display will show decimal 10, meaning that the unknown input frequency is 10 pps. Similarly, the tens counter will reset-and-carry into the hundreds counter, and the hundreds counter will reset and carry into the thousands counter, as is expected when counting in decimal.

The BCD outputs of the four-stage counter are applied to a set of registers, and the output of these registers drives a four-digit seven-segment display via BCD-to-seven-segment display decoders. Referring to the divide-by-two and monostable block, you can see that the 7476 $\overline{Q}$ output is used to trigger a 74121 one-shot, whose RC network has been chosen to generate a 1 μs pulse. Referring again to the timing waveforms in the inset in Figure 12-18, you can see that the positive edge of the 7476 $\overline{Q}$ output triggers the 74121 which generates a positive 1 μs pulse at its Q output and a negative 1 μs pulse at its $\overline{Q}$ output. The positive edge of the 74121 Q output clocks the four 74173 4-bit registers, which store the new frequency count. This stored value appears at the Q outputs of the 74173s, where it is decoded by the 7447s and displayed on the seven-segment displays. The 74173s are included to hold the display value constant, except for the few microseconds when the display is being updated to a new value. The negative 1 μs pulse at the 74121 $\overline{Q}$ output is used to clear the four-stage counter once the new frequency count has been loaded into the 74173 registers.

To explain this circuit in a little more detail, let us examine each of the frequency counter's different range selections (controlled by SW_1), which are shown in the table in Figure 12-18.

1 Hz or 1 s Range: When SW_1 selects this frequency, a count window pulse of one second is generated. Using this window of time and a 4-digit display, we can measure any input frequency from 0001 Hz to 9999 Hz. If this maximum of 9999 Hz is exceeded, the counter will reset all four digits. For example, an input of 9999 Hz will be displayed as 9999, whereas an input of 10000 Hz will be displayed as 0000. To measure an input frequency in excess of 9999 Hz, therefore, we will have to use a smaller window so that fewer pulses are counted.

10 Hz or 100 ms Range: When SW_1 selects this frequency, a count window pulse of 1⁄10 of a second is generated. This means that if the unknown input frequency is 10,000 Hz, only 1⁄10 of these pulses will be switched through the AND gate to the counter (1⁄10 of 10,000 = 1000). This value of 1000 is smaller and can be displayed on a 4-digit display; however, a display of 1000 Hz is not accurate since the input frequency is 10,000 Hz. Since 10,000 Hz = 10 kHz, on this range we will add a decimal point in the mid position and add a prefix of kHz after the value so that the readout is correct, as shown in the table in Figure 12-18. An input of 10,000 Hz, therefore, will be displayed as 10.00 kHz.

100 Hz or 10 ms Range: When SW_1 selects this frequency, a count window pulse of 1⁄100 of a second is generated. This means that if the unknown input frequency is 100,000 Hz, only 1/100 of these pulses will be switched through the AND gate to the counter (1/100 of 100,000 = 1000). This value of 1000 is smaller and can be displayed on a 4-digit display; however, a display of 1000 Hz is not accurate since the input frequency is 100,000 Hz. Since 100,000 Hz = 100 kHz, on this range we will move the position of the decimal point and keep the prefix of kHz after the value so that the readout is correct, as shown in the table in Figure 12-18. An input of 100,000 Hz, therefore, will be displayed as 100.0 kHz.

1 kHz or 1 ms Range: When SW_1 selects this frequency, a count window pulse of 1/1000 of a second is generated. This means that if the unknown input frequency is 1,000,000 Hz, only 1/1000 of these pulses will be switched through the AND gate to the counter (1/1000 of 1,000,000 = 1000). This value of 1000 is smaller and can be displayed on a 4-digit display; however, a display of 1000 Hz is not accurate since the input frequency is 1,000,000 Hz. Since 1,000,000 Hz = 1 MHz, on this range we will again move the position of the decimal point and add a prefix of MHz after the value so that the readout is correct, as shown in the table in Figure 12-18. An input of 1,000,000 Hz, therefore, will be displayed as 1.000 MHz.

12-3-3 ***The Multiplexed Display***

Multiplexed Display Circuit
A display circuit in which the drive circuitry is time-shared.

The display circuitry used for the previously described digital clock circuit and the frequency counter circuit has two basic disadvantages: it uses a large number of ICs and it uses a large amount of power. Both of these disadvantages can be solved by using a **multiplexed display circuit** similar to the one shown in Figure 12-19. This circuit needs only one 7447 decoder IC instead of the six that were needed for the digital clock circuit and four which were needed for the frequency counter circuit. Second, since a multiplexed display has only one digit ON at a time, the total circuit current, and therefore power consumption, is dramatically less. For example, each LED in each seven-segment display draws approximately 20 mA, and therefore each seven-segment display can draw a maximum of 140 mA. If all six digits in the digital clock circuit are at their maximum, a total current of 6 × 140 mA = 840 mA (0.84 A) will be needed. With a multiplexed display, only one seven-segment display is enabled at a time, and therefore the maximum display current at any one time is only 140 mA. If you combine this with the fact that each 7447 draws 260 mA, and only one 7447 (instead of four or six) is needed, you can see that the current savings are dramatic.

The multiplexed display circuit shown in Figure 12-19 operates in the following way. To *multiplex* is to simultaneously transmit two or more signals over a single channel. With the multiplexed display circuit in Figure 12-19, the 74293 MOD-8 counter controls the switching operation by controlling both the 74138 1-of-8 decoder and the four 74151 1-of-8 multiplexers. Referring to the table below the circuit in Figure 12-19, you can see that when the 74293 reaches a count of 001 (decimal 1), the 74138 will make LOW its active-LOW 1 output which will turn ON Q_1 and switch the +5 V supply to only the first seven-segment display. The 001 output from the 74293 is also applied to the four 74151 multiplexers, which will respond by switching all of their 1 inputs through to their Y outputs. This will connect the A_0, A_1, A_2, and A_3 4-bit word from the

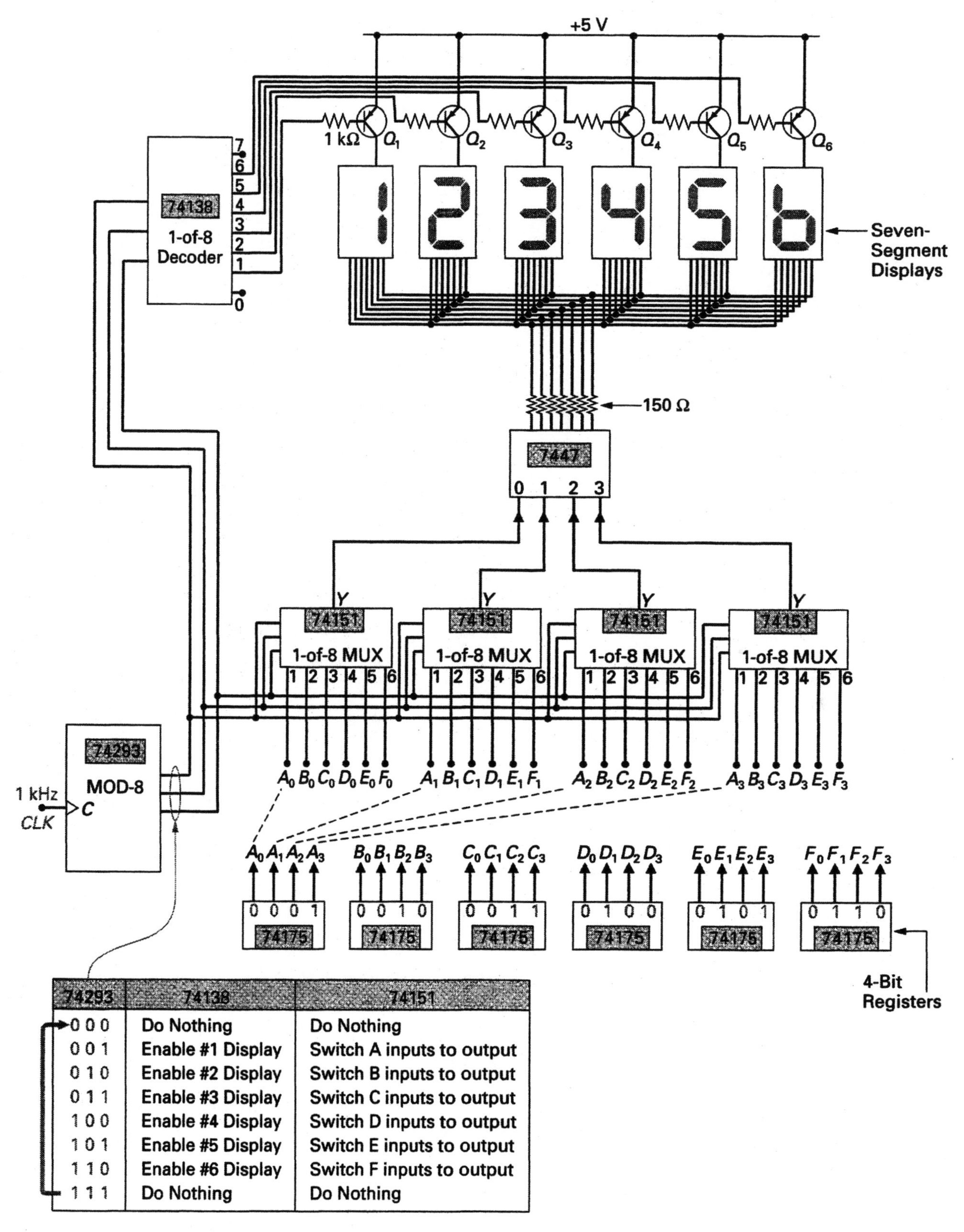

FIGURE 12-19 A 6-Digit Multiplexed Display Circuit.

left-most 74175 register through to the 7447, where it will be decoded from BCD to a corresponding seven-segment code. This seven-segment code is applied to all six seven-segment displays; however, since only one of the displays is receiving a +5 V supply, the data will only be displayed on the left-most seven-segment display.

Referring again to the table in Figure 12-19, you can see that as the count of the 74293 advances, the displays are enabled in turn by the 74138, and the corresponding input words from the registers are switched through by the 74151 multiplexers. If the cycle is repeated at a fast enough rate, your eyes will not see the ON/OFF switching of each display, only a continuous display. **Time division multiplexing (TDM),** or the timesharing of several signals over one set of wires or a single channel is a method frequently used in many electronic circuits and systems.

Time Division Multiplexing (TDM) A time-sharing technique in which several signals can be transmitted within a single channel.

SELF-TEST REVIEW QUESTIONS FOR SECTION 12-3

1. When a counter is used as a ____________ it provides a parallel output that represents the number of input clock pulses applied. As a ____________, a counter provides a serial output that is a fraction of the input frequency.
2. Do the counters in the seconds counter, minutes counter, and hours counter circuits in Figure 12-16 function as counters or frequency dividers?
3. Do the MOD-10 counters in the count window generator circuit in Figure 12-18 function as counters or frequency dividers?
4. Do the MOD-10 counters in the counter, register, decoder, and display circuit in Figure 12-18 function as counters or frequency dividers?

12-4 TROUBLESHOOTING COUNTER CIRCUITS

To review, the procedure for fixing a failure can be broken down into three steps—diagnose, isolate, and repair. Since this chapter has been devoted to counters, let us first examine the operation of a typical counter circuit, and then apply our three-step troubleshooting process to this circuit.

12-4-1 *A Counter Circuit*

As an example, we will apply our three-step troubleshooting procedure to the frequency counter circuit previously discussed and shown in Figure 12-18.

Step 1: Diagnose

As before, it is extremely important that you first understand the operation of a circuit and how all of the devices within it are supposed to work, so that you are able to determine whether or not a circuit malfunction really exists. If you were preparing to troubleshoot the circuit in Figure 12-18, your first step should be to read through the circuit description and review the operation of each integrated circuit until you feel completely confident with the correct operation of the entire circuit. Once you are fully familiar with the operation of the circuit, you will easily be able to diagnose the problem as either an *operator error* or a *circuit malfunction.*

Distinguishing an operator error from an actual circuit malfunction is an important first step, and a wrong diagnosis can waste much time and effort. For example, the following could be interpreted as circuit malfunctions, when in fact they are simply operator errors.

Example 1.
Symptom: SW_1 has selected the 1 ms range, input frequency = 100 Hz, display shows 0000.
Diagnosis: Operator Error—A 1 ms window will not sample any of the input since each cycle of the input lasts for 10 ms (0.01 Hz). To determine low input frequencies, the operator will need to use a wider window.

Example 2.
Symptom: SW_1 has selected the 100 ms range, input frequency = 12.3 kHz, display shows 1230.
Diagnosis: Operator Error—Since no circuitry is included to turn ON the decimal point in the correct position and display kHz, the display has to be interpreted based on the range selected. For example, 1230 on the 100 ms range = 12.30 kHz.

Once you have determined that the problem is not an operator error, but is in fact a circuit malfunction, proceed to Step 2 and isolate the circuit failure.

Step 2: Isolate

No matter what circuit or system failure has occurred, you should always follow a logical and sequential troubleshooting procedure. Let us review some of the isolating techniques, and then apply them to our example circuit in Figure 12-18.

a. Use a cause-and-effect troubleshooting process, which means study the effects you are getting from the faulty circuit and then logically reason out what could be the cause.

b. Check first for obvious errors before leaping into a detailed testing procedure. Is the power OFF or not connected to the circuit? Are there wiring errors? Are all of the ICs correctly oriented?

c. Using a logic probe or voltmeter, test that power and ground are connected to the circuit and are present at all points requiring power and ground. If the whole circuit, or a large section of the circuit, is not operating, the problem is normally power. Using a multimeter, check that all of the dc voltages for the circuit are present at all IC pins that should have power or a HIGH input, and are within tolerance. Second, check that 0 V, or ground, is connected to all IC ground pins and to all inputs that should be tied LOW.

d. Use your senses to check for broken wires, loose connectors, overheating or smoking components, pins not making contact, and so on.

e. Test the clock signals using the logic probe or the oscilloscope. Although clock signals can be easily checked with a logic probe, the oscilloscope is best for checking the signal's amplitude, frequency, pulse width, and so on. The oscilloscope will also display timing problems and noise spikes (glitches) which could be false-triggering a circuit into operation at the wrong time. With the frequency counter circuit in Figure 12-18, you should first test the 100 kHz output from the crystal oscillator and then test in sequence the 10 kHz, 1 kHz, 100 Hz, 10 Hz, and 1 Hz outputs from the count window generator circuit. Problems in frequency divider chains are generally easily detected since an incorrect frequency output normally indicates that the previous stage has malfunctioned.

f. Perform a static test on the circuit. With our circuit example in Figure 12-18, you should disconnect the inputs to the counter, register, decoder, and display circuit, and then use a logic pulser to clock the counter, and a logic probe to test that the valid count is appearing at the outputs of each stage of the counter. For example, after nineteen clock pulse inputs from the logic pulser, the logic probe should indicate the following logic levels at the counter outputs.

$$\begin{array}{cccc} 0000 & 0000 & 1000 & 1001_{BCD} \\ 0 & 0 & 1 & 9 \end{array}$$

By pulsing the register clock input, this value should appear at the output of the 74173s, be decoded by the 7447s, and be displayed. By pulsing the counter clear input, you should be able to clear all of the outputs of the counter. The divide-by-two and monostable circuit can also be tested in this way by disconnecting the inputs and using the logic pulser as a signal source and the logic probe as a signal detector.

g. With a dynamic test, the circuit is tested while it is operating at its normal clock frequency (which in Figure 12-18 is 100 kHz), and therefore all of the inputs and outputs are continually changing. Although a logic probe can detect a pulse waveform, the oscilloscope will display more signal detail, and since it can display more than one signal at a time, it is ideal for making signal comparisons and looking for timing errors. If the clock section is suspected, you can check the rest of the circuit by disconnecting the clock input to the circuit and then using a function generator to generate a 100 kHz clock.

In some instances, you will discover that some circuits will operate normally when undergoing a static test, and yet fail a dynamic test. This effect usually points to a timing problem involving the clock signals and/or the propagation delay times of the ICs used within the circuit.

h. Noise due to electromagnetic interference (EMI) can false-trigger an IC, such as the counters in Figure 12-18. This problem can be overcome by not leaving any of the IC's inputs unconnected and, therefore, floating. Connect unneeded active-HIGH inputs to ground and active-LOW inputs to $+V_{CC}$.

i. Apply the half-split method of troubleshooting first to a circuit, and then to a section of a circuit, to help speed up the isolation process. With our circuit example in Figure 12-18, a good mid-point check would be to test the different frequency outputs from the count window generator circuit.

Also remember that a load problem can make a source appear at fault. If an output is incorrect, it would be safer to disconnect the source from the rest of the circuit and then recheck the output. If the output is still incorrect, the problem is definitely within the source; however, if the output is correct, the load is probably shorting the line either HIGH or LOW.

j. Substitution can be used to help speed up your troubleshooting process. Once the problem is localized to an area containing only a few ICs, substitute suspect ICs with known working ICs (one at a time) to see if the problem can be quickly remedied.

k. Many electronic system manufacturers provide troubleshooting trees in the system technical manual. These charts are a graphical means to show the sequence of tests to be performed on a suspected circuit or system. Figure 12-20 shows a simple troubleshooting tree for our frequency counter circuit.

Step 3: Repair

Once the fault has been found, the final step is to repair the circuit and then perform a final test to see that the circuit and the system is now fully operational.

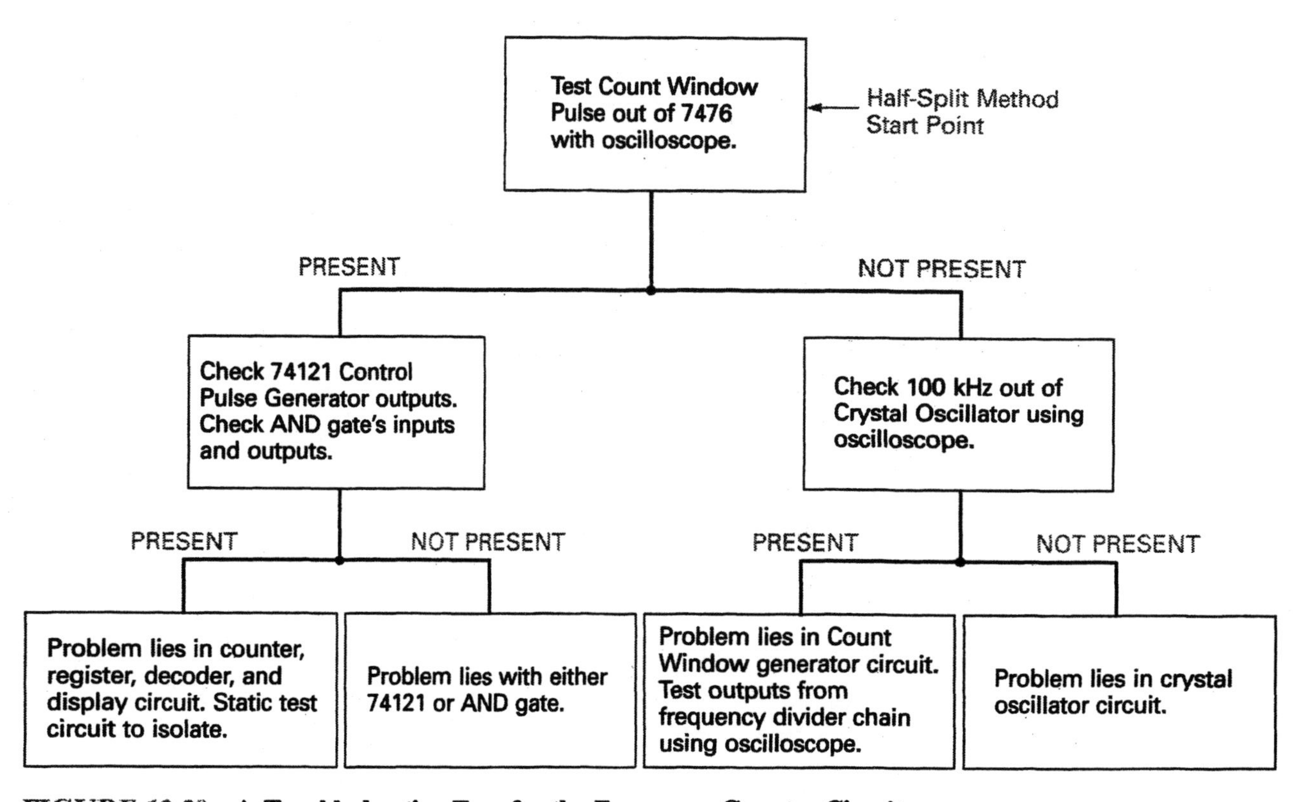

FIGURE 12-20 A Troubleshooting Tree for the Frequency Counter Circuit.

12-4-2 *Sample Problems*

Once you have constructed a circuit like the circuit in Figure 12-18, introduce a few errors to see what effects or symptoms they produce. Then logically reason out why a particular error or cause has a particular effect on the circuit. Never short any two points together unless you have carefully thought out the consequences, but generally it is safe to open a path and see the results. Here are some problems for our example circuit in Figure 12-18.

a. Remove the 100 kHz crystal.

b. Disconnect one of the *RCO* outputs in the count window generator circuit.

c. Disconnect the clock input to one of the 74173s.

d. Disconnect the $\overline{Q}$ output of the divide-by-two circuit.

e. Disconnect the Q output of the 74121.

f. Disconnect the *ENT* and *ENP* inputs of the first 74LS160A counter from the +5 V and connect these lines to ground.

g. Disconnect one of the inputs to a 7447.

SELF-TEST REVIEW QUESTIONS FOR SECTION 12-4

---

1. What are the three basic troubleshooting steps?
2. Which of the 74LS160As in Figure 12-18 function as frequency dividers, and which function as counters?
3. To test the count window generator circuit in Figure 12-18, the clock oscillator input is disconnected and a logic pulser is used to inject a 500 kHz input. Does a problem exist if outputs of 5 kHz, 500 Hz, 50 Hz, and 5 Hz are measured with the oscilloscope at each of the SW_1 positions?
4. What symptom would you get from the circuit in Figure 12-18 if you were to apply a permanent HIGH to the unknown frequency input?

REVIEW QUESTIONS

Multiple-Choice Questions

1. The ____________ of a counter describes the number of different output combinations generated by the counter.

a. division factor
b. modulus
c. maximum count
d. propagation delay time

2. The ____________ of a counter describes the time it will take for all of the outputs to be updated.

   a. division factor **b.** modulus
   c. maximum count **d.** propagation delay time

3. The ____________ of a counter describes how much lower the final flip-flop's output frequency will be compared to the input clock frequency.

   a. division factor **b.** modulus
   c. maximum count **d.** propagation delay time

4. The ____________ of a counter describes the highest binary value that will appear at the parallel outputs.

   a. division factor **b.** modulus
   c. maximum count **d.** propagation delay time

5. A MOD-10 or decade counter will divide the input clock frequency by a factor of ____________.

   a. 16 **b.** 8 **c.** 10 **d.** 100

6. A four-stage binary counter will have a natural modulus of ____________.

   a. 16 **b.** 8 **c.** 10 **d.** 100

7. A(an) ____________ counter will start counting from a loaded value up to its natural modulus.

   a. synchronous **b.** asynchronous
   c. presettable **d.** cascaded

8. The maximum counting speed of a counter is limited by the counter's ____________.

   a. division factor **b.** modulus
   c. maximum count **d.** propagation delay time

9. What is the primary advantage of a synchronous counter?

   a. High counting speed **b.** Long propagation delay
   c. Circuit simplicity **d.** Ripple action

10. What is the primary advantage of an asynchronous counter?

   a. High counting speed **b.** Long propagation delay
   c. Circuit simplicity **d.** Ripple action

11. Cascaded up counters generate a ____________ output to the next significant counter.

   a. carry output **b.** borrow output
   c. *RCO* **d.** both a. and c. are true

12. Cascaded down counters generate a ____________ output to the next significant counter.

   a. carry output **b.** borrow output
   c. *RCO* **d.** both a. and c. are true

13. A state transition diagram shows a counter's ____________.

   a. output states
   b. division factor
   c. propagation delay
   d. modulus

14. An up/down counter is also referred to as a ____________ counter.

   a. presettable
   b. synchronous
   c. bidirectional
   d. asynchronous

15. A multiplexed display makes use of ____________ division multiplexing.

   a. time b. frequency c. phase d. pulse

Essay Questions

16. What is the difference between an asynchronous counter and a synchronous counter? (Introduction)

17. Define the following terms as they relate to counters (12-1)

   a. Maximum count
   b. Modulus
   c. Division factor
   d. Programmable modulus

18. Sketch a simple three-stage asynchronous up counter circuit. (12-1-1)

19. Sketch a simple divide-by-eight asynchronous frequency divider circuit. (12-1-1)

20. Briefly describe how an asynchronous counter can be modified to make it function as a MOD-10 counter. (12-1-4)

21. What is a state transition diagram? (12-1-4)

22. Briefly describe the operation of a presettable asynchronous counter. (12-1-5)

23. Sketch a simple synchronous up counter and briefly describe its operation. (12-2-1)

24. What advantages does the synchronous counter have over the asynchronous counter? (12-2-2)

25. Why are counters cascaded? (12-2-3)

26. What is a carry bit? (12-2-3)

27. What is a bidirectional counter? (12-2-5)

28. What is a borrow bit? (12-2-5)

29. Why do counters in some applications supply a parallel output, while in other applications they supply a serial output? (12-3)

30. Draw a schematic diagram showing how to upgrade the digital clock circuit in Figure 12-16 to include a multiplexed display like the one shown in Figure 12-19. (12-3)

Practice Problems

31. Calculate the frequency division factor of the following counters.
    a. MOD-12 counter
    b. MOD-4096 counter
    c. A counter with 8,192 different outputs states
    d. A counter with a maximum count of 15
32. Calculate the maximum count of a
    a. 4-bit counter.
    b. 6-bit counter.
    c. 24-bit counter.
33. Calculate the modulus of a
    a. 4-bit counter.
    b. 6-bit counter.
    c. 24-bit counter.
34. Assuming a flip-flop propagation delay of 25 ns, calculate the total asynchronous counter propagation delay time and the upper clock pulse frequency limit for the following:
    a. 4-bit counter
    b. 6-bit counter
    c. 24-bit counter

To practice your circuit recognition and operation skills, refer to Figure 12-21 and answer the following seven questions.

35. Is this circuit a synchronous or an asynchronous counter?
36. What is this counter's natural modulus?
37. What is this counter's maximum count?
38. What is this counter's frequency division between the clock input and Q_4?
39. What would be the Q_4 output frequency if the clock input were 2 kHz?
40. What would be this counter's upper clock pulse frequency limit if each flip-flop had a propagation delay of 22 ns?
41. Assuming an 8-bit asynchronous presettable counter, calculate
    a. its natural modulus.
    b. its preset modulus if 0011 00100 is loaded.

To practice your circuit recognition and operation skills, refer to Figure 12-22 and answer the following question.

42. Briefly describe the operation of the
    a. 74193 IC in this circuit including the NAND gate's function at the clock input and the function of SW_1.
    b. 74154 IC in this circuit.
    c. If the alligator clip is connected to the 74154's pin 14 output and SW_1 is pressed, how will this circuit operate?

Troubleshooting Questions

43. How would you apply the half-split troubleshooting method to the circuit in Figure 12-22?

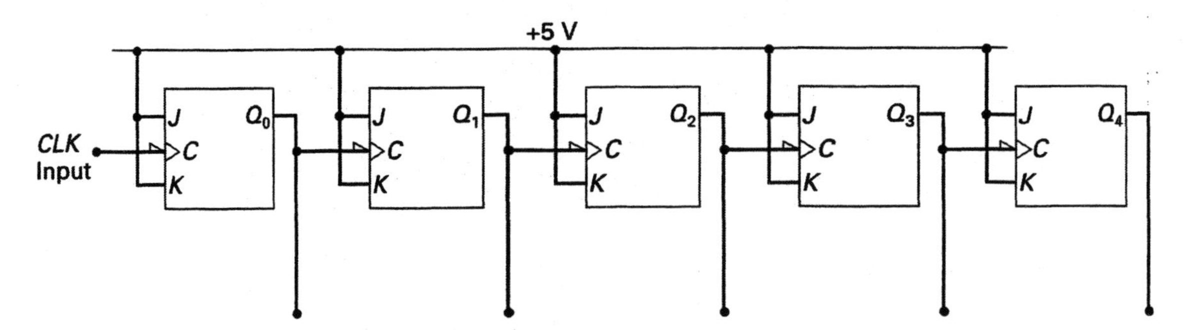

FIGURE 12-21 A 5 Stage Up Counter.

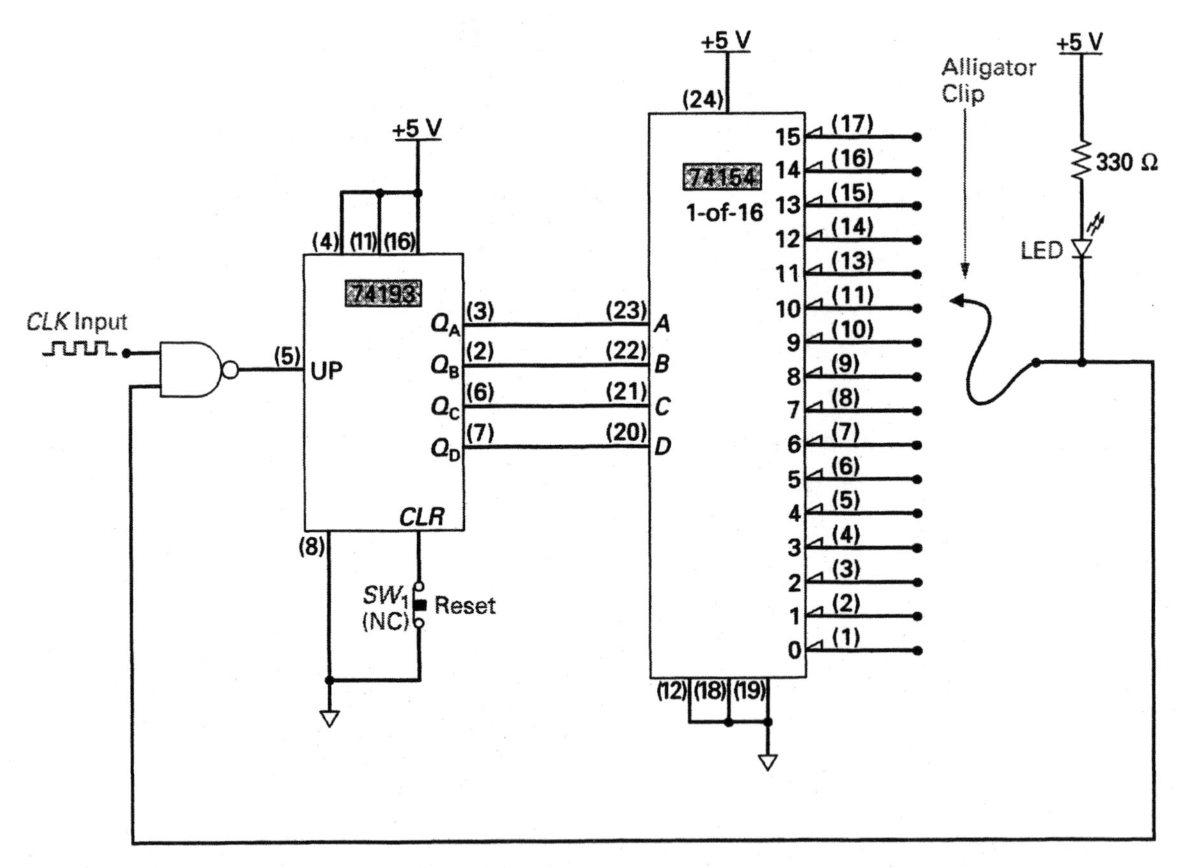

FIGURE 12-22 Count Up to a Selected Value and Stop Circuit.

44. What symptoms would you expect from the circuit shown in Figure 12-22 if the following malfunctions were to occur?

a. The NAND gate's output was permanently HIGH.

b. The line between 74193 (pin 7) and 74154 (pin 20) was open.

c. The LED was connected in the opposite direction.
d. The NAND gate's output was connected to the DOWN input instead of the UP input.

45. Briefly describe how to static test the circuit shown in Figure 12-22.

Chapter 13

OBJECTIVES

After completing this chapter, you will be able to:

1. Describe how to complete the four basic arithmetic operations—binary addition, binary subtraction, binary multiplication, and binary division.
2. Explain how the sign-magnitude, one's complement, and two's complement number systems are used to represent positive and negative numbers.
3. Describe how an adder circuit can add, subtract, multiply, and divide two's complement numbers.
4. Describe how the floating-point number system can be used to represent a large range of values.
5. Define the differences between a half-adder circuit, a full-adder circuit, and a parallel-adder circuit.
6. Explain the operation of a typical parallel-adder IC.
7. List and describe some arithmetic circuit applications such as
   a. a two's complement adder/subtractor circuit.
   b. an arithmetic-logic unit (ALU) IC.
   c. the ALU within a microprocessor unit.
8. Describe how to troubleshoot digital logic circuits containing arithmetic circuits using the three-step method.

Arithmetic Operations and Circuits

OUTLINE

VIGNETTE

The Wizard of Menlo Park

Thomas Alva Edison was born to Samuel and Nancy Edison on February 11, 1847. As a young boy he had a keen and inquisitive mind, yet he did not do well at school, so his mother, a former school teacher, withdrew him from school and tutored him at home. In later life he said that his mother taught him to read well and instilled a love for books that lasted the rest of his life. In fact, the inventor's personal library of more than ten thousand volumes is preserved at the Edison Laboratory National Monument in West Orange, New Jersey.

At the age of twenty-nine, after several successful inventions, Edison put into effect what is probably his greatest idea—the first industrial research laboratory. Choosing Menlo Park in New Jersey, which was then a small rural village, Edison had a small building converted into a laboratory for his fifteen-member staff, and a house built for his wife and two small daughters. When asked to explain the point of this lab, Edison boldly stated that it would produce "a minor invention every ten days and a big thing every six months or so." At the time, most of the scientific community viewed his prediction as preposterous; however, in the next ten years Edison would be granted 420 patents, including those for the electric light bulb, the motion picture, the phonograph, the universal electric motor, the fluorescent lamp, and the medical fluoroscope.

Over one thousand patents were granted to Edison during his lifetime; and his achievements at what he called his "invention factory" earned him the nickname "the wizard of Menlo Park." When asked about his genius he said, "Genius is two percent inspiration and ninety-eight percent perspiration."

INTRODUCTION

At the heart of every digital electronic computer system is a circuit called a microprocessor unit (MPU). To perform its function, the MPU contains several internal circuits, and one of these circuits is called the arithmetic-logic unit, or ALU. This unit is the number-crunching circuit of every digital system and, as its name implies, it can be controlled to perform either arithmetic operations (such as addition and subtraction) or logic operations (such as AND and OR).

In order to understand the operation of an ALU, we must begin by discussing how binary numbers can represent any value, whether it is large, small, positive, or negative. We must also discuss how these numbers can be manipulated to perform arithmetic operations such as addition, subtraction, multiplication, and division. All of these topics will be covered in the first section of this chapter.

In the second section of this chapter, we will discuss how logic circuits can be connected to form arithmetic circuits, and how these circuits can perform arithmetic operations.

In the third section of this chapter, we will combine all of the topics covered in this chapter and see how an ALU can perform either arithmetic operations or logic operations.

In the fourth section of this chapter, we will practice circuit troubleshooting techniques by applying our three-step troubleshooting procedure to a typical arithmetic circuit.

13-1 ARITHMETIC OPERATIONS

Before we discuss the operation of arithmetic circuits, we must first understand the arithmetic operations these circuits perform. In the first section we will see how two binary numbers are added together, how one binary number can be subtracted from another, how one binary number is multiplied by another, and how one binary number is divided into another—in short, the principles of binary arithmetic.

13-1-1 ***Binary Arithmetic***

The addition, subtraction, multiplication, and division of binary numbers is performed in exactly the same way as the addition, subtraction, multiplication, and division of decimal numbers. The only difference is that the decimal number system has ten digits, whereas the binary number system has two.

As we discuss each of these arithmetic operations, we will use decimal arithmetic as a guide to binary arithmetic so that we can compare the known to the unknown.

Binary Addition

Figure 13-1(a) illustrates how the decimal values 29,164 and 63,729 are added together. First the units 4 and 9 are added producing a sum of 13, or 3 carry 1. The carry is then added to the 6 and 2 in the tens column producing a sum of 9 with no carry. This process is then repeated for each column, moving right to left, until the total sum of the addend, augend, and their carries has been obtained.

When adding two decimal values, many combinations are available since decimal has ten different digits. With binary, however, only two digits are available and therefore only four basic combinations are possible. These four rules of binary addition are shown in Figure 13-1(b). To help understand these rules, let us apply them to the example in Figure 13-1(c) in which we will add 11010 (decimal 26) to 11100 (decimal 28). Like decimal addition, we will begin at the right in the ones column and proceed to the left. In the ones column 0 + 0 = 0 with a carry of 0 (Rule 1). In the twos column 1 + 0 and the 0 carry from

		(10,000)	(1,000)	(100)	(10)	(1)
Carry	0	1	0	0	1	
Addend:		2	9,	1	6	4
Avgend:	+	6	3,	7	2	9
Sum:		9	2,	8	9	3

$4 + 9 = 13$ or 1 carry 3

(a)

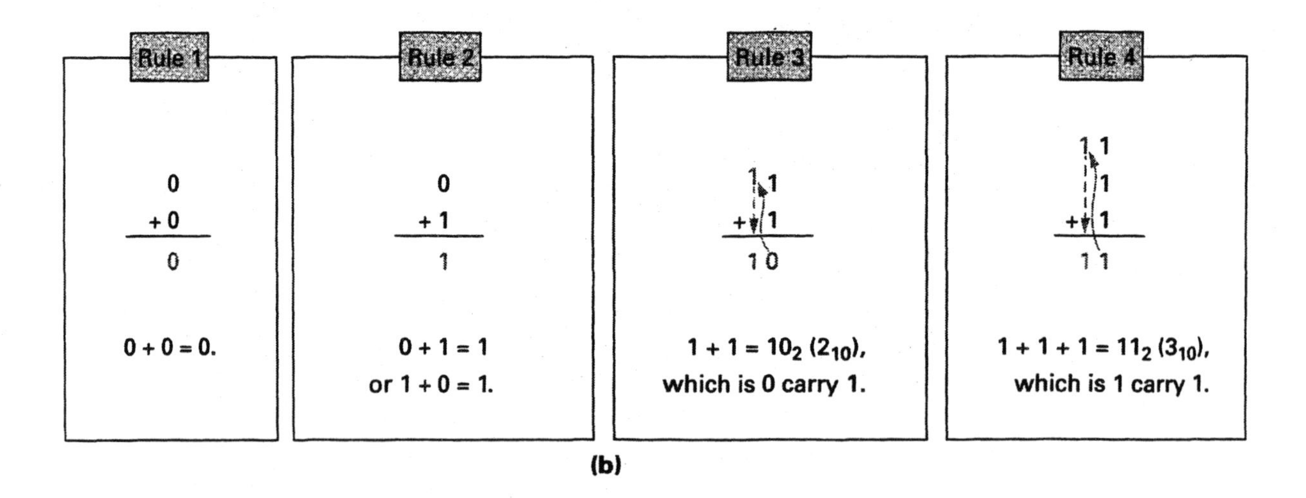

(b)

	(32)	(16)	(8)	(4)	(2)	(1)		(10)	(1)
Carry:	1	1	0	0	0			1	
Addend:		1	1	0	1	0_2	=	2	6_{10}
Avgend:	+	1	1	1	0	0_2		+ 2	8_{10}
Sum:	1	1	0	1	1	0_2		5	4_{10}

(c)

FIGURE 13-1 Binary Addition.

the ones column = 1 with no carry (Rule 2). In the fours column we have 0 + 1 plus a carry of 0 from the twos column. This results in a fours column total of 1, with a carry of 0 (Rule 2). In the eights column we have 1 + 1 plus a 0 carry from the fours column, producing a total of 10_2 (1 + 1 = decimal 2, which is 10 in binary). In the eights column this 10 (decimal 2) is written down as a total of 0 with a carry of 1 (Rule 3). In the sixteens column, we have 1 + 1, plus a carry of 1 from the eights column, producing a total of 11_2 (1 + 1 + 1 = decimal 3, which is 11 in binary). In the sixteens column, this 11 (decimal 3) is written down as a total of 1 with a carry of 1 (Rule 4). Since there is only a carry of 1 in the thirty-twos column, the total in this column will be 1.

Converting the binary addend and augend to their decimal equivalents as seen in Figure 13-1(c), you can see that 11010 (decimal 26) plus 11100 (decimal 28) will yield a sum or total of 110110 (decimal 54).

■ **EXAMPLE:**

Find the sum of the following binary numbers.

a. 1011 + 1101 = ?   b. 1000110 + 1100111 = ?

■ ***Solution:***

a.

Carry:		1	1	1	1		
Augend:			1	0	1	1_2	(11_{10})
Addend:	+		1	1	0	1_2	(13_{10})
Sum:		1	1	0	0	0_2	(24_{10})

b.

Carry:		1	0	0	0	1	1	0		
Augend:			1	0	0	0	1	1	0_2	(70_{10})
Addend:	+		1	1	0	0	1	1	1_2	(103_{10})
Sum:		1	0	1	0	1	1	0	1_2	(173_{10})

Binary Subtraction

Figure 13-2(a) reviews the decimal subtraction procedure by subtracting 4,615 from 7,003. Starting in the units column, you can see that since we cannot take 5 away from 3, we will have to go to a higher order minuend unit and borrow. Since the minuend contains no tens or hundreds, we will have to go to the thousands column. From this point a chain of borrowing occurs as a thousand is borrowed and placed in the hundreds column (leaving 6 thousands), one of the hundreds is borrowed and placed in the tens column (leaving 9 hundreds), and one of the tens is borrowed and placed in the units column (leaving 9 tens). After borrowing 10, the minuend units digit has a value of 13, and if we now perform the subtraction, the result or difference of 13 − 5 = 8. Since all of the other minuend digits are now greater than their respective subtrahend digits, there will be no need for any further borrowing to obtain the difference.

When subtracting one decimal value from another, many combinations are available since decimal has ten different digits. With binary, however, only two digits are available and therefore only four basic combinations are possible. These four rules of binary subtraction are shown in Figure 13-2(b). To help understand these rules, let us apply them to the example in Figure 13-2(c) in which we will subtract 11100 (decimal 28) from 1010110 (decimal 86). Like decimal subtraction, we will begin at the right in the ones column and proceed to the left. In the ones column 0 − 0 = 0 (Rule 1). In the twos column 1 − 0 = 1 (Rule 2), and in the fours column 1 − 1 = 0 (Rule 3). In the eights column we cannot subtract 1 from nothing or 0, so therefore we must borrow 1 from the sixteens column, making the minuend in the eights column 10 (decimal 2). The difference can now be calculated in the eights column since 10 − 1 = 1 (decimal 2 − 1 = 1, Rule 4). Due to the previous borrow, the minuend in the

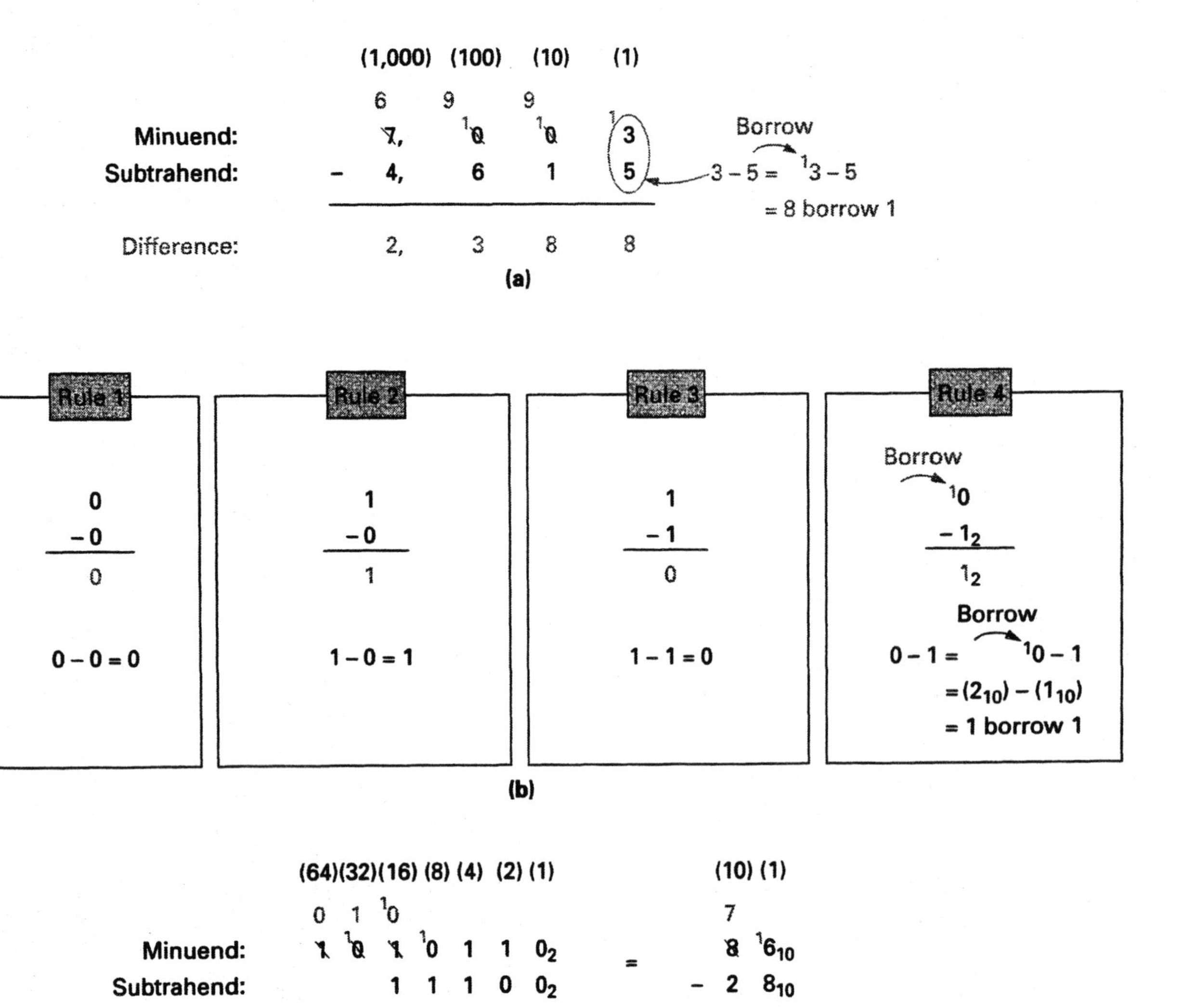

FIGURE 13-2 Binary Subtraction.

sixteens column is now 0 and therefore we need to borrow a 1 from the thirty-twos column. Since the thirty-twos column is also 0, we will need to borrow from the sixty-fours column. Borrowing 1 from the sixty-fours column will leave a minuend of 0 and make the minuend in the thirty-twos column equal to 10 (decimal 2). Borrowing 1 from the 10 (borrowing decimal 1 from 2) in the thirty-twos column will leave a minuend of 1 and make the minuend in the sixteens column equal to 10 (decimal 2). We can now subtract the subtrahend of 1 from the minuend of 10 (decimal 2) in the sixteens column to obtain a difference of 1. Due to the previous borrow, the minuend in the thirty-twos column is equal to 1, and $1 - 0 = 1$ (Rule 2). Finally, since the minuend and subtrahend are both 0 in the sixty-fours column, the subtraction is complete.

Converting the binary minuend and subtrahend to their decimal equivalents, as seen in Figure 13-2(c), you can see that 1010110 (decimal 86) minus 11100 (decimal 28) will result in a difference of 111010 (decimal 58).

■ **EXAMPLE:**

Find the difference between the following binary numbers.

a. 100111 − 11101 = ? b. 101110111 − 1011100 = ?

■ ***Solution:***

Minuend:		${}^{0}\not{1}$	${}^{1}\not{0}$	${}^{1}0$	1	1	1_2	(39_{10})
a. *Subtrahend:*	−		1	1	1	0	1_2	(29_{10})
Difference:		0	0	1	0	1	0_2	(10_{10})

Minuend:		1	0	1	${}^{0}\not{1}$	${}^{10}\not{1}$	${}^{1}0$	1	1	1_2	(375_{10})
b. *Subtrahend:*	−			1	0	1	1	1	0	0_2	(92_{10})
Difference:		1	0	0	0	1	1	0	1	1_2	(283_{10})

Binary Multiplication

Figure 13-3(a) reviews the decimal multiplication procedure by multiplying 463 by 23. To perform this operation, you would begin by multiplying each digit of the multiplicand by the units digit of the multiplier to obtain the first partial product. Second, you would multiply each digit of the multiplicand by the tens digit of the multiplier to obtain the second partial product. Finally, all of the partial products are added to obtain the final product.

When multiplying one decimal value by another, many combinations are available since decimal has ten different digits. With binary multiplication, however, only two digits are available and therefore only four basic combinations are possible. These four rules of binary multiplication are shown in Figure 13-3(b). To help understand these rules, let us apply them to the example in Figure 13-3(c) in which we will multiply 1011 (decimal 11) by 101 (decimal 5). Like decimal multiplication, we begin by multiplying each bit of the multiplicand by the ones column multiplier bit to obtain the first partial product. Second, we multiply each bit of the multiplicand by the twos column multiplier bit to obtain the second partial product. Remember that the LSB of the partial product must always be directly under its respective multiplier bit, and therefore the LSB of the second partial product should be under the twos column multiplier bit. In the next step, we add together the first and second partial products to obtain a sum of the partial products. With decimal multiplication we usually calculate all of the partial products, and then add them together to obtain the final product. This procedure can also be used with binary multiplication; however, it is better to add only two partial products at a time since the many carries in binary become hard to keep track of and easily lead to errors. In the next step, we multiply each bit of the multiplicand by the fours column multiplier bit to obtain the third partial product. Remember that the LSB of this third partial product must be placed directly under its respective multiplier bit. The final step is to add the third partial product to the previous partial sum to obtain the final product or result.

If you study the partial products obtained in this example, you will notice that they were either exactly equal to the multiplicand (when the multiplier was

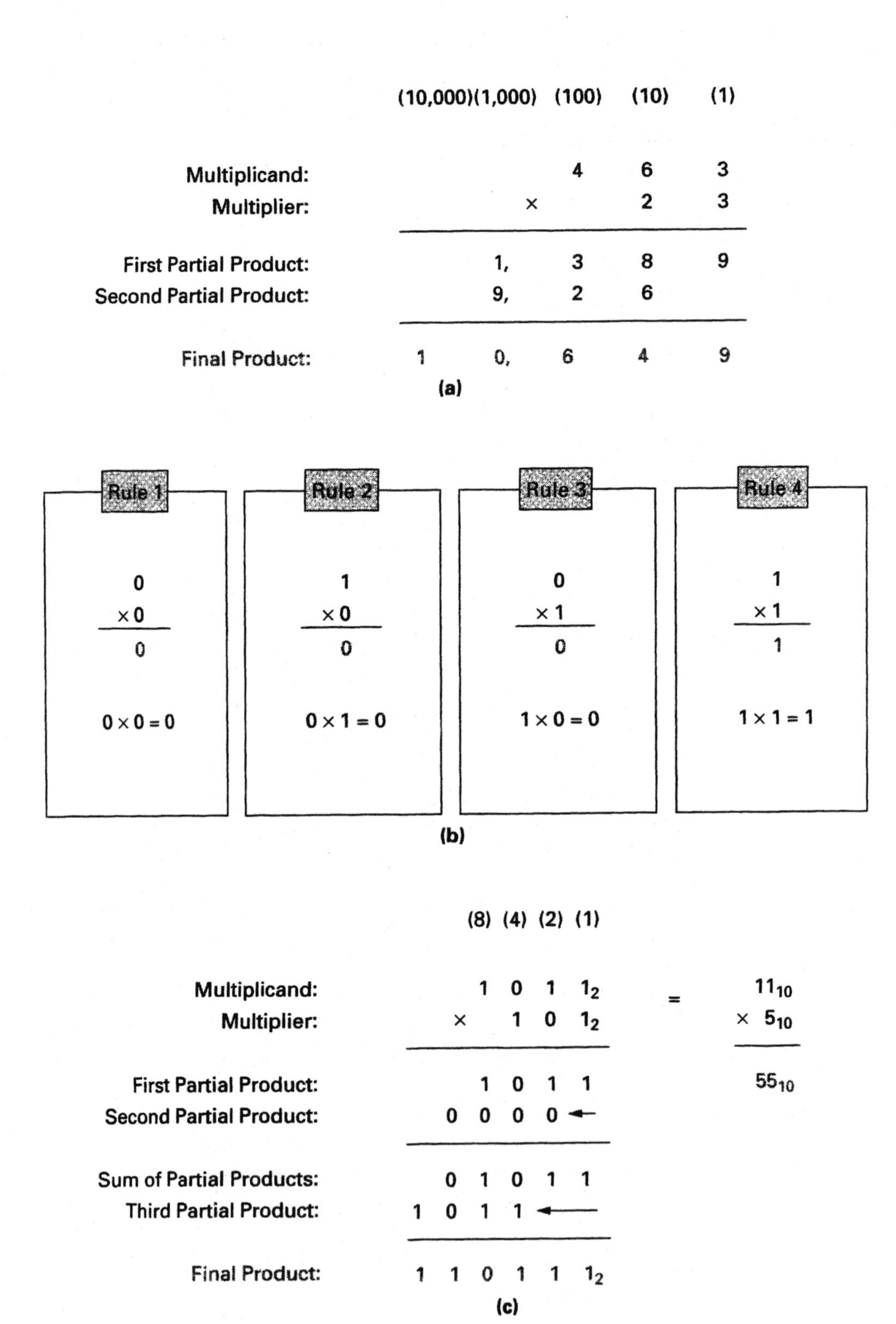

FIGURE 13-3 Binary Multiplication.

1) or all 0s (when the multiplier was 0). You can use this shortcut in future examples; however, be sure to always place the LSB of the partial products directly below their respective multiplier bits.

■ **EXAMPLE:**

Multiply 1101_2 by 1011_2.

■ ***Solution:***

Multiplicand:	1 1 0 1_2	13_{10}
Multiplier:	× 1 0 1 1_2	× 11_{10}
First Partial Product:	1 1 0 1	143_{10}
Second Partial Product:	1 1 0 1←	
Sum of Partial Product:	1 0 0 1 1 1	
Third Partial Product:	0 0 0 0 ←←	
Sum of Partial Product:	1 0 0 1 1 1	
Fourth Partial Product:	1 1 0 1 ←←←	
Final Product:	1 0 0 0 1 1 1 1_2	

Binary Division

Figure 13-4(a) reviews the decimal division procedure by dividing 830 by 23. This long division procedure begins by determining how many times the divisor (23) can be subtracted from the first digit of the dividend (8). Since the dividend is smaller, the quotient is 0. Next, we see how many times the divisor (23) can be subtracted from the first two digits of the dividend (83). Since 23 can be subtracted three times from 83, a quotient of 3 results, and the product of 3 × 23, or 69, is subtracted from the first two digits of the dividend, giving a remainder of 14. Then the next digit of the dividend (0) is brought down to the remainder (14), and we see how many times the divisor (23) can be subtracted from the new remainder (140). In this example, 23 can be subtracted six times from 140, a quotient of 6 results, and the product of 6 × 23, or 138, is subtracted from the remainder, giving a new remainder of 2. Therefore, 830 ÷ 23 = 36, remainder 2.

Figure 13-4(b) illustrates an example of binary division, which is generally much simpler than decimal division. To perform this operation, we must first determine how many times the divisor (101_2) can be subtracted from the first bit of the dividend (1_2). Since the dividend is smaller than the divisor, a quotient of 0_2 is placed above the first digit of the dividend. Next, we see how many times the divisor (101_2) can be subtracted from the first two bits of the dividend (10_2), and this again results in a quotient of 0_2. Similarly, the divisor (101_2) cannot be subtracted from the first three bits of the dividend, so the quotient is once again 0_2. Finally, we find a dividend (1001_2) that is greater than the divisor (101_2), and therefore a 1_2 is placed in the quotient, and the divisor is subtracted from the first four bits of the dividend, resulting in a remainder of

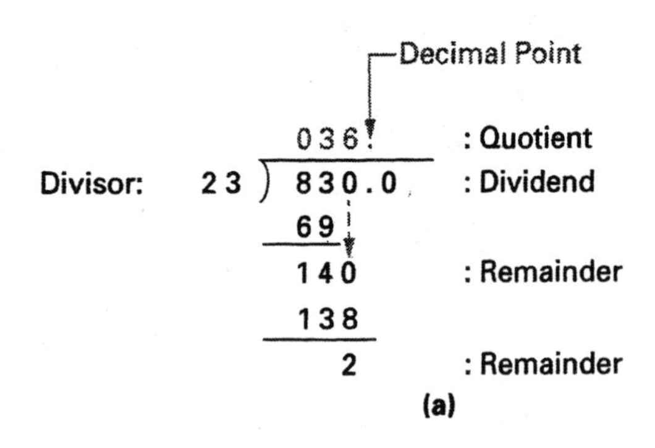

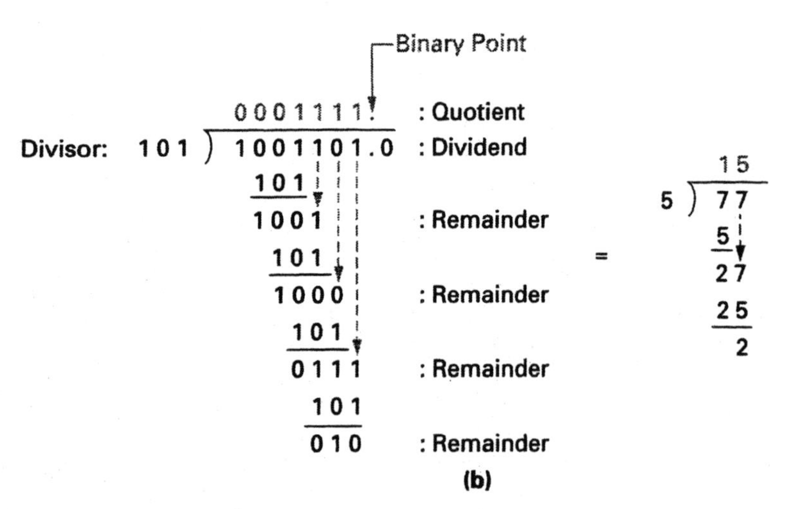

FIGURE 13-4 Binary Division.

100_2. Then the next bit of the dividend (1_2) is brought down to the remainder (100_2), and we see how many times the divisor (101_2) can be subtracted from the new remainder (1001_2). Since 101_2 is smaller than 1001_2, a quotient of 1_2 results. The divisor is subtracted from the remainder, resulting in a new remainder of 100_2. Then the next bit of the dividend (0_2) is brought down to the remainder (100_2), and we see how many times the divisor (101_2) can be subtracted from the new remainder (1000_2). Since 101_2 is smaller than 1000_2, a quotient of 1 results. The divisor is subtracted from the remainder, resulting in a new remainder of 11_2. Then the next bit of the dividend (1_2) is brought down to the remainder (11_2), and we see how many times the divisor (101_2) can be subtracted from the new remainder (111_2). Since 101_2 is smaller than 111_2, a quotient of 1 results. The divisor is subtracted from the remainder, resulting in a final remainder of 10. Therefore, $1001101_2 \div 101_2 = 1111_2$, remainder 10 ($77 \div 5 = 15$, remainder 2).

■ **EXAMPLE:**

Divide 1110101_2 by 110_2.

■ ***Solution:***

	Binary		Decimal
	0 0 1 0 0 1 1_2	:*Quotient*	019_{10}
Divisor:	110)1 1 1 0 1 0 1_2	:*Dividend*	$6_{10})117_{10}$
	1 1 0 ↓ ↓ ↓ ↓		
	0 0 1 0 1 0 ↓	:*Remainder*	
	1 1 0 ↓		6↓
	1 0 0 1	:*Remainder*	57
	1 1 0		54
	0 0 1 1_2	:*Remainder*	3_{10}

13-1-2 Representing Positive and Negative Numbers

Many digital systems, such as calculators and computers, are used to perform mathematical functions on a wide range of numbers. Some of these numbers are positive and negative signed numbers; therefore, a binary code is needed to represent these numbers.

The Sign and Magnitude Number System

The **sign and magnitude number system** is a binary code system used to represent positive and negative numbers. A sign magnitude number contains a *sign bit* (0 for positive, 1 for negative) followed by the *magnitude bits.*

Sign and Magnitude Number System A binary code used to represent positive and negative numbers in which the MSB signifies sign and the following bits indicate magnitude.

As an example, Figure 13-5(a) shows some 4-bit sign-magnitude positive numbers. All of these sign-magnitude binary numbers—0001, 0010, 0011, and 0100—have an MSB, or sign bit, of 0, and therefore they are all positive numbers. The remaining three bits in these 4-bit words indicate the magnitude of the number based on the standard 421 binary column weight ($001_2 = 1_{10}$, $010_2 = 2_{10}$, $011_2 = 3_{10}$, $100_2 = 4_{10}$).

As another example, Figure 13-5(b) shows some 4-bit sign-magnitude negative numbers. All of these sign-magnitude binary numbers—1100, 1101, 1110, and 1111—have an MSB, or sign bit, of 1, and therefore they are all negative numbers. The remaining three bits in these 4-bit words indicate the magnitude of the number based on the standard 421 binary column weight ($100_2 = 4_{10}$, $101_2 = 5_{10}$, $110_2 = 6_{10}$, $111_2 = 7_{10}$).

If you need to represent larger decimal numbers, you simply use more bits, as shown in Figure 13-5(c) and (d). The principle still remains the same in that the MSB indicates the sign of the number, and the remaining bits indicate the magnitude of the number.

The sign-magnitude number system is an ideal example of how binary numbers could be coded to represent positive and negative numbers. This code system, however, requires complex digital hardware for addition and subtraction, and is therefore seldom used.

Sign Bit	Magnitude Bits
0 = + 1 = −	= Binary Equivalent

4-Bit Sign-Magnitude Numbers

$0\ 0\ 0\ 1_2$	$= +1_{10}$
$0\ 0\ 1\ 0_2$	$= +2_{10}$
$0\ 0\ 1\ 1_2$	$= +3_{10}$
$0\ 1\ 0\ 0_2$	$= +4_{10}$

Sign Bit   Magnitude Bits

(a)

$1\ 1\ 0\ 0_2$	$= -4_{10}$
$1\ 1\ 0\ 1_2$	$= -5_{10}$
$1\ 1\ 1\ 0_2$	$= -6_{10}$
$1\ 1\ 1\ 1_2$	$= -7_{10}$

Sign Bit   Magnitude Bits

(b)

8-Bit Sign-Magnitude Numbers

$0\ 0\ 0\ 0\quad 1\ 1\ 1\ 1_2$	$= +15_{10}$
$0\ 1\ 0\ 1\quad 0\ 0\ 0\ 0_2$	$= +80_{10}$
$1\ 0\ 0\ 0\quad 1\ 0\ 1\ 0_2$	$= -10_{10}$
$1\ 1\ 1\ 1\quad 1\ 1\ 1\ 1_2$	$= -127_{10}$

(c)

16-Bit Sign-Magnitude Numbers

$0\ 0\ 0\ 0\quad 0\ 0\ 0\ 0\quad 0\ 0\ 0\ 0\quad 0\ 0\ 0\ 1_2$	$= +1_{10}$
$0\ 0\ 0\ 0\quad 0\ 0\ 0\ 0\quad 1\ 0\ 1\ 0\quad 0\ 1\ 0\ 0_2$	$= +164_{10}$
$1\ 0\ 0\ 0\quad 0\ 0\ 0\ 0\quad 0\ 0\ 1\ 1\quad 0\ 0\ 1\ 0_2$	$= -50_{10}$
$1\ 0\ 0\ 0\quad 1\ 1\ 1\ 1\quad 0\ 0\ 0\ 0\quad 0\ 0\ 0\ 0_2$	$= -3840_{10}$

(d)

FIGURE 13-5 The Sign-Magnitude Number System.

■ EXAMPLE:

What are the decimal equivalents of the following sign-magnitude number system codes?

a. 0111 b. 1010 c. 0001 1111 d. 1000 0001

■ *Solution:*

a. +7 b. −2 c. +31 d. −1

■ EXAMPLE:

What are the 8-bit sign-magnitude number system codes for the following decimal values?

a. +8 b. −65 c. +127 d. −127

■ *Solution:*

a. 0000 1000 b. 1100 0001 c. 0111 1111 d. 1111 1111

One's Complement Number System
A binary code used to represent positive and negative numbers, in which negative values are determined by inverting all of the binary digits of an equivalent positive value.

The One's Complement (1's Complement) Number System

The **one's complement number system** became popular in early digital systems. It also used an MSB sign bit, that was either a 0 (indicating a positive number) or a 1 (indicating a negative number).

Figure 13-6(a) shows some examples of 8-bit one's complement positive numbers. As you can see, one's complement positive numbers are no different from sign-magnitude positive numbers. The only difference between the one's complement number system and the sign-magnitude number system is the way in which negative numbers are represented. To understand the code used to represent negative numbers, we must first understand the meaning of the term "one's complement." To one's complement a binary number means to invert or change all of the 1s in the binary word to 0s, and all of the 0s in the binary word to 1s. Performing the one's complement operation on a positive number will change the number from a one's complement positive number to a one's complement negative number. For example, by one's complementing the number 00000101, or +5, we will get 11111010, which is −5. In Figure 13-6(b), you will see the negative one's complement codes of the positive one's complement codes shown in Figure 13-6(a).

The next question you may have is, how can we determine the decimal value of a negative one's complement code? The answer is to do the complete opposite, or complement, to that of the positive one's complement code. This means count the 1s in a positive one's complement code, and count the 0s in a negative one's complement code. For example, with a positive one's complement code (MSB = 0), the value is determined by simply adding together all of the column weights that contain a 1 (00000110 has a 1 in the fours column and twos column, so this number is 4 + 2, or +6). On the other hand, for a negative one's complement code (MSB = 1), the value is determined by doing the opposite and adding together all of the column weights that contain a 0 (11110101 has a 0 in the eights column and twos column, so this number is 8 + 2, or −10).

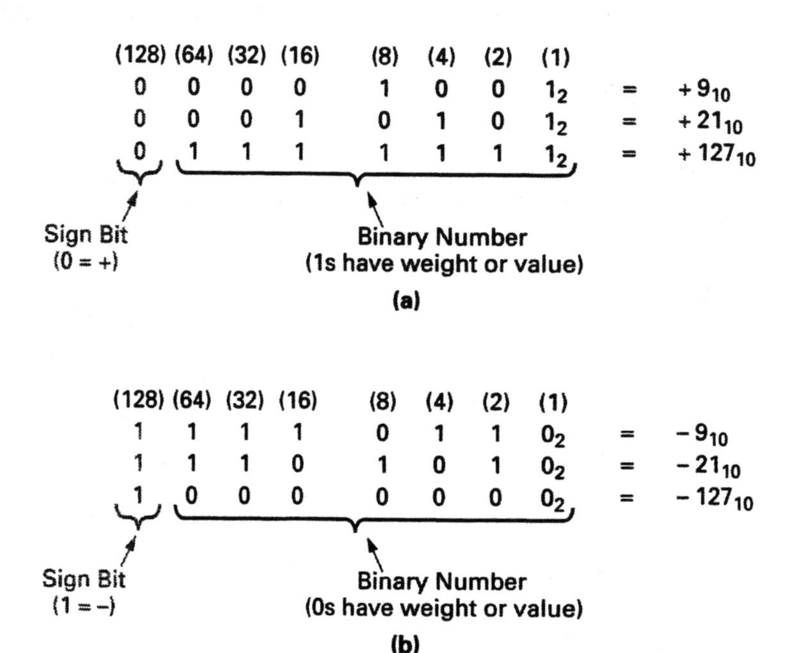

FIGURE 13-6 The One's Complement Number System.

■ EXAMPLE:

What are the decimal equivalents of the following one's complement number system codes?

a. 0010 0001 b. 1101 1101

■ *Solution:*

a. MSB = 0 (number is positive, so add together all of the column weights that contain a 1). The number 0010 0001 has a 1 in the thirty-twos column and a one in the units column, so 0010 0001 = +33.
b. MSB = 1 (number is negative, so add together all of the column weights that contain a 0). The number 1101 1101 has a 0 in the thirty-twos column and a 0 in the twos column, so 1101 1101 = −34.

Since the term "one's complement" can be used to describe both an operation (change all of the 1s to 0s, and all of the 0s to 1s), and a positive and negative number system, some confusion can occur. To avoid this, remember that if you are asked to "one's complement a number," it means to change all of the 1s in the binary word to 0s and all the 0s to 1s. If, on the other hand, you are asked to "find the decimal equivalent of a one's complement number," you should decode the sign bit and binary value to determine the decimal equivalent value.

■ EXAMPLE:

One's complement the following numbers.

a. 0101 1111 b. 1110 0101

■ *Solution:*

To one's complement a number means to change all of the 1s to 0s and all of the 0s to 1s.

a. 1010 0000 b. 0001 1010

■ EXAMPLE:

Find the decimal equivalent of the following one's complement numbers.

a. 0101 1111 b. 1110 0101

■ *Solution:*

a. MSB = 0 (number is positive, so add together all of the column weights that contain a 1). The number 0101 1111 = +95.
b. MSB = 1 (number is negative, so add together all of the column weights that contain a 0). The number 1110 0101 = −26.

Two's Complement Number System
A binary code used to represent positive and negative numbers, in which negative values are determined by inverting all of the binary digits of an equivalent positive value and then adding 1.

The Two's Complement (2's Complement) Number System

The **two's complement number system** is used almost exclusively in digital systems to represent positive and negative numbers.

Positive two's complement numbers are no different from positive sign-magnitude and one's complement numbers in that an MSB sign bit of 0 indicates a positive number and the remaining bits of the word indicate the value of the number. Figure 13-7(a) shows two examples of positive two's complement numbers.

Negative two's complement numbers are determined by two's complementing a positive two's complement number code. To two's complement a number first one's complement it (change all the 1s to 0s and 0s to 1s), and then add 1. For example, if we were to two's complement the number 0000 0101 (+5), we would obtain the two's complement number code for −5. The procedure to follow to two's complement 0000 0101 (+5), would be as follows:

0000 0101	Original Number (+5)
1111 1010	One's Complement (inverted)
+ 1	Plus 1
1111 1011	Two's Complement Code for −5

The new two's complement code obtained after this two's complement operation is the negative equivalent of the original number. In our example, therefore:

Original Number:	0000 0101 = +5
After Two's Complement Operation:	1111 1011 = −5

Any negative equivalent two's complement code therefore can be obtained by simply two's complementing the same positive two's complement number. For example, to find the negative two's complement code for −125, we would perform the following operation.

	0111 1101	Original Number (+125)
Two's Complement Operation →	1000 0010	One's Complement (inverted)
	+ 1	Plus 1
	1000 0011	Two's Complement Code for −125

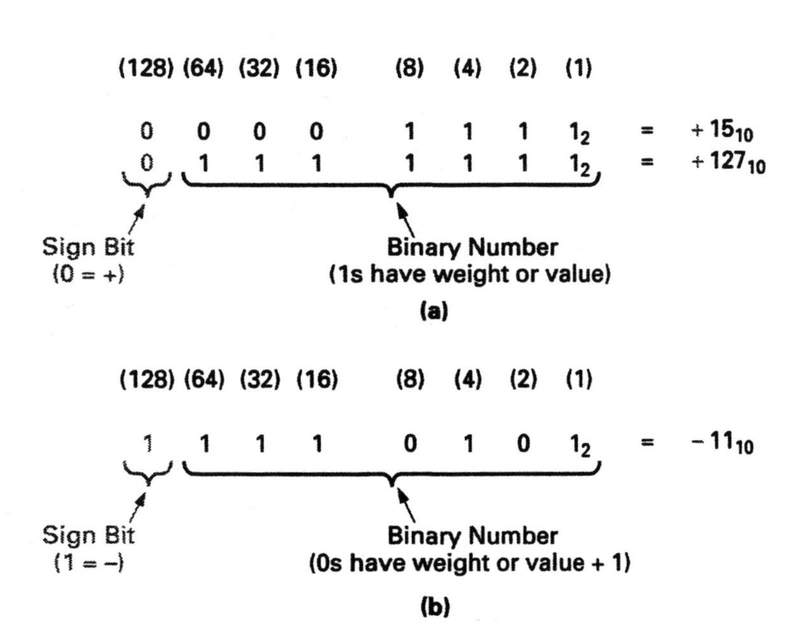

FIGURE 13-7 The Two's Complement Number System.

■ **EXAMPLE:**

Find the negative two's complement codes for

a. −26 b. −67

■ ***Solution:***

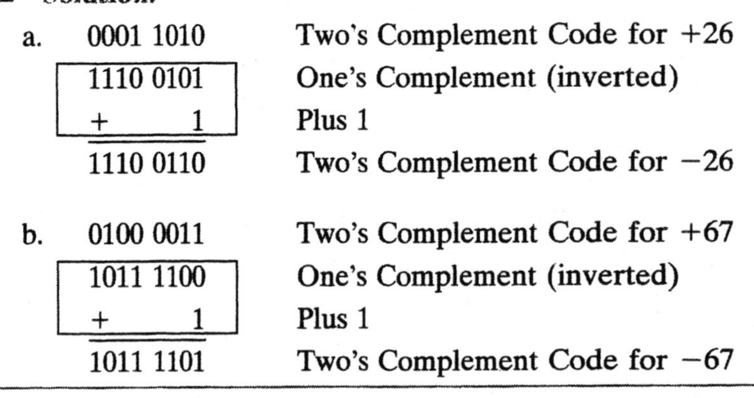

Performing the two's complement operation on a number twice will simply take the number back to its original value and polarity. For example, let's take the two's complement code for +6 (0000 0110) and two's complement it twice.

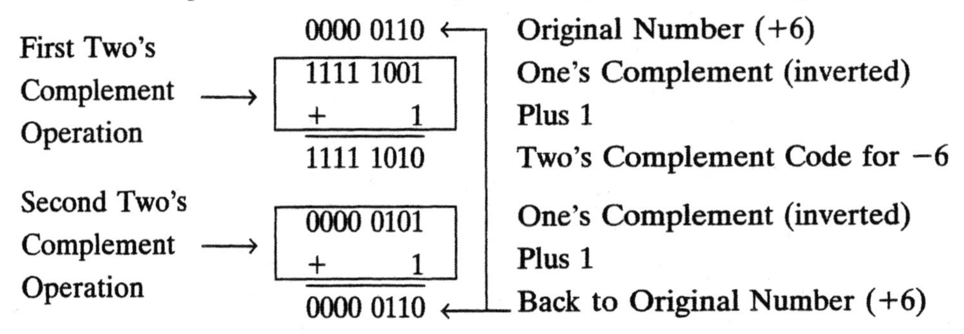

The decimal equivalent of a positive two's complement number (MSB = 0) is easy to determine, since the binary value following the sign bit applies the same as any binary number, as shown in Figure 13-7(a). On the other hand, the decimal equivalent of a negative two's complement number (MSB = 1) is not determined in the same way. To determine the value of a negative two's complement number, refer to the remaining bits following the sign bit, add the weight of all columns containing a 0, and then add 1 to the result. For example, the negative two's complement number in Figure 13-7(b) has 0s in the eights and twos columns, and therefore its value will equal the sum of 8 and 2, plus the additional 1 (due to two's complement). As a result:

$$\overset{}{1\,1\,1\,1}\;\overset{(8)(4)(2)(1)}{0\,1\,0\,1} = (8 + 2) + 1 = -11$$

■ **EXAMPLE:**

Determine the decimal equivalent of the following negative two's complement numbers.

a. 1110 0001 b. 1011 1011

c. 1000 0000 d. 1111 1111

■ ***Solution:***

(±)(64)(32)(16) (8) (4) (2) (1)
a. 1 1 1 0 0 0 0 1_2 = (16 + 8 + 4 + 2) + 1 = -31_{10}
(±)(64)(32)(16) (8) (4) (2) (1)
b. 1 0 1 1 1 0 1 1_2 = (64 + 4) + 1 = -69_{10}
(±)(64)(32)(16) (8) (4) (2) (1)
c. 1 0 0 0 0 0 0 0_2 = (64 + 32 + 16 + 8 + 4 + 2 + 1) + 1 = -128_{10}
(±)(64)(32)(16) (8) (4) (2) (1)
d. 1 1 1 1 1 1 1 1_2 = (0) + 1 = -1_{10}

As with the sign-magnitude and one's complement number systems, if you need to represent larger decimal numbers with a two's complement code, you will have to use a word containing more bits. For example, a total of 16 different combinations is available when we use a 4-bit word ($2^4 = 16$), as shown in Figure 13-8. With standard binary, these 16 different combinations give us a count from 0 to 15, as shown in the center column in Figure 13-8. As far as two's complement is concerned, we split these 16 different codes in half, and ended up with 8 positive two's complement codes and 8 negative two's complement codes, as shown in the right column in Figure 13-8. Since decimal zero uses one of the positive two's complement codes, the 16 different combinations give us a range of -8_{10} to $+7_{10}$ (the maximum positive value, +7, is always one less than the maximum negative value, −8, since decimal zero uses one of the positive codes).

Using an 8-bit word will give us 256 different codes ($2^8 = 256$), as shown in Figure 13-9. Dividing this total in half will give us 128 positive two's complement codes and 128 negative two's complement codes. Since decimal zero uses one of the positive two's complement codes, the 256 different combinations gives us a range of -128_{10} to $+127_{10}$.

	4-Bit Word	Standard Binary Value	2's Complement Value	
2^4 = 16 Words	0 0 0 0	0	0	8 Positive Numbers
	0 0 0 1	1	+1	
	0 0 1 0	2	+2	
	0 0 1 1	3	+3	
	0 1 0 0	4	+4	
	0 1 0 1	5	+5	
	0 1 1 0	6	+6	
	0 1 1 1	7	+7	
	1 0 0 0	8	−8	8 Negative Numbers
	1 0 0 1	9	−7	
	1 0 1 0	10	−6	
	1 0 1 1	11	−5	
	1 1 0 0	12	−4	
	1 1 0 1	13	−3	
	1 1 1 0	14	−2	
	1 1 1 1	15	−1	

FIGURE 13-8 4-Bit Two's Complement Codes.

8-Bit Word	Standard Binary Value	2's Complement Value
0 0 0 0 0 0 0 0	0	0
0 0 0 0 0 0 0 1	1	+1
0 0 0 0 0 0 1 0	2	+2
0 0 0 0 0 0 1 1	3	+3
0 0 0 0 0 1 0 0	4	+4
0 0 0 0 0 1 0 1	5	+5
0 0 0 0 0 1 1 0	6	+6
0 0 0 0 0 1 1 1	7	+7
⋮	⋮	⋮ (128 Positive Numbers)
0 1 1 1 1 0 1 1	123	+123
0 1 1 1 1 1 0 0	124	+124
0 1 1 1 1 1 0 1	125	+125
0 1 1 1 1 1 1 0	126	+126
0 1 1 1 1 1 1 1	127	+127
1 0 0 0 0 0 0 0	128	−128
1 0 0 0 0 0 0 1	129	−127
1 0 0 0 0 0 1 0	130	−126
1 0 0 0 0 0 1 1	131	−125
1 0 0 0 0 1 0 0	132	−124
1 0 0 0 0 1 0 1	133	−123
⋮	⋮	⋮ (128 Negative Numbers)
1 1 1 1 1 0 1 1	251	−5
1 1 1 1 1 1 0 0	252	−4
1 1 1 1 1 1 0 1	253	−3
1 1 1 1 1 1 1 0	254	−2
1 1 1 1 1 1 1 1	255	−1

$2^8 =$ 256 Words

FIGURE 13-9 8-Bit Two's Complement Codes.

■ **EXAMPLE:**

What would be the two's complement range for a 16-bit word?

■ ***Solution:***

$2^{16} = 65{,}536$; half of $65{,}536 = 32{,}768$; therefore, the range will be $-32{,}768$ to $+32{,}767$.

Arithmetic-Logic Unit (ALU)
The section of a computer that performs all arithmetic and logic operations.

13-1-3 Two's Complement Arithmetic

To perform its function, a computer's MPU contains several internal circuits, and one of these circuits is called the **arithmetic-logic unit,** or **ALU.** This unit is the number-crunching circuit of every digital system and, as its name implies, it can be controlled to perform either arithmetic operations (such as addition or subtraction) or logic operations (such as AND and OR). The arithmetic-logic unit has two parallel inputs, one parallel output, and a set of function select control lines, as shown in Figure 13-10.

As far as arithmetic operations are concerned, the ALU has to be able to perform the four basic operations which, as you know, are addition, subtraction, multiplication, and division. One would expect, therefore, that an ALU would contain a separate circuit for each of these operations. This is not so, since all four operations can be performed using a binary-adder circuit. To explain this point, because the binary-adder circuit's function is to perform addition, and since multiplication is simply repeated addition, we can also use the binary-adder circuit to perform any multiplication operations. For example, 3×4 means that you are to take the number 4 and add it three times ($4 + 4 + 4 = 12$, therefore $3 \times 4 = 12$). But what about subtraction and division? The answer to this problem is the two's complement number system. As you will see in this section, by representing all values as two's complement numbers, we can perform a subtraction operation by slightly modifying the second number, and then adding the two numbers together. This means that our binary-adder circuit can also be used to perform subtraction operations, and since division is simply repeated subtraction, we can also use the subtraction operation of the binary-adder circuit to perform any division operation. For example, $12 \div 4$ means that you are to take the number 12 and see how many times you can subtract 4 from it ($12 - 4 - 4 - 4 = 0$, therefore there are three 4s in 12, $12 \div 4 = 3$).

By using two's complement numbers, therefore, we can use a binary-adder circuit within an ALU to add, subtract, multiply, and divide. Having one circuit

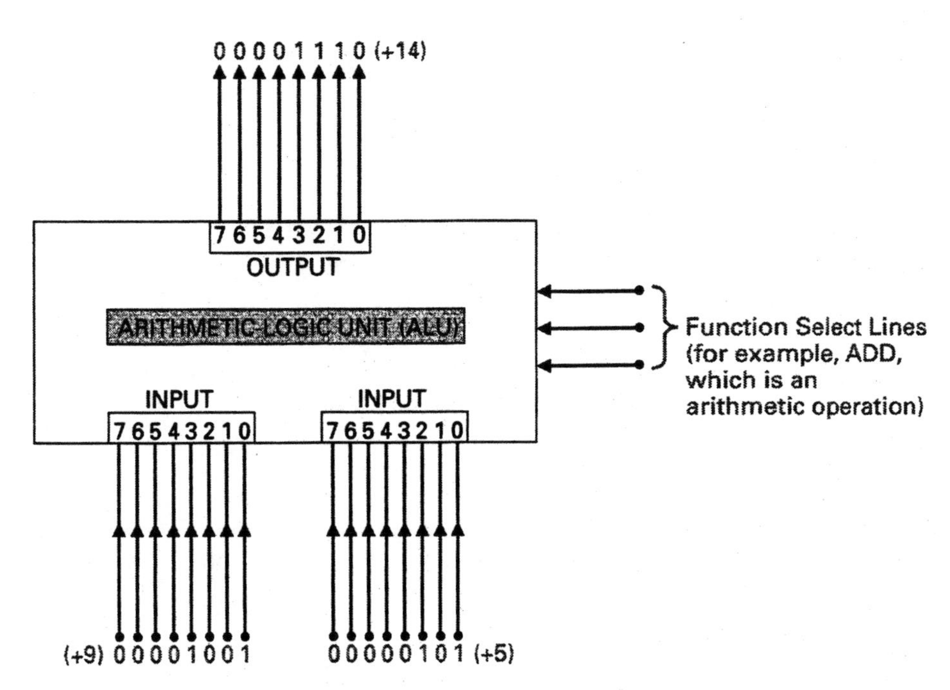

FIGURE 13-10 Block Diagram of the Arithmetic-Logic Unit (ALU).

able to perform all four mathematical operations means that the circuit will be small in size, fast to operate, consume very little power, and be cheap to manufacture. In this section, we will see how we can perform all four mathematical operations by always adding two's complement numbers.

Adding Positive Numbers

To begin with, let us prove that the addition of two positive two's-complement numbers will render the correct result.

■ **EXAMPLE:**

Add +9 (0000 1001) and +5 (0000 0101).

■ ***Solution:***

```
                                        0000 1001
                                      + 0000 0101
                                      -----------
MSB of 0 = positive number. ----->      0000 1110
```

When both of the numbers to be added are positive two's complement numbers (both have an MSB of 0), it makes sense that the sum will have an MSB of 0, indicating a positive result. This is always true, unless the range of the two's complement word is exceeded. For example, the maximum positive number that can be represented by an 8-bit word is +127 (0111 1111). If the sum of the two positive two's complement numbers to be added exceeds the upper positive range limit, there will be a *two's complement overflow* into the sign bit. For instance:

```
                                 (±)(64)(32)(16) (8)(4)(2)(1)
                                   0  1  1  0    0  1  0  0      (+100)
                               +   0  0  0  1    1  1  1  0    + (+30)
                                 ----------------------------  --------
MSB of 1 = negative number. --->   1  0  0  0    0  0  1  0      (−126)
```

If two positive numbers are applied to an arithmetic circuit, and it is told to add the two, the result should also be a positive number. Most arithmetic circuits are able to detect this problem if it occurs by simply monitoring the MSB of the two input words and the MSB of the output word, as shown in Figure 13-11. If the two input sign bits are the same, but the output sign bit is different, then a two's complement overflow has occurred.

Adding Positive and Negative Numbers

By using the two's complement number system, we can also add numbers with unlike signs and obtain the correct result. As an example, let us add +9 (0000 1001) and −5 (1111 1011).

```
                                   0000 1001       (+9)
                                 + 1111 1011     + (−5)
                                 -----------     ------
Ignore the final carry. ----> 1    0000 0100       (+4)
```

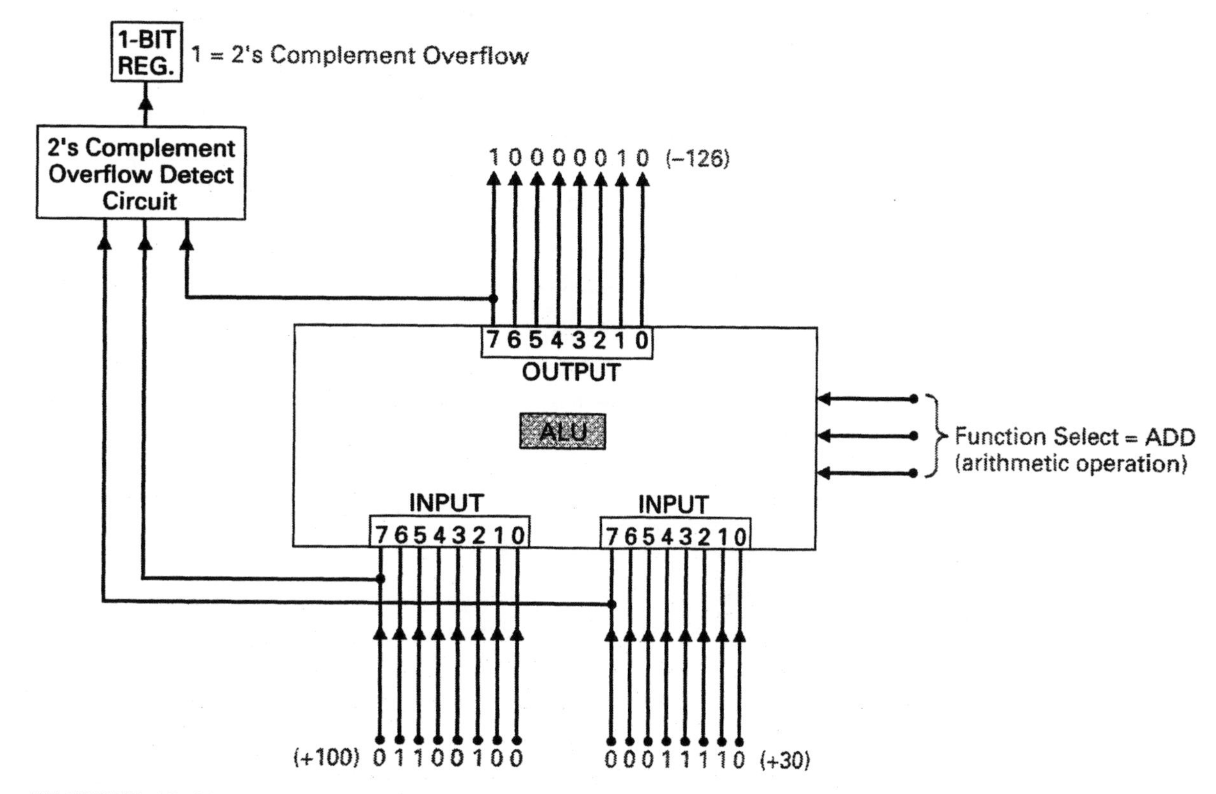

FIGURE 13-11 Two's Complement Overflow Detection.

A final carry will always be generated whenever the sum is a positive number; however, if this carry is ignored because it is beyond the 8-bit two's complement word, the answer will be correct.

■ **EXAMPLE:**

Add +10 (0000 1010) and −12 (1111 0100).

■ ***Solution:***

0000 1010	(+10)
+ 1111 0100	+ (−12)
1111 1110	(−2)

Adding Negative Numbers

As a final addition test for the two's complement number system, let us add two negative numbers to see if we can obtain the correct negative sum.

■ **EXAMPLE:**

Add −5 (1111 1011) and −4 (1111 1100).

■ ***Solution:***

	1111 1011	(−5)
	+ 1111 1100	+ (−4)
Ignore the final carry. ⟶ 1	1111 0111	(−9)

The sum in this example is the correct two's complement code for −9.

Two's complement overflow can also occur in this condition if the sum of the two negative numbers exceeds the negative range of the word. For example, with an 8-bit two's complement word, the largest negative number is −128, and if we add −100 (1001 1100) and −30 (1110 0010), we will get an error, as follows:

	1001 1100	(−100)
	+ 1110 0010	+ (−30)
Ignore the final carry. ⟶ 1	0111 1110	(+126)

Once again, the two's complement overflow detect circuit, shown in Figure 13-11, will indicate such an error since two negative numbers were applied to the input and a positive number appeared at the output.

In all of the addition examples mentioned so far (adding positive numbers, adding positive and negative numbers, and adding negative numbers) you can see that an ALU added the bits of the input bytes without any regard for whether they were signed or unsigned numbers. This brings up a very important point, which is often misunderstood. The ALU adds the bits of the input words using the four basic rules of binary addition, and it does not know if these input and output bit patterns are coded two's complement words or standard unsigned binary words. It is therefore up to us to maintain a consistency, which means that if we are dealing with coded two's complement words at the input of an ALU, we must interpret the output of an ALU as a two-coded two's complement word. On the other hand, if we are dealing with standard unsigned binary words at the input of an ALU, we must interpret the output of an ALU as a standard unsigned binary word.

Subtraction Through Addition

The key advantage of the two's complement number system is that it enables us to perform a subtraction operation using the adder circuit within the ALU. To explain how this is done, let us take an example and assume that we want to subtract 3 (0000 0011) from 8 (0000 1000), as follows:

Minuend:	0000 1000	(+8)
Subtrahend:	+ 0000 0011	− (+3)
Difference:	0000 0101	+5

If you subtract the subtrahend of +3 from the minuend of +8, you will obtain the correct result of +5 (0000 0101). However, we must find the difference to this problem by using addition and not subtraction. The solution to this is achieved by *two's complementing the subtrahend, and then adding the result to the minuend.* The following shows how this is done.

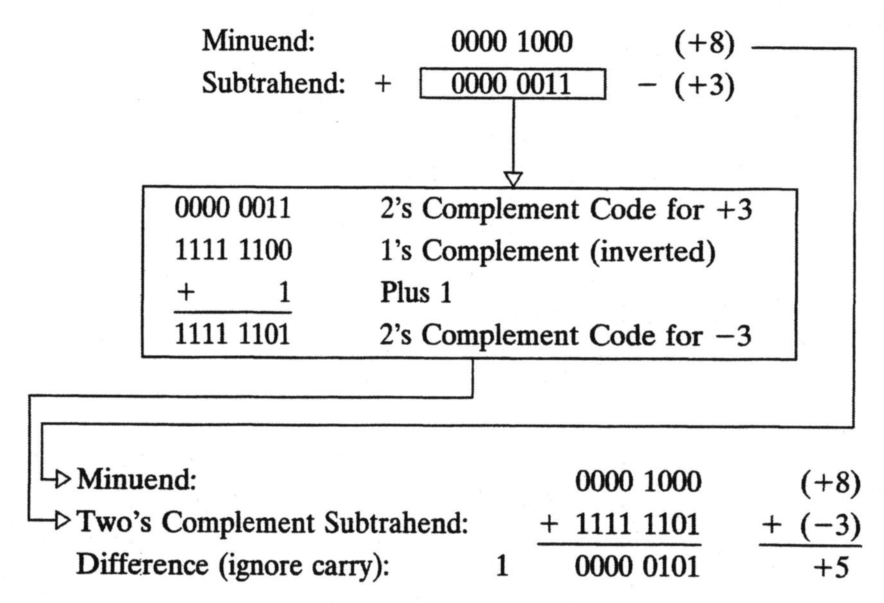

As you can see from this operation, adding the −3 to +8 achieves exactly the same result as subtracting +3 from +8. We can therefore perform a subtraction by two's complementing the subtrahend of a subtraction problem, and then adding the result to the minuend.

Digital electronic calculators and computer systems perform the previously described steps automatically. That is, they first instruct the ALU to one's complement the subtrahend input (function select control inputs = invert, which is a logic operation), then add 1 to the result (function select control inputs = +1, which is an arithmetic operation), and finally add the result of the two previous operations (which is the two's complemented subtrahend) to the minuend (function select = ADD, arithmetic operation). Although this step-by-step procedure or **algorithm** seems complicated, it is a lot simpler than having to include a separate digital electronic circuit to perform all subtraction operations. Having one circuit able to perform both addition and subtraction means that the circuit will be small in size, fast to operate, consume very little power, and be cheap to manufacture.

Algorithm
A set of rules for the solution of a problem.

In Figure 13-12, you can see a **flow chart** summarizing the subtraction through addition procedure. Flow charts provide a graphic way of describing a step-by-step procedure (algorithm) or a program of instructions.

Flow Chart
A graphic representation of a step-by-step procedure or program of instructions.

Multiplication Through Repeated Addition

Most digital electronic calculators and computers do not include a logic circuit that can multiply, since the adder circuit can also be used to perform this arithmetic operation. For example, to multiply 6 × 3, you simply add 6 three times.

Multiplicand:	6		6
Multiplier:	× 3	=	+ 6
Product:	18		+ 6
			18

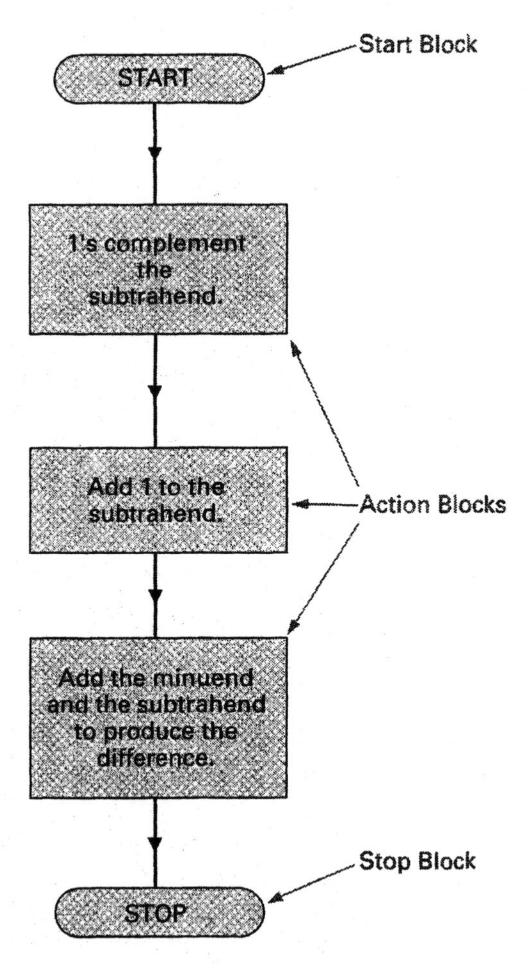

FIGURE 13-12 Flow Chart for Subtraction through Addition.

Like subtraction through addition, an algorithm must be followed when a digital system needs to perform a multiplication operation. Figure 13-13 shows the flow chart for multiplication through repeated addition. Let us apply this procedure to our previous example of 6 × 3 and see how the multiplicand 6 is added three times to obtain a final product of 18. The steps would be as follows:

Start:	Multiplicand = 6, Multiplier = 3, Product = 0.
After First Pass:	Multiplicand = 6, Multiplier = 2, Product = 6.
After Second Pass:	Multiplicand = 6, Multiplier = 1, Product = 12.
After Third Pass:	Multiplicand = 6, Multiplier = 0, Product = 18
Stop.	

In this example we cycled through the flow chart three times (until the multiplier was 0) and then finally obtained a product of 18.

Although this algorithm seems complicated, it is a lot simpler than having to include a separate digital circuit for multiplication. Having one circuit able to perform addition, subtraction, and multiplication means that the circuit will

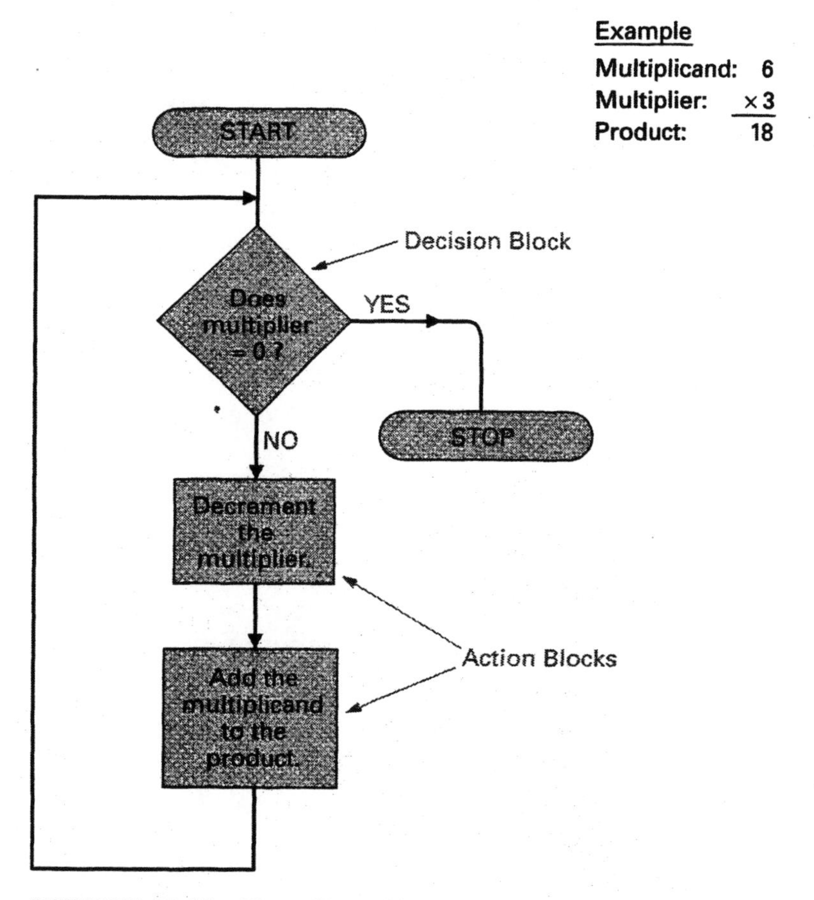

FIGURE 13-13 Flow Chart for Multiplication through Repeated Addition.

be small in size, fast to operate, consume very little power, and be cheap to manufacture.

Division Through Repeated Subtraction

Up to this point, we have seen how we can use the two's complement number system and the adder circuit within an ALU for addition, subtraction, and multiplication. Just as multiplication can be achieved through repeated addition, division can be achieved through repeated subtraction. For example, to divide 31 by 9 you simply see how many times 9 can be subtracted from 31, as follows:

	3	:Quotient		31	
Divisor:	9)31	:Dividend		−9	:First Subtraction
	27		=	22	
	4	:Remainder		−9	:Second Subtraction
				13	
				−9	:Third Subtraction
				4	

Since we were able to subtract 9 from 31 three times with a remainder of 4, then 31 ÷ 9 = 3, remainder 4.

Like multiplication, an algorithm must be followed when a digital system needs to perform a division. Figure 13-14 shows the flow chart for division through repeated subtraction. Let us apply this procedure to our previous example of 31 ÷ 3, and see how the quotient is incremented to 3, leaving a remainder of 4. The steps would be as follows:

Start: Dividend = 31, Divisor = 9, Quotient = 0, Remainder = 31.
After First Pass: Dividend = 22, Divisor = 9, Quotient = 1, Remainder = 22.
After Second Pass: Dividend = 13, Divisor = 9, Quotient = 2, Remainder = 13.
After Third Pass: Dividend = 4, Divisor = 9, Quotient = 3, Remainder = 4.
Stop.

When this program of instructions cycles or loops back for the third pass, the divisor of 9 is subtracted from the remaining dividend of 4, resulting in −5. When the question Is the dividend negative? is asked, the answer will be yes, so the branch on the right is followed. This branch of the flow chart will add the divisor of 9 to the dividend of −5 to restore it to the correct remainder of

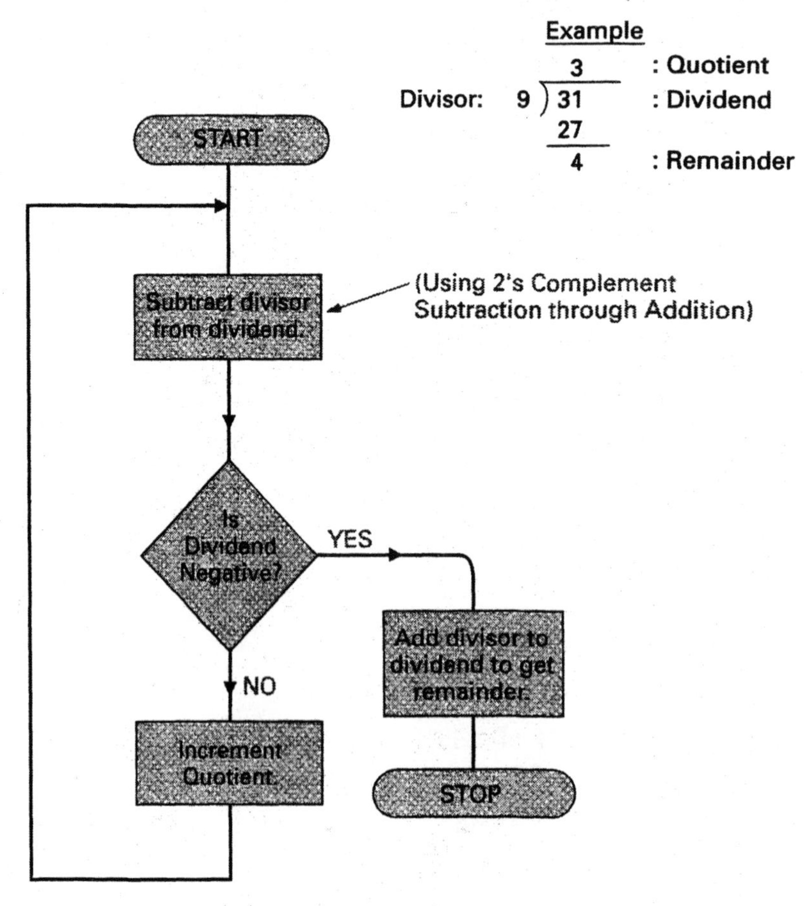

FIGURE 13-14 Flow Chart for Division through Repeated Subtraction.

4 before the program stops. To summarize, these are the steps followed by this program:

```
  (+31)
− (+ 9)
  (+22)   :Quotient = 1
− (+ 9)
  (+13)   :Quotient = 2
− (+ 9)
  (+ 4)   :Quotient = 3
− (+ 9)
  (− 5)
− (+ 9)   :Add Divisor to Dividend
  (+ 4)   :Remainder = 4
```

When a digital system is instructed to perform a division, it will use the ALU for all arithmetic and logic operations. Since division is achieved through repeated subtraction, and subtraction is achieved through addition, the ALU's adder circuit can also be used for division.

Although this algorithm seems complicated, it is a lot simpler than having to include a separate digital circuit for division. As previously mentioned, having one circuit able to perform addition, subtraction, multiplication, and division means that the circuit will be small in size, fast to operate, consume very little power, and be cheap to manufacture.

13-1-4 Representing Large and Small Numbers

While two's complementing enables us to represent positive and negative numbers, the range is limited to the number of bits in the word. For example, the range of an 8-bit two's complement word is from −128 to +127, while the range of a 16-bit two's complement word is from −32,768 to +32,767. To extend this range, we can keep increasing the number of bits; however, these large words become difficult to work with. Another problem arises when we wish to represent fractional numbers such as 1.57.

These problems are overcome by using the **floating-point number system,** which uses scientific notation to represent a wider range of values. Figure 13-15 explains the two-byte (16-bit) floating-point number system. The first byte is called the **mantissa,** and it is always an unsigned binary number that is interpreted as a value between 0 and 1. The second byte is called the **exponent,** and it is always a two's complement number and therefore represents a range from 10^{-128} to 10^{+127}. As an example, let us see how the floating point number system can be used to represent a large value number such as 27,000,000,000,000 and a small value number such as 0.0000000015.

Floating-Point Number System
A number system in which a value is represented by a mantissa and an exponent.

Mantissa
The fractional part of a logarithm.

Exponent
A small number placed to the right and above a value to indicate the number of times that symbol is a factor.

■ **EXAMPLE:**

Give the two-byte floating-point binary values for 27,000,000,000,000.

■ ***Solution:***

The value 27,000,000,000,000 would be written in scientific notation as 0.27×10^{14}, and when coded as a floating-point number, would appear as follows:

$$27{,}000{,}000{,}000{,}000$$
$$= 0.27 \times 10^{14}$$
$$= \boxed{0001\ 1011} \quad \boxed{0000\ 1110}$$

Mantissa (unsigned binary) = 27, which is interpreted as 0.27.	Exponent (two's complement) = +14, which is interpreted as 10^{14}.

In this example, the mantissa byte (standard binary number) is equal to 27 (0001 1011) and is therefore interpreted as 0.27, and the exponent byte (two's complement number) is equal to +14 (0000 1110) and is therefore interpreted as 10^{14}.

■ **EXAMPLE:**

Give the two-byte floating-point binary values for 0.0000000015.

■ ***Solution:***

The value 0.0000000015 would be written in scientific notation as 0.15×10^{-8}, and when coded as a floating-point number, would appear as follows:

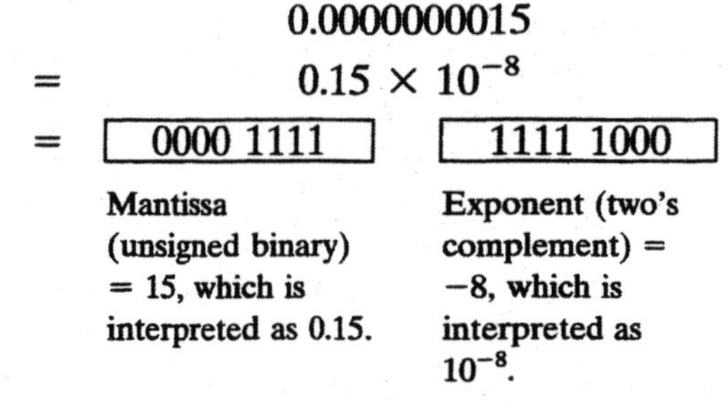

In this example, the mantissa byte (standard binary number) is equal to 15 (0000 1111) and is therefore interpreted as 0.15, and the exponent byte (two's complement number) is equal to −8 (1111 1000) and is therefore interpreted as 10^{-8}.

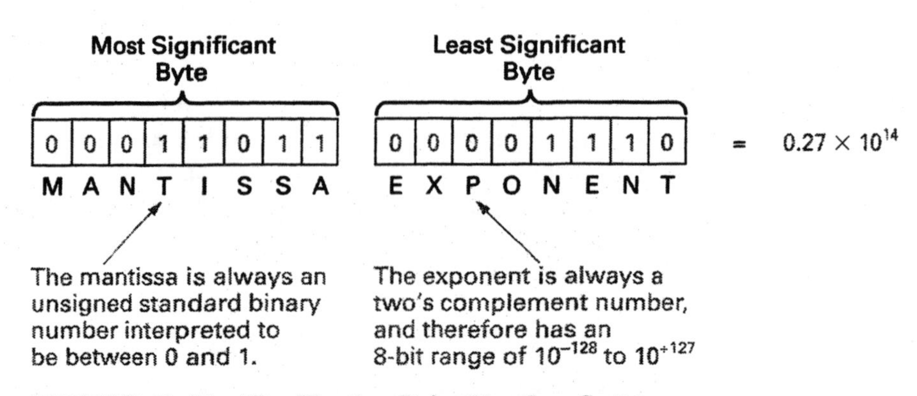

FIGURE 13-15 The Floating-Point Number System.

As with all binary number coding methods, we must know the number representation system being used if the bit pattern is to be accurately deciphered. For instance, the 16-bit word in Figure 13-15 would be very different if it were first interpreted as an unsigned binary number and then compared to a two's complement number and to a floating-point number.

SELF-TEST REVIEW QUESTIONS FOR SECTION 13-1

1. Perform the following binary arithmetic operations.
   a. 1011 + 11101 = ? **b.** 1010 − 1011 = ?
   c. 101 × 10 = ? **d.** 10111 ÷ 10 = ?
2. Two's complement the following binary numbers.
   a. 1011 **b.** 0110 1011 **c.** 0011 **d.** 1000 0111
3. What is the decimal equivalent of the following two's complement numbers?
   a. 1011 **b.** 0110 **c.** 1111 1011 **d.** 0000 0110
4. The two's complement number system makes it possible for us to perform addition, subtraction, multiplication, and division all with a/an __________ circuit.

13-2 ARITHMETIC CIRCUITS

In the previous section, we discovered that there is no need to have a separate circuit to add, another circuit to subtract, another circuit to multiply, and yet another circuit to divide. If we use two's complement numbers, these four mathematical operations can be performed with a simple binary-adder circuit. In this section, we will examine the different types of binary-adder circuits and a typical binary-adder IC.

13-2-1 ***Half-Adder Circuit***

Figure 13-16(a) reviews the four rules of binary addition. Studying the sum column of the truth table, you may recognize that the output generated for each input combination is exactly the same as that generated by an XOR gate (odd number of 1s at the input gives a 1 at the output). Referring now to the carry column in the truth table, you may recognize that in this case the output generated for each input combination is exactly the same as that generated by an AND gate (any 0 in gives a 0 out). Figure 13-16(b) shows how these two gates could be connected to sense and add an *A* and *B* input, and generate both a sum and carry output that follows the truth table in Figure 13-16(a).

The circuit in Figure 13-16(b) is called a **half-adder circuit,** and if we ever need to add only two input bits, this circuit would be ideal. In most binary addition operations, however, we will need to add three input bits at a time. To explain this, let us add 11_2 (3_{10}) and 01_2 (1_{10}).

Half-Adder Circuit
A digital circuit that can sum a 2-bit binary input and generate a SUM and CARRY output.

A	B	Sum	Carry	
0	0	0	0	0 + 0 = 0 with no carry.
0	1	1	0	0 + 1 = 1 with no carry.
1	0	1	0	1 + 0 = 1 with no carry.
1	1	0	1	1 + 1 = 2_{10} or 10_2, which is 0 carry 1.

(a)

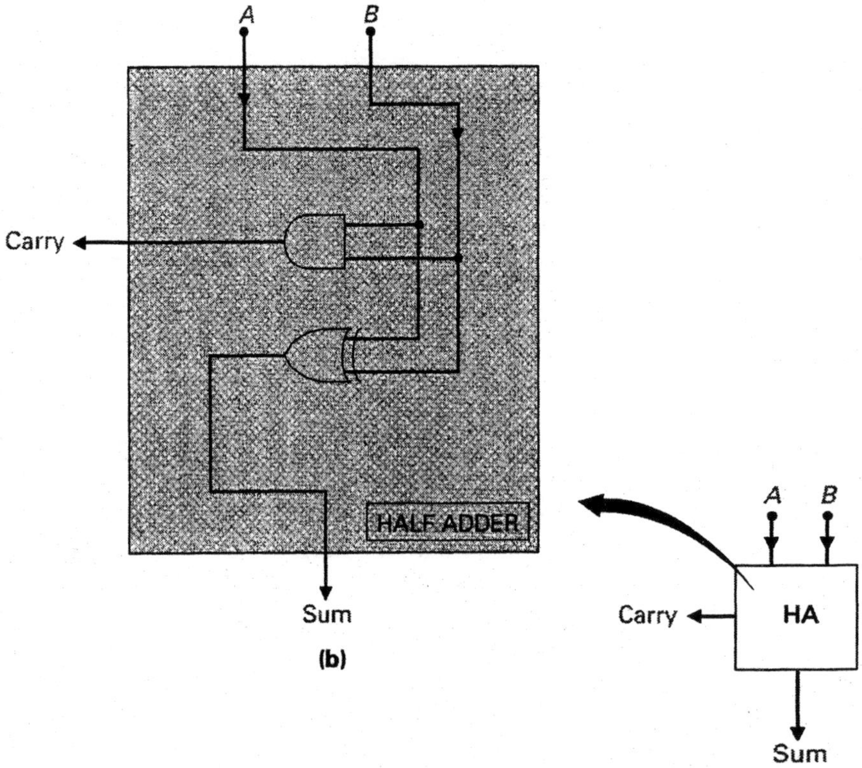

A B

Carry

HALF ADDER

Sum

(b)

A B

Carry HA

Sum

FIGURE 13-16 The Half Adder (HA). (a) Truth Table. (b) Logic Circuit.

$$
\begin{array}{lccc}
\text{Carry:} & 1 & 1 & \\
 & & 1 & 1_2 \\
+ & & 0 & 1_2 \\
\hline
\text{Sum:} & 1 & 0 & 0_2
\end{array}
\qquad
\begin{array}{r}
3_{10} \\
+\ 1_{10} \\
\hline
4_{10}
\end{array}
$$

In the first column (units column) we only need to add two bits, 1 + 1, and therefore a half-adder circuit could be used to perform this addition. In the second column, however, we need to add three bits, 1 + 0 plus the 1 carry from the first column. A different logic circuit that can add a 3-bit input (*A*, *B* and carry) and then generate a sum and carry output is needed.

13-2-2 *Full-Adder Circuit*

A **full-adder circuit** is able to add a 3-bit input and generate a sum and carry output. Figure 13-17(a) summarizes the addition operation that must be performed by a full-adder circuit. The full-adder circuit shown in Figure 13-17(b) has the three inputs *A*, *B*, and carry input (CI), and the two outputs sum and carry output (CO). The full-adder circuit contains two half adders, which in combination will generate the correct sum and carry output logic levels for any of the input combinations listed in Figure 13-17(a).

Full-Adder Circuit
A digital circuit that can sum a 3-bit binary input and generate a SUM and CARRY output.

If we ever need to add only three input bits, the full-adder circuit would be ideal. In most binary addition operations, however, we will need to add several 3-bit columns. To explain this, let us add 0111_2 (7_{10}) and 0110_2 (6_{10}).

	(8)	(4)	(2)	(1)		
Carry:	1	1	0			
	0	1	1	1_2		7_{10}
	+ 0	1	1	0_2		+ 6_{10}
Sum:	1	1	0	1_2		13_{10}

In the first column (units) we only need to add two bits; therefore, a half-adder circuit could be used to find the sum of this 2-bit column. In the remaining columns, however, we will always have a carry input from the previous column, and therefore a parallel-adder circuit will be needed to sum these 3-bit columns.

13-2-3 *Parallel-Adder Circuit*

Figure 13-18(a) shows a 4-bit **parallel-adder circuit.** The half adder (HA) on the right side will add A_1 and B_1, and produce at its output S_1 (sum 1 output) and C_1 (carry 1 output). All of the remaining adders will have to be full adders (FA) since these columns will always need to add three input bits: the *A* input, *B* input, and the carry input from the previous column. This 4-bit parallel-adder circuit will therefore perform the following addition.

Parallel-Adder Circuit
A digital circuit that can simultaneously add the corresponding parts of two binary input words.

Carry:	C_4	C_3	C_2	C_1	
Addend:		A_4	A_3	A_2	A_1
Augend: +		B_4	B_3	B_2	B_1
Sum:	C_4	S_4	S_3	S_2	S_1

Figure 13-18(b) shows how this 4-bit adder circuit will add an A_4, A_3, A_2, and A_1 input of 0111_2 (7_{10}) and a B_4, B_3, B_2, and B_1 input of 0110_2 (6_{10}). The result will be as follows:

Carry:	0	1	1	0	
Addend:		0	1	1	1
Augend: +		0	1	1	0
Sum:	0	1	1	0	1

By adding more full adders to the left side of a parallel-adder circuit, you can construct a parallel-adder circuit of any size. For example, an 8-bit parallel-

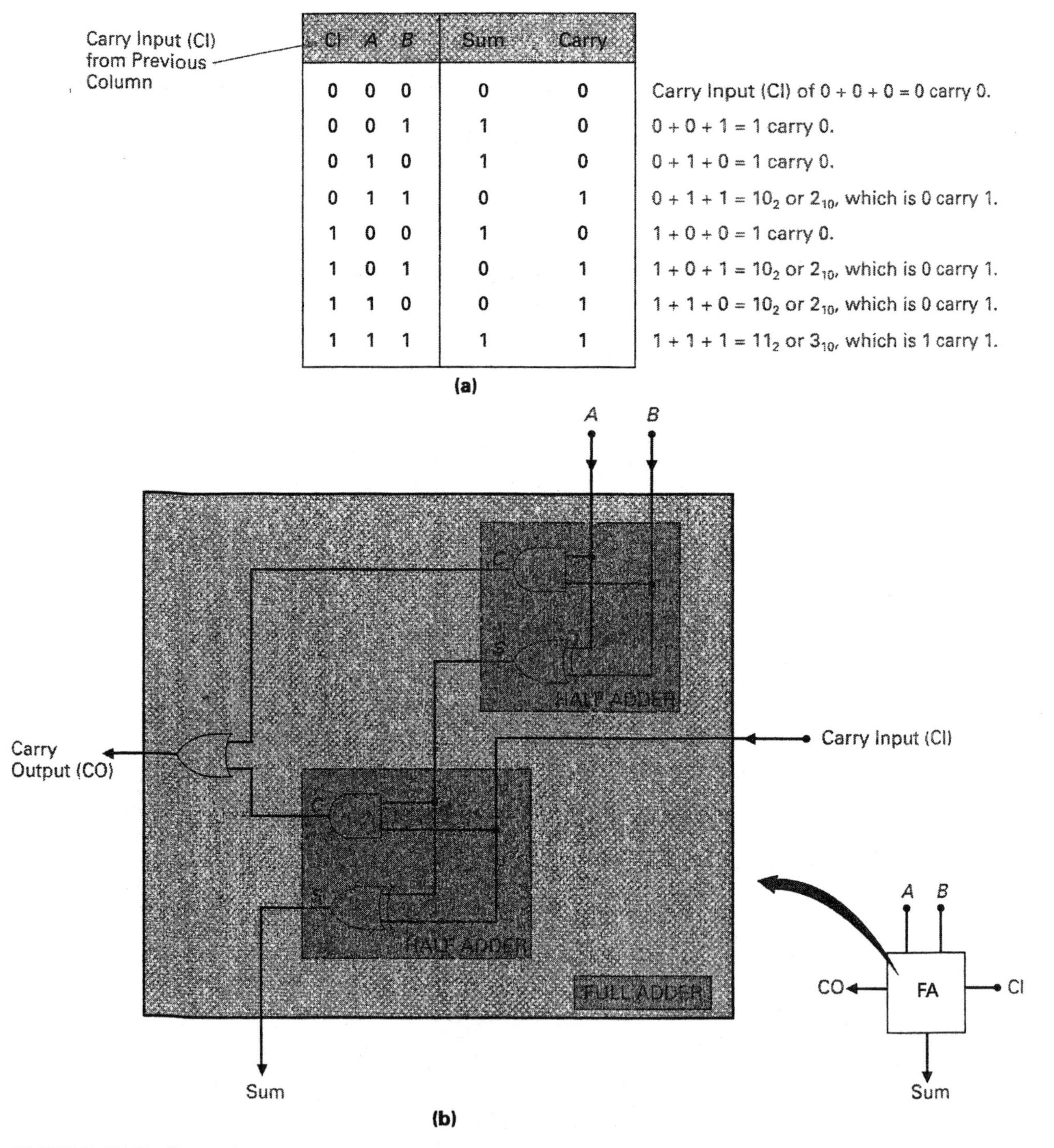

CI	A	B	Sum	Carry	
0	0	0	0	0	Carry Input (CI) of 0 + 0 + 0 = 0 carry 0.
0	0	1	1	0	0 + 0 + 1 = 1 carry 0.
0	1	0	1	0	0 + 1 + 0 = 1 carry 0.
0	1	1	0	1	0 + 1 + 1 = 10_2 or 2_{10}, which is 0 carry 1.
1	0	0	1	0	1 + 0 + 0 = 1 carry 0.
1	0	1	0	1	1 + 0 + 1 = 10_2 or 2_{10}, which is 0 carry 1.
1	1	0	0	1	1 + 1 + 0 = 10_2 or 2_{10}, which is 0 carry 1.
1	1	1	1	1	1 + 1 + 1 = 11_2 or 3_{10}, which is 1 carry 1.

FIGURE 13-17 The Full Adder (FA). (a) Truth Table. (b) Logic Circuit.

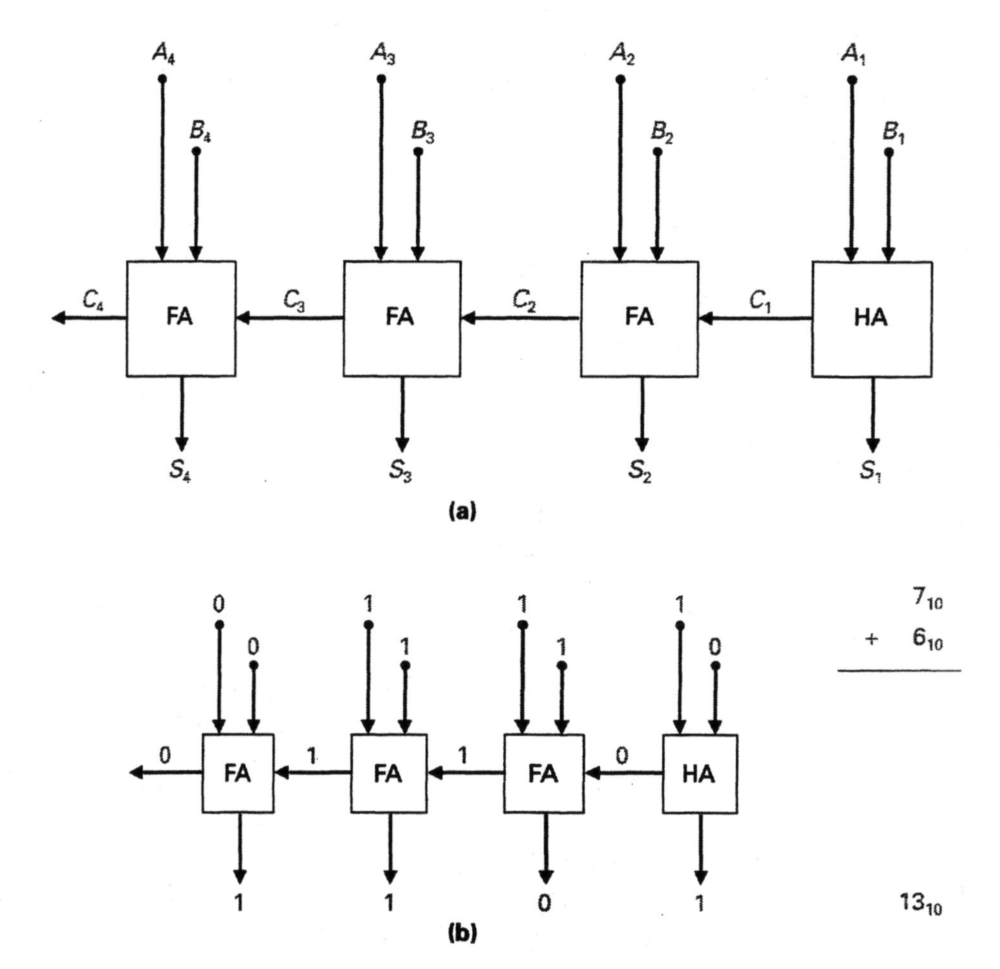

FIGURE 13-18 The Parallel Adder.

adder circuit would contain one half adder (in the units position) and seven full adders, while a 16-bit parallel-adder circuit would need one half adder and fifteen full adders.

Figure 13-19 shows the data sheet details for the 7483A, 74LS83, 74LS283, and 74HC283, which are examples of commonly used 4-bit parallel-adder ICs. The only difference between the 283s and the 83s is that the V_{CC} and ground pins are at the corners of the chip for 283s. The inputs to the 7483 IC are A_4, A_3, A_2, and A_1, which is collectively called the P input; B_4, B_3, B_2, and B_1, which is collectively called the Q input; and an LSB C_0 (pin 13), which is the carry input. The outputs are an MSB carry (C_4) and the sum bits, which are symbolized with the Greek capital letter *sigma* (Σ) and are labeled Σ_4, Σ_3, Σ_2, and Σ_1. Referring to the logic symbol for this 7483, you can see that the sigma symbol is also used to denote an adder IC.

The 7483 IC is called a *fast carry* or *look-ahead carry* IC because of the time delay that occurs as the carry ripples from the LSB column to the second column, from the second to the third, and from the third to the fourth. For example, consider the following addition.

IC TYPE: SN7483A– 4-Bit Binary Full Adders with Fast Carry

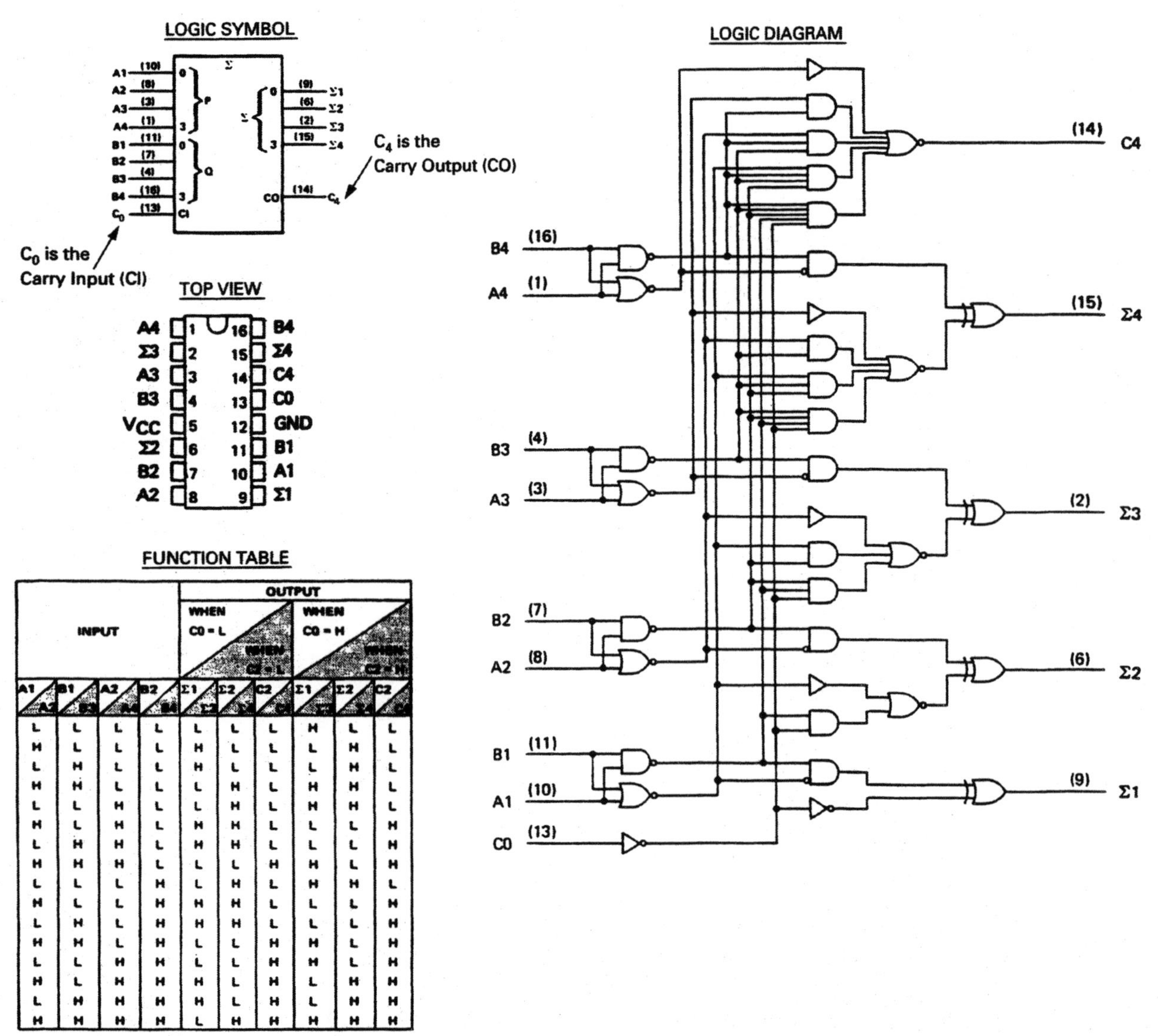

INPUT				OUTPUT WHEN C0 = L / WHEN C2 = L			OUTPUT WHEN C0 = H / WHEN C2 = H		
A1 / A3	B1 / B3	A2 / A4	B2 / B4	Σ1 / Σ3	Σ2 / Σ4	C2 / C4	Σ1 / Σ3	Σ2 / Σ4	C2 / C4
L	L	L	L	L	L	L	H	L	L
H	L	L	L	H	L	L	L	H	L
L	H	L	L	H	L	L	L	H	L
H	H	L	L	L	H	L	H	H	L
L	L	H	L	L	H	L	H	H	L
H	L	H	L	H	H	L	L	L	H
L	H	H	L	H	H	L	L	L	H
H	H	H	L	L	L	H	H	L	H
L	L	L	H	L	H	L	H	H	L
H	L	L	H	H	H	L	L	L	H
L	H	L	H	H	H	L	L	L	H
H	H	L	H	L	L	H	H	L	H
L	L	H	H	L	L	H	H	L	H
H	L	H	H	H	L	H	L	H	H
L	H	H	H	H	L	H	L	H	H
H	H	H	H	L	H	H	H	H	H

H = high level, L = low level

NOTE: Input conditions at A1, B1, A2, B2, and C0 are used to determine outputs Σ1 and Σ2 and the value of the internal carry C2. The values at C2, A3, B3, A4, and B4 are then used to determine outputs Σ3, Σ4, and C4.

FIGURE 13-19 An IC Containing a 4-Bit Binary Adder. (Courtesy of Texas Instruments)

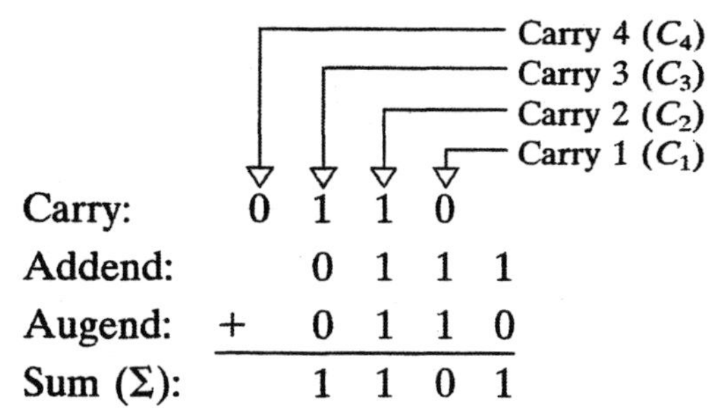

If each full adder within the 7483 had a propagation delay of 30 ns, it would take 30 ns for the first FA to generate C_1, another 30 ns for the second FA to generate C_2, another 30 ns for the third FA to generate C_3, and another 30 ns for the fourth FA to generate C_4. This means that the sum will not be available for 120 ns (4×30 ns) after the 4-bit A input and B input have been applied. This *carry propagation delay,* or *carry ripple delay,* becomes a real problem with 16-bit and 32-bit adder circuits. To overcome this problem, IC logic designers include several logic gates that look at the lower-order bits of the two inputs and generate higher-order carry bits. Most commercially available adder ICs feature full internal look-ahead across all 4-bits, generating the carry in typically 10 ns.

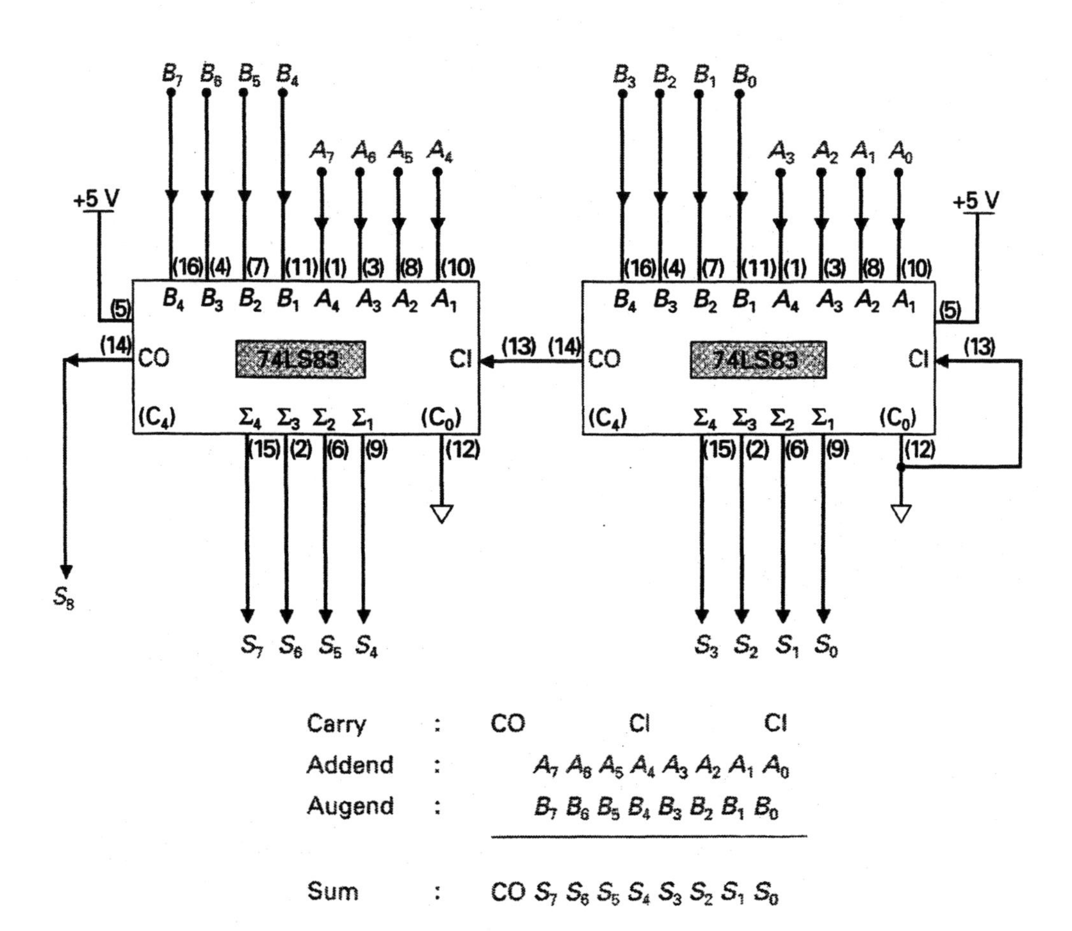

FIGURE 13-20 An 8-Bit Parallel-Adder Circuit.

Application—An 8-Bit Parallel Circuit

Figure 13-20 shows how two 74LS83 4-bit adders could be cascaded to form an 8-bit parallel-adder circuit. The carry input of the first 74LS83 is connected LOW, while the carry output of the first 74LS83 is connected to the carry input of the second 74LS83. This circuit will add the two 8-bit A and B inputs and produce an 8-bit sum output. In the following section we will be using this circuit to form a complete two's complement adder/subtractor circuit.

SELF-TEST REVIEW QUESTIONS FOR SECTION 13-2

1. A half adder can be used to add __________ bits, while a full adder can add __________ bits.
2. Which of the following adder circuits could be used to add two binary words?
   a. Half adder **b.** Full adder **c.** Parallel adder **c.** Diamond head adder
3. How would you describe the 7483 IC?
4. The sum outputs of an adder circuit are normally symbolized with the Greek capital letter __________.

13-3 ARITHMETIC CIRCUIT APPLICATIONS

In this section we will be examining some of the applications for arithmetic circuits.

13-3-1 Basic Two's Complement Adder/Subtractor Circuit

Figure 13-21 shows how two 7483 ICs could be connected to form an 8-bit two's complement adder/subtractor. In this circuit, there is also an 8-bit input switch circuit, an 8-bit output LED display circuit, and two 8-bit registers (A and B). This circuit operates as follows.

A set of eight input switches and two 74126 three-state buffers (controlled by the active-HIGH input enable, E_I) are used to connect any 8-bit word (set by the switches) onto the 8-bit data bus (D_0 through D_7). The 8-bit input data can be loaded into the A register (by pulsing LOW, $\overline{L_A}$), the B register (by pulsing LOW, $\overline{L_B}$), or the output register (by pulsing LOW, $\overline{L_O}$).

Referring to the two 74LS83s (which form an 8-bit adder circuit) you can see that the A register supplies the 8-bit A input (A_7 through A_0), while the B register supplies the 8-bit B input (B_7 through B_0). The 8-bit sum output from the two 74LS83s (Σ_7 through Σ_0) can be switched onto the 8-bit data bus by pulsing the sum enable control line (E_Σ) HIGH.

The two 7486s connected to the B input of the adder/subtractor contain eight XOR gates which are included to function as controlled inverters under

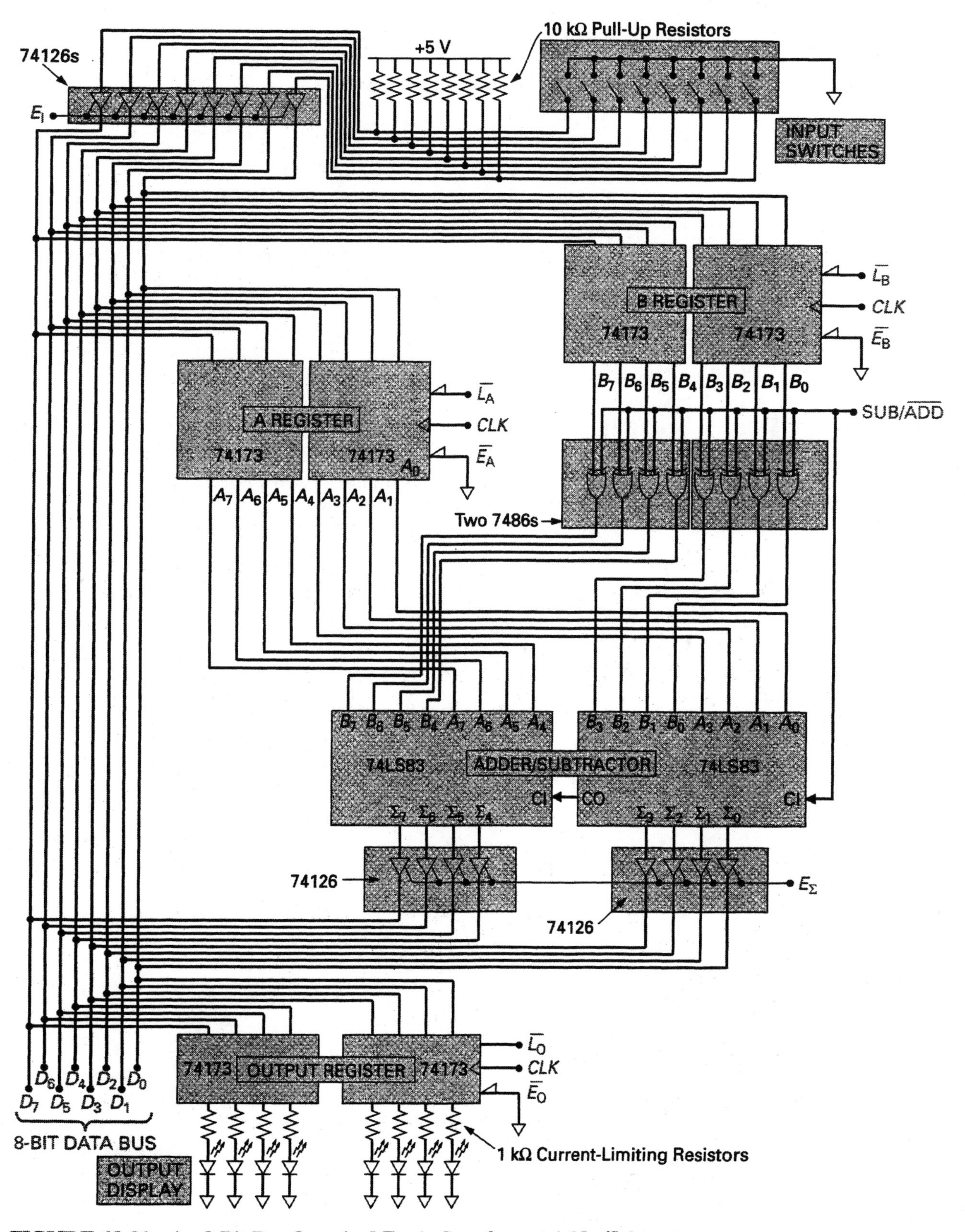

FIGURE 13-21 An 8-Bit Bus-Organized Two's Complement Adder/Subtractor Circuit.

the control of the $SUB/\overline{ADD}$ input. When the $SUB/\overline{ADD}$ control line is LOW, inputs B_7 through B_0 will pass through the XOR gates without inversion, and therefore the sum output of the 74LS83s will be

$$\Sigma_7 \text{ through } \Sigma_0 = A_7 \text{ through } A_0 + B_7 \text{ through } B_0$$

The LSB carry input (CI) is also connected to the $SUB/\overline{ADD}$ control line, and since this input is LOW in this instance, it will have no effect on the adder circuit's result.

On the other hand, when the $SUB/\overline{ADD}$ control line is HIGH, the XOR gates will act as inverters and invert the output from the B register. The B input to the 74LS83Ss will therefore be the one's complement of the original word stored in the B register. Since the $SUB/\overline{ADD}$ control line is HIGH in this instance, the LSB carry input will also be HIGH, and therefore a 1 will be added to the LSB of the adder. As a result, when the $SUB/\overline{ADD}$ control line is HIGH, the B input word is first one's complemented, and then a 1 is added to the LSB, so the overall effect is that the B input is two's complemented. Since the B input word is two's complemented and added to the A input word, the result will be a two's complement subtraction through addition operation or

$$\Sigma_7 \text{ through } \Sigma_0 = A_7 \text{ through } A_0 - B_7 \text{ through } B_0$$

By using the input switches, the operator is able to load an 8-bit two's complement word into the A register (E_I = HIGH, $\overline{L_A}$ = LOW), and then load a different 8-bit two's complement word into the B register (E_I = HIGH, $\overline{L_B}$ = LOW). Then, by controlling the logic level on the $SUB/\overline{ADD}$ control line (HIGH = subtract, LOW = add) you can control whether the sum or the difference appears at the output of the 74LS83s. To display the output of the 74LS83s on the 8-bit LED display, simply enable the 74126 three-state output buffers (E_Σ = HIGH) and then load the result from the data bus into the output register ($\overline{L_O}$ = LOW).

13-3-2 An Arithmetic-Logic Unit (ALU) IC

The 74181 IC is a good example of an arithmetic-logic unit (ALU) that can perform sixteen different arithmetic operations and sixteen different logic operations on two 4-bit input words.

Figure 13-22 shows some of the data sheet details for this IC. Referring to the Function Select Table, you can see that the mode control input M, pin 8, determines whether the ALU performs a logic operation (M = HIGH) or arithmetic operation (M = LOW). The four select inputs (S_3, S_2, S_1, and S_0) are used to select which one of the sixteen different logic functions or arithmetic operations are performed on the A input (A_3, A_2, A_1, or A_0) and B input (B_3, B_2, B_1, or B_0) words. For example, if M = HIGH (logic function) and S_3, S_2, S_1, and S_0 = HHHL ($F = A + B$, F = A OR B), then the output word F (F_3, 210) will be the result of ORing input word A with input word B. As another example, if M = LOW (arithmetic operation), S_3, S_2, S_1, and S_0 = HLLH (F =

IC TYPE: SN74S181– 4-Bit Arithmetic-Logic Unit (ALU)

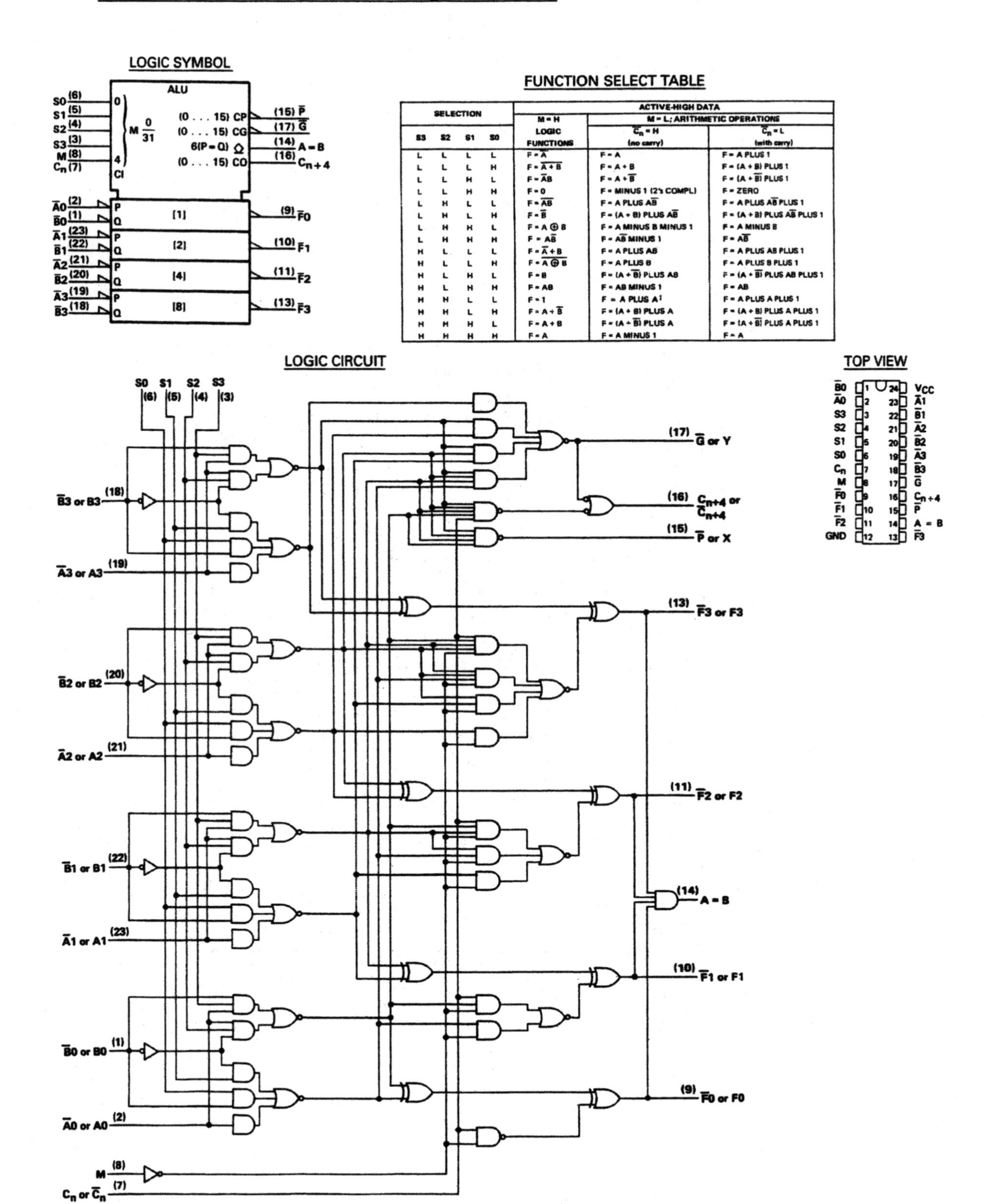

FUNCTION SELECT TABLE

SELECTION				ACTIVE-HIGH DATA		
				M = H	M = L; ARITHMETIC OPERATIONS	
S3	S2	S1	S0	LOGIC FUNCTIONS	$\overline{C}_n$ = H (no carry)	$\overline{C}_n$ = L (with carry)
L	L	L	L	$F = \overline{A}$	F = A	F = A PLUS 1
L	L	L	H	$F = \overline{A + B}$	F = A + B	F = (A + B) PLUS 1
L	L	H	L	$F = \overline{A}B$	$F = A + \overline{B}$	$F = (A + \overline{B})$ PLUS 1
L	L	H	H	F = 0	F = MINUS 1 (2's COMPL)	F = ZERO
L	H	L	L	$F = \overline{AB}$	$F = A$ PLUS $A\overline{B}$	$F = A$ PLUS $A\overline{B}$ PLUS 1
L	H	L	H	$F = \overline{B}$	$F = (A + B)$ PLUS $A\overline{B}$	$F = (A + B)$ PLUS $A\overline{B}$ PLUS 1
L	H	H	L	$F = A \oplus B$	F = A MINUS B MINUS 1	F = A MINUS B
L	H	H	H	$F = A\overline{B}$	$F = A\overline{B}$ MINUS 1	$F = A\overline{B}$
H	L	L	L	$F = \overline{A} + B$	F = A PLUS AB	F = A PLUS AB PLUS 1
H	L	L	H	$F = \overline{A \oplus B}$	F = A PLUS B	F = A PLUS B PLUS 1
H	L	H	L	F = B	$F = (A + \overline{B})$ PLUS AB	$F = (A + \overline{B})$ PLUS AB PLUS 1
H	L	H	H	F = AB	F = AB MINUS 1	F = AB
H	H	L	L	F = 1	F = A PLUS A[†]	F = A PLUS A PLUS 1
H	H	L	H	$F = A + \overline{B}$	F = (A + B) PLUS A	F = (A + B) PLUS A PLUS 1
H	H	H	L	F = A + B	$F = (A + \overline{B})$ PLUS A	$F = (A + \overline{B})$ PLUS A PLUS 1
H	H	H	H	F = A	F = A MINUS 1	F = A

FIGURE 13-22 An IC Containing an Arithmetic-Logic Unit (ALU). (Courtesy of Texas Instruments)

A plus B) when the active-LOW carry input is HIGH ($\overline{C_n}$ = HIGH, no carry), then the output word F will be the sum of input word A plus input word B.

Application—Single Chip Arithmetic-Logic Unit (ALU Circuit)

Figure 13-23 shows how a 74181 4-bit ALU IC could be connected so that its sixteen different logic functions and sixteen different arithmetic operations could be tested.

An ALU can be found within the microprocessor unit of every digital electronic computer system. In Figure 13-24 you can see the three basic blocks of a computer system which are the MPU, memory, and input/output units. It is the MPU that controls the operation of a computer system by fetching binary coded instructions from memory, and then decoding and executing these instructions. Referring to the inset in Figure 13-24, you can see the ALU within the MPU. It is, in fact, the calculations and logic comparisons performed by the ALU that enable the MPU to make decisions and therefore process data or information. In the final chapter of this text, we will examine the operation of the microprocessor unit and discuss in more detail how it interfaces with the rest of the computer system.

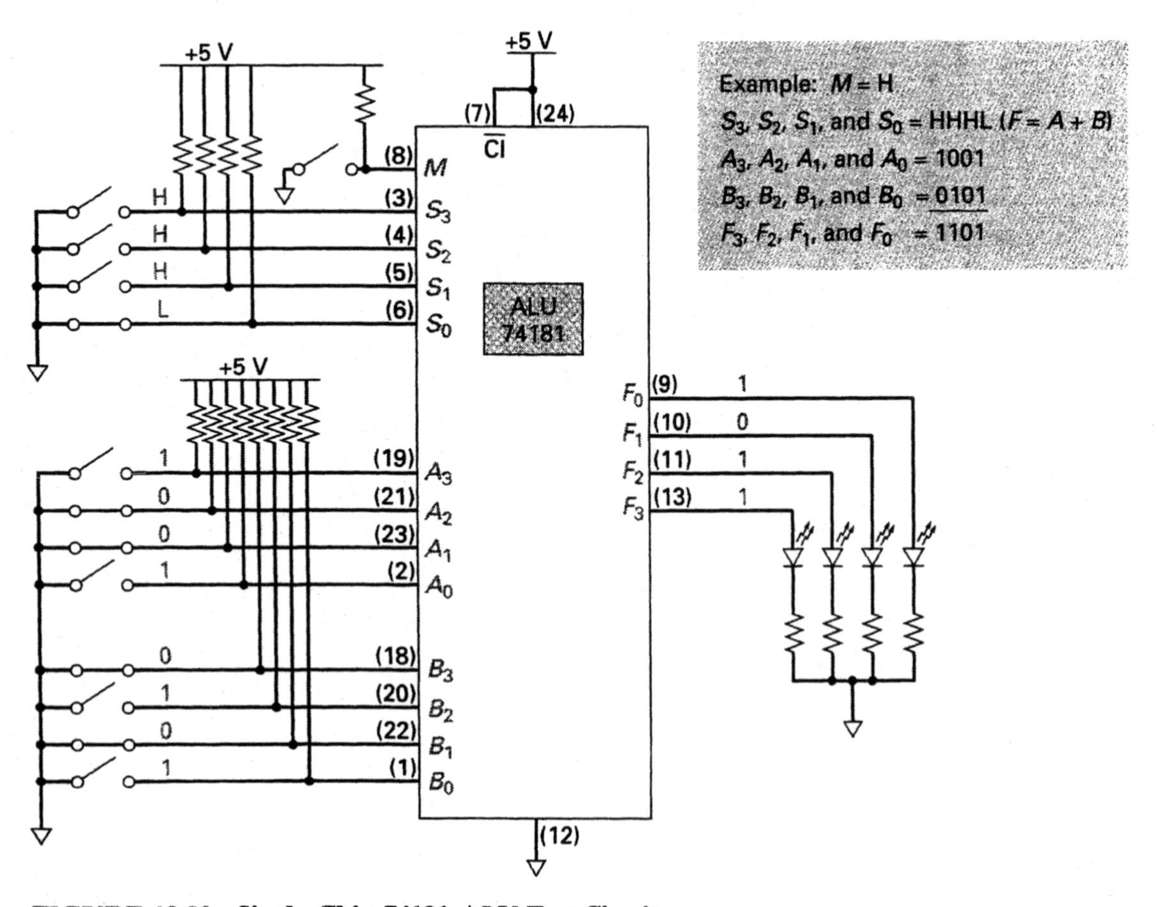

FIGURE 13-23 Single Chip 74181 ALU Test Circuit.

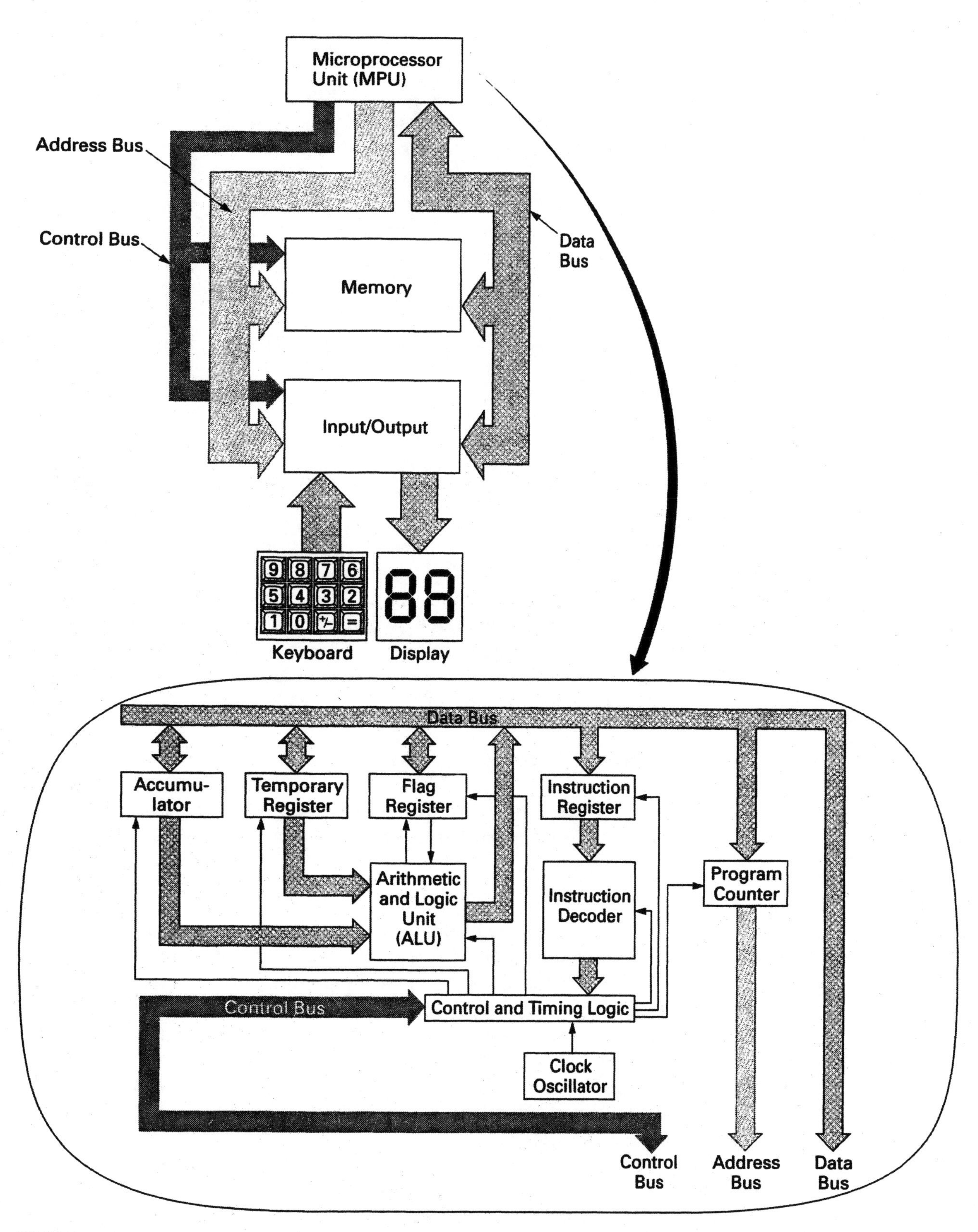

FIGURE 13-24 The Basic Computer Block Diagram.

SELF-TEST REVIEW QUESTIONS FOR SECTION 13-3

1. With the two's complement adder/subtractor circuit in Figure 13-21, which ICs are responsible for one's complementing during a subtraction operation?
2. With the two's complement adder/subtractor circuit in Figure 13-21, how is the additional 1 added so that we can two's complement the B input word?
3. What operation would a 74181 IC perform if its control inputs were
   a. M = H, S = LHLH? **b.** M = L, S = LHHL, $\overline{C_n}$ = L?
4. What would appear at the output of a 74181 if A = 0011, B = 0110, M = L, S = HLLH?

13-4 TROUBLESHOOTING ARITHMETIC CIRCUITS

To review, the procedure for fixing a failure can be broken down into three steps—diagnose, isolate, and repair. Since this chapter has been devoted to arithmetic operations and circuits, let us first examine the operation of a typical arithmetic circuit, and then apply our three-step troubleshooting process to this circuit.

13-4-1 *An Arithmetic Circuit*

As an example, we will apply our three-step troubleshooting procedure to the two's complement adder/subtractor circuit shown in Figure 13-21.

Step 1: Diagnose

As before, it is extremely important that you first understand the operation of a circuit, and how all of the devices within it are supposed to work, so that you are able to determine whether or not a circuit malfunction really exists. If you were preparing to troubleshoot the circuit in Figure 13-21, your first step should be to read through the circuit description and review the operation of each integrated circuit until you feel completely confident with the correct operation of the entire circuit. Once you are fully familiar with the operation of the circuit, you will easily be able to diagnose the problem as either an *Operator Error* or a *Circuit Malfunction.*

Distinguishing an operator error from an actual circuit malfunction is an important first step, and a wrong diagnosis can waste a lot of time and effort. For example, the following could be interpreted as circuit malfunctions, when in fact they are simply operator errors.

Example 1.
Symptom: Register A will not load the word 01011110. Control lines are as follows:

E_I = HIGH

$\overline{L_A}$ = HIGH

Input Switches = 01011110

Diagnosis: Operator Error—The enable control lines are active-HIGH while the load control lines are active-LOW. Therefore, to load register A with the input word set on the switches, E_I must be HIGH and $\overline{L_A}$ must be LOW.

Example 2.
Symptom: When SUB/$\overline{\text{ADD}}$ = LOW, A = 0111 1111 (+127), and B = 0000 0010 (+2). The result (Σ) = 1000 0001 (−127) when it should be (+127) + (+2) = +129.
Diagnosis: Operator Error—The 8-bit two's complement range of −128 to +127 has been exceeded.

Example 3.
Symptom: Circuit seems to be malfunctioning since it will not display the difference between register A and register B. It always gives a result of 0000 0000.

E_I = HIGH

$\overline{L_A}$ and $\overline{L_B}$ = LOW

Input Switches = 01011100

SUB/$\overline{\text{ADD}}$ = HIGH

Σ = 00000000

Diagnosis: Operator Error—If both $\overline{L_A}$ and $\overline{L_B}$ are made LOW simultaneously, the same value from the switches will be loaded into both registers via the data bus. The difference between the same value will therefore always be zero.

Once you have determined that the problem is not an operator error, but is in fact a circuit malfunction, proceed to Step 2 and isolate the circuit failure.

Step 2: Isolate

No matter what circuit or system failure has occurred, you should always follow a logical and sequential troubleshooting procedure. Let us review some of the isolating techniques, and then apply them to our example circuit in Figure 13-21.

a. Use a cause-and-effect troubleshooting process, which means study the effects you are getting from the faulty circuit and then logically reason out what could be the cause.

b. Check first for obvious errors before leaping into a detailed testing procedure. Is the power OFF or not connected to the circuit? Are there wiring errors? Are all of the ICs correctly oriented?

c. Using a logic probe or voltmeter, test that power and ground are connected to the circuit and are present at all points requiring power and ground. If the whole circuit, or a large section of the circuit, is not operating, the problem is normally power. Using a multimeter, check that all of the dc voltages for the circuit are present at all IC pins that should have power or a HIGH input, and are within tolerance. Secondly, check that 0 V, or ground, is connected to all IC ground pins and all inputs that should be tied LOW.

d. Use your senses to check for broken wires, loose connectors, overheating or smoking components, pins not making contact, and so on.

e. Test the clock signals using the logic probe or the oscilloscope. Although clock signals can be easily checked with a logic probe, the oscilloscope is best for checking the signal's amplitude, frequency, pulse width, and so on. The oscilloscope will also display timing problems and noise spikes (glitches) which could be false-triggering a circuit into operation at the wrong time.

f. Perform a static test on the circuit. With our example circuit in Figure 13-21, you could use the logic pulser to supply single-step clock pulses, and then test each data bus transfer. For example:

Input switches to register A (E_I = HIGH, $\overline{L_A}$ = LOW, CLK = ↑)

Input switches to register B (E_I = HIGH, $\overline{L_B}$ = LOW, CLK = ↑)

Adder/subtractor to output register (E_Σ = HIGH, $\overline{L_O}$ = LOW, CLK = ↑)

During each of these transfers, you can use the logic probe to test the logic levels at all points between source and destination.

By holding the circuit in a static condition, you can also use the logic probe to test the add and subtract action of the 74LS83s.

g. With a dynamic test, the circuit is tested while it is operating at its normal clock frequency, and therefore all of the inputs and outputs are continually changing. Although a logic probe can detect a pulse waveform, the oscilloscope will display more signal detail, and since it can display more than one signal at a time, it is ideal for making signal comparisons and looking for timing errors. If the clock section is suspected, you can check the rest of the circuit by disconnecting the clock input to the circuit, and then using a function generator to generate the clock signal.

In some instances, you will discover that some circuits will operate normally when undergoing a static test, and yet fail a dynamic test. This effect usually points to a timing problem involving the clock signal, control signals, and/or propagation delay times of the ICs used within the circuit.

h. Noise due to electromagnetic interference (EMI) can false-trigger an IC. This problem can be overcome by not leaving any of the IC's inputs unconnected and, therefore, floating. Connect unneeded active-HIGH inputs to ground and active-LOW inputs to $+V_{CC}$.

i. Apply the half-split method of troubleshooting first to a circuit, and then to a section of a circuit, to help speed up the isolation process. Also remember that a load problem can make a source appear at fault. If an output is incorrect, it would be safer to disconnect the source from the rest of the circuit and then recheck the output. If the output is still incorrect, the problem is definitely within the source; however, if the output is correct, the load is probably shorting the line either HIGH or LOW.

j. Substitution can be used to help speed up your troubleshooting process. Once the problem is localized to an area containing only a few ICs, substitute suspect ICs with known working ICs (one at a time) to see if the problem can be quickly remedied.

Step 3: Repair

Once the fault has been found, the final step is to repair the circuit and then perform a final test to see that the circuit and the system are now fully operational.

13-4-2 Sample Problems

Once you have constructed a circuit like the circuit in Figure 13-21, introduce a few errors to see what effects or symptoms they produce. Then logically reason out why a particular error or cause has a particular effect on the circuit. Never short any two points together unless you have carefully thought out the consequences, but generally it is safe to open a path and see the results. Here are some examples for our example circuit in Figure 13-21.

a. Open a data bus line at a point in the circuit. Pick several other points in the circuit to see the different symptoms.

b. Disconnect the SUB/$\overline{\text{ADD}}$ input to the 74LS83.

c. Disconnect the $\overline{E_O}$ control line from ground.

d. Disconnect one of the E_I control lines to one of the 74126 input ICs.

e. Disconnect the *CLK* input to the A register.

f. Open the SUB/$\overline{\text{ADD}}$ control line connection to the two 7486s.

g. Open the CO to CI connection between the 74LS83s.

h. Connect $\overline{E_A}$ HIGH instead of LOW.

SELF-TEST REVIEW QUESTIONS FOR SECTION 13-4

1. What are the three basic troubleshooting steps?
2. How would you use a logic probe to test the input circuit in Figure 13-21?
3. How would you static test the adder/subtractor section of the circuit in Figure 13-21?
4. How could you quickly test the input and output circuit in Figure 13-21?

REVIEW QUESTIONS

Multiple-Choice Questions

1. The sum of 101_2 and 011_2 is

   a. 0111_2. **b.** 8_{10}. **c.** 1001_2. **d.** 6_{10}.

2. Which digital logic unit is used to perform arithmetic operations and logic functions?

   a. MPU **b.** ALU **c.** CPU **d.** ICU

3. With binary addition, $1 + 1 = ?$

   a. 10_2 **b.** 0 carry 1 **c.** 2_{10} **d.** all of the above

4. With binary subtraction, $0 - 1 = ?$

   a. 10_2 **b.** 1 borrow 0
   c. 1 borrow 1 **d.** all of the above

5. What would be the result of $1011101_2 \times 01_2$?

   a. 1011101_2 **b.** 1110101_2 **c.** 000111_2 **d.** 1011100_2

6. The four rules of binary multiplication are $0 \times 0 =$ ________, $1 \times 0 =$ ________, $0 \times 1 =$ ________, and $1 \times 1 =$ ________.

   a. 0, 1, 0, 1. **b.** 0, 0, 1, 1. **c.** 0, 0, 0, 0. **d.** 0, 0, 0, 1.

7. What would be the result of $1001_2 \div 0011_2$?

   a. 3_{10} **b.** 3, remainder zero **c.** 0011_2 **d.** All of the above

8. The one's complement of $0101\ 0001_2$ would be

   a. $0101\ 0010_2$. **b.** $1010\ 1110_2$.
   c. $1010\ 1111_2$. **d.** all of the above.

9. What would be the decimal equivalent of 1001_2 if it were a two's complement number?

   a. +9 **b.** −9 **c.** +7 **d.** −7

10. Which number system makes it possible for us to add, subtract, multiply, and divide binary numbers all with an adder circuit?

    a. Sign-magnitude **b.** One's complement
    c. Two's complement **d.** Floating point

11. Which binary code system enables us to represent extremely large numbers or small fractions with fewer bits?

    a. Sign-magnitude **b.** One's complement
    c. Two's complement **d.** Floating point

12. With a half-adder circuit, the sum output of the 2-bit input is generated by a/an ________ gate, while the carry output of the 2-bit input is generated by a/an ________ gate.

    a. XOR, AND **b.** XNOR, NAND
    c. NOR, NAND **d.** AND, XOR

13. Which circuit is used to add a 2-bit input and a carry from the previous column?

    a. Half-adder
    b. Full-adder
    c. Parallel-adder
    c. All of the above

14. The 7483 IC is a ____________-bit adder circuit.

    a. 2 b. 6 c. 4 d. 3

15. The 74181 IC could be described as a/an ____________ .

    a. 4-bit FA b. 8-bit FA c. 4-bit PA c. 8-bit PA

Essay Questions

16. List the rules of each of the following: (13-1-1)

    a. Binary addition
    b. Binary subtraction
    c. Binary multiplication
    d. Binary division

17. Using an example, briefly describe how to perform the following operations. (13-1-1)

    a. Binary addition
    b. Binary subtraction
    c. Binary multiplication
    d. Binary division

18. Briefly describe how the following number systems are used to represent positive and negative binary numbers. (13-1-2)

    a. Sign-magnitude
    b. One's complement
    c. Two's complement

19. Why is the two's complement number system most frequently used to represent positive and negative numbers? (13-1-3)

20. How would you determine the maximum positive and maximum negative value of a 10-bit two's complement number? (13-1-2)

21. Describe how to two's complement a binary number. (13-1-2)

22. Using the two's complement number system, briefly describe how to perform each of the following operations. (13-1-3)

    a. Add two positive numbers
    b. Add two negative numbers
    c. Add a positive and a negative number
    d. Subtract through addition
    e. Multiply through repeated addition
    f. Divide through repeated subtraction

23. What is the floating-point number system? (13-1-4)

24. Sketch and briefly describe the operation of the following circuits. (13-2)

    a. Half adder b. Full adder c. Parallel adder

25. Sketch a 6-bit parallel-adder circuit using HA and FA blocks. (13-2)

26. What is an ALU? (Introduction, 13-1-3)

27. Sketch a circuit to show how one 7486 and one 7483 could be connected to form a 4-bit adder/subtractor circuit. (13-3-1)

28. Briefly describe the operation of the following ICs. (13-2, 13-3)

    a. 7483 **b.** 74181

29. What is a bus-organized adder/subtractor circuit? (13-3)

30. Using a flow chart, first describe how subtraction is achieved through addition, and then incorporate this flow chart within another flow chart showing how division is achieved through repeated subtraction. (13-1-3)

Practice Problems

31. Add the following 8-bit numbers.

    a. 0110 1000 and 0001 1111
    b. 0111 1111 and 0100 0011
    c. 0001 1010 and 0011 1010

32. Subtract the following 8-bit numbers.

    a. 0001 1010 from 1010 1010
    b. 0001 1111 from 0011 0000
    c. 0010 0000 from 1000 0000

33. Multiply the following binary numbers.

    a. 0101 1011 by 011
    b. 0001 1111 by 1011
    c. 1110 by 010

34. Divide the following binary numbers.

    a. 0110 into 0111 0110
    b. 101 into 1010 1010
    c. 1001 into 1011 1110

35. Convert each of the following decimal numbers to an 8-bit sign-magnitude number.

    a. +34 **b.** −83 **c.** +83

36. One's complement the following numbers.

    a. 0110 1001 **b.** 1010 1010 **c.** 1111 0000

37. Convert all of the following two's complement numbers to their decimal equivalents.

    a. 0001 0000 **b.** 1011 1110 **c.** 1111 0101

38. Convert each of the following decimal numbers to an 8-bit two's complement number.

    a. +119 **b.** −119 **c.** −34

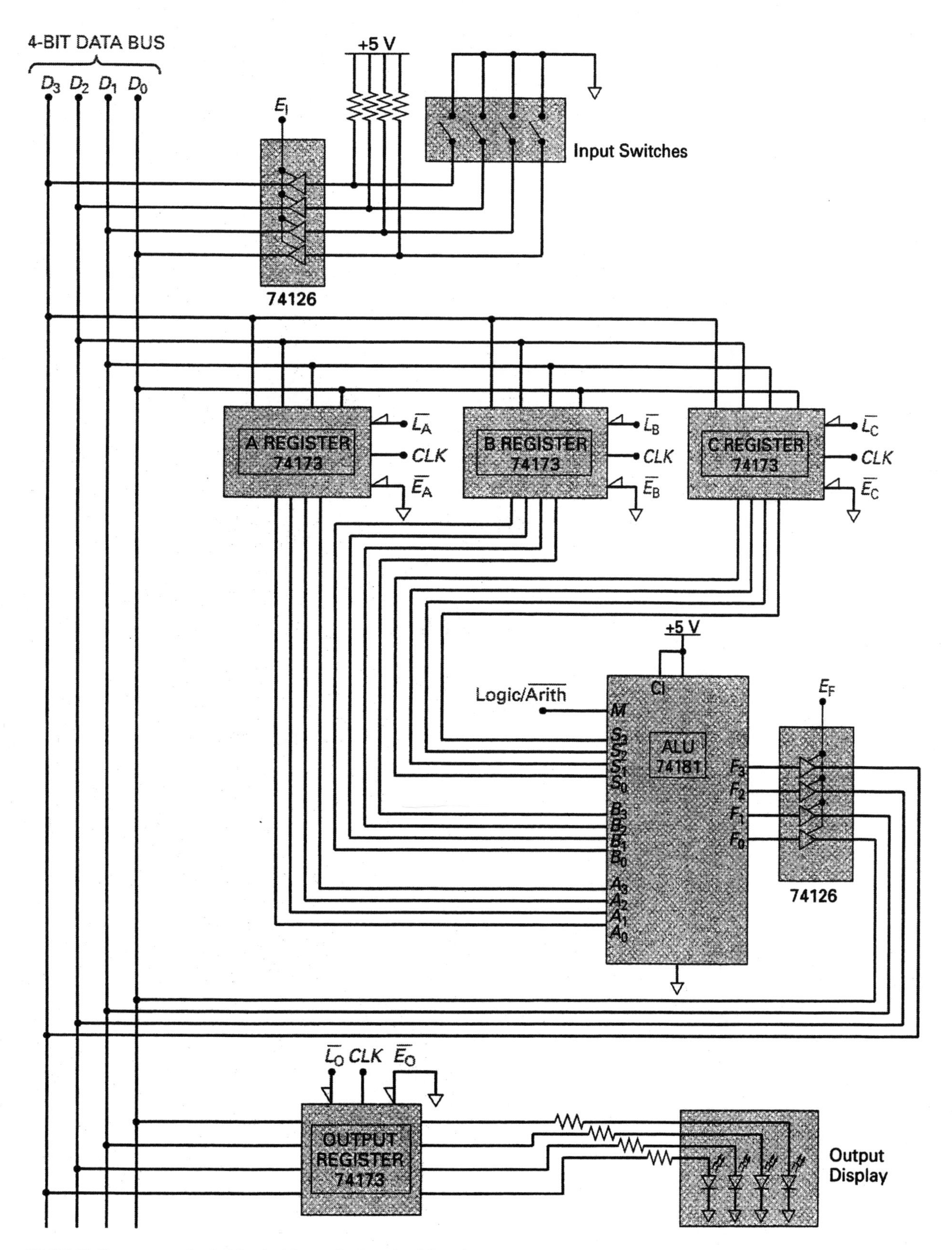

FIGURE 13-25 A 4-Bit Arithmetic-Logic Circuit.

39. Two's complement the following numbers.

    a. 0101 1010 b. 1000 0001 c. 1111 1101

40. What would be the two's complement range of a 6-bit binary number?

To practice your circuit recognition and operation skills, refer to Figure 13-25 and answer the following five questions.

41. Give a one-line definition of each of the ICs used in this circuit.
42. What is the purpose of the C register in this circuit?
43. Describe the sequence of events that would have to be followed if we wanted to
    a. input a word into the A register.
    b. input a word into the B register.
    c. display the output of the ALU on the LED display.
    d. connect a 4-bit word from the input switches directly to the output display.
44. Describe the sequence of events that would have to be followed if we wanted to
    a. input and add two 4-bit words and display the result.
    b. input and OR two 4-bit words and display the result.
    c. input and AND two 4-bit words and display the result.
45. Describe the sequence of events that would have to be followed if we wanted to input and add two 4-bit words, and then store the result in the A register.

Troubleshooting Questions

46. What operator errors would you expect from the circuit in Figure 13-25?
47. How could you use a logic pulser and logic probe to static test the circuit in Figure 13-25?
48. How would you apply the half-split troubleshooting method to the circuit in Figure 13-25?
49. What symptoms would you expect from the circuit shown in Figure 13-25 if the following malfunctions were to occur?
    a. The control line E_F is permanently HIGH.
    b. The LEDs are connected in the opposite direction.
    c. Control line $\overline{L_O}$ is permanently LOW.
    d. One of the data bus lines is open between the 4-bit data bus and the 74126 output of the ALU.
50. Sketch a simple troubleshooting tree for the circuit shown in Figure 13-25.

Chapter 14

OBJECTIVES

After completing this chapter, you will be able to:

1. List the different types of primary memory devices and secondary memory devices.
2. Describe the operation of a basic semiconductor read-only memory (ROM).
3. Explain the following memory characteristics.
   a. Number of address lines
   b. Memory size
   c. Storage density
   d. Memory configuration
   e. Read-only operation
   f. Bipolar and MOS ROMs
   g. Access time
4. Describe the operation and characteristics of the following ROM types.
   a. Mask programmable ROMs
   b. Programmable ROMs
   c. Erasable programmable ROMs
5. Describe many of the applications for ROMs.
6. Explain how ROMs are tested.
7. Describe the difference between a read-only memory and a read/write memory.
8. Describe the operation of a basic sequential-access memory (SAM) and a basic random-access memory (RAM).
9. Describe the operation, characteristics, read and write timing, and advantages of static RAMs.
10. Describe the operation, characteristics, read and write timing, and advantages of dynamic RAMs.
11. Describe in what applications static RAMs are used and in what applications dynamic RAMs are used.
12. Explain the operation of a typical memory-mapped RAM circuit.
13. Explain how RAMs are tested.
14. Describe how to troubleshoot digital logic circuits containing memory circuits using the three-step method.

Semiconductor Memories

OUTLINE

VIGNETTE

The Turning Enigma

During World War II, the Germans developed a cipher-generating apparatus called Enigma. This electromechanical teleprinter would scramble messages with several randomly spinning rotors that could be set to a predetermined pattern by the sender. This key-and-plug pattern was changed three times a day by the Germans. To British intelligence, cracking the secrets of Enigma became of the utmost importance. With this objective in mind, every brilliant professor and eccentric researcher was gathered at a Victorian estate near London called Bletchley Park. They specialized in everything from engineering to literature and were collectively called the "backroom boys."

By far the strangest and definitely most gifted was an unconventional theoretician from Cambridge University named Alan Turing. He wore rumpled clothes and had a shrill stammer and a crowing laugh that aggravated even his closest friends. He had other legendary idiosyncrasies that included not setting his watch by checking a clock but instead sighting on a certain star from a specific spot and then mentally calculating the time of day. He also insisted on wearing his gas mask whenever he was out, not for fear of a gas attack, but simply because it helped his hay fever.

Turing's eccentricities may have been strange; however, his genius was indisputable. At the age of twenty-six he wrote a paper in which he outlined what he called his "universal machine," since it could solve any mathematical or logical problem. The data, or in this case the intercepted enemy messages, could be entered into the machine on paper tape and then compared with known Enigma codes until a match was found.

In 1943 Turing's ideas began to take shape as the backroom boys began developing a machine that used two thousand vacuum tubes and incorporated five photoelectric readers that could process twenty-five thousand characters per second. It was named Colossus and incorporated the stored program and other ideas from a paper Turing had written seven years earlier.

Turing could have gone on to accomplish much more; however, his idiosyncrasies kept getting in his way. He became totally preoccupied with abstract questions concerning machine intelligence. His unconventional personal lifestyle led to his arrest in 1952 and—after a sentence of psychoanalysis—his suicide two years later.

Before joining the backroom boys at Bletchley Park, Turing's genius was clearly apparent at Cambridge; however, how much of a role he played in the development of Colossus is still unknown and remains a secret guarded by the British Official Secrets Act. Turing was never fully recognized for his important role in the development of this innovative machine, except by one of his Bletchley Park colleagues at his funeral who said, "I won't say what Turing did made us win the war, but I daresay we might have lost it without him."

INTRODUCTION

In Chapter 11 we saw how several flip-flops could be connected to form a register circuit, and how a register could be used to store a binary word. When a register is used to store binary data, it is called a memory register, and in digital systems, memory storage is a very important function.

In a digital electronic computer system, the microprocessor unit (MPU) decodes and executes binary instruction codes and directs the flow of binary data codes in order to perform its function. The instruction codes and the data codes within a digital computer system are stored in either a *primary memory* or a *secondary memory*. These two basic memory categories are shown in Figure 14-1.

The computer's primary memory is made up of two semiconductor memory IC types called ROM (read-only memory) and RAM (random-access memory). The MPU has the binary data that it needs to quickly access data stored in these high-speed ROM and RAM ICs. The ROM and RAM ICs, however, have a high cost per bit of storage, which means that their high speed of operation comes at a high cost. Secondary memory devices, such as floppy disks, hard disks, and optical disks, will store data at a cheaper cost per bit; therefore these devices are used to store the large volume of data that is not presently being used by the MPU.

In this chapter, we will be discussing in detail the semiconductor ROM and RAM primary memory devices.

14-1 SEMICONDUCTOR READ-ONLY MEMORIES (ROMs)

A **read-only memory,** or **ROM,** is an electronic circuit that permanently stores binary data. In this section we will examine the operation, characteristics, applications, and testing of ROM ICs, the different types of which are shown shaded in Figure 14-2.

Read-Only Memory (ROM)
A non-volatile random-access memory in which the stored binary data is loaded into the device before it is installed into a system. Once installed into a system, data can only be read out of ROM.

14-1-1 A Basic Diode ROM

Figure 14-3(a) shows how a simple ROM circuit could be constructed using an 8-position rotary switch, a set of diodes, and four pull-up resistors. Each of the eight rows (row 0 through 7) is equivalent to a 4-bit register, and when selected, the 4-bit binary word stored will be present at the 4-bit output (D_3, D_2, D_1, and D_0).

When the rotary switch is placed in position 0, all three diodes in row 0 will be turned ON, producing a LOW output at D_3, D_2, and D_1. The 4-bit word stored at memory location 0 is therefore 0001.

When the rotary switch is placed in position 1, a different pattern of diodes will be turned ON, producing a LOW output at D_3, D_1, and D_0. The 4-bit output word stored at memory location 1 is therefore 0100.

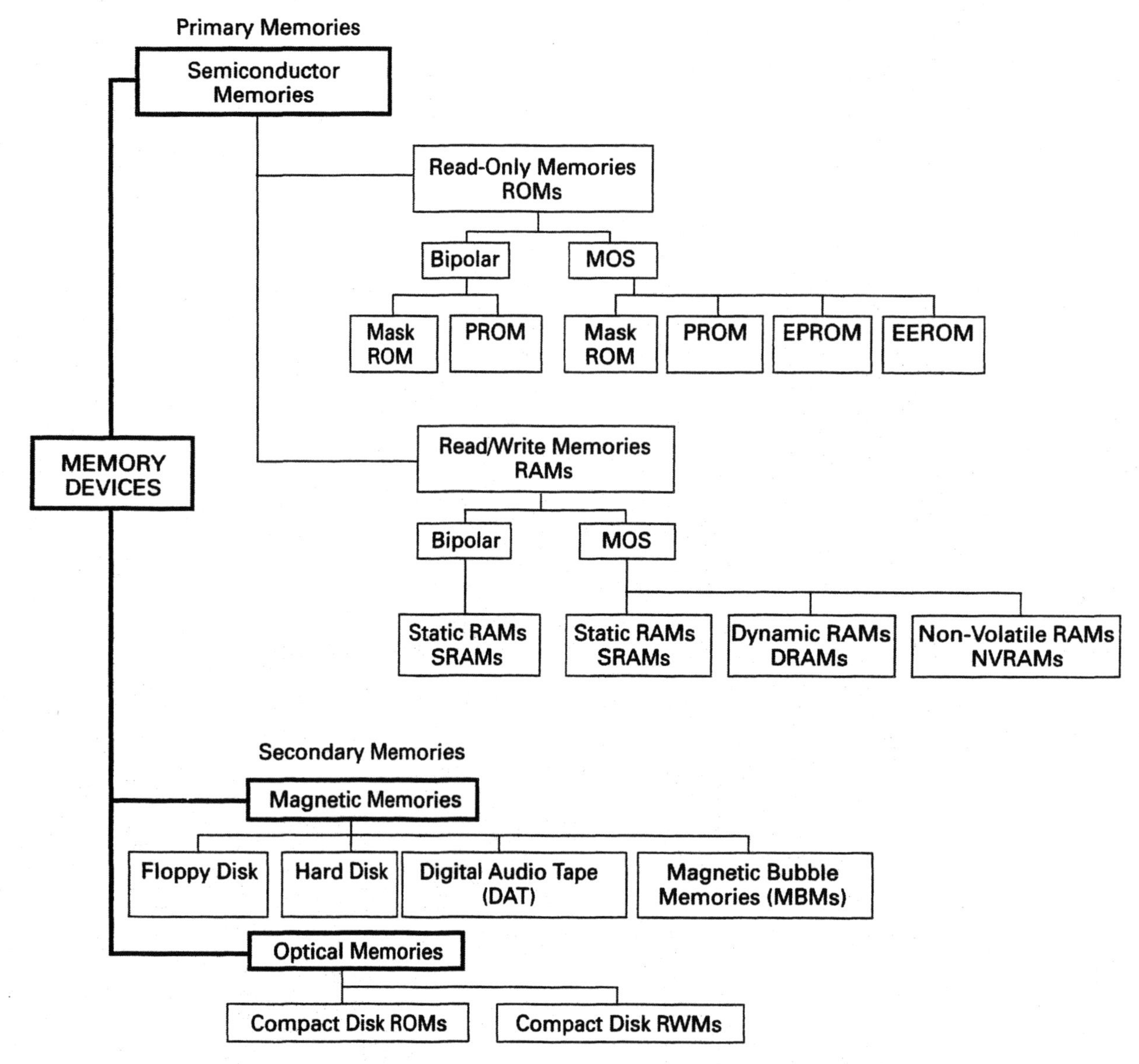

FIGURE 14-1 Memory Devices.

Address
A binary code that designates a particular location in either a storage, input or output device.

Data
A general term used for meaningful information.

Figure 14-3(b) shows all of the 4-bit words stored in each of the eight memory locations. By adding or removing diodes from the matrix, you can change the 4-bit word stored in any of the eight memory locations.

Referring to the bottom of the table in Figure 14-3(b), you can see that the switch position is called the **address,** while the contents of each memory location are called the **data.** Each and every memory location is assigned its own unique address, so we can access the data in any one of the desired memory locations. This method is similar to the postal system, which gains access to us through our unique address. The address of a memory location and the data stored in that memory location are completely unrelated. For example, at address 2 in Figure 14-3, we have a stored data value of 1011 (decimal 11). Sim-

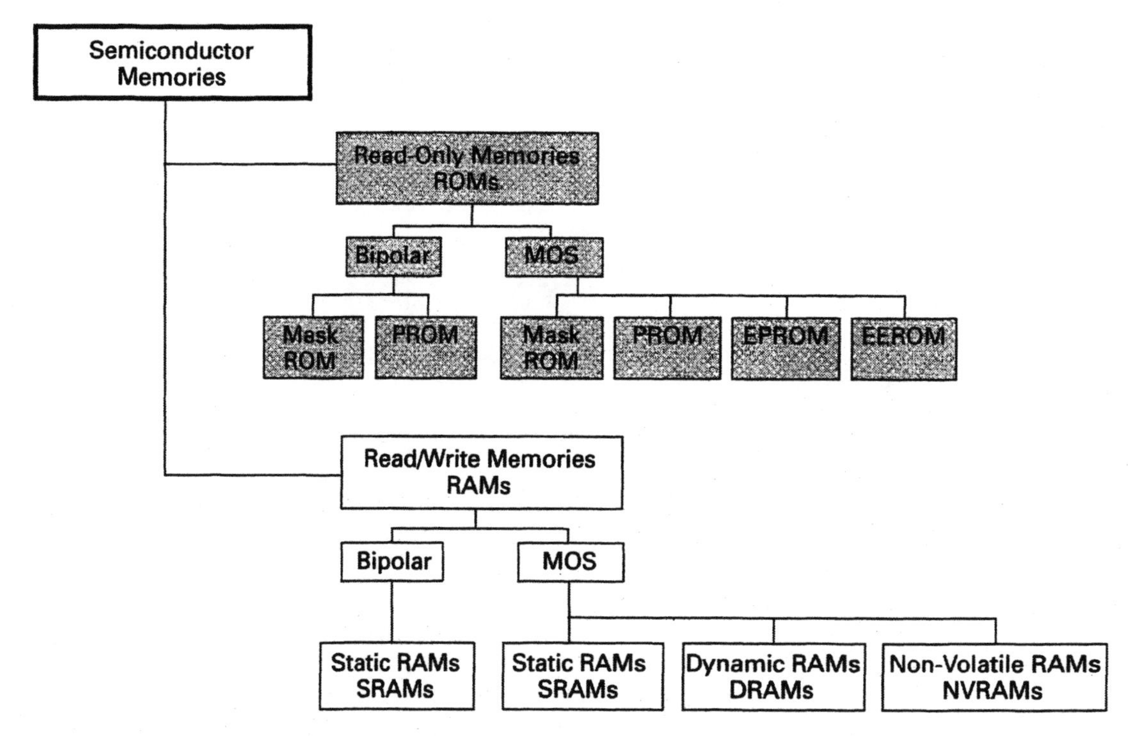

FIGURE 14-2 Semiconductor Memories: The ROM Family.

ilarly, the postal system address and data at that address are also unrelated, in that address 10752 A Street can have any number of letters in its mail box. In a digital electronic computer system, the MPU will send out a specific address so that it can gain access to the data stored at that address.

14-1-2 A Diode ROM With Internal Decoding

Figure 14-4(a) shows how a one-of-eight decoder could be used to select the desired memory location instead of using an eight-position rotary switch.

The three address inputs A_2, A_1, and A_0 drive the one-of-eight decoder, which will make one of the eight rows LOW based on the binary address applied to the address inputs. For example, if an address of 010 (decimal 2) is applied to the address inputs A_2, A_1, and A_0, only NAND gate C will have a HIGH on all three of its inputs. This will generate a LOW out on row 2, producing a data output of 1011 at D_3, D_2, D_1, and D_0, respectively. The data stored at each address is listed in the table in Figure 14-4(b).

14-1-3 Semiconductor ROM Characteristics

In this section we will examine characteristics that are common to all semiconductor ROMs. These include number of address lines, memory size and organization, read-only operation, volatility, access time, and input/output timing.

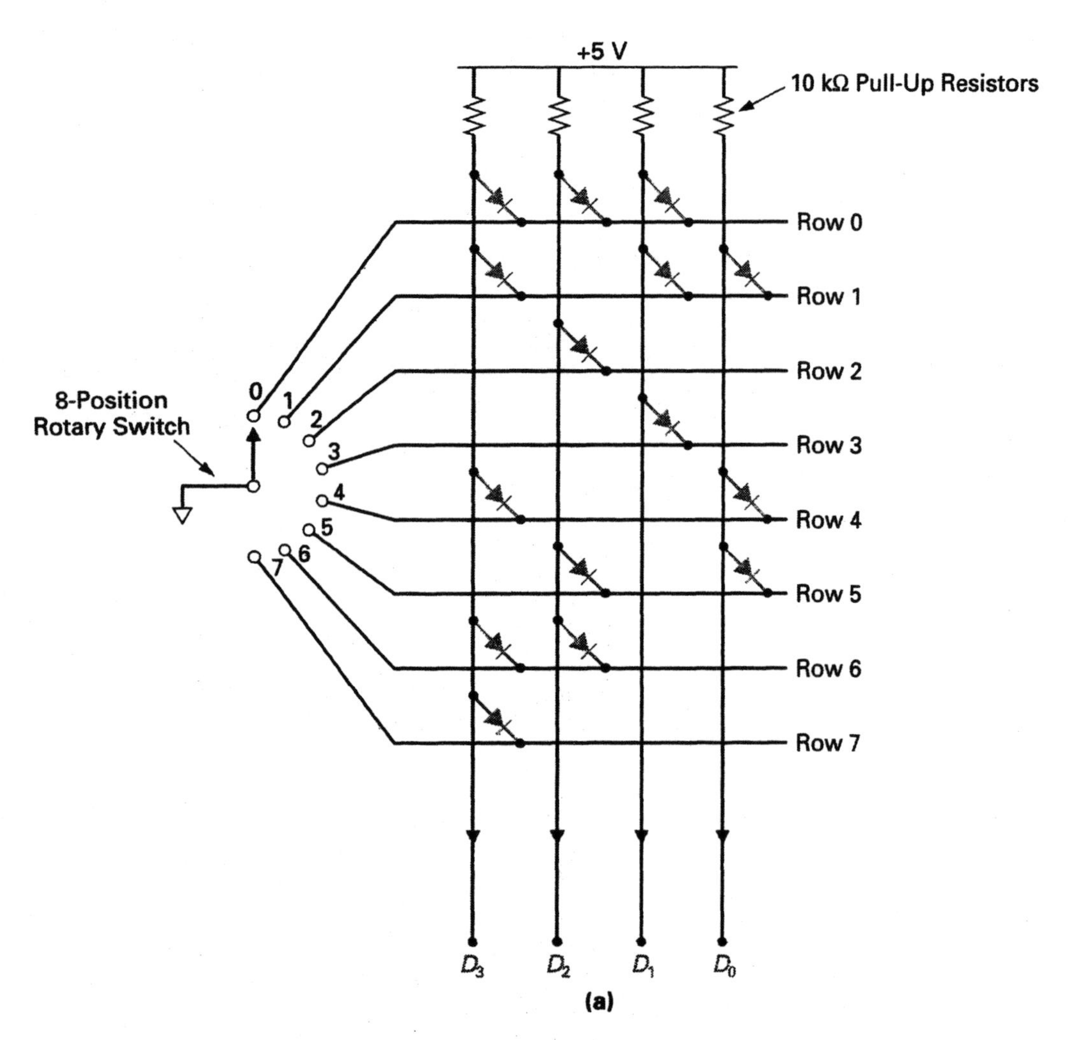

Switch Position	Output Word
0	0001
1	0100
2	1011
3	1101
4	0110
5	1010
6	0011
7	0111
Address	Data

(b)

FIGURE 14-3 A Basic ROM Circuit.

Number of Address Lines

Although the ROM circuit in Figure 14-4(a) is more complex than the ROM circuit in Figure 14-3(a), the decoder enables the circuit to be electronically accessed using three address lines instead of manually accessed by eight switched input lines. When an on-chip decoder is included with the ROM mem-

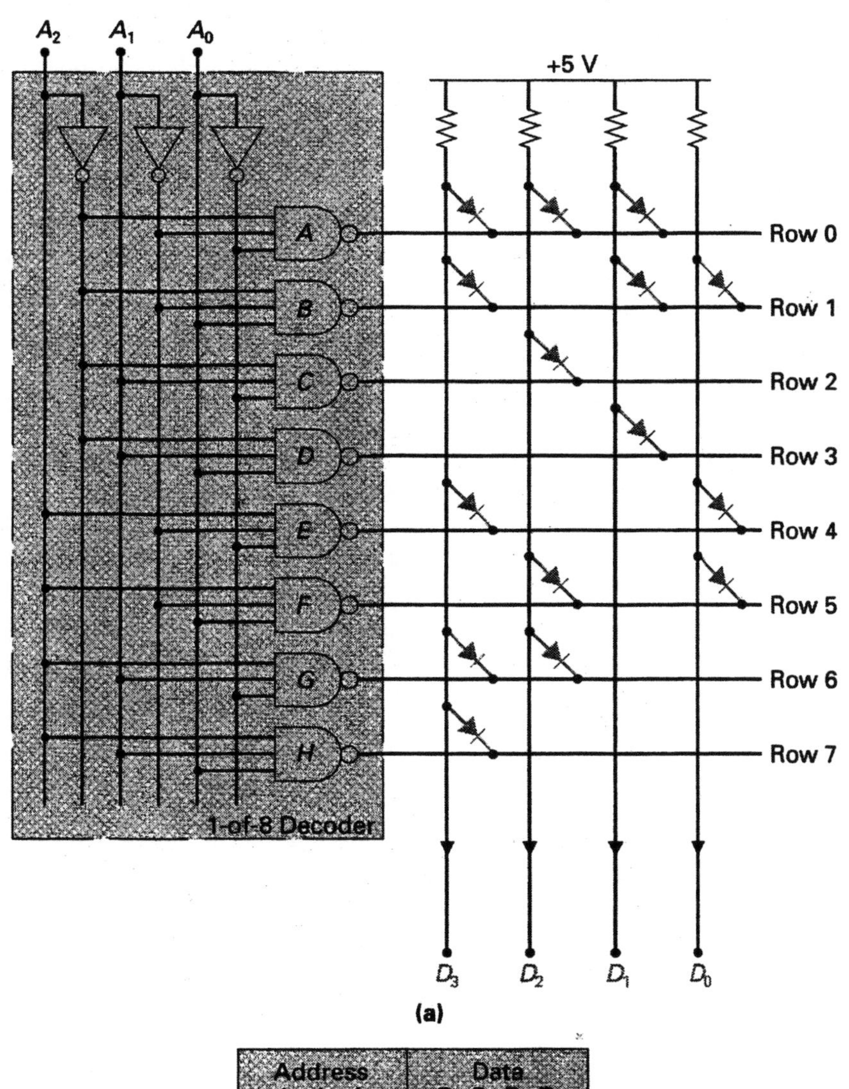

(a)

Address A_2 A_1 A_0	Data D_3 D_2 D_1 D_0
0 0 0	0 0 0 1
0 0 1	0 1 0 0
0 1 0	1 0 1 1
0 1 1	1 1 0 1
1 0 0	0 1 1 0
1 0 1	1 0 1 0
1 1 0	0 0 1 1
1 1 1	0 1 1 1

(b)

FIGURE 14-4 A Basic ROM Circuit with Internal Decoding.

ory circuit, the number of address lines (n) applied to the IC will indicate how many memory locations are available within the IC (2^n). For example, the three address lines applied to the ROM circuit in Figure 14-4(a) could address one of eight memory locations ($2^n = 2^3 = 8$ memory locations).

■ EXAMPLE:

Calculate the number of memory locations that can be accessed by a ROM memory IC, if it has

a. four address lines applied
b. eight address lines applied
c. sixteen address lines applied

■ ***Solution:***

a. Four address lines can address $2^4 = 16$ memory locations.
b. Eight address lines can address $2^8 = 256$ memory locations.
c. Sixteen address lines can address $2^{16} = 65{,}536$ memory locations.

As you can see from these examples, there is a direct relationship between the number of address lines applied to a memory circuit and the number of memory locations available within that memory circuit. The left and center columns in Table 14-1 show the correlation between the two. The right column in Table 14-1 shows the abbreviation that is used to represent these large values. The label "k" indicates 1000 in decimal; however, here we are dealing with binary (base-2), where k indicates 1,024. A 2k memory, therefore, has 2,048 memory locations (2 × 1,024), and a 1M (one meg) memory has 1,048,576 memory locations (1,024 × 1,024).

Table 14-1 The Relationship Between Address Bits and Memory Locations.

NUMBER OF ADDRESS BITS (n)	EXACT NUMBER OF MEMORY LOCATIONS (2^n)	ABBREVIATED NUMBER OF MEMORY LOCATIONS
8	256	256
9	512	512
10	1,024	1k
11	2,048	2k
12	4,096	4k
13	8,192	8k
14	16,384	16k
15	32,768	32k
16	65,536	64k
17	131,072	128k
18	262,144	256k
19	524,288	256k
20	1,048,576	1M
21	2,097,152	2M
22	3,194,304	4M
23	8,388,608	8M
24	16,777,216	16M
25	33,554,432	32M
26	67,108,864	64M

Memory Size
A term used to describe the total number of binary digits stored in a memory.

Memory Size and Memory Configuration

The term **memory size** describes the number of bits of data stored in the memory. For example, the basic ROM in Figure 14-4 is a 32-bit memory since it can store thirty-two binary digits of data. The term **memory configuration** describes how the memory circuit is organized into groups of bits or words. For example, the thirty-two storage bits in the ROM circuit in Figure 14-4 are organized into eight 4-bit words (8 by 4, 8 × 4).

The term **storage density** is used to compare one memory to another. For example, if memory A IC can store thirty-two bits and memory B IC can store sixty-four bits, then memory B is said to contain more storage bits per chip.

Memory Configuration A term used to describe the organization of the storage bits within a memory.

Storage Density A relative term used to compare the memory storage ability of one device to another.

■ **EXAMPLE:**

The 8355 ROM IC shown in Figure 14-5 has eleven address inputs and eight data outputs. What is the size and configuration of this read-only memory?

■ ***Solution:***

The eleven address line inputs to this IC indicate that there are 2^{11}, or 2,048, different memory locations. The eight data lines to this IC indicate that an 8-bit word is stored at each of the memory locations. Therefore, the 8355 ROM IC has a memory size of 16,384 bits (2,048 × 8) and is configured as 2,048 by 8.

The Read-Only Operation of a ROM

When a binary word is **stored** in a memory location, the process is called a **memory-write operation.** The term "write" is used, since writing a word into memory is similar to writing a word onto a page.

When a binary word is **retrieved** from a memory location, the process is called a **memory-read operation.** The term "read" is used, since reading a word out of memory is similar to reading a word from a page.

With a ROM, the binary information contained in each memory location is stored or written into the ROM only once. This process of entering data into the ROM is called **programming** or *burning in* the ROM. Once a ROM IC has been programmed, it is installed in a circuit within a digital system, and from that point on the MPU only reads or *fetches* data out of the ROM. Once it has

Stored A term used to describe when data is placed in a memory location or output device.

Write Operation To store information into a location in memory or an output device.

Retrieved A term used to describe when data is extracted from a memory location or input device.

Read Operation To extract information from a location in memory or an input device.

Programming A process of preparing a sequence of operating instructions for a computer system, and then loading these instructions in the computer's memory.

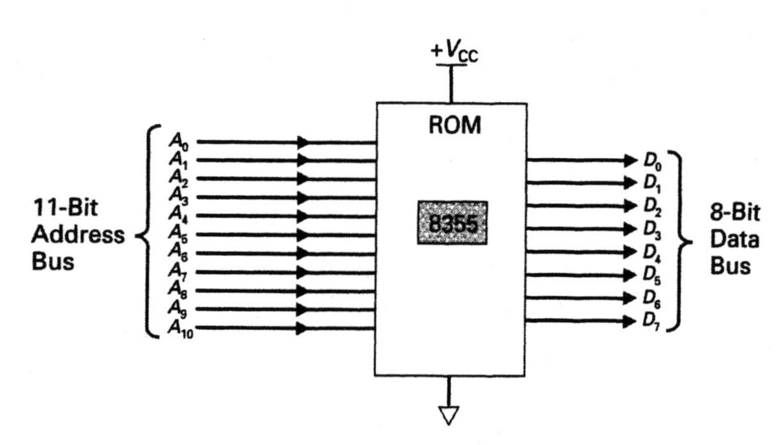

FIGURE 14-5 A Read-Only Memory IC.

been installed in a circuit, therefore, the ROM will only ever operate as a read-only memory.

Bipolar and MOS ROMs

Architecture The basic structure of a device, circuit or system.

Figure 14-6(a) shows the basic internal structure of a ROM IC. This **architecture** is essentially the same, irrespective of whether the device is a bipolar ROM or MOS ROM.

As far as operation, the *chip enable* ($\overline{CE}$) or *chip select* ($\overline{CS}$) is normally an active-LOW control input that either enables ($\overline{CE}$ = LOW) or disables ($\overline{CE}$ = HIGH) the ROM. It achieves this by disabling the address decoder's AND gates and the three-state output buffers.

When enabled, the one-of-eight decoder selects one of the eight rows depending on the address applied to the three address inputs A_2, A_1, and A_0 ($2^3 = 8$). The decoder will give a HIGH output at only one of its AND gate outputs, and this HIGH will be applied to four blocks in the ROM matrix. Each of these blocks contains a transistor and a *programmable link,* as shown in the inset in Figure 14-6(a). It is the presence or absence of the programmable link that determines whether a 1 or 0 is stored in each of these blocks. If this were a bipolar ROM, the programmable link would connect the row line from the decoder to the base of an NPN bipolar transistor. If this were a MOS ROM, the programmable link would connect the row line from the decoder to the gate of a field-effect transistor. In both cases, the presence of a link will allow the HIGH on the row line from the decoder to turn ON the transistor and connect a HIGH onto the column line and to the data output line below. On the other hand, the absence of a link will keep the transistor OFF, and therefore maintain the column line and the data output line below as LOW.

The logic symbol for this 32-bit ROM, which is organized as an 8 by 4 (eight 4-bit words), is shown in Figure 14-6(b).

Most semiconductor ROM ICs use MOSFETs because their relatively small storage cell size makes it possible for a greater number of storage bits per IC. MOS ROMs also cost less per bit, and consume less power than bipolar ROMs. Bipolar TTL and ECL ROM ICs are available; however, their small storage capacity, high cost per bit, and high power consumption makes them undesirable except for high-speed applications.

Access Time and Input/Output ROM Timing

Access Time The time interval between a device being addressed, and the data at that address being available at the device's output.

The **access time** of a memory is the time interval between the memory receiving a new address input and the data stored at that address being available at the data outputs. To explain this in more detail, Figure 14-6(c) shows the typical timing waveforms for a ROM read operation. The upper waveform represents the address inputs, the center waveform represents the active-LOW chip-enable ($\overline{CE}$) input, and the lower waveform represents the data outputs. Since the address and data waveforms represent more than one line, they are shown as a block on the timing waveform since some of the lines are HIGH and some are LOW. Whenever these blocks criss-cross, as can be seen with the address waveform at time t_1, a change in logic levels has occurred on the lines.

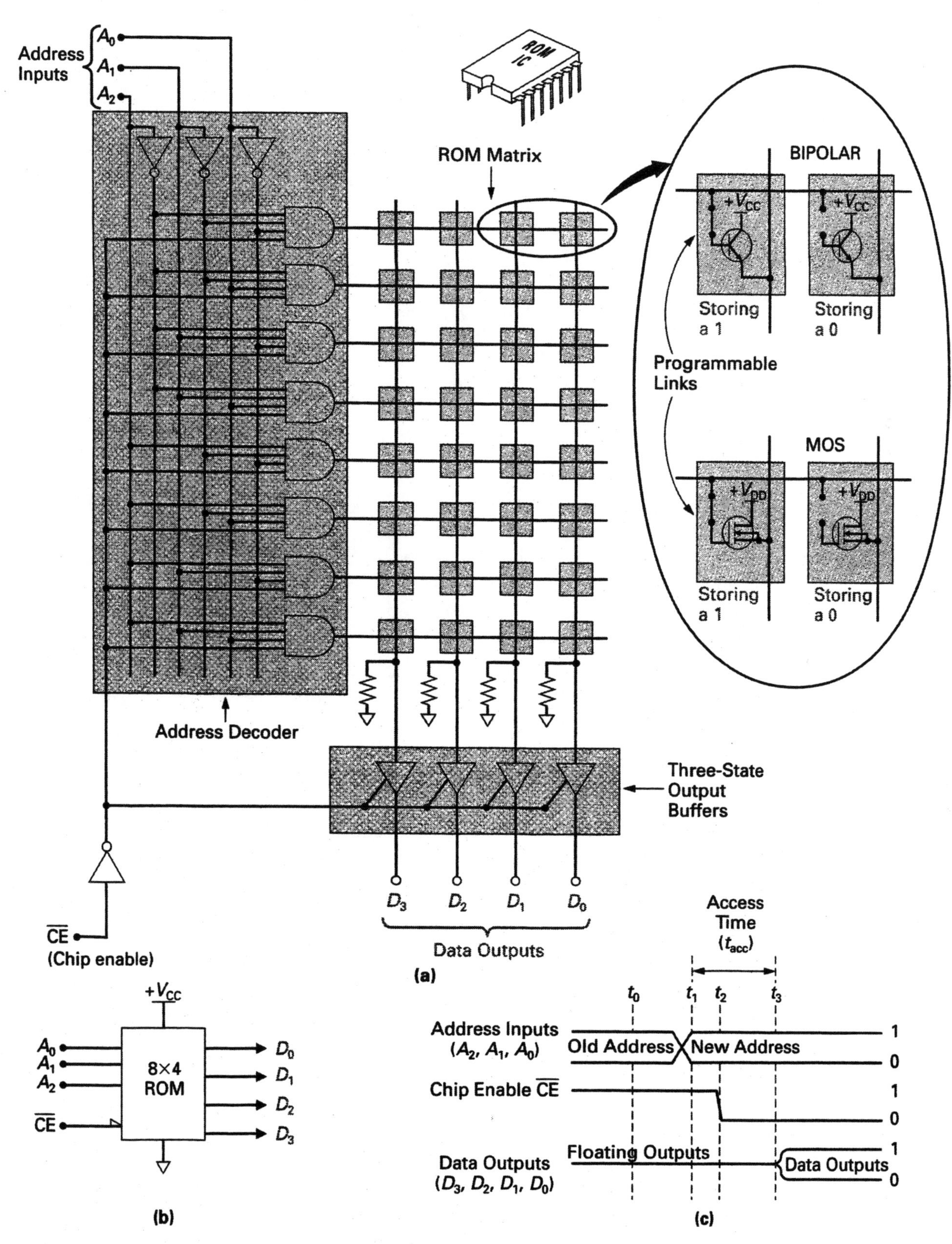

FIGURE 14-6 Typical ROM IC. (a) Basic Architecture. (b) Logic Symbol. (c) Timing Waveform.

Referring to the timing points shown in Figure 14-6(c), you can see that at time t_0, the chip-enable input is HIGH and therefore the ROM IC is disabled. At time t_1, a new address is placed on the 3-bit address bus (A_2, A_1, and A_0), and the ROM's internal address decoder will begin to decode the new address input to determine which memory location is to be enabled. At time t_2, the $\overline{CE}$ input is made active, enabling the final stage of the address decoder and the output three-state buffers. At time t_3, the data outputs (D_3, D_2, D_1, and D_0) change from a floating output state to a logic level that reflects the data stored in the addressed memory location.

Since the access time of a memory is the time interval between the memory receiving a new address input and the data being available at the data output, the access time (t_{acc}) is the time interval between t_1 and t_3 in Figure 14-6(c). Typically, bipolar ROMs have access times in the 10 ns to 50 ns range, while MOS ROMs will have access times from 35 ns to 500 ns.

14-1-4 ROM Types

There are three basic types of ROM—mask programmable ROMs (MROMs), programmable ROMs (PROMs), and erasable programmable ROMs (EPROMs, EEPROMs, and flash memories). Let us now examine each of these types and see how they differ.

Mask Programmable ROMs (MROMs)

Mask Programmable ROM (MROM)
A read-only memory in which the manufacturer generates a photographic negative or mask designed to store the customer-specified data requirements.

Mask programmable ROMs, or **MROMs,** are permanently programmed by the manufacturer by simply adding or leaving out diodes or transistors, as previously shown. A customer sends the manufacturer a truth table stating what data should be stored at each address, and then the manufacturer generates a photographic negative called a *mask* which is used to produce the interconnections on the ROM matrix. These masks are normally expensive to develop and therefore this type of ROM only becomes economical if a large number of ROMs are required.

The disadvantage of these mask programmed ROMs is that they cannot be reprogrammed with any design modifications since programming is permanent.

As an example, Figure 14-7 shows the logic symbol and pin assignment for a 7488 IC, which is a 256-bit mask programmable ROM. The 256 bits of storage in this IC are organized as thirty-two 8-bit words. The 5-bit address input (A_0 through A_4) will select one of the thirty-two memory locations. The active-LOW select ($\overline{S}$) input will connect the addressed 8-bit word onto the data outputs (D_0 through D_7). This ROM can be programmed to have any 8-bit data word stored in its thirty-two memory locations by sending the manufacturer a truth table of the desired contents.

As another example, Figure 14-8 shows the logic symbol and pin assignment for a TMS47256 IC, which is a 262,144-bit mask programmable ROM. The 262,144 bits of storage in this IC are organized as 32,768 8-bit words. The 15-bit address input is used to select one of the 32,768 memory locations. The 8-bit data output is under three-state control and is only enabled when both $\overline{E}$

DIGITAL DEVICE DATA SHEET

IC TYPE: 7488 – A 256-BIT MASK PROGRAMMABLE BIPOLAR ROM (32 × 8)

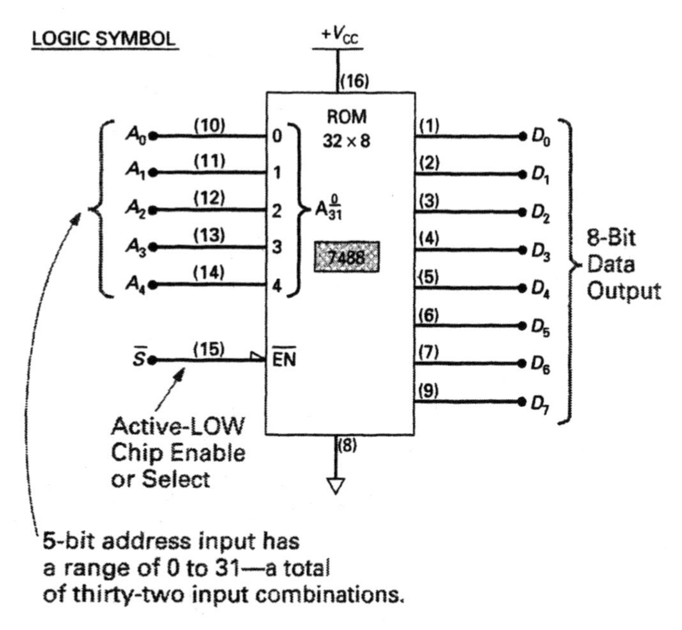

FIGURE 14-7 An IC Containing a 256-Bit Mask Programmable ROM That is Organized to Store 32 8-Bit Words.

and $\overline{S}$ are LOW. The $\overline{E}$ input also controls a power-down function. When $\overline{E}$ is HIGH, the IC's internal circuits are put in a low-power standby mode, and the IC will only draw approximately one-quarter of the normal supply current. This feature is ideal for portable battery-operated systems. This MROM is available from Texas Instruments as an NMOS version (TMS47256) with an access time of 200 ns and a standby power dissipation of 82.5 mW, or as a CMOS version (TMS47C256) with an access time of 150 ns and a standby power dissipation of 2.8 mW.

Programmable ROMs (PROMs)

Although the MROM is extremely expensive, the cost per IC can become minimal when a large volume of ROMs is needed (thousands). On the other hand, if only a small number of ROMs is needed, one option is to use a fusible-link **programmable ROM,** or **PROM.** Unlike the MROM whose data is etched in stone by the manufacturer, so to speak, the PROM comes with a clean slate and can be programmed by the user. Like the MROM, however, once the PROM has been programmed, the etched data cannot be erased.

Programmable ROM (PROM) A read-only memory in which the customer can store any desired data codes by controlling fusible links.

Figure 14-9(a) shows how the architecture of a PROM is similar to that of the MROM shown in Figure 14-6(a). The difference with the PROM, however, is that each of its programmable links is fusible. All of these fusible links are

DIGITAL DEVICE DATA SHEET

IC TYPE: TMS47256 – A 256,144-BIT MASK PROGRAMMABLE ROM (32,768 × 8)

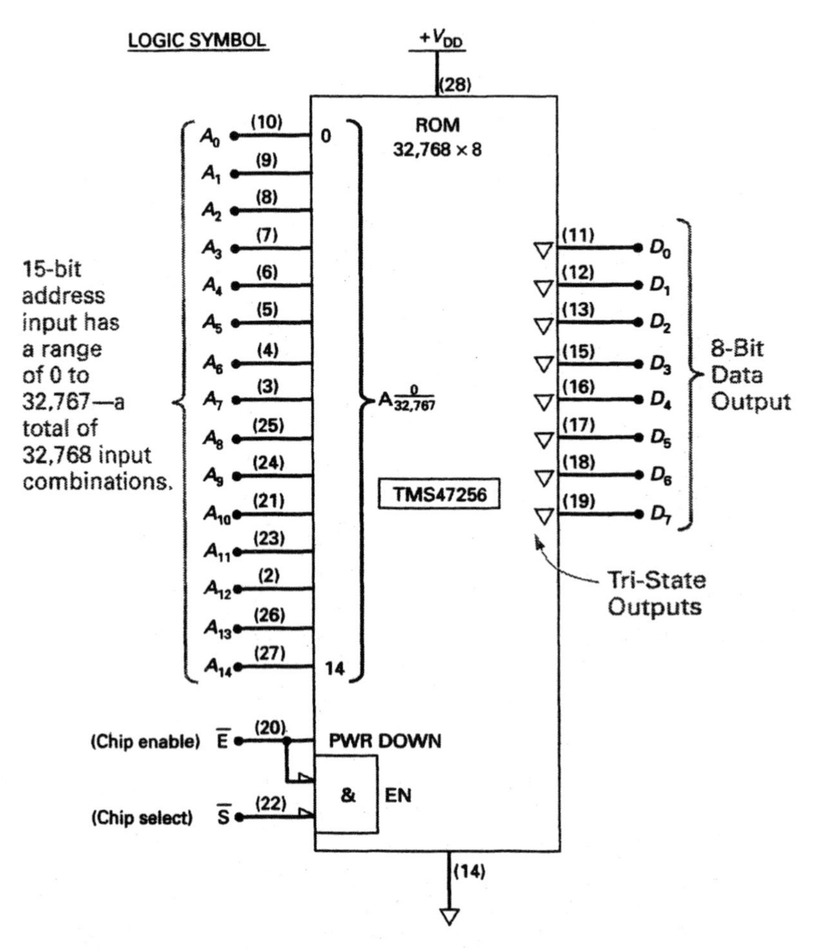

FIGURE 14-8 An IC Containing a 262,144-Bit Mask Programmable ROM That is Organized to Store 32,768 8-Bit Words.

initially intact when the PROM arrives from the manufacturer, and it is up to the user to either blow a fuse link (to store a 0) or leave it intact (to store a 1).

The programming or burning-in process is performed by a *PROM programmer unit,* similar to the one shown in Figure 14-9(b). By placing the PROM into the IC socket on the unit and entering the desired truth table using the keyboard, the PROM programmer unit will then automatically step through each address and blow or leave the fuse intact in each memory location. Once the data has been written into every memory location, the programmer can read back the stored data to verify that it matches the desired truth table.

Both bipolar (low-density, high-speed, high power dissipation) and MOS (high-density, low-speed, low power dissipation) PROMs are available. For example, the 74186 is a 512-bit bipolar PROM, which is organized as sixty-four words of 8 bits each, has an access time of 50 ns. As another example, the

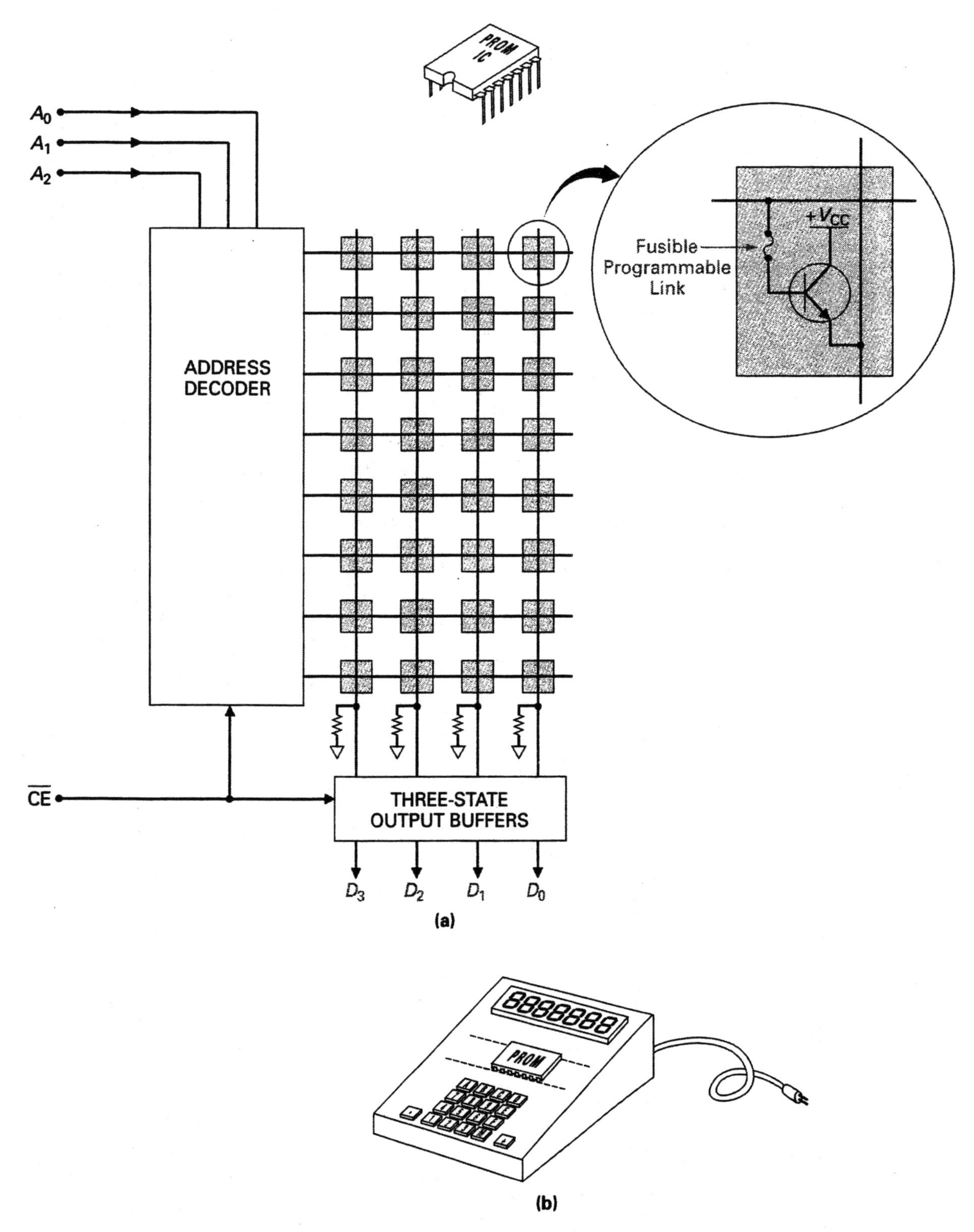

FIGURE 14-9 Typical PROM IC. (a) Basic Architecture. (b) Programming.

TMS27PC256 is a 256k-bit CMOS PROM which is organized as 32k words of eight bits each and has an access time of 120 ns.

Erasable Programmable ROMs (EPROMs, EEPROMs, and Flash)

Erasable PROM (EPROM) A PROM device that can have its contents erased using an ultraviolet light source, and then be re-programmed with new data.

Electrically Erasable PROM (EEPROM) A PROM device that can have its contents erased using an electrical pulse, and then be re-programmed with new data.

The ultraviolet-light **erasable PROM (EPROM),** the **electrically erasable PROM (EEPROM),** and *flash memories* are MOS circuits that enable the user to store data in the same way as the standard PROM, using the PROM programmer unit. The important difference with these devices is that if a modification to the stored data needs to be made at a later time, the contents of the IC can be erased and then reprogrammed with the new data.

The EPROM. With the ultraviolet erasable PROM, or EPROM, the data is erased by shining a high-intensity ultraviolet light through a quartz window on the top of the IC for about twenty minutes. Data in this IC type is stored as a charge or no charge, and the ultraviolet light causes these charges to leak off. The disadvantage with the EPROM is that the light erases the entire contents of the IC, and therefore if you only wish to change one byte or one bit, you will still have to erase everything and then reprogram the IC with the entire truth table.

As an example, Figure 14-10 shows the logic symbol, and physical appearance of a 2732 IC, which can store 4,096 (4k) 8-bit words. Referring to

DIGITAL DEVICE DATA SHEET

IC TYPE: 2732 – 4k × 8 NMOS EPROM

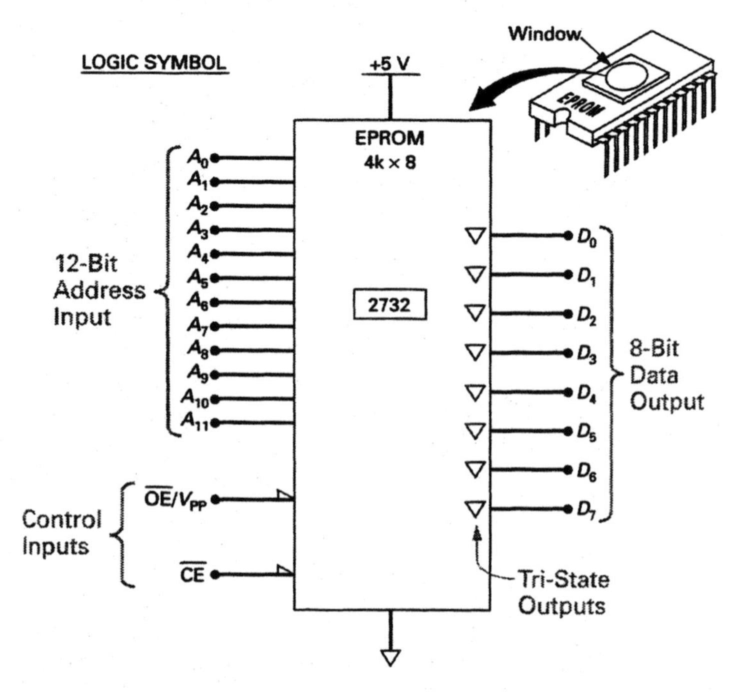

FIGURE 14-10 An IC Containing an NMOS EPROM That Is Able to Store 4,096 8-Bit Words.

the physical appearance of this IC, you can see the EPROM IC's quartz window, which is generally covered with a sticker once it has been programmed so that the circuit is shielded from ultraviolet rays present in sunlight and fluorescent lighting. The 2732 has twelve address inputs and eight data outputs.

To load or program the 2732 EPROM, the address input code is first applied to select the desired memory location (A_0 through A_{11}), then the data to be stored in that memory location is applied to the data outputs (D_0 through D_7). A +21 V programming pulse voltage (V_{PP}) is applied to the V_{PP} control input at the same time the chip-enable control input ($\overline{CE}$) is pulsed LOW. These two control inputs are typically pulsed active (V_{PP} = +21 V, $\overline{CE}$ = LOW) for 50 ms. The process is then repeated to store the next byte. Once programmed, the 2732 is placed in a circuit and operated as a read-only memory.

To operate the 2732 IC as a ROM, the $\overline{CE}$ must first be enabled, the address of the desired memory location must be applied to the address inputs, and then the output enable must be made active ($\overline{OE}$ = LOW) in order for the selected data to be switched through the three-state buffers to the data outputs. When the $\overline{CE}$ of the 2732 is inactive, the IC is in a standby mode and it consumes 175 mW compared to 500 mW when enabled.

To erase the data stored in the 2732, simply expose the circuit through the IC window to an ultraviolet light source for twenty minutes.

The EEPROM. The electrically erasable PROM, or EEPROM, uses an electrical pulse to program data (like the EPROM); however, unlike the EPROM, it also makes use of an electrical pulse for erasing the stored data. This means that an ultraviolet light source is not needed for this type of PROM, since programming and erasing can be performed without removing the IC package from the circuit.

When programming an EEPROM, a 21 V pulse of typically 10 ms is needed to store a byte (an EPROM requires a 21 V pulse of typically 50 ms). When many thousands of bytes are involved, this makes the EEPROM programming and verifying process a lot less time consuming. The key advantage of the EEPROM, however, is its ability to erase only a single byte of stored data, or to erase the entire device with a single 21 V/10 ms pulse. This means that a single byte modification can be performed without having to erase the entire device and then completely reprogram, as was the case with the EPROM. The erase time of 10 ms for the EEPROM is also a lot faster than that of the ultraviolet EPROM, which requires twenty minutes of UV exposure to erase its contents.

The EPROM is normally cheaper than the EEPROM, and both are generally used to develop a product. Once the design is finalized, however, a mask ROM, or MPROM, will be ordered in bulk for mass production.

As an example, Figure 14-11 shows the logic symbol and function table of a 2864 IC, which can store 8,192 (8k) 8-bit words. The 2864 contains high-voltage circuitry on the chip along with the 8k by 8 of memory. This high-voltage circuit generates the high-voltage pulse needed to either program or erase the memory. As a result, only a $+V_{CC}$ supply voltage is needed. Address inputs A_0 through A_{12} are used to access one of the 8,192 memory locations. The data

DIGITAL DEVICE DATA SHEET

IC TYPE: 2864 – 8k × 8 EEPROM WITH ON-CHIP HV CIRCUITRY

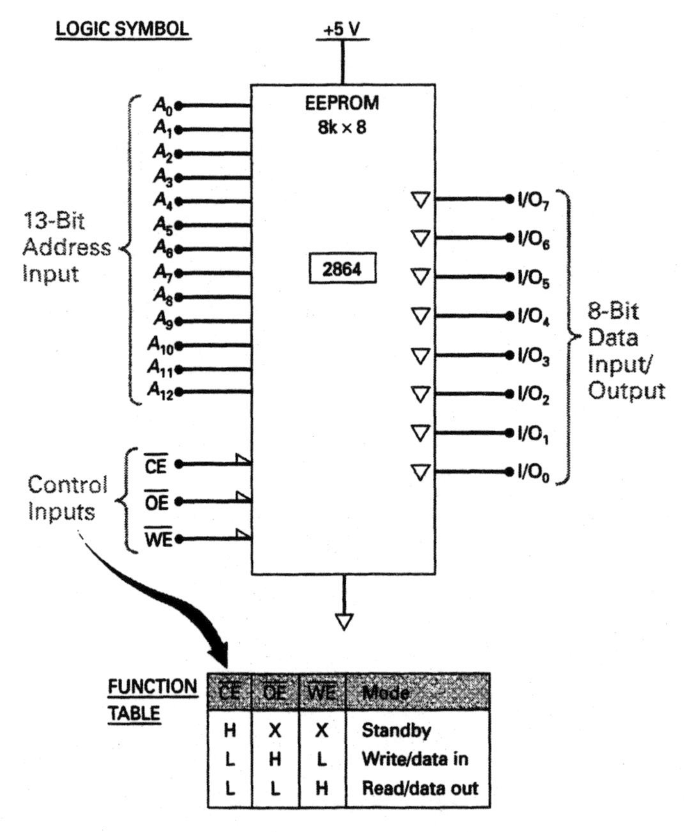

$\overline{CE}$	$\overline{OE}$	$\overline{WE}$	Mode
H	X	X	Standby
L	H	L	Write/data in
L	L	H	Read/data out

FIGURE 14-11 An IC Containing an EEPROM That Is Able to Store 8,192 8-Bit Words.

lines are labeled I/O (input/output) since they are used to write data into a memory location during the programming operation or read data out of a memory location during normal read-only memory operation. The function of the three active-LOW control inputs—chip-enable, output-enable, and write-enable—are shown in the IC's function table. The chip-enable input is used to either enable or disable the chip, and when HIGH, will place the chip in its low power consumption, standby mode. When the chip-enable input is LOW, the condition of the output-enable and write-enable will control whether data is written into the memory ($\overline{CE}$ = LOW, $\overline{OE}$ = HIGH, $\overline{WE}$ = LOW) or data is read out of the memory ($\overline{CE}$ = LOW, $\overline{OE}$ = LOW, $\overline{WE}$ = HIGH).

The flash memory. To compare, EPROMs have fast access times, a high storage density, and a low cost per bit of storage. The EPROM's disadvantage, however, is that it has to be removed from its circuit to be erased, which takes twenty minutes. On the other hand, EEPROMs have fast access times and can be erased and reprogrammed without being removed from the circuit. The

EEPROM's disadvantage, however, is that it has a low storage density and a much higher cost per bit of storage.

A **flash memory** has the advantages of both the EPROM and the EEPROM by combining the in-circuit erasability of the EEPROM with the high storage density and low cost of the EPROM. It is called a flash memory because of its high-speed erase and write data times. This new technology is rapidly improving, and at present the cost of a flash memory is much less than an EEPROM, but not as low-cost as an EPROM. As far as write time is concerned, a typical flash memory will have a write time of 10 μs per byte, compared to 100 μs per byte for a fast EPROM, and 5 ms for an EEPROM.

Flash Memory
A re-programmable ROM device that has extremely fast erase-data and write-data speeds.

14-1-5 *ROM Applications*

As you study all of the ROM applications discussed in this section, you will notice that all the ROM is really doing is converting a coded input word at the address inputs into a desired output word at the data outputs.

Code Conversion

Figure 14-12 shows how a ROM could be used as a code converter. The input code is applied to the address input pins of the ROM. This code will be decoded by the ROM's internal decoder, which will select one of its internal memory locations and then apply the data content or desired output code to the data output pins.

Some standard ROM code converters are available. For example, there is a ROM code converter to convert the 7-bit alphanumeric ASCII (American Standard Code for Information Interchange) code into another code such as the 7-bit alphanumeric BCDIC (Binary Coded Decimal Interchange Code) code, which is used by some computers instead of ASCII. There is also a ROM code converter available that will achieve the opposite by converting BCDIC codes to ASCII codes.

Arithmetic Operations

As we discovered in the previous chapter, a simple adder circuit and the two's complement number system can be used to perform addition, subtraction, multiplication, and division. More complex mathematical functions, such as trigonometric and logarithmic operations, would require a more complex circuit

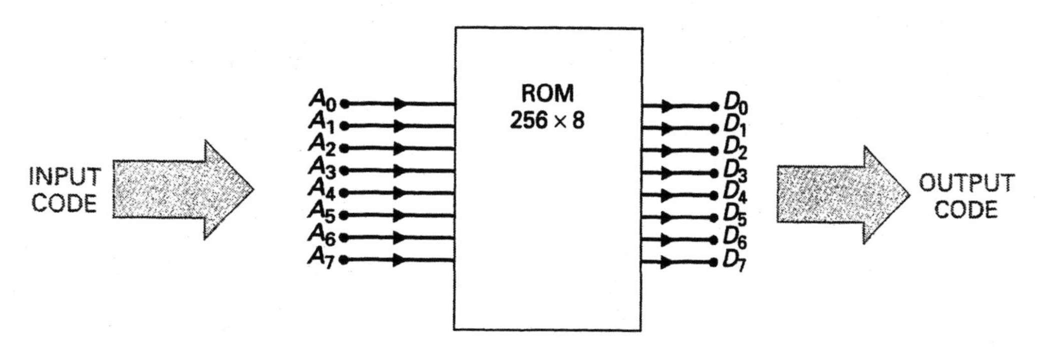

FIGURE 14-12 The ROM as a Code Converter.

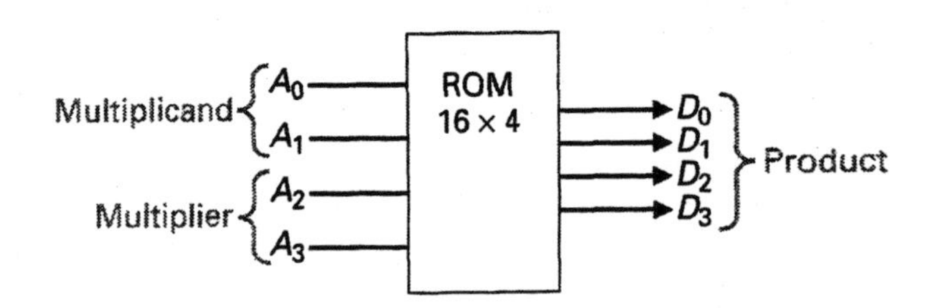

A_3 A_2	A_1 A_0	D_3 D_2 D_1 D_0	
0 0 ×	0 0 =	0 0 0 0	0 × 0 = 0
0 0 ×	0 1 =	0 0 0 0	0 × 1 = 0
0 0 ×	1 0 =	0 0 0 0	0 × 2 = 0
0 0 ×	1 1 =	0 0 0 0	0 × 3 = 0
0 1 ×	0 0 =	0 0 0 0	1 × 0 = 0
0 1 ×	0 1 =	0 0 0 1	1 × 1 = 1
0 1 ×	1 0 =	0 0 1 0	1 × 2 = 2
0 1 ×	1 1 =	0 0 1 1	1 × 3 = 3
1 0 ×	0 0 =	0 0 0 0	2 × 0 = 0
1 0 ×	0 1 =	0 0 1 0	2 × 1 = 2
1 0 ×	1 0 =	0 1 0 0	2 × 2 = 4
1 0 ×	1 1 =	0 1 1 0	2 × 3 = 6
1 1 ×	0 0 =	0 0 0 0	3 × 0 = 0
1 1 ×	0 1 =	0 0 1 1	3 × 1 = 3
1 1 ×	1 0 =	0 1 1 0	3 × 2 = 6
1 1 ×	1 1 =	1 0 0 1	3 × 3 = 9

FIGURE 14-13 The ROM as a Multiplier.

and algorithm. In many of these instances, a ROM is used to implement the arithmetic operation. To use a simple example, Figure 14-13 shows how a ROM could be used to multiply two 2-bit words. In this circuit, the 2-bit multiplicand and 2-bit multiplier are applied to the ROM's 4-bit address input. Since the correct product for every possible input combination is stored in the ROM, the correct result will always be present at the data outputs.

Look-Up Tables

A look-up table is not that different from a code converter. For example, the MM4232 ROM will convert a 9-bit binary input representing an angle between 0° and 90° to an 8-bit binary output that is equivalent to the sine of the input angle.

Logic Circuit Replacement

A ROM can also be used to replace a logic circuit. For example, an eight-input/four-output logic circuit could be replaced by an 8-bit address input/4-bit data output ROM, which could be programmed to give the desired output for any of the 8-bit input combinations.

Character Generator

Another typical application for ROMs is as a character generator. For example, the Motorola MC6571 ROM is programmed with all of the 7 by 9 dot matrix codes used to display a number, letter, or symbol on a video display monitor screen.

Keyboard Encoder

Keyboard encoders are also not that different from code converters. For example, the MM5740 ROM is used within some computer keyboards to convert the keyboard switch matrix code generated when a key is pressed, to a corresponding ASCII code that can be recognized by the computer.

Non-Volatile Computer Memory

The term **non-volatile** describes a device that will retain its information in the absence of power. ROMs are mainly used in computer systems to store instruction codes and data codes that must not be lost when the system power is turned off. Several ROMs are often used to store information, such as system start-up instructions, keyboard encoding codes, character generator codes, and look-up tables.

Non-Volatile
A memory storage device that will retain its stored data in the absence of power.

14-1-6 *ROM Testing*

A **ROM listing** sheet lists the data values that should be stored in every memory location of a ROM. A ROM can be tested, therefore, by simply accessing or addressing each memory location, and then checking that the stored data in that memory location matches what is listed in the ROM listing.

ROM Listing
A table showing the binary codes stored in every memory location of a ROM.

Figure 14-14 shows a simple ROM test circuit. In this circuit, a push-button switch is used to clock a counter, causing it to advance its count. The counter's

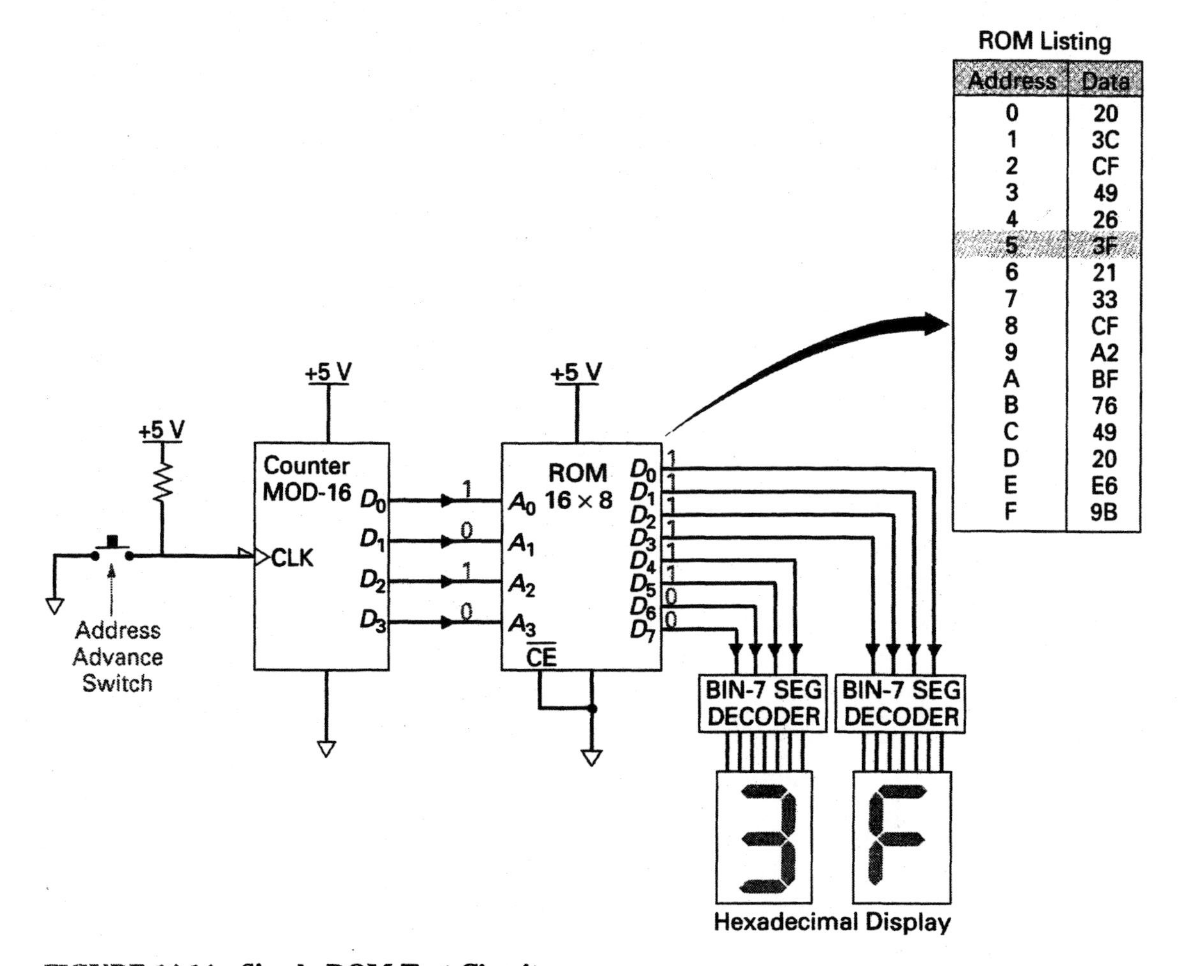

Address	Data
0	20
1	3C
2	CF
3	49
4	26
5	3F
6	21
7	33
8	CF
9	A2
A	BF
B	76
C	49
D	20
E	E6
F	9B

FIGURE 14-14 Simple ROM Test Circuit.

output is used to supply a sequential address to the ROM being tested. The data output of the ROM is displayed on a seven-segment hexadecimal display. This circuit, therefore, will allow us to step through each address and check the stored data value against the value shown in the ROM listing. This check will verify the operation and contents of the ROM.

The test circuit in Figure 14-14 would be ideal for testing a small 16 by 8 ROM; however, most digital system ROMs will be much larger, having thousands of memory locations, making a complete test with this circuit long and laborious. In these cases, a ROM tester can be used to check the memory's contents. This test instrument automatically reads the contents of the ROM being tested and compares it to the stored data within a known good reference ROM to check for any false operation or storage errors.

Most ROMs within digital electronic computer systems are tested by a built-in **self-diagnostic program.** This test program will generally use the **check-sum method** to verify the ROM's contents. With the check-sum method, the computer reads out and adds all of the data stored in every one of the ROM's memory locations to obtain a sum (ignoring carries). The resultant sum is compared with a sum value that was stored in a specific address in the ROM when it was programmed. If a difference exists between the present check-sum and the previously stored check-sum, then there is definitely a fault with the programmed data.

Self-Diagnostic Program
A test program stored within a computer to test itself.

Check-Sum Method
A ROM testing method in which the computer reads out and sums the entire contents of the memory and then compares the total to a known good value.

SELF-TEST REVIEW QUESTIONS FOR SECTION 14-1

1. Is a ROM a volatile or non-volatile device?
2. What is a memory's address input, and how are the number of address input lines related to the number of memory locations?
3. Define the following terms.
   a. Storage density
   b. Memory configuration
   c. Burning in
   d. Access time
4. Give the full names for the following abbreviations.
   a. MROM b. PROM c. EPROM d. EEPROM
5. How is a ROM used as a code converter?
6. What is a ROM listing, and how can it be used to test a ROM?

14-2 SEMICONDUCTOR READ/WRITE MEMORIES (RWMs)

As its name implies, the previously discussed read-only memory, or ROM, can only read out or retrieve data from its memory locations once it is installed within a digital system. Data, such as ASCII coded text, numerical codes, or microprocessor instruction codes, are permanently stored in the ROM's memory locations before the device is installed in a digital circuit.

Read/write memories (RWMs) can perform either a read or a write operation after they are installed in a digital circuit. Unlike the ROM which contains permanent storage locations, a RWM acts as a temporary storage location for binary data. When data is stored in a RWM, the data is said to be written into the memory, whereas when data is retrieved from an RWM, it is said that data is being read from memory. In this section, we will be concentrating on the semiconductor read/write memories shown shaded in the chart in Figure 14-15. To begin with, let us discuss why read/write memories (RWMs) are more commonly referred to as **random-access memories (RAMs).**

Read/Write Memory A memory that can have data written into or read out from any addressable storage location.

Random Access Memory A memory circuit characteristic in which any storage location can be directly accessed.

14-2-1 *SAMs Versus RAMs*

The shift register discussed in Chapter 11 is an example of a **sequential access memory (SAM),** or *serial memory.* To explain this, Figure 14-16 shows a 512-bit shift register which can store sixty-four bytes (a 64 by 8 SAM). These sixty-four bytes of data are numbered byte 0 through byte 63, and are stored serially in 512 adjacent shift register storage elements. To store an 8-bit word in this serial memory, you must first apply a 6-bit address to specify the storage location for the data byte. This 6-bit address is loaded into the address register and then applied to the left input of the comparator.

Sequential Access Memory A memory circuit characteristic in which any storage location can only be accessed in sequence or one after the other.

For every clock pulse at the input, the data within the shift register is shifted one position to the right. The clock input is also applied to a MOD-8 counter, which clocks the byte counter each time a complete word or eight bits

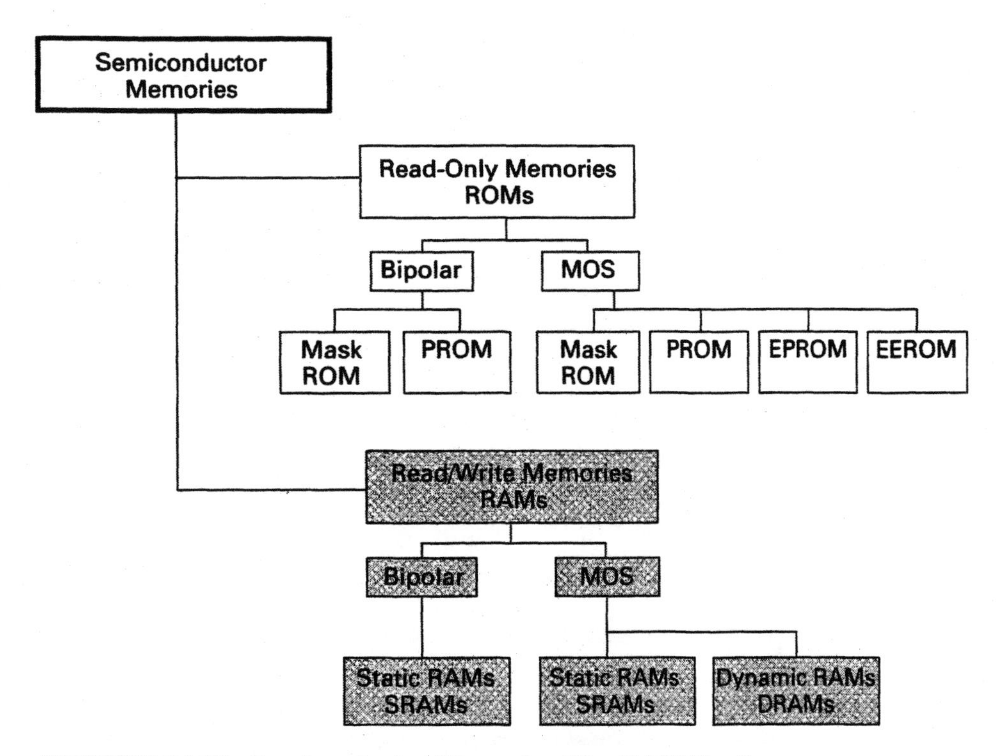

FIGURE 14-15 Semiconductor Memories: The RAM Family.

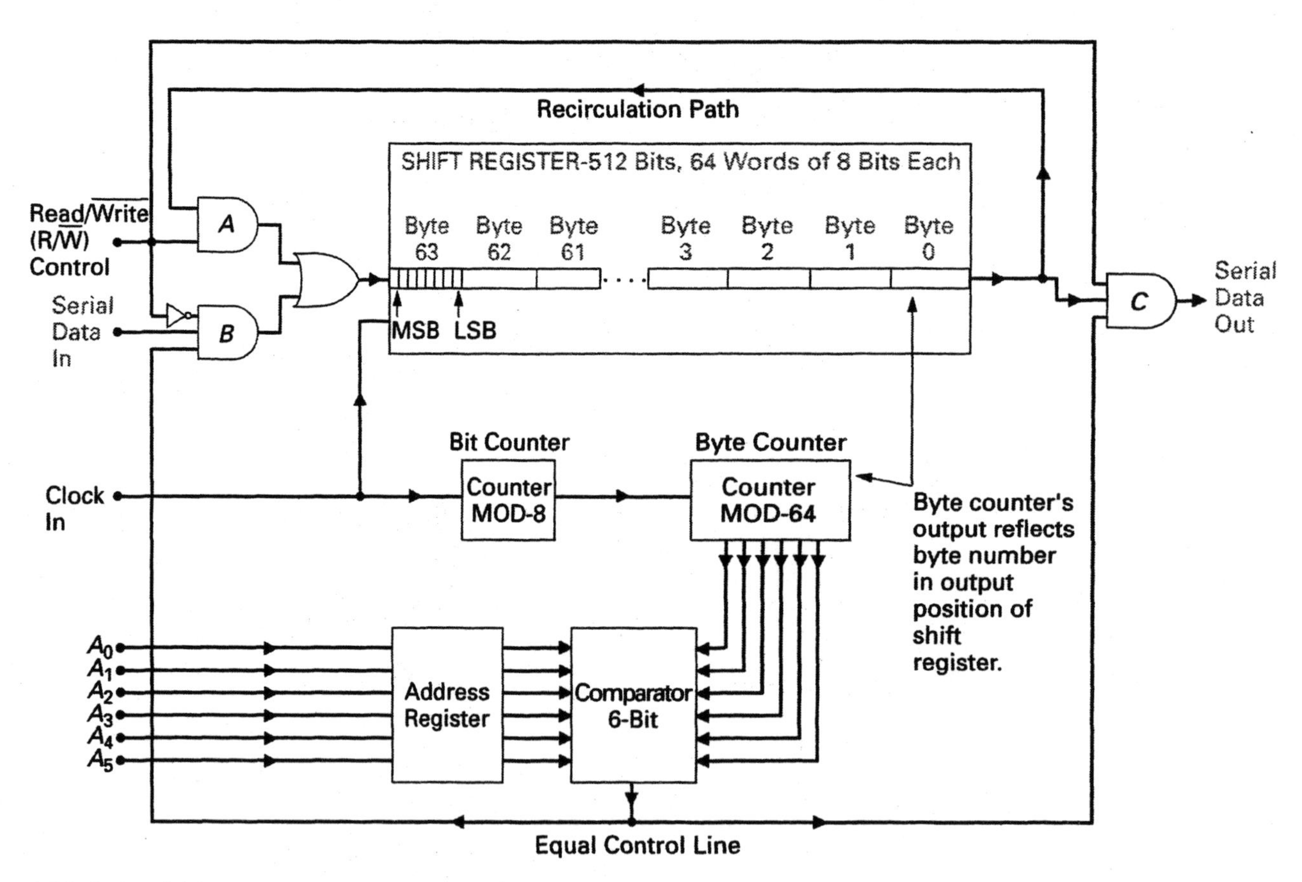

FIGURE 14-16 A 64 × 8 Sequential Access Read/Write Memory.

have been shifted right. The byte counter, therefore, keeps track of the number of bytes shifted and which byte is presently being applied to the output.

To access memory location 9, for example, the address 001001 (decimal 9) is loaded into the address register and applied to the comparator. To read or retrieve the contents from memory location 9, the read/write control line is made HIGH, enabling AND gates *A* and *C*. The data bytes stored in the shift register are shifted right and recirculated back to the input via AND gate *A*, until byte 9 is in the output position of the shift register. Since the byte counter's contents always reflect the number of the byte in the output position, the byte counter will also contain 9, and since both inputs to the comparator are equal, its output will go active or HIGH. This signal will enable AND gate *C*, allowing the 8-bit data contents in memory location 9 to be serially shifted out of the memory. After eight clock pulses have occurred and the entire contents of memory location 9 have been applied to the output, the byte counter will increment and the comparator's output will go inactive, completing the read-from-memory-9 operation.

To write or store an 8-bit data value into memory location 9, for example, we must once again load 9 into the address register and then shift and recirculate the contents of the shift register until byte 9 is in the output position. At this time, the byte counter will contain 9, the equal control line from the comparator will go HIGH, and this signal will be applied to AND gates *B* and *C*. Since the read/write control line is LOW in this instance, AND gates *A* and *C* will be disabled while AND gate *B* will be enabled. The 8-bit serial word to be stored will be shifted and stored in the byte 9 memory location, while the old contents of memory location 9 are lost since the recirculation path is blocked by the disabled AND gate *A*. After eight clock pulses have been applied, and all eight bits of the serial input have been stored, the byte counter will increment and disable the equal control line and AND gate *B*, completing the write operation.

To access a memory location in a serial memory, therefore, you must step or sequence through all of the bits until the desired word is located. This is why this memory type is called a sequential access memory (SAM). As you can imagine, the access times for SAMs memory are very slow, and this is their major disadvantage. They are, however, still used today in serial data, low-cost applications where high-speed access is not required. Other sequential access memory devices that are not widely used, except in specialized applications, are *charge coupled devices (CCDs)* and *magnetic bubble memories (MBMs)*.

Unlike the serial memory which has its memory locations end-to-end, a *parallel memory* has all of its memory locations stacked, as seen in Figure 14-17. This means that there is no need to sequence through every memory location, since the address decoder will enable only one memory location and therefore allow us to randomly access any byte. For example, the 512-bit register

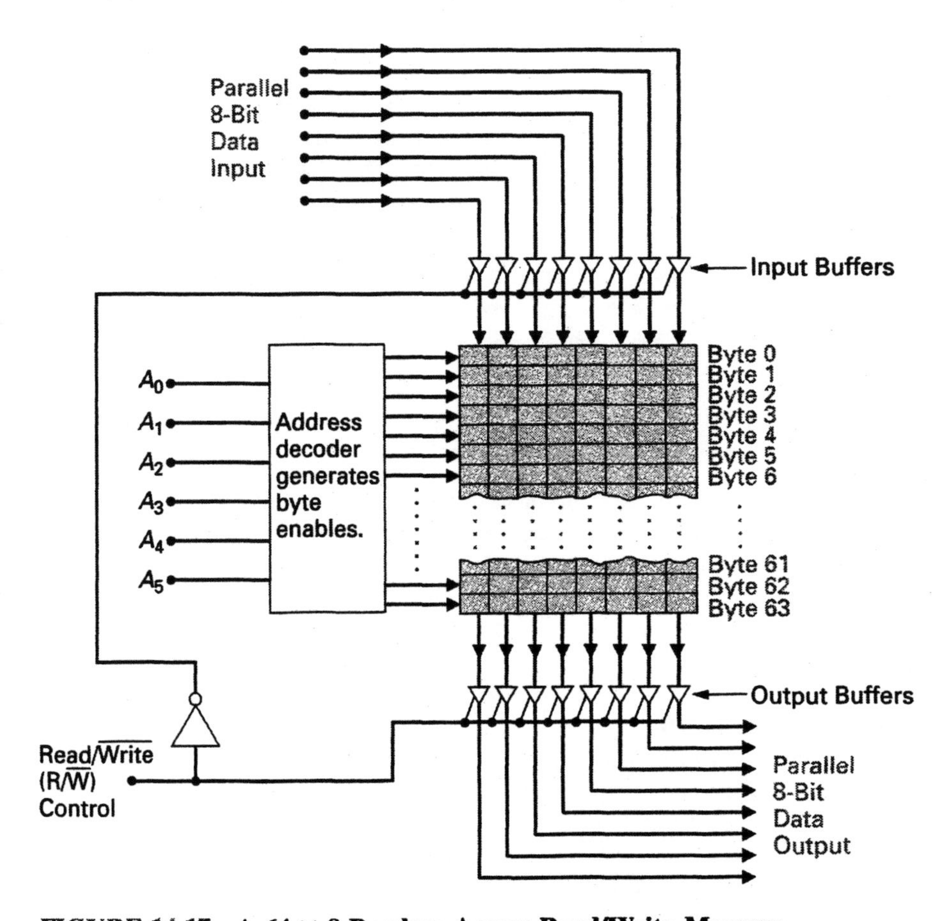

FIGURE 14-17 A 64 × 8 Random Access Read/Write Memory.

block in Figure 14-17 is organized as sixty-four words of eight bits each. These sixty-four bytes are numbered byte 0 through byte 63 and are stacked in parallel. To access a memory location in this RAM, first apply a 6-bit address word to the decoder, then take the read/write control line HIGH (read) or LOW (write). If the read/write control line is made HIGH, the output buffers are enabled and the contents of the addressed memory location is read out in parallel onto the 8-bit data output. On the other hand, if the read/write control line is made LOW, the input buffers are enabled and the parallel 8-bit data word being applied to the data input lines is stored in the memory location enabled by the address decoder.

The RAM is more widely used than the SAM, because its parallel format results in faster memory access times. Before we leave this topic of random access, it is important to point out that ROMs are random access in nature, even though the term RAM is now only associated with read/write memories.

14-2-2 RAM Types

Volatile
A memory storage device that will not retain its stored data in the absence of power.

As was shown previously in Figure 4-15, there are two basic types of RAM—the static RAM (SRAM) and the dynamic RAM (DRAM). To begin with, the SRAM and DRAM are both **volatile,** which means that they will lose their data if power is interrupted. As far as storing data is concerned, both of these RAM types use a different technique. Static RAMs, for instance, latch bits of data in high-speed bistable flip-flops, whereas dynamic RAMs store each bit of data as a charge level in a capacitor. The DRAM capacitor circuit is smaller than the SRAM bistable flip-flop circuit; however, the DRAM needs additional refresh circuitry to replenish the charge lost from the capacitors due to capacitor leakage. Each RAM type, therefore, has its own advantage that makes it more suitable for certain applications. For example, the DRAM has a higher storage density but requires refresh circuitry, while the SRAM operates at a faster speed but is larger in size. Let us now examine these two types in more detail, beginning with the static RAM.

Static Random-Access Memories (SRAMs)

Static RAM
A RAM device in which data is stored in a bistable flip-flop circuit.

There are basically two types of **static RAMs** available. *Bipolar SRAMs* use TTL devices to construct bistable flip-flop memory circuits, which have a low storage density (small amount of storage per chip) but operate at a high speed. *MOS SRAMs* use either MOS or CMOS technology along with a capacitor to construct bistable flip-flop storage cells that have a high storage density but operate at a slower speed.

SRAM storage cell. Figure 14-18(a) shows the basic storage cell for a TTL static RAM. This storage cell, which could also be constructed using *n*-channel E-type MOSFETs, is capable of storing one single bit of data.

Write operation: To latch a binary 1 or binary 0 into the bistable flip-flop, the word select control line must be pulsed HIGH to turn ON Q_3 and Q_4. If a HIGH is to be stored, it will appear on the data input line (Data Input = 1, $\overline{\text{Data Input}}$ = 0), and this positive voltage will be applied through Q_3 to the

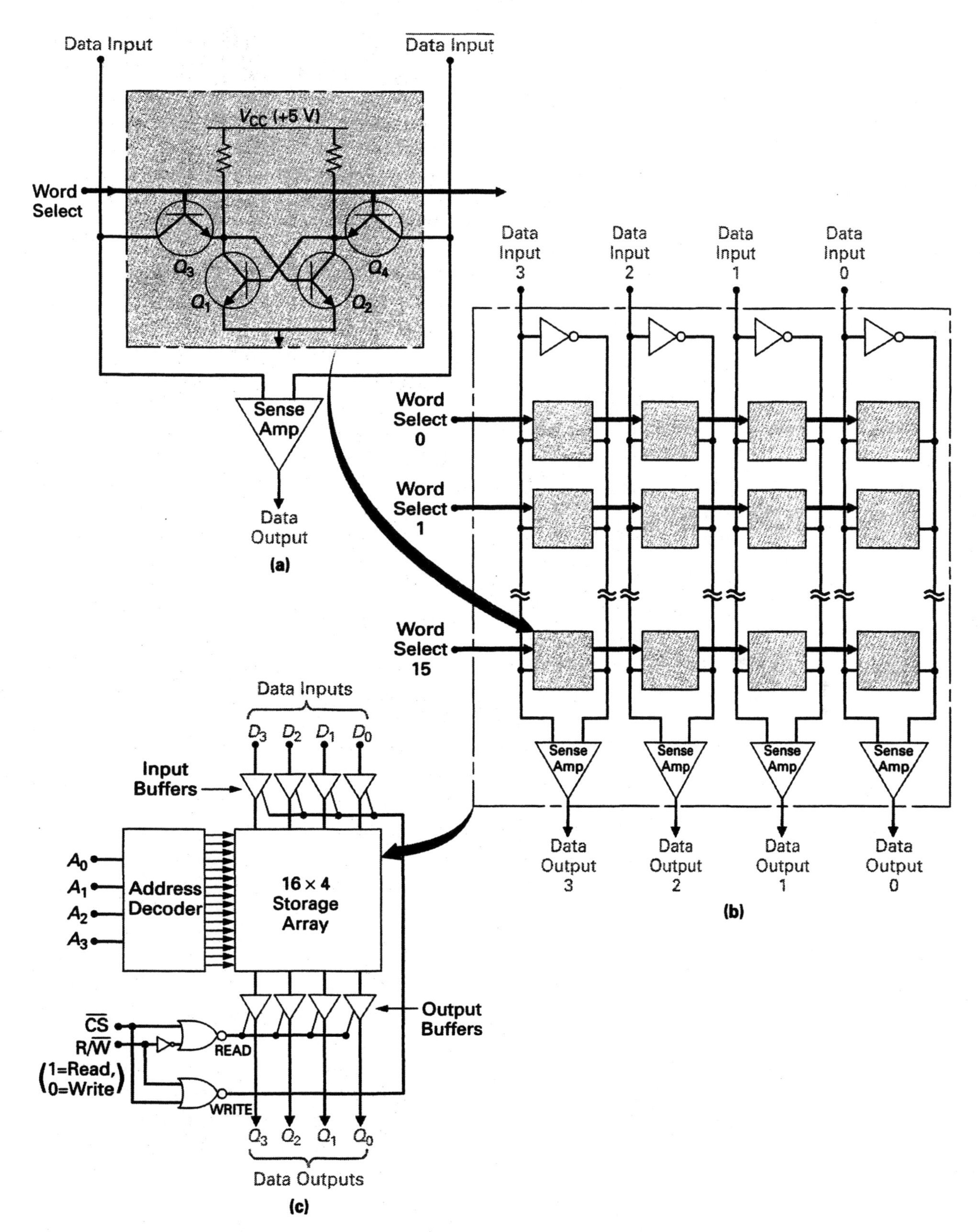

FIGURE 14-18 Static RAM. (a) Storage Cell. (b) Storage Array. (c) Internal Architecture.

base of Q_2. This HIGH will force Q_2 to conduct and its dropping collector voltage will be applied to the base of Q_1, turning Q_1 OFF. With Q_1 OFF, its collector voltage will go HIGH, maintaining Q_2 ON. When the word-select control line goes inactive or LOW, the storage cell will remain in the same condition (Q_2 ON, Q_1 OFF) with a binary 1 latched in the flip-flop due to the cross-coupling between Q_2 and Q_1. On the other hand, if we wanted to store or write a binary 0 into this storage cell, the data input line is made LOW (Data Input = 0, $\overline{\text{Data Input}}$ = 1), and then the word-select control line will again be pulsed HIGH.

When in the read mode, the data inputs are both floating and therefore no data is being applied to the flip-flop. To read or retrieve data out of the storage cell, the word-select line is pulsed HIGH in exactly the same way as the write operation, turning ON Q_3 and Q_4 and switching the stored logic level to the output. For example, if the flip-flop has a logic 1 stored (Q_1 OFF, Q_2 ON), the left input of the sense amplifier (a comparator) will receive a HIGH voltage, while the right input of the sense amplifier will receive a LOW voltage. In this instance, the sense amplifier will produce a HIGH data output. On the other hand, if the flip-flop has a logic 0 stored (Q_1 ON, Q_2 OFF), the left input of the sense amplifier will receive a LOW voltage, while the right input of the sense amplifier will receive a HIGH voltage. In this instance, the sense amplifier will produce a LOW data output.

Figure 14-18(b) shows how sixty-four storage cells could be arranged to form a 16 by 4 (sixteen 4-bit words) static RAM. In this illustration, each of the squares represents one of the storage cells shown in Figure 14-18(a). For simplicity, only three of the sixteen 4-bit sets have been shown in Figure 14-18(b). Each of the data inputs is inverted to provide a *Data Input* and a $\overline{\textit{Data Input}}$. A sense amplifier is required for each of the four data output lines, so four sense comparator amplifiers are included at the bottom of the storage array.

SRAM storage array. Figure 14-18(c) shows all of the basic blocks in a 16 by 4 static RAM. The four-line to sixteen-line address decoder will decode the 4-bit address and make active one of the sixteen word select control lines. The three-state input buffers are enabled by a HIGH signal on the write control line, while the three-state output buffers are enabled by a HIGH on the read control line. These write and read control lines are controlled by the chip-select ($\overline{CS}$) and read/write ($R/\overline{W}$) inputs. When the $\overline{CS}$ and $R/\overline{W}$ control lines are both LOW, the input buffers are enabled and data is written into the addressed memory location. On the other hand, when the $\overline{CS}$ is LOW and $R/\overline{W}$ control line is HIGH, the output buffers are enabled and the data at the addressed memory location is read out and applied to the Q outputs. Therefore, when the $\overline{CS}$ line is LOW (SRAM chip is selected or enabled), the logic level on the $R/\overline{W}$ control line will govern what operation is performed (1 = read, 0 = write). On the other hand, if the $\overline{CS}$ line is switched HIGH, both the input buffers and output buffers are disabled, and the SRAM chip is effectively disconnected from the data input and data output bus lines.

SRAM read and write timing. The logic symbol for the 16 by 4 SRAM is shown in Figure 14-19(a). Now that we have covered the operation of the SRAM cell and the operation of an SRAM's internal architecture, let us see

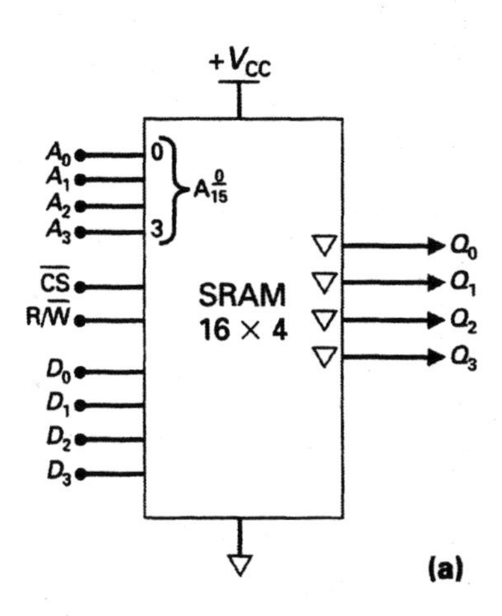

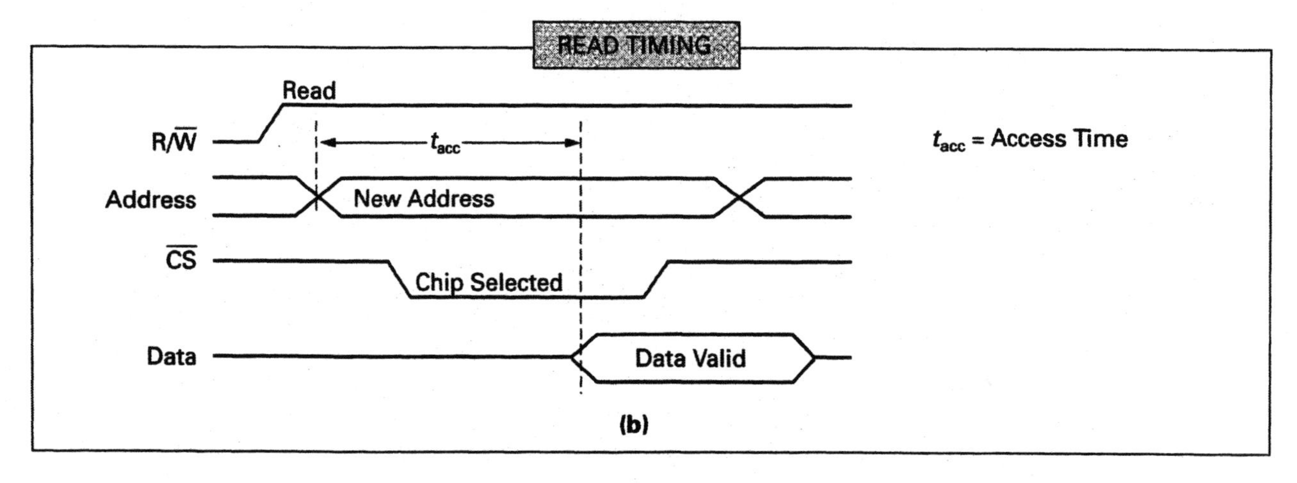

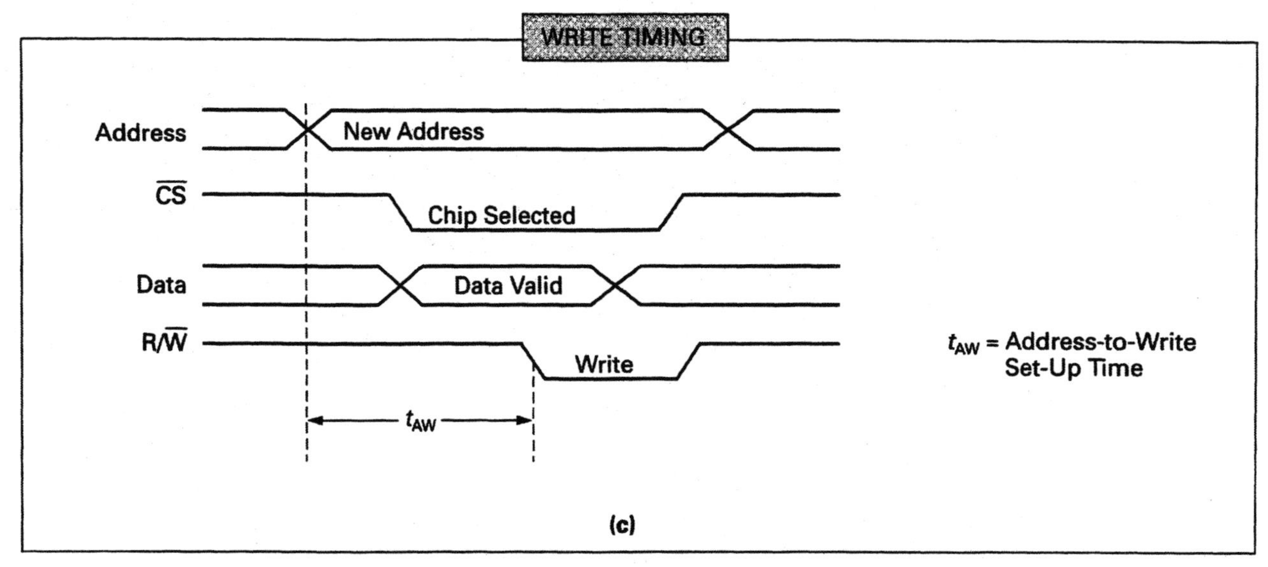

FIGURE 14-19 Static RAM (SRAM) Logic Symbol and Read/Write Timing.

how all of the control and input/output lines are used to perform a memory-read and a memory-write operation.

Figure 14-19(b) shows the typical timing waveforms for an SRAM read operation. Working from the top waveform to the bottom, you can see that the $R/\overline{W}$ control line input is first made HIGH (read operation). The desired address is then applied to the 4-bit address inputs, and then the chip select control input is activated ($\overline{CS}$ = LOW). After a short delay, the data stored at the addressed memory location will appear at the Q outputs. The *access time* (t_{acc}) of a memory IC is governed by the IC's internal address decoder. Assuming no delays are introduced by the $R/\overline{W}$ or $\overline{CS}$ control lines, the read-access time of a memory IC is the time it takes for a new address to be decoded and the contents of the memory location to appear at the output.

Figure 14-19(c) shows the typical timing waveforms for an SRAM write operation. Working from the top waveform to the bottom, you can see that the desired address is first applied to the 4-bit address inputs, and then the chip select control input is activated ($\overline{CS}$ = LOW). The data to be stored is then applied to the data inputs, and when the $R/\overline{W}$ control line input is brought LOW, the data is written into the addressed memory location. The *address-to-write set-up time* (t_{AW}), is a time delay that must be introduced between a new address being applied and the $R/\overline{W}$ control line going LOW. This time delay is needed so that the memory IC's internal decoder has enough time to decode and activate the new address before it is signaled to perform a write operation.

SRAM data sheets. Figure 14-20(a) shows the data sheet details for the 2114A, which is a typical static RAM IC. The ten address input lines (A_0 through A_9) are used to select one of the 1,024 internal memory locations. The active-LOW chip-select ($\overline{CS}$) control line input is used to enable the IC, and the active-LOW write-enable ($\overline{WE}$) control line input operates in exactly the same way as the previously discussed $R/\overline{W}$ input. Each memory location in this IC can store a 4-bit word, and a set of four data lines is used for both input (write) and output (read) operations. During a read operation, these four data input/output (I/O) lines become output lines, and the 4-bit word from the addressed memory location appears on these four I/O lines. During a write operation, these four data input/output lines become input lines, and the 4-bit word applied to these four I/O inputs is written into the addressed memory location. It is the $\overline{WE}$ control line that determines whether these four I/O data lines are being used for input or output operations.

It is probably safe to say that a single RAM IC chip is not normally large enough for a digital system's memory requirements. In most applications, several RAM ICs will be interconnected to form a larger RAM memory circuit. Figure 14-20(b) shows how two 2114A ICs could be connected to form a 1k by 8 RAM (1,024 8-bit words). The 10-bit address input is applied to both ICs along with the $\overline{CS}$ and $\overline{WE}$ control line inputs. IC5 connects to data lines 0 through 3, and IC6 connects to data lines 4 through 7. These two 2114As are therefore effectively connected in parallel, providing an 8-bit read/write data capability.

In Chapter 17 you will be introduced to "SAM," which is a "simple all-purpose microcomputer" system. Referring to the SAM schematic diagram,

DIGITAL DEVICE DATA SHEET

IC TYPE: 2114A – 1k × 4 STATIC RAM

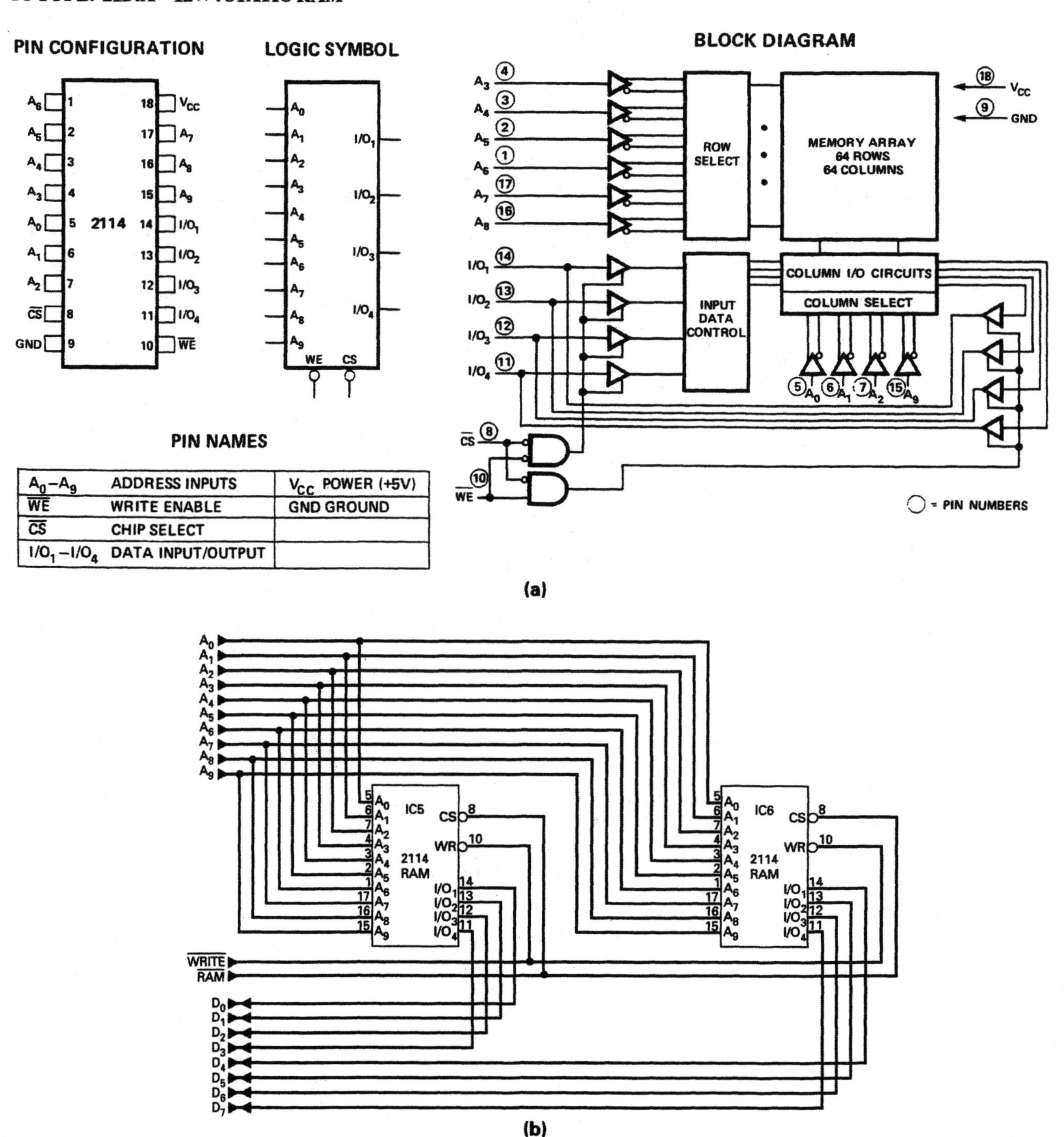

FIGURE 14-20 An IC Containing a 1,024 by 4-Bit Bipolar Static RAM. (a) Data Sheet. (b) Application Circuit. (Part (a) courtesy of Intel Corporation)

which is shown on pages 689–692, you should be able to find the 2114 RAM memory circuit shown in Figure 14-20. The IC numbers are the same (IC5 = U5, IC6 = U6), and the control signals are also the same ($\overline{\text{WRITE}}$ = NWR or not write, $\overline{\text{RAM}}$ = NRAM or not ram).

As another example, Figure 14-21 shows the logic symbol and function table for a TMS4016, which is a 2k by 8 MOS static RAM IC. Eleven address line inputs are used to select one of the 2,048 memory locations, and eight input/output data lines are available.

Dynamic Random-Access Memories (DRAMs)

Dynamic RAM
A RAM device in which data is stored as a charge which needs to be refreshed.

As mentioned previously, there are static RAMs and **dynamic RAMs.** As we saw in the previous section, memory systems using static RAMs appear to be ideal since they are fast, there are many types to choose from, and their circuitry is simple. The question, therefore, is Why bother having another type of RAM? There are two good reasons why dynamic RAMs are ideal in certain applications. The first is that you can package more storage cells into a dynamic RAM IC than you can into a static RAM IC. This can be easily seen by comparing the single static RAM cell shown in Figure 14-22(a), to the single dynamic RAM cell shown in Figure 14-22(b). A dynamic RAM storage cell has only one MOSFET and a capacitor compared to a static RAM storage cell which normally contains six MOSFETs. Therefore, approximately three dynamic storage cells could fit into the same space as one static storage cell, giving the

DIGITAL DEVICE DATA SHEET

IC TYPE: TMS4016 – 2,048 × 8 STATIC RAM

LOGIC SYMBOL

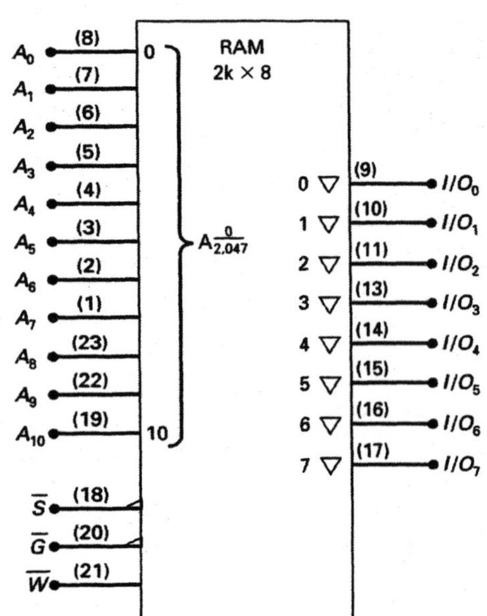

FUNCTION TABLE

$\overline{W}$	$\overline{S}$	$\overline{G}$	I/O_0 – I/O_7	Mode
0	0	0	Valid Data	Write
1	0	0	Data Output	Read
X	1	X	High Impedance	Three-State
X	X	1	High Impedance	Three-State

FIGURE 14-21 An IC Containing a 2,048 by 8 MOS Static RAM.

DRAM a large space-saving advantage. The second reason for using dynamic RAMs is power dissipation. Since power dissipation is dependent on the number of devices per storage cell, it is easy to imagine why DRAMs have a lower power dissipation since fewer MOSFETs are used for each storage cell.

DRAM storage cell. The DRAM storage cell shown in Figure 14-22(b) operates in the following way.

To write data into this storage cell, the data bit to be stored is first applied to the data-in line, and then the bit select control line is pulsed HIGH. A HIGH on the bit select control line will cause Q_1 to conduct (*n*-channel MOSFET), allowing either the HIGH or LOW data input to be applied directly to the capacitor, causing it to either charge (a 1 is stored) or discharge (a 0 is stored).

To read data out of this storage cell, the bit select control line is switched HIGH to turn ON Q_1, allowing either the stored binary 1 (capacitor charged) or binary 0 (capacitor discharged) to pass through to the data-out line.

Now let us see how these dynamic storage cells are connected to form a complete dynamic RAM circuit.

DRAM storage array. Nearly all dynamic RAM ICs are organized as multiple 1-bit read/write memories. This means that they have multiple memory locations; however, each memory location has only a 1-bit word (for example, 16k by 1-bit, 64k by 1-bit, 256k by 1-bit, and so on).

Figure 14-22(c) shows the internal architecture of a typical dynamic RAM circuit. This dynamic RAM circuit has 16,384 1-bit memory locations, arranged into a 128 by 128 matrix. Each of the dynamic cells in this matrix has its own unique row and column position in the matrix. To address each of these 16,384 1-bit memory locations will require fourteen address inputs ($2^{14} = 16{,}384$). Looking at the dynamic RAM circuit in Figure 14-22(c), you will notice that this circuit only has seven address input pins. To reduce the number of pins on the IC package, manufacturers often use an **address multiplexing** technique, in which an address input pin is used to input two different address bits. With this technique, the 14-bit address is applied to the dynamic RAM address decoder in two 7-bit groups. First, the seven lower-order address bits (A_0 through A_6) are applied to the seven address inputs, and are stored in the row address register and decoded by the row address decoder by activating the *row address strobe* ($\overline{RAS}$). Second, the seven higher-order address bits (A_7 through A_{13}) are applied to the seven address inputs, and are stored in the column address register and decoded by the column address decoder by activating the *column address strobe* ($\overline{CAS}$). The row address strobe and column address strobe control line inputs, therefore, allow us to input a 14-bit address on seven address input pins, which decreases the size of the IC package, decreasing the size of memory circuits and the final digital system. The outputs of the row and column address registers drive the row and column address decoders, and these decoders will bit-select one of the 16,384 dynamic RAM storage cells. Once activated, the read/write control line input will determine whether a read operation (storage cell → data-out pin) or write operation (data-in pin → storage cell) is performed.

Address Multiplexing
A time-division technique in which two or more address codes are switched through a single input at different times.

The dynamic RAM, therefore, uses an on-chip capacitor for each storage element. A charge is stored on the capacitor to indicate a binary 1, and no

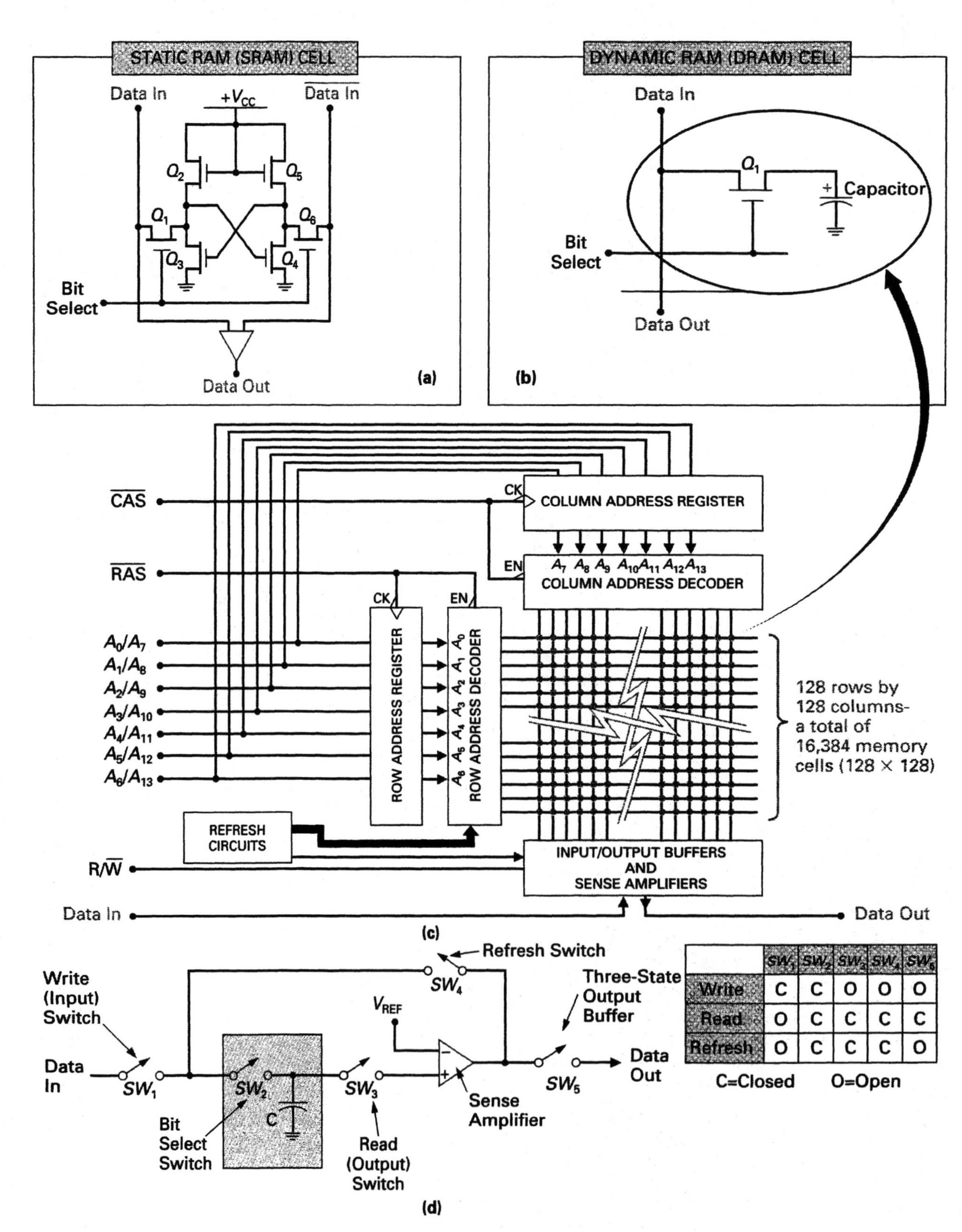

	SW_1	SW_2	SW_3	SW_4	SW_5
Write	C	C	O	O	O
Read	O	C	C	C	C
Refresh	O	C	C	C	O

C=Closed O=Open

FIGURE 14-22 Dynamic RAMs. (a) Static Cell. (b) Dynamic Cell. (c) Internal Architecture. (d) Simplified DRAM Cell Operation.

charge on the capacitor is used to indicate a binary 0. This technique simplifies the storage cell, permitting denser memory chips that dissipate less power. There is a problem, however, in that the charge leaks off the capacitor due to capacitor leakage, and therefore a binary 1 can become a binary 0 in just a few milliseconds. When a binary 1 is stored in a DRAM, therefore, it must be *refreshed* continually so that the data stored within the DRAM is accurate. Most DRAM ICs today have on-chip refresh circuits, as shown in Figure 14-22(c), to automatically rewrite the same data back into each memory location. This refresh action will typically take place every two ms. After a refresh, all of the binary 1s are restored to full charge and all of the binary 0s are left as no-charge.

DRAM write, read, and refresh. To explain the write, read, and refresh operations in more detail, Figure 14-22(d) shows a simplified DRAM cell and its input/output circuitry. In this diagram the circuit's three-state input buffer, three-state output buffer, and the internal cell MOSFETs have been represented as switches, which are under the control of the read/write input. The table associated with this illustration indicates which of the switches are open and closed for each of the three operations.

To write data to the cell, SW_1 and SW_2 are closed so that the data input can be applied directly to the capacitor. At the same time, SW_3, SW_4, and SW_5 are opened to disconnect the storage capacitor from the rest of the circuit.

To read data from the cell, SW_3 is closed to connect the stored capacitor voltage to the sense amplifier. The sense amplifier compares the stored capacitor voltage to a reference voltage in order to determine whether a logic 0 or logic 1 was stored, and then produces a valid 0 V or 5 V (binary 0 or 1) at its output. SW_5 switches the stored bit from the sense amplifier through to the data output. In this mode, SW_4 is also closed, along with SW_2, to refresh the capacitor voltage (recharge to reinforce the HIGH, or discharge to reinforce the LOW). During each read operation, therefore, a refresh action is performed.

To only refresh the cell, SW_2, SW_3, and SW_4 are closed, so that the logic level stored in the capacitor can be fed back to the capacitor to reinforce the stored HIGH or LOW.

DRAM read and write timing. The logic symbol for the previously discussed 16,384 by 1-bit DRAM is shown in Figure 14-23(a). Now that we have covered the operation of the DRAM cell and the operation of a DRAM's internal architecture, let us see how all of the control and input/output lines are used to perform a memory-read and a memory-write operation.

Figure 14-23(b) shows the typical timing waveforms for a DRAM read operation. Working from the top waveform to the bottom, you can see that first the seven lower-order address bits (row address) are applied to the seven address input pins. This row address input is strobed, or latched, into the row address register when the row address strobe control line ($\overline{RAS}$) is brought LOW. Then the seven higher-order address bits (column address) are applied to the seven address input pins. This column address input is strobed, or latched, into the column address register when the column address strobe control line ($\overline{CAS}$) is brought LOW. The $R/\overline{W}$ control line is used to control the input and output buffers, and therefore control whether a read or write operation is performed. In this example, the $R/\overline{W}$ control line goes HIGH after the $\overline{CAS}$ goes

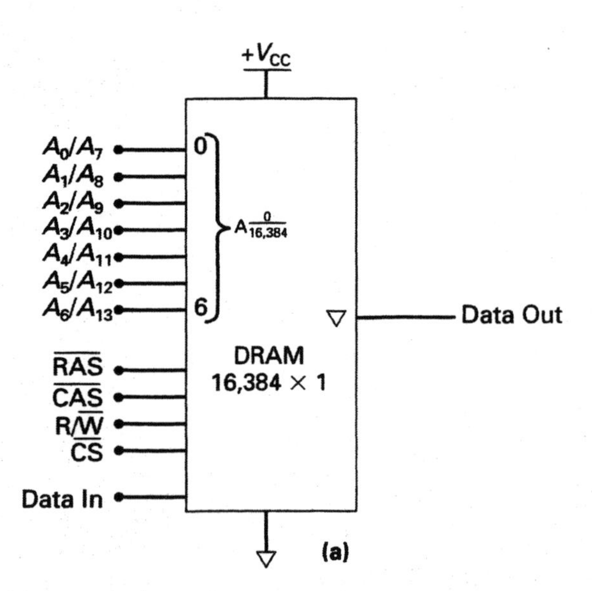

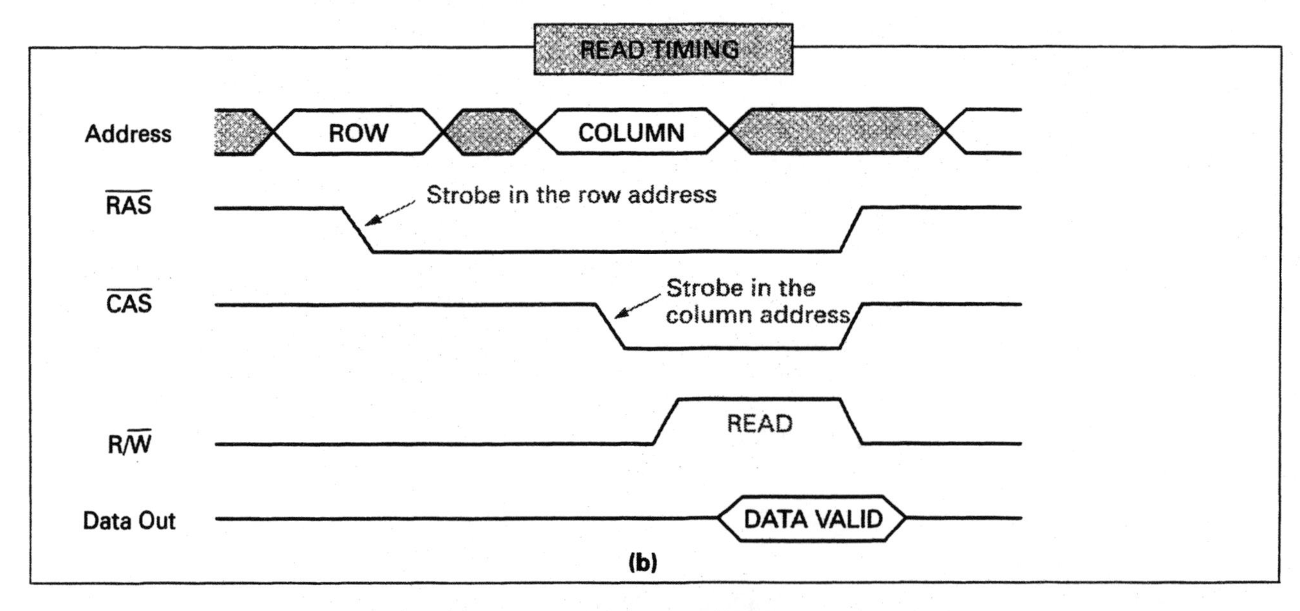

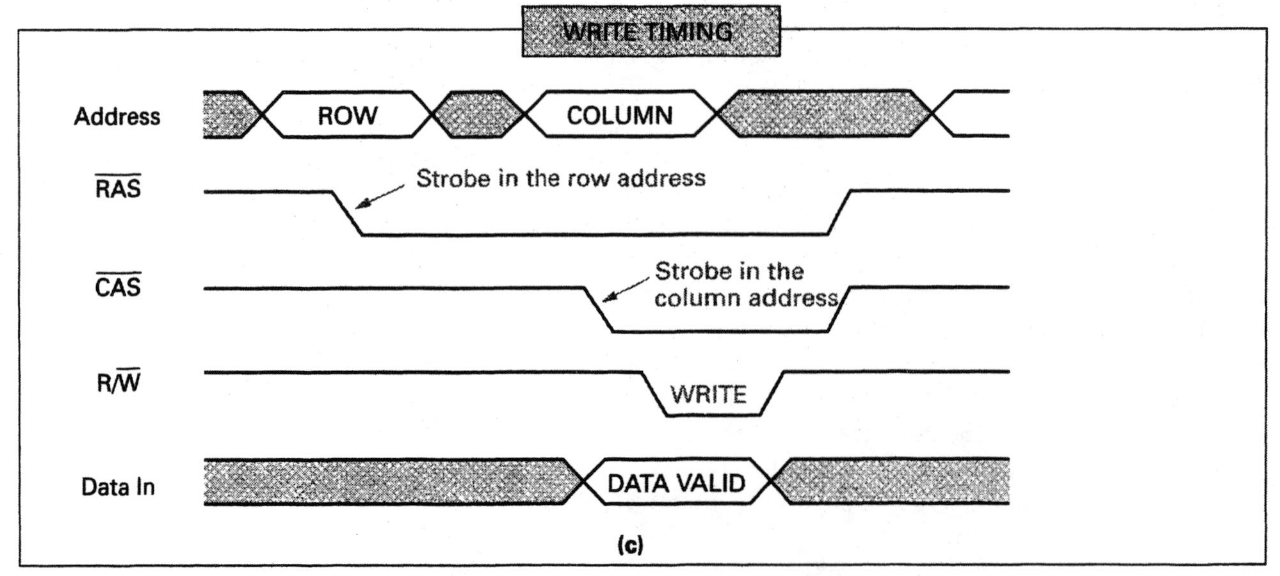

FIGURE 14-23 Dynamic RAM (DRAM) Logic Symbol and Read/Write Timing.

LOW, to enable a read operation. The three-state output buffers will therefore be taken out of their high impedance state and the stored binary 1 or binary 0 will appear at the data-out pin.

Figure 14-23(c) shows the typical timing waveforms for a DRAM write operation. Working from the top waveform to the bottom, you can see that the row address and column address are latched into the DRAM in exactly the same way as the read operation. Once the 14-bit address has been inputted, the $R/\overline{W}$ control line will control whether a read or write operation is performed. In this example, the $R/\overline{W}$ control line goes LOW after the $\overline{CAS}$ goes LOW to enable a write operation. The input switch will therefore be enabled, and the binary 1 or binary 0 input being applied to the data-in pin will be stored or written into the addressed memory location.

As mentioned previously, nearly all dynamic RAM ICs are organized as multiple 1-bit read/write memories. This means that they have multiple memory locations; however, each memory location has only a 1-bit word (for example, 16k by 1-bit, 64k by 1-bit, 256k by 1-bit, and so on). Figure 14-24 shows how

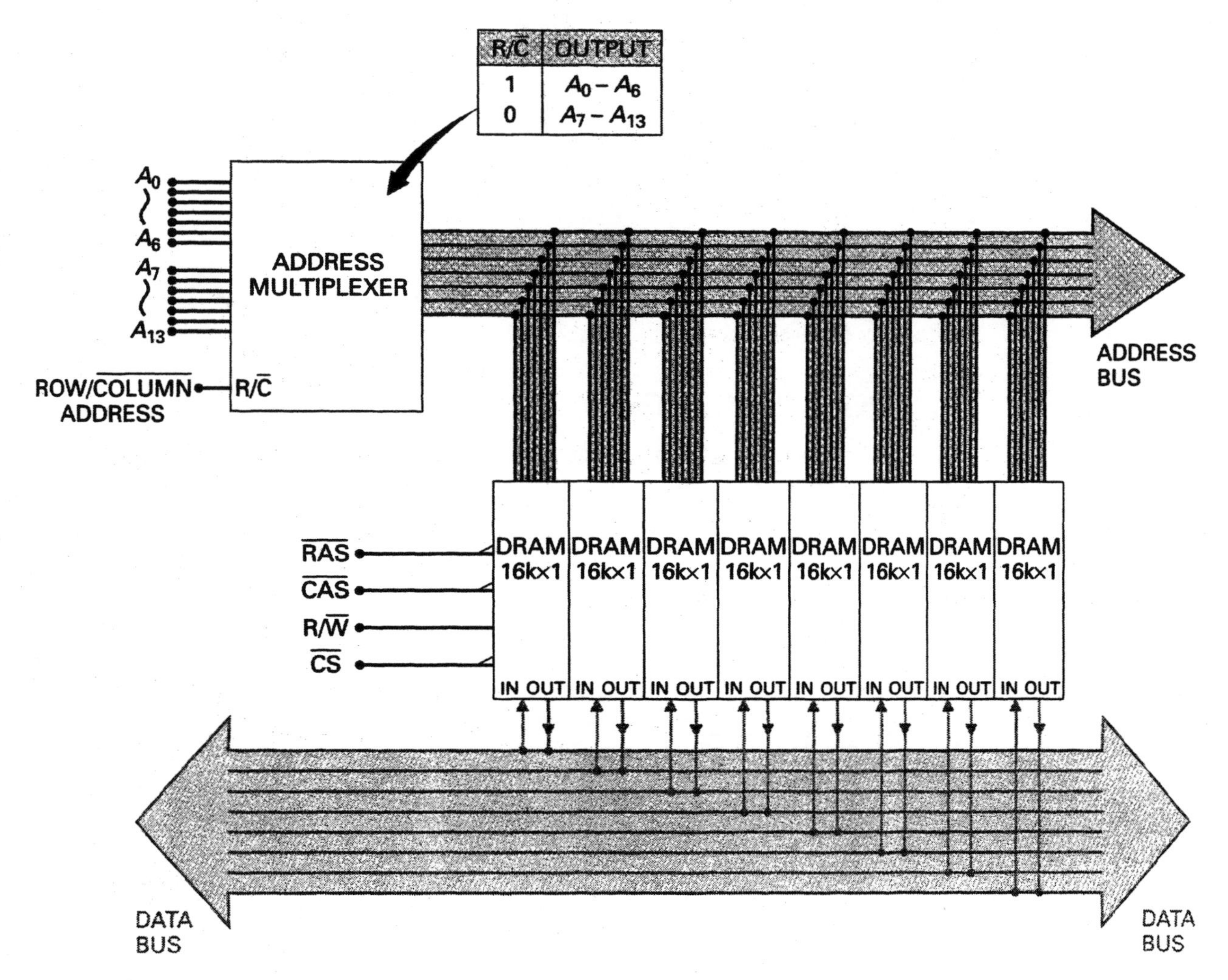

FIGURE 14-24 A 16k by 8-Bit DRAM Memory Circuit.

eight 16k by 1-bit DRAM ICs can be connected to form a 16k by 8-bit DRAM read/write memory circuit. An address multiplexer is used to switch the seven lower-order address bits through to the DRAMs just before $\overline{RAS}$ goes active, and then switch the seven higher-order address bits through to the DRAMs just before the $\overline{CAS}$ goes active. The $\overline{RAS}$, $\overline{CAS}$, $R/\overline{W}$, and $\overline{CS}$ control line inputs are connected to all eight DRAM ICs so that they are all activated simultaneously.

DRAM data sheets. Only MOSFET dynamic RAM ICs are available, and their sizes range from 16k by 1-bit (16kbit) to 16M by 1-bit (16Mbit). Figure 14-25 shows the data sheet logic symbol for the TMS4164, which is a typical 64k by 1-bit (64kbit) dynamic RAM IC which, like most DRAM ICs today, has a built-in refresh circuit. The operation of this IC is not that different than the previously described 16k by 1-bit DRAM, except that it is four times larger. Address multiplexing is again used to reduce the IC's pin count, and therefore the $\overline{RAS}$ control input is used to latch the eight lower-order address bits into the IC when they are being applied to the eight address input pins, and the $\overline{CAS}$ control input is used to latch the eight higher-order address bits into the IC when they are being applied to the eight address input pins. As can be seen in the table in Figure 14-25, the level and edge of the $\overline{RAS}$ control is a multi-function input. It is used to power down the DRAM (low-power standby mode), initiate a refresh operation, and strobe in the row address.

14-2-3 *RAM Applications*

As far as applications are concerned, dynamic RAMs are more widely used than static RAMs because they are smaller in size, lower in cost, and dissipate less power than an equivalent static RAM. Static RAMs, however, have faster

DIGITAL DEVICE DATA SHEET

IC TYPE: TMS4164 – 64k × 1-BIT DYNAMIC RAM (DRAM)

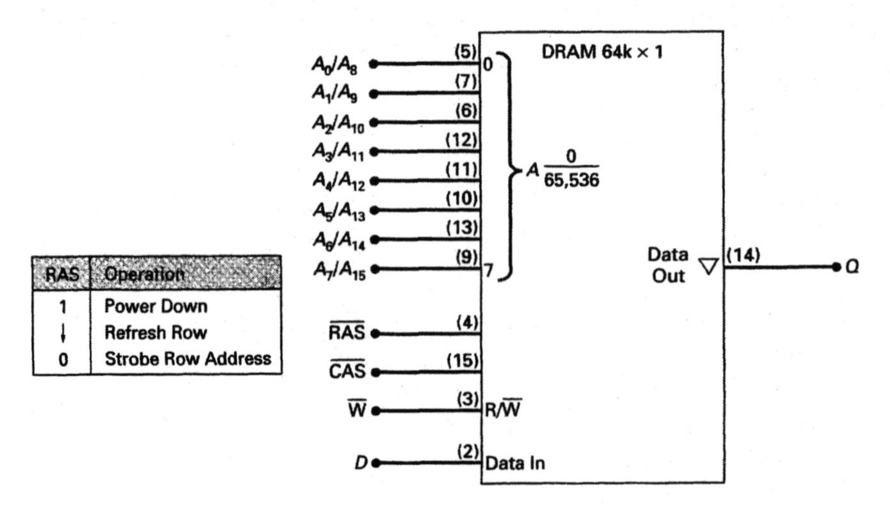

RAS	Operation
1	Power Down
↓	Refresh Row
0	Strobe Row Address

FIGURE 14-25 An IC Containing a 65,536 by 1-Bit DRAM.

access times and do not require refresh circuitry and the special timing requirements that go along with refreshing the stored data (you must refresh at least every 2 ms). Semiconductor memory manufacturers, however, are continually improving the DRAM, making its disadvantages easier to deal with.

SRAMs or DRAMs

In general, static RAMs are used when only a small amount of read/write memory is required (64k or less), or when high-speed access is needed. Dynamic RAMs, on the other hand, are used in applications where a large amount of read/write memory is needed (megabytes). Some systems employ a small amount of SRAM for high-speed read/write memory requirements and a large amount of DRAM for the bulk of the system's read/write memory needs.

A Typical RAM Memory Circuit

Whether the application is a personal computer or a digital oscilloscope, all digital computer systems use RAM. As discussed previously, a microprocessor unit (MPU) is at the heart of every digital computer system. The MPU is a complex integrated circuit that contains control logic circuits, instruction decoding circuits, temporary storage circuits, and arithmetic processing circuits. To be useful, the microprocessor IC must be able to exchange information with memory ICs (both ROM and RAM) input devices, such as a keyboard, and output devices, such as a display.

Figure 14-26(a) shows how a typical RAM memory circuit could be connected in a digital computer system. At this time, we will simply think of the MPU as a system controller that can select or address any memory location in RAM, and then either read data from that memory location, or write data to that memory location. It achieves this by controlling the address bus, the data bus, and generating control signals. To review, the data bus is used for data transfers, and since all devices share this bus, the microprocessor will only activate one device at a time, and then read data from that device via the data bus, or write data to that device via the data bus (under the control of the $R/\overline{W}$ control line from the MPU). Before a data transfer can take place on the data bus, however, the microprocessor must output an address onto the address bus. This address specifies which memory location the microprocessor wishes to access.

In Figure 14-26(a), four 256 by 8-bit RAM ICs have been connected to form a 1,024 (4 × 256) by 8-bit RAM memory. Since most computer systems use more than one memory IC, the address from the MPU must indicate which memory IC should be selected, and which word within that IC should be accessed. This is achieved by assigning a unique set of addresses to each memory IC.

The **memory map** shown in Figure 14-26(b) shows how each of the four RAM ICs has its own unique set of 256 addresses. For example, addresses 0_{10} through 255_{10} (0000_{16} through $00FF_{16}$) are assigned to RAM 0, addresses 256_{10} through 511_{10} (0100_{16} through $01FF_{16}$) are assigned to RAM 1, addresses 512_{10} through 766_{10} (0200_{16} through $02FF_{16}$) are assigned to RAM 2, and addresses 767_{10} through 1023_{10} (0300_{16} through $03FF_{16}$) are assigned to RAM 3.

Memory Map
An address listing diagram showing the boundaries of the address space.

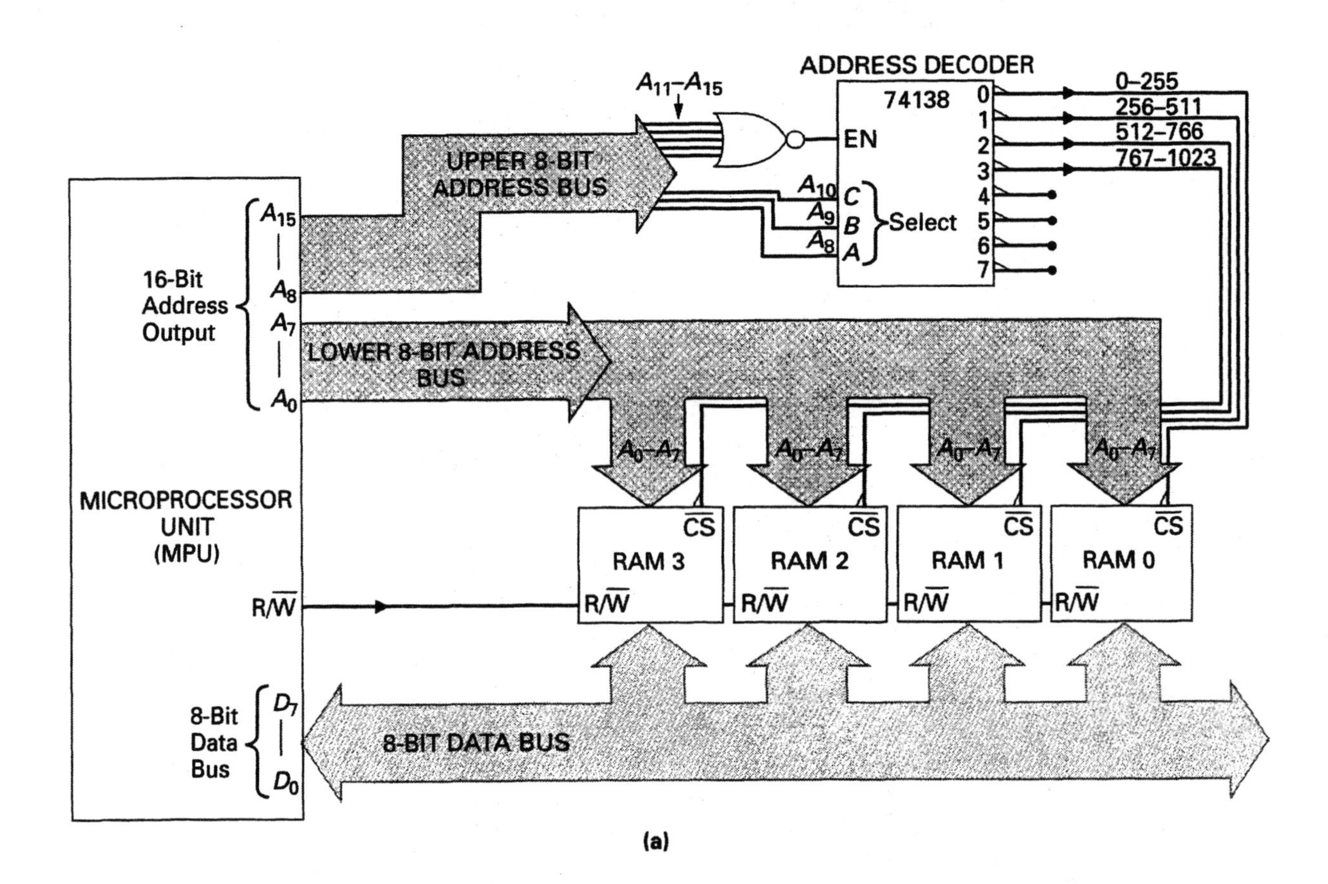

(a)

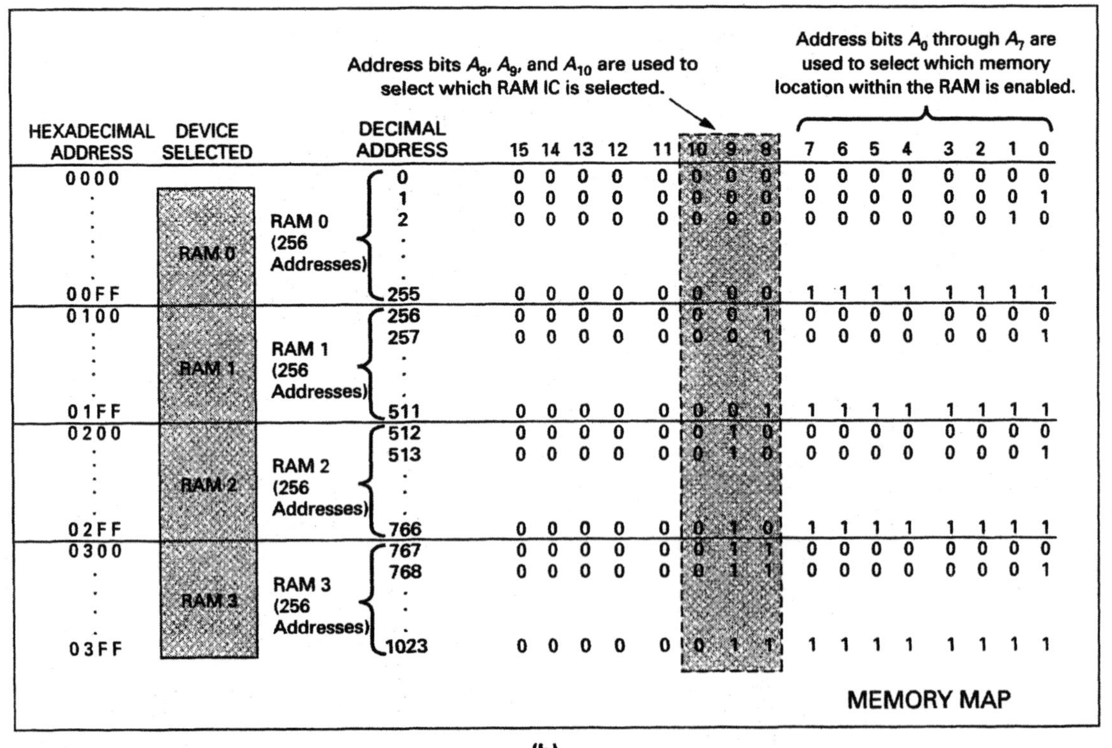

(b)

FIGURE 14-26 A Typical RAM Memory Circuit.

Referring to the 16-bit address bus from the MPU in Figure 14-26(a), you can see that the upper eight bits of the address bus are applied to an **address decoder,** and the lower eight bits of the address bus are applied to all four RAMs. By decoding address bits A_8 through A_{15}, the 74138 address decoder can determine which one of the four RAM ICs should be enabled. The eight lower-order address bits from the MPU (A_0 through A_7) will then be decoded by the selected RAM IC to determine which word within the RAM IC is accessed. In the circuit in Figure 14-26(a), you can see that address bits A_8, A_9, and A_{10} are applied to the *ABC* select inputs of the 74138.

Address Decoder
A circuit that decodes the applied addresses and activates certain outputs based on which zone the address is within.

In the memory map in Figure 14-26(b), you can see how address bits A_8, A_9, and A_{10} will advance by one for every 256 addresses. For example, address bits A_8, A_9, and A_{10} will remain at 000 (decimal 0) for the first 256 addresses, advance to 001 (decimal 1) for the next 256 addresses, advance to 010 (decimal 2) for the next 256 addresses, and so on. Since these three bits are applied to the 74138s *ABC* select inputs, output 0 will go active (LOW) for the first 256 addresses, output 1 will go active (LOW) for the next 256 addresses, output 2 will go active (LOW) for the next 256 addresses, and so on. Therefore, by connecting each of the 74138's active-LOW outputs to each of the active-LOW chip-select inputs of the RAM ICs, as shown in Figure 14-26(a), we will ensure that: RAM 0 is only enabled whenever any address from 0000_{16} to $00FF_{16}$ is generated by the MPU; RAM 1 is only enabled whenever any address from 0100_{16} to $01FF_{16}$ is generated by the MPU; RAM 2 is only enabled whenever any address from 0200_{16} to $02FF_{16}$ is generated by the MPU; and RAM 3 is only enabled whenever any address from 0300_{16} to $03FF_{16}$ is generated by the MPU. The rest of the higher-order address bits (A_{11} through A_{15}) are used to enable (EN) the decoder only when they are all LOW.

To summarize, with the address decoding technique, the lower-order address bits are sent directly to the memory's address line, and the higher-order address bits are decoded to generate the chip selects, with only one IC selected at any given time. In other circuit designs, you may see that the number of bits fed directly to the memory devices may change (depending on the number of words that need to be addressed, 2^n), and this will change which higher-order address bits are decoded to generate chip selects. In the following troubleshooting memory circuits section, you will see another example of how an address decoder is used to create 1k (1,024) memory map blocks, instead of the 256 memory map blocks shown in Figure 14-26(b).

14-2-4 *RAM Testing*

When testing a ROM IC, all we had to do was read out the contents of each memory location and compare it to the ROM listing to see if a memory cell error was present. With a RAM IC, we have to test both the read and the write ability of all the memory cells, making the testing of a RAM a little more lengthy.

Figure 14-27 shows a simple RAM test circuit. In this circuit, a 10-bit counter is triggered to sequence through the 1,024 addresses needed for a 2114 IC being tested. Four data switches can be used to generate any 4-bit word

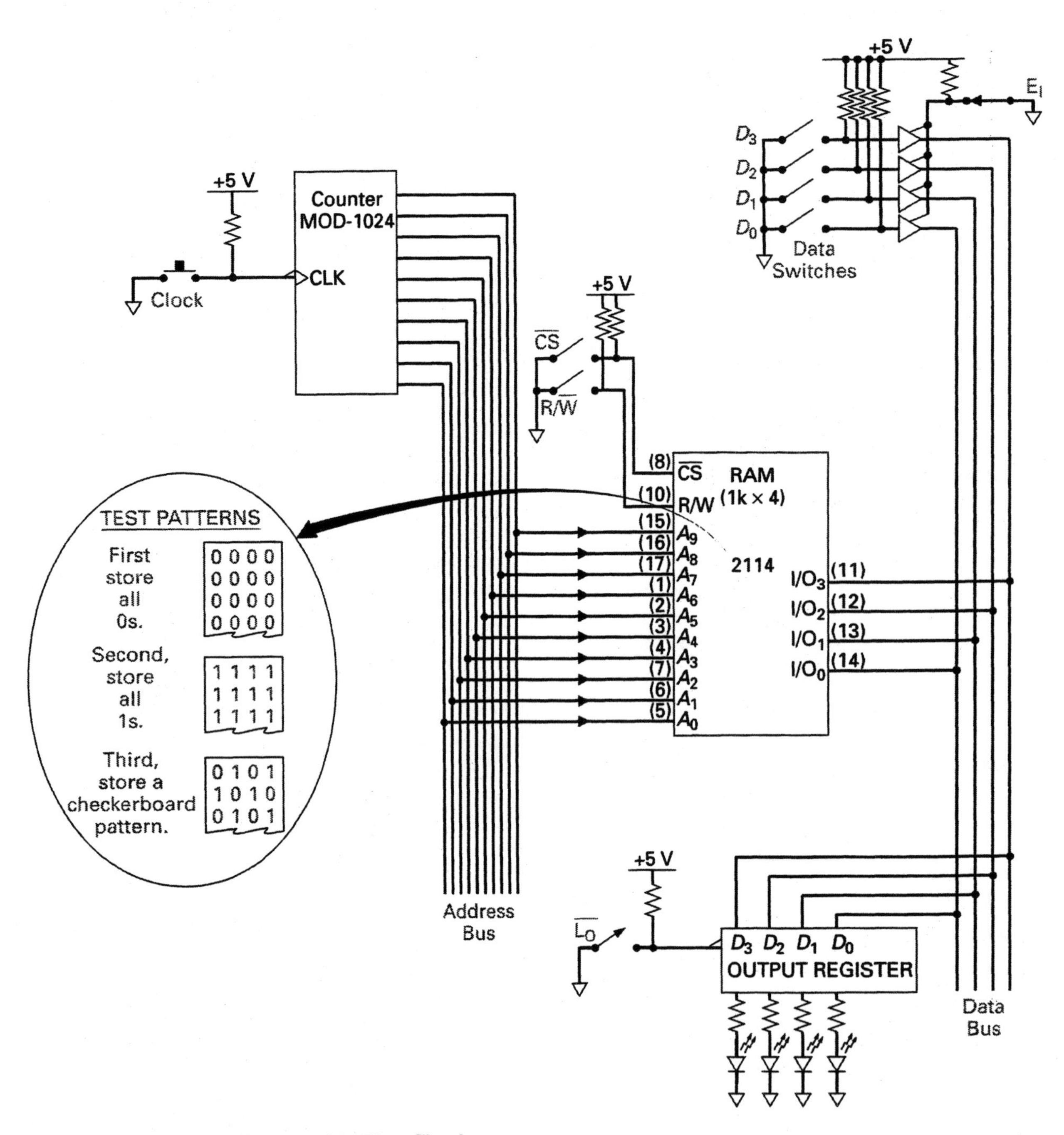

FIGURE 14-27 A Simple RAM Test Circuit.

combination, which can then be applied to the 2114 and stored in any memory location (write operation). A 4-bit output register and LED display is included so that a 4-bit word retrieved from any memory location can be viewed (read operation). RAM ICs are normally tested by first storing 0s in every memory cell and then reading out all of these 0s and checking that no 1s are present. The second step is to store 1s in every memory cell and then read out all of these 1s and check that no 0s are present. These first two checks can be seen in the inset in Figure 14-27. These two checks, however, do not test for a memory cell short between two adjacent memory cells. For example, if one memory cell is shorted to an adjacent memory cell, storing all 1s or storing all 0s will not expose this fault since the short will cause both memory cells to be at the same logic level anyway.

To determine whether a memory cell short exists within a RAM IC, a **checkerboard test pattern** is normally written into the RAM IC, as shown in the inset in Figure 14-27. With the checkerboard test pattern, adjacent cells will have opposite bits stored, and if the logic levels do not alternate when the pattern is read out of the RAM, a memory cell problem exists. After the RAM is checked with this checkerboard pattern, the pattern is normally reversed to further test the RAM, and then read out and checked again. Since there is no way to repair an internal RAM IC failure, the IC must be discarded and replaced if a problem is found.

Checkerboard Test
A RAM testing method in which the computer first stores zeros in every memory location, then stores ones, and then alternates the storage of ones and zeros.

The test circuit in Figure 14-27 would be ideal for testing a small 1k by 4 RAM; however, most digital system RAMs will be much larger, having possibly millions of memory locations, making a complete test with this circuit long and laborious. In these cases, a RAM tester can be used to check the RAM's read/write ability. This test instrument automatically writes several test patterns into the RAM and then reads back the data and compares it to check for any operation or storage error.

Most RAMs within digital electronic computer systems are tested by a built-in self-diagnostic program. This test program will generally use the checkerboard test pattern method to verify the RAM's operation.

SELF-TEST REVIEW QUESTIONS FOR SECTION 14-2

1. What is the difference between a read/write memory and a read-only memory?
2. What is the difference between a SAM and a RAM?
3. Are ROMs RAMs?
4. Is a RAM a volatile or non-volatile device?
5. Describe the difference between a static RAM and a dynamic RAM.
6. What are the advantages and disadvantages of SRAMs and DRAMs?
7. What is a memory map?
8. How are RAMs usually tested?

14-3 TROUBLESHOOTING MEMORY CIRCUITS

To review, the procedure for fixing a failure can be broken down into three steps—diagnose, isolate, and repair. Since this chapter has been devoted to memories, let us first examine the operation of a typical memory circuit, and then apply our three-step troubleshooting process to this circuit.

14-3-1 A Memory Circuit

As an example, we will apply our three-step troubleshooting procedure to the 8-bit digital circuit shown in Figure 14-28. This circuit has eight data input switches, which can be connected through to the 8-bit data bus by making the $\overline{INPUT}$ control line active. The eight-LED display is controlled by the output register, which can be loaded with the contents of the data bus by making the $\overline{OUTPUT}$ control line active. The 1k by 8-bit ROM IC is selected only when the $\overline{ROM}$ control line is active and when the $R/\overline{W}$ switch is HIGH (read-only). The two 2114 ICs in this circuit provide a 1k by 8-bit RAM that is enabled by making the $\overline{RAM}$ control line active and is controlled by the $R/\overline{W}$ switch.

A 13-bit address is supplied by thirteen address switches, shown in the upper left side of Figure 14-28. The upper three bits of the address are applied to an address decoder, and the lower ten bits of the address are connected to an address bus and applied to this circuit's ROM and RAM. The 10-bit lower-order address will be decoded by the selected memory IC to determine which word within the memory is accessed. Address bits A_{10}, A_{11}, and A_{12} are applied to the *ABC* select inputs of the 74138 address decoder, which will determine whether the ROM, RAM, input port, or output port is enabled.

In the memory map in Figure 14-28, you can see how address bits A_{10}, A_{11}, and A_{12} will advance by one for every 1,024 addresses. For example, address bits A_{10}, A_{11}, and A_{12} will remain at 000 (decimal 0) for the first 1,024 addresses, advance to 001 (decimal 1) for the next 1,024 addresses, advance to 010 (decimal 2) for the next 1,024 addresses, and advance to 011 (decimal 3) for the next 1,024 addresses. Since these three bits are applied to the 74138's *ABC* select inputs, output 0 will go active (LOW) for the first 1,024 addresses, output 1 will go active (LOW) for the next 1,024 addresses, output 2 will go active (LOW) for the next 1,024 addresses, and output 3 will go active (LOW) for the next 1,024 addresses. By connecting the 74138's active-LOW 0 through 3 outputs to the active-LOW enable inputs of the ROM, RAM, input, and output, as shown in Figure 14-28, we can ensure that: the ROM will be enabled whenever any address from 0000_{16} to $03FF_{16}$ is generated; the RAM will be enabled whenever any address from 0400_{16} to $07FF_{16}$ is generated; the input port will be enabled whenever any address from 0800_{16} to $0BFF_{16}$ is generated; and the output port will be enabled whenever any address from $0C00_{16}$ to $0FFF_{16}$ is generated.

Memory Mapped Input/Output
An addressing technique in which input and output devices are assigned a zone of addresses in the same way as a memory.

This technique of assigning the input port and the output port their own block or range of addresses in a memory map is called **memory mapped input/output.** Although it seems a waste to assign 1,024 addresses to a single input port and 1,024 addresses to a single output port, the hardware design of the circuit is simplified, since one decoder does all of the decoding (input, output, and memory). In addition, it is easier to have a 74138 decode only three address inputs, rather than have a thirteen-input gate circuit decode thirteen address inputs so that only one address will enable only one port.

Since only the input, output, ROM, or RAM can be enabled at any given time, a temporary register has been included to act like one of the temporary registers in a MPU. The temporary register in Figure 14-28 can be loaded with the data present on the data bus by pressing the $\overline{LOAD}$ control switch, or apply

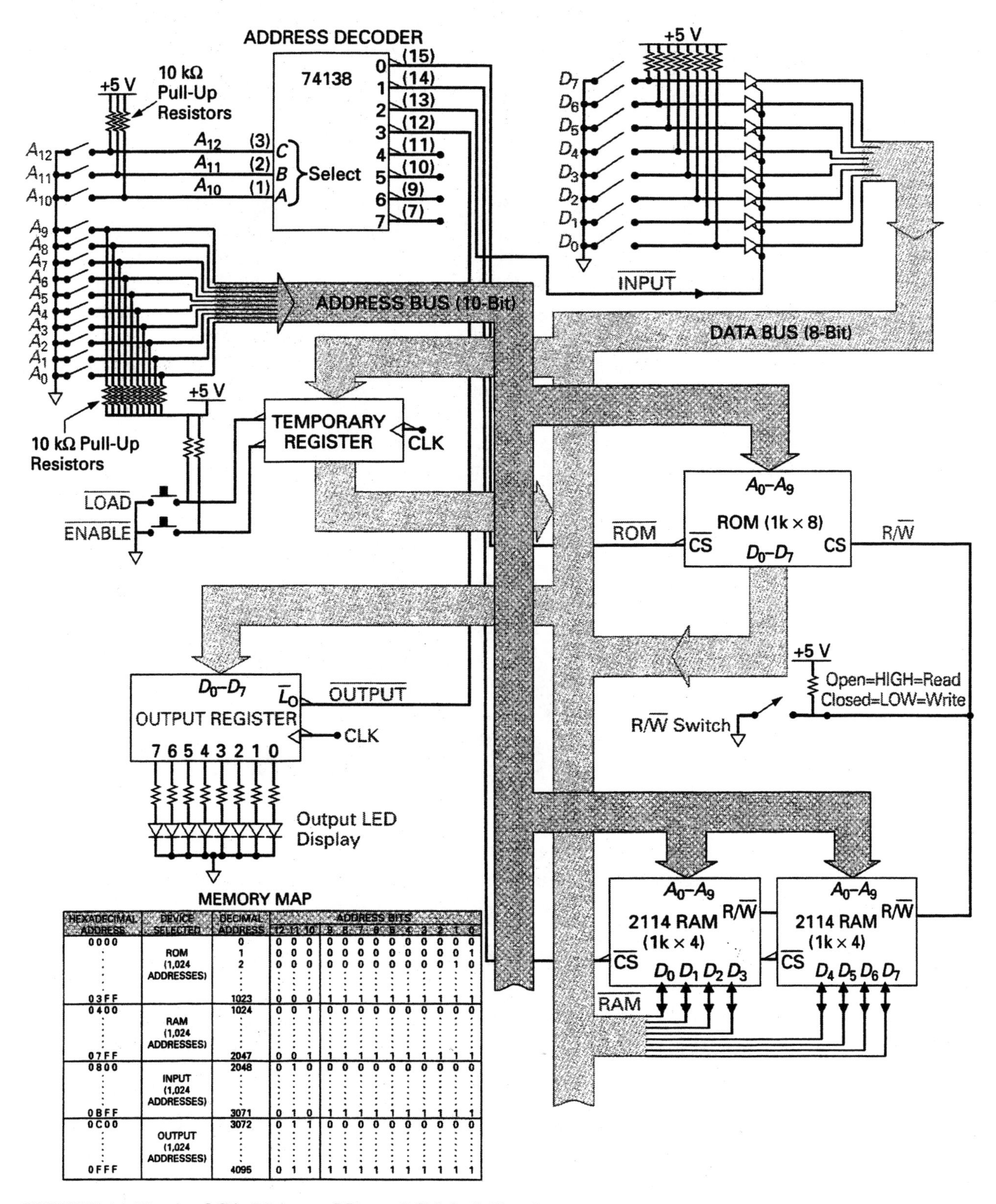

HEXADECIMAL ADDRESS	DEVICE SELECTED	DECIMAL ADDRESS	12	11	10	9	8	7	6	5	4	3	2	1	0
0000	ROM (1,024 ADDRESSES)	0	0	0	0	0	0	0	0	0	0	0	0	0	0
:		1	0	0	0	0	0	0	0	0	0	0	0	0	1
:		2	0	0	0	0	0	0	0	0	0	0	0	1	0
:		:	:	:	:	:	:	:	:	:	:	:	:	:	:
03FF		1023	0	0	0	1	1	1	1	1	1	1	1	1	1
0400	RAM (1,024 ADDRESSES)	1024	0	0	1	0	0	0	0	0	0	0	0	0	0
:		:	:	:	:	:	:	:	:	:	:	:	:	:	:
07FF		2047	0	0	1	1	1	1	1	1	1	1	1	1	1
0800	INPUT (1,024 ADDRESSES)	2048	0	1	0	0	0	0	0	0	0	0	0	0	0
:		:	:	:	:	:	:	:	:	:	:	:	:	:	:
0BFF		3071	0	1	0	1	1	1	1	1	1	1	1	1	1
0C00	OUTPUT (1,024 ADDRESSES)	3072	0	1	1	0	0	0	0	0	0	0	0	0	0
:		:	:	:	:	:	:	:	:	:	:	:	:	:	:
0FFF		4095	0	1	1	1	1	1	1	1	1	1	1	1	1

FIGURE 14-28 An 8-Bit Memory Mapped Digital Circuit.

its contents to the data bus by pressing the $\overline{ENABLE}$ control switch. This register is controlled by the operator, and can have data loaded from the input port, ROM, or RAM, and can apply its contents to either the output port or RAM.

Step 1: Diagnose

As before, it is extremely important that you first understand the operation of a circuit, and how all of the devices within it are supposed to work, so that you are able to determine whether or not a circuit malfunction really exists. If you were preparing to troubleshoot the circuit in Figure 14-28, your first step should be to read through the circuit description and review the operation of each integrated circuit until you feel completely confident with the correct operation of the entire circuit. Once you are fully familiar with the operation of the circuit, you will easily be able to diagnose the problem as either an *operator error* or a *circuit malfunction.*

Distinguishing an operator error from an actual circuit malfunction is an important first step, and a wrong diagnosis can waste a lot of time and effort. For example, the following could be interpreted as circuit malfunctions, when in fact they are simply operator errors.

Example 1.
Symptom: The circuit will not enable the input and output port at the same time for a direct data transfer.
Diagnosis: Operator Error—The 74138 address decoder ensures that only one device is enabled at any one time (either ROM, RAM, input, or output). To transfer the contents of the input port to the output port, therefore, first the input port must be enabled and the inputted data loaded in the temporary register. Then the output port should be enabled and the temporary register's contents transferred to the output port.

Example 2.
Symptom: There is no specific input port address, only a range of addresses. Which one should be used?
Diagnosis: Operator Error—With a memory mapped I/O system, the input and output port are both assigned a block of addresses to simplify address decoding. To enable the input port, address bits A_{10}, A_{11}, and A_{12} must be 010, and to enable the output port, address bits A_{10}, A_{11}, and A_{12} must be 011.

Once you have determined that the problem is not an operator error, but is in fact a circuit malfunction, proceed to Step 2 and isolate the circuit failure.

Step 2: Isolate

No matter what circuit or system failure has occurred, you should always follow a logical and sequential troubleshooting procedure. Let us review some of the isolating techniques, and then apply them to our example circuit in Figure 14-28.

a. Use a cause-and-effect troubleshooting process, which means study the effects you are getting from the faulty circuit and then logically reason out what could be the cause.

b. Check first for obvious errors before leaping into a detailed testing procedure. Is the power OFF or not connected to the circuit? Are there wiring errors? Are all of the ICs correctly oriented?

c. Using a logic probe or voltmeter, test that power and ground are connected to the circuit and are present at all points requiring power and ground. If the whole circuit, or a large section of the circuit, is not operating, the problem is normally power. Using a multimeter, check that all of the dc voltages for the circuit are present at all IC pins that should have power or a HIGH input and are within tolerance. Secondly, check that 0 V or ground is connected to all IC ground pins and all inputs that should be tied LOW.

d. Use your senses to check for broken wires, loose connectors, overheating or smoking components, pins not making contact, and so on.

e. Test the clock signals using the logic probe or the oscilloscope. Although clock signals can be easily checked with a logic probe, the oscilloscope is best for checking the signal's amplitude, frequency, pulse width, and so on. The oscilloscope will also display timing problems and noise spikes (glitches) which could be false-triggering a circuit into operation at the wrong time.

f. Perform a static test on the circuit. With our circuit example in Figure 14-28, it is easy to perform a static test since this circuit operates at the speed at which we control the address bus inputs and control signals. Each step in a procedure can therefore be verified by checking logic levels with a logic probe.

g. Noise due to electromagnetic interference (EMI) can interfere with the data present on bus lines and control lines. This problem can be overcome by including pull-up resistors and by not leaving any of the IC's inputs unconnected and, therefore, floating. Connect unneeded active-HIGH inputs to ground and active-LOW inputs to $+V_{CC}$.

h. Apply the half-split method of troubleshooting first to a circuit, and then to a section of a circuit, to help speed up the isolation process. With our circuit example in Figure 14-28, a good mid-point check would be to first test the outputs of the address decoder to see if address bits A_{10}, A_{11}, and A_{12} are switching these chip-select lines LOW.

Also remember that a load problem can make a source appear at fault. If an output is incorrect, it would be safer to disconnect the source from the rest of the circuit and then recheck the output from the source. If the output is still incorrect, the problem is definitely within the source; however, if the output is correct, the load is probably shorting the line either HIGH or LOW. With common bus lines, an open is easy to isolate when checked with the logic probe since the data is present before the open but not present after the open. A short on a bus line will reflect a problem back to every device driving that shorted bus line. By disconnecting each source device in turn, you will be able to determine which of the sources is functioning normally. If the problem seems to be a short in one of the loads, turn all sources OFF, and then use the logic pulser to inject a signal onto the suspected line, and then use the current tracer to find the shorted path.

i. Substitution can be used to help speed up your troubleshooting process. Once the problem is localized to an area containing only a few ICs, substitute suspect ICs with known working ICs (one at a time) to see if the problem can be quickly remedied.

Step 3: Repair

Once the fault has been found, the final step is to repair the circuit and then perform a final test to see that the circuit and the system are now fully operational.

14-3-2 Sample Problems

Once you have constructed a circuit like the circuit in Figure 14-28, introduce a few errors to see what effects or symptoms they produce. Then logically reason out why a particular error or cause has a particular effect on the circuit. Never short any two points together unless you have carefully thought out the consequences, but generally it is safe to open a path and see the results. Here are some examples for our example circuit in Figure 14-28.

a. Open the ground connection to all address switches.

b. Open one of the address inputs to the 74138.

c. Open one of the data bus lines at any point.

d. Disconnect one of the outputs of the 74138.

e. Connect the $R/\overline{W}$ control line permanently LOW.

f. Disconnect one of the address inputs to the ROM or RAM ICs.

g. Disconnect the chip-select input to the high-order 2114.

SELF-TEST REVIEW QUESTIONS FOR SECTION 14-3

---

1. Design a simple troubleshooting tree for the circuit in Figure 14-28.
2. If a circuit problem were isolated to the ROM in Figure 14-28, what tests would you perform on the ROM to verify that it was indeed faulty?
3. If a circuit problem were isolated to the RAM in Figure 14-28, what tests would you perform on that RAM to verify that it was indeed faulty?
4. What symptoms would you get from the circuit in Figure 14-28 if the address inputs to the 74138 were reversed so that A_{10} was applied to pin 3, A_{11} was applied to pin 2, and A_{12} was applied to pin 1?

REVIEW QUESTIONS

Multiple-Choice Questions

1. How many memory locations can be accessed by a ROM IC if it has ten address line inputs?

   a. 16 **b.** 1,024 **c.** 256 **d.** 2,048

2. A memory is said to be ____________ if it loses its data whenever power is removed.

   a. non-volatile **b.** accessible **c.** volatile **d.** sequential

3. If a RAM IC has twelve address inputs and eight data input/outputs, what is its size and configuration?

   a. 2k by 8 **b.** 4k by 8 **c.** 1,024 × 8 **d.** 8,192 by 8

4. The ______________ of a memory is the time interval between the memory receiving a new address input and the data at that address being available at the data outputs.

   a. storage density
   b. programmable link
   c. access time
   d. memory configuration

5. The total storage capacity of an 8k by 8 RAM is ______________ bits.

   a. 16,384 **b.** 65,536 **c.** 64,000 **d.** 49,152

6. Which of the following ROM types is the cheapest when ordered in large quantities?

   a. MROM **b.** PROM **c.** EPROM **d.** EEPROM

7. Which of the following ROM types can easily be erased and reprogrammed?

   a. MROM **b.** PROM **c.** EPROM **d.** EEPROM

8. Which of the following ROM types can be erased and is low in cost?

   a. MROM **b.** PROM **c.** EPROM **d.** EEPROM

9. A read/write memory is normally always referred to as a ______________.

   a. ROM **b.** EEPROM **c.** RAM **d.** MBM

10. The storage cell in a static RAM is a ______________.

    a. bistable flip-flop
    b. FET and capacitor
    c. monostable flip-flop
    d. Both a. and c. are true

11. Which of the RAM types has a higher storage density but operates slower?

    a. SRAM **b.** DRAM **c.** SAM **d.** All of the above

12. Which of the RAM types has a lower storage density but operates faster?

    a. SRAM **b.** DRAM **c.** SAM **d.** All of the above

13. Which of the following RAM types has a lower power consumption?

    a. SRAM **b.** DRAM **c.** SAM **d.** All of the above

14. Static RAMs and dynamic RAMs are both non-volatile.

    a. True **b.** False

15. Dynamic RAM ICs will normally use an address multiplexing technique to ______________.

    a. refresh
    b. reduce the number of data inputs
    c. enable an address decoder
    d. reduce the number of address inputs

16. The notebook computer used to develop the manuscript for this textbook has 32 megabytes of RAM. Which RAM type do you think is being used in this application?

    a. SRAM **b.** DRAM **c.** SAM **d.** ERAM

17. If address bits 5, 6, and 7 are applied to the *ABC* select inputs of an address decoder, and address bits 0 through 4 are applied directly to the memory, what is the size of the memory map blocks?

    a. 16 **b.** 32 **c.** 8 **d.** 64

18. If address bits 8, 9, and 10 are applied to the *ABC* select inputs of an address decoder, and address bits 0 through 7 are applied directly to the memory, what is the size of the memory map blocks?

    a. 256 **b.** 1,024 **c.** 512 **d.** 128

19. If address bits 13, 14, and 15 are applied to the *ABC* select inputs of an address decoder, and address bits 0 through 12 are applied directly to the memory, what is the size of the memory map blocks?

    a. 4k **b.** 32k **c.** 8k **d.** 16k

20. The memory mapped input/output technique assigns memory, and input ports and output ports, a block of addresses.

    a. True **b.** False

Essay Questions

21. What is the difference between a primary main memory and a secondary mass storage memory? (Introduction)
22. Give the full names for the following abbreviations. (Chapter 14)

    a. ROM **b.** RAM **c.** DRAM **d.** SRAM **e.** RAS **f.** CAS **g.** MROM **h.** EPROM **i.** EEPROM **j.** RAM **k.** CCD **l.** MBM **m.** MPU

23. What is the difference between a memory IC's address input pins and data pins? (14-1-1)
24. Why is an address decoder included in a memory circuit? (14-1-2)
25. How can a memory's size and configuration be determined? (14-1-3)
26. Define the following terms. (14-1-3)

    a. Memory write operation **b.** Memory read operation
    c. Access time **d.** Storage density

27. What is a mask programmable ROM? (14-1-4)
28. What is the difference between an MROM and a PROM? (14-1-4)
29. What is the difference between an EPROM and a PROM? (14-1-4)
30. What is the difference between an EPROM, EEPROM, and flash memory? (14-1-4)
31. What is the main application for ROMs? (14-1-5)
32. How could a ROM be used to replace a logic gate circuit? (14-1-5)
33. What is meant by the term "non-volatile"? (14-1-5)
34. What is a ROM listing? (14-1-6)
35. What is the difference between an RWM and RAM? (14-2)
36. Are ROMs RAMs? (14-2)

37. Briefly describe the operation of a sequential access memory. (14-2-1)
38. What is the key disadvantage with a sequential access memory? (14-2-1)
39. What is the difference between a sequential access memory and a random access memory? (14-2-1)
40. Define the terms serial memory and parallel memory. (14-2-1)
41. What are the two basic RAM types? (14-2-2)
42. What is meant by the term "volatile"? (14-2-2)
43. What is the difference between an SRAM storage cell and a DRAM storage cell? (14-2-2)
44. What advantages do SRAMs have over DRAMs? (14-2-2)
45. What advantages do DRAMs have over SRAMs? (14-2-2)
46. Why do DRAMs need refresh circuitry? (14-2-2)
47. What is address multiplexing? (14-2-2)
48. Why are SRAMs used in small memory applications and DRAMs used in large memory applications? (14-2-3)
49. What is a memory map? (14-2-3)
50. What is an address decoder? (14-2-3)

Practice Problems

To practice your data sheet interpretation skills, refer to Figure 14-29 and answer the following six questions.

51. What type of memory is the 28F256?
52. What is the size of this memory IC, and how are the memory cells organized?
53. How many data pins and address pins does this IC have?
54. What is the maximum access time of this IC?
55. Define the function of the following control pins:

    a. CE **b.** OE **c.** WE **d.** V_{PP}

56. If this memory is a read-only device, why are the data pins called input/output lines?

To practice your circuit recognition and operation skills, refer to Figure 14-30 and answer the following four questions.

57. What is the purpose of the $\overline{ROM}$ control signal input?
58. What is the purpose of the $\overline{READ}$ control signal input?
59. How is data written into this IC?
60. How is data read from this IC?

To practice your circuit recognition and operation skills, refer to the SAM schematic diagram on pages 689–692 and answer the following five questions.

61. Where is the 2316 ROM circuit shown in Figure 14-30 on the SAM schematic diagram?

DEVICE: A 28F256A—256k (32k × 8) CMOS Flash Memory

The 28F256A is a 256-kilobit non-volatile memory organized as 32768 bytes of 8 bits.

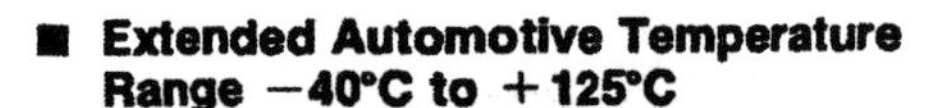

- **Extended Automotive Temperature Range −40°C to +125°C**
- **Flash Electrical Chip-Erase**
  - **—1 Second Typical Chip-Erase**
- **Quick-Pulse Programming Algorithm**
  - **—10 μs Typical Byte-Program**
  - **—0.5 Second Chip-Program**
- **1,000 Erase/Program Cycles Minimum Over Automotive Temperature Range**
- **12.0V ±5% V_{PP}**
- **High-Performance Read**
  - **—120/150 ns Maximum Access Time**
- **CMOS Low Power Consumption**
  - **—30 mA Maximum Active Current**
  - **—100 μA Maximum Standby Current**
- **Integrated Program/Erase Stop Timer**

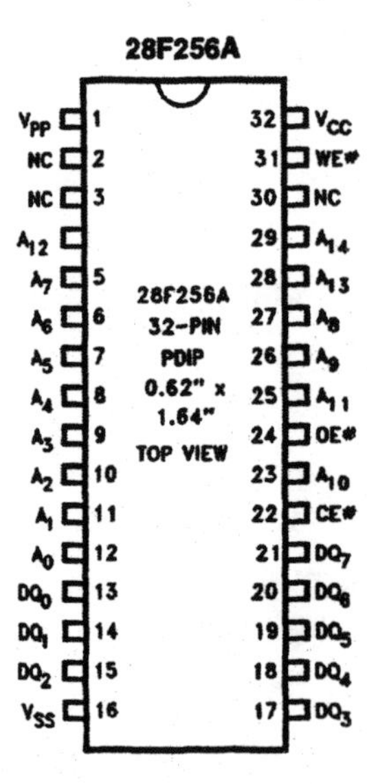

Table 1. Pin Description

Symbol	Type	Name and Function
A_0–A_{14}	INPUT	**ADDRESS INPUTS** for memory addresses. Addresses are internally latched during a write cycle.
DQ_0–DQ_7	INPUT/OUTPUT	**DATA INPUT/OUTPUT:** Inputs data during memory write cycles; outputs data during memory read cycles. The data pins are active high and float to tri-state OFF when the chip is deselected or the outputs are disabled. Data is internally latched during a write cycle.
CE#	INPUT	**CHIP ENABLE:** Activates the device's control logic, input buffers, decoders and sense amplifiers. CE# is active low; CE high deselects the memory device and reduces power consumption to standby levels.
OE#	INPUT	**OUTPUT ENABLE:** Gates the devices output through the data buffers during a read cycle. OE# is active low.
WE#	INPUT	**WRITE ENABLE:** Controls writes to the control register and the array. Write enable is active low. Addresses are latched on the falling edge and data is latched on the rising edge of the WE# pulse. **Note:** With $V_{PP} \leq 6.5V$, memory contents cannot be altered.
V_{PP}		**ERASE/PROGRAM POWER SUPPLY** for writing the command register, erasing the entire array, or programming bytes in the array.
V_{CC}		**DEVICE POWER SUPPLY** (5V ±10%)
V_{SS}		**GROUND**
NC		**NO INTERNAL CONNECTION** to device. Pin may be driven or left floating.

FIGURE 14-29 A ROM Data Sheet. (Courtesy of Intel Corporation)

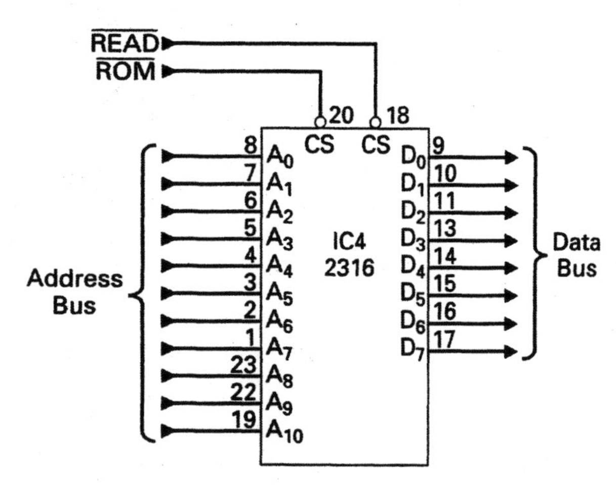

FIGURE 14-30 The ROM Circuit from SAM (Simple All-Purpose Microcomputer).

62. The NRD (not read) control line input to the 2316 ROM IC comes from what IC?

63. The NROM (not ROM) control line input to the 2316 ROM IC comes from what IC?

64. Which two ICs output the 11-bit address that is applied to the 2316 IC?

65. The 8-bit data output of the 2316 ROM IC should connect to the data bus, and then back to the MPU IC. Which IC do you think is the MPU?

To practice your data sheet interpretation skills, refer to Figure 14-31 and answer the following six questions.

66. What type of RAM is the 2114?

67. How many memory cells does this RAM have?

68. How many data pins and address pins does this IC have?

69. What is the access time and power dissipation for a 2114L?

70. How many chip-select inputs does this IC have?

71. What is the purpose of the $\overline{\text{WE}}$ control line input?

To practice your circuit recognition and operation skills, refer to Figure 14-32 and answer the following four questions.

72. What is the purpose of the $\overline{RAM}$ control signal input?

73. What is the purpose of the $\overline{WRITE}$ control signal input?

74. How is data written into this IC?

75. How is data read from this IC?

To practice your circuit recognition and operation skills, refer again to the SAM schematic diagram on pages 689–692 and answer the following three questions.

76. Where is the 2114 RAM circuit shown in Figure 14-32 on the SAM schematic diagram?

DIGITAL DEVICE DATA SHEET

IC TYPE: 2114 – 1024 × 4 STATIC RAM

	2114-2	2114-3	2114	2114L2	2114L3	2114L
Max. Access Time (ns)	200	300	450	200	300	450
Max. Power Dissipation (mw)	525	525	525	370	370	370

- **High Density 18 Pin Package**
- **Identical Cycle and Access Times**
- **Single +5V Supply**
- **No Clock or Timing Strobe Required**
- **Completely Static Memory**
- **Directly TTL Compatible: All Inputs and Outputs**
- **Common Data Input and Output Using Three-State Outputs**
- **Pin-Out Compatible with 3605 and 3625 Bipolar PROMs**

The Intel® 2114 is a 4096-bit static Random Access Memory organized as 1024 words by 4-bits using N-channel Silicon-Gate MOS technology. It uses fully DC stable (static) circuitry throughout — in both the array and the decoding — and therefore requires no clocks or refreshing to operate. Data access is particularly simple since address setup times are not required. The data is read out nondestructively and has the same polarity as the input data. Common input/output pins are provided.

The 2114 is designed for memory applications where high performance, low cost, large bit storage, and simple interfacing are important design objectives. The 2114 is placed in an 18-pin package for the highest possible density.

It is directly TTL compatible in all respects: inputs, outputs, and a single +5V supply. A separate Chip Select ($\overline{CS}$) lead allows easy selection of an individual package when outputs are or-tied.

The 2114 is fabricated with Intel's N-channel Silicon-Gate technology — a technology providing excellent protection against contamination permitting the use of low cost plastic packaging.

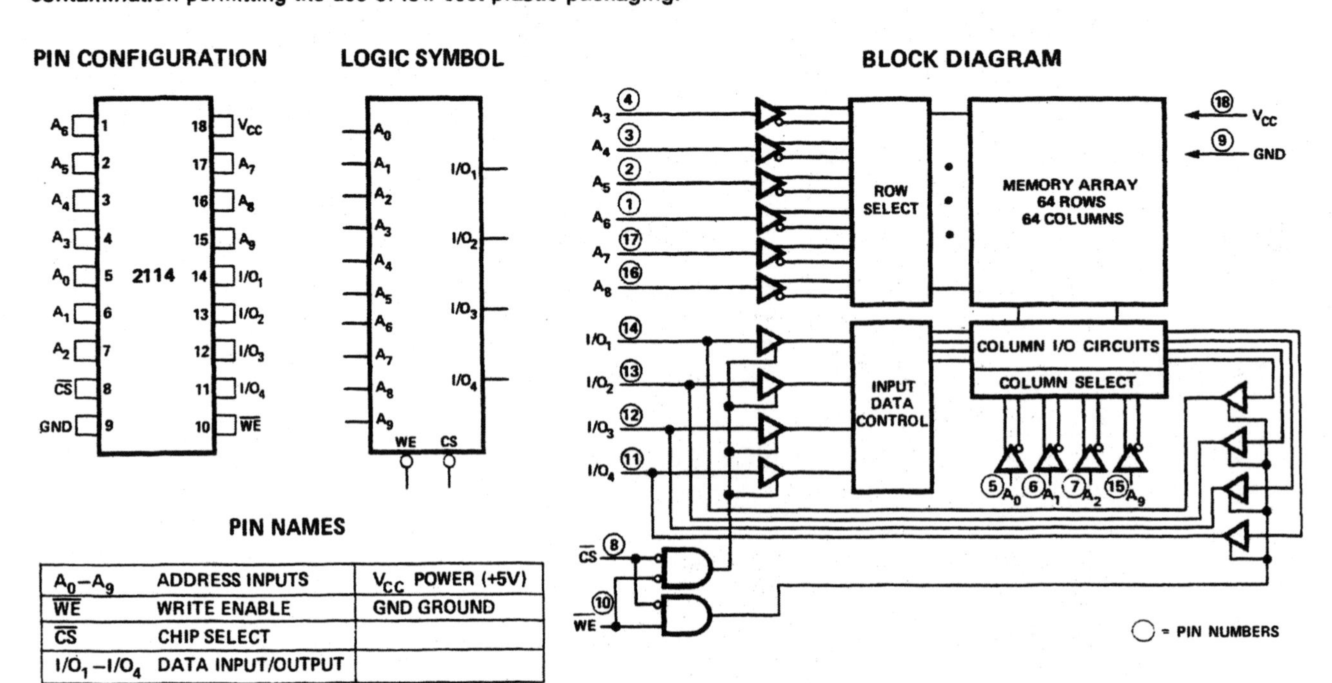

A_0–A_9	ADDRESS INPUTS	V_{CC} POWER (+5V)
$\overline{WE}$	WRITE ENABLE	GND GROUND
$\overline{CS}$	CHIP SELECT	
I/O_1–I/O_4	DATA INPUT/OUTPUT	

FIGURE 14-31 A RAM Data Sheet. (Courtesy of Intel Corporation)

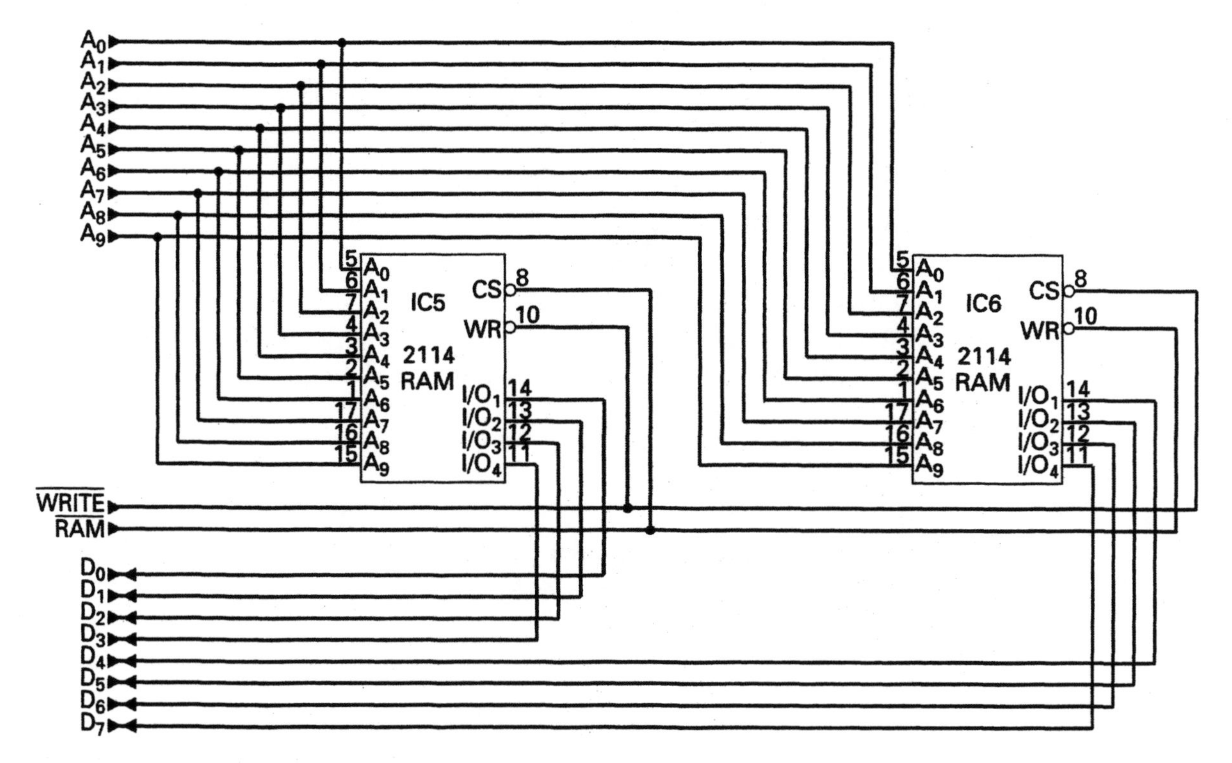

FIGURE 14-32 The RAM Circuit from SAM (Simple All-Purpose Microcomputer).

77. The NWR (not write) control line input to the 2114 RAMs comes from what IC?

78. The NRAM (not RAM) control line input to the 2114 RAMs comes from what IC?

To practice your circuit recognition and operation skills, refer to Figure 14-33 and answer the following three questions.

79. Which address bits are applied to the address decoder's *ABC* select inputs?

80. What is the size of each block in the memory map?

81. What device is selected when A_{11}, A_{12}, and A_{13} equal:

a. 000 **b.** 001 **c.** 010 **d.** 011 **e.** 100 **f.** 101 **g.** 110 **h.** 111

To practice your circuit recognition and operation skills, refer again to the SAM schematic diagram on pages 689–692 and answer the following four questions.

82. Where is the address decoder shown in Figure 14-33 located on the SAM schematic diagram?

83. What is the purpose of the four-LED displays DS1, DS2, DS3, and DS4?

84. What is the purpose of the displays DS7 and DS8?

85. Is the SAM system using memory mapped input/output?

Troubleshooting Questions

86. What is the key difference between testing a ROM IC and testing a RAM IC?

87. What is the check-sum testing method?

88. What is the checkerboard test pattern?

To practice your circuit troubleshooting skills, refer to the SAM schematic diagram included on pages 689–692 and answer the following two questions.

89. At the data output of the 2316 ROM is a jumper which can be removed and reinstalled so that a fault is introduced to the SAM system. What symptoms would you expect if jumper 2 were removed and reinstalled to open data line 5 from the ROM?

90. At the address input of U6 is a jumper which can be removed and reinstalled so that a fault is introduced to the SAM system. What symptoms would you expect if jumper 4 were removed and reinstalled to open the address line 7 input to U6?

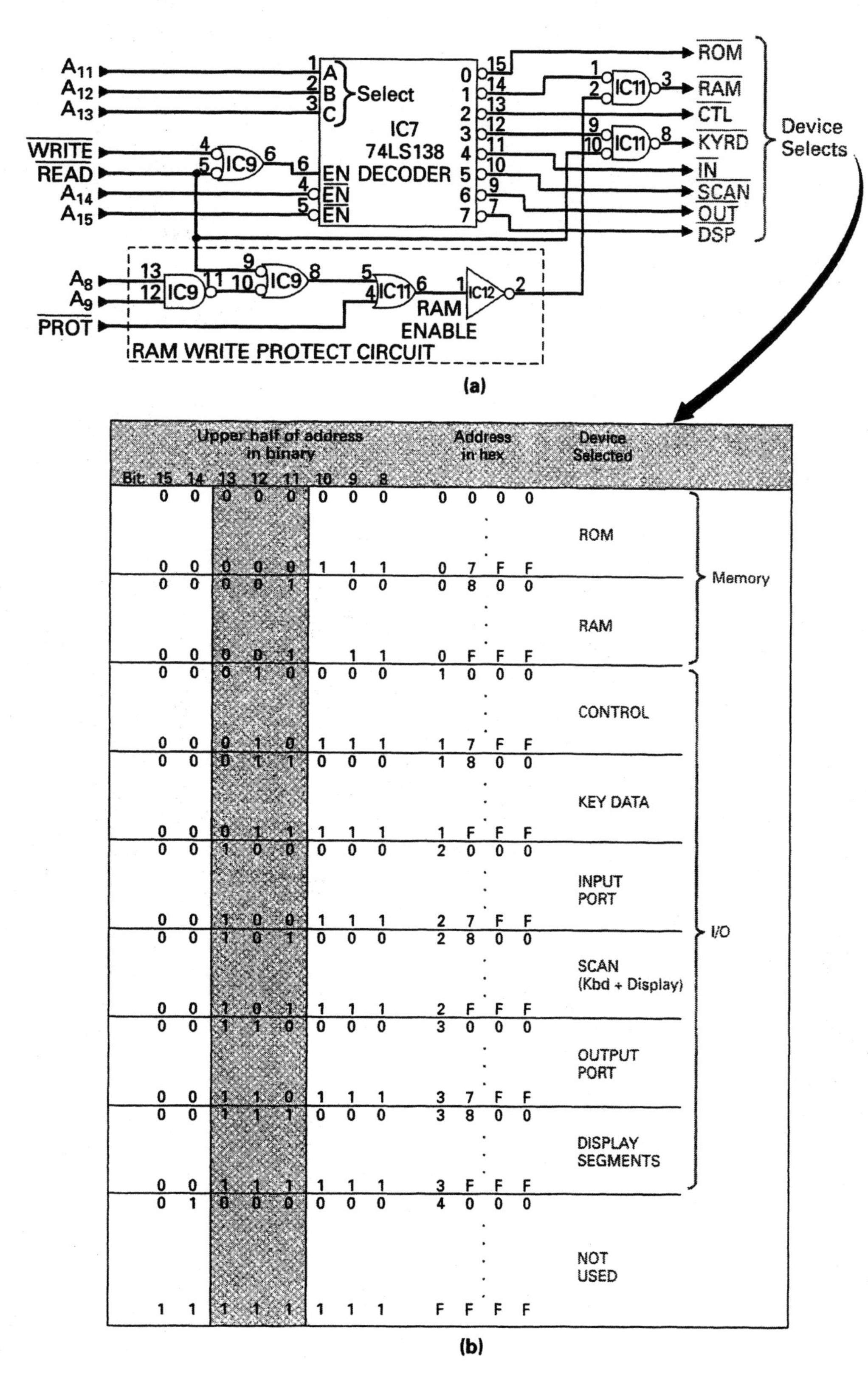

Upper half of address in binary								Address in hex				Device Selected	
Bit: 15	14	13	12	11	10	9	8						
0	0	0	0	0	0	0	0	0	0	0	0	ROM	Memory
0	0	0	0	0	1	1	1	0	7	F	F		
0	0	0	0	1		0	0	0	8	0	0	RAM	
0	0	0	0	1		1	1	0	F	F	F		
0	0	0	1	0	0	0	0	1	0	0	0	CONTROL	I/O
0	0	0	1	0	1	1	1	1	7	F	F		
0	0	0	1	1	0	0	0	1	8	0	0	KEY DATA	
0	0	0	1	1	1	1	1	1	F	F	F		
0	0	1	0	0	0	0	0	2	0	0	0	INPUT PORT	
0	0	1	0	0	1	1	1	2	7	F	F		
0	0	1	0	1	0	0	0	2	8	0	0	SCAN (Kbd + Display)	
0	0	1	0	1	1	1	1	2	F	F	F		
0	0	1	1	0	0	0	0	3	0	0	0	OUTPUT PORT	
0	0	1	1	0	1	1	1	3	7	F	F		
0	0	1	1	1	0	0	0	3	8	0	0	DISPLAY SEGMENTS	
0	0	1	1	1	1	1	1	3	F	F	F		
0	1	0	0	0	0	0	0	4	0	0	0	NOT USED	
1	1	1	1	1	1	1	1	F	F	F	F		

(b)

FIGURE 14-33 The Address Decoder and Memory Map For a SAM (Simple All-Purpose Microcomputer).

Chapter 15

OBJECTIVES

After completing this chapter, you will be able to:

1. Describe the characteristics, operation, programming, and application of a basic programmable logic device (PLD).
2. Explain how a PROM can be made to function as a PLD.
3. Describe the internal architecture, basic operation, programming, and application of a programmable logic array (PLA).
4. Explain the data sheet details for a typical PLA IC.
5. Describe the internal architecture, basic operation, programming, and application of a programmable array logic (PAL).
6. Explain the data sheet details for a typical PAL IC.
7. Describe the internal architecture, basic operation, and programming of a programmable logic sequencer (PLS).
8. Explain the data sheet details for a typical PLS IC.
9. Describe how a universal programmer unit is used to program a PLD.
10. Describe the internal architecture, basic operation, and programming of an erasable programmable logic device (EPLD).
11. Explain the data sheet details for a typical EPLD IC.
12. Describe how to troubleshoot digital logic circuits containing programmable logic devices using the three-step method.

Programmable Logic Devices

OUTLINE

VIGNETTE

A Problem with Early Mornings

René Descartes was born in Brittany, France, in 1596. At the age of eight he had surpassed most of his teachers at school and was therefore sent on to the Jèsuit college in La Flèche, one of the best in Europe. It was here that his genius in mathematics became apparent; however, due to his extremely delicate health, his professors allowed him to study in bed until midday.

In 1616 he had an urge to see the world, so he joined the army, which made use of Descartes' mathematical genius in military engineering. While traveling, Descartes met Dutch philosopher Isaac Beekman who convinced him to leave the army and, in his words, "turn his mind back to science and more worthy occupations."

After leaving the army, Descartes traveled looking for some purpose to his life. Then on November 10, 1619, it happened. Descartes was in Neuberg, Germany, where he had shut himself in a well-heated room for the winter. It was on the eve of St. Martin's that a freezing blizzard forced Descartes to retire early. That night Descartes describes having an extremely vivid dream that clarified his purpose and showed him that physics and all sciences could be reduced to geometry, and therefore could all be interconnected like a chain.

In his time, and to this day, he is heralded as an analytical genius. In fact, Descartes' problem-solving ability can still be used as a guide to solving any problem.

Descartes's four-step procedure for solving a problem is as follows:

1. Never accept anything as true unless it is clear and distinct enough to exclude doubt from your mind.
2. Divide the problem in as many parts as necessary to reach a solution.
3. Start with the simplest things and proceed step by step towards the complex.
4. Review the solution so completely and generally that you are sure nothing was omitted.

INTRODUCTION

In the previous chapter, it was shown that a ROM simply converts a coded input word at the address inputs into a desired output word at the data outputs. A ROM, therefore, can be used to replace a logic circuit made up of several logic gates. For example, an eight-input/four-output IC logic circuit could be replaced by an 8-bit address input/4-bit data output PROM, which when programmed would give the desired 4-bit output for any 8-bit input combination. Replacing several logic gate ICs with a single PROM IC will save printed circuit board space, cut inventory and assembly costs, and reduce the board's power consumption.

In this chapter, you will be introduced to the programmable logic device or PLD. Like the PROM, the PLD is a fuse-programmable device that can be programmed by the customer and can be used to replace a complete logic circuit. Unlike the PROM, however, the fusible links within a PLD are used to interconnect the logic gates, flip-flops, and registers within the PLD IC to achieve any desired logic function.

The four different types of programmable logic devices are the Programmable Logic Array (PLA), the Programmable Array Logic (PAL), the Programmable Logic Sequencer (PLS), and the Erasable Programmable Logic Device (EPLD). The abbreviations for these devices are sometimes preceded by an "F" (FPLA, FPAL, and so on), to indicate that the PLD is customer- or field-programmable. Before discussing each of the PLD types in detail, let us begin by examining the operation and application of a basic programmable logic device.

15-1 PROGRAMMABLE LOGIC DEVICES

Figure 15-1(a) shows a simple **programmable logic device,** or **PLD.** Basically, a PLD consists of a set of uncommitted logic gates connected by an array of fuses. Most PLDs contain an AND array decoder which drives an OR array. Using these AND/OR arrays, it is possible to implement any logic function expressed in the sum-of-products form. During the programming of a PLD, some of the fuse links are blown or broken, leaving only the connections required to form the desired logic circuit. The PLD is used instead of a collection of SSI devices to build a control logic circuit, since a single flexible PLD can be custom programmed to exactly fit any desired need.

Programmable Logic Device (PLD)
A device that can be programmed to perform a desired logic function.

15-1-1 A Basic Programmable Logic Device

Figure 15-1(a) shows the internal circuit diagram of a basic two-input/four-output PLD. Two noninverting buffers, two inverting buffers, and four AND gates form a one-of-four decoder. This decoder will make HIGH only one of its four outputs based on the binary value applied to inputs A and B, as shown in the table in Figure 15-1(a). The four AND gate outputs from the decoder are called *product lines,* and since each AND gate has two different inputs, each of these product lines is unique ($\overline{A}\overline{B}$, $\overline{A}B$, $A\overline{B}$, and AB). Initially, all of these product lines are connected to each of the OR gates via a fusible link, as shown in the inset in Figure 15-1(a). These fusible links are usually shown as an "X" on a PLD schematic diagram.

The circuit diagram shown in Figure 15-1(a) has only two inputs, four outputs, and four products, and even this small circuit is quite elaborate. Most of the commercially available PLDs are much larger in size and therefore a simpler schematic representation is generally used. Figure 15-1(b) shows the PLD from

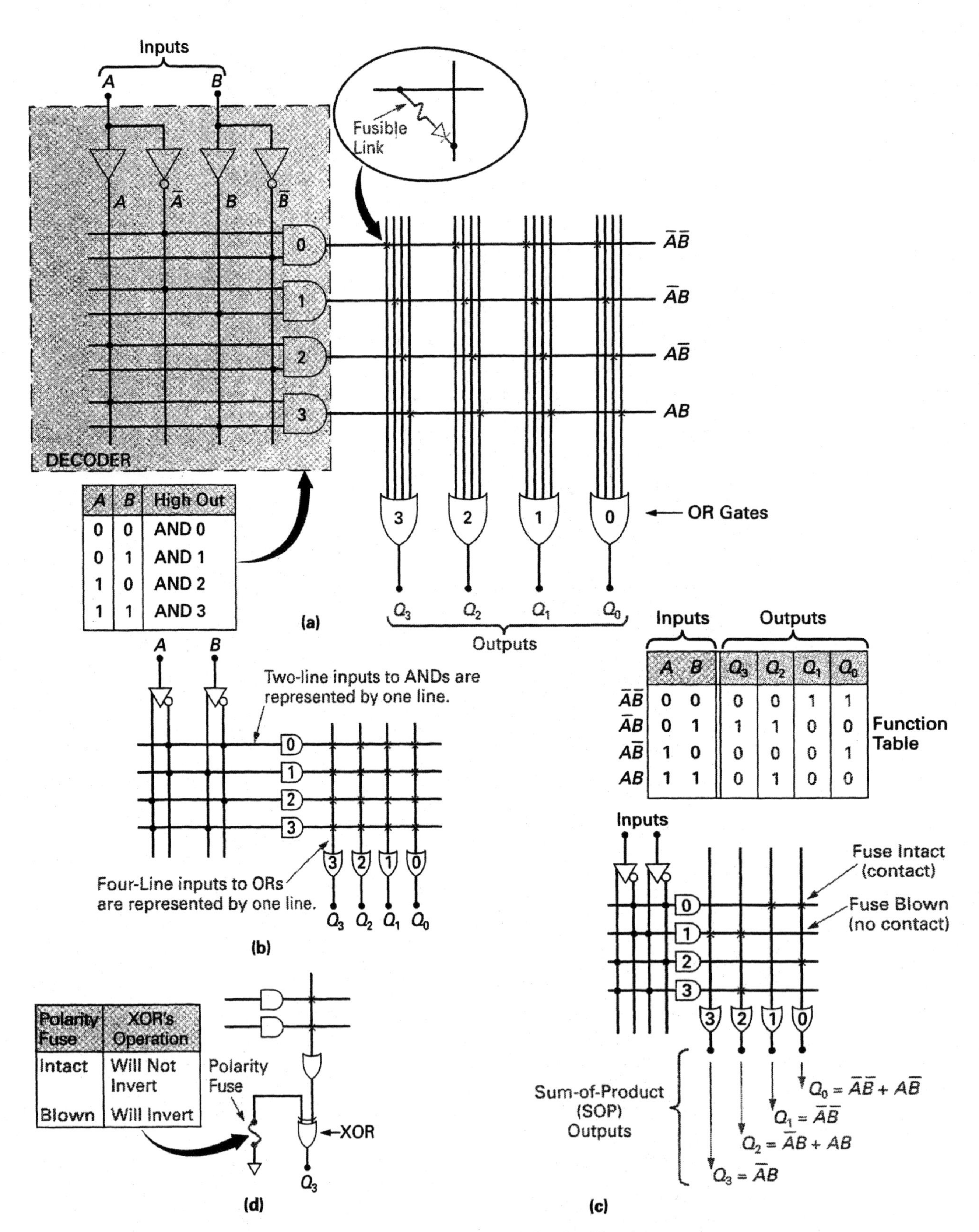

FIGURE 15-1 A Programmable Logic Device (PLD). (a) Basic Circuit. (b) Simplified Circuit Diagram. (c) Application. (d) Programmable Output.

Figure 15-1(a), redrawn using the simple schematic representation. In this reduced schematic, the inverting buffers and noninverting buffers are shown as one with a bubble indicating the inverted output. The two inputs to each AND gate are now shown as a single line with a dot indicating a hard-wired connection (a connection that cannot be changed). The absence of a dot, therefore, indicates that no connection exists. The four inputs to each OR gate are also shown as a single line, but in this case the "Xs" are used to represent a fusible link. To determine the number of input lines to an AND gate or OR gate, therefore, you simply count the number of dots or "Xs" shown on the input line.

Let us now see how we can use this simple PLD to implement a desired logic function. Figure 15-1(c) shows an example application, in which the simple two-input/four-output-product PLD can be programmed so that its Q_3, Q_2, Q_1, and Q_0 outputs will generate the desired HIGH or LOW outputs for each A and B input combination. By blowing fuses 1, 3, and 4 at the input of OR gate 3, the Q_3 output will only go HIGH when $AB = 01$ ($Q_3 = \overline{A}B$). By blowing fuses 1 and 3 at the input of OR gate 2, the Q_2 output will only go HIGH when $AB = 01$ or 11 ($Q_2 = \overline{A}B + AB$). By blowing fuses 2, 3, and 4 at the input of OR gate 1, the Q_1 output will only go HIGH when $AB = 00$ ($Q_1 = \overline{A}\overline{B}$). By blowing fuses 2 and 4 at the input of OR gate 0, the Q_0 output will only go HIGH when $AB = 00$ or 10 ($Q_0 = \overline{A}\overline{B} + A\overline{B}$).

Most of the commercially available PLDs have a *programmable output polarity fuse* at each of the outputs along with an XOR gate, as shown in Figure 15-1(d). This feature is added so that the output can be inverted or not, based on whether the polarity fuse is blown or left intact. When the fuse is left intact, a LOW is applied to the second input of the XOR gate so no inversion occurs between input and final output. When the fuse is blown, a HIGH is applied to the second input of the exclusive OR gate so an inversion occurs between input and final output. This action is summarized in the table in Figure 15-1(d).

■ **EXAMPLE:**

Show how a two-input/four-output/four-product PLD can be used to implement the logic function shown in the truth table in Figure 15-2(a). Also, list the sum-of-products (SOP) Boolean expression for each of the four outputs.

■ ***Solution:***

Figure 15-2(b) shows how a simple two-input/four-output/four-product PLD can be programmed so that its Q_3, Q_2, Q_1, and Q_0 outputs will generate the desired HIGH or LOW outputs for each A and B input combination shown in Figure 15-2(a). By blowing fuses 3 and 4 at the input of OR gate 3, the Q_3 output will only go HIGH when $AB = 00$ or 01 ($Q_3 = \overline{A}\overline{B} + \overline{A}B$). By blowing fuses 2 and 3 at the input of OR gate 2, the Q_2 output will only go HIGH when $AB = 00$ or 11 ($Q_2 = \overline{A}\overline{B} + AB$). By blowing fuses 1, 2, and 4 at the input of OR gate 1, the Q_1 output will only go HIGH when $AB = 10$ ($Q_1 = A\overline{B}$). By blowing fuses 1 and 2 at the input of OR gate 0, the Q_0 output will only go HIGH when $AB = 10$ or 11 ($Q_0 = A\overline{B} + AB$). The sum-of-products Boolean expressions for these outputs are also listed in Figure 15-2(c).

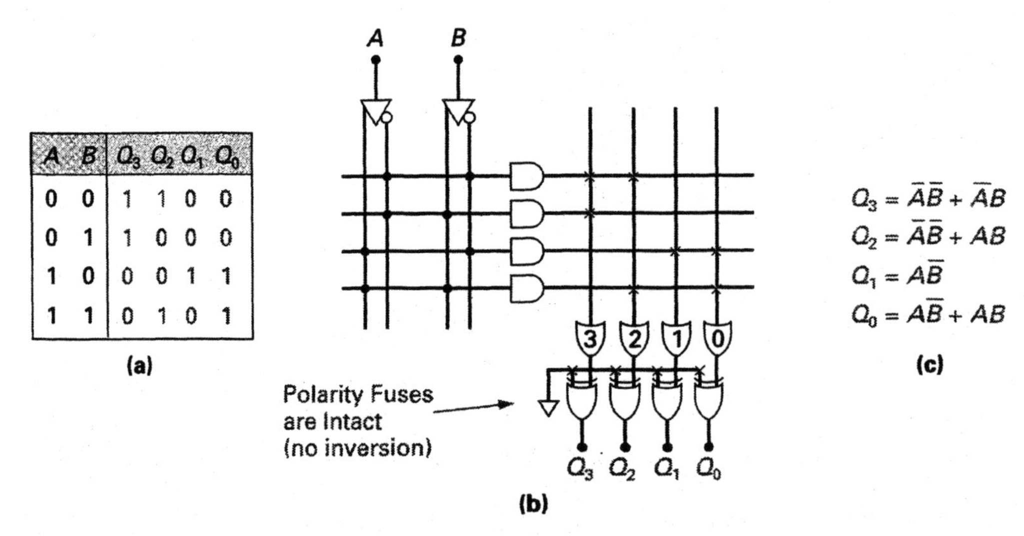

A	B	Q_3	Q_2	Q_1	Q_0
0	0	1	1	0	0
0	1	1	0	0	0
1	0	0	0	1	1
1	1	0	1	0	1

FIGURE 15-2 Using a PLD to Implement a Desired Function.

15-1-2 *Using the PROM as a PLD*

In the previous chapter, we discussed the operation and internal logic structure of a PROM or programmable read-only memory. The internal schematic of a sixteen-words-by-four-bits PROM is shown in Figure 15-3(a). The hard-wired *fixed array* (not programmable) of AND gates forms an address decoder which feeds into a *programmable array* of OR gates. The inputs applied to the PROM form an address which propagates through the AND array and will turn ON (make HIGH) only one of the sixteen product lines or AND gate outputs. The programming of the OR array by the customer will govern which of the outputs will be HIGH or LOW based on whether a fuse has been blown or left intact. A PROM, therefore, is a programmable logic device since it can be programmed to implement a logic function.

15-1-3 *The Programmable Logic Array (PLA)*

The PROM is ideal in applications where we need to generate an output code for every possible input combination. In some applications, however, we may not want to generate an output code when some input codes are applied. For example, we may want a PLD to have an 8-bit input and generate 2-bit output codes for only 12 of the possible 256 input combinations. To meet this need, manufacturers developed the **programmable logic array (PLA).** Referring to the internal schematic of the PLA shown in Figure 15-3(b), you can see that this PLD has a programmable AND array and a programmable OR array, making it an extremely versatile logic device.

Programmable Logic Array (PLA)
A PLD that has both a programmable AND array and a programmable OR array.

A Programmable Logic Array (PLA) Application

Figure 15-4 shows how a programmable logic array can be used as a BCD to seven-segment code converter. In this application, only ten of the possible sixteen AND output products are used. Figure 15-4(a) reviews the BCD to

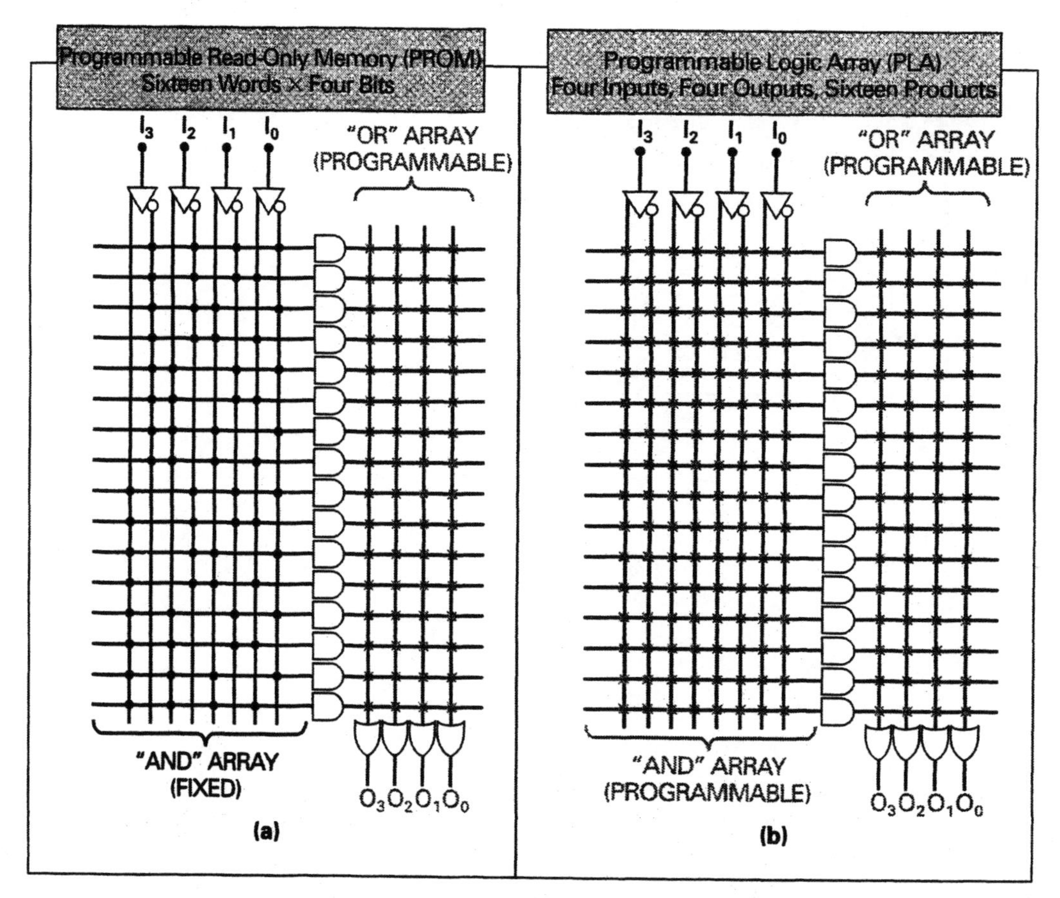

FIGURE 15-3 Programmable Logic Array (PLA).

seven-segment code conversion function table and shows which of the segments in a seven-segment display are activated in order to display the digits 0 through 9.

Figure 15-4(b) shows the fuse maps for both the AND array and OR array of a four-input/seven-output PLA. The fuses left intact in these arrays will ensure that the correct seven-segment codes will be generated for each of the BCD inputs. For example, when the code 0000_{BCD} is applied to the 4-bit input, AND gate 0 will generate a HIGH (product line $\overline{A}\overline{B}\overline{C}\overline{D}$). Since six fuses connected to this product line were left intact during programming, outputs *a, b, c, d, e,* and *f* will go HIGH. Since the last fuse connected to this product line was blown during programming, output *g* will be LOW. Therefore, the seven-segment code 1111110 will be generated at the output of this PLA whenever a 0000_{BCD} code is applied to the input.

The polarity fuses have all been left intact so that active-HIGH outputs can be obtained to drive a common-cathode display. To drive a common-anode display, the polarity fuses should all be blown so that the PLA will generate active-HIGH outputs.

A Programmable Logic Array (PLA) Data Sheet

Figure 15-5 shows the data sheet details for the PLS153A field-programmable logic array. This device has eight inputs (I_0 through I_7), ten bidirectional

INPUTS				Digit Displayed	OUTPUTS						
D	C	B	A		a	b	c	d	e	f	g
L	L	L	L	0	ON	ON	ON	ON	ON	ON	OFF
L	L	L	H	1	OFF	ON	ON	OFF	OFF	OFF	OFF
L	L	H	L	2	ON	ON	OFF	ON	ON	OFF	ON
L	L	H	H	3	ON	ON	ON	ON	OFF	OFF	ON
L	H	L	L	4	OFF	ON	ON	OFF	OFF	ON	ON
L	H	L	H	5	ON	OFF	ON	ON	OFF	ON	ON
L	H	H	L	6	ON	OFF	ON	ON	ON	ON	ON
L	H	H	H	7	ON	ON	ON	OFF	OFF	OFF	OFF
H	L	L	L	8	ON	ON	ON	ON	ON	ON	ON
H	L	L	H	9	ON	ON	ON	OFF	OFF	ON	ON

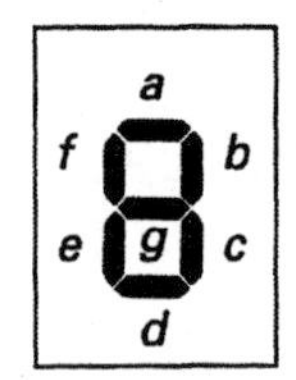

(a)

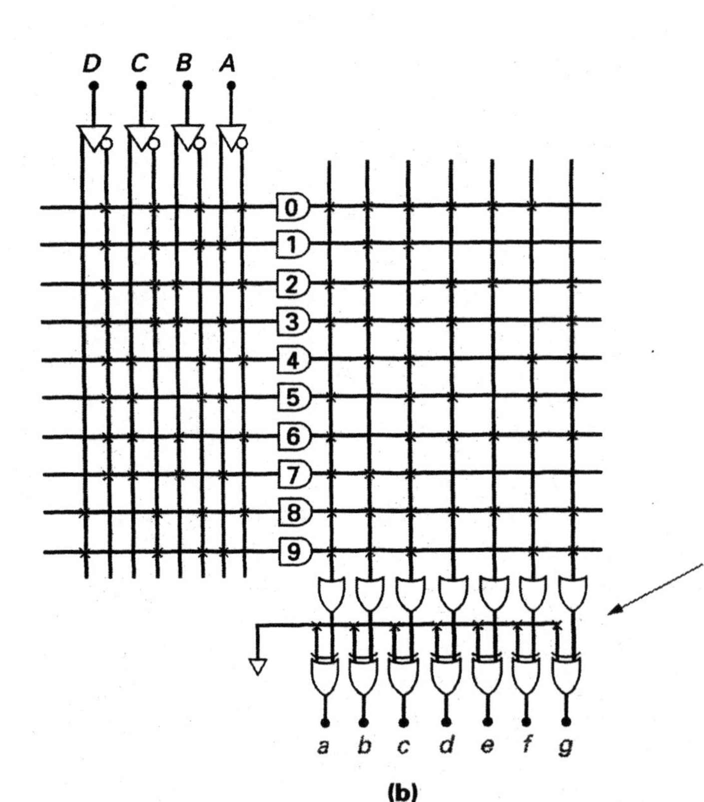

(b)

FIGURE 15-4 Application—Using a PLA as a BCD to Seven-Segment Decoder.

input/output line (B_0 through B_9), and forty-two product terms. The upper AND array, the lower OR array, and the output XOR gates are programmable, even though fuses ("Xs") are not shown.

Later in Figures 15-11 and 15-12, you will see how the PLS153A can be used in a circuit application; you will also see how the programming of this device is documented in a program table.

DIGITAL DEVICE DATA SHEET

IC TYPE: PLS153A – Programmable Logic Array (18 × 42 × 10)

DESCRIPTION

The PLS153 and PLS153A are two-level logic elements, consisting of 42 AND gates and 10 OR gates with fusible link connections for programming I/O polarity and direction.

All AND gates are linked to 8 inputs (I) and 10 bidirectional I/O lines (B). These yield variable I/O gate configurations via 10 direction control gates (D), ranging from 18 inputs to 10 outputs.

On-chip T/C buffers couple either True (I, B) or Complement ($\bar{I}$, $\bar{B}$) input polarities to all AND gates, whose outputs can be optionally linked to all OR gates. Their output polarity, in turn, is individually programmable through a set of EX-OR gates for implementing AND/OR or AND/NOR logic functions.

The PLS153 and PLS153A are field-programmable, enabling the user to quickly generate custom patterns using standard programming equipment.

FEATURES

- Field-Programmable (Ni-Cr links)
- 8 inputs
- 42 AND gates
- 10 OR gates
- 10 bidirectional I/O lines
- Active-High or -Low outputs
- 42 product terms:
  - 32 logic terms
  - 10 control terms
- I/O propagation delay:
  - PLS153: 40ns (max)
  - PLS153A: 30ns (max)
- Input loading: –100μA (max)
- Power dissipation: 650mW (typ)
- 3-State outputs
- TTL compatible

APPLICATIONS

- Random logic
- Code converters
- Fault detectors
- Function generators
- Address mapping
- Multiplexing

N Package

Pin	Signal	Pin	Signal
1	I_0	20	V_{CC}
2	I_1	19	B_9
3	I_2	18	B_8
4	I_3	17	B_7
5	I_4	16	B_6
6	I_5	15	B_5
7	I_6	14	B_4
8	I_7	13	B_3
9	B_0	12	B_2
10	GND	11	B_1

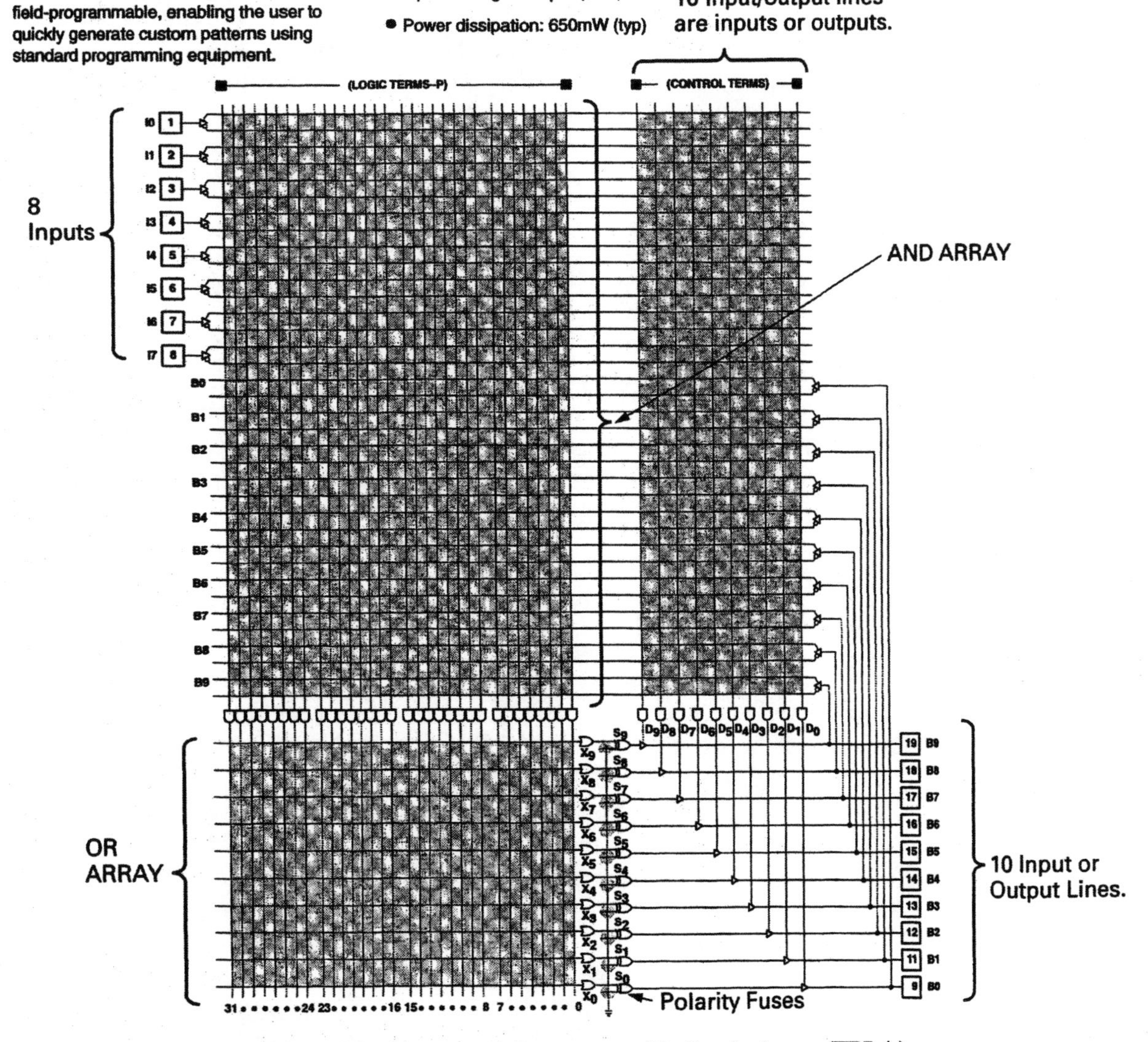

FIGURE 15-5 An IC Containing a Field-Programmable Logic Array (FPLA) Circuit. (Permission granted by Philips Semiconductors)

15-1-4 *The Programmable Array Logic (PAL)*

Programmable Array Logic (PAL) A PLD that has a programmable AND array driving a hard-wired OR array.

Figure 15-6(a) compares the programmable read-only memory (PROM) PLD, the programmable logic array (PLA) PLD, and the **programmable array logic (PAL)** PLD. With the PROM, the AND array is hard-wired while the OR array is programmable. With the PLA, both the AND array and the OR array can be programmed, making this PLD type the most versatile of the three. The disadvantage with the PLA, however, is that its complex internal circuit is more difficult to manufacture, making it a more expensive device. In addition, its complex internal circuit also makes it more difficult to program and to test.

The PAL has a programmable AND array driving a hard-wired OR array, as shown on the right of Figure 15-6(a). In Figure 15-6(b), the PAL's internal schematic has been redrawn so that the circuit arrangement can be seen more easily. This circuit arrangement is not as versatile as the PLA, but is easier to manufacture, and therefore costs less. Programming a PAL is also easier because unused product terms stay at a LOW logic level, and have no effect on the outputs. This means that the programmer only needs to consider the input addresses which will have to activate HIGH outputs.

A Programmable Array Logic (PAL) Data Sheet

Figure 15-7 shows the data sheet details for the PAL12H6, which has twelve inputs and six outputs. This twenty-pin PAL logic array has sixteen product-term lines with a programmable fuse link at each intersection.

Security fuses are included within this type of PAL to prevent direct copying of proprietary logic patterns. Blowing the security fuses disables the address decoders to make it impossible to read the fuse map. If you attempt to read the fuse map after the security fuses have been blown, the fuse map will be scrambled or the array will appear unprogrammed. Once the security fuses have been blown, therefore, it is impossible to program the fuse array.

A Programmable Array Logic (PAL) Application

Figure 15-8 shows an application for the PAL12H6 PLD. In this application, the fusible logic device has been programmed to implement the basic inverter gate, AND gate, OR gate, NAND gate, NOR gate and XOR gate. Figure 15-8(a) shows the programmed PAL12H6 internal schematic, and Figure 15-8(b) shows the pin-out for this application.

To follow one of the gates from input to output, let us see how the NOT function is achieved between the input at pin 19 and the output at pin 18. Referring to Figure 15-8(a) and following the input at pin 19, you can see that the inverted input at pin 19 is connected by a fuse to the input of the OR gate driving the output at pin 18. This arrangement, therefore, ensures that the input at pin 19 is inverted and then connected directly to the output at pin 18, so the basic inverter gate function is performed. Each of these logic functions can be verified in this way by tracing the signal path from input to output.

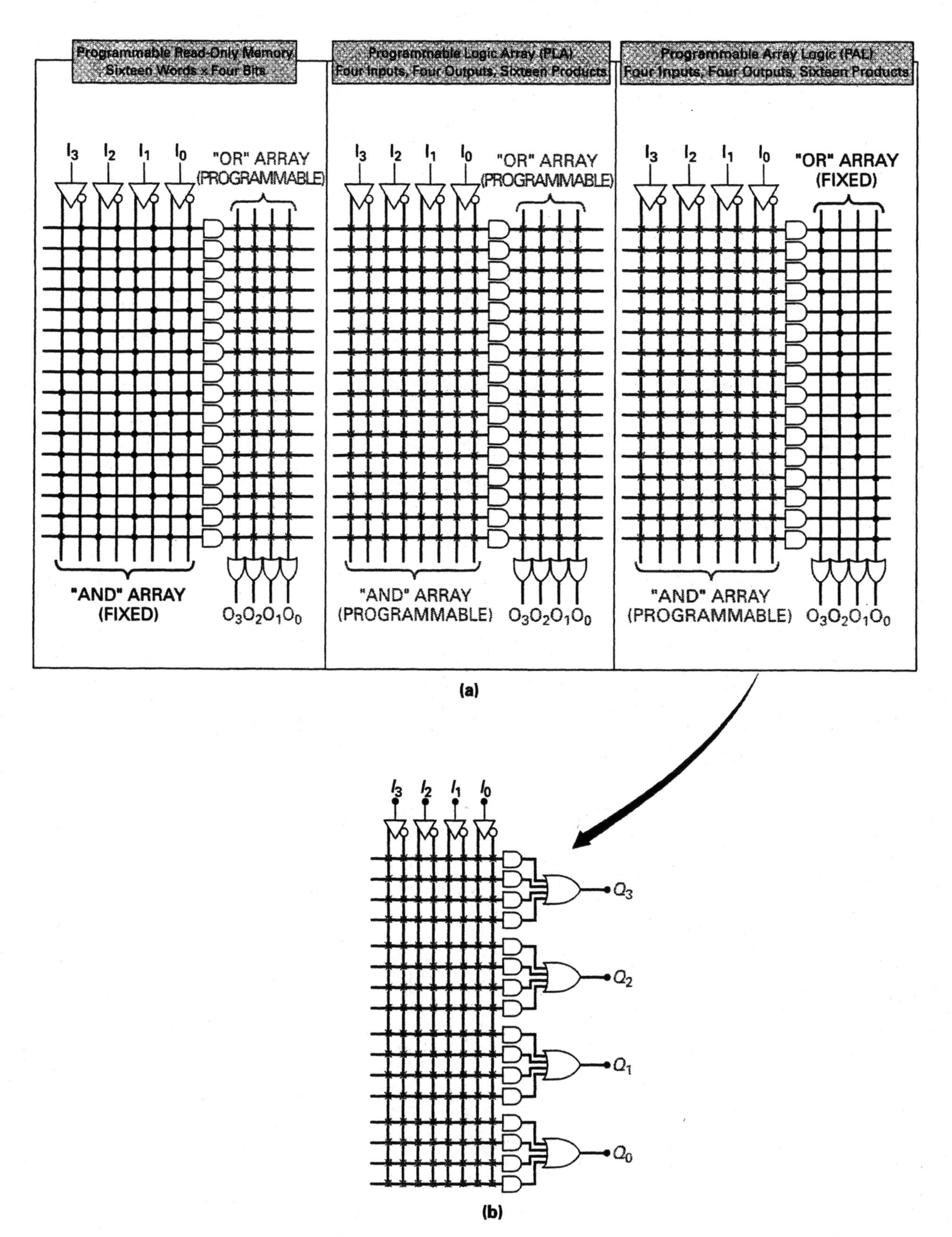

FIGURE 15-6 Programmable Array Logic (PAL).

DIGITAL DEVICE DATA SHEET

IC TYPE: PAL 12H6 (Twelve Inputs, Six Outputs)

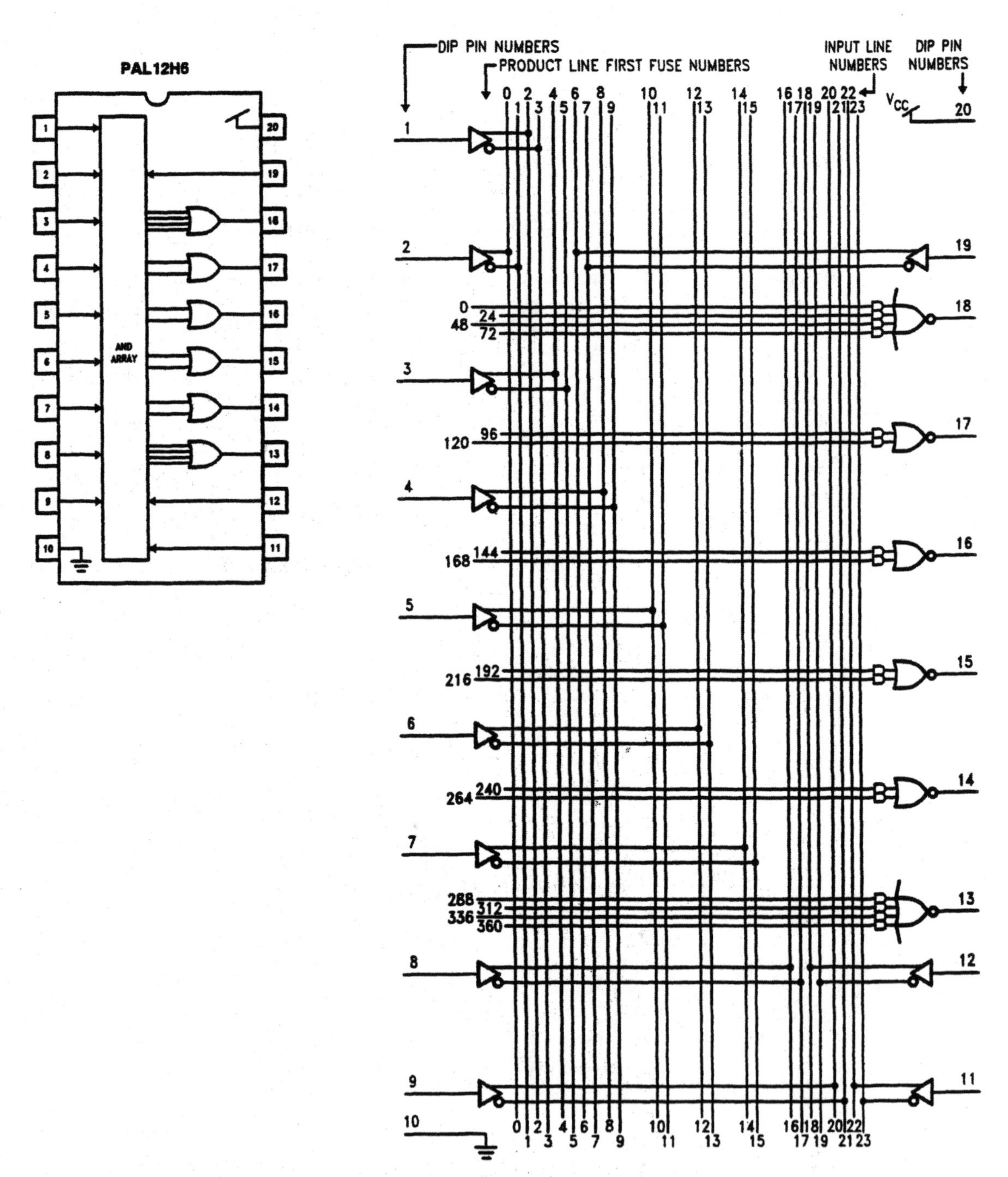

FIGURE 15-7 An IC Containing a Field-Programmable Array Logic (FPAL) Circuit. (Courtesy of National Semiconductor)

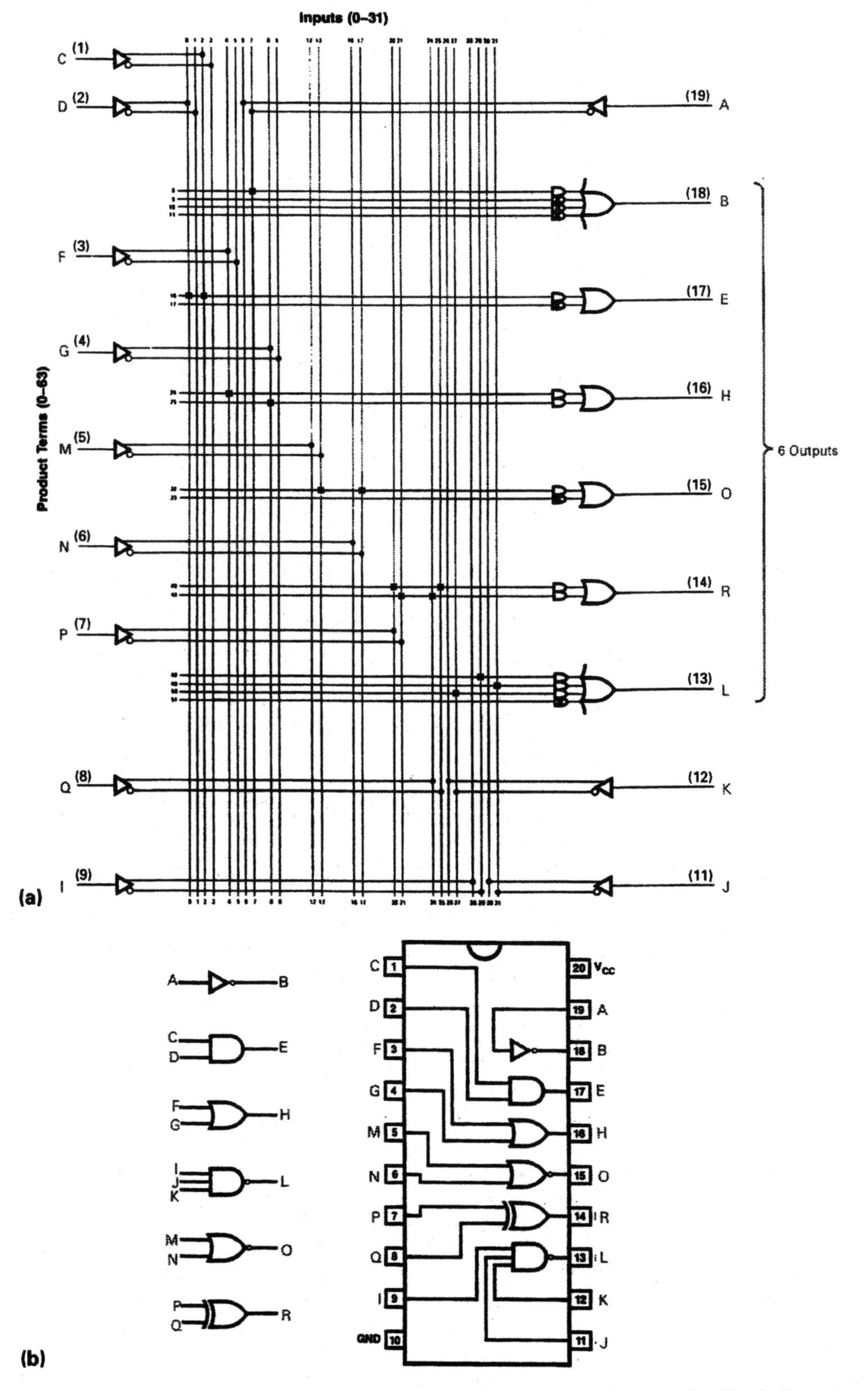

FIGURE 15-8 Application—Using a PAL12H6 to Implement the Basic Inverter, AND, OR, NAND, NOR, and XOR Functions. (Courtesy of National Semiconductor)

15-1-5 The Programmable Logic Sequencer (PLS)

Programmable Logic Sequencer (PLS)
A PLD that contains flip-flops, latches, and registers making it able to perform sequential logic and combinational logic functions.

The **programmable logic sequencer (PLS)** gives the circuit designer even more versatility. These devices contain flip-flops, latches, input registers, and output registers, making it possible to program these devices to perform sequential logic and combinational logic functions.

A Programmable Logic Sequencer (PLS) Data Sheet

Figure 15-9 shows the data sheet details for the PLS16R8, which has eight inputs and eight registered outputs. Device outputs are either taken directly from the AND-OR functions (combinatorial) or passed through D-type flip-flops (registered). It is the registers that allow this device to implement sequential logic circuits. Pin 1 provides register clocking, pin 11 provides three-state output control, and on power-up all registers are reset automatically.

15-1-6 Programming PLDs

A *universal programming unit,* like the PROM programmer shown previously in Figure 14-9(b), can also be used to program field-programmable logic devices (FPLDs) that are shipped to the customer with all fuses intact. To program a FPLD, you will first need to input the *fuse map* into the universal programmer. The fuse map describes which fuses within the FPLD should be blown and which should be left intact. Most of the FPLD manufacturers supply free programming software, which can be used on any personal computer to input the fuse map. Once this data has been entered, the personal computer can be instructed to transfer the fuse map data to the universal programming unit. The next step is to place a new FPLD into the socket on the universal programmer and then instruct the universal programmer to begin programming. The universal programmer will then automatically address each fuse in the array sequentially, and either blow the fuse or leave it intact based on the fuse map. Once the programming of the FPLD is complete, the universal programmer will test the programmed FPLD against the fuse map and indicate whether the device has passed or failed.

15-1-7 The Erasable Programmable Logic Device (EPLD)

All of the PLDs discussed so far are programmed by blowing their internal fuses. Once a fuse has been blown, it cannot be reset, and therefore any mistake during programming will mean that the device will have to be discarded. This disadvantage has now been overcome by the **erasable programmable logic device (EPLD).** These PLDs are similar to EEPROMs in that they can be electrically erased and reprogrammed because their fuses are electronic switches instead of fusible links.

Erasable Programmable Logic Device (EPLD)
A PLD that can be electrically erased and then reprogrammed.

An Erasable Programmable Logic Device (EPLD) Data Sheet

Figure 15-10 shows the data sheet details for the GAL16V8, which has eight dedicated inputs (pins 2 through 9), two special function inputs (pins 1

DIGITAL DEVICE DATA SHEET

IC TYPE – PLS 16R8 (Eight Inputs, Eight Registered Outputs)

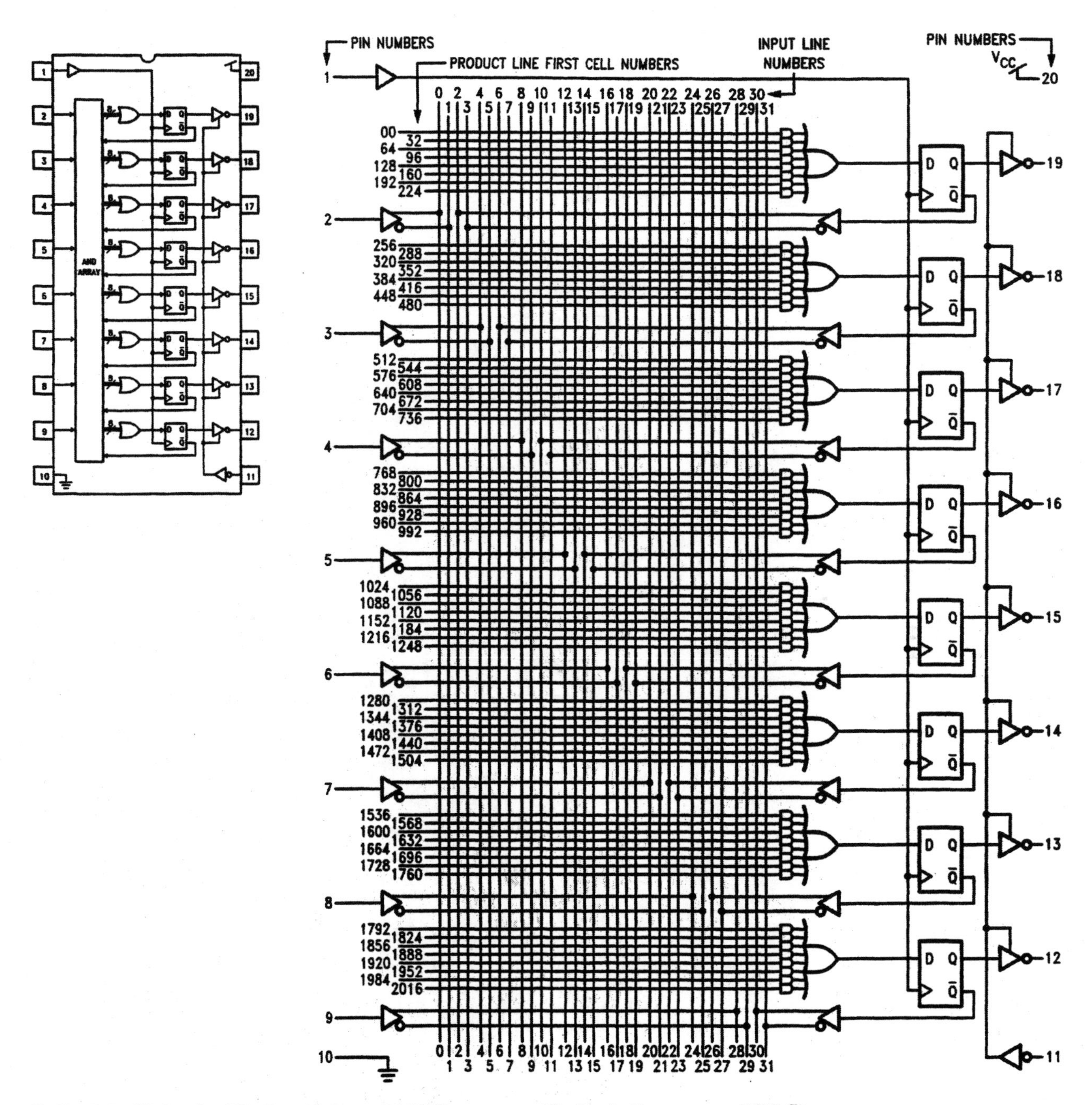

FIGURE 15-9 An IC Containing a Field-Programmable Logic Sequencer (FPLS) Circuit. (Courtesy of National Semiconductor)

and 11), and eight pins that can be used as either inputs or outputs (pins 12 through 19). The versatility of this device lies in the programmable *output logic macro cell (OLMC)*. The eight different products from the AND gates are applied to each of the eight OLMCs. These OLMCs will OR the products and then either apply the output to the output pin if a combinational output is required, or clock a D-type flip-flop if a registered sequential output is required.

DIGITAL DEVICE DATA SHEET

IC TYPE: GAL 16V8 – Generic Array Logic (Erasable PLD)

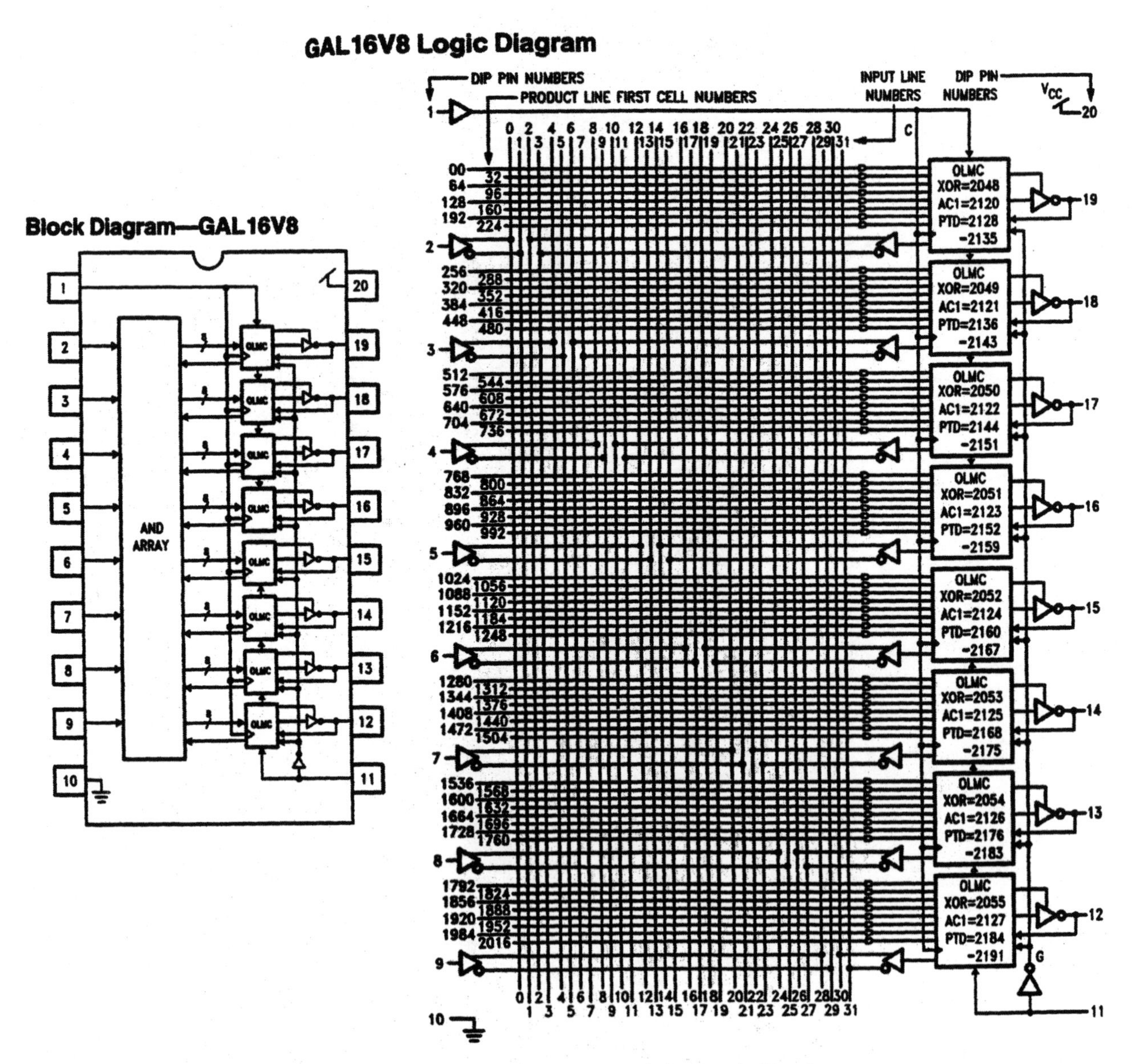

FIGURE 15-10 An IC Containing an Erasable Programmable Logic Device (EPLD). (Courtesy of National Semiconductor)

SELF-TEST REVIEW QUESTIONS FOR SECTION 15-1

1. What are the four basic types of programmable logic devices?
2. Is a PROM a PLD?
3. What are the differences among a PROM, a PLA, and a PAL?
4. What advantages does the PAL have over the PLA?
5. What is the advantage of the programmable logic sequencer?
6. What is the advantage of the EPLD?

15-2 TROUBLESHOOTING PLD CIRCUITS

To review, the procedure for fixing a failure can be broken down into three steps—diagnose, isolate, and repair. Since this chapter has been devoted to programmable logic devices, let us first examine the operation of a typical PLD circuit, and then apply our three-step troubleshooting process to this circuit.

15-2-1 A PLD Circuit

As an example, we will apply our three-step troubleshooting procedure to the PLD circuit shown in Figure 15-11(a). In this circuit, a PLS153A (shown previously in Figure 15-5) has been programmed with the fuse map shown in Figure 15-11(b) so that it will function as a BCD to seven-segment decoder. To explain the internal operation of this PLD, let us see how the circuit in Figure 15-11(b) will operate when 0000_{BCD} (decimal 0) is applied to the I_0 through I_3 input (ABCD). A 0000 input will be inverted to 1111, and via the fuses on the far right of the AND array, apply all HIGH inputs to the right-most AND gate, giving a HIGH out. The HIGH from this AND gate will give a 1111110 output at S_9 through S_3 via the fuses. Since all of the polarity fuses have been blown, 1111110 will be inverted to 0000001 and applied to the output and cause all of the segments within a common-anode display to turn ON except for segment *g*. A 0000 BCD input, therefore, to the PLS153A will be converted to a 0000001 seven-segment code, which will cause a "0" to be displayed.

All of the seven-segment codes for displaying decimal digits 0 through 9 have been programmed into this PLS153A, and each can be accessed by applying a corresponding 4-bit BCD input code.

Step 1: Diagnose

As before, it is extremely important that you first understand the operation of a circuit, and how all of the devices within it are supposed to work, so that you are able to determine whether or not a circuit malfunction really exists. If you were preparing to troubleshoot the circuit in Figure 15-11, your first step should be to read through the circuit description and review the operation of each integrated circuit until you feel completely confident with the correct operation of the entire circuit. When troubleshooting circuits with PLDs, you are

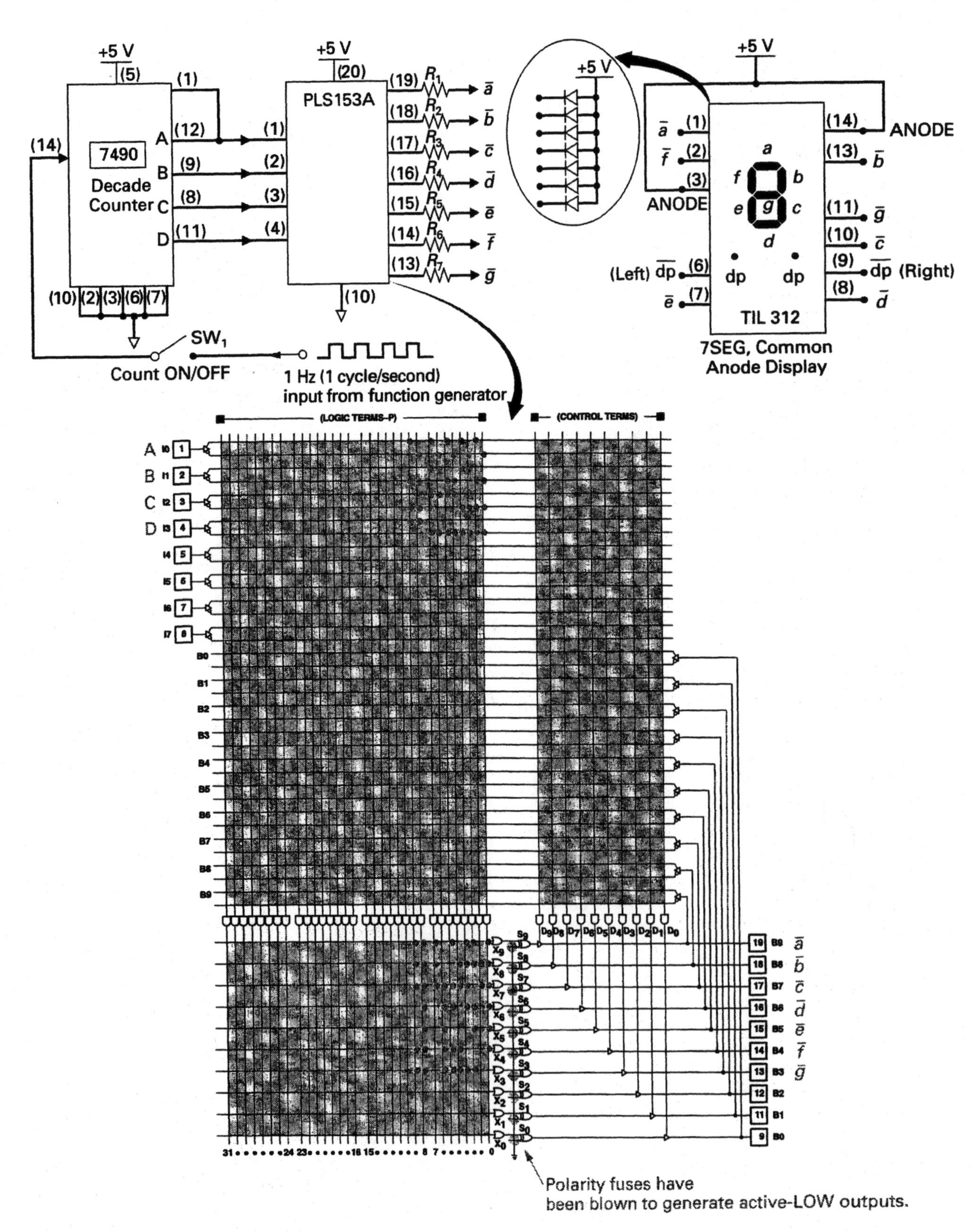

FIGURE 15-11 Troubleshooting PLD Circuits. (Permission granted by Philips Semiconductors)

first going to have to find out how the PLD has been programmed so that you can determine whether or not the device is acting as it should. This documentation will either be in the form of a fuse map like the one shown in Figure 15-11(b), or a program table like the one shown in Figure 15-12. Once you are fully familiar with the operation of the circuit, you will easily be able to diagnose the problem as either an *operator error* or a *circuit malfunction.*

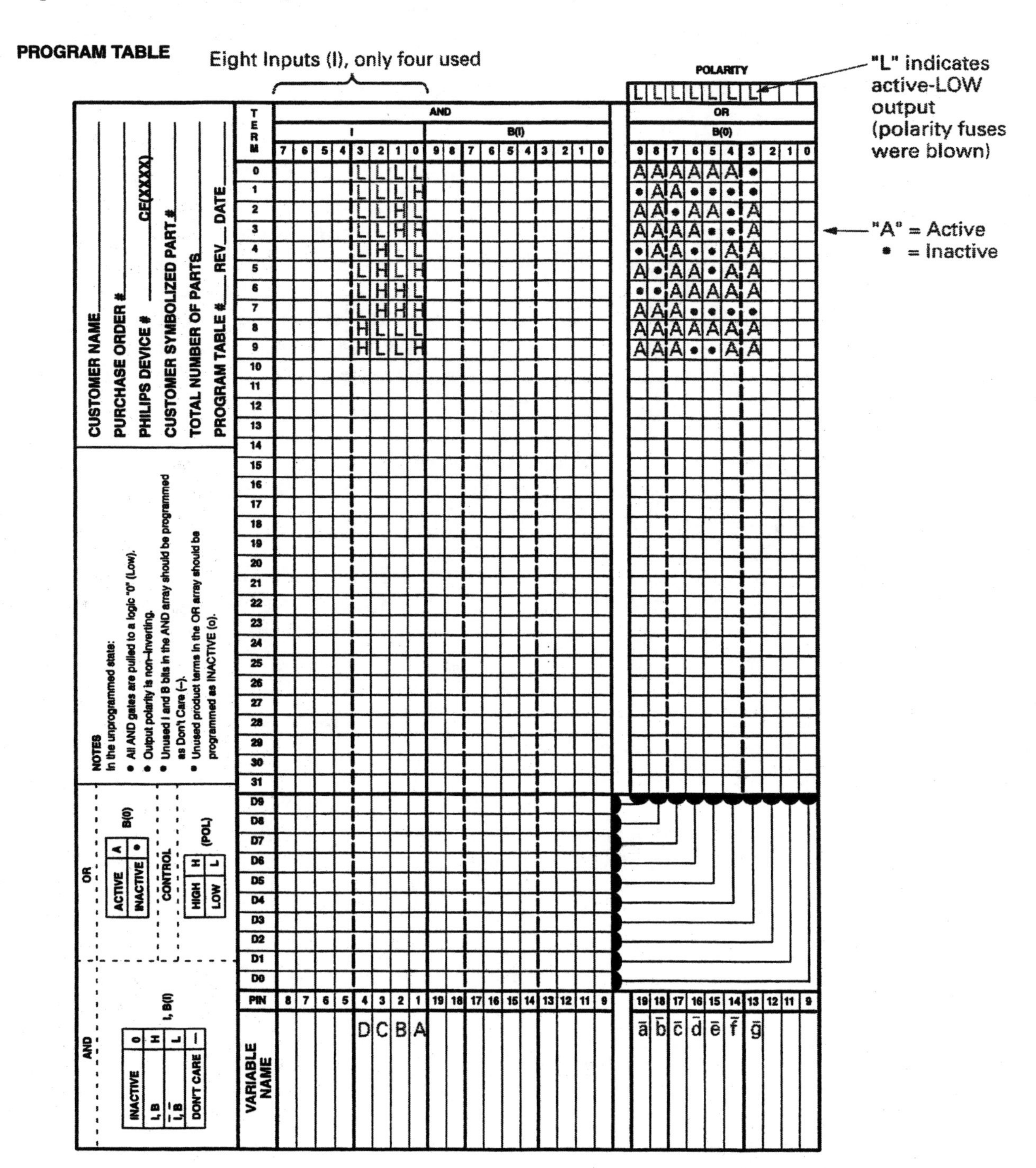

FIGURE 15-12 An FPLA Program Table. (Permission granted by Philips Semiconductor)

Step 2: Isolate

No matter what circuit or system failure has occurred, you should always follow a logical and sequential troubleshooting procedure. Let us review some of the isolating techniques and apply them to our example circuit in Figure 15-11.

a. Use a cause-and-effect troubleshooting process, which means study the effects you are getting from the faulty circuit and then logically reason out what could be the cause.

b. Check first for obvious errors before leaping into a detailed testing procedure. Is the power OFF or not connected to the circuit? Are there wiring errors? Are all of the ICs correctly oriented?

c. Using a logic probe or voltmeter test that power and ground are connected to the circuit and are present at all points requiring power and ground.

d. Use your senses to check for broken wires, loose connectors, overheating or smoking components, pins not making contact, and so on.

e. Test the clock signals using the logic probe or the oscilloscope. Although clock signals can be easily checked with a logic probe, the oscilloscope is best for checking the signal's amplitude, frequency, pulse width, and so on. The oscilloscope will also display timing problems and noise spikes (glitches) which could be false-triggering a circuit into operation at the wrong time.

f. Perform a static test on the circuit. With our circuit example in Figure 15-11, you can use a logic pulser to single-step the circuit from 0 through 9.

g. With a dynamic test, the circuit is tested while it is operating at its normal clock frequency (which in Figure 15-11 is 1 Hz), and all of the inputs and outputs are continually changing.

h. Noise due to electromagnetic interference (EMI) can false-trigger an IC, such as the counter in Figure 15-11. This problem can be overcome by not leaving any of the IC's inputs unconnected and, therefore, floating. Connect unneeded active-HIGH inputs to ground and active-LOW inputs to $+V_{CC}$.

i. Apply the half-split method of troubleshooting first to a circuit, and then to a section of a circuit, to help speed up the isolation process. With our circuit example in Figure 15-11, a good midpoint check would be to test the seven-segment codes out of the PLD.

j. Substitution can be used to help speed up your troubleshooting process. Once the problem is localized to an area containing only a few ICs, substitute suspect ICs with known working ICs (one at a time) to see if the problem can be quickly remedied.

Step 3: Repair

Once the fault has been found, the final step is to repair the circuit and then perform a final test to see that the circuit and the system are now fully operational.

SELF-TEST REVIEW QUESTIONS FOR SECTION 15-2

1. What is a fuse map?

2. What is a program table?

3. How can a fuse map or program table help you to troubleshoot a circuit containing a PLD?
4. Should a BCD to seven-segment PLD decoder/driver be tested differently than a standard BCD to seven-segment decoder/driver?

REVIEW QUESTIONS

Multiple-Choice Questions

1. Which of the following is a programmable logic device?
   a. ROM **b.** RAM **c.** PROM **d.** RWM
2. How many product lines would a four-input PROM have?
   a. Thirty-two **b.** Sixteen **c.** Twenty-four **d.** Eight
3. A PROM has a __________ AND array and a __________ OR array.
   a. programmable, fixed **b.** fixed, programmable
   c. programmable, programmable **d.** fixed, fixed
4. A PLA has a __________ AND array and a __________ OR array.
   a. programmable, fixed **b.** fixed, programmable
   c. programmable, programmable **d.** fixed, fixed
5. A PAL has a __________ AND array and a __________ OR array.
   a. programmable, fixed **b.** fixed, programmable
   c. programmable, programmable **d.** fixed, fixed
6. Which of the following PLD types contains flip-flops, latches, input registers, and output registers?
   a. PROM **b.** PLA **c.** PAL **d.** PLS
7. Which of the following PLD types is the most versatile?
   a. PROM **b.** PAL **c.** PLA **d.** Both a. and b.
8. Which of the following PLD types uses electronic switches instead of fusible links?
   a. EPLD **b.** RAM **c.** PROM **d.** PAL
9. Which of the following PLD types has a fixed OR array?
   a. PAL **b.** PLA **c.** EPLD **d.** PLS
10. Which of the following PLD types can be programmed by the customer using a universal programming unit?
    a. PROM **b.** FPLA **c.** FPAL **d.** All of the above

Essay Questions

11. Give the full names of the following abbreviations. (Chapter 15)
    a. PAL **b.** PLA **c.** PROM **d.** EPLD **e.** PLS **f.** PLD
12. What is a programmable logic device? (Introduction)
13. What are the four different types of PLDs? (Introduction)
14. Why is an "F" used as a prefix for a PLD? (Introduction)
15. Sketch a simple two-input/three-output PROM PLD. (15-1-1)
16. How many product lines will a four-input/four-output PROM PLD have? (15-1-1)

17. What is a fusible link? (15-1)
18. Why are PLD circuits generally redrawn? (15-1-1)
19. What is a programmable output polarity fuse? (15-1-1)
20. Describe how a PLD can be used to implement a sum-of-products Boolean expression. (15-1-1)
21. Can a PROM be used as a PLD? (15-1-2)
22. Using a sketch, describe the internal schematic of a typical PLA. (15-1-3)
23. Using a sketch, describe the internal schematic of a typical PAL. (15-1-4)
24. What advantage(s) does the PAL have over the PLA? (15-1-4)
25. What is a PLS? (15-1-5)
26. What advantage does the PLS have over the PLA and PAL? (15-1-5)
27. What is a universal programming unit? (15-1-6)
28. What is a fuse map? (15-1-6)
29. What is an EPLD and what is its key advantage? (15-1-7)
30. What is a program table? (15-2-2)

Practice Problems

To practice your device recognition and operation skills, refer to Figure 15-13 and answer the following nine questions.

31. Which type of PLD is being used in this application?
32. How many inputs are being applied to this IC, and how many outputs are generated by this IC?
33. What is the key function of this PLD?
34. Which outputs are active for each of the following input codes?

    a. 0001 **b.** 1010 **c.** 1111 **d.** 0101
35. If the outputs were connected to a seven-segment display, what would be displayed if the following inputs were applied?

    a. 0011 **b.** 1011 **c.** 1111 **d.** 1100
36. The ____________ input is used to suppress leading 0s in a multi-digit display. The signal is propagated from the most-significant digit to the least-significant digit. If the digit input is 0, and the ____________ input is LOW (indicating that the previous digit is also 0), all segments are left blank and this digit position's ____________ output is set LOW.

    a. intensity control (IC) **b.** ripple-blanking (RB)
    c. lamp test (LT) **d.** NC
37. The ____________ input controls the duty cycle of this display driver. When this input is HIGH, all segments are turned OFF. Pulsing this pin with a duty-cycled signal will control the display's apparent brightness.

    a. intensity control (IC) **b.** ripple-blanking (RB)
    c. lamp test (LT) **d.** NC
38. When the ____________ input is made HIGH, all of the segment outputs are made HIGH. This input allows you to check to see if all segments can be energized.

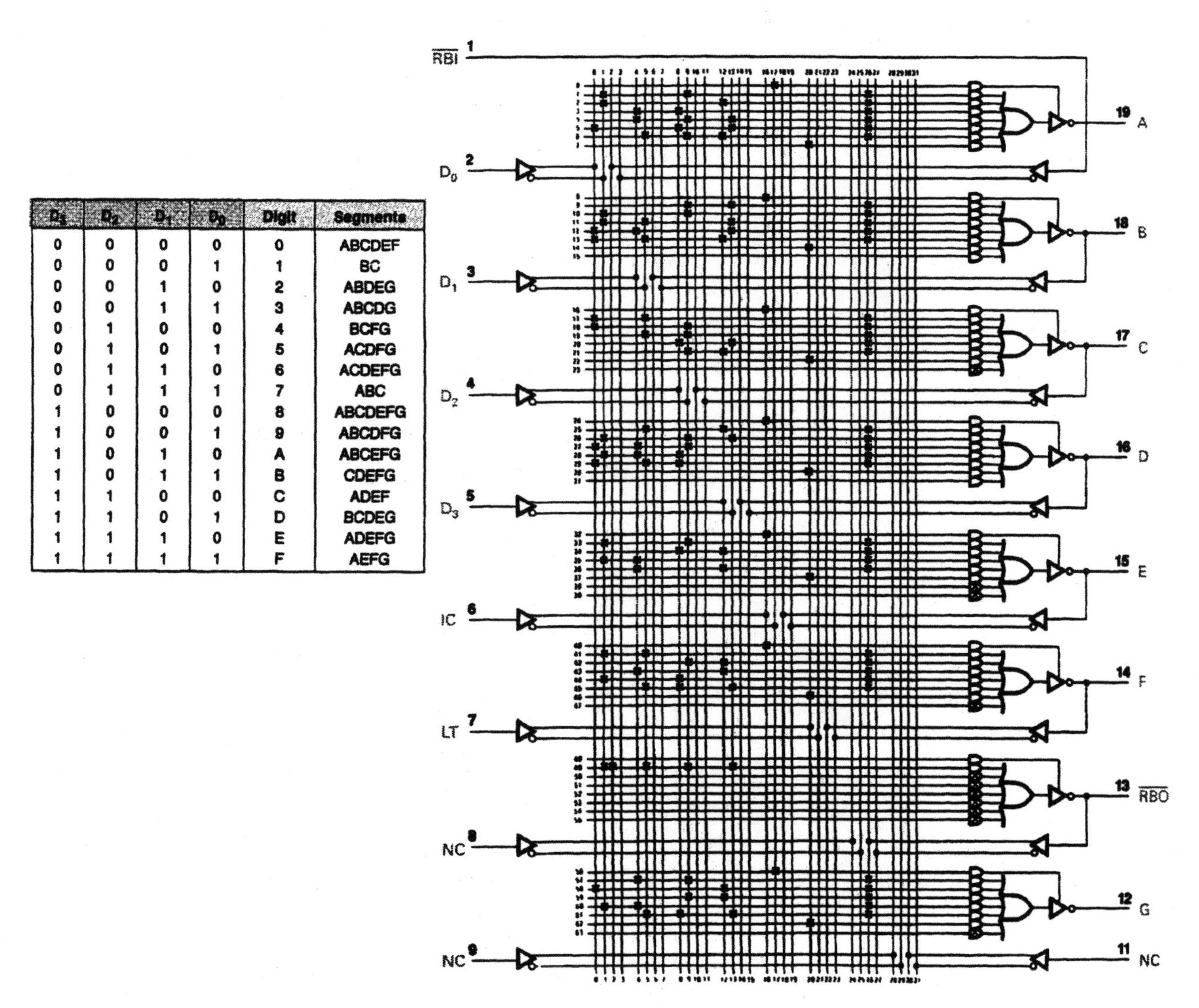

D_3	D_2	D_1	D_0	Digit	Segments
0	0	0	0	0	ABCDEF
0	0	0	1	1	BC
0	0	1	0	2	ABDEG
0	0	1	1	3	ABCDG
0	1	0	0	4	BCFG
0	1	0	1	5	ACDFG
0	1	1	0	6	ACDEFG
0	1	1	1	7	ABC
1	0	0	0	8	ABCDEFG
1	0	0	1	9	ABCDFG
1	0	1	0	A	ABCEFG
1	0	1	1	B	CDEFG
1	1	0	0	C	ADEF
1	1	0	1	D	BCDEG
1	1	1	0	E	ADEFG
1	1	1	1	F	AEFG

FIGURE 15-13 Using a PLD as a Hexadecimal to Seven-Segment Decoder. (Courtesy of National Semiconductor)

39. How many inputs to this PLD are not used?

40. Figure 15-14 shows how three PLDs could be used to display the data present on a 12-bit data bus. What would be the minimum to maximum hexadecimal value displayed?

To practice your device recognition and operation skills, refer to Figure 15-15 and answer the following five questions.

41. Which type of PLD is being used in this application?

42. How many inputs are being applied to this IC, and how many outputs are generated by this IC?

43. What is the key function of this PLD?

44. What is an address decoder?

45. Are the outputs active-HIGH or active-LOW?

Troubleshooting Questions

46. How can a PLD fuse map help you to test a PLD?
47. How can a PLD program table help you to test a PLD?
48. Can a universal programming unit help to test a PLD?
49. How would you test the PLD shown in Figure 15-13?
50. How would you test the PLD shown in Figure 15-15?

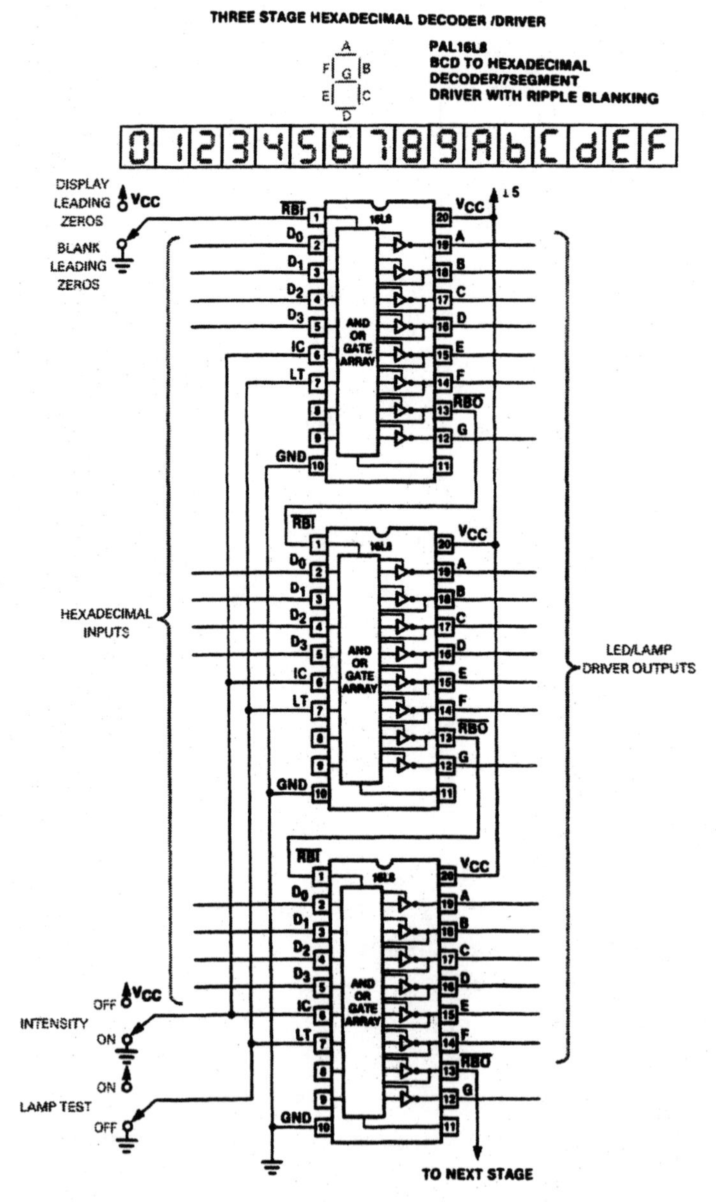

FIGURE 15-14 Three-Stage Hexadecimal Decoder/Driver Using PLDs. (Courtesy of National Semiconductor)

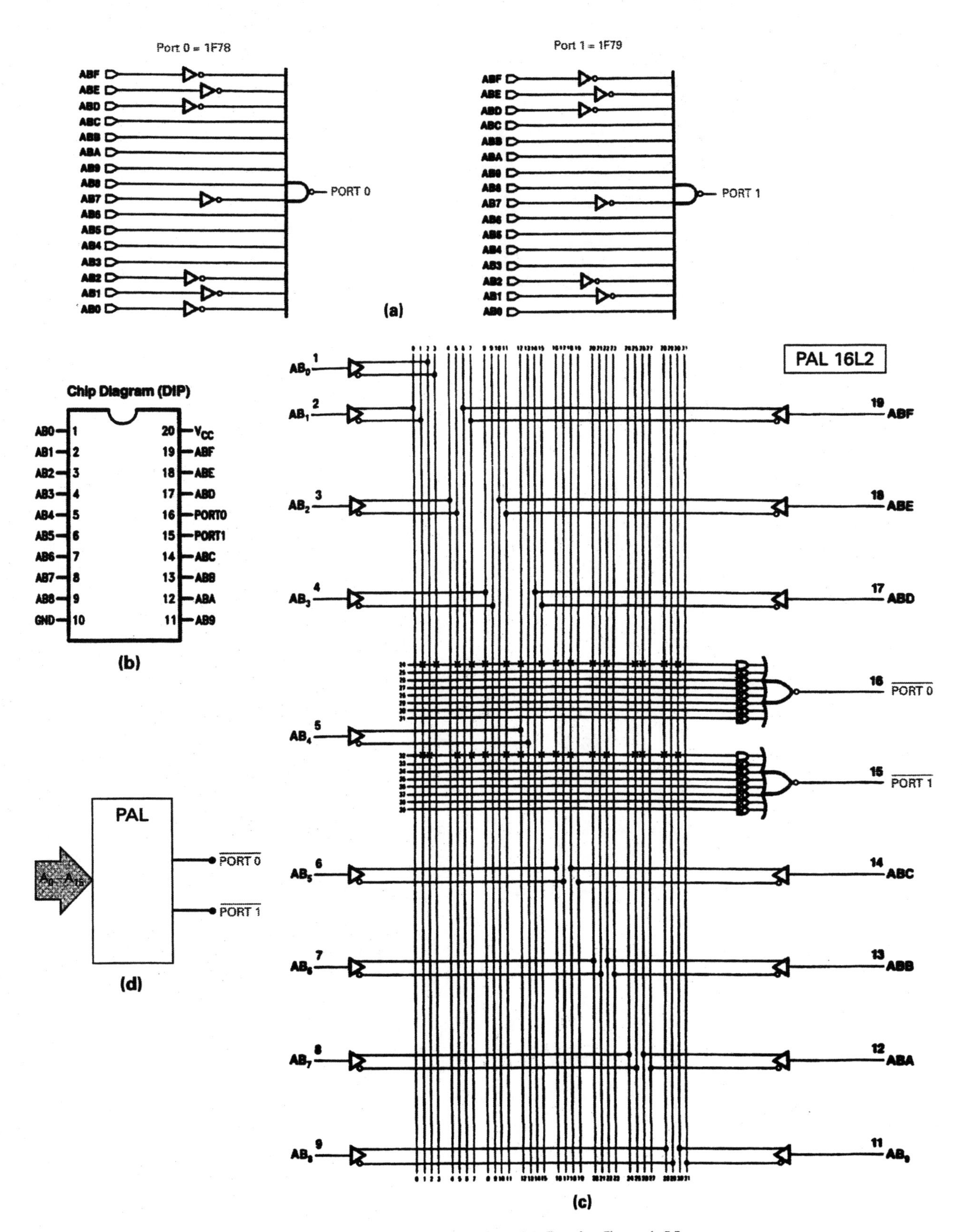

FIGURE 15-15 Using a PLD as an Address Decoder. (a) Logic Gate Address Decoders. (b) PLD Pin Out. (c) Internal IC Circuit with Five Connections. (d) Block Symbol. (Courtesy of National Semiconductor)

Chapter 16

OBJECTIVES

After completing this chapter, you will be able to:

1. Describe how analog and digital devices are connected to a microcomputer.
2. Explain the process of converting analog data to digital data, and digital data to analog data.
3. Name two methods of digital-to-analog conversion.
4. Describe the operation of a binary-weighted resistor digital-to-analog converter (DAC) circuit.
5. Describe the operation of an *R*/2*R* digital-to-analog converter circuit.
6. Define the following digital-to-analog converter characteristics.
   a. Resolution
   b. Monotonicity
   c. Settling time
   d. Accuracy
7. Interpret specifications from a typical DAC data sheet and describe how the device would typically be connected in an application.
8. Describe how DAC ICs are tested.
9. Name two methods of analog-to-digital conversion.
10. Describe the operation of a staircase analog-to-digital converter (ADC) circuit.
11. Describe the operation of a successive approximation analog-to-digital converter circuit.
12. Describe the operation of a flash analog-to-digital converter circuit.
13. Interpret specifications from a typical ADC data sheet and describe how the device would typically be connected in an application.
14. Describe how ADC ICs are tested.
15. Describe how to troubleshoot data converter circuits containing DACs and ADCs using the three-step method.

Analog and Digital Signal Converters

OUTLINE

VIGNETTE

The First Pocket Calculator

During the seventeenth century, European thinkers were obsessed with any device that could help them with mathematical calculation. Scottish mathematician John Napier decided to meet this need, and in 1614 he published his new discovery of logarithms. In this book, consisting mostly of tediously computed tables, Napier stated that a logarithm is the exponent of a base number. For example:

The common logarithm (base-10) of 100 is 2 ($100 = 10^2$).
The common logarithm of 10 is 1 ($10 = 10^1$).
The common logarithm of 27 is 1.43136 ($27 = 10^{1.43136}$).
The common logarithm of 6 is 0.77815 ($6 = 10^{0.77815}$).

Any number therefore, no matter how large or small, can be represented by or converted to a logarithm. He also outlined how the multiplication of two numbers could be a achieved by simply adding the numbers' logarithms. For example, if the logarithm of 2 (which is 0.30103), is added to the logarithm of 4 (which is 0.60206), the result will be 0.90309, which is the logarithm of the number 8 ($0.30103 + 0.60206 = 0.90309$, $2 \times 4 = 8$). Therefore, the multiplication of two large numbers can be achieved by looking up the logarithm of the two numbers in a log table, adding the logarithms together, and then finding the number that corresponds to the sum in an antilog (reverse log) table. In this example, the antilog of 0.90309 is 8.

Napier's logarithm and antilogarithm tables are stored today in every scientific calculator. Just ten years after Napier's death in 1617, William Oughtred developed a handy mechanical device that could be used for rapid calculation. This device, considered the first pocket calculator, was the slide rule.

INTRODUCTION

As discussed in Chapter 1, electronic information exists as either analog data or as digital data. With analog data, the signal voltage change is an analogy of the physical information it is representing. For example, if a person were to speak into a microphone, the pitch and loudness of the sound wave would determine the frequency and amplitude of the voltage signal generated by the microphone. With a digital signal, information is first converted into a group of HIGH and LOW voltage pulses that represent the binary digits 1 and 0. For example, the binary digit code 1101001 is generated whenever the lower case "i" key on an ASCII computer keyboard is pressed. In this case, therefore, the information "i" is encoded into a digital code, or a code that makes use of binary digits.

Whenever you have two different forms, there is always a need to convert back and forth between these two different forms. For example, in previous chapters we have seen how we have needed circuits to convert between binary and decimal, bipolar logic levels and MOS logic levels, serial data and parallel data, and so on. In this chapter, we will examine data conversion techniques and the circuits that are used to perform digital-to-analog (D/A) data conversion and analog-to-digital (A/D) data conversion. We will also discuss the application, testing, and troubleshooting of these data conversion circuits.

16-1 ANALOG AND DIGITAL SIGNAL CONVERSION

Today, microcomputers are used in almost every electronic system because of their ability to quickly process and store large amounts of data, make systems more versatile, and perform more functions. Digital processing within electronic systems has therefore become more and more predominant. Many of the input signals applied to a microcomputer for storage or processing are analog in nature, and therefore a data conversion circuit is needed to interface these analog inputs to a digital system. Similarly, a data conversion circuit may also be needed to interface the digital microcomputer system to an analog output device.

16-1-1 Connecting Analog and Digital Devices to a Computer

Figure 16-1 shows a simplified block diagram of a microcomputer. No matter what the application, a microcomputer has three basic blocks—a microprocessor unit, a memory unit, and an input/output unit. Getting information into the computer is the job of input devices, and as you can see in Figure 16-1, there are two input paths into a computer—one path from analog input devices, and the other path from digital input devices. Information in the form of light, sound, heat, pressure, or any other real world quantity is analog in nature and has an infinite number of input levels. Sensors or transducers of this type, such as photoelectric cells for light or microphones for sound, will generate analog signals. Since the computer operates on information in only digital or binary form, a translation or conversion is needed if the computer is going to be able to understand or interpret an analog signal. This signal processing is achieved by an analog-to-digital converter (ADC) which transforms the varying analog input voltage into digital codes which the computer can understand. Once the analog information has been encoded into digital form, the information can then enter the computer through an electronic doorway called the **input port.** Some input devices, such as a computer keyboard, automatically generate a digital code for each key that is pressed (ASCII codes). The codes generated by these digital input devices can be connected directly to the computer without having to pass through a converter.

Input Port
An electronic doorway used for transmitting information into a computer.

Output Port
An electronic doorway used for transmitting information out of a computer.

Similarly, the digital output information from a microcomputer's **output port** can be used to drive either an analog device or a digital device. If an analog

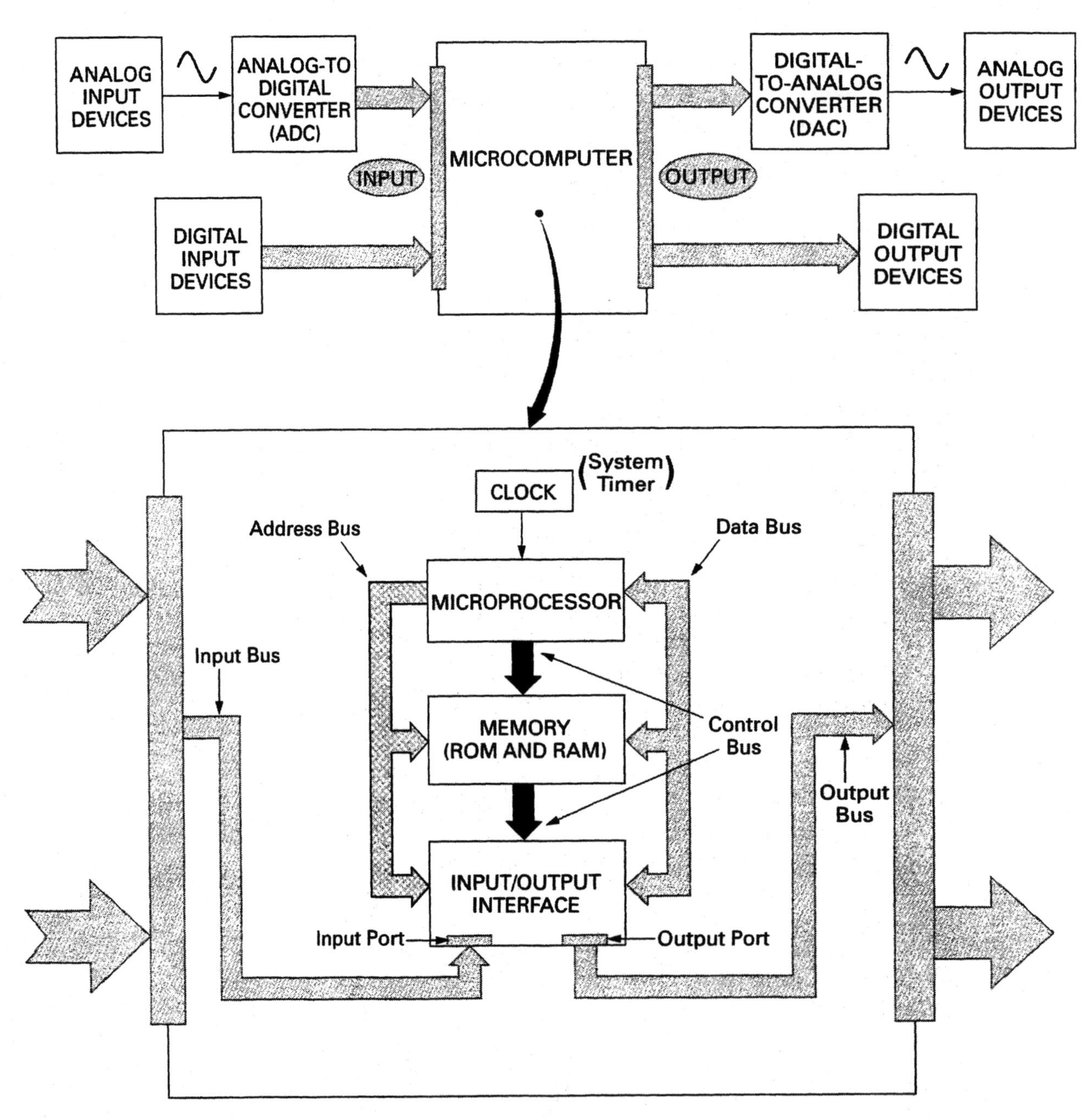

FIGURE 16-1 Connecting Analog and Digital Devices to a Computer.

output is desired, as in the case of a voice signal for a loudspeaker, the digital computer's output will have to be converted into a corresponding analog signal by a digital-to-analog converter (DAC). If only a digital output is desired, as in the case of a printer which translates the binary codes into printed characters, the digital computer output can be connected directly to the output device without the need for a converter.

16-1-2 *Converting Information Signals*

Figure 16-2(a) shows how analog data can be converted into digital data, and Figure 16-2(b) shows how digital data can be converted into analog data. To give this process purpose, let us consider how the music information stored on a compact disc (CD) is first recorded and then played back. Like all sound, music is made up of waves of compressed air. When these waves strike the diaphragm of a microphone, an analog voltage signal is generated. In the past, this analog data was recorded on magnetic tapes or as a grooved track on a record. These data storage devices were susceptible to wear and tear, temperature, noise, and age. Using digital recording techniques, binary codes are stored on a compact disc to achieve near-perfect fidelity to live sound.

Figure 16-2(a) shows how an **analog-to-digital converter (ADC)** is used during the recording process to convert the analog music input into a series of digital output codes. These digital codes are used to control the light beam of a recording laser so that it will engrave the binary 0s and 1s onto a compact disk (CD) in the form of pits or spaces. The ADC is triggered into operation by a sampling pulse that causes it to measure the input voltage of the analog signal at that particular time, and generate an equivalent digital output code. For example, on the active edge of sampling pulse 1, the analog input voltage is at 2 V so the binary code 010 (decimal 2) is generated at the output of the ADC. On the active edge of sampling pulse 2, the analog voltage has risen to 4 V, so the binary code 100 (decimal 4) is generated at the output of the ADC. On the active edge of sampling pulses 3, 4, and 5, the binary codes 110 (decimal 6), 111 (decimal 7), and 110 (decimal 6) are generated at the output, representing the analog voltages of 6 V, 7 V, and 6 V, respectively, and so on.

Analog-to-Digital Converter (ADC)
A circuit that converts analog input signals into equivalent digital output signals.

Figure 16-2(b) shows how a **digital-to-analog converter (DAC)** is used during the music playback process to convert the digital codes stored on a CD to an analog output signal. Another laser is used during playback to read the pits and spaces on the compact disc as 0s and 1s. These codes are then applied to the DAC, which converts the digital input codes into discrete voltages. The DAC is triggered into operation by a strobe pulse that causes it to convert the code currently being applied at the input into an equivalent output voltage.

Digital-to-Analog Converter (DAC)
A circuit that converts digital input signals into equivalent analog output signals.

For example, on the active edge of strobe pulse 1, the digital code 010 (decimal 2) is being applied to the DAC, so it will generate 2 V at its output. On the active edge of strobe pulse 2, the digital code 100 (decimal 4) is being applied to the DAC, so it will generate 4 V at its output. On the active edge of strobe pulse 3, the digital code 110 (decimal 6) is being applied to the DAC, so it will generate 6 V at its output. On the active edge of strobe pulse 4, the digital code 111 (decimal 7) is being applied to the DAC, so it will generate 7 V at its output. On the active edge of strobe pulse 5, the digital code 110 (decimal 6) is being applied to the DAC, so it will generate 6 V at its output, and so on. If the output of a DAC is then applied to a low-pass filter, the discrete voltage steps can be blended into a smooth wave that closely approximates the original analog wave, as shown by the dashed line in Figure 16-2(b).

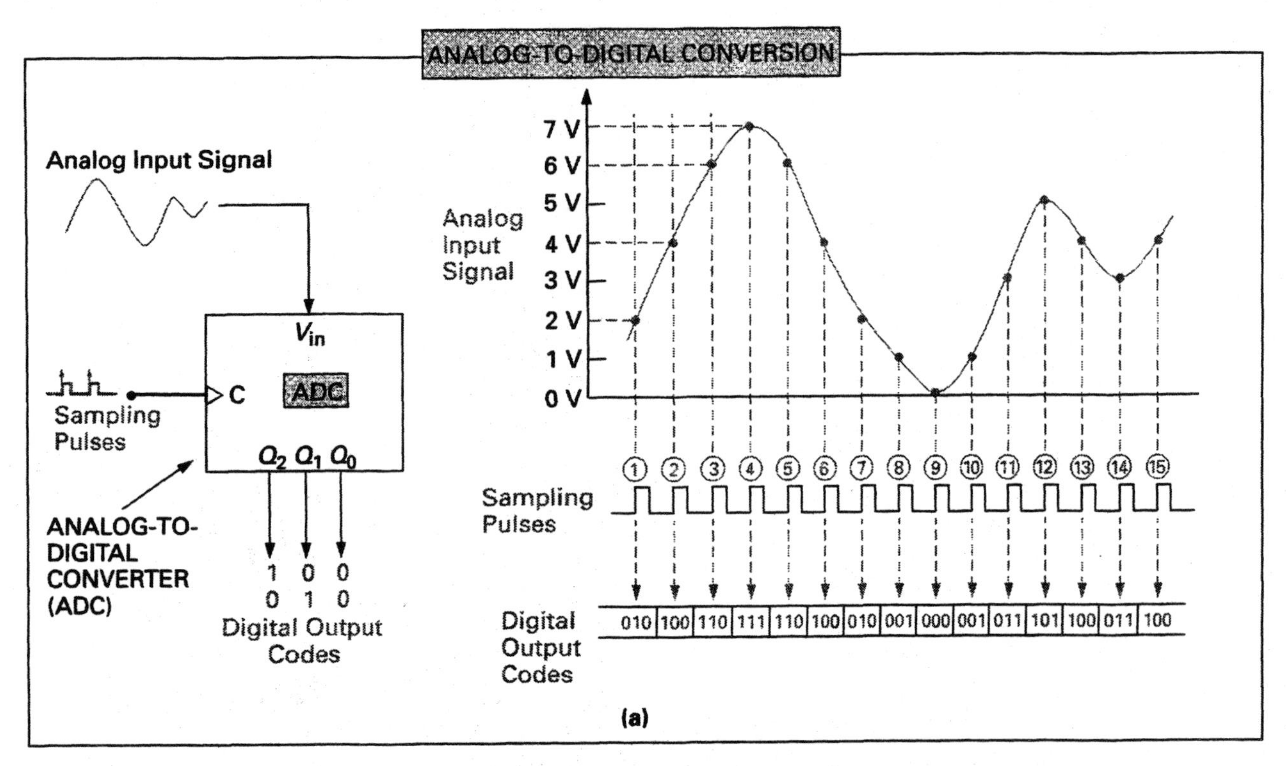

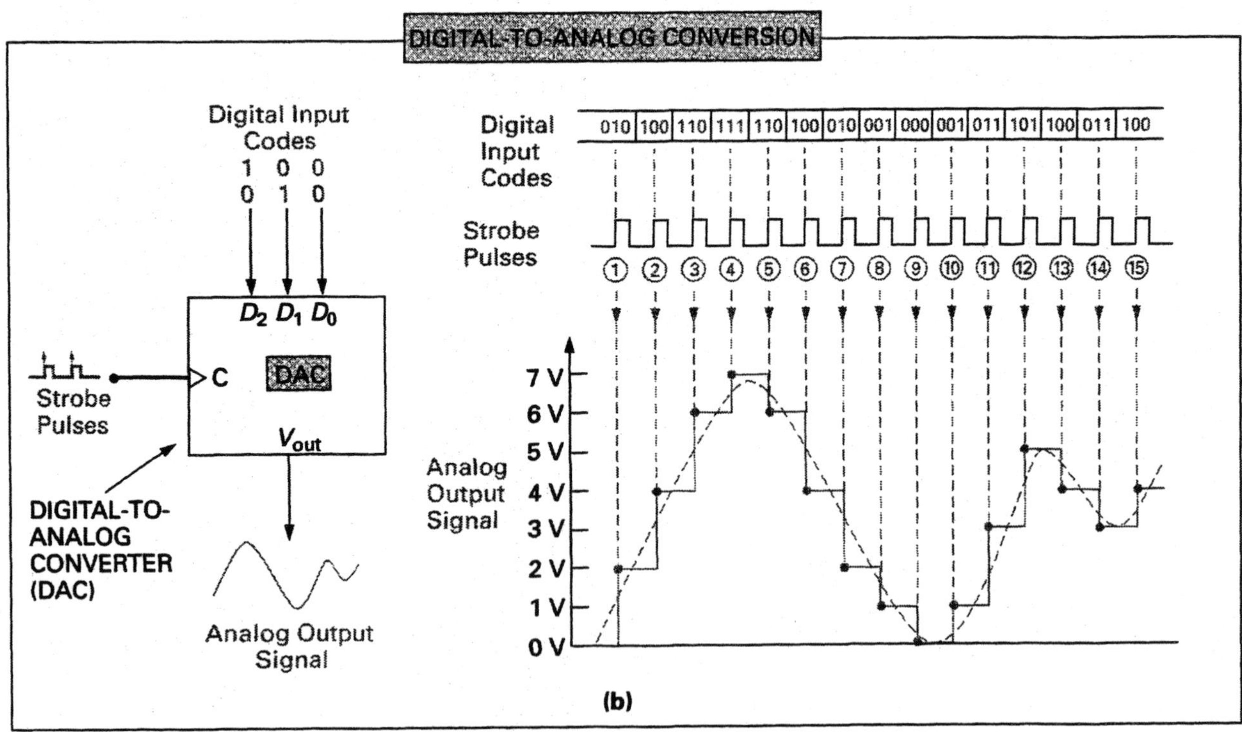

FIGURE 16-2 Converting Information Signals.

SELF-TEST REVIEW QUESTIONS FOR SECTION 16-1

1. Give the full names of the following abbreviations.
   a. ADC   b. DAC
2. To interface a microcomputer with analog devices, a ________ would be connected to the input port, and a ________ would be connected to the output port.
3. An ADC will convert a/an ________ input into an equivalent ________ output.
4. A DAC will convert a/an ________ input into an equivalent ________ output.

16-2 DIGITAL-TO-ANALOG CONVERTERS (DACs)

A DAC (pronounced "dak") is a circuit that generates an analog output voltage that is proportional to the value of the binary input code. There are a variety of digital-to-analog converter ICs available with varying characteristics. In this section, we will be discussing digital-to-analog conversion techniques and characteristics, a typical DAC data sheet, and DAC applications and testing.

16-2-1 ***Binary-Weighted Resistor DAC***

Binary-Weighted Resistor Circuit
A DAC circuit that uses a power-of-2 weighted resistor network to convert a digital signal input into an equivalent analog signal output.

One of the simplest digital-to-analog converter circuits is the **binary-weighted resistor DAC,** shown in Figure 16-3(a). In this circuit, the 4-bit digital input to this circuit is used to control four HIGH-LOW input switches (SW_1 through SW_4). The setting of these input switches will determine the analog output voltage generated at the output, as listed in the table in Figure 16-3(b). A -10 V reference voltage is applied to a binary-weighted input resistor network, and an inverting op-amp is used to convert a current input into a positive analog output voltage.

Let us now examine the operation of this circuit in more detail. In the example in Figure 16-3(a), a digital input of 1011 (decimal 11) is being applied to the DAC circuit, so SW_1 is set HIGH, SW_2 is set HIGH, SW_3 is set LOW, and SW_4 is set HIGH. This means that input resistors $8R$, $4R$ and R will contribute to the final output, whereas input resistor $2R$ will not contribute to the final output. A negative reference voltage has been used because the op-amp is connected as an inverting amplifier, so a negative reference voltage input will generate positive analog output voltages, as listed in the table in Figure 16-3(b). The final analog output voltage from this circuit is determined by the reference

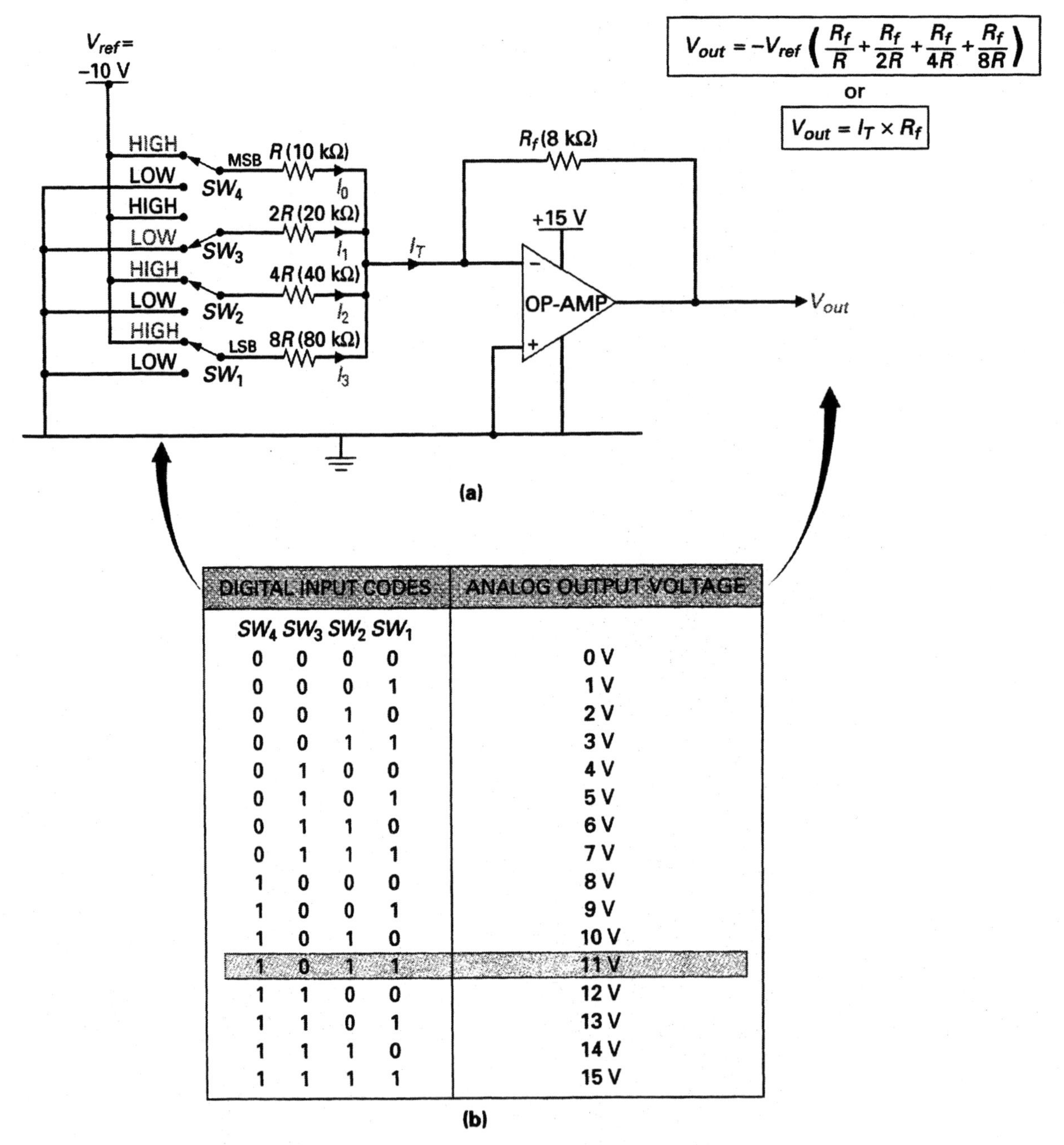

DIGITAL INPUT CODES SW_4 SW_3 SW_2 SW_1	ANALOG OUTPUT VOLTAGE
0 0 0 0	0 V
0 0 0 1	1 V
0 0 1 0	2 V
0 0 1 1	3 V
0 1 0 0	4 V
0 1 0 1	5 V
0 1 1 0	6 V
0 1 1 1	7 V
1 0 0 0	8 V
1 0 0 1	9 V
1 0 1 0	10 V
1 0 1 1	11 V
1 1 0 0	12 V
1 1 0 1	13 V
1 1 1 0	14 V
1 1 1 1	15 V

(b)

FIGURE 16-3 Binary-Weighted Resistor Digital-to-Analog Converter (DAC).

voltage, and the ratio of the feedback resistor R_f to the contributing input resistors. The formula is as follows:

$$V_{out} = -V_{ref}\left(\frac{R_f}{R} + \frac{R_f}{2R} + \frac{R_f}{4R} + \frac{R_f}{8R}\right)$$

For the 1101 (decimal 11) digital input example shown in Figure 16-3(a), the analog output voltage will be

$$V_{out} = -V_{ref}\left(\frac{R_f}{R} + \frac{R_f}{2R} + \frac{R_f}{4R} + \frac{R_f}{8R}\right)$$
$$= -(-10\text{ V})\left(\frac{8\text{ k}\Omega}{10\text{ k}\Omega} + 0 + \frac{8\text{ k}\Omega}{40\text{ k}\Omega} + \frac{8\text{ k}\Omega}{80\text{ k}\Omega}\right)$$
$$= 10\text{ V }(0.8 + 0 + 0.2 + 0.1)$$
$$= 10\text{ V} \times 1.1 = 11\text{ V}$$

■ **EXAMPLE:**

What analog output voltage will be generated by the circuit in Figure 16-3(a) if a digital input of 1001 is applied?

■ ***Solution:***

If a digital input of 1001 (decimal 9) is applied to the DAC circuit, SW_4 will be set HIGH, SW_3 will be set LOW, SW_2 will be set LOW, and SW_1 will be set HIGH. This means that input resistors $8R$ and R will contribute to the final output, whereas input resistors $4R$ and $2R$ will not contribute to the final output. The circuit will therefore generate the following analog output voltage.

$$V_{out} = -V_{ref}\left(\frac{R_f}{R} + \frac{R_f}{2R} + \frac{R_f}{4R} + \frac{R_f}{8R}\right)$$
$$= -(-10\text{ V})\left(\frac{8\text{ k}\Omega}{10\text{ k}\Omega} + 0 + 0 + \frac{8\text{ k}\Omega}{80\text{ k}\Omega}\right)$$
$$= 10\text{ V }(0.8 + 0 + 0 + 0.1)$$
$$= 10\text{ V} \times 0.9 = 9\text{ V}$$

The analog output voltage generated by the circuit in Figure 16-3(a) can also be calculated using the formula

$$V_{out} = I_T \times R_f$$

For a 1101 (decimal 11) input, therefore, the circuit in Figure 16-3(a) will generate the following analog output voltage.

$$V_{out} = I_T \times R_f$$
$$I_0 = \frac{-V_{ref}}{R} = \frac{-(-10\text{ V})}{10\text{ k}\Omega} = \frac{10\text{ V}}{10\text{ k}\Omega} = 1\text{ mA}$$
$$I_1 = 0\text{ mA}$$
$$I_2 = \frac{-V_{ref}}{4R} = \frac{-(-10\text{ V})}{40\text{ k}\Omega} = \frac{10\text{ V}}{40\text{ k}\Omega} = 0.25\text{ mA}$$
$$I_3 = \frac{-V_{ref}}{8R} = \frac{-(-10\text{ V})}{80\text{ k}\Omega} = \frac{10\text{ V}}{80\text{ k}\Omega} = 0.125\text{ mA}$$
$$I_T = I_1 + I_2 + I_3 + I_4 = 1\text{ mA} + 0\text{ mA} + 0.25\text{ mA} + 0.125\text{ mA}$$
$$= 1.375\text{ mA}$$
$$V_{out} = I_T \times R_f = 1.375\text{ mA} \times 8\text{ k}\Omega = 11\text{ V}$$

■ **EXAMPLE:**

Using the $V_{out} = I_T \times R_f$ formula, calculate the analog output voltage generated by the circuit in Figure 16-3(a) if a digital input of 1001 were applied.

■ ***Solution:***

$$V_{out} = I_T \times R_f$$

$$I_0 = \frac{-V_{ref}}{R} = \frac{-(-10 \text{ V})}{10 \text{ k}\Omega} = \frac{10 \text{ V}}{10 \text{ k}\Omega} = 1 \text{ mA}$$

$$I_1 = 0 \text{ mA}$$

$$I_2 = 0 \text{ mA}$$

$$I_3 = \frac{-V_{ref}}{8R} = \frac{-(-10 \text{ V})}{80 \text{ k}\Omega} = \frac{10 \text{ V}}{80 \text{ k}\Omega} = 0.125 \text{ mA}$$

$$I_T = I_1 + I_2 + I_3 + I_4 = 1 \text{ mA} + 0 \text{ mA} + 0 \text{ mA} + 0.125 \text{ mA} = 1.125 \text{ mA}$$

$$V_{out} = I_T \times R_f = 1.125 \text{ mA} \times 8 \text{ k}\Omega = 9 \text{ V}$$

16-2-2 ***R/2R Ladder DAC***

One of the key disadvantages of the binary-weighted resistor DAC is that the weighted resistors create difficulties as the number of digital input bits increases. For example, an 8-bit DAC will need eight weighted resistors of *R*, 2*R*, 4*R*, 8*R*, 16*R*, 32*R*, 64*R*, and 128*R*. A 16-bit DAC would need sixteen weighted resistors, with the largest resistance being 32,768 times larger than the smallest resistor. The large difference between the LSB resistor value and the MSB resistor value creates several problems. Another disadvantage is the extreme difference in current in each branch of the binary-weighted resistor network. For example, having a very large LSB resistor value will result in an extremely small current value that is susceptible to noise. In addition, having an extremely small MSB resistor value will result in an extremely large current value that will cause temperature and load changes.

***R/2R* Ladder Circuit** A DAC circuit that uses a resistor network having only two different values—R and 2R.

The problems of the binary-weighted resistor DAC can be overcome by using the ***R2/R* ladder DAC,** shown in Figure 16-4(a). This *R*/2*R* resistor network does not have the resistance range problems of the weighted-resistor network because it only makes use of two resistance values, *R* and 2*R*. Like the binary-weighted resistor DAC, the *R*/2*R* ladder DAC circuit uses binary-controlled switches to switch binary weighted currents through to the summing junction of the op-amp. The op-amp is included, as in the binary-weighted resistor DAC, to convert the input current into a proportional output voltage.

The R/2R ladder resistor network operates as a current divider. Figures 16-4(b) through (j) show how this circuit operates as a current divider. In Figure 16-4(b), you can see that below junction *D* we have a 2*R* resistor in parallel (||) with a 2*R* resistor, which is equivalent to *R*, as redrawn in Figure 16-4(c). Moving on to the redrawn circuit in Figure 16-4(c), you can see that we now have *R* in series with *R* directly below junction *C*, which is equivalent to 2*R*, as

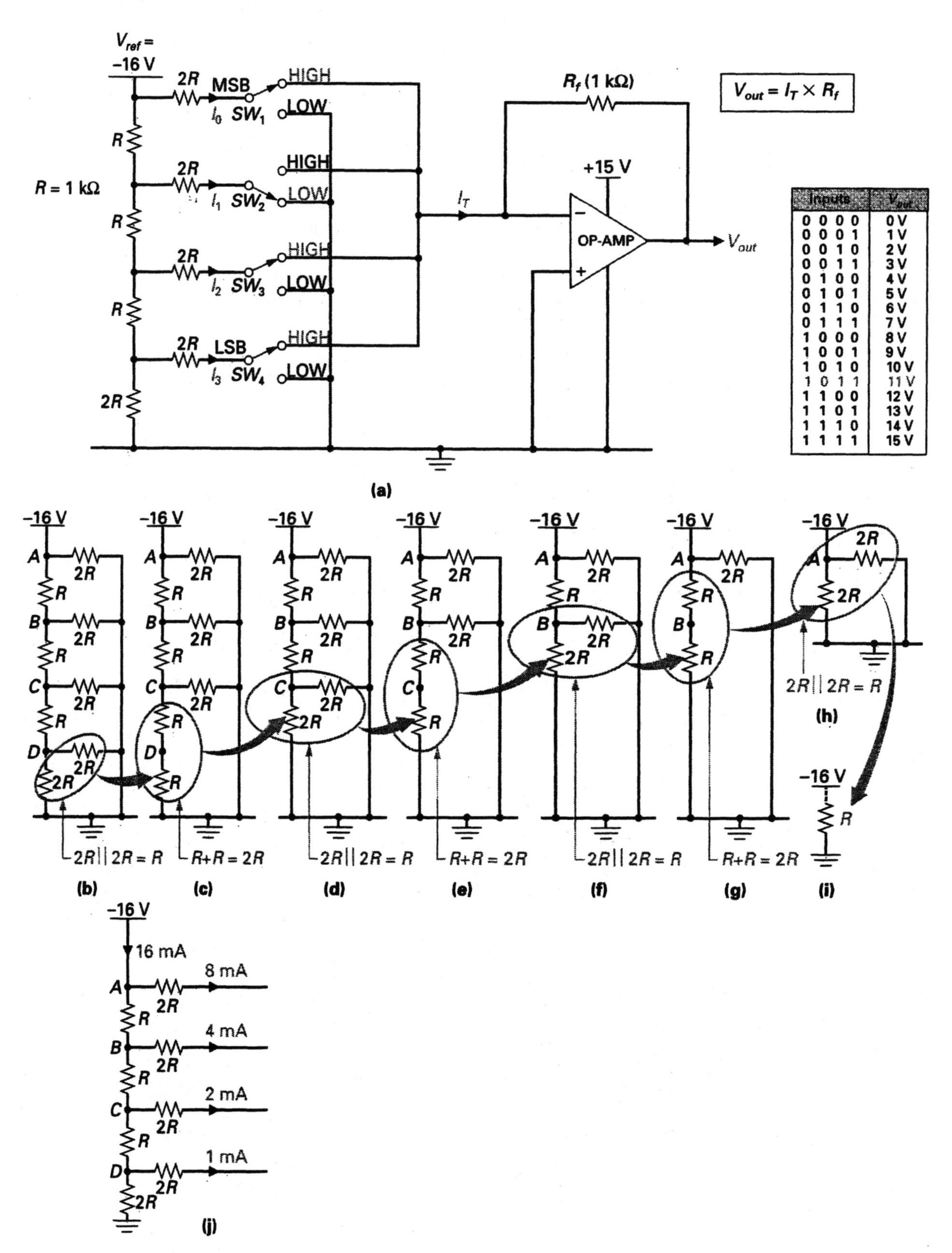

Inputs	V_{out}
0 0 0 0	0 V
0 0 0 1	1 V
0 0 1 0	2 V
0 0 1 1	3 V
0 1 0 0	4 V
0 1 0 1	5 V
0 1 1 0	6 V
0 1 1 1	7 V
1 0 0 0	8 V
1 0 0 1	9 V
1 0 1 0	10 V
1 0 1 1	11 V
1 1 0 0	12 V
1 1 0 1	13 V
1 1 1 0	14 V
1 1 1 1	15 V

FIGURE 16-4 ***R/2R*** **Digital-to-Analog Converter (DAC).**

redrawn in Figure 16-4(d). Moving on to the redrawn circuit in Figure 16-4(d), you can see that we now have $2R$ in parallel with $2R$ below junction C, which is equivalent to R, as redrawn in Figure 16-4(e). Moving on to the redrawn circuit in Figure 16-4(e), you can see that we now have R in series with R directly below junction B, which is equivalent to $2R$, as redrawn in Figure 16-4(f). Moving on to the redrawn circuit in Figure 16-4(f), you can see that we now have $2R$ in parallel with $2R$ below junction B, which is equivalent to R, as redrawn in Figure 16-4(g). Moving on to the redrawn circuit in Figure 16-4(g), you can see that we now have R in series with R directly below junction A, which is equivalent to $2R$ as redrawn in Figure 16-4(h). Moving on to the redrawn circuit in Figure 16-4(h) you can see that we now have $2R$ in parallel with $2R$ below junction A, which is equivalent to R, as redrawn in Figure 16-4(i). The $R/2R$ ladder resistor network is therefore equivalent to R, so the total current can be calculated with Ohm's Law as follows:

$$I_T = \frac{V_{ref}}{R}$$

In the example circuit in Figure 16-4(a), $V_{ref} = -16$ V and $R = 1$ kΩ, so the total current will be

$$I_T = \frac{V_{ref}}{R} = \frac{-16}{1\text{ k}\Omega} = 16\text{ mA}$$

The $R/2R$ ladder circuit will divide this total current into binary levels, as shown in Figure 16-4(j). The 16 mA arriving at junction A will see a parallel path made up of $2R$ and $2R$, as was shown in Figure 16-4(h), so the 16 mA will be halved, with 8 mA flowing through each branch, as seen in Figure 16-4(j). The 8 mA arriving at junction B will see a parallel path made up of $2R$ and $2R$, as was shown in Figure 16-4(f), so the 8 mA will be halved, with 4 mA flowing through each branch, as seen in Figure 16-4(j). The 4 mA arriving at junction C will see a parallel path made up of $2R$ and $2R$, as was shown in Figure 16-4(d), so the 4 mA will be halved, with 2 mA flowing through each branch, as seen in Figure 16-4(j). The 2 mA arriving at junction D will see a parallel path made up of $2R$ and $2R$, as was shown in Figure 16-4(b), so the 2 mA will be halved, with 1 mA flowing through each branch, as seen in Figure 16-4(j). These binary-weighted current values are switched through to the summing junction of the op-amp based on the condition of the binary-controlled electronic switches. For example, if a 1011 digital input were applied to this $R/2R$ DAC circuit, SW_1 would be in its HIGH position, SW_2 would be set LOW, SW_3 would be set HIGH, and SW_4 would be set HIGH, as shown in Figure 16-4(a). This means that the 8 mA, 2 mA, and 1 mA current values will be switched through to the summing junction of the op-amp, so I_T will equal 11 mA (8 mA + 2 mA + 1 mA = 11 mA). Stated as a formula, I_T is equal to

$$I_{out} = I_{ref}\left(\frac{D_3}{2} + \frac{D_2}{4} + \frac{D_1}{8} + \frac{D_0}{16}\right)$$

$$I_{out} = I_{ref}\left(\frac{D_3}{2} + \frac{D_2}{4} + \frac{D_1}{8} + \frac{D_0}{16}\right)$$
$$= 16\text{ mA}\left(\frac{1}{2} + \frac{0}{4} + \frac{1}{8} + \frac{1}{16}\right)$$
$$= 16\text{ mA}\left(\frac{8 + 0 + 2 + 1}{16}\right)$$
$$= 16\text{ mA} \times \left(\frac{11}{16}\right)$$
$$= 11\text{ mA}$$

For a 1011 (decimal 11) digital input, therefore, this *R*/2*R* ladder DAC will generate an analog output voltage of

$$V_{out} = I_T \times R_f = 11\text{ mA} \times 1\text{ k}\Omega = 11\text{ V}$$

■ **EXAMPLE:**

Calculate the analog output voltage generated by the circuit in Figure 16-4(a) if a digital input of 0110 were applied.

■ ***Solution:***

If a digital input of 0110 (decimal 6) were applied to the *R*/2*R* ladder DAC circuit, SW_4 would be set LOW, SW_3 would be set HIGH, SW_2 would be set HIGH, and SW_1 would be LOW. The total current applied to the summing junction of the op-amp would be

$$I_T = I_0 + I_1 + I_2 + I_3 = 0 + 4\text{ mA} + 2\text{ mA} + 0 = 6\text{ mA}$$

The analog output voltage will therefore be

$$V_{out} = I_T \times R_f = 6\text{ mA} \times 1\text{ k}\Omega = 6\text{ V}$$

16-2-3 DAC Characteristics

The characteristics of a device determine its quality and its suitability for different applications. Digital-to-analog converters are normally classified as being either a current output or voltage output IC, based on whether a voltage-to-current op-amp is included within the IC's internal circuitry. The other key specification to look for is the number of digital input bits, and the output voltage range or current range. For example, Figure 16-5(a) shows a 0–5 V 8-bit DAC. In this test circuit, the DAC is being driven by an 8-bit free-running counter. This means that the DAC will receive a continually incrementing digital count input, and will therefore generate a continually increasing analog output voltage waveform. This **staircase waveform** output will begin at 0 V when the digital input is 0000 0000 (decimal 0), will rise to a maximum analog output voltage when the digital input is 1111 1111 (decimal 255), and will then repeat the cycle. The table in Figure 16-5(b) explains the staircase output waveform in

Staircase Waveform
A waveform that continually steps up in value from zero to a maximum value, and then repeats.

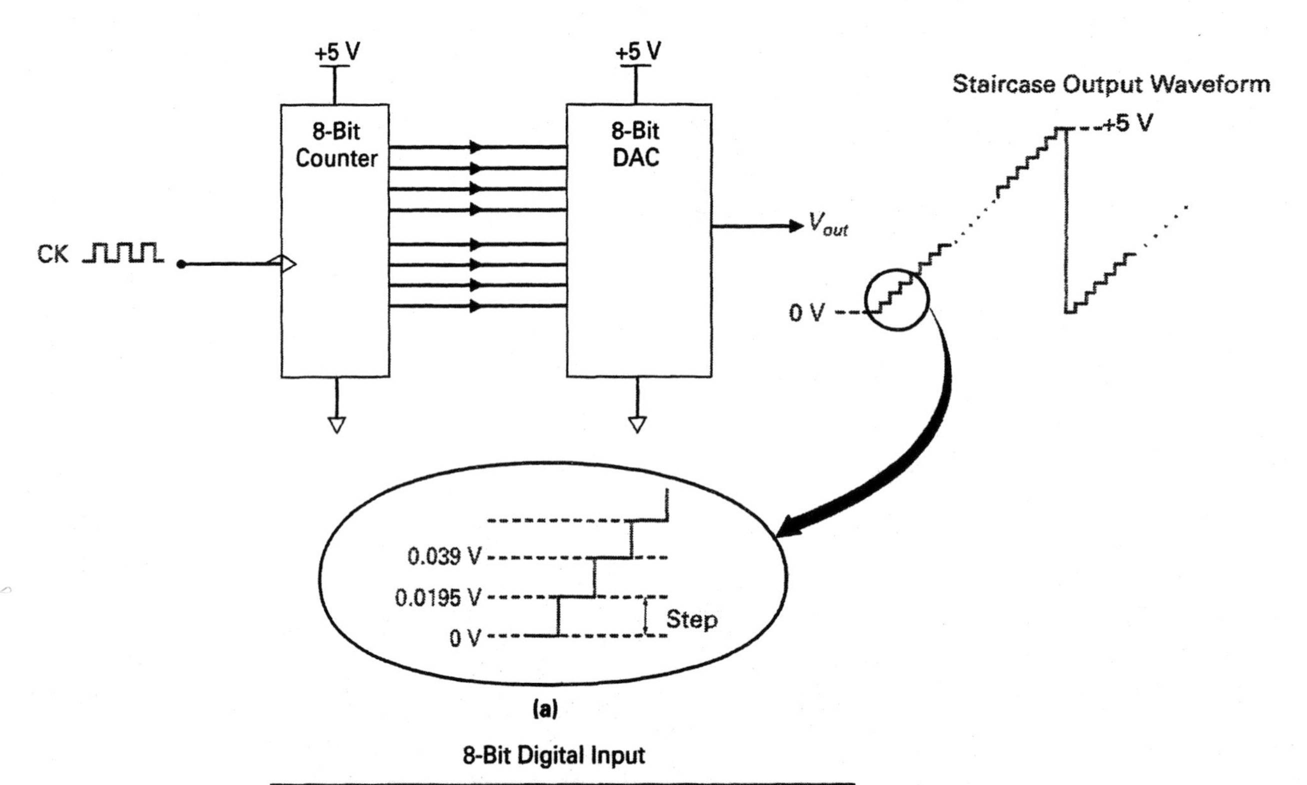

(a)

8-Bit Digital Input

	MSB							LSB		Analog Output Voltage (V_{out})
	$\frac{5\text{ V}}{2}=$	$\frac{5\text{ V}}{4}=$	$\frac{5\text{ V}}{8}=$	$\frac{5\text{ V}}{16}=$	$\frac{5\text{ V}}{32}=$	$\frac{5\text{ V}}{64}=$	$\frac{5\text{ V}}{128}=$	$\frac{5\text{ V}}{256}=$		
	2.5 V	1.25 V	625 mV	312 mV	156 mV	78 mV	39 mV	19.5 mV		
0	0	0	0	0	0	0	0	0	=	0 V
1	0	0	0	0	0	0	0	1	=	19.5 mV
2	0	0	0	0	0	0	1	0	=	39 mV
3	0	0	0	0	0	0	1	1	=	58.5 mV
.	.	.	.	.	.	.	.	.	.	
.	.	.	.	.	.	.	.	.	.	
.	.	.	.	.	.	.	.	.	.	
.	.	.	.	.	.	.	.	.	.	
.	.	.	.	.	.	.	.	.	.	
.	.	.	.	.	.	.	.	.	.	
254	1	1	1	1	1	1	1	0	=	4.96 V
255	1	1	1	1	1	1	1	1	=	4.98 V

(b)

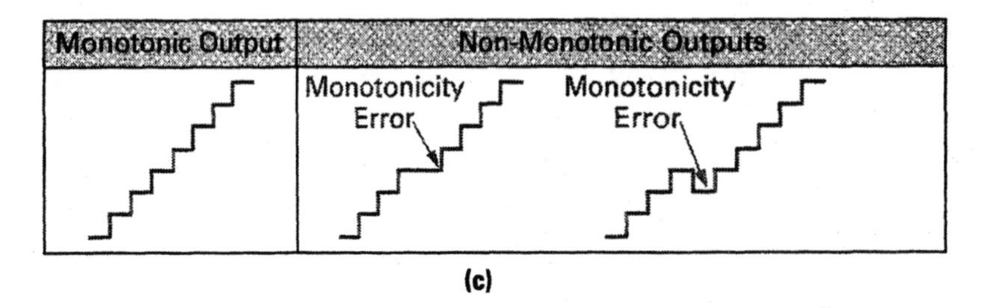

(c)

FIGURE 16-5 DAC Characteristics.

more detail. As you know, each of the digital input bits is weighted, and therefore:

the LSB will have a weight of $5\text{ V}/256 = 19.5$ mV,
the next bit will have a weight of $5\text{ V}/128 = 39$ mV,
the next bit will have a weight of $5\text{ V}/64 = 78$ mV,
the next bit will have a weight of $5\text{ V}/32 = 156$ mV,
the next bit will have a weight of $5\text{ V}/16 = 312$ mV,
the next bit will have a weight of $5\text{ V}/8 = 625$ mV,
the next bit will have a weight of $5\text{ V}/4 = 1.25$ V,
and the MSB will have a weight of $5\text{ V}/2 = 2.5$ V.

As the digital input count advances by 1, the analog output of the DAC will increase in 19.5 mV steps, as shown in the inset in Figure 16-5(a). The DAC's analog output, therefore, will generate 255 steps, and have a **step size** of 19.5 mV.

$$\text{Step Size} = \frac{\text{Max. Rated Output Voltage}}{2^n}$$

$$\text{Step Size} = \frac{\text{Max. Rated Output Voltage}}{2^n} = \frac{5\text{ V}}{2^8} = \frac{5\text{ V}}{256} = 19.5\text{ mV}$$

Adding up all of the individual bit weights, you can see that when the digital input is 1111 1111 (decimal 255), the analog output will be its maximum, which is 4.98 V.

Resolution

Resolution means smallest detail, and the resolution of a DAC is its step size, or the smallest analog output change that can occur as a result of an increment in the digital input. A DAC's resolution is normally expressed as a percentage of the maximum rated output, and can be calculated with the following formula.

Resolution
A term used to describe the amount of detail present.

$$\text{Percent Resolution} = \frac{\text{Step Size}}{\text{Max. Rated Output Voltage}} \times 100$$

To apply this formula to an example, the DAC in Figure 16-5(a) will have a resolution of

$$\text{Percent Resolution} = \frac{\text{Step Size}}{\text{Max. Rated Output Voltage}} \times 100$$

$$= \frac{0.0195\text{ V}}{5\text{ V}} \times 100$$

$$= 0.0039 \times 100 = 0.39\%$$

■ EXAMPLE:

Calculate the resolution of a 0–5 V 10-bit DAC.

■ *Solution:*

First calculate the step size, and then calculate the DAC's resolution.

$$Step\ Size = \frac{Max.\ Rated\ Output\ Voltage}{2^n} = \frac{5\text{ V}}{2^{10}} = \frac{5\text{ V}}{1{,}024} = 4.9\text{ mV}$$

$$Percent\ Resolution = \frac{Step\ Size}{Max.\ Rated\ Output\ Voltage} \times 100$$

$$= \frac{4.9\text{ mV}}{5\text{ V}} \times 100$$

$$= 0.0009765 \times 100 = 0.098\%$$

As you can see from this example, a 0–5 V 10-bit DAC has a much better resolution than a 0–5 V 8-bit DAC.

Monotonicity

Monotonic
A term used to describe the condition of a DAC's output. A monotonic output from a DAC will have an increasing analog output for every increase in the digital input.

When a free-running counter is connected to the input of a DAC, its output voltage should continually step up, producing a staircase waveform. A DAC is said to be **monotonic** if its output increases for each successive increase in the digital input, as shown in the left portion of Figure 16-5(c). If an increase in the digital input does not cause the output of the DAC to step up, or if the output steps backwards, as shown in the right portion of Figure 16-5(c), the DAC is said to be non-monotonic. Monotonicity errors can be caused by circuit malfunctions within the DAC IC, and they can also be caused by circuit problems external to the IC. As always, you will need to isolate the device to determine whether the problem is source or load. We will be discussing the testing of DACs and the troubleshooting of data conversion circuits in a later section in this chapter.

Settling Time

Settling Time
The time it takes for a DAC's analog output to settle to 99.95% of its new value.

The **settling time** of a DAC is the time it takes for the DAC's analog output to settle to 99.95% of its new value after a digital input has been applied. This DAC characteristic describes how fast a conversion can be made, and so it will determine the maximum frequency at which the DAC can operate. As an example, the MC1408 current DAC shown in Figure 16-6 has a settling time of about 300 ns, whereas most voltage output DACs will have settling times ranging from 50 ns to 20 μs.

Accuracy

Relative Accuracy
The amount a DAC's output level has deviated relative to an ideal theoretical output value.

The **relative accuracy** of a DAC describes how much the output level has deviated from its ideal theoretical output value. This accuracy specification is usually expressed as a percentage, and describes the percentage of error rather

DIGITAL DEVICE DATA SHEET

IC TYPE: MC1408 – 8-BIT DIGITAL-TO-ANALOG CONVERTER (DAC)

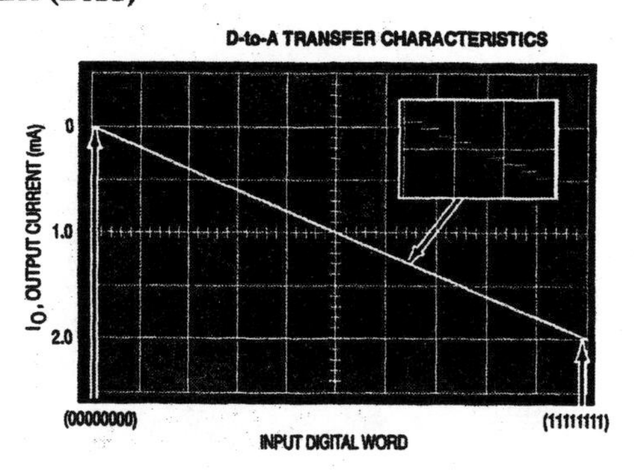

EIGHT-BIT MULTIPLYING DIGITAL-TO-ANALOG CONVERTER

. . . designed for use where the output current is a linear product of an eight-bit digital word and an analog input voltage.

- Eight-Bit Accuracy Available in Both Temperature Ranges
  Relative Accuracy: ±0.19% Error maximum
  (MC1408L8, MC1408P8, MC1508L8)
- Seven and Six-Bit Accuracy Available with MC1408 Designated by 7 or 6 Suffix after Package Suffix
- Fast Settling Time — 300 ns typical
- Noninverting Digital Inputs are MTTL and CMOS Compatible
- Output Voltage Swing — +0.4 V to –5.0 V
- High-Speed Multiplying Input
  Slew Rate 4.0 mA/µs
- Standard Supply Voltages: + 5.0 V and –5.0 V to – 15 V

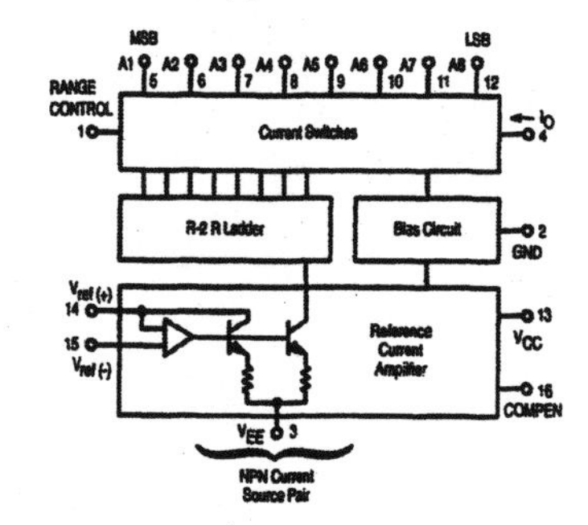

MAXIMUM RATINGS (T_A = + 25°C unless otherwise noted.)

Rating	Symbol	Value	Unit
Power Supply Voltage	V_{CC} V_{EE}	+5.5 –16.5	Vdc
Digital Input Voltage	V_5 thru V_{12}	0 to +5.5	Vdc
Applied Output Voltage	V_O	+0.5, –5.2	Vdc
Reference Current	I_{14}	5.0	mA
Reference Amplifier Inputs	V_{14},V_{15}	V_{CC},V_{EE}	Vdc
Operating Temperature Range MC1508 MC1408 Series	T_A	 –55 to + 125 0 to + 75	°C
Storage Temperature Range	T_{stg}	–65 to +150	°C

FIGURE 16-6 An IC Containing an 8-Bit Digital-to-Analog (D/A) Converter. (Copyright of Motorola, Used by Permission)

than the percentage of accuracy. As an example, the MC1408 DAC shown in Figure 16-6 could have a ±0.19% error relative to the full-scale output.

16-2-4 A DAC Data Sheet and Application Circuit

Figure 16-6 shows data sheet details for an MC1408 8-bit current DAC. This IC contains an *R*/2*R* ladder circuit, a reference current circuit, and eight binary current steering switches.

Figure 16-7 shows how this IC can be connected to function as a 0 to +5 V 8-bit digital-to-analog converter. A +5 V reference voltage ($+V_{ref}$), in conjunction with rheostat R_1, sets up a reference current of 2 mA for the DAC's internal ladder circuit. When all of the digital inputs are HIGH, therefore, the analog output current (I_{out}) will be

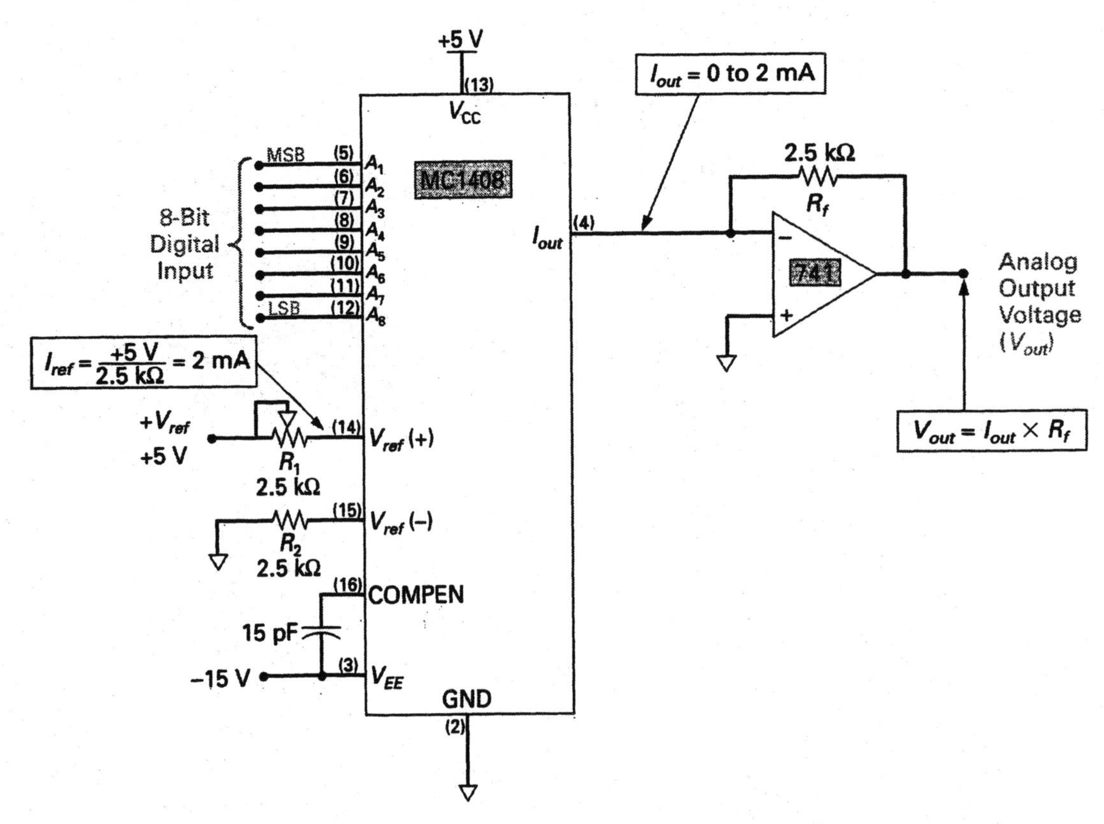

FIGURE 16-7 Application—An 8-Bit Digital-to-Analog Converter Circuit.

$$
\begin{aligned}
I_{out} &= I_{ref}\left(\frac{A_1}{2} + \frac{A_2}{4} + \frac{A_3}{8} + \frac{A_4}{16} + \frac{A_5}{32} + \frac{A_6}{64} + \frac{A_7}{128} + \frac{A_8}{256}\right) \\
&= 2\text{ mA}\left(\frac{1}{2} + \frac{1}{4} + \frac{1}{8} + \frac{1}{16} + \frac{1}{32} + \frac{1}{64} + \frac{1}{128} + \frac{1}{256}\right) \\
&= 2\text{ mA} \times \left(\frac{255}{256}\right) \\
&= 1.99\text{ mA}
\end{aligned}
$$

The analog output current is applied to a 741 op-amp, which will develop an output voltage that is equal to

$$V_{out} = I_T \times R_f$$

■ **EXAMPLE:**

What will be the analog output voltage from the circuit in Figure 16-7 if a digital input of 1100 0001 is applied?

■ ***Solution:***

The first step should be to calculate I_{out}, and then calculate V_{out}.

$$
\begin{aligned}
I_{out} &= I_{ref}\left(\frac{A_1}{2}+\frac{A_2}{4}+\frac{A_3}{8}+\frac{A_4}{16}+\frac{A_5}{32}+\frac{A_6}{64}+\frac{A_7}{128}+\frac{A_8}{256}\right)\\
&= 2\text{ mA}\left(\frac{1}{2}+\frac{1}{4}+\frac{0}{8}+\frac{0}{16}+\frac{0}{32}+\frac{0}{64}+\frac{0}{128}+\frac{1}{256}\right)\\
&= 2\text{ mA}\times\left(\frac{128+64+1}{256}\right)\\
&= 2\text{ mA}\times\left(\frac{193}{256}\right) = 1.51\text{ mA}\\
V_{out} &= I_T \times R_f = 1.51\text{ mA}\times 2.5\text{ k}\Omega = 3.8\text{ V}
\end{aligned}
$$

16-2-5 *Testing DACs*

One of the easiest ways to test a DAC is to have a free-running counter generate the digital input and have an oscilloscope monitor the analog output, as shown in Figure 16-8. Ideally, the output should be a straight-line monotonic staircase,

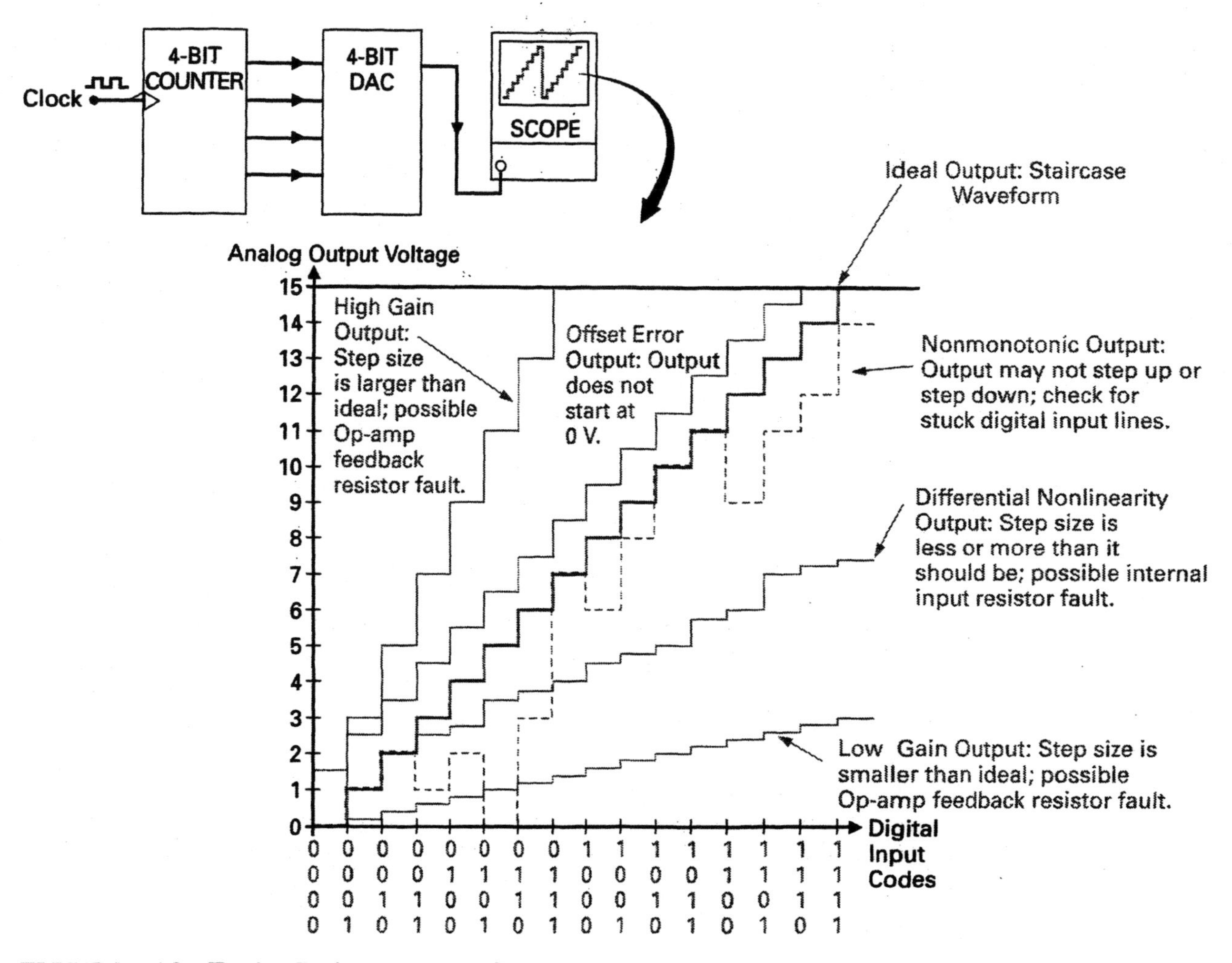

FIGURE 16-8 Testing Digital-to-Analog Converters.

as shown in Figure 16-8. If a fault exists, you may obtain an incorrect output similar to one of the waveform examples also shown in Figure 16-8. In this situation, you will need to isolate whether the problem is internal or external to the DAC IC, and then replace the faulty device.

SELF-TEST REVIEW QUESTIONS FOR SECTION 16-2

1. What is the main disadvantage of the weighted-resistor DAC?
2. What is the purpose of the op-amp in a DAC circuit?
3. Define the following terms.
   - **a.** Staircase output
   - **b.** Resolution
   - **c.** Step size
   - **d.** Monotonicity
   - **e.** Settling time
   - **f.** Accuracy
4. How can a DAC be tested?

16-3 ANALOG-TO-DIGITAL CONVERTERS (ADCs)

An ADC (pronounced "aye-dee-see") is a circuit that generates a binary output code that is proportional to an analog input voltage. A variety of analog-to-digital converter ICs are available with varying characteristics. In this section, we will be discussing analog-to-digital conversion techniques and characteristics, a typical ADC data sheet, and ADC applications and testing.

16-3-1 Staircase ADC

Staircase ADC
An ADC circuit that uses a staircase waveform and comparator to determine the value of the analog input voltage, and generate an equivalent digital output.

One of the simplest analog-to-digital converter circuits is the basic **staircase ADC,** shown in Figure 16-9(a). To convert an analog input into a digital output, this ADC circuit has a counter, comparator, AND gate, and a DAC. The analog input voltage is applied to the negative input of the comparator. The positive input of the comparator is a staircase voltage waveform from the DAC. The DAC generates a staircase voltage waveform because its digital input is connected to a counter incremented by a clock signal. This clock signal is either switched through the AND gate or blocked by the AND gate, based on the logic level output from the comparator.

To explain the operation of the circuit in Figure 16-9(a), let us step through the timing waveforms shown in Figure 16-9(b), which show how an analog input voltage of 5.9 V is converted to the equivalent 4-bit digital code 0110 (decimal 6). To begin with, the active-LOW start conversion ($\overline{SC}$) input is used to RESET the counter to 0 V. This zero count will be applied to the comparator's positive input. Since the comparator's positive input of 0 V is less than the negative analog input of 5.9 V at this time, the comparator will give a HIGH output.

The HIGH from the comparator will enable the AND gate, allowing the clock signal to pass through the AND gate and increment the counter to 0001

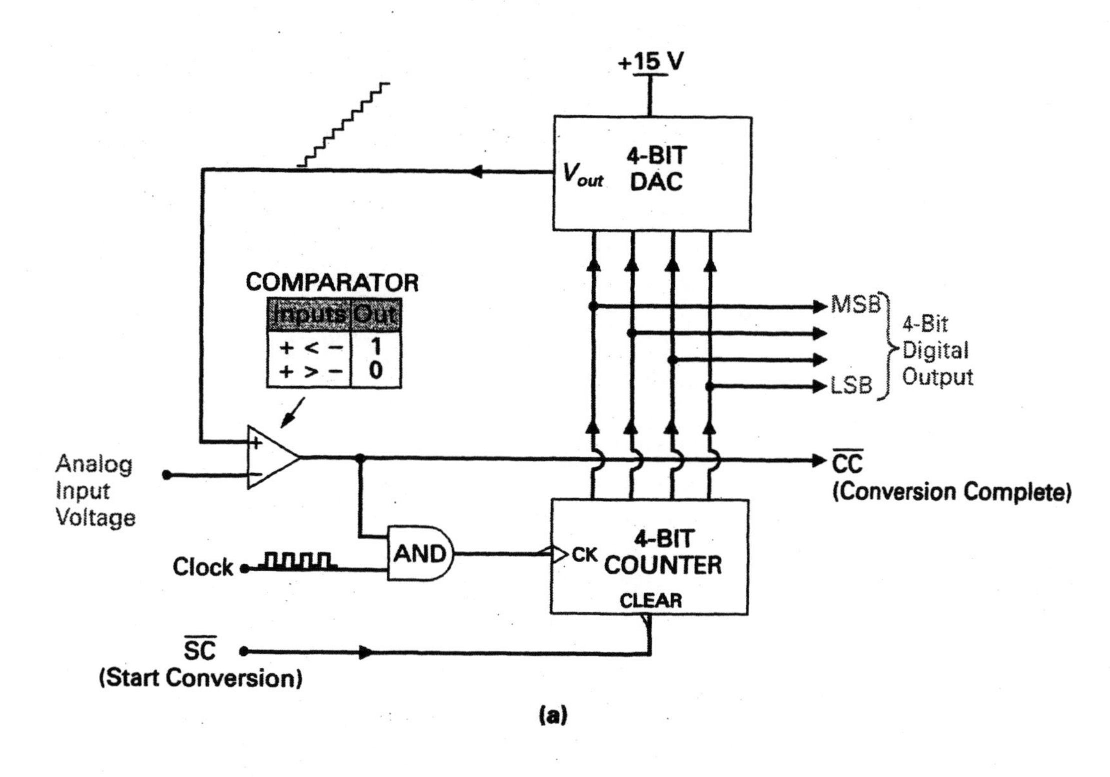

(a)

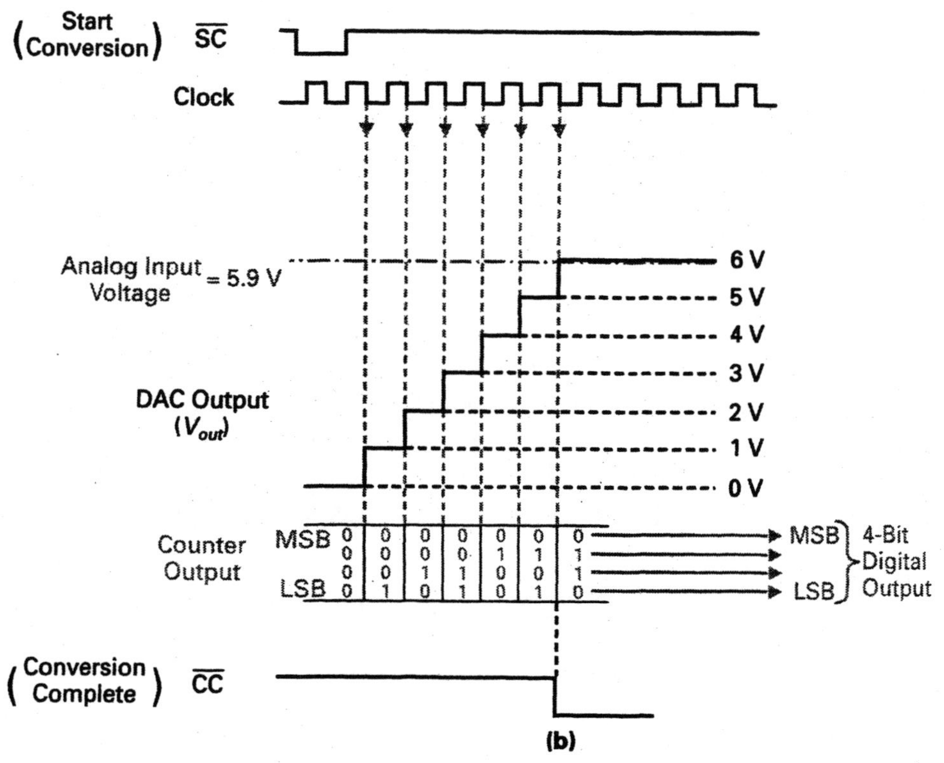

(b)

FIGURE 16-9 A Basic Staircase Analog-to-Digital Converter (ADC).

(decimal 1). The 0001 count will cause the DAC to generate a 1 V output, which is still less than the analog input voltage of 5.9 V, so the comparator will continue to enable the AND gate and allow the counter to be incremented further to 0010 (decimal 2). The 0010 count will cause the DAC to generate a 2 V output, which is still less than the analog input voltage of 5.9 V, so the comparator will continue to enable the AND gate and allow the counter to be incremented further to 0011 (decimal 3). This process will continue until the counter has been incremented to 0110 (decimal 6), and the DAC is therefore generating an output of 6 V. At this time, the comparator's positive input is greater than the analog input voltage of 5.9 V, so the comparator's output will go LOW.

The LOW output will disable the AND gate, preventing the clock signal from passing through to the counter, so the counter will hold its count of 0110. In addition to controlling the AND gate, the output of the comparator is also used to generate an active-LOW conversion complete ($\overline{CC}$) output. This $\overline{CC}$ output is used to indicate that the present 4-bit code appearing at the digital output is the binary equivalent of the analog input voltage.

■ **EXAMPLE:**

What digital output will be generated by the circuit in Figure 16-9(a) if an analog input voltage of 11.8 V were applied?

■ ***Solution:***

If an analog input of 11.8 V were applied to the ADC circuit, the counter would be incremented until it reached a count of 1100 (decimal 12).

16-3-2 ***Successive Approximation ADC***

One of the key disadvantages with the staircase ADC is that its conversion speed is dependent on the analog input voltage. For example, with the circuit in Figure 16-9, it would take fifteen clock pulses to convert an analog input voltage of 15 V to a digital output of 1111. For an 8-bit ADC, in the worst case, it would take 255 clock pulses to convert a maximum analog input voltage to 1111 1111. The worst case conversion time, therefore, is equal to

$$t_{conv(\max)} = (2^n - 1) \times t_{ck}$$

$t_{conv(\max)}$ = maximum conversion time.
n = number of bits.
t_{ck} = clock pulse period.

■ **EXAMPLE:**

Calculate the maximum conversion time of a 12-bit staircase ADC converter with a 20 kHz clock pulse applied.

■ ***Solution:***

First, we will need to determine the clock pulse input period, and then we can use the maximum conversion time formula.

$$t_{ck} = \frac{1}{f} = \frac{1}{20 \text{ kHz}} = 50 \ \mu\text{s}$$

$$\begin{aligned} t_{conv(\max)} &= (2^n - 1) \times t_{ck} \\ &= (2^{12} - 1) \times 50 \ \mu\text{s} \\ &= (4{,}095) \times 50 \ \mu\text{s} \\ &= 0.2 \text{ s} \end{aligned}$$

The slow conversion time of the staircase ADC can be overcome by using the **successive approximation ADC,** shown in Figure 16-10(b). Before we discuss the operation of this circuit, let us first examine the successive approximation (SA) method. As an example, Figure 16-10(a) shows how the SA method can be used to determine the digital equivalent of the analog input by making only four comparisons. In the example shown highlighted in Figure 16-10(a), an analog input voltage of 11.2 V has been applied. Following the comparisons shown in binary (and decimal in the inset), you can see that the SA method first sets the MSB HIGH (1000, or decimal 8) and then tests to see whether this value is greater than or less than the analog input voltage. Since the analog input voltage of 11.2 V is greater than the first SA value of 1000, the MSB is kept HIGH, the next MSB is set HIGH (1100, or decimal 12), and another comparison is made. Since the analog input voltage of 11.2 V is less than the second SA value of 1100, the second MSB is reset LOW, the next MSB is set HIGH (1010, or decimal 10), and another comparison is made. Since the analog input voltage of 11.2 V is greater than the third SA value of 1010, the third MSB is kept HIGH, the LSB is set HIGH (1011, or decimal 11), and a final comparison is made. Since the analog input voltage of 11.2 V is greater than the fourth SA value of 1011, the LSB is kept HIGH, and an equivalent digital code has been found.

Successive Approximation ADC
An ADC circuit tests and sets each of the digital output word's bits in sequence to find the digital output code that is equivalent to the analog input.

Referring to the successive approximation ADC block diagram in Figure 16-10(b), you can see that this circuit also contains a DAC and a comparator. This circuit, however, does not contain a counter since the *successive approximation register (SAR)* and control circuits will generate the special sequence of binary codes needed to converge on the digital equivalent of the analog input voltage. The start conversion ($\overline{SC}$) input is used to reset the SAR, and initiate the series of four comparisons. The SA values at the output of the SAR register are applied to the DAC, and as with the staircase ADC, the DAC output controls a comparator, which in turn controls an AND gate, which in turn controls the clock input. After each of the bits has been set and tested, the conversion complete output ($\overline{CC}$) is used to indicate that the present 4-bit code appearing at the digital output is the binary equivalent of the analog input voltage.

The key advantage of the successive approximation ADC is its conversion speed. Unlike the staircase ADC, the SA ADC does not need to sequence through every possible binary combination to find a match. An SA ADC will only need to set and test each of the digital output bits, so a 4-bit ADC will

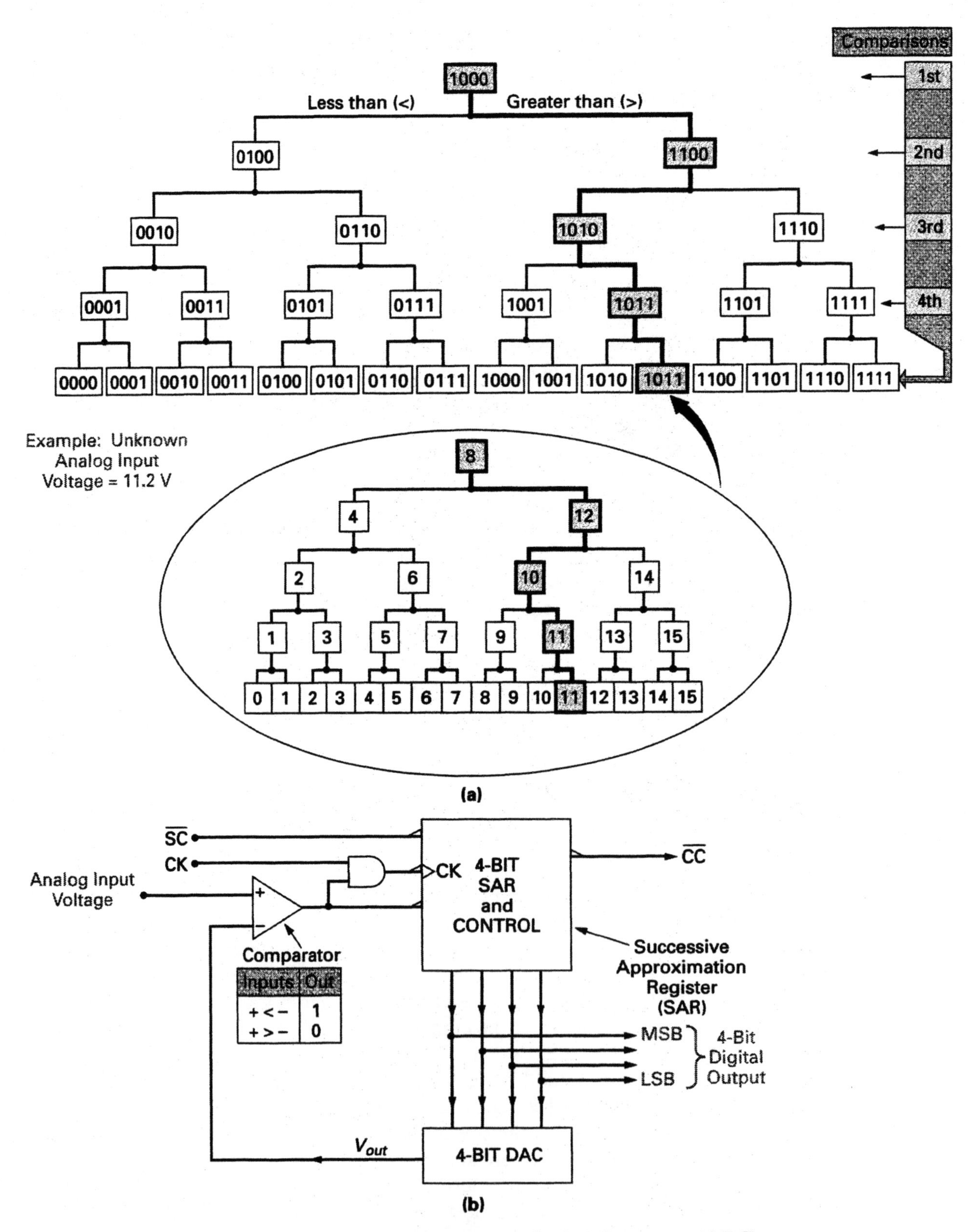

FIGURE 16-10 A Successive Approximation Analog-to-Digital Converter (ADC).

only need to make four comparisons, an 8-bit ADC will only need to make eight comparisons, and so on. This means that an SA ADC's conversion time will be the same for any input value. Having a fixed conversion time for any input value will make system timing a lot easier.

16-3-3 *Flash ADC*

Flash ADC
An ADC circuit that uses a resistive voltage divider and comparators for a fast conversion time.

One of simplest ways to convert an analog input voltage into an equivalent digital output is to use a **flash ADC.** To explain the operation of this ADC type, Figure 16-11(a) shows a 3-bit flash converter circuit. Flash ADCs use comparators to compare the analog input voltage to successively smaller equally spaced reference voltages provided by a resistive voltage-divider network. A comparator will generate a HIGH output whenever the analog input voltage is greater than the applied reference voltage. The outputs from the comparators are applied to a 3-bit priority encoder. The 3-bit binary output code generated by the priority encoder will be based on which of the highest-order inputs is HIGH, as shown in Figure 16-11(b). For example, when an analog input of 4.1 V is applied, the priority encoder inputs 4, 3, 2, and 1 will be HIGH, so the binary output code generated will be 100 (decimal 4) since 4 is the highest-order input.

The advantage of this analog-to-digital conversion method is its fast conversion time which is typically in the nanoseconds (hence the name "flash"). The disadvantage of this ADC type is the large number of comparators that are needed. In most cases, $2^n - 1$ comparators are needed, so an 8-bit flash ADC will need 255 comparators, a 16-bit flash ADC will need 65,535 comparators, and so on.

16-3-4 *An ADC Data Sheet and Application Circuit*

Figure 16-12 shows data sheet details for an ADC0803 8-bit successive approximation ADC which contains an on-chip clock circuit.

Figure 16-13 shows how this IC can be connected to function as an 8-bit analog-to-digital converter. In this example circuit, the 8-bit digital output from the ADC0803 (pins 11 through 18) have been applied to an LED display. In most digital system applications, the parallel digital output will be under three-state control, based on the active-LOW $\overline{RD}$ control input. The active-LOW $\overline{WR}$ control input is used to start a conversion, and the active-LOW $\overline{INTR}$ output is used to indicate when a conversion is complete. The $\overline{INTR}$ output is connected directly to the $\overline{WR}$ input to ensure that the conversion is continuous. The resistor, capacitor, and buffer circuit are included to ensure that the $\overline{WR}$ input is taken LOW when power is first applied to the circuit.

The *RC* circuit connected to the CLK-R and CLK-IN pins will determine the converter's internal clock circuit frequency based on the following formula.

$$f = \frac{1}{1.1\,RC}$$

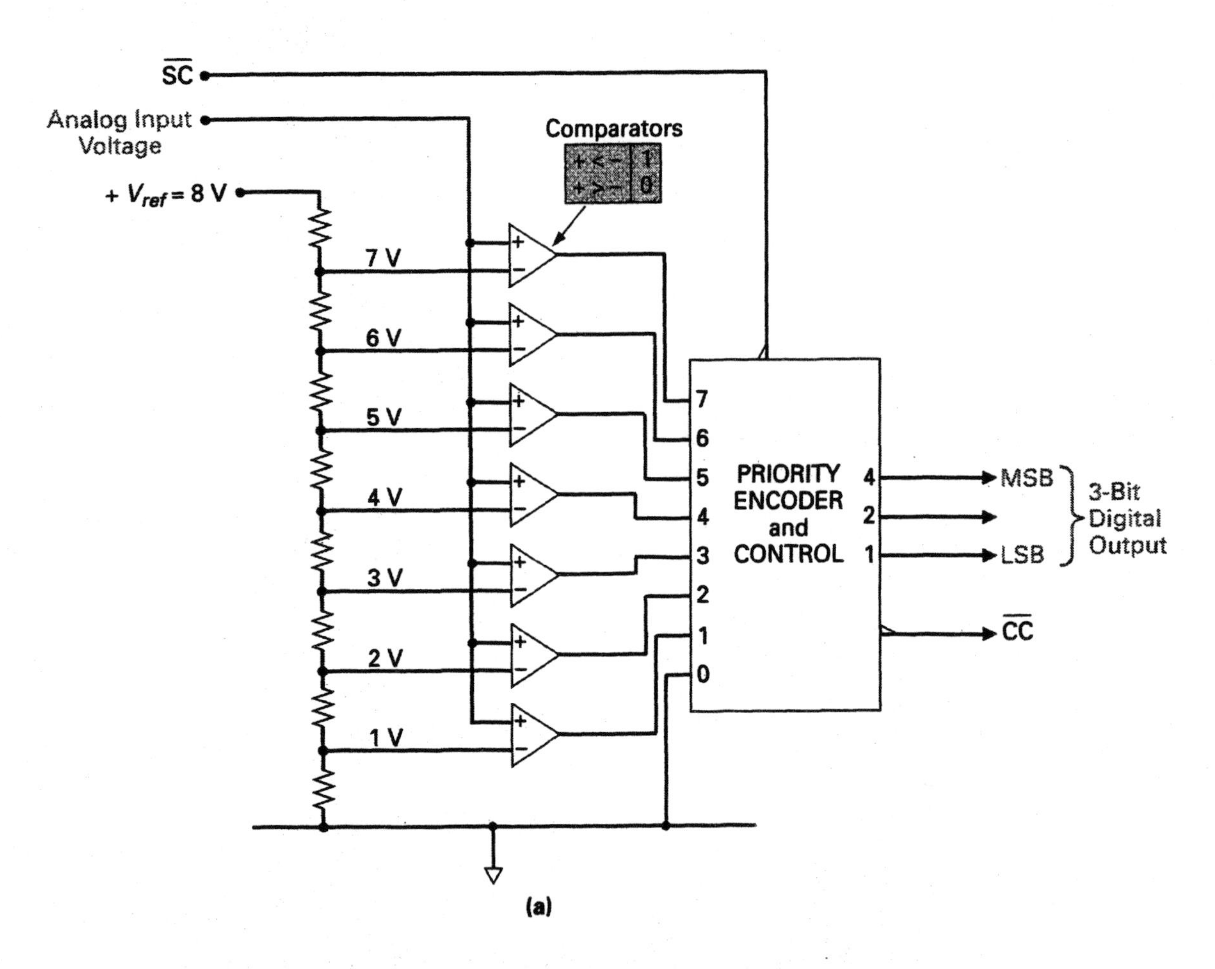

Analog Input Voltage	Priority Encoder Inputs								Priority Encoder Outputs		
	7	6	5	4	3	2	1	0	4	2	1
0 V to 0.9 V	0	0	0	0	0	0	0	0	0	0	0
1 V to 1.9 V	0	0	0	0	0	0	1	0	0	0	1
2 V to 2.9 V	0	0	0	0	0	1	1	0	0	1	0
3 V to 3.9 V	0	0	0	0	1	1	1	0	0	1	1
4 V to 4.9 V	0	0	0	1	1	1	1	0	1	0	0
5 V to 5.9 V	0	0	1	1	1	1	1	0	1	0	1
6 V to 6.9 V	0	1	1	1	1	1	1	0	1	1	0
7 V or Greater	1	1	1	1	1	1	1	0	1	1	1

(b)

FIGURE 16-11 A Flash Analog-to-Digital Converter.

In the example circuit in Figure 16-13, the converter clock frequency will be

$$f = \frac{1}{1.1\ RC} = \frac{1}{1.1 \times 10\ \text{k}\Omega \times 150\ \text{pF}} = 606\ \text{kHz}$$

The ADC0803 has both a positive analog input (pin 6) and a negative analog input (pin 7). This versatility means that you can ground pin 7 and have a positive input, ground pin 6 and have a negative input, or have the ADC

DIGITAL DEVICE DATA SHEET

IC TYPE: ADC0803/4-1 – 8-BIT ANALOG-TO-DIGITAL CONVERTER (ADC)

DESCRIPTION

The ADC0803 family is a series of three CMOS 8-bit successive approximation A/D converters using a resistive ladder and capacitive array together with an auto-zero comparator. These converters are designed to operate with microprocessor-controlled buses using a minimum of external circuitry. The 3-State output data lines can be connected directly to the data bus.

The differential analog voltage input allows for increased common-mode rejection and provides a means to adjust the zero-scale offset. Additionally, the voltage reference input provides a means of encoding small analog voltages to the full 8 bits of resolution.

FEATURES

- Compatible with most microprocessors
- Differential inputs
- 3-State outputs
- Logic levels TTL and MOS compatible
- Can be used with internal or external clock
- Analog input range 0V to V_{CC}
- Single 5V supply
- Guaranteed specification with 1MHz clock

APPLICATIONS

- Transducer-to-microprocessor interface
- Digital thermometer
- Digitally-controlled thermostat
- Microprocessor-based monitoring and control systems

ABSOLUTE MAXIMUM RATINGS

SYMBOL	PARAMETER	RATING	UNIT
V_{CC}	Supply voltage	6.5	V
	Logic control input voltages	-0.3 to +16	V
	All other input voltages	-0.3 to (V_{CC} +0.3)	V
T_A	Operating temperature range		
	ADC0803/04-1 LCD	-40 to +85	°C
	ADC0803/04-1 LCN	-40 to +85	°C
	ADC0803/04-1 CD	0 to +70	°C
	ADC0803/04-1 CN	0 to +70	°C
T_{STG}	Storage temperature	-65 to +150	°C
T_{SOLD}	Lead soldering temperature (10 seconds)	300	°C
P_D	Maximum power dissipation T_A=25°C (still air)[1]		
	N package	1690	mW
	D package	1390	mW

BLOCK DIAGRAM

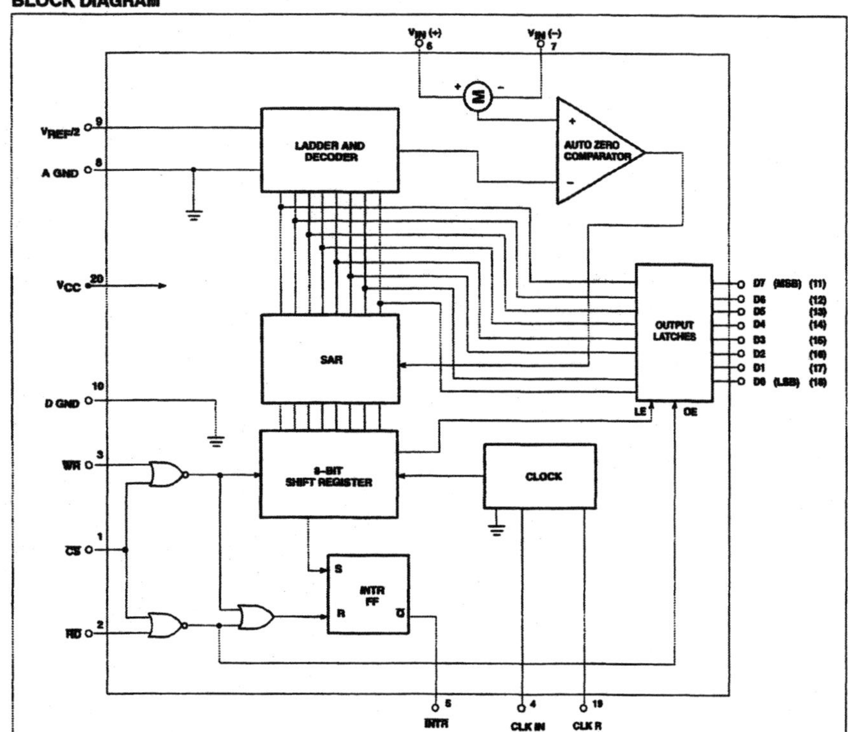

PIN CONFIGURATION

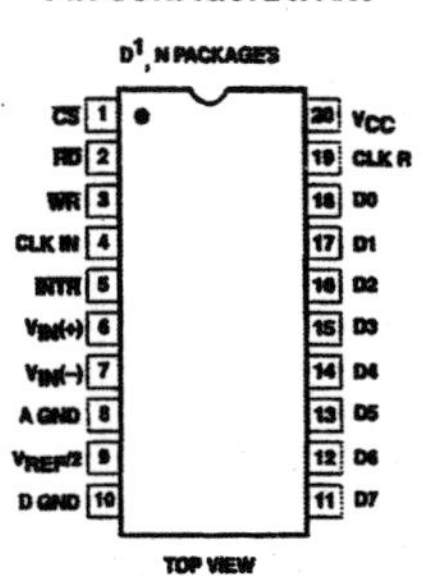

FIGURE 16-12 **An IC Containing an 8-Bit Analog-to-Digital (A/D) Converter.**

generate an output code based on the difference in voltage between pins 6 and 7 (differential input).

The +5 V supply voltage is applied to pin 20, and both the analog and digital ground pins must be grounded. The $V_{ref}/2$ input to the ADC0803 is used to determine which analog input voltage will generate the maximum digital

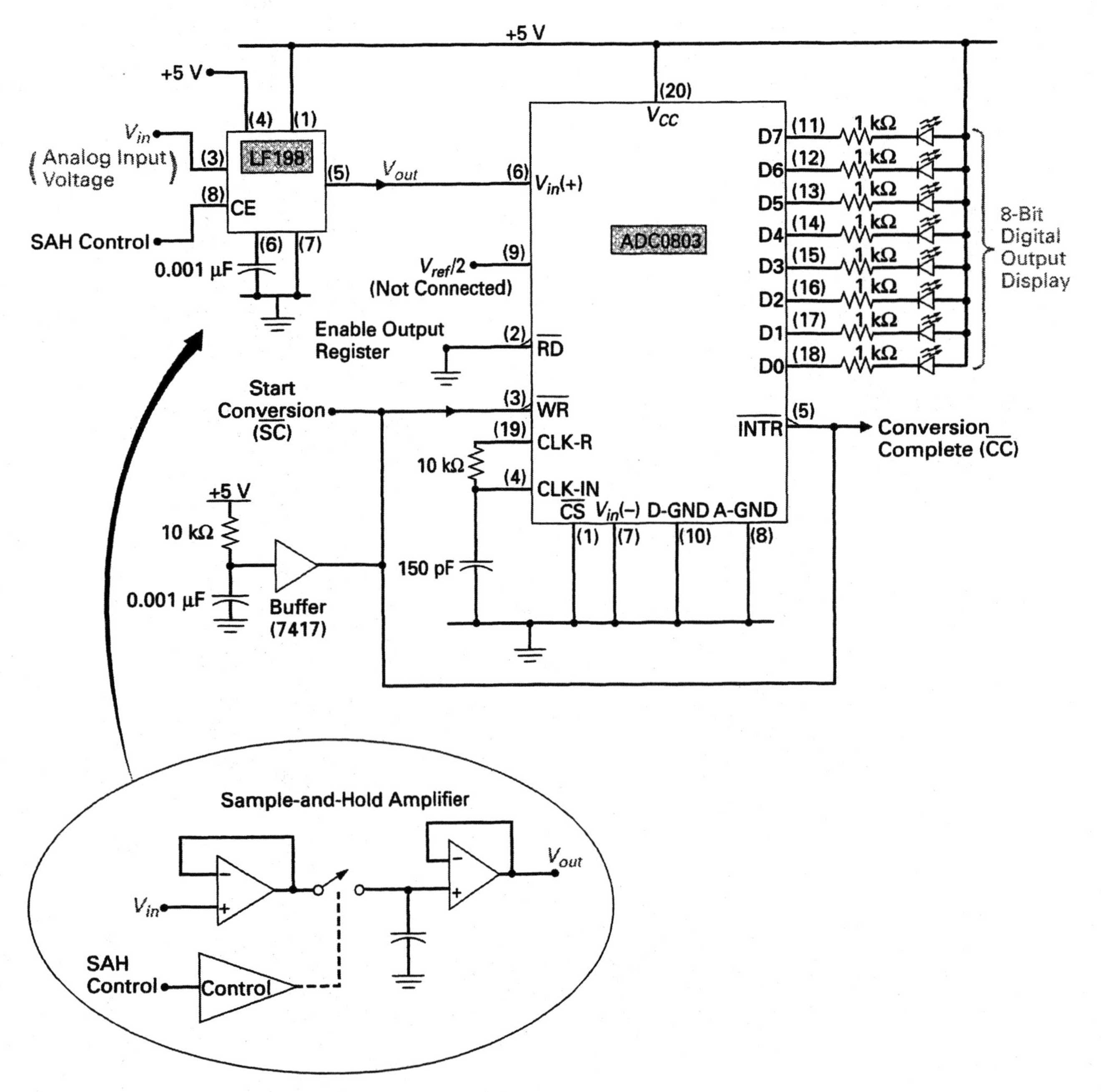

FIGURE 16-13 Application—An 8-Bit Analog-to-Digital Converter Circuit.

Sample-and-Hold Amplifier
A circuit that will sample the value of a changing analog input signal when triggered, and hold this value constant at its output.

output of 1111 1111 (decimal 255). If this input is not connected, the supply voltage of +5 V will mean that the analog input range will be from 0 V to +5 V. When a voltage is applied to the $V_{ref}/2$ input, a different analog input range can be obtained. For example, to generate a maximum digital output code of 1111 1111 when an analog input of +4 V is applied, connect +2 V to the $V_{ref}/2$ input (+4 V/2 = 2 V).

The application circuit in Figure 16-13 includes an LF198 **sample-and-hold amplifier** at the analog input. This circuit has been included to prevent any error due to analog input voltage changes during conversion. Up until this time, we have assumed that the analog input voltage remains constant while the converter is making the conversion. This does not always happen, and if a change in the analog input voltage were to occur while the converter was performing the conversion, the digital output would be inaccurate. To prevent this problem, a sample-and-hold (SAH) amplifier can be included at the input of an ADC to sample the analog input voltage and hold this sample constant while the ADC makes the conversion. The sample-and-hold amplifier contains two unity-gain noninverting amplifiers ($V_{out} = V_{in}$), a logic-controlled switch, and an external capacitor, as shown in the inset in Figure 16-13. When the SAH control input is made HIGH, the logic-controlled switch is closed, and the external capacitor charges quickly to equal the analog input voltage. When the logic-controlled switch is made LOW, the capacitor will retain the analog input voltage level, and this unchanging voltage will be applied to the ADC. Using a capacitor of 0.001 μF, the LF198 must have its switch closed for at least 4 μs in order to get an accurate sample (acquisition time), and when in the hold condition the output voltage will decrease due to leakage paths at a rate of 30 mV/s (droop rate).

16-3-5 *Testing ADCs*

One of the easiest ways to test an ADC is to apply a linear ramp to the analog input and have a display monitor the digital output, as shown in Figure 16-14. If operating correctly, the output should be an incrementing digital code, as listed in the table in Figure 16-14 in the ideal output column. If a fault exists, you may obtain incorrect output codes, missing codes, or an offset output, similar to the other columns listed in the table in Figure 16-14. In this situation, you will need to isolate whether the problem is internal or external to the ADC IC, and then replace the faulty device.

SELF-TEST REVIEW QUESTIONS FOR SECTION 16-3

1. What is the main disadvantage of the staircase ADC?
2. What two types of ADCs make use of a DAC?
3. How does a successive approximation ADC operate?
4. What is the fastest type of ADC?
5. How can an ADC be tested?

16-4 TROUBLESHOOTING DATA CONVERTER CIRCUITS

To review, the procedure for fixing a failure can be broken down into three steps—diagnose, isolate, and repair. Since this chapter has been devoted to data converter circuits, let us first examine the operation of a typical converter circuit, and then apply our three-step troubleshooting process to this circuit.

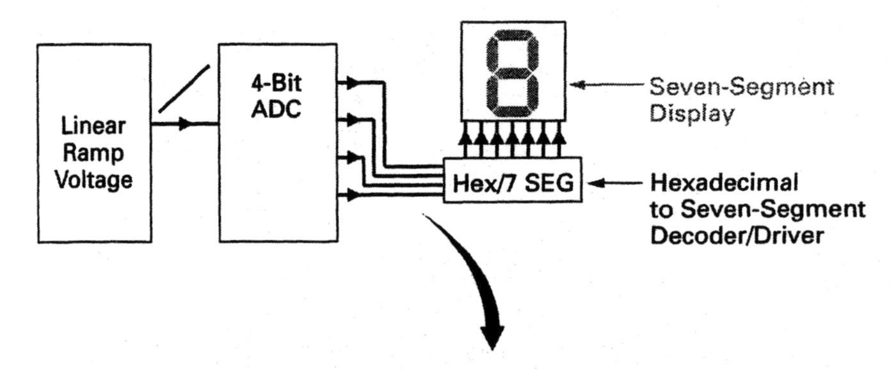

Input Voltage	Ideal Output		Incorrect Codes		Missing Codes		Offset Output	
	8 4 2 1	Display	8 4 2 1	Display	8 4 2 1	Display	8 4 2 1	Display
0 V to 0.9 V	0 0 0 0	0	0 0 0 0	0	0 0 0 0	0	0 0 1 1	3
1 V to 1.9 V	0 0 0 1	1	0 0 0 1	1	0 0 0 1	1	0 1 0 0	4
2 V to 2.9 V	0 0 1 0	2	0 0 0 1	1	0 0 1 0	2	0 1 0 1	5
3 V to 3.9 V	0 0 1 1	3	0 0 0 0	0	0 0 1 1	3	0 1 1 0	6
4 V to 4.9 V	0 1 0 0	4	0 1 0 0	4	0 1 0 0	4	0 1 1 1	7
5 V to 5.9 V	0 1 0 1	5	0 1 0 1	5	0 1 0 0	4	1 0 0 0	8
6 V to 6.9 V	0 1 1 0	6	0 1 0 1	5	0 1 1 0	6	1 0 0 1	9
7 V to 7.9 V	0 1 1 1	7	0 1 0 1	5	0 1 1 1	7	1 0 1 0	A
8 V to 8.9 V	1 0 0 0	8	0 1 1 0	6	1 0 0 0	8	1 0 1 1	b
9 V to 9.9 V	1 0 0 1	9	1 0 0 1	9	1 0 0 1	9	1 1 0 0	C
10 V to 10.9 V	1 0 1 0	A	1 0 1 0	A	1 0 1 0	A	1 1 0 1	d
11 V to 11.9 V	1 0 1 1	b	1 0 1 1	b	1 0 1 0	A	1 1 1 0	E
12 V to 12.9 V	1 1 0 0	C	1 0 1 1	b	1 1 0 0	C	1 1 1 1	F
13 V to 13.9 V	1 1 0 1	d	1 1 0 1	d	1 1 0 1	d	1 1 1 1	F
14 V to 14.9 V	1 1 1 0	E	1 1 1 0	E	1 1 1 0	E	1 1 1 1	F
15 V or Greater	1 1 1 1	F	1 1 1 0	E	1 1 1 1	F	1 1 1 1	F

FIGURE 16-14 Testing Analog-to-Digital Converters.

16-4-1 A Data Converter Circuit

As an example, we will apply our three-step troubleshooting procedure to the ADC/DAC circuit shown in Figure 16-15. In this circuit, the output from an ADC0803 analog-to-digital converter is used to drive the input of an MC1408 digital-to-analog converter. These application circuits were shown previously in Figure 16-7 and Figure 16-13.

Step 1: Diagnose

As before, it is extremely important that you first understand the operation of a circuit, and how all of the devices within it are supposed to work, so that you are able to determine whether or not a circuit malfunction really exists. If you were preparing to troubleshoot the circuit in Figure 16-15, your first step should be to read through the circuit description and review the operation of each integrated circuit until you feel completely confident with the correct

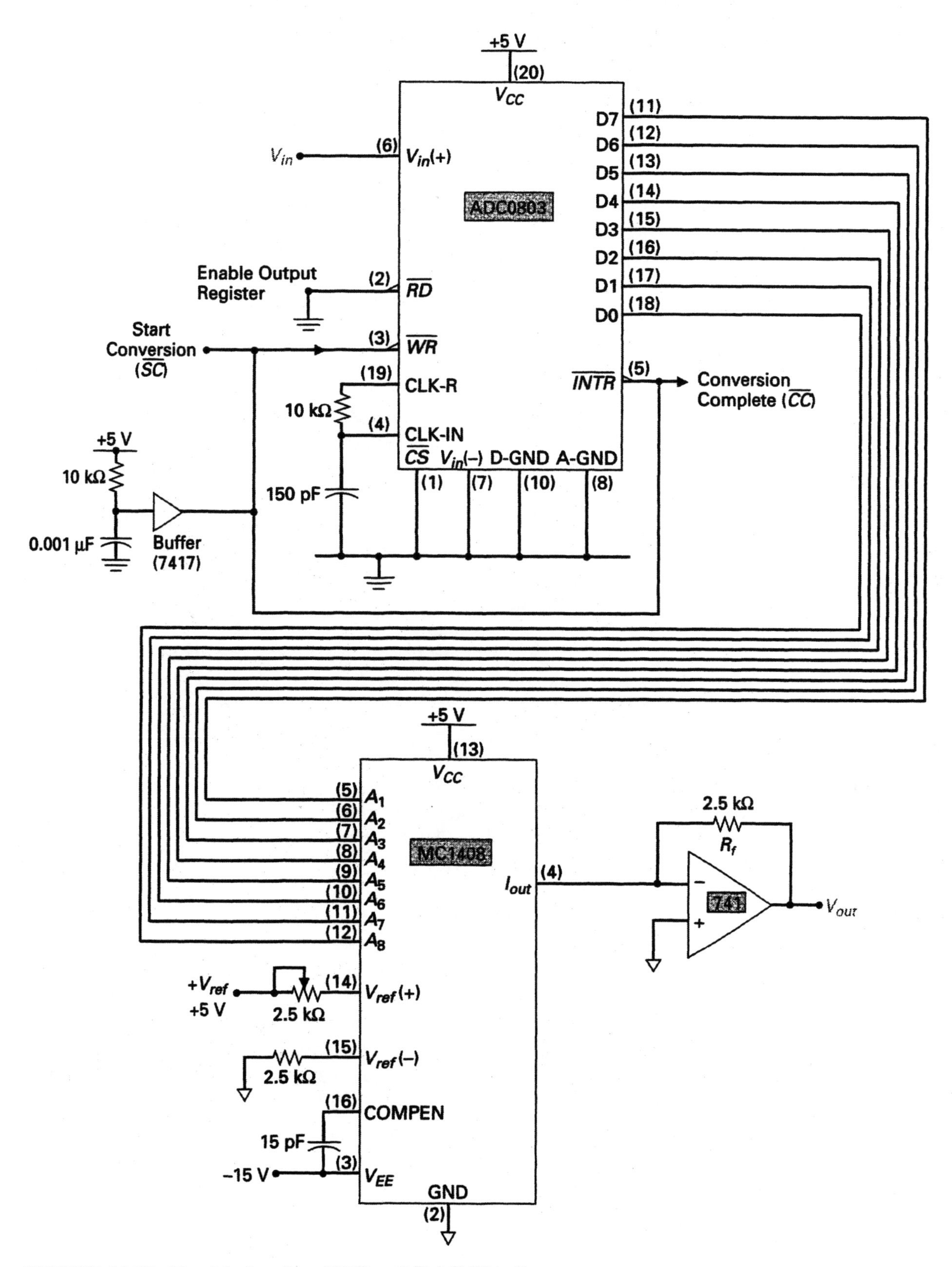

FIGURE 16-15 Troubleshooting ADC and DAC Circuits.

operation of the entire circuit. Once you are fully familiar with the operation of the circuit, you will easily be able to diagnose the problem as either an *operator error* or a *circuit malfunction.*

Step 2: Isolate

No matter what circuit or system failure has occurred, you should always follow a logical and sequential troubleshooting procedure. Let us review some of the isolating techniques and then apply them to our example circuit in Figure 16-15.

a. Use a cause-and-effect troubleshooting process, which means study the effects you are getting from the faulty circuit and then logically reason out what could be the cause.

b. Check first for obvious errors before leaping into a detailed testing procedure. Is the power OFF or not connected to the circuit? Are there wiring errors? Are all of the ICs correctly oriented?

c. Using a logic probe or voltmeter, test that power and ground are connected to the circuit and are present at all points requiring power and ground. If the whole circuit, or a large section of the circuit, is not operating, the problem is normally power. Using a digital multimeter, check that all of the dc voltages for the circuit are present at all IC pins that should have power or a HIGH input and are within tolerance. Secondly, check that 0 V or ground is connected to all IC ground pins and all inputs that should be tied LOW.

d. Use your senses to check for broken wires, loose connectors, overheating or smoking components, pins not making contact, and so on.

e. Perform a static test on the circuit. With our example circuit in Figure 16-15, you could apply a calibrated dc voltage to the analog input, use a logic probe to monitor the digital code generated by the ADC, and then use a DVM to compare the analog input voltage with the analog output voltage.

f. Perform a dynamic test on the circuit. With our example circuit in Figure 16-15, you could apply a linear 0 V to +5 V ramp input voltage to the analog input and then use a dual trace oscilloscope to compare the analog input ramp with the analog output ramp.

g. Noise due to electromagnetic interference (EMI) can interfere with the operation of a circuit. This problem can be overcome by not leaving any of the IC's inputs unconnected and, therefore, floating. Connect unneeded active-HIGH inputs to ground and active-LOW inputs to $+V_{CC}$.

h. Apply the half-split method of troubleshooting first to a circuit, and then to a section of a circuit, to help speed up the isolation process. With our example circuit in Figure 16-15, the obvious midpoint check would be to test the digital code generated by the ADC. Remember that a load problem can make a source appear at fault. If an output is incorrect, it would be safer to disconnect the source from the rest of the circuit and then recheck the output. If the output is still incorrect, the problem is definitely within the source; however, if the output is correct, the load is probably shorting the line either HIGH or LOW.

i. Substitution can be used to help speed up your troubleshooting process. Once the problem is localized to an area, substitute suspect ICs with known working ICs (one at a time) to see if the problem can be quickly remedied.

Step 3: Repair

Once the fault has been found, the final step is to repair the circuit and then perform a final test to see that the circuit and the system are now fully operational.

16-4-2 Sample Problems

Once you have constructed a circuit like the circuit in Figure 16-15, introduce a few errors to see what effects or symptoms they produce. Then logically reason out why a particular error or cause has a particular effect on the circuit. Never short any two points together unless you have carefully thought out the consequences, but generally it is safe to open a path and see the results. Here are some examples for our circuit example in Figure 16-15.

a. Connect pin 4 of the ADC0803 HIGH.

b. Open the connection between pin 5 and pin 3 of the ADC0803.

c. Connect pin 1 of the ADC0803 HIGH instead of LOW.

d. Connect one of the eight data line inputs to the MC1408 HIGH instead of connecting it from the ADC0803.

e. Connect one of the eight data line inputs to the MC1408 LOW instead of connecting it from the ADC0803.

f. Open the connection between pin 4 of the MC1408 and the input of the 741 op-amp.

g. Change the rheostat connected to the MC1408 so that $I_{ref} = 1.5$ mA.

SELF-TEST REVIEW QUESTIONS FOR SECTION 16-4

1. How would you test an MC1408 DAC?
2. What are some of the more typical DAC output errors?
3. How would you test an ADC0803 ADC?
4. What are some of the more typical ADC output errors?

REVIEW QUESTIONS

Multiple-Choice Questions

1. A ____________ converts an analog input voltage into a proportional digital output code.

a. ROM **b.** DAC **c.** ADC **d.** RWM

2. A ____________ converts a digital input code into a proportional analog output voltage.

a. ROM **b.** DAC **c.** ADC **d.** RWM

3. Which of the following signal converters is the most frequently used DAC?

   a. Binary-weighted **b.** Staircase **c.** *R*/2*R* ladder
   d. Successive approximation **e.** Flash

4. Which of the following signal converters is the fastest ADC?

   a. Binary-weighted **b.** Staircase **c.** *R*/2*R* ladder
   d. Successive approximation **e.** Flash

5. Most DACs give a/an __________ output.

   a. parallel binary **b.** analog current **c.** serial binary **d.** analog voltage

6. Most ADCs give a/an __________ output.

   a. parallel binary **b.** analog current **c.** serial binary **d.** analog voltage

7. The op-amp in a DAC circuit is used to convert a/an __________ input into a __________ output.

   a. parallel binary, current **b.** serial binary, voltage
   b. current, voltage **d.** analog voltage, current

8. Which of the following data converter types can be used to generate a staircase output waveform?

   a. Staircase **b.** Flash **c.** R/2R **d.** Successive approximation

9. The __________ of a DAC describes the smallest analog output change.

   a. monotonicity **b.** settling time **c.** accuracy **d.** resolution

10. What circuit can be used to hold the input to an ADC constant during the conversion cycle?

    a. R/2R circuit **b.** Flash circuit
    c. V-to-I circuit **d.** Sample-and-hold circuit

Essay Questions

11. What are the three basic blocks of a microcomputer? (16-1-1)
12. Briefly describe how a microcomputer is able to store and process information that originates in analog form. (16-1-1)
13. Briefly describe how a microcomputer generates an analog output signal. (16-1-2)
14. Using a sketch, briefly describe the conversion process performed by an ADC. (16-1-2)
15. Using a sketch, briefly describe the conversion process performed by a DAC. (16-1-2)
16. How does a binary-weighted resistor DAC achieve digital-to-analog conversion? (16-2-1)
17. What is the key disadvantage of a binary-weighted resistor DAC? (16-2-2)
18. How does an R/2R ladder DAC achieve digital-to-analog conversion? (16-2-2)

19. Define the following DAC characteristics. (16-2-3)

  a. Resolution
  b. Monotonicity
  c. Settling time
  d. Accuracy

20. Using the data sheet shown in Figure 16-6, determine the typical settling time and the maximum reference current for the MC1408. (16-2-3)

21. What is the purpose of the op-amp in Figure 16-7? (16-2-4)

22. In regards to a DAC, what is the relationship between the reference current input and the analog current output? (16-2-4)

23. How does a staircase ADC achieve analog-to-digital conversion? (16-3-1)

24. What is the key disadvantage of a staircase ADC? (16-3-2)

25. How does a successive approximation ADC achieve analog-to-digital conversion? (16-3-2)

26. How does a flash ADC achieve analog-to-digital conversion? (16-3-3)

27. What is the key advantage of the flash ADC? (16-3-3)

28. Referring to Figure 16-13, describe the function of the interrupt output and the write input. (16-3-4)

29. Why would the ADC0803 shown in Figure 16-12 have a read input? (16-3-4)

30. What is a sample-and-hold amplifier, and in what application is it typically used? (16-3-4)

Practice Problems

31. If a 4-bit DAC has a 12 V reference applied, calculate each of the following:

  a. Step size
  b. Percent resolution
  c. The weight of each binary column
  d. The output voltage when 0010 is applied
  e. The output voltage when 1011 is applied

32. If an 8-bit DAC has a 5 V reference applied, calculate each of the following:

  a. Step size
  b. Percent resolution
  c. The weight of each binary column
  d. The output voltage when 0000 0101 is applied
  e. The output voltage when 1011 0111 is applied

33. If an MC1408 has a 5 mA reference current applied, what will be the output current for the following digital inputs?

  a. 1010 0001 b. 0001 0101 c. 1111 1111 d. 0001 0001

34. Calculate the maximum conversion time of a 16-bit staircase ADC converter with a 200 kHz clock.

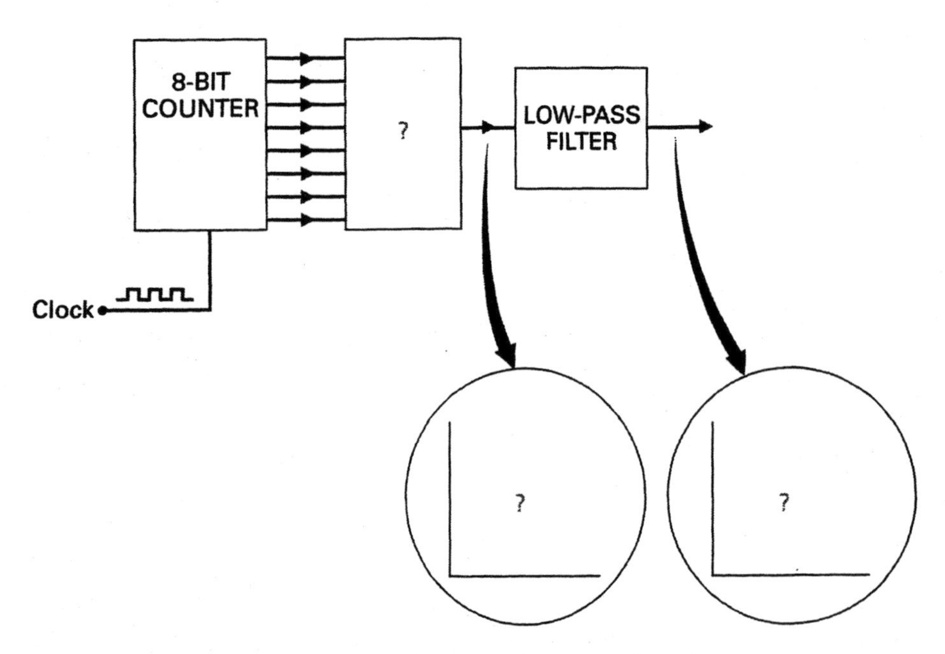

FIGURE 16-16 Application—A Ramp Generator Circuit.

To practice your circuit recognition and operation skills, refer to Figure 16-16 and answer the following five questions.

35. Would the circuit indicated by the ? in this circuit be a DAC or an ADC?

36. Sketch the output you would expect from the ? circuit.

37. Sketch the output you would expect from the low-pass filter circuit.

38. Why was the low-pass filter included in this circuit?

39. If the converter in this circuit had a 3 V reference applied, calculate each of the following:

a. Step size
b. Percent resolution
c. The weight of each binary column
d. The output voltage when 0000 0101 is applied
e. The output voltage when 1011 0111 is applied

To practice your circuit recognition and operation skills, refer to Figure 16-17 and answer the following four questions.

40. What is the purpose of this circuit?

41. Would the circuit indicated by the ? in this circuit be a DAC or an ADC?

42. What is the function of the PLD in this circuit?

43. Briefly describe the operation of this circuit.

To practice your circuit recognition and operation skills, refer to Figure 16-18 and answer the following four questions.

44. Is this circuit a DAC or an ADC?

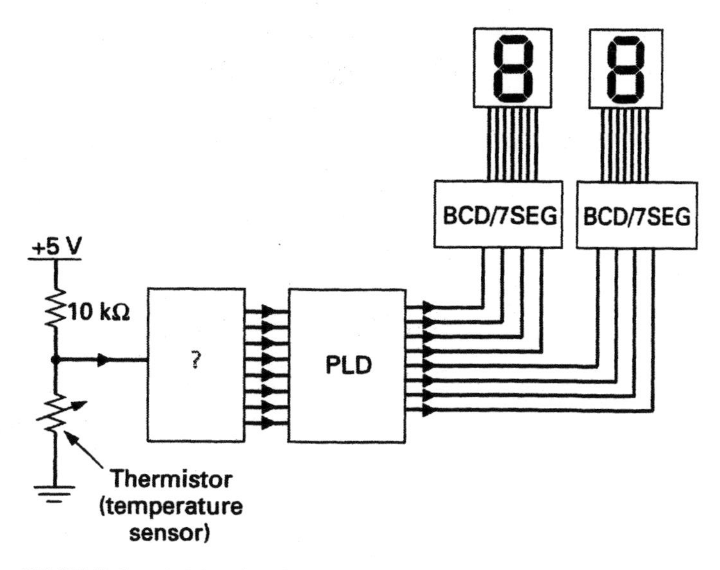

FIGURE 16-17 **Application—A Digital Thermometer.**

45. What circuit name is given to this type of converter?

46. Indicate what logic level will appear at test points 1, 2, and 3 for each of the four input voltage ranges listed in the table in this figure.

47. Complete the table shown in this figure to show what logic levels will be present at the MSB and LSB outputs for the four input voltage ranges listed in the table.

To practice your circuit recognition and operation skills, refer to Figure 16-19 and answer the following three questions.

48. Which of the blocks in this system will perform the following conversions?

- **a.** Parallel digital input to a serial digital output
- **b.** Serial digital input to a parallel digital output
- **c.** Digital input to an analog output
- **d.** Analog input to a digital output

49. Sketch the output for the input shown in Figure 16-19(b).

50. Sketch the output for the input shown in Figure 16-19(c).

Troubleshooting Questions

51. How would you test a DAC?

52. How would you test an ADC?

53. How would you test the application circuit shown previously in Figure 16-7?

54. How would you test the application circuit shown previously in Figure 16-13?

55. How would you perform a static test and a dynamic test on the circuit in Figure 16-18?

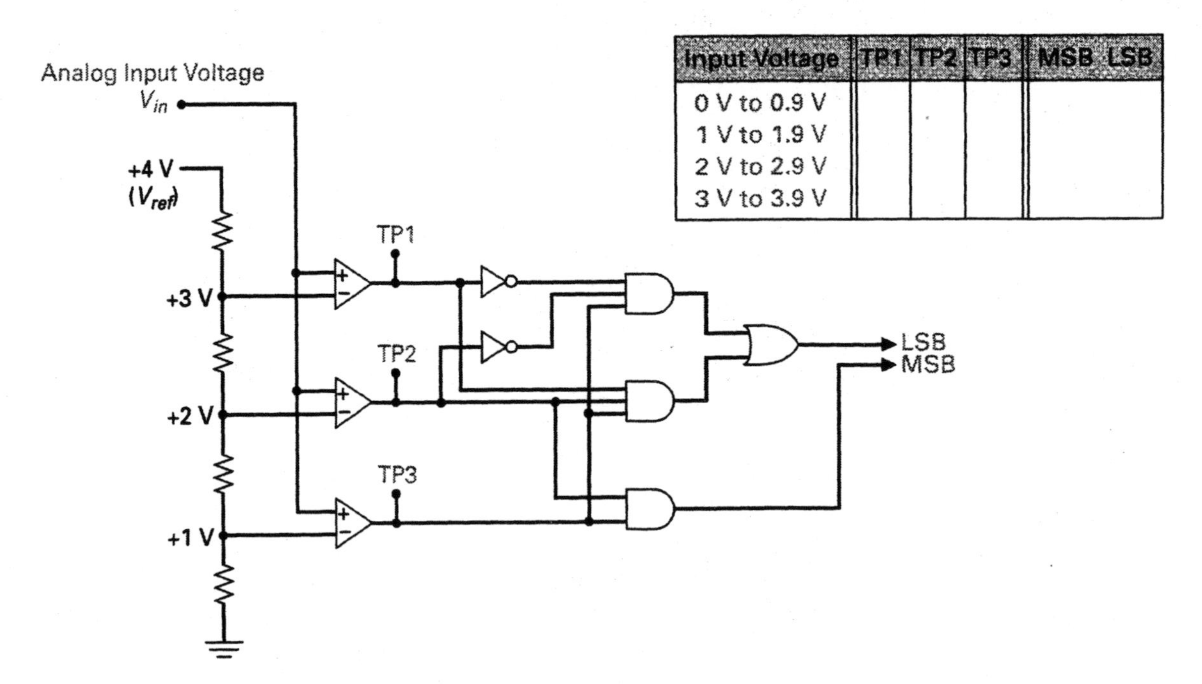

Input Voltage	TP1	TP2	TP3	MSB LSB
0 V to 0.9 V				
1 V to 1.9 V				
2 V to 2.9 V				
3 V to 3.9 V				

FIGURE 16-18 A Converter.

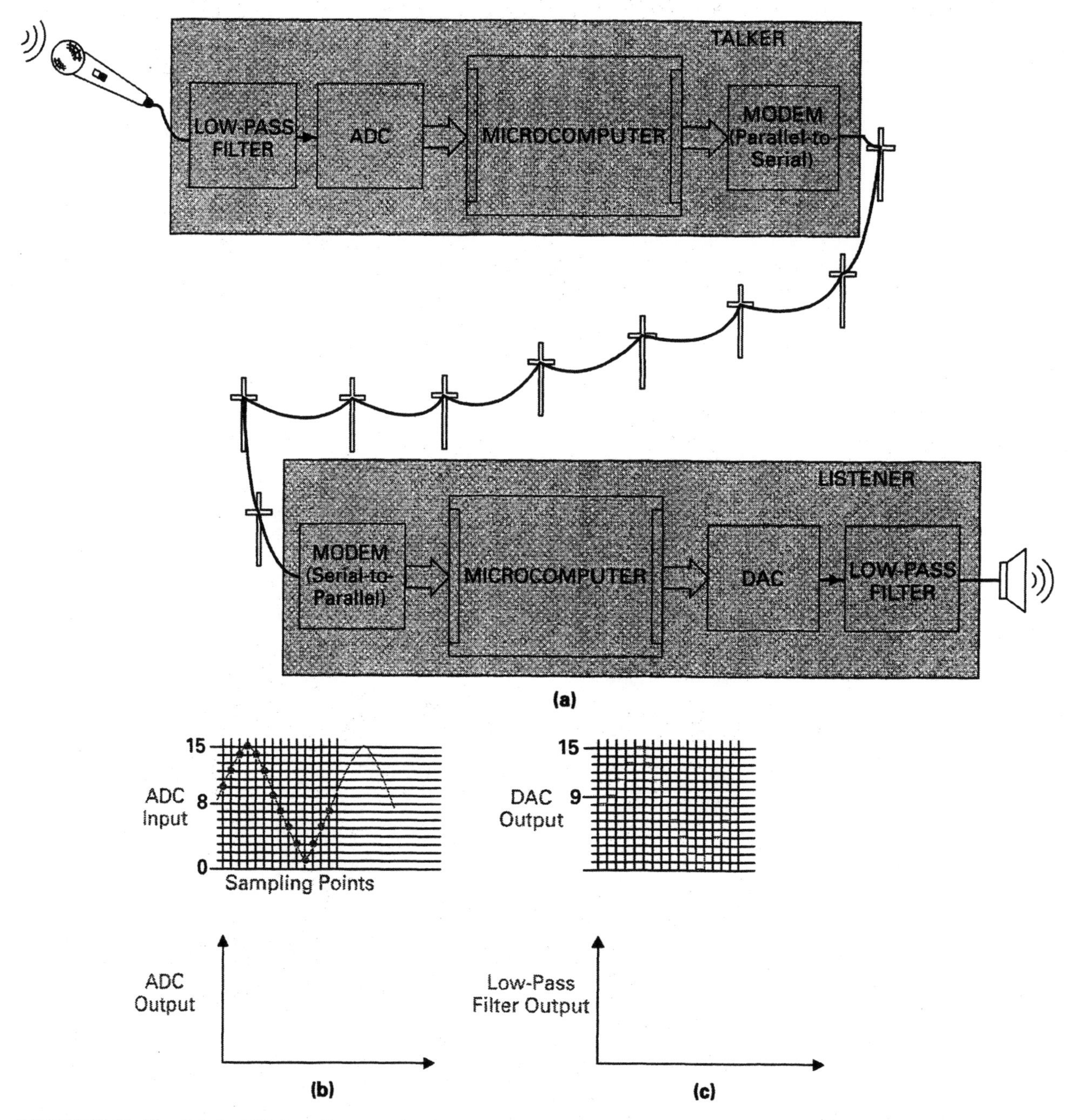

FIGURE 16-19 Digital Voice Communication.

Chapter 17

OBJECTIVES

After completing this chapter, you will be able to:

1. Define the term microprocessor.
2. Describe the difference between a hard-wired digital system and a microprocessor-based digital system.
3. List and describe the functions of the three basic blocks that make up a microcomputer.
4. Define the difference between computer hardware and software.
5. Describe how flowcharts can help a programmer write programs.
6. Describe the meaning, structure, and differences between programs that are written in machine language, assembly language, and a high-level language.
7. Describe the operation of a microprocessor-based educational microcomputer down to the component level.
8. Understand and be able to interpret a typical microprocessor-based system schematic diagram.
9. Describe the basic pin-out, internal operation, and instruction set of the 8085 microprocessor IC.
10. Explain the differences between microprocessor-based circuit troubleshooting and digital logic circuit troubleshooting.
11. Step through a troubleshooting chart and explain the testing procedures for a typical microprocessor-based system.
12. Describe the operation of the signature analyzer and logic analyzer and explain how these test instruments can be used to troubleshoot microprocessor-based systems.

Introduction to Microprocessors

OUTLINE

Note: Material in this chapter relating to the SAM microcomputer (5036A) courtesy of Hewlett-Packard Company.

VIGNETTE

Making an Impact

John Von Neuman, a mathematics professor at the Institute of Advanced Studies, delighted in amazing his students by performing complex computations in his head faster than they could with pencil, paper, and reference books. He possessed a photographic memory, and at his frequently held lavish parties in his home in Princeton, New Jersey, he gladly occupied center stage to recall from memory entire pages of books read years previously, the lineage of European royal families, and a store of controversial limericks. His memory, however, failed him in his search for basic items in a house he had lived in for seventeen years. On many occasions when traveling, he would become so completely absorbed in mathematics that he would have to call his office to find out where he was going and why.

Born in Hungary, he was quick to demonstrate his genius. At the age of six he would joke with his father in classical Greek. At the age of eight, he had mastered calculus, and in his mid-twenties he was teaching and making distinct contributions to the science of quantum mechanics, which is the cornerstone of nuclear physics.

Next to fine clothes, expensive restaurants, and automobiles, which he had to replace annually due to smash-ups, was his love of his work. His interest in computers began when he became involved in the top secret Manhattan Project at Los Alamos, New Mexico, where he proved mathematically the implosive method of detonating an atom bomb. Working with the then-available computers, he became aware that they could become much more than a high-speed calculator. He believed that they could be an all-purpose scientific research tool, and he published these ideas in a paper. This was the first document to outline the logical organization of the electronic digital computer, and was widely circulated to all scientists throughout the world. In fact, even to this day, scientists still refer to computers as "Von Neuman machines."

Von Neuman collaborated on a number of computers of advanced design for military applications, such as the development of the hydrogen bomb and ballistic missiles.

In 1957, at the age of 54, he lay in a hospital dying of bone cancer. Under the stress of excruciating pain, his brilliant mind began to break down. Since Von Neuman had been privy to so much highly classified information, the Pentagon had him surrounded with only medical orderlies specially cleared for security for fear he might, in pain or sleep, give out military secrets.

INTRODUCTION

The microprocessor is a large, complex integrated circuit containing all the computation and control circuitry for a small computer. The introduction of the microprocessor has caused a dramatic change in the design of digital electronic systems. Before microprocessors, random or hard-wired digital circuits were designed using individual logic blocks, such as gates, flip-flops, registers, counters, and so on. These building blocks were interconnected to achieve the desired end, as required by the application. Using random logic, each application required a unique design and there was little similarity between one system and another. This approach was very similar to analog circuit design, in that the structure of the circuit was governed by the function that needed to be performed. Once constructed, the function of the circuit was difficult to change.

The microprocessor, on the other hand, provides a general-purpose control system which can be adapted to a wide variety of applications with only slight circuit modification. The individuality of a microprocessor-based system is provided by a list of instructions (called the program) which controls the system's operation. A microprocessor-based system therefore has two main elements—the actual components or hardware, and the programs or software.

In the first section of this chapter, you will be introduced to the hardware that makes up a microprocessor system and the software that controls it. In the second section of this chapter, you will see how all of the circuits discussed in previous chapters come together to form a complete microprocessor-based system. The system is called "SAM," which is an acronym for "simple all-purpose microcomputer." We will be examining SAM's schematic diagram down to the component level to remove any mystery associated with microcomputers and to give you the confidence that comes from understanding a complete digital electronic microprocessor-based system. In addition, we will discuss general microprocessor system troubleshooting techniques and the detailed troubleshooting procedures for the SAM system.

17-1 MICROCOMPUTER BASICS

The earliest electronic computers were constructed using thousands of vacuum tubes. These machines were extremely large and unreliable, and were mostly a laboratory curiosity. Figure 17-1 compares the electronic numeric integrator and computer, or ENIAC, which was unveiled in 1946, to today's personal computer. The ENIAC weighed 38 tons, measured 18 feet wide and 88 feet long and used 17,486 vacuum tubes. These vacuum tubes produced a great deal of heat and developed frequent faults requiring constant maintenance. Today's personal computer, on the other hand, makes use of semiconductor integrated circuits

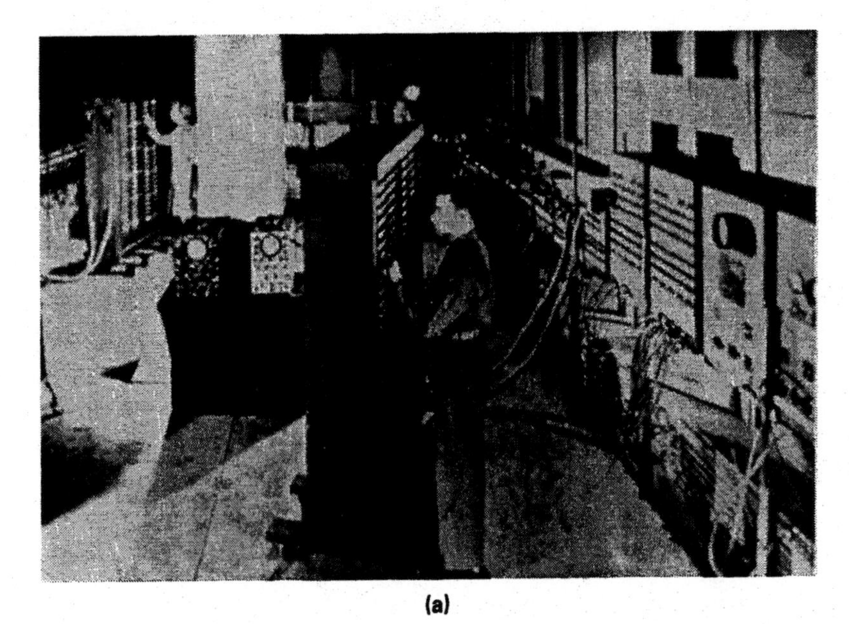

(a)

(b)

FIGURE 17-1 Microcomputer Systems—Past and Present. (Photo (b) Courtesy of Apple Computer, Inc.)

and is far more powerful, versatile, portable, and reliable than the ENIAC. Another advantage is cost: in 1946 the ENIAC calculator cost $400,000 to produce, whereas a present-day high-end personal computer can be purchased for about $4,000.

17-1-1 *Hardware*

Figure 17-2 shows the block diagram of a basic microcomputer system. The microprocessor (also called central processor unit or CPU) is the "brains" of the system. It contains all of the logic circuitry needed to recognize and execute the program of instructions stored in memory. The input port connects the processor to the keyboard so that data can be read from this input device. The output port connects the processor to the display so that we can write data to this output device. Combined in this way, the microprocessor unit, memory unit, and input/output unit form a **microcomputer.**

Microcomputer Complete system, including CPU, memory, and I/O interfaces.

The blocks within the microcomputer are interconnected by three buses. The microprocessor uses the address bus to select which memory location, input port, or output port it wishes to put information into or take information out of. Once the microprocessor has selected the location using the address bus, data or information is transferred via the data bus. In most cases, data will travel either from the processor to memory, from memory to the processor, from the input port to the processor, or from the processor to the output port. The control bus is a group of control signal lines that are used by the processor to coordinate the transfer of data within the microcomputer.

A list of instructions is needed to direct a microcomputer system so that it will perform a desired task. For example, if we wanted the system in Figure

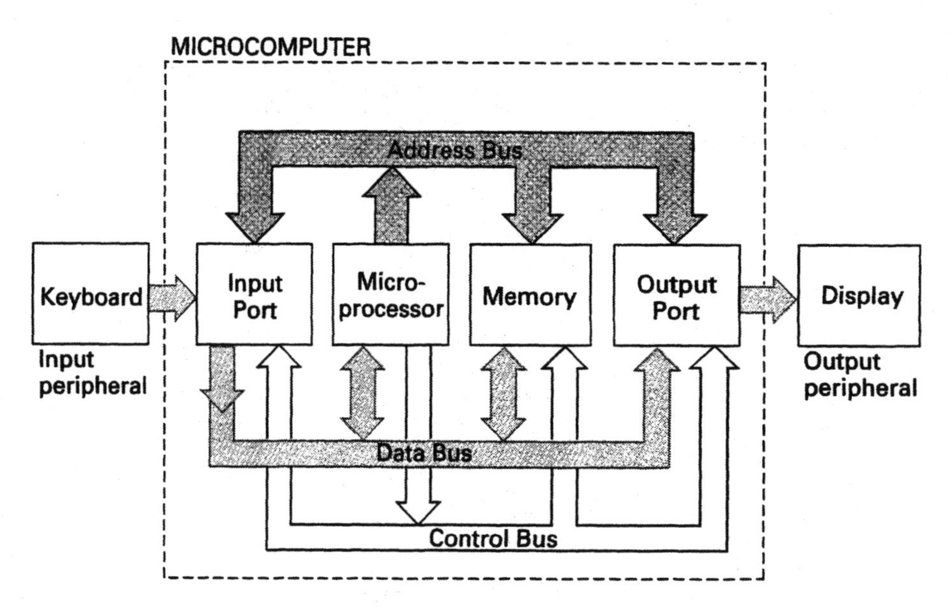

FIGURE 17-2 Basic Microcomputer System.

17-2 to display the number of any key pressed on the keyboard, the program would be as follows:

1. Read the data from the keyboard.
2. Write the data to the display.
3. Repeat (go to Step 1).

For a microprocessor to perform a task from a list of instructions, the instructions must first be converted into codes that the microprocessor can understand. These instruction codes are stored, or programmed, into the system's memory. When the program is run, the microprocessor begins by reading the first coded instruction from memory, decoding its meaning, and then performing the indicated operation. The processor then reads the instruction from the next location in memory, decodes the meaning of this next instruction, and then performs the indicated operation. This process is then repeated, one memory location after another, until all of the instructions within the program have been fetched, decoded and then executed.

The input/output devices connected to the input and output ports (the keyboard and display, for example) are called the **peripherals.** Peripherals are the system's interface with the user, or the system's interface with other equipment such as printers or data storage devices. As an example, Figure 17-3 shows how a microcomputer could be made to operate as a microprocessor-based digital voltmeter. Its input peripherals are an analog-to-digital converter (ADC), and a range and function switches. The output peripheral is a digital display.

Peripheral
Any interface device connected to a computer. Also, a mass storage or communications device connected to a computer.

As you come across different microcomputer applications, you will see that the basic microcomputer system is always the same. The only differences between one system and another are in the program and peripherals. Whether the system is a personal computer or a digital voltmeter, the only differences are in the program of instructions that controls the microcomputer and the peripherals connected to the microcomputer.

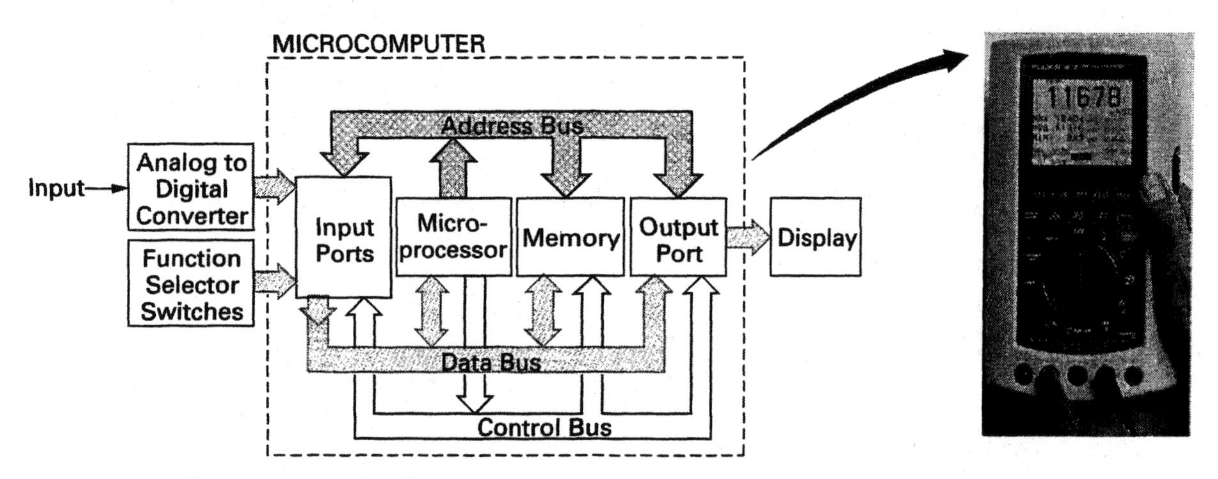

FIGURE 17-3 A Microprocessor-Based Digital Voltmeter. (Photo courtesy of Fluke Corporation. Reproduced with permission.)

17-1-2 Software

Microcomputer programs are first written in a way that is convenient for the person writing the program, or *programmer.* Once written, the program must then be converted and stored as a code that can be understood by the microprocessor. When writing a computer program, the programmer must tell the computer what to do down to the most minute detail. Computers can act with tremendous precision and speed for long periods of time, but they must be told exactly what to do. A computer can respond to a change in conditions, but only if it contains a program that says "if this condition occurs, do this."

The Microcomputer as a Logic Device

To understand how a program controls a microcomputer, let us see how we could make a microcomputer function as a simple two-input AND gate. As can be seen in Figure 17-4, the two AND gate inputs will be applied to the input port, and the single AND gate output will appear at the output port. To make the microcomputer operate as an AND gate, a program of instructions will first have to be loaded into the system's memory. This list of instructions will be as follows:

1. Read the input port.
2. Go to Step 5 if all inputs are HIGH; otherwise, continue.
3. Set output LOW.
4. Go to Step 1.
5. Set output HIGH.
6. Go to Step 1.

Studying the steps in this program, you can see that first the input port is read. Then the inputs are examined to see if they are all HIGH, since that is the

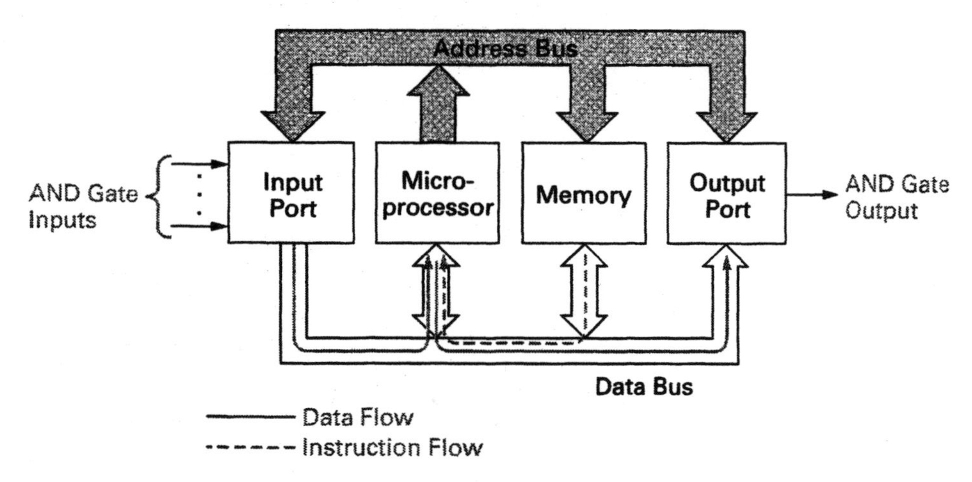

FIGURE 17-4 **A Microprocessor-Based AND Gate.**

function of an AND gate. If the inputs are all HIGH, the output is set HIGH. If the inputs are not all HIGH, the output is set LOW. Once the procedure has been completed, the program jumps back to Step 1 and repeats indefinitely, so that the output will continuously follow any changes at the input.

You may be wondering why we would use this complex system to perform a simple AND gate function. If this were the only function that we wanted the microcomputer to perform, then it would definitely be easier to use a simple AND logic gate; however, the microcomputer provides tremendous flexibility. It allows the function of the gate to be arbitrarily redefined just by changing the program. You could easily add more inputs, and the gate function could be extremely complex. For example, we could have the microcomputer function as an eight-input electronic lock. The output of this lock will only go HIGH if the inputs are turned ON in a specific order. Using traditional logic, this would require a complex circuit; however, this function could be easily implemented using the same microprocessor system used for the simple AND gate. Of course, a new program would be needed, more complex than the simple AND gate program, but the hardware will not change. Additionally, the combination of this electronic lock could easily be changed by simply modifying the program.

Flowcharts

Flowcharts provide a graphic way of describing the operation of a program. They are composed of different types of blocks interconnected with lines. The three basic types of blocks used for flowcharts are shown in Figure 17-5. A rectangular block describes each action that the program takes. A diamond-shaped block is used for each decision, such as testing the value of a variable. An oval block is used at the beginning of the flowchart with the name of the program placed inside it. This oval block can also be used to mark the end of the flowchart.

As an example, Figure 17-6 shows a flowchart for the AND gate program discussed previously. For each line of the program there is a block, except for the two "go to" instructions. These are represented as lines indicating how the program will flow from one block to another. While the flowchart contains the

Flowchart or Flow Diagram
Graphical representation of program logic. Flowcharts enable the designer to visualize the procedure necessary for each item in the program. A complete flowchart leads directly to the final code.

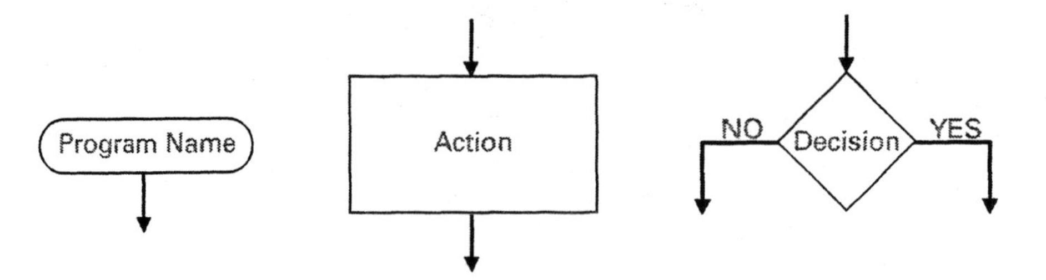

FIGURE 17-5 **Flowchart Symbols.**

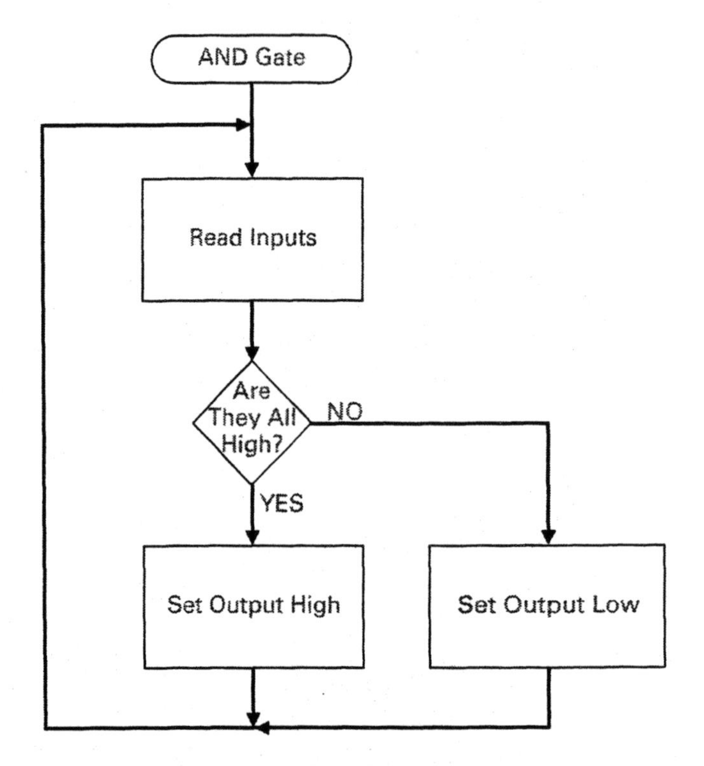

FIGURE 17-6 **AND Gate Flowchart.**

same information as the program list, it is in a more graphic form. When you first set out to write a program, a flowchart is a good way to organize your thoughts and document what the program must do. By going through the flowchart manually, you can check the logic of the flowchart and then write the actual program from it. Flowcharts are also useful for understanding a program that has been written in the past.

Machine Language Binary language (often represented in hex) that is directly understood by the processor. All other programming languages must be translated into binary code before they can be entered into the processor.

Programming Languages

Writing programs in English is convenient since it is the language most people understand. Unfortunately, English is meaningless to a microprocessor. The language understood by the microprocessor is called **machine language,** or machine code. Since microprocessors deal directly with digital signals, machine language instructions are binary codes, such as 00111100, 11100111, and so on.

Machine language is not easy for people to use, since 00111100 has no obvious meaning. It can be made easier to work with by using the hexadecimal representation of 0011 1100, which is 3C; however, this still does not provide the user with any clue as to the meaning of this instruction. To counteract this problem, microprocessor manufacturers replace each instruction code with a short name called a **mnemonic,** or memory aid. As an example, Intel's 8085 microprocessor uses the mnemonic "INR A" for the code 3C, since this code instructs the 8085 microprocessor to "increment its internal A register." The mnemonics are much easier to remember than the machine codes. By assigning a mnemonic to each instruction code, you can write programs using the mnemonics instead of machine codes. Once the program is written, the mnemonics can easily be converted to their equivalent machine codes. Programs written using mnemonics are called **assembly language** programs.

Mnemonic Code
To assist the human memory, the binary numbered codes are assigned groups of letters (or mnemonic symbols) that suggest the definition of the instruction.

Assembly Language
A program written as a series of statements using mnemonic symbols that suggest the definition of the instruction. It is then translated into machine language by an assembler program.

The machine language is generally determined by the design of the microprocessor chip and cannot be modified. The assembly language mnemonics are made up by the microprocessor's manufacturer as a convenience for programmers, and are not set by the microprocessor's design. For example, you could write INC A instead of INR A, as long as both were translated to the machine code 3C. A microprocessor is designed to recognize a specific list or group of codes called the **instruction set,** and each microprocessor type has its own set of instructions.

Instruction Set
Total group of instructions that can be executed by a given microprocessor.

While assembly language is a vast improvement over machine language, it is still difficult to use for complex programs. To make programming easier, **high-level languages** have been developed. These are similar to English and are generally independent of any particular microprocessor. For example, a typical instruction might be "LET COUNT = 10" or "PRINT COUNT". These instructions give a more complicated command than those that the microprocessor can understand. Therefore, microcomputers using high-level languages also contain long, complex programs (permanently stored in their memory) that translate the high-level language program into a machine language program. A single high-level instruction may translate into dozens of machine language instructions. Such translator programs are called **compilers.**

High-Level Language
A language closer to the needs of the problem to be handled than to the language of the machine on which it is to be implemented.

Compiler
Translation program that converts high-level instructions into a set of binary instructions (machine code) for execution.

A Programming Example

To show the difference between machine language, assembly language, and a high-level language, let us use a simple programming example. Figure 17-7(a) shows the flow chart for a program that counts to ten. There are no input or output operations in this program, since all we will be doing is having the contents of a designated memory location count from zero to ten, and then repeat.

The translation from the flow chart to a high-level language is fairly simple, as seen in Figure 17-7(b). The high-level language used in this example is called **BASIC,** which stands for Beginner's All-purpose Symbolic Instruction Code and has the advantage of being simple and similar to English. Following the program listing shown in Figure 17-7(b), you can see that the first two lines of the program correspond exactly to the first two action blocks of the count-to-ten flow chart. In the first line, the memory location called COUNT is set to 0. In the

BASIC
An easy-to-learn and use language available on most microcomputer systems.

Line No.	Instruction	Description
1	LET COUNT = 0	Set Count to 0
2	LET COUNT = COUNT + 1	Increment Count
3	IF COUNT = 10 THEN 1	Go to 1 if Count = 10
4	GO TO 2	Otherwise go to 2

(b)

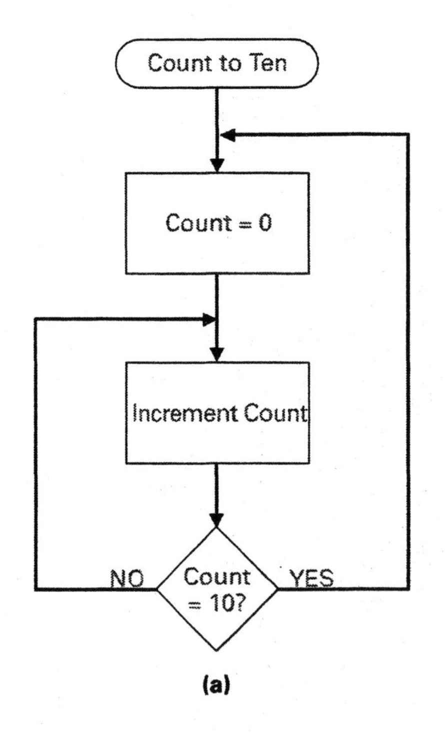

(a)

Label	Instruction	Comments
START:	MVI A,0	;Set A register to 0
LOOP:	INR A	;Increment A register
	CPI 10	;Compare A register to 10
	JZ START	;Go to beginning if A = 10
	JMP LOOP	;Repeat

(c)

Memory Address (Hex)	Memory Contents (Hex)	Memory Contents (Binary)
07F0	3E	00111110
07F1	00	00000000
07F2	3C	00111100
07F3	FE	11111110
07F4	0A	00001010
07F5	CA	11001010
07F6	F0	11110000
07F7	07	00000111
07F8	C3	11000011
07F9	F2	11110010
07FA	07	00000111

(d)

BASIC Language		8085 Assembly Language			8085 Machine Language	
Line No.	Instruction	Label	Instruction		Address	Contents
1	LET COUNT = 0	START:	MVI	A,0	07F0 07F1	3E *Opcode* 00 *Data*
2	LET COUNT = COUNT + 1	LOOP:	INR	A	07F2	3C *Opcode*
3	IF COUNT = 10 THEN 1		CPI	10_{10}	07F3 07F4	FE *Opcode* 0A *Data*
			JZ	START	07F5 07F6 07F7	CA *Opcode* F0 } *Address* 07 }
4	GO TO 2		JMP	LOOP	07F8 07F9 07FA	C3 *Opcode* F2 } *Address* 07 }

(e)

FIGURE 17-7 A Count-to-Ten Programming Example. (a) Flowchart. (b) High-Level Language Program. (c) Assembly Language Program. (d) Machine Language Program. (e) Summary of Program Languages.

second line, LET COUNT = COUNT + 1 is simply a way of saying "increment the count." Lines 3 and 4 perform the function of the decision block in the count-to-ten flow chart. Line 3 specifies that if COUNT = 10, then the next instruction executed should be line 1. If the count is not equal to ten, the COUNT = 10 instruction has no effect, and the program continues with line 4, which says "go to line 2." To test whether this BASIC program will count to ten and then repeat, try following the program step by step to see if it works, starting with a value of zero.

Assembly language is not one specific language, but a class of languages. Each microprocessor has its own machine language and therefore its own assembly language, which is defined by the manufacturer. Figure 17-7(c) shows the assembly language listing for the count-to-ten program. This program is certainly more cryptic than the BASIC language program, but it performs exactly the same function.

The three columns in Figure 17-7(c) are for labels, instructions, and comments. The label provides the same function as the line number in the BASIC program. Instead of numbering every line, you simply make up a name (called a label) for each line to which you need to refer. A colon (:) is used to identify the label. A line only needs a label if there is another instruction in the program that refers to that line. The comments are an aid to understanding the program, and a semicolon (;) is used to identify the beginning of a comment. High-level language programs do not always need many comments because the instructions themselves are more descriptive. For assembly language programs, however, comments are an invaluable aid for people other than the programmer, or when the programmer returns to a program after some time.

The first instruction is MVI A, 0, which means "move immediately to the accumulator the data zero." The **accumulator** is also called the A register, and is a storage location inside the microprocessor. This assembly language instruction is equivalent to the BASIC instruction LET COUNT = 0, except instead of making up a name for the variable (COUNT), we used a pre-assigned name (A) for a register inside the microprocessor. This MVI A, 0 instruction will therefore load the data 0 into the microprocessor's internal A register. The next instruction, INR A, means "increment the value in the accumulator," and this assembly language instruction is equivalent to the BASIC instruction LET COUNT = COUNT + 1. The next three instructions implement the decision function. The instruction CPI 10 means "compare the value in the accumulator with the value 10." The result of this comparison will determine whether a special flip-flop within the microprocessor called the *zero flag* is SET or RESET. If the value in the accumulator is not equal to 10, the zero flag is RESET LOW, whereas if the value in the accumulator is equal to 10, the zero flag is SET HIGH. The next instruction, JZ START, means "jump if the zero flag is SET to the line with the label START." This instruction tests the zero flag, and if it is SET (accumulator = 10), it will cause the program to jump to the line with the label START. Together, these two instructions (CPI 10 and JZ START) perform the function of the BASIC statement IF COUNT = 0 THEN 1. The last instruction, JMP LOOP, means "jump to the line with the label LOOP." This instruction simply causes the program to jump to the line with the label LOOP, and is equivalent to the BASIC statement GO TO 2.

Accumulator
One or more registers associated with the Arithmetic and Logic Unit (ALU), which temporarily store sums and other arithmetical and logical results of the ALU.

Opcode
The first part of a machine-language instruction that specifies the operation to be performed. The other parts specify the data, address, or port. For the 8085, the first byte of each instruction is the opcode.

Figure 17-7(d) shows the machine language listing for the count-to-ten program. Although this language looks the most alien to us, its sequence of 1s and 0s is the only language that the microprocessor understands. In this example, each memory location holds eight bits of data. To program the microcomputer, we will have to store these 8-bit codes in the microcomputer's memory. Each instruction begins with an **opcode,** or operation code, that specifies the operation to be performed, and since all 8085 opcodes are eight bits, each opcode will occupy one memory location. An 8085 opcode may be followed by one byte of data, two bytes of data, or no bytes, depending upon the instruction type used.

Stepping through the machine language program shown in Figure 17-7(d), you can see that the first byte ($3E_{16}$) at address $07F0_{16}$ is the opcode for the instruction MVI A. The MVI A instruction is made up of two-bytes. The first byte is the opcode specifying that you want to move some data into the accumulator, and the second byte stored in the very next memory location (address $07F1_{16}$) contains the data 00_{16} to be stored in the accumulator. The third memory location (address 07F2) contains the opcode for the second instruction, INR A. This opcode (3C) tells the microprocessor to increment the accumulator. The INR A instruction has no additional data, and therefore the instruction only occupies one memory location. The next code, FE, is the opcode for the compare instruction, CPI. Like the MVI A, 0 instruction, the memory location following the opcode contains the data required by the instruction. Since the machine language program is shown in hexadecimal notation, the data 10 (decimal) appears as 0A (hex). This instruction compares the accumulator with the value 10 and sets the zero flag if they are equal, as described earlier. The next instruction, JZ, has the opcode CA and appears at address 07F5. This opcode tells the microprocessor to jump if the zero flag is set. The next two memory locations contain the address to jump to. Since addresses in an 8085 system are sixteen bits long, it takes two memory locations (eight bits each) to store an address. The two parts of the address are stored in the reverse order than you might expect. The least significant half of the address is stored first and then the most significant half of the address is stored next. The address 07F0 is therefore stored as F0 07. The assembly language instruction JZ START means that the processor should jump to the instruction labeled START. The machine code must therefore use the actual address that corresponds to the label START, which in this case is address 07F0. The last instruction, JMP LOOP, is coded in the same way as the previous jump instruction. The only difference is that this jump instruction is independent of any flag condition, and therefore no flags will be tested to determine whether this jump should occur. When the program reaches the C3 jump opcode, it will always jump to the address specified in the two bytes following the opcode, which in this example is address 07F2. The machine language program, therefore, contains a series of bytes, some of which are opcodes, some are data, and some are addresses.

You must know the size and format of an instruction if you want to be able to interpret the operation of the program correctly. To compare the differences and show the equivalents, Figure 17-7(e) combines the high-level language, assembly language, and machine language listings for the count-to-ten program.

SELF-TEST REVIEW QUESTIONS FOR SECTION 17-1

1. What three blocks are interconnected to form a microcomputer, and what is the function of each block?
2. What three buses are needed to interconnect the three blocks within a microcomputer, and what is the function of each bus?
3. What is the difference between hardware and software?
4. Briefly describe the three following languages.
   a. Machine language b. Assembly language c. High-level language
5. What is a flowchart?
6. Define the following terms.
   a. Instruction set b. Peripheral c. Programmer d. Mnemonic
   e. Compiler f. Accumulator g. Flag register h. Opcode

17-2 A MICROCOMPUTER SYSTEM

In this section, you will see how all of the circuits discussed in the previous chapters come together to form a complete microprocessor-based system. The system is called "SAM," which is an acronym for "simple all-purpose microcomputer." We will be examining SAM's schematic diagram down to the component level to remove any mystery associated with microcomputers and give you the confidence that comes from understanding a complete digital electronic microprocessor-based system. In addition, we will discuss general microprocessor system troubleshooting techniques and the detailed troubleshooting procedures for the SAM system.

This section of the textbook will read very similar to a manufacturer's technical service manual. This has been done intentionally to introduce you to the format and style of a typical tech manual and so you can see how this resource will provide you with the detailed data that you need to have before attempting to troubleshoot any electronic system.

17-2-1 Theory of Operation

On pages 689 through 692 you will find the schematic diagram for the SAM system. You may wish to copy and combine this schematic, since we will be referring to it constantly throughout the circuit description. You may also want to color code your copy of the schematic to help you more easily see the circuit interconnections. For example, using highlighters, you could make all address lines yellow, all data lines blue, and all control signal lines green.

Introduction

The simple all-purpose microcomputer system was designed specifically for educational use. It uses Intel's 8085 microprocessor and has 2kB (kilobytes) of ROM and 1kB of RAM. The system contains two input ports that are connected to a keyboard from which you can enter programs, store data, and give commands to control the operation of the microcomputer; and a set of slide switches. The output port display allows you to view the contents of the memory and registers. In addition, there is an output port using LEDs and a speaker that is controlled by the processor. There are also LEDs on the address bus, the data bus, and the major control lines so you can monitor their activity. The ROM in the SAM contains programs to read the keyboard, execute keyboard commands, and send data to the display. The entire operation of the system is governed by a built-in software program called the *monitor*. Using the keyboard and the display, a program can be loaded into RAM, and then it can be run at full speed or executed one step at a time to allow you to follow the operation of the program in detail. Special switches are also provided to establish loop operations for test purposes, and moveable jumpers are included to insert faults in the circuits for training in troubleshooting microprocessor systems.

Block Diagram Description

A simplified block diagram of the SAM system is shown in Figure 17-8. The three main functional blocks of this microcomputer are the microprocessor, the memory, and the input/output ports. These blocks within the microcomputer are interconnected by three groups of signals—address, data, and control.

Address bus. The microprocessor is the only address bus "talker," which means that it is the only device that ever sends out addresses onto the 16-bit

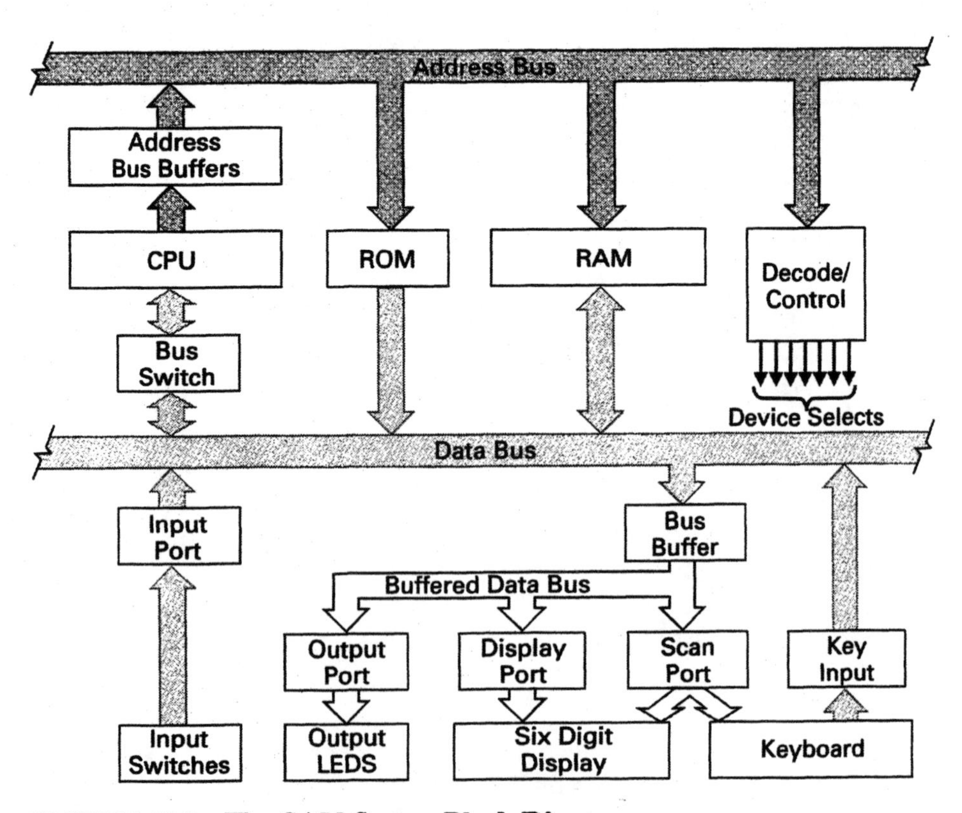

FIGURE 17-8 The SAM System Block Diagram.

address bus. Every other device connected to the address bus, therefore, is a "listener" since these devices only receive addresses from the microprocessor. This one-direction flow of address information on this bus accounts for why the address bus is called a **unidirectional bus.** Before any data transfer can take place via the data bus, the microprocessor must first output an address onto the address bus to specify which memory location or input/output port the processor wishes to access. In this way, the processor can select any part of the system it wishes to communicate with.

Unidirectional Bus Wire or group of wires in which data flows in only one direction. Each device connected to a unidirectional bus is either a transmitter, or a receiver, but not both.

Bidirectional Bus Indicates that signal flow may be in either direction. Common bidirectional buses are three-state or open collector TTL.

Data bus. The 8-bit data bus is a **bidirectional bus** (two-direction), and therefore any device that talks on this bus must have eight driver outputs, and any device that listens must have eight data inputs. The microprocessor and the RAM are data bus talkers and listeners, while the ROM is only a data bus talker. The input ports are talkers since they input data from outside the system and put it on the data bus, whereas the output ports are listeners since they take data off the data bus and send it outside the system. The arrows on the data bus lines in Figure 17-8 summarize how each of the devices communicates with the data bus.

The microprocessor, RAM, ROM, and input ports contain three-state drivers on their data outputs. The select input to the RAM, ROM, and input ports enables the three-state drivers and causes the data from the selected device to appear on the data bus. The microprocessor is the controller of the system, so it will ensure that no more than one device is trying to use the data bus at any given time. For example, if the microprocessor wants to read data from the ROM, it would three-state its own data lines and generate the control signals needed to cause the ROM's select input to be true. The ROM's outputs would then appear on the data bus, and the microprocessor would read the data off the bus. Reading the RAM or an input port is done in a similar manner. On the other hand, if the microprocessor wanted to write data to the RAM or an output port, the microprocessor would first place the data to be written on its data lines. It would then generate the control signals needed to cause a write pulse to be sent to the appropriate device, which would then input the data from the bus.

Control bus. The address bus is used to select a particular memory location or input/output port. The data bus is then used to carry the data, and the control bus is used to control this process. The control bus consists of a number of control signals, most of which are generated by the microprocessor.

The three main control signals generated by the microprocessor are $\overline{RD}$ (not read or NRD), $\overline{WR}$ (not write or NWR), and IO/$\overline{M}$ (input/output/not memory or IO/NM). If NRD is LOW (active), it indicates that a read is in progress and the microprocessor will be expecting the device which is being addressed to put data on the data bus. If NWR is LOW (active), a write is in progress and the microprocessor will be putting data on the data bus, expecting the addressed device to store this data. If IO/NM is LOW, the operation in progress (which may be a read or write) is a memory operation. If IO/NM is HIGH, the operation in progress (which may be an output or input) is an input/output port operation.

Inside the 8085

At the heart of the SAM system is an 8085 microprocessor (U3 on the SAM schematic). In most microcomputer circuit descriptions, the microprocessor is treated as a device with known characteristics, but whose internal structure is of no concern. Since a knowledge of the microprocessor's internal operation will help you gain a clearer understanding of the operation of the SAM system, we will take a look at the internal structure and operation of the 8085.

Figure 17-9 shows the basic signals that connect to a microprocessor. There are sixteen address outputs which drive the address bus, and eight data lines connected to the data bus. The $\overline{RD}$ and $\overline{WR}$ output control signals coordinate the movement of data on the data bus, and the RESET input is used to initialize the microprocessor's internal circuitry. The two connections at the top of the microprocessor are for an external crystal, which is used as the frequency-determining device for the microprocessor's internal clock oscillator. The output from the microprocessor's internal clock oscillator is called the *system clock*, and it is used to synchronize all devices in the system and set the speed at which instructions are executed.

The inset in Figure 17-9 shows a simplified internal block diagram of the 8085 microprocessor. The accumulator, or A register, connects to the data bus and the arithmetic and logic unit (ALU). The ALU performs all data manipulation, such as incrementing a value or adding two numbers. The temporary register is automatically controlled by the microprocessor's control circuitry, and this register feeds the ALU's other input. The flags in the flag register are a collection of flip-flops that indicate certain characteristics about the result of the most recent operation performed by the ALU. For example, the zero flag is set if the result from an operation is 0, and as discussed earlier, this flag is tested by the JZ (jump if 0) instruction. The instruction register, instruction decoder, program counter, and control and timing logic are used for fetching instructions from memory and directing their execution.

Let us now take a closer look at the microprocessor's *fetch-execute cycle.* As an example, let us imagine that we need to read an MVI A instruction from ROM at address 0200. The first step is to retrieve the opcode for this instruction. This process is called the *instruction fetch,* and is shown in Figure 17-10(a). As its name implies, the program counter is used to keep count by pointing to the current address in the program. In this example, the program counter will contain the address 0200, and this will be placed on the address bus and cause memory location 0200 to be selected. The ROM will then place the contents of memory location 0200 on the data bus, and the microprocessor will store this data, which is the opcode for MVI A (3E) in the instruction register.

Once the opcode is stored, the *instruction execute* process begins, as shown in Figure 17-10(b). The instruction register feeds the instruction decoder, which recognizes the opcode and provides control signals to the control and timing circuitry. For example, for an MVI A instruction, the control and timing logic circuit first reads the opcode 3E, and then increments the address in the program counter. The instruction decoder determines that this opcode is always followed by a byte of data, so the contents of the memory location pointed to by the program counter is read, and the second byte of this instruction is placed into

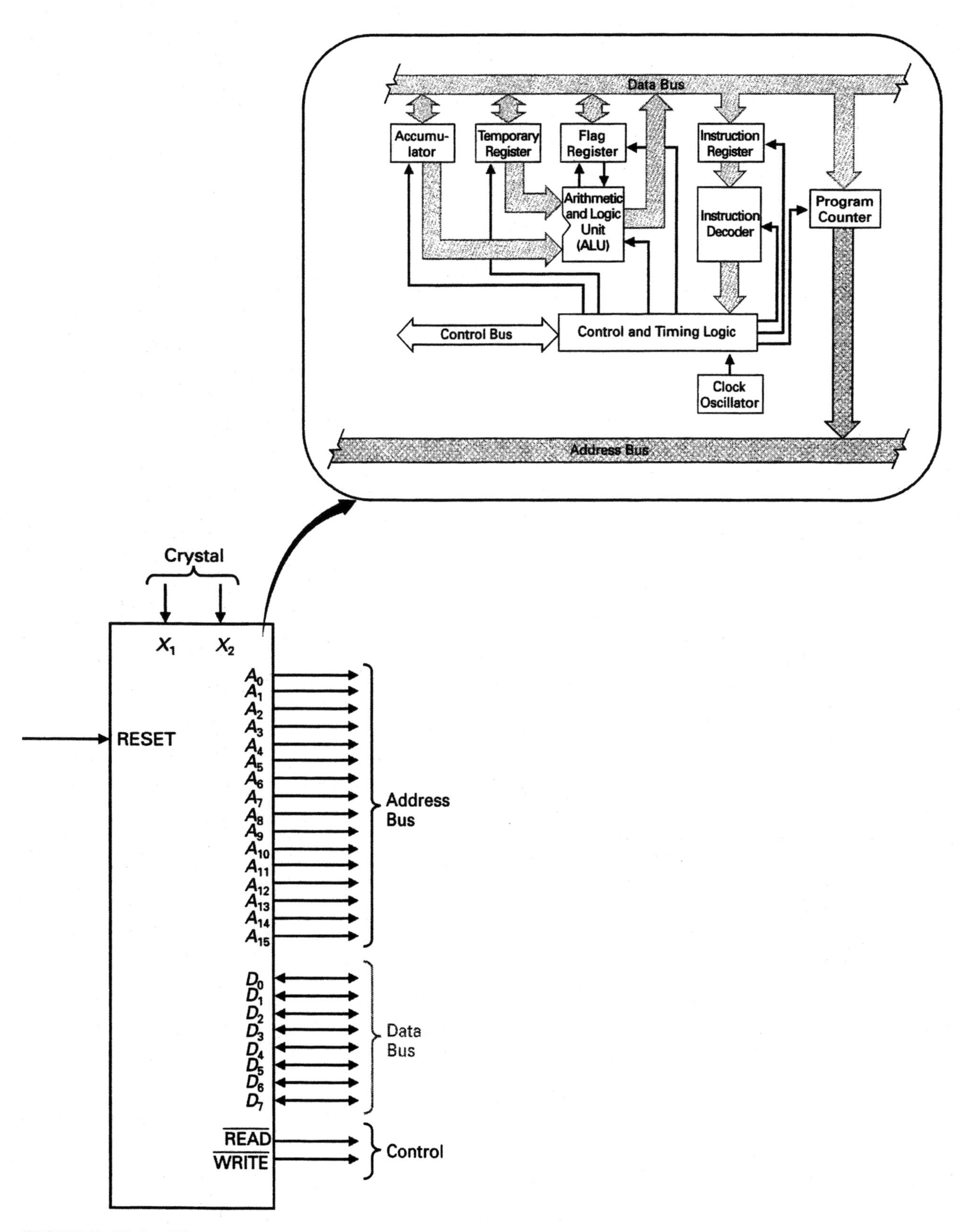

FIGURE 17-9 The Microprocessor.

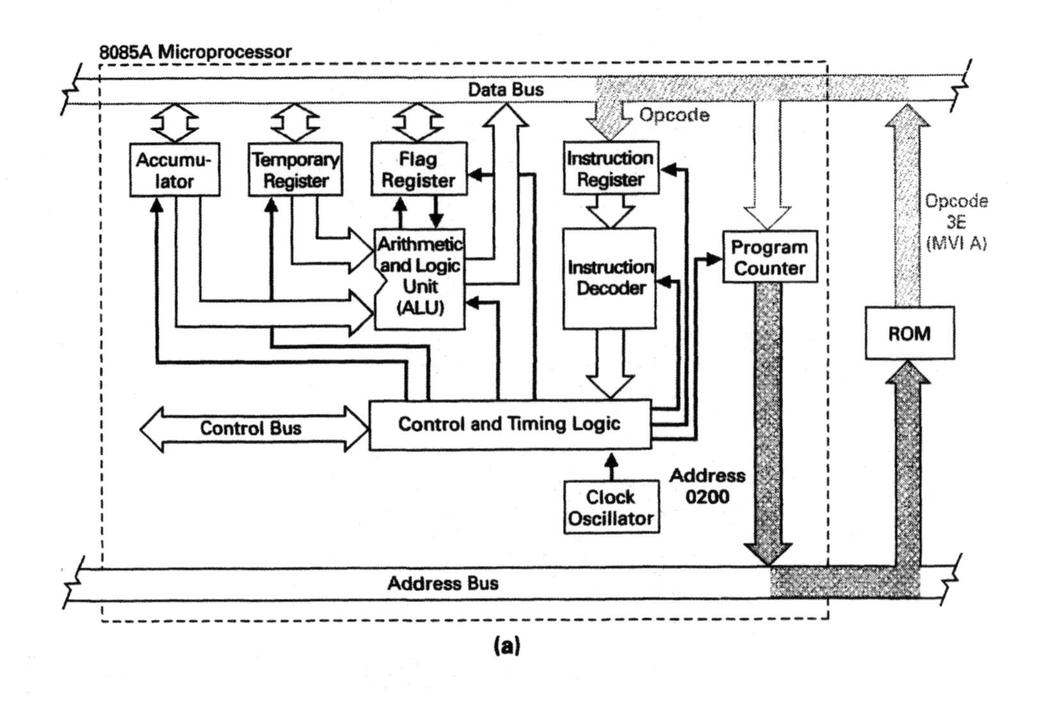

(a)

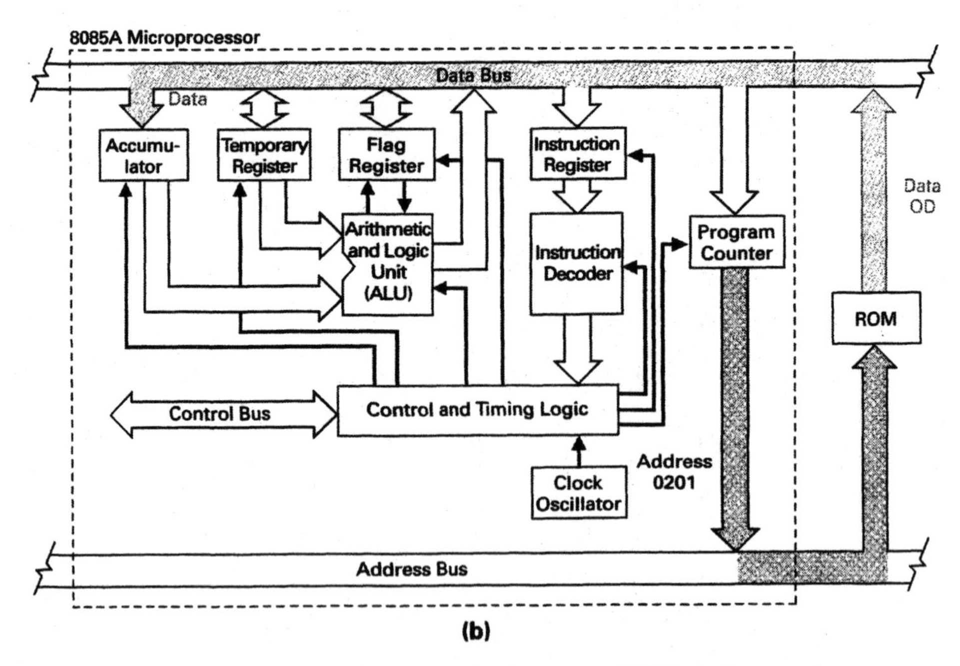

(b)

FIGURE 17-10 **The Fetch-Execute Cycle for an MVI A Instruction.**

the accumulator. The program counter is again incremented, and the next byte of the main program (which will be the next instruction's opcode) is read into the instruction register, and then the execution of this instruction begins.

The real work is done in the execute phase of the instruction. Four basic types of operations can be performed by the 8085 microprocessor:

1. Read data from memory or an input port.

2. Write date to memory or an output port.

3. Perform an operation internal to the microprocessor.

4. Transfer control to another memory location.

The first two operation types are self-explanatory. The third involves manipulating the 8085's internal registers (such as the accumulator) without accessing the external memory or input/output ports. For example, the contents of one register may be incremented or decremented. The fourth operation type includes instructions such as JMP, which causes the program counter and therefore the program to shift to point to a new memory location.

To give you a little more detail on the 8085 microprocessor, Figure 17-11 shows the data sheet for this microprocessor. This data sheet includes a more detailed internal block diagram, a pin-out diagram, and a description of the function of each pin.

To summarize the operation of the microprocessor, we can say that the microprocessor keeps reading sequentially through the memory, one location after another, performing the indicated operations. Exceptions to this occur when a jump instruction is executed or at the occurrence of an **interrupt.** The 8085 has a few interrupt input pins which are used to tell the microprocessor to stop its current task and perform another task immediately. Jump instructions or interrupts will cause the microprocessor to stop its sequential flow through a program and begin executing instructions from another address.

Interrupt Involves suspension of the normal program that the microprocessor is executing in order to handle a sudden request for service (interrupt). The processor then jumps from the program it was executing to the interrupt service routine. When the interrupt service routine is completed, control returns to the interrupted program.

Remember that opcodes and data are intermixed in memory. One address might contain an opcode, the next two addresses a jump address, the next address an opcode, and the next address a piece of data. It is the programmer's responsibility to be sure that the memory contains a valid sequence of opcodes and data. The microprocessor cannot distinguish between opcodes, jump addresses, and data since they are all simply bit patterns stored in the memory. All such information is read in exactly the same way and it travels over the same data bus. The microprocessor must always keep track of whether it is reading an opcode or data and treat each appropriately. The processor always assumes that the first location it reads contains an opcode, and goes from there. If the opcode requires a second byte of data, the microprocessor is aware of this (because of the instruction decoder), and treats the next byte accordingly. It then assumes that the byte following the last byte in an instruction is the next opcode. The microprocessor will not know if a programmer has placed data in a position that should have an opcode. Such errors will cause the system to "crash" since the microprocessor would decode mispositioned data as opcodes.

SAM Circuit Description

As we proceed through this section, you should have your SAM schematic open, since we will be referring to devices on the schematic along with simplified SAM circuit diagrams shown in this textbook.

DIGITAL DEVICE DATA SHEET

IC: 8085A – SINGLE CHIP 8-BIT *n*-CHANNEL MICROPROCESSOR

- **0.8 μs Instruction Cycle; 5 MHz Internal Clock**
- **Single +5V Power Supply**
- **On-Chip Clock Generator (with External Crystal or RC Network)**
- **On-Chip System Controller; Advanced Cycle Status Information Available for Large System Control**
- **Four Vectored Interrupts (One is non-Maskable)**
- **Serial In/Serial Out Port**
- **Decimal, Binary and Double Precision Arithmetic**
- **Direct Addressing Capability to 64K Bytes of Memory**

The Intel® 8085A is a new generation, complete 8-bit parallel central processing unit (CPU). Its instruction set is 100% software compatible with the 8080A microprocessor, and it is designed to improve the present 8080's performance by higher system speed. Its high level of system integration allows a minimum system of three ICs: 8085A (CPU), 8156 (RAM), and 8355/8755A (ROM/PROM).

The 8085A incorporates all of the features that the 8224 (clock generator) and 8228 (system controller) provided for the 8080, thereby offering a high level of system integration.

The 8085A uses a multiplexed Data Bus. The address is split between the 8-bit address bus and the 8-bit data bus. The on-chip address latches of 8155/8355/8755 memory products allows a direct interface with 8085A.

The 8085A-2 is a higher performance version of the 8085A.

BLOCK DIAGRAM

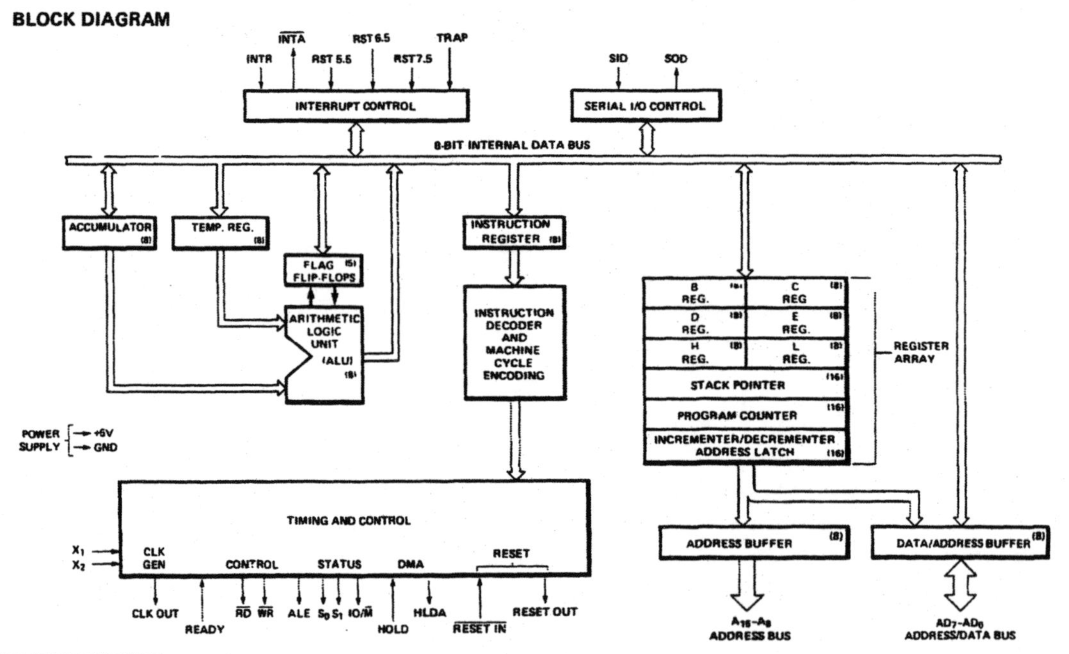

PIN DESCRIPTION

The following describes the function of each pin:

A_8-A_{15} (Output 3-State)

Address Bus; The most significant 8-bits of the memory address or the 8-bits of the I/O address, 3-stated during Hold and Halt modes.

AD_{0-7} (Input/Output 3-State)

Multiplexed Address/Data Bus; Lower 8-bits of the memory address (or I/O address) appear on the bus during the first clock cycle of a machine state. It then becomes the data bus during the second and third clock cycles.

3-stated during Hold and Halt modes.

ALE (Output)

Address Latch Enable: It occurs during the first clock cycle of a machine state and enables the address to get latched into the on-chip latch of peripherals. The falling edge of ALE is set to guarantee setup and hold times for the address information ALE can also be used to strobe the status information. ALE is never 3-stated.

FIGURE 17-11 An IC Containing a Single-Chip Microprocessor. (Courtesy of Intel Corporation)

S0, S1 (Output)

Data Bus Status. Encoded status of the bus cycle:

S_1	S_0	
0	0	HALT
0	1	WRITE
1	0	READ
1	1	FETCH

S_1 can be used as an advanced R/$\overline{W}$ status.

$\overline{RD}$ (Output 3-State)

READ; indicates the selected memory or I/O device is to be read and that the Data Bus is available for the data transfer. 3-stated during Hold and Halt.

$\overline{WR}$ (Output 3-State)

WRITE: indicates the data on the Data Bus is to be written into the selected memory or I/O location. Data is set up at the trailing edge of $\overline{WR}$. 3-stated during Hold and Halt modes.

READY (Input)

If Ready is high during a read or write cycle, it indicates that the memory or peripheral is ready to send or receive data. If Ready is low, the CPU will wait for Ready to go high before completing the read or write cycle.

HOLD (Input)

HOLD; indicates that another Master is requesting the use of the Address and Data Buses. The CPU, upon receiving the Hold request, will relinquish the use of buses as soon as the completion of the current machine cycle. Internal processing can continue. The processor can regain the buses only after the Hold is removed. When the Hold is acknowledged, the Address, Data, $\overline{RD}$, $\overline{WR}$, and IO/$\overline{M}$ lines are 3-stated.

HLDA (Output)

HOLD ACKNOWLEDGE; indicates that the CPU has received the Hold request and that it will relinquish the buses in the next clock cycle. HLDA goes low after the Hold request is removed. The CPU takes the buses one half clock cycle after HLDA goes low.

INTR (Input)

INTERRUPT REQUEST; is used as a general purpose interrupt. It is sampled only during the next to the last clock cycle of the instruction. If it is active, the Program Counter (PC) will be inhibited from incrementing and an $\overline{INTA}$ will be issued. During this cycle a RESTART or CALL instruction can be inserted to jump to the interrupt service routine. The INTR is enabled and disabled by software. It is disabled by Reset and immediately after an interrupt is accepted.

$\overline{INTA}$ (Output)

INTERRUPT ACKNOWLEDGE; is used instead of (and has the same timing as) $\overline{RD}$ during the instruction cycle after an INTR is accepted. It can be used to activate the 8259 Interrupt chip or some other interrupt port.

RST 5.5 }
RST 6.5 } (Inputs)
RST 7.5 }

RESTART INTERRUPTS; These three inputs have the same timing as INTR except they cause an internal RESTART to be automatically inserted.

PIN CONFIGURATION

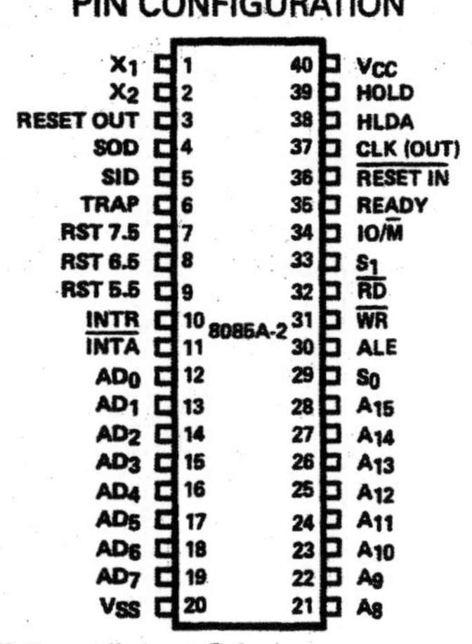

RST 7.5 —Highest Priority
RST 6.5
RST 5.5 —Lowest Priority

The priority of these interrupts is ordered as shown above. These interrupts have a higher priority than the INTR.

TRAP (Input)

Trap interrupt is a nonmaskable restart interrupt. It is recognized at the same time as INTR. It is unaffected by and mask or interrupt Enable. It has the highest priority of any interrupt.

$\overline{RESET\ IN}$ (Input)

Reset sets the Program Counter to zero and resets the Interrupt Enable and HLDA flip-flops. None of the other flags or registers (except the instruction register) are affected. The CPU is held in the reset condition as long as Reset is applied.

RESET OUT (Output)

Indicates CPU is being reset. Can be used as a system RESET. The signal is synchronized to the processor clock.

X_1, X_2 (Input)

Crystal or R, C network connections to set the internal clock generator. X_1 can also be an external clock input instead of a crystal. The input frequency is divided by 2 to give the internal operating frequency.

CLK (Output)

Clock Output for use as a system clock when a crystal or R/C network is used as an input to the CPU. The period of CLK is twice the X_1, X_2 input period.

IO/$\overline{M}$ (Output)

IO/$\overline{M}$ indicates whether the Read/Write is to memory or I/O. Tri-stated during Hold and Halt modes.

SID (Input)

Serial input data line. The data on this line is loaded into accumulator bit 7 whenever a RIM instruction is executed.

SOD (Output)

Serial output data line. The output SOD is set or reset as specified by the SIM instruction.

V_{CC}	V_{SS}
+5 volt supply.	Ground Reference

FIGURE 17-11 **(Continued).**

Decoding. The SAM system does not use I/O mapping for its I/O ports, so the IO/NM output from the microprocessor is not used for decoding. The SAM system uses memory-mapped input/output, which means that the I/O ports are treated as addressable devices similar to memory space.

Figure 17-12(a) shows the SAM address decoder (U7 on the SAM schematic), and Figure 17-12(b) shows the memory map or address map for the SAM system. Only the first quarter of the address space is used, so address bits 14 and 15 are always 0. The address space used in the SAM system is divided into eight equal sections, each of which has 2k, or 2,048, locations. The ROM occupies the first 2k addresses (0000 to 07FF hex), and the RAM has been assigned the next 2k addresses (0800 to 0FFF). The six address sections following the RAM are used for I/O ports. The control port is used by the monitor program (system control program in ROM) to provide some special functions, which will be described later. The key data, scan, and display segment ports control the keyboard and display, and the input and output ports are used for the switches and LEDs.

To simplify the address decoding hardware, 2,048 addresses have been assigned to each memory type and each of the input and output ports. The SAM system makes use of 16k addresses for 2k of ROM, 1k of RAM, and six I/O ports. The SAM system could have used only 2,048 + 1,024 + 6 = 3,078 addresses, which would result in more unused address space and a much more complicated address decoding circuit. Referring to the simplified SAM address decoder circuit shown in Figure 17-12(a) and U7 on the SAM schematic, let us now see how this circuit operates. There are two things to note in this decoding arrangement:

1. Some of the read/write control is mixed with the decoding.
2. There is a special circuit for RAM write protect.

The binary addresses listed in the memory map in Figure 17-12(b) show that the A_{11}, A_{12}, and A_{13} lines specify which section is to be addressed. Therefore, these lines are used to provide the binary-select inputs to U7 (a 74138 binary to one-of-eight decoder IC). This provides eight separate active-LOW outputs, one for each of the 2k byte blocks in the SAM memory map. Address decoder U7 has three enable inputs, two that are active-LOW and one that is active-HIGH. All three enable inputs must be true to allow any of the outputs to be true. This is achieved by connecting the A_{14} and A_{15} lines to the two active-LOW enables. This will prevent any of the outputs from being true unless both A_{14} and A_{15} are LOW.

The read/write control gating is distributed throughout the circuitry using the enables of the decoder, the memory devices, and the I/O ports to reduce the number of gates required. The decoder's third enable input is connected to a gate which generates the OR of NRD and NWR. This has the effect of allowing the device-select output of U7 to be true only when either a read or a write is in progress. This is necessary because the address bus will not contain meaningful information if either NRD or NWR are true. In addition, the ROM and the input ports are to be selected only if a read is being performed. If they respond to both a read or a write, a bus conflict could occur. For example, if a write to the ROM is attempted, the microprocessor would put data on the

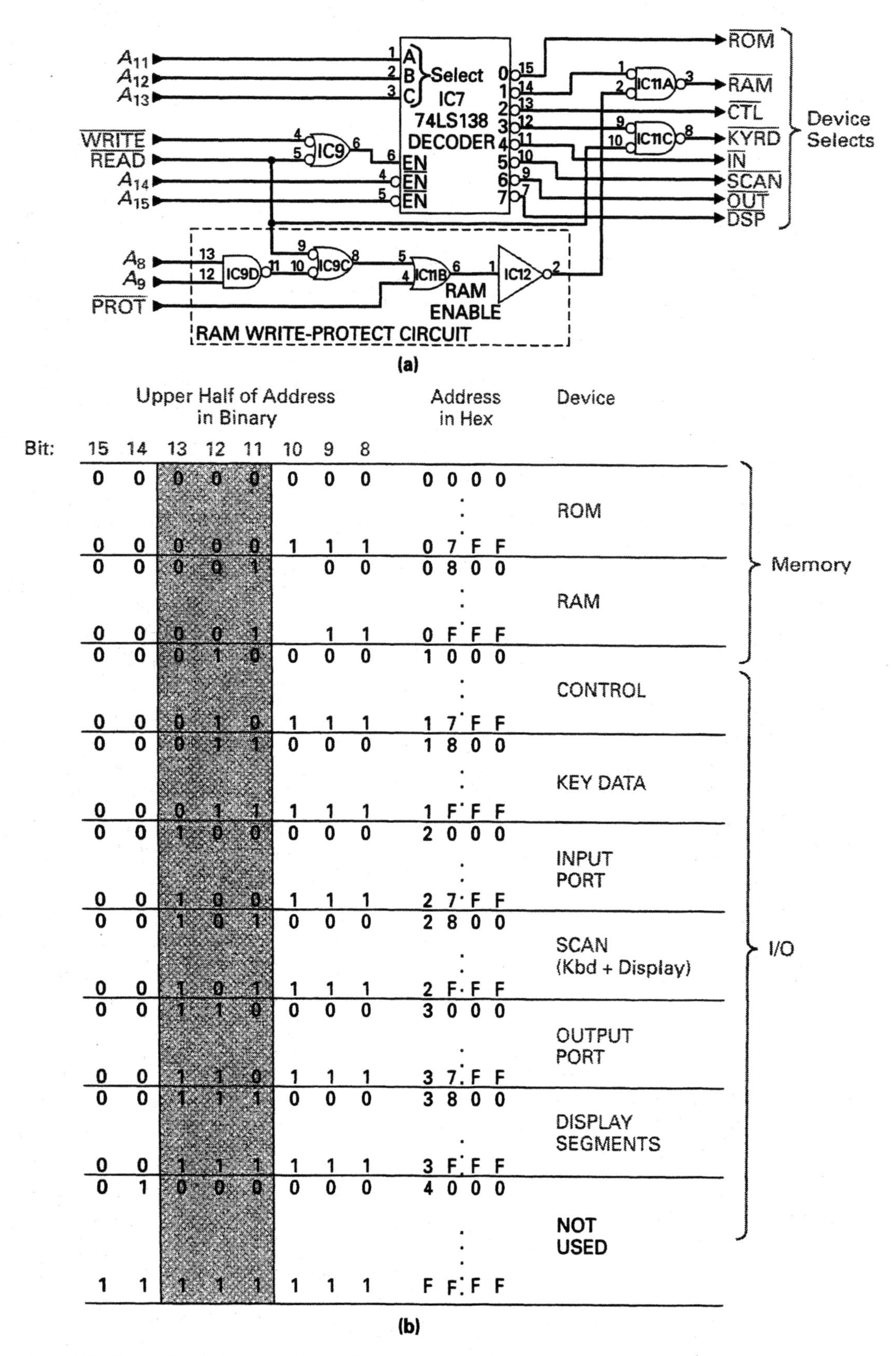

Bit:	15	14	13	12	11	10	9	8	Address in Hex	Device	
	0	0	0	0	0	0	0	0	0000	ROM	Memory
	0	0	0	0	0	1	1	1	07FF		
	0	0	0	0	1		0	0	0800	RAM	
	0	0	0	0	1		1	1	0FFF		
	0	0	0	1	0	0	0	0	1000	CONTROL	I/O
	0	0	0	1	0	1	1	1	17FF		
	0	0	0	1	1	0	0	0	1800	KEY DATA	
	0	0	0	1	1	1	1	1	1FFF		
	0	0	1	0	0	0	0	0	2000	INPUT PORT	
	0	0	1	0	0	1	1	1	27FF		
	0	0	1	0	1	0	0	0	2800	SCAN (Kbd + Display)	
	0	0	1	0	1	1	1	1	2FFF		
	0	0	1	1	0	0	0	0	3000	OUTPUT PORT	
	0	0	1	1	0	1	1	1	37FF		
	0	0	1	1	1	0	0	0	3800	DISPLAY SEGMENTS	
	0	0	1	1	1	1	1	1	3FFF		
	0	1	0	0	0	0	0	0	4000	NOT USED	
	1	1	1	1	1	1	1	1	FFFF		

Column headings: Upper Half of Address in Binary (bits 15–8); Address in Hex; Device.

(b)

FIGURE 17-12 SAM's Address Decoder and Address Map.

data bus to be written to the ROM. If the ROM were allowed to be enabled by a write, then it would also put data on the data bus, which is an unacceptable situation.

To solve this problem, U11C ANDs the NRD signal with the device select. This is shown for the KYRD port, but not for the ROM and IN ports. The ROM and IN port chips each have two enables, so one is used for the device select and one for NRD. This effectively ANDs the NRD signal with the device select. For the output ports, the situation is slightly different. In an attempt to read an output port (which is not a valid operation), a write will be performed instead, and the port will be loaded with invalid data. This is acceptable, since the software should know not to do this and even if it does, no real damage will be caused. This is in contrast to the situation of writing to an input port, which causes a hardware conflict and must not be allowed. Therefore, it is not necessary to AND the NWR signal with the device select for the output port U15.

The RAM's device select should be true when a read or write to the RAM's address space is in progress. The gate (U11A) on the RAM's device-select line is for the write protect circuit. The write-protect circuitry helps prevent the RAM's contents from being accidentally lost. A programming error may result in the microprocessor running wild (usually by interpreting data as an opcode), and often this will result in incorrect data being written into the RAM. To prevent this, the SAM system contains control latch U8, which provides an NPROT input. When this latch is SET, the RAM will be protected. The monitor program will set this protect latch whenever a program is running and reset the protect latch at all other times so that data can be stored in the RAM so SAM can be programmed. Since the program may want to use the RAM to store data during program execution, only the first three-quarters of the RAM is protected. Address lines A_8 and A_9 indicate which quarter of the RAM is being addressed, and if they are both HIGH, then the last quarter is being addressed and the memory will not be protected. To achieve this, A_8 and A_9 are ANDed together in U9D and the result is then ORed with NRD and NPRT in U11B. This produces the RAM enable signal, which will be true if A_8 and A_9 are HIGH if a read is in progress or if the protect latch is not set. If the protect latch is set, then the RAM will be disabled unless a read is in progress or if A_8 and A_9 are HIGH.

RAMs. Figure 17-13 shows the SAM system's RAM circuit (U5 and U6 on the SAM schematic). The SAM system uses two 1k by 4-bit SRAMs connected to form a 1k by 8-bit RAM memory circuit. The address and control pins for both chips are parallel connected together, while U5 connects to data lines 0 through 3 and U6 connects to data lines 4 through 7.

ROM. Figure 17-14 shows the SAM system's ROM circuit (U4 on the SAM schematic). The eleven low-order address lines (A_0 through A_{10}) supply an address to this 2k by 8-bit masked-program ROM. The 8-bit data output from this IC is connected to the system's data bus. The two selects for this IC must be true for the ROM's three-state output drivers to be enabled. The ROM will therefore only drive the data bus if the ROM select is true and the operation is a read.

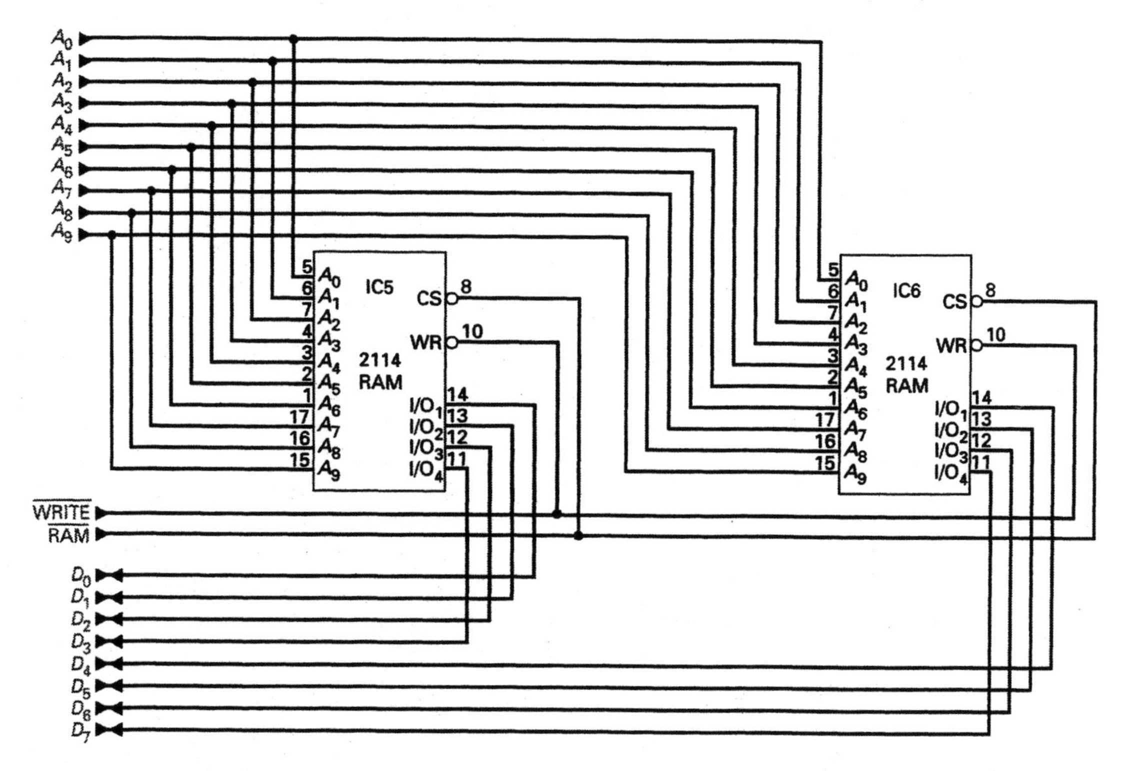

FIGURE 17-13 SAM's RAM Circuit.

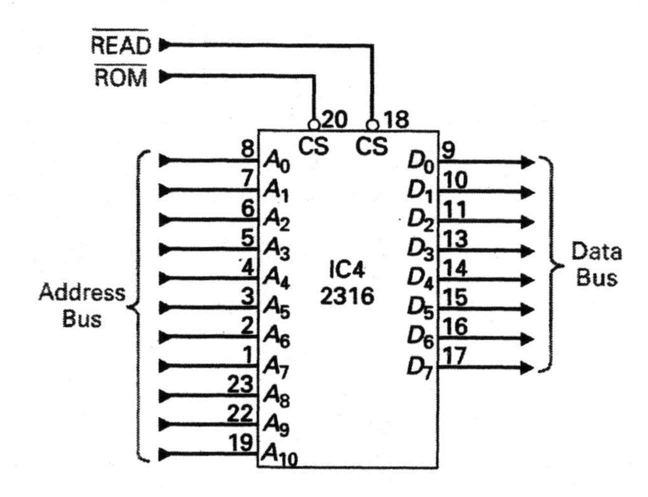

FIGURE 17-14 SAM's ROM Circuit.

Speaker. Figure 17-15 shows the SAM system's serial output circuit, which will have to be traced carefully on the SAM schematic. The speaker is driven by the microprocessor's serial output, so this is in effect a 1-bit output port. The U3 (8085) SOD output is buffered by U18A and sent to the edge connector for use by external hardware, as desired. It is then buffered again before driving the speaker. The speaker draws so much current that the edge connector would not have valid logic levels if the speaker buffer were not included. A 100 Ω resistor in series with the speaker limits the current to levels which will not

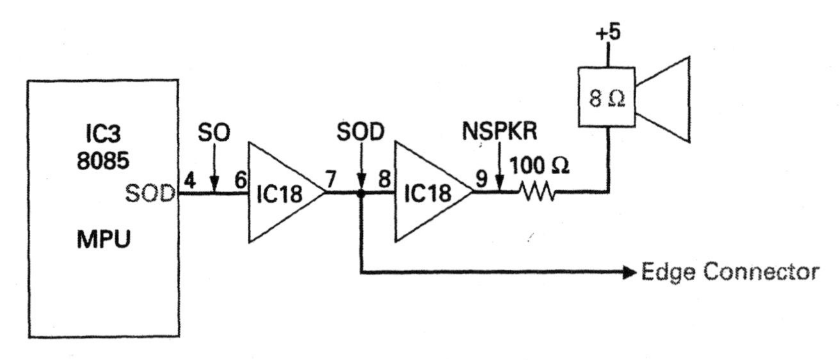

FIGURE 17-15 SAM's Serial Output Circuit.

damage the buffer. The other connection to the speaker is tied to +5 V because the TTL buffer can sink more current than it can source (it can pull more current through the speaker than it can push through the speaker). All of the actual tone generation is controlled by the software. A beep program within the ROM turns the serial output ON and OFF several hundred times a second. This several-hundred-hertz square wave is applied to the speaker for a few seconds.

The control port. Figure 17-16 shows the SAM system's control port register (U8 on the SAM schematic). The control port is used by the microprocessor to send signals to special circuits. The control port IC (U8) is a 4-bit register clocked by the control port select signal, which is generated by address decoder U7. This is similar to the other output ports; however, it is unusual because the data inputs are connected to the address bus instead of the data bus. Therefore, the data written to the port is independent of the data on the data bus. The control port will be selected by any address from 1000 to 17FF. This allows the eleven low-order address lines to contain any value and still select the port. Note that A_0, A_1, and A_2 provide the data inputs to this register, and the data sent to the port is determined by the address used to select the port. For example, an address of 1000 would RESET or clear all of the bits, whereas an address of 1001 would SET the Q output and therefore RESET the

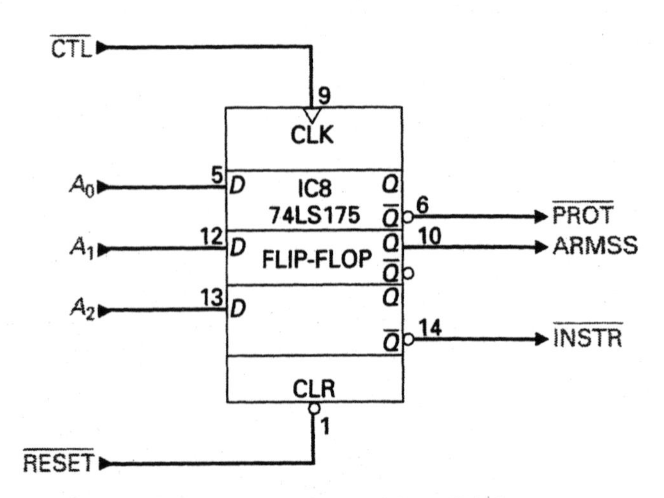

FIGURE 17-16 SAM's Control Port Register.

$\overline{\text{PROT}}$ bit. As another example, an address of 1004 would SET the Q output and therefore RESET the $\overline{\text{INSTR}}$ bit. This technique simplifies the software. Since it does not matter what data is sent to the port (only the address matters), the software does not need to set up a value within the microprocessor before it writes the value to the port. The $\overline{\text{PROT}}$ bit of this port is used to control the memory protect circuit, and if active, the first three-fourths of the RAM will be write protected as described earlier. The other two control bits from this port are used to control the single-step circuitry.

The multiplexed bus. Figure 17-17(a) shows the SAM system demultiplexing circuit (U1 and U2 on the SAM circuit diagram). To keep the number of IC pins to a minimum, the 8085 microprocessor chip (U3) multiplexes the data bus with the lower half of the address bus. An 8-bit address bus from the 8085 carries the upper half of the address bus, and an 8-bit address/data bus carries the data and the lower half of the address. The address latch enable (ALE) signal is generated by the microprocessor to indicate when the address/data bus contains address information. This signal is used to latch the address/data bus contents into the address latch IC (U2) when the lower half of the address is being generated by the 8085. The 8-bit latch U2 latches the address information off the address/data bus on the negative edge of the ALE signal (the inverter U12 is needed to select this edge). The three-state buffer U1 is used to boost the current level of the upper half of the address bus for the address bus LED display, and is not part of the demultiplexing circuit.

The waveforms in Figure 17-17(b) show the 8085 microprocessor's bus timing. The A_8 through A_{15} address lines always contain the high-order address byte. At the beginning of each memory cycle, the low-order address byte is placed on the address/data bus. The trailing edge of the ALE indicates that a low-order address byte is present on the address/data bus and causes the demultiplexing latch to store the low-order address byte. The address information is then removed from the address/data bus to allow for data transfer. If a read operation is in progress, the microprocessor will generate a read signal, and the addressed memory or I/O device will place the data on the address/data bus. On the trailing edge of the NRD signal, the microprocessor will read the data off the bus.

The write cycle is similar, except that the direction of the data transfer is reversed. At the beginning of the cycle, the low-order address byte is placed on the address data bus and the ALE line is pulsed. The microprocessor then issues a write pulse and places the data on the address/data bus. On the trailing edge of the NWR, the addressed memory device or I/O device will store the data from the bus.

Interrupts and reset. Figure 17-18 shows the SAM interrupt and reset circuit. Interrupts provide a means for hardware external to the microprocessor to request immediate action by the processor. They allow the usual program flow to be interrupted and cause control to be transferred to a special program routine. There are two groups of interrupts available on the 8085 microprocessor: TRAP, RST 5.5, 6.5, and 7.5, which are controlled by the individual pins on the microprocessor; and RST 1, 2, 3, 4, 5, 6, and 7, which are controlled by INTR (interrupt request) and INTA (interrupt acknowledge). Only the TRAP, RST

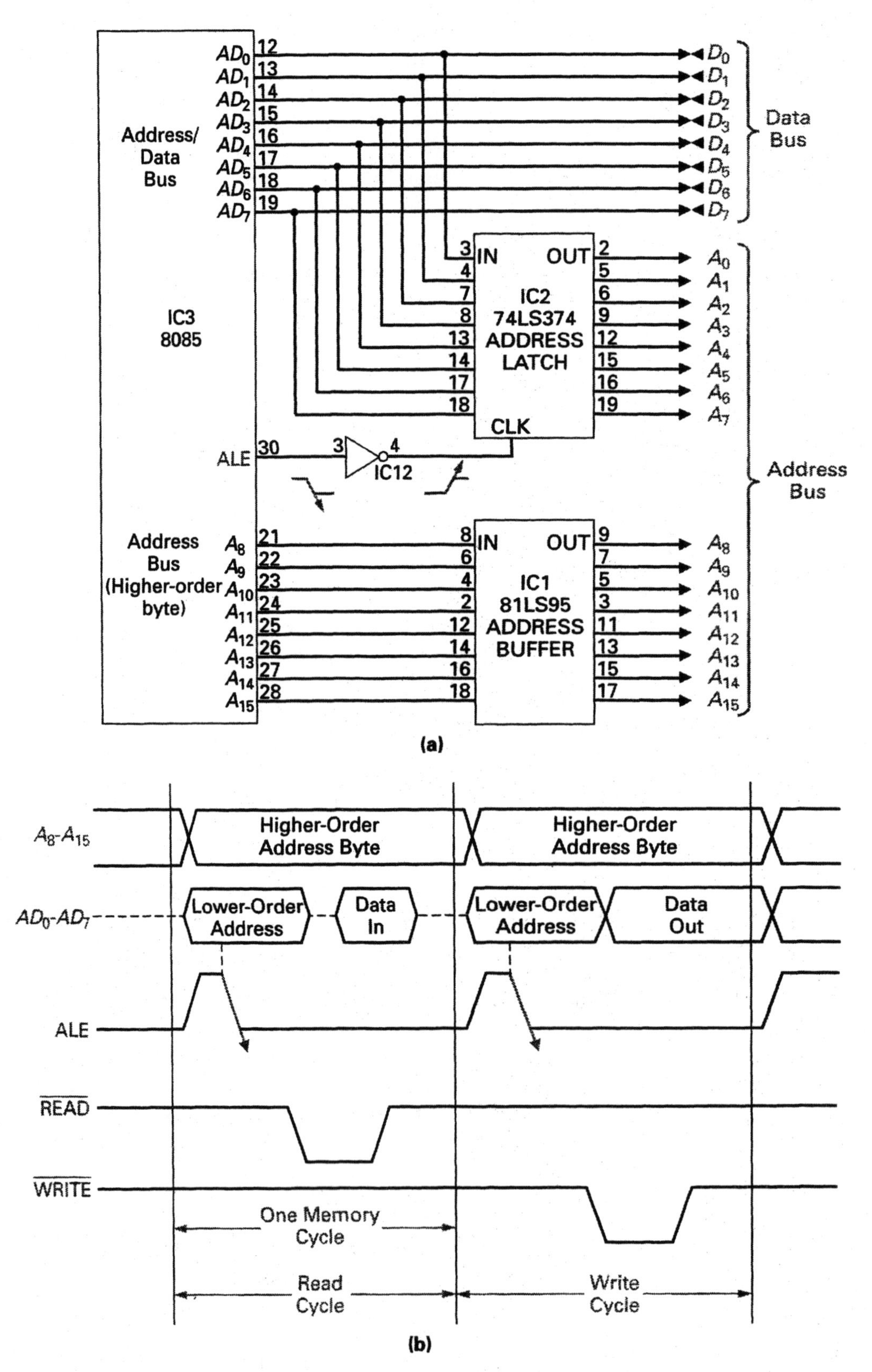

FIGURE 17-17 SAM's Multiplexed Bus.

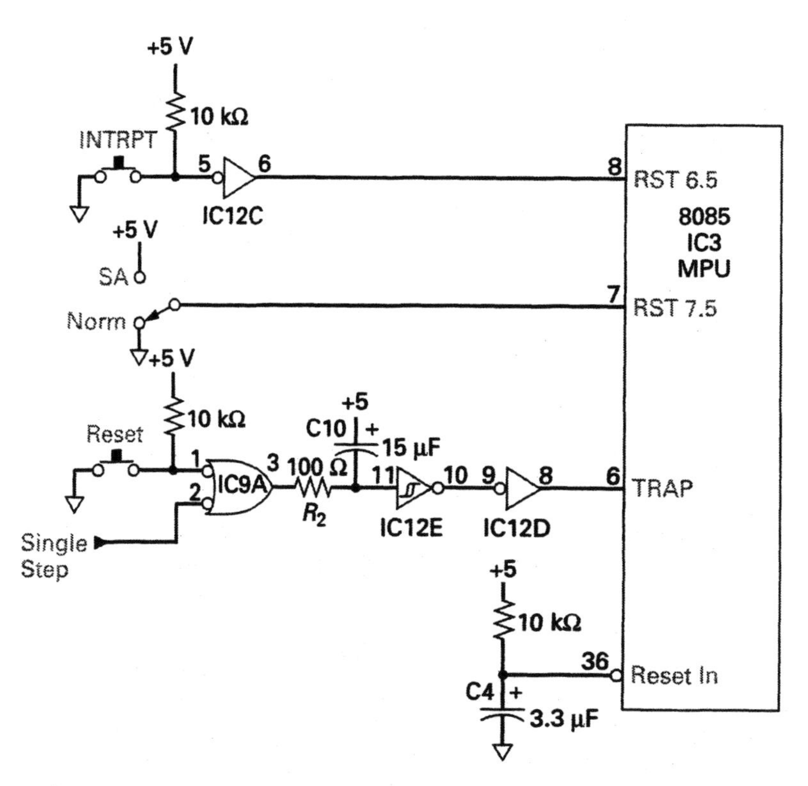

FIGURE 17-18 **SAM's Interrupt and Reset Circuitry.**

7.5, and RST 6.5 interrupts are used in the SAM system. All that is required to initiate one of the interrupts is to apply a signal to the corresponding pin on the microprocessor. The RST 5.5 and 6.5 interrupt inputs will respond to a HIGH logic level, whereas the RST 7.5 interrupt input will only respond to a positive edge. The SAM system uses the TRAP input for the keyboard's RESET key, the RST 6.5 input for the INTRPT key on the keyboard, and the RST 7.5 input for the SA (signature analysis) switch. Signature analysis is a troubleshooting method which will be described later in the SAM troubleshooting section. The NAND gate U9A (which is equivalent to a bubbled-input OR) is used to allow the single-step circuitry to access the TRAP input. The 100 Ω resistor and the capacitor are used to debounce the RESET switch, which is necessary to cause only one interrupt. With the other interrupts, switch debouncing is not necessary because the software disables the interrupt as soon as it is acknowledged, and so prevents a second interrupt. The TRAP input cannot be disabled, so it must be debounced using hardware.

Some allowances must be made for the fact that more than one interrupt may occur at the same time. To accommodate this, each interrupt is assigned a priority, and the interrupt with the highest priority will be acknowledged first. The 8085 TRAP interrupt has the highest priority, followed by the RST 7.5, 6.5, and 5.5, in that order, with INTR having the lowest priority.

The reset-in pin on U3 (pin 36) is used for power-up initialization. When a LOW level is applied to this pin, the microprocessor's internal circuitry is cleared. The program counter is set to 0000 (hex), so program execution begins from this address. This address is in ROM and contains a start-up program that initializes the complete SAM system. The slow-start circuit connected to the reset-in line will provide an automatic power-ON pulse.

Ready. Figure 17-19 shows the SAM system's single-step circuit and, as before, the IC numbers on this circuit correspond to the "U" numbers on the SAM schematic. When the ready input to the microprocessor (pin 35) is brought LOW, the microprocessor enters a wait state. The buses are not put in a three-state position, but remain at their current status until the ready input is brought HIGH again. This input is used for the hardware (HDWR) single-step mode, which allows the observation of the address bus, data bus, and status LEDs for each step of a program.

Data bus buffer. The 8085 microprocessor will not generate enough current at its address and data bus outputs to drive all of the devices connected to the buses plus a set of address bus monitoring LEDs (DS1 through DS4) and data bus monitoring LEDs (DS5 and DS6). Bus buffers are therefore included to boost current. Figure 17-20 shows how a data bus buffer IC14 (U14 on the SAM schematic) is included in this circuit. Since data flow can only travel in one direction (from the data bus to the buffered data bus), only output devices can be connected to the buffered data bus. Any input devices must be connected to the data bus since they will be sending data back to the microprocessor.

Timing. In order for any microprocessor-based system to operate correctly, many timing relationships must be satisfied. To give you an example of the timing involved, Figure 17-21 shows a write-and-read timing operation.

Referring to the write timing waveforms shown in Figure 17-21(a), you can see that the address must be stable for some period of time (called the access time), before any operation can be performed. This allows the memory's inter-

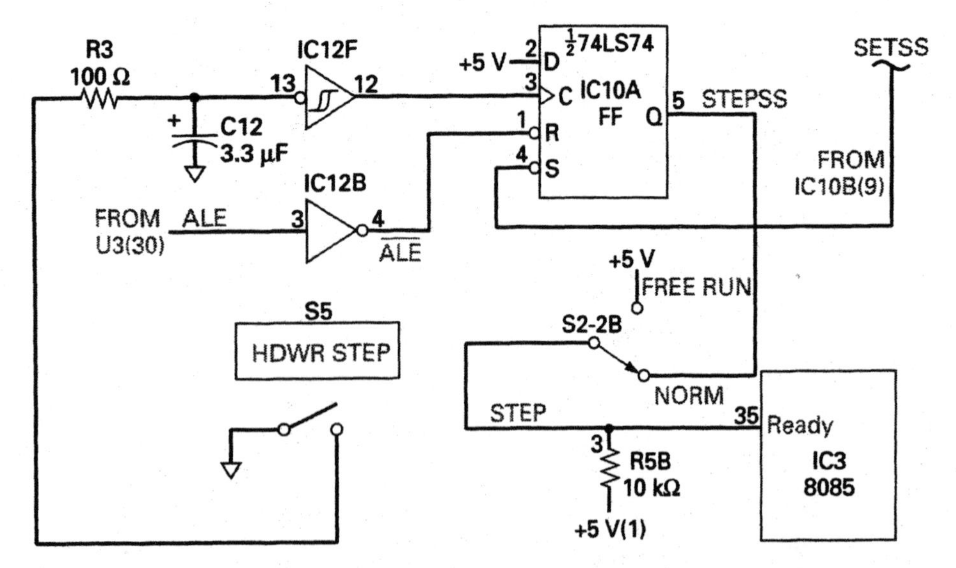

FIGURE 17-19 SAM's Single-Step Circuit (Advances Microprocessor One Machine Cycle).

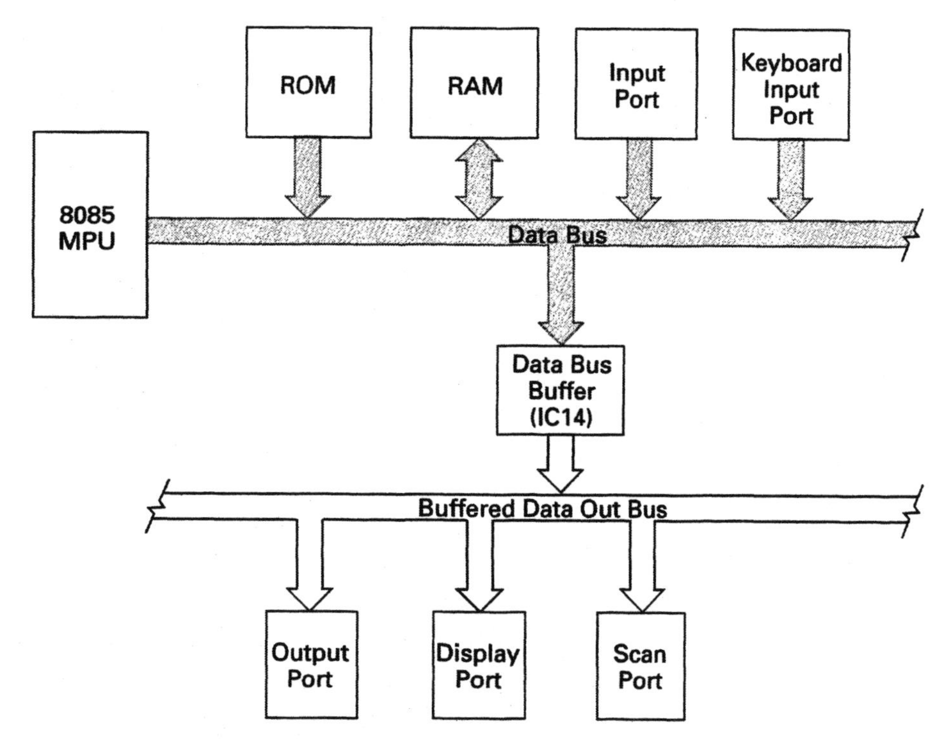

FIGURE 17-20 SAM's Data Bus Buffer.

nal address decoders to select the specified memory location. The data must then be stable for some time before the write pulse (called setup time) and after the write pulse (called hold time).

Referring to the read timing waveforms shown in Figure 17-21(b), you can see that, as with the write operation, the address must be stable for some time to allow the memory's internal decoders to settle. A read pulse is then generated, and after a short time (equal to the memory IC's access time), the accessed data will be placed on the data bus. This data must be allowed to stabilize (setup time) before the data is read into the microprocessor on the rising edge of the NRD pulse, and the data must also be stable for some time after the read pulse (hold time).

Figure 17-22(a) shows the microprocessor timing for a typical instruction. In microprocessor-based systems, the basic unit of time is called the *state,* or timing state (T-state), and is equal to one clock period. The 8085 microprocessor in SAM generates a 2 MHz system clock signal, and so each T-state is equal to 500 ns (1/2 MHz = 500 ns). A **machine cycle** consists of between three and six T-states, and it is the time it takes the machine or microprocessor to complete an operation. An **instruction cycle** is the time it takes a microprocessor to fetch and execute a complete instruction. Most simple instructions have an instruction cycle that has only one machine cycle, whereas more complex instructions may consist of five machine cycles.

Machine Cycle
Basic period of time required to manipulate data in a system.

Instruction Cycle
All of the machine cycles necessary to fully execute an instruction.

As an example, Figure 17-22(b) shows the complete system timing for the two-byte OUT instruction. Timing states in this diagram use the prefix T (T_1, T_2, and so on), and machine cycles use the prefix M (M_1, M_2, and so on). The

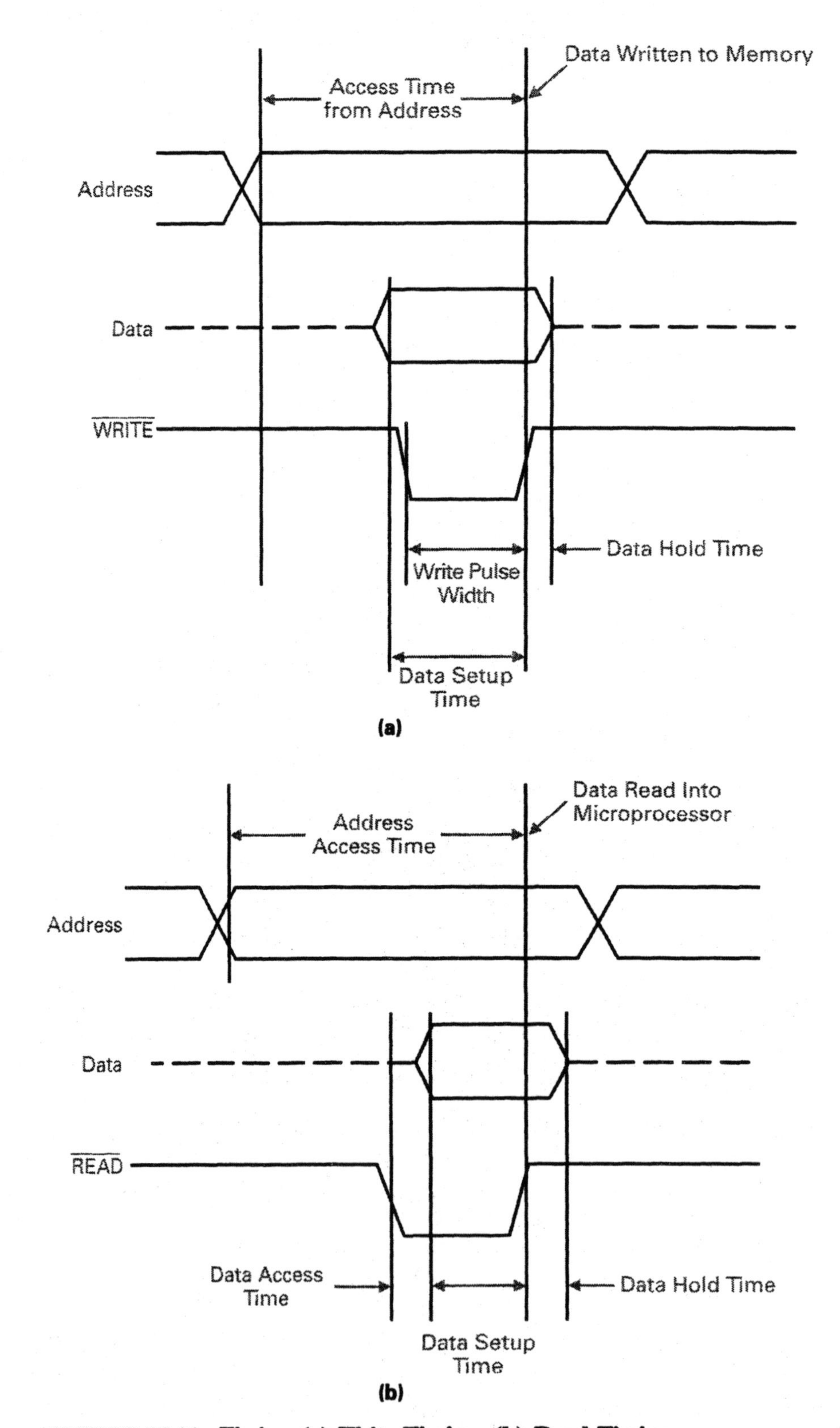

FIGURE 17-21 Timing. (a) Write Timing. (b) Read Timing.

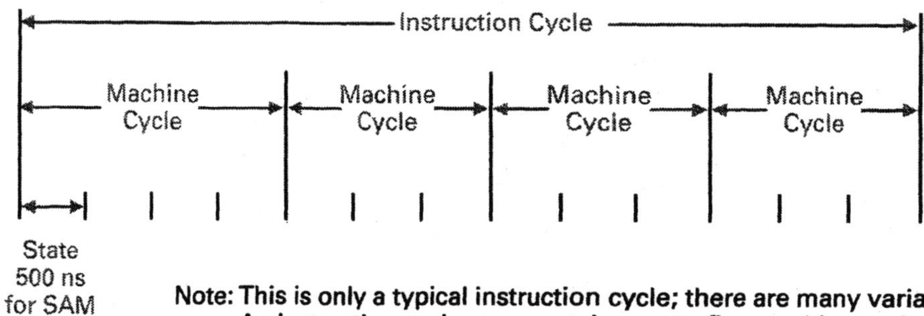

Note: This is only a typical instruction cycle; there are many variations.
An instruction cycle may contain one to five machine cycles.
A machine cycle may contain three to six states.

(a)

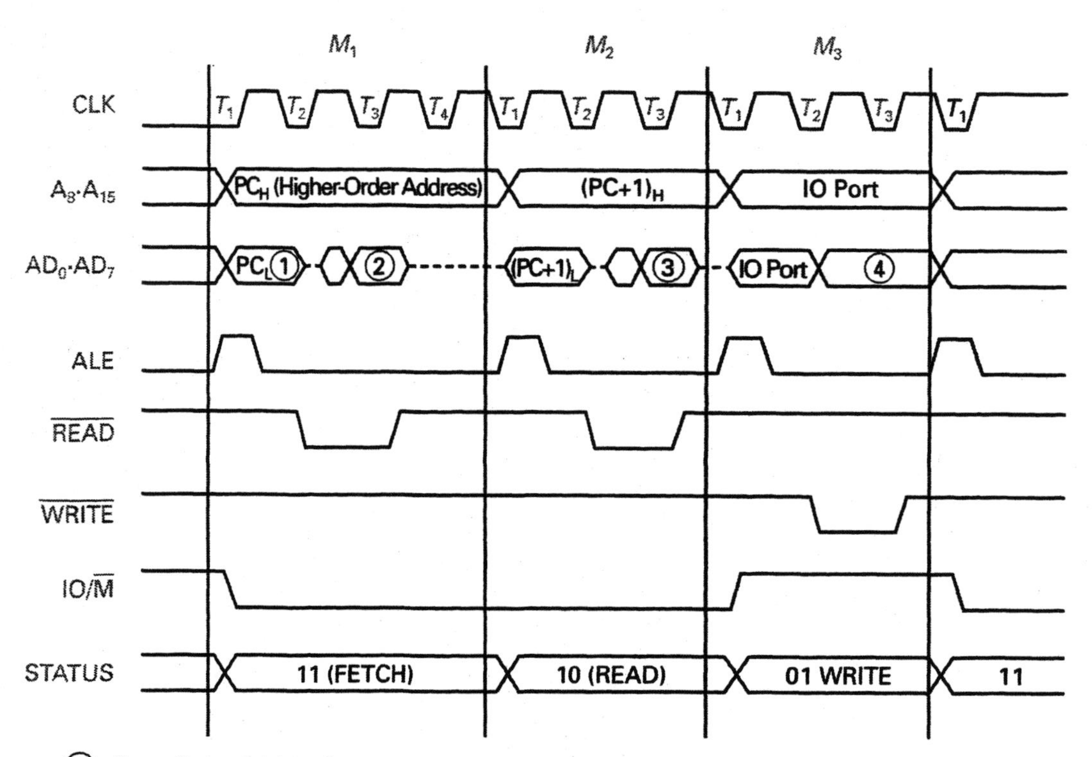

① (Low-Order Address)

② Data from Memory (Instruction)

③ Data from Memory (I/O Port Address)

④ Data to Memory or Peripheral

(b)

FIGURE 17-22 Instruction Cycle. (a) Summary of Microprocessor Timing. (b) Fetch and Execution of an OUT Instruction.

first byte of this two-byte instruction (the opcode) tells the microprocessor to send the contents of the accumulator to the output port specified in the second byte of the instruction. In the first machine cycle (M_1), the opcode is fetched from memory. In the second byte of the instruction (M_2), the output port address is read from memory. In the third machine cycle (M_3), the instruction is executed and the contents of the accumulator is written to the output port.

All instructions require one machine cycle to fetch the opcode. Single-byte simple instructions (such as the MOV A,B instruction, which transfers the data from one of the microprocessor's internal registers to another) can be performed in only one machine cycle. On the other hand, more complex multiple-byte instructions (such as STA 0837, which tells the microprocessor to store the contents of the accumulator at address 0837) will require four machine cycles since three bytes need to be retrieved from memory before the instruction can be executed.

Input switches. Figure 17-23 shows SAM's eight-switch input circuit (S3 and U13 on the SAM schematic). The switches at the inputs of the input port will cause the input to be LOW if the switch is closed. The input port's three-state buffer will only be enabled when both the NRD and the NIN control lines are active.

Output LEDs. Figure 17-24 shows SAM's eight-LED output circuit (DS12 through DS19—U15 on the SAM schematic). Since the LEDs form a common-anode display, a LOW stored in the output register will turn ON an LED. The output port's register is clocked by the NOUT control line.

Input keyboard and output display. Figure 17-25 shows a simplified diagram of SAM's keyboard circuit (S4 through S29, U17, and U18B on the SAM schematic). If we were to connect each of the twenty-six keys to its own input port bit, we would need to have four 8-bit input ports ($4 \times 8 = 32$). Using a multiplexed keyboard, however, we can interface up to 256 keys using only two ports. A multiplexed keyboard has the keys arranged in a matrix that has intersecting column lines and row lines. An output port (U17, address 2800) drives

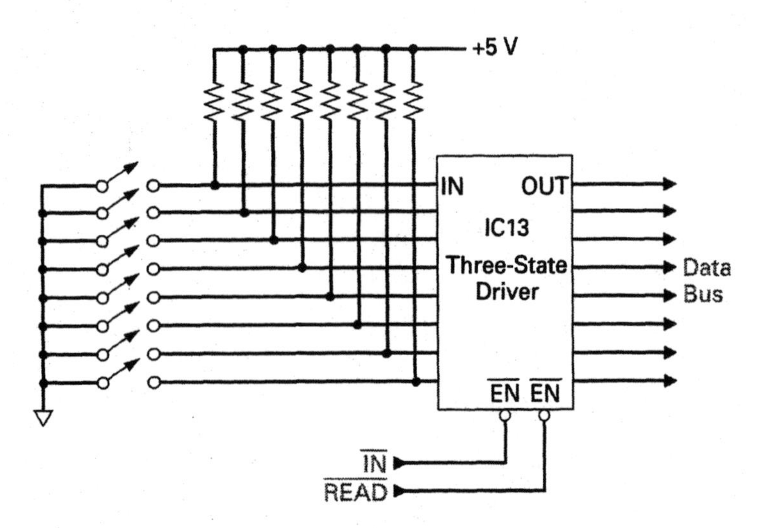

FIGURE 17-23 SAM's Eight-Switch Input Port.

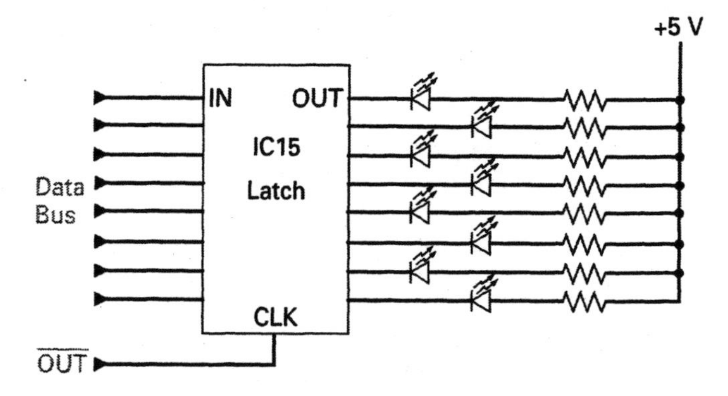

FIGURE 17-24 SAM's Eight-LED Output Port.

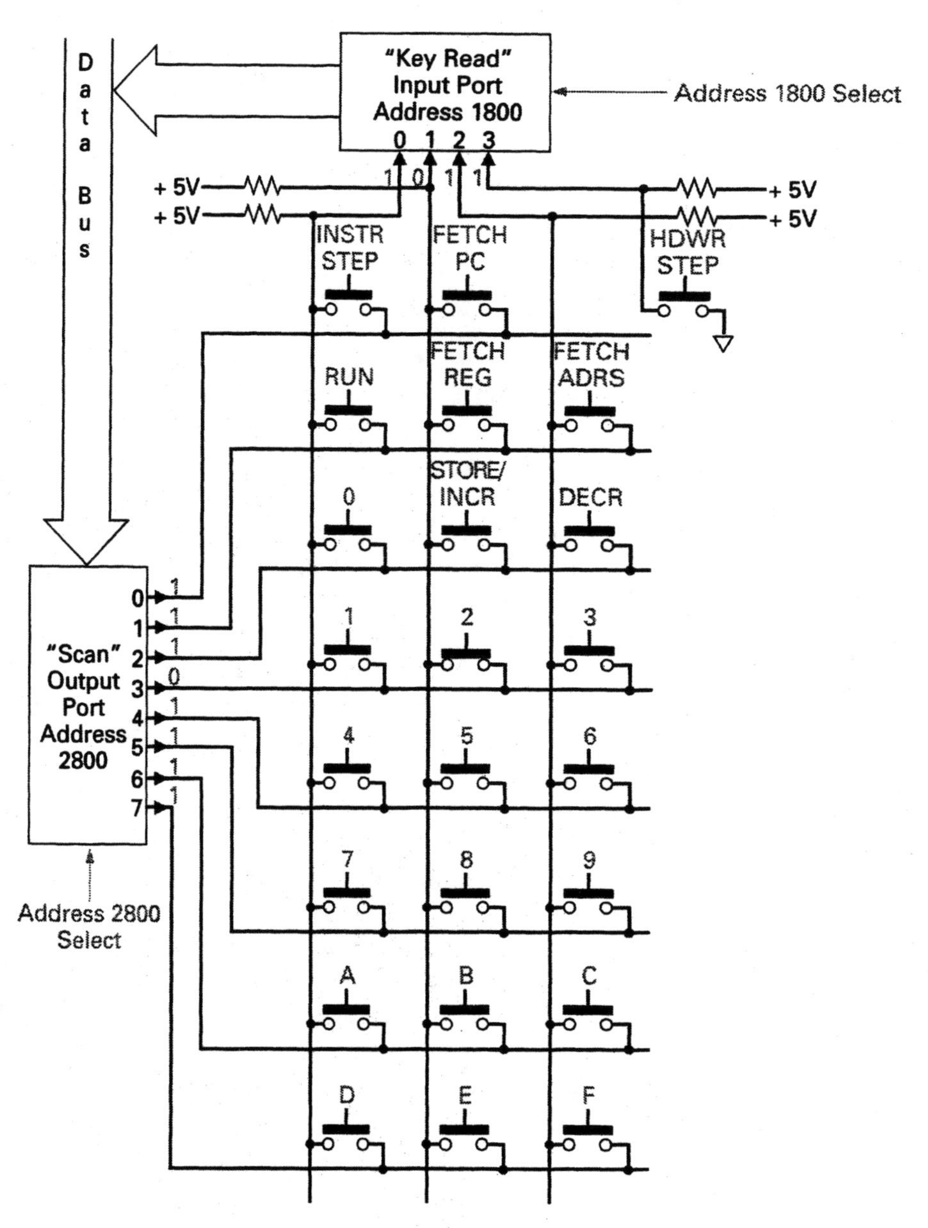

FIGURE 17-25 SAM's Keyboard Interface.

the rows, and an input port (U18B, address 1800) reads the columns. The SAM monitor program scans the keyboard by making LOW only one row at a time, and then reading in the contents of the input port. If any of the keys are pressed in the row that is LOW, then the column line which that key is on will be forced LOW, and so that input port bit will be LOW. If no keys are pressed, the input port lines will all be pulled up to a HIGH level. The monitor program knows which row the activated key is in based on which output port bit is set LOW. It also knows which column the activated key is in based on which port bit is LOW (as seen in the example in Figure 17-25 in which key 2 has been pressed). To check all the keys, each output port bit must be set LOW in turn or scanned. At each step, only four keys are checked, and the process is so fast that the entire keyboard can be checked in less time than it takes a person to do the fastest possible press of a key.

Figure 17-26 shows SAM's display circuit (DS9 through DS11, U19, U17, and U20 on the SAM schematic). Like the keyboard, the seven-segment display also uses a scanning technique. The SAM system has six seven-segment display digits; however, only one display is on at any instant. They are each turned on in sequence, but this happens so fast that they appear to all be on at the same time. A character is displayed by putting a HIGH level on the common connection, and a LOW on the segment inputs based on which of the segments we wish to turn ON. In this multiplexed display, one 8-bit output port (U17 and U20, address 2800) is used to select one of the six digits, and another output port (U19, address 3800) supplies the segment information to the selected digit. A control program within the SAM monitor program can be run to control the display and display data stored in a set of six preselected RAM addresses. The control program operates in the following way. First, the segment information for digit 1 is sent to the segment port and the digit port is loaded with a value that will activate only digit 1. Then, the segment information for digit 2 is sent to the segment port and the digit port is loaded with a value that will activate only digit 2. Then, the segment information for digit 3 is sent to the segment port and the digit port is loaded with a value that will activate only digit 3. This continues until all six digits have been driven with segment data. The program then jumps back to its start point and the process is repeated.

17-2-2 Troubleshooting Microprocessor Systems

Although the troubleshooting of microprocessor-based systems is very similar to standard digital logic circuits, there are several testing procedures and problems that are unique to microcomputer systems. For one thing, most of the control is in the software, so signal flow is hard to trace. Another difficulty is that since the microprocessor cannot be stopped, measurements must be taken while the processor is running. This requirement reduces the effectiveness of the logic probe and logic pulser but enhances the usefulness of the current tracer, oscilloscope, and two specialized test instruments called a signature analyzer and a logic analyzer. Before we begin our three-step troubleshooting procedure, let us see how these two microprocessor analyzers operate.

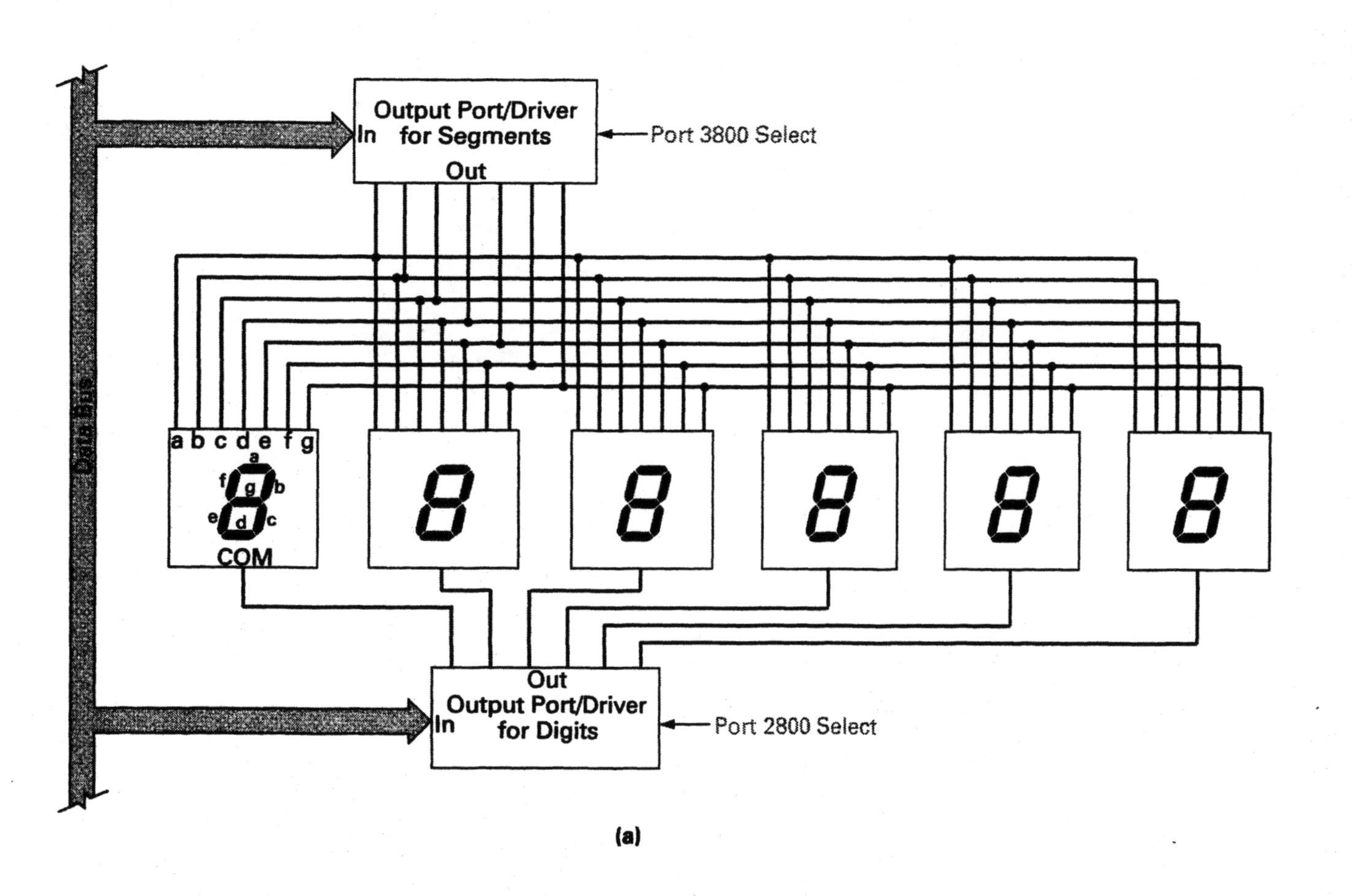

(a)

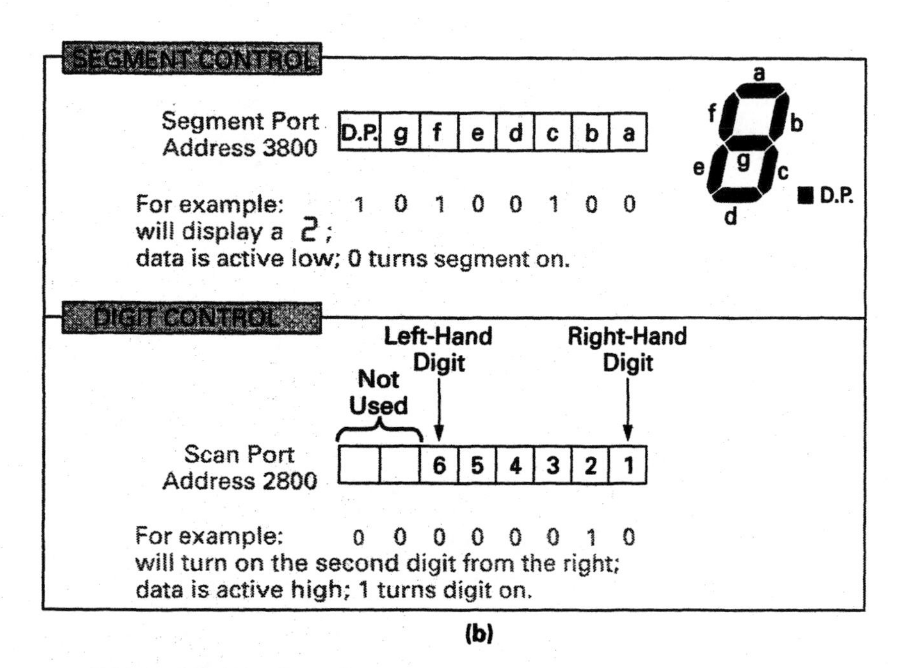

(b)

FIGURE 17-26 SAM's Display Interface.

The Signature Analyzer

Signature Analyzer Instrument used to convert the long, complex serial data streams present on microprocessor system nodes into four-digit signatures.

The **signature analyzer** (SA) is an easy-to-use and highly accurate test instrument that converts the long, complex serial data streams on microprocessor system logic nodes (junction points) into 4-digit signatures. Figure 17-27 shows a typical signature analyzer. The character set used for signature analysis contains sixteen characters: 0, 1, 2, 3, 4, 5, 6, 7, 8, 9, A, C, F, H, P, and U. These characters were chosen so that when they are displayed on the signature analyzer's seven-segment display, they can easily be distinguished from one another.

The signature analyzer has DATA, START, STOP, and CLOCK signal inputs. The DATA input receives the data from the node being tested. The START signal input is connected to the system being tested and tells the signature analyzer when to begin looking at the data, and the STOP signal input tells it when to stop. Between the START and STOP signals, data is processed every time a new CLOCK input occurs. The connection for these inputs is usually specified in the system technical manual.

To test the SAM system using a signature analyzer, the following steps would be followed.

a. Turn OFF the SAM system and perform the following.

b. The next step is to get the SAM system to free run. This is achieved by changing the position of the switches shown in Figure 17-28(a), so that the eight data bus switches are opened, and the free-run switch is in the up position. Opening the data bus in this way isolates the microprocessor from the rest of the system, and therefore no instructions can be sent to the microprocessor. The free-run switch hard-wires the data bus with 0111 1111 whenever a read operation is performed (due to the pull-up

FIGURE 17-27 A Signature Analyzer. (Courtesy of Hewlett-Packard Company)

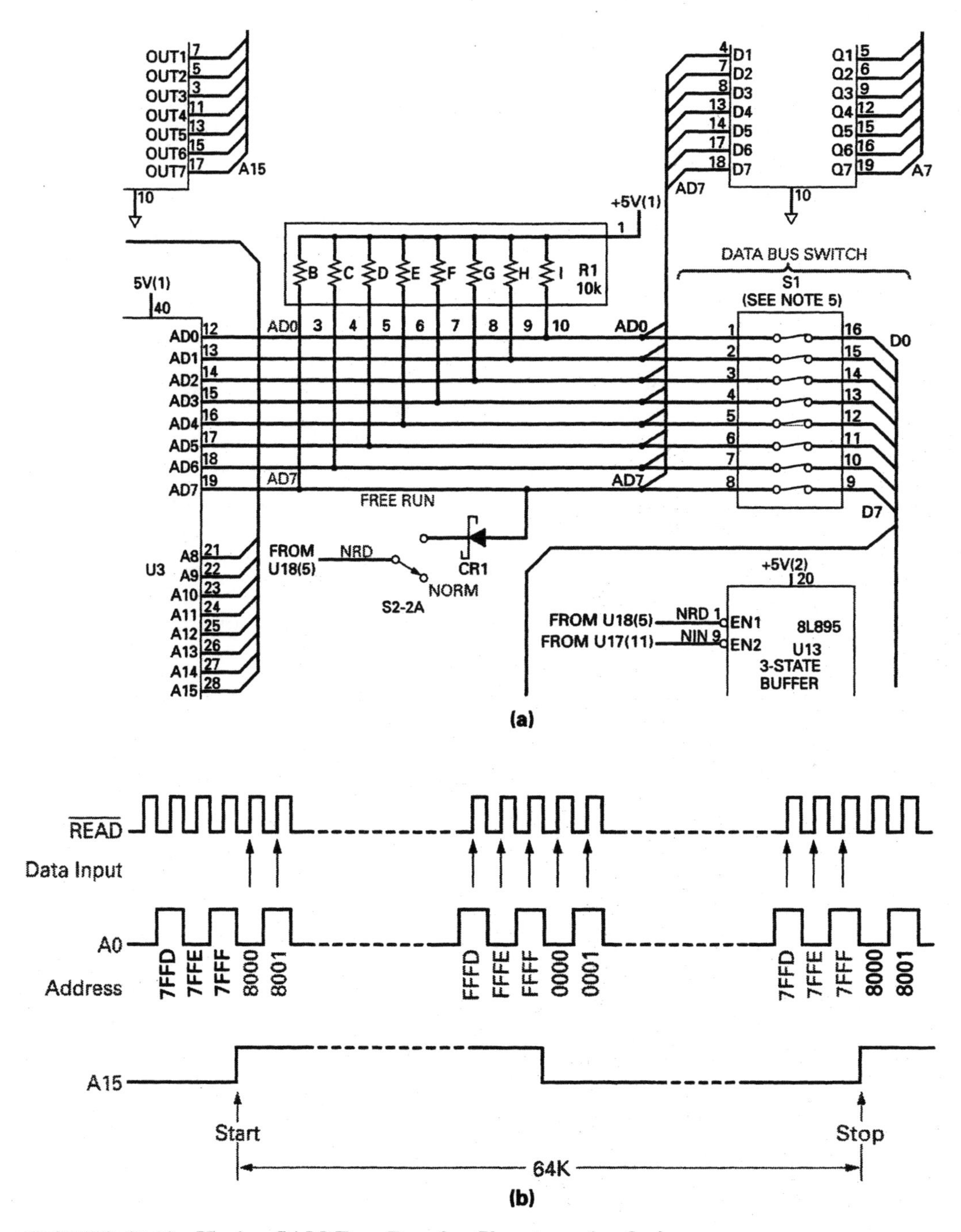

FIGURE 17-28 Having SAM Free-Run for Signature Analysis.

resistors and the NRD signal). This means that as the microprocessor steps through every address from 0000 to FFFF, it will read in the opcode 7F at every one of the addresses, as shown in Figure 17-28(b). Since this is the opcode for MOV A,A (move the contents of the accumulator into the accumulator), the microprocessor will basically do nothing but address the entire system (decoders, ROM, RAM, and all the I/O ports), and then repeat the process.

c. Connect the signature analyzer's test leads to the SAM system. Clip the GND lead of the SA to SAM's ground, the START lead and STOP leads of the SA to SAM's A15 address line, and the CLOCK lead of the SA to SAM's NRD line.

d. Turn ON the SAM system and use the DATA probe of the signature analyzer to check the signatures at different nodes. Figure 17-29 shows a sample of the signature analysis table for the SAM system when it is in the free-run test loop.

Node signatures are meaningful because they are generated by a routine that exercises portions of the circuit in a controlled, repeatable manner. The correct signatures are provided by the system manufacturer and reflect signatures gathered from a known good product. Remember that there is no such thing as an "almost" right signature. Signatures are either right or wrong, and a signature of 5H22 is no more related to 5H23 than to C7FP.

As well as the free-run test loop, the SAM system also has a signature analysis test loop. The SA test loop is a program permanently stored in SAM's ROM for the sole purpose of providing a more detailed, controlled stimulus for signature analysis. The signature analysis switch (shown previously in Figure 17-18) is connected to an interrupt line on the microprocessor, and when closed will cause the microprocessor to run the SA test loop program. This program provides stimulus signals to the devices that the microprocessor talks to, such as the scan and display ports, the speaker, and the control logic. This program or routine also reads the devices that talk to the microprocessor such as the input port switches and the keyboard. This simple repeating program, therefore, exercises and stimulates the system's devices so that their repeating patterns can be monitored by a signature analyzer.

Having two test loops (the free-run and SA test loops) for the SAM system enables us to apply the half-split method. When the SA test loop program runs, it exercises the entire system to as great an extent as possible. If a fault exists in the system and the SA test loop program cannot operate, the solution is to free-run the processor. The decoders and the memories do not need to be functional for this test loop to work, only the processor itself and its immediate circuitry needs to be operating. So even if the processor refuses to free-run, the problem has been narrowed down significantly.

The Logic Analyzer

Logic Analyzer
Test system capable of displaying 0's and 1's, as well as performing complex test functions. Logic analyzers typically have 16 to 32 input lines and can store sequences of sixteen or more bits on each of the input lines.

Figure 17-30 shows a **logic analyzer** connected to a personal computer. The logic analyzer is a specialized instrument used for examining a number of logic signals or data sequences present in microprocessor systems. A logic analyzer is similar to an oscilloscope in many ways since it has signal inputs, a trigger, timing circuitry, and a CRT (cathode ray tube) display. Its signal inputs, however, differ from those of an oscilloscope in that it can monitor up to sixty-four threshold-sensitive logic inputs (1s and 0s). As well as a display of multiple waveforms, the logic analyzer can also display a tabular listing of digital data, as shown in the examples in Figure 17-31.

SA SWITCHES	SA CONNECTIONS
START ⎍	A15
STOP ⎍	A15
CLOCK ⎍	READ

VCC SIGNATURE 0001

SA FR

NORM

FREERUN

NORM

BUS SWITCH

SAM SWITCHES

LOGIC 1

LOGIC 0

7 6 5 4 3 2 1 0

INPUT SWITCHES

SIGNATURES

Data Lines		Address Lines			
D0	X	A0	UUUU	A8	HC89
D1	X	A1	5555	A9	2H70
D2	X	A2	CCCC	A10	HPP0
D3	X	A3	7F7F	A11	1293
D4	X	A4	5H21	A12	HAP7
D5	X	A5	0AFA	A13	3C96
D6	X	A6	UPFH	A14	3827
D7	X	A7	52F8	A15	755P

U1

Signature	Pin	Pin	Signature
GND	1	20	VCC
1293	2	19	0000
1293	3	18	755P
HPP0	4	17	755P
HPP0	5	16	3827
2H70	6	15	3827
2H70	7	14	3C96
HC89	8	13	3C96
HC89	9	12	HAP7
GND	10	11	HAP7

U2

Signature	Pin	Pin	Signature
0000	1	20	VCC
UUUU	2	19	52F8
0001-B	3	18	0000-B
0001-B	4	17	0001-B
5555	5	16	UPFH
CCCC	6	15	0AFA
0001-B	7	14	0001-B
0001-B	8	13	0001-B
7F7F	9	12	5H21
GND	10	11	0001-B

U3

Signature	Pin	Pin	Signature
X	1	40	VCC
X	2	39	0000
0000	3	38	0000
0000 or 0001	4	37	0001-B
0000	5	36	0001
0000	6	35	0001
0000	7	34	0000
0000	8	33	0001
0000	9	32	0001-B
0000	10	31	0001
0001	11	30	0000-B
0001-B	12	29	0001
0001-B	13	28	755P
0001-B	14	27	3827
0001-B	15	26	3C96
0001-B	16	25	HAP7
0001-B	17	24	1293
0001-B	18	23	HPP0
0000-B	19	22	2H70
GND	20	21	HC89

U4

Signature	Pin	Pin	Signature
52F8	1	24	VCC
UPFH	2	23	HC89
0AFA	3	22	2H70
5H21	4	21	0001
7F7F	5	20	3PCF
CCCC	6	19	HPP0
5555	7	18	0000-B
UUUU	8	17	X
X	9	16	X
X	10	15	X
X	11	14	X
GND	12	13	X

U5

Signature	Pin	Pin	Signature
UPFH	1	18	VCC
0AFA	2	17	52F8
5H21	3	16	HC89
7F7F	4	15	2H70
UUUU	5	14	X
5555	6	13	X
CCCC	7	12	X
84AF	8	11	X
GND	9	10	0001

U6

Signature	Pin	Pin	Signature
UPFH	1	18	VCC
0AFA	2	17	52F8
5H21	3	16	HC89
7F7F	4	15	2H70
UUUU	5	14	X
5555	6	13	X
CCCC	7	12	X
84AF	8	11	X
GND	9	10	0001

U7

Signature	Pin	Pin	Signature
1293	1	16	VCC
HAP7	2	15	3PCF
3C96	3	14	84AF
3827	4	13	960F
755P	5	12	4154
0001-B	6	11	UA87
1920	7	10	597C
GND	8	9	C34C

U8

Signature	Pin	Pin	Signature
0001	1	16	VCC
0000	2	15	F770
0001	3	14	F771
GND	4	13	CCCC
UUUU	5	12	5555
8HUC	6	11	U6AH
8HUA	7	10	U6AF
GND	8	9	960F

X = Don't Care Signature B = Blinking GND or VCC Signature

FIGURE 17-29 A Signature Analysis Table.

FIGURE 17-30 A Logic Analyzer Connected to a Personal Computer. (Courtesy of Hewlett-Packard Company)

Troubleshooting the SAM System

To review, the procedure for fixing a failure can be broken down into three steps—diagnose, isolate, and repair. Since this chapter has been devoted to microprocessor-based systems, we will be applying our three-step troubleshooting procedure to the SAM system.

Step 1: DIAGNOSE

As before, it is extremely important that you first understand the operation of a circuit, and how all of the devices within it are supposed to work, so that you are able to determine whether or not a circuit malfunction really exists. If you were preparing to troubleshoot the SAM system, your first step should be to study the block diagram, circuit description, and schematic, and review the operation of each integrated circuit until you feel completely confident with the correct operation of the entire circuit. Once you are fully familiar with the operation of the circuit, you will easily be able to diagnose the problem as either an operator error or a circuit malfunction. Few things are more frustrating than trying to fix something that is not broken.

A great deal of diagnostic information can often be obtained from a system without removing the product's covers. Use the input devices (such as the keyboard) and watch the output devices (such as the display) to see what

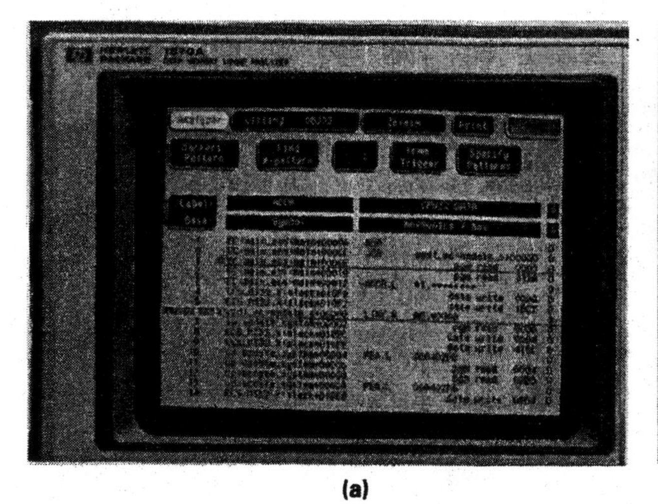

(a)

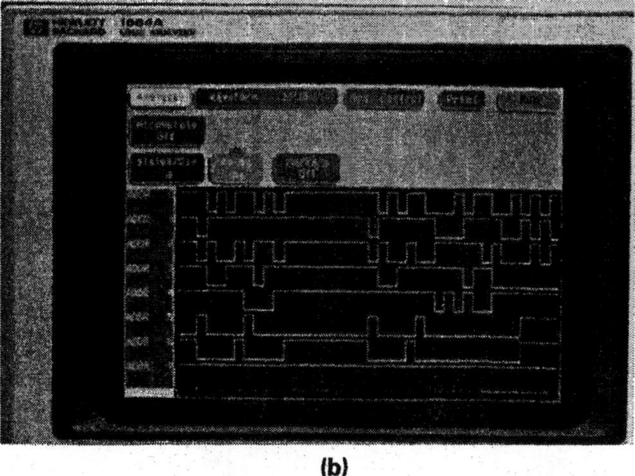

(b)

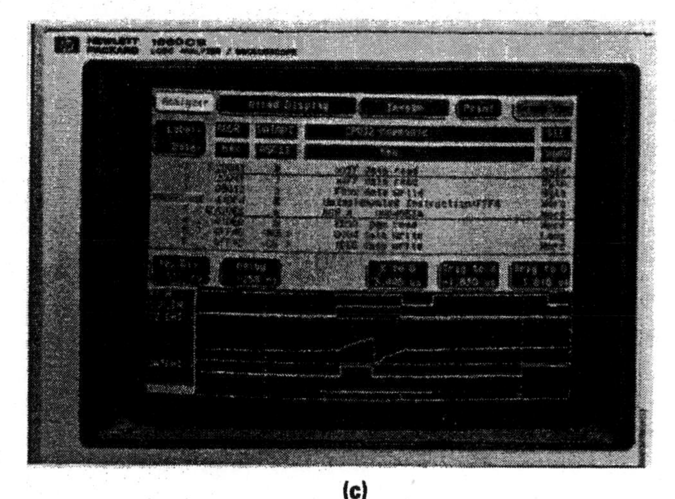

(c)

FIGURE 17-31 Logic Analyzer Display Examples. (a) Data in Tabular Form So That Program Sequence Can Be Followed. (b) Address Lines Displayed in Timing Analysis Mode. (c) Mixed Display Showing Address, Data, and Mnemonics, along with Waveform Timing. (Courtesy of Hewlett-Packard Company)

responses you get. For example, if the system seems to be dead when the power is turned on, the problem is more than likely power. But if a certain part of the display is dead, the problem is probably in the display drive circuit or display. If the system is operating to some extent, take advantage of the built-in self-diagnostic programs detailed in the tech manual.

Step 2: ISOLATE

No matter what circuit or system failure has occurred, always follow a logical and sequential troubleshooting procedure. A troubleshooting tree is a useful guide for directing the sequence of tests you should follow to locate a system fault. Most system manufacturers include these charts in their technical manuals. As an example, let us step through the troubleshooting flowchart for the SAM system, which is shown in Figure 17-32.

Power-Up Does the SAM system power up when power is first turned ON? This can be checked by looking at a system's display, which normally comes up with a start-up message (micro-lab up). If this message appears, you can assume that all is reasonably well since the SAM system has executed the start-up program and is able to control the display (microprocessor,

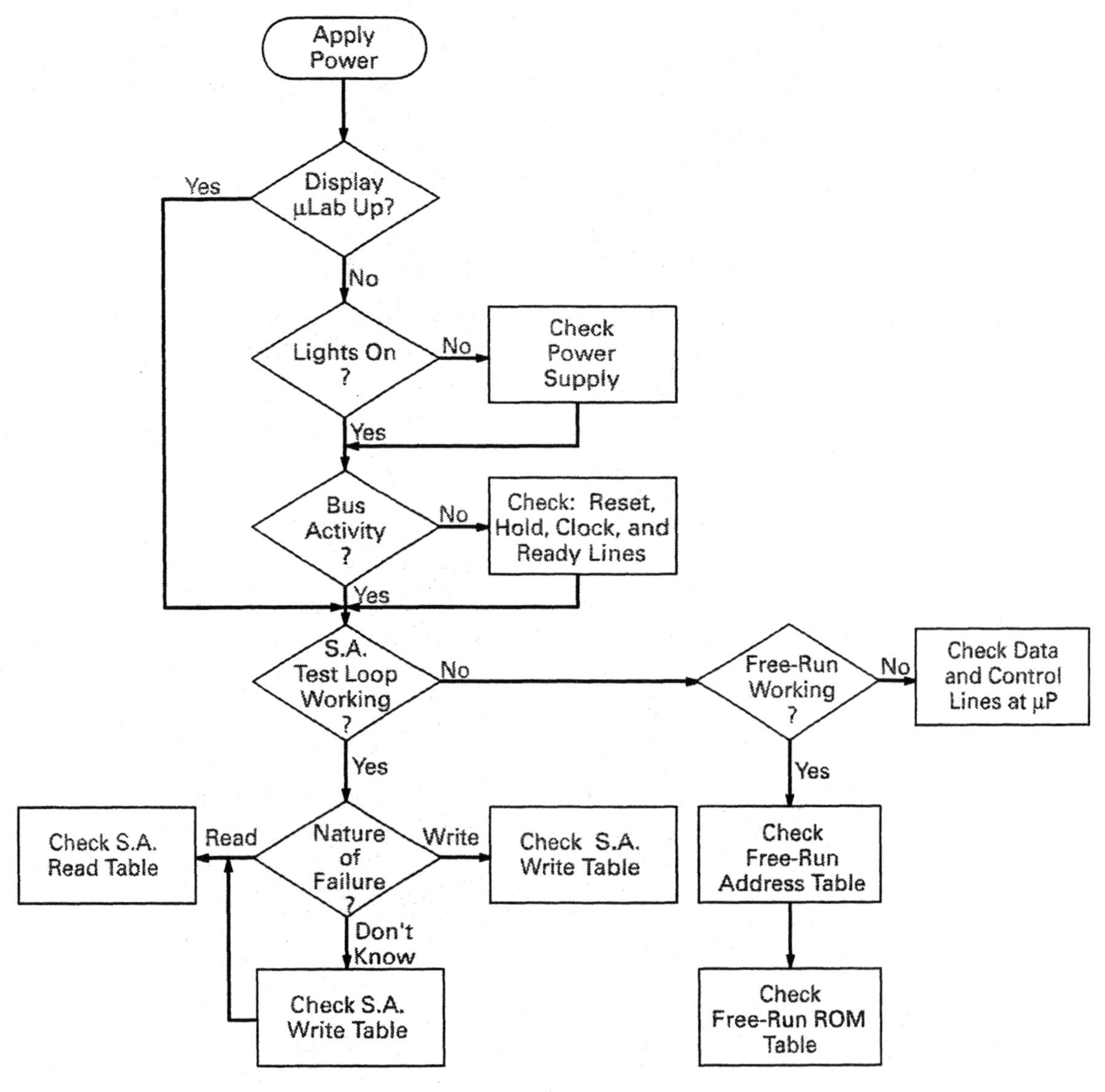

FIGURE 17-32 Troubleshooting Flowchart for the SAM System.

ROM, RAM, decoding, control circuits, and buses must be working to some degree).

Lights On If the SAM system fails to power up and none of the LED indicators light, the power supply or primary source should be suspected.

Bus Activity If the power supply is good, then the next logical test is to determine the presence of activity on the data and address bus lines using a logic probe. A lack of activity would explain the failure to power up and could be caused by a faulty clock or an incorrect logic state could be present on the reset, hold, or ready inputs to the microprocessor. A faulty ROM could also cause a halt instruction to be executed, which would cause all bus activity to stop. If these lines are good, continue on to the next step.

SA Test Loop Working This point in the procedure is basically a half-split. The signature analysis test loop requires more of the circuit to operate than the free-run test loop. Try to get the SAM system into the SA test loop by pressing the RESET key and then sliding the SA switch up and down once. If it is working, all of the output LEDs and display segments will light and the speaker will beep once. If the SA test loop runs, you know that the essential portions of the system are operating satisfactorily. If only portions of the system seem to be operating correctly, use the signature analyzer to locate faulty signatures and remember that these incorrect signatures often point to the portion of the circuit that has the problem.

Free-Run Test Loop Working When the SA test loop will not run, you need to test smaller portions of the circuit. In the free-run mode, the microprocessor stimulates portions of the circuit through the address bus. This address bus stimulus is unsophisticated (open-loop) compared to the well-controlled data patterns used in the SA test loop; however, the free-run mode does exercise the address bus drive circuits, the decoding and control circuits, and the ROM. The advantage of this test mode is that, in order to use it, little more than the microprocessor chip has to work. If the microprocessor will not free-run, check the reset, hold, ready and interrupt control line inputs, the clock, V_{CC} and ground, and the power-on reset connections to the microprocessor. You should also check that the microprocessor is getting the free-run instruction. If all of these appear to be correct, the microprocessor is probably defective since there are few other factors that could keep a properly powered and controlled microprocessor from running. If the free-run test mode is working (but the SA test mode is not), then you should start testing signatures beginning with those on the address bus lines, and then proceeding to the chip-select pins, address decoder and other control circuits, and the microprocessor.

Step 3: REPAIR

Once the fault has been found, the final step is to repair the circuit and then perform a final test to see that the circuit and the system are again fully operational.

SELF-TEST REVIEW QUESTIONS FOR SECTION 17-2

1. What is the difference between a unidirectional and bidirectional bus?
2. Define the function of the following 8085 microprocessor pins.
   a. NWR **b.** NRD **c.** IO/NM **d.** RESET
   e. READY **f.** ALE **g.** HOLD **h.** TRAP
3. Briefly describe a microprocessor's fetch-execute cycle.
4. What is a microprocessor
   a. interrupt? **b.** T-state?
   c. machine cycle? **d.** instruction cycle?

5. What is a signature analyzer, and how is it used to test a microprocessor-based system?
6. Briefly describe why a logic analyzer is useful in troubleshooting digital electronic systems.

REVIEW QUESTIONS

Multiple-Choice Questions

1. Microprocessor-based systems are more flexible than hard-wired logic designs because ________.
   a. they are faster
   b. they use LSI devices
   c. their operation is controlled by software
   d. the hardware is specialized
2. The peripherals of a microcomputer system are ________.
   a. the memory devices
   b. the microprocessor
   c. the software
   d. the I/O devices
3. The personality of a microprocessor-based system is determined primarily by ________.
   a. the microprocessor used
   b. the program and peripherals
   c. the number of data bus lines
   d. the type of memory ICs used
4. The language that the microprocessor understands directly is called ________.
   a. assembly language
   b. high-level language
   c. English language
   d. machine language
5. The main purpose of a microprocessor's accumulator is ________.
   a. keeping track of the next instruction to be executed
   b. temporary data storage
   c. selecting which interrupts should be enabled
   d. storing opcodes
6. The microprocessor's program counter is used for ________.
   a. keeping track of the next instruction to be executed
   b. temporary data storage
   c. selecting which interrupts should be enabled
   d. storing opcodes
7. Interrupts are used primarily for ________.
   a. breaking a program into modular segments
   b. responding quickly to unpredictable events
   c. speeding up program execution
   d. halting the system

8. When an interrupt occurs, the microprocessor will ___________.

   a. jump to the interrupt service routine
   b. halt until another request is made
   c. continue executing the main program
   d. complete the current program and then stop

9. The microprocessor knows which bytes to interpret as opcodes because ___________.

   a. every byte is an opcode
   b. every third byte is an opcode
   c. each opcode indicates the number of information bytes that follow
   d. the programmer has highlighted the opcodes

10. The purpose of the ALU in a microprocessor is to ___________.

    a. interpret the opcodes
    b. perform arithmetic and logic operations
    c. control the address bus
    d. calculate the number of machine cycles required

11. In a microprocessor system, the data bus is ___________.

    a. unidirectional and three-state
    b. bidirectional and three-state
    c. unidirectional and bidirectional
    d. all of the above are true

12. In a system using several 1 kB memory devices, which address lines are connected to the memory chips?

    a. A0–A9 b. A10–A15 c. A0–A15 d. A0–A7

13. In a system with a 16-bit address bus, what is the maximum number of 2 kByte memory devices it could contain?

    a. 16 b. 32 c. 64 d. 128

14. Memory mapped input/output ___________.

    a. treats I/O ports as memory locations
    b. tends to be more wasteful of address space than I/O mapped decoding
    c. is used in the SAM system
    d. is all of the above

15. A bus conflict will occur when ___________.

    a. an output port is enabled by a NRD signal
    b. more than one device is reading from the data bus
    c. more than one device is writing to the data bus
    d. MOS and TTL devices are both connected to a data bus

16. The main advantage of keypad and display scanning is ___________.

    a. faster I/O operations
    b. a reduction in software costs
    c. a reduction in hardware costs
    d. all of the above

17. The primary purpose of assigning priorities to interrupt lines is to ______________.

   a. select the interrupt routine address
   b. determine which interrupts are most often used
   c. specify which interrupt is to be selected when more than one occurs
   d. all of the above

18. In the free-run test mode, microprocessor circuit stimulus patterns are generated by ______________.

   a. a program stored in the ROM
   b. a program stored in the RAM
   c. random action of the circuits
   d. sequential cycling of the microprocessor

19. Signatures can ______________.

   a. provide information about the nature of the fault
   b. tell whether a node is operating correctly or not
   c. verify the test setup
   d. all of the above

20. Compared to oscilloscopes, logic analyzers have ______________.

   a. multi-line sequential trigger circuits
   b. fast internal clocks and memory
   c. high resolution CRTs
   d. all of the above

Essay Questions

21. What is a microprocessor? (Introduction)
22. What three blocks form a microcomputer? (17-1-1)
23. What is the difference between programming a microcomputer and running a program in a microcomputer? (17-1-1)
24. What is the difference between software and hardware? (17-1)
25. What are peripherals? (17-1-1)
26. What is a flowchart? (17-1-2)
27. Briefly describe each of the following programming languages. (17-1-2)

   a. Machine language **b.** Assembly language **c.** High-level language

28. Define the following terms. (17-1)

   a. Opcode **b.** Instruction **c.** Flag **d.** Accumulator **e.** BASIC **f.** Mnemonic **g.** Program **h.** Peripheral **i.** Flowchart

29. What is the difference between a unidirectional bus and a bidirectional bus? (17-2-1)
30. Describe the function of the address bus, data bus, and control bus. (17-2-1)
31. In relation to buses, what is the difference between a talker and listener? (17-2-1)

32. Briefly describe the microprocessor's fetch-execute cycle. (17-2-1)
33. What is an interrupt? (17-2-1)
34. In relation to microprocessor timing, define the following terms. (17-2-1)
    a. Timing state **b.** Machine cycle **c.** Instruction cycle
35. Why does the 8085 microprocessor have a multiplexed address/data bus? (17-2-1)

Practice Problems

36. Describe the function of the SAM system program shown in Figure 17-33.

To practice your circuit recognition and operation skills, refer to U4 on the SAM schematic and answer the following three questions.

37. What is the size and organization of this IC?
38. How is this IC enabled, and where do the enables originate from?
39. Is this IC a data bus talker or listener or both?

To practice your circuit recognition and operation skills, refer to U5 and U6 on the SAM schematic and answer the following four questions.

40. What type of ICs are these?
41. What is the size and organization of these ICs?
42. How are these ICs enabled, and where do the enables originate from?
43. Are these ICs data bus talkers or listeners or both?

To practice your circuit recognition and operation skills, refer to U3 on the SAM schematic and answer the following seven questions.

44. Where does the signal input to pin 8 originate from?
45. What is connected between pins 1 and 2, and how is this related to the output on pin 37?

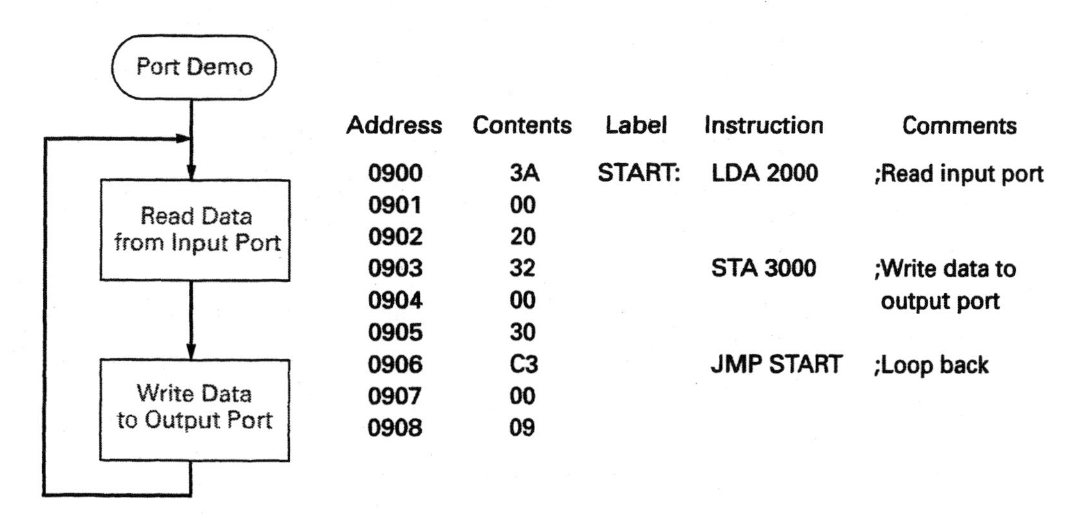

Address	Contents	Label	Instruction	Comments
0900	3A	START:	LDA 2000	;Read input port
0901	00			
0902	20			
0903	32		STA 3000	;Write data to output port
0904	00			
0905	30			
0906	C3		JMP START	;Loop back
0907	00			
0908	09			

FIGURE 17-33 **An Echo Program for the SAM System.**

46. How is the address and data information separated if both appear on pins 12 through 19?
47. Where does the signal output on pin 4 go to?
48. How does the *RC* connected to pin 36 input operate?
49. What is the function of the signal outputs on pins 30, 31, and 32?
50. Are pins 21 through 28 multiplexed?

To practice your circuit recognition and operation skills, refer to the SAM schematic and answer the following ten questions.

51. What is the function of DS1 through DS4?
52. Describe the operation of U7 and how it governs the system memory map.
53. What is the function of DS5 and DS6?
54. Describe the operation of the eight-switch input port.
55. Describe the operation of the eight-LED output port.
56. How are the scanning of the keyboard circuit and the scanning of the seven-segment display circuit related?
57. Describe the operation of the keyboard circuit.
58. Describe the operation of the seven-segment display circuit.
59. What is the function of DS7 and DS8?
60. Why does this circuit have a buffered data bus?

Troubleshooting Questions

61. Briefly describe the operation and function of the following digital test instruments.

    a. Logic probe **b.** Logic pulser **c.** Current tracer
    d. Signature analyzer **e.** Logic analyzer

62. When troubleshooting microprocessor systems, what advantage(s) does the logic analyzer have over the oscilloscope?
63. In regards to the SAM troubleshooting tree shown in Figure 17-32, describe the meaning of the following blocks:

    a. Apply power **b.** Lights on **c.** Bus activity

64. To practice troubleshooting skills, let us introduce some faults into the SAM system, and then use the SAM troubleshooting chart to logically isolate the fault. On the SAM schematic you will find twelve troubleshooting jumper plugs (J_1–J_{12}). One jumper position is normal (as shown in the schematic), and the other position introduces a circuit fault. These fault jumpers simulate real-world circuit malfunctions as shorted and open traces, stuck and open outputs, and functional failures within the ICs. Studying your SAM schematic, describe the fault, the symptoms you would expect, and the sequence of tests you would make to lead to the problem.
65. What is the purpose of a test loop, and what is the difference between a free-run and program test loop?

SAM Schematic Diagram (pages 689–692)

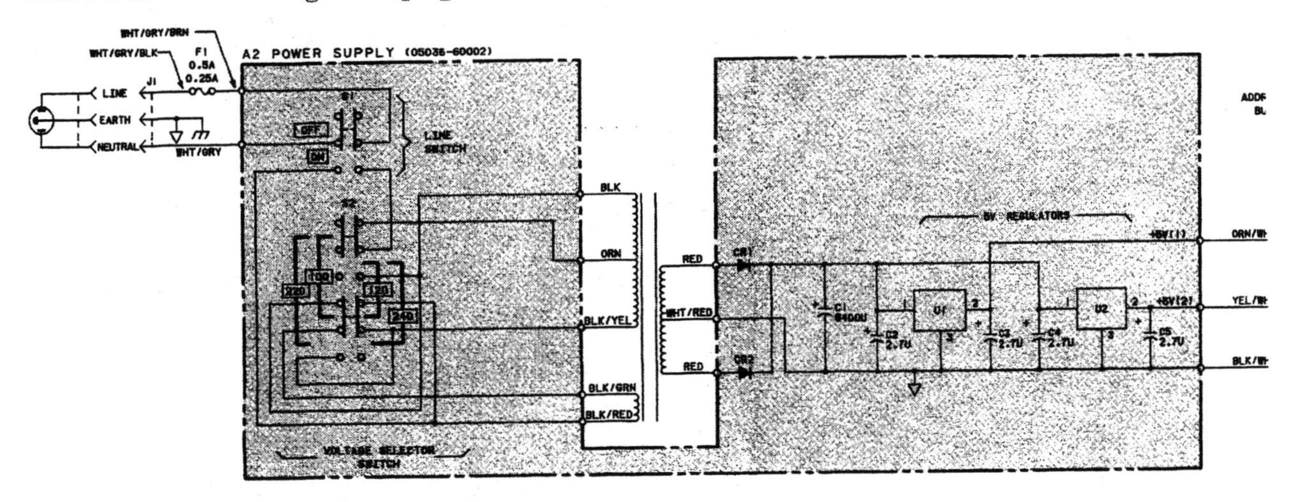

REFERENCE DESIGNATIONS
A1
C1-C16 CR1, CR2 DS1-DS19 LS1 R1-R8 S1-S29 U1-U20 W1-W14 XU4 Y1
A2
C1-C5 CR1, CR2 S1, S2 U1, U2

TABLE OF ACTIVE ELEMENTS

A1		
Reference Designation	HP Part Number	Mfg or Industry Part Number
CR1	1901-0518	1901-0518
CR2	1901-0731	1901-0731
DS1-DS6, DS8	1990-0652	HLMP-6220:1X4:
DS7A, DS7B	1990-0685	1990-0685
DS9-DS11	1990-0667	1990-0667
DS12, DS16	1990-0673	5082-4690
DS13, DS15, DS17, DS19	1990-0675	5082-4590
DS14, DS18	1990-0674	5082-4990
U1, U13, U14	1820-1794	DM81LS95N
U2	1820-1997	SN74LS374PC
U3	1820-2074	P8085
U4	1818-0773	1818-0773
U5, U6	1818-0436	P2114
U7	1820-1216	SN74LS138N
U8	1820-1195	SN74LS175N
U9	1820-1197	SN74LS00N
U10	1820-1112	SN74LS74N
U11	1820-1208	SN74LS32N
U12	1820-1416	SN74LS14N
U15, U16, U17	1820-1730	SN74LS273N
U18	1820-1759	DM81LS97N
U19	1820-2138	DS8871N
U20	1820-1231	SN75492N
Y1	0410-1142	0410-1142
A2		
CR1, CR2	1901-0662	MR751
U1, U2	1826-0122	7805UC

NOTES

1. THE TRACE BETWEEN THE TERMINALS AT J1,J2,J3 AND J4 MUST BE CUT IF THESE LINES ARE TO BE USED WITH PERIPHERALS.
2. S3 SWITCHES ARE SHOWN IN CLOSED (LOGIC Ø) POSITION.
3. S1 SWITCHES ARE SHOWN IN CLOSED (NORM) POSITION. ALL SECTIONS OF S1 MUST BE OPEN FOR FREERUN MODE.
4. SIGNAL MNEMONICS PRECEDED BY THE LETTER 'N' INDICATE LOW POLARITY.

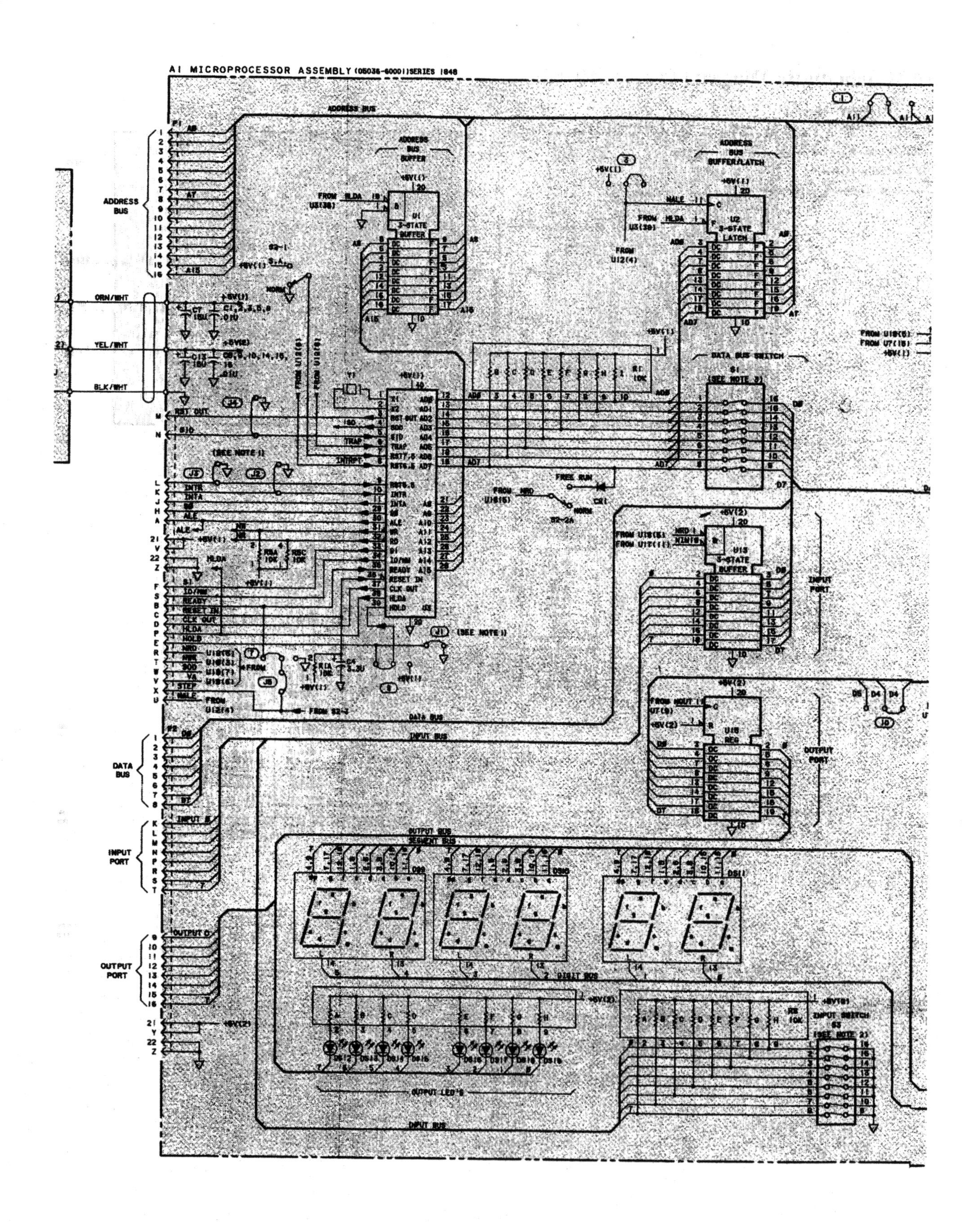
A1 MICROPROCESSOR ASSEMBLY (05036-60001) SERIES 1848
ADDRESS BUS
DATA BUS
INPUT PORT
OUTPUT PORT
ADDRESS BUS BUFFER
ADDRESS BUS BUFFER/LATCH
DATA BUS SWITCH
INPUT BUS
OUTPUT BUS
SEGMENT BUS
DIGIT BUS
OUTPUT LED'S
INPUT SWITCH

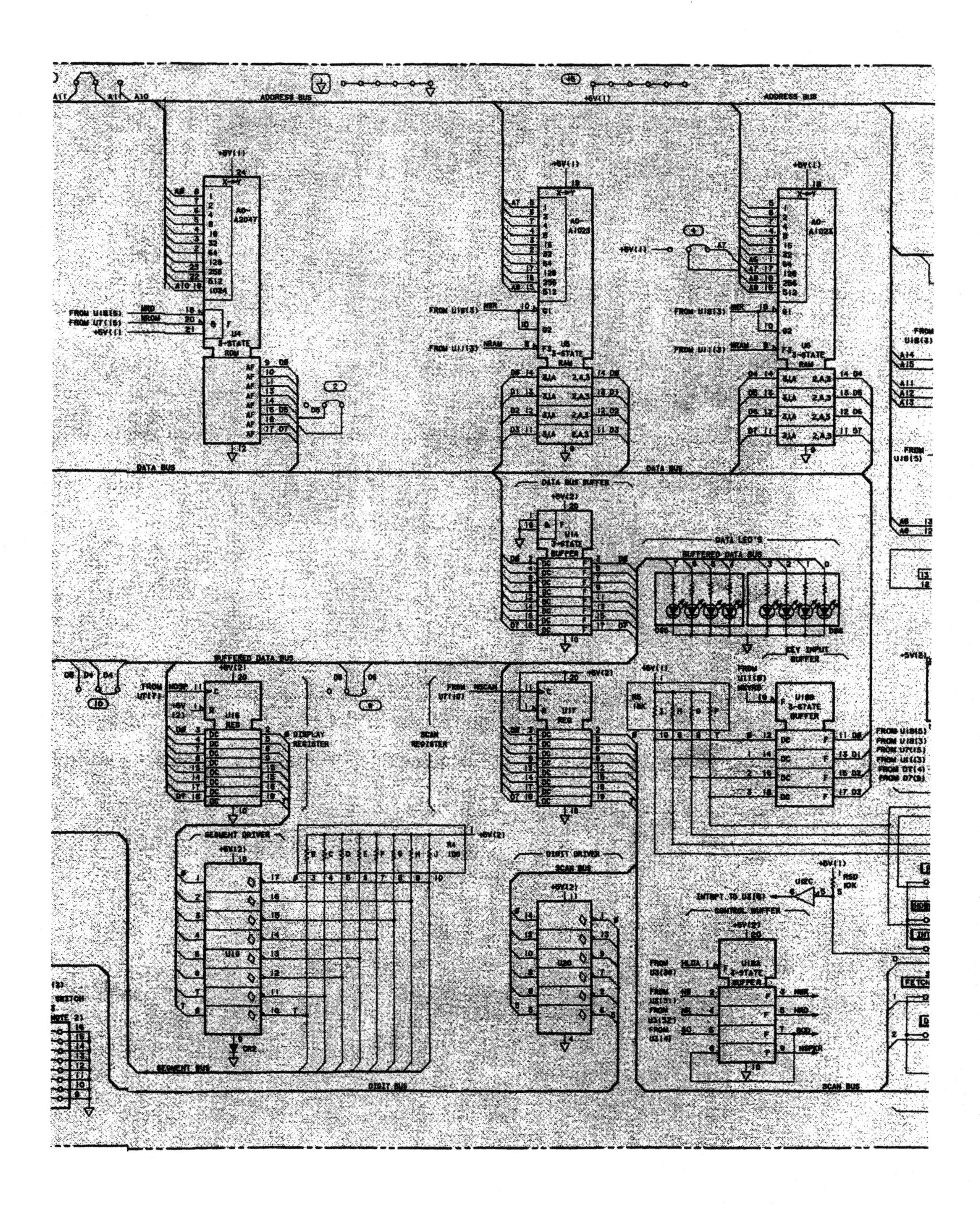

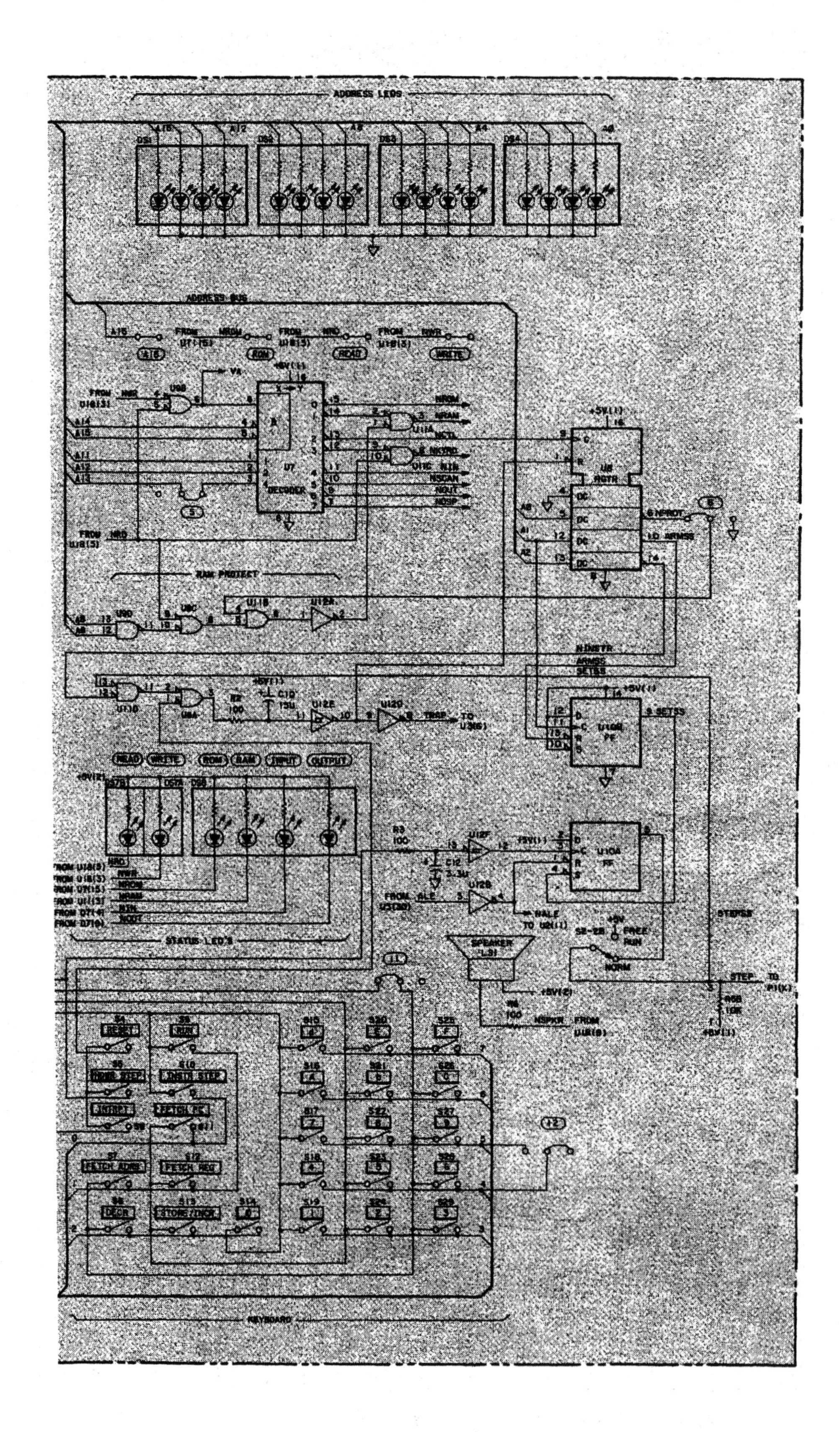
ADDRESS LEDS
ADDRESS BUS
RAM PROTECT
DECODER
STATUS LED'S
SPEAKER
KEYBOARD

APPENDIX

A

Answers to Self-Test Review Questions

STR 1-1

1. (a) Analog electronic signal
2. (d) Digital electrical signal
3. Because their output varies in direct proportion to the input—a linear function
4. Analog readout

STR 1-2

1. (a) Calculator
2. (b) Computer

STR 2-1

1. No difference
2. Base 10
3. 2 = 1000, 6 = 100, 3 = 10, 9 = 1
4. Reset and carry

STR 2-2

1. 26_{10}
2. 10111_2
3. LSB = Least Significant Bit, MSB = Most Significant Bit
4. 1101110_2

STR 2-3

1. (a) Base 10 (b) Base 2 (c) Base 16
2. 21_{10} 15_{16}
3. BF7A
4. Binary equivalent = 100001
   Hexadecimal equivalent = 21

STR 2-4

1. 8
2. 232_8
3. 4
4. 4267_8

STR 2-5

1. $0111\ 0110\ 0010\ 1001_{BCD}$
2. 11_{10}
3. Only one digit changes as you step to the next code group.
4. B a b b a g e

STR 3-1

1. Hardware
2. Software
3. Vacuum tube
4. Diode, transistor

STR 3-2

1. OR, AND
2. Binary 1
3. AND gate
4. OR
5. Binary 0
6. AND

STR 3-3

1. NOT gate
2. Binary 0
3. NAND
4. NOR
5. Binary 1
6. NAND

STR 3-4

1. OR
2. XOR
3. XNOR
4. XNOR

STR 3-5

1. IEEE/ANSI
2. Instead of using distinctive shapes to represent logic gates, it uses a special dependency notation system to indicate how the output is dependent on the input.

STR 4-1

1. Transistor to transistor logic
2. Ten
3. HIGH
4. LOW, HIGH, and high impedance
5. High speed applications
6. Higher packing density

STR 4-2

1. True
2. Voltage control, high input impedance, and low power dissipation
3. P channel and N channel
4. Low power

STR 4-3

1. (a) Transistor Outline
   (b) Dual-in-line package
   (c) Surface-mount technology
2. SMT
3. (a) SSI
4. E-MOSFET, MOS

STR 4-4

1. Lower cost
2. CMOS
3. Signal voltage and current incompatibility
4. Provides interfacing by shifting signal voltage for compatibility
5. BiCMOS family
6. TTL series

STR 5-1

1. Voltmeter
2. Oscilloscope
3. Pulser
4. Current tracer, logic probe

STR 5-2

1. New chip substitution
2. Open, short
3. Midpoint start during isolation
4. 2-state timing signal
5. Logic pulser and current tracer, logic probe

STR 5-3

1. a. Record orientation of IC
   b. De-solder using solder wick or vacuum bulb (grounded iron is required with MOS devices)
   c. Avoid overheating board or component
2. A final test is the application of power and clock to ensure all aspects of the system operate properly.

STR 6-1

1. a. AND b. XOR c. OR
   d. NAND e. NOR f. NAND
2. $\overline{A} \cdot \overline{B} = \overline{A + B}$
3. a. $\overline{AB}$ b. $(\overline{A} \cdot \overline{B}) \cdot C$
4. a. A, B, C → Y

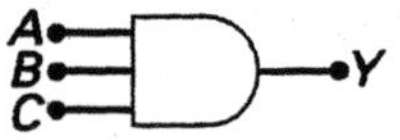

   b. L, D → Y

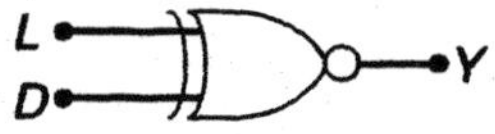

   c. A, B → Y

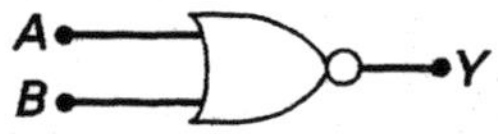

   d. A, B, C → Y

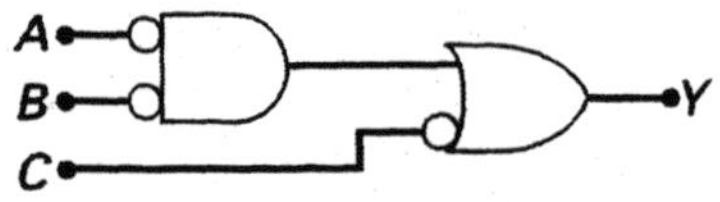

   e. A, L, D → Y

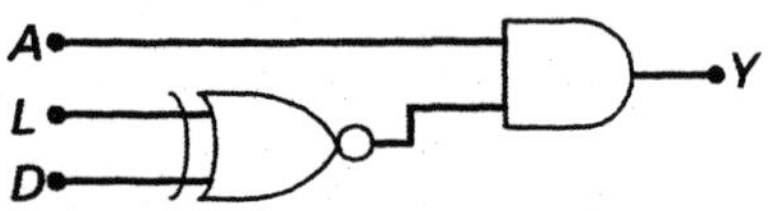

   f. A, B, C → Y

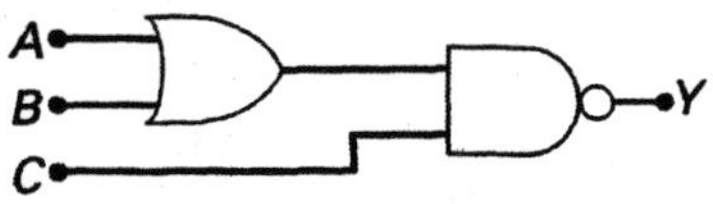

STR 6-2

1. a. 1 b. 1 c. 0 d. A
2. $A(B + C) = (AB) + (AC)$
3. $(AB)C$ or $(AC)B$
4. $A(B + C + D)$

STR 6-3

1. Product, sum
2. a. $A\,\overline{B}\,C\,D$ b. $\overline{A}\,B\,C\,\overline{D}$
3. SOP
4. Fundamental products are ORed and then the circuit is developed.

STR 6-4

1. 4
2. $y = B(A + \overline{A}) = B(1) = B$
3. Larger
4. With 1s for larger grouping

STR 7-1

1. All HIGH except line 9 which would be LOW
2. a. 10 b. 16
3. Output "*b*" would be the only HIGH
4. Matching low power consumption

STR 7-2

1. Ten to four line
2. Inverter gates wired to each 74147 output
3. 0111 (7 has priority)
4. a. 1101 b. 0001 c. HLHH d. OFF, ON, OFF, ON

STR 7-3

1. Diagnose, Isolate, and Repair
2. An operator error is an invalid operating procedure or visual misunderstanding of display sequencing that may be corrected by proper operating procedure; a circuit malfunction is a valid defect that must be isolated and then repaired.
3. A midpoint chosen to begin the isolation process
4. Only the display of the number 2 would not be affected.

STR 8-1

1. (b) One of several, one
2. 4
3. False
4. Parallel
5. Parallel
6. 74157

STR 8-2

1. (a) One, one of several
2. Decoder
3. $C = 1, B = 0, A = 0$
4. True

STR 8-3

1. XNOR
2. The three outputs are labeled $A < B$, $A > B$, and $A = B$
3. For additional chips to handle larger word size
4. $A < B = 1, A > B = 0, A = B$ is 0

STR 8-4

1.

Decimal	Gray Code	Odd-Parity Bit
0	0000	1
1	0001	0
2	0011	1
3	0010	0
4	0110	1
5	0111	0

2. True
3. Both, as a parity generator and a parity checker
4. 8-bits (7-bit ASCII code plus 1-bit for parity)

STR 8-5

1. Diagnose, Isolate, and Repair
2. A single pulse condition with no clock
3. Midpoint selection to begin isolation
4. Both displays would be OFF at the same time and ON at the same time. Displays would therefore show same data.

STR 9-1

1. Latch
2. Level triggered, edge triggered, and pulse triggered
3. No change, race
4. a. Edge triggering
   b. Pulse triggering
5. A latch to eliminate switch noise
6. Input delay is accomplished by capacitor charge time

STR 9-2

1. Data
2. No
3. D flip-flop has a single bit input.
4. (d) All of the above

STR 9-3

1. Diagnose, Isolate, and Repair
2. A single pulse condition with no clock
3. Midpoint selection to begin isolation
4. Constant continuous clock

STR 10-1

1. (d)
2. HIGH, toggle
3. (b) A 63 kHz square wave
4. (c) J-K, D-type, S-R

STR 10-2

1. Free running
2. Monostable
3. Bistable
4. Pulse generation, pulse delay, and timing
5. An internal 5 kOhm voltage divider
6. Depending upon external wiring, it may be used as a monostable or astable multivibrator, a frequency divider, or as a modulator.

STR 10-3

1. Diagnose, Isolate, and Repair
2. Usually a single pulse condition with no clock
3. Midpoint selection to begin isolation
4. The counter's Q outputs would all be PRESET HIGH when SW_1 was pressed, instead of all being cleared LOW.

STR 11-1

1. (c) PIPO
2. Data storage

STR 11-2

1. 8 pulses
2. SIPO, PISO
3. 1 pulse
4. Bidirectional

STR 11-3

1. High impedance (floating)
2. BUS
3. Bus conflict occurs when more than one signal seeks control of the bus due to lack of signal isolation. It is avoided through the use of tristate devices.
4. HIGH, LOW, and control inputs

STR 11-4

1. LOAD = bus data enters register. ENABLE = data read (output) is allowed
2. a. A BUS is a group of parallel wires designed to carry binary data.
   b. A BUS Organized Digital system occurs when one set of wires is connected to all devices—as long as those that transmit data onto the bus have three-state output control.
   c. BUS conflict occurs when more than one signal seeks control of the bus due to lack of signal isolation.
   d. A 16-bit BUS utilizes 16 wires.
3. Universal Asynchronous Receiver and Transmitter. It can be used to interface the parallel data format within a digital electronics system to the serial data format of external equipment.
4. 64 times larger
5. The feedback path is twisted with the johnson counter.

STR 11-5

1. Diagnose, Isolate, and Repair
2. Ground "G" inputs. Use test circuit for data input changes. Substitute logic pulser for clock. Monitor outputs with logic probe.
3. Use zero detect gate as the midpoint
4. No initialization

STR 12-1

1. Are not, do not
2. Modulus = 256, maximum count = 255.
3. 30 ns $\times$ 8 = 240 ns, 1 $\times$ 10^9 $\div$ 240 = 4.2 MHz.
4. Natural Modulus = 256, preset modulus = 226

STR 12-2

1. False
2. Asynchronous has fewer gates, synchronous counter is faster

3. Carry, borrow
4. Count

STR 12-3

1. Binary counter, frequency divider
2. They function as counters
3. They function as frequency dividers.
4. They function as counters.

STR 12-4

1. Diagnose, Isolate, and Repair
2. Count window 74LS160s function as frequency dividers. MOD-10 74LS160s function as counters.
3. No problem exists within dividers
4. Display should read 0000 Hz

STR 13-1

1. (a) 1011 + 11101 = 101000 (b) 1010 − 1011 = 1111 0001 (c) 101 × 10 = 1010 (d) 10111 ÷ 10 = 1011
2. (a) 0101 (b) 1001 0101 (c) 1101 (d) 0111 1001
3. (a) −5 (b) +6 (c) −5 (d) +6
4. Adder circuit

STR 13-2

1. 2, 3 bits
2. (c) Parallel adder
3. The 7483 is an 8-bit, parallel, 2-word adder or a 4-bit binary full-adder with fast carry.
4. SIGMA

STR 13-3

1. ICs are the two 7486 XOR chips.
2. A 1 is added via the carry-in input of the 7483
3. (a) $F = \overline{B}$ (b) $F = A$ minus B
4. 1001 ($F = A$ plus B)

STR 13-4

1. Diagnose, Isolate, and Repair
2. Logic probe each register input as corresponding input switch is taken HIGH.
3. Probe the input and output of the 7483s and the 7486s to see that they are performing the function selected.
4. Enable the input 74126s and the output 74173s so that the input is transferred directly to the output.

STR 14-1

1. Non-volatile
2. A memory's address input is a gating system designed to access a certain cell. The number of address lines is related to these memory locations as function of 2^n.
3. (a) Storage density is a comparison of one memory content to another.
   (b) Memory configuration describes how the circuit is organized into groups of bits or words.
   (c) Burning-in is the process of entering data into a ROM.
   (d) Access time is the interval between a new address input, and the data stored at that address being available at the output.
4. (a) MROM = Masked Read Only Memory
   (b) PROM = Programmable Read Only Memory
   (c) EPROM = Erasable Programmable Read Only Memory
   (d) EEPROM = Electrically Erasable PROM
5. Input code is applied to the address inputs of a ROM where it is decoded internally to select that memory location which, in turn, supplies the desired output code.
6. A ROM listing is a sheet that lists the data values that should be stored in each memory location within the ROM. ROM tests may be made by accessing each word location to check stored content with the listing.

STR 14-2

1. Once programmed, the contents of a ROM cannot be changed.
2. With a SAM, one must sequence through all bits until the desired word is located. This is accomplished directly with a RAM due to its random access capabilities.
3. No. Although access is the same, a ROM has no "write" capability.
4. A volatile device
5. A dynamic RAM MOS cell capacitor requires refreshing whereas a static RAM has bipolar cells.
6. A SRAM has fast, simple circuitry whereas the DRAM has more packing density, and lower power consumption due to less MOSFETs per cell.
7. A memory map is an address-listing diagram showing which address blocks are assigned to each part of the system.
8. RAMs are tested by writing a test pattern into every memory location and then reading out and testing the stored contents: the stored patterns could be all zeros, then all ones, and then a checkerboard pattern.

STR 14-3

1. A simple troubleshooting tree for Figure 14-28:

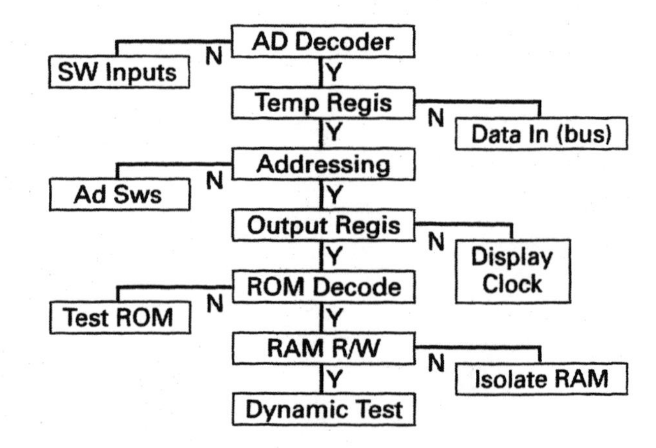

2. Compare contents of ROM with its listing.
3. Use a test circuit to ensure all test patterns would be accepted and delivered by the RAM.

4. ROM would be selected when A_{10}, A_{11}, and A_{12} are 000; RAM would be selected when inputs are 100; INPUT would be selected when inputs are 010; and OUTPUT would be selected when inputs are 110.

STR 15-1

1. The four types of PLDs are:
   PLA = Programmable Logic Array
   PAL = Programmable Array Logic
   PLS = Programmable Logic Sequencer
   EPLD = Erasable Programmable Logic Device
2. A PROM may be used as a PLD.
3. With a PROM, the AND array is fixed while the OR array is programmable. PLA AND and OR arrays are both programmable. A PAL AND array is programmable while the OR array is fixed.
4. A PAL circuit arrangement is not as versatile as the PLA, but it is easier to manufacture and costs less.
5. The advantage is the PLS register outputs that allow the device to perform sequential logic.
6. An advantage of the EPLD is it may be electrically erased and reprogrammed.

STR 15-2

1. A fuse map indicates which fuses within the FPLD should be blown and which should be left intact.
2. A program table is a paper listing of how a PLD has been programmed.
3. Understanding the contents of a fuse map or program table will help the technician determine the differences between operator error and valid defects.
4. No, but a new PLD must be programmed prior to chip substitution.

STR 16-1

1. (a) ADC = Analog to Digital Converter
   (b) DAC = Digital to Analog Converter
2. ADC, DAC
3. Analog, digital
4. Digital, analog

STR 16-2

1. A weighted resistor DAC disadvantage is that as bits increase, large branch current differences may introduce noise or cause output load change.
2. The OP-AMP within a DAC is used to convert a current input into an equivalent voltage output.
3. (a) A staircase is a continually increasing analog output.
   (b) Resolution of a DAC is the smallest analog output change that may occur as a result of an increment in the digital input.
   (c) Step size is the DAC resolution.
   (d) Monotonicity describes correct staircase waveform behavior.
   (e) Settling time is the time it takes for a change in a DAC's analog output to settle to 99.95% of the new value.
   (f) The relative accuracy of a DAC describes how much the output level has deviated from its ideal theoretical output value.
4. DACs may be tested with an input counter to produce an output that may be viewed on an oscilloscope to determine high gain, offset, non-monotonic, and other problems.

STR 16-3

1. The main disadvantage of the staircase ADC is that its conversion speed is dependent on the analog input voltage.
2. The staircase and successive approximation ADC circuits make use of a DAC.
3. A successive approximation ADC operates by first setting and then comparing bits to obtain an equivalent digital output.
4. The flash converter is the fastest type of ADC.
5. An ADC may be tested by applying a linear ramp to the analog input while monitoring the digital output with a display.

STR 16-4

1. An MC1408 DAC may be tested with a free running counter generating the digital input while an oscilloscope monitors the analog output.
2. DAC output errors are high gain, offset, poor resolution, non-monotonic output, non-linear, or low gain.
3. An ADC0801 may be tested by applying a linear ramp signal to the analog input while a display monitors the digital output.
4. ADC output faults are offset, output, missing digital codes, and incorrect code errors.

STR 17-1

1. The three blocks are the microprocessor unit (MPU), the memory unit, and the input/output unit. Functions are:
   Microprocessor—logic circuitry recognizes and executes the program stored in memory
   Memory—contains the program of instruction that is fetched and executed by the MPU
   Input/output—allows data to be read in from, or sent out to peripheral devices
2. The three buses are address, data, and control. Functions are:
   Address—used by the MPU to select memory locations or input/output ports
   Data—used to transfer data or information between devices
   Control—carries signals used by the processor to coordinate the transfer of data within the microcomputer
3. Hardware—the units and components of the system.
   Software—written programs stored as code which are understood by the MPU.
4. a. Machine language—the digital code understood by the MPU

b. Assembly language—programs written using mnemonics which are easier to remember
c. High-level language—an independent language designed to make programming easier that later is compiled into machine language for the MPU

5. A flow chart is a graphic form that describes the operation of a program using oval, rectangular, and diamond shaped blocks.
6. a. Instruction set—contains a list of the machine codes that are recognized by a specific type of microprocessor
b. Peripherals—input and output devices connected to microprocessor system ports
c. Programmer—a person who writes a program in any language which will be understood by the MPU
d. Mnemonic—a short name, or memory aid, used in ASSEMBLY language programming.
e. Compiler—a program that translates high level language into machine code.
f. Accumulator—an MPU internal register (register A), used in conjunction with the ALU; a storage location within an MPU chip
g. Flag Register—an internal register within an MPU chip that is a collection of flip-flops used to indicate the results of certain instructions
h. Op-code—the operation code or "do" portion of an instruction

STR 17-2

1. A unidirectional bus receives data from only one transmitter chip, or "talker." Data transfers are therefore one-way only. A bidirectional bus has several "talkers"; therefore, data flow may be reversed. However, those transmitters must have three-state output control to avoid bus conflict.
2. 8085 pin out signals are:
a. WR—indicates data bus data is to be written into the selected memory or I/O location
b. RD—indicates selected memory or I/O location is to be read. Data bus is ready for transfer.
c. IO/NM—high indicates I/O read/write is to I/O; low indicates read/write is to memory
d. RESET—resets CPU. Once the program counter, interrupt enable, and HLDA flip-flops are reset, it allows a peripheral read.
e. READY—high allows completion of read/write cycle while a low creates a wait state
f. ALE—the address-latch enable signal clocks the lower order address bits for storage
g. HOLD—causes CPU to relinquish the use of buses at the end of the present machine cycle for use by another master, usually a peripheral
h. TRAP—highest priority signal of any interrupt request; recognized as INTR occurs
3. During FETCH, the OP-CODE is read from memory and transferred to the instruction register. During EXECUTE, control/timing logic decodes the Op-code, increments the program register, and retrieves the bytes necessary to complete the task.
4. a. Interrupt—provides a means for devices external to the microprocessor to request immediate action by the processor
b. T-state—the basic unit of time for a microprocessor system equal to one clock period
c. Machine cycle—the time it takes an operation to be completed; from three to six states
d. Instruction cycle—the time it takes to fetch and execute a complete instruction; usually from one to five machine cycles
5. Test equipment which converts long and complex serial data streams at certain points within an MPU system into four-digit displayed "signatures"
6. Specialized test equipment used for examining multiple logic signals or data sequences present within a MPU system.

APPENDIX

B

Answers to Odd-Numbered Problems

Chapter 1

1. b **3.** a **5.** b **7.** d **9.** c

(The answers to essay questions 11 through 15 can be found in the indicated sections that follow the questions.)

Chapter 2

1. b **3.** c **5.** d **7.** a **9.** d

(The answers to essay questions 11 through 24 can be found in the indicated sections that follow the questions.)

25. 1001, Decimal 9

27. Decimal equivalents are:
a. 57 **b.** 39 **c.** 06

29. BCD Equivalents are:
a. 0010 0011 0110 0101 **b.** 0010 0100

31. Codes in order are: BCD, Gray, Excess-3, and ASCII.

33. Stx practice makes perfect etx.

35. 0100010 1000101 1100001 1110011 1101000 1101100 1111001 0100000 1110011 1100001 1101001 1100100 0101100 0100000 1101110 1101111 1110100 0100000 1000101 1100001 1110011 1101000 1101100 1111001 0100000 1100100 1101111 1101110 1100101 0100010 0101110

Chapter 3

1. a **3.** b **5.** b **7.** b **9.** b **11.** b **13.** d **15.** c **17.** b **19.** a

(The answers to essay questions 21 through 30 can be found in the indicated sections that follow the questions.)

31. OR gate

33. Light, $Y = 1$; dark, $Y = 0$

35. OR

37. Truth table outputs (Y) should be:

a. 0 (AND)	**b.** 0 (OR)	**c.** 0 (XOR)
0	1	1
0	1	1
1	1	0

d. 1 (XNOR)	**e.** 1 (NOR)	**f.** 0 (NOT)
0	0	1
0	0	
1	0	

39. C = HIGH

Y_0	0
Y_1	1
Y_2	1
Y_3	0

Chapter 4

1. c **3.** c **5.** a **7.** a **9.** d **11.** a **13.** d **15.** a

(The answers to essay questions 16 through 35 can be found in the indicated sections that follow the questions.)

37. NOR logic, any HIGH in = LOW out

39. Symbol means device will have open collector output

41. Yes, 24 mA needed, 33 mA is maximum

43. 74H05

45. Truth table output is:
a. 0
1
Float
Float

47. The 74LS125 has a LOW activated control input while the 72LS126 has a HIGH activated control input.

49. Circuit clips, or rectifies, an ac signal which is then formed into a positive pulse waveshape within the trigger circuit.

51. N type E-MOSFET

53. NOR Function

55. $t_{pd} = 24$ ns

Chapter 5

1. b **3.** a **5.** b **7.** b **9.** b

(The answers to essay questions 11 through 25 can be found in the indicated sections that follow the questions.)

27. Oscilloscope

29. PRF = 2000 pps
PRT = 500 μs

31. Eliminates external faults

33. Input internally grounded

35. **a.** No **b.** No **c.** No

Chapter 6

1. c **3.** c **5.** b **7.** c **9.** a **11.** c **13.** a **15.** b **17.** c **19.** c

(The answers to essay questions 21 through 30 can be found in the indicated sections that follow the questions.)

31. **a.** $Y = (\overline{A} \cdot B) + (B \cdot C) + D$

b. $Y = \overline{(A \cdot B)} + \overline{(\overline{\overline{A}} \cdot \overline{\overline{B}})}$

c. $Y = \overline{\overline{\overline{(\overline{A} \cdot \overline{B} \cdot \overline{C})} + \overline{(A \cdot B \cdot \overline{C})}}}$

d. Y = (LIT + NLT) · (SLW + FST)

33. a.

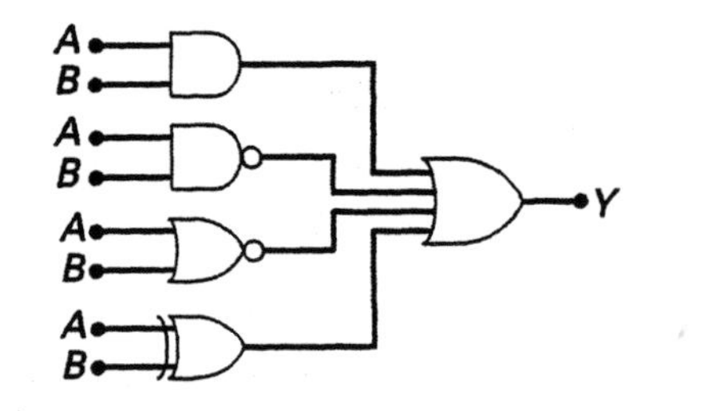

b.

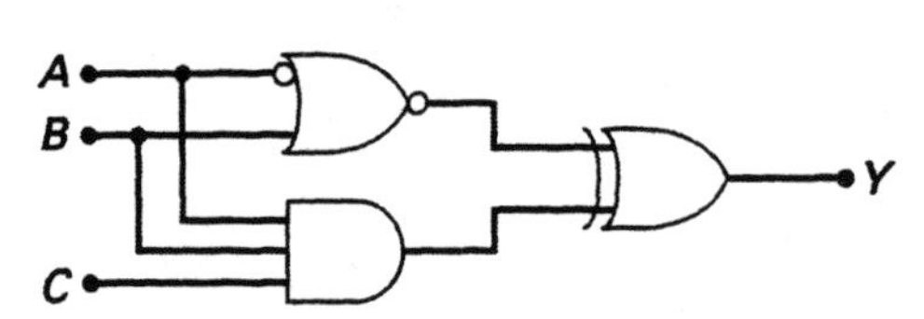

35.

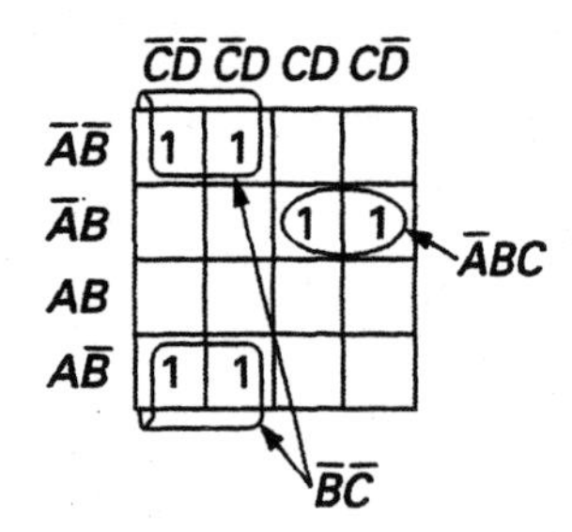

$Y_2 = \overline{B}\,\overline{C} + \overline{A}BC$

37. $Y_1 = \overline{A}\,\overline{B}\,\overline{C}\,\overline{D} + \overline{A}\,\overline{B}\,\overline{C}D + \overline{A}\,\overline{B}C\overline{D} + \overline{A}\,\overline{B}CD + A\overline{B}\,\overline{C}D + A\overline{B}CD + AB\overline{C}\,\overline{D} + AB\overline{C}D + ABC\overline{D} + ABCD$

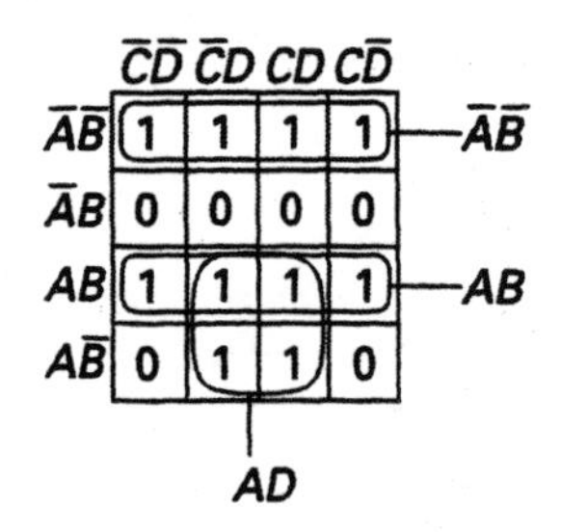

$Y_1 = \overline{A}\,\overline{B} + AB + AD$
(50 Inputs to 9 Inputs)

39. (Question 35, Y_1)

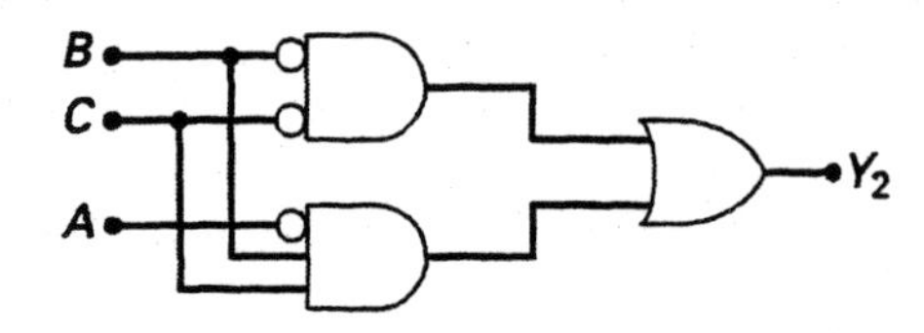

(Question 37, Y_2)

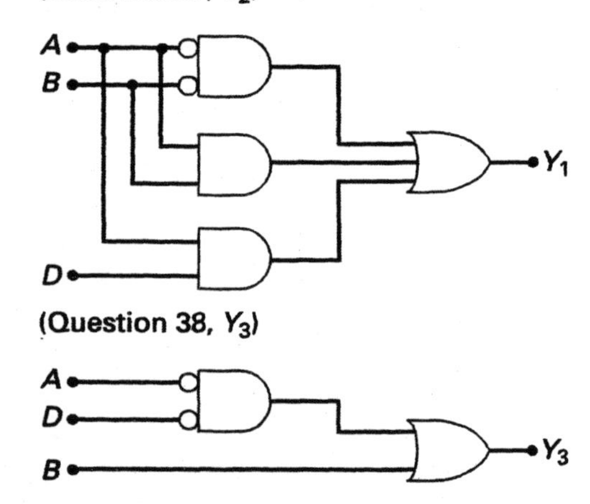

(Question 38, Y_3)

Chapter 7

1. a **3.** a **5.** c **7.** b **9.** d

(The answers to essay questions 11 through 25 can be found in the indicated sections that follow the questions.)

27. a.

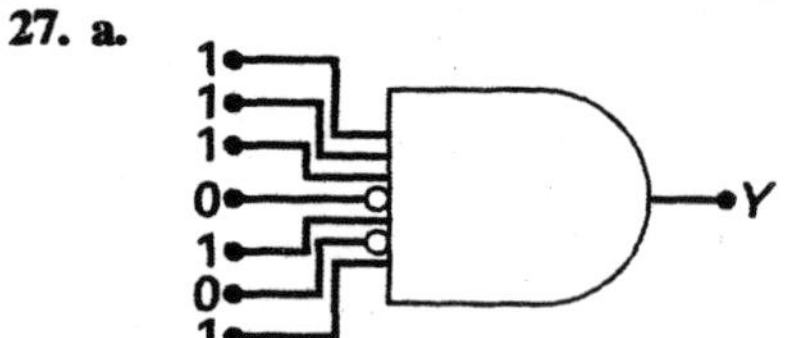

b.

1 1 0 0 1 1 → Y

c.

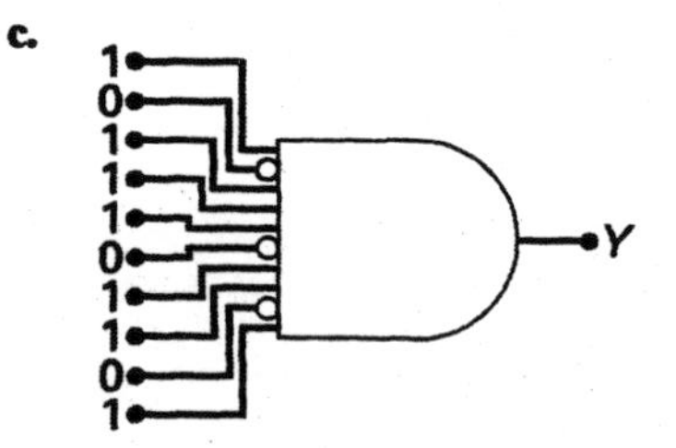

29. Pull-up resistors for input switches
31. Active-LOW
33. $D\ C\ B\ A = 0100$
35. Common-cathode display
37. Controls the counting of the 7490
39. $DCBA$ input = 1000, output = 1111111
41. Switch position determines output code (diode matrix)
43. A priority encoder circuit (9-line decimal to 4-line BCD)
45. Active-LOW inputs
47. Negative, Anode, ON
49. 1000 (Negative Logic code for 7)
51. See section 7-3

53. **a.** No display
b. The g segment will not light. Affects display of 2, 3, 4, 5, 6, 8, and 9.
c. Constant display of 9
55. **a.** Output code will not be accurate whenever burned-out LED should be ON.
b. Pin 14 (LED 4) output from 74147 will always be LOW (only codes 8 and 9 will be displayed).
c. The corresponding LED will constantly be OFF.
d. The corresponding LED will constantly be ON.

Chapter 8

1. c **3.** a **5.** b **7.** b **9.** a **11.** d **13.** d **15.** a
(The answers to essay questions 16 through 35 can be found in the indicated sections that follow the questions.)
37. Parallel-to-Serial
39. Clock increments counter and negative-logic display until switch is pressed. MUX LOW output stops clock from reaching counter.
41. Anode, LOW
43. Continuous count
45. The 7485 is used as a comparator.
47. Word selector
49. $A = B$ output would be HIGH, A word would be selected because $\overline{A}/B$ would be LOW.
51. 74138 acts as demultiplexer (decoder) of 74193 count input
53. Count 0 to 4, and then repeat.
55. Count 0 to 2, and then repeat.
57. When strobe $\overline{G_1}$ is HIGH, outputs are: Y = LOW, W = HIGH.
59. **a.** B word selected unless $A = B$
b. B word selected when $A > B$
c. No input action, Y = LOW
d. The 7485 would still function and so would display. No word would be selected by disabled 74157.

Chapter 9

1. d **3.** a **5.** c **7.** b **9.** a **11.** c **13.** c **15.** d
(The answers to essay questions 16 through 35 can be found in the indicated sections that follow the questions.)
37. NOR

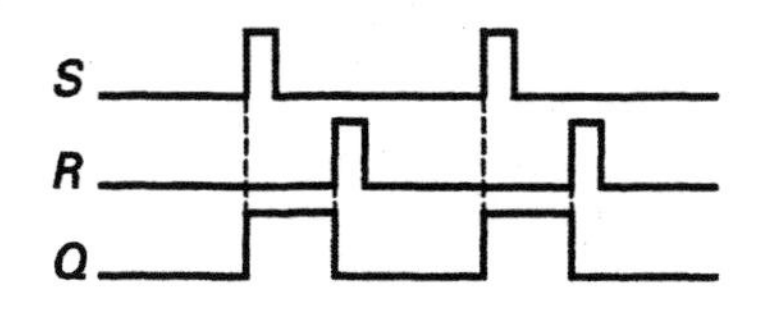

39. NAND

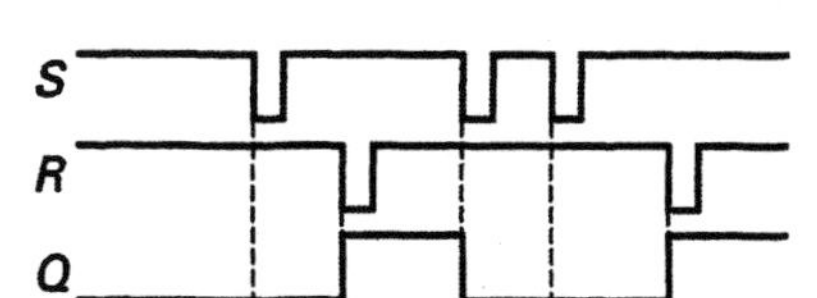

41.

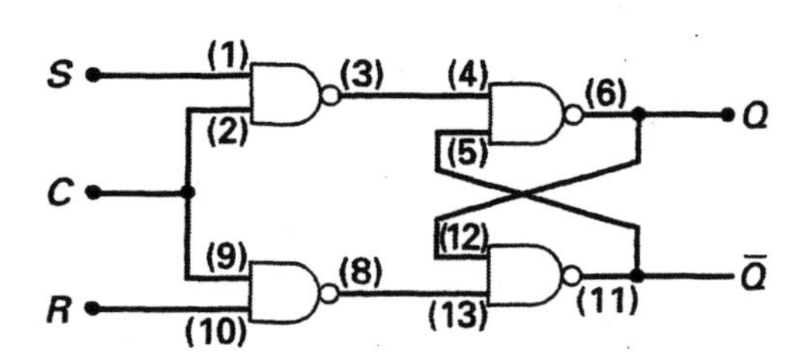

43. Use art from answer 41 plus the following:

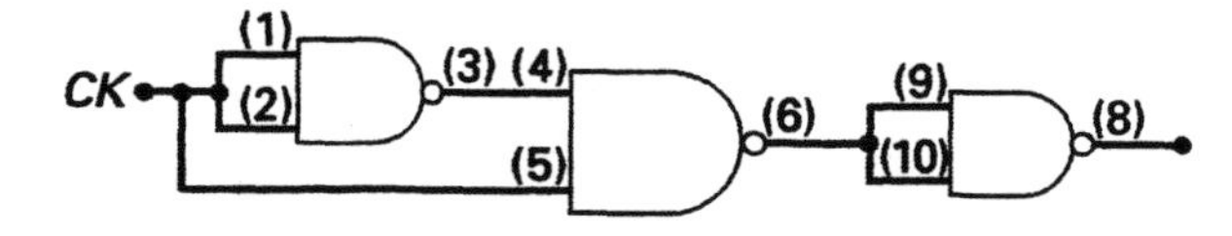

45. Negative edge-triggered D-type f-f
47. 0110 (six)
49. LOAD = HIGH, first negative edge of clock will load f-fs, LOAD = LOW
51. NAND S-R latch
53. OFF; $Y = 0$ ON; $Y = 1$ kHz clock
55. **a.** 100 kHz **b.** 50 kHz
57. Positive edge-triggered D-type f-f
59. Q will go LOW, and LED will turn ON.
61. Disconnect clock, pulse LOAD, and probe gates. Pulse clock, probe Q outputs. Pulse CLR, probe Q outputs.
63. **a.** X will either remain latched HIGH or LOW, therefore no action.
b. Y will always be LOW
65. **a.** Same action, out-of-phase = LED ON, in-phase = LED OFF
b. LED will display reverse indications.
c. Constant clear, LED will always be ON

Chapter 10

1. c **3.** d **5.** c **7.** b **9.** d **11.** b **13.** c **15.** d
17. d **19.** d
(The answers to essay questions 21 through 30 can be found in the indicated sections that follow the questions.)
31. Negative edge
33. **a.** 2 MHz **b.** 0 Hz
35.

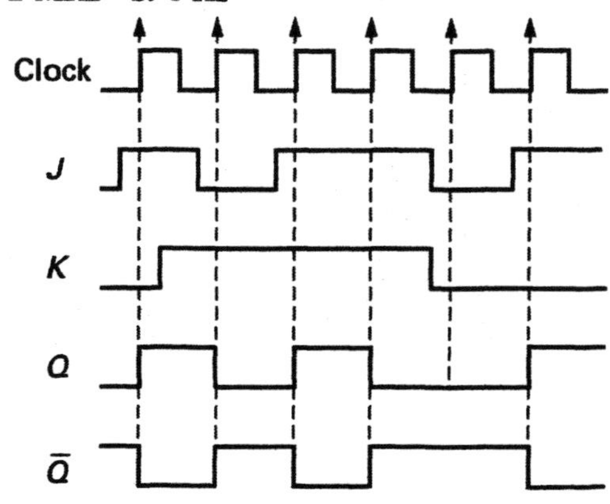

37. Pulse

39.

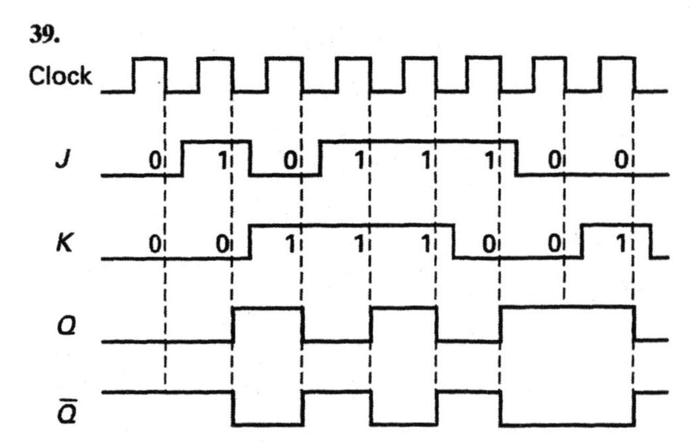

41. $f_{max} = \dfrac{1.43}{R_1 + (2 \times R_2) \times C_1} =$

$\dfrac{1.43}{1\ \Omega + (2 \times 1\ k\Omega) \times 0.1\ \mu F} = \dfrac{1.43}{0.0002} = 7.15\ kHz$

$f_{min} = \dfrac{1.43}{R_1 + (2 \times R_2) \times C_1} =$

$\dfrac{1.43}{100\ k\Omega + (2 \times 1\ k\Omega) \times 0.1\ \mu F} = \dfrac{1.43}{0.0102} = 140\ Hz$

43. $P_w = 0.7 \times RC = 0.7 \times 1\ \Omega \times 0.01\ \mu F = 0.007\ \mu s$
$P_w = 0.7 \times RC = 0.7 \times 100\ k\Omega \times 0.01\ \mu F = 0.7\ ms$

45. 10-27(c)

47. 10-27(b)

49. $Q = 1$ (preset is constantly active)

51. First probe latch in OFF and ON state. Disconnect 4 MHz input, pulse clock input, and probe *JK* f-f.

53. **a.** Q output would not change.
b. Q output will toggle for each clock pulse.
c. Q output would not change.
d. Q output would not change.
e. *JK* f-f would switch from toggle to no-change condition as switch bounced.

55. **a.** 555 timer output will be at a frequency determined only by R_2 and C_1.
b. 555 output will always be LOW (reset).
c. Frequency range would be changed.
d. Tone pitch would vary in the opposite direction (as ON time of 74121 is increased at Q, it is decreased at $\bar{Q}$).
e. No change
f. PW could not be varied.

Chapter 11

1. a **3.** d **5.** a **7.** b **9.** a **11.** b **13.** d **15.** b
17. a **19.** d

(The answers to essay questions 21 through 30 can be found in the indicated sections that follow the questions.)

31.

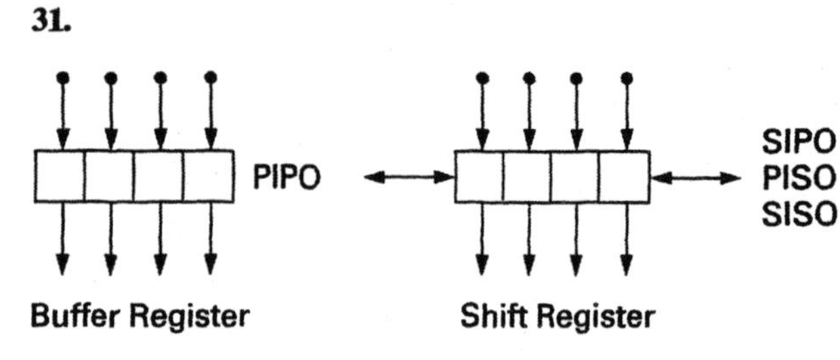

33. The 74175 is a quad D-type flip-flop, PIPO buffer register.

35. SISO

37. **a.** $\bar{K}$ serial input **b.** J serial input
c. A parallel load input **d.** Q_A parallel output
e. SHIFT/$\overline{\text{LOAD}}$; HIGH = shift, LOW = load

39. When SHIFT/$\overline{\text{LOAD}}$ is HIGH, the contents in the register is shifted by an active clock input. When SHIFT/$\overline{\text{LOAD}}$ is LOW, the 8-bit parallel input word is loaded into the register after an active clock input.

41. Synchronous

43. Active-LOW

45. 10001111

47. Ring counter, rotating zero state

49. **a.** When power-up is stable (slow-start circuit action), zero is loaded into register A for rotation with clock, as controlled by NAND A.
b. When a key is pressed, clock will be inhibited by NAND A only when zero from ring counter has located row with key closure. The ROW code will be available at 74147 A output and the column code will be available at 74147 B output.

51. Disconnect clock and use pulser to trigger circuit. Probe should reveal if ring counter, row and column encoders, zero detect gate, and matrix are operating as they should.

53. **a.** Shift register would be locked in load condition—no zero rotate.
b. Same symptoms as answer **a.**
c. SH/$\overline{\text{LD}}$ line would go HIGH immediately at power up—zero may not be loaded into A register.
d. Associated ROW would not have zero scan—keys on row would not generate codes or stop clock.
e. Clock will be shorted to ground—74199 will not load or shift.
f. This will have no effect—pin 1 is normally connected to pin 2 in this application.
g. Zero would shift through 74199 once and then be replaced by a register filled with HIGHs (open would give a HIGH input to 74199 input).

55. 74147 A input = 11101111, 74147 B input = 11101111. 74147 A and B outputs = HLHH HLHH.

Chapter 12

1. b **3.** a **5.** c **7.** c **9.** a **11.** d **13.** a **15.** a

(The answers to essay questions 16 through 30 can be found in the indicated sections that follow the questions.)

31. a. ÷ 12 **b.** ÷ 4096 **c.** ÷ 8192 **d.** ÷ 16
33. a. 16 **b.** 64 **c.** 16,777,200
35. Asynchronous
37. 31
39. 62.5 Hz
41. a. 256 **b.** 256 − 52 = 204
43. Disconnect clip, and then check counter inputs to 74154.
45. Disconnect clock input to NAND gate and connect pulser. Single step circuit and use logic probe to check 74193's count and 74194's LOW output sequence. Reset counter and connect clip to see if circuit operates as it should.

Chapter 13

1. b **3.** d **5.** a **7.** d **9.** d **11.** d **13.** b **15.** c
(The answers to essay questions 16 through 30 can be found in the indicated sections that follow the questions.)
31. a. 1000 0111 **b.** 1100 0010 **c.** 0101 0100
33. a. 0001 0001 0001 **b.** 0001 0101 0101 **c.** 0001 1100
35. +34 = 0010 0010
−83 = 1010 1101
+83 = 0101 0011
37. a. +16 **b.** −66 **c.** −11
39. a. 1010 0110 **b.** 0111 1111 **c.** 0000 0011
41. 74126 is a quad bus buffer with 3-state outputs. 74173 is a 4-bit D-type register with 3-state outputs. 74181 is an ALU.
43. a. Set input switches to desired value, pulse E_I HIGH, pulse $\overline{L_A}$ LOW.
b. Set input switches to desired value, pulse E_I HIGH, pulse $\overline{L_B}$ LOW.
c. Pulse E_F HIGH, pulse $\overline{L_0}$ LOW.
d. Set input switches to desired value, pulse E_I HIGH, pulse $\overline{L_0}$ LOW.
45. Follow same procedure as 44**a.** until output of 74181 has result. Pulse E_F HIGH, pulse $\overline{L_A}$ LOW.
47. Disconnect clock and then use logic pulser to pulse enable, load and clock inputs in sequence to see if input, output, registers, and ALU are operating.
49. a. Bus conflict
b. If anodes were connected to +5 V, display would be complement of result.
c. Output register contents and LED display with change with each bus change.
d. When E_F is HIGH, broken F-Bus line will FLOAT associated data bus line.

Chapter 14

1. b **3.** b **5.** b **7.** d **9.** c **11.** b **13.** b **15.** d
17. a **19.** a
(The answers to essay questions 21 through 50 can be found in the indicated sections that follow the questions.)
51. CMOS Flash memory
53. 8 Data, 15 Address
55. a. Chip Enable: Activates the IC
b. Output Enable: Enables the IC's output data buffers
c. Write Enable: Controls writes to the control register and the array
d. V_{pp}: Erase/Program power supply for programming bytes in the array
57. Enable ROM IC (Chip Select)
59. When installed in a circuit, this MROM is a read-only device.
61. 2316 is designated as U4 on schematic.
63. U7, pin 15
65. U3, 8085
67. 4096
69. 450 ns, 370 mW
71. $\overline{WE}$ is a read/$\overline{\text{write}}$ control
73. Control read/$\overline{\text{write}}$ control
75. $\overline{CS}$ should be active, $\overline{WE}$ should be HIGH
77. U18 pin 3
79. A_{11}, A_{12}, and A_{13}
81. a. ROM **b.** RAM **c.** CONTROL **d.** KEY DATA **e.** INPUT **f.** SCAN **g.** OUTPUT **h.** DISPLAY
83. Display address bus logic levels
85. Yes
87. See section 14-1-6
89. Data bus line 5 would be pulled constantly HIGH, resulting in ROM data errors.

Chapter 15

1. c **3.** b **5.** a **7.** c **9.** a
(The answers to essay questions 11 through 30 can be found in the indicated sections that follow the questions.)
31. PAL
33. Hexadecimal to seven-segment decoder
35. a. 3 **b.** B **c.** F **d.** C
37. Intensity control
39. 3 (pins 8, 9, and 11)
41. PAL
43. Address decoding for Port $1F78_{16}$ and Port $1F79_{16}$
45. Active-LOW
47. A program table also indicates how a PLD has been programmed, including whether outputs are active-LOW or active-HIGH.
49. Static test the circuit by single-stepping a binary count applied to the input and probing the outputs to see that the correct outputs are made active.

Chapter 16

1. c **3.** c **5.** b **7.** b **9.** d
(The answers to essay questions 11 through 30 can be found in the indicated sections that follow the questions.)
31.
a. $\text{Stepsize} = \frac{V_0}{2^n} = \frac{12\text{ V}}{16} = 0.75\text{ V}$
b. $\%\text{ Res.} = \frac{\text{s/size}}{V_0} = \frac{0.75\text{ V}}{12\text{ V}} = 6.25\%$
c. LSB = 12 V/16 = 0.75 V
1 = 12 V/8 = 1.5 V
2 = 12 V/4 = 3.0 V
MSB = 12 V/2 = 6.0 V
d. 0010 = 1.5 V
e. 1011 = 8.25 V

33.
a. $5\left(\frac{64 + 32 + 1}{256}\right) = 1.9$ mA

b. $5\left(\frac{16 + 5}{256}\right) = 0.4$ mA

c. $5\left(\frac{255}{256}\right) = 4.9$ mA

d. $5\left(\frac{16 + 1}{256}\right) = 0.33$ mA

35. DAC

37.

39. **a.** 0.012 mV **b.** 0.0004%
c. LSB = 12 mV, 24 mV, 48 mV, 96 mV, 384 mV, 768 mV, MSB = 1.54 V
d. 60 mV **e.** 2.2 V

41. ADC

43. Temperature change will vary dc input voltage to ADC. Equivalent digital output is decoded by PLD and displayed in decimal on 7-segment displays.

45. Flash converter

47.

MSB	LSB
0	0
0	1
1	0
1	1

49. 1010, 1100, 1110, 1111, 1110, 1100, 1001, 0111, 0101, 0011, 0001, 0011, 0101, 0111.

51. Apply free-running counter to MC1408 input, monitor analog output with an oscilloscope.

53. Test DAC as described in question 51, disconnect op-amp if output loading is suspected.

55. Static test digital gates with logic pulser and probe. Dynamic test converter using linear ramp input.

Chapter 17

1. c **3.** b **5.** b **7.** b **9.** c **11.** b **13.** b **15.** c **17.** c **19.** d

(The answers to essay questions 21 through 35 can be found in the indicated sections that follow the questions.)

37. U4 is a 2 k × 8 ROM.

39. A talker (Read-only)

41. Together U5 and U6 form a 1 k × 8 RAM.

43. Both talkers (read) and listeners (write)

45. External crystal, determines clock signal frequency output (pin 37)

47. The serial output data goes to U18A (6), and then onto the speaker.

49. Pin 30 (address latch enable, ALE): when active it indicates an address is present at the microprocessor's address and address/data output pins (active-LOW).
Pin 31 (write control output): indicates to external devices that data on data bus should be written into addressed location (active-LOW).
Pin 32 (read control output): indicates to external devices that addressed device should place addressed data on data bus (active-LOW).

51. Displays contents of address bus.

53. Displays contents of buffered data bus

55. When U15 is enabled by NOUT, the contents on the data bus will be latched and displayed on the eight common-anode LED display.

57. See section 17-2-1

59. These LEDs show which device has been enabled by the address decoder.

61. See sections 5-1 and 17-2-2

63. See section 17-2-2

65. See section 17-2-2.

Index

B

C

D

E

F

G

H

I

N

O

P

R

S

U

V